a LANGE medical book

Jawetz, Melnick, & Adelberg's
Medical Microbiology

Twenty-Fifth Edition

Geo. F. Brooks, MD
*Professor Emeritus of Laboratory Medicine and
Microbiology and Immunology
University of California
San Francisco*

Karen C. Carroll, MD
*Professor of Pathology
The Johns Hopkins University School of Medicine
Director, Division Medical Microbiology
The Johns Hopkins Hospital
Baltimore*

Janet S. Butel, PhD
*Distinguished Service Professor
Chair, Department of Molecular Virology and
Microbiology
Baylor College of Medicine
Houston*

Stephen A. Morse, PhD
*Associate Director for Science
Bioterrorism Preparedness and Response Program
National Center for Infectious Diseases
Centers for Disease Control and Prevention
Atlanta*

Timothy A. Mietzner, PhD
*Associate Professor
Department of Microbiology and Molecular
Genetics
University of Pittsburgh School of Medicine Pittsburgh
Adjunct Associate Professor of Microbiology
Arizona School of Dentistry and Oral Health
Mesa*

New York Chicago San Francisco Lisbon London Madrid Mexico City Milan
New Delhi San Juan Seoul Singapore Sydney Toronto

ISBN 978-0-07-162496-1
MHID 0-07162496-1
ISSN 1054-2744

Notice

Medicine is an ever-changing science. As new research and clinical experience broaden our knowledge, changes in treatment and drug therapy are required. The authors and the publisher of this work have checked with sources believed to be reliable in their efforts to provide information that is complete and generally in accord with the standards accepted at the time of publication. However, in view of the possibility of human error or changes in medical sciences, neither the authors nor the publisher nor any other party who has been involved in the preparation or publication of this work warrants that the information contained herein is in every respect accurate or complete, and they disclaim all responsibility for any errors or omissions or for the results obtained from use of the information contained in this work. Readers are encouraged to confirm the information contained herein with other sources. For example and in particular, readers are advised to check the product information sheet included in the package of each drug they plan to administer to be certain that the information contained in this work is accurate and that changes have not been made in the recommended dose or in the contraindications for administration. This recommendation is of particular importance in connection with new or infrequently used drugs.

This book was set in Minion Pro by Newgen Publishing and Imaging Services.
The editors were Michael Weitz and Harriet Lebowitz.
The production supervisor was Catherine Saggese.
Production management was provided by Newgen Publishing and Imaging Services.
The designer was Elise Lansdon; the cover designer was Margaret Webster-Shapiro.
Cover Photo: Scanning electron micrograph of *Streptococcus*. Magnification x6, 200 (at 4x4'); Credit: David M. Phillips/Photo Researchers, Inc.
RR Donnelley was the printer and binder.

This book is printed on acid-free paper.

International Edition ISBN 978-0-07-174271-9; MHID 0-07-174271-9

Contents

* Chapters 1–7 were edited by Stephen A. Morse in his private capacity. No official support or endorsement by CDC is intended or should be inferred.

SECTION IV
VIROLOGY 373

Janet S. Butel, PhD

SECTION **V**

MYCOLOGY   625

Thomas G. Mitchell, PhD

viii   Contents

Preface

The twenty-fifth edition of *Jawetz, Melnick, & Adelberg's Medical Microbiology* remains true to the goals of the first edition published in 1954 "to provide a brief, accurate and up-to-date presentation of those aspects of medical microbiology that are of particular significance to the fields of clinical infections and chemotherapy." The current edition has some new features that include a change in the format, the addition of color photographs, and an increased number of new and revised review questions at the end of each chapter. All chapters have been revised extensively consistent with the tremendous expansion of medical knowledge afforded by molecular mechanisms, advances in our understanding of microbial pathogenesis and the discovery of novel pathogens.

New also to this edition is Timothy Mietzner, PhD, Associate Professor in the Department of Microbiology and Molecular Genetics at the University of Pittsburgh School of Medicine. Dr. Mietzner's extensive expertise in molecular microbial pathogenesis will add significantly to the current and future editions and we welcome his participation.

The change to full color has provided the opportunity to prepare many new photographs and photomicrographs, and their new uniformity has resulted in a distinct benefit in the Bacteriology Section (III). The Gram stains were photographed using the same microscope, camera, and software settings. The pictures were cropped for printing as square images one column wide. The result is that the relative size of the bacteria can be compared from one image to another. Thus, *Escherichia coli* (Figure 15–1) appear larger than *Haemophilus influenzae* (Figure 18–1) and *Franciscella tularensis* (Figure 18–2), much as would be seen when observing the bacteria through the microscope. The authors hope that comparing the relative size of bacteria in photographs will be useful to students of microbiology.

Geo. F. Brooks
San Francisco
Karen C. Carroll
Baltimore
Janet S. Butel
Houston
Stephen A. Morse
Atlanta
March 2010

C H A P T E R

1

The Science of Microbiology

INTRODUCTION

Microbiology is the study of microorganisms, a large and diverse group of microscopic organisms that exist as single cells or cell clusters; it also includes viruses, which are microscopic but not cellular. Microorganisms have a tremendous impact on all life and the physical and chemical makeup of our planet. They are responsible for cycling the chemical elements essential for life, including carbon, nitrogen, sulfur, hydrogen, and oxygen; more photosynthesis is carried out by microorganisms than by green plants. It has been estimated that 5×10^{30} microbial cells exist on earth; excluding cellulose, these cells constitute about 90% of the biomass of the entire biosphere. Humans also have an intimate relationship with microorganisms; more than 90% of the cells in our bodies are microbes.

BIOLOGIC PRINCIPLES ILLUSTRATED BY MICROBIOLOGY

Nowhere is **biologic diversity** demonstrated more dramatically than by microorganisms, creatures that are not directly visible to the unaided eye. In form and function, be it biochemical property or genetic mechanism, analysis of microorganisms takes us to the limits of biologic understanding. Thus, the need for **originality**—one test of the merit of a scientific **hypothesis**—can be fully met in microbiology. A useful hypothesis should provide a basis for **generalization**, and microbial diversity provides an arena in which this challenge is ever-present.

Prediction, the practical outgrowth of science, is a product created by a blend of technique and theory. **Biochemistry**, **molecular biology**, and **genetics** provide the tools required for analysis of microorganisms. **Microbiology**, in turn, extends the horizons of these scientific disciplines. A biologist might describe such an exchange as **mutualism**, ie, one that benefits all of the contributing parties. Lichens are an example of microbial mutualism. Lichens consist of a fungus and phototropic partner, either an alga (a eukaryote) or a cyanobacterium (a prokaryote). The phototropic component is the primary producer, whereas the fungus provides the phototroph with an anchor and protection from the elements. In biology, mutualism is called **symbiosis**, a continuing association of different organisms. If the exchange operates primarily to the benefit of one party, the association is described as **parasitism**, a relationship in which a **host** provides the primary benefit to the parasite. Isolation and characterization of a parasite—eg, a pathogenic bacterium or virus—often require effective mimicry in the laboratory of the growth environment provided by host cells. This demand sometimes represents a major challenge to the investigator.

The terms "mutualism," "symbiosis," and "parasitism" relate to the science of **ecology**, and the principles of environmental biology are implicit in microbiology. Microorganisms are the products of **evolution**, the biologic consequence of **natural selection** operating upon a vast array of genetically diverse organisms. It is useful to keep the complexity of natural history in mind before generalizing about microorganisms, the most heterogeneous subset of all living creatures.

A major biologic division separates the eukaryotes, organisms containing a membrane-bound nucleus, from prokaryotes, organisms in which DNA is not physically separated from the cytoplasm. As described below and in Chapter 2, further major distinctions can be made between eukaryotes

1

and prokaryotes. Eukaryotes, for example, are distinguished by their relatively large size and by the presence of specialized membrane-bound organelles such as mitochondria.

As described more fully below, eukaryotic microorganisms—or, phylogenetically speaking, the Eukarya—are unified by their distinct cell structure and phylogenetic history. Among the groups of eukaryotic microorganisms are the **algae**, the **protozoa**, the **fungi**, and the **slime molds**.

The unique properties of viruses set them apart from living creatures. Eukaryotes and prokaryotes are organisms because they contain all of the enzymes required for their replication and possess the biologic equipment necessary for the production of metabolic energy. Thus, eukaryotes and prokaryotes stand distinguished from **viruses**, which depend upon host cells for these necessary functions.

VIRUSES

Viruses lack many of the attributes of cells, including the ability to replicate. Only when it infects a cell does a virus acquire the key attribute of a living system: reproduction. Viruses are known to infect all cells, including microbial cells. Host–virus interactions tend to be highly specific, and the biologic range of viruses mirrors the diversity of potential host cells. Further diversity of viruses is exhibited by their broad array of strategies for replication and survival.

A viral particle consists of a nucleic acid molecule, either DNA or RNA, enclosed in a protein coat, or capsid (sometimes itself enclosed by an envelope of lipids, proteins, and carbohydrates). Proteins—frequently glycoproteins—in the capsid determine the specificity of interaction of a virus with its host cell. The capsid protects the nucleic acid and facilitates attachment and penetration of the host cell by the virus. Inside the cell, viral nucleic acid redirects the host's enzymatic machinery to functions associated with replication of the virus. In some cases, genetic information from the virus can be incorporated as DNA into a host chromosome. In other instances, the viral genetic information can serve as a basis for cellular manufacture and release of copies of the virus. This process calls for replication of the viral nucleic acid and production of specific viral proteins. **Maturation** consists of assembling newly synthesized nucleic acid and protein subunits into mature viral particles, which are then liberated into the extracellular environment. Some very small viruses require the assistance of another virus in the host cell for their duplication. The delta agent, also known as hepatitis D virus, is too small to code for even a single capsid protein and needs help from hepatitis B virus for transmission. Viruses are known to infect a wide variety of plant and animal hosts as well as protists, fungi, and bacteria. However, most viruses are able to infect specific types of cells of only one host species.

A number of transmissible plant diseases are caused by **viroids**—small, single-stranded, covalently closed circular RNA molecules existing as highly base-paired rod-like structures. They range in size from 246 to 375 nucleotides in length. The extracellular form of the viroid is naked RNA—there is no capsid of any kind. The RNA molecule contains no protein-encoding genes, and the viroid is therefore totally dependent on host functions for its replication. Viroid RNA is replicated by the DNA-dependent RNA polymerase of the plant host; preemption of this enzyme may contribute to viroid pathogenicity.

The RNAs of viroids have been shown to contain inverted repeated base sequences at their 3' and 5' ends, a characteristic of transposable elements (see Chapter 7) and retroviruses. Thus, it is likely that they have evolved from transposable elements or retroviruses by the deletion of internal sequences.

The general properties of animal viruses pathogenic for humans are described in Chapter 29. Bacterial viruses are described in Chapter 7.

PRIONS

A number of remarkable discoveries in the past 3 decades have led to the molecular and genetic characterization of the transmissible agent causing **scrapie**, a degenerative central nervous system disease of sheep. Studies have identified a scrapie-specific protein in preparations from scrapie-infected brains of sheep that is capable of reproducing the symptoms of scrapie in previously uninfected sheep (Figure 1–1). Attempts to identify additional components, such as nucleic acid, have been unsuccessful. To distinguish this agent from viruses and viroids, the term **prion** was introduced to

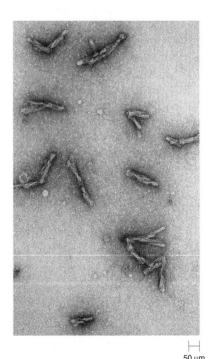

⊢——⊣
50 μm

FIGURE 1–1 Prion. Prions isolated from the brain of a scrapie-infected hamster. This neurodegenerative disease is caused by a prion. (Reproduced with permission from Stanley B. Prusiner/Visuals Unlimited.)

emphasize its proteinaceous and infectious nature. The cellular form of the prion protein (PrP^c) is encoded by the host's chromosomal DNA. PrP^c is a sialoglycoprotein with a molecular weight of 33,000–35,000 and a high content of α-helical secondary structure that is sensitive to proteases and soluble in detergent. PrP^c is expressed on the surface of neurons via a glycosylphosphatidyl inositol anchor in both infected and uninfected brains. An abnormal isoform of this protein (PrP^{res}) is the only known component of the prion and is associated with transmissibility. It has the same amino acid sequence as PrP^c, but differs physically from the normal cellular isoform by its high β-sheet content, its insolubility in detergents, its propensity to aggregate, and its partial resistance to proteolysis. It is believed that PrP^{res} induces PrP^c to fold or refold into the prion form.

There are additional prion diseases of importance (Table 1–1). Kuru, Creutzfeldt-Jakob disease (CJD), Gerstmann-Sträussler-Scheinker disease, and fatal familial insomnia affect humans. Bovine spongiform encephalopathy, which is thought to result from the ingestion of feeds and bone meal prepared from rendered sheep offal, has been responsible for the deaths of more than 184,000 cattle in Great Britain since its discovery in 1985. A new variant of CJD (vCJD) has been associated with human ingestion of prion-infected beef in the UK and France. A common feature of all of these diseases is the conversion of a host-encoded sialoglycoprotein to a protease-resistant form as a consequence of infection.

Human prion diseases are unique in that they manifest as sporadic, genetic, and infectious diseases. The study of prion biology is an important emerging area of biomedical investigation, and much remains to be learned.

PROKARYOTES

The primary distinguishing characteristics of the prokaryotes are their relatively small size, usually on the order of 1 μm in diameter, and the absence of a nuclear membrane. The DNA of almost all bacteria is a circle with a length of about 1 mm; this is the prokaryotic chromosome. Most prokaryotes have only a single chromosome. The chromosomal DNA must be folded more than 1000-fold just to fit within the prokaryotic cell membrane. Substantial evidence suggests that the folding may be orderly and may bring specified regions of the DNA into proximity. The specialized region of the cell containing DNA is termed the **nucleoid** and can be visualized by electron microscopy as well as by light microscopy after treatment of the cell to make the nucleoid visible. Thus, it would be a mistake to conclude that subcellular differentiation, clearly demarcated by membranes in eukaryotes, is lacking

TABLE 1–1   Common Human & Animal Prion Diseases

Type	Name	Etiology
Human prion diseases		
Acquired	Variant Creutzfeldt-Jakob disease[a]	Associated with ingestion or inoculation of prion-infected material
	Kuru	
	Iatrogenic Creutzfeldt-Jakob disease[b]	
Sporadic	Creutzfeldt-Jakob disease	Source of infection unknown
Familial	Gerstmann-Sträussler-Scheinker	Associated with specific mutations within the gene encoding PrP
	Fatal familial insomnia	
	Creutzfeldt-Jakob disease	
Animal prion diseases		
Cattle	Bovine spongiform encephalopathy	Exposure to prion-contaminated meat and bone meal
Sheep	Scrapie	Ingestion of scrapie-contaminated material
Deer, elk	Chronic wasting disease	Ingestion of prion-contaminated material
Mink	Transmissible mink encephalopathy	Source of infection unknown
Cats	Feline spongiform encephalopathy[a]	Exposure to prion-contaminated meat and bone meal

[a]Associated with exposure to BSE-contaminated materials.

[b]Associated with prion-contaminated biologic materials, such as dura mater grafts, corneal transplants, and cadaver-derived human growth hormone, or prion-contaminated surgical instruments.

Reproduced with permission from ASM News 3:570, Dec, 2008.

in prokaryotes. Indeed, some prokaryotes form membrane-bound subcellular structures with specialized function such as the chromatophores of photosynthetic bacteria (see Chapter 2).

Prokaryotic Diversity

The small size of the prokaryotic chromosome limits the amount of genetic information it can contain. Recent data based on genome sequencing indicate that the number of genes within a prokaryote may vary from 468 in *Mycoplasma genitalium* to 7825 in *Streptomyces coelicolor*, and many of these genes must be dedicated to essential functions such as energy generation, macromolecular synthesis, and cellular replication. Any one prokaryote carries relatively few genes that allow physiologic accommodation of the organism to its environment. The range of potential prokaryotic environments is unimaginably broad, and it follows that the prokaryotic group encompasses a heterogeneous range of specialists, each adapted to a rather narrowly circumscribed niche.

The range of prokaryotic niches is illustrated by consideration of strategies used for generation of metabolic energy. Light from the sun is the chief source of energy for life. Some prokaryotes such as the purple bacteria convert light energy to metabolic energy in the absence of oxygen production. Other prokaryotes, exemplified by the blue-green bacteria (**cyanobacteria**), produce oxygen that can provide energy through respiration in the absence of light. **Aerobic organisms** depend upon respiration with oxygen for their energy. Some **anaerobic organisms** can use electron acceptors other than oxygen in respiration. Many anaerobes carry out **fermentations** in which energy is derived by metabolic rearrangement of chemical growth substrates. The tremendous chemical range of potential growth substrates for aerobic or anaerobic growth is mirrored in the diversity of prokaryotes that have adapted to their utilization.

Prokaryotic Communities

A useful survival strategy for specialists is to enter into **consortia**, arrangements in which the physiologic characteristics of different organisms contribute to survival of the group as a whole. If the organisms within a physically interconnected community are directly derived from a single cell, the community is a **clone** that may contain up to 10^8 cells. The biology of such a community differs substantially from that of a single cell. For example, the high cell number virtually assures the presence within the clone of at least one cell carrying a variant of any gene on the chromosome. Thus, genetic variability—the wellspring of the evolutionary process called natural selection—is assured within a clone. The high number of cells within clones also is likely to provide physiologic protection to at least some members of the group. Extracellular polysaccharides, for example, may afford protection against potentially lethal agents such as antibiotics or heavy metal ions. Large amounts of polysaccharides produced by the high number of cells within a clone may allow cells within the interior to survive exposure to a lethal agent at a concentration that might kill single cells.

Many bacteria exploit a cell-cell communication mechanism called **quorum sensing** to regulate the transcription of genes involved in diverse physiologic processes, including bioluminescence, plasmid conjugal transfer, and the production of virulence determinants. Quorum sensing depends on the production of one or more diffusible signal molecules termed, **autoinducers** or **pheromones**, which enable a bacterium to monitor its own cell population density. It is an example of multicellular behavior in prokaryotes.

A distinguishing characteristic of prokaryotes is their capacity to exchange small packets of genetic information. This information may be carried on **plasmids**, small and specialized genetic elements that are capable of replication within at least one prokaryotic cell line. In some cases, plasmids may be transferred from one cell to another and thus may carry sets of specialized genetic information through a population. Some plasmids exhibit a **broad host range** that allows them to convey sets of genes to diverse organisms. Of particular concern are **drug resistance plasmids** that may render diverse bacteria resistant to antibiotic treatment.

The survival strategy of a single prokaryotic cell line may lead to a range of interactions with other organisms. These may include symbiotic relationships illustrated by complex nutritional exchanges among organisms within the human gut. These exchanges benefit both the microorganisms and their human host. Parasitic interactions can be quite deleterious to the host. Advanced symbiosis or parasitism can lead to loss of functions that may not allow growth of the symbiont or parasite independent of its host.

The **mycoplasmas**, for example, are parasitic prokaryotes that have lost the ability to form a cell wall. Adaptation of these organisms to their parasitic environment has resulted in incorporation of a substantial quantity of cholesterol into their cell membranes. Cholesterol, not found in other prokaryotes, is assimilated from the metabolic environment provided by the host. Loss of function is exemplified also by obligate intracellular parasites, the **chlamydiae** and **rickettsiae**. These bacteria are extremely small (0.2–0.5 µm in diameter) and depend upon the host cell for many essential metabolites and coenzymes. This loss of function is reflected by the presence of a smaller genome with fewer genes (see Table 7–1).

The most widely distributed examples of bacterial symbionts appear to be chloroplasts and mitochondria, the energy-yielding organelles of eukaryotes. A substantial body of evidence points to the conclusion that ancestors of these organelles were **endosymbionts**, prokaryotes that established symbiosis within the cell membrane of the ancestral eukaryotic host. The presence of multiple copies of the organelles may have contributed to the relatively large size of eukaryotic cells and to their capacity for specialization, a trait ultimately reflected in the evolution of differentiated multicellular organisms.

Classification of the Prokaryotes

An understanding of any group of organisms requires their **classification**. An appropriate classification system allows a scientist to choose characteristics that allow swift and accurate categorization of a newly encountered organism. The categorization allows prediction of many additional traits shared by other members of the category. In a hospital setting, successful classification of a pathogenic organism may provide the most direct route to its elimination. Classification may also provide a broad understanding of relationships among different organisms, and such information may have great practical value. For example, elimination of a pathogenic organism will be relatively long-lasting if its habitat is occupied by a nonpathogenic variant.

The principles of prokaryotic classification are discussed in Chapter 3. At the outset it should be recognized that any prokaryotic characteristic might serve as a potential criterion for classification. However, not all criteria are equally effective in grouping organisms. Possession of DNA, for example, is a useless criterion for distinguishing organisms because all cells contain DNA. The presence of a broad host range plasmid is not a useful criterion because such plasmids may be found in diverse hosts and need not be present all of the time. Useful criteria may be structural, physiologic, biochemical, or genetic. **Spores**—specialized cell structures that may allow survival in extreme environments—are useful structural criteria for classification because well-characterized subsets of bacteria form spores. Some bacterial groups can be effectively subdivided on the basis of their ability to ferment specified carbohydrates. Such criteria may be ineffective when applied to other bacterial groups that may lack any fermentative capability. A biochemical test, the **Gram stain**, is an effective criterion for classification because response to the stain reflects fundamental and complex differences in the bacterial cell surface that divide most bacteria into two major groups.

Genetic criteria are increasingly employed in bacterial classification, and many of these advances are made possible by the development of recombinant DNA technology. It is now possible to design DNA probes that swiftly identify organisms carrying specified genetic regions with common ancestry. Comparison of DNA sequences for some genes led to the elucidation of **phylogenetic relationships** among prokaryotes. Ancestral cell lines can be traced, and organisms can be grouped on the basis of their evolutionary affinities. These investigations have led to some striking conclusions. For example, comparison of cytochrome c sequences suggests that all eukaryotes, including humans, arose from one of three different groups of purple photosynthetic bacteria. This conclusion in part explains the evolutionary origin of eukaryotes, but it does not fully take into account the generally accepted view that the eukaryotic cell was derived from the evolutionary merger of different prokaryotic cell lines.

Bacteria & Archaebacteria: The Major Subdivisions within the Prokaryotes

A major success in molecular phylogeny has been the demonstration that prokaryotes fall into two major groups. Most investigations have been directed to one group, the bacteria. The other group, the archaebacteria, has received relatively little attention until recently, in part because many of its representatives are difficult to study in the laboratory. Some archaebacteria, for example, are killed by contact with oxygen, and others grow at temperatures exceeding that of boiling water. Before molecular evidence became available, the major subgroupings of archaebacteria seemed disparate. The methanogens carry out an anaerobic respiration that gives rise to methane; the halophiles demand extremely high salt concentrations for growth; and the thermoacidophiles require high temperature and acidity. It has now been established that these prokaryotes share biochemical traits such as cell wall or membrane components that set the group entirely apart from all other living organisms. An intriguing trait shared by archaebacteria and eukaryotes is the presence of **introns** within genes. The function of introns—segments of DNA that interrupt informational DNA within genes—is not established. What is known is that introns represent a fundamental characteristic shared by the DNA of archaebacteria and eukaryotes. This common trait has led to the suggestion that—just as mitochondria and chloroplasts appear to be evolutionary derivatives of the bacteria—the eukaryotic nucleus may have arisen from an archaebacterial ancestor.

PROTISTS

The "true nucleus" of eukaryotes (from Gr *karyon* "nucleus") is only one of their distinguishing features. The membrane-bound organelles, the microtubules, and the microfilaments of eukaryotes form a complex intracellular structure unlike that found in prokaryotes. The agents of motility for eukaryotic cells are flagella or cilia—complex multistranded structures that do not resemble the flagella of prokaryotes. Gene expression in eukaryotes takes place through a series of events achieving physiologic integration of the nucleus with the endoplasmic reticulum, a structure that has no counterpart in prokaryotes. Eukaryotes are set apart by the organization of their cellular DNA in chromosomes separated by a distinctive mitotic apparatus during cell division.

In general, genetic transfer among eukaryotes depends upon fusion of **haploid gametes** to form a **diploid** cell containing a full set of genes derived from each gamete. The life cycle of many eukaryotes is almost entirely in the diploid state, a form not encountered in prokaryotes. Fusion of gametes to form reproductive progeny is a highly specific event and establishes the basis for eukaryotic **species**. This term can be applied only metaphorically to the prokaryotes, which exchange fragments of DNA through recombination. Taxonomic groupings of eukaryotes frequently are based on shared **morphologic properties**, and it is noteworthy that many taxonomically useful determinants are those associated with reproduction. Almost all successful eukaryotic species are those in which closely related cells, members of the same species, can recombine to form viable offspring. Structures that contribute directly or indirectly to the reproductive event tend to be highly developed and—with minor modifications among closely related species—extensively conserved.

Microbial eukaryotes—**protists**—are members of the four following major groups: algae, protozoa, fungi, and slime molds. It should be noted that these groupings are not necessarily phylogenetic: Closely related organisms may have been categorized separately because underlying biochemical and genetic similarities may not have been recognized.

Algae

The term "algae" has long been used to denote all organisms that produce O_2 as a product of photosynthesis. One major subgroup of these organisms—the blue-green bacteria, or cyanobacteria—are prokaryotic and no longer are termed algae. This classification is reserved exclusively for photosynthetic eukaryotic organisms. All algae contain chlorophyll in the photosynthetic membrane of their subcellular chloroplast. Many algal species are unicellular microorganisms. Other algae may form extremely large multicellular structures. Kelps of brown algae sometimes are several hundred meters in length. A number of algae produce toxins that are poisonous to humans and other animals. Dinoflagellates, a unicellular algae, cause algal blooms, or red tides, in the ocean (Figure 1–2). Red tides caused by the dinoflagellate *Gonyaulax* species are serious as this organism produces neurotoxins such as **saxitoxin** and **gonyautoxins**, which accumulate in shellfish (eg, clams, mussels, scallops, and oysters) that feed on this organism. Ingestion of these shellfish by

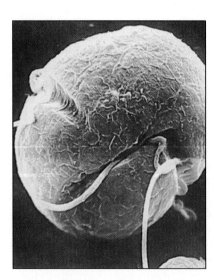

FIGURE 1–2 The dinoflagellate *Gymnodinium* scanning electron micrograph (4000×). (Reproduced with permission from David M. Phillips/Visuals Unlimited.)

humans results in symptoms of **paralytic shellfish poisoning** and can lead to death.

Protozoa

Protozoa are unicellular nonphotosynthetic protists. The most primitive protozoa appear to be flagellated forms that in many respects resemble representatives of the algae. It seems likely that the ancestors of these protozoa were algae that became **heterotrophs**—the nutritional requirements of such organisms are met by organic compounds. Adaptation to a heterotrophic mode of life was sometimes accompanied by loss of chloroplasts, and algae thus gave rise to the closely related protozoa. Similar events have been observed in the laboratory to be the result of either mutation or physiologic adaptation.

From flagellated protozoa appear to have evolved the ameboid and the ciliated types; intermediate forms are known that have flagella at one stage in the life cycle and pseudopodia (characteristic of the ameba) at another stage. A fourth major group of protozoa, the sporozoa, are strict parasites that are usually immobile; most of which reproduce sexually and asexually in alternate generations by means of spores. Protozoan parasites of humans are discussed in Chapter 46.

Fungi

The fungi are nonphotosynthetic protists growing as a mass of branching, interlacing filaments ("hyphae") known as a **mycelium**. Although the hyphae exhibit cross walls, the cross walls are perforated and allow free passage of nuclei and cytoplasm. The entire organism is thus a coenocyte (a multinucleated mass of continuous cytoplasm) confined within a series of branching tubes. These tubes, made of polysaccharides such as chitin, are homologous with cell walls. The mycelial forms are called **molds**; a few types, **yeasts**, do not form a mycelium but are easily recognized as fungi by the nature of their sexual reproductive processes and by the presence of transitional forms.

The fungi probably represent an evolutionary offshoot of the protozoa; they are unrelated to the actinomycetes, mycelial bacteria that they superficially resemble. The major subdivisions (phyla) of fungi are: Chytridiomycota, Zygomycota (the zygomycetes), Ascomycota (the ascomycetes), Basidiomycota (the basidiomycetes), and the "deuteromycetes" (or imperfect fungi).

The evolution of the ascomycetes from the phycomycetes is seen in a transitional group, members of which form a zygote but then transform this directly into an ascus. The basidiomycetes are believed to have evolved in turn from the ascomycetes. The classification of fungi and their medical significance are discussed further in Chapter 45.

Slime Molds

These organisms are characterized by the presence, as a stage in their life cycle, of an ameboid multinucleate mass of

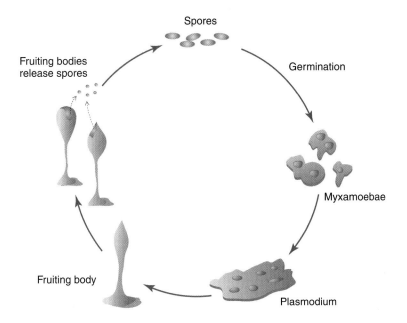

A

B

FIGURE 1–3 Slime molds. **A**: Life cycle of an acellular slime mold. **B**: Fruiting body of a cellular slime mold. (Reproduced with permission from Carolina Biological Supply/Phototake.)

cytoplasm called a **plasmodium**. The plasmodium of a slime mold is analogous to the mycelium of a true fungus. Both are coenocytic. In the latter, cytoplasmic flow is confined to the branching network of chitinous tubes, whereas in the former the cytoplasm can flow in all directions. This flow causes the plasmodium to migrate in the direction of its food source, frequently bacteria. In response to a chemical signal, 3′, 5′-cyclic AMP (see Chapter 7), the plasmodium, which reaches macroscopic size, differentiates into a stalked body that can produce individual motile cells. These cells, flagellated or ameboid, initiate a new round in the life cycle of the slime mold (Figure 1–3). The cycle frequently is initiated by sexual fusion of single cells.

The life cycle of the slime molds illustrates a central theme of this chapter: the interdependency of living forms. The growth of slime molds depends upon nutrients provided by bacterial or, in some cases, plant cells. Reproduction of the slime molds via plasmodia can depend upon intercellular recognition and fusion of cells from the same species. Full understanding of a microorganism requires both knowledge of the other organisms with which it coevolved and an appreciation of the range of physiologic responses that may contribute to survival.

REVIEW QUESTIONS

1. Which one of the following terms characterizes the interaction between a fungus and algae in a lichen?
   (A) Parasitism
   (B) Symbiosis
   (C) Endosymbiosis
   (D) Endoparasitism
   (E) Consortia

2. Which one of the following agents lacks nucleic acid?
   (A) Bacteria
   (B) Viruses
   (C) Viroids
   (D) Prions
   (E) Protozoa

3. Which one of the following is not a protist?
   (A) Bacteria
   (B) Algae
   (C) Protozoa
   (D) Fungi
   (E) Slime molds

4. Which one of the following agents simultaneously contains both DNA and RNA?
   (A) Bacteria
   (B) Viruses
   (C) Viroids
   (D) Prions
   (E) Plasmids

5. A 65-year-old man develops dementia, progressive over several months, along with ataxia and somnolence. An electroencephalographic pattern shows paroxysms with high voltages and slow waves, suggestive of Creutzfeldt–Jakob disease. This disease is caused by which of the following agents?
   (A) Bacterium
   (B) Virus
   (C) Viroid
   (D) Prion
   (E) Plasmid

Answers

1. B	3. A	5. D
2. D	4. A	

REFERENCES

Belay ED: Transmissible spongiform encephalopathies in humans. Annu Rev Microbiol 1999;53:283. [PMID: 10547693]

Diener TO: Viroids and the nature of viroid diseases. Arch Virol 1999;15(Suppl):203.

Lederberg J (editor): *Encyclopedia of Microbiology,* 4 vols. Academic Press, 1992.

Olsen GJ, Woese CR: The winds of (evolutionary) change: Breathing new life into microbiology. J Bacteriol 1994;176:1. [PMID: 8282683]

Pelczar MJ Jr, Chan ECS, Krieg NR: *Microbiology: Concepts and Applications.* McGraw-Hill, 1993.

Priola SA: How animal prions cause disease in humans. Microbe 2008;3:568.

Prusiner SB: Biology and genetics of prion diseases. Annu Rev Microbiol 1994;48:655.

Reisser W (editor): *Algae and Symbiosis: Plants, Animals, Fungi, Viruses, Interactions Explored.* Biopress, 1992.

Schloss PD, Handlesman J: Status of the microbial census. Microbiol Mol Biol Rev 2004;68:686.

Sleigh MA: *Protozoa and Other Protists.* Chapman & Hall, 1990.

Whitman WB, Coleman DC, Wiebe WJ: Prokaryotes: The unseen majority. Proc Natl Acad Sci USA 1998;95:6578. [PMID: 7826022]

Cell Structure

INTRODUCTION

In this chapter we discuss the basic structure and function of the components that make up eukaryotic and prokaryotic cells. The chapter begins with a discussion of the microscope. Historically, it was the microscope that first revealed the presence of bacteria and later, the secrets of cell structure. Today, it remains a powerful tool in cell biology.

OPTICAL METHODS

The Light Microscope

The resolving power of the light microscope under ideal conditions is about half the wavelength of the light being used. (**Resolving power** is the distance that must separate two point sources of light if they are to be seen as two distinct images.) With yellow light of a wavelength of 0.4 µm, the smallest separable diameters are thus about 0.2 µm, ie, one-third the width of a typical prokaryotic cell. The useful magnification of a microscope is the magnification that makes visible the smallest resolvable particles. Several types of light microscopes are commonly used in microbiology:

A. Bright-Field Microscope

The bright-field microscope is most commonly used in microbiology courses and consists of two series of lenses (**objective** and **ocular lens**), which function together to resolve the image. These microscopes generally employ a 100-power objective lens with a 10-power ocular lens, thus magnifying the specimen 1000 times. Particles 0.2 µm in diameter are therefore magnified to about 0.2 mm and so become clearly visible. Further magnification would give no greater resolution of detail and would reduce the visible area (**field**).

With this microscope, specimens are rendered visible because of the differences in **contrast** between them and the surrounding medium. Many bacteria are difficult to see well because of their lack of contrast with the surrounding medium. Dyes (stains) can be used to stain cells or their organelles and increase their contrast so that they can be more easily seen in the bright-field microscope.

B. Phase Contrast Microscope

The phase contrast microscope was developed to improve contrast differences between cells and the surrounding medium, making it possible to see living cells without staining them; with bright-field microscopes, killed and stained preparations must be used. The phase contrast microscope takes advantage of the fact that light waves passing through transparent objects, such as cells, emerge in different phases depending on the properties of the materials through which they pass. This effect is amplified by a special ring in the objective lens of a phase contrast microscope, leading to the formation of a dark image on a light background.

C. Dark-Field Microscope

The dark-field microscope is a light microscope in which the lighting system has been modified to reach the specimen from the sides only. This is accomplished through the use of a special condenser that both blocks direct light rays and deflects light off a mirror on the side of the condenser at an oblique angle. This creates a "dark field" that contrasts against the highlighted edge of the specimens and results when the oblique rays are reflected from the edge of the specimen upward into the objective of the microscope. Resolution by dark-field microscopy is quite high. Thus, this technique has been particularly useful for observing organisms such as *Treponema pallidum*, a spirochete which is less than 0.2 µm in diameter and therefore cannot be observed with a bright-field or phase contrast microscope (Figure 2–1A).

D. Fluorescence Microscope

The fluorescence microscope is used to visualize specimens that **fluoresce**, which is the ability to absorb short wavelengths of light (ultraviolet) and give off light at a longer wavelength (visible). Some organisms fluoresce naturally because of the presence within the cells of naturally fluorescent substances such as chlorophyll. Those that do not naturally fluoresce may be stained with a group of fluorescent dyes called **fluorochromes**. Fluorescence microscopy is widely used in clinical diagnostic microbiology. For example, the fluorochrome auramine O, which glows yellow when exposed to ultraviolet light, is strongly absorbed by *Mycobacterium tuberculosis*, the bacterium that

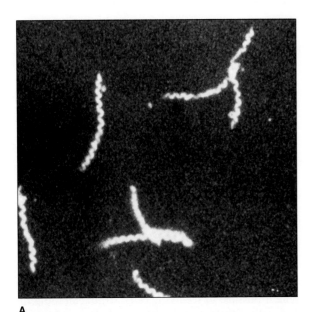

A

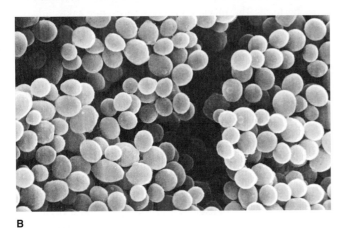

B

FIGURE 2–1 **A**: Positive dark-field examination. Treponemes are recognizable by their characteristic corkscrew shape and deliberate forward and backward movement with rotation about the longitudinal axis. (Reproduced with permission from Charles Stratton/Visuals Unlimited.) **B**: Scanning electron microscope of bacteria—*Staphylococcus aureus* (32,000×). (Reproduced with permission from David M. Phillips/Photo Researchers, Inc.)

causes tuberculosis. When the dye is applied to a specimen suspected of containing *M tuberculosis* and exposed to ultraviolet light, the bacterium can be detected by the appearance of bright yellow organisms against a dark background.

The principal use of fluorescence microscopy is a diagnostic technique called the **fluorescent-antibody (FA) technique** or **immunofluorescence**. In this technique, specific antibodies (eg, antibodies to *Legionella pneumophila*) are chemically labeled with a fluorochrome such as **fluorescein isothiocyanate (FITC)**. These fluorescent antibodies are then added to a microscope slide containing a clinical specimen. If the specimen contains *L pneumophila*, the fluorescent antibodies will bind to antigens on the surface of the bacterium, causing it to fluoresce when exposed to ultraviolet light.

E. Differential Interference Contrast (DIC) Microscope

Differential interference contrast microscopes employ a polarizer to produce polarized light. The polarized light beam passes through a prism that generates two distinct beams; these beams pass through the specimen and enter the objective lens where they are recombined into a single beam. Because of slight differences in refractive index of the substances each beam passed through, the combined beams are not totally in phase but instead create an interference effect, which intensifies subtle differences in cell structure. Structures such as spores, vacuoles, and granules appear three dimensional. DIC microscopy is particularly useful for observing unstained cells because of its ability to generate images that reveal internal cell structures that are less apparent by bright-field techniques.

The Electron Microscope

The high resolving power of the electron microscope has enabled scientists to observe the detailed structures of prokaryotic and eukaryotic cells. The superior resolution of the electron microscope is due to the fact that electrons have a much shorter wavelength than the photons of white light.

There are two types of electron microscopes in general use: the **transmission electron microscope (TEM)**, which has many features in common with the light microscope, and the **scanning electron microscope (SEM)**. The TEM was the first to be developed and employs a beam of electrons projected from an electron gun and directed or focused by an electromagnetic condenser lens onto a thin specimen. As the electrons strike the specimen, they are differentially scattered by the number and mass of atoms in the specimen; some electrons pass through the specimen and are gathered and focused by an electromagnetic objective lens, which presents an image of the specimen to the projector lens system for further enlargement. The image is visualized by allowing it to impinge on a screen that fluoresces when struck with the electrons. The image can be recorded on photographic film. TEM can resolve particles 0.001 μm apart. Viruses, with diameters of 0.01–0.2 μm, can be easily resolved.

The SEM generally has a lower resolving power than the TEM; however, it is particularly useful for providing three-dimensional images of the surface of microscopic objects. Electrons are focused by means of lenses into a very fine point. The interaction of electrons with the specimen results in the release of different forms of radiation (eg, secondary electrons) from the surface of the material, which can be captured by an appropriate detector, amplified, and then imaged on a television screen (Figure 2–1B).

An important technique in electron microscopy is the use of "shadowing." This involves depositing a thin layer of heavy metal (such as platinum) on the specimen by placing it in the path of a beam of metal ions in a vacuum. The beam is

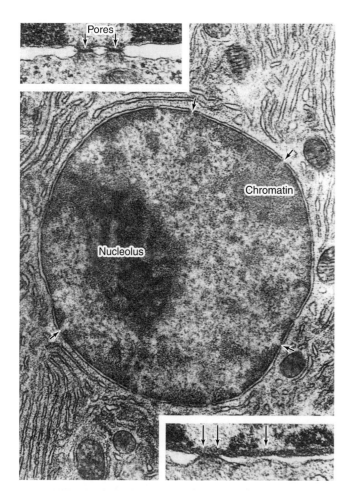

FIGURE 2–2 Electron micrograph of a thin section of a typical eukaryotic nucleus showing a prominent nucleolus and large aggregations of heterochromatin against the nuclear membrane, which is traversed by pores (at arrows). **Inset upper left**: Two nuclear pores and their pore diaphragms. **Inset lower right**: The fibrous lamina present in the inner aspect of the nuclear envelope. Several mitochondria are visible in the cytoplasm. (Reproduced with permission from Fawcett DW: *Bloom and Fawcett, A Textbook of Histology,* 12th ed. Copyright © 1994. By permission of Chapman & Hall, New York, NY.)

directed at a low angle to the specimen, so that it acquires a "shadow" in the form of an uncoated area on the other side. When an electron beam is then passed through the coated preparation in the electron microscope and a positive print is made from the "negative" image, a three-dimensional effect is achieved (eg, see Figure 2–22).

Other important techniques in electron microscopy include the use of ultrathin sections of embedded material, a method of freeze-drying specimens that prevents the distortion caused by conventional drying procedures, and the use of negative staining with an electron-dense material such as phosphotungstic acid or uranyl salts (eg, see Figure 42–1). Without these heavy metal salts, there would not be enough contrast to detect the details of the specimen.

Confocal Scanning Laser Microscope

The **confocal scanning laser microscope (CSLM)** couples a laser light source to a light microscope. In confocal scanning laser microscopy, a laser beam is bounced off a mirror that directs the beam through a scanning device. Then the laser beam is directed through a pinhole that precisely adjusts the plane of focus of the beam to a given vertical layer within the specimen. By precisely illuminating only a single plane of the specimen, illumination intensity drops off rapidly above and below the plane of focus, and stray light from other planes of focus are minimized. Thus, in a relatively thick specimen, various layers can be observed by adjusting the plane of focus of the laser beam.

Cells are often stained with fluorescent dyes to make them more visible. Alternatively, false color images can be generated by adjusting the microscope in such a way as to make different layers take on different colors. The CSLM is equipped with computer software to assemble digital images for subsequent image processing. Thus, images obtained from different layers can be stored and then digitally overlaid to reconstruct a three-dimensional image of the entire specimen.

Scanning Probe Microscopes

A new class of microscopes, called **scanning probe microscopes**, measure surface features by moving a sharp probe over the object's surface. The **scanning tunneling microscope** and the **atomic force microscope** are examples of this new class of microscopes, which enable scientists to view atoms or molecules on the surface of a specimen. For example, interactions between proteins of the bacterium *Escherichia coli* can be studied with the atomic force microscope.

EUKARYOTIC CELL STRUCTURE

The Nucleus

The **nucleus** contains the cell's genome. It is bounded by a membrane that consists of a pair of unit membranes separated by a space of variable thickness. The inner membrane is usually a simple sac, but the outermost membrane is, in many places, continuous with the endoplasmic reticulum. The **nuclear membrane** exhibits selective permeability due to pores, which consist of a complex of several proteins whose function is to import substances into and export substances out of the nucleus. The chromosomes of eukaryotic cells contain linear DNA macromolecules arranged as a double helix. They are only visible with a light microscope when the cell is undergoing division and the DNA is in a highly condensed form; at other times, the chromosomes are not condensed and appear as in Figure 2–2. Eukaryotic DNA macromolecules are associated with basic proteins called **histones** that bind to the DNA by ionic interactions.

A structure often visible within the nucleus is the **nucleolus**, an area rich in RNA that is the site of ribosomal RNA synthesis (Figure 2–2). Ribosomal proteins synthesized in the cytoplasm are transported into the nucleolus and combine with ribosomal RNA to form the small and large subunits of the eukaryotic ribosome. These are then exported to the cytoplasm where they associate to form an intact ribosome that can function in protein synthesis.

Cytoplasmic Structures

The cytoplasm of eukaryotic cells is characterized by the presence of an endoplasmic reticulum, vacuoles, self-reproducing plastids, and an elaborate cytoskeleton composed of microtubules, microfilaments, and intermediate filaments.

The **endoplasmic reticulum (ER)** is a network of membrane-bound channels continuous with the nuclear membrane. Two types of endoplasmic reticulum are recognized: **rough**, which contains attached 80S ribosomes, and **smooth**, which does not (Figure 2–2). Rough ER is a major producer of glycoproteins and also produces new membrane material that is transported throughout the cell; smooth ER participates in the synthesis of lipids and in some aspects of carbohydrate metabolism. The **Golgi apparatus** consists of a stack of membranes that function in concert with the ER to chemically modify and sort products of the ER into those destined to be secreted and those that function in other membranous structures of the cell.

The plastids include **mitochondria** and **chloroplasts**. Several lines of evidence suggest that mitochondria and chloroplasts were descendents of ancient prokaryotic organisms and arose from the engulfment of a prokaryotic cell by a larger cell (**endosymbiosis**). Mitochondria are of prokaryotic size, and its membrane, which lacks sterols, is much less rigid than the eukaryotic cell's cytoplasmic membrane, which does contain sterols. Mitochondria contain two sets of membranes. The outermost membrane is rather permeable having numerous minute channels that allow passage of ions and small molecules (eg, ATP). Invagination of the outer membrane forms a system of inner folded membranes called **cristae**. The cristae are the sites of enzymes involved in respiration and ATP production. Cristae also contain specific transport proteins that regulate passage of metabolites into and out of the mitochondrial **matrix**. The matrix contains a number of enzymes, in particular those of the citric acid cycle. Chloroplasts are photosynthetic cell organelles that are capable of converting the energy of sunlight into chemical energy through photosynthesis. Chlorophyll and all other components needed for photosynthesis are located in a series of flattened membrane discs called **thylakoids**. The size, shape, and number of chloroplasts per cell vary markedly; in contrast to mitochondria, chloroplasts are generally much larger than prokaryotes. Mitochondria and chloroplasts contain their own DNA, which exists in a covalently closed circular form and codes for some (not all) of their constituent proteins and transfer RNAs. Mitochondria and chloroplasts also contain 70S ribosomes, the same as those of prokaryotes.

Some eukaryotic microorganisms (eg, *Trichomonas vaginalis*) lack mitochondria and contain instead a membrane-enclosed respiratory organelle called the **hydrogenosome**. Hydrogenosomes may have arisen by endosymbiosis and some have been identified that contain DNA and ribosomes. The hydrogenosome, while similar in size to mitochondria, lacks cristae and the enzymes of the tricarboxylic acid cycle. Pyruvate is taken up by the hydrogenosome and H_2, CO_2, acetate, and ATP are produced.

Lysosomes are membrane-enclosed sacs that contain various digestive enzymes that the cell uses to digest macromolecules such as proteins, fats, and polysaccharides. The lysosome allows these enzymes to be partitioned away from the cytoplasm proper where they could destroy key cellular macromolecules if not contained. Following the hydrolysis of macromolecules in the lysosome, the resulting monomers pass from the lysosome into the cytoplasm where they serve as nutrients.

The **peroxisome** is a membrane-enclosed structure whose function is to produce H_2O_2 from the reduction of O_2 by various hydrogen donors. The H_2O_2 produced in the peroxisome is subsequently degraded to H_2O and O_2 by the enzyme **catalase**.

The **cytoskeleton** is a three-dimensional structure that fills the cytoplasm. The primary types of fibers comprising the cytoskeleton are **microfilaments**, **intermediate filaments**, and **microtubules**. Microfilaments are about 3–6 nm in diameter and are polymers composed of subunits of the protein **actin**. These fibers form scaffolds throughout the cell defining and maintaining the shape of the cell. Microfilaments can also carry out cellular movements including gliding, contraction, and cytokinesis.

Microtubules are cylindrical tubes, 20–25 nm in diameter and are composed of subunits of the protein **tubulin**. Microtubules assist microfilaments in maintaining cell structure, form the spindle fibers for separating chromosomes during mitosis, and also play an important role in cell motility. Intermediate filaments are about 10 nm in diameter and provide tensile strength for the cell.

Surface Layers

The cytoplasm is enclosed within a plasma membrane composed of protein and phospholipid, similar to the prokaryotic cell membrane illustrated later (see Figure 2–10). Most animal cells have no other surface layers; however, plant cells have an outer cell wall composed of cellulose. Many eukaryotic microorganisms also have an outer **cell wall**, which may be composed of a polysaccharide such as cellulose or chitin or may be inorganic, eg, the silica wall of diatoms.

Motility Organelles

Many eukaryotic microorganisms have organelles called **flagella** (eg, *Trichomonas vaginalis*) or **cilia** (eg, *Balantidium coli*) that move with a wave-like motion to propel the cell through

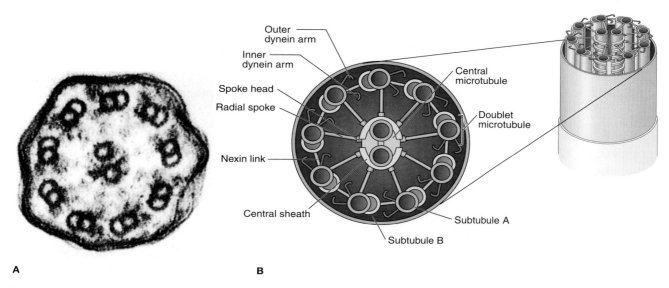

A **B**

FIGURE 2–3 Cilia and Flagella Structure. A: An electron micrograph of a cilium cross section. Note the two central microtubles surrounded by nine microtubule doublets (160,000x). (Reproduced with permission from KG Murti/Visuals Unlimited.) B: A diagram of cilia and flagella structure. (Reproduced with permission from Willey JM, Sherwood LM, Woolverton CJ (eds): Prescott, Harley, and Klein's Microbiology, 7th edt. New York: Mcgraw-Hill; 2008.)

water. Eukaryotic flagella emanate from the polar region of the cell, whereas cilia, which are shorter than flagella, surround the cell. Both the flagella and the cilia of eukaryotic cells have the same basic structure and biochemical composition. Both consist of a series of microtubules, hollow protein cylinders composed of a protein called **tubulin**, surrounded by a membrane. The arrangement of the microtubules is called the "9 + 2 system" because it consists of nine peripheral pairs of microtubules surrounding two single central microtubules (Figure 2–3).

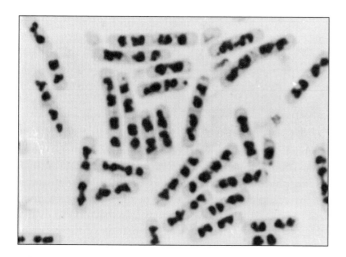

FIGURE 2–4 Nucleoids of *Bacillus cereus* (2500×). (Reproduced with permission from Robinow C: Bacteriol Rev 1956;20:207.)

PROKARYOTIC CELL STRUCTURE

The prokaryotic cell is simpler than the eukaryotic cell at every level, with one exception: The cell envelope is more complex.

The Nucleoid

Prokaryotes have no true nuclei; instead they package their DNA in a structure known as the **nucleoid**. The nucleoid can be seen with the light microscope in stained material (Figure 2–4). It is Feulgen-positive, indicating the presence of DNA. The negatively charged DNA is at least partially neutralized by small polyamines and magnesium ions, but histone-like proteins exist in bacteria and presumably play a role similar to that of histones in eukaryotic chromatin.

Electron micrographs of a typical prokaryotic cell such as Figure 2–5 reveal the absence of a nuclear membrane and a mitotic apparatus. The exception to this rule is the planctomycetes, a divergent group of aquatic bacteria, which have a nucleoid surrounded by a nuclear envelope consisting of two membranes. The distinction between prokaryotes and eukaryotes that still holds is that prokaryotes have no eukaryotic-type mitotic apparatus. The nuclear region (Figure 2–5) is filled with DNA fibrils. The nucleoid of most bacterial cells consists of a single continuous circular molecule ranging in size from 0.58 to almost 10 million base pairs. However, a few bacteria have been shown to have two, three, or even four dissimilar chromosomes. For example, *Vibrio cholerae* and *Brucella melitensis* have two dissimilar chromosomes. There are exceptions to this rule of circularity because some prokaryotes (eg, *Borrelia burgdorferi* and

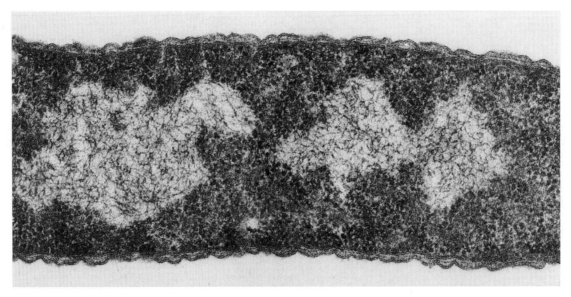

FIGURE 2–5 Thin section of *Escherichia coli* cell fixed with osmium tetroxide and postfixed with aqueous uranyl acetate showing two nuclear regions filled with DNA fibrils. (Reproduced with permission from Robinow C, Kellenberger E: Microbiol Rev 1994;58:211.)

Streptomyces coelicolor) have been shown to have a linear chromosome.

In bacteria, the number of nucleoids, and therefore the number of chromosomes, depends on the growth conditions (Figure 2–4). Rapidly growing bacteria have more nucleoids per cell than slowly growing ones; however, when multiple copies are present they are all the same (ie, prokaryotic cells are **haploid**).

Cytoplasmic Structures

Prokaryotic cells lack autonomous plastids, such as mitochondria and chloroplasts; the electron transport enzymes are localized instead in the cytoplasmic membrane. The photosynthetic pigments (carotenoids, bacteriochlorophyll) of photosynthetic bacteria are contained in intracytoplasmic membrane systems of various morphologies. Membrane vesicles (**chromatophores**) or lamellae are commonly observed membrane types. Some photosynthetic bacteria have specialized nonunit membrane-enclosed structures called **chlorosomes**. In some cyanobacteria (formerly known as blue-green algae), the photosynthetic membranes often form multilayered structures known as thylakoids (Figure 2–6). The major accessory pigments used for light harvesting are the phycobilins found on the outer surface of the thylakoid membranes.

Bacteria often store reserve materials in the form of insoluble granules, which appear as refractile bodies in the cytoplasm when viewed in a phase contrast microscope. These so-called inclusion bodies almost always function in the storage of energy or as a reservoir of structural building blocks. Most cellular inclusions are bounded by a thin non-unit membrane consisting of lipid, which serves to separate the inclusion from the cytoplasm proper. One of the most common inclusion bodies consists of **poly-β-hydroxybutyric acid (PHB)**, a lipid-like compound consisting of chains of β-hydroxybutyric acid units connected through ester linkages. PHB is produced when the source of nitrogen, sulfur, or phosphorous is limited and there is excess carbon in the medium (Figure 2–7A). Another storage product formed by prokaryotes when carbon is in excess is **glycogen**, which is a polymer of glucose. PHB and glycogen are used as carbon sources when protein and nucleic acid synthesis are resumed. A variety of prokaryotes are capable of oxidizing reduced sulfur compounds such as hydrogen sulfide and thiosulfate, producing intracellular granules of elemental **sulfur (Figure 2–C)**. As the reduced sulfur source becomes limiting, the sulfur in the granules is oxidized, usually to sulfate, and the granules slowly disappear. Many bacteria accumulate large reserves of inorganic phosphate in the form of granules of **polyphosphate (Figure 2–7B)**. These granules can be degraded and used as sources of phosphate for nucleic acid and phospholipid synthesis to support growth. These granules are sometimes termed **volutin granules** or **metachromatic granules** because they stain red with a blue dye. They are characteristic features of the corynebacteria (Chapter 13).

Certain groups of autotrophic bacteria that fix carbon dioxide to make their biochemical building blocks contain polyhedral bodies surrounded by a protein shell (**carboxysomes**) containing the key enzyme of CO_2 fixation, **ribulose-bisphosphate carboxylase (Figure 2–6B)**. **Magnetosomes** are intracellular crystal particles of the iron mineral magnetite (Fe_3O_4) that allow certain aquatic bacteria to exhibit magnetotaxis (ie, migration or orientation of the cell with respect to the earth's magnetic field). Magnetosomes are surrounded by a nonunit membrane containing phospholipids, proteins, and glycoproteins. **Gas vesicles** are found almost

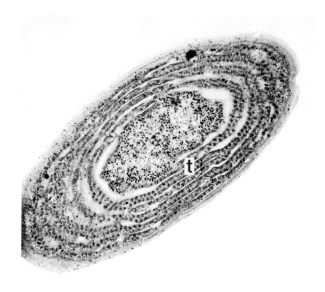

A

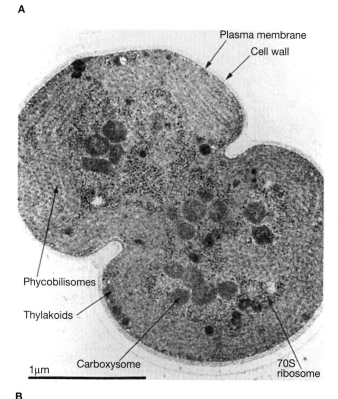

Plasma membrane
Cell wall

Phycobilisomes

Thylakoids

1μm   Carboxysome   70S
                                  ribosome

B

FIGURE 2–6 **A**: Thin section of *Synechococcus lividus* showing an extensive thylakoid system. The phycobillisomes lining these thylakoids are clearly visible as granules at location t (85,000x). (Reproduced with permission from Elizabeth Gentt/Visuals Unlimited.) **B**: Thin section of *Synechocystis* during division. Many structures are visible. (Reproduced with permission from Carlsberg Research Communications 42:77-98, 1977, Carlsberg Laboratories.)

exclusively in microorganisms from aquatic habitats, where they provide buoyancy. The gas vesicle membrane is a 2 nm thick layer of protein, impermeable to water and solutes but permeable to gases; thus, gas vesicles exist as gas-filled

structures surrounded by the constituents of the cytoplasm (Figure 2–8).

Bacteria contain proteins resembling both the actin and nonactin cytoskeletal proteins of eukaryotic cells as additional proteins that play cytoskeletal roles (Figure 2–9). Actin homologs (eg, MreB, Mbl) perform a variety of functions, helping to determine cell shape, segregate chromosomes, and localize proteins with the cell. Nonactin homologs (eg, FtsZ) and unique bacterial cytoskeletal proteins (eg, SecY, MinD) are involved in determining cell shape and in regulation of cell division and chromosome segregation.

The Cell Envelope

Prokaryotic cells are surrounded by complex envelope layers that differ in composition among the major groups. These structures protect the organisms from hostile environments, such as extreme osmolarity, harsh chemicals, and even antibiotics.

The Cell Membrane

A. Structure

The bacterial cell membrane, also called the cytoplasmic membrane, is visible in electron micrographs of thin sections (Figure 2–8). It is a typical "unit membrane" composed of phospholipids and upward of 200 different kinds of proteins. Proteins account for approximately 70% of the mass of the membrane, which is a considerably higher proportion than that of mammalian cell membranes. Figure 2–10 illustrates a model of membrane organization. The membranes of prokaryotes are distinguished from those of eukaryotic cells by the absence of sterols, the only exception being mycoplasmas that incorporate sterols, such as cholesterol, into their membranes when growing in sterol-containing media.

The cell membranes of the *Archaea* (see Chapter 1) differ from those of the *Bacteria*. Some Archaeal cell membranes contain unique lipids, **isoprenoids**, rather than fatty acids, linked to glycerol by an ether rather than an ester linkage. Some of these lipids have no phosphate groups, and therefore, they are not phospholipids. In other species the cell membrane is made up of a lipid monolayer consisting of long lipids (about twice as long as a phospholipid) with glycerol ethers at both ends (diglycerol tetraethers). The molecules orient themselves with the polar glycerol groups on the surfaces and the nonpolar hydrocarbon chain in the interior. These unusual lipids contribute to the ability of many Archaea to grow under environmental conditions such as high salt, low pH, or very high temperature.

B. Function

The major functions of the cytoplasmic membrane are (1) selective permeability and transport of solutes; (2) electron transport and oxidative phosphorylation, in aerobic species; (3) excretion of hydrolytic exoenzymes; (4) bearing the

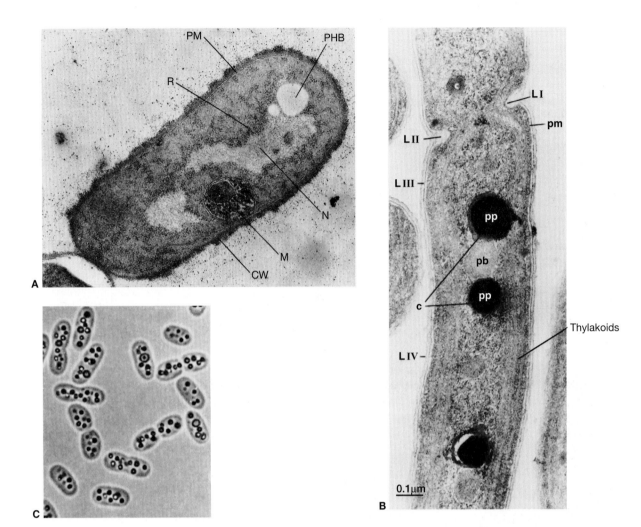

FIGURE 2–7 Inclusion Bodies in Bacteria. A: Electron micrograph of *Bacillus megaterium* (30,500x) showing poly-β-hydroxybutyric acid inclusion body, PHB; cell wall, CW; nucleoid, N; plasma membrane, PM; "mesosome," M; and ribosomes, R. B: Ultrastructure of the cyanobacterium *Anacystis nidulans*. The bacterium is dividing and a septum is partially formed, LI and LII. Several structural features can be seen, including cell wall layers, LIII and LIV; polyphosphate granules, pp; a polyhedral body, pb; cyanophycin material, c; and plasma membrane, pm. C: *Cromatium vinosum*, a purple sulfur bacterium, with intracellular sulfur granules, bright field microscopy (2,000x). **A**: (Reproduced with permission from Ralph A. Slepecky/Visuals Unlimited.) **B**: (Reproduced with permission from National Research Council of Canada.) **C**: (Reproduced with permission from *The Shorter Bergey's Manual of Determinative Bacteriology*, 8th ed, John Holt, editor, 1977 Copyright Bergey's Manual Trust. Published by Williams & Wilkins.)

enzymes and carrier molecules that function in the biosynthesis of DNA, cell wall polymers, and membrane lipids; and (5) bearing the receptors and other proteins of the chemotactic and other sensory transduction systems.

At least 50% of the cytoplasmic membrane must be in the semifluid state in order for cell growth to occur. At low temperatures, this is achieved by greatly increased synthesis and incorporation of unsaturated fatty acids into the phospholipids of the cell membrane.

1. Permeability and transport—The cytoplasmic membrane forms a hydrophobic barrier impermeable to most hydrophilic molecules. However, several mechanisms (**transport systems**) exist that enable the cell to transport nutrients

into and waste products out of the cell. These transport systems work against a concentration gradient to increase the concentration of nutrients inside the cell, a function that requires energy in some form. There are three general transport mechanisms involved in membrane transport: **passive transport**, **active transport**, and **group translocation**.

a. Passive transport—This mechanism relies on diffusion, uses no energy, and operates only when the solute is at higher concentration outside than inside the cell. **Simple diffusion** accounts for the entry of very few nutrients including dissolved oxygen, carbon dioxide, and water itself. Simple diffusion provides neither speed nor selectivity. **Facilitated diffusion** also uses no energy so the solute never achieves an internal concentration greater than what

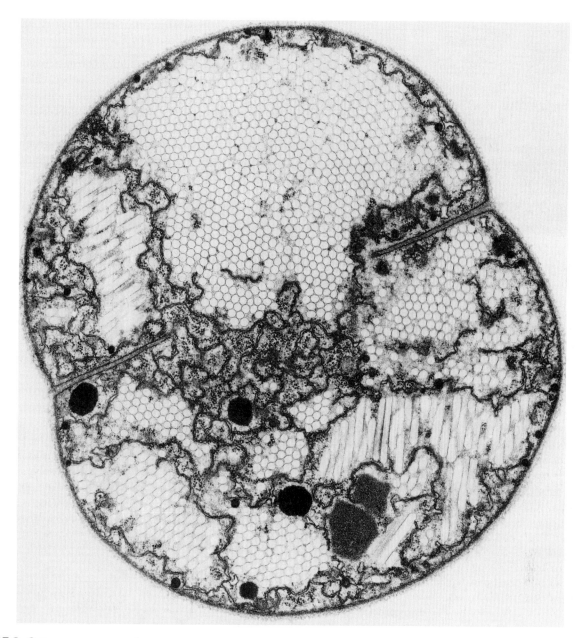

FIGURE 2–8 Transverse section of a dividing cell of the cyanobacterium *Microcystis* species showing hexagonal stacking of the cylindric gas vesicles (31,500×). (Micrograph by HS Pankratz. Reproduced with permission from Walsby AE: Gas vesicles. Microbiol Rev 1994;58:94.)

exists outside the cell. However, facilitated diffusion is selective. **Channel proteins** form selective channels that facilitate the passage of specific molecules. Facilitated diffusion is common in eukaryotic microorganisms (eg, yeast), but is rare in prokaryotes. Glycerol is one of the few compounds that enters prokaryotic cells by facilitated diffusion.

b. Active transport—Many nutrients are concentrated more than a thousand fold as a result of active transport. There are two types of active transport mechanisms depending upon the source of energy employed: **ion-coupled transport** and **ATP-binding cassette (ABC) transport**.

1) *Ion-coupled transport*—These systems move a molecule across the cell membrane at the expense of a previously established ion gradient such as **protonmotive** or **sodium-motive force**. There are three basic types: **uniport**, **symport**, and **antiport** (Figure 2–11). Ion-coupled transport is particularly common in aerobic organisms, which have an easier time generating an ion-motive force than do anaerobes. Uniporters catalyze the transport of a substrate independent of any coupled ion. Symporters catalyze the simultaneous transport of two substrates in the same direction by a single carrier; for example, an H^+ gradient can permit symport of an oppositely charged ion (eg, glycine) or a neutral molecule (eg, galactose). Antiporters catalyze the simultaneous transport of two like-charged compounds in opposite directions by a common carrier (eg, H^+:Na^+). Approximately 40% of the substrates transported by *E coli* utilize this mechanism.

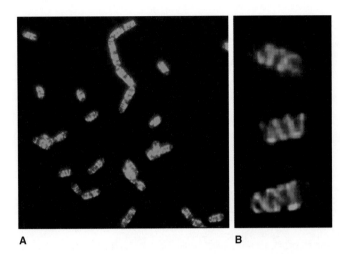

A   **B**

FIGURE 2–9 The Prokaryotic Cytoskeleton. Visualization of the MreB-like cytoskeletal protein (Mbl) of *Bacillus subtilis*. The Mbl protein has been fused with green fluorescent protein and live cells have been examined by fluorescence microscopy. A: Arrows point to the helical cytoskeleton cables that extend the length of the cells. B: Three of the cells from (A) are shown at a higher magnification. (Courtesy of Rut Carballido-Lopez and Jeff Errington.)

2) *ABC transport*—This mechanism employs ATP directly to transport solutes into the cell. In gram-negative bacteria, the transport of many nutrients is facilitated by specific **binding proteins** located in the periplasmic space; in gram-positive cells the binding proteins are attached to the outer surface of the cell membrane. These proteins function by transferring the bound substrate to a membrane-bound protein complex. Hydrolysis of ATP is then triggered, and the energy is used to open the membrane pore and allow the unidirectional movement of the substrate into the cell. Approximately 40% of the substrates transported by *E coli* utilize this mechanism.

c. Group translocation—In addition to true transport, in which a solute is moved across the membrane without change in structure, bacteria use a process called group translocation (**vectorial metabolism**) to effect the net uptake of certain sugars (eg, glucose and mannose), the substrate becoming phosphorylated during the transport process. In a strict sense, group translocation is not active transport because no concentration gradient is involved. This process allows bacteria to utilize their energy resources efficiently by coupling transport with metabolism. In this process, a membrane carrier protein is first phosphorylated in the cytoplasm at the expense of phosphoenolpyruvate; the phosphorylated carrier protein then binds the free sugar at the exterior membrane face and transports it into the cytoplasm, releasing it as sugar-phosphate. Such systems of sugar transport are called **phosphotransferase** systems. Phosphotransferase systems are also involved in movement toward these carbon sources (**chemotaxis**) and in the regulation of several other metabolic pathways (**catabolite repression**).

d. Special transport processes—Iron (Fe) is an essential nutrient for the growth of almost all bacteria. Under anaerobic conditions, Fe is generally in the +2 oxidation state and soluble. However, under aerobic conditions, Fe is generally in the +3 oxidation state and insoluble. The internal compartments of animals contain virtually no free Fe; it is sequestered in complexes with such proteins as **transferrin** and **lactoferrin**. Some bacteria solve this problem by secreting **siderophores**—compounds that chelate Fe and promote its transport as a soluble complex. One major group of siderophores consists of derivatives of hydroxamic acid

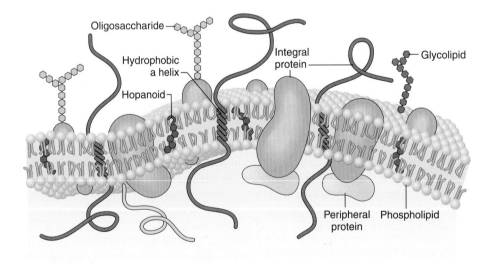

FIGURE 2–10 Bacterial Plasma Membrane Structure. This diagram of the fluid mosaic model of bacterial membrane structure shown the integral proteins (green and red) floating in a lipid bilayer. Peripheral proteins (yellow) are associated loosely with the inner membrane surface. Small spheres represent the hydrophilic ends of membrane phospholipids and wiggly tails, the hydrophobic fatty acid chains. Other membrane lipids such as hopanoids (purple) may be present. For the sake of clarity, phospholipids are shown proportionately much larger size than in real membranes. (Reproduced with permission from Willey JM, Sherwood LM, Woolverton CJ (eds): Prescott, Harley, and Klein's Microbiology, 7th edt. New York: Mcgraw-Hill; 2008.)

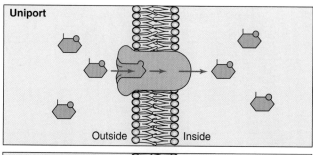

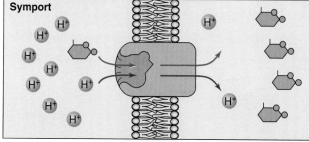

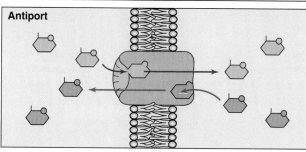

FIGURE 2–11 Three types of porters: **A**: uniporters, **B**: symporters, and **C**: antiporters. Uniporters catalyze the transport of a single species independently of any other, symporters catalyze the cotransport of two dissimilar species (usually a solute and a positively charged ion, H⁺) in the same direction, and antiporters catalyze the exchange transport of two similar solutes in opposite directions. A single transport protein may catalyze just one of these processes, two of these processes, or even all three of these processes, depending on conditions. Uniporters, symporters, and antiporters have been found to be structurally similar and evolutionarily related, and they function by similar mechanisms. (Reproduced with permission from Saier MH Jr: Peter Mitchell and his chemiosmotic theories. ASM News 1997;63:13.)

(–CONH₂OH), which chelate Fe^{3+} very strongly. The iron-hydroxamate complex is actively transported into the cell by the cooperative action of a group of proteins that span the outer membrane, periplasm, and inner membrane. The iron is released, and the hydroxamate can exit the cell and be used again for iron transport.

Some pathogenic bacteria use a fundamentally different mechanism involving specific receptors that bind host transferrin and lactoferrin (as well as other iron-containing host proteins). The Fe is removed and transported into the cell by an energy-dependent process.

2. Electron transport & oxidative phosphorylation—The cytochromes and other enzymes and components

of the respiratory chain, including certain dehydrogenases, are located in the cell membrane. The bacterial cell membrane is thus a functional analog of the mitochondrial membrane—a relationship which has been taken by many biologists to support the theory that mitochondria have evolved from symbiotic bacteria. The mechanism by which ATP generation is coupled to electron transport is discussed in Chapter 6.

3. Excretion of hydrolytic exoenzymes & pathogenicity proteins—All organisms that rely on macromolecular organic polymers as a source of nutrients (eg, proteins, polysaccharides, lipids) excrete hydrolytic enzymes that degrade the polymers to subunits small enough to penetrate the cell membrane. Higher animals secrete such enzymes into the lumen of the digestive tract; bacteria (both gram-positive and gram-negative) secrete them directly into the external medium or into the periplasmic space between the peptidoglycan layer and the outer membrane of the cell wall in the case of gram-negative bacteria (see The Cell Wall, below).

In gram-positive bacteria, proteins are secreted directly, but proteins secreted by gram-negative bacteria must traverse the outer membrane as well. Six pathways of protein secretion have been described in bacteria: the type I, type II, type III, type IV, type V, and type VI secretion systems. A schematic overview of the type I–V systems is presented in Figure 2–12. The type I and IV secretion systems have been described in both gram-negative and gram-positive bacteria, while the type II, III, V, and VI secretion systems have been found only in gram-negative bacteria. Proteins secreted by the type I and type III pathways traverse the inner membrane (IM) and outer membrane (OM) in one step, whereas proteins secreted by the type II and type V pathways cross the IM and OM in separate steps. Proteins secreted by the type II and type V pathways are synthesized on cytoplasmic ribosomes as preproteins containing an extra **leader** or **signal sequence** of 15 to 40 amino acids—most commonly about 30 amino acids—at the amino terminal and require the sec system for transport across the IM. In *E coli*, the sec pathway comprises a number of IM proteins (SecD to SecF, SecY), a cell membrane-associated ATPase (SecA) that provides energy for export, a **chaperone** (SecB) that binds to the preprotein, and the periplasmic signal peptidase. Following translocation, the leader sequence is cleaved off by the membrane-bound signal peptidase and the mature protein is released into the periplasmic space. In contrast, proteins secreted by the type I and type III systems do not have a leader sequence and are exported intact.

In gram-negative and gram-positive bacteria, another plasma membrane translocation system, called the ***tat* pathway**, can move proteins across the plasma membrane. In gram-negative bacteria, these proteins are then delivered to the type II system (Figure 2–12). The *tat* pathway is distinct from the *sec* system in that it translocates already folded proteins.

Although proteins secreted by the type II and type V systems are similar in the mechanism by which they cross the

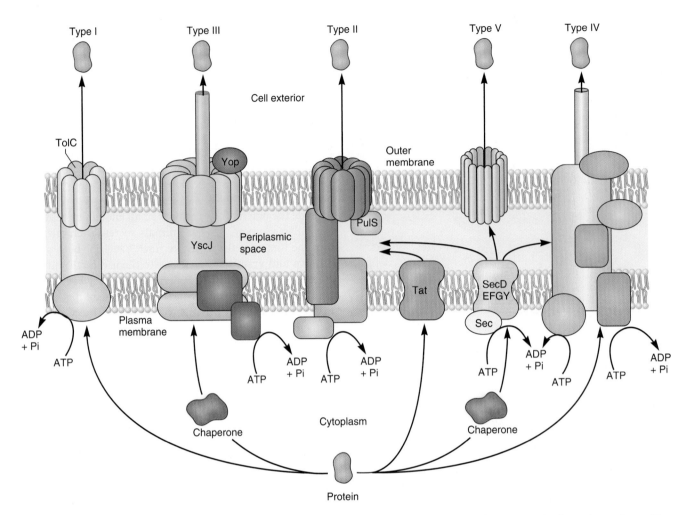

FIGURE 2–12 The Protein Secretion Systems of Gram-Negative Bacteria. Five secretion systems of gram-negative bacteria are shown. The Sec-dependent and Tat pathways deliver proteins from the cytoplasm to the periplasmic space. The type II, type V, and sometimes type IV systems complete the secretion process begun by the Sec-dependent pathway. The Tat system appears to deliver proteins only to the type II pathway. The type I and type III systems bypass the Sec-dependent and Tat pathways, moving proteins directly from the cytoplasm, through the outer membrane, to the extracellular space. The type IV system can work either with the Sec-dependent pathway or can work alone to transport proteins to the extracellular space. Proteins translocated by the Sec-dependent pathway and the type III pathway are delivered to those systems by chaperone proteins. (Reproduced with permission from Willey JM, Sherwood LM, Woolverton CJ (eds): Prescott, Harley, and Klein's Microbiology, 7th edt. New York: Mcgraw-Hill; 2008.)

IM, differences exist in how they traverse the OM. Proteins secreted by the type II system are transported across the OM by a multiprotein complex (Figure 2–12). This is the primary pathway for the secretion of extracellular degradative enzymes by gram-negative bacteria. Elastase, phospholipase C, and exotoxin A are secreted by this system in *Pseudomonas aeruginosa*. However, proteins secreted by the type V system autotransport across the outer membrane by virtue of a carboxyl terminal sequence which is enzymatically removed upon release of the protein from the OM. Some extracellular proteins—eg, the IgA protease of *Neisseria gonorrhoeae* and the vacuolating cytotoxin of *Helicobacter pylori*—are secreted by this system.

The type I and type III secretion pathways are *sec*-independent and thus do not involve amino terminal processing of the secreted proteins. Protein secretion by these pathways

occurs in a continuous process without the presence of a cytoplasmic intermediate. Type I secretion is exemplified by the α-hemolysin of *E coli* and the adenylyl cyclase of *Bordetella pertussis*. Type I secretion requires three secretory proteins: an IM ATP-binding cassette (ABC transporter), which provides energy for protein secretion; an OM protein; and a membrane fusion protein, which is anchored in the inner membrane and spans the periplasmic space (Figure 2–12). Instead of a signal peptide, the information is located within the carboxyl terminal 60 amino acids of the secreted protein.

The type III secretion pathway is a **contact-dependent** system. It is activated by contact with a host cell, and then injects a toxin protein into the host cell directly. The type III secretion apparatus is composed of approximately 20 proteins, most of which are located in the IM. Most of these IM components are homologous to the flagellar biosynthesis apparatus of both gram-negative and gram-positive bacteria. As in type I secretion, the proteins secreted via the type III pathway are not subject to amino terminal processing during secretion.

Type IV pathways secrete either polypeptide toxins (directed against eukaryotic cells) or protein-DNA complexes either between two bacterial cells or between a bacterial and a eukaryotic cell. Type IV secretion is exemplified by the protein-DNA complex delivered by *Agrobacterium tumefaciens* into a plant cell. Additionally, *B pertussis* and *H pylori* possess type IV secretion systems that mediate secretion of pertussis toxin and interleukin-8-inducing factor, respectively. The *sec*-independent type VI secretion was recently described in *P aeruginosa*, where it contributes to pathogenicity in patients with cystic fibrosis. This secretion system is composed of 15–20 proteins whose biochemical functions are not well understood. However, recent studies suggest that some of these proteins share homology with bacteriophage tail proteins.

The characteristics of the protein secretion systems of bacteria are summarized in Table 9–6.

4. Biosynthetic functions—The cell membrane is the site of the carrier lipids on which the subunits of the cell wall are assembled (see the discussion of synthesis of cell wall substances in Chapter 6) as well as of the enzymes of cell wall biosynthesis. The enzymes of phospholipid synthesis are also localized in the cell membrane.

5. Chemotactic systems—Attractants and repellents bind to specific receptors in the bacterial membrane (see Flagella below). There are at least 20 different chemoreceptors in the membrane of *E coli*, some of which also function as a first step in the transport process.

The Cell Wall

The internal osmotic pressure of most bacteria ranges from 5 to 20 atm as a result of solute concentration via active transport. In most environments, this pressure would be sufficient to burst the cell were it not for the presence of a high-tensile-strength cell wall (Figure 2–13). The bacterial cell wall owes its strength to a layer composed of a substance variously referred to as murein, mucopeptide, or **peptidoglycan** (all are synonyms). The structure of peptidoglycan will be discussed below.

Most bacteria are classified as gram-positive or gram-negative according to their response to the Gram-staining procedure. This procedure was named for the histologist Hans Christian Gram, who developed this differential staining procedure in an attempt to stain bacteria in infected tissues. The Gram stain depends on the ability of certain bacteria (the gram-positive bacteria) to retain a complex of crystal violet (a purple dye) and iodine after a brief wash with alcohol or acetone. Gram-negative bacteria do not retain the dye-iodine complex and become translucent, but they can then be counterstained with safranin (a red dye). Thus, gram-positive bacteria look purple under the microscope, and gram-negative

FIGURE 2–13 Isolated Gram-Positive Cell Wall. The peptidoglycan wall from *Bacillus megaterium*, a gram-positive bacterium. The latex spheres have a diameter of 0.25 μm (Reproduced with permission from Willey JM, Sherwood LM, Woolverton CJ (eds): Prescott, Harley, and Klein's Microbiology, 7th edt. New York: Mcgraw-Hill; 2008.)

bacteria look red. The distinction between these two groups turns out to reflect fundamental differences in their cell envelopes (Figure 2–14).

In addition to giving osmotic protection, the cell wall plays an essential role in cell division as well as serving as a primer for its own biosynthesis. Various layers of the wall are the sites of major antigenic determinants of the cell surface, and one component—the lipopolysaccharide of gram-negative cell walls—is responsible for the nonspecific endotoxin activity of gram-negative bacteria. The cell wall is, in general, nonselectively permeable; one layer of the gram-negative wall, however—the outer membrane—hinders the passage of relatively large molecules (see below).

The biosynthesis of the cell wall and the antibiotics that interfere with this process are discussed in Chapter 6.

A. The Peptidoglycan Layer

Peptidoglycan is a complex polymer consisting, for the purposes of description, of three parts: a backbone, composed of alternating *N*-acetylglucosamine and *N*-acetylmuramic acid; a set of identical tetrapeptide side chains attached to *N*-acetylmuramic acid; and a set of identical peptide cross-bridges (Figure 2–15). The backbone is the same in all bacterial species; the tetrapeptide side chains and the peptide cross-bridges vary from species to species, those of *Staphylococcus aureus* being illustrated in Figure 2–15. In many gram-negative cell walls, the cross-bridge consists of a direct peptide linkage between the diaminopimelic acid (DAP) amino group of one side chain and the carboxyl group of the terminal D-alanine of a second side chain.

The tetrapeptide side chains of all species, however, have certain important features in common. Most have L-alanine at position 1 (attached to *N*-acetylmuramic acid), D-glutamate

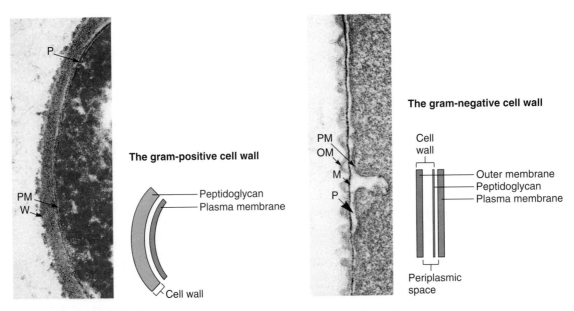

FIGURE 2–14 Gram-Positive and Gram-Negative Cell Walls. The gram-positive envelope is from *Bacillus licheniiformis* (**left**), and the gram-negative micrograph is of *Aquaspirillum serpens* (**right**). M; peptidoglycan or murein layer; OM, outer membrane; IM, plasma membrane; P, periplasmic space; W, gram-positive peptidoglycan wall. (Reproduced with permission from T. J. Beveridge/Biological Photo Service.)

or substituted D-glutamate at position 2, and D-alanine at position 4. Position 3 is the most variable one: Most gram-negative bacteria have diaminopimelic acid at this position, to which is linked the lipoprotein cell wall component discussed below. Gram-positive bacteria usually have L-lysine at position 3; however, some may have diaminopimelic acid or another amino acid at this position.

Diaminopimelic acid is a unique element of bacterial cell walls. It is never found in the cell walls of *Archaea* or eukaryotes. Diaminopimelic acid is the immediate precursor of lysine in the bacterial biosynthesis of that amino acid (see Figure 6–18). Bacterial mutants that are blocked prior to diaminopimelic acid in the biosynthetic pathway grow normally when provided with diaminopimelic acid in the medium; when given L-lysine alone, however, they lyse, since they continue to grow but are specifically unable to make new cell wall peptidoglycan.

The fact that all peptidoglycan chains are cross-linked means that each peptidoglycan layer is a single giant molecule. In gram-positive bacteria, there are as many as 40 sheets of peptidoglycan, comprising up to 50% of the cell wall material; in gram-negative bacteria, there appears to be only one or two sheets, comprising 5–10% of the wall material. Bacteria owe their shapes, which are characteristic of particular species, to their cell wall structure.

B. Special Components of Gram-Positive Cell Walls

Most gram-positive cell walls contain considerable amounts of **teichoic** and **teichuronic acids**, which may account for up to 50% of the dry weight of the wall and 10% of the dry weight of the total cell. In addition, some gram-positive walls may contain polysaccharide molecules.

1. Teichoic and teichuronic acids—The term teichoic acids encompasses all wall, membrane, or capsular polymers containing glycerophosphate or ribitol phosphate residues. These polyalcohols are connected by phosphodiester linkages and usually have other sugars and D-alanine attached (Figure 2–16A). Because they are negatively charged, teichoic acids are partially responsible for the negative charge of the cell surface as a whole. There are two types of teichoic acids: **wall teichoic acid (WTA)**, covalently linked to peptidoglycan, and **membrane teichoic acid**, covalently linked to membrane glycolipid. Because the latter are intimately associated with lipids, they have been called **lipoteichoic acids (LTA)**. Together with peptidoglycan, WTA and LTA make up a polyanionic network or matrix that provides functions relating to the elasticity, porosity, tensile strength, and electrostatic properties of the envelope. Although not all gram-positive bacteria have conventional LTA and WTA, those that lack these polymers generally have functionally similar ones.

Most teichoic acids contain large amounts of D-alanine, usually attached to position 2 or 3 of glycerol or position 3 or 4 of ribitol. In some of the more complex teichoic acids, however, D-alanine is attached to one of the sugar residues. In addition to D-alanine, other substituents may be attached to the free hydroxyl groups of glycerol and ribitol, eg, glucose, galactose, *N*-acetylglucosamine, *N*-acetylgalactosamine, or succinate. A given species may have more than one type of sugar substituent in addition to D-alanine; in such cases, it is not certain whether the different sugars occur on the same or on separate teichoic acid molecules. The composition of the teichoic acid formed by a given bacterial species can vary with the composition of the growth medium.

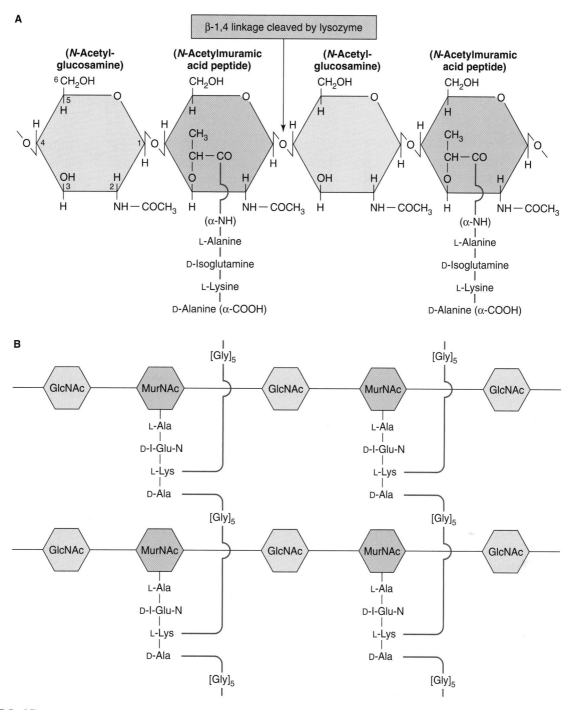

FIGURE 2–15 A: A segment of the peptidoglycan of *Staphylococcus aureus*. The backbone of the polymer consists of alternating subunits of *N*-acetylglucosamine and *N*-acetylmuramic acid connected by β1→4 linkages. The muramic acid residues are linked to short peptides, the composition of which varies from one bacterial species to another. In some species, the L-lysine residues are replaced by diaminopimelic acid, an amino acid that is found in nature only in prokaryotic cell walls. Note the D-amino acids, which are also characteristic constituents of prokaryotic cell walls. The peptide chains of the peptidoglycan are cross-linked between parallel polysaccharide backbones, as shown in Figure 2–15B. **B**: Schematic representation of the peptidoglycan lattice that is formed by cross-linking. Bridges composed of pentaglycine peptide chains connect the α-carboxyl of the terminal D-alanine residue of one chain with the ε-amino group of the L-lysine residue of the next chain. The nature of the cross-linking bridge varies among different species.

The teichoic acids constitute major surface antigens of those gram-positive species that possess them, and their accessibility to antibodies has been taken as evidence that they lie on the outside surface of the peptidoglycan. Their activity is often increased, however, by partial digestion of the peptidoglycan; thus, much of the teichoic acid may lie between the cytoplasmic membrane and the peptidoglycan layer, possibly extending upward through pores in the latter (Figure 2–16B).

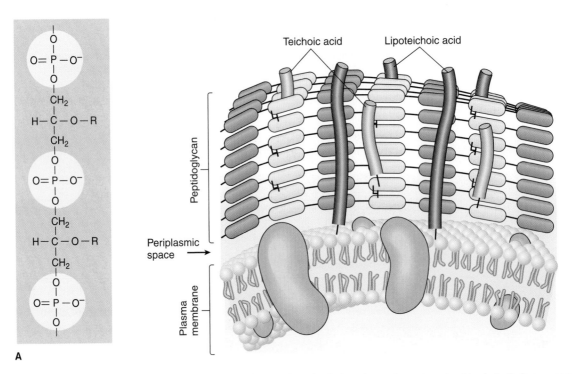

FIGURE 2–16 **A**: Teichoic Acid Structure. The segment of a teichoic acid made of phosphate, glycerol, and a side chain, R. R may represent D-alanine, glucose, or other molecules. **B**: Teichoic and lipoteichoic acids of the gram-positive envelope. (Reproduced with permission from Willey JM, Sherwood LM, Woolverton CJ (eds): Prescott, Harley, and Klein's Microbiology, 7th edt. New York: Mcgraw-Hill; 2008.)

In the pneumococcus (*Streptococcus pneumoniae*), the teichoic acids bear the antigenic determinants called Forssman antigen. In *Streptococcus pyogenes*, LTA is associated with the M protein that protrudes from the cell membrane through the peptidoglycan layer. The long M protein molecules together with the LTA form microfibrils that facilitate the attachment of *S pyogenes* to animal cells.

The **teichuronic acids** are similar polymers, but the repeat units include sugar acids (such as *N*-acetylmannosuronic or D-glucosuronic acid) instead of phosphoric acids. They are synthesized in place of teichoic acids when phosphate is limiting.

2. Polysaccharides—The hydrolysis of gram-positive walls has yielded, from certain species, neutral sugars such as mannose, arabinose, rhamnose, and glucosamine and acidic sugars such as glucuronic acid and mannuronic acid. It has been proposed that these sugars exist as subunits of polysaccharides in the cell wall; the discovery, however, that teichoic and teichuronic acids may contain a variety of sugars (Figure 2–16A) leaves the true origin of these sugars uncertain.

C. Special Components of Gram-Negative Cell Walls

Gram-negative cell walls contain three components that lie outside of the peptidoglycan layer: lipoprotein, outer membrane, and lipopolysaccharide (Figure 2–17).

1. Outer membrane—The outer membrane is chemically distinct from all other biological membranes. It is a bilayered structure; its inner leaflet resembles in composition that of the cell membrane while its outer leaflet contains a distinctive component, a **lipopolysaccharide (LPS)** (see below). As a result, the leaflets of this membrane are asymmetrical, and the properties of this bilayer differ considerably from those of a symmetrical biologic membrane such as the cell membrane.

The ability of the outer membrane to exclude hydrophobic molecules is an unusual feature among biologic membranes and serves to protect the cell (in the case of enteric bacteria) from deleterious substances such as bile salts. Because of its lipid nature, the outer membrane would be expected to exclude hydrophilic molecules as well. However, the outer membrane has special channels, consisting of protein molecules called **porins**, that permit the passive diffusion of low-molecular-weight hydrophilic compounds like sugars, amino acids, and certain ions. Large antibiotic molecules penetrate the outer membrane relatively slowly, which accounts for the relatively high antibiotic resistance of gram-negative bacteria. The permeability of the outer membrane varies widely from one gram-negative species to another; in *P aeruginosa*, for example, which is extremely resistant to antibacterial agents, the outer membrane is 100 times less permeable than that of *E coli*.

The major proteins of the outer membrane, named according to the genes that code for them, have been placed into several functional categories on the basis of mutants in which they are lacking and on the basis of experiments in which

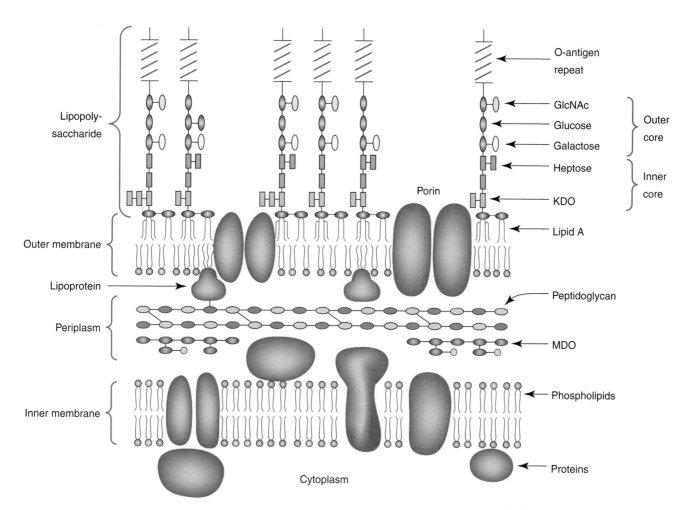

FIGURE 2–17 Molecular representation of the envelope of a gram-negative bacterium. Ovals and rectangles represent sugar residues, and circles depict the polar head groups of the glycerophospholipids (phosphatidylethanolamine and phosphatidylglycerol). (MDO, membrane-derived oligosaccharides.) The core region shown is that of *Escherichia coli* K-12, a strain that does not normally contain an O-antigen repeat unless transformed with an appropriate plasmid. (Reproduced with permission from Raetz CRH: Bacterial endotoxins: Extraordinary lipids that activate eucaryotic signal transduction. J Bacteriol 1993;175:5745.)

purified proteins have been reconstituted into artificial membranes. Porins, exemplified by OmpC, D, and F and PhoE of *E coli* and *Salmonella typhimurium*, are trimeric proteins that penetrate both faces of the outer membrane (Figure 2–18). They form relatively nonspecific pores that permit the free diffusion of small hydrophilic solutes across the membrane. The porins of different species have different exclusion limits, ranging from molecular weights of about 600 in *E coli* and *S typhimurium* to more than 3000 in *P aeruginosa*.

Members of a second group of outer membrane proteins, which resemble porins in many ways, are exemplified by LamB and Tsx. LamB, an inducible porin that is also the receptor for lambda bacteriophage, is responsible for most of the transmembrane diffusion of maltose and maltodextrins; Tsx, the receptor for T6 bacteriophage, is responsible for the transmembrane diffusion of nucleosides and some amino acids. LamB allows some passage of other solutes; however, its relative specificity may reflect weak interactions of solutes with configuration-specific sites within the channel.

The OmpA protein is an abundant protein in the outer membrane. The OmpA protein participates in the anchoring of the outer membrane to the peptidoglycan layer; it is also the sex pilus receptor in F-mediated bacterial conjugation (Chapter 7).

The outer membrane also contains a set of less abundant proteins that are involved in the transport of specific molecules such as vitamin B_{12} and iron-siderophore complexes. They show high affinity for their substrates and probably function like the classic carrier transport systems of the cytoplasmic membrane. The proper function of these proteins requires energy coupled through a protein called TonB. Additional minor proteins include a limited number of enzymes, among them phospholipases and proteases.

The topology of the major proteins of the outer membrane, based on cross-linking studies and analyses of functional relationships, is shown in Figure 2–17. The outer membrane is connected to both the peptidoglycan layer and the cytoplasmic membrane. The connection with the peptidoglycan layer

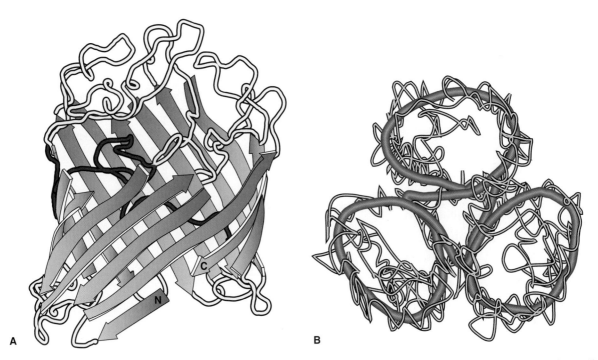

FIGURE 2–18 **A**: General fold of a porin monomer (OmpF porin from *Escherichia coli*). The large hollow β-barrel structure is formed by antiparallel arrangement of 16 β-strands. The strands are connected by short loops or regular turns on the periplasmic rim (bottom), and long irregular loops face the cell exterior (top). The internal loop, which connects β-strands 5 and 6 and extends inside the barrel, is highlighted in dark. The chain terminals are marked. The surface closest to the viewer is involved in subunit contacts. **B**: Schematic representation of the OmpF trimer. The view is from the extracellular space along the molecular threefold symmetry axis. (Reproduced with permission from Schirmer T: General and specific porins from bacterial outer membranes. J Struct Biol 1998;121:101.)

is primarily mediated by the outer membrane lipoprotein (see below). About one-third of the lipoprotein molecules are covalently linked to peptidoglycan and help hold the two structures together. A noncovalent association of some of the porins with the peptidoglycan layer plays a lesser role in connecting the outer membrane with this structure. Outer membrane proteins are synthesized on ribosomes bound to the cytoplasmic surface of the cell membrane; how they are transferred to the outer membrane is still uncertain, but one hypothesis suggests that transfer occurs at zones of adhesion between the cytoplasmic and outer membranes, which are visible in the electron microscope. Unfortunately, firm evidence for such areas of adhesion has proven hard to come by.

2. Lipopolysaccharide (LPS)—The LPS of gram-negative cell walls consists of a complex glycolipid, called lipid A, to which is attached a polysaccharide made up of a core and a terminal series of repeat units (Figure 2–19A). The lipid A component is embedded in the outer leaflet of the membrane anchoring the LPS. LPS is synthesized on the cytoplasmic membrane and transported to its final exterior position. The presence of LPS is required for the function of many outer membrane proteins.

Lipid A consists of phosphorylated glucosamine disaccharide units to which are attached a number of long-chain fatty acids (Figure 2–19). β-Hydroxymyristic acid, a C14 fatty acid, is always present and is unique to this lipid; the other

fatty acids, along with substituent groups on the phosphates, vary according to the bacterial species.

The polysaccharide **core**, shown in Figure 2–19A and B is similar in all gram-negative species that have LPS and includes two characteristic sugars, **ketodeoxyoctanoic acid (KDO)** and a heptose. Each species, however, contains a unique repeat unit, that of *Salmonella* being shown in Figure 2–19A. The repeat units are usually linear trisaccharides or branched tetra- or pentasaccharides. The repeat unit is referred to as the O antigen. The hydrophilic carbohydrate chains of the O antigen cover the bacterial surface and exclude hydrophobic compounds.

The negatively charged LPS molecules are noncovalently cross-bridged by divalent cations (ie, Ca^{2+} and Mg^{2+}); this stabilizes the membrane and provides a barrier to hydrophobic molecules. Removal of the divalent cations with chelating agents or their displacement by polycationic antibiotics such as polymyxins and aminoglycosides renders the outer membrane permeable to large hydrophobic molecules.

LPS, which is extremely toxic to animals, has been called the **endotoxin** of gram-negative bacteria because it is firmly bound to the cell surface and is released only when the cells are lysed. When LPS is split into lipid A and polysaccharide, all of the toxicity is associated with the former. The O antigen is highly immunogenic in a vertebrate animal. Antigenic specificity is conferred by the O antigen as this antigen is highly variable among species and even in strains within a

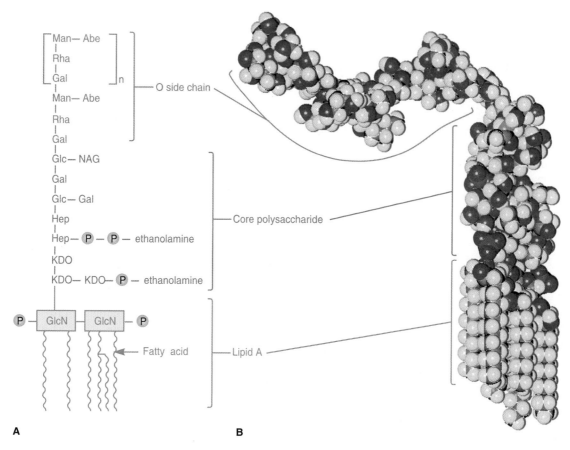

FIGURE 2–19 Lipopolysaccharide structure. **A:** The lipopolysaccharide from *Salmonella*. This slightly simplified diagram illustrates one form of the LPS. (Abe, abequose; Gal, galactose; GlcN, glucosamine; Hep, heptulose; KDO, 2-keto-3-deoxyoctonate; Man, mannose; NAG, *N*-acetylglucosamine; P, phosphate; Rha, L-rhamnose.) Lipid A is buried in the outer membrane. **B:** Molecular model of an *Escherichia coli* lipopolysaccharide. The lipid A and core polysaccharide are straight; the O side chain is bent at an angle in this model. (Reproduced with permission from Willey VM, Sherwood LM, Woolverton CJ: Prescott, Harley, & Klein's *Microbiology*, 7th edition, McGraw-Hill, 2008.)

species. The number of possible antigenic types is very great: Over 1000 have been recognized in *Salmonella* alone. Not all gram-negative bacteria have outer membrane LPS composed of a variable number of repeated oligosaccharide units (see Figure 2–19); the outer membrane glycolipids of bacteria that colonize mucosal surfaces (eg, *Neisseria meningitidis*, *N gonorrhoeae*, *Haemophilus influenzae*, and *Haemophilus ducreyi*) have relatively short, multiantennary (ie, branched) glycans. These smaller glycolipids have been compared with the "R-type" truncated LPS structures, which lack O antigens and are produced by rough mutants of enteric bacteria such as *E coli*. However, their structures more closely resemble those of the glycosphingolipids of mammalian cell membranes, and they are more properly termed **lipooligosaccharides (LOS)**. These molecules exhibit extensive antigenic and structural diversity even within a single strain. LOS is an important virulence factor. Epitopes have been identified on LOS which mimic host structures and may enable these organisms to evade the immune response of the host. Some LOS (eg, those from *N gonorrhoeae, N meningitidis,* and *H ducreyi*) have a terminal *N*-acetyllactosamine (Galβ-1→4-GlcNAc) residue which is immunochemically similar to the precursor of the human

erythrocyte i antigen. In the presence of a bacterial enzyme called sialyltransferase and a host or bacterial substrate (cytidine monophospho-*N*-acetylneuraminic acid, CMP-NANA), the *N*-acetyllactosamine residue is sialylated. This sialylation, which occurs in vivo, provides the organism with the environmental advantages of molecular mimicry of a host antigen and the biologic masking thought to be provided by sialic acids.

3. Lipoprotein—Molecules of an unusual **lipoprotein** cross-link the outer membrane and peptidoglycan layers (Figure 2–17). The lipoprotein contains 57 amino acids, representing repeats of a 15-amino-acid sequence; it is peptide-linked to DAP residues of the peptidoglycan tetrapeptide side chains. The lipid component, consisting of a diglyceride thioether linked to a terminal cysteine, is noncovalently inserted in the outer membrane. Lipoprotein is numerically the most abundant protein of gram-negative cells (ca 700,000 molecules per cell). Its function (inferred from the behavior of mutants that lack it) is to stabilize the outer membrane and anchor it to the peptidoglycan layer.

4. The periplasmic space—The space between the inner and outer membranes, called the **periplasmic space**, contains the peptidoglycan layer and a gel-like solution of proteins. The periplasmic space is approximately 20–40% of the cell volume, which is far from insignificant. The periplasmic proteins include binding proteins for specific substrates (eg, amino acids, sugars, vitamins, and ions), hydrolytic enzymes (eg, alkaline phosphatase and 5′-nucleotidase) that break down nontransportable substrates into transportable ones, and detoxifying enzymes (eg, β-lactamase and aminoglycoside-phosphorylase) that inactivate certain antibiotics. The periplasm also contains high concentrations of highly branched polymers of D-glucose, eight to ten residues long, which are variously substituted with glycerol phosphate and phosphatidylethanolamine residues; some contain O-succinyl esters. These so-called **membrane-derived oligosaccharides** appear to play a role in osmoregulation, since cells grown in media of low osmolarity increase their synthesis of these compounds 16-fold.

D. The Acid-Fast Cell Wall

Some bacteria, notably the tubercle bacillus (*M tuberculosis*) and its relatives have cell walls that contain large amounts of **waxes**, complex branched hydrocarbons (70–90 carbons long) known as **mycolic acids**. The cell wall is composed of peptidoglycan and an external asymmetric lipid bilayer; the inner leaflet contains mycolic acids linked to an arabinoglycan and the outer leaflet contains other extractable lipids. This is a highly ordered lipid bilayer in which proteins are embedded forming water-filled pores through which nutrients and certain drugs can pass slowly. Some compounds can also penetrate the lipid domains of the cell wall albeit slowly. This hydrophobic structure renders these bacteria resistant to many harsh chemicals including detergents and strong acids. If a dye is introduced into these cells by brief heating or treatment with detergents, it cannot be removed by dilute hydrochloric acid, as in other bacteria. These organisms are therefore called **acid-fast.** The permeability of the cell wall to hydrophilic molecules is 100- to 1000-fold lower than for *E coli* and may be responsible for the slow growth rate of mycobacteria.

E. Cell Walls of the Archaea

The *Archaea* do not have cell walls like the *Bacteria*. Some have a simple S-layer (see below) often comprised of glycoproteins. Some *Archaea* have a rigid cell wall composed of polysaccharides or a peptidoglycan called **pseudomurein**. The pseudomurein differs from the peptidoglycan of bacteria by having L-amino acids rather than D-amino acids and disaccharide units with an α-1→3 rather than a β-1→4 linkage. *Archaea* that have a pseudomurein cell wall are gram-positive.

F. Crystalline Surface Layers

Many bacteria, both gram-positive and gram-negative bacteria as well as archaebacteria, possess a two-dimensional crystalline, subunit-type layer lattice of protein or glycoprotein molecules (**S-layer**) as the outermost component of the cell envelope. In both gram-positive and gram-negative bacteria, this structure is sometimes several molecules thick. In some *Archaea*, they are the only layer external to the cell membrane.

S-layers are generally composed of a single kind of protein molecule, sometimes with carbohydrates attached. The isolated molecules are capable of self-assembly, ie, they make sheets similar or identical to those present on the cells. S-layer proteins are resistant to proteolytic enzymes and protein-denaturing agents. The function of the S-layer is uncertain but is probably protective. In some cases, it has been shown to protect the cell from wall-degrading enzymes, from invasion by *Bdellovibrio bacteriovorous* (a predatory bacterium), and from bacteriophages. It also plays a role in the maintenance of cell shape in some species of archaebacteria, and it may be involved in cell adhesion to host epidermal surfaces.

G. Enzymes that Attack Cell Walls

The β1→4 linkage of the peptidoglycan backbone is hydrolyzed by the enzyme **lysozyme**, which is found in animal secretions (tears, saliva, nasal secretions) as well as in egg white. Gram-positive bacteria treated with lysozyme in low-osmotic-strength media lyse; if the osmotic strength of the medium is raised to balance the internal osmotic pressure of the cell, free spherical bodies called **protoplasts** are liberated. The outer membrane of the gram-negative cell wall prevents access of lysozyme unless disrupted by an agent such as ethylene-diaminetetraacetic acid (EDTA), a compound that chelates divalent cations; in osmotically protected media, cells treated with EDTA-lysozyme form **spheroplasts** that still possess remnants of the complex gram-negative wall, including the outer membrane.

Bacteria themselves possess a number of **autolysins**, hydrolytic enzymes that attack peptidoglycan, including muramidases, glucosaminidases, endopeptidases, and carboxypeptidases. These enzymes catalyze the turnover or degradation of peptidoglycan in bacteria. These enzymes presumably participate in cell wall growth and turnover and in cell separation, but their activity is most apparent during the dissolution of dead cells (autolysis).

Enzymes that degrade bacterial cell walls are also found in cells that digest whole bacteria, eg, protozoa and the phagocytic cells of higher animals.

H. Cell Wall Growth

Cell wall synthesis is necessary for cell division; however, the incorporation of new cell wall material varies with the shape of the bacterium. Rod-shaped bacteria (eg, *E coli, Bacillus subtilis*) have two modes of cell wall synthesis; new peptidoglycan is inserted along a helical path leading to elongation of the cell, and is inserted in a closing ring around the future division site, leading to the formation of the division septum. Coccoid cells such as *S aureus* do not seem to have an elongation mode of cell wall synthesis. Instead, new peptidoglycan is inserted only at the division site. A third form of cell wall growth is exemplified by *S pneumoniae,* which are not true cocci, as their shape is not totally round, but instead have the shape of a rugby ball. *S pneumoniae* synthesize cell wall not only at the septum but also at the so-called equatorial rings (Figure 2–20).

I. Protoplasts, Spheroplasts, & L Forms

Removal of the bacterial wall may be accomplished by hydrolysis with lysozyme or by blocking peptidoglycan synthesis with an antibiotic such as penicillin. In osmotically protective media, such treatments liberate **protoplasts** from gram-positive cells and **spheroplasts** (which retain outer membrane and entrapped peptidoglycan) from gram-negative cells.

If such cells are able to grow and divide, they are called **L forms.** L forms are difficult to cultivate and usually require a medium that is solidified with agar as well as having the right osmotic strength. L forms are produced more readily with penicillin than with lysozyme, suggesting the need for residual peptidoglycan.

Some L forms can revert to the normal bacillary form upon removal of the inducing stimulus. Thus, they are able to resume normal cell wall synthesis. Others are stable and never revert. The factor that determines their capacity to revert may again be the presence of residual peptidoglycan, which normally acts as a primer in its own biosynthesis.

Some bacterial species produce L forms spontaneously. The spontaneous or antibiotic-induced formation of L forms

in the host may produce chronic infections, the organisms persisting by becoming sequestered in protective regions of the body. Since L-form infections are relatively resistant to antibiotic treatment, they present special problems in chemotherapy. Their reversion to the bacillary form can produce relapses of the overt infection.

J. The Mycoplasmas

The **mycoplasmas** are cell wall-lacking bacteria containing no peptidoglycan. There are also wall-less *Archaea,* but they have been less well studied. Genomic analysis places the mycoplasmas close to the gram-positive bacteria from which they may have been derived. Mycoplasmas lack a target for cell wall-inhibiting antimicrobial agents (eg, penicillins and cephalosporins) and are therefore resistant to these drugs. Some, like *Mycoplasma pneumoniae,* an agent of pneumonia, contain sterols in their membranes. The difference between L forms and mycoplasmas is that when the murein is allowed to reform, L forms revert to their original bacteria shape, but mycoplasmas never do.

Capsule & Glycocalyx

Many bacteria synthesize large amounts of extracellular polymer when growing in their natural environments. With one known exception (the poly-D-glutamic acid capsules of *Bacillus anthracis* and *Bacillus licheniformis*), the extracellular material is polysaccharide (Table 2–1). The terms **capsule**

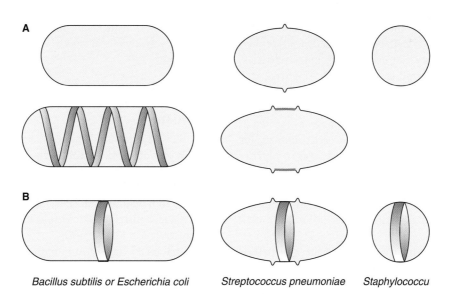

Bacillus subtilis or Escherichia coli   *Streptococcus pneumoniae*   *Staphylococcu*

FIGURE 2–20 Incorporation of new cell wall in differently shaped bacteria. Rod-shaped bacteria such as *Bacillus subtilis* or *Escherichia coli* have two modes of cell wall synthesis: new peptidoglycan is inserted along a helical path **(A)**, leading to elongation of the lateral wall, and is inserted in a closing ring around the future division site, leading to the formation of the division septum **(B)**. *Streptococcus pneumoniae* cells have the shape of a rugby ball and elongate by inserting new cell wall material at the so-called equatorial rings **(A)**, which correspond to an outgrowth of the cell wall that encircles the cell. An initial ring is duplicated, and the two resultant rings are progressively separated, marking the future division sites of the daughter cells. The division septum is then synthesized in the middle of the cell **(B)**. Round cells such as *Staphylococcus aureus* do not seem to have an elongation mode of cell wall synthesis. Instead, new peptidoglycan is inserted only at the division septum **(B)**. (Reproduced with permission from Scheffers DJ and Pinho MG: Microbiol Mol Biol Rev 2005;69:585.)

TABLE 2–1   Chemical Composition of the Extracellular Polymer in Selected Bacteria

Organism	Polymer	Chemical Subunits
Bacillus anthracis	Polypeptide	D-Glutamic acid
Enterobacter aerogenes	Complex polysaccharide	Glucose, fucose, glucuronic acid
Haemophilus influenzae	Serogroup b	Ribose, ribitol, phosphate
Neisseria meningitidis	Homopolymers and heteropolymers, eg,	
	Serogroup A	Partially O-acetylated N-acetylmannosaminephosphate
	Serogroup B	N-Acetylneuraminic acid (sialic acid)
	Serogroup C	Acetylated sialic acid
	Serogroup 135	Galactose, sialic acid
Pseudomonas aeruginosa	Alginate	D-Manuronic acid, L-glucuronic acid
Streptococcus pneumoniae	Complex polysaccharide (many types), eg,	
(pneumococcus)	Type II	Rhamnose, glucose, glucuronic acid
	Type III	Glucose, glucuronic acid
	Type VI	Galactose, glucose, rhamnose
	Type XIV	Galactose, glucose, N-acetylglucosamine
	Type XVIII	Rhamnose, glucose
Streptococcus pyogenes (group A)	Hyaluronic acid	N-Acetylglucosamine, glucuronic acid
Streptococcus salivarius	Levan	Fructose

and **slime layer** are frequently used to describe polysaccharide layers; the more inclusive term **glycocalyx** is also used. Glycocalyx is defined as the polysaccharide-containing material lying outside the cell. A condensed, well-defined layer closely surrounding the cell that excludes particles, such as India ink, is referred to as a capsule (Figure 2–21). If the glycocalyx is loosely associated with the cell and does not exclude particles, it is referred to as a slime layer. Extracellular polymer is synthesized by enzymes located at the surface of the bacterial cell. *Streptococcus mutans*, for example, uses two enzymes—glucosyl transferase and fructosyl transferase—to synthesize long-chain dextrans (poly-D-glucose) and levans (poly-D-fructose) from sucrose. These polymers are called **homopolymers**. Polymers containing more than one kind of monosaccharide are called **heteropolymers**.

The capsule contributes to the invasiveness of pathogenic bacteria—encapsulated cells are protected from phagocytosis unless they are coated with anticapsular antibody. The glycocalyx plays a role in the adherence of bacteria to surfaces in their environment, including the cells of plant and animal hosts. *S mutans*, for example, owes its capacity to adhere tightly to tooth enamel to its glycocalyx. Bacterial cells of the same or different species become entrapped in the glycocalyx, which forms the layer known as plaque on the tooth surface; acidic products excreted by these bacteria cause dental caries (Chapter 11). The essential role of the glycocalyx in this process—and its formation

from sucrose—explains the correlation of dental caries with sucrose consumption by the human population. Because outer polysaccharide layers bind a significant amount of water, the glycocalyx layer may also play a role in resistance to desiccation.

Flagella

A. Structure

Bacterial flagella are thread-like appendages composed entirely of protein, 12–30 nm in diameter. They are the organs of locomotion for the forms that possess them. Three types of arrangement are known: **monotrichous** (single polar flagellum), **lophotrichous** (multiple polar flagella), and **peritrichous** (flagella distributed over the entire cell). The three types are illustrated in Figure 2–22.

A bacterial flagellum is made up of several thousand molecules of a protein subunit called **flagellin**. In a few organisms (eg, caulobacter), flagella are composed of two types of flagellin, but in most only a single type is found. The flagellum is formed by the aggregation of subunits to form a helical structure. If flagella are removed by mechanically agitating a suspension of bacteria, new flagella are rapidly formed by the synthesis, aggregation, and extrusion of flagellin subunits; motility is restored within 3–6 minutes. The flagellins of different bacterial species presumably differ from one another in primary structure. They are highly antigenic (**H antigens**), and some of the

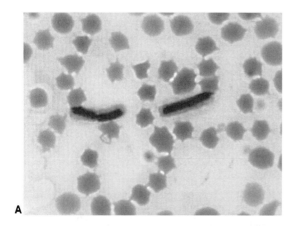

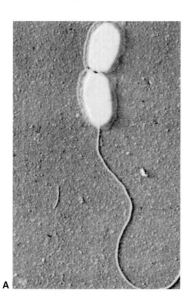

FIGURE 2–21 Bacterial capsules. **A**: *Bacillus anthracis* M'Faydean capsule stain, grown at 35°C, in defibrinated horse blood. **B**: Demonstration of the presence of a capsule in *Bacillus anthracis* by negative staining with India ink. This method is useful for improving visualization of encapsulated bacteria in clinical samples such as blood, blood culture bottles, or cerebrospinal fluid. (CDC, courtesy of Larry Stauffer, Oregon State Public Health Laboratory.)

immune responses to infection are directed against these proteins.

The flagellum is attached to the bacterial cell body by a complex structure consisting of a hook and a basal body. The hook is a short curved structure that appears to act as the universal joint between the motor in the basal structure and the flagellum. The basal body bears a set of rings, one pair in gram-positive bacteria and two pairs in gram-negative bacteria. An electron micrograph and interpretative diagrams of the gram-negative structure are shown in Figures 2–23 and 2–24; the rings labeled L and P are absent in gram-positive cells. The complexity of the bacterial flagellum is revealed by genetic studies, which show that over 40 gene products are involved in its assembly and function.

Flagella are made stepwise (Figure 2–24). First the basal body is assembled and inserted into the cell envelope. Then the hook is added, and finally, the filament is assembled progressively by the addition of flagellin subunits to its growing tip. The flagellin subunits are extruded through a hollow central channel in the flagella; when it reaches the tip it condenses with its predecessors, and thus the filament elongates.

B. Motility

Bacterial flagella are semirigid helical rotors to which the cell imparts a spinning movement. Rotation is driven by the flow of protons into the cell down the gradient produced by the primary proton pump (see above); in the absence of a metabolic energy source, it can be driven by a proton motive force generated by ionophores. Bacteria living in alkaline environments (alkalophiles) use the energy of the sodium ion gradient—rather than the proton gradient—to drive the flagellar motor (Figure 2–25).

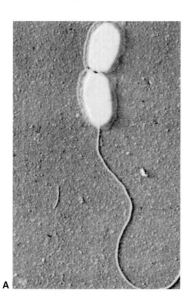

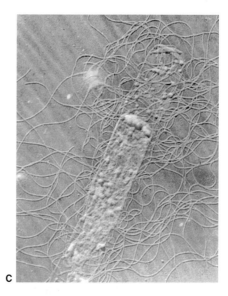

FIGURE 2–22 Bacterial flagellation. **A**: *Vibrio metchnikovii,* a monotrichous bacterium (7500×). (Reproduced with permission from van Iterson W: Biochim Biophys Acta 1947;1:527.) **B**: Electron micrograph of *Spirillum serpens,* showing lophotrichous flagellation (9000×). (Reproduced with permission from van Iterson W: Biochim Biophys Acta 1947;1:527.) **C**: Electron micrograph of *Proteus vulgaris,* showing peritrichous flagellation (9000×). Note basal granules. (Reproduced with permission from Houwink A, van Iterson W: Biochim Biophys Acta 1950;5:10.)

FIGURE 2-23 Electron micrograph of a negatively stained lysate of *Rhodospirillum molischianum,* showing the basal structure of an isolated flagellum. (Reproduced with permission from Cohen-Bazire G, London L: Basal organelles of bacterial flagella. J Bacteriol 1967;94:458.)

All of the components of the flagellar motor are located in the cell envelope. Flagella attached to isolated, sealed cell envelopes rotate normally when the medium contains a suitable substrate for respiration or when a proton gradient is artificially established.

When a peritrichous bacterium swims, its flagella associate to form a posterior bundle that drives the cell forward in a straight line by counterclockwise rotation. At intervals, the flagella reverse their direction of rotation and momentarily dissociate, causing the cell to tumble until swimming resumes in a new, randomly determined direction. This behavior makes possible the property of **chemotaxis:** A cell that is moving away from the source of a chemical attractant tumbles and reorients itself more frequently than one that is moving toward the attractant, the result being the net movement of the cell toward the source. The presence of a chemical attractant (such as a sugar or an amino acid) is sensed by specific receptors located in the cell membrane (in many cases, the same receptor also participates in membrane transport of that molecule). The bacterial cell is too small to be able to detect the existence of a spatial chemical gradient (ie, a gradient between its two poles); rather, experiments show that it detects temporal gradients, ie, concentrations that decrease with time during which the cell is moving away from the attractant source and increase with time during which the cell is moving toward it.

Some compounds act as repellants rather than attractants. One mechanism by which cells respond to attractants and repellents involves a cGMP-mediated methylation and demethylation of specific proteins in the membrane. Attractants cause a transient inhibition of demethylation of these proteins, while repellents stimulate their demethylation.

The mechanism by which a change in cell behavior is brought about in response to a change in the environment is called **sensory transduction**. Sensory transduction is responsible not only for chemotaxis but also for **aerotaxis** (movement toward the optimal oxygen concentration), **phototaxis** (movement of photosynthetic bacteria toward the light), and **electron acceptor taxis** (movement of respiratory bacteria toward alternative electron acceptors, such as nitrate

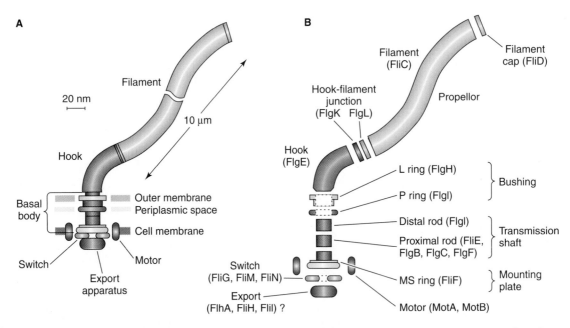

FIGURE 2-24 **A:** General structure of the flagellum of a gram-negative bacterium, such as *Escherichia coli* or *Salmonella typhimurium.* The filament-hook-basal body complex has been isolated and extensively characterized. The location of the export apparatus has not been demonstrated. **B:** An exploded diagram of the flagellum showing the substructures and the proteins from which they are constructed. The FliF protein is responsible for the M-ring feature, S-ring feature, and collar feature of the substructure shown, which is collectively termed the MS ring. The location of FliE with respect to the MS ring and the rod—and the order of the FlgB, FlgC, and FlgF proteins within the proximal rod—is not known. (From Macnab RM: Genetics and biogenesis of bacterial flagella. Annu Rev Genet 1992;26:131. Reproduced with permission from *Annual Review of Genetics,* Volume 26, © 1992 by Annual Reviews.)

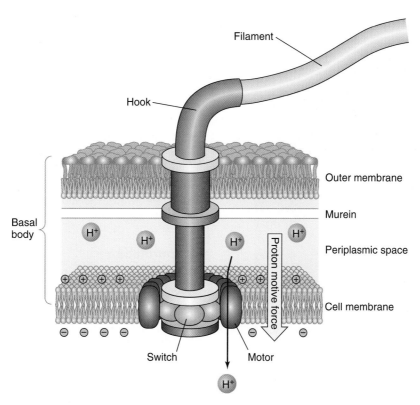

Filament

Hook

Basal body

H⁺ H⁺ H⁺ H⁺

Outer membrane

Murein

Periplasmic space

Proton motive force

Cell membrane

Switch   Motor

H⁺

FIGURE 2–25 Structural components within the basal body of the flagellum allow the inner portion of this structure, the rods of the basal body, and the attached hook-filament complex to rotate. The outer rings remain statically in contact with the inner and outer cell membranes and cell wall (murein), anchoring the flagellum complex to the bacterial cell envelope. Rotation is driven by the flow of protons through the motor from the periplasmic space, outside the cell membrane, into the cytoplasm in response to the electric field and proton gradient across the membrane, which together constitute the proton motive force. A switch determines the direction of rotation, which in turn determines whether the bacteria swim forward (due to counterclockwise rotation of the flagellum) or tumble (due to clockwise rotation of the flagellum). (Reproduced with permission from Saier MH Jr: Peter Mitchell and his chemiosmotic theories. ASM News 1997;63:13.)

and fumarate). In these three responses, as in chemotaxis, net movement is determined by regulation of the tumbling response.

Pili (Fimbriae)

Many gram-negative bacteria possess rigid surface appendages called **pili** (L "hairs") or **fimbriae** (L "fringes"). They are shorter and finer than flagella; like flagella, they are composed of structural protein subunits termed **pilins**. Some pili contain a single type of pilin, others more than one. Minor proteins termed **adhesins** are located at the tips of pili and are responsible for the attachment properties. Two classes can be distinguished: ordinary pili, which play a role in the adherence of symbiotic and pathogenic bacteria to host cells, and sex pili, which are responsible for the attachment of donor and recipient cells in bacterial conjugation (see Chapter 7). Pili are illustrated in Figure 2–26, in which the sex pili have been coated with phage particles for which they serve as specific receptors.

Motility via pili is completely different from flagellar motion. Pilin molecules are arranged helically to form a straight cylinder that does not rotate and lacks a complete basal body. Their tips strongly adhere to surfaces at a distance from the cells. Pili then depolymerize from the inner end, thus retracting inside the cell. The result is that the bacterium moves in the direction of the adhering tip. This kind of surface motility is called **twitching** and is widespread among piliated bacteria. Unlike flagella, pili grow from the inside of the cell outward.

The virulence of certain pathogenic bacteria depends on the production not only of toxins but also of "colonization antigens," which are ordinary pili that provide the cells with adherent properties. In enteropathogenic *E coli* strains, both the enterotoxins and the colonization antigens (pili) are genetically determined by transmissible plasmids, as discussed in Chapter 7.

In one group of gram-positive cocci, the streptococci, fimbriae are the site of the main surface antigen, the M protein. Lipoteichoic acid, associated with these fimbriae, is responsible for the adherence of group A streptococci to epithelial cells of their hosts.

Pili of different bacteria are antigenically distinct and elicit the formation of antibodies by the host. Antibodies against the pili of one bacterial species will not prevent the attachment of another species. Some bacteria (see Chapter 21), such as *N gonorrhoeae*, are able to make pili of different

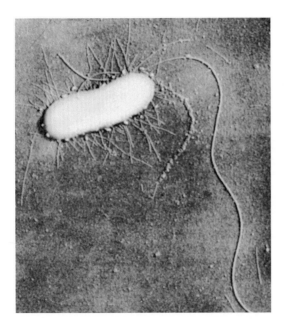

FIGURE 2–26 Surface appendages of bacteria. Electron micrograph of a cell of *Escherichia coli* possessing three types of appendages: ordinary pili (short, straight bristles), a sex pilus (longer, flexible, with phage particles attached), and several flagella (longest, thickest). Diameters: ordinary pili: 7 nm; sex pili: 8.5 nm; flagella: 25 nm. (Courtesy of J Carnahan and C Brinton.)

antigenic types (**antigenic variation**) and thus can still adhere to cells in the presence of antibodies to their original type of pili. Like capsules, pili inhibit the phagocytic ability of leukocytes.

Endospores

Members of several bacterial genera are capable of forming **endospores** (see Figure 2–27). The two most common are gram-positive rods: the obligately aerobic genus *Bacillus*

and the obligately anaerobic genus *Clostridium*. The other bacteria known to form endospores are *Thermoactinomyces, Sporolactobacillus, Sporosarcina, Sporotomaculum, Sporomusa,* and *Sporohalobacter*. These organisms undergo a cycle of differentiation in response to environmental conditions: The process, **sporulation**, is triggered by near depletion of any of several nutrients (carbon, nitrogen, or phosphorous). Each cell forms a single internal spore that is liberated when the mother cell undergoes autolysis. The spore is a resting cell, highly resistant to desiccation, heat, and chemical agents; when returned to favorable nutritional conditions and activated (see below), the spore **germinates** to produce a single vegetative cell.

A. Sporulation

The sporulation process begins when nutritional conditions become unfavorable, near depletion of the nitrogen or carbon source (or both) being the most significant factor. Sporulation occurs massively in cultures that have terminated exponential growth as a result of this near depletion.

Sporulation involves the production of many new structures, enzymes, and metabolites along with the disappearance of many vegetative cell components. These changes represent a true process of **differentiation**: A series of genes whose products determine the formation and final composition of the spore are activated. These changes involve alterations in the transcriptional specificity of RNA polymerase, which is determined by the association of the polymerase core protein with one or another promoter-specific protein called a **sigma factor**. During vegetative growth, a sigma factor designated σ^A predominates. Then, during sporulation, five other sigma factors are formed that cause various spore genes to be expressed at various times in specific locations.

The sequence of events in sporulation is highly complex: Differentiation of a vegetative cell of *B subtilis* into an endospore takes about 7 hours under laboratory conditions.

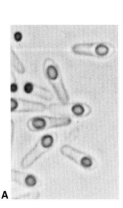
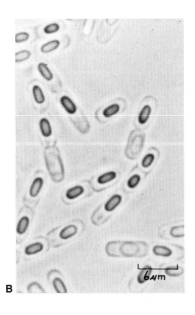
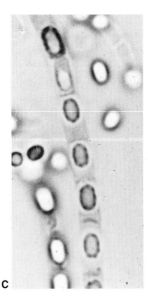

FIGURE 2–27 Sporulating cells of bacillus species. **A:** Unidentified bacillus from soil. **B:** *Bacillus cereus.* **C:** *Bacillus megaterium.* (Reproduced with permission from Robinow CF, in: Structure. Vol 1 of: *The Bacteria: A Treatise on Structure and Function.* Gunsalus IC, Stanier RY [editors]. Academic Press, 1960.)

A   B   6 μm   C

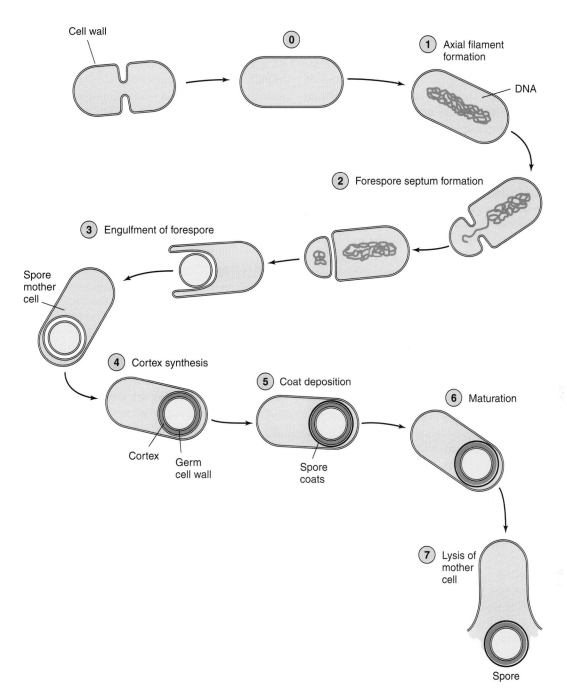

FIGURE 2–28 The stages of endospore formation. (Reproduced with permission from Merrick MJ: Streptomyces. In: *Developmental Biology of Procaryotes*. Parish JH [editor]. Univ California Press, 1979.)

Different morphologic and chemical events occur at sequential stages of the process. Seven different stages have been identified.

Morphologically, sporulation begins with the formation of an axial filament (Figure 2–28). The process continues with an infolding of the membrane so as to produce a double membrane structure whose facing surfaces correspond to the cell wall-synthesizing surface of the cell envelope. The growing points move progressively toward the pole of the cell so as to engulf the developing spore.

The two spore membranes now engage in the active synthesis of special layers that will form the cell envelope: the **spore wall** and the **cortex**, lying outside the facing membranes. In the newly isolated cytoplasm, or core, many vegetative cell enzymes are degraded and are replaced by a set of unique spore constituents.

B. Properties of Endospores

1. Core—The core is the spore protoplast. It contains a complete nucleus (chromosome), all of the components of the protein-synthesizing apparatus, and an energy-generating system based on glycolysis. Cytochromes are lacking even in aerobic species, the spores of which rely on a shortened electron transport pathway involving flavoproteins. A number of vegetative cell enzymes are increased in amount (eg, alanine racemase), and a number of unique enzymes are formed (eg, dipicolinic acid synthetase). Spores contain no reduced pyridine nucleotides or ATP. The energy for germination is stored as 3-phosphoglycerate rather than as ATP.

The heat resistance of spores is due in part to their dehydrated state and in part to the presence in the core of large amounts (5–15% of the spore dry weight) of **calcium dipicolinate**, which is formed from an intermediate of the lysine biosynthetic pathway (see Figure 6–18). In some way not yet understood, these properties result in the stabilization of the spore enzymes, most of which exhibit normal heat lability when isolated in soluble form.

2. Spore wall—The innermost layer surrounding the inner spore membrane is called the spore wall. It contains normal peptidoglycan and becomes the cell wall of the germinating vegetative cell.

3. Cortex—The cortex is the thickest layer of the spore envelope. It contains an unusual type of peptidoglycan, with many fewer cross-links than are found in cell wall peptidoglycan. Cortex peptidoglycan is extremely sensitive to lysozyme, and its autolysis plays a role in spore germination.

4. Coat—The coat is composed of a keratin-like protein containing many intramolecular disulfide bonds. The impermeability of this layer confers on spores their relative resistance to antibacterial chemical agents.

5. Exosporium—The exosporium is composed of proteins, lipids, and carbohydrates. It consists of a paracrystalline basal layer and a hair-like outer region. The function of the exosporium is unclear. Spores of some *Bacillus* species (eg, *B anthracis* and *B cereus*) possess an exosporium, while other species (eg, *B atrophaeus*) have spores that lack this structure.

C. Germination

The germination process occurs in three stages: activation, initiation, and outgrowth.

1. Activation—Most endospores cannot germinate immediately after they have formed. But they can germinate after they have rested for several days or are first activated, in a nutritionally rich medium, by one or another agent that damages the spore coat. Among the agents that can overcome spore dormancy are heat, abrasion, acidity, and compounds containing free sulfhydryl groups.

2. Initiation—Once activated, a spore will initiate germination if the environmental conditions are favorable. Different species have evolved receptors that recognize different effectors as signaling a rich medium: Thus, initiation is triggered by L-alanine in one species and by adenosine in another. Binding of the effector activates an autolysin that rapidly degrades the cortex peptidoglycan. Water is taken up, calcium dipicolinate is released, and a variety of spore constituents are degraded by hydrolytic enzymes.

3. Outgrowth—Degradation of the cortex and outer layers results in the emergence of a new vegetative cell consisting of the spore protoplast with its surrounding wall. A period of active biosynthesis follows; this period, which terminates in cell division, is called outgrowth. Outgrowth requires a supply of all nutrients essential for cell growth.

STAINING

Stains combine chemically with the bacterial protoplasm; if the cell is not already dead, the staining process itself will kill it. The process is thus a drastic one and may produce artifacts.

The commonly used stains are salts. **Basic stains** consist of a colored cation with a colorless anion (eg, methylene blue$^+$ chloride$^-$); **acidic stains** are the reverse (eg, sodium$^+$ eosinate$^-$). Bacterial cells are rich in nucleic acid, bearing negative charges as phosphate groups. These combine with the positively charged basic dyes. Acidic dyes do not stain bacterial cells and hence can be used to stain background material a contrasting color (see Negative Staining below).

The basic dyes stain bacterial cells uniformly unless the cytoplasmic RNA is destroyed first. Special staining techniques can be used, however, to differentiate flagella, capsules, cell walls, cell membranes, granules, nucleoids, and spores.

The Gram Stain

An important taxonomic characteristic of bacteria is their response to Gram stain. The Gram-staining property appears to be a fundamental one, since the Gram reaction is correlated with many other morphologic properties in phylogenetically related forms (Chapter 3). An organism that is potentially gram-positive may appear so only under a particular set of environmental conditions and in a young culture.

The Gram-staining procedure (see Chapter 47 for details) begins with the application of a basic dye, crystal violet. A solution of iodine is then applied; all bacteria will be stained blue at this point in the procedure. The cells are then treated with alcohol. Gram-positive cells retain the crystal violet-iodine complex, remaining blue; gram-negative cells are completely decolorized by alcohol. As a last step, a counterstain (such as the red dye safranin) is applied so that the decolorized gram-negative cells will take on a contrasting color; the gram-positive cells now appear purple.

The basis of the differential Gram reaction is the structure of the cell wall, as discussed earlier in this chapter.

The Acid-Fast Stain

Acid-fast bacteria are those that retain carbolfuchsin (basic fuchsin dissolved in a phenol-alcohol-water mixture) even when decolorized with hydrochloric acid in alcohol. A smear of cells on a slide is flooded with carbolfuchsin and heated on a steam bath. Following this, the discolorization with acid-alcohol is carried out, and finally a contrasting (blue or green) counterstain is applied (see Chapter 47). Acid-fast bacteria (mycobacteria and some of the related actinomycetes) appear red; others take on the color of the counterstain.

Negative Staining

This procedure involves staining the background with an acidic dye, leaving the cells contrastingly colorless. The black dye nigrosin is commonly used. This method is used for those cells or structures difficult to stain directly (Figure 2–21B).

The Flagella Stain

Flagella are too fine (12–30 nm in diameter) to be visible in the light microscope. However, their presence and arrangement can be demonstrated by treating the cells with an unstable colloidal suspension of tannic acid salts, causing a heavy precipitate to form on the cell walls and flagella. In this manner, the apparent diameter of the flagella is increased to such an extent that subsequent staining with basic fuchsin makes the flagella visible in the light microscope. Figure 2–29 shows cells stained by this method.

In peritrichous bacteria, the flagella form into bundles during movement, and such bundles may be thick enough to be observed on living cells by dark-field or phase contrast microscopy.

The Capsule Stain

Capsules are usually demonstrated by the negative staining procedure or a modification of it (Figure 2–21). One such "capsule stain" (Welch method) involves treatment with hot

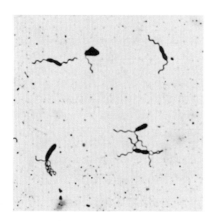

FIGURE 2–29 Flagella stain of *Pseudomonas* species. (Reproduced with permission from Leifson E: J Bacteriol 1951;62:377.)

crystal violet solution followed by a rinsing with copper sulfate solution. The latter is used to remove excess stain because the conventional washing with water would dissolve the capsule. The copper salt also gives color to the background, with the result that the cell and background appear dark blue and the capsule a much paler blue.

Staining of Nucleoids

Nucleoids are stainable with the Feulgen stain, which is specific for DNA (see Figure 2–4).

The Spore Stain

Spores are most simply observed as intracellular refractile bodies in unstained cell suspensions or as colorless areas in cells stained by conventional methods (Figure 2–27). The spore wall is relatively impermeable, but dyes can be made to penetrate it by heating the preparation. The same impermeability then serves to prevent decolorization of the spore by a period of alcohol treatment sufficient to decolorize vegetative cells. The latter can finally be counterstained. Spores are commonly stained with malachite green or carbolfuchsin.

MORPHOLOGIC CHANGES DURING GROWTH

Cell Division

Most bacteria divide by **binary fission** into two equal progeny cells. In a growing culture of a rod-shaped bacterium such as *E coli*, cells elongate and then form a partition that eventually separates the cell into two daughter cells. The partition is referred to as a **septum** and is a result of the inward growth of the cytoplasmic membrane and cell wall from opposing directions until the two daughter cells are pinched off. The chromosomes, which have doubled in number preceding the division, are distributed equally to the two daughter cells.

Although bacteria lack a mitotic spindle, the septum is formed in such a way as to separate the two sister chromosomes formed by chromosomal replication. This is accomplished by the attachment of the chromosome to the cell membrane. According to one model, completion of a cycle of DNA replication triggers active membrane synthesis between the sites of attachment of the two sister chromosomes. The chromosomes are then pushed apart by the inward growth of the septum, one copy going to each daughter cell.

Cell Groupings

If the cells remain temporarily attached following division, certain characteristic groupings result. Depending on the plane of division and the number of divisions through which the cells remain attached, the following may occur in the coccal forms: chains (streptococci), pairs (diplococci), cubical bundles (sarcinae), or flat plates. Rods may form pairs or chains.

Following fission of some bacteria, characteristic post-division movements occur. For example, a "whipping" motion can bring the cells into parallel positions; repeated division and whipping result in the "palisade" arrangement characteristic of diphtheria bacilli.

REVIEW QUESTIONS

1. A 22-year-old man presents with a painless 1-cm ulcer on the shaft of his penis. Inguinal lymphadenopathy is present. The patient admits trading drugs for sex and has several sexual partners. An RPR test is positive, and syphilis is suspected; however, a Gram stain of a swab specimen from the ulcer shows no bacteria. *Treponema pallidum*, the causative agent of syphilis, cannot be visualized by light microscopy because

   (A) It is transparent
   (B) It cannot be stained by ordinary stains
   (C) It has a diameter of <0.2 μm
   (D) The wavelength of white light is too long
   (E) Rapid movement of the organism prevents visualization

2. Chloramphenicol, an antibiotic that inhibits bacterial protein synthesis, will also affect which of the following eukaryotic organelles?

   (A) Mitochondria
   (B) Golgi apparatus
   (C) Microtubules
   (D) Endoplasmic reticulum
   (E) Nuclear membrane

3. Which of the following structures is not part of the bacterial cell envelope?

   (A) Peptidoglycan
   (B) Lipopolysaccharide
   (C) Capsule
   (D) Gas vacuole
   (E) S-layer

4. Which of the following transport mechanisms functions without the requirement for energy?

   (A) Binding protein-dependent
   (B) Group translocation
   (C) Symport
   (D) Uniport
   (E) Facilitated diffusion

5. Which of the following components is present in gram-negative bacteria but not in gram-positive bacteria?

   (A) Peptidoglycan
   (B) Lipid A
   (C) Capsule
   (D) Flagella
   (E) Pili

6. Which of the following components is present in gram-positive bacteria but not in gram-negative bacteria?

   (A) Peptidoglycan
   (B) Capsule
   (C) Flagella
   (D) Teichoic acid
   (E) Diaminopimelic acid

7. In the fall of 2001, a series of letters containing spores of *Bacillus anthracis* were mailed to members of the media and to U.S. Senate offices. The result was 22 cases of anthrax, with 5 deaths. The heat resistance of bacterial spores, such as those of *Bacillus anthracis,* is due in part to their dehydrated state and in part to the presence of large amounts of

   (A) Diaminopimelic acid
   (B) D-Glutamic acid
   (C) Calcium dipicolinate
   (D) Sulfhydryl-containing proteins
   (E) Lipid A

Answers

1. C	3. D	5. B	7. C
2. A	4. E	6. D	

REFERENCES

Balows A et al (editors): The Prokaryotes, A Handbook on the Biology of Bacteria: Ecophysiology, Isolation, Identification, Applications, 2nd ed, 4 vols. Springer, 1992.

Barreteau H, Kovac A, Boniface A, Sova M, Gobec S, Blanot D: Cytoplasmic steps of peptidoglycan biosynthesis. FEMS Microbiol Rev 2008;32:168.

Barton LL: Structural and Functional Relationships in Prokaryotes. Springer, 2005.

Bermudes D, Hinkle G, Margulis L: Do prokaryotes contain microtubules? Microbiol Rev 1994;58:387. [PMID: 7968920]

Blair DF: How bacteria sense and swim. Annu Rev Microbiol 1995;49:489. [PMID: 8561469]

Craig L, Pique ME, Tainer JA: Type IV pilus structure and bacterial pathogenicity. Nature Rev Microbiol 2004;2:363.

Dautin N, Bernstein HD: Protein secretion in gram-negative bacteria via the autotransporter pathway. Annu Rev Microbiol 2007;61:89.

Drlica K, Riley M (editors): The Bacterial Chromosome. American Society for Microbiology, 1990.

Economou A, Christie PJ, Fernandez RC, Palmer T, Plano GV, Pugsley AP: Secretion by the numbers: Protein traffic in prokaryotes. Mol Microbiol 2006;62:308.

Henriques AO, Moran CP Jr: Structure, assembly, and function of the spore surface layers. Annu Rev Microbiol 2007;61:555.

Hinnebusch J, Tilly K: Linear plasmids and chromosomes in bacteria. Mol Microbiol 1993;10:917. [PMID: 7934868]

Hueck CJ: Type III protein secretion systems in bacterial pathogens of animals and plants. Microbiol Mol Biol Rev 1998;62:379. [PMID: 9618447]

Leiman PG et al: Type VI secretion apparatus and phage tail-associated protein complexes share a common evolutionary origin. Proc Natl Acad Sci USA 2009;106:4154.

Liu J, Barry CE III, Besra GS, Nikaido H: Mycolic acid structure determines the fluidity of the mycobacterial cell wall. J Biol Chem 1996;271:29545.

Messner P et al: III. Biochemistry of S-layers. FEMS Microbiol Rev 1997;20:25. [PMID: 9276927]

Moat AG, Foster JW: *Microbial Physiology,* 3rd ed. Wiley-Liss, 1995.

Naroninga N: Morphogenesis of *Escherichia coli.* Microbiol Mol Biol Rev 1998;62:110.

Neuhaus FC, Baddiley J: A continuum of anionic charge: Structures and functions of D-alanyl-teichoic acids in gram-positive bacteria. Microbiol Mol Biol Rev 2003;67:686.

Nikaido H: Molecular basis of bacterial outer membrane permeability revisited. Microbiol Mol Biol Rev 2003;67:593.

Rachel R et al: Fine structure of S-layers. FEMS Microbiol Rev 1997;20:13.

Sauvage E, Kerff F, Terrak M, Ayala JA, Charlier P: The penicillin-binding proteins: Structure and role in peptidoglycan biosynthesis. FEMS Microbiol Rev 2008;32:234.

Schaechter M, Ingraham JL, Neidhardt FC: *Microbe*. American Society for Microbiology, 2006.

Scheffers DJ, Pinho MG: Bacterial cell wall synthesis: New insights from localization studies. Microbiol Mol Biol Rev 2005;69:585.

Schirmer T: General and specific porins from bacterial outer membranes. J Struct Biol 1998;121:101. [PMID: 9615433]

Scott JR, Barnett TC: Surface proteins of gram-positive bacteria and how they get there. Annu Rev Microbiol 2006;60:397.

Sonenshein AL, Hoch JA, Losick R: Bacillus subtilis *and Its Closest Relatives*. American Society for Microbiology, 2002.

Vaara M: Agents that increase the permeability of the outer membrane. Microbiol Rev 1992;56:395. [PMID: 1406489]

Vollmer W, Blanot D, de Pedro MA: Peptidoglycan structure and architecture. FEMS Microbiol Rev 2008;32:149.

Walsby AE: Gas vesicles. Microbiol Rev 1994;58:94. [PMID: 8177173]

Whittaker CJ, Klier CM, Kolenbrander PE: Mechanisms of adhesion by oral bacteria. Annu Rev Microbiol 1996;50:513. [PMID: 8905090]

Classification of Bacteria

TAXONOMY—THE VOCABULARY OF MEDICAL MICROBIOLOGY

One has only to peruse the table of contents of this book to appreciate the diversity of medical pathogens that are associated with infectious diseases. It has been estimated that we currently have the capacity to identify less than ten percent of the pathogens responsible for causing human disease because of our current inability to culture or target these organisms using molecular probes. Yet the diversity of even these identifiable pathogens is so great that it is important to understand the subtle differences associated with each infectious agent. The reason for understanding these subtleties is significant is that each of these infectious agents has specifically adapted to a mode(s) of transmission, a mechanism to infect the human host (colonization), and a mechanism to cause disease (pathology). As such, a vocabulary that consistently communicates the unique characteristics of infectious organisms to students, microbiologists, and healthcare workers is critical to avoid the chaos that would ensue without the organizational restraints of bacterial **taxonomy** (*Gk. taxon* = arrangement; eg, the classification of organisms in an ordered system that indicates a natural relationship).

Classification, **nomenclature**, and **identification** are three separate but interrelated areas of bacterial taxonomy. **Classification** is the categorization of organisms into taxonomic groups. Classification of bacteria () requires experimental and observational techniques; this is because biochemical, physiologic, genetic, and morphologic properties are often necessary for an adequate description of a taxon. **Nomenclature** refers to the naming of an organism by international rules (established by a recognized group of medical professionals) according to its characteristics. **Identification** refers to the practical use of a classification scheme to: (1) isolate and distinguish desirable organisms from undesirable ones, (2) verify the authenticity or special properties of a culture in a clinical setting, and (3) isolate and identify the causative agent of a disease. The latter may lead to the selection of specific pharmacologic treatments directed toward their eradication (Chapter 28), a vaccine mitigating their pathology, or a public health measure (eg, hand washing or use of a condom) that prevents their transmission.

Identification schemes are not classification schemes, though there may be some superficial similarity. An identification scheme for a group of organisms can be devised only after that group has first been classified, ie, recognized as being different from other organisms. For example, the popular literature has reported *Escherichia coli* as being a cause of hemolytic uremic syndrome (HUS) in infants. There are hundreds of different strains that are classified as *E coli* but only a few that are associated with HUS. These strains can be distinguished from the many *E coli* strains by antibody reactivity with their O- and H-antigens, as described in Chapter 2 (for example, *E coli* O157:H7). Taxonomy, and the nomenclature that accompanies it, is an imprecise and evolving science. Just as our societal vocabulary evolves, so does the vocabulary of medical microbiology. Any professional associated with infectious disease should be aware of the evolving taxonomy of infectious microorganisms.

Taxonomic ranks form the basis for the organization of bacteria. Linnaean taxonomy is the system most familiar to biologists. It uses the formal taxonomic ranks (in order) kingdom, phylum, class, order, family, genus, and species. The lower ranks are approved by a consensus of experts in the scientific community (see Table 3–1). Of these ranks the family, genus, and species are the most useful.

TABLE 3–1 Taxonomic Ranks

Formal Rank	Example
Kingdom	Prokaryotae
Division	Gracilicutes
Class	Scotobacteria
Order	Eubacteriales
Family	Enterobacteriaceae
Genus	*Escherichia*
Species	*coli*
Subtype	*Escherichia coli* O157: H7

CRITERIA FOR CLASSIFICATION OF BACTERIA

Growth on Media

Suitable criteria for purposes of bacterial classification include many of the properties that were described in the preceding chapter. One criterion is the growth on bacteriologic media. In contrast to viruses and parasites, many bacterial pathogens can be isolated on solid agar-containing media. The general cultivation of most bacteria requires media rich in metabolic nutrients. These media generally include agar, a carbon source, and an acid hydrolysate or enzymatically degraded source of biologic material (eg, casein). Because of the undefined composition of the latter, these types of media are referred to as **complex media**.

Clinical samples from normally nonsterile sites (eg, the throat or the colon) contain more than one type of organism, including potential pathogens and resident microbial flora. Media can be **nonselective** or **selective** and are used to distinguish among the various bacteria in a clinical sample containing many different organisms.

1. Nonselective media

Blood agar and chocolate agar are examples of complex, nonselective media, which support the growth of many different bacteria. Nonselective media are important for the isolation of unknown bacteria from a specimen. Numerous types of bacterial colonies are commonly observed when clinical specimens are inoculated onto nonselective media.

2. Selective media

Because of the diversity of microorganisms that typically reside at some sampling sites (eg, the intestinal tract), selective media are used to eliminate (or reduce) the large numbers of irrelevant bacteria in these specimens. The basis for selective media is the incorporation of an inhibitory agent that specifically selects against the growth of irrelevant bacteria. Examples of such agents are:

- Sodium azide—selects for gram-positive bacteria over gram-negative bacteria
- Bile salts (eg, sodium deoxycholate)—select for gram-negative enteric bacteria, and inhibit gram-negative mucosal and most gram-positive bacteria
- Colistin and nalidixic acid—inhibit the growth of many gram-negative bacteria

Examples of selective media are MacConkey agar (contains bile) that selects for the Enterobacteriaceae and CNA blood agar (contains colistin and nalidixic acid) that selects for Staphylococci and Streptococci.

3. Differential media

Upon culture, some bacteria produce characteristic pigments, and others can be differentiated on the basis of their complement of extracellular enzymes; the activity of these enzymes often can be detected as zones of clearing surrounding colonies grown in the presence of insoluble substrates (eg, zones of **hemolysis** in agar medium containing red blood cells). Many of the members of the Enterobacteriaceae can be differentiated on the basis of their ability to metabolize lactose. For example, pathogenic salmonellae and shigellae do not ferment lactose and on a MacConkey plate form clear colonies, whereas lactose-fermenting members of the Enterobacteriaceae (eg, *E coli*) form red or pink colonies. The number of differential media employed in today's clinical laboratories is far beyond the scope of this chapter. However, it should be noted that biochemical identification is an important means to classify microbial pathogens.

Bacterial microscopy

Appropriately associated bacteria on allows them to be examined using appropriately stained samples. Historically, the Gram stain, together with visualization by light microscopy, has been among the most informative methods for classifying the eubacteria. This staining technique is generally the first step in broadly dividing bacteria on the basis of fundamental differences in the structure of their cell walls (see Chapter 2). Gram-positive bacteria have a thick mesh-like cell wall made of peptidoglycan (50–90% by weight of the cell envelope), while gram-negative bacteria have a thinner layer (10% by weight of the cell envelope). The Gram stain procedure is described elsewhere in this text.

Biochemical tests

Tests such as the **oxidase test**, which uses an artificial electron acceptor, can be used to distinguish organisms on the basis of the presence or absence of a respiratory enzyme, cytochrome C, the lack of which differentiates the Enterobacteriaceae from other gram-negative rods. Similarly, **catalase** activity can be used, for example, to differentiate between the gram-positive cocci. Antimicrobial sensitivity (eg, colistin and/or nalidixic acid) may also be used. Ultimately there are many examples of biochemical tests that can ascertain the presence of characteristic metabolic functions and be used to group bacteria into a specific **taxon**.

Immunologic Tests—Serotypes, serogroups, and serovars

The designation "sero" simply indicates the use of antibodies (polyclonal or monoclonal) that react with specific bacterial cell surface structures such as lipopolysaccharide (LPS), flagella, or capsular antigens. The terms "serotype," "serogroups," and "serovars" are, for all practical purposes, identical—they all utilize the specificity of these antibodies to subdivide strains of a particular bacterial species.

Genetic Instability

The value of a taxonomic criterion depends upon the biologic group being compared. Traits shared by all or none of the members of a group cannot be used to distinguish its members, but they may define a group (eg, all staphylococci produce the enzyme catalase). Developments in molecular biology now make it possible to investigate the relatedness of genes or genomes by comparing sequences among different bacteria (Chapter 7). For these cases genetic instability can cause some traits to be highly variable within a biologic group or even within a specific taxonomic group. For example, antibiotic resistance genes or genes encoding enzymes (lactose utilization, etc.) may be carried on **plasmids** or **bacteriophages** (Chapter 7), extrachromosomal genetic elements that may be transferred among unrelated bacteria or that may be lost from a subset of bacterial strains identical in all other respects. Many organisms are difficult to cultivate, and in these instances techniques that reveal relatedness by measurement of nucleic acid hybridization or by DNA sequence analysis may be of particular value.

CLASSIFICATION SYSTEMS

Keys

Keys organize bacterial traits in a manner that permits efficient identification of organisms. The ideal system should contain the minimum number of features required for a correct categorization. Groups are split into smaller subgroups based upon the presence (+) or absence (–) of a diagnostic character. Continuation of the process with different characteristics guides the investigator to the smallest defined subgroup containing the analyzed organism. In the early stages of this process, organisms may be assigned to subgroups on the basis of characteristics that do not reflect genetic relatedness. It would be perfectly reasonable, for example, for a bacterial key to include a group such as "bacteria forming red pigments when propagated on a defined medium," even though this would include such unrelated forms as *Serratia marcescens* (Chapter 15) and purple photosynthetic bacteria (Chapter 6). These two disparate bacterial assemblages occupy distinct niches and depend upon entirely different forms of energy metabolism. Nevertheless, preliminary grouping of the assemblages would be useful because it would immediately make it possible for an investigator having to identify a red-pigmented culture to narrow the range of possibilities to relatively groups.

Numerical Taxonomy

Numerical taxonomy (also referred to as phenetics or taxometrics) became widely used in the 1970s. These classification schemes use a large number (frequently greater than 100) of unweighted taxonomically useful characteristics. The Analytical Profile Index (API) is a method commonly used to identify a wide range of microorganisms. APIs consist of a number of plastic strips, each of which has about 20 miniature compartments containing biochemical reagents. Almost all cultivatable bacterial groups and more than 550 different species can be identified using the results of these API tests. These identification systems have extensive databases of microbial biochemical reactions. The numerical clusters derived from these tests identify different strains at selected levels of overall similarity (usually >80% at the species level) on the basis of the frequency with which they share traits. In addition, numerical classification provides percentage frequencies of positive character states for all strains within each cluster. The limitation of this approach is that it is a **static system**, which does not allow for the evolution of bacteria and routine discovery of new bacterial pathogens.

Phylogenetic Classification: Toward an Understanding of Evolutionary Relationships among Bacteria

Phylogenetic classifications are measures between two organisms and imply that they share a common ancestor. The fossil record has made such inferences relatively easy to draw for many representatives of plants and animals. However, no such record exists for bacteria, and in the absence of molecular evidence, the distinction between **convergent** and **divergent evolution** for bacterial traits can be difficult to establish.

The genetic properties of bacteria allow genes to be exchanged among distantly related organisms. Furthermore, multiplication of bacteria is almost entirely vegetative, and their mechanisms of genetic exchange rarely involve recombination among large portions of their genomes (Chapter 7). Therefore, the concept of a **species**—the fundamental unit of eukaryotic phylogeny—has an entirely different meaning when applied to bacteria. A eukaryotic species is a biologic group capable of interbreeding, with the ultimate intent to produce variable offspring. Because bacteria replicate clonally by binary fission, they do not require a complementary set of chromosomes in order to reproduce. Consequently, the definition of a bacterial species is necessarily pragmatic and is operationally defined. For the purposes of categorizing bacteria, a **species** is a genomically coherent group of individual isolates or strains sharing a high degree of similarity in many independent features, when comparably tested under highly standardized conditions. The decision to circumscribe clusters of organisms within a bacterial species is made by the taxonomist, who may choose to subdivide the group into **biotypes** and to cluster species within genera. Broader groupings such as families may be proposed. The formal ranks used in the taxonomy of bacteria are listed in Table 3–1. For practical purposes, only the ranks of the family, genus, and species are commonly used. There is considerable genetic diversity among bacteria. Chemical characterization of bacterial genomic DNA reveals a wide range of nucleotide base compositions among different bacterial strains. The guanine + cytosine (G + C) content of closely related bacteria is similar, indicating that genetic relatedness of DNA from similar organisms can be used as a measure of taxonomic relatedness. The parameter of DNA–DNA similarity based on the difference in thermal denaturation midpoint has been a gross method for species delineation.

A more precise method is DNA sequencing. This method has become a routine procedure, and comparison of the DNA sequences of divergent genes can give a measure of their relatedness. Genes for different functions, such as those encoding surface antigens to escape immune surveillance, diverge at different rates relative to "housekeeping" genes such as those that encode cytochromes. Thus, DNA sequence differences among rapidly diverging genes can be used to ascertain the genetic distance of closely related groups of bacteria, and sequence differences among housekeeping genes can be used to measure the relatedness of widely divergent groups of bacteria.

Ribosomal RNA

Ribosomes have an essential role in protein synthesis for all organisms. Genetic sequences encoding both ribosomal RNAs (rRNA) and proteins (both of which are required to comprise a functional ribosome) have been highly conserved throughout evolution and have diverged more slowly than other chromosomal genes. Comparison of the nucleotide sequence of 16S ribosomal RNA from a range of biologic sources revealed evolutionary relationships among widely divergent organisms and has led to the elucidation of a new kingdom, the **Archaebacteria**. The phylogenetic tree based on ribosomal RNA (rRNA) data, showing the separation of bacteria, archaea, and eukaryote families, is depicted in Figure 3–1 which shows the three major domains of biological life as they are currently understood. From this diagram two Kingdoms, the Eubacteria (true bacteria) and the Archeabacteria are distinct from the Eukaryotic branch.

DESCRIPTION OF THE MAJOR CATEGORIES & GROUPS OF BACTERIA

Bergey's Manual of Systematic Bacteriology

The definitive work on the taxonomic organization of bacteria is the latest edition of *Bergey's Manual of Systematic Bacteriology.* First published in 1923, this publication taxonomically classifies known bacteria that have or have not been cultured or well-described, in the form of a key. A companion volume, *Bergey's Manual of Determinative Bacteriology,* serves as an aid in the identification of those bacteria that have been described and cultured. The major bacteria that cause infectious diseases, as categorized in *Bergey's Manual,* are listed in Table 3–2. Because it is likely that emerging information concerning phylogenetic relationships will lead to further modifications in the organization of bacterial groups within *Bergey's Manual,* its designations must be regarded as a work in progress.

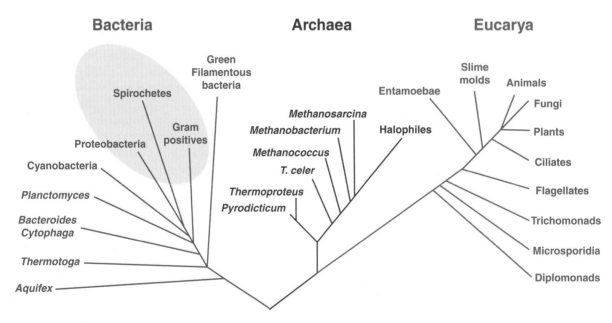

FIGURE 3–1 A phylogenetic tree based on rRNA data, showing the separation of bacteria, archaea, and eukaryotes families. The groups of the major known pathogenic bacteria are designated in grey. The only group of pathogenic bacteria that does not cluster in this shaded area is the *Bacteroides* group.

TABLE 3–2 Major Categories and Groups of Bacteria That Cause Disease in Humans as Part of an Identification Scheme Described in *Bergey's Manual of Determinative Bacteriology*, 9th Ed.

Bergey's Manual of Systematic Bacteriology	
I. Gram-negative eubacteria that have cell walls	
Group 1: The spirochetes	*Treponema*
	Borrelia
	Leptospira
Group 2: Aerobic/microaerophilic, motile helical/vibroid gram-negative bacteria	*Campylobacter*
	Helicobacter
	Spirillum
Group 3: Nonmotile (or rarely motile) curved bacteria	None
Group 4: Gram-negative aerobic/microaerophilic rods and cocci	*Alcaligenes*
	Bordetella
	Brucella
	Francisella
	Legionella
	Moraxella
	Neisseria
	Pseudomonas
	Rochalimaea
	Bacteroides (some species)
Group 5: Facultatively anaerobic gram-negative rods	*Escherichia* (and related coliform bacteria)
	Klebsiella
	Proteus
	Providencia
	Salmonella
	Shigella
	Yersinia
	Vibrio
	Haemophilus
	Pasteurella
Group 6: Gram-negative, anaerobic, straight, curved, and helical rods	*Bacteroides*
	Fusobacterium
	Prevotella
Group 7: Dissimilatory sulfate- or sulfur-reducing bacteria	None
Group 8: Anaerobic gram-negative cocci	None
Group 9: The rickettsiae and chlamydiae	*Rickettsia*
	Coxiella
	Chlamydia
Group 10: Anoxygenic phototrophic bacteria	None
Group 11: Oxygenic phototrophic bacteria	None
Group 12: Aerobic chemolithotrophic bacteria and assorted organisms	None
Group 13: Budding or appendaged bacteria	None
Group 14: Sheathed bacteria	None
Group 15: Nonphotosynthetic, nonfruiting gliding bacteria	*Capnocytophaga*
Group 16: Fruiting gliding bacteria: the myxobacteria	None
II. Gram-positive bacteria that have cell walls	
Group 17: Gram-positive cocci	*Enterococcus*
	Peptostreptococcus
	Staphylococcus
	Streptococcus
Group 18: Endospore-forming gram-positive rods and cocci	*Bacillus*
	Clostridium
Group 19: Regular, nonsporing gram-positive rods	*Erysipelothrix*
	Listeria
Group 20: Irregular, nonsporing gram-positive rods	*Actinomyces*
	Corynebacterium
	Mobiluncus
Group 21: The mycobacteria	*Mycobacterium*
Group 22–29: Actinomycetes	*Nocardia*
	Streptomyces
	Rhodococcus
III. Cell wall-less eubacteria: The mycoplasmas or mollicutes	
Group 30: Mycoplasmas	*Mycoplasma*
	Ureaplasma
IV. Archaebacteria	
Group 31: The methanogens	None
Group 32: Archaeal sulfate reducers	None
Group 33: Extremely halophilic archaebacteria	None
Group 34: Cell wall-less archaebacteria	None
Group 35: Extremely thermophilic and hyperthermophilic sulfur metabolizers	None

As discussed in Chapter 2, there are two different groups of prokaryotic organisms: eubacteria and archaebacteria. Both are small unicellular organisms that replicate asexually. Eubacteria refer to classic bacteria as science has historically understood them. They lack a true nucleus, have characteristic lipids that make up their membranes, possess a peptidoglycan cell wall, and have a protein and nucleic acid synthesis machinery that can be selectively inhibited by antimicrobial agents. In contrast, archaebacteria do not have a classic peptidoglycan cell wall and have many characteristics (eg, protein synthesis and nucleic acid replication machinery) that are similar to those of eukaryotic cells (Table 3–3).

The Eubacteria

A. Gram-Negative Eubacteria

This is a heterogeneous group of bacteria that have a complex (gram-negative type) cell envelope consisting of an outer membrane, a periplasmic space containing a thin peptidoglycan layer and a cytoplasmic membrane. The cell shape (Figure 3–2) may be spherical, oval, straight or curved rods, helical, or filamentous; some of these forms may be sheathed or encapsulated. Reproduction is by binary fission, but some groups reproduce by budding. Fruiting bodies and myxospores may be formed by the myxobacteria. Motility, if present, occurs by means of flagella or by gliding motility. Members of this category may be **phototrophic** or **nonphototrophic** (Chapter 5) bacteria and include **aerobic, anaerobic, facultatively anaerobic,** and **microaerophilic** species; some members are obligate intracellular parasites.

B. Gram-Positive Eubacteria

These bacteria have a cell wall profile of the gram-positive type; cells generally, but not always, stain gram-positive. The cell envelope of gram-positive organisms consists of a thick cell wall that determines cellular shape and a cytoplasmic membrane. These cells may be encapsulated and can exhibit flagellar-mediated motility. Cells may be spherical, rods, or filaments; the rods and filaments may be nonbranching or may show true branching. Reproduction is generally by binary fission. Some bacteria in this category produce **spores** (eg, *Bacillus* and *Clostridium spp.*) as resting forms that are highly resistant to disinfection. The gram-positive eubacteria are generally **chemosynthetic**

TABLE 3–3   Key Characteristics Shared by Archaebacteria and Eukaryotic Cells That are Absent from Eubacteria

Characteristic	Eubacteria	Archaebacteria, Eukaryotes
Elongation factor-2 (EF-2) contains the amino acid diphthamide and is therefore ADP-ribosylable by diphtheria toxin	No	Yes
The methionyl initiator tRNA is not formylated	No	Yes
Some tRNA genes contain introns	No	Yes in eukaryotes
Protein synthesis is inhibited by anisomycin but not by chloramphenicol	No	Yes
DNA-dependent RNA polymerases are multicomponent enzymes insensitive to the antibiotics rifampin and streptomycin	No	Yes
DNA-dependent RNA polymerases are multicomponent enzymes and are insensitive to the antibiotics rifampin and streptolydigin	No	Yes

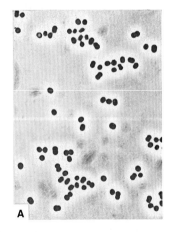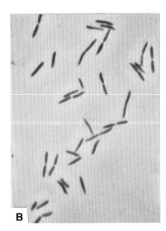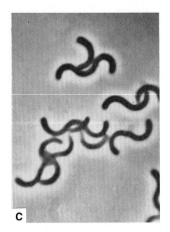

FIGURE 3–2 The cell shapes that occur among unicellular true bacteria. (A) Coccus. (B) Rod. (C) Spiral. (Phase contrast, 1500 ×.) (Reproduced with permission from Stanier RY, Doudoroff M, Adelberg EA: *The Microbial World*, 3rd ed. Copyright © 1970. By permission of Prentice-Hall, Inc., Englewood Cliffs, NJ.)

heterotrophs (Chapter 5) and include aerobic, anaerobic, and facultatively anaerobic species. The groups within this category include simple asporogenous and sporogenous bacteria as well as the structurally complex actinomycetes and their relatives.

C. Eubacteria Lacking Cell Walls

These are microorganisms that lack cell walls (commonly called **mycoplasmas** and making up the class Mollicutes) and do not synthesize the precursors of peptidoglycan. They are enclosed by a unit membrane, the plasma membrane (Figure 3–3). They resemble the **L-forms** (Chapter 25) that can be generated from many species of bacteria (notably gram-positive eubacteria); unlike L-forms, however, mycoplasmas never revert to the walled state, and there are no antigenic relationships between mycoplasmas and eubacterial L-forms.

Six genera have been designated as mycoplasmas (Chapter 25) on the basis of their habitat; however, only two genera contain animal pathogens. Mycoplasmas are highly pleomorphic organisms and range in size from vesicle-like forms to very small (0.2 μm), filterable forms (meaning that they are too small to be captured on filters that routinely trap most bacteria). Reproduction may be by budding, fragmentation, or binary fission, singly or in combination. Most species require a complex medium for growth and tend to form characteristic "fried egg" colonies on a solid medium. A unique characteristic of the Mollicutes is that some genera require cholesterol for growth; unesterified cholesterol is a unique component of the membranes of both sterol-requiring and non-sterol-requiring species if present in the medium.

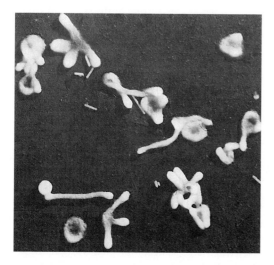

FIGURE 3–3 Electron micrograph of cells of a member of the mycoplasma group, the agent of bronchopneumonia in the rat (1960 ×). (Reproduced with permission from Klieneberger-Nobel E, Cuckow FW: A study of organisms of the pleuropneumonia group by electron microscopy. J Gen Microbiol 1955;12:99.)

The Archaebacteria

These organisms are predominantly inhabitants of extreme terrestrial and aquatic environments (high salt, high temperature, anaerobic) and are often referred to as "extremeophiles; some are symbionts in the digestive tract of animals. The archaebacteria consist of aerobic, anaerobic, and facultatively anaerobic organisms that are **chemolithotrophs**, **heterotrophs**, or **facultative heterotrophs** (Chapter 5). Some species are **mesophiles**, while others are capable of growing at temperatures above 100°C. These hyperthermophilic archaebacteria are uniquely adapted for growth and multiplication at high temperatures. With few exceptions enzymes isolated from these organisms are intrinsically more thermostable than their counterparts from mesophilic organisms. Some of these thermostable enzymes, such as the DNA polymerase from *Thermus aquaticus* (Taq polymerase), are important components of DNA amplification methods such as the polymerase chain reaction (PCR).

Archaebacteria can be distinguished from eubacteria in part by their lack of a peptidoglycan cell wall, possession of isoprenoid diether or diglycerol tetraether lipids, and characteristic ribosomal RNA sequences. Archaebacteria also share some molecular features with eukaryotes (see Table 3–3). Cells may have a diversity of shapes, including spherical, spiral, and plate or rod shaped; unicellular and multicellular forms in filaments or aggregates also occur. Multiplication occurs by either binary fission, budding, constriction, fragmentation, or other unknown mechanisms.

SUBTYPING & ITS APPLICATION

Under certain circumstances (such as an epidemic) it is important to distinguish between strains of a given species or to identify a particular strain. This is called **subtyping** and is accomplished by examining bacterial isolates for characteristics that allow discrimination below the species level. Classically, subtyping has been accomplished by biotyping, serotyping, antimicrobial susceptibility testing and bacteriophage typing. For example, more than 130 serogroups of *Vibrio cholerae* have been identified on the basis of antigenic differences in the O-polysaccharide of their LPS; however, only the O1 and O139 serogroups are associated with epidemic and pandemic cholera. Within these serogroups, only strains that produce a particular bundle-forming pilus and cholera toxin are virulent and cause the disease cholera. By contrast, nontoxigenic *V cholerae* strains, which are not associated with epidemic cholera, have been isolated from environmental specimens, from food, and from patients with sporadic diarrhea.

Serological Typing

Clonality with respect to isolates of microorganisms from a common source outbreak (**point source spread**) is an important concept in the epidemiology of infectious diseases. Etiologic agents associated with these outbreaks are generally **clonal;** in other words, they are the progeny of a single cell and thus, for all practical purposes, are genetically identical. Thus, subtyping plays an important role in discriminating these particular microorganisms. Recent advances in biotechnology have dramatically improved our ability to subtype microorganisms. Hybridoma technology has resulted in the development of monoclonal antibodies against cell surface antigens, which have been used to create highly standardized antibody-based subtyping systems that describe bacterial **serotypes.** This is an important tool for defining the epidemiologic spread of a bacterial infection.

Other organisms cannot be identified as unique serotypes. For example, some pathogens (eg, *Neisseria gonorrhoeae*) are transmitted as an inoculum composed of **quasispecies** (meaning that there is extensive antigenic variation among the bacteria present in the inoculum). In these cases, groups of hybridomas that recognize variants of the original organisms are used to categorize serovariants or **serovars. Genotyping Multilocus enzyme electrophoresis** (MLEE), which has been a standard method for investigating eukaryotic population genetics, has also been used to study the genetic diversity and clonal structure of pathogenic microorganisms. MLEE involves the determination of the mobilities of a set of soluble enzymes (usually 15–25 enzymes) by starch gel electrophoresis. Because the rate of migration of a protein during electrophoresis and its net electrostatic charge are determined by its amino acid sequence, mobility variants (referred to as **electromorphs** or **allozymes**) of an enzyme reflect amino acid substitutions in the protein sequence, which reflect changes in the DNA sequence encoding the protein. The enzyme-encoding structural genes of *E coli* exhibit extensive genetic diversity; however, by using MLEE, investigators at the Centers for Disease Control and Prevention were able to ascertain that strains of *E coli* serotype O157:H7, a pathogen associated with outbreaks of hemorrhagic colitis and hemolytic uremic syndrome (HUS) in children (Chapter 15), were descended from a clone that is widely distributed in North America.

Chemical Fingerprinting

The characterization or identification of isolates has been improved by applying physical methods to prokaryotic cells, such as Fourier transform infrared spectroscopy (FTIR), pyrolysis/mass spectrometry, and matrix-assisted laser desorption/ionization with time-of-flight (Maldi/Tof) or spray ionization mass spectrometry. The equipment required for these powerful techniques is expensive and is not routinely available to a clinical laboratory.

NUCLEIC ACID-BASED TAXONOMY

Since 1975, developments in nucleic acid isolation, amplification, and sequencing spurred the evolution of nucleic acid-based subtyping systems. These include plasmid profile analysis, restriction endonuclease analysis, ribotyping, pulsed field gel electrophoresis, PCR amplification and restriction endonuclease digestion of specific genes, arbitrarily primed PCR, and nucleic acid sequence analysis.

Plasmid Analysis

Plasmid profile analysis was the first, and the simplest, nucleic acid-based technique applied to epidemiologic studies. Plasmids, which are extrachromosomal genetic elements (Chapter 7), are isolated from each bacterium and then separated by agarose gel electrophoresis to determine their number and size. However, plasmids of identical size with different sequences can exist in many bacteria. Thus, digesting the plasmids with restriction endonucleases and then comparing the number and size of the resulting restriction fragments often provides additional useful information. Plasmid analysis has been shown to be most useful for examining outbreaks that are restricted in time and place (eg, an outbreak in a hospital), particularly when they are combined with other identification methods.

Restriction Endonucleases Analysis

The use of restriction enzymes to cleave DNA into discrete fragments is one of the most basic procedures in molecular biology. Restriction endonucleases recognize short DNA sequences (restriction sequence), and they cleave double-stranded DNA within or adjacent to this sequence. Restriction sequences range from four to more than 12 bases in length and occur throughout the bacterial chromosome. Restriction enzymes that recognize short sequences (eg, four base pairs) occur more frequently than those that recognize longer sequences (eg, 12 base pairs). Thus, enzymes that recognize short DNA sequences will produce more fragments than enzymes that recognize infrequently occurring long DNA sequences. Several subtyping methods employ restriction endonuclease-digested DNA. The basic method involves digesting DNA with an enzyme that recognizes a frequently occurring restriction site and separating the fragments, which generally range from 0.5 kb to 50 kb in length, by agarose gel electrophoresis followed by visualization under ultraviolet light after staining with ethidium bromide. One of the major limitations of this technique is the difficulty in interpreting the complex profiles consisting of hundreds of bands that may be unresolved and overlapping. The use of restriction endonucleases that cut at infrequently occurring restriction sites has circumvented this problem. Digestion of DNA with these enzymes generally results in five to 20 fragments ranging from approximately 10 kb to 800 kb in length. Separation of these large DNA fragments is accomplished by a technique called **pulsed field gel electrophoresis** (PFGE), which requires specialized equipment. Theoretically, all bacterial isolates can be typed by this method. Its advantage is that the restriction profile consists of a few well-resolved bands representing the entire bacterial chromosome in a single gel.

Southern Blot Analysis

This analysis was named after its inventor, Edwin Mellor Southern, and has been used as a subtyping method to identify isolates associated with outbreaks. For this analysis, DNA preparations from bacterial isolates are subjected to restriction endonuclease digestion. Following agarose gel electrophoresis, the separated restriction fragments are transferred to a nitrocellulose or nylon membrane. These double-stranded DNA fragments are first converted into single-stranded linear sequences. Using a labeled fragment of DNA as a probe, it is possible to identify the restriction fragments containing sequences (loci) that are homologous to the probe by complementation to the bound single-stranded fragments (Figure 3–4). **Restriction**

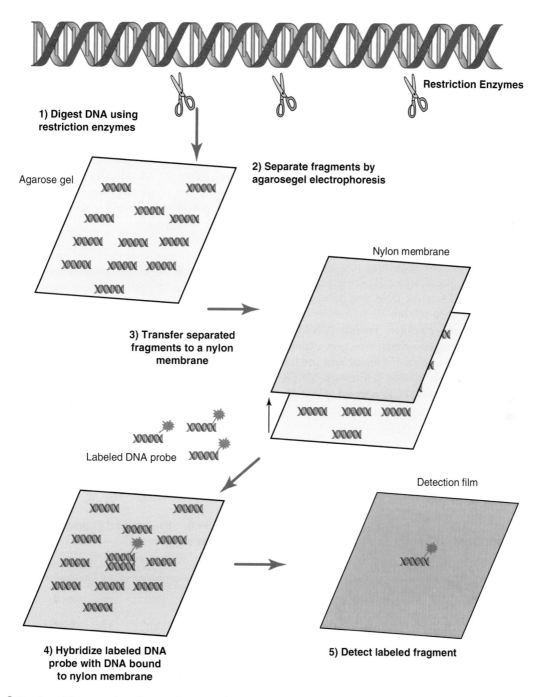

FIGURE 3–4 Southern blot procedure showing how specific loci on separated DNA fragments can be detected with a labeled DNA probe. This procedure in essence allows for the discrimination of DNA at three levels: (1) at the level of restriction enzyme recognition, (2) by the size of the DNA fragment, and (3) by the hybridization of a DNA probe to a specific locus defined by a specific band at a specific position of the membrane.

fragment length polymorphisms (RFLPs) refer to variations in both the number of loci that are homologous to the probe and the location of restriction sites that are within or flanking those loci.

Ribotyping

This method uses Southern blot analysis to detect polymorphisms of rRNA genes, which are present in all bacteria. Because ribosomal sequences are highly conserved, they can be detected with a common probe prepared from the 16S and 23S rRNA of a eubacterium *E coli*. Many organisms have multiple copies (five to seven) of these genes, resulting in patterns with a sufficient number of bands to provide good discriminatory power; however, ribotyping will be of limited value for some microorganisms, such as mycobacteria, which have only a single copy of these genes.

Repetitive Sequences

In the current genomic era of molecular medicine hundreds of microbial genomes have now been sequenced. With this era have come bioinformatical tools to mine this wealth of DNA sequence information to identify novel targets for pathogen subtyping, such as the **repetitive sequences** that have been found in different species (see Chapter 7). These repetitive sequences have been termed **satellite DNA** and have repeating units that range from 10 to 100 bp. They are commonly referred to as **variable number tandem repeats (VNTRs)**. VNTRs have been found in regions controlling gene expression and within open reading frames. The repeat unit and the number of copies repeated side by side define each VNTR locus. A genotyping approach using PCR, referred to as **multiple-locus VNTR analysis** (MLVA), takes advantage of the levels of diversity generated by both repeat unit size variation and copy number among a number of characterized loci. It has proved especially useful in subtyping monomorphic species such as *Bacillus anthracis*, *Yersinia pestis*, and *Francisella tularensis*.

Microbial Forensics

Genotyping methods are progressing toward the identification of **single nucleotide polymorphisms** (SNPs) in both open reading frames and intergenic regions to address a diverse range of epidemiologic and evolutionary questions. The field of microbial forensics developed in the wake of bioterrorist attacks with spores of *Bacillus anthracis* (**Anthrax**) in the fall of 2001. Microbial forensics was part of the criminal investigation and involved the use of many of the techniques described above to identify the precise strain and substrain of the microorganism used in a biocrime and its forensically meaningful source.

NONCULTURE METHODS FOR THE IDENTIFICATION OF PATHOGENIC MICROORGANISMS

Attempts to estimate total numbers of eubacteria, archaebacteria, and viruses are problematic because of difficulties such as detection in and recovery from the environment, our incomplete knowledge of obligate microbial associations, and the problem of the species concept in these groups. Nevertheless, estimates indicate that the numbers of uncultured microbial taxa greatly exceed those of the cultured organisms (Table 3–4). However, more recent estimates suggest that the number of bacterial species in the world is between 10^7 and 10^9. Until very recently, microbial identification required the isolation of pure cultures (or in some instances defined co-cultures) followed by testing for multiple physiologic and biochemical traits. Clinicians have long been aware of human diseases that are associated with visible but nonculturable microorganisms. Scientists are now employing a PCR-assisted approach using rRNA to identify pathogenic microorganisms in situ. The first phase of this approach involves the extraction of DNA from a suitable specimen, the use of standard molecular techniques to obtain a clone library, the retrieval of rRNA sequence information, and a comparative analysis of the retrieved sequences. This yields information on the identity or relatedness of the sequences in comparison with the available data base. In the second phase, proof that the sequences are from cells in the original specimen is obtained by in situ hybridization using sequence-specific probes. This approach has been used in the identification of pathogenic microorganisms. For example, a previously uncharacterized pathogen has been identified as the Whipple-disease-associated rod-shaped bacterium now designated *Tropheryma whipplei*. This rRNA approach has also been used to identify the etiologic agent of bacillary angiomatosis as *Bartonella henselae* and to show that the opportunistic pathogen *Pneumocystis jiroveci* is a member of the fungi. Undoubtedly these and other techniques will identify additional etiologic agents in the future.

TABLE 3–4   Known and Estimated Numbers of Biologic Species[a]

Group	Known Species	Estimated Total Species	Percentage of Known Species
Viruses	5,000	130,000	4
Bacteria	4,760	40,000	12
Fungi	69,000	1,500,000	5
Algae	40,000	60,000	67
Protozoa	30,800	100,000	31

[a]Modified from Bull AT et al: Biodiversity as a source of innovation in biotechnology. Ann Rev Microbiol 1992;46:219.

REVIEW QUESTIONS

1. Eubacteria that lack cell walls and do not synthesize the precursors of peptidoglycan are called

   (A) Gram negative bacteria
   (B) Viruses
   (C) Mycoplasmas
   (D) Serovar variant
   (E) Bacilli

2. Archaebacteria can be distinguished from eubacteria by their lack of

   (A) DNA
   (B) RNA
   (C) Ribosomes
   (D) Peptidoglycan
   (E) A nucleus

3. A 16-year-old cystic fibrosis patient is admitted to the hospital. A sputum culture yields *Burkholderia cepacia*. Subsequently, there are two other patients with *Burkholderia cepacia* bacteremia, and the organism is cultured from the sputum of four additional patients. During this nosocomial outbreak of *Burkholderia cepacia,* 50 environmental and seven patient isolates are being subtyped to identify the source of the outbreak. Which of the following techniques would be most useful in this endeavor?

   (A) Culture
   (B) Ribotyping
   (C) 16S rRNA sequencing
   (D) Antimicrobial susceptibility testing
   (E) Nucleic acid sequencing

4. An unculturable gram-positive microorganism has been visualized in tissue specimens obtained from patients with a previously undescribed disease. Which of the following techniques would be most useful in identifying this organism

   (A) Serology
   (B) PCR amplification and sequencing of rRNA genes
   (C) Multilocus enzyme electrophoresis
   (D) SDS-polyacrylamide gel electrophoresis
   (E) Pulsed field gel electrophoresis

5. The DNA polymerase from *Thermus aquaticus*is is an important component of DNA amplification methods such as the polymerase chain reaction. This organism is capable of growing at temperatures above 100°C. Organisms that are capable of growth at these temperatures are referred to as

   (A) Mesophiles
   (B) Psychrophiles
   (C) Halophiles
   (D) Thermophiles
   (E) Chemolithotrophs

Answers

1. C	3. E	5. D
2. D	4. B	

REFERENCES

Achtman M, Wagner M: Microbial diversity and the genetic nature of microbial species. Nat Rev Microbiol 2008;6:431.

Amann RI, Ludwig W, Schleiffer K-H: Phylogenetic identification and in situ detection of individual microbial cells without culture. Microbiol Rev 1995;59:143.

Boone DR, Castenholz RW (editors): *Bergey's Manual of Systematic Bacteriology.* Vol. 1. *The Archaea and the Deeply Branching and Phototrophic Bacteria,* 2nd ed. Springer, 2001.

Breeze RG, Budowle B, Schutzer SE (editors): *Microbial Forensics.* Elsevier, 2005.

Brenner DJ, Krieg NR, Staley JT (editors): *Bergey's Manual of Systematic Bacteriology.* Vol. 2. *The Proteobacteria.* Part A. *Introductory Essays.* Springer, 2005.

Brenner DJ, Krieg NR, Staley JT (editors): *Bergey's Manual of Systematic Bacteriology.* Vol. 2. *The Proteobacteria.* Part B. *The Gammaproteobacteria.* Springer, 2005.

Brenner DJ, Krieg NR, Staley JT (editors): *Bergey's Manual of Systematic Bacteriology.* Vol. 3. *The Proteobacteria.* Part C. *The Alpha-, Beta-, Delta-, and Episilonproteobacteria.* Springer, 2005

Colwell RR, Grimes DJ (editors): *Nonculturable microorganisms in the Environment.* ASM Press, 2000.

Curtis TP, Sloan WT, Scannell JW: Estimating prokaryotic diversity and its limits. Proc Natl Acad Sci U S A 2002;99:10494.

Edman JC et al: Ribosomal RNA sequence shows *Pneumocystis carinii* to be a member of the fungi. Nature (London) 1988;334:519.

Fernandez LA: Exploring prokaryotic diversity: There are other molecular worlds. Molec Microbiol 2005;55:5–15.

Fredericks DN, Relman DA: Sequence-based identification of microbial pathogens: A reconsideration of Koch's postulates. Clin Microbiol Rev 1996;9:18.

Holt JG et al (editors): *Bergey's Manual of Determinative Bacteriology,* 9th ed. Williams & Wilkins, 1994.

Medini D et al: Microbiology in the post-genomic era. Nat Rev Microbiol 2008;6:429.

Persing DH et al (editors): *Molecular Microbiology. Diagnostic Principles and Practice.* ASM Press, 2004.

Riley LW: *Molecular Epidemiology of Infectious Diseases. Principles and Practices.* ASM Press, 2004.

Rosello-Mora R, Amann R: The species concept for prokaryotes. FEMS Microbiol Rev 2001;25:39.

Schloss PD, Handelsman J: Status of the microbial census. Microbiol Molec Biol Rev 2004;68:686.

Stringer JR et al: A new name *(Pneumocystis jiroveci)* for Pneumocystis from humans. Emerg Infect Dis 2002;8:891.

Whitman WB, Coleman DC, Wiebe WJ: Prokaryotes: The unseen majority. Proc Natl Acad Sci U S A 1998;95:6578.

The Growth, Survival, & Death of Microorganisms

SURVIVAL OF MICROORGANISMS IN THE NATURAL ENVIRONMENT

The population of microorganisms in the biosphere is roughly constant: Growth is counterbalanced by death. The survival of any microbial group within its niche is determined in large part by successful competition for nutrients and by maintenance of a pool of living cells during nutritional deprivation. It is increasingly evident that many microorganisms exist in consortia formed by representatives of different genera. Other microorganisms, often characterized as single cells in the laboratory, form cohesive colonies in the natural environment.

Most of our understanding of microbial physiology has come from the study of isolated cell lines growing under optimal conditions, and this knowledge forms the basis for this section. Nevertheless, it should be remembered that many microorganisms compete in the natural environment while under nutritional stress, a circumstance that may lead to a physiologic state quite unlike that observed in the laboratory. Furthermore, it should be recognized that a vacant microbial niche in the environment will soon be filled. Public health procedures that eliminate pathogenic microorganisms by clearing their niche are likely to be less successful than methods that leave the niche occupied by successful nonpathogenic competitors.

THE MEANING OF GROWTH

Growth is the orderly increase in the sum of all the components of an organism. Thus, the increase in size that results when a cell takes up water or deposits lipid or polysaccharide is not true growth. Cell multiplication is a consequence of growth; in unicellular organisms, growth leads to an increase in the number of individuals making up a population or culture.

The Measurement of Microbial Concentrations

Microbial concentrations can be measured in terms of cell concentration (the number of viable cells per unit volume

of culture) or of biomass concentration (dry weight of cells per unit volume of culture). These two parameters are not always equivalent, because the average dry weight of the cell varies at different stages in the history of a culture. Nor are they of equal significance: In studies of microbial genetics or the inactivation of cells, cell concentration is the significant quantity; in studies on microbial biochemistry or nutrition, biomass concentration is the significant quantity.

A. Cell Concentration

The viable cell count (Table 4–1) is usually considered the measure of cell concentration. However, for many purposes the turbidity of a culture, measured by photoelectric means, may be related to the viable count in the form of a **standard curve**. A rough visual estimate is sometimes possible: A barely turbid suspension of *Escherichia coli* contains about 10^7 cells per milliliter, and a fairly turbid suspension contains about 10^8 cells per milliliter. In using turbidimetric measurements, it must be remembered that the correlation between turbidity and viable count can vary during the growth and death of a culture; cells may lose viability without producing a loss in turbidity of the culture.

B. Biomass Density

In principle, biomass can be measured directly by determining the dry weight of a microbial culture after it has been washed with distilled water. In practice, this procedure is cumbersome, and the investigator customarily prepares a standard curve that correlates dry weight with turbidity. Alternatively, the concentration of biomass can be estimated indirectly by measuring an important cellular component such as protein or by determining the volume occupied by cells that have settled out of suspension.

EXPONENTIAL GROWTH

The Growth Rate Constant

The growth rate of cells unlimited by nutrient is first order: The rate of growth (measured in grams of biomass produced

TABLE 4–1　**Example of a Viable Count**

Dilution	Plate Count[a]
Undiluted	Too crowded to count
10^{-1}	Too crowded to count
10^{-2}	510
10^{-3}	72
10^{-4}	6
10^{-5}	1

[a]Each count is the average of three replicate plates.

per hour) is the product of the **growth rate constant**, k, and the biomass concentration, B:

$$\frac{dB}{dt} = kB \qquad (1)$$

Rearrangement of equation (1) demonstrates that the growth rate constant is the rate at which cells are producing more cells:

$$k = \frac{B\,dt}{dB} \qquad (2)$$

A growth rate constant of 4.3 h^{-1}, one of the highest recorded, means that each gram of cells produces 4.3 g of cells per hour during this period of growth. Slowly growing organisms may have growth rate constants as low as 0.02 h^{-1}. With this growth rate constant, each gram of cells in the culture produces 0.02 g of cells per hour.

Integration of equation (1) yields

$$\text{In}\frac{B_1}{B_0} = 2.3\log_{10}\frac{B_1}{B_0} = k(t_1 - t_0) \qquad (3)$$

The natural logarithm of the ratio of B_1 (the biomass at time 1 [t_1]) to B_0 (the biomass at time zero [t_0]) is equal to the product of the growth rate constant (k) and the difference in time ($t_1 - t_0$). Growth obeying equation (3) is termed exponential because biomass increases exponentially with respect to time. Linear plots of exponential growth can be produced by plotting the logarithm of biomass concentration (B) as a function of time (t).

Calculation of the Growth Rate Constant & Prediction of the Amount of Growth

Many bacteria reproduce by binary fission, and the average time required for the population, or the biomass, to double is known as the **generation time** or **doubling time** (t_d). Usually the t_d is determined by plotting the amount of growth on a semi-logarithmic scale as a function of time; the time

required for doubling the biomass is t_d (Figure 4–1). The growth rate constant can be calculated from the doubling time by substituting the value 2 for B_1/B_0 and t_d for $t_1 - t_0$ in equation (3), which yields

$$\text{In}2 = kt_d$$
$$k = \frac{\text{In}2}{t_d} \qquad (4)$$

A rapid doubling time corresponds to a high growth rate constant. For example, a doubling time of 10 minutes (0.17 hour) corresponds to a growth rate constant of 4.1 h^{-1}. The relatively long doubling time of 35 hours corresponds to a growth rate constant of 0.02 h^{-1}.

The calculated growth rate constant can be used either to determine the amount of growth that will occur in a specified period of time or to calculate the amount of time required for a specified amount of growth.

The amount of growth within a specified period of time can be predicted on the basis of the following rearrangement of equation (3):

$$\log_{10}\frac{B_1}{B_0} = \frac{k(t_1 - t_0)}{2.3} \qquad (5)$$

For example, it is possible to determine the amount of growth that would occur if a culture with a growth rate constant of 4.1 h^{-1} grew exponentially for 5 hours:

$$\log_{10}\frac{B_1}{B_0} = \frac{4.1h^{-1} \times 5h}{2.3} \qquad (6)$$

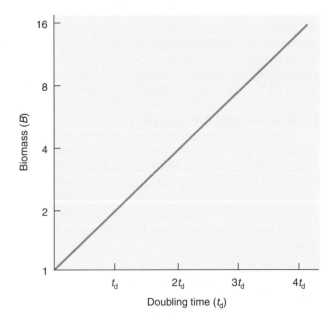

FIGURE 4–1 Exponential growth. The biomass (B) doubles with each doubling time (t_d).

In this example, the increase in biomass is 10^{-9}; a single bacterial cell with a dry weight of 2×10^{-13} g would give rise to 0.2 mg of biomass, a quantity that would densely populate a 5-mL culture. Clearly, this rate of growth cannot be sustained for a long period of time. Another 5 hours of growth at this rate would produce 200 kg dry weight of biomass, or roughly a ton of cells.

Another rearrangement of equation (3) allows calculation of the amount of time required for a specified amount of growth to take place. In equation (7), shown below, N, cell concentration, is substituted for B, biomass concentration, to permit calculation of the time required for a specified increase in cell number.

$$t_1 - t_0 = \frac{2.3 \log_{10}(N_1 / N_0)}{k} \quad (7)$$

Using equation (7), it is possible, for example, to determine the time required for a slowly growing organism with a growth rate constant of 0.02 h^{-1} to grow from a single cell into a barely turbid cell suspension with a concentration of 10^7 cells per milliliter.

$$t_1 - t_0 = \frac{2.3 \times 7}{0.02 h^{-1}} \quad (8)$$

Solution of equation (8) reveals that about 800 hours—slightly more than a month—would be required for this amount of growth to occur. The survival of slowly growing organisms implies that the race for biologic survival is not always to the swift—those species flourish that compete successfully for nutrients and avoid annihilation by predators and other environmental hazards.

THE GROWTH CURVE

If a fixed volume of liquid medium is inoculated with microbial cells taken from a culture that has previously been grown to saturation and the number of viable cells per milliliter is determined periodically and plotted, a curve of the type shown in Figure 4–2 is usually obtained. The phases of the bacterial growth curve shown in Figure 4–2 are reflections of the events in a population of cells, not in individual cells. This type of culture is referred to as a **batch culture**. The typical growth curve may be discussed in terms of four phases (Table 4–2).

The Lag Phase

The lag phase represents a period during which the cells, depleted of metabolites and enzymes as the result of the unfavorable conditions that existed at the end of their previous culture history, adapt to their new environment. Enzymes and intermediates are formed and accumulate until they are present in concentrations that permit growth to resume.

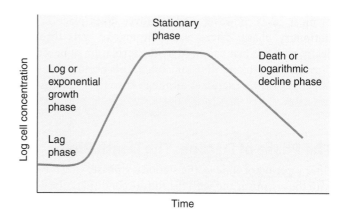

FIGURE 4–2 A bacterial growth curve.

TABLE 4–2   Phases of the Microbial Growth Curve

Phase	Growth Rate
Lag	Zero
Exponential	Constant
Maximum stationary	Zero
Decline	Negative (death)

If the cells are taken from an entirely different medium, it often happens that they are genetically incapable of growth in the new medium. In such cases a long lag may occur, representing the period necessary for a few mutants in the inoculum to multiply sufficiently for a net increase in cell number to be apparent.

The Exponential Phase

During the exponential phase, the mathematics of which has already been discussed, the cells are in a steady state. New cell material is being synthesized at a constant rate, but the new material is itself catalytic, and the mass increases in an exponential manner. This continues until one of two things happens: either one or more nutrients in the medium become exhausted, or toxic metabolic products accumulate and inhibit growth. For aerobic organisms, the nutrient that becomes limiting is usually oxygen. When the cell concentration exceeds about 1×10^7/mL (in the case of bacteria), the growth rate will decrease unless oxygen is forced into the medium by agitation or by bubbling in air. When the bacterial concentration reaches 4–5×10^9/mL, the rate of oxygen diffusion cannot meet the demand even in an aerated medium, and growth is progressively slowed.

The Maximum Stationary Phase

Eventually, the exhaustion of nutrients or the accumulation of toxic products causes growth to cease completely.

In most cases, however, cell turnover takes place in the stationary phase: There is a slow loss of cells through death, which is just balanced by the formation of new cells through growth and division. When this occurs, the total cell count slowly increases although the viable count stays constant.

The Phase of Decline: The Death Phase

After a period of time in the stationary phase, which varies with the organism and with the culture conditions, the death rate increases until it reaches a steady level. The mathematics of steady-state death is discussed below. In most cases the rate of cell death is much slower than that of exponential growth. Frequently, after the majority of cells have died, the death rate decreases drastically, so that a small number of survivors may persist for months or even years. This persistence may in some cases reflect cell turnover, a few cells growing at the expense of nutrients released from cells that die and lyse.

A phenomenon, in which cells are called **viable but not culturable (VBNC)**, is thought to be the result of a genetic response triggered in starving, stationary phase cells. Just as some bacteria form spores as a survival mechanism, others are able to become dormant without changes in morphology. Once the appropriate conditions are available (eg, passage through an animal), VNBC microbes resume growth.

MAINTENANCE OF CELLS IN THE EXPONENTIAL PHASE

Cells can be maintained in the exponential phase by transferring them repeatedly into fresh medium of identical composition while they are still growing exponentially. This is referred to as **continuous culture**; the most common type of continuous culture device used is a chemostat.

The Chemostat

This device consists of a culture vessel equipped with an overflow siphon and a mechanism for dripping in fresh medium from a reservoir at a regulated rate. The medium in the culture vessel is stirred by a stream of sterile air; each drop of fresh medium that enters causes a drop of culture to siphon out.

The medium is prepared so that one nutrient limits growth yield. The vessel is inoculated, and the cells grow until the limiting nutrient is exhausted; fresh medium from the reservoir is then allowed to flow in at such a rate that the cells use up the limiting nutrient as fast as it is supplied. Under these conditions, the cell concentration remains constant and the growth rate is directly proportionate to the flow rate of the medium.

DEFINITION & MEASUREMENT OF DEATH

The Meaning of Death

For a microbial cell, death means the irreversible loss of the ability to reproduce (grow and divide). The empirical test of death is the culture of cells on solid media: A cell is considered dead if it fails to give rise to a colony on any medium. Obviously, then, the reliability of the test depends upon choice of medium and conditions: A culture in which 99% of the cells appear "dead" in terms of ability to form colonies on one medium may prove to be 100% viable if tested on another medium. Furthermore, the detection of a few viable cells in a large clinical specimen may not be possible by directly plating a sample, as the sample fluid itself may be inhibitory to microbial growth. In such cases, the sample may have to be diluted first into liquid medium, permitting the outgrowth of viable cells before plating.

The conditions of incubation in the first hour following treatment are also critical in the determination of "killing." For example, if bacterial cells are irradiated with ultraviolet light and plated immediately on any medium, it may appear that 99.99% of the cells have been killed. If such irradiated cells are first incubated in a suitable buffer for 20 minutes, however, plating will indicate only 10% killing. In other words, irradiation determines that a cell will "die" if plated immediately but will live if allowed to repair radiation damage before plating.

A microbial cell that is not physically disrupted is thus "dead" only in terms of the conditions used to test viability.

The Measurement of Death

When dealing with microorganisms, one does not customarily measure the death of an individual cell, but the death of a population. This is a statistical problem: Under any condition that may lead to cell death, the probability of a given cell's dying is constant per unit time. For example, if a condition is employed that causes 90% of the cells to die in the first 10 minutes, the probability of any one cell dying in a 10-minute interval is 0.9. Thus, it may be expected that 90% of the surviving cells will die in each succeeding 10-minute interval, and a death curve similar to those shown in Figure 4–3 will be obtained.

The number of cells dying in each time interval is thus a function of the number of survivors present, so that death of a population proceeds as an exponential process according to the general formula

$$S = S_0 e^{-kt} \qquad (9)$$

where S_0 is the number of survivors at time zero and S is the number of survivors at any later time t. As in the case of exponential growth, $-k$ represents the rate of exponential death when the fraction $\ln (S/S_0)$ is plotted against time.

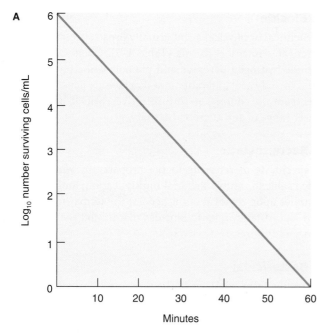

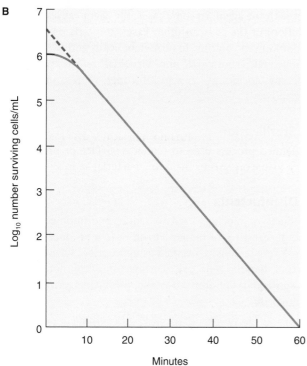

FIGURE 4–3 Death curve of a suspension of 10^6 viable microorganisms per milliliter. **A**: Single-hit curve. **B**: Multihit curve. The straight-line portion extrapolates to 6.5, corresponding to 4×10^6 cells. The number of targets is thus 4×10^6, or four per cell.

The one-hit curve shown in Figure 4–3A is typical of the kinetics of inactivation observed with many antimicrobial agents. The fact that it is a straight line from time zero (dose zero)—rather than exhibiting an initial shoulder—means that a single "hit" by the inactivating agent is sufficient to kill the cell; ie, only a single target must be damaged in order for the entire cell to be inactivated. Such a target might be the chromosome of a uninucleate bacterium or the cell membrane; conversely, it could not be an enzyme or other cell constituent that is present in multiple copies.

A cell that contains several copies of the target to be inactivated exhibits a multihit curve of the type shown in Figure 4–3B. Extrapolation of the straight-line portion of the curve to the ordinate permits an estimate of the number of targets (eg, 4 in Figure 4–3B).

Sterilization

In practice, we speak of "sterilization" as the process of killing all of the organisms in a preparation. From the above considerations, however, we see that no set of conditions is guaranteed to sterilize a preparation. Consider Figure 4–3, for example. At 60 minutes, there is one organism (10^0) left per milliliter. At 70 minutes there would be 10^{-1}, at 80 minutes 10^{-2}, etc. By 10^{-2} organisms per milliliter we mean that in a total volume of 100 mL, one organism would survive. How long, then, does it take to "sterilize" the culture? All we can say is that after any given time of treatment, the probability of having any surviving organisms in 1 mL is that given by the curve in Figure 4–3. After 2 hours, in the above example, the probability is 1×10^{-6}. This would usually be considered a safe sterilization time, but a 1000-L lot might still contain one viable organism.

Note that such calculations depend upon the curve's remaining unchanged in slope over the entire time range. Unfortunately, it is very common for the curve to bend upward after a certain period, as a result of the population being heterogeneous with respect to sensitivity to the inactivation agent. Extrapolations are dangerous and can lead to errors such as those encountered in early preparations of sterile polio vaccine.

The Effect of Drug Concentration

When antimicrobial substances (drugs) are used to inactivate microbial cells, it is commonly observed that the concentration of drug employed is related to the time required to kill a given fraction of the population by the following expression:

$$C^n t = K \qquad (10)$$

In this equation, C is the drug concentration, t is the time required to kill a given fraction of the cells, and n and K are constants.

This expression says that, for example, if $n = 6$ (as it is for phenol), then doubling the concentration of the drug will reduce the time required to achieve the same extent of inactivation 64-fold. That the effectiveness of a drug varies with the sixth power of the concentration suggests that six molecules of the drug are required to inactivate a cell, although there is no direct chemical evidence for this conclusion.

In order to determine the value of n for any drug, inactivation curves are obtained for each of several concentrations, and the time required at each concentration to inactivate a fixed fraction of the population is determined. For example, let the first concentration used be C_1 and the time required to inactivate 99% of the cells be t_1. Similarly, let C_2 and t_2 be the second concentration and time required to inactivate 99% of the cells. From equation (10), we see that

$$C_1^n\ t_1 = C_2^n\ t_2 \qquad (11)$$

Solving for n gives

$$n = \frac{\log t_2 - \log t_1}{\log C_1 - \log C_2} \qquad (12)$$

Thus, n can be determined by measuring the slope of the line that results when $\log t$ is plotted against $\log C$ (Figure 4–4). If n is experimentally determined in this manner, K can be determined by substituting observed values for C, t, and n in equation (10).

ANTIMICROBIAL AGENTS

Definitions

The following terms are commonly employed in connection with antimicrobial agents and their uses.

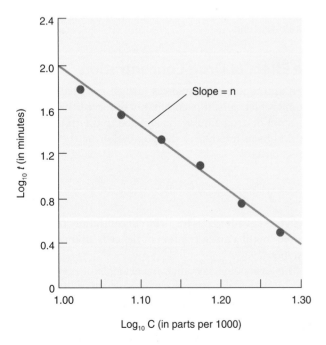

FIGURE 4–4 Relationship between drug concentration and time required to kill a given fraction of a cell population.

A. Biocide

A chemical or physical agent, usually broad spectrum, which inactivates microorganisms (Table 4–3). Chemical biocides include hydrogen peroxide and phenols while physical biocides include heat and radiation. Biocides are generally broad spectrum, in contrast to anti-infectives, which have a narrower range of antimicrobial activity.

B. Bacteriostatic

A specific term referring to the property by which a biocide is able to inhibit bacterial multiplication; multiplication resumes upon removal of the agent. (The terms "fungistatic" and "sporostatic" refer to biocides that inhibit the growth of fungi and spores, respectively.)

C. Bactericidal

A specific term referring to the property by which a biocide is able to kill bacteria. Bactericidal action differs from bacteriostasis only in being irreversible; ie, the "killed" organism can no longer reproduce, even after being removed from contact with the agent. In some cases, the agent causes lysis (dissolution) of the cells; in other cases, the cells remain intact and may even continue to be metabolically active. (The terms "fungicidal," "sporicidal," and "virucidal" refer to the property whereby biocides are able to kill fungi, spores, and viruses, respectively.)

D. Sterilization

A defined process used to render a surface or product free from viable organisms including bacterial spores.

E. Disinfectants

Products or biocides used to reduce the number of viable microorganisms, or bioburden, on or in a product or surface to a level previously specified as appropriate for its intended further handling or use. Disinfectants are not necessarily sporicidal, but are sporostatic, inhibiting germination or outgrowth.

F. Septic

Characterized by the presence of pathogenic microbes in living tissue.

G. Antiseptic

A biocide or product that destroys or inhibits the growth of microorganisms in or on living tissue (eg, skin).

H. Aseptic

Free of, or using methods to keep free of, microorganisms.

I. Preservation

The prevention of multiplication of microorganisms in formulated products, including pharmaceuticals and foods.

TABLE 4–3   Some Common Biocides Used for Antisepsis, Disinfection, Preservation, and Other Purposes

Agent	Formula	Uses
Alcohols Ethanol	CH_3-CHOH	Antisepsis, disinfection, preservation
Isopropanol	$\begin{array}{c} CH_3 \\ CH_3 \end{array}\!\!\!\!>\!CHOH$	
Aldehydes Glutaraldehyde	$O = \overset{H}{C}CH_2CH_2CH_2\overset{H}{C} = O$	Disinfection, sterilization, preservation
Formaldehyde	$\overset{H}{\underset{H}{}}\!\!\!>\!C = O$	

Biguanides
Chlorhexidine
	Cl—⟨ ⟩—N(HCN)$_2$H(CH$_2$)$_6$N(HCN)$_2$H—⟨ ⟩—Cl
	        ‖ ‖
	       NH NH
		Antisepsis, antiplaque activity, preservation, disinfection

Bisphenols Triclosan		Antisepsis, antiplaque activity
Hexachlorophene		Deodorant, preservation
Halogen-releasing agents Chlorine compounds	$\rightarrow OCl^-, HOCl, Cl_2$	Disinfection, antisepsis
Iodine compounds	$\rightarrow I_2$	
Heavy metal derivatives Silver compounds	Ag	Preservation, antisepsis
Mercury compounds	Hg	Disinfection
Organic acids Benzoic acid	⟨ ⟩—COOH	Preservation
Propionic acid	CH_3-CH_2-COOH	Sodium or calcium salt used for preservation
Peroxygens Hydrogen peroxide	H_2O_2	Disinfection, sterilization
Ozone	O_3	
Peracetic acid	CH_3COOOH	
Phenols and cresols Phenol		Disinfection, preservation
Cresol		
Quaternary ammonium compounds	$\left[\begin{array}{c} R^1 \quad R^3 \\ N \\ R^2 \quad R^4 \end{array}\right]^+ X^-$	Disinfection, antisepsis, preservation

(continued)

TABLE 4–3   Some Common Biocides Used for Antisepsis, Disinfection, Preservation, and Other Purposes (continued)

Agent	Formula	Uses
Cetrimide		Disinfection, antisepsis, preservation
Benzalkonium chloride		
Vapor phase Ethylene oxide		Sterilization, disinfection
Formaldehyde		
Hydrogen peroxide	H_2O_2	

J. Antibiotics

Naturally occurring or synthetic organic compounds which inhibit or destroy selective bacteria, generally at low concentrations.

Modes of Action

A. Damage to DNA

A number of physical and chemical agents act by damaging DNA; these include ionizing radiations, ultraviolet light, and DNA-reactive chemicals. Among the last category are alkylating agents and other compounds that react covalently with purine and pyrimidine bases to form DNA adducts or interstrand cross-links. Radiations damage DNA in several ways: Ultraviolet light, for example, induces cross-linking between adjacent pyrimidines on one or the other of the two polynucleotide strands, forming pyrimidine dimers; ionizing radiations produce breaks in single and double strands. Radiation-induced and chemically induced DNA lesions kill the cell mainly by interfering with DNA replication. See Chapter 7 for a discussion of DNA repair systems.

B. Protein Denaturation

Proteins exist in a folded, three-dimensional state determined by intramolecular covalent disulfide linkages and a number of noncovalent linkages such as ionic, hydrophobic, and hydrogen bonds. This state is called the tertiary structure of the protein; it is readily disrupted by a number of physical or chemical agents, causing the protein to become nonfunctional. The disruption of the tertiary structure of a protein is called protein denaturation.

C. Disruption of Cell Membrane or Wall

The cell membrane acts as a selective barrier, allowing some solutes to pass through and excluding others. Many compounds are actively transported through the membrane, becoming concentrated within the cell. The membrane is also the site of enzymes involved in the biosynthesis of components of the cell envelope. Substances that concentrate at the cell surface may alter the physical and chemical properties of the membrane, preventing its normal functions and therefore killing or inhibiting the cell.

The cell wall acts as a corseting structure, protecting the cell against osmotic lysis. Thus, agents that destroy the wall (eg, lysozyme) or prevent its normal synthesis (eg, penicillin) may bring about lysis of the cell.

D. Removal of Free Sulfhydryl Groups

Enzyme proteins containing cysteine have side chains terminating in sulfhydryl groups. In addition to these, coenzymes such as coenzyme A and dihydrolipoate contain free sulfhydryl groups. Such enzymes and coenzymes cannot function unless the sulfhydryl groups remain free and reduced. Oxidizing agents thus interfere with metabolism by forming disulfide linkages between neighboring sulfhydryl groups:

$$R-SH + HS-R \xrightarrow{-2H} R-S-S-R$$

Many metals such as mercuric ion likewise interfere by combining with sulfhydryls:

$$R - SH \atop R - SH + Hg \genfrac{}{}{0pt}{}{Cl}{Cl} \longrightarrow \genfrac{}{}{0pt}{}{R - S}{R - S} \diagdown\diagup Hg + 2HCl$$

There are many sulfhydryl enzymes in the cell; therefore, oxidizing agents and heavy metals do widespread damage.

E. Chemical Antagonism

The interference by a chemical agent with the normal reaction between a specific enzyme and its substrate is known as "chemical antagonism." The antagonist acts by combining with some part of the holoenzyme (either the protein apoenzyme, the mineral activator, or the coenzyme), thereby preventing attachment of the normal substrate. ("Substrate" is here used in the broad sense to include cases in which the inhibitor combines with the apoenzyme, thereby preventing attachment to it of coenzyme.)

An antagonist combines with an enzyme because of its chemical affinity for an essential site on that enzyme. Enzymes perform their catalytic function by virtue of their affinity for their natural substrates; hence any compound structurally resembling a substrate in essential aspects may also have an affinity for the enzyme. If this affinity is great enough, the "analog" will displace the normal substrate and prevent the proper reaction from taking place.

Many holoenzymes include a mineral ion as a bridge either between enzyme and coenzyme or between enzyme and substrate. Chemicals that combine readily with these minerals will again prevent attachment of coenzyme or substrate; eg, carbon monoxide and cyanide combine with the iron atom in heme-containing enzymes and prevent their function in respiration.

Chemical antagonists can be conveniently discussed under two headings: antagonists of energy-yielding processes, and antagonists of biosynthetic processes. The former include poisons of respiratory enzymes (carbon monoxide, cyanide) and of oxidative phosphorylation (dinitrophenol); the latter include analogs of the building blocks of proteins (amino acids) and of nucleic acids (nucleotides). In some cases the analog simply prevents incorporation of the normal metabolite (eg, 5-methyltryptophan prevents incorporation of tryptophan into protein), and in other cases the analog replaces the normal metabolite in the macromolecule, causing it to be nonfunctional. The incorporation of *p*-fluorophenylalanine in place of phenylalanine in proteins is an example of the latter type of antagonism.

Reversal of Antibacterial Action

In the section on definitions, the point was made that bacteriostatic action is, by definition, reversible. Reversal can be brought about in several ways.

A. Removal of Agent

When cells that are inhibited by the presence of a bacteriostatic agent are removed by centrifugation, washed thoroughly in the centrifuge, and resuspended in fresh growth medium, they will resume normal multiplication.

B. Reversal by Substrate

When a chemical antagonist of the analog type binds reversibly with the enzyme, it is possible to displace it by adding a high concentration of the normal substrate. Such cases are termed "competitive inhibition." The ratio of inhibitor concentration to concentration of substrate reversing the inhibition is called the **antimicrobial index**; it is usually very high (100–10,000), indicating a much greater affinity of enzyme for its normal substrate.

C. Inactivation of Agent

An agent can often be inactivated by adding to the medium a substance that combines with it, preventing its combination with cellular constituents. For example, mercuric ion can be inactivated by addition to the medium of sulfhydryl compounds such as thioglycolic acid.

D. Protection Against Lysis

Osmotic lysis can be prevented by making the medium isotonic for naked bacterial protoplasts. Concentrations of 10–20% sucrose are required. Under such conditions penicillin-induced protoplasts remain viable and continue to grow as L forms.

Resistance to Antibacterial Agents

The ability of bacteria to become resistant to antibacterial agents is an important factor in their control. The mechanisms by which resistance is acquired are discussed in Chapters 7: Microbial Genetics and 28: Antimicrobial Chemotherapy.

Physical Agents

A. Heat

Application of heat is the simplest means of sterilizing materials, provided the material is itself resistant to heat damage. A temperature of 100°C will kill all but spore forms of bacteria within 2–3 minutes in laboratory-scale cultures; a temperature of 121°C for 15 minutes is utilized to kill spores. Steam is generally used, both because bacteria are more quickly killed when moist and because steam provides a means for distributing heat to all parts of the sterilizing vessel. At sea level, steam must be kept at a pressure of 15 lb/sq in. (psi) in excess of atmospheric pressure to obtain a temperature of 121°C; autoclaves or pressure cookers are used for this purpose. At higher altitudes, the pressure would need to be higher than 15 psi to reach 121°C. For sterilizing materials that must remain dry, circulating hot air electric ovens are available; since heat is less effective on dry material, it is customary to apply a temperature of 160–170°C for 1 hour or more.

Under the conditions described above (ie, excessive temperatures applied for long periods of time), heat acts by denaturing cell proteins and nucleic acids and by disrupting cell membranes.

B. Radiation

Ultraviolet light and ionizing radiations have various applications as sterilizing agents. Their modes of action are discussed above.

Chemical Agents

The chemical structures and uses of biocides are shown in Table 4–3.

A. Alcohols

Ethyl alcohol, isopropyl alcohol, and *n*-propanol exhibit rapid, broad-spectrum antimicrobial activity against vegetative bacteria, viruses, and fungi but are not sporicidal. Activity is optimal when they are diluted to a concentration of 60–90% with water.

B. Aldehydes

Glutaraldehyde is used for low-temperature disinfection and sterilization of endoscopes and surgical equipment. It is normally used as a 2% solution to achieve sporicidal activity. Formaldehyde is bactericidal, sporicidal, and virucidal.

C. Biguanides

Chlorhexidine is widely used in hand washing and oral products and as a disinfectant and preservative. Mycobacteria are generally highly resistant.

D. Bisphenols

The bisphenols are widely used in antiseptic soaps and hand rinses. In general, they are broad-spectrum but have little activity against *Pseudomonas aeruginosa* and molds. Triclosan and hexachlorophene are bactericidal and sporostatic.

E. Halogen-Releasing Agents

The most important types of chlorine-releasing agents are sodium hypochlorite, chlorine dioxide, and sodium dichloroisocyanurate, which are oxidizing agents that destroy the cellular activity of proteins. Hypochlorous acid is the active compound responsible for the bactericidal and virucidal effect of these compounds. At higher concentrations, these compounds are sporicidal. Iodine is rapidly bactericidal, fungicidal, tuberculocidal, virucidal, and sporicidal. Iodophors (eg, povidone-iodine) are complexes of iodine and a solubilizing agent or carrier, which acts as a reservoir of the active I_2.

F. Heavy Metal Derivatives

Silver sulfadiazine, a combination of two antibacterial agents, Ag^+ and sulfadiazine, has a broad spectrum of activity. Binding to cell components such as DNA may be responsible for its inhibitory properties.

G. Organic Acids

Organic acids are used as preservatives in the pharmaceutical and food industries. Benzoic acid is fungistatic; propionic acid is both bacteriostatic and fungistatic.

H. Peroxygens

Hydrogen peroxide has broad-spectrum activity against viruses, bacteria, yeasts, and bacterial spores. Sporicidal activity requires higher concentrations (10–30%) of H_2O_2 and longer contact times.

I. Phenols

Phenol and many phenolic compounds have antiseptic, disinfectant, or preservative properties.

J. Quaternary Ammonium Compounds

These compounds have two regions in their molecular structures, one a water-repelling (hydrophobic) group and the other a water-attracting (hydrophilic) group. Cationic detergents, as exemplified by quaternary ammonium compounds (QACs), are useful antiseptics and disinfectants. QACs have been used for a variety of clinical purposes (eg, preoperative disinfection of unbroken skin) as well as for cleaning hard surfaces. They are sporostatic; they inhibit the outgrowth of spores but not the actual germination process. QACs are also mycobacteriostatic and have an effect on lipid-enveloped but not lipid-nonenveloped viruses.

K. Vapor-Phase Sterilants

Heat-sensitive medical devices and surgical supplies can be effectively sterilized by vapor-phase systems employing ethylene oxide, formaldehyde, hydrogen peroxide, or peracetic acid.

Chemotherapeutic Agents

The natures and modes of action of these drugs are discussed in Chapter 28.

REVIEW QUESTIONS

1. A 23-year-old woman has 10 *Escherichia coli* inoculated into her bladder while having sex. These *Escherichia coli* have a generation time of 20 minutes. After a lag of 20 minutes, the *Escherichia coli* enter the logarithmic phase of growth. After 3 hours of logarithmic growth, the total number of cells is

   (A) 2560
   (B) 5012

(C) 90

(D) 1028

(E) 1,000,000

2. A 73-year-old woman is admitted to the hospital for intravenous treatment of an abscess caused by *Staphylococcus aureus*. Subsequent to her treatment and discharge from the hospital, it is necessary to disinfect the hospital room. One thousand of the *Staphylococcus aureus* cells are exposed to a disinfectant. After 10 minutes, 90% of the cells are killed. How many cells remain viable after 20 minutes?

(A) 500

(B) 100

(C) 10

(D) 1

(E) 0

3. The action of which of the following agents or processes on bacteria can be reversed?

(A) A disinfectant

(B) A bactericidal agent

(C) A bacteriostatic agent

(D) Autoclaving at 121°C for 15 minutes

(E) Dry heat at 160–170°C for 1 hour

4. The growth rate of bacteria during the exponential phase of growth is

(A) Zero

(B) Increasing

(C) Constant

(D) Decreasing

(E) Negative

5. The growth rate of bacteria during the maximum stationary phase of growth is

(A) Zero

(B) Increasing

(C) Constant

(D) Decreasing

(E) Negative

Answers

1. A	3. C	5. A
2. C	4. C	

REFERENCES

Block SS (editor): *Disinfection, Sterilization, and Preservation,* 5th ed. Lippincott Williams & Wilkins, 2001.

Colwell RR, Grimes DJ (editors): *Nonculturable Microorganisms in the Environment.* ASM Press, 2000.

Donohue WD: The cell cycle of *Escherichia coli.* Annu Rev Microbiol 1993;47:199.

Gerhardt P et al (editors): *Manual of Methods for General Bacteriology.* American Society for Microbiology, 1981.

Kjelleberg S (editor): *Starvation in Bacteria.* Plenum Press, 1993.

Kjelleberg S et al: The transient phase between growth and nongrowth of heterotrophic bacteria. Annu Rev Microbiol 1987;41:25. [PMID: 3318670]

Kolter R, Siegels DA, Tormo A: The stationary phase of the bacterial life cycle. J Bacteriol 1992;174:345. [PMID: 1729229]

McDonnell GE: *Antisepsis, Disinfection, and Sterilization: Types, Action, and Resistance.* ASM Press, 2007.

McDonnell G, Russell AD: Antiseptics and disinfectants: Activity, action, and resistance. Clin Microbiol Rev 1999;12:147. [PMID: 9880479]

Olmstad RN (editor): *APIC Infection Control and Applied Epidemiology: Principles and Practices.* Mosby Year Book, 1996.

Russell AD, Hugo WB, Ayliffe GAJ (editors): *Principles and Practice of Disinfection, Preservation and Sterilization,* 3rd ed. Blackwell Scientific Publications, 1999.

Sancar A, Sancar GB: DNA repair enzymes. Annu Rev Biochem 1988;57:29. [PMID: 3052275]

Siegels DA, Kolter R: Life after log. J Bacteriol 1992;174:345.

Cultivation of Microorganisms

Cultivation is the process of propagating organisms by providing the proper environmental conditions. Growing microorganisms are making replicas of themselves, and they require the elements present in their chemical composition. Nutrients must provide these elements in metabolically accessible form. In addition, the organisms require metabolic energy in order to synthesize macromolecules and maintain essential chemical gradients across their membranes. Factors that must be controlled during growth include the nutrients, pH, temperature, aeration, salt concentration, and ionic strength of the medium.

REQUIREMENTS FOR GROWTH

Most of the dry weight of microorganisms is organic matter containing the elements carbon, hydrogen, nitrogen, oxygen, phosphorus, and sulfur. In addition, inorganic ions such as potassium, sodium, iron, magnesium, calcium, and chloride are required to facilitate enzymatic catalysis and to maintain chemical gradients across the cell membrane.

For the most part, the organic matter is in macromolecules formed by **anhydride bonds** between building blocks. Synthesis of the anhydride bonds requires chemical energy, which is provided by the two phosphodiester bonds in ATP (adenosine triphosphate; see Chapter 6). Additional energy required to maintain a relatively constant cytoplasmic composition during growth in a range of extracellular chemical environments is derived from the **proton motive force**. The proton motive force is the potential energy that can be derived by passage of a proton across a membrane. In eukaryotes, the membrane may be part of the mitochondrion or the chloroplast. In prokaryotes, the membrane is the cytoplasmic membrane of the cell.

The proton motive force is an electrochemical gradient with two components: a difference in pH (hydrogen ion concentration) and a difference in ionic charge. The charge on the outside of the bacterial membrane is more positive than the charge on the inside, and the difference in charge contributes to the free energy released when a proton enters the cytoplasm from outside the membrane. Metabolic processes that generate the proton motive force are discussed in Chapter 6. The free energy may be used to move the cell, to maintain ionic or molecular gradients across the membrane, to synthesize anhydride bonds in ATP, or for a combination of these purposes. Alternatively, cells given a source of ATP may use its anhydride bond energy to create a proton motive force that in turn may be used to move the cell and to maintain chemical gradients.

In order to grow, an organism requires all of the elements in its organic matter and the full complement of ions required for energetics and catalysis. In addition, there must be a source of energy to establish the proton motive force and to allow macromolecular synthesis. Microorganisms vary widely in their nutritional demands and their sources of metabolic energy.

SOURCES OF METABOLIC ENERGY

The three major mechanisms for generating metabolic energy are fermentation, respiration, and photosynthesis. At least one of these mechanisms must be employed if an organism is to grow.

Fermentation

The formation of ATP in fermentation is not coupled to the transfer of electrons. Fermentation is characterized by **substrate phosphorylation**, an enzymatic process in which a pyrophosphate bond is donated directly to ADP (adenosine diphosphate) by a phosphorylated metabolic intermediate. The phosphorylated intermediates are formed by metabolic rearrangement of a fermentable substrate such as glucose, lactose, or arginine. Because fermentations are not accompanied by a change in the overall oxidation-reduction state of the fermentable substrate, the elemental composition of the products of fermentation must be identical to those of the substrates. For example, fermentation of a molecule of glucose ($C_6H_{12}O_6$) by the Embden-Meyerhof pathway (see Chapter 6) yields a net gain of two pyrophosphate bonds in ATP and produces two molecules of lactic acid ($C_3H_6O_3$).

Respiration

Respiration is analogous to the coupling of an energy-dependent process to the discharge of a battery. Chemical reduction of an oxidant (electron acceptor) through a specific series of electron carriers in the membrane establishes the proton motive force across the bacterial membrane. The reductant (electron donor) may be organic or inorganic: eg, lactic acid serves as a reductant for some organisms, and hydrogen gas is a reductant for other organisms. Gaseous oxygen (O_2) often is employed as an oxidant, but alternative oxidants that are employed by some organisms include carbon dioxide (CO_2), sulfate (SO_4^{2-}), and nitrate (NO_3^-).

Photosynthesis

Photosynthesis is similar to respiration in that the reduction of an oxidant via a specific series of electron carriers establishes the proton motive force. The difference in the two processes is that in photosynthesis the reductant and oxidant are created photochemically by light energy absorbed by pigments in the membrane; thus, photosynthesis can continue only as long as there is a source of light energy. Plants and some bacteria are able to invest a substantial amount of light energy in making water a reductant for carbon dioxide. Oxygen is evolved in this process, and organic matter is produced. Respiration, the energetically favorable oxidation of organic matter by an electron acceptor such as oxygen, can provide photosynthetic organisms with energy in the absence of light.

NUTRITION

Nutrients in growth media must contain all the elements necessary for the biologic synthesis of new organisms. In the following discussion, nutrients are classified according to the elements they supply.

Carbon Source

As mentioned above, plants and some bacteria are able to use photosynthetic energy to reduce carbon dioxide at the expense of water. These organisms belong to the group of **autotrophs**, creatures that do not require organic nutrients for growth. Other autotrophs are the **chemolithotrophs**, organisms that use an inorganic substrate such as hydrogen or thiosulfate as a reductant and carbon dioxide as a carbon source.

Heterotrophs require organic carbon for growth, and the organic carbon must be in a form that can be assimilated. Naphthalene, eg, can provide all the carbon and energy required for respiratory heterotrophic growth, but very few organisms possess the metabolic pathway necessary for naphthalene assimilation. Glucose, on the other hand, can support the fermentative or respiratory growth of many organisms. It is important that growth substrates be supplied at levels appropriate for the microbial strain that is being grown: Levels that will support the growth of one organism may inhibit the growth of another organism.

Carbon dioxide is required for a number of biosynthetic reactions. Many respiratory organisms produce more than enough carbon dioxide to meet this requirement, but others require a source of carbon dioxide in their growth medium.

Nitrogen Source

Nitrogen is a major component of proteins, nucleic acids, and other compounds, accounting for approximately 5% of the dry weight of a typical bacterial cell. Inorganic dinitrogen (N_2) is very prevalent, as it comprises 80% of the earth's atmosphere. It is also a very stable compound, primarily because of the high activation energy required to break the nitrogen–nitrogen triple bond. However, nitrogen may be supplied in a number of different forms, and microorganisms vary in their abilities to assimilate nitrogen (Table 5–1). The end product of all pathways for nitrogen assimilation is the most reduced form of the element, ammonia (NH_3). When NH_3 is available, it diffuses into most bacteria through transmembrane channels as dissolved gaseous NH_3 rather than ionic ammonium ion (NH_4^+).

The ability to assimilate N_2 reductively via NH_3, which is called **nitrogen fixation**, is a property unique to prokaryotes, and relatively few bacteria are capable of breaking the nitrogen–nitrogen triple bond. This process (see Chapter 6) requires a large amount of metabolic energy and is readily inactivated by oxygen. The capacity for nitrogen fixation is found in widely divergent bacteria that have evolved quite different biochemical strategies to protect their nitrogen-fixing enzymes from oxygen.

Most microorganisms can use NH_3 as a sole nitrogen source, and many organisms possess the ability to produce NH_3 from amines ($R—NH_2$) or from amino acids ($RCHNH_2COOH$), generally intracellularly. Production of NH_3 from the deamination of amino acids is called **ammonification**. Ammonia is introduced into organic matter by biochemical pathways involving glutamate and glutamine. These pathways are discussed in Chapter 6.

Many microorganisms possess the ability to assimilate nitrate (NO_3^-) and nitrite (NO_2^-) reductively by conversion of these ions into NH_3. These processes are termed **assimilatory nitrate reduction** and **assimilatory nitrite reduction**, respectively. These pathways for assimilation differ from pathways used for **dissimilation** of nitrate and nitrite. The dissimilatory pathways are used by organisms that employ these ions as terminal electron acceptors in respiration. Some autotrophic bacteria (eg, *Nitrosomonas*, *Nitrobacter*) are able to convert NH_3 to gaseous N_2 under anaerobic conditions; this process is known as **denitrification**. Our understanding of the nitrogen cycle continues to evolve. In the mid-1990s, the **anammox** reaction was discovered. The reaction

$$NH_4^+ + NO_2^- \rightarrow N_2 + 2H_2O$$

in which ammonia is oxidized by nitrite, is a microbial process that occurs in anoxic waters of the ocean and is a major pathway by which nitrogen is returned to the atmosphere.

TABLE 5–1 Sources of Nitrogen in Microbial Nutrition

Compound	Valence of N
NO_3^-	+5
NO_2^-	+3
N_2	0
NH_4^+	−3
R-NH_2[a]	−3

[a]R, organic radical.

Sulfur Source

Like nitrogen, sulfur is a component of many organic cell substances. It forms part of the structure of several coenzymes and is found in the cysteinyl and methionyl side chains of proteins. Sulfur in its elemental form cannot be used by plants or animals. However, some autotrophic bacteria can oxidize it to sulfate (SO_4^{2-}). Most microorganisms can use sulfate as a sulfur source, reducing the sulfate to the level of hydrogen sulfide (H_2S). Some microorganisms can assimilate H_2S directly from the growth medium, but this compound can be toxic to many organisms.

Phosphorus Source

Phosphate (PO_4^{3-}) is required as a component of ATP, nucleic acids, and such coenzymes as NAD, NADP, and flavins. In addition, many metabolites, lipids (phospholipids, lipid A), cell wall components (teichoic acid), some capsular polysaccharides, and some proteins are phosphorylated. Phosphate is always assimilated as free inorganic phosphate (P_i).

Mineral Sources

Numerous minerals are required for enzyme function. Magnesium ion (Mg^{2+}) and ferrous ion (Fe^{2+}) are also found in porphyrin derivatives: magnesium in the chlorophyll molecule, and iron as part of the coenzymes of the cytochromes and peroxidases. Mg^{2+} and K^+ are both essential for the function and integrity of ribosomes. Ca^{2+} is required as a constituent of gram-positive cell walls, though it is dispensable for gram-negative bacteria. Many marine organisms require Na^+ for growth. In formulating a medium for the cultivation of most microorganisms, it is necessary to provide sources of potassium, magnesium, calcium, and iron, usually as their ions (K^+, Mg^{2+}, Ca^{2+}, and Fe^{2+}). Many other minerals (eg, Mn^{2+}, Mo^{2+}, Co^{2+}, Cu^{2+}, and Zn^{2+}) are required; these frequently can be provided in tap water or as contaminants of other medium ingredients.

The uptake of iron, which forms insoluble hydroxides at neutral pH, is facilitated in many bacteria and fungi by their production of **siderophores**—compounds that chelate iron and promote its transport as a soluble complex. These include hydroxamates (–$CONH_2OH$) called sideramines, and derivatives of catechol (eg, 2,3-dihydroxy-benzoylserine). Plasmid-determined siderophores play a major role in the invasiveness of some bacterial pathogens (see Chapter 7). Siderophore- and nonsiderophore-dependent mechanisms of iron uptake by bacteria are discussed in Chapter 9.

Growth Factors

A growth factor is an organic compound which a cell must contain in order to grow but which it is unable to synthesize. Many microorganisms, when provided with the nutrients listed above, are able to synthesize all of the building blocks for macromolecules (Figure 5–1): amino acids; purines, pyrimidines, and pentoses (the metabolic precursors of nucleic acids); additional carbohydrates (precursors of polysaccharides); and fatty acids and isoprenoid compounds. In addition, free-living organisms must be able to synthesize the complex vitamins that serve as precursors of coenzymes.

Each of these essential compounds is synthesized by a discrete sequence of enzymatic reactions; each enzyme is produced under the control of a specific gene. When an organism undergoes a gene mutation resulting in failure of one of these enzymes to function, the chain is broken and the end product is no longer produced. The organism must then obtain that compound from the environment: The compound has become a **growth factor** for the organism. This type of mutation can be readily induced in the laboratory.

Different microbial species vary widely in their growth factor requirements. The compounds involved are found in and are essential to all organisms; the differences in requirements reflect differences in synthetic abilities. Some species require no growth factors, while others—like some of the lactobacilli—have lost, during evolution, the ability to synthesize as many as 30–40 essential compounds and hence require them in the medium.

ENVIRONMENTAL FACTORS AFFECTING GROWTH

A suitable growth medium must contain all the nutrients required by the organism to be cultivated, and such factors as pH, temperature, and aeration must be carefully controlled. A liquid medium is used; the medium can be gelled for special purposes by adding agar or silica gel. Agar, a polysaccharide extract of a marine alga, is uniquely suitable for microbial cultivation because it is resistant to microbial action and because it dissolves at 100°C but does not gel until cooled below 45°C; cells can be suspended in the medium at 45°C and the medium quickly cooled to a gel without harming them.

Nutrients

On the previous pages, the function of each type of nutrient is described and a list of suitable substances presented. In general, the following must be provided: (1) hydrogen donors and

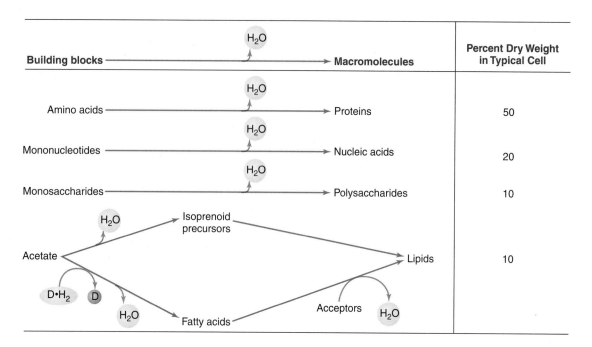

FIGURE 5–1 Macromolecular synthesis. Polymerization of building blocks into macromolecules is achieved largely by the introduction of anhydride bonds. Formation of fatty acids from acetate requires several steps of biochemical reduction using organic hydrogen donors (D · H_2).

acceptors: about 2 g/L; (2) carbon source: about 1 g/L; (3) nitrogen source: about 1 g/L; (4) minerals: sulfur and phosphorus, about 50 mg/L of each, and trace elements, 0.1–1 mg/L of each; (5) growth factors: amino acids, purines, and pyrimidines, about 50 mg/L of each, and vitamins, 0.1–1 mg/L of each.

For studies of microbial metabolism, it is usually necessary to prepare a completely synthetic medium in which the exact characteristics and concentration of every ingredient are known. Otherwise, it is much cheaper and simpler to use natural materials such as yeast extract, protein digest, or similar substances. Most free-living microbes will grow well on yeast extract; parasitic forms may require special substances found only in blood or in extracts of animal tissues. Nevertheless, there are parasitic microbes (eg, *Treponema pallidum*) that cannot be grown in vitro or that grow inside eukaryotic cells (eg, *Chlamydia trachomatis*).

For many organisms, a single compound (such as an amino acid) may serve as energy source, carbon source, and nitrogen source; others require a separate compound for each. If natural materials for nonsynthetic media are deficient in any particular nutrient, they must be supplemented.

Hydrogen Ion Concentration (pH)

Most organisms have a fairly narrow optimal pH range. The optimal pH must be empirically determined for each species. Most organisms (**neutralophiles**) grow best at a pH of 6.0–8.0, although some forms (**acidophiles**) have optima as low as pH 3.0 and others (**alkaliphiles**) have optima as high as pH 10.5.

Microorganisms regulate their internal pH over a wide range of external pH values by pumping protons in or out of their cells. Acidophiles maintain an internal pH of about 6.5 over an external range of 1.0–5.0; neutralophiles maintain an internal pH of about 7.5 over an external range of 5.5–8.5; and alkaliphiles maintain an internal pH of about 9.5 over an external range of 9.0–11.0. Internal pH is regulated by a set of proton transport systems in the cytoplasmic membrane, including a primary, ATP-driven proton pump and a Na^+/H^+ exchanger. A K^+/H^+ exchange system has also been proposed to contribute to internal pH regulation in neutralophiles.

Temperature

Different microbial species vary widely in their optimal temperature ranges for growth: **Psychrophilic** forms grow best at low temperatures (15–20°C); **mesophilic** forms grow best at 30–37°C; and most **thermophilic** forms grow best at 50–60°C. Some organisms are **hyperthermophilic** and can grow at well above the temperature of boiling water, which exists under high pressure in the depths of the ocean. Most organisms are mesophilic; 30°C is optimal for many free-living forms, and the body temperature of the host is optimal for symbionts of warm-blooded animals.

The upper end of the temperature range tolerated by any given species correlates well with the general thermal stability of that species' proteins as measured in cell extracts. Microorganisms share with plants and animals the **heat-shock response**, a transient synthesis of a set of "heat-shock proteins," when exposed to a sudden rise in temperature above the growth optimum. These proteins appear to be unusually heat-resistant and to stabilize the heat-sensitive proteins of the cell.

The relationship of growth rate to temperature for any given microorganism is seen in a typical Arrhenius plot (Figure 5–2). Arrhenius showed that the logarithm of the velocity of any chemical reaction (log k) is a linear function of the reciprocal of the temperature ($1/T$); since cell growth is the result of a set of chemical reactions, it might be expected to show this relationship. Figure 5–2 shows this to be the case over the normal range of temperatures for a given species; log k decreases linearly with $1/T$. Above and below the normal range, however, log k drops rapidly, so that maximum and minimum temperature values are defined.

Beyond their effects on growth rate, extremes of temperature kill microorganisms. Extreme heat is used to sterilize preparations (see Chapter 4); extreme cold also kills microbial cells, although it cannot be used safely for sterilization. Bacteria also exhibit a phenomenon called **cold shock**: the killing of cells by rapid—as opposed to slow—cooling. For example, the rapid cooling of *Escherichia coli* from 37°C to 5°C can kill 90% of the cells. A number of compounds protect cells from either freezing or cold shock; glycerol and dimethyl sulfoxide are most commonly used.

Aeration

The role of oxygen as hydrogen acceptor is discussed in Chapter 6. Many organisms are obligate aerobes, specifically requiring oxygen as hydrogen acceptor; some are facultative, able to live aerobically or anaerobically; and others are obligate anaerobes, requiring a substance other than oxygen as hydrogen acceptor and being sensitive to oxygen inhibition.

The natural by-products of aerobic metabolism are the reactive compounds hydrogen peroxide (H_2O_2) and superoxide (O_2^-). In the presence of iron, these two species can generate hydroxyl radicals ($\bullet OH$), which can damage any biologic macromolecule:

$$O_2^- + H_2O_2 \xrightarrow{Fe^{3+}/Fe^{2+}} O_2 + OH^- + \bullet OH$$

Many aerobes and aerotolerant anaerobes are protected from these products by the presence of superoxide dismutase, an enzyme that catalyzes the reaction

$$2O_2^- + 2H^+ \rightarrow O_2 + H_2O_2$$

and by the presence of catalase, an enzyme that catalyzes the reaction

$$2H_2O_2 \rightarrow 2H_2O + O_2$$

Some fermentative organisms (eg, *Lactobacillus plantarum*) are aerotolerant but do not contain catalase or superoxide dismutase. Oxygen is not reduced, and therefore H_2O_2 and O_2^- are not produced. All strict anaerobes lack both superoxide dismutase and catalase. Some anaerobic organisms (eg, *Peptococcus anaerobius*) have considerable tolerance to oxygen as a result of their ability to produce high levels of an enzyme (NADH oxidase) that reduces oxygen to water according to the reaction

$$NADH + H^+ + \tfrac{1}{2}O_2 \longrightarrow NAD^+ + H_2O$$

Hydrogen peroxide owes much of its toxicity to the damage it causes to DNA. DNA repair-deficient mutants are exceptionally sensitive to hydrogen peroxide; the *recA* gene product, which functions in both genetic recombination and repair, has been shown to be more important than either catalase or superoxide dismutase in protecting *E coli* cells against hydrogen peroxide toxicity.

The supply of air to cultures of aerobes is a major technical problem. Vessels are usually shaken mechanically to introduce oxygen into the medium, or air is forced through the medium by pressure. The diffusion of oxygen often becomes the limiting factor in growing aerobic bacteria; when a cell concentration of $4–5 \times 10^9$/mL is reached, the rate of diffusion of oxygen to the cells sharply limits the rate of further growth.

Obligate anaerobes, on the other hand, present the problem of oxygen exclusion. Many methods are available for this: Reducing agents such as sodium thioglycolate can be added

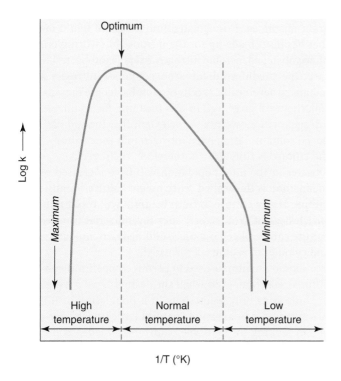

FIGURE 5–2 General form of an Arrhenius plot of bacterial growth. (Reproduced with permission from Ingraham JL: Growth of psychrophilic bacteria, J Bacteriol July 76(1):75–80,1958.)

to liquid cultures; tubes of agar can be sealed with a layer of petrolatum and paraffin; the culture vessel can be placed in a container from which the oxygen is removed by evacuation or by chemical means; or the organism can be handled within an anaerobic glove-box.

Ionic Strength & Osmotic Pressure

To a lesser extent, such factors as osmotic pressure and salt concentration may have to be controlled. For most organisms, the properties of ordinary media are satisfactory; however, for marine forms and organisms adapted to growth in strong sugar solutions, for example, these factors must be considered. Organisms requiring high salt concentrations are called **halophilic**; those requiring high osmotic pressures are called **osmophilic**.

Most bacteria are able to tolerate a wide range of external osmotic pressures and ionic strengths because of their ability to regulate internal osmolality and ion concentration. Osmolality is regulated by the active transport of K^+ ions into the cell; internal ionic strength is kept constant by a compensating excretion of the positively charged organic polyamine putrescine. Since putrescine carries several positive charges per molecule, a large drop in ionic strength is effected at only a small cost in osmotic strength.

CULTIVATION METHODS

Two problems will be considered: the choice of a suitable medium and the isolation of a bacterial organism in pure culture.

The Medium

The technique used and the type of medium selected depend upon the nature of the investigation. In general, three situations may be encountered: (1) One may need to raise a crop of cells of a particular species that is on hand; (2) one may need to determine the numbers and types of organisms present in a given material; or (3) one may wish to isolate a particular type of microorganism from a natural source.

A. Growing Cells of a Given Species

Microorganisms observed microscopically to be growing in a natural environment may prove exceedingly difficult to grow in pure culture in an artificial medium. Certain parasitic forms, eg, have never been cultivated outside the host. In general, however, a suitable medium can be devised by carefully reproducing the conditions found in the organism's natural environment. The pH, temperature, and aeration are easy to duplicate; the nutrients present the major problem. The contribution made by the living environment is important and difficult to analyze; a parasite may require an extract of the host tissue, and a free-living form may require a substance excreted by a microorganism with which it is associated in

nature. Considerable experimentation may be necessary in order to determine the requirements of the organism, and success depends upon providing a suitable source of each category of nutrient listed at the beginning of this chapter. The cultivation of obligate parasites such as chlamydiae is discussed in Chapter 27.

B. Microbiologic Examination of Natural Materials

A given natural material may contain many different microenvironments, each providing a niche for a different species. Plating a sample of the material under one set of conditions will allow a selected group of forms to produce colonies but will cause many other types to be overlooked. For this reason, it is customary to plate out samples of the material using as many different media and conditions of incubation as is practicable. Six to eight different culture conditions are not an unreasonable number if most of the forms present are to be discovered.

Since every type of organism present must have a chance to grow, solid media are used and crowding of colonies is avoided. Otherwise, competition will prevent some types from forming colonies.

C. Isolation of a Particular Type of Microorganism

A small sample of soil, if handled properly, will yield a different type of organism for every microenvironment present. For fertile soil (moist, aerated, rich in minerals and organic matter) this means that hundreds or even thousands of types can be isolated. This is done by selecting for the desired type. One gram of soil, eg, is inoculated into a flask of liquid medium that has been made up for the purpose of favoring one type of organism, eg, aerobic nitrogen fixers (azotobacter). In this case, the medium contains no combined nitrogen and is incubated aerobically. If cells of azotobacter are present in the soil, they will grow well in this medium; forms unable to fix nitrogen will grow only to the extent that the soil has introduced contaminating fixed nitrogen into the medium. When the culture is fully grown, therefore, the percentage of azotobacter in the total population will have increased greatly; the method is thus called "enrichment culture." Transfer of a sample of this culture to fresh medium will result in further enrichment of azotobacter; after several serial transfers, the culture can be plated out on a solidified enrichment medium and colonies of azotobacter isolated.

Liquid medium is used to permit competition and hence optimal selection, even when the desired type is represented in the soil as only a few cells in a population of millions. Advantage can be taken of "natural enrichment." For example, in looking for kerosene oxidizers, oil-laden soil is chosen, since it is already an enrichment environment for such forms.

Enrichment culture, then, is a procedure whereby the medium is prepared so as to duplicate the natural environment ("niche") of the desired microorganism, thereby selecting for it. An important principle involved in such selection is the following: The organism selected for will be the type whose nutritional requirements are barely satisfied. Azotobacter, eg, grows best in a medium containing organic nitrogen, but its minimum requirement is the presence of N_2; hence it is selected for in a medium containing N_2 as the sole nitrogen source. If organic nitrogen is added to the medium, the conditions no longer select for azotobacter but rather for a form for which organic nitrogen is the minimum requirement.

When searching for a particular type of organism in a natural material, it is advantageous to plate the organisms obtained on a **differential medium** if available. A differential medium is one that will cause the colonies of a particular type of organism to have a distinctive appearance. For example, colonies of *E coli* have a characteristic iridescent sheen on agar containing the dyes eosin and methylene blue (EMB agar). EMB agar containing a high concentration of one sugar will also cause organisms which ferment that sugar to form reddish colonies. Differential media are used for such purposes as recognizing the presence of enteric bacteria in water or milk and the presence of certain pathogens in clinical specimens.

Table 5–2 presents examples of enrichment culture conditions and the types of bacteria they will select. However, in spite of our best efforts, many environments contain numerous uncultured bacteria.

Isolation of Microorganisms in Pure Culture

In order to study the properties of a given organism, it is necessary to handle it in pure culture free of all other types of organisms. To do this, a single cell must be isolated from all other cells and cultivated in such a manner that its collective progeny also remain isolated. Several methods are available.

A. Plating

Unlike cells in a liquid medium, cells in or on a gelled medium are immobilized. Therefore, if few enough cells are placed in or on a gelled medium, each cell will grow into an isolated colony. The ideal gelling agent for most microbiologic media is **agar**, an acidic polysaccharide extracted from certain red algae. A 1.5–2% suspension in water dissolves at 100°C, forming a clear solution that gels at 45°C. Thus, a sterile agar solution can be cooled to 50°C, bacteria or other microbial cells added, and then the solution quickly cooled below 45°C to form a gel. (Although most microbial cells are killed at 50°C, the time-course of the killing process is sufficiently slow at this temperature to permit this procedure; see Figure 4–3.) Once gelled, agar will not again liquefy until it is heated above 80°C, so that any temperature suitable for the

TABLE 5–2 Some Enrichment Cultures

Nitrogen Source	Carbon Source	Atmosphere	Illumination	Predominant Organism Initially Enriched
N_2	CO_2	Aerobic or anaerobic	Dark	None
			Light	Cyanobacteria
	Alcohol, fatty acids, etc	Anaerobic	Dark	None
		Air	Dark	Azotobacter
	Glucose	Anaerobic	Dark	*Clostridium pasteurianum*
		Air	Dark	Azotobacter
$NaNO_3$	CO_2	Aerobic or anaerobic	Dark	None
			Light	Green algae and cyanobacteria
	Alcohol, fatty acids, etc	Anaerobic	Dark	Denitrifiers
		Air	Dark	Aerobes
	Glucose	Anaerobic	Dark	Fermenters
		Air	Dark	Aerobes
NH_4Cl	CO_2	Anaerobic	Dark	None
		Aerobic	Dark	Nitrosomonas
		Aerobic or anaerobic	Light	Green algae and cyanobacteria
	Alcohol, fatty acids, etc	Anaerobic	Dark	Sulfate or carbonate reducers
		Aerobic	Dark	Aerobes
	Glucose	Anaerobic	Dark	Fermenters
		Aerobic	Dark	Aerobes

Note: Constituents of all media: $MgSO_4$, K_2HPO_4, $FeCl_3$, $CaCl_2$, $CaCO_3$, and trace elements.

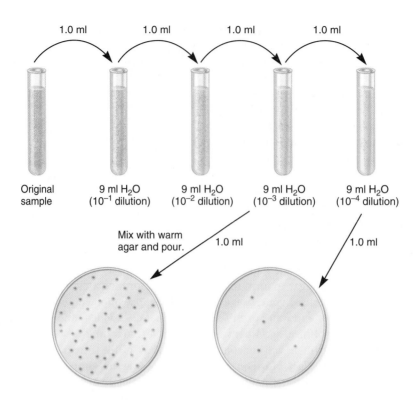

FIGURE 5–3 The pour-plate technique. The original sample is diluted several times to thin out the population sufficiently. The most diluted samples are then mixed with warm agar and poured into Petri dishes. Isolated cells grow into colonies and are used to establish pure cultures. The surface colonies are circular; subsurface colonies are lenticular (lens shaped). (Reproduced with permission from Willey JM, Sherwood LM, Woolverton CJ: Prescott, Harley, & Klein's *Microbiology*, 7th edition, McGraw-Hill, 2008.)

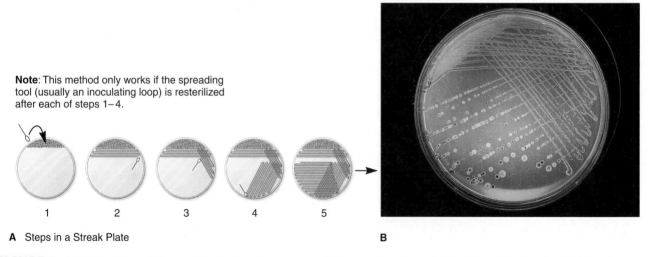

Note: This method only works if the spreading tool (usually an inoculating loop) is resterilized after each of steps 1–4.

A Steps in a Streak Plate

B

FIGURE 5–4 Streak-plate technique. **A**: A typical streaking pattern. **B**: An example of a streak plate. (Reproduced with permission from Kathy Park Talaro.)

incubation of a microbial culture can subsequently be used. In the **pour-plate method**, a suspension of cells is mixed with melted agar at 50°C and poured into a **Petri dish**. When the agar solidifies, the cells are immobilized in the agar and grow into colonies. If the cell suspension is sufficiently dilute, the colonies will be well separated, so that each has a high probability of being derived from a single cell (Figure 5–3). To make certain of this, however, it is necessary to pick a colony of the desired type, suspend it in water, and replate. Repeating

this procedure several times ensures that a pure culture will be obtained.

Alternatively, the original suspension can be streaked on an agar plate with a wire loop (**streak-plate technique**). As the streaking continues, fewer and fewer cells are left on the loop, and finally the loop may deposit single cells on the agar (Figure 5–4). The plate is incubated, and any well-isolated colony is then removed, resuspended in water, and again streaked on agar. If a suspension (and not just a bit of growth from a colony or slant) is streaked, this method is just as reliable as and much faster than the pour-plate method.

In the **spread plate technique**, a small volume of dilute microbial suspension containing ca 30–300 cells is transferred to the center of an agar plate and spread evenly over the surface with a sterile bent-glass rod. The dispersed cells develop into isolated colonies. Because the number of colonies should equal the number of viable organisms in a sample, spread plates can be used to count the microbial population.

B. Dilution

A much less reliable method is that of extinction dilution. The suspension is serially diluted, and samples of each dilution are plated. If only a few samples of a particular dilution exhibit growth, it is presumed that some of the colonies started from single cells. This method is not used unless plating is for some reason impossible. An undesirable feature of this method is that it can only be used to isolate the predominant type of organism in a mixed population.

REVIEW QUESTIONS

1. Most microorganisms pathogenic for humans grow best in the laboratory when cultures are incubated at
   (A) 15–20°C
   (B) 20–30°C
   (C) 30–37°C
   (D) 38–50°C
   (E) 50–55°C

2. The process by which microorganisms form ATP during the fermentation of glucose is characterized by
   (A) Coupling of ATP production with the transfer of electrons
   (B) Denitrification
   (C) The reduction of oxygen
   (D) Substrate phosphorylation
   (E) Anaerobic respiration

3. Which of the following culture techniques and media would enumerate the greatest number of microbial species in a soil sample?
   (A) Enrichment culture
   (B) A plate of selective medium
   (C) A plate of differential medium
   (D) A tube of nutrient broth
   (E) A number of different media and conditions of incubation

4. Polymerization of building blocks (eg, amino acids) into macromolecules (eg, proteins) is achieved largely by
   (A) Dehydration
   (B) Reduction
   (C) Oxidation
   (D) Assimilation
   (E) Hydrolysis

5. A strain of *E coli* undergoes a mutation such that it can no longer grow in a defined medium consisting of glucose, mineral salts, and ammonium chloride. However, it is capable of growth in this medium if methionine is added. The methionine is referred to as
   (A) An inorganic salt
   (B) A sulfur source
   (C) A growth factor
   (D) An energy source
   (E) A nitrogen source

Answers

1. C   3. E   5. C
2. D   4. A

REFERENCES

Adams MW: Enzymes and proteins from organisms that grow near or above 100 degrees C. Annu Rev Med 1993;47:627. [PMID: 8257111]

Koch AL: Microbial physiology and ecology of slow growth. Microbiol Molec Biol Rev 1997;61:305. [PMID: 9293184]

Lederberg J (editor): *Encyclopedia of Microbiology,* 4 vols. Academic Press, 1992.

Maier RM, Pepper IL, Gerba CP: *Environmental Microbiology.* Academic Press, 1992.

Marzlut GA: Regulation of sulfur and nitrogen metabolism in filamentary fungi. Annu Rev Microbiol 1993;42:89.

Pelczar MJ Jr, Chan ECS, Krieg NR: *Microbiology: Concepts and Applications.* McGraw-Hill, 1993.

Schloss PD, Handelsman J: Status of the microbial census. Microbiol Molec Biol Rev 2004;68:686.

Microbial Metabolism

ROLE OF METABOLISM IN BIOSYNTHESIS & GROWTH

Microbial growth requires the polymerization of biochemical building blocks into proteins, nucleic acids, polysaccharides, and lipids. The building blocks must come preformed in the growth medium or must be synthesized by the growing cells. Additional biosynthetic demands are placed by the requirement for coenzymes that participate in enzymatic catalysis. Biosynthetic polymerization reactions demand the transfer of anhydride bonds from ATP. Growth demands a source of metabolic energy for the synthesis of anhydride bonds and for the maintenance of transmembrane gradients of ions and metabolites.

The biosynthetic origins of building blocks and coenzymes can be traced to relatively few precursors, called **focal metabolites**. Figures 6–1 through 6–4 illustrate how the respective focal metabolites glucose 6-phosphate, phosphoenolpyruvate, oxaloacetate, and α-ketoglutarate give rise to most biosynthetic end products. Microbial metabolism can be divided into four general categories: (1) pathways for the interconversion of focal metabolites, (2) assimilatory pathways for the formation of focal metabolites, (3) biosynthetic sequences for the conversion of focal metabolites to end products, and (4) pathways that yield metabolic energy for growth and maintenance.

When provided with building blocks and a source of metabolic energy, a cell synthesizes macromolecules. The sequence of building blocks within a macromolecule is determined in one of two ways. In nucleic acids and proteins, it is **template-directed**: DNA serves as the template for its own synthesis and for the synthesis of the various types of RNA; messenger RNA serves as the template for the synthesis of proteins. In carbohydrates and lipids, on the other hand, the arrangement of building blocks is determined entirely by enzyme specificities. Once the macromolecules have been synthesized, they self-assemble to form the supramolecular structures of the cell, eg, ribosomes, membranes, cell wall, flagella, and pili.

The rate of macromolecular synthesis and the activity of metabolic pathways must be regulated so that biosynthesis is balanced. All of the components required for macromolecular synthesis must be present for orderly growth, and control must be exerted so that the resources of the cell are not expended on products that do not contribute to growth or survival.

This chapter contains a review of microbial metabolism and its regulation. Microorganisms represent extremes of evolutionary divergence, and a vast array of metabolic pathways are found within the group. For example, any of more than half a dozen different metabolic pathways may be used for assimilation of a relatively simple compound, benzoate, and a single pathway for benzoate assimilation may be regulated by any of more than half a dozen control mechanisms. Our goal will be to illustrate the principles that underlie metabolic pathways and their regulation. The primary principle that determines metabolic pathways is that they are achieved by organizing relatively few biochemical type reactions in a specific order. Many biosynthetic pathways can be deduced by examining the chemical structures of the starting material, the end product, and, perhaps, one or two metabolic intermediates. The primary principle underlying metabolic regulation is that enzymes tend to be called into play only when their catalytic activity is demanded. The activity of an enzyme may be changed by varying either the amount of enzyme or the amount of substrate. In some cases, the activity of enzymes may be altered by the binding of specific **effectors**, metabolites that modulate enzyme activity.

FOCAL METABOLITES & THEIR INTERCONVERSION

Glucose 6-Phosphate & Carbohydrate Interconversions

Figure 6–1 illustrates how glucose 6-phosphate is converted to a range of biosynthetic end products via phosphate esters of carbohydrates with different chain lengths. Carbohydrates possess the empirical formula $(CH_2O)_n$, and the primary objective of carbohydrate metabolism is to change n, the length of the carbon chain. Mechanisms by which the chain lengths of carbohydrate phosphates are interconverted are summarized in Figure 6–5. In one case, oxidative reactions are used to remove a single carbon from glucose 6-phosphate, producing the pentose derivative ribulose

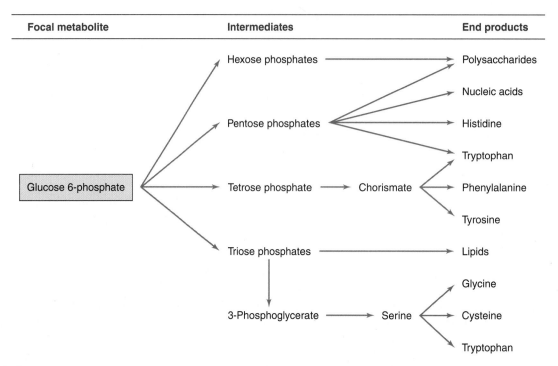

FIGURE 6–1 Biosynthetic end products formed from glucose 6-phosphate. Carbohydrate phosphate esters of varying chain length serve as intermediates in the biosynthetic pathways.

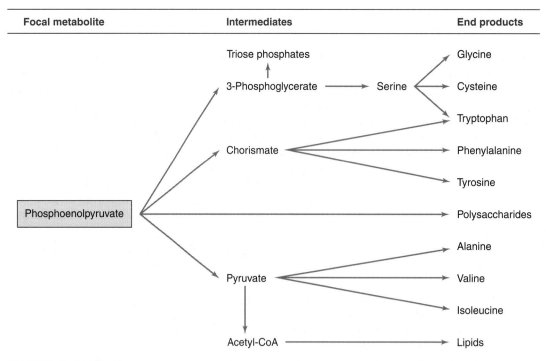

FIGURE 6–2 Biosynthetic end products formed from phosphoenolpyruvate.

5-phosphate. Isomerase and epimerase reactions interconvert the most common biochemical forms of the pentoses: ribulose 5-phosphate, ribose 5-phosphate, and xylulose 5-phosphate. Transketolases transfer a two-carbon fragment from a donor to an acceptor molecule. These reactions allow pentoses to form or to be formed from carbohydrates of varying chain lengths. As shown in Figure 6–5, two pentose 5-phosphates (n = 5) are interconvertible with triose 3-phosphate (n = 3) and heptose 7-phosphate (n = 7); pentose 5-phosphate (n = 5) and tetrose 4-phosphate (n = 4) are interconvertible with triose 3-phosphate (n = 3) and hexose 6-phosphate (n = 6).

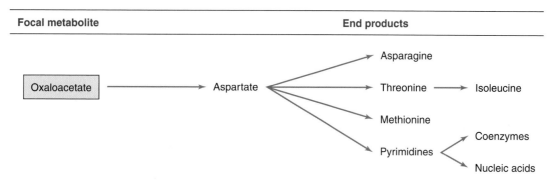

FIGURE 6–3 Biosynthetic end products formed from oxaloacetate. The end products aspartate, threonine, and pyrimidines serve as intermediates in the synthesis of additional compounds.

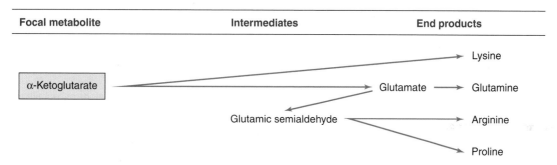

FIGURE 6–4 Biosynthetic end products formed from α-ketoglutarate.

The six-carbon hexose chain of fructose 6-phosphate can be converted to two three-carbon triose derivatives by the consecutive action of a kinase and an aldolase on fructose 6-phosphate. Alternatively, aldolases, acting in conjunction with phosphatases, can be used to lengthen carbohydrate molecules: Triose phosphates give rise to fructose 6-phosphate; a triose phosphate and tetrose 4-phosphate form heptose 7-phosphate. The final form of carbohydrate chain length interconversion is the transaldolase reaction, which interconverts heptose 7-phosphate and triose 3-phosphate with tetrose 4-phosphate and hexose 6-phosphate.

The coordination of different carbohydrate rearrangement reactions to achieve an overall metabolic goal is illustrated by the hexose monophosphate shunt (Figure 6–6). This metabolic cycle is used by cyanobacteria for the reduction of NAD^+ to NADH, which serves as a reductant for respiration in the dark. Many organisms use the hexose monophosphate shunt to reduce $NADP^+$ to NADPH, which is used for biosynthetic reduction reactions. The first steps in the hexose monophosphate shunt are the oxidative reactions that shorten six hexose 6-phosphates (abbreviated as six C_6 in Figure 6–6) to six pentose 5-phosphates (abbreviated six C_5). Carbohydrate rearrangement reactions convert the six C_5 molecules to five C_6 molecules so that the oxidative cycle may continue.

Clearly, all reactions for interconversion of carbohydrate chain lengths are not called into play at the same time. Selection of specific sets of enzymes, essentially the determination of the metabolic pathway taken, is dictated by the source of carbon and the biosynthetic demands of the cell. For example, a cell given triose phosphate as a source of carbohydrate will use the aldolase-phosphatase combination to form fructose 6-phosphate; the kinase that acts on fructose 6-phosphate in its conversion to triose phosphate would not be expected to be active under these circumstances. If demands for pentose 5-phosphate are high, as in the case of photosynthetic carbon dioxide assimilation, transketolases that can give rise to pentose 5-phosphates are very active.

In sum, glucose 6-phosphate can be regarded as a focal metabolite because it serves both as a direct precursor for metabolic building blocks and as a source of carbohydrates of varying length that are used for biosynthetic purposes. Glucose 6-phosphate itself may be generated from other phosphorylated carbohydrates by selection of pathways from a set of reactions for chain length interconversion. The reactions chosen are determined by the genetic potential of the cell, the primary carbon source, and the biosynthetic demands of the organism. Metabolic regulation is required to ensure that reactions which meet the requirements of the organism are selected.

Formation & Utilization of Phosphoenolpyruvate

Triose phosphates, formed by the interconversion of carbohydrate phosphoesters, are converted to phosphoenolpyruvate by the series of reactions shown in Figure 6–7. Oxidation of

Dehydrogenases

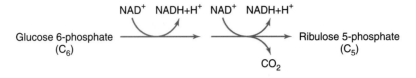

Transketolases

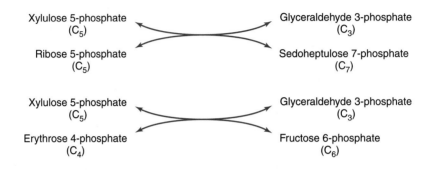

Kinase, Aldolase

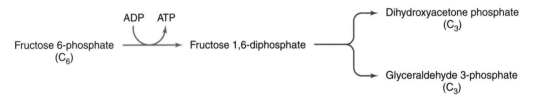

Aldolase, Phosphatase

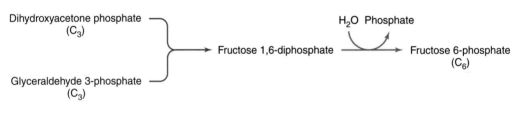

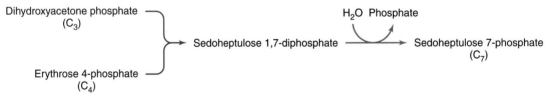

Transaldolase

FIGURE 6–5 Biochemical mechanisms for changing the length of carbohydrate molecules. The general empirical formula for carbohydrate phosphate esters, $(C_nH_{2n}O_n)$-N-phosphate, is abbreviated (C_n) in order to emphasize changes in chain length.

glyceraldehyde 3-phosphate by NAD^+ is accompanied by the formation of the acid anhydride bond on the one carbon of 1,3-diphosphoglycerate. This phosphate anhydride is transferred in a **substrate phosphorylation** to ADP, yielding an energy-rich bond in ATP. Another energy-rich phosphate bond is formed by dehydration of 2-phosphoglycerate to phosphoenolpyruvate; via another substrate phosphorylation, phosphoenolpyruvate can donate the energy-rich bond

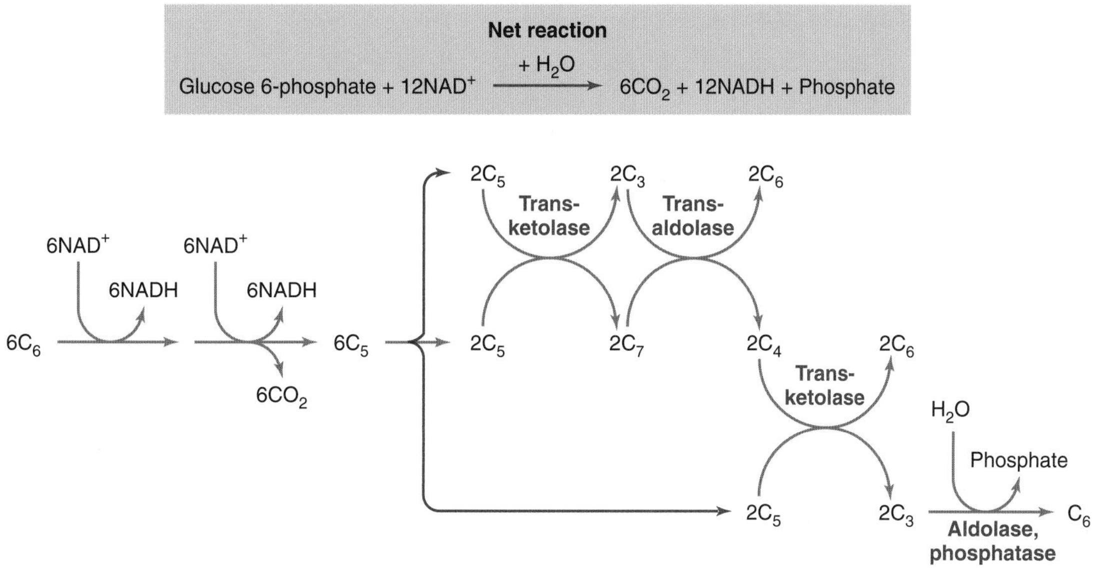

Net reaction

$$\text{Glucose 6-phosphate} + 12\text{NAD}^+ \xrightarrow{+\text{H}_2\text{O}} 6\text{CO}_2 + 12\text{NADH} + \text{Phosphate}$$

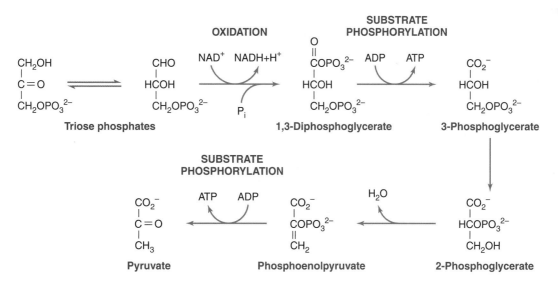

FIGURE 6–6 The hexose monophosphate shunt. Oxidative reactions (Figure 6–5) reduce NAD^+ and produce CO_2, resulting in the shortening of the six hexose phosphates (abbreviated C_6) to six pentose phosphates (abbreviated C_5). Carbohydrate rearrangements (Figure 6–5) convert the pentose phosphates to hexose phosphates so that the oxidative cycle may continue.

FIGURE 6–7 Formation of phosphoenolpyruvate and pyruvate from triose phosphate. The figure draws attention to two sites of substrate phosphorylation and to the oxidative step that results in the reduction of NAD^+ to NADH. Repetition of this energy-yielding pathway demands a mechanism for oxidizing NADH to NAD^+. Fermentative organisms achieve this goal by using pyruvate or metabolites derived from pyruvate as oxidants.

to ADP, yielding ATP and pyruvate. Thus, two energy-rich bonds in ATP can be obtained by the metabolic conversion of triose phosphate to pyruvate. This is an oxidative process, and in the absence of an exogenous electron acceptor, the NADH generated by oxidation of glyceraldehyde 3-phosphate must be oxidized to NAD^+ by pyruvate or by metabolites derived from pyruvate. The products formed as a result of this process vary and, as described later in this chapter, can be used in the identification of clinically significant bacteria.

Formation of phosphoenolpyruvate from pyruvate requires a substantial amount of metabolic energy, and two anhydride ATP bonds invariably are invested in the process. Some organisms—*Escherichia coli,* for example—directly phosphorylate pyruvate with ATP, yielding AMP and inorganic phosphate (P_i). Other organisms use two metabolic steps: One ATP pyrophosphate bond is invested in the carboxylation of pyruvate to oxaloacetate, and a second pyrophosphate bond (often carried by GTP rather than ATP) is used to generate phosphoenolpyruvate from oxaloacetate.

Formation & Utilization of Oxaloacetate

As described above, many organisms form oxaloacetate by the ATP-dependent carboxylation of pyruvate. Other organisms, such as *E coli*, which form phosphoenolpyruvate directly from pyruvate, synthesize oxaloacetate by carboxylation of phosphoenolpyruvate.

Succinyl-CoA is a required biosynthetic precursor for the synthesis of porphyrins and other essential compounds. Some organisms form succinyl-CoA by reduction of oxaloacetate via malate and fumarate. These reactions represent a reversal of the metabolic flow observed in the conventional tricarboxylic acid cycle (see Figure 6–10).

Formation of α-Ketoglutarate From Pyruvate

Conversion of pyruvate to α-ketoglutarate requires a metabolic pathway that diverges and then converges (Figure 6–8). In one branch, oxaloacetate is formed by carboxylation of pyruvate or phosphoenolpyruvate. In the other branch, pyruvate is oxidized to acetyl-CoA. It is noteworthy that, regardless of

the enzymatic mechanism used for the formation of oxaloacetate, acetyl-CoA is required as a positive metabolic effector for this process. Thus, the synthesis of oxaloacetate is balanced with the production of acetyl-CoA. Condensation of oxaloacetate with acetyl-CoA yields citrate. Isomerization of the citrate molecule produces isocitrate, which is oxidatively decarboxylated to α-ketoglutarate.

ASSIMILATORY PATHWAYS

Growth With Acetate

Acetate is metabolized via acetyl-CoA, and many organisms possess the ability to form acetyl-CoA (Figure 6–9). Acetyl-CoA is used in the biosynthesis of α-ketoglutarate, and in most respiratory organisms, the acetyl fragment in acetyl-CoA is oxidized completely to carbon dioxide via the tricarboxylic acid cycle (Figure 6–10). The ability to utilize acetate as a net source of carbon, however, is limited to relatively few microorganisms and plants. Net synthesis of biosynthetic precursors from acetate is achieved by

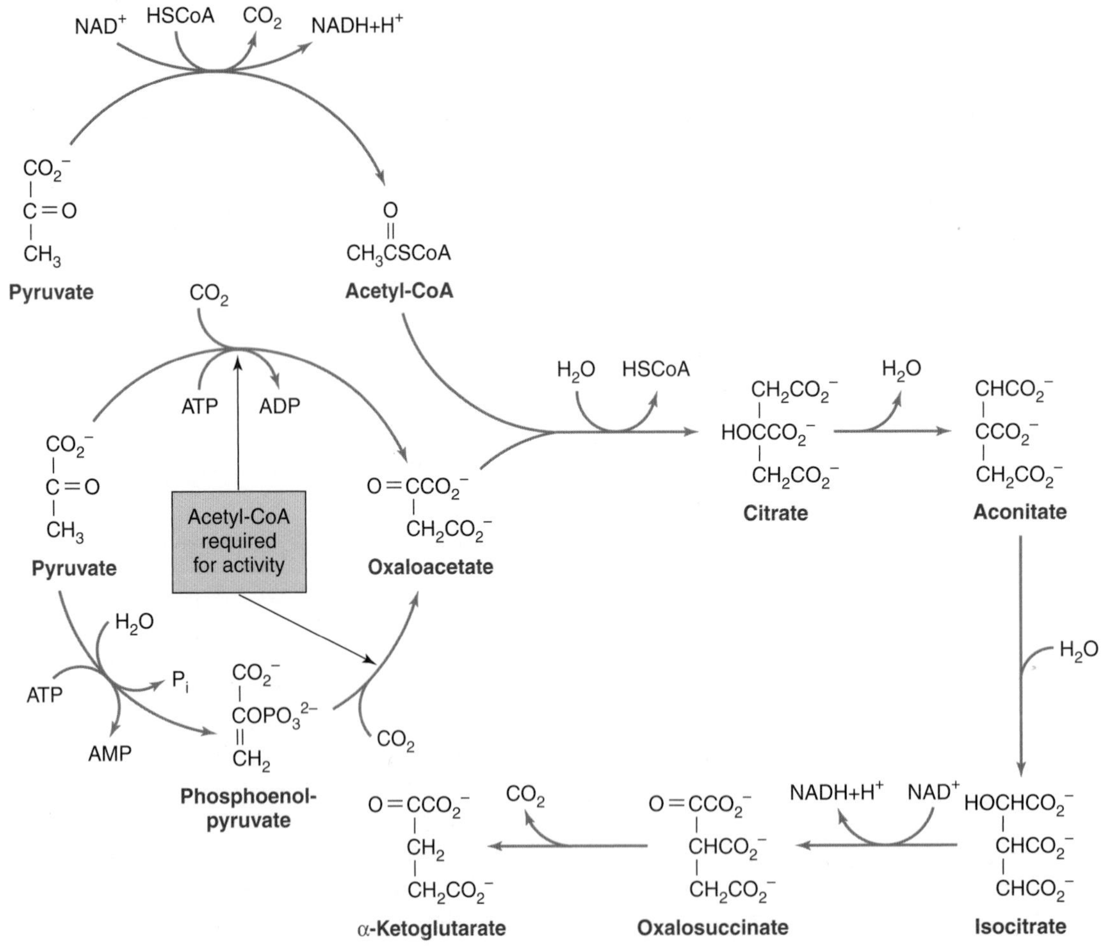

FIGURE 6–8 Conversion of pyruvate to α-ketoglutarate. Pyruvate is converted to α-ketoglutarate by a branched biosynthetic pathway. In one branch, pyruvate is oxidized to acetyl-CoA; in the other, pyruvate is carboxylated to oxaloacetate.

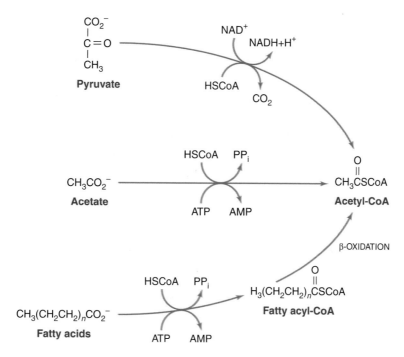

FIGURE 6–9 Biochemical sources of acetyl-CoA.

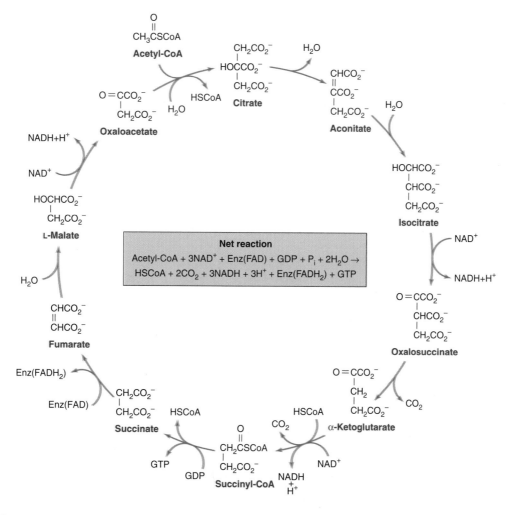

FIGURE 6–10 The tricarboxylic acid cycle. There are four oxidative steps, three giving rise to NADH and one giving rise to a reduced flavoprotein, Enz(FADH$_2$). The cycle can continue only if electron acceptors are available to oxidize the NADH and reduced flavoprotein.

coupling reactions of the tricarboxylic acid cycle with two additional reactions catalyzed by isocitrate lyase and malate synthase. As shown in Figure 6–11, these reactions allow the *net* oxidative conversion of two acetyl moieties from acetyl-CoA to one molecule of succinate. Succinate may be used for biosynthetic purposes after its conversion to oxaloacetate, α-ketoglutarate, phosphoenolpyruvate, or glucose 6-phosphate.

Growth With Carbon Dioxide: The Calvin Cycle

Like plants and algae, a number of microbial species can use carbon dioxide as a sole source of carbon. In almost all of these organisms, the primary route of carbon assimilation is via the Calvin cycle, in which carbon dioxide and ribulose diphosphate combine to form two molecules of 3-phosphoglycerate (Figure 6–12A). 3-Phosphoglycerate is phosphorylated to 1,3-diphosphoglycerate, and this compound is reduced to the triose derivative, glyceraldehyde 3-phosphate. Carbohydrate rearrangement reactions (Figure 6–5) allow triose phosphate to be converted to the pentose derivative ribulose 5-phosphate, which is phosphorylated to regenerate the acceptor molecule, ribulose 1,5-diphosphate (Figure 6–12B). Additional reduced carbon, formed by the reductive assimilation of carbon dioxide, is converted to focal metabolites for biosynthetic pathways.

Cells that can use carbon dioxide as a sole source of carbon are termed **autotrophic**, and the demands for this pattern of carbon assimilation can be summarized briefly as follows: In addition to the primary assimilatory reaction giving rise to 3-phosphoglycerate, there must be a mechanism for

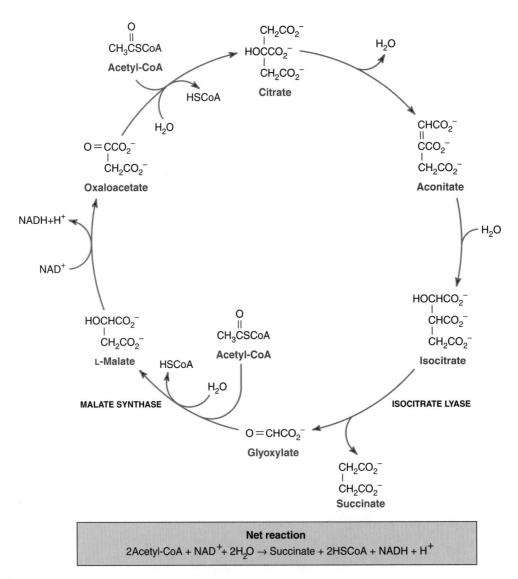

Net reaction

$$2\text{Acetyl-CoA} + \text{NAD}^+ + 2\text{H}_2\text{O} \rightarrow \text{Succinate} + 2\text{HSCoA} + \text{NADH} + \text{H}^+$$

FIGURE 6–11   The glyoxylate cycle. Note that the reactions which convert malate to isocitrate are shared with the tricarboxylic acid cycle (Figure 6–10). Metabolic divergence at the level of isocitrate and the action of two enzymes, isocitrate lyase and malate synthase, modify the tricarboxylic acid cycle so that it reductively converts two molecules of acetyl-CoA to succinate.

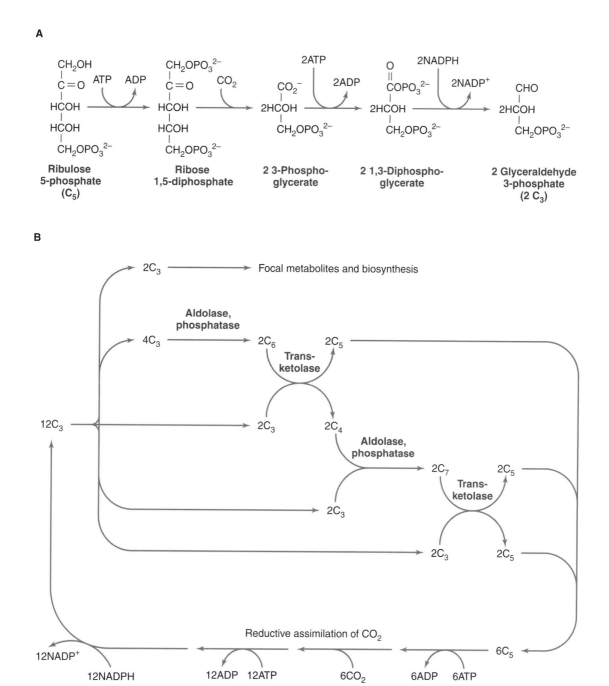

FIGURE 6–12 The Calvin cycle. **A:** Reductive assimilation of CO_2. ATP and NADPH are used to reductively convert pentose 5-phosphate (C_5) to two molecules of triose phosphate (C_3). **B:** The Calvin cycle is completed by carbohydrate rearrangement reactions (Figure 6–5) that allow the net synthesis of carbohydrate and the regeneration of pentose phosphate so that the cycle may continue.

regenerating the acceptor molecule, ribulose 1,5-diphosphate. This process demands the energy-dependent reduction of 3-phosphoglycerate to the level of carbohydrate. Thus, autotrophy requires carbon dioxide, ATP, NADPH, and a specific set of enzymes.

Depolymerases

Many potential growth substrates occur as building blocks within the structure of biologic polymers. These large molecules are not readily transported across the cell membrane and often are affixed to even larger cellular structures. Many microorganisms elaborate extracellular depolymerases that hydrolyze proteins, nucleic acids, polysaccharides, and lipids. The pattern of depolymerase activities can be useful in the identification of microorganisms.

Oxygenases

Many compounds in the environment are relatively resistant to enzymatic modification, and utilization of these compounds as growth substrates demands a special class of enzymes, oxygenases. These enzymes directly employ the potent oxidant molecular oxygen as a substrate in reactions that convert a relatively intractable compound to a form in which it can be assimilated by thermodynamically favored reactions. The action of oxygenases is illustrated in Figure 6–13, which shows the role of two different oxygenases in the utilization of benzoate.

Reductive Pathways

Some microorganisms live in extremely reducing environments that favor chemical reactions which would not occur in organisms using oxygen as an electron acceptor. In these organisms, powerful reductants can be used to drive reactions that allow the assimilation of relatively intractable compounds. An example is the reductive assimilation of benzoate, a process in which the aromatic ring is reduced and opened to form the dicarboxylic acid pimelate. Further metabolic reactions convert pimelate to focal metabolites.

Nitrogen Assimilation

The reductive assimilation of molecular nitrogen, also referred to as **nitrogen fixation**, is required for continuation of life on our planet. Nitrogen fixation is accomplished by a variety of bacteria and cyanobacteria using a multicomponent **nitrogenase enzyme complex**. Despite the variety of organisms capable of fixing nitrogen, the nitrogenase complex is similar in most of them (Figure 6–14). Nitrogenase is a complex of two enzymes—one enzyme (dinitrogenase reductase) contains iron and the other (dinitrogenase) contains iron and molybdenum. Together, these enzymes catalyze the following reaction:

$$N_2 + 6H^+ + 6e^- + 12ATP \longrightarrow 2NH_3 + 12ADP + 12P_i$$

Because of the high activation energy of breaking the very strong triple bond that joins two nitrogen atoms, this reductive assimilation of nitrogen demands a substantial amount of metabolic energy. Somewhere between 20 and 24 molecules of ATP are hydrolyzed as a single N_2 molecule is reduced to two molecules of NH_3.

Additional physiologic demands are placed by the fact that nitrogenase is readily inactivated by oxygen. Aerobic organisms that employ nitrogenase have developed elaborate mechanisms to protect the enzyme against inactivation. Some form specialized cells in which nitrogen fixation takes place, and others have developed elaborate electron transport chains to protect nitrogenase against inactivation by oxygen. The most significant of these bacteria in agriculture are the Rhizobiaceae, organisms that fix nitrogen symbiotically in the root nodules of leguminous plants.

The capacity to use ammonia as a nitrogen source is widely distributed among organisms. The primary portal of entry of nitrogen into carbon metabolism is glutamate, which is formed by reductive amination of α-ketoglutarate. As shown in Figure 6–15, there are two biochemical mechanisms by which this can be achieved. One, the single-step reduction catalyzed by glutamate dehydrogenase (Figure 6–15A), is effective in environments in which there is an ample supply of ammonia. The other, a two-step process in which glutamine is an intermediate (Figure 6–15B), is employed in environments in which ammonia is in short supply. The latter mechanism allows cells to invest the free energy formed by hydrolysis of a pyrophosphate bond in ATP into the assimilation of ammonia from the environment.

The amide nitrogen of glutamine, an intermediate in the two-step assimilation of ammonia into glutamate (Figure 6–15B), is also transferred directly into organic nitrogen appearing in the structures of purines, pyrimidines, arginine, tryptophan, and glucosamine. The activity and synthesis of glutamine synthase are regulated by the ammonia supply

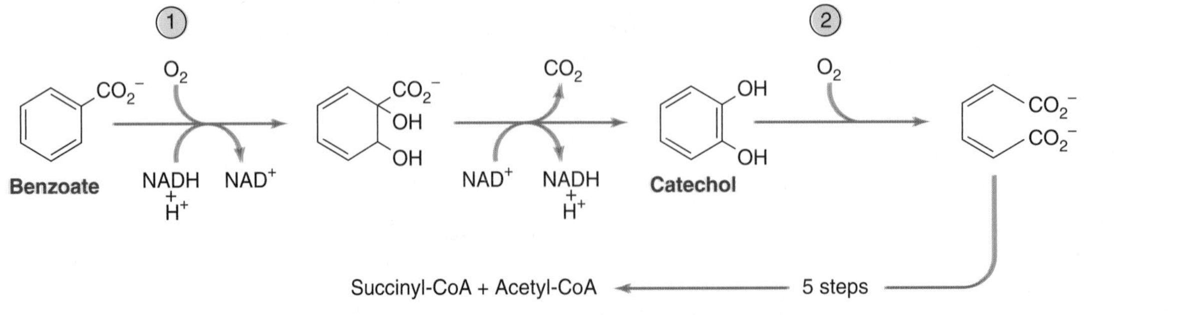

FIGURE 6–13 The role of oxygenases in aerobic utilization of benzoate as a carbon source. Molecular oxygen participates directly in the reactions that disrupt the aromaticity of benzoate and catechol.

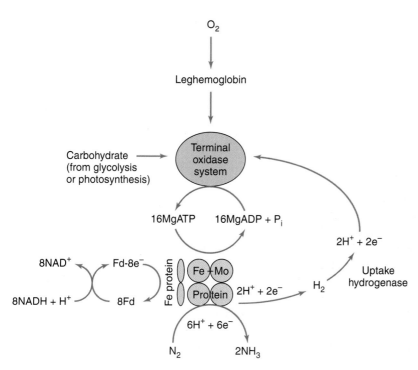

FIGURE 6–14 Reduction of N_2 to two molecules of NH_3. In addition to reductant, the nitrogenase reaction requires a substantial amount of metabolic energy. The number of ATP molecules required for reduction of a single nitrogen molecule to ammonia is uncertain; the value appears to lie between 12 and 16. The overall reaction requires $8NADH + H^+$. Six of these are used to reduce N_2 to $2NH_3$, and two are used to form H_2. The uptake hydrogenase returns H_2 to the system, thus conserving energy. (Redrawn and reproduced, with permission, from Moat AG, Foster JW: *Microbial Physiology*, 4th ed. Wiley-Liss, 2002. Reprinted by permission of John Wiley & Sons, Inc.)

and by the availability of metabolites containing nitrogen derived directly from the amide nitrogen of glutamine.

Most of the organic nitrogen in cells is derived from the α-amino group of glutamate, and the primary mechanism by which the nitrogen is transferred is **transamination**. The usual acceptor in these reactions is an α-keto acid, which is transformed to the corresponding α-amino acid. α-Ketoglutarate, the other product of the transamination reaction, may be converted to glutamate by reductive amination (Figure 6–15).

BIOSYNTHETIC PATHWAYS

Tracing the Structures of Biosynthetic Precursors: Glutamate & Aspartate

In many cases, the carbon skeleton of a metabolic end product may be traced to its biosynthetic origins. Glutamine, an obvious example, clearly is derived from glutamate (Figure 6–16). The glutamate skeleton in the structures of arginine and proline (Figure 6–16) is less obvious but readily discernible. Similarly, the carbon skeleton of aspartate, directly derived from the focal metabolite oxaloacetate, is evident in the structures of asparagine, threonine, methionine, and pyrimidines (Figure 6–17). In some cases, different carbon skeletons combine in a biosynthetic pathway. For example, aspartate semialdehyde and pyruvate combine to form the metabolic precursors of lysine, diaminopimelic acid, and dipicolinic acid (Figure 6–18). The latter two compounds are found only in prokaryotes. Diaminopimelic acid is a component of peptidoglycan in the cell wall, and dipicolinic acid represents a major component of endospores.

Synthesis of Cell Wall Peptidoglycan

The structure of peptidoglycan is shown in Figure 2–19; the pathway by which it is synthesized is shown in simplified form in Figure 6–19A. The synthesis of peptidoglycan begins with the stepwise synthesis in the cytoplasm of UDP-*N*-acetylmuramic acid-pentapeptide. *N*-acetylglucosamine is first attached to UDP and then converted to UDP-*N*-acetylmuramic acid by condensation with phosphoenolpyruvate and reduction. The amino acids of the pentapeptide are sequentially added, each addition catalyzed by a different enzyme and each involving the split of ATP to ADP + P_i.

The UDP-*N*-acetylmuramic acid-pentapeptide is attached to bactoprenol (a lipid of the cell membrane) and receives a molecule of *N*-acetylglucosamine from UDP. Some bacteria (eg, *Staphylococcus aureus*) form a pentaglycine derivative in a series of reactions using glycyl-tRNA as the donor; the completed disaccharide is polymerized to an oligomeric intermediate before being transferred to the growing end of a glycopeptide polymer in the cell wall.

Final cross-linking (Figure 6–19B) is accomplished by a transpeptidation reaction in which the free amino group of a pentaglycine residue displaces the terminal D-alanine residue of a neighboring pentapeptide. Transpeptidation is catalyzed by one of a set of enzymes called penicillin-binding proteins (PBPs). PBPs bind penicillin and other β-lactam antibiotics covalently due, in part, to a structural similarity between these antibiotics and the pentapeptide precursor. Some PBPs have transpeptidase or carboxypeptidase activities, their relative rates perhaps controlling the degree of cross-linking in peptidoglycan (a factor important in cell septation).

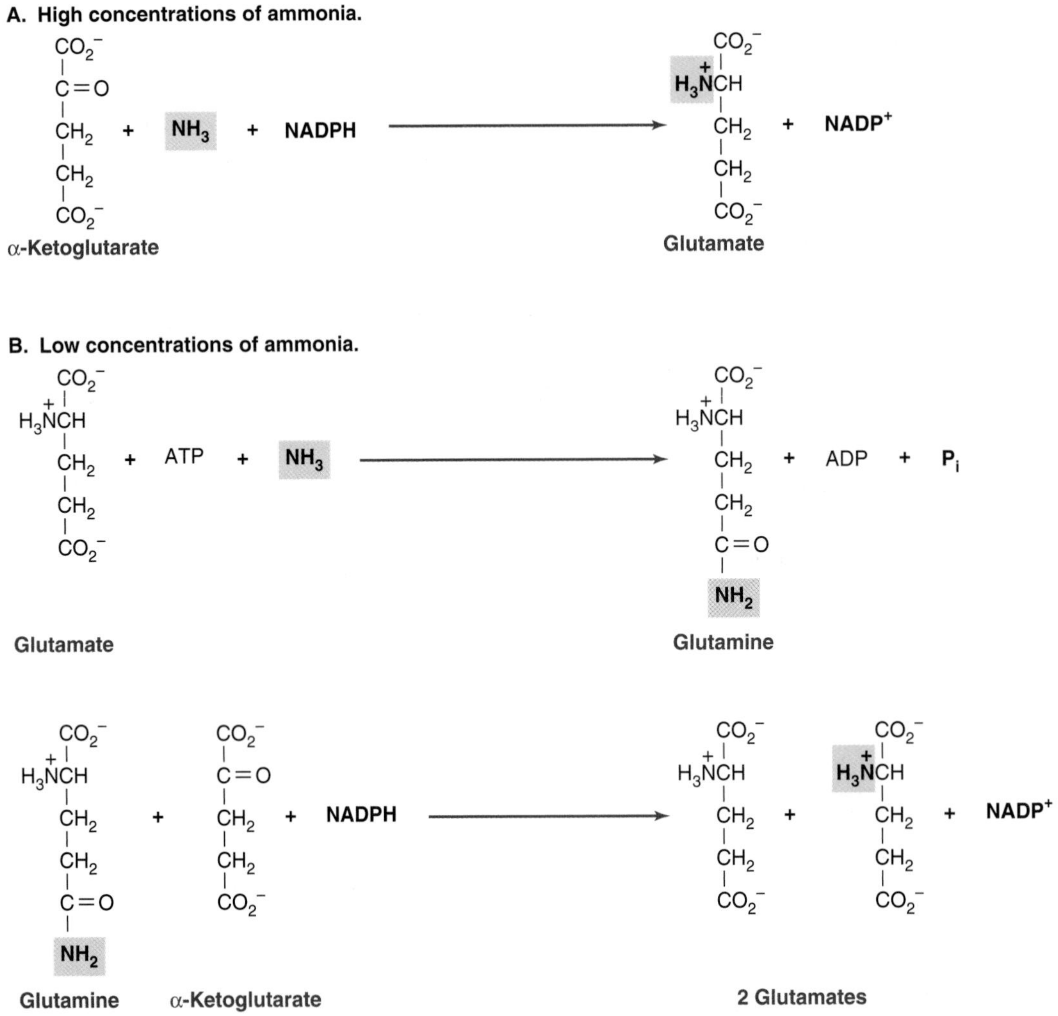

A. High concentrations of ammonia.

$$\alpha\text{-Ketoglutarate} + NH_3 + NADPH \longrightarrow Glutamate + NADP^+$$

B. Low concentrations of ammonia.

$$Glutamate + ATP + NH_3 \longrightarrow Glutamine + ADP + P_i$$

$$Glutamine + \alpha\text{-Ketoglutarate} + NADPH \longrightarrow 2\ Glutamates + NADP^+$$

FIGURE 6–15 Mechanisms for the assimilation of NH_3. **A:** When the NH_3 concentration is high, cells are able to assimilate the compound via the glutamate dehydrogenase reaction. **B:** When, as most often is the case, the NH_3 concentration is low, cells couple the glutamine synthase and glutamate synthase reactions in order to invest the energy produced by hydrolysis of a pyrophosphate bond into ammonia assimilation.

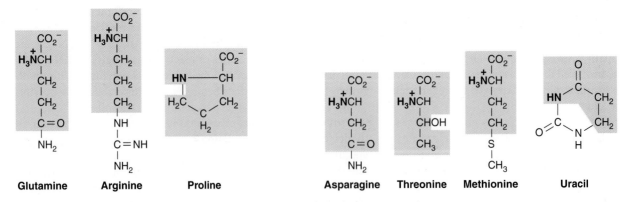

Glutamine   Arginine   Proline   Asparagine   Threonine   Methionine   Uracil

FIGURE 6–16 Amino acids formed from glutamate.   **FIGURE 6–17** Biosynthetic end products formed from aspartate.

The biosynthetic pathway is of particular importance in medicine, as it provides a basis for the selective antibacterial action of several chemotherapeutic agents. Unlike their host cells, bacteria are not isotonic with the body fluids. Their contents are under high osmotic pressure, and their viability depends on the integrity of the peptidoglycan lattice in the cell wall being maintained throughout the growth cycle. Any compound that inhibits any step in the biosynthesis of

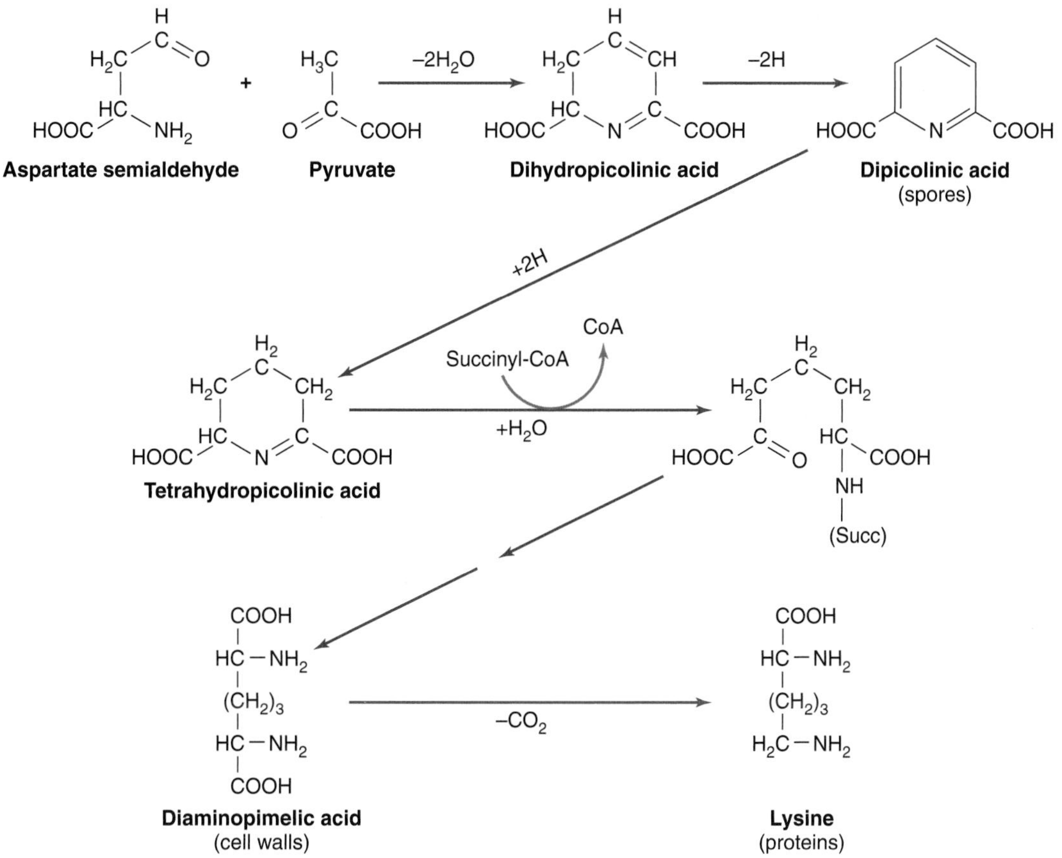

FIGURE 6–18
Biosynthetic end products formed from aspartate semialdehyde and pyruvate.

peptidoglycan causes the wall of the growing bacterial cell to be weakened and the cell to lyse. The sites of action of several antibiotics are shown in Figure 6–19A and B.

Synthesis of Cell Envelope Lipopolysaccharide

The general structure of the antigenic lipopolysaccharide of gram-negative cell envelopes is shown in Figure 2–20. The biosynthesis of the repeating end-group, which gives the cell envelope its antigenic specificity, is shown in Figure 6–20. Note the resemblance to peptidoglycan synthesis: In both cases, a series of subunits is assembled on a lipid carrier in the membrane and then transferred to open ends of the growing polymer.

Synthesis of Extracellular Capsular Polymers

The capsular polymers, a few examples of which are listed in Table 2–1, are enzymatically synthesized from activated subunits. No membrane-bound lipid carriers have been implicated in this process. The presence of a capsule is often environmentally determined: Dextrans and levans, for example, can only be synthesized using the disaccharide sucrose (fructose–glucose) as the source of the appropriate subunit, and their synthesis thus depends on the presence of sucrose in the medium.

Synthesis of Reserve Food Granules

When nutrients are present in excess of the requirements for growth, bacteria convert certain of them to intracellular reserve food granules. The principal ones are starch, glycogen, poly-β-hydroxybutyrate, and volutin, which consists mainly of inorganic polyphosphate (see Chapter 2). The type of granule formed is species-specific. The granules are degraded when exogenous nutrients are depleted.

PATTERNS OF MICROBIAL ENERGY-YIELDING METABOLISM

As described in Chapter 5, there are two major metabolic mechanisms for generating the energy-rich acid pyrophosphate bonds in ATP: **substrate phosphorylation** (the direct transfer of a phosphate anhydride bond from an organic donor to ADP) and phosphorylation of ADP by inorganic phosphate. The latter reaction is energetically unfavorable and must be driven by a transmembrane electrochemical gradient, the **proton motive force**. In respiration, the electrochemical gradient is created from externally supplied reductant and oxidant. Energy released by transfer of electrons from the reductant to the oxidant through membrane-bound carriers is coupled to the formation of the transmembrane electrochemical gradient. In photosynthesis, light energy

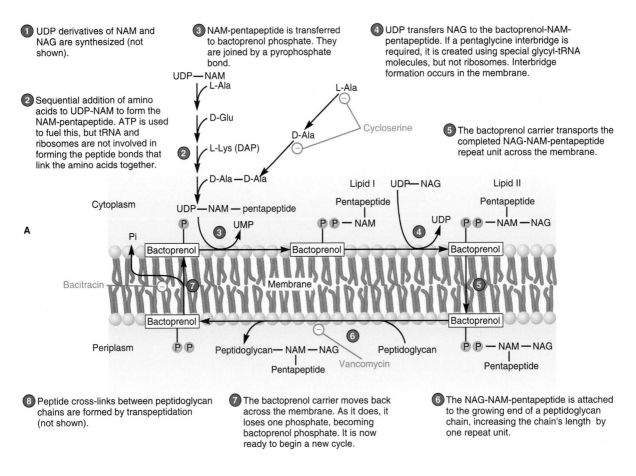

① UDP derivatives of NAM and NAG are synthesized (not shown).

② Sequential addition of amino acids to UDP-NAM to form the NAM-pentapeptide. ATP is used to fuel this, but tRNA and ribosomes are not involved in forming the peptide bonds that link the amino acids together.

③ NAM-pentapeptide is transferred to bactoprenol phosphate. They are joined by a pyrophosphate bond.

④ UDP transfers NAG to the bactoprenol-NAM-pentapeptide. If a pentaglycine interbridge is required, it is created using special glycyl-tRNA molecules, but not ribosomes. Interbridge formation occurs in the membrane.

⑤ The bactoprenol carrier transports the completed NAG-NAM-pentapeptide repeat unit across the membrane.

⑧ Peptide cross-links between peptidoglycan chains are formed by transpeptidation (not shown).

⑦ The bactoprenol carrier moves back across the membrane. As it does, it loses one phosphate, becoming bactoprenol phosphate. It is now ready to begin a new cycle.

⑥ The NAG-NAM-pentapeptide is attached to the growing end of a peptidoglycan chain, increasing the chain's length by one repeat unit.

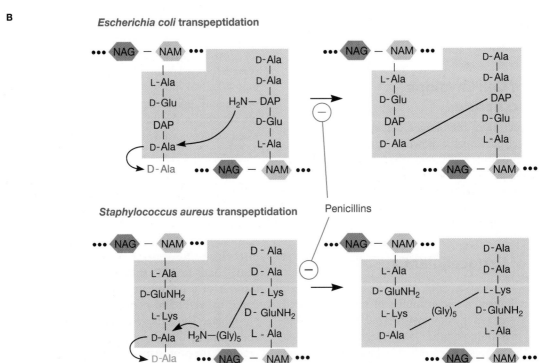

FIGURE 6–19   A: Peptidoglycan synthesis. NAM is *N*-acetylmuramic acid and NAG is *N*-acetylglucosamine. The pentapeptide contains L-lysine in *Staphylococcus aureus* peptidoglycan, and diaminopimelic acid (DAP) in *Escherichia coli*. Inhibition by bacitracin, cycloserine, and vancomycin also is shown. The numbers correspond to six of the eight stages discussed in the text. Stage eight is depicted in Figure 6-19B. **B:** Transpeptidation. The transpeptidation reactions in the formation of the peptidoglycans of *Escherichia coli* and *Staphylococcus aureus*.

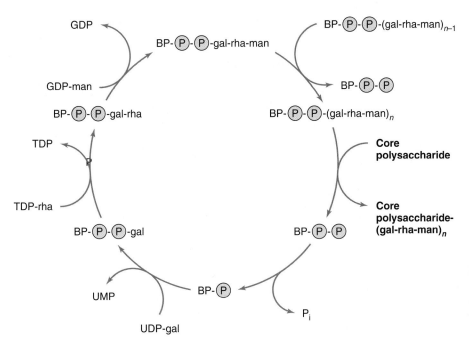

FIGURE 6–20 Synthesis of the repeating unit of the polysaccharide side chain of *Salmonella newington* and its transfer to the lipopolysaccharide core. BP, bactoprenol.

generates membrane-associated reductants and oxidants; the proton motive force is generated as these electron carriers return to the ground state. These processes are discussed below.

Pathways of Fermentation

A. Strategies for Substrate Phosphorylation

In the absence of respiration or photosynthesis, cells are entirely dependent upon substrate phosphorylation for their energy: Generation of ATP must be coupled to chemical rearrangement of organic compounds. Many compounds can serve as fermentable growth substrates, and many pathways for their fermentation have evolved. These pathways have the following three general stages: (1) Conversion of the fermentable compound to the phosphate donor for substrate phosphorylation. This stage often contains metabolic reactions in which NAD⁺ is reduced to NADH. (2) Phosphorylation of ADP by the energy-rich phosphate donor. (3) Metabolic steps that bring the products of the fermentation into chemical balance with the starting materials. The most frequent requirement in the last stage is a mechanism for oxidation of NADH, generated in the first stage of fermentation, to NAD⁺ so that the fermentation may proceed. In the following sections, examples of each of the three stages of fermentation are considered.

B. Fermentation of Glucose

The diversity of fermentative pathways is illustrated by consideration of some of the mechanisms used by microorganisms to achieve substrate phosphorylation at the expense of glucose. In principle, the phosphorylation of ADP to

ATP can be coupled to either of two chemically balanced transformations:

$$\underset{(C_6H_{12}O_6)}{\text{Glucose}} \longrightarrow \underset{(C_3H_6O_3)}{2\,\text{Lactic acid}}$$

or

$$\underset{(C_6H_{12}O_6)}{\text{Glucose}} \longrightarrow \underset{C_2H_6O)}{2\,\text{Ethanol}} + \underset{(CO_2)}{2\,\text{Carbon dioxide}}$$

The biochemical mechanisms by which these transformations are achieved vary considerably.

In general, the fermentation of glucose is initiated by its phosphorylation to glucose 6-phosphate. There are two mechanisms by which this can be achieved: (1) Extracellular glucose may be transported across the cytoplasmic membrane into the cell and then phosphorylated by ATP to yield glucose 6-phosphate and ADP. (2) In many microorganisms, extracellular glucose is phosphorylated as it is being transported across the cytoplasmic membrane by an enzyme system in the cytoplasmic membrane that phosphorylates extracellular glucose at the expense of phosphoenolpyruvate, producing intracellular glucose 6-phosphate and pyruvate. The latter process is an example of **vectorial metabolism**, a set of biochemical reactions in which both the structure and the location of a substrate are altered. It should be noted that the choice of ATP or phosphoenolpyruvate as a phosphorylating agent does not alter the ATP yield of fermentation, because phosphoenolpyruvate is used as a source of ATP in the later stages of fermentation (Figure 6–7).

C. The Embden-Meyerhof Pathway

This pathway (Figure 6–21), a commonly encountered mechanism for the fermentation of glucose, uses a kinase and an aldolase (Figure 6–5) to transform the hexose (C_6) phosphate to two molecules of triose (C_3) phosphate. Four substrate phosphorylation reactions accompany the conversion of the triose phosphate to two molecules of pyruvate. Thus, taking into account the two ATP pyrophosphate bonds required to form triose phosphate from glucose, the Embden-Meyerhof pathway produces a net yield of two ATP pyrophosphate bonds. Formation of pyruvate from triose phosphate is an oxidative process, and the NADH formed in the first metabolic step (Figure 6–21) must be converted to NAD^+ for the fermentation to proceed; two of the simpler mechanisms for achieving this goal are illustrated in Figure 6–22. Direct reduction of pyruvate by NADH produces lactate as the end product of fermentation and thus results in acidification of the medium. Alternatively, pyruvate may be decarboxylated to acetaldehyde, which is then used to oxidize NADH, resulting in production of the neutral product ethanol. The pathway

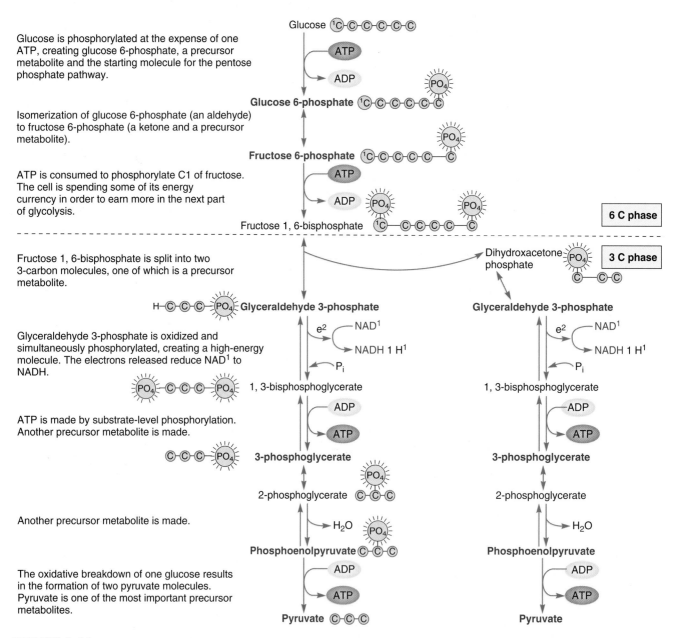

FIGURE 6–21 The Embden-Meyerhof pathway. This is one of three glycolytic pathways used to catabolize glucose to pyruvate and it can function during aerobic respiration, anaerobic respiration, and fermentation. When used during a respiratory process, the electrons accepted by NAD^+ are transferred to an electron transport chain and are ultimately accepted by an exogenous electron acceptor. When used during fermentation, the electrons accepted by NAD^+ are donated to an endogenous electron acceptor (eg, pyruvate). The Embden-Meyerhof pathway is also an important amphibolic pathway, as it generates several precursor metabolites (shown in blue).

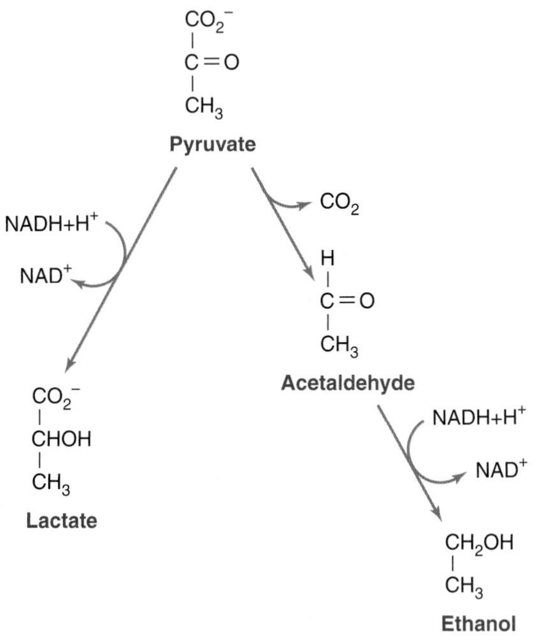

taken is determined by the evolutionary history of the organism and, in some microorganisms, by the growth conditions.

D. The Entner-Doudoroff & Heterolactate Fermentations

Alternative pathways for glucose fermentation include some specialized enzyme reactions, and these are shown in Figure 6–23. The Entner-Doudoroff pathway diverges from other pathways of carbohydrate metabolism by a dehydration of 6-phosphogluconate followed by an aldolase reaction that produces pyruvate and triose phosphate (Figure 6–23A). The heterolactate fermentation and some other fermentative pathways depend upon a phosphoketolase reaction (Figure 6–23B) that phosphorolytically cleaves a ketosephosphate to produce acetyl phosphate and triose phosphate. The acid anhydride acetyl phosphate may be used to synthesize ATP or may allow the oxidation of two NADH molecules to NAD^+ as it is reduced to ethanol.

The overall outlines of the respective Entner-Doudoroff and heterolactate pathways are shown in Figures 6–24 and 6–25. The pathways yield only a single molecule of triose phosphate from glucose, and the energy yield is correspondingly low: Unlike the Embden-Meyerhof pathway, the Entner-Doudoroff and heterolactate pathways yield only a single net substrate phosphorylation of ADP per molecule of glucose fermented. Why have the alternative pathways

FIGURE 6–22 Two biochemical mechanisms by which pyruvate can oxidize NADH. **Left**: Direct formation of lactate, which results in net production of lactic acid from glucose. **Right**: Formation of the neutral products carbon dioxide and ethanol.

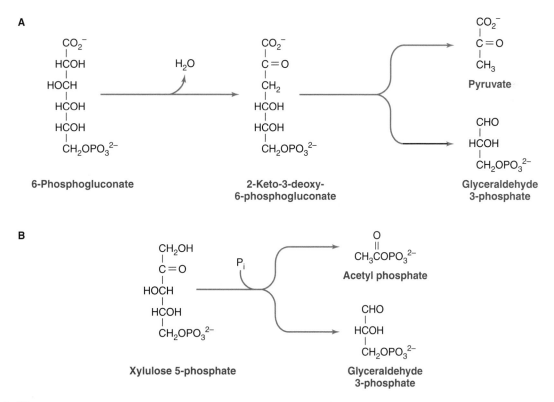

FIGURE 6–23 Reactions associated with specific pathways of carbohydrate fermentation. **A**: Dehydratase and aldolase reactions used in the Entner-Doudoroff pathway. **B**: The phosphoketolase reaction. This reaction, found in several pathways for fermentation of carbohydrates, generates the mixed acid anhydride acetyl phosphate, which can be used for substrate phosphorylation of ADP.

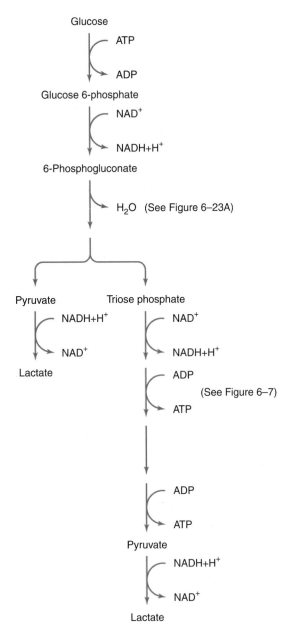

FIGURE 6–24 The Entner-Doudoroff pathway.

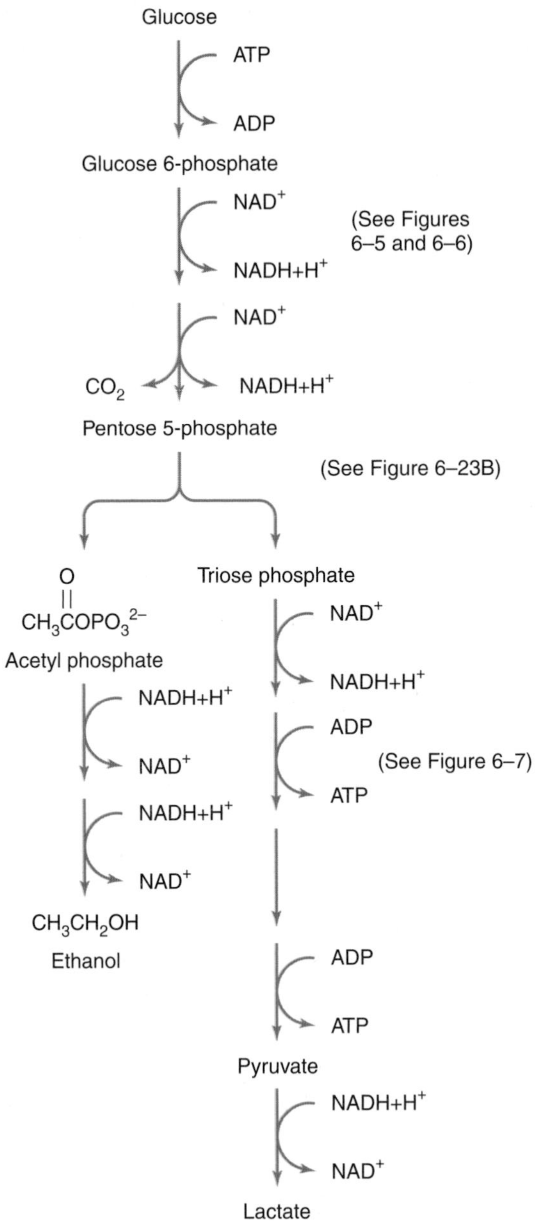

FIGURE 6–25 The heterolactic fermentation of glucose.

for glucose fermentation been selected in the natural environment? In answering this question, two facts should be kept in mind. First, in direct growth competition between two microbial species, the rate of substrate utilization can be more important than the amount of growth. Second, glucose is but one of many carbohydrates encountered by microorganisms in their natural environment. Pentoses, for example, can be fermented quite efficiently by the heterolactate pathway.

E. Additional Variations in Carbohydrate Fermentations

Pathways for carbohydrate fermentation can accommodate many more substrates than described here, and the end

products may be far more diverse than suggested thus far. For example, there are numerous mechanisms for oxidation of NADH at the expense of pyruvate. One such pathway is the reductive formation of succinate. Many clinically significant bacteria form pyruvate from glucose via the Embden-Meyerhof pathway, and they may be distinguished on the basis of reduction products formed from pyruvate, reflecting the enzymatic constitution of different species. The major products of fermentation, listed in Table 6–1, form the basis for many diagnostic tests.

F. Fermentation of Other Substrates

Carbohydrates are by no means the only fermentable substrates. Metabolism of amino acids, purines, and pyrimidines

TABLE 6–1   Microbial Fermentations Based on the Embden-Meyerhof Pathway

Fermentation	Organisms	Products
Ethanol	Some fungi (notably some yeasts)	Ethanol, CO_2
Lactate (homofermentation)	*Streptococcus* Some species of *Lactobacillus*	Lactate (accounting for at least 90% of the energy source carbon)
Lactate (heterofermentation)	*Enterobacter, Aeromonas, Bacillus polymyxa*	Ethanol, acetoin, 2,3-butylene glycol, CO_2, lactate, acetate, formate (total acids = 21 mol[a])
Propionate	*Clostridium propionicum, Propionibacterium, Corynebacterium diphtheriae,* Some species of *Neisseria, Veillonella, Micromonospora*	Propionate, acetate, succinate, CO_2
Mixed acid	*Escherichia, Salmonella, Shigella, Proteus*	Lactate, acetate, formate, succinate, H_2, CO_2, ethanol (total acids = 159 mol[a])
Butanol-butyrate	*Butyribacterium, Zymosarcina maxima* Some species of *Clostridium*	Butanol, butyrate, acetone, isopropanol, acetate, ethanol, H_2, CO_2

[a]Per 100 mol of glucose fermented.

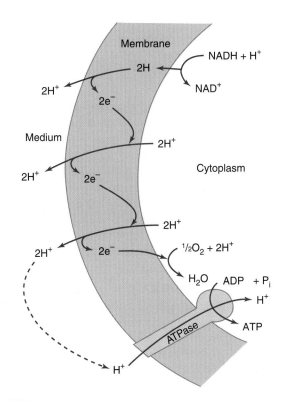

FIGURE 6–26 The coupling of electron transport in respiration to the generation of ATP. The indicated movements of protons and electrons are mediated by carriers (flavoprotein, quinone, cytochromes) associated with the membrane. The flow of protons down their electrochemical gradient, via the membrane ATPase, furnishes the energy for the generation of ATP from ADP and P_i. See text for explanation.

may allow substrate phosphorylations to occur. For example, arginine may serve as an energy source by giving rise to carbamoyl phosphate, which can be used to phosphorylate ADP to ATP. Some organisms ferment pairs of amino acids,

using one as an electron donor and the other as an electron acceptor:

Patterns of Respiration

Respiration requires a closed membrane. In bacteria, the membrane is the cell membrane. Electrons are passed from a chemical reductant to a chemical oxidant through a specific set of electron carriers within the membrane, and as a result, the proton motive force is established (Figure 6–26); return of protons across the membrane is coupled to the synthesis of ATP. As suggested in Figure 6–26, the biologic reductant for respiration frequently is NADH, and the oxidant often is oxygen.

Tremendous microbial diversity is exhibited in the sources of reductant used to generate NADH, and many microorganisms can use electron acceptors other than oxygen. Organic growth substrates are converted to focal metabolites that may reduce NAD^+ to NADH either by the hexose monophosphate shunt (Figure 6–6) or by the tricarboxylic acid cycle (Figure 6–10). Additional reductant may be generated during the breakdown of some growth substrates, eg, fatty acids (Figure 6–9).

Some bacteria, called **chemolithotrophs**, are able to use inorganic reductants for respiration. These energy sources include hydrogen, ferrous iron, and several reduced forms of sulfur and nitrogen. ATP derived from respiration and NADPH generated from the reductants can be used to drive the Calvin cycle (Figure 6–12).

Compounds and ions other than O_2 may be used as terminal oxidants in respiration. This ability, the capacity for **anaerobic respiration**, is a widespread microbial trait. Suitable electron acceptors include nitrate, sulfate, and carbon dioxide. Respiratory metabolism dependent upon carbon dioxide as an electron acceptor is a property found among representatives of a large microbial group, the **archaebacteria**.

Representatives of this group possess, for example, the ability to reduce carbon dioxide to acetate as a mechanism for generating metabolic energy.

Bacterial Photosynthesis

Photosynthetic organisms use light energy to separate electronic charge, to create membrane-associated reductants and oxidants as a result of a photochemical event. Transfer of electrons from the reductant to the oxidant creates a proton motive force. Many bacteria carry out a photosynthetic metabolism that is entirely independent of oxygen. Light is used as a source of metabolic energy, and carbon for growth is derived either from organic compounds (**photoheterotroph**) or from a combination of an inorganic reductant (eg, thiosulfate) and carbon dioxide (**photolithotroph**). These bacteria possess a single photosystem that, although sufficient to provide energy for the synthesis of ATP and for the generation of essential transmembrane ionic gradients, does not allow the highly exergonic reduction of NADP$^+$ at the expense of water. This process, essential for oxygen-evolving photosynthesis, rests upon additive energy derived from the coupling of two different photochemical events, driven by two independent photochemical systems. Among prokaryotes, this trait is found solely in the cyanobacteria (blue-green bacteria). Among eukaryotic organisms, the trait is shared by algae and plants in which the essential energy-providing organelle is the chloroplast.

REGULATION OF METABOLIC PATHWAYS

In their normal environment, microbial cells generally regulate their metabolic pathways so that no intermediate is made in excess. Each metabolic reaction is regulated not only with respect to all others in the cell but also with respect to the concentrations of nutrients in the environment. Thus, when a sporadically available carbon source suddenly becomes abundant, the enzymes required for its catabolism increase in both amount and activity; conversely, when a building block (such as an amino acid) suddenly becomes abundant, the enzymes required for its biosynthesis decrease in both amount and activity.

The regulation of enzyme activity as well as enzyme synthesis provides both **fine control** and **coarse control** of metabolic pathways. For example, the inhibition of enzyme activity by the end product of a pathway constitutes a mechanism of fine control, since the flow of carbon through that pathway is instantly and precisely regulated. The inhibition of enzyme synthesis by the same end product, on the other hand, constitutes a mechanism of coarse control. The preexisting enzyme molecules continue to function until they are diluted out by further cell growth, although unnecessary protein synthesis ceases immediately.

The mechanisms by which the cell regulates enzyme activity are discussed in the following section. The regulation of enzyme synthesis is discussed in Chapter 7.

The Regulation of Enzyme Activity
A. Enzymes as Allosteric Proteins

In many cases, the activity of an enzyme catalyzing an early step in a metabolic pathway is inhibited by the end product of that pathway. Such inhibition cannot depend on competition for the enzyme's substrate binding site, however, because the structures of the end product and the early intermediate (substrate) are usually quite different. Instead, inhibition depends on the fact that regulated enzymes are **allosteric**: Each enzyme possesses not only a catalytic site, which binds substrate, but also one or more other sites that bind small regulatory molecules, or **effectors**. The binding of an effector to its site causes a conformational change in the enzyme such that the affinity of the catalytic site for the substrate is reduced (allosteric inhibition) or increased (allosteric activation).

Allosteric proteins are usually oligomeric. In some cases, the subunits are identical, each subunit possessing both a catalytic site and an effector site; in other cases, the subunits are different, one type possessing only a catalytic site and the other only an effector site.

B. Feedback Inhibition

The general mechanism which has evolved in microorganisms for regulating the flow of carbon through biosynthetic pathways is the most efficient that one can imagine. The end product in each case allosterically inhibits the activity of the first—and only the first—enzyme in the pathway. For example, the first step in the biosynthesis of isoleucine not involving any other pathway is the conversion of L-threonine to α-ketobutyric acid, catalyzed by threonine deaminase. Threonine deaminase is allosterically and specifically inhibited by L-isoleucine and by no other compound (Figure 6–27); the other four enzymes of the pathway are not affected (although their synthesis is repressed).

C. Allosteric Activation

In some cases, it is advantageous to the cell for an end product or an intermediate to activate rather than inhibit a particular enzyme. In the breakdown of glucose by E coli, for example, overproduction of the intermediates glucose 6-phosphate and phosphoenolpyruvate signals the diversion of some glucose to the pathway of glycogen synthesis; this is accomplished by the allosteric activation of the enzyme converting glucose 1-phosphate to ADP-glucose (Figure 6–28).

D. Cooperativity

Many oligomeric enzymes, possessing more than one substrate binding site, show cooperative interactions of substrate molecules. The binding of substrate by one catalytic site increases the affinity of the other sites for additional substrate molecules. The net effect of this interaction is to produce an exponential increase in catalytic activity in response to an arithmetic increase in substrate concentration.

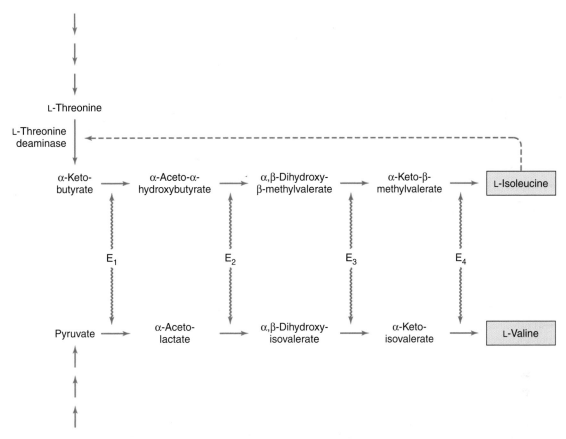

FIGURE 6–27 Feedback inhibition of L-threonine deaminase by L-isoleucine (dashed line). The pathways for the biosynthesis of isoleucine and valine are mediated by a common set of four enzymes, as shown.

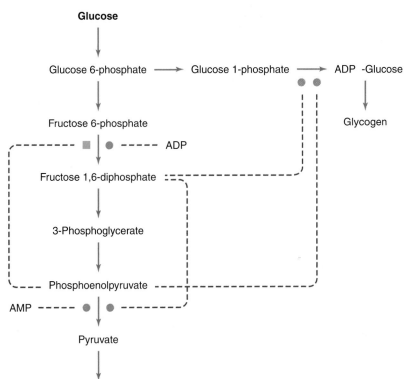

FIGURE 6–28 Regulation of glucose utilization by a combination of allosteric activation ● and allosteric inhibition ■ (Reproduced with permission from Stanier RY, Adelberg EA, Ingraham JL: *The Microbial World,* 4th ed. Prentice-Hall, 1976.)

E. Covalent Modification of Enzymes

The regulatory properties of some enzymes are altered by covalent modification of the protein. For example, the response of glutamine synthetase to metabolic effectors is altered by adenylylation, the covalent attachment of ADP to a specific tyrosyl side chain within each enzyme subunit. The enzymes controlling adenylylation also are controlled by covalent modification. The activity of other enzymes is altered by their phosphorylation.

F. Enzyme Inactivation

The activity of some enzymes is removed by their hydrolysis. This process can be regulated and sometimes is signaled by covalent modification of the enzyme targeted for removal.

REVIEW QUESTIONS

1. The synthesis of which of the following cell components is dependent upon a template?
   (A) Lipopolysaccharide
   (B) Peptidoglycan
   (C) Capsular polysaccharide
   (D) Deoxyribonucleic acid
   (E) Phospholipids

2. The synthesis of which of the following cell components is determined entirely by enzyme specificities?
   (A) DNA
   (B) Ribosomal RNA
   (C) Flagella
   (D) Lipopolysaccharide
   (E) Protein

3. The steps leading to the synthesis of peptidoglycan occur in the cytoplasm, on the cytoplasmic membrane, and extracellularly. Which antibiotic inhibits an extracellular step in peptidoglycan biosynthesis?
   (A) Cycloserine
   (B) Rifampin
   (C) Penicillin
   (D) Bacitracin
   (E) Streptomycin

4. Amino acids are found in the protein, peptidoglycan, and capsule of bacteria. Which of the following amino acids is found only in peptidoglycan?
   (A) L-Lysine
   (B) Diaminopimelic acid
   (C) D-Glutamate
   (D) L-Alanine
   (E) None of the above

5. The ability to use compounds and ions other than oxygen as terminal oxidants in respiration is a widespread microbial trait. This capacity is called
   (A) Photosynthesis
   (B) Fermentation
   (C) Anaerobic respiration
   (D) Substrate phosphorylation
   (E) Nitrogen fixation

Answers

1. D   3. C   5. C
2. D   4. B

REFERENCES

Atlas RM, Bartha R: *Microbial Ecology: Fundamentals and Applications,* 4th ed. Benjamin Cummings, 1998.

Downs DM: Understanding microbial metabolism. Annu Rev Microbiol 2006;60:533

Gibson J, Harwood CS: Metabolic diversity in aromatic compound utilization by anaerobic microbes. Annu Rev Microbiol 2002; 56:345. [PMID: 12142480]

Hillen W, Stülke: Regulation of carbon catabolism in *Bacillus* species. Annu Rev Microbiol 2000;54:849. [PMID: 11018147]

Hurst CJ et al (editors): *Manual of Environmental Microbiology,* 2nd ed. ASM Press, 2002.

Ishihama A: Functional modulation of *Escherichia coli* RNA polymerase. Annu Rev Microbiol 2000;54:499. [PMID: 11018136]

Leigh JA, Dodsworth JA: Nitrogen regulation in bacteria and archaea. Annu Rev Microbiol 2007;61:349

Maier RM, Pepper IL, Gerba CP: *Environmental Microbiology.* Academic Press, 2000.

Moat AG, Foster JW: *Microbial Physiology,* 4th ed. Wiley-Liss, 2002.

Neidhardt FC et al (editors): *Escherichia coli* and *Salmonella. Cellular and Molecular Biology,* 2nd ed. Vols 1 and 2. ASM Press, 1996.

Peters JW, Fisher K, Dean DR: Nitrogenase structure and function. Annu Rev Microbiol 1995;49:335. [PMID: 8561464]

Roberts IS: The biochemistry and genetics of capsular polysaccharide production in bacteria. Annu Rev Microbiol 1996;50:285. [PMID: 8905082]

Russell JB, Cook GM: Energetics of bacterial growth: Balance of anabolic and catabolic reactions. Microbiol Rev 1995;59:48. [PMID: 7708012] Schaechter M, Ingraham JL, Neidhardt FC: *Microbe.* ASM Press, 2006.

Microbial Genetics

The science of **genetics** defines and analyzes **heredity**, or constancy and change in the vast array of physiologic functions that form the properties of organisms. The basic unit of heredity is the **gene**, a segment of deoxyribonucleic acid **(DNA)** that encodes in its nucleotide sequence information for a specific physiologic property. The traditional approach to genetics has been to identify genes on the basis of their contribution to **phenotype**, or the collective structural and physiologic properties of an organism. A phenotypic property, be it eye color in humans or resistance to antibiotics in a bacterium, is generally observed at the level of the organism. The chemical basis for variation in phenotype is change in **genotype**, or alteration in the DNA sequence, within a gene or within the organization of genes.

DNA as the fundamental element of heredity was suggested in the 1930s from a seminal experiment performed by Frederick Griffith. In this experiment (Figure 7–1), killed virulent *Streptococcus pneumoniae* type III-S (possessing a capsule), when injected into mice along with living but non-virulent type II-R pneumococci (lacking a capsule), resulted in a lethal infection from which viable type III-S pneumococci were recovered. The implication was that some chemical entity transformed the live, nonvirulent strain to the virulent phenotype. A decade later Avery, MacLeod, and McCarty discovered that DNA was the transforming agent forming the foundation for molecular biology as we understand it today. Subsequent investigations with bacteria revealed the presence of **restriction enzymes**, proteins that cleave DNA at specific sites, giving rise to DNA **restriction fragments**. **Plasmids** were identified as small genetic elements carrying genes and capable of independent replication in bacteria and yeasts. The introduction of a DNA restriction fragment into a plasmid allows the fragment to be amplified many times. Amplification of specific regions of DNA also can be achieved with bacterial enzymes using **polymerase chain reaction (PCR)** or other enzyme-based method of nucleic acid amplification. DNA amplified by these sources and digested with appropriate restriction enzymes can be inserted into plasmids. Genes can be placed under control of high-expression bacterial **promoters** that allow encoded proteins to be expressed at increased levels. Bacterial genetics have fostered development of **genetic engineering** not only in prokaryotes but also in eukaryotes.

This technology is responsible for the tremendous advances in the field of medicine realized today.

ORGANIZATION OF GENES

The Structure of DNA & RNA

Genetic information in bacteria is stored as a sequence of **DNA bases** (Figure 7–2). In bacteriophages and viruses genetic information can be stored as sequences of **ribonucleic acid** (RNA) (Chapter 29). Most DNA molecules are double-stranded, with **complementary bases** (A-T; G-C) paired by hydrogen bonding in the center of the molecule (Figure 7–3). The orientation of the two DNA strands is **antiparallel:** One strand is chemically oriented in a $5'\rightarrow3'$ direction, and its complementary strand runs $3'\rightarrow5'$. The complementarity of the bases enables one strand (**template strand**) to provide the information for copying or expression of information in the other strand (**coding strand**). The base pairs are stacked within the center of the DNA double helix (Figure 7–2), and they determine its genetic information. Each turn of the helix has one major groove and one minor groove. Many proteins with the capacity to bind DNA and regulate gene expression interact predominately with the major groove, where atoms comprising the bases are more exposed. Each of the four bases is bonded to phospho-2'-deoxyribose to form a **nucleotide**. The negatively charged phosphodiester backbone of DNA faces the solvent. The length of a DNA molecule is usually expressed in thousands of base pairs, or **kilobase pairs (kbp)**. A small virus may contain a single DNA molecule of lessthan 0.5 kbp, whereas the single DNA genome that ecodes *Escherichia coli* is >4000 Kbp. In either case,each base pair is separated from the next by about 0.34 nm, or 3.4×10^{-7} mm, so that the total length of the *E coli* chromosome is roughly 1 mm. Since the overall dimensions of the bacterial cell are roughly 1000-fold smaller than this length, it is evident that a substantial amount of folding, or **supercoiling**, contributes to the physical structure of the molecule in vivo.

RNA most frequently occurs in single-stranded form. The base uracil (U) replaces thymine (T) in DNA, so the complementary bases that determine the structure of RNA are

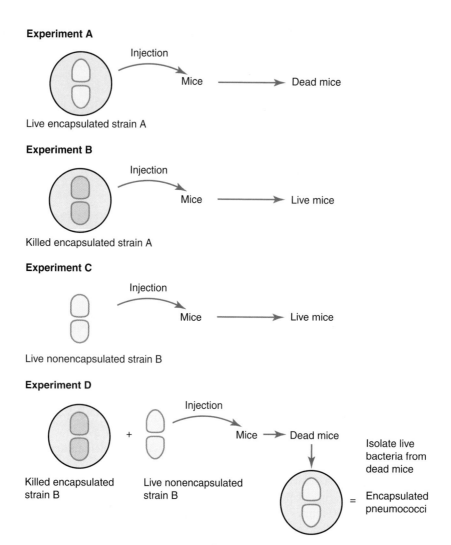

FIGURE 7–1 Griffiths' experiment demonstrating evidence for a transforming factor, later identified as DNA. In a series of experiments, mice were injected with live or killed encapsulated or nonencapsulated *Streptococcus pneumoniae*, as indicated in experiments A through D. The key experiment is D, showing that the killed encapsulated bacteria could supply a factor that allowed the nonencapsulated bacteria to kill mice. Besides providing key support for the importance of the capsule for pneumococcal virulence, experiment D also illustrates the principle of DNA as the fundamental basis of genetic transformation. (Reproduced by permission from Mietner & WcClane, Microbial Pathogenesis: A Principles-Oriented Approach, Fence Creek Publishing, 1999.)

A-U and C-G. The overall structure of single-stranded RNA molecules is determined by pairing between bases within the strand-forming loops, with the result that single-stranded RNA molecules assume a compact structure capable of expressing genetic information contained in DNA.

The most general function of RNA is communication of DNA gene sequences in the form of **messenger RNA (mRNA)** to **ribosomes**. These processes are referred to as **transcription** and **translation**. mRNA is transcribed as the RNA complement to the coding DNA strand. This mRNA is then translated by ribosomes. The ribosomes, which contain both **ribosomal RNA (rRNA)** and proteins, translate this message into the primary structure of proteins via amino-acyl-**transfer RNAs (tRNAs)**. RNA molecules range in size from the small tRNAs, which contain fewer than 100 bases, to mRNAs, which may carry genetic messages extending to

several thousand bases. Bacterial ribosomes contain three kinds of rRNA, with respective sizes of 120, 1540, and 2900 bases and a number of proteins (Figure 7–4). Corresponding rRNA molecules in eukaryotic ribosomes are somewhat larger. The need for expression of individual gene changes in response to physiologic demand, and requirements for flexible gene expression are reflected in the rapid metabolic turnover of most mRNAs. On the other hand, tRNAs and rRNAs—which are associated with the universally required function of protein synthesis—tend to be stable and together account for more than 95% of the total RNA in a bacterial cell. A few RNA molecules have been shown to function as enzymes (**ribozymes**). For example, the 23S RNA in the 50S ribosomal subunit (Figure 7–4) catalyzes the formation of the peptide bond during protein synthesis. Recently, a new class of RNA molecules called **small interfering RNA** (siRNA)

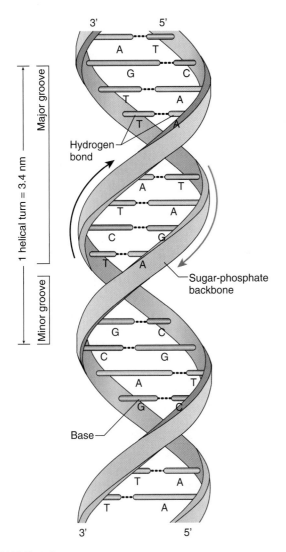

FIGURE 7–2 A schematic drawing of the Watson-Crick structure of DNA, showing helical sugar-phosphate backbones of the two strands held together by hydrogen bonding between the bases. (Redrawn with permission from Snyder L, Champness W: *Molecular Genetics of Bacteria,* 2nd ed. ASM Press, 2002.)

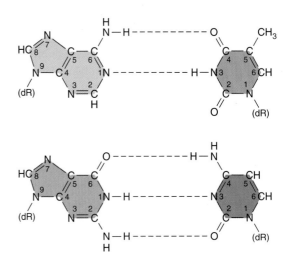

FIGURE 7–3 Normal base-pairing in DNA. **Top**: adenine-thymidine (A-T) pairing; **bottom**: guanine-cytosine (G-C) pair. Hydrogen bonds are indicated by dotted lines. Not that the G-C pairing shares three sets of hydrogen bonds whereas the A-T pairing has only two. Consequently, a G-C interaction is stronger than an A-T interaction. (dR, deoxyribose of the sugar-phosphate DNA backbone.)

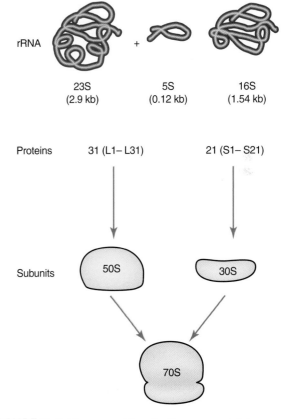

FIGURE 7–4 The composition of a ribosome containing one copy each of the 16S, 23S, and 5S RNAs as well as many proteins. The proteins of the large 50S subunit are designated L1 to L31. The proteins of the small 30S subunit are designated S1 to S21. (Redrawn with permission from Snyder L, Champness W: *Molecular Genetics of Bacteria,* 2nd ed. ASM Press, 2002.)

was described in plants. siRNAs are double-stranded RNA molecules, 20–25 nucleotides in length, that play a variety of roles in biology. Some have been shown to function as regulators by either binding near the 5′ end of an mRNA, preventing ribosomes from translating that message, or base pairing directly with a strand of DNA near the promoter, preventing transcription.

The Eukaryotic Genome

The **genome** is the totality of genetic information in an organism. Almost all of the eukaryotic genome is carried on two or more linear chromosomes separated from the cytoplasm within the membrane of the nucleus. **Diploid** eukaryotic cells contain two **homologues** (divergent evolutionary copies) of each chromosome. **Mutations**, or genetic changes, frequently cannot be detected in diploid cells because the contribution of one gene copy compensates for changes in the function of its homologue. A gene that does not achieve phenotypic expression in the presence of its homologue is **recessive**, whereas a gene that overrides the effect of its homologue is **dominant**. The effects of mutations can be most readily discerned in **haploid** cells, which carry only a single copy of most genes. Yeast cells (which are eukaryotic) are frequently investigated because they can be maintained and analyzed in the haploid state.

Eukaryotic cells contain mitochondria and, in some cases, chloroplasts. Within each of these organelles is a circular molecule of DNA that contains a few genes whose function relates to that particular organelle. Most genes associated with organelle function, however, are carried on eukaryotic chromosomes. Many yeast contain an additional genetic element, an independently replicating 2 μm circle containing about 6.3 kbp of DNA. Such small circles of DNA, termed **plasmids**, are frequently encountered in the genetics of prokaryotes. The small size of plasmids renders them amenable to genetic manipulation and, after their alteration, may allow their introduction into cells. Therefore, plasmids are commonly used in genetic engineering.

Repetitive DNA, which occurs in large quantities in eukaryotic cells, has been increasingly identified in prokaryotes. In eukaryotic genomes, repetitive DNA is infrequently associated with coding regions and is located primarily in extragenic regions. These short-sequence repeats (SSRs) or short tandemly repeated sequences (STRs) occur in several to thousands of copies dispersed throughout the genome. The presence of prokaryotic SSRs and STRs is well-documented, and some show extensive-length polymorphisms. This variability is thought to be caused by slipped-strand mispairing and is an important prerequisite for bacterial phase variation and adaptation. Many eukaryotic genes are interrupted by **introns**, intervening sequences of DNA that are missing in processed mRNA when it is translated. Introns have been observed in archaebacterial genes but with a few rare exceptions are not found in eubacteria (see Table 3–3).

The Prokaryotic Genome

Most prokaryotic genes are carried on the bacterial chromosome. And with few exceptions, bacterial genes are haploid. Genome sequence data from more than 340 microbial genomes have indicated that most prokaryotic genomes (>90%) consist of a single circular DNA molecule containing from 580 kbp to more than 5220 kbp of DNA (Table 7–1). A few bacteria (eg, *Brucella melitensis*, *Burkholderia pseudomallei*, and *Vibrio cholerae*) have genomes consisting of two circular DNA molecules. Many bacteria contain additional genes on plasmids that range in size from several to 100 kbp.

Covalently closed DNA circles (bacterial chromosomes and plasmids), which contain genetic information necessary for their own replication, are called **replicons**. Because prokaryotes do not contain a nucleus, a membrane does not separate bacterial genes from cytoplasm as in eukaryotes.

Some bacterial species are efficient at causing disease in higher organisms because they possess specific genes for pathogenic determinants. These genes are often clustered together in the DNA and are referred to as **pathogenicity islands**. These gene segments can be quite large (up to 200 kbp) and encode a collection of virulence genes. Pathogenicity islands (1) have a different G + C content from the rest of the genome, (2) are closely linked on the chromosome to tRNA genes, (3) are flanked by direct repeats, and (4) contain diverse genes important for pathogenesis—including adhesins, invasins, and exotoxins—as well as those that can be involved in mobilization.

Genes essential for bacterial growth (often referred to as "housekeeping genes") are carried on the chromosome, and plasmids carry genes associated with specialized functions (Table 7–2). Many plasmids encode genes that mediate their transfer from one organism to another as well as other genes associated with genetic acquisition or rearrangement of DNA. Therefore, genes with independent evolutionary origins may be assimilated by plasmids that are widely disseminated among bacterial populations. A consequence of such genetic events has been observed in the swift spread among bacterial populations of plasmid-borne resistance to antibiotics after their liberal use in hospitals.

Transposons are genetic elements that contain several genes, including those necessary for their migration from one genetic locus to another. In doing so, they create **insertion mutations**. The involvement of relatively short transposons (0.75–2.0 kbp long), known as **insertion elements**, produce the majority of insertion mutations. These insertion elements (also known as insertion sequence [IS] elements) carry only the genes for enzymes needed to promote their own transposition to another genetic locus, but cannot replicate on their own. Almost all bacteria carry IS elements, with each species harboring its own characteristic ones. Related IS elements can sometimes be found in different bacteria, implying that at some point in evolution they have crossed species barriers. Plasmids also carry IS elements, which are important in the formation of high-frequency recombinant **(Hfr)** strains (see below). Complex transposons carry genes for specialized functions such as antibiotic resistance and are flanked by insertion sequences. Unlike plasmids, transposons do not contain genetic information necessary for their own replication. Selection of transposons depends upon their replication as part of a replicon. Detection or genetic exploitation of transposons is achieved by selection of the specialized genetic information (normally, resistance to an antibiotic) that they carry.

The Viral Genome

Viruses are capable of survival, but not growth, in the absence of a cell host. Replication of the viral genome

TABLE 7–1 Comparison of Genome Sizes in Selected Prokaryotes, Bacteriophages, and Viruses

	Organism	Size (kbp)
Prokaryotes		
Archae	*Methanococcus jannaschii*	1660
	Archaeoglobus fulgidus	2180
Eubacteria	*Mycoplasma genitalium*	580
	Mycoplasma pneumoniae	820
	Borrelia burgdorferi	910
	Chlamydia trachomatis	1040
	Rickettsia prowazekii	1112
	Treponema pallidum	1140
	Chlamydia pneumoniae	1230
	Helicobacter pylori	1670
	Haemophilus influenzae	1830
	Francisella tularensis	1893
	Coxiella burnetii	1995
	Neisseria meningitides serogroup A	2180
	Neisseria meningitides serogroup B	2270
	Brucella melitensis[a]	2117 + 1178
	Mycobacterium tuberculosis	4410
	Escherichia coli	4640
	Bacillus anthracis	5227
	Burkholderia pseudomallei[a]	4126 + 3182
Bacteriophage	Lambda	48
Viruses	Ebola	19
	Variola major	186
	Vaccinia	192
	Cytomegalovirus	229

[a]Organisms with two different circular chromosomes.

TABLE 7–2 Examples of Metabolic Activities Determined by Plasmids

Organism	Activity
Pseudomonas species	Degradation of camphor, toluene, octane, salicylic acid
Bacillus stearothermophilus	α-Amylase
Alcaligenes eutrophus	Utilization of H_2 as oxidizable energy source
Escherichia coli	Sucrose uptake and metabolism, citrate uptake
Klebsiella species	Nitrogen fixation
Streptococcus (group N)	Lactose utilization, galactose phosphotransferase system, citrate metabolism
Rhodospirillum rubrum	Synthesis of photosynthetic pigment
Flavobacterium species	Nylon degradation

depends upon the metabolic energy and the macromolecular synthetic machinery of the host. Frequently, this form of genetic parasitism results in debilitation or death of the host cell. Therefore, successful propagation of the virus requires (1) a stable form that allows the virus to survive in the absence of its host, (2) a mechanism for invasion of a host cell, (3) genetic information required for replication of the viral components within the cell, and (4) additional information that may be required for packaging the viral components and liberating the resulting virus from the host cell.

Distinctions are frequently made between viruses associated with eukaryotes and viruses associated with prokaryotes, the latter being termed **bacteriophage or phage**. With more than 5000 isolates of known morphology, phages constitute the largest of all viral groups. Much of our understanding of viruses—indeed, many fundamental concepts of molecular biology—has emerged from investigation of the bacteriophage, and it is this group of viruses that is discussed in this chapter.

Bacteriophages occur in more than 140 bacterial genera and many different habitats. The nucleic acid molecule of bacteriophages is surrounded by a protein coat. Some phages also contain lipid. Considerable variability is found in the nucleic acid of phages. Many phages contain double-stranded DNA; others contain double-stranded RNA, single-stranded RNA, or single-stranded DNA. Unusual bases such as hydroxymethylcytosine are sometimes found in the phage nucleic acid. Bacteriophages exhibit a wide variety of morphologies. Many phages contain specialized syringe-like structures (tails) that bind to receptors on the cell surface and inject the phage nucleic acid into a host cell (Figure 7–5); other phages appear cubic, filamentous, or pleomorphic.

Phages can be distinguished on the basis of their mode of propagation. **Lytic phages** produce many copies of themselves as they kill their host cell. The most thoroughly studied lytic phages, the T-even (eg, T2, T4) phages of *E coli*, have demonstrated the need for precisely timed expression of viral genes in order to coordinate events associated with phage formation. **Temperate phages** are able to enter a nonlytic **prophage** state in which replication of their nucleic acid is linked to replication of host cell DNA. Bacteria carrying prophages are termed **lysogenic** because a physiologic signal can trigger a lytic cycle resulting in death of the host cell and liberation of many copies of the phage. The best characterized temperate phage is the *E coli* phage λ (lambda). Genes that determine the lytic or lysogenic response to λ infection have been identified and their complex interactions explored in detail.

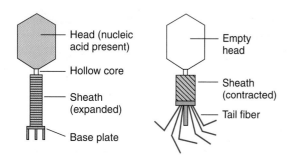

FIGURE 7–5 Illustrations of phage T2 with or without nucleic acid. Note that when the phage is loaded with nucleic acid it takes on a different form than when the nucleic acid is absent. These diagrams are redrawn from electron micrographic observations

Filamentous phages, exemplified by the well-studied *E coli* phage M13, are exceptional in several respects. Their filaments contain single-stranded DNA complexed with protein and are extruded from their hosts, which are debilitated but not killed by the phage infection. Engineering of DNA into phage M13 has provided single strands that are valuable sources for DNA analysis and manipulation.

REPLICATION

Double-stranded DNA is synthesized by **semiconservative replication**. As the parental duplex unwinds, each strand serves as a template (ie, the source of sequence information) for DNA replication. New strands are synthesized with their bases in an order complementary to that in the preexisting strands. When synthesis is complete, each daughter molecule contains one parental strand and one newly synthesized strand.

Eukaryotic DNA

Replication of eukaryotic DNA begins at several growing points along the linear chromosome. Accurate replication of the ends of linear chromosomes requires enzymatic activities different from the normal functions associated with DNA replication. These activities may involve **telomeres**, specialized DNA sequences (carried on the ends of eukaryotic chromosomes) that seem to be associated with accurate replication of chromosome ends. Eukaryotes have evolved specialized machinery, called a **spindle**, which pulls daughter chromosomes into separate nuclei newly formed by the process of **mitosis**. More extensive division of nuclei by **meiosis** halves the chromosomal number of diploid cells to form haploid cells. Accurate segregation of chromosomes during the reductive divisions of meiosis is an important factor in maintaining chromosomal structure within a species. Frequently, the haploid cells are **gametes**. Formation of gametes followed by their fusion to form diploid **zygotes** is the primary source of genetic variability via recombination in eukaryotes.

Bacterial DNA

Bacteria lack anything resembling the complex structures associated with the segregation of eukaryotic chromosomes into different daughter nuclei. The replication of bacterial DNA begins at one point and moves in both directions (ie, **bidirectional replication**). In the process, the two old strands of DNA are separated and used as templates to synthesize new strands (**semiconservative replication**). The structure where the two strands are separated and the new synthesis is occurring is referred to as the **replication fork**. Replication of the bacterial chromosome is tightly controlled, and the number of each chromosome (when more than one is present) per growing cell falls between one and four. Some bacterial plasmids may have as many as 30 copies in one bacterial cell, and mutations causing relaxed control of plasmid replication can result in even higher copy numbers.

The replication of circular double-stranded bacterial DNA begins at the *ori* locus and involves interactions with several proteins. In *E coli,* chromosome replication terminates in a region called *ter.* The **origin** (*ori*) and **termination sites** (*ter*) for replication are located at opposite points on the circular DNA chromosome. The two daughter chromosomes are separated, or resolved, before cell division, so that each progeny gets one of the daughter DNAs. This is accomplished with the aid of recombination and **topoisomerases**, enzymes that alter the supercoiling of double-stranded DNA. (In supercoiling the DNA molecule coils up like a telephone cord, which shortens the molecule.) The topoisomerases act by transiently cutting one or both strands of the DNA to relax the coil and extend the DNA molecule. Because bacterial topoisomerases are essential and unique, they are targets of **antibiotics** (eg, fluoroquinolones). Similar processes lead to the replication of plasmid DNA, except that in some cases replication is unidirectional.

Transposons

Transposons do not carry the genetic information required to couple their own replication to cell division, and therefore their propagation depends on their physical integration with a bacterial replicon. This association is fostered by enzymes that confer the ability of transposons to form copies of themselves; these enzymes may allow the transposons to integrate within the same replicon or an independent replicon. The specificity of sequence at the insertion site is generally low, so that transposons often seem to insert in a random pattern, but they tend to favor regions encoding tRNAs. Many plasmids are transferred among bacterial cells, and insertion of a transposon into such a plasmid is a vehicle that leads to the transposon's dissemination throughout a bacterial population.

Phage

Bacteriophages exhibit considerable diversity in the nature of their nucleic acid, and this diversity is reflected in different modes of replication. Fundamentally different propagation strategies are exhibited by lytic and temperate phages. Lytic phages produce many copies of themselves in a single burst of growth. Temperate phages establish themselves as prophages either by becoming part of an established replicon (chromosome or plasmid), or by forming an independent replicon.

The dsDNA of many lytic phages is linear, and the first stage in their replication is the formation of circular DNA. This process depends upon **cohesive ends**, complementary single-stranded tails of DNA that hybridize. **Ligation**, formation of a phosphodiester bond between the 5′ and 3′ DNA ends, gives rise to covalently closed circular DNA that may undergo replication in a manner similar to that used for other replicons. Cleavage of the circles produces linear DNA that is packaged inside protein coats to form daughter phages.

The ssDNA of filamentous phages is converted to a circular double-stranded replicative form. One strand of the replicative form is used as a template in a continuous process that produces single-stranded DNA. The template is a rolling circle, and the ssDNA it produces is cleaved and packaged with protein for extracellular extrusion.

ssRNA phages are among the smallest extracellular particles containing information that allows for their own replication. The RNA of phage MS2, for example, contains (in fewer than 4000 nucleotides) three genes that can act as mRNA following infection. One gene encodes the coat protein, and another encodes an RNA polymerase that forms a dsRNA replicative form. ssRNA produced from the replicative form is the core of new infective particles. The mechanism of propagation of RNA bacteriophage via RNA intermediates contrasts strongly with propagation of **retroviruses**, animal RNA viruses that use RNA as a template for DNA synthesis.

Some temperate bacteriophages, exemplified by *E coli* phage P1, can be established in a prophage state as a plasmid. The dsDNA of other temperate bacteriophages is established as a prophage by its insertion into the host chromosome. The site of insertion may be quite specific, as exemplified by integration of *E coli* phage λ at a single *int* locus on the bacterial chromosome. The specificity of integration is determined by identity of the shared DNA sequence by the *int* chromosomal locus and a corresponding region of the phage genome. Other temperate phages, such as *E coli* phage Mu, integrate in any of a wide range of chromosomal sites and in this aspect resemble transposons.

Prophages contain genes required for lytic replication (also called vegetative replication), and expression of these genes is repressed during maintenance of the prophage state. A manifestation of repression is that an established prophage frequently confers cellular immunity against lytic infection by similar phage. A cascade of molecular interactions triggers **derepression** (release from repression), so that a prophage undergoes vegetative replication, leading to formation of a burst of infectious particles. Artificial stimuli such as ultraviolet light may cause derepression of the prophage. The switch between lysogeny—propagation of the phage genome with the host—and vegetative phage growth at the expense of the cell may be determined in part by the cell's physiologic state. A nonreplicating bacterium will not support vegetative growth of phage, whereas a vigorously growing cell contains sufficient energy and building blocks to support rapid phage replication.

TRANSFER OF DNA

The haploid nature of the bacterial genome might be presumed to limit the genomic plasticity of a bacterium. However, the ubiquity of diverse bacteria in the environment provides a fertile gene pool that contributes to their remarkable genetic diversity through mechanisms of genetic exchange. Bacterial genetic exchange is typified by transfer of a relatively small fragment of a donor genome to a recipient cell followed by genetic recombination. Bacterial genetic recombination is quite unlike the fusion of gametes observed with eukaryotes; it demands that this donor DNA be replicated in the recombinant organism. Replication can be achieved either by integration of the donor DNA into the recipient's chromosome or by establishment of donor DNA as an independent replicon.

Restriction & Other Constraints on Gene Transfer

Restriction enzymes (restriction endonucleases) provide bacteria with a mechanism to distinguish between their own DNA and DNA from other biologic sources. These enzymes hydrolyze (cleave) DNA at restriction sites determined by specific DNA sequences ranging from four to 13 bases. DNA fragments can be selectively prepared because of this selectivity; this is the foundation of genetic engineering. Each bacterial strain that possesses a restriction system is able also to disguise these recognition sites in its own DNA by modifying them through methylation of adenine or cytosine residues within the site. These restriction-modification systems fall into two broad classes: type I systems, in which the restriction and modification activities are combined in a single multisubunit protein, and type II systems, which consist of separate endonucleases and methylases. A direct biologic consequence of restriction can be cleavage of donor DNA before it has an opportunity to become established as part of a recombinant replicon. Therefore, many recipients used in genetic engineering are dysfunctional in the *res* genes associated with restriction-modification.

Some plasmids exhibit a narrow host range and are able to replicate only in a closely related set of bacteria. Other plasmids, exemplified by some drug resistance plasmids, replicate in a wide range of bacterial recombinants. However, not all types of plasmids can stably coexist in a cell. Some types will interfere with the replication or partitioning of another type, so if two such plasmids are introduced into the same cell, one or the other will be lost at a higher than normal rate when the cell divides. The phenomenon is called **plasmid incompatibility;** two plasmids that cannot stably coexist belong to the same **incompatibility (Inc) group**, and two plasmids that can stably coexist belong to different Inc groups.

Mechanisms of Recombination

Donor DNA that does not carry information necessary for its own replication must recombine with recipient DNA in order to become established in a recipient strain. The recombination may be **homologous**, a consequence of close similarity in the sequences of donor and recipient DNA, or **nonhomologous**, the result of enzyme-catalyzed recombination between dissimilar DNA sequences. Homologous recombination almost always involves exchange between genes that share common ancestry. The process requires a set of genes designated *rec,* and dysfunctions in these genes give rise to bacteria that can maintain closely homologous genes in the absence of recombination. Nonhomologous recombination depends on enzymes encoded by the integrated DNA and is most clearly exemplified by the insertion of DNA into a recipient to form a copy of a donor transposon.

The mechanism of recombination mediated by *rec* gene products is reciprocal: Introduction of a donor sequence into a recipient is mirrored by transfer of the homologous recipient sequence into the donor DNA. Increasing scientific attention is being paid to the role of **gene conversion**—the nonreciprocal transfer of DNA sequences from donor to recipient—in the acquisition of genetic diversity.

Mechanisms of Gene Transfer

The DNA composition of microorganisms can be remarkably fluid. DNA can be transferred from one organism to another, and that DNA can be stably incorporated in the recipient, permanently changing its genetic composition. This process is called **lateral** or **horizontal** gene transfer to differentiate it from the inheritance of parental genes, a process called **vertical** inheritance. Three broad mechanisms mediate efficient movement of DNA between cells—**conjugation, transduction**, and **transformation**.

Conjugation requires donor cell-to-recipient cell contact to transfer only one strand of DNA (Figure 7–6). The recipient completes the structure of double-stranded DNA by synthesizing the strand that complements the strand acquired from the donor. In **transduction**, donor DNA is carried in a phage coat and is transferred into the recipient by the mechanism used for phage infection. **Transformation**, the direct uptake of "naked" donor DNA by the recipient cell, may be natural or forced. Relatively few bacterial species are naturally competent for transformation; these species assimilate donor DNA in linear form. Forced transformation is induced in the laboratory, where, after treatment with high salt and temperature shock, many

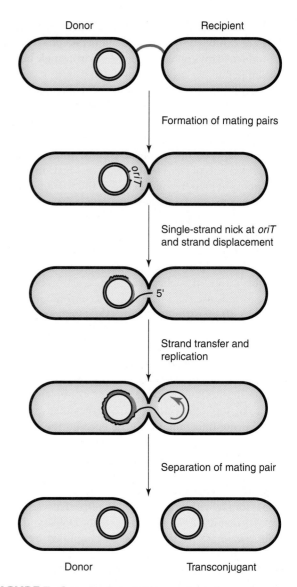

FIGURE 7–6 Mechanism of DNA transfer during conjugation. The donor cell produces a pilus, which is encoded by the plasmid and contacts a potential recipient cell that does not contain the plasmid. Retraction of the pilus brings the cells into close contact, and a pore forms in the adjoining cell membranes. Formation of the mating pair signals the plasmid to begin transfer from a single-stranded nick at *oriT*. The nick is made by plasmid encoded *tra* functions. The 5′ end of a single strand of the plasmid is transferred to the recipient through the pore. During transfer, the plasmid in the donor is replicated, its DNA synthesis being primed by the 3′ OH of the *oriT* nick. Replication of the single strand in the recipient proceeds by a different mechanism with RNA primers. Both cells now contain double-stranded plasmids, and the mating pair separates. (Redrawn with permission from Snyder L, Champness W: *Molecular Genetics of Bacteria.* ASM Press, 1997.)

bacteria are rendered competent for the uptake of extracellular plasmids. The capacity to force bacteria to incorporate extracellular plasmids by transformation is fundamental to genetic engineering.

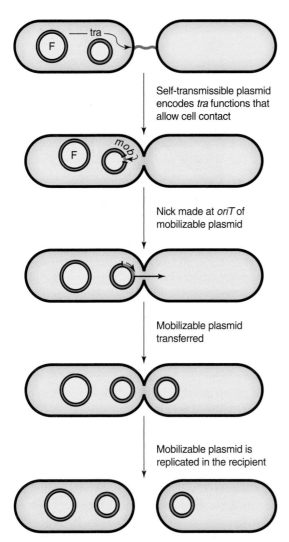

Self-transmissible plasmid encodes *tra* functions that allow cell contact

Nick made at *oriT* of mobilizable plasmid

Mobilizable plasmid transferred

Mobilizable plasmid is replicated in the recipient

FIGURE 7–7 Mechanism of plasmid mobilization. The donor cell carries two plasmids, a self-transmissible plasmid, F, which encodes the *tra* functions that promote cell contact and plasmid transfer, and a mobilizable plasmid. The *mob* functions encoded by the mobilizable plasmid make a single-stranded nick at *oriT* in the *mob* region. Transfer and replication of the mobilizable plasmid then occur. The self-transmissible plasmid may also transfer. (Redrawn with permission from Snyder L, Champness W: *Molecular Genetics of Bacteria.* ASM Press, 2nd ed. 2002.)

A. Conjugation

Plasmids are the genetic elements most frequently transferred by conjugation. Genetic functions required for transfer are encoded by the *tra* genes, which are carried by self-transmissible **plasmids**. Some self-transmissible plasmids can mobilize other plasmids or portions of the chromosome for transfer. In some cases mobilization is achieved because the *tra* genes provide functions necessary for transfer of an otherwise nontransmissible plasmid (Figure 7–7 and Figure 7–8). In other cases, the self-transmissible plasmid integrates with the DNA of another replicon and, as an extension of itself, carries a strand of this DNA into a recipient cell.

Genetic analysis of *E coli* was greatly advanced by elucidation of **fertility** factors carried on a plasmid designated F⁺. This plasmid confers certain donor characteristics upon cells; these characteristics include a sex pilus, an extracellular multimeric protein extrusion that attaches donor cells to recipient organisms lacking the fertility factor. A bridge between the cells allows a strand of the F⁺ plasmid, synthesized by the donor, to pass into the recipient, where the complementary strand of DNA is formed. The F⁺ fertility factor can integrate into numerous loci in the chromosome of donor cells. The integrated fertility factor creates high frequency recombination (HFR) donors from which chromosomal DNA is transferred (from the site of insertion) in a direction determined by the orientation of insertion (Figure 7–9).

The rate of chromosomal transfer from Hfr cells is constant, and compilation of results from many conjugation experiments has allowed preparation of an *E coli* **genetic map** in which distances between loci are measured in number of minutes required for transfer in conjugation. A similar map has been constructed for the related coliform (*E coli*-like) bacterium *Salmonella typhimurium*, and comparison of the two maps shows related patterns of gene organizationcies.

Analogous procedures with other plasmids have enabled researchers to map the circular chromosomes of members of distant bacterial genera; eg, drug resistance plasmids, termed **R factors**, can promote chromosomal transfer from diverse bacteria, including *Pseudomonas* spp. Comparison of chromosomal maps of *Pseudomonas aeruginosa* and *Pseudomonas putida* shows that few, albeit significant, genetic rearrangements accompanied divergence of these two closely related species. *Pseudomonas* maps have little in common with those of the biologically distant coliform bacteria.

Integration of chromosomal DNA into a conjugal plasmid can produce a recombinant replicon—an **F** (fertility) prime, or **R** (resistance) prime, depending on the plasmid—in which the integrated chromosomal DNA can be replicated on the plasmid independently of the chromosome. This occurs when the integrated plasmid (eg, F) is bracketed by two copies of an IS element. Bacteria carrying gene copies, a full set on the chromosome and a partial set on a prime, are **partial diploids**, or **merodiploids**, and are useful for complementation studies. A wild-type gene frequently complements its mutant homologue, and selection for the wild-type phenotype can allow maintenance of merodiploids in the laboratory. Such strains can allow analysis of interactions between different **alleles**, genetic variants of the same gene. Merodiploids frequently are genetically unstable because recombination between the plasmid and the homologous chromosome can result in loss or exchange of mutant or wild-type alleles. This problem can frequently be circumvented by maintenance of merodiploids in a genetic background in which *recA*, a gene required for recombination between homologous segments of DNA, has been inactivated by mutation.

Homologous genes from different organisms may have diverged to an extent that prevents homologous

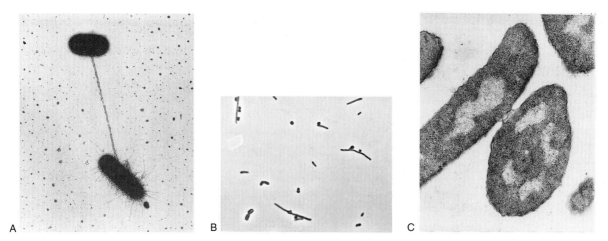

FIGURE 7–8 **A:** A male and a female cell joined by an F pilus (sex pilus). **B:** Mating pairs of *E coli* cells. Hfr cells are elongated. **C:** Electron micrograph of a thin section of a mating pair. The cell walls of the mating partners are in intimate contact in the "bridge" area. (Photographs [B] and [C] reproduced with permission from Gross JD and Caro LG: DNA transfer in bacterial conjugation. J Mol Biol 1966;16:269.)

recombination between them but does not alter the capacity of one gene to complement the missing activity of another. For example, the genetic origin of an enzyme required for amino acid biosynthesis is unlikely to influence catalytic activity in the cytoplasm of a biologically distant host. A merodiploid carrying a gene for such an enzyme would also carry flanking genes derived from the donor organism. Therefore, conventional microbial genetics, based on selection of prime plasmids, can be used to isolate genes from fastidious organisms in *E coli* or *P aeruginosa*. The significance of this technology lies in its ability to simplify or to circumvent the relatively expensive procedures demanded by genetic engineering.

B. Transduction

Transduction is phage-mediated genetic recombination in bacteria. In simplest terms, a transducing particle might be regarded as bacterial nucleic acid in a phage coat. Even a lytic phage population may contain some particles in which the phage coat surrounds DNA derived from the bacterium rather than from the phage. Such populations have been used to transfer genes from one bacterium to another. Temperate phages are preferred vehicles for gene transfer because infection of recipient bacteria under conditions that favor lysogeny minimizes cell lysis and thus favors survival of recombinant strains. Indeed, a recipient bacterium carrying an appropriate prophage may form a repressor that renders the cell immune to lytic infection; such cells may still take up bacterial DNA from transducing particles. Transducing mixtures carrying donor DNA can be prepared under conditions that favor the lytic phage cycle.

The size of DNA in transducing particles is usually no more than several percent of the bacterial chromosome, and therefore **co-transduction**—transfer of more than one gene at a time—is limited to linked bacterial genes. This process is of particular value in mapping genes that lie too close together to be placed in map order using the gross method of conjugal transfer. Mutant phages can be identified on the basis of the morphology of the **plaque** they form by lysis of a lawn of bacteria growing on solidified agar medium.

Pathogenicity islands are often transported by phages. For example, two phages transport pathogenicity islands responsible for converting a benign form of *Vibrio cholerae* into the pathogenic form responsible for epidemic cholera (see Chapter 17). These phage encode genes for cholera toxin and bundle forming pili which function in attachment.

The speed with which phages recombine and replicate has made them central subjects for study of these processes, and many generalizations concerning the underlying mechanisms have emerged from phage genetics. The capacity of phages to make rapid replicas of their DNA makes them valuable to genetic engineering. Of particular value are recombinant phages engineered so that they contain DNA inserts from another biologic source. Inserted DNA can be replicated with the swiftness that characterizes phage DNA and regained in a form useful for manipulation. Single-stranded DNA, produced by phage M13 and its derivatives, serves as a template for sequencing and site-directed mutagenesis.

C. Transformation

Direct uptake of donor DNA by recipient bacteria depends on their **competence** for transformation. Natural competence is unusual among bacteria, and some of these strains are transformable only in the presence of **competence factors**, produced only at a specific point in the growth cycle. Other strains readily undergo natural transformation, and these organisms offer promise for genetic engineering because of the ease with which they incorporate modified DNA into their chromosomes. Naturally competent transformable

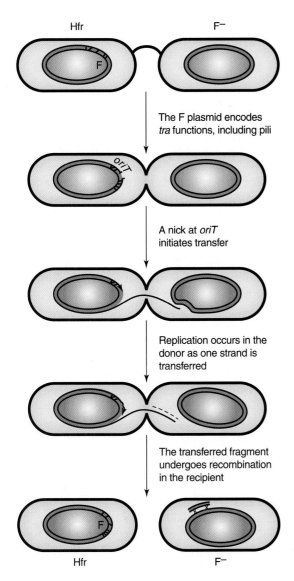

FIGURE 7–9 Transfer of chromosomal DNA by an integrated plasmid. Formation of mating pairs, nicking of the F *oriT* sequence, and transfer of the 5′ end of a single strand of F DNA proceed as in transfer of the F plasmid. Transfer of a covalently linked chromosomal DNA will also occur as long as the mating pair is stable. Complete chromosome transfer rarely occurs, and so the recipient cell remains F⁻, even after mating. Replication in the donor usually accompanies DNA transfer. Some replication of the transferred single strand may also occur. Once in the recipient cell, the transferred DNA may recombine with homologous sequences in the recipient chromosome. (Redrawn with permission from Snyder L, Champness W: *Molecular Genetics of Bacteria,* 2nd ed. ASM Press, 2002.)

bacteria are found in several genera and include *Bacillus subtilis, Haemophilus influenzae, Neisseria gonorrhoeae, Neisseria meningitidis,* and *Streptococcus pneumoniae.* DNA fragments containing genes from such organisms can be readily identified on the basis of their ability to transform mutant cells to the wild type. These techniques represent a substantial advance over the laborious procedures used by Avery and his

colleagues to demonstrate that the pneumococcal transforming principle was DNA (Figure 7–1).

Genetic transformation is recognized as a major force in microbial evolution. Natural transformation is an active process demanding specific proteins produced by the recipient cell. For *Neisseria* and *Haemophilus* spp. specific DNA sequences (**uptake sequences**) are required for uptake of the DNA. These uptake sequences are species specific, thus restricting genetic exchange to a single species. The DNA that is not incorporated can be degraded and used as a source of nutrients to support microbial growth.

Most bacteria are unable to undergo natural transformation. In these cases transformation can be forced by treatment with calcium chloride and temperature shock. Transformation with engineered recombinant plasmids by this procedure is a cornerstone of modern molecular biology because it enables DNA from diverse biologic sources to be established as part of well-characterized bacterial replicons.

MUTATION & GENE REARRANGEMENT

Spontaneous Mutations

Mutations are changes in DNA sequence. Spontaneous mutations for a given gene in a wild-type background generally occur with a frequency of 10^{-6}–10^{-8} in a population derived from a single bacterium (depending on the bacterial species and conditions used to identify the mutation). The mutations include **base substitutions**, **deletions**, **insertions**, and **rearrangements**. Base substitutions can arise as a consequence of mispairing between complementary bases during replication. In *E coli*, this occurs about once every 10^{10} times it incorporates a nucleotide; a remarkably accurate process. Occurance of a mispaired base is minimized by enzymes associated with **mismatch repair**, a mechanism that essentially proofreads a newly synthesized strand to ensure that it perfectly complements its template. Mismatch repair enzymes distinguish the newly synthesized strand from the preexisting strand on the basis of methylation of adenine in GATC sequences of the preexisting strand. When DNA damage is too extensive, a special DNA repair system, the **SOS response**, rescues cells in which DNA has been damaged. The **SOS response** is a postreplication DNA repair system that allows DNA replication to bypass lesions or errors in the DNA.

Many base substitutions escape detection at the phenotypic level because they do not significantly disrupt the function of the gene product. For example, **missense mutations**, which result in substitution of one amino acid for another, may be without discernible phenotypic effect. **Nonsense mutations** terminate synthesis of proteins and thus result in a protein truncated at the site of mutation. The gene products of nonsense mutations are usually inactive.

The consequences of deletion or insertion mutations also are severe because they can drastically alter the amino acid sequence of gene products. As described below, accurate expression of DNA sequences depends on translation of nucleotide triplet codons in perfect phase. Insertion or deletion of a single nucleotide disrupts the phase of translation and thus introduces an entirely different protein sequence distal to the amino acid codon altered by the mutation.

Many spontaneous mutations are the result of deletions that remove large portions of genes or even sets of genes. These large deletions involve recombination between directly repeated sequences (eg, IS elements) and almost never revert. Other spontaneous mutations cause duplication, frequently in tandem, of comparable lengths of DNA. Such mutations usually are unstable and readily revert. Still other mutations can invert lengthy DNA sequences or transpose such sequences to new loci. Comparative gene maps of related bacterial strains have shown that such rearrangements can be fixed in natural populations. These observations point to the fact that linear separation of DNA fragments does not completely disrupt possibilities for physical and chemical interaction among them.

Mutagens

The frequency of mutation is greatly enhanced by exposure of cells to mutagens. Ultraviolet (UV) light is a **physical mutagen** that damages DNA by linking neighboring thymine bases to form dimers (Chapter 4). Sequence errors can be introduced during enzymatic repair of this genetic damage. **Chemical mutagens** may act by altering either the chemical or the physical structure of DNA. Reactive chemicals alter the structure of bases in DNA. For example, nitrous acid (HNO_2) substitutes hydroxyl groups for amino groups. The resulting DNA has altered template activity during subsequent rounds of replication. **Frameshift mutations**—introduction or removal of a single base pair from DNA—are caused by slight slippage of DNA strands. This slippage is favored by acridine dyes (eg, acridine orange), which can intercalate between bases.

In general, the direct effect of chemical or physical mutagens is damage to DNA. The resulting mutations are introduced by the replication process and escape the repair enzymes described above. Mutations that change the activity of replication or repair enzymes can make a bacterium more susceptible to biologic mutagens and are referred to a mutator strains.

Reversion & Suppression

Regaining an activity lost as a consequence of mutation, termed **phenotypic reversion**, may or may not result from restoration of the original DNA sequence, as would be demanded by **genotypic reversion**. Frequently, a mutation at a second locus, called a **suppressor mutation**, restores the lost activity. In **intragenic suppression**, after a primary mutation has changed an enzyme's structure so that its activity has been lost, a second mutation, at a different site in the enzyme's gene, restores the structure required for activity. **Extragenic suppression** is caused by a second mutation lying outside the originally affected gene.

GENE EXPRESSION

The tremendous evolutionary separation of eukaryotic and prokaryotic genomes is illustrated by comparing their mechanisms of gene expression, which share only a small subset of properties. In both groups, genetic information is encoded in DNA, transcribed into mRNA, and translated on ribosomes through tRNA into the structure of proteins. The triplet nucleotide codons used in translation are generally shared, and many enzymes associated with macromolecular synthesis in the two biologic groups have similar properties. The mechanism by which the sequence of nucleotides in a gene determines the sequence of amino acids in a protein is largely similar in prokaryotes and eukaryotes and is as follows:

(1) RNA polymerase forms a single polyribonucleotide strand, called "messenger RNA" (mRNA), using DNA as a template; this process is called **transcription**. The mRNA has a nucleotide sequence complementary to a template strand in the DNA double helix if read in the 3′ to 5′ direction. Thus, an mRNA is oriented in a 5′ to 3′ direction.

(2) Amino acids are enzymatically activated and transferred to specific adapter molecules of RNA, called "transfer RNA" (tRNA). Each adapter molecule has a triplet of bases (**anticodon**) complementary to a triplet of bases on mRNA, and at one end its specific amino acid. The triplet of bases on mRNA is called the **codon** for that amino acid.

(3) mRNA and tRNA come together on the surface of the ribosome. As each tRNA finds its complementary nucleotide triplet on mRNA, the amino acid that it carries is put into peptide linkage with the amino acid of the preceding tRNA molecule. The enzyme **peptidyltransferase** (which is actually the 23S rRNA, ie, a **ribozyme**) catalyzes the formation of the peptide bond. The ribosome moves along the mRNA, the polypeptide growing sequentially until the entire mRNA molecule has been translated into a corresponding sequence of amino acids. This process, called **translation**, is diagrammed in Figure 7–10.

In prokaryotes, genes associated with related functions are typically clustered or occur in **operons**. Because there is no nucleus, transcription/translation is coupled, meaning that the nascent mRNA attaches to a ribosome and is translated at the same time it is transcribed. This coupled transcription/translation allows for the rapid response to changes in the environment. Likewise the mRNA is rapidly turned over, having a half-life on the order of seconds to minutes.

In eukaryotes, clustering of related genes is unusual. **Enhancer sequences** are regions of eukaryotic DNA that increase transcription and may lie distantly upstream from the transcribed gene. Eukaryotic genes carry **introns**, DNA

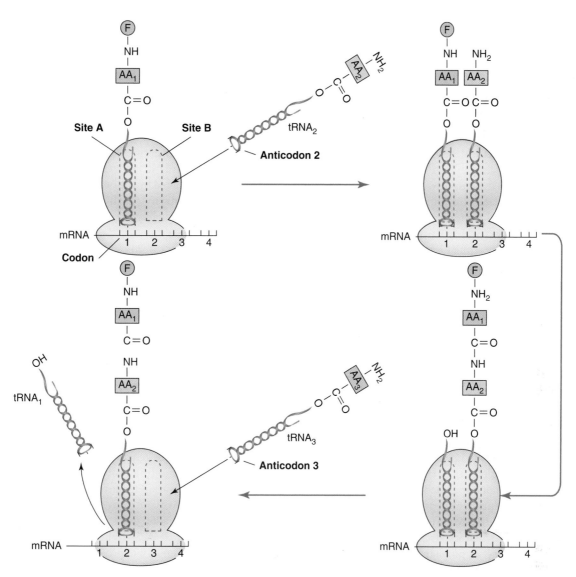

FIGURE 7–10 Four stages in the lengthening of a polypeptide chain on the surface of a 70S ribosome. **Top left**: A tRNA molecule bearing the anticodon complementary to codon 1 at one end and AA$_1$ at the other binds to site A. AA$_1$ is attached to the tRNA through its carboxyl group; its amino nitrogen bears a formyl group (F). **Top right**: A tRNA molecule bearing AA$_2$ binds to site B; its anticodon is complementary to codon 2. **Bottom right**: An enzyme complex catalyzes the transfer of AA$_1$ to the amino group of AA$_2$, forming a peptide bond. (Note that transfer in the opposite direction is blocked by the prior formylation of the amino group of AA$_1$.) **Bottom left**: The ribosome moves to the right, so that sites A and B are now opposite codons 2 and 3; in the process, tRNA$_1$ is displaced and tRNA$_2$ moves to site A. Site B is again vacant and is ready to accept tRNA$_3$ bearing AA$_3$. (When the polypeptide is completed and released, the formyl group is enzymatically removed.) (Redrawn and reproduced by permission of Stanier RY, Doudoroff M, Adelberg EA: *The Microbial World*, 3rd ed. Copyright © 1970. Prentice-Hall, Inc., Englewood Cliffs, NJ.)

insertions that are not found in prokaryotic genes. Introns separate **exons**, the coding regions of eukaryotic genes. Transcribed introns are removed from eukaryotic transcripts during RNA processing, a series of enzymatic reactions that take place in the nucleus. The mRNA of eukaryotes is polyadenylated at the 3′ end, protecting it from exonucleases so that it can traverse the nuclear membrane into the cytosol, where the ribosomes are located; in this case, translation is uncoupled from transcription. Because of this polyadenylation, eukaryotic mRNAs have half-lives on the order of hours to days.

Eukaryotic and prokaryotic ribosomes differ in many respects. Eukaryotic ribosomes are larger and have a sedimentation coefficient of 80S compared with the 70S sedimentation coefficient of prokaryotic ribosomes. The 40S and 60S eukaryotic ribosomal subunits are larger than the corresponding 30S and 50S ribosomal subunits of prokaryotes, and the eukaryotic ribosomes are relatively rich in protein. Significant differences are inherent in the sensitivity of the ribosomal activities to antibiotics (eg, tetracycline), many of which selectively inhibit protein synthesis in prokaryotic but not in eukaryotic cytoplasms (see Chapter 9). It should be remembered, however, that **mitochondrial** ribosomes in eukaryotes resemble those from prokaryotes.

Regulation of Prokaryotic Gene Expression

Specific proteins, the products of regulatory genes, govern expression of structural genes that encode enzymes. Transcription of DNA into mRNA begins at the **promoter**, the DNA sequence that binds RNA polymerase. The level of gene expression is determined in part by the ability of a promoter to bind the polymerase, and the intrinsic effectiveness of promoters differs widely. Further controls over gene expression are exerted by regulatory proteins that can bind to regions of DNA near promoters.

Many prokaryotic structural genes that encode a series of metabolic reactions are clustered on **operons**. This contiguous series of genes are expressed as a single mRNA transcript, and expression of the transcript may be governed by a single regulatory gene. For example, five genes associated with tryptophan biosynthesis are clustered in the *trp* operon of *E coli*. Gene expression is governed by attenuation, as described below, and is also controlled by repression: Binding of tryptophan by a **repressor protein** gives it a conformation that allows it to attach to the *trp* **operator**, a short DNA sequence that helps to regulate gene expression. Binding of the repressor protein to the operator prevents transcription of the *trp* genes. Repression can be viewed as a course-control mechanism, an all-or-none approach to gene regulation. This form of control is independent of attenuation, a fine-tuning mechanism that also is used to govern *trp* gene expression.

Attenuation is a regulatory mechanism of some biosynthetic pathways (eg, the tryptophan biosynthetic pathway) that controls the efficiency of transcription after transcription has been initiated, but before mRNA synthesis of the operon's genes takes place, especially when the end product of the pathway is in short supply. For example, under normal growth conditions, most *trp* mRNA transcripts terminate before they reach the structural genes of the *trp* operon. However, during conditions of severe tryptophan starvation, the premature termination of transcription is abolished, allowing expression of the operon at 10-fold higher levels than under normal conditions. The explanation for this phenomenon resides in the 162 bp regulatory sequence in front of the *trp* structural genes (Figure 7–11) referred to as the **leader sequence** or *trpL*. The *trp* leader sequence can be transcribed into mRNA and subsequently translated into a 14 amino acid polypeptide with two adjacent tryptophan residues, a very rare occurrence. At the end of *trpL*, and upstream of the regulatory signals that control translation of the *trp* structural genes, is a **Rho-independent terminator**. The DNA sequence of this region suggests that the encoded mRNA has a high probability of forming **stem loop secondary structures**. These have been named the **pause loop** (1:2), the **terminator loop** (3:4), and the **antiterminator loop** (2:3). Attenuation of the *trp* operon uses the secondary structure of the mRNA to sense the amount of tryptophan in the cell (as trp-tRNA) according to the model shown in Figure 7–11.

Prevention of transcription by a repressor protein is called **negative control**. The opposite form of transcriptional regulation—initiation of transcription in response to binding of an **activator protein**—is termed **positive control**. Both forms of control are exerted over expression of the *lac* operon, genes associated with fermentation of lactose in *E coli*. The operon contains three structural genes. Transport of lactose into the cell is mediated by the product of the *lacY* gene. Beta-galactosidase, the enzyme that hydrolyzes lactose to galactose and glucose, is encoded by the *lacZ* gene. The product of the third gene (*lacA*) is a transacetylase; the physiologic function of this enzyme has not been clearly elucidated.

As a by-product of its normal function, β-galactosidase produces allolactose, a structural isomer of lactose. Lactose itself does not influence transcriptional regulation; rather this function is served by allolactose, which is the **inducer** of the *lac* operon because it is the metabolite that most directly elicits gene expression. In the absence of allolactose, the *lac* repressor, a product of the independently controlled *lacI* gene, exerts negative control over transcription of the *lac* operon by binding to the *lac* operator. In the presence of the inducer, the repressor is released from the operator, and transcription takes place.

Expression of the *lac* operon and many other operons associated with energy generation is enhanced by the binding of **cyclic AMP-binding protein (CAP)** to a specific DNA sequence near the promoter for the regulated operon. The protein exerts positive control by enhancing RNA polymerase activity. The metabolite that triggers the positive control by binding to CAP is 3′,5′-cyclic AMP (cAMP). This compound, formed in energy-deprived cells, acts through CAP to enhance expression of catabolic enzymes that give rise to metabolic energy.

Cyclic AMP is not alone in its ability to exert control over unlinked genes in *E coli*. A number of different genes respond to the nucleotide ppGpp (in which "p" denotes phosphodiester and "G" denotes guanine) as a signal of amino acid starvation, and unlinked genes are expressed as part of the SOS response to DNA damage. Yet another set of unlinked genes is called into play in response to heat shock. This response is found in both prokaryotes and eukaryotes.

Elucidation of prokaryotic systems of transcriptional control has proved to have both conceptual and technical value. The *lac* operon has provided a useful model for comparative studies of gene expression. For example, the phenomenon of repression, first clearly described for the *lac* operon, accounts for the lysogenic response to infection by a temperate phage such as λ. Studies of the *E coli lac* system has provided many genetic derivatives that are useful in genetic engineering. Insertion of foreign DNA into plasmids to form **recombinant vectors** is frequently monitored phenotypically by use of a color test to monitor insertional inactivation of the *lacY* gene, and the *lac* promoter is often used to achieve controlled expression of inserted genes.

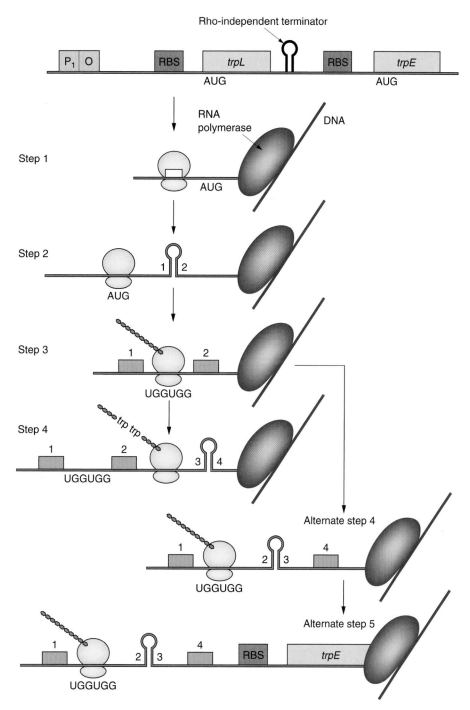

FIGURE 7–11 The predictions of the attenuation model. (Step 1) Coupled transcription/translation takes place as for any bacterial gene. (Step 2) RNA polymerase pauses and a 1:2 stem loop forms. (Step 3) The ribosome disrupts the 1:2 stem loop and encounters the two *trp* codons. (Step 4) If enough tryptophan is present, charged trp-tRNAs will be present and the ribosomes will translate *trpL*. This causes the RNA polymerase to stop at the Rho-independent terminator composed of a 3:4 stem loop. (Alternate Step 4) If tryptophan is limiting (no trpt-RNA), the ribosome stalls at the two *trp* codons, while RNA polymerase continues. The 2:3 stem loop forms. (Alternate Step 5) The 3:4 terminator cannot form and the RNA polymerase continues transcribing into the *trp* structural genes. This exposes the ribosome binding site (RBS) upstream of *trpE*, allowing translation. (Reproduced with permission from Trun N, Trempy J: *Fundamental Bacterial Genetics*. Blackwell Science Ltd, 2004.)

GENETIC ENGINEERING

Engineering is the application of science to social needs. In recent years, engineering based on bacterial genetics has transformed biology. Specified DNA fragments can be isolated and amplified, and their genes can be expressed at high levels. The nucleotide specificity required for cleavage by restriction enzymes allows fragments containing genes or parts of genes to be ligated (incorporated) into plasmids ("vectors") that can in turn be used to transform bacterial cells. Bacterial colonies or **clones** carrying specified genes can be identified by **hybridization** of DNA or RNA with labeled **probes** (similar to that shown in Figure 3-4). Alternatively, protein products encoded by the genes can be recognized either by enzyme activity or by immunologic techniques. The latter procedures have been greatly enhanced by the remarkable selectivity with which **monoclonal antibodies** (see Chapter 8) bind to specific antigenic determinants in proteins. Thus, genetic engineering techniques can be used to isolate virtually any gene and many of these genes can be expressed so that a biochemically recognizable property can be studied or exploited.

Isolated genes can be used for a variety of purposes. **Site-directed mutagenesis** can identify and alter the DNA sequence of a gene. Nucleotide residues essential for gene function can thus be determined and, if desired, altered. With hybridization techniques, DNA can be used as a probe that recognizes nucleic acids corresponding to the complementary sequence of its own DNA. For example, a latent virus in animal tissue can be detected with a DNA probe even in the absence of overt viral infection. The protein products of isolated viral genes offer great promise as vaccines because they can be prepared without genes that encode the replication of viral nucleic acid. Moreover, proteins such as insulin that have useful functions can be prepared in large quantities from bacteria that express cloned genes.

Preparation of DNA Fragments with Restriction Enzymes

The genetic diversity of bacteria is reflected in their extensive range of **restriction enzymes**, which possess remarkable selectivity that allows them to recognize specific regions of DNA for cleavage. DNA sequences recognized by restriction enzymes are predominantly palindromes (inverted sequence repetitions). A typical sequence palindrome, recognized by the frequently used restriction enzyme *Eco*R1, is GAATTC; the inverted repetition, inherent in the complementarity of the G-C and A-T base pairs, results in the 5′ sequence TTC being reflected as AAG in the 3′ strand.

The length of DNA fragments produced by restriction enzymes varies tremendously because of the individuality of DNA sequences. The average length of the DNA fragment is determined in large part by the number of specific bases recognized by an enzyme. Most restriction enzymes recognize four, six, or eight base sequences; however, other restriction enzymes recognize 10, 11, 12, or 15 base sequences. Recognition of four bases yields fragments with an average length of 250 base pairs and therefore is generally useful for analysis or manipulation of gene fragments. Complete genes are frequently encompassed by restriction enzymes that recognize six bases and produce fragments with an average size of about 4kbp. Restriction enzymes that recognize eight bases produce fragments with a typical size of 64kbp and are useful for analysis of large genetic regions. Restriction enzymes that recognize more than 10 bases are useful for construction

of a physical map and for molecular typing by pulse-field gel electrophoresis.

Physical Separation of Differently Sized DNA Fragments

Much of the simplicity underlying genetic engineering techniques lies in the fact that **gel electrophoresis** permits DNA fragments to be separated on the basis of size (Figure 7–12): The smaller the fragment, the more rapid the rate of migration. Overall rate of migration and optimal range of size for separation are determined by the chemical nature of the gel and by the degree of its cross-linking. Highly cross-linked gels optimize the separation of small DNA fragments. The dye **ethidium bromide** forms a brightly fluorescent adduct as it binds to DNA, so that small amounts of separated DNA fragments can be visualized on gels (Figure 7–12A). Specific DNA fragments can be recognized by probes containing complementary sequences (Figure 7–12B and C).

Pulsed-field gel electrophoresis allows the separation of DNA fragments containing up to 100 kbp that are separated on high resolution polyacrylamide gels. Characterizations of such large fragments have allowed construction of a physical map for the chromosomes from several bacterial species and have been invaluable in fingerprinting bacterial isolates associated with infectious disease outbreaks.

Cloning of DNA Restriction Fragments

Many restriction enzymes cleave asymmetrically and produce DNA fragments with **cohesive (sticky) ends** that may hybridize with one another. This DNA can be used as a donor with plasmid recipients to form genetically engineered recombinant plasmids. For example, cleavage of DNA with *Eco*R1 produces DNA containing the 5′ tail sequence AATT and the complementary 3′ tail sequence TTAA (Figure 7–13). Cleavage of a plasmid (a circular piece of DNA) with the same restriction enzyme produces a linear fragment with cohesive ends that are identical to one another. Enzymatic removal of the free phosphate groups from these ends ensures that they will not be ligated to form the original circular plasmid. Ligation in the presence of other DNA fragments containing free phosphate groups produces **recombinant plasmids**, which contain DNA fragments as inserts in covalently closed circular DNA. Plasmids must be in a circular form in order to replicate in a bacterial host.

Recombinant plasmids may be introduced into a bacterial host, frequently *E coli*, by transformation. Alternatively, **electroporation** is a recently developed procedure that introduces DNA into bacteria using an electrical gradient. Transformed cells may be selected on the basis of one or more drug resistance factors encoded by plasmid genes. The resulting bacterial population contains a **library** of recombinant

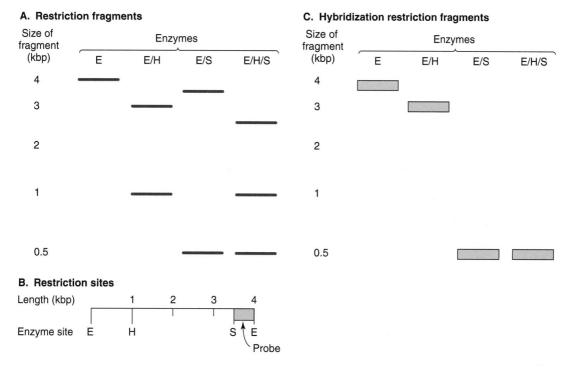

A. Restriction fragments

C. Hybridization restriction fragments

B. Restriction sites

FIGURE 7–12 **A**: Separation of DNA fragments on the basis of size by electrophoresis through a gel. Smaller fragments migrate more rapidly than large fragments, and, over a range determined by the properties of the gel, the distance migrated is roughly proportionate to the logarithm of the size of the fragment. DNA fragments can be visualized on the basis of their fluorescence after staining with a dye. **B**: The size of restriction fragments is determined by the location of restriction sites within the DNA. In this example, a 4.0 kilobase pair (kbp) fragment formed by restriction enzyme *Eco*R1 (E) contains respective sites for restriction enzymes *Hind*III (H) and *Sal*I (S) at positions corresponding to 1.0 and 3.5 kbp. The electrophoretic pattern in **A** reveals that restriction enzyme E does not cut the 4.0 kbp fragment (first lane); cleavage with restriction enzyme H produces fragments of 3.0 and 1.0 kbp (second lane); cleavage with restriction enzyme S yields fragments of 3.5 and 0.5 kbp (third lane); and cleavage with both H and S forms fragments of 2.5, 1.0, and 0.5 kbp (fourth lane). The 0.5 kbp fragment lying between the S and E sites was selected as a probe to determine DNA with hybridizing sequences as shown in C. **C**: Identification of hybridizing fragments. Restriction fragments were separated as in A. The hybridization procedure reveals those fragments that hybridized with the 0.5 kbp probe. These are the 4.0 kbp fragment formed by restriction enzyme E, the 3.0 kbp fragment lying between the E and H sites, and the 0.5 kbp fragment lying between the S and H sites.

plasmids carrying various cloned inserted restriction fragments derived from the donor DNA. Hybridization techniques may be used to identify bacterial colonies carrying specific DNA fragments, or, if the plasmid expresses the inserted gene, colonies can be screened for the gene product (Figure 7–14).

CHARACTERIZATION OF CLONED DNA

Restriction Mapping

Manipulation of cloned DNA requires an understanding of its nucleic acid sequence. Preparation of a **restriction map** is the first step in gaining this understanding. A restriction map is constructed much like a jigsaw puzzle from fragment sizes produced by **single digests**, which are prepared with individual restriction enzymes, and by **double digests**, which are formed with pairs of restriction enzymes. Restriction maps are also the initial step toward DNA sequencing, because they identify fragments that will provide **subclones** (relatively

small fragments of DNA) that may be subjected to more rigorous analysis, which may involve DNA sequencing. In addition, restriction maps provide a highly specific information base that allows DNA fragments, identified on the basis of size, to be associated with specific gene function.

Sequencing

DNA sequencing displays gene structure and enables researchers to deduce the structure of gene products. In turn, this information makes it possible to manipulate genes in order to understand or alter their function. In addition, DNA sequence analysis reveals regulatory regions that control gene expression and genetic "hot spots" particularly susceptible to mutation. Comparison of DNA sequences reveals evolutionary relationships that provide a framework for unambiguous classification of organisms and viruses. Such comparisons may facilitate identification of conserved regions that may prove particularly useful as specific hybridization probes to detect the organisms or viruses in clinical samples.

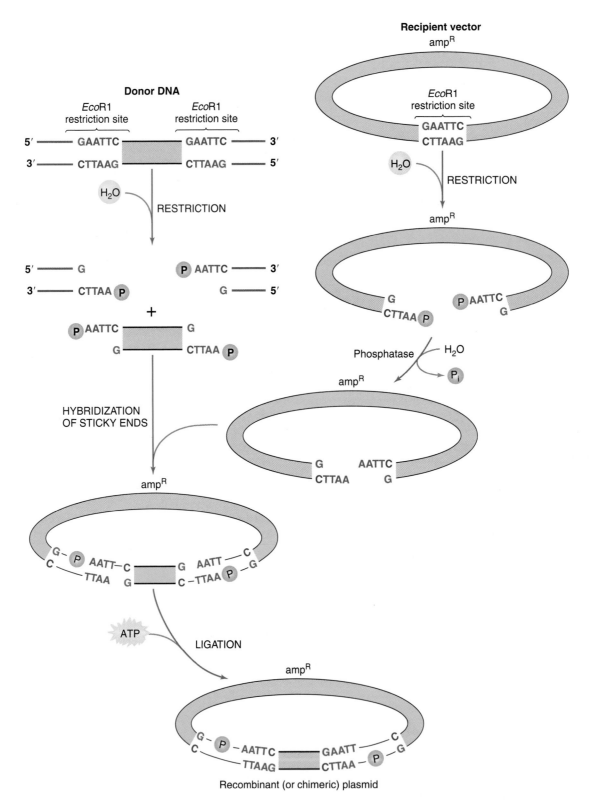

FIGURE 7–13 Formation of a recombinant, or chimeric, plasmid from donor DNA and a recipient vector. The vector, a plasmid that carries an *Eco*R1 restriction site, is cleaved by the enzyme and prepared for ligation by removal of the terminal phosphate groups. This step prevents the sticky ends of the plasmid from being ligated in the absence of an insert. The donor DNA is treated with the same restriction enzyme, and covalently bound circles are formed by ligation. A drug resistance marker, shown as amp[R] on the plasmid, can be used to select the recombinant plasmids after their transformation into *Escherichia coli*. Enzymes of the host bacterium complete covalent bonding of the circular DNA and mediate its replication.

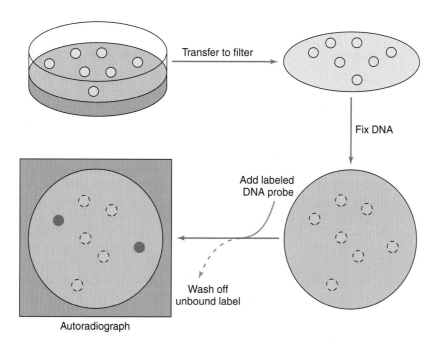

FIGURE 7–14 Use of probes to identify clones containing a specific fragment of DNA. Colonies may be transferred to a filter and baked so that the cells lyse and the DNA adheres to the filter. The filter can then be treated with a solution containing a suitably labeled DNA probe, which specifically hybridizes to the desired clones. Subsequent autoradiography of the filter identifies these clones (dark circles). Alternatively, the clones may be probed with antibodies to determine whether they have synthesized a specific protein product.

The two generally employed methods of DNA sequence determination are the **Maxam-Gilbert technique**, which relies on the relative chemical liability of different nucleotide bonds, and the **Sanger (dideoxy termination) method**, which interrupts elongation of DNA sequences by incorporating dideoxynucleotides into the sequences. Both techniques produce a nested set of oligonucleotides starting from a single origin and entail separation on a sequencing gel of DNA strands that differ by the increment of a single nucleotide. A polyacrylamide sequencing gel separates strands that differ in length from one to several hundred nucleotides and reveals DNA sequences of varying lengths.

Four parallel lanes on the same gel reveal the relative length of strands undergoing dideoxy termination at adenine, cytidine, guanidine, and thymidine. Comparison of four lanes containing reaction mixes that differ solely in the method of chain termination makes it possible to determine DNA sequence by the Sanger method (Figure 7–15). The relative simplicity of the Sanger method has led to its more general use, but the Maxam-Gilbert technique is widely employed because it can expose regions of DNA that are protected by specific binding proteins against chemical modification.

DNA sequencing is greatly facilitated by genetic manipulation of *E coli* bacteriophage M13, which contains single-stranded DNA. The replicative form of the phage DNA is a covalently closed circle of double-stranded DNA that has been engineered so that it contains a multiple cloning site that permits integration of specific DNA fragments that have been previously identified by restriction mapping. Bacteria

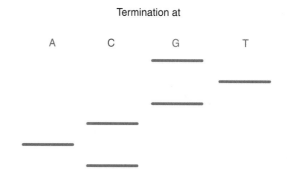

FIGURE 7–15 Determination of a DNA sequence by the Sanger (dideoxy termination) method. Enzymatic elongation of DNA is interrupted by inclusion of dideoxy analogs of the trinucleotides corresponding to A, C, G, and T separately in parallel reaction mixes. The resulting sets of interrupted elongated strands are separated on a sequencing gel, and the sequence can be deduced by noting the base corresponding to each increment of chain length. The sequencing gel is read from the bottom up; each band corresponds to an increase of one base.

infected with the replicative form secrete modified phages containing, within their protein coat, single-stranded DNA that includes the inserted sequence. This DNA serves as the **template** for elongation reactions. The origin for elongation is determined by a DNA **primer**, which can be synthesized by highly automated machines for **chemical oligonucleotide synthesis**. Such machines, which can produce DNA strands containing 75 or more oligonucleotides in a predetermined sequence, are essential for sequencing and for the modification of DNA by site-directed mutagenesis.

Chemically synthesized oligonucleotides can serve as primers for the **PCR**, a procedure that allows amplification and sequencing of DNA lying between the primers. Thus, in many instances, DNA need not be cloned in order to be sequenced or to be made available for engineering.

The study of biology has been revolutionized by the development of technology that allows sequencing and analysis of entire genomes, ranging from viruses to unicellular prokaryotic and eukaryotic microorganisms to humans. This has been facilitated by use of the procedure known as **shotgunning**. In this procedure, the DNA is broken into random smaller fragments to create a fragment library. These unordered fragments are sequenced by automated DNA sequencers and reassembled in the correct order using powerful computer software. A sufficient number of fragments are sequenced to ensure adequate coverage of the genome so that when they are assembled, most of the genome is represented without leaving too many gaps. (To achieve this, the entire genome is usually covered five- to eightfold, leaving about 0.1% of the total DNA unsequenced.) After the random fragments have been assembled by areas of overlapping sequence, any remaining gaps can be identified and closed. Advanced data processing permits annotation of the sequence data in which putative coding regions, operons, and regulatory sequences are identified. Already, the genomes of a number of important microorganisms have been sequenced. The continued analysis of sequence data from important human pathogens combined with studies on molecular pathogenesis will facilitate our understanding of how these organisms cause disease and, ultimately, will lead to better vaccines and therapeutic strategies.

SITE-DIRECTED MUTAGENESIS

Chemical synthesis of oligonucleotides enables researchers to perform controlled introduction of base substitutions into a DNA sequence. The specified substitution may be used to explore the effect of a predesigned mutation on gene expression, to examine the contribution of a substituted amino acid to protein function, or—on the basis of prior information about residues essential for function—to inactivate a gene. Single-stranded oligonucleotides containing the specified mutation are synthesized chemically and hybridized to single-stranded phage DNA, which carries the wild-type sequence as an insert (Figure 7–16). The resulting partially dsDNA is enzymatically converted to the fully double-stranded replicative form. This DNA, which contains the wild-type sequence on one strand and the mutant sequence on the other, is used to infect a bacterial host by transformation. Replication results in segregation of wild-type and mutant DNA, and the double-stranded mutant gene can be

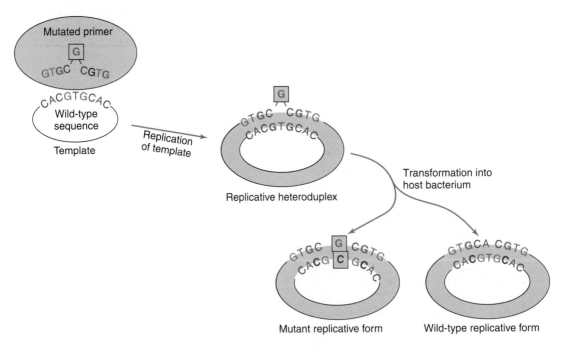

FIGURE 7–16 Site-directed mutagenesis. A chemically synthesized primer containing mutation G (in box) is hybridized to a wild-type sequence inserted in DNA from a single-stranded phage. Polymerization reactions are used to form the double-stranded heteroduplex carrying the mutation on one strand. Introduction of the heteroduplex into a host bacterium followed by segregation produces derivation strains carrying replicative forms with either the wild-type insert or an insert that has acquired the chemically designed mutation.

isolated and subsequently cloned from the replicative form of the phage.

ANALYSIS WITH CLONED DNA: HYBRIDIZATION PROBES

Hybridization probes (southern blotting, see Figure 3-4) are used routinely in the cloning of DNA. The amino acid sequence of a protein can be used to deduce the DNA sequence from which a probe may be constructed and employed to detect a bacterial colony containing the cloned gene. **Complementary DNA**, or **cDNA**, encoded by mRNA, can be used to detect the gene that encoded that mRNA. Hybridization of DNA to RNA by **Northern blots** can provide quantitative information about RNA synthesis. Specific DNA sequences in restriction fragments separated on gels can be revealed by **Southern blots**, a method that uses hybridization of DNA to DNA. These blots can be used to detect overlapping restriction fragments. Cloning of these fragments makes it possible to isolate flanking regions of DNA by a technique known as **chromosomal walking**. With **Western blots**, another frequently employed detection technique, antibodies are used to detect cloned genes by binding to their protein products.

Probes can be used in a broad range of analytic procedures. Some regions of human DNA exhibit substantial variability in the distribution of restriction sites. This variability is termed **restriction fragment length polymorphism (RFLP)**. Oligonucleotide probes that hybridize with RFLP DNA fragments can be used to trace DNA from a small sample to its human donor. Thus, the technique is valuable to forensic science. Applications of RFLP to medicine include identification of genetic regions that are closely linked to human genes with dysfunctions coupled to genetic disease. This information has been and will continue to be a valuable aid in **genetic counseling**.

DNA probes offer the promise of techniques for rapidly identifying fastidious organisms in clinical specimens that are difficult to grow in a microbiology laboratory. Furthermore, extensions of the technique afford opportunities to identify pathogenic agents rapidly and directly in infected tissue. Kits for identification of many bacterial and viral pathogens are commercially available.

Application of diagnostic DNA probes requires an appreciation of (1) the probes themselves, (2) systems used to detect the probes, (3) targets (the DNA to which the probes hybridize), and (4) the conditions of hybridization. Probes may be relatively large restriction fragments derived from cloned DNA or oligonucleotides corresponding to a specific region of DNA. Larger probes may provide greater accuracy because they are less sensitive to single base changes in target DNA. On the other hand, hybridization reactions occur more rapidly with small probes, and they can be designed against conserved regions of DNA in which base substitutions are unlikely to have occurred. Amplification of a target by PCR followed by a detection of the amplified product after hybridization to a probe has proved more sensitive than direct detection methods.

Recently, significant improvements have occurred in molecular diagnostic testing methods, especially those that incorporate nucleic acid amplification technologies such as PCR. Several commercial instruments have become available that combine PCR amplification of target DNA with detection of amplicons in the same closed vessel. This technology has been referred to as **real-time PCR**, implying that PCR amplicons can be detected in real time. In actuality, "real time" refers to the detection of amplicons after each PCR cycle. Probe detection formats involve detecting fluorophores. Results are semi-quantitative and can be obtained in considerably less time than it takes to perform a conventional PCR assay.

MANIPULATION OF CLONED DNA

Genetic engineering techniques permit separation and entirely independent expression of genes associated with pathogens. Vaccines prepared with engineered genes afford previously unattainable measures of safety. For example, a vaccine might be prepared against a viral coat protein that was produced in the absence of any genes associated with replicative viral functions; inoculation with such a vaccine would therefore entail no risk of introducing functional virus. Potential difficulties in the development of such vaccines stem from the ease with which viral mutations may produce genetic variants that are not recognized by the immune defense system of a vaccinated individual. Ultimately, vaccines now (and in the future will) contain a range of proteins that anticipate the genetic response of pathogens.

Recombinant Strains in the Environment

Major scientific advances have sometimes elicited adverse public reactions, so it is prudent to consider the potential consequences of genetic engineering. Of most immediate concern are known pathogens that have undergone relatively slight genetic modification. These have been and should be investigated in laboratories specially designed to contain them. The need for containment diminishes after genes for specific functions, such as protein coats, are separated from genes associated with replication or toxicity of a pathogen. For the most part, standard precautions associated with microbiology laboratories should be observed, if for no other reason than they foster habits that are valuable if a potential pathogen should enter the laboratory.

Interesting exceptions to this general rule are engineered organisms that may provide a social benefit if introduced into the environment. Many such organisms derive from non-pathogenic bacteria that occur naturally with a frequency as high as 10^5/g of soil. The available evidence suggests that predation and competition rapidly eliminate engineered bacterial strains after they are introduced into the environment. The primary challenge would ideally be to maintain biologically beneficial, engineered organisms in the environment rather than to eliminate them. However, this is not without social consequence. Among the examples of engineered organisms are *Pseudomonas* strains that produce a protein that favors formation of ice crystals. The value of these wild-type organisms is appreciated by ski slope owners, who have deliberately introduced the bacteria into the environment without arousing any public concern. An unfortunate side effect of the introduction of these organisms is that the ice crystals they promote can injure sensitive crops such as lettuce during seasons in which light frost is likely. Mutant bacteria that do not form ice crystals were designed by microbiologists who hoped that the mutant organisms might protect lettuce crops by temporarily occupying the niche normally inhabited by the ice-forming strains; however, attempts to use the mutant organisms in field studies were met with substantial protest, and studies were conducted only after lengthy and expensive legal delays. The legal precedents that have emerged from this and related applications will establish guidelines for the progressive and beneficial use of genetic engineering techniques and facilitate determination of situations in which extreme caution is justified.

REVIEW QUESTIONS

1. Mutations in bacteria can occur by which of the following mechanisms?
   (A) Base substitutions
   (B) Deletions
   (C) Insertions
   (D) Rearrangements
   (E) All of the above

2. The form of genetic exchange in which donor DNA is introduced to the recipient by a bacterial virus is
   (A) Transformation
   (B) Conjugation
   (C) Transfection
   (D) Transduction
   (E) Horizontal transfer

3. The form of genetic exchange in bacteria that is most susceptible to the activity of deoxyribonuclease during the process of DNA uptake is
   (A) Transformation
   (B) Conjugation
   (C) Transfection
   (D) Transduction
   (E) All of the above

4. Replication of which of the following requires physical integration with a bacterial replicon?
   (A) Single-stranded DNA bacteriophage
   (B) Double-stranded DNA bacteriophage
   (C) Single-stranded RNA bacteriophage
   (D) Plasmid
   (E) Transposon

5. The formation of a mating pair during the process of conjugation in *Escherichia coli* requires
   (A) Lysis of the donor
   (B) A sex pilus
   (C) Transfer of both strands of DNA
   (D) A restriction endonuclease
   (E) Integration of a transposon

Answers

1. E	3. A	5. B
2. D	4. E	

REFERENCES

Alberts B et al: *Molecular Biology of the Cell,* 4th ed. Garland, 2002.

Ausubel FM et al: *Current Protocols in Molecular Biology.* Wiley, 1987.

Avery O., Mcleod C, McCarty M. Studies on the chemical nature of the substance inducing transformation of pneumococcal types: Induction of transformation by a desoxyribonucleic acid fraction isolated from pneumococcus type III. J Exp Med 1944;79(2):137. [PMID: 19871359]

Bushman F: *Lateral DNA Transfer. Mechanisms and Consequences.* Cold Spring Harbor Laboratory Press, 2002.

Charlebois RL (editor): *Organization of the Prokaryotic Genome.* American Society for Microbiology, 1999.

Condon C: RNA processing and degradation in *Bacillus subtilis.* Microbiol Mol Biol Rev 2003;67:157. [PMID: 12794188]

Drlica K, Riley M (editors): *The Bacterial Chromosome.* American Society for Microbiology, 1990.

Fraser CM, Read TD, Nelson KE (editors): *Microbial Genomes.* Humana Press, 2004.

Grohmann E, Muuth G, Espinosa M: Conjugative plasmid transfer in gram-positive bacteria. Microbiol Mol Biol Rev 2003;67:277. [PMID: 12794193]

Hatfull GF: Bacteriophage genomics. Curr Opin Microbiol. 2008;5:447.

Koonin EV, Makarova KS, Aravind L: Horizontal gene transfer in prokaryotes: Quantification and classification. Annu Rev Microbiol 2001;55:709. [PMID: 11544372]

Kornberg A, Baker T: *DNA Replication,* 2nd ed. Freeman, 1992.

Lengler JW, Drews G, Schlegel HG (editors): *Biology of the Prokaryotes.* Blackwell Science, 1999.

Liebert CA, Hall RM, Summers AO: Transposon Tn*21,* flagship of the floating genome. Microbiol Mol Biol Rev 1999;63:507. [PMID: 10477306]

Murray NE: Type I restriction systems: Sophisticated molecular machines (a legacy of Bertani and Weigle). Microbiol Mol Biol Rev 2000;64:412. [PMID: 10839821]

Ptashne M: *A Genetic Switch: Phage Lambda and Higher Organisms,* 2nd ed. Blackwell, 1992.

Rawlings DE, Tietze E: Comparative biology of IncQ and IncQ-like plasmids. Microbiol Mol Biol Rev 2001;65:481. [PMID: 11729261]

Reischl U, Witter C, Cockerill F (editors): *Rapid Cycle Real-Time PCR—Methods and Applications.* Springer, 2001.

Rhodius V et al: Impact of genomic technologies on studies of bacterial gene expression. Annu Rev Microbiol 2002;56:599. [PMID: 12142487]

Riley MA, Wertz JE: Bacteriocins: Evolution, ecology, and application. Annu Rev Microbiol 2002;56:117. [PMID: 12142491]

Sambrook J, Russell NO: *Molecular Cloning: A Laboratory Manual,* 3rd ed. Cold Spring Harbor Laboratory, 2001.

Singleton P, Sainsbury D: *A Dictionary of Microbiology and Molecular Biology,* 3rd ed. Wiley, 2002.

Snyder L, Champness W: *Molecular Genetics of Bacteria.* ASM Press, 1997.

Trun N, Trempy J: *Fundamental Bacterial Genetics.* Blackwell Science Ltd, 2004.

van Belkum A et al: Short-sequence DNA repeats in prokaryotic genomes. Microbiol Mol Biol Rev 1998;62:275.

Zimmer C, Störl K, Störl J: Microbial DNA topoisomerases and their inhibition by antibiotics. J Basic Microbiol 1990;30:209–224.

C H A P T E R

8

Immunology

Roderick Nairn, PhD*

The study of immunology, a broad field encompassing both basic research and clinical applications, deals with host defense reactions to foreign (nonself) entities known as antigens, antigen recognition molecules, and cell-mediated host defense functions, especially as they relate to immunity to disease, hypersensitivity (including allergy), autoimmunity, immunodeficiency, and transplantation. This chapter presents the basic principles of immunology, particularly as they relate to response to infection. The reader is referred to texts on immunology for more detailed discussions.

IMMUNITY & THE IMMUNE RESPONSE

Immune responses can be innate (nonadaptive) or adaptive (acquired) (see Figure 8–1).

Innate Immunity

Innate immunity is resistance that is preexisting and is not acquired through contact with a nonself (foreign) entity known as an **antigen**. It is nonspecific and includes barriers to infectious agents—eg, skin and mucous membranes, phagocytic cells, inflammatory mediators, and complement components.

Adaptive Immunity

Adaptive immunity, which occurs after exposure to an antigen (eg, an infectious agent) is specific and is mediated by either antibody or lymphocytes. It can be passive or active.

A. Passive Immunity

Passive immunity is transmitted by antibodies or lymphocytes preformed in another host. The passive administration of antibody (in antisera) against certain viruses (eg, hepatitis B) can be useful during the incubation period to limit viral multiplication, eg, after a needle-stick injury to someone who has not been vaccinated. The main advantage of passive immunization with preformed antibodies is the prompt availability of large amounts of antibody; disadvantages are the short life span of these antibodies and possible hypersensitivity reactions if antibodies (immunoglobulins) from another species are administered.

B. Active Immunity

Active immunity is induced after contact with foreign antigens (eg, microorganisms or their products). This contact may consist of clinical or subclinical infection, immunization with live or killed infectious agents or their antigens, exposure to microbial products (eg, toxins, toxoids), or transplantation of foreign cells. In all these instances the host actively produces antibodies, and lymphocytes acquire the ability to respond to the antigens. Advantages of active immunity include long-term protection (based on memory of prior contact with antigen and the capacity to respond faster and to a greater extent on subsequent contact with the same antigen); disadvantages include the slow onset of protection and the need for prolonged or repeated contact with the antigen.

*Provost & Vice Chancellor for Academic & Student Affairs, University of Colorado Denver, Colorado.

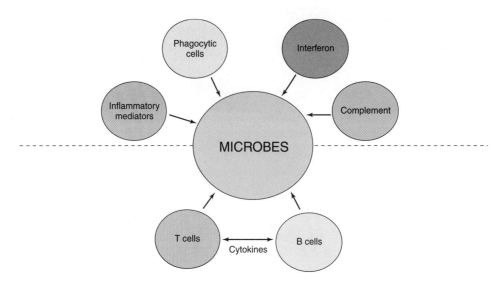

FIGURE 8–1 Top: The innate immune system is characterized by physiologic barriers to entry of pathogenic organisms and very fast host defense responses. **Bottom**: The adaptive immune system consists of cells displaying antigen recognition molecules and has the capacity for long-term memory.

GLOSSARY[1]

Alleles: Variants of a single genetic locus.

Anaphylatoxins: Fragments of complement proteins released during activation. Result in increased vascular permeability and attract leukocytes.

Antibody (Ab): A protein produced as a result of interaction with an antigen. The protein has the ability to combine with the antigen that stimulated its production.

Antigen (Ag): A substance that can react with an antibody. Not all antigens can induce antibody production; those that can are also called immunogens.

B cell (also B lymphocyte): Strictly, a bursa-derived cell in avian species and, by analogy, a cell derived from the equivalent of the bursa (bone marrow) in nonavian species. B cells are the precursors of plasma cells that produce antibody.

Cell Adhesion Molecules (CAMs): For example, the integrins and selectins. These are molecules that mediate the binding of cells to other cells or to extracellular matrix molecules such as fibronectin.

Cell-mediated (cellular) immunity: Immunity in which the participation of lymphocytes and macrophages is predominant. Cell-mediated immunity is a term generally applied to the type IV hypersensitivity reaction (see below).

Chemokines: Low-molecular-weight proteins that stimulate leukocyte movement.

Chemotaxis: A process whereby phagocytic cells are attracted to the vicinity of invading pathogens.

Complement: A set of plasma proteins that is the primary mediator of antigen–antibody reactions.

Cytolysis: The lysis of bacteria or of cells such as tumor or red blood cells by insertion of the membrane attack complex derived from complement activation.

Cytotoxic T cell: T cells that can kill other cells, eg, cells infected with intracellular pathogens.

Endotoxins: Bacterial toxins released from damaged cells.

Epitope: Site on an antigen recognized by an antibody. Also known as an antigenic determinant.

Hapten: A molecule that is not immunogenic by itself but can react with specific antibody after being joined to a suitable carrier molecule.

Histocompatible: Sharing major histocompatibility complex (transplantation) antigens.

Humoral immunity: Pertaining to immunity in a body fluid and used to denote immunity mediated by antibody and complement.

Hypersensitivity reactions:

  (1) **Antibody-mediated hypersensitivity:**

    Type I. Immediate: IgE antibody is induced by allergen and binds via its Fc receptor to mast cells and eosinophils. After encountering the antigen

[1] Modified and reproduced, with permission, from Stites DP, Stobo JD, Wells JV (editors): *Basic & Clinical Immunology*, 6th ed. Originally published by Appleton & Lange. Copyright © 1987 by the McGraw-Hill Companies, Inc.

again, the fixed IgE becomes cross-linked, inducing degranulation and release of mediators, especially histamine.

Type II. Antigens on a cell surface combine with antibody, which leads to complement-mediated lysis (eg, transfusion or Rh reactions) or other cytotoxic membrane damage (eg, autoimmune hemolytic anemia).

Type III. Immune complex: Antigen–antibody immune complexes are deposited in tissues, complement is activated, and poly-morphonuclear cells are attracted to the site, causing tissue damage.

(2) **Cell-mediated hypersensitivity:**

Type IV. Delayed: T lymphocytes, sensitized by an antigen, release cytokines upon second contact with the same antigen. The cytokines induce inflammation and activate macrophages.

Immune response: Development of resistance (immunity) to a foreign substance (eg, infectious agent). It can be antibody-mediated (humoral), cell-mediated (cellular), or both.

Immunity:

(1) **Innate immunity:** Nonspecific host defense not acquired through contact with an antigen. It includes skin and mucous membrane barriers to infectious agents and a variety of nonspecific immunologic factors.

(2) **Adaptive immunity:** Protection acquired by deliberate introduction of an antigen into a responsive host. Active immunity is specific and is mediated by either antibody or lymphoid cells (or both).

Immunoglobulin: A glycoprotein, composed of H and L chains, that functions as antibody. All antibodies are immunoglobulins, but not all immunoglobulins have antibody function.

Immunoglobulin class: A subdivision of immunoglobulin molecules based on structural (amino acid sequence) differences. In humans there are five immunoglobulin classes: IgG, IgM, IgA, IgE, and IgD.

Immunoglobulin subclass: A subdivision of the classes of immunoglobulins based on structural differences in the H chains. For human IgG there are four subclasses: IgG1, IgG2, IgG3, and IgG4.

Inflammation: Local accumulation of fluid and cells after injury or infection.

Interferon: One of a heterogeneous group of low-molecular-weight proteins elaborated by infected host cells that protect noninfected cells from viral infection. Interferons, which are cytokines, also have immunomodulating functions.

Leukocyte: General term for a white blood cell.

Lymphocyte: A mononuclear cell 7–12 μm in diameter containing a nucleus with densely packed chromatin and a small rim of cytoplasm. Lymphocytes include the T cells and B cells, which have primary roles in immunity.

Macrophage: A phagocytic mononuclear cell derived from bone marrow monocytes and found in tissues and at the site of inflammation. Macrophages serve accessory roles in immunity, particularly as antigen-presenting cells (APCs).

Major histocompatibility complex (MHC): A cluster of genes located in close proximity, eg, on human chromosome 6, that encode the histocompatibility antigens (MHC molecules).

Membrane attack complex: The end product of activation of the complement cascade, which contains C5, C6, C7, and C8 (and C9). The membrane attack complex makes holes in the membranes of gram-negative bacteria, killing them and, in red blood or other cells, resulting in lysis.

Monoclonal antibodies: Each B lymphocyte produces antibody of a single specificity. However, normal B cells do not grow indefinitely. If B cells are fused to a myeloma cell by somatic cell hybridization and fused cells that secrete the desired antibody specificity are selected, an immortalized antibody-producing cell line, known as a hybridoma, is obtained, and these hybrid cells produce monoclonal antibodies.

Monocyte: A circulating phagocytic blood cell that develops into tissue macrophages.

Natural killer (NK) cells: Large granular lymphoid cells with no known antigen-specific receptors. They are able to recognize and kill certain virally infected cells, and also activate the innate response.

Opsonin: A substance capable of enhancing phagocytosis. Antibodies and complement are the two main opsonins.

Opsonization: The coating of an antigen or particle (eg, infectious agent) by substances, such as antibodies, complement components, fibronectin, and so forth, that facilitate uptake of the foreign particle into a phagocytic cell.

Plasma cell: A terminally differentiated B cell that secretes antibody.

Polymorphonuclear cell (PMN): Also known as a neutrophil or granulocyte, a PMN is characterized by a multilobed nucleus. PMNs migrate from the circulation to a site of inflammation by chemotaxis and are phagocytic for bacteria and other particles.

T cell (also T lymphocyte): A thymus-derived cell that participates in a variety of cell-mediated immune reactions.

Thymocytes: Developing T cells found in the thymus.

Vaccination: Induction of immunity by injecting a dead or attenuated form of a pathogen.

MECHANISMS OF INNATE IMMUNITY

Physiologic Barriers at the Portal of Entry

A. The Skin

Few microorganisms are capable of penetrating intact skin, but many can enter sweat or sebaceous glands and hair follicles and establish themselves there. Sweat and sebaceous secretions—by virtue of their acid pH and certain chemical substances (especially fatty acids)—have antimicrobial properties that tend to eliminate pathogenic organisms. Lysozyme, an enzyme that dissolves some bacterial cell walls, is present on the skin and can help provide protection against some microorganisms. Lysozyme is also present in tears and in respiratory and cervical secretions.

The skin produces a variety of antimicrobial agents including a protein with antibacterial properties known as psoriasin.

B. Mucous Membranes

In the respiratory tract, a film of mucus covers the surface and is constantly being driven upward by ciliated cells toward the natural orifices. Bacteria tend to stick to this film. In addition, mucus and tears contain lysozyme and other substances with antimicrobial properties. For some microorganisms, the first step in infection is their attachment to surface epithelial cells by means of adhesive bacterial surface proteins (eg, the pili of gonococci and *Escherichia coli*). If such cells have IgA antibody on their surfaces—a host resistance mechanism—attachment may be prevented. (The organism can overcome this resistance mechanism by breaking down the antibody with a protease.)

When organisms enter the body via mucous membranes, they tend to be taken up by phagocytes and are transported into regional lymphatic vessels that carry them to lymph nodes. The phagocytes act as barriers to further spread of large numbers of bacteria. The mucociliary apparatus for removal of bacteria in the respiratory tract is aided by pulmonary macrophages. Special protective mechanisms in the respiratory tract include the hairs at the nares and the cough reflex, which prevents aspiration.

In the gastrointestinal tract, several systems function to inactivate bacteria: Saliva contains numerous hydrolytic enzymes; the acidity of the stomach kills many ingested bacteria (eg, *V cholerae*); and the small intestine contains many proteolytic enzymes and active macrophages.

It must be remembered that most mucous membranes of the body carry a constant normal microbial flora that itself opposes establishment of pathogenic microorganisms ("bacterial interference") and has important physiologic functions. For example, in the adult vagina, an acid pH is maintained by normal lactobacilli, inhibiting establishment of yeasts, anaerobes, and gram-negative bacteria.

Innate Immunologic Mechanisms

The innate immune system uses both soluble and membrane-bound **pattern recognition receptors** (PRRs) to "sense" the presence of invading microbes. These receptors include molecules such as the **toll-like receptors** (TLRs) that recognize highly conserved pathogen molecules such as lipopolysaccharide on gram-negative bacteria. Very early in the response to infection (first few hours), the engulfment of microorganisms by macrophages (phagocytosis) and the activation of complement by the alternative pathway (see Figure 8–9 and discussion later in this chapter) are the important nonspecific host responses. The next line of defense includes some responses that are still nonadaptive—eg, release of cytokines from macrophages—and the release of other mediators that trigger the **inflammatory response**. The inflammatory response occurs rapidly and generally serves to hold the spread of pathogen until a specific adaptive response is initiated. However, some microorganisms have found ways to evade these nonspecific host responses—eg, bacteria (pneumococci) with polysaccharide-rich capsules can evade phagocytosis, and some viruses (such as poxviruses) produce cytokine receptor homologs that function as competitive antagonists of the cytokines. These evasion mechanisms slow the immune response long enough for the microorganism to establish a niche.

A. Phagocytic Cells

Mononuclear phagocytic cells are present in blood, lymphoid tissue, liver, spleen, lung, and other tissues that are efficient in uptake and removal of particulate matter from lymph vessels and the bloodstream. They include cells lining blood and lymph sinuses (Kupffer cells in the liver) and macrophages.

An important function of the spleen is filtering microorganisms from the bloodstream. Patients whose spleens have been removed or are nonfunctional (eg, in sickle cell disease) often suffer from bacterial sepsis, particularly with pneumococci and salmonellae. Phagocytosis is greatly enhanced by opsonins. When macrophages recognize microbial constituents, they are stimulated to release cytokines that cause the recruitment of more phagocytic cells to the site of infection.

B. Phagocytosis

During bacterial infection, the number of circulating phagocytic cells often increases. The main functions of phagocytic cells include migration, chemotaxis, ingestion, and microbial killing. Microorganisms (and other particles) that enter the lymphatics, lung, or bloodstream are engulfed by any of a variety of phagocytic cells. Among them are polymorphonuclear leukocytes (granulocytes), phagocytic monocytes (macrophages), and fixed macrophages of the reticuloendothelial system (see above). Many microorganisms elaborate chemotactic factors that attract phagocytic cells. Defects in chemotaxis may account for hypersusceptibility to certain infections; the defects may be acquired or inherited. Phagocytosis can occur in the absence of serum antibodies early in the infectious process.

1. Factors affecting phagocytosis—Phagocytosis is made more efficient by the presence of antibodies (opsonins) that coat the surface of bacteria and facilitate their ingestion by phagocytes. Opsonization can occur by three mechanisms: (1) Antibody alone can act as opsonin; (2) antibody plus antigen can activate complement via the classic pathway to yield opsonin; and (3) opsonin may be produced by C3 via the alternative pathway (see Figure 8–9). Macrophages have receptors on their membranes for the Fc portion of antibody and for the C3 component of complement. These receptors aid the phagocytosis of antibody-coated particles.

Ingestion of foreign particles (eg, microorganisms) has the following effects on phagocytic granulocytes: (1) Oxygen consumption increases, and there is increased generation of superoxide anion (O_2^-) and increased release of H_2O_2; (2) glycolysis increases via the hexose monophosphate shunt; and (3) lysosomes rupture, and their hydrolytic enzymes are discharged into the phagocytic vacuole to form a digestive vacuole, or "phagolysosome." Morphologically, this process appears as "degranulation" of granulocytes. Inhibition of these mechanisms is an important part of the infectious process, or pathogenesis, of **legionella pneumonia**. In Chédiak-Higashi syndrome, most microorganisms are phagocytosed normally, but intracellular killing is impaired, because a defect in a cytoplasmic protein causes abnormal granule membrane fusion, leading to lysosomal dysfunction and recurrent infection by pyogenic bacteria.

2. Granulocytes (polymorphonuclear leukocytes, or neutrophils)—Granulocytes contain granules composed of lysozyme, other hydrolytic enzymes, several cationic proteins, the defensins (antimicrobial components), lactoferrin, and toxic nitrogen oxides.

The mechanisms of intracellular killing of microorganisms in phagocytic granulocytes include nonoxidative mechanisms (eg, activation of hydrolytic enzymes in contact with microorganisms, action of antimicrobial peptides) and oxidative mechanisms. Among the latter, the following have been implicated:

a. Increased oxidative activity results in accumulation of H_2O_2. In the presence of oxidizable cofactors (halides such as chlorine), an acid pH, and the enzyme myeloperoxidase, H_2O_2 is converted to HOCl, which is an effective antimicrobial agent.

b. In normal granulocytes, superoxide anion (O_2^-) is generated when particles are phagocytosed. The superoxide radical may be directly lethal for many microorganisms. Children suffering from chronic granulomatous disease have granulocytes that ingest microbes normally, but they have a genetic deficiency of the NADPH oxidase system required to produce the superoxide anion so important in the antimicrobial activity of phagocytes. This defect may be responsible for the impaired killing ability of granulocytes associated with this disease and explains the susceptibility of these patients to infections, especially staphylococcal infections.

When the bone marrow of patients is suppressed by disease, drugs, or radiation, the number of functional granulocytes falls. If the granulocyte level drops below protective levels, the patient is highly susceptible to opportunistic infection by bacteria.

3. Macrophages (circulating phagocytic monocytes)—Macrophages are derived from monocyte stem cells in bone marrow, have a longer life span than circulating granulocytic phagocytes, and continue their activity at a lower pH.

Macrophages in blood can be activated by various stimulants, or "activators," including microbes and their products, antigen–antibody complexes, inflammation, activated T lymphocytes, cytokines (see below), and injury. Activated macrophages have an increased number of lysosomes and produce and release interleukin-1 (IL-1), which has a wide range of activity in inflammation. IL-1 participates in fever production and in activation of lymphoid cells, resulting in the release of other cytokines.

Intracellular killing in macrophages probably includes mechanisms similar to those described above for granulocytes.

C. Alternative Pathway of Complement Activation

The complement system, a set of proteins that enhance the function of both the adaptive and innate responses to infection, is discussed later in this chapter. One pathway of complement activation, the alternative pathway, is very important as a first line of defense against infection by microorganisms. As shown in Figure 8–9, the alternative complement pathway can be activated by microbial surfaces and proceeds in the absence of antibody. There are several antimicrobial properties of complement proteins that contribute to host defense, including opsonization, lysis of bacteria, and amplification of inflammatory responses through the anaphylatoxins C5a and C3a.

Some microorganisms have developed mechanisms to interfere with the complement system and in that way evade the immune response. For example, vaccinia virus encodes a soluble protein that functions as a complement control protein by blocking both major pathways of complement activation through binding to C3b and C4b.

D. Inflammatory Response

Any injury to tissue, such as that following establishment and multiplication of microorganisms, elicits an inflammatory response. The innate immune response of macrophages includes the release of **cytokines**, including IL-1 and tumor necrosis factor-α (TNF-α). The other mediators released from activated macrophages include prostaglandins and leukotrienes. These inflammatory mediators begin to elicit changes in local blood vessels. This begins with dilation of local arterioles and capillaries, from which plasma escapes. Edema fluid accumulates in the area of injury, and fibrin forms a network and occludes the lymphatic channels, limiting the spread of organisms. A second effect of the mediators is to induce changes in expression of various adhesion molecules on endothelial cells and on leukocytes. Adhesion molecules such as the selectins and integrins cause leukocytes to attach to the endothelial cells of the blood vessels and thereby promote their movement across the vessel wall. Thus, polymorphonuclear leukocytes in the capillaries stick to the walls and then migrate out (extravasation) of the capillaries toward the irritant. This migration (chemotaxis) is stimulated by substances in the inflammatory exudate, including some small polypeptides called **chemokines**. Chemokines are synthesized by macrophages and by endothelial cells. IL-8 is an example of a chemokine (see later discussion and Table 8–3). These compounds function mainly to recruit monocytes and neutrophils from the blood into sites of infection. Phagocytes engulf the microorganisms, and intracellular digestion begins. Soon the pH of the inflamed area becomes more acid, and cellular proteases induce lysis of the leukocytes. Large mononuclear macrophages arrive on the site and, in turn, engulf leukocytic debris as well as microorganisms and pave the way for resolution of the local inflammatory process.

Cytokines and derivatives of arachidonic acid, including prostaglandins and leukotrienes, are mediators of the inflammatory response. Drugs that inhibit synthesis of prostaglandins (by blocking the enzyme cyclooxygenase) act as antiinflammatory agents.

E. Fever

Fever is the most common systemic manifestation of the inflammatory response and a cardinal symptom of infectious disease.

The ultimate regulator of body temperature is the thermoregulatory center in the hypothalamus. Among the substances capable of inducing fever (pyrogens) are endotoxins of gram-negative bacteria and cytokines released from lymphoid cells, such as IL-1.

Various activators can act upon mononuclear phagocytes and other cells and induce them to release IL-1. Among these activators are microbes and their products; toxins, including endotoxins; antigen–antibody complexes; inflammatory processes; and many others. IL-1 is carried by the bloodstream to the thermoregulatory center in the hypothalamus, where physiologic responses are initiated that result in fever (eg, increased heat production, reduced heat loss). Other effects of IL-1 are mentioned below.

Cytokines are small soluble proteins that are produced by one cell and influence other cells. These molecules have a variety of properties—eg, IL-1 promotes lymphocyte proliferation in addition to inducing fever and interleukin-2, produced by T cells, causes T cell proliferation and has numerous other immunomodulating functions. These molecules are described further, later in this chapter.

F. Interferons

Viral infection induces the expression of antiviral proteins known as **interferons**. These proteins, called interferon-α (IFN-α) and interferon-β (IFN-β), are distinct from the interferon-γ (IFN-γ) produced by activated T lymphocytes. The alpha and beta interferons help control viral replication by inhibiting protein synthesis in cells.

G. Natural Killer (NK) Cells

Natural killer cells represent a distinct functional population of lymphocytes. They play a role in antibody-dependent cellular cytotoxicity (ADCC) and have a role in the early phases of infection with certain viruses such as herpesviruses and some intracellular bacteria. They resemble large, granular lymphocytes morphologically related to T cells. They do not express antigen-specific receptors. They do have two types of surface receptors: (1) lectin-like NK-cell receptors that bind proteins not carbohydrates and (2) killer immunoglobulin-like receptors (KIRs) which recognize HLA-B or HLA-C molecules. These NK-cell receptors have both activation and inhibition properties. They can lyse target cells that have undergone malignant transformation and may play a role in immune surveillance against tumor establishment. They can kill certain virus-infected cells with altered levels of MHC class I molecules. The lytic activity of NK cells is enhanced by high levels of alpha and beta interferons.

MECHANISMS OF SPECIFIC HOST DEFENSE

Adaptive Response

The adaptive response can be antibody-mediated (humoral), cell-mediated (cellular), or both. An encounter with a microbial or viral agent usually elicits a complex variety of responses. An overview of these is given here, and details are presented later in this chapter.

Upon entry of a potential pathogen into the host and after interaction with the nonadaptive defense system just described, it or its major antigens are taken up by antigen-presenting cells (APCs), eg, macrophages, dendritic cells, etc. These nonself antigens reappear on the APC surface complexed with proteins encoded by the major histocompatibility complex (MHC) and are presented to clones of T lymphocytes. The MHC-antigen complexes are recognized by specific receptors on the surface of T cells, and these cells then produce a variety of cytokines that induce lymphocyte proliferation. The two arms of the immune response—cell-mediated and antibody-mediated—develop concurrently.

In the **antibody-mediated** arm, helper (CD4) T lymphocytes recognize the pathogen's antigens complexed with class II MHC proteins on the surface of an antigen-presenting cell (macrophage or B cell) and produce cytokines that activate B cells expressing antibodies that specifically match the antigen. The B cells undergo clonal proliferation and differentiate to form plasma cells, which then produce specific immunoglobulins (antibodies). Major host defense functions of antibodies include neutralization of toxins and viruses and opsonization (coating) of the pathogen, which aids its uptake by phagocytic cells. Antibody-mediated defense is important against pathogens that produce toxins (eg, *Clostridium tetani*) or have polysaccharide capsules that interfere with phagocytosis (eg, the pneumococci). It applies mainly to extracellular pathogens and their toxins.

In the **cell-mediated** arm, the antigen-MHC class II complex is recognized by helper (CD4) T lymphocytes, while the antigen-MHC class I complex is recognized by cytotoxic (CD8) T lymphocytes. Each class of T cells produces cytokines, becomes activated, and expands by clonal proliferation.

Helper T cell activity, in addition to stimulating B cells to produce antibodies, promotes the development of delayed hypersensitivity and thereby also serves in the defense against intracellular agents, including intracellular bacteria (eg, mycobacteria), fungi, protozoa, and viruses. **Cytotoxic** T cell activity is aimed mainly at the destruction of cells in tissue grafts, tumor cells, or cells infected by some viruses. Thus, T cells are mainly utilized to activate B cell responses and to cope with intracellular pathogens.

Figure 8–1 summarizes the adaptive and innate host defense mechanisms used to combat microorganisms. The net result of effective immunity is the host's resistance to microbial and other pathogens and foreign cells. By contrast, impaired immunity manifests itself as excessive susceptibility to such pathogens or tumors. Specific examples are presented below.

Antigens

The features of antigens that largely determine immunogenicity in the immune response are as follows.

A. Foreignness (Difference From "Self")

In general, molecules recognized as "self" are not immunogenic; for immunogenicity, molecules must be recognized as "nonself."

B. Molecular Size

The most potent immunogens are usually large proteins. Generally, molecules with a molecular weight less than 10,000 are weakly immunogenic, and very small ones (eg, amino acids) are nonimmunogenic. Certain small molecules (eg, haptens) become immunogenic only when linked to a carrier protein.

C. Chemical and Structural Complexity

A certain amount of chemical complexity is required—eg, amino acid homopolymers are less immunogenic than heteropolymers containing two or three different amino acids.

D. Genetic Constitution of the Host

Two strains of the same species of animal may respond differently to the same antigen because of a different composition of genes involved in the immune response, eg, different MHC alleles.

E. Dosage, Route, and Timing of Antigen Administration

Since the degree of the immune response depends on the amount of antigen given, the immune response can be optimized by carefully defining the dosage (including number of doses), route of administration, and timing of administration (including intervals between doses).

It is possible to enhance the immunogenicity of a substance by mixing it with an **adjuvant**. Adjuvants are substances that stimulate the immune response—eg, by facilitating uptake into antigen-presenting cells.

Cellular Basis of the Immune Response

During embryonic development, blood cell precursors (hematopoietic stem cells) are found in fetal liver and other tissues; in postnatal life, the stem cells reside in bone marrow. They can differentiate in several ways. Stem cells may differentiate into cells of the myeloid series or into cells of the lymphoid series. Lymphoid progenitor cells evolve into two main lymphocyte populations, B cells and T cells. NK cells are also derived from the lymphoid progenitor.

A. B Cells

B cells are lymphocytes that develop in the bone marrow in mammals. In birds they develop in the bursa of Fabricius, a gut appendage. They rearrange their immunoglobulin genes and express a unique receptor for antigen on their cell surface. At this point, they migrate to a secondary lymphoid organ—eg, spleen—and may be activated by an encounter with antigen to become antibody-secreting plasma cells.

B. T Cells

T cells are lymphocytes that require maturation in the thymus and form several subclasses with specific functions. They are the source of cell-mediated immunity, discussed below.

Some lymphocytic cells (eg, natural killer cells; see above) lack features of B or T cells but have significant immunologic roles. Figure 8–2 presents an overview of immunologically active lymphocytes and their interactions.

ANTIGEN RECOGNITION MOLECULES

In order for the immune system to respond to nonself, ie, foreign antigen, a recognition system capable of precisely distinguishing self from nonself had to evolve. The next section of this chapter deals with the molecules used to recognize foreign antigens. First, we shall review the structure and function of **antibodies**, the soluble recognition products of B lymphocytes. Then we shall review some membrane-bound receptors for antigen, the B cell receptor for antigen, the T cell receptor for antigen, and the products of the **major histocompatibility complex (MHC)**.

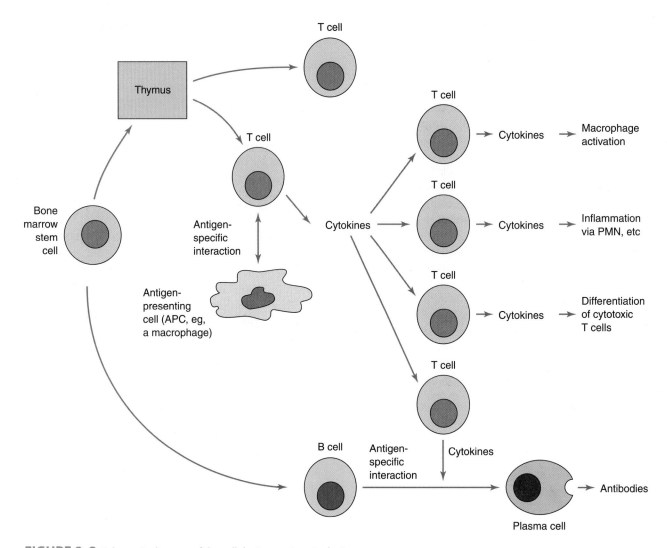

FIGURE 8–2 Schematic diagram of the cellular interactions in the immune response.

ANTIBODIES

Antibodies **(immunoglobulins)** are formed by B lymphocytes. Each individual has a large pool of different B lymphocytes (about 10^{11}) that have a life span of days or weeks and are found in the bone marrow, lymph nodes, and gut-associated lymphoid tissues (eg, tonsils or appendix).

B cells display immunoglobulin molecules (10^5/cell) on their surface. These immunoglobulins serve as receptors for a specific antigen, so that each B cell can respond to only one antigen or a closely related group of antigens. All immature B cells carry IgM immunoglobulins on their surface, and most also carry IgD. B cells also have surface receptors for the Fc portion of immunoglobulins and for several complement components.

An antigen interacts with the B lymphocyte that shows the best "fit" by virtue of its immunoglobulin surface receptor. The antigen binds to this receptor, and the B cell is stimulated to divide and form a clone **(clonal selection)**. Such selected B cells soon become plasma cells and secrete antibody. Since

each person can make about 10^{11} different antibody molecules, there is an antigen-binding site on a B cell to fit almost any antigenic determinant.

The initial step in antibody formation is phagocytosis of the antigen, usually by antigen-presenting cells (chiefly macrophages or B cells) that process and present the antigen to T cells. These activated T cells then interact with B cells. B cells that carry the surface immunoglobulin which best fits the antigen are stimulated to proliferate and differentiate into plasma cells (Figure 8–2), which form the specific antibody proteins or differentiate into long-lived memory cells. The plasma cells synthesize an immunoglobulin of the same specificity as that carried by the B precursor cells.

Antibody Structure & Function

Antibodies are immunoglobulins which react specifically with the antigen that stimulated their production. They make up about 20% of plasma proteins.

Antibodies that arise in an animal in response to a single complex antigen are heterogeneous because they are formed by several different clones of cells, each expressing an antibody capable of reacting with a different antigenic determinant on the complex antigen. These antibodies are said to be **polyclonal**. Antibodies that arise from a single clone of cells, eg, in a plasma cell tumor (myeloma), are homogeneous and are referred to as **monoclonal**. Monoclonal antibodies can be produced by fusing a myeloma cell with an antibody-producing lymphocyte. Such **hybridomas** produce virtually unlimited quantities of monoclonal antibodies in vitro. Important information about the structure and function of antibodies has been derived from the study of monoclonal antibodies.

All immunoglobulin molecules are made up of light and heavy polypeptide chains. The terms light and heavy refer to molecular weight—ie, light chains have a molecular weight of approximately 25,000, whereas heavy chains have a molecular weight of approximately 50,000. **Light (L) chains** are of one of two types, κ (kappa) or λ (lambda); classification is made based on amino acid differences in their constant regions (see Figure 8–3). Both types occur in all classes of immunoglobulins (IgG, IgM, IgA, IgE, and IgD), but any one immunoglobulin molecule contains only one type of L chain. The amino terminal portion of each L chain contains part of the antigen-binding site. **Heavy (H) chains** are distinct for each of the five immunoglobulin classes and are designated γ (gamma), μ (mu), α (alpha), δ (delta), and ε (epsilon) (Table 8–1). The amino terminal portion of each H chain participates in the antigen-binding site; the other (carboxyl) terminal forms the Fc fragment (see Figure 8–3), which has various biologic activities (eg, complement activation and binding to cell surface receptors).

An individual antibody molecule always consists of identical H chains and identical L chains. The simplest antibody molecule has a Y shape (Figure 8–3) and consists of four polypeptide chains: two H chains and two L chains. The four chains are covalently linked by disulfide bonds.

If such an antibody molecule is treated with a proteolytic enzyme (eg, papain), peptide bonds in the **hinge region** are broken. This breakage produces two identical Fab fragments, which carry the antigen-binding sites, and one Fc fragment, which is involved in placental transfer, complement fixation, attachment for various cells, and other biologic activities.

L and H chains are subdivided into **variable regions** and **constant regions**. The regions are composed of three-dimensionally folded, repeating segments called domains. The structure of these domains has been determined at high resolution by x-ray crystallography. An L chain consists of one variable domain (V_L) and one constant domain (C_L). Most H chains consist of one variable domain (V_H) and three or more constant domains (C_H). Each domain is approximately 110 amino acids long. Variable regions are responsible for antigen binding; constant regions are responsible for the biologic functions described below.

Within the variable regions of both L and H chains are subregions consisting of extremely variable (**hypervariable**) amino acid sequences that form the antigen-binding site. The

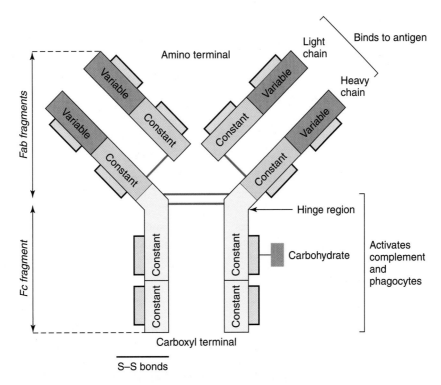

FIGURE 8–3 Schematic representation of an IgG molecule, indicating the location of the constant and the variable regions on the light and heavy chains. Fab fragment is fragment antigen binding, Fc fragment is fragment crystallizable.

TABLE 8–1   Properties of Human Immunoglobulins

	IgG	IgA	IgM	IgE	IgD
Heavy chain symbol	γ	α	μ	ε	δ
Molecular weight (×1000)	150	170–600[a]	900	190	150
Serum concentration (mg/mL)	7–18	0.8–4	0.4–2.5	< 0.0005	< 0.003
Serum half-life (days)	21	7	7	2	2
Activates complement	Yes (+)	No	Yes (++)	No	No
Percentage of total immunoglobulins in serum	80	13	6	< 1	< 1

[a]In secretions, eg, saliva, milk, and tears and in respiratory, intestinal, and genital tract secretions IgA is generally found as a dimer or a tetramer but in serum IgA exists primarily as a monomer.

hypervariable regions form the area of the antibody molecule complementary in structure to the antigenic determinant or epitope and are therefore also known as complementarity-determining regions (CDRs). Only five to ten amino acids in each hypervariable region constitute the antigen-binding site. Antigen binding is noncovalent, involving van der Waals, electrostatic, and other weak forces as well as hydrogen and other bonds.

Small molecules such as haptens bind to antibodies in a cleft formed by the heavy and light chain variable domains. The interaction of an antibody with a large native protein (eg, a viral protein) occurs, by contrast, with a conformational or discontinuous epitope that represents a surface area of the protein antigen. Most or all of the CDRs of the antibody molecule are involved in this binding.

Immunoglobulin Classes

A. IgG

Each IgG molecule consists of two L chains and two H chains linked by disulfide bonds (molecular formula H_2L_2). Because it has two identical antigen-binding sites, it is said to be divalent. There are four subclasses (IgG1 to IgG4), based on amino acid sequence differences in the H chains and on the number and location of disulfide bonds. IgG1 is 65% of the total IgG. IgG2 is directed against polysaccharide antigens and may be an important host defense against encapsulated bacteria.

IgG is the predominant antibody in secondary responses and constitutes an important defense against bacteria and viruses. It is the only antibody to pass the placenta and is therefore the most abundant immunoglobulin in newborns.

B. IgM

IgM is the main immunoglobulin produced early in the *primary* immune response. IgM is present on the surface of virtually all uncommitted B cells. It is composed of five H_2L_2 units (each similar to one IgG unit) and one molecule of J (joining) chain (Figure 8–4). The pentamer (MW 900,000) has a total of ten identical antigen-binding sites and thus

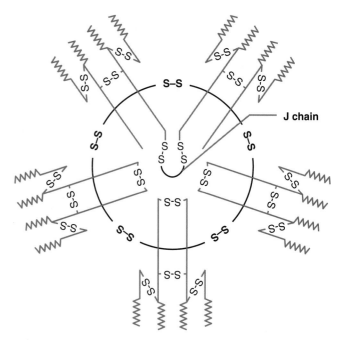

FIGURE 8–4 Schematic diagram of the pentameric structure of human IgM. The IgM monomers are connected to each other and the J chain by disulphide bonds.

a valence of 10. It is the most efficient immunoglobulin in agglutination, complement fixation, and other antigen–antibody reactions and is important also in defense against bacteria and viruses. It can be produced by a fetus undergoing an infection. Since its interaction with antigen can involve all ten binding sites, it has the highest avidity of all immunoglobulins.

C. IgA

IgA is the main immunoglobulin in secretions such as milk, saliva, and tears and in secretions of the respiratory, intestinal, and genital tracts. It protects mucous membranes from attack by bacteria and viruses.

Each secretory IgA molecule (MW 400,000) consists of two H_2L_2 units and one molecule each of J chain and secretory component. The latter is a protein derived from cleavage of the poly-Ig receptor. This receptor binds IgA dimers and facilitates their transport across mucosal epithelial cells. Some IgA exists in serum as a monomer H_2L_2 (MW 170,000). There are at least two subclasses, IgA1 and IgA2. Some bacteria (eg, neisseriae) can destroy IgA1 by producing a protease and can thus overcome antibody-mediated resistance on mucosal surfaces.

D. IgE

The Fc region of IgE binds to a receptor on the surface of mast cells, basophils, and eosinophils. This bound IgE acts as a receptor for the antigen that stimulated its production, and the resulting antigen–antibody complex triggers allergic responses of the immediate (anaphylactic) type through the release of mediators. In persons with such antibody-mediated allergic hypersensitivity, IgE concentration is greatly increased, and IgE may appear in external secretions. Serum IgE is also typically increased during helminth infections.

E. IgD

IgD acts as an antigen receptor when present on the surface of certain B lymphocytes. In serum it is present only in trace amounts.

Immunoglobulin Genes & Generation of Diversity

Special genetic mechanisms have evolved to produce the very large number of immunoglobulin molecules (about 10^{11}) that develop in the host in response to antigenic stimulation without requiring excessive numbers of genes. Thus, immunoglobulin genes (and, as we shall see later, T cell receptor genes) undergo somatic recombination to produce the enormous diversity of antibody specificities.

Each immunoglobulin chain consists of a variable (V) and a constant (C) region. For each type of immunoglobulin chain—ie, kappa light chain (κ), lambda light chain (λ), and the five heavy chains (γH, μH, αH, εH, and δH)—there is a separate pool of gene segments located on different chromosomes. In humans the multigene families are found on the following chromosomes: λ, chromosome 22; κ, chromosome 2; and the heavy chain family, chromosome 14. Each of the three gene loci contains a set of different V gene segments widely separated from C gene segments. During B cell differentiation, the DNA is rearranged to bring the selected gene segments adjacent to each other in the genome. A family of enzymes known as the V(D)J recombinases are responsible for this gene rearrangement process.

The variable region of each L chain is encoded by two gene segments: V and J. The variable region of each H chain is encoded by three gene segments: V, D, and J. The segments are united into one functional V-variable gene by DNA rearrangement. Each assembled V-variable gene is then transcribed with the appropriate C-constant gene to produce a messenger RNA (mRNA) that encodes for the complete peptide chain. L and H chains are synthesized separately on polysomes and finally assembled in the cytoplasm to form H_2L_2 units by means of disulfide bonds. The carbohydrate moiety is then added during progress through the membrane components of the cell (eg, Golgi apparatus), and the immunoglobulin molecule is released from the cell.

This gene rearrangement mechanism permits the assembly of an enormous variety of immunoglobulin molecules. Antibody diversity depends on (1) multiple V, D, and J gene segments; (2) combinatorial association, ie, the association of any V gene segment with any D or J segment; (3) the random combining of different L and H chains; (4) somatic hypermutation; and (5) junctional diversity, created by imprecise joining during rearrangement with the addition of nucleotides by the enzyme terminal deoxynucleotidyl transferase to form a complete joint.

Immunoglobulin Class Switching

Initially, all B cells matched to an antigen carry IgM specific for that antigen and produce IgM in response to this exposure to antigen. Later, gene rearrangement permits elaboration of antibodies of the same antigenic specificity but of different immunoglobulin classes. In **class switching**, the same assembled V_H gene can sequentially associate with different C_H genes, so that the immunoglobulin produced later (IgG, IgA, or IgE) has the same specificity as the original IgM but different biologic characteristics. Class switching is dependent on cytokines released from T cells and also happens after antigenic stimulation.

CELL SURFACE RECEPTORS FOR ANTIGEN

B Cell Receptor for Antigen

B cells express a form of IgM that is located on the cell surface. Cell surface IgM has the same antigen specificity as the secreted IgM antibody molecule. This is achieved by a differential RNA splicing mechanism. The μ-chain RNA transcript can include a sequence that encodes about 25 hydrophobic amino acids, which enables the IgM molecule to localize in the cell membrane as a transmembrane receptor. Later in development of the B cell, regulation of RNA processing allows expression of a membrane-bound form of IgD, again with the same antigen-binding specificity. Throughout this process, the same V region segment is being expressed with different C region segments.

As a membrane-bound receptor, IgM or IgD interacts with other cell surface molecules, known as Igα and Igβ, that can transduce signals subsequent to antigen binding by interacting with tyrosine kinase molecules, and the other components of the signal transduction machinery. These signals result in biochemical events involving intracellular phosphatases, kinases, GTP-binding proteins, lipid mediators, calcium ions, and other intermediates, eventually leading to cell activation.

T Cell Receptor for Antigen

The T cell receptor is a transmembrane heterodimeric protein composed of two disulfide-linked chains. This receptor resembles a membrane-bound Fab fragment of immunoglobulin. There are two different classes of T cell receptor. The two chains are known as α and β in one class and as γ and δ in the other. γδ-Expressing T cells are relatively infrequent in humans and seem to be predisposed toward recognition of frequently encountered bacterial antigens—eg, certain glycolipid and phosphorylated lipid moieties. αβ T cells make up the predominant T cell phenotype and are subdivided by their expression of other cell surface markers, the proteins known as CD4 and CD8, into helper and cytotoxic functional classes, respectively.

The T cell receptor proteins have variable and constant regions similar to antibodies. The variable regions are located at the amino terminals of the polypeptide chain farthest away from the cell membrane. Both chains contribute to the variable domain that has been shown to interact with antigen presented by self proteins encoded in the major histocompatibility complex (MHC).

The T cell receptor genes closely resemble immunoglobulin genes, and the generation of diversity in the T cell receptor is accomplished in a fashion largely analogous to that described earlier for immunoglobulins. Thus, there are multiple variable region segments, contributing a repertoire of different antigen specificities; multiple V, D, and J segments that can combine in different ways just as for antibodies; and random combination of a large number of α and β chains. There are two differences from the situation described earlier for antibodies: (1) No evidence for somatic mutation in T cell receptors has been obtained, and (2) the potential for increasing the repertoire of potential antigen specificities by junctional diversity is much greater for T cell receptors than for antibodies. There are more J and D segments for T cell receptor genes than for immunoglobulin genes. In essence, however, the encoding of T cell receptors is very much like that described for immunoglobulins. For example, the variable regions of the α and γ chains of the T cell receptor are like the variable regions of immunoglobulin light chains in having V and J segments, whereas the β and δ chains are like immunoglobulin heavy chains in being encoded by V, D, and J segments.

In all functional antigen-specific T cells, the two T cell receptor chains are noncovalently associated with six other polypeptide chains composed of four different proteins that make up the CD3 complex. The invariant proteins of the CD3 complex are responsible for transducing the signal received by the T cell receptor on recognition of antigen to the inside of the cell. All four different proteins of the CD3 complex are transmembrane proteins that can interact with cytosolic tyrosine kinases on the inside of the membrane. It is this interaction that begins the biochemical events of signal transduction leading to gene transcription, cell activation, and initiation of the functional activities of T cells.

The CD4 and CD8 molecules that differentiate the two major functional classes of T cell function as co-receptor molecules on the T cell surface. During recognition of antigen, the CD4 and CD8 molecules interact with the T cell receptor complex and with MHC molecules. CD4 binds to MHC class II molecules, and CD8 binds to MHC class I molecules. This greatly increases the sensitivity of antigen recognition by T cells.

The Major Histocompatibility Complex

The major histocompatibility complex (MHC) was first detected as the genetic locus encoding the glycoprotein molecules (transplantation antigens) responsible for the rapid rejection of tissue grafts transplanted between genetically nonidentical individuals. It is now known that MHC molecules bind peptide antigens and present them to T cells. Thus, these transplantation antigens are responsible for antigen recognition by the T cell receptor. In this respect, the T cell receptor is different from antibody. Antibody molecules interact with antigen directly; the T cell receptor only recognizes antigen presented by MHC molecules on another cell, the antigen-presenting cell. The T cell receptor is specific for antigen, but the antigen must be presented on a self MHC molecule. The T cell receptor is also specific for the MHC molecule. If the antigen is presented by another allelic form of the MHC molecule in vitro (normally only in an experimental situation), there is no recognition by the T cell receptor. This phenomenon is known as MHC restriction.

In humans, the MHC is a cluster of extensively studied genes located on chromosome 6. Among the many important genes in the human MHC, also known as HLA (human leukocyte antigens), are those that encode the class I, class II, and class III MHC proteins. As outlined in Table 8–2, class I proteins are encoded by the HLA-A, -B, and -C genes. These proteins are made up of two chains: (1) a transmembrane glycoprotein of MW 45,000, noncovalently associated with (2) a non-MHC-encoded polypeptide of MW 12,000 that is known as β_2-microglobulin. Class I molecules are to be found on virtually all nucleated cells in the body.

Class II proteins are encoded by the HLA-D region. As shown in Table 8–2, there are three main sets: the DP-, DQ-, and DR-encoded molecules. This locus retains control of immune responsiveness, and different allelic forms of these genes confer striking differences in the ability to mount an immune response against a given antigen.

The HLA-D locus-encoded proteins are made up of two noncovalently associated transmembrane glycoproteins of about MW 33,000 and MW 29,000. Unlike class I proteins, they have a restricted tissue distribution and are chiefly found on macrophages, B cells, and other antigen-presenting cells. Their expression on other cells—eg, endothelial cells—can be induced by interferon-gamma.

TABLE 8-2 **Important Features of Some Human MHC Gene Products**

	Class I	Class II
Genetic loci (partial list)	HLA-A, -B, and -C	HLA-DP, -DQ, and -DR
Polypeptide composition	MW 45,000 + β$_2$M (MW 12,000)	α chain (MW 33,000), β chain (MW 29,000), Ii chain (MW 30,000)
Cell distribution	All nucleated somatic cells	Antigen-presenting cells (macrophages, B cells, etc), activated human T cells
Present peptide antigens to	CD8 T cells	CD4 T cells
Size of peptide bound	8–11 residues	10–30 or more residues

The class II MHC locus also includes genes encoding proteins involved in antigen processing, eg, TAP (see Figure 8–7). The class III MHC locus encodes complement proteins and several cytokines.

The genes of the MHC exhibit a remarkable genetic variability. The MHC is **polygenic** in that there are several genes for each class of molecule. The MHC is also **polymorphic**. Thus, a large number of alleles exist in the population for each of the genes. Each individual inherits a restricted set of alleles from its parents. Sets of MHC genes tend to be inherited as a block or **haplotype**, as there are relatively infrequent crossover events at this locus.

Much is known about the structural organization and sequence of MHC genes and proteins. Perhaps the most important information, however, has come from the x-ray analysis of crystals of MHC proteins. It was these studies that helped to clearly explain the function of the MHC proteins. The x-ray analysis (Figure 8–5) shows that the domains of the class I MHC molecule farthest away from the membrane are composed of two parallel α helices above a platform created by a β-pleated sheet. The whole structure undoubtedly looks like a **cleft** whose sides are formed by the α helices and floored by the β sheets. The x-ray analysis also showed that the cleft was occupied by a peptide. In essence, then, the T cell receptor sees the peptide antigen bound in a cleft provided by the MHC protein. A simplified diagram of this interaction is provided in Figure 8–6A.

MHC proteins show a broad specificity for peptide antigens, and many different peptides can be presented by any given MHC allele (one peptide is bound at a time). The α helices that form the binding cleft are the site of the amino acid residues that are polymorphic in MHC proteins (ie, those that vary between alleles). This means that different alleles can bind and present different peptide antigens. For all these reasons, MHC polymorphism has a major effect on antigen recognition.

Analysis of the function of T cells with respect to interaction with MHC molecules reveals that peptide antigens associated with class I MHC molecules are recognized by CD8-positive cytotoxic T lymphocytes, whereas class II-associated peptide antigens are recognized by CD4-positive helper T cells.

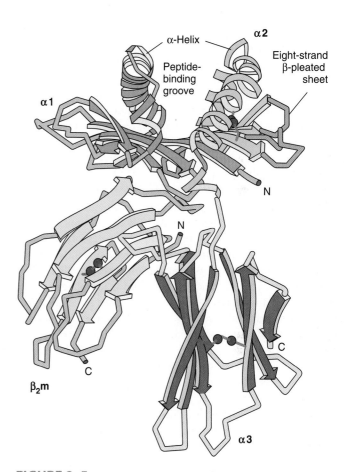

FIGURE 8–5 Diagrammatic structure of a class I HLA molecule. (Reproduced with permission from Bjorkman PJ et al: Structure of the human class I histocompatibility antigen, HLA-A2. Nature 1987;329:506.)

The Immunoglobulin Supergene Family

All the molecules discussed—antibodies, the T cell receptor, and MHC proteins—have structural features in common. All of these molecules—and a long list of other immunologically relevant molecules, including the T cell subpopulation markers CD4 and CD8—have a domain structure built on the three-dimensional feature known as the **immunoglobulin fold**. Undoubtedly, the members of this family evolved in

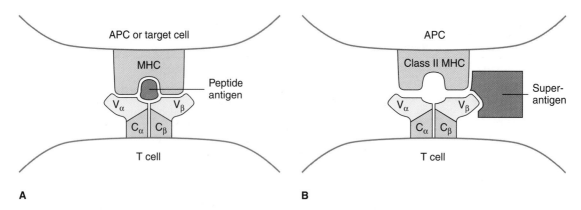

FIGURE 8–6 Binding of antigen by MHC and T cell receptor. In **panel A**, a model of the interaction between peptide antigen, MHC, and the T cell receptor is shown. The V_α and V_β regions of the TCR are shown interacting with the α helices that form the peptide binding groove of MHC. In **panel B**, a model of the interaction between a superantigen, MHC, and the T cell receptor is shown. The superantigen interacts with the V_β region of the TCR and with class II MHC outside the peptide binding groove. (Adapted from Stites DG et al [editors]: *Medical Immunology*, 9th ed. McGraw-Hill, 1997.)

such a way as to supply a common function to the organism. One part of this function is to act as a recognition unit or receptor at the cell surface.

Antigen Processing & Presentation

Antigen processing and presentation are the means by which antigens become associated with self MHC molecules for presentation to T cells with appropriate receptors. Proteins from exogenous antigens, such as bacteria, are internalized via endocytic vesicles into antigen-presenting cells such as the various types of dendritic cells and macrophages. Then, as illustrated in Figure 8–7, they are exposed to cellular proteases in intracellular vesicles. Peptides, approximately 10 to 30 amino acid residues in length, are generated in endosomal vesicles. The endosomal vesicles can then fuse with exocytic vesicles containing class II MHC molecules.

The class II MHC molecules are synthesized, as for other membrane glycoproteins, in the rough endoplasmic reticulum and then proceed out through the Golgi apparatus. A third polypeptide, **the invariant chain (Ii)**, protects the binding site of the class II αβ dimer until the lowered pH of the compartment created after fusion with an endosomal vesicle causes a dissociation of the Ii chain. The MHC class II-peptide antigen complex is then transported to the cell surface for display and recognition by a T cell receptor of a CD4 T cell.

Endogenous antigens—eg, cytosolic viral proteins synthesized in an infected cell—are processed for presentation by class I MHC molecules. Some of the steps involved are diagrammed in Figure 8–7. In brief, cytosolic proteins are broken down by a proteolytic complex known as the **proteasome**. The cytosolic peptides gain access to nascent MHC class I molecules in the rough endoplasmic reticulum via peptide transporter systems (transporters associated with antigen processing; TAPs). The TAP genes are also encoded in the MHC. Within the lumen of the endoplasmic reticulum, peptide antigens approximately 8 to 11 residues in length complex with nascent

MHC class I proteins and cooperate with β_2-microglobulin to create a stable, fully folded MHC class I-peptide antigen complex that is then transported to the cell surface for display and recognition by CD8 cytotoxic T cells.

The binding groove of the class I molecule is more constrained than that of the class II molecule, and for that reason shorter peptides are found in class I than in class II MHC molecules.

Understanding the details of antigen processing has clarified our thinking about T cell function. Thus, it is now understood why T cells do not respond to carbohydrate antigens (they would not fit in the groove) and why T cells recognize only linear antigenic determinants (they respond only to proteolytically processed antigen). Whether an antigen is destined for class I or class II presentation depends only on the intracellular compartments it traverses.

Several viruses attempt to defeat the immune response by interfering with the antigen-processing pathways. For example, an HIV Tat protein is able to inhibit expression of class I MHC molecules. A herpesvirus protein binds to the transporter proteins (TAPs), preventing transport of viral peptides into the endoplasmic reticulum, where class I molecules are being synthesized. A consequence of these inhibitory mechanisms is that these viruses can evade the immune response because the cells they infect are not recognized by effector lymphocytes.

Some **superantigens** are able to bind to MHC molecules outside the peptide-binding cleft. One consequence is that whereas an individual peptide complexed to an MHC molecule will normally stimulate only a small percentage of the T cells in an individual, superantigens cause up to 10% of T cells to be nonspecifically activated. Examples of superantigens include certain bacterial toxins, including the staphylococcal enterotoxins, toxic shock syndrome toxin, and group A streptococcal pyrogenic exotoxin A. These antigens bind to the "outside" of the MHC protein and to the T cell receptor (Figure 8–6B). They are active at very low

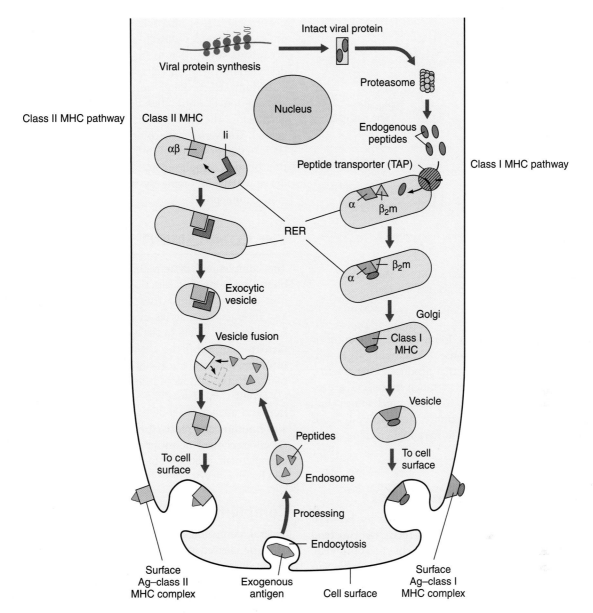

FIGURE 8–7 Antigen-processing pathways. (Modified and reproduced with permission from Parslow TG et al [editors]: *Medical Immunology,* 10th ed. McGraw-Hill, 2001.)

concentrations (10^{-9} mol/L) and cause T cells expressing particular Vβ sequences to be stimulated and to release large amounts of cytokines, including IL-1 and tumor necrosis factor (TNF). It is the release of large amounts of cytokines from stimulation of a high percentage of the pool of T lymphocytes that explains to a large extent the pathogenesis of diseases caused by organisms expressing superantigens.

ANTIBODY-MEDIATED (HUMORAL) IMMUNITY

The Primary Response

When an individual encounters an antigen for the first time, antibody to that antigen is detectable in the serum within days or weeks depending on the nature and dose of the antigen and the route of administration (eg, oral, parenteral). The serum antibody concentration continues to rise for several weeks and then declines; it may drop to very low levels (Figure 8–8). The first antibodies formed are IgM, followed by IgG, IgA, or both. IgM levels tend to decline sooner than IgG levels.

The Secondary Response

In the event of a second encounter with the same antigen (or a closely related "cross-reacting" one) months or years after the primary response, the antibody response is more rapid and rises to higher levels than during the primary response. This change in response is attributed to the persistence of

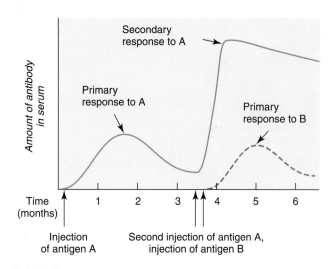

FIGURE 8–8 Rate of antibody production following initial antigen administration and a second "booster" injection.

antigen-sensitive "memory cells" following the first immune response. In the secondary response, the amount of IgM produced is qualitatively similar to that produced after the first contact with the antigen; however, much more IgG is produced, and the level of IgG tends to persist much longer than in the primary response. Furthermore, such antibody tends to bind antigen more firmly (ie, to have higher affinity) and thus to dissociate less easily.

Protective Functions of Antibodies

Because of the close structural complementarity between antibodies and the antigen that elicited them, the two tend to bind to each other whenever they meet, in vitro or in vivo. This binding is noncovalent and involves electrostatic, van der Waals, and other weak forces as well as hydrogen and other bonds. Antibodies can produce resistance to infection by opsonizing (coating) organisms, which makes them more readily ingested by phagocytes; antibodies can bind to viruses and reduce their ability to bind to cellular receptor molecules and invade host cells; and most importantly, antibodies can neutralize toxins of microorganisms (eg, diphtheria, tetanus, and botulism) and inactivate their harmful effects.

Antibodies can be induced actively in the host by administering appropriate antigens or preparations containing them (toxoids of diphtheria, tetanus), but protection is delayed until the antibodies reach helpful concentrations. In contrast, antibodies can be administered passively (ie, preformed in another host), which makes them immediately available for preventive or therapeutic purposes. The latter approach (passive immunization) has been used in the management of "needlestick" injuries in individuals not vaccinated against hepatitis B.

Antibody-mediated immunity against bacteria is most effective when directed against microbial infections in which virulence is related to polysaccharide capsules (eg,

pneumococcus, haemophilus, neisseria). In such infections, antibodies complex with the capsular antigens and make the organisms susceptible to ingestion by phagocytic cells and destruction within the cells.

Many cell-mediated immune responses also require the cooperation of antibodies directed against offending antigens before the latter can be inactivated or eliminated (see below). Conversely, the binding of antibodies to antigens leads to the formation of immune complexes, and the deposition of such complexes may be an important feature in the development of organ dysfunction, eg, poststreptococcal glomerulonephritis.

THE COMPLEMENT SYSTEM

The complement system includes serum and membrane-bound proteins that function in both adaptive and innate host defense systems. These proteins are highly regulated and interact via a series of proteolytic cascades. The term "complement" refers to the ability of these proteins to complement (augment) the effects of other components of the immune system (eg, antibody). Complement has several main effects: (1) lysis of cells (eg, bacteria and tumor cells), (2) production of mediators that participate in inflammation and attract phagocytes, (3) opsonization of organisms and immune complexes for clearance by phagocytosis, and (4) enhancement of antibody-mediated immune responses. Complement proteins are synthesized mainly by the liver and by phagocytic cells.

Complement Activation

Several complement components are proenzymes, which must be cleaved to form active enzymes. The components of the classic pathway are numbered from C1 to C9, and the reaction sequence is C1-C4-C2-C3-C5-C6-C7-C8-C9. Up to C5, activation involves proteolytic cleavage, liberating smaller fragments from C2 through C5. The smaller fragments are by convention denoted by the letter a (eg, C4a) and the larger fragments by b (eg, C5b). Activation of the complement system can be initiated either by antigen–antibody complexes or by a variety of nonimmunologic molecules.

Sequential activation of complement components (Figure 8–9) occurs via three main pathways.

A. The Classic Pathway

Only IgM and IgG activate or fix complement via the classic pathway. Of the IgGs, only IgG subclasses 1, 2, and 3 fix complement; IgG4 does not. C1, which is bound to a site in the Fc region, is composed of three proteins: C1q, C1r, and C1s. C1q is an aggregate of polypeptides that bind to the Fc portion of IgG and IgM. The antibody-antigen immune complex bound to C1 activates C1s, which cleaves C4 and C2 to form C4b2b. The latter is an active C3 convertase, which cleaves C3 molecules into two fragments: C3a and C3b. C3a, an anaphylatoxin, is discussed below. C3b forms a complex with C4b2b,

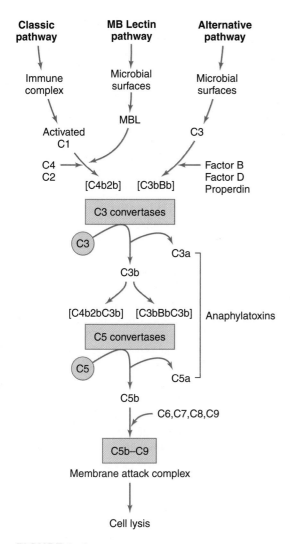

FIGURE 8–9 Complement reaction sequence.

producing a new enzyme, C5 convertase, which cleaves C5 to form C5a and C5b. C5a is an anaphylatoxin and a chemotactic factor (see below). C5b binds to C6 and C7 to form a complex that inserts into the membrane bilayer. C8 then binds to the C5b/C6/C7 complex, followed by the polymerization of up to sixteen C9 molecules to produce the membrane attack complex that generates a channel or pore in the membrane and causes cytolysis by allowing free passage of water across the cell membrane.

B. The Alternative Pathway

Many unrelated substances, from complex chemicals (eg, endotoxin) to infectious agents (eg, parasites), activate a different pathway. C3 is cleaved, and a C3 convertase is generated via the action of factors B, D, and properdin. The alternative C3 convertase (C3bBb) generates more C3b. The additional C3b binds to the C3 convertase to form C3bBbC3b, which is the alternative pathway C5 convertase that generates C5b, leading to production of the membrane attack complex described above.

C. Mannan-Binding Lectin Pathway

In recent years, the concept of an additional pathway of complement activation has emerged—the MB lectin pathway. Its main constituent is a plasma protein termed MBL, which is short for mannan-binding lectin. MBL binds to sugar residues like mannose found in microbial surface polysaccharides such as LPS. The MBL complex, when bound to microbial surfaces, can activate C4 and C2. The rest of this pathway is the same as the classic pathway of complement activation.

Regulation of the Complement System

Several serum proteins regulate the complement system at different stages: (1) C1 inhibitor binds to and inactivates the serine protease activity of C1r and C1s; (2) factor I cleaves C3b and C4b, thereby reducing the amount of C5 convertase available; (3) factor H enhances the effect of factor I on C3b; and (4) factor P (properdin) protects C3b and stabilizes the C3 convertase of the alternative pathway. Regulation is also provided by proteins that have the ability to accelerate the decay of the complement proteins—eg, decay-accelerating factor, a membrane-bound protein found on most blood cell surfaces that can act to accelerate dissociation of the C3 convertases of both pathways.

Major Biologic Effects of Complement

A. Opsonization

Cells, antigen–antibody complexes, and other particles are phagocytosed much more efficiently in the presence of C3b because of the presence of C3b receptors on the surface of many phagocytes.

B. Chemotaxis

C5a stimulates movement of neutrophils and monocytes toward sites of antigen deposition.

C. Anaphylatoxins

C3a and C5a can produce increased vascular permeability and smooth muscle contraction. C3a and C5a also stimulate mast cells to release histamine.

D. Cytolysis

Insertion of the C5b6789 complex into the cell surface leads to killing or lysis of many types of cells, including erythrocytes, bacteria, and tumor cells.

Clinical Consequences of Complement Deficiencies

Many genetic deficiencies of complement proteins have been described, and these generally lead to enhanced susceptibility to infectious disease—eg, C2 deficiency frequently leads to serious pyogenic bacterial infections.

Deficiency in components of the membrane attack complex greatly enhances susceptibility to neisserial infections. Deficiencies in components of the alternative pathway are also known—eg, properdin deficiency is associated with greater susceptibility to meningococcal disease. There are also deficiencies in complement regulating proteins. For example, lack of the C1 inhibitor protein leads to hereditary angioedema.

CELL-MEDIATED IMMUNITY

Antibody-mediated immunity is most important in toxin-induced disorders, in microbial infections in which polysaccharide capsules determine virulence, and as a part of the host defense response to some viral infections. However, in most microbial infections, it is cell-mediated immunity that imparts resistance and aids in recovery, though the cooperation of antibodies may be required. Furthermore, cell-mediated immunity is central in host defense against intracellular pathogens such as viruses and in combating tumor cells. The important role of cell-mediated immunity is underlined in clinical situations in which its suppression (eg, AIDS) results in overwhelming infections or tumors.

The cell-mediated immune system includes several cell types and their products. APCs present antigen to T lymphocytes via their cell surface-situated MHC proteins. T cell receptors recognize the antigen, and a specific T cell clone becomes activated and begins to proliferate. Because there are a number of T cell subpopulations and because their interactions (either directly or through the production of soluble cytokines) result in an intricate response system, selected aspects of the system are discussed separately below.

Development of T Cells

Within the thymus, T cell progenitor cells undergo differentiation (under the influence of thymic hormones) into T cell subpopulations. Much has been learned about this process in recent years, and the reader is referred to specialty texts for details. T cells differentiate in the thymus into committed cells expressing a specific T cell receptor and become positive for the expression of either the CD4 or CD8 coreceptor molecules. After differentiation in the thymus, T cells undergo positive and negative selection processes that result in the retention of only those cells with the most useful antigen receptors, ie, those that are nonself antigen-specific and self MHC-restricted. Those clones that are potentially antiself are either deleted or functionally inactivated (made *anergic*). A consequence of the selection processes is that about 95% of thymocytes die in the thymus. Only a minority of developing T cells express the appropriate receptors to be retained and to exit into the periphery where they may mature into effective T cells.

T Cell Proliferation & Differentiation

T cell proliferation depends on a variety of events. Naive T cells are activated when they encounter antigen on APCs. Antigen alone, however, is insufficient. Resting T cells must receive two signals for activation to occur. One signal comes from the T cell receptor interacting with an MHC-antigen complex presented on another cell. Cell adhesion molecules are important in the interaction between the two cell types. Recognition of antigen triggers a set of biochemical pathways in the cell that result eventually in DNA synthesis and mitosis. As described earlier, critical to the signaling event are the proteins of the CD3 complex associated with the T cell receptor chains. CD3 transduces the signal to the cytoplasm that results eventually in the transcription of, eg, IL-2 and IL-2 receptor genes. Release of IL-2 results in activation of other T cells bearing IL-2 receptors. Another **costimulatory signal** required for T cell activation comes from interaction between a molecule known as CD80 (B7), which is found on professional antigen-presenting cells such as B cells and macrophages, and its receptor partner, CD28, on the T cell. Without this second signal, exposure of T cells to antigen may lead to their functional inactivation (anergy) or death. Once naive T cells are activated by antigen-MHC complexes plus a costimulatory signal, they secrete the cytokine IL-2 and express IL-2 receptors. T cell proliferation can now be induced in an autocrine fashion. The proliferating T cells may then differentiate into **effector cells**.

T cells fall into two broad categories: CD4-expressing cells and CD8-expressing cells.

Proliferating CD4 T cells can become one of two main categories of effector T cell: TH1 cells or TH2 cells. The control of this differentiation lies largely in the cytokines elicited by interaction of the pathogen with the nonadaptive immune system. In an environment of IFN-γ, TH1 cells dominate and either activate macrophages or cause B cells to switch to IgG synthesis. In either case, this can promote bacterial clearance either by direct destruction in the macrophage or by destruction after phagocytosis of opsonized particles. In an environment where IL-4 is being produced, TH2 cells predominate and activate mast cells and eosinophils and cause B cells to synthesize IgE. This aids in the response to infection by worms, etc.

CD8 T cells can become fully activated effector cells either by interacting with MHC-antigen complexes on professional antigen-presenting cells that express high densities of costimulatory molecules (CD80; B7) or via the help of CD4 T cells that are interacting with the same antigen on a cell that expresses only low levels of costimulatory molecules. In the latter case, cytokines released from the helper T cell may help drive the final stages of activation of the CD8 T cell: (1) Cytotoxic (killer) T cells express CD8 and recognize foreign peptides generated from cytotoxic pathogens, such as viruses, associated with class I MHC molecules. (2) TH1 T cells express CD4 and recognize foreign peptides generated in the endocytic pathway that are associated with class II MHC molecules. These cells activate macrophages, or induce B cells to make IgG, enabling removal of infecting bacteria. (3) TH2 T cells express CD4 and recognize foreign peptides generated as above in association with class II MHC molecules. These cells activate B cells to secrete immunoglobulin E and activate mast cells and eosinophils, enabling removal of parasites such as worms.

T Cell Functions

T cells have both effector and regulatory functions.

A. Effector Functions

Cell-mediated immunity and delayed hypersensitivity reactions are produced mainly against antigens of intracellular parasites, including viruses, fungi, some protozoa, and bacteria (eg, mycobacteria). A deficiency in cell-mediated immunity manifests itself primarily as marked susceptibility to infection by such microorganisms and to certain tumors.

In the response to allografts or tumors, CD4-positive cells can recognize foreign class II MHC molecules in addition to specific antigens and are activated. CD8-positive cytotoxic T cells then respond to the production of cytokines by CD4 cells, recognize class I MHC molecules on the "foreign" cells, and proceed to destroy those cells. In the case of virus-infected cells, the CD8 lymphocytes must recognize both virus-determined antigens and class I MHC molecules on infected cells.

B. Regulatory Functions

T cells play a central role in regulating both humoral (antibody-mediated) and cellular (cell-mediated) immunity. Antibody production by B cells usually requires the participation of T helper cells (T cell-dependent response), but antibodies to some antigens (eg, polymerized macromolecules such as bacterial capsular polysaccharide) are the result of a T cell-independent response.

In the T cell-dependent B cell response to antigen, both B and T cells must have the same class II MHC specificity. In such T cell-dependent responses, the antigen interacts with IgM on the B cell surface. It is then internalized and processed. Fragments of the antigen are returned to the B cell surface in association with class II MHC molecules. These interact with the T cell receptor on the T helper cell, which produces cytokines that enhance division of the B cells and also help them to differentiate into antibody-producing plasma cells, expressing immunoglobulin of other classes (eg, IgG, IgA). Just as for T cells, B cells require two signals for activation. One signal is from the B cell receptor for antigen, and the

TABLE 8–3 Selected Important Cytokines

Name	Major Cellular Source	Selected Biologic Effects
IFN-α, -β	Macrophages (IFN-α), fibroblasts (IFN-β)	Antiviral
IFN-γ (interferon)	T cells, NK cells	Activates macrophages, B cells, T cells and NK cells
TNF-α (tumor necrosis factor alpha)	Macrophages, T cells	Cell activation, fever, cachexia, antitumor
TNF-β (tumor necrosis factor beta), LT (lymphotoxin)	T cells	Activates PMNs and macrophages
IL-1 (interleukin-1)	Macrophages	Cell activation, fever
IL-2 (interleukin-2)	T cells	T cell growth and activation
IL-3 (interleukin-3)	T cells	Hematopoiesis
IL-4 (interleukin-4)	T cells, mast cells	B cell proliferation and switching to IgE, TH2 differentiation
IL-5 (interleukin-5)	T cells	Differentiation of eosinophils, activates B cells
IL-7 (interleukin-7)	Bone marrow stromal cells	T cell progenitor differentiation
IL-8 (interleukin-8)	Macrophages, T cells	Chemotactic for neutrophils
IL-10 (interleukin-10)	T cells	Inhibits activated macrophages and dendritic cells
IL-12 (interleukin-12)	Macrophages	Differentiation of T cells, activation of NK cells
GM-CSF (granulocyte-macrophage colony-stimulating factor)	T cells, macrophages, monocytes	Differentiation of myeloid progenitor cells
M-CSF (monocyte-macrophage colony-stimulating factor)	Macrophages, monocytes, fibroblasts	Differentiation of monocytes and macrophages
G-CSF (granulocyte colony-stimulating factor)	Fibroblasts, monocytes, macrophages	Stimulates neutrophil production in bone marrow

second costimulatory signal results from interaction of CD40 on the B cell with CD154 (CD40 ligand) on the helper T cell.

In other cell-mediated responses, antigen is processed by APCs, and fragments are presented in conjunction with class II MHC molecules on the APC surface. These interact with the T cell receptor on T helper cells, which produce cytokines to stimulate growth of appropriate CD4 (T helper) cells. Important cytokines are briefly described below and in Table 8–3.

When an imbalance exists in the number of active CD4 and CD8 cells, cellular immune mechanisms are grossly impaired. Thus, in AIDS, the normal ratio of CD4 to CD8 cells (>1.5) is lost. Some CD4 cells are destroyed by HIV. This results in a CD4:CD8 ratio of less than 1, leading to extreme susceptibility to development of many opportunistic infections and certain tumors.

CYTOKINES

Cytokines are the soluble mediators of host defense responses, both specific and nonspecific. As such they have a critically important role in the effector mechanisms involved in eliminating foreign antigens such as microorganisms. Table 8–3 lists a small number of important cytokines.

Many different cytokines are produced during immune responses. The same cytokine can be produced by multiple cell types and can have multiple effects on the same cell and can also act on many different cell types. Their effects are mediated by binding to specific receptors on target cells. Thus, cytokines are like other hormones in that their effects are mediated through receptors that signal target cells respond to. As can be appreciated from Table 8–3, cytokines (like other hormones) frequently act as growth factors.

HYPERSENSITIVITY

The term "hypersensitivity" denotes a condition in which an immune response results in exaggerated or inappropriate reactions that are harmful to the host. In a given individual, such reactions typically occur after the second contact with a specific antigen (allergen). The first contact is a necessary preliminary event that induces sensitization to that allergen.

There are four main types of hypersensitivity reactions. Types I, II, and III are antibody-mediated; type IV is T cell-mediated.

Type I: Immediate Hypersensitivity (Allergy)

Type I hypersensitivity manifests itself in tissue reactions occurring within seconds after the antigen combines with the matching antibody. It may take place as a systemic anaphylaxis (eg, after administration of heterologous proteins) or as a local reaction (eg, an atopic allergy such as hay fever).

The general mechanism of immediate hypersensitivity involves the following steps. An antigen induces the formation of IgE antibody, which binds firmly by its Fc portion to a receptor on mast cells, basophils, and eosinophils. Some time later, a second contact of the individual with the same antigen results in the antigen's fixation to cellbound IgE, crosslinking of IgE molecules, and release of pharmacologically active mediators from cells within seconds to minutes. Cyclic nucleotides and calcium are essential in the release of mediators. There may also be a second "late phase" that lasts for several days and involves infiltration of tissues with leukocytes, particularly eosinophils.

A. Type 1 Hypersensitivity Mediators
Some important mediators and their main effects are listed below.

1. Histamine—Histamine exists in a preformed state in platelets and in granules of mast cells, basophils, and eosinophils. Its release causes vasodilation, increased capillary permeability, and smooth muscle contraction (eg, bronchospasm). Antihistamine drugs can block histamine receptor sites and are relatively effective in allergic rhinitis. Histamine is one of the primary mediators of a Type I reaction.

2. Prostaglandins & leukotrienes—Prostaglandins and leukotrienes are derived from arachidonic acid via the cyclooxygenase pathway. Prostaglandins chiefly produce bronchoconstriction. Leukotrienes chiefly cause increased permeability of capillaries.

These mediators, along with cytokines such as TNF-α and IL-4, are referred to as secondary mediators of a Type I reaction.

B. Treatment and Prevention of Anaphylactic Reactions
Treatment aims to reverse the action of mediators by maintaining the airway, providing artificial ventilation if necessary, and supporting cardiac function. One or more of the following may be given: epinephrine, antihistamines, and corticosteroids.

Prevention relies on identification of the allergen (often by skin test) and subsequent avoidance.

C. Atopy
Atopic hypersensitivity disorders exhibit a strong familial predisposition and are associated with elevated IgE levels. Predisposition to atopy is clearly genetic, but symptoms are induced by exposure to specific allergens. These antigens are typically environmental (eg, respiratory allergy to pollens, ragweed, or house dust) or foods (eg, intestinal allergy to shellfish). Common clinical manifestations include hay fever, asthma, eczema, and urticaria. Many sufferers give immediate

type reactions to skin tests (injection, patch, scratch) using the offending antigen.

Type II: Hypersensitivity

Type II hypersensitivity involves the binding of IgG antibodies to cell surface antigens or extracellular matrix molecules. Antibody directed at cell surface antigens can activate complement (or other effectors) to damage the cells. The result may be complement-mediated lysis, as occurs in hemolytic anemias, ABO transfusion reactions, and Rh hemolytic disease.

Drugs such as penicillin can attach to surface proteins on red blood cells and initiate antibody formation. Such autoimmune antibodies may then combine with the cell surface, with resulting hemolysis. In Goodpasture syndrome, antibody forms against basement membranes of kidney and lung, resulting in severe damage to the membranes through activity of complement-attracted leukocytes. In some cases, antibodies to cell surface receptors alter function without cell injury—eg, in Graves disease, an autoantibody binds to the thyroid-stimulating hormone (TSH) receptor and by stimulating the thyroid causes hyperthyroidism.

Type III: Immune Complex Hypersensitivity

When antibody combines with its specific antigen, immune complexes are formed. Normally, they are promptly removed, but occasionally they persist and are deposited in tissues, resulting in several disorders. In persistent microbial or viral infections, immune complexes may be deposited in organs (eg, the kidneys), resulting in dysfunction. In autoimmune disorders, "self" antigens may elicit antibodies that bind to organ antigens or are deposited in organs and tissues as complexes, especially in joints (arthritis), kidneys (nephritis), and blood vessels (vasculitis). Finally, environmental antigens such as fungal spores and certain drugs can cause immune complex formation with disease.

Wherever immune complexes are deposited, they activate the complement system, and macrophages and neutrophils are attracted to the site, where they cause inflammation and tissue injury. There are two major forms of immune complex-mediated hypersensitivity. One is local (**Arthus reaction**) and typically elicited in the skin when a low dose of antigen is injected and immune complexes form locally. IgG antibodies are involved, and the resulting activation of complement leads to activation of mast cells and neutrophils, mediator release, and enhanced vascular permeability. This typically occurs in about 12 hours. A second form of type III hypersensitivity involves systemic immune complex disease. There are several examples, including diseases such as acute poststreptococcal glomerulonephritis.

Acute poststreptococcal glomerulonephritis is a well-known immune complex disease. Its onset occurs several weeks after a group A β-hemolytic streptococcal infection, particularly of the skin, and often occurs with infection due to nephritogenic types of streptococci. The complement level is typically low, suggesting an antigen–antibody reaction with consumption of complement. Lumpy deposits of immunoglobulin and complement component C3 are seen along glomerular basement membranes stained by immunofluorescence, suggesting antigen–antibody complexes. It is likely that streptococcal antigen–antibody complexes are filtered out by glomeruli, that they fix complement and attract neutrophils, and that the resulting inflammatory process damages the kidney.

Type IV: Cell-Mediated (Delayed) Hypersensitivity

Cell-mediated hypersensitivity is a function not of antibody but of specifically sensitized T lymphocytes that activate macrophages to cause an inflammatory response. The response is delayed—ie, it usually starts 2–3 days after contact with the antigen and often lasts for days.

A. Contact Hypersensitivity

Contact hypersensitivity occurs after sensitization with simple chemicals (eg, nickel, formaldehyde), plant materials (poison ivy, poison oak), topically applied drugs (eg, sulfonamides, neomycin), some cosmetics, soaps, and other substances. In all cases, small molecules enter the skin and then, acting as haptens, attach to body proteins to serve as complete antigen. Cell-mediated hypersensitivity is induced, particularly in skin. When the skin again comes in contact with the offending agent, the sensitized person develops erythema, itching, vesication, eczema, or necrosis of skin within 12–48 hours. Patch testing on a small area of skin can sometimes identify the offending antigen. Subsequent avoidance of the material will prevent recurrences. The antigen-presenting cell in contact sensitivity is probably the Langerhans cell in the epidermis, which interacts with CD4 TH1 cells that drive the response.

B. Tuberculin-Type Hypersensitivity

Delayed hypersensitivity to antigens of microorganisms occurs in many infectious diseases and has been used as an aid in diagnosis. It is typified by the tuberculin reaction. When a small amount of tuberculin is injected into the epidermis of a patient previously exposed to *Mycobacterium tuberculosis*, there is little immediate reaction; gradually, however, induration and redness develop and reach a peak in 24–72 hours. Mononuclear cells accumulate in the subcutaneous tissue, and there are CD4 TH1 cells in abundance. A positive skin test indicates that the person has been infected with the agent but does not imply the presence of current disease. However, a recent change of skin test response from negative to positive suggests recent infection and possible current activity.

A positive skin test response assists in diagnosis. For example, in leprosy, a positive skin test indicates tuberculoid disease, with active cell-mediated immunity, whereas a negative test suggests lepromatous leprosy, with weak cell-mediated immunity.

INADEQUATE IMMUNE RESPONSES TO INFECTIOUS AGENTS

There are a considerable number of inherited immune deficiency diseases that can affect the host response to infection. The reader is referred to other texts for details, but in brief, these defects can result in a variety of immune system changes, including reduced levels of antibody, phagocytic cell alterations, and lack of effector cells. Any of these changes can create a situation where the host is highly susceptible to infections. Some of these situations have been mentioned earlier in the chapter—eg, defects such as **chronic granulomatous** disease that reduce antibacterial activity.

In some cases, the pathogen ultimately causes immune suppression—an example is infection with HIV, which alters T cell immunity and allows further infection with **opportunistic pathogens**. In other situations, certain bacteria release toxins that function as superantigens, initially stimulating large numbers of T cells to proliferate but, because of the release of cytokines from T cells, ultimately suppressing the immune response and allowing the pathogen to multiply.

The pathogen itself may have mechanisms to actively avoid the immune response. For example, several pathogens alter their antigenic structure by mutation to evade the immune defenses. Influenza virus undergoes antigenic variation by two mutational mechanisms called **antigenic shift** and **antigenic drift** that create new antigenic phenotypes which evade the host's current immunity and allow reinfection with the virus. Several other pathogens have similar evasion strategies—eg, trypanosomes alter their surface glycoproteins and streptococci alter their surface carbohydrate antigens.

Examples of other avoidance strategies were discussed earlier, eg, viral proteins that inhibit the development of an effective immune response. A strategy used by a few pathogens is to become inactive; eg, herpes simplex virus becomes transcriptionally inactive in a state referred to as **latency** in certain nerve cells after infection and may stay in this state until the immune response declines, whereupon a new cycle of viral replication may be initiated.

IMMUNOLOGIC DIAGNOSTIC TESTS

Reactions of antigens and antibodies are highly specific. An antigen will react only with antibody elicited by that antigen or by a closely related antigen. Because of this high specificity, reactions between an antigen and an antibody can be used to identify one by means of the other.

Antigen–antibody reactions are used to identify specific components in mixtures of either one. Microorganisms and other cells possess a variety of antigens and may thus react with many different antibodies. Monoclonal antibodies are excellent tools for the identification of antigens because they have a single known specificity and are homogeneous. Antisera generated as part of an immune response contain complex mixtures of antibodies and are heterogeneous. This makes them less useful for specific tests. Possible cross-reactions between related antigens can limit the test's specificity.

Enzyme-Linked Immunosorbent Assay (ELISA)

Enzyme immunoassay, which has many variations, depends on the conjugation of an enzyme to an antibody. The enzyme is detected by assaying for enzyme activity with its substrate.

To measure antibody, known antigens are fixed to a solid phase (eg, plastic microdilution plate), incubated with test antibody dilutions, washed, and re-incubated with an anti-immunoglobulin labeled with an enzyme (eg, horseradish peroxidase). Enzyme activity, measured by adding the specific substrate and estimating the color reaction, is a direct function of the amount of antibody bound. This type of assay is used, eg, to detect antibodies to HIV proteins in blood samples.

Immunofluorescence

Fluorescent dyes (eg, fluorescein, rhodamine) can be covalently attached to antibody molecules and made visible by ultraviolet light in the fluorescence microscope. Such labeled antibody can be used to identify antigens (eg, on the surfaces of bacteria such as streptococci or treponemes) or in cells in histologic section or other specimens. A **direct immunofluorescence** reaction occurs when known labeled antibody interacts directly with unknown antigen. An **indirect immunofluorescence** reaction occurs when a two-stage process is used—eg, a known antigen is attached to a slide, unknown serum is added, and the preparation is washed. If the unknown serum antibody matches the antigen, it will remain fixed to it on the slide and can be detected by adding a fluorescent-labeled anti-immunoglobulin or other antibody-specific reagent such as staphylococcus protein A and examining the slide by ultraviolet microscopy.

Another use of fluorescent-tagged antibody molecules is to count and classify cells by **flow cytometry** using a fluorescence-activated cell sorter (FACS).

Flow cytometry analyzes a single-cell suspension flowing through a set of laser beams to measure the relative amount of light scattered by microscopic particles (providing information on relative size and granularity) and the relative fluorescence of those particles. For a mixture of white blood cells, it is relatively easy to separate the cells in this mixture into major classes—eg, small lymphocytes separated from granulocytes that are larger and contain more granules (scatter more light). With the availability of panels of monoclonal antibodies (that can be detected by fluorescent anti-immunoglobulin) to cell surface proteins, it is also possible to count subpopulations of cells—eg, CD4 expressing helper T cells from CD8 expressing cytotoxic T cells, or antibody expressing B cells from T cells. This technology is widely used both in clinical medicine and in biomedical research—eg, to enumerate CD4 T cells in HIV-positive patients or to distinguish tumor cells from normal white blood cells.

Immunoblotting

Immunoblotting (sometimes called Western blotting) is a method for identifying a particular antigen in a complex mixture of proteins. The complex mixture of proteins is subjected to sodium dodecyl sulfate (SDS)-polyacrylamide gel electrophoresis (PAGE). This separates the proteins according to molecular size. The gel is then covered with a membrane (often a sheet of nitrocellulose), and the proteins are "transferred" by electrophoresis to the membrane. The nitrocellulose membrane (blot) acquires a replica of the proteins separated by SDS-PAGE. During the transfer, the SDS is largely removed from the proteins and, at least for some proteins, there is refolding and enough conformation is restored so that antibodies can react with the proteins on the membrane.

The nitrocellulose membrane is then reacted with an enzyme-labeled antibody in a direct test or in an indirect test, with antibody followed by an enzyme-labeled anti-immunoglobulin. The protein antigen then becomes visible as a band on the membrane. None of the other proteins in the mixture are detected. This technique is used, eg, to confirm an HIV-positive ELISA test by demonstrating the presence of antibodies to specific HIV proteins in a patient's serum.

REVIEW QUESTIONS

1. What is a characteristic of the adaptive immune response and not of the innate response?

   (A) Physical barriers
   (B) Chemical barriers
   (C) Clonal expansion of effector cells
   (D) Inflammatory mediators
   (E) Phagocytosis

2. The antibodies that arise in the serum of an individual immunized with a single complex antigen containing many epitopes (antigenic determinants) are subjected to electrophoresis and then structural analyses (amino acid sequencing, etc). Which of the following best describes the antibodies that are produced?

   (A) Homogeneous in chemical structure
   (B) Monoclonal
   (C) Heterogeneous in chemical structure
   (D) A single sharp spike on gel electrophoresis
   (E) A pure single protein

3. How is generation of diversity in the immune response principally achieved?

   (A) Recombination among sets of gene segments to generate a repertoire of antigen recognition molecules
   (B) A massive amount of DNA given over to genes for antigen recognition molecules
   (C) Only germline diversification of genes encoding antigen recognition molecules
   (D) Only somatic diversification of genes encoding antigen recognition molecules
   (E) Antigen "instruction" of a small set of lymphocytes to produce an infinite array of antigen recognition molecules

4. Which genetic mechanism increases the number of different antibody molecules during an immune response without increasing the diversity of the pool of antigen receptor specificities?

   (A) V gene segment recombination
   (B) Class switching
   (C) Somatic hypermutation
   (D) Junctional variability due to imprecise V, D, J joining
   (E) Gene duplication, ie, multiple V, D, and J gene segments

5. What is the principal function of the class I and class II MHC molecules?

   (A) They frustrate the efforts of transplant surgeons
   (B) They bind peptide antigens for presentation to antigen-specific receptors on B cells
   (C) They help in endocytosis of antigens by phagocytic cells
   (D) They bind carbohydrate antigens directly for presentation on T cells
   (E) They display peptide antigens for review by antigen-specific receptors on T cells

6. MHC class I molecules need to bind peptide antigens to fold properly and to be expressed at the cell surface. What would you expect to be the most common health problem in a child with a defect in the function of the peptide transporter (TAP) found in the endoplasmic reticulum?

   (A) Chronic upper respiratory viral infections
   (B) Parasitic infections
   (C) Infections with encapsulated bacteria
   (D) Pronounced allergies to household pets
   (E) Autoimmune disease

7. What is the presentation of a particular peptide antigen in association with either class I or class II MHC molecules at the cell surface most dependent on?

   (A) The intrinsic binding specificity differences of class I and class II MHC molecules
   (B) Differences in the amino acid sequence of the peptide antigens, particularly with respect to certain anchor residues
   (C) Polymorphic differences between the MHC class I and class II molecules
   (D) The initial route the peptide antigen takes through the cell, ie, endocytic vesicle versus cytoplasm
   (E) The source of the peptide antigen, ie, bacterial protein or viral protein

8. Superantigens are peptide toxins released from virulent strains of staphylococcal or streptococcal bacteria. To what entity do these superantigens bind in order to elicit their effect?

   (A) TCR γ chains
   (B) TCR β chain and MHC Class II chain complex.
   (C) MHC class I molecules
   (D) Peptide antagonists
   (E) Only γ/δ T cell receptors

9. Which class of antibody molecule has the ability to cross the placenta?

   (A) IgG
   (B) IgA
   (C) IgM
   (D) IgE
   (E) IgD

10. A 2-year-old child with recurrent bacterial infections causing otitis media and pneumonia is most likely deficient in

    (A) T cells
    (B) Phagocytes
    (C) B cells

(D) TAP-1/TAP-2 molecules

(E) C1 esterase inhibitor

11. What initiates the classical pathway of complement activation most efficiently?

(A) IgG

(B) Mannose-containing bacterial glycolipids

(C) Microbial surfaces

(D) IgM-antigen immune complexes

(E) Endotoxin

12. Positive and negative selection processes in the thymus generate a T cell repertoire that is both self-tolerant and self-restricted. This means that the mature T cells which develop have a repertoire of receptors that are specific for

(A) Nonself antigenic peptides bound to nonself MHC molecules

(B) Self antigenic peptides bound to self MHC molecules

(C) Self antigenic peptides bound to nonself MHC molecules

(D) Nonself antigenic peptides bound to self MHC molecules

(E) Any peptide antigen bound to any MHC molecule

13. Which "receptor-ligand" pair supplies the crucial second signal for activation of the naive T cell by an APC in a secondary lymphoid organ?

(A) CD80 (B7)/CD28

(B) MHC Class II/CD4

(C) MHC Class I/CD8

(D) MHC Class II/TCR

(E) ICAM-1/LFA-1

14. A man in his twenties presents in the emergency room with shortness of breath and fatigue. He is also very pale. Two days earlier he was given penicillin for an infection. He had had penicillin previously without problems and stated that he had no "allergy" to penicillin. Laboratory testing shows that antibodies to penicillin are present in the patient's serum and that he is breaking down his own red blood cells. He is diagnosed with immune hemolytic anemia. This patient is suffering from which type of hypersensitivity reaction?

(A) Type I

(B) Type II

(C) Type III

(D) Type IV (DTH)

15. Which one of the following cell types expresses receptors for IgE on its cell surface that stimulate the cell to mount a response to parasites such as worms?

(A) T cells

(B) B cells

(C) Promonocytes

(D) NK cells

(E) Mast cells

16. Which immunologic test is widely used to precisely enumerate and collect cells expressing an antigen bound by a fluorescence-tagged monoclonal antibody?

(A) ELISA

(B) Direct Immunofluorescence

(C) Western blotting

(D) Fluorescence-activated cell sorting

(E) Indirect Immunofluorescence

17. In any given immunoglobulin molecule, the light chains are

(A) Identical to each other except in their antigenic determinants

(B) Identical to each other

(C) Identical to each other except in their hypervariable regions

(D) Of related but different amino acid sequences

(E) Identical to each other except in their overall domain structure

18. Antigen–antibody complexes are phagocytosed more effectively in the presence of which complement component?

(A) C3a and C5a

(B) C3b

(C) C56789 complex

(D) MBL

(E) Properdin

19. A 2-year-old boy presented with a high fever (102°F) and pain in the ears. A diagnosis of otitis media was established. The fever subsided after treatment with a nonsteroidal antiinflammatory drug. Which of the following cytokines is most likely involved in the development of the boy's fever?

(A) Interleukin-1

(B) Interleukin-2

(C) Interleukin-4

(D) Interleukin-10

(E) Interleukin-12

20. A primary immune response requires approximately how much time to produce detectable levels of antibody in the serum?

(A) Immediate

(B) 7–10 hours

(C) 24 hours

(D) 5–7 days

(E) 21 days

21. NK cells express a killer immunoglobulin-like receptor which recognizes:

(A) MHC Class I molecules

(B) MHC Class II molecules

(C) Cell adhesion molecules

(D) Glycophospholipid molecules

(E) CD40 molecules

Answers

1. C	7. D	13. A	19. A
2. C	8. B	14. B	20. D
3. A	9. A	15. E	21. A
4. B	10. C	16. D	
5. E	11. D	17. B	
6. A	12. D	18. B	

REFERENCES

Abbas AK, Lichtman AH, Pillai S: *Cellular and Molecular Immunology,* 6th ed. Updated. Saunders Elsevier, 2007.

Kindt TJ, Goldsby RA, Osborne BA: *Kuby Immunology,* 6th ed. W.H. Freeman, 2007.

Murphy K, Travers P, Wolport M: *Janeway's Immunobiology* 7th ed. Garland Science, 2008.

Nairn R, Helbert M: *Immunology for Medical Students.* 2nd ed. Mosby/Elsevier, 2007.

Paul WE (editor): *Fundamental Immunology,* 6th ed. Wolters Kluwer/Lippincott Williams & Wilkins, 2008.

C H A P T E R

9

Pathogenesis of Bacterial Infection

The pathogenesis of bacterial infection includes initiation of the infectious process and the mechanisms that lead to the development of signs and symptoms of disease. Characteristics of bacteria that are pathogens include transmissibility, adherence to host cells, invasion of host cells and tissues, toxigenicity, and ability to evade the host's immune system. Many infections caused by bacteria that are commonly considered to be pathogens are inapparent or asymptomatic. Disease occurs if the bacteria or immunologic reactions to their presence cause sufficient harm to the person.

Terms frequently used in describing aspects of pathogenesis are defined in the Glossary (see below). Refer to the Glossary in Chapter 8 for definitions of terms used in immunology and in describing aspects of the host's response to infection.

GLOSSARY

Adherence (adhesion, attachment): The process by which bacteria stick to the surfaces of host cells. Once bacteria have entered the body, adherence is a major initial step in the infection process. The terms adherence, adhesion, and attachment are often used interchangeably.

Carrier: A person or animal with asymptomatic infection that can be transmitted to another susceptible person or animal.

Infection: Multiplication of an infectious agent within the body. Multiplication of the bacteria that are part of the normal flora of the gastrointestinal tract, skin, etc, is generally not considered an infection; on the other hand, multiplication of pathogenic bacteria (eg, *Salmonella* species)—even if the person is asymptomatic—is deemed an infection.

Invasion: The process whereby bacteria, animal parasites, fungi, and viruses enter host cells or tissues and spread in the body.

Nonpathogen: A microorganism that does not cause disease; may be part of the normal flora.

Opportunistic pathogen: An agent capable of causing disease only when the host's resistance is impaired (ie, when the patient is "immunocompromised").

Pathogen: A microorganism capable of causing disease.

Pathogenicity: The ability of an infectious agent to cause disease. (See also virulence.)

Toxigenicity: The ability of a microorganism to produce a toxin that contributes to the development of disease.

Virulence: The quantitative ability of an agent to cause disease. Virulent agents cause disease when introduced into the host in small numbers. Virulence involves adherence, invasion, and toxigenicity (see above).

IDENTIFYING BACTERIA THAT CAUSE DISEASE

Humans and animals have abundant normal flora that usually do not produce disease (see Chapter 10) but achieve a balance that ensures the survival, growth, and propagation of both the bacteria and the host. Some bacteria that are important causes of disease are cultured commonly with the normal flora (eg, *Streptococcus pneumoniae, Staphylococcus aureus*). Sometimes bacteria that are clearly pathogens (eg, *Salmonella typhi*) are present, but infection remains latent or subclinical and the host is a "carrier" of the bacteria.

It can be difficult to show that a specific bacterial species is the cause of a particular disease. In 1884, Robert Koch proposed a series of postulates that have been applied broadly to link many specific bacterial species with particular diseases. **Koch's postulates** are summarized in Table 9–1.

Koch's postulates have remained a mainstay of microbiology; however, since the late 19th century, many microorganisms that do not meet the criteria of the postulates have been shown to cause disease. For example, *Treponema pallidum* (syphilis) and *Mycobacterium leprae* (leprosy) cannot be grown in vitro; however, there are animal models of infection with these agents. In another example, *Neisseria gonorrhoeae* (gonorrhea), there is no animal model of infection even though the bacteria can readily be cultured in vitro; experimental infection in humans has been produced, which substitutes for an animal model.

In other instances, Koch's postulates have been at least partially satisfied by showing bacterial pathogenicity in an in vitro model of infection rather than in an animal model. For example, some forms of *Escherichia coli*-induced diarrhea (Chapter 15) have been defined by the interaction of the *E coli* with host cells in culture.

The host's immune responses also should be considered when an organism is being investigated as the possible cause of a disease. Thus, development of a rise in specific antibody during recovery from disease is an important adjunct to Koch's postulates.

Modern-day microbial genetics has opened new frontiers to study pathogenic bacteria and differentiate them from nonpathogens. Molecular cloning has allowed investigators to isolate and modify specific virulence genes and study them with models of infection. The ability to study genes associated with virulence has led to a proposed form of **molecular Koch's postulates**. These postulates are summarized in Table 9–1.

Some pathogens are difficult or impossible to grow in culture, and for that reason it is not possible with Koch's postulates or the molecular Koch's postulates to establish the cause of their associated diseases. The polymerase chain reaction is used to amplify microorganism-specific nucleic acid sequences from host tissues or fluids. The sequences are used to identify the infecting organisms. The molecular guidelines for establishing microbial disease causation are listed in Table 9–1. This approach has been used to establish the causes of several diseases, including Whipple disease (*Tropheryma*

TABLE 9–1   Guidelines for Establishing the Causes of Infectious Diseases

Koch's Postulates	Molecular Koch's Postulates	Molecular Guidelines for Establishing Microbial Disease Causation
1. The microorganism should be found in all cases of the disease in question, and its distribution in the body should be in accordance with the lesions observed	1. The phenotype or property under investigation should be significantly associated with pathogenic strains of a species and not with nonpathogenic strains	1. The nucleic acid sequence of a putative pathogen should be present in most cases of an infectious disease, and preferentially in anatomic sites where pathology is evident
2. The microorganism should be grown in pure culture in vitro (or outside the body of the host) for several generations	2. Specific inactivation of the gene or genes associated with the suspected virulence trait should lead to a measurable decrease in pathogenicity or virulence	2. The nucleic acid sequence of a putative pathogen should be absent from most healthy controls. If the sequence is detected in healthy controls, it should be present with a lower prevalence as compared with patients with disease, and in lower copy numbers
3. When such a pure culture is inoculated into susceptible animal species, the typical disease must result	3. Reversion or replacement of the mutated gene with the wild-type gene should lead to restoration of pathogenicity or virulence	3. The copy number of a pathogen-associated nucleic acid sequence should decrease or become undetectable with resolution of the disease (eg, with effective treatment) and should increase with relapse or recurrence of disease
4. The microorganism must again be isolated from the lesions of such experimentally produced disease		4. The presence of a pathogen-associated nucleic acid sequence in healthy subjects should help predict the subsequent development of disease
		5. The nature of the pathogen inferred from analysis of its nucleic acid sequence should be consistent with the known biologic characteristics of closely related organisms and the nature of the disease. The significance of a detected microbial sequence is increased when microbial genotype predicts microbial morphology, pathology, clinical features of disease, and host response

whipplei), bacillary angiomatosis (*Bartonella henselae*), human monocytic ehrlichiosis (*Ehrlichia chaffeensis*), hantavirus pulmonary syndrome (Sin Nombre virus), and Kaposi sarcoma (human herpesvirus 8).

Analysis of infection and disease through the application of principles such as Koch's postulates leads to classification of bacteria as pathogens, opportunistic pathogens, or non-pathogens. Some bacterial species are always considered to be pathogens, and their presence is abnormal; examples include *Mycobacterium tuberculosis* (tuberculosis) and *Yersinia pestis* (plague). Such bacteria readily meet the criteria of Koch's postulates. Other species are commonly part of the normal flora of humans (and animals) but also can frequently cause disease. For example, *E coli* is part of the gastrointestinal flora of normal humans but is also a common cause of urinary tract infections, traveler's diarrhea, and other diseases. Strains of *E coli* that cause disease are differentiated from those that do not by determining (1) whether they are virulent in animals or in vitro models of infection and (2) whether they have a genetic makeup that is significantly associated with production of disease. Other bacteria (eg, *Pseudomonas* species, *Stenotrophomonas maltophilia,* and many yeasts and molds) only cause disease in immunosuppressed and debilitated persons and are **opportunistic pathogens**.

TRANSMISSION OF INFECTION

Bacteria (and other microorganisms) adapt to the environment, including animals and humans, where they normally reside and subsist. In doing so, the bacteria ensure their survival and enhance the possibility of transmission. By producing asymptomatic infection or mild disease, rather than death of the host, microorganisms that normally live in people enhance the possibility of transmission from one person to another.

Some bacteria that commonly cause disease in humans exist primarily in animals and incidentally infect humans. For example, *Salmonella* and *Campylobacter* species typically infect animals and are transmitted in food products to humans. Other bacteria produce infection of humans that is inadvertent, a mistake in the normal life cycle of the organism; the organisms have not adapted to humans, and the disease they produce may be severe. For example, *Y pestis* (plague) has a well-established life cycle in rodents and rodent fleas, and transmission by the fleas to humans is inadvertent; *Bacillus anthracis* (anthrax) lives in the environment, occasionally infects animals, and is transmitted to humans by products such as raw hair from infected animals. The *Clostridium* species are ubiquitous in the environment and are transmitted to humans by ingestion (eg, *C perfringens* gastroenteritis and *C botulinum* [botulism]) or when wounds are contaminated by soil (eg, *C perfringens* [gas gangrene] and *C tetani* [tetanus]).

The clinical manifestations of diseases (eg, diarrhea, cough, genital discharge) produced by microorganisms often promote transmission of the agents. Examples of clinical syndromes and how they enhance transmission of the causative bacteria are as follows: *Vibrio cholerae* can cause voluminous diarrhea which may contaminate salt and fresh water; drinking water or seafood such as oysters and crabs may be contaminated; ingestion of contaminated water or seafood can produce infection and disease. Similarly, contamination of food products with sewage containing *E coli* that cause diarrhea results in transmission of the bacteria. *M tuberculosis* (tuberculosis) naturally infects only humans; it produces respiratory disease with cough and production of aerosols, resulting in transmission of the bacteria from one person to another.

Many bacteria are transmitted from one person to another on hands. A person with *S aureus* carriage in the anterior nares may rub his nose, pick up the staphylococci on the hands, and spread the bacteria to other parts of the body or to another person, where infection results. Many opportunistic pathogens that cause nosocomial infections are transmitted from one patient to another on the hands of hospital personnel. Hand washing is thus an important component of infection control.

The most frequent **portals of entry of pathogenic bacteria** into the body are the sites where mucous membranes meet with the skin: respiratory (upper and lower airways), gastrointestinal (primarily mouth), genital, and urinary tracts. Abnormal areas of mucous membranes and skin (eg, cuts, burns, and other injuries) are also frequent sites of entry. Normal skin and mucous membranes provide the primary defense against infection. To cause disease, pathogens must overcome these barriers.

THE INFECTIOUS PROCESS

Once in the body, bacteria must attach or adhere to host cells, usually epithelial cells. After the bacteria have established a primary site of infection, they multiply and spread directly through tissues or via the lymphatic system to the bloodstream. This infection (bacteremia) can be transient or persistent. Bacteremia allows bacteria to spread widely in the body and permits them to reach tissues particularly suitable for their multiplication.

Pneumococcal pneumonia is an example of the infectious process. *S pneumoniae* can be cultured from the nasopharynx of 5–40% of healthy people. Occasionally, pneumococci from the nasopharynx are aspirated into the lungs; aspiration occurs most commonly in debilitated people and in settings such as coma when normal gag and cough reflexes are diminished. Infection develops in the terminal air spaces of the lungs in persons who do not have protective antibodies against that capsular polysaccharide type of the pneumococci. Multiplication of the pneumococci and resultant inflammation lead to pneumonia. The pneumococci enter the lymphatics of the lung and move to the bloodstream. Between 10% and 20% of persons with pneumococcal pneumonia have bacteremia at the time the diagnosis of pneumonia is made. Once bacteremia occurs, the pneumococci can spread to secondary sites of infection (eg, cerebrospinal fluid, heart valves, joint spaces). The major complications of pneumococcal pneumonia are meningitis, endocarditis, and septic arthritis.

The infectious process in cholera involves ingestion of *V cholerae,* chemotactic attraction of the bacteria to the gut epithelium, motility of the bacteria by a single polar flagellum, and penetration of the mucous layer on the intestinal surface. The *V cholerae* adherence to the epithelial cell surface is mediated by pili and possibly other adhesins. Production of cholera toxin results in flow of chloride and water into the lumen of the gut, causing diarrhea and electrolyte imbalance.

GENOMICS & BACTERIAL PATHOGENICITY

Bacteria are haploid (Chapter 7) and limit genetic interactions that might change their chromosomes and potentially disrupt their adaptation and survival in specific environmental niches.

The Clonal Nature of Bacterial Pathogens

One important result of the conservation of chromosomal genes in bacteria is that the organisms are clonal. For most pathogens there are only one or a few clonal types that are spread in the world during a period of time. For example, epidemic serogroup A meningococcal meningitis occurs in Asia, the Middle East, and Africa, and occasionally spreads into Northern Europe and the Americas. On several occasions, over a period of decades, single clonal types of serogroup A *Neisseria meningitidis* have been observed to appear in one geographic area and subsequently spread to others with resultant epidemic disease. There are many types of *Haemophilus influenzae,* but only clonal *H influenzae* type b is commonly associated with disease. There are two clonal types of *Bordetella pertussis,* both associated with disease. Similarly, *S typhi* (typhoid fever) from patients is of two clonal types. There are, however, mechanisms that bacteria use, or have used a long time in the past, to transmit virulence genes from one to another.

Mobile Genetic Elements

A primary mechanism for exchange of genetic information between bacteria is transfer of extrachromosomal mobile genetic elements: plasmids or phages. The genes that code for many bacterial virulence factors commonly are on plasmids or are carried by phages. Transfer of these mobile genetic elements between members of one species or, less commonly, between species can result in transfer of virulence factors. Sometimes the genetic elements are part of highly mobile DNA (transposons; Chapter 7) and there is recombination between the extrachromosomal DNA and the chromosome (illegitimate or nonhomologous recombination; Chapter 7).

If this recombination occurs, the genes coding for virulence factors may become chromosomal. A few examples of plasmid- and phage-encoded virulence factors are in Table 9–2.

Pathogenicity Islands

Large groups of genes that are associated with pathogenicity and are located on the bacterial chromosome are termed pathogenicity islands (PAIs). They are large organized groups of genes, usually 10 to 200 kb in size. The major properties of PAIs are as follows: they have one or more virulence genes; they are present in the genome of pathogenic members of a species, but absent in the nonpathogenic members; they are large; they typically have a different guanine plus cytosine (G + C) content than the rest of the bacterial genome; they are commonly associated with tRNA genes; they are often found with parts of the genome associated with mobile genetic elements; they often have genetic instability; and they often represent mosaic structures with components acquired at different times. Collectively, the properties of PAIs suggest that they originate from gene transfer from foreign species. A few examples of PAI virulence factors are in Table 9–3.

REGULATION OF BACTERIAL VIRULENCE FACTORS

Pathogenic bacteria (and other pathogens) have adapted both to saprophytic or free-living states, possibly environments outside of the body, and to the human host. In the adaptive process, pathogens husband their metabolic needs

TABLE 9–2   Examples of Virulence Factors Encoded by Genes on Mobile Genetic Elements

Genus/Species	Virulence Factor and Disease
Plasmid encoded	
Escherichia coli	Heat-labile and heat-stable enterotoxins that cause diarrhea
Escherichia coli	Hemolysin (cytotoxin) of invasive disease and urinary tract infections
Escherichia coli and *Shigella* species	Adherence factors and gene products involved in mucosal invasion
Bacillus anthracis	Capsule essential for virulence (on one plasmid)
	Edema factor, lethal factor, protective antigen all essential for virulence (on another plasmid)
Phage encoded	
Clostridium botulinum	Botulinum toxin that causes paralysis
Corynebacterium diphtheriae	Diphtheria toxin that inhibits human protein synthesis
Vibrio cholerae	Cholera toxin that can cause a severe watery diarrhea

TABLE 9–3 **A Few Examples of the Very Large Number of Pathogenicity Islands of Human Pathogens**

Genus/Species	PAI Name	Virulence Characteristics
Escherichia coli	PAI I$_{536}$	Alpha hemolysin, fimbriae, adhesions, in urinary tract infections
Escherichia coli	PAI I$_{J96}$	Alpha hemolysin, P-pilus in urinary tract infections
Escherichia coli (EHEC)	O1#7	Macrophage toxin of enterohemorrhagic Escherichia coli (EHEC)
Salmonella typhimurium	SPI-1	Invasion and damage of host cells; diarrhea
Yersinia pestis	HPI/pgm	Genes that enhance iron uptake
Vibrio cholerae El Tor O1	VPI-1	Neuraminidase, utilization of amino sugars
Staphylococcus aureus	SCC mec	Methicillin and other antibiotic resistance
Staphylococcus aureus	SaPI1	Toxic shock syndrome toxin-1, enterotoxin
Enterococcus faecalis	NP^m	Cytolysin, biofilm formation

and products. They have evolved complex signal transduction systems to regulate the genes important for virulence. Environmental signals often control the expression of the virulence genes. Common signals include temperature, iron availability, osmolality, growth phase, pH, and specific ions (eg, Ca^{2+}) or nutrient factors. A few examples are presented in the following paragraphs.

The gene for diphtheria toxin from *Corynebacterium diphtheriae* is carried on temperate bacteriophages. Toxin is produced only by strains lysogenized by the phages. Toxin production is greatly enhanced when *C diphtheriae* is grown in a medium with low iron.

Expression of virulence genes of *B pertussis* is enhanced when the bacteria are grown at 37°C and suppressed when they are grown at lower temperatures or in the presence of high concentrations of magnesium sulfate or nicotinic acid.

The virulence factors of *V cholerae* are regulated on multiple levels and by many environmental factors. Expression of the cholera toxin is higher at pH 6.0 than at pH 8.5 and higher also at 30°C than at 37°C.

Osmolality and amino acid composition also are important. As many as 20 other genes of *V cholerae* are similarly regulated.

Y pestis produces a series of virulence plasmid-encoded proteins. One of these is an antiphagocytic fraction 1 capsular protein that results in antiphagocytic function. This protein is expressed maximally at 35–37°C, the host temperature, and minimally at 20–28°C, the flea temperature at which antiphagocytic activity is not needed. The regulation of other virulence factors in *Yersinia* species also is influenced by environmental factors.

Motility of bacteria enables them to spread and multiply in their environmental niches or in patients. *Yersinia enterocolitica* and *Listeria monocytogenes* are common in the environment where motility is important to them. Presumably, motility is not important in the pathogenesis of the diseases caused by these bacteria. *Y enterocolitica* is motile when grown at 25°C but not when grown at 37°C. Similarly, *Listeria*

is motile when grown at 25°C and not motile or minimally motile when grown at 37°C.

BACTERIAL VIRULENCE FACTORS

Many factors determine bacterial virulence or ability to cause infection and disease.

Adherence Factors

Once bacteria enter the body of the host, they must adhere to cells of a tissue surface. If they did not adhere, they would be swept away by mucus and other fluids that bathe the tissue surface. Adherence, which is only one step in the infectious process, is followed by development of microcolonies and subsequent steps in the pathogenesis of infection.

The interactions between bacteria and tissue cell surfaces in the adhesion process are complex. Several factors play important roles: surface hydrophobicity and net surface charge, binding molecules on bacteria (ligands), and host cell receptor interactions. Bacteria and host cells commonly have net negative surface charges and, therefore, repulsive electrostatic forces. These forces are overcome by hydrophobic and other more specific interactions between bacteria and host cells. In general, the more hydrophobic the bacterial cell surface, the greater the adherence to the host cell. Different strains of bacteria within a species may vary widely in their hydrophobic surface properties and ability to adhere to host cells.

Bacteria also have specific surface molecules that interact with host cells. Many bacteria have **pili**, hair-like appendages that extend from the bacterial cell surface and help mediate adherence of the bacteria to host cell surfaces. For example, some *E coli* strains have type 1 pili, which adhere to epithelial cell receptors; adherence can be blocked in vitro by addition of d-mannose to the medium. *E coli* organisms that cause urinary tract infections commonly do not have d-mannose-mediated adherence but have P-pili, which attach to a portion of the P blood group antigen; the minimal recognition structure is the disaccharide α-d-galactopyranosyl-(1–4)-β-d-galactopyranoside (GAL–GAL binding adhesion). The *E coli* that cause diarrheal diseases (see Chapter 15) have pilus-mediated adherence to intestinal epithelial cells, though the pili and specific molecular mechanisms of adherence appear to be different depending upon the form of the *E coli* that induce the diarrhea.

Other specific ligand-receptor mechanisms have evolved to promote bacterial adherence to host cells, illustrating the diverse mechanisms employed by bacteria. Group A streptococci (*Streptococcus pyogenes*) (see Chapter 14) also have hair-like appendages, termed fimbriae, that extend from the cell surface. **Lipoteichoic acid**, protein F, and M protein are found on the fimbriae. The lipoteichoic acid and protein F cause adherence of the streptococci to buccal epithelial cells; this adherence is mediated by fibronectin, which acts as the host cell receptor molecule. M protein acts as an antiphagocytic molecule and is a major virulence factor.

Antibodies that act against the specific bacterial ligands that promote adherence (eg, pili and lipoteichoic acid) can block adherence to host cells and protect the host from infection.

Invasion of Host Cells & Tissues

For many disease-causing bacteria, invasion of the host's epithelium is central to the infectious process. Some bacteria (eg, *Salmonella* species) invade tissues through the junctions between epithelial cells. Other bacteria (eg, *Yersinia* species, *N gonorrhoeae*, *Chlamydia trachomatis*) invade specific types of the host's epithelial cells and may subsequently enter the tissue. Once inside the host cell, bacteria may remain enclosed in a vacuole composed of the host cell membrane, or the vacuole membrane may be dissolved and bacteria may be dispersed in the cytoplasm. Some bacteria (eg, *Shigella* species) multiply within host cells, whereas other bacteria do not.

"Invasion" is the term commonly used to describe the entry of bacteria into host cells, implying an active role for the organisms and a passive role for the host cells. In many infections, the bacteria produce virulence factors that influence the host cells, causing them to engulf (ingest) the bacteria. The host cells play a very active role in the process.

Toxin production and other virulence properties are generally independent of the ability of bacteria to invade cells and tissues. For example, *C diphtheriae* is able to invade the epithelium of the nasopharynx and cause symptomatic sore throat even when the *C diphtheriae* strains are nontoxigenic.

In vitro studies with cells in tissue culture have helped characterize the mechanisms of invasion for some pathogens; however, the in vitro models have not necessarily provided a complete picture of the invasion process. Full understanding of the invasion process, as it occurs in naturally acquired infection, has required study of genetically engineered mutants and their ability to infect susceptible animals and humans. Thus, understanding of eukaryotic cell invasion by bacteria requires satisfying much of Koch's postulates and the molecular Koch's postulates. The following paragraphs contain examples of bacterial invasion of host cells as part of the infectious process.

Shigella species adhere to host cells in vitro. Commonly, HeLa cells are used; these undifferentiated unpolarized cells were derived from a cervical carcinoma. The adherence causes actin polymerization in the nearby portion of the HeLa cell, which induces the formation of pseudopods by the HeLa cells and engulfment of the bacteria. Adherence and invasion are mediated at least in part by products of genes located on a large plasmid common to many shigellae. There are multiple proteins, including the **invasion plasmid antigens** (IpA-D), that contribute to the process. Once inside the HeLa cells, the shigellae either are released or escape from the phagocytic vesicle, where they multiply in the cytoplasm. Actin polymerization propels the shigellae within a HeLa cell and from one cell into another. In vivo the shigellae adhere to integrins on the surface of M cells in Peyer's patches and not to the polarized absorptive cells of the mucosa. M cells normally sample antigens and present them to macrophages in the submucosa. The shigellae are phagocytosed by the M cells and pass through the M cells into the underlying collection of macrophages. Shigellae inside the M cells and macrophages can cause these cells to die by activating the normal cell death process (apoptosis). The shigellae spread to adjacent mucosal cells in a manner similar to the in vitro model, by actin polymerization that propels the bacteria.

From studies using cells in vitro, it appears that the adherence-invasion process with *Y enterocolitica* is similar to that of *Shigella*. Yersiniae adhere to the host cell membrane and cause it to extrude protoplasmic projections. The bacteria are then engulfed by the host cell with vacuole formation. Invasion is enhanced when the bacteria are grown at 22°C rather than at 37°C. Once yersiniae have entered the cell, the vacuolar membrane dissolves and the bacteria are released into the cytoplasm. In vivo, the yersiniae are thought to adhere to and invade the M cells of Peyer's patches rather than the polarized absorptive mucosal cells, much like shigellae.

L monocytogenes from the environment is ingested in food. Presumably, the bacteria adhere to and invade the intestinal mucosa, reach the bloodstream, and disseminate. The pathogenesis of this process has been studied in vitro. *L monocytogenes* adheres to and readily invades macrophages and cultured undifferentiated intestinal cells. The listeriae induce engulfment by the host cells. A protein, **internalin**, has a primary role in this process. The engulfment process, movement within a cell and movement between cells, requires actin polymerization to propel the bacteria, as with shigellae.

Legionella pneumophila infects pulmonary macrophages and causes pneumonia. Adherence of the legionellae to the macrophage induces formation of a long, thin pseudopod which then coils around the bacteria, forming a vesicle

(coiling phagocytosis). The vesicle remains intact, phago-lysosome fusion is inhibited, and the bacteria multiply within the vesicle.

N gonorrhoeae uses pili as primary adhesins and **opacity associated proteins (Opa)** as secondary adhesins to host cells. Certain Opa proteins mediate adherence to polymorphonuclear cells. Some gonococci survive after phagocytosis by these cells. Pili and Opa together enhance the invasion of cells cultured in vitro. In uterine (fallopian) tube organ cultures, the gonococci adhere to the microvilli of nonciliated cells and appear to induce engulfment by these cells. The gonococci multiply intracellularly and migrate to the subepithelial space by an unknown mechanism.

Toxins

Toxins produced by bacteria are generally classified into two groups: extracellular toxins often called exotoxins and endotoxins. The primary features of the two groups are listed in Table 9–4.

A. Exotoxins

Many gram-positive and gram-negative bacteria produce exotoxins of considerable medical importance. Some of these toxins have had major roles in world history. For example, tetanus caused by the toxin of *C tetani* killed as many as 50,000 soldiers of the Axis powers in World War II; the Allied forces, however, immunized military personnel against tetanus, and very few died of that disease. Vaccines have been developed for some of the exotoxin-mediated diseases and continue to be important in the prevention of disease. These

vaccines—called **toxoids**—are made from exotoxins, which are modified so that they are no longer toxic. Many exotoxins consist of A and B subunits. The B subunit generally mediates adherence of the toxin complex to a host cell and aids entrance of the exotoxin into the host cell. The A subunit provides the toxic activity. Examples of some pathogenetic mechanisms associated with exotoxins are given below. Other toxins of specific bacteria are discussed in the chapters covering those bacteria.

C diphtheriae is a gram-positive rod that can grow on the mucous membranes of the upper respiratory tract or in minor skin wounds (Chapter 12). Strains of *C diphtheriae* that carry a temperate bacteriophage with the structural gene for the toxin are toxigenic and produce **diphtheria toxin** and cause **diphtheria**. Many factors regulate toxin production; when the availability of inorganic iron is the factor limiting the growth rate, then maximal toxin production occurs. The toxin molecule is secreted as a single polypeptide molecule (MW 62,000). This native toxin is enzymatically degraded into two fragments, A and B, linked together by a disulfide bond. Fragment B (MW 40,700) binds to specific host cell receptors and facilitates the entry of fragment A (MW 21,150) into the cytoplasm. Fragment A inhibits peptide chain elongation factor EF-2 by catalyzing a reaction that yields free nicotinamide plus an inactive adenosine diphosphate-ribose-EF-2 complex. Arrest of protein synthesis disrupts normal cellular physiologic functions. Diphtheria toxin is very potent.

C tetani is an anaerobic gram-positive rod that causes tetanus (Chapter 11). *C tetani* from the environment contaminates wounds, and the spores germinate in the anaerobic environment of the devitalized tissue. Infection often is

TABLE 9–4 Characteristics of Exotoxins and Endotoxins (Lipopolysaccharides)

Exotoxins	Endotoxins
Excreted by living cell; high concentrations in liquid medium	Integral part of the cell wall of gram-negative bacteria. Released on bacterial death and in part during growth. May not need to be released to have biologic activity
Produced by both gram-positive and gram-negative bacteria	Found only in gram-negative bacteria
Polypeptides with a molecular weight of 10,000–900,000	Lipopolysaccharide complexes. Lipid A portion probably responsible for toxicity
Relatively unstable; toxicity often destroyed rapidly by heating at temperatures above 60°C	Relatively stable; withstand heating at temperatures above 60°C for hours without loss of toxicity
Highly antigenic; stimulate formation of high-titer antitoxin. Antitoxin neutralizes toxin	Weakly immunogenic; antibodies are antitoxic and protective. Relationship between antibody titers and protection from disease is less clear than with exotoxins
Converted to antigenic, nontoxic toxoids by formalin, acid, heat, etc. Toxoids are used to immunize (eg, tetanus toxoid)	Not converted to toxoids
Highly toxic; fatal to animals in microgram quantities or less	Moderately toxic; fatal for animals in tens to hundreds of micrograms
Usually bind to specific receptors on cells	Specific receptors not found on cells
Usually do not produce fever in the host	Usually produce fever in the host by release of interleukin-1 and other mediators
Frequently controlled by extrachromosomal genes (eg, plasmids)	Synthesis directed by chromosomal genes

minor and not clinically apparent. The vegetative forms of *C tetani* produce the toxin **tetanospasmin** (MW 150,000) that is cleaved by a bacterial protease into two peptides (MW 50,000 and MW 100,000) linked by a disulfide bond. The toxin initially binds to receptors on the presynaptic membranes of motor neurons. It then migrates by the retrograde axonal transport system to the cell bodies of these neurons to the spinal cord and brain stem. The toxin diffuses to terminals of inhibitory cells including both glycinergic interneurons and γ-aminobutyric acid-secreting neurons from the brain stem. The toxin degrades synaptobrevin, a protein required for docking of neurotransmitter vesicles on the presynaptic membrane. Release of the inhibitory glycine and γ-aminobutyric acid is blocked, and the motor neurons are not inhibited. Spastic paralysis results. Extremely small amounts of toxin can be lethal for humans. Tetanus is totally preventable in immunologically normal people by immunization with tetanus toxoid.

C botulinum causes botulism. It is found in soil or water and may grow in foods (canned, vacuum-packed, etc) if the environment is appropriately anaerobic. An exceedingly potent toxin (the most potent toxin known) is produced. It is heat-labile and is destroyed by sufficient heating. There are multiple distinct serologic types of toxin. Types A, B, and E are most commonly associated with human disease. The toxin is very similar to tetanus toxin, with a 150,000 MW protein that is cleaved into 100,000 MW and 50,000 MW proteins linked by a disulfide bond. Botulinum toxin is absorbed from the gut and binds to receptors of presynaptic membranes of motor neurons of the peripheral nervous system and cranial nerves. Proteolysis, by the light chain of botulinum toxin, of target proteins in the neurons inhibits the release of acetylcholine at the synapse, resulting in lack of muscle contraction and paralysis.

Spores of *C perfringens* are introduced into wounds by contamination with soil or feces. In the presence of necrotic tissue (an anaerobic environment), spores germinate and vegetative cells can produce several different toxins. Many of these are necrotizing and hemolytic and—together with distention of tissue by gas formed from carbohydrates and interference with blood supply—favor the spread of **gas gangrene**. The **alpha toxin** of *C perfringens* is a **lecithinase** that damages cell membranes by splitting lecithin to phosphorylcholine and diglyceride. Theta toxin also has a necrotizing effect. Collagenases and DNAses are produced by clostridiae as well.

Some *S aureus* strains growing on mucous membranes (eg, the vagina in association with menstruation), or in wounds, elaborate **toxic shock syndrome toxin-1 (TSST-1)**, which causes **toxic shock syndrome** (Chapter 13). The illness is characterized by shock, high fever, and a diffuse red rash that later desquamates; multiple other organ systems are involved as well. TSST-1 is a super antigen and stimulates lymphocytes to produce large amounts of IL-1 and TNF (Chapter 8). The major clinical manifestations of the disease appear to be secondary to the effects of the cytokines. TSST-1 may act synergistically

with low levels of lipopolysaccharide to yield the toxic effect. Many of the systemic effects of TSST-1 are similar to those of toxicity due to lipopolysaccharide (below).

Some strains of group A β-hemolytic streptococci produce **pyrogenic exotoxin A** that is similar to or the same as streptococcal erythrogenic toxin, which results in scarlet fever. Rapidly progressive soft tissue infection by streptococci that produce the pyrogenic exotoxin A has many clinical manifestations similar to those of staphylococcal toxic shock syndrome. The pyrogenic exotoxin A also is a super antigen that acts in a manner similar to TSST-1.

B. Exotoxins Associated with Diarrheal Diseases & Food Poisoning

Exotoxins associated with diarrheal diseases are frequently called enterotoxins. (See also Table 48–3.) Characteristics of some important enterotoxins are discussed below.

V cholerae has produced epidemic diarrheal disease (**cholera**) in many parts of the world (Chapter 17) and is another toxin-produced disease of historical and current importance. After entering the host via contaminated food or drink, *V cholerae* penetrates the intestinal mucosa and attaches to microvilli of the brush border of gut epithelial cells. *V cholerae,* usually of the serotype O1 (and O139), can produce an enterotoxin with a molecular weight of 84,000. The toxin consists of two subunits—A, which is split into two peptides, A_1 and A_2, linked by a disulfide bond, and B. Subunit B has five identical peptides and rapidly binds the toxin to cell membrane ganglioside molecules. Subunit A enters the cell membrane and causes a large increase in adenylate cyclase activity and in the concentration of cAMP. The net effect is rapid secretion of electrolytes into the small bowel lumen, with impairment of sodium and chloride absorption and loss of bicarbonate. Life-threatening massive diarrhea (eg, 20–30 L/day) can occur, and acidosis develops. The deleterious effects of cholera are due to fluid loss and acid-base imbalance; treatment, therefore, is by electrolyte and fluid replacement.

Some strains of *S aureus* produce enterotoxins while growing in meat, dairy products, or other foods. In typical cases, the food has been recently prepared but not properly refrigerated. There are at least six distinct types of the **staphylococcal enterotoxin**. After the preformed toxin is ingested, it is absorbed in the gut, where it stimulates neural receptors. The stimulus is transmitted to the vomiting center in the central nervous system. Vomiting, often projectile, results within hours. Diarrhea is less frequent. Staphylococcal food poisoning is the most common form of food poisoning. *S aureus* enterotoxins are super antigens.

Enterotoxins are also produced by some strains of *Y enterocolitica* (Chapter 19), *Vibrio parahaemolyticus* (Chapter 17), *Aeromonas* species (Chapter 17), and other bacteria, but the role of these toxins in pathogenesis is not as well defined. The enterotoxin produced by *C perfringens* is discussed in Chapter 11.

C. Lipopolysaccharides of Gram-Negative Bacteria

The lipopolysaccharides (LPS, endotoxin) of gram-negative bacteria are derived from cell walls and are often liberated when the bacteria lyse. The substances are heat-stable, have molecular weights between 3000 and 5000 (**lipooligosaccharides, LOS**) and several million (**lipopolysaccharides**), and can be extracted (eg, with phenol-water). They have three main regions (Table 9–5; see Figure 2–19).

The **pathophysiologic effects of LPS** are similar regardless of their bacterial origin except for those of *Bacteroides* species, which have a different structure and are less toxic (Chapter 21). LPS in the bloodstream is initially bound to circulating proteins which then interact with receptors on macrophages and monocytes and other cells of the reticuloendothelial system. IL-1, TNF, and other cytokines are released, and the complement and coagulation cascades are activated. The following can be observed clinically or experimentally: fever, leukopenia, and hypoglycemia; hypotension and shock resulting in impaired perfusion of essential organs (eg, brain, heart, kidney); intravascular coagulation; and death from massive organ dysfunction.

Injection of LPS produces **fever** after 60–90 minutes, the time needed for the body to release IL-1. Injection of IL-1 produces fever within 30 minutes. Repeated injection of IL-1 produces the same fever response each time, but repeated injection of LPS causes a steadily diminishing fever response because of tolerance due in part to reticuloendothelial blockade and in part to IgM antibodies to LPS.

Injection of LPS produces early **leukopenia**, as does bacteremia with gram-negative organisms. Secondary leukocytosis occurs later. The early leukopenia coincides with the onset of fever due to liberation of IL-1. LPS enhances glycolysis in many cell types and can lead to **hypoglycemia**.

Hypotension occurs early in gram-negative bacteremia or following injection of LPS. There may be widespread arteriolar and venular constriction followed by peripheral vascular dilatation, increased vascular permeability, decrease in venous return, lowered cardiac output, stagnation in the microcirculation, peripheral vasoconstriction, shock, and impaired organ perfusion and its consequences. Disseminated intravascular coagulation also contributes to these vascular changes.

LPS is among the many different agents that can activate the alternative pathway of the **complement cascade**, precipitating a variety of complement-mediated reactions (anaphylatoxins, chemotactic responses, membrane damage, etc) and a drop in serum levels of complement components (C3, C5–9).

Disseminated intravascular coagulation (DIC) is a frequent complication of gram-negative bacteremia and can also occur in other infections. LPS activates factor XII (Hageman factor)—the first step of the intrinsic clotting system—and sets into motion the coagulation cascade, which culminates in the conversion of fibrinogen to fibrin. At the same time, plasminogen can be activated by LPS to plasmin (a proteolytic enzyme), which can attack fibrin with the formation of fibrin split products. Reduction in platelet and fibrinogen levels and detection of fibrin split products are evidence of DIC. Heparin can sometimes prevent the lesions associated with DIC.

LPS causes platelets to adhere to vascular endothelium and occlusion of small blood vessels, causing ischemic or hemorrhagic necrosis in various organs.

Endotoxin levels can be assayed by the limulus test: A lysate of amebocytes from the horseshoe crab (limulus) gels or coagulates in the presence of 0.0001 μg/mL of endotoxin.

D. Peptidoglycan of Gram-Positive Bacteria

The peptidoglycan of gram-positive bacteria is made up of cross-linked macromolecules that surround the bacterial cells (Chapter 2 and Figure 2–15). Vascular changes leading to shock may also occur in infections due to gram-positive bacteria that contain no LPS. Gram-positive bacteria have considerably more cell wall-associated peptidoglycan than do gram-negative bacteria. Peptidoglycan released during infection may yield many of the same biologic activities as LPS, though peptidoglycan is invariably much less potent than LPS.

Enzymes

Many species of bacteria produce enzymes that are not intrinsically toxic but do play important roles in the infectious process. Some of these enzymes are discussed below.

A. Tissue-Degrading Enzymes

Many bacteria produce tissue-degrading enzymes. The best-characterized are enzymes from *C perfringens* (Chapter 11), *S aureus* (Chapter 13), group A streptococci (Chapter 14), and, to a lesser extent, anaerobic bacteria (Chapter 21). The roles of tissue-degrading enzymes in the pathogenesis of infections appear obvious but have been difficult to prove, especially those of individual enzymes. For example, antibodies against the tissue-degrading enzymes of streptococci do not modify the features of streptococcal disease.

TABLE 9–5   Composition of Lipopolysaccharide "Endotoxins" in the Cell Walls of Gram-Negative Bacteria

Chemistry	Common Name
(a) Repeating oligosaccharide (eg, man-rha-gal) combinations make up type-specific haptenic determinants (outermost in cell wall)	(a) O-specific polysaccharide. Induces specific immunity
(b) (*N*-acetylglucosamine, glucose, galactose, heptose). Same in all gram-negative bacteria	(b) Common core polysaccharide
(c) Backbone of alternating heptose and phosphate groups linked through KDO (2-keto-3-deoxyoctonic acid) to lipid. Lipid is linked to peptidoglycan (by glycoside bonds)	(c) Lipid A with KDO responsible for primary toxicity

In addition to **lecithinase**, *C perfringens* produces the proteolytic enzyme **collagenase**, which degrades collagen, the major protein of fibrous connective tissue, and promotes spread of infection in tissue.

S aureus produces **coagulase**, which works in conjunction with blood factors to coagulate plasma. Coagulase contributes to the formation of fibrin walls around staphylococcal lesions, which helps them persist in tissues. Coagulase also causes deposition of fibrin on the surfaces of individual staphylococci, which may help protect them from phagocytosis or from destruction within phagocytic cells.

Hyaluronidases are enzymes that hydrolyze hyaluronic acid, a constituent of the ground substance of connective tissue. They are produced by many bacteria (eg, staphylococci, streptococci, and anaerobes) and aid in their spread through tissues.

Many hemolytic streptococci produce **streptokinase (fibrinolysin)**, a substance that activates a proteolytic enzyme of plasma. This enzyme is then able to dissolve coagulated plasma and probably aids in the rapid spread of streptococci through tissues. Streptokinase has been used in treatment of acute myocardial infarction to dissolve fibrin clots.

Many bacteria produce substances that are **cytolysins**—ie, they dissolve red blood cells (**hemolysins**) or kill tissue cells or leukocytes (**leukocidins**). **Streptolysin O**, for example, is produced by group A streptococci and is lethal for mice and hemolytic for red blood cells from many animals. Streptolysin O is oxygen-labile and can therefore be oxidized and inactivated, but it is reactivated by reducing agents. It is antigenic. The same streptococci also produce oxygen-stable, serum-inducible **streptolysin S**, which is not antigenic. Clostridia produce various hemolysins, including the lecithinase described above. Hemolysins are produced by most strains of *S aureus*; staphylococci also produce leukocidins. Most gram-negative rods isolated from sites of disease produce hemolysins. For example, *E coli* strains that cause urinary tract infections typically produce hemolysins, whereas those strains that are part of the normal gastrointestinal flora may or may not produce hemolysins.

B. IgA1 Proteases

Immunoglobulin A is the secretory antibody on mucosal surfaces. It has two primary forms, IgA1 and IgA2, that differ near the center, or hinge region of the heavy chains of the molecules (Chapter 8). IgA1 has a series of amino acids in the hinge region that are not present in IgA2. Some bacteria that cause disease produce enzymes, **IgA1 proteases**, that split IgA1 at specific proline-threonine or proline-serine bonds in the hinge region and inactivate its antibody activity. IgA1 protease is an important virulence factor of the pathogens *N gonorrhoeae*, *N meningitidis*, *H influenzae*, and *S pneumoniae*. The enzymes are also produced by some strains of *Prevotella melaninogenica*, some streptococci associated with dental disease, and a few strains of other species that occasionally cause disease. Nonpathogenic species of the same genera do not have genes coding for the enzyme and do not produce it. Production of IgA1 protease allows pathogens to inactivate the primary antibody found on mucosal surfaces and thereby eliminate protection of the host by the antibody.

Antiphagocytic Factors

Many bacterial pathogens are rapidly killed once they are ingested by polymorphonuclear cells or macrophages. Some pathogens evade phagocytosis or leukocyte microbicidal mechanisms by adsorbing normal host components to their surfaces. For example, *S aureus* has surface protein A, which binds to the Fc portion of IgG. Other pathogens have surface factors that impede phagocytosis—eg, *S pneumoniae*, *N meningitidis*; many other bacteria have polysaccharide capsules. *S pyogenes* (group A streptococci) has M protein. *N gonorrhoeae* (gonococci) has pili. Most of these antiphagocytic surface structures show much antigenic heterogeneity. For example, there are more than 90 pneumococcal capsular polysaccharide types and more than 150 M protein types of group A streptococci. Antibodies against one type of the antiphagocytic factor (eg, capsular polysaccharide, M protein) protect the host from disease caused by bacteria of that type but not from those with other antigenic types of the same factor.

A few bacteria (eg, *Capnocytophaga* and *Bordetella*) produce soluble factors or toxins that inhibit chemotaxis by leukocytes and thus evade phagocytosis by a different mechanism.

Intracellular Pathogenicity

Some bacteria (eg, *M tuberculosis*, *Brucella* species, and *Legionella* species) live and grow in the hostile environment within polymorphonuclear cells, macrophages, or monocytes. The bacteria accomplish this feat by several mechanisms: they may avoid entry into phagolysosomes and live within the cytosol of the phagocyte; they may prevent phagosome–lysosome fusion and live within the phagosome; or they may be resistant to lysosomal enzymes and survive within the phagolysosome.

Many bacteria can live within nonphagocytic cells (see previous section, Invasion of Host Cells and Tissues).

Antigenic Heterogeneity

The surface structures of bacteria (and of many other microorganisms) have considerable antigenic heterogeneity. Often these antigens are used as part of a serologic classification system for the bacteria. The classification of the 2000 or so different salmonellae is based principally on the types of the O (lipopolysaccharide side chain) and H (flagellar) antigens. Similarly, there are more than 150 *E coli* O types and more than 100 *E coli* K (capsule) types. The antigenic type of the bacteria may be a marker for virulence, related to the clonal nature of pathogens, though it may not actually be the virulence factor (or factors). *V cholerae* O antigen type 1 and O antigen type 139 typically produce cholera toxin, whereas very few of the many other O types produce the toxin. Only some of the group A streptococcal M protein types are associated with a high incidence of poststreptococcal glomerulonephritis. *N meningitidis* capsular polysaccharide types A and C are associated with epidemic meningitis. In the examples cited above and in other typing systems that use surface antigens in serologic classification, antigenic types for a given isolate of the species remain constant during infection and on subculture of the bacteria.

Some bacteria and other microorganisms have the ability to make frequent shifts in the antigenic form of their surface structures in vitro and presumably in vivo. One well-known example is *Borrelia recurrentis,* which causes relapsing fever. A second widely studied example is *N gonorrhoeae* (see Chapter 20). The gonococcus has three surface-exposed antigens that switch forms at very high rates of about one in every 1000: lipooligosaccharide, 6–8 types; pili, innumerable types; and Opa, 10–12 types for each strain. The number of antigenic forms is so large that each strain of *N gonorrhoeae* appears to be antigenically distinct from every other strain. Switching of forms for each of the three antigens appears to be under the control of different genetic mechanisms. It is presumed that frequent switching of antigenic forms allows gonococci to evade the host's immune system; gonococci that are not attacked by the immune system survive and cause disease.

Bacterial Secretion Systems

Bacterial secretion systems are important in the pathogenesis of infection and are essential for the interaction of bacteria with the eukaryotic cells of the host. The gram-negative bacteria have cell walls with cytoplasmic membranes and outer membranes; a thin layer of peptidoglycan is present. Gram-positive bacteria have a cytoplasmic membrane and a very thick layer of peptidoglycan (Chapter 2). Some gram-negative bacteria and some gram-positive bacteria have capsules as well. The complexity and rigidity of the cell wall structures necessitate mechanisms for the translocation of proteins across the membranes. These secretion systems are involved in cellular functions such as the transport of proteins that make pili or flagella and in the secretion of enzymes or toxins into the extracellular environment. The differences in cell wall structure between gram-negative and gram-positive bacteria result in some differences in the secretion systems. The basic mechanisms of the different bacterial secretion systems are discussed in Chapter 2. (Note: The specific bacterial secretion systems were named in the order of their discovery and not by their mechanisms of action.)

Both gram-negative and gram-positive bacteria have a general secretion pathway (*sec*) as the major mechanism for protein secretion. This pathway is involved in the insertion of most of the bacterial membrane proteins and provides the major pathway for proteins crossing the bacterial cytoplasmic membrane. Some bacterial secretion systems are *sec*-dependent including the **type V secretion system**, which have autotransporters, two partner secretion pathways, and chaperone/usher pathways and the complex **type II secretion system**. The type V and type II pathways function in gram-negative bacteria to transport proteins across the outer membrane after the proteins have been translocated across the

cytoplasmic membrane by the general secretion pathway. The *sec*-independent pathways include the **type I secretion system** or **ABC secretion system** (ATP binding cassette) and the **type III secretion system**. The type I and type III pathways do not interact with proteins that have been transported across the cytoplasmic membrane by the Sec system. Instead these systems translocate proteins across both the cytoplasmic and outer membranes. The type III, which is activated upon contact with a eukaryotic host cell, promotes transport of proteins directly from inside the bacterium to the inside of the host cell using a needle-like structure; once in the host cell cytoplasm, the transported proteins can manipulate host cell function. The **type IV secretion pathway** can be either *sec*-dependent or *sec*-independent and is able to transport proteins or DNA. Some examples of the secretion systems and their roles in pathogenesis are shown in Table 9–6. These examples are but a small sample designed to illustrate the roles of the large number of molecular secretion activities used by bacteria to provide nutrients and facilitate their pathogenesis.

The Requirement for Iron

Iron is an essential nutrient for the growth and metabolism of nearly all microorganisms and is an essential cofactor of numerous metabolic and enzymatic processes. The availability of iron in humans for microbial assimilation is limited because the iron is sequestered by the high-affinity iron-binding proteins transferrin in serum and lactoferrin on mucosal surfaces. The ability of a microbial pathogen to efficiently obtain iron from the host environment is critical to its ability to cause disease. The requirement for iron, how bacteria acquire iron, and bacterial iron metabolism are discussed in Chapter 5.

Iron availability affects the virulence of many pathogens. For example, iron is an essential virulence factor in *Pseudomonas aeruginosa.* The use of animal models in *Listeria monocytogenes* infection has demonstrated that increased iron results in enhanced susceptibility to infection while iron depletion results in prolonged survival; iron supplementation therapy yields an increase in lethal infections.

Decreased iron availability can also be important in pathogenesis. For example, the gene for diphtheria toxin resides on a lysogenic bacteriophage and only strains of *C diphtheriae* that carry the lysogenic bacteriophage are toxigenic. In the presence of low iron availability there is increased production of diphtheria toxin and potentially more severe disease. The virulence of *N meningitidis* for mice is increased 1000-fold or more when the bacteria are grown under iron-limited conditions.

Human iron deficiency also plays a role in the infectious process. Iron deficiency affects hundreds of millions of people worldwide. Iron deficiency can affect multiple organ systems including the immune system and can result in impaired cell-mediated immunity and decreased polymorphonuclear cell function. Providing iron therapy during an active infection probably should be delayed because many pathogenic microorganisms can utilize the small amounts of supplemental iron resulting in an increase in virulence.

TABLE 9-6 Examples of Molecules Translocated by Bacterial Secretion Systems and Their Relevance to Pathogenesis

Secretion System	Genus Species	Substrate/Role in Pathogenesis
Type I (sec-independent)	Escherichia coli	α Hemolysin makes holes in cell membranes
	Proteus vulgarus	Hemolysin
	Morganella morganii	Hemolysin
	Bordetella pertussis	Adenylate cyclase which catalyzes synthesis of cAMP
	Pseudomonas aeruginosa	Alkaline protease
	Serratia marcescens	Zn protease yields host cell damage
Type II (can utilize sec or tat)	Pseudomonas aeruginosa	Elastase, exotoxin A, phospholipase C, others
	Legionella pneumophila	Acid phosphatase, lipase, phospholipase, protease, RNAse
	Vibrio cholera	Cholera toxin
	Serratia marcescens	Hemolysin
Type III (sec-independent; contact-dependent)	Yersinia species	Ysc-Yop system; toxins that block phagocytosis and induce apoptosis
	Pseudomonas aeruginosa	Cytotoxin
	Shigella species	Controls host cell signaling, invasion, and death
	Salmonella enterica subspecies enterica serotypes Choleraesuis, Dublin, Paratyphi, Typhi, Typhimurium, etc	Effectors from Salmonella pathogenicity Islands I and II (SPI1 and SPI2), which promote attachment to and invasion of host cells
	Escherichia coli	Enterohemorrhagic (EHEC) and enteropathogenic (EPEC); disruption of epithelial barriers and tight junctions
	Vibrio parahaemolyticus	Direct cytotoxicity
Type IV (sec-dependent and sec-independent)		
Protein substrates	Bordetella pertussis	Pertussis toxin
	Helicobacter pylori	Cytotoxin
DNA substrates	Neisseria gonorrhoeae	DNA export system
	Helicobacter pylori	DNA uptake and release system
Type V	Neisseria gonorrhoeae	IgA1 protease splits IgA1 in hinge region and destroys antibody activity (sec-dependent)
	Haemophilus influenzae	IgA1 protease, adhesins
	Escherichia coli	Serine protease, adhesins, type 1 pili, P-pili
	Shigella flexneri	Serine protease
	Serratia marscesens	Proteases
	Bordetella species	Adhesins
	Bordetella pertussis	Filamentous hemagglutinin
	Yersinia pestis	Capsular antigen

The Role of Bacterial Biofilms

A biofilm is an aggregate of interactive bacteria attached to a solid surface or to each other and encased in an exopolysaccharide matrix. This is distinct from planktonic or free-living bacterial, in which interactions of the microorganisms do not occur in the same way. Biofilms form a slimy coat on solid surfaces and occur throughout nature. A single species of bacteria may be involved or more than one species may coaggregate to form a biofilm. Fungi—including yeasts—are occasionally involved. Once a biofilm is formed quorum sensing molecules produced by the bacteria in the biofilm accumulate resulting in a modification of the metabolic activity of the bacteria. The basic biology of biofilm exopolysaccharide (glycocalyx) is discussed in Chapter 2; the quorum sensing molecules are discussed in Chapter 1.

The bacteria in the exopolysaccharide matrix may be protected from the host's immune mechanisms. This matrix also functions as a diffusion barrier for some antimicrobials, while other antimicrobials may bind to it. Some of the bacteria within the biofilm show marked resistance to antimicrobials in comparison to the same strain of bacteria grown free-living in broth, which helps to explain why it is so difficult to treat infections associated with biofilms.

Biofilms are important in human infections that are persistent and difficult to treat. A few examples include *Staphylococcus epidermidis* and *S aureus* infections of central venous catheters, eye infections such as that occur with contact lenses and intraocular lenses, in dental plaque, and in prosthetic joint infections. Perhaps the most profound example of a biofilm in human infection is in *P aeruginosa* airway infections in cystic fibrosis patients. There are many other examples.

REVIEW QUESTIONS

1. A 22-year-old woman who works in a plant nursery presents with a history of fever and cough for 2 months. Over this period of time she has lost 5 kg. Chest x-ray shows bilateral upper lobe infiltrates with cavities. A stain of her sputum shows acid-fast bacilli. The likely means by which the patient acquired her infection is

   (A) Sexual activity
   (B) Ingesting the microorganisms in her food
   (C) Holding onto contaminated hand rails when she takes public transportation
   (D) Handling potting soil
   (E) Breathing aerosolized droplets containing the microorganism

2. During a pandemic of a well-characterized disease, a group of 175 airline passengers flew from Lima, Peru, to Los Angeles. Lunch on the plane included crab salad, which was eaten by about two-thirds of the passengers. After landing in Los Angeles, many of the passengers transferred to other flights with destinations in other parts of California and other Western states. Two of the passengers who stayed in Los Angeles developed severe watery diarrhea. The status of the other passengers was unknown. The likely cause of the diarrhea in the two passengers is

   (A) *Escherichia coli* O157:H7 (lipopolysaccharide O antigen 157; flagellar antigen 7)
   (B) *Vibrio cholerae* type O139 (lipopolysaccharide O antigen 139)
   (C) *Shigella dysenteriae* type 1
   (D) *Campylobacter jejuni*
   (E) *Entamoeba histolytica*

3. A 65-year-old woman has a long-term central venous catheter for intravenous therapy. She develops fever and subsequently has multiple blood cultures positive for *Staphylococcus epidermidis*. All the *Staphylococcus epidermidis* isolates have the same colony morphology and antimicrobial susceptibility pattern suggesting they are the same strain. A *Staphylococcus epidermidis* biofilm is thought to be present on the catheter. Which one of the following statements about such an infection is correct?

   (A) The biofilm containing the *Staphylococcus epidermidis* is likely to wash off the catheter.
   (B) Production of an extracellular polysaccharide inhibits growth of the *Staphylococcus epidermidis,* limiting the infection
   (C) The *Staphylococcus epidermidis* in the biofilm are likely to be more susceptible to antimicrobial therapy because the bacteria have decreased rates of metabolism
   (D) The quorum sensing ability of *Staphylococcus epidermidis* results in increased susceptibility to antimicrobial therapy
   (E) The complex molecular interactions within the biofilm make it difficult to provide effective antimicrobial therapy and it is likely the catheter will have to be removed to cure the infection

4. The first microorganism to satisfy Koch's postulates (in the late 19th century) was

   (A) *Treponema pallidum*
   (B) *Stenotrophomonas maltophilia*
   (C) *Mycobacterium leprae*
   (D) *Bacillus anthracis*
   (E) *Neisseria gonorrhoeae*

5. Which of the following statements about lipopolysaccharide is correct?

   (A) It interacts with macrophages and monocytes yielding release of cytokines
   (B) The toxic component is the O side chain
   (C) It forms holes in red blood cell membranes yielding hemolysis
   (D) It causes hypothermia
   (E) It causes paralysis

6. A 27-year-old man had a rhinoplasty. A nasal tampon was placed to control the bleeding. Approximately 4 hours later, he developed headache, muscle aches, and abdominal cramps with diarrhea. He then developed an erythematous rash (resembling sunburn) over much of his body, including the palms and soles. His blood pressure is 80/50 mm Hg. The nasal tampon remained in place. His liver enzyme tests were elevated, and there was evidence of moderate renal failure. This patient's illness was likely to be caused by which of the following?

   (A) Lipopolysaccharide
   (B) Peptidoglycan
   (C) A toxin that is a superantigen
   (D) A toxin that has A and B subunits
   (E) Lecithinase (alpha toxin)

7. The organism most likely to be responsible for the patient's disease (Question 6) is

   (A) *Escherichia coli*
   (B) *Corynebacterium diphtheriae*
   (C) *Clostridium perfringens*
   (D) *Neisseria meningitidis*
   (E) *Staphylococcus aureus*

8. Which of the following is most likely to be associated with the formation of a bacterial biofilm?

    (A) Airway colonization in a cystic fibrosis patient with a mucoid (alginate-producing) strain of *Pseudomonas aeruginosa*
    (B) Urinary tract infection with *Escherichia coli*
    (C) Meningitis with *Neisseria meningitidis*
    (D) Tetanus
    (E) Impetigo caused by *Staphylococcus aureus*

9. Regarding bacterial type III secretions systems, which of the following statements is correct?

    (A) They are commonly found in gram-positive commensal bacteria.
    (B) They play an important role in the pathogenesis of toxin-induced diseases of *Clostridium* species, tetanus, botulism, gas gangrene, and pseudomembranous colitis.
    (C) They cause release of effectors of pathogenesis into the extracellular environment promoting bacterial colonization and multiplication.
    (D) They directly inject bacterial proteins into host cells across bacterial and host cell membranes promoting pathogenesis of infections.
    (E) Mutations which prevent the bacterial type III secretion from functioning enhance pathogenesis

10. Which of the following statements is correct?

    (A) Lipopolysaccharide is part of the cell wall of *Escherichia coli*
    (B) Cholera toxin is attached to the flagella of *Vibrio cholerae*
    (C) The lecithinase of *Clostridium perfringens* causes diarrhea
    (D) Toxic shock syndrome toxin-1 is produced by hemolytic stains of *Staphylococcus epidermidis*

11. A 15-year-old Bangladeshi girl develops severe watery diarrhea. The stool looks like "rice water." It is voluminous—more than 1 L in the last 90 minutes. She has no fever and seems otherwise normal except for the effects of loss of fluid and electrolytes. The most likely cause of her illness is

    (A) *Clostridium difficile* enterotoxin
    (B) A toxin with A and B subunits
    (C) *Shigella dysenteriae* type 1 that produces Shiga toxin
    (D) Enterotoxigenic *Escherichia coli* that produces heat-labile and heat-stable toxins
    (E) Staphylococcal enterotoxin F

12. The most important thing that can be done to treat the patient (Question 11) is

    (A) To give her ciprofloxacin
    (B) To give her a toxoid vaccine
    (C) To give her the appropriate antitoxin
    (D) To treat her with fluid and electrolyte replacement
    (E) To culture her stool to make the correct diagnosis, then treat specifically

13. A 23-year-old woman has a history of recurrent urinary tract infections, including at least one episode of pyelonephritis. Blood typing shows the P blood group antigen. Which of the following is likely to be the primary cause of her infections?

    (A) *Escherichia coli* that produce heat-stable toxin
    (B) *Escherichia coli* with K1 (capsular type 1) antigen
    (C) *Escherichia coli* O139 (lipopolysaccharide O antigen 139)
    (D) *Escherichia coli* with P-pili (fimbriae)
    (E) *Escherichia coli* O157:H7 (lipopolysaccharide O antigen 157; flagellar antigen 7)

14. A 55-year-old man presents with gradually increasing weight loss, abdominal pain, diarrhea, and arthropathy. During the evaluation process, a small bowel biopsy is done. After processing, examination of the specimen by light microscopy reveals periodic acid-Schiff (PAS)-positive inclusions in the bowel wall. Which of the following tests could be done to confirm the diagnosis of Whipple disease, caused by *Tropheryma whipplei*?

    (A) Culture on agar media
    (B) Polymerase chain reaction (PCR) amplification and sequencing of an appropriate segment of DNA
    (C) Cocultivation with *Escherichia coli*
    (D) In situ hybridization
    (E) Direct fluorescent antibody test

15. Which of the following best describes the mechanism of action of diphtheria toxin?

    (A) Forms pores in red blood cells causing hemolysis
    (B) Degrades lecithin in eukaryotic cell membranes (C) Causes release of tumor necrosis factor
    (D) Inhibits elongation factor 2
    (E) Causes increased adenylate cyclase activity

Answers

1. E	5. A	9. D	13. D
2. B	6. C	10. A	14. B
3. E	7. E	11. B	15. D
4. D	8. A	12. D	

REFERENCES

Barton LL: *Structural and Functional Relationships in Prokaryotes.* Springer, 2005.

Coburn B, Sekirov, Finlay BB: Type III secretion systems and disease. Clin Microbiol Rev 2007;20:535.

Costerton JW et al: Bacterial biofilms: A common cause of persistent infections. Science 1999;284:1318.

Falkow S: Molecular Koch's postulates applied to microbial pathogenicity. Rev Infect Dis 1988;10(Suppl 3):S274.

Fredricks DN, Relman DA: Sequence-based identification of microbial pathogens: A reconsideration of Koch's postulates. Clin Microbiol Rev 1996;9:18.

Götz F: MicroReview: *Staphylococcus* and biofilms. Mol Microbiol 2002;43:1367.

Nickerson CA, Schurr MJ (eds): *Molecular Paradigms of Infectious Disease: A Bacterial Perspective.* Springer, 2006.

Relman DA, Falkow S: A molecular perspective of microbial pathogenicity. In: *Mandell, Douglas and Bennett's Principles and Practice of Infectious Diseases*, 7th ed. Mandell GL, et al (editors). Elsevier, 2010.

Salyers AA, Whitt DD: *Bacterial Pathogenesis*, 2nd ed. American Society for Microbiology, 2002.

Schmidt H, Hensel M: Pathogenicity islands in bacterial pathogenesis. Clin Microbiol Rev 2004;17:14.

Schroeder GN, Hilbi H: Molecular pathogenesis of *Shigella* spp.: controlling host cell signaling, invasion, and death by type III secretion. Clin Microbiol Rev 2008;21:134.

Normal Human Microbiota

The term "normal microbial flora or microbiota" denotes the population of microorganisms that inhabit the skin and mucous membranes of healthy normal persons. Research has shown that these "normal flora" now referred to as "normal microbiota" provide a first line of defense against microbial pathogens, assist in digestion, play a role in toxin-degradation, and contribute to maturation of the immune system. Shifts in the normal microbiota or stimulation of inflammation by these commensals may cause diseases such as inflammatory bowel disease. In a broad attempt to understand the role played by resident microbial ecosystems in human health and disease, in 2007, the National Institutes of Health launched the Human Microbiome Project. One aspect of this project involves having several research groups simultaneously embark upon surveying the microbial communities on human skin and in mucosal areas such as the mouth, esophagus, stomach, colon, and vagina using small-subunit (16S) ribosomal RNA gene sequencing. Numerous observations have already been made. For example, it has been determined that there are large differences between individuals in terms of the numbers and types of species of microorganisms inhabiting the colon and that obesity may be correlated with the types of microbes involved in specific metabolic pathways in the gastrointestinal tract. The reader should be aware that this field is rapidly evolving and our understanding of the human microbiota will necessarily change as more information about resident microbial communities becomes available through the Human Microbiome Project.

The skin and mucous membranes always harbor a variety of microorganisms that can be arranged into two groups: (1) The resident flora consists of relatively fixed types of microorganisms regularly found in a given area at a given age; if disturbed, it promptly reestablishes itself. (2) The transient flora consists of nonpathogenic or potentially pathogenic microorganisms that inhabit the skin or mucous membranes for hours, days, or weeks; it is derived from the environment, does not produce disease, and does not establish itself permanently on the surface. Members of the transient flora are generally of little significance so long as the normal resident flora remains intact. However, if the resident flora is disturbed, transient microorganisms may colonize, proliferate, and produce disease.

Organisms frequently encountered in specimens obtained from various areas of the human body—and considered normal flora—are listed in Table 10–1. The classification of anaerobic normal bacterial flora is discussed in Chapter 21.

It is likely that microorganisms that can be cultured in the laboratory represent only a fraction of those that are part of the normal resident or transient microbial flora. When the broad range polymerase chain reaction is used to amplify bacterial 16S rDNA many previously unidentified bacteria can be detected, as in secretions from patients with bacterial vaginosis. The number of species that make up the normal microbiota has been shown to be much greater than is recognized. Thus, the understanding of normal microbiota is in transition. As mentioned above, the relationship of previously unidentified microorganisms, which are potentially part of the normal microbiota, to disease is likely to change.

ROLE OF THE RESIDENT MICROBIOTA

The microorganisms that are constantly present on body surfaces are commensals. Their flourishing in a given area depends upon physiologic factors of temperature, moisture, and the presence of certain nutrients and inhibitory substances. Their presence is not essential to life, because "germ-free" animals can be reared in the complete absence of a normal microbial flora. Yet the resident flora of certain areas plays a definite role in maintaining health and normal function. Members of the resident flora in the intestinal tract synthesize vitamin K and aid in the absorption of nutrients. On mucous membranes and skin, the resident flora may prevent colonization by pathogens and possible disease through "bacterial interference." The mechanism of bacterial interference may involve competition for receptors or binding sites on host cells, competition for nutrients, mutual inhibition by metabolic or toxic products, mutual inhibition by antibiotic materials or bacteriocins, or other mechanisms. Suppression of the normal microbiota clearly creates a partial local void that tends to be filled by organisms from the environment or from other parts of the body. Such organisms behave as opportunists and may become pathogens.

TABLE 10–1   Normal Bacterial Microbiota

Skin

Staphylococcus epidermidis

Staphylococcus aureus (in small numbers)

Micrococcus species

α-Hemolytic and nonhemolytic streptococci (eg *Streptococcus mitis*)

Corynebacterium species

Propionibacterium species

Peptostreptococcus species

Acinetobacter species

Small numbers of other organisms (*Candida* species, *Pseudomonas aeruginosa*, etc)

Nasopharynx

Any amount of the following: diphtheroids, nonpathogenic *Neisseria* species, α-hemolytic streptococci; *S epidermidis*, nonhemolytic streptococci, anaerobes (too many species to list; varying amounts of *Prevotella* species, anaerobic cocci, *Fusobacterium* species, etc)

Lesser amounts of the following when accompanied by organisms listed above: yeasts, *Haemophilus* species, pneumococci, *S aureus*, gram-negative rods, *Neisseria meningitidis*

Gastrointestinal tract and rectum

Various Enterobacteriaceae except *Salmonella, Shigella, Yersinia, Vibrio*, and *Campylobacter* species

Glucose non-fermenting gram-negative rods

Enterococci

α-Hemolytic and nonhemolytic streptococci

Diphtheroids

Staphylococcus aureus in small numbers

Yeasts in small numbers

Anaerobes in large numbers (too many species to list)

Genitalia

Any amount of the following: *Corynebacterium* species, *Lactobacillus* species, α-hemolytic and nonhemolytic streptococci, nonpathogenic *Neisseria* species

The following when mixed and not predominant: enterococci, Enterobacteriaceae and other gram-negative rods, *Staphylococcus epidermidis*, *Candida albicans*, and other yeasts

Anaerobes (too many to list); the following may be important when in pure growth or clearly predominant: *Prevotella, Clostridium*, and *Peptostreptococcus* species

On the other hand, members of the normal microbiota may themselves produce disease under certain circumstances. These organisms are adapted to the noninvasive mode of life defined by the limitations of the environment. If forcefully removed from the restrictions of that environment and introduced into the bloodstream or tissues, these organisms may become pathogenic. For example, streptococci of the viridans group are the most common resident organisms of the upper respiratory tract. If large numbers of them are introduced into the bloodstream (eg, following tooth extraction or oral surgery), they may settle on deformed or prosthetic heart valves and produce infective endocarditis. Small numbers occur transiently in the bloodstream with minor trauma (eg, dental scaling or vigorous brushing). *Bacteroides* species are the most common resident bacteria of the large intestine and are quite harmless in that location. If introduced into the free peritoneal cavity or into pelvic tissues along with other bacteria as a result of trauma, they cause suppuration and bacteremia. There are many other examples, but the important point is that the normal resident microbiota are harmless and may be beneficial in their normal location in the host and in the absence of coincident abnormalities. They may produce disease if introduced into foreign locations in large numbers and if predisposing factors are present.

NORMAL MICROBIOTA OF THE SKIN

Because of its constant exposure to and contact with the environment, the skin is particularly apt to contain transient microorganisms. Nevertheless, there is a constant and well-defined resident flora, modified in different anatomic areas by secretions, habitual wearing of clothing, or proximity to mucous membranes (mouth, nose, and perineal areas).

The predominant resident microorganisms of the skin are aerobic and anaerobic diphtheroid bacilli (eg, *Corynebacterium, Propionibacterium*); nonhemolytic aerobic and anaerobic staphylococci (*S epidermidis* and other coagulase-negative staphylococci, occasionally *S aureus,* and *Peptostreptococcus* species); gram-positive, aerobic, spore-forming bacilli that are ubiquitous in air, water, and soil; α-hemolytic streptococci (viridans streptococci) and enterococci (*Enterococcus* species); and gram-negative coliform bacilli and *Acinetobacter*. Fungi and yeasts are often present in skin folds; acid-fast, nonpathogenic mycobacteria occur in areas rich in sebaceous secretions (genitalia, external ear).

Among the factors that may be important in eliminating nonresident microorganisms from the skin are the low pH, the fatty acids in sebaceous secretions, and the presence of lysozyme. Neither profuse sweating nor washing and bathing can eliminate or significantly modify the normal resident flora. The number of superficial microorganisms may be diminished by vigorous daily scrubbing with soap containing hexachlorophene or other disinfectants, but the flora is rapidly replenished from sebaceous and sweat glands even when contact with other skin areas or with the environment is completely excluded. Placement of an occlusive dressing on skin tends to result in a large increase in the total microbial population and may also produce qualitative alterations in the flora.

Anaerobes and aerobic bacteria often join to form synergistic infections (gangrene, necrotizing fasciitis, cellulitis) of skin and soft tissues. The bacteria are frequently part of the normal microbial flora. It is usually difficult to pinpoint one specific organism as being responsible for the progressive lesion, since mixtures of organisms are usually involved.

NORMAL MICROBIOTA OF THE MOUTH & UPPER RESPIRATORY TRACT

The flora of the nose consists of prominent corynebacteria, staphylococci *(S epidermidis, S aureus),* and streptococci.

The mucous membranes of the mouth and pharynx are often sterile at birth but may be contaminated by passage through the birth canal. Within 4–12 hours after birth, viridans streptococci become established as the most prominent members of the resident flora and remain so for life. They probably originate in the respiratory tracts of the mother and attendants. Early in life, aerobic and anaerobic staphylococci, gram-negative diplococci (neisseriae, *Moraxella catarrhalis*), diphtheroids, and occasional lactobacilli are added. When teeth begin to erupt, the anaerobic spirochetes, *Prevotella* species (especially *Prevotella melaninogenica*), *Fusobacterium* species, *Rothia* species, and *Capnocytophaga* species (see below) establish themselves, along with some anaerobic vibrios and lactobacilli. *Actinomyces* species are normally present in tonsillar tissue and on the gingivae in adults, and various protozoa may also be present. Yeasts (*Candida* species) occur in the mouth.

In the pharynx and trachea, a similar flora establishes itself, whereas few bacteria are found in normal bronchi. Small bronchi and alveoli are normally sterile. The predominant organisms in the upper respiratory tract, particularly the pharynx, are nonhemolytic and α-hemolytic streptococci and neisseriae. Staphylococci, diphtheroids, haemophili, pneumococci, mycoplasmas, and prevotellae are also encountered.

Infections of the mouth and respiratory tract are usually caused by mixed oronasal flora, including anaerobes. Periodontal infections, perioral abscesses, sinusitis, and mastoiditis may involve predominantly *P melaninogenica, Fusobacteria,* and *Peptostreptococci.* Aspiration of saliva (containing up to 10^2 of these organisms and aerobes) may result in necrotizing pneumonia, lung abscess, and empyema.

The Role of the Normal Mouth Microbiota in Dental Caries

Caries is a disintegration of the teeth beginning at the surface and progressing inward. First the surface enamel, which is entirely noncellular, is demineralized. This has been attributed to the effect of acid products of bacterial fermentation. Subsequent decomposition of the dentin and cement involves bacterial digestion of the protein matrix.

Dental plaque has come to be viewed and managed as a complex biofilm, which can be defined simplistically as an accumulation of microbes within a matrix. The advantages for the microbes in the biofilm include protection from environmental hazards (including antimicrobials) and optimization of spatial arrangements that maximize energy through movement of nutrients. Organisms within the biofilm interact dynamically at multiple metabolic and molecular levels. In dental plaque, the initial colonizing organisms that lead to a slime layer are mainly gram-positive cocci and gram-positive rods that form microcolonies on the hard, smooth enamel surface. The plaque or biofilm consists mainly of gelatinous deposits of high-molecular-weight glucans in which acid-producing bacteria adhere to the enamel. The carbohydrate polymers (glucans) are produced mainly by streptococci (*Streptococcus mutans,* peptostreptococci), perhaps in association with actinomycetes. There appears to be a strong correlation between the presence of *S mutans* and caries on specific enamel areas. The essential second step in caries production appears to be the formation of large amounts of acid (pH < 5.0) from carbohydrates by streptococci and lactobacilli in the plaque. High concentrations of acid demineralize the adjoining enamel and initiate caries.

In experimental "germ-free" animals, cariogenic streptococci can induce the formation of plaque and caries. Adherence to smooth surfaces requires both the synthesis of water-insoluble glucan polymers by glucosyltransferases and the participation of binding sites on the surface of microbial cells. (Perhaps carbohydrate polymers also aid the attachment of some streptococci to endocardial surfaces.) Other members of the oral microflora, eg, veillonellae, may complex with glucosyltransferase of *Streptococcus salivarius* in saliva and then synthesize water-insoluble carbohydrate polymers to adhere to tooth surfaces. Adherence may be initiated by salivary IgA antibody to *S mutans*. Certain diphtheroids and streptococci that produce levans can induce specific soft tissue damage and bone resorption typical of periodontal disease. Proteolytic organisms, including actinomycetes and bacilli, play a role in the microbial action on dentin that follows damage to the enamel. As the biofilm matures in the absence of good oral hygiene, there is cross-linking with *Fusobacterium* species followed by, in the third stage, predominantly gram-negative bacteria. The development of caries also depends on genetic, hormonal, nutritional, and many other factors.

Control of caries involves physical removal of plaque, limitation of sucrose intake, good nutrition with adequate protein intake, and reduction of acid production in the mouth by limitation of available carbohydrates and frequent cleansing. The application of fluoride to teeth or its ingestion in water results in enhancement of acid resistance of the enamel. Control of periodontal disease requires removal of calculus (calcified deposit) and good mouth hygiene.

Periodontal pockets in the gingiva are particularly rich sources of organisms, including anaerobes, that are rarely encountered elsewhere. While they may participate in periodontal disease and tissue destruction, attention is drawn to them when they are implanted elsewhere, eg, producing infective endocarditis or bacteremia in a granulocytopenic host. Examples are *Capnocytophaga* species and *Rothia dentocariosa*. *Capnocytophaga* species are fusiform, gram-negative, gliding anaerobes; *Rothia* species are pleomorphic, aerobic, gram-positive rods. Both probably participate in the complex microbial flora of periodontal disease with prominent bone destruction. In granulocytopenic immunodeficient patients, they can lead to serious opportunistic lesions in other organs.

NORMAL MICROBIOTA OF THE INTESTINAL TRACT

At birth the intestine is sterile, but organisms are soon introduced with food. In breast-fed children, the intestine contains large numbers of lactic acid streptococci and lactobacilli. These aerobic and anaerobic, gram-positive, nonmotile organisms (eg, *Bifidobacterium* species) produce acid from carbohydrates and tolerate pH 5.0. In bottle-fed children, a more mixed flora exists in the bowel, and lactobacilli are less prominent. As food habits develop toward the adult pattern, the bowel flora changes. Diet has a marked influence on the relative composition of the intestinal and fecal flora. Bowels of newborns in intensive care nurseries tend to be colonized by Enterobacteriaceae, eg, *Klebsiella*, *Citrobacter*, and *Enterobacter*.

In the normal adult, the esophagus contains microorganisms arriving with saliva and food. The stomach's acidity keeps the number of microorganisms at a minimum (10^3–10^5/g of contents) unless obstruction at the pylorus favors the proliferation of gram-positive cocci and bacilli. The normal acid pH of the stomach markedly protects against infection with some enteric pathogens, eg, cholera. Administration of antacids, H_2-receptor antagonists, and proton pump inhibitors for peptic ulcer disease and gastroesophageal reflux disease leads to a great increase in microbial flora of the stomach, including many organisms usually prevalent in feces. As the pH of intestinal contents becomes alkaline, the resident flora gradually increases. In the adult duodenum, there are 10^3–10^6 bacteria per gram of contents; in the jejunum and ileum, 10^5–10^8 bacteria per gram; and in the cecum and transverse colon, 10^8–10^{10} bacteria per gram. In the upper intestine, lactobacilli and enterococci predominate, but in the lower ileum and cecum, the flora is fecal. In the sigmoid colon and rectum, there are about 10^{11} bacteria per gram of contents, constituting 60% of the fecal mass. Anaerobes outnumber facultative organisms by 1000-fold. In diarrhea, the bacterial content may diminish greatly, whereas in intestinal stasis the count rises.

In the normal adult colon, 96–99% of the resident bacterial flora consists of anaerobes: *Bacteroides* species, especially *Bacteroides fragilis*; *Fusobacterium* species; anaerobic lactobacilli, eg, bifidobacteria; clostridia (*Clostridium perfringens*, 10^3–10^5/g); and anaerobic gram-positive cocci (*Peptostreptococcus* species). Only 1–4% are facultative aerobes (gram-negative coliform bacteria, enterococci, and small numbers of protei, pseudomonads, lactobacilli, candidae, and other organisms). More than 100 distinct types of organisms, which can be cultured routinely in the laboratory, occur regularly in normal fecal flora. There probably are more than 500 species of bacteria in the colon including many that are likely unidentified. Minor trauma (eg, sigmoidoscopy, barium enema) may induce transient bacteremia in about 10% of procedures.

The important functions of intestinal microbiota can be divided into three major categories (see review by O'Hara et al in the reference section). The first of these are protective functions in which the resident bacteria displace and inhibit potential pathogens indirectly by competing for nutrients and receptors or directly through the production of antimicrobial factors, such as bacteriocins and lactic acid. Second, commensal organisms are important for the development and function of the mucosal immune system. They induce the secretion of IgA, influence the development of the intestinal humoral immune system, and modulate local T-cell responses and cytokine profiles. The third category consists of a broad range of metabolic functions. Intestinal bacteria produce short-chain fatty acids that control intestinal epithelial cell differentiation. They synthesize vitamin K, biotin, and folate and enhance ion absorption. Certain bacteria metabolize dietary carcinogens and assist with fermentation of nondigestible dietary residue. There is now evidence that gut bacteria can influence fat deposition in the host leading to obesity.

Antimicrobial drugs taken orally can, in humans, temporarily suppress the drug-susceptible components of the fecal flora. This is commonly done by the preoperative oral administration of insoluble drugs. For example, neomycin plus erythromycin can in 1–2 days suppress part of the bowel flora, especially aerobes. Metronidazole accomplishes that for anaerobes. If lower bowel surgery is performed when the counts are at their lowest, some protection against infection by accidental spill can be achieved. However, soon thereafter the counts of fecal flora rise again to normal or higher than normal levels, principally of organisms selected out because of relative resistance to the drugs employed. The drug-susceptible microorganisms are replaced by drug-resistant ones, particularly staphylococci, enterobacters, enterococci, protei, pseudomonads, *Clostridium difficile,* and yeasts.

The feeding of large quantities of *Lactobacillus acidophilus* may result in the temporary establishment of this organism in the gut and the concomitant partial suppression of other gut microflora.

The anaerobic flora of the colon, including *B fragilis,* clostridia, and peptostreptococci, plays a main role in abscess formation originating in perforation of the bowel. *Prevotella bivia* and *Prevotella disiens* are important in abscesses of the pelvis originating in the female genital organs. Like *B fragilis,* these species are penicillin-resistant; therefore, another agent should be used.

While the intestinal microbiota are normally an asset for the host, in genetically susceptible individuals, some components of the flora can result in disease. For example, inflammatory bowel diseases are felt to be associated with a loss of immune tolerance to bacterial antigens. This leads to intense inflammation caused by an exuberant immune response. Similar mechanisms may be important in intestinal malignancy such as colon cancer.

NORMAL MICROBIOTA OF THE URETHRA

The anterior urethra of both sexes contains small numbers of the same types of organisms found on the skin and perineum. These organisms regularly appear in normal voided urine in numbers of 10^2–10^4/mL.

NORMAL MICROBIOTA OF THE VAGINA

Soon after birth, aerobic lactobacilli appear in the vagina and persist as long as the pH remains acid (several weeks). When the pH becomes neutral (remaining so until puberty), a mixed flora of cocci and bacilli is present. At puberty, aerobic and anaerobic lactobacilli reappear in large numbers and contribute to the maintenance of acid pH through the production of acid from carbohydrates, particularly glycogen. This appears to be an important mechanism in preventing the establishment of other, possibly harmful microorganisms in the vagina. If lactobacilli are suppressed by the administration of antimicrobial drugs, yeasts or various bacteria increase in numbers and cause irritation and inflammation. After menopause, lactobacilli again diminish in number and a mixed flora returns.

The normal vaginal flora includes group B streptococci in as many as 25% of women of childbearing age. During the birth process, a baby can acquire group B streptococci, which subsequently may cause neonatal sepsis and meningitis. The normal vaginal flora often includes also α-hemolytic streptococci, anaerobic streptococci (peptostreptococci), *Prevotella* species, clostridia, *Gardnerella vaginalis, Ureaplasma urealyticum,* and sometimes *Listeria* or *Mobiluncus* species. The cervical mucus has antibacterial activity and contains lysozyme. In some women, the vaginal introitus contains a heavy flora resembling that of the perineum and perianal area. This may be a predisposing factor in recurrent urinary tract infections. Vaginal organisms present at time of delivery may infect the newborn (eg, group B streptococci).

NORMAL MICROBIOTA OF THE CONJUNCTIVA

The predominant organisms of the conjunctiva are diphtheroid*s, S epidermidis,* and nonhemolytic streptococci. Neisseriae and gram-negative bacilli resembling haemophili (*Moraxella* species) are also frequently present. The conjunctival flora is normally held in check by the flow of tears, which contain antibacterial lysozyme.

REVIEW QUESTIONS

1. A 26-year-old woman visits her physician because of an unusual vaginal discharge. On examination the physician observes a thin, homogeneous, white-gray discharge that adheres to the vaginal wall. The pH of the discharge is 5.5 (normal: <4.3). On Gram stain, many epithelial cells covered with gram-variable rods are seen. Bacterial vaginosis is diagnosed. Which one of the following normal genital flora microorganisms is greatly decreased in bacterial vaginosis?

   (A) *Corynebacterium* species
   (B) *Staphylococcus epidermidis*
   (C) *Prevotella* species
   (D) *Candida albicans*
   (E) *Lactobacillus* species

2. Certain microorganisms are never considered to be members of the normal flora. They are always considered to be pathogens. Which one of the following organisms fits into that category?

   (A) *Streptococcus pneumoniae*
   (B) *Escherichia coli*
   (C) *Mycobacterium tuberculosis*
   (D) *Staphylococcus aureus*
   (E) *Neisseria meningitidis*

3. A 9-year-old girl develops fever and severe pain on the right side of her throat. On examination, redness and swelling in the right peritonsillar area are seen. A peritonsillar abscess is diagnosed. The most likely organisms to be cultured from this abscess are

   (A) *Staphylococcus aureus*
   (B) *Streptococcus pneumoniae*
   (C) *Corynebacterium* species and *Prevotella melaninogenica*
   (D) Normal oral nasal flora
   (E) Viridans streptococci and *Candida albicans*

4. A 70-year-old man with a history of diverticulosis of the sigmoid colon experiences a sudden onset of severe left lower quadrant abdominal pain. Fever develops. The severe pain gradually subsides and is replaced by a constant aching pain and marked abdominal tenderness. A diagnosis of probable ruptured diverticulum is made and the patient is taken to the operating room. The diagnosis of ruptured diverticulum is confirmed and an abscess next to the sigmoid colon is found. The most likely bacteria to be found in the abscess are

   (A) Mixed normal gastrointestinal flora
   (B) *Bacteroides fragilis* alone
   (C) *Escherichia coli* alone
   (D) *Clostridium perfringens* alone
   (E) *Enterococcus* species alone

5. Antimicrobial therapy can decrease the amount of susceptible bowel flora and allow proliferation of relatively resistant colonic bacteria. Which one of the following species can proliferate and produce a toxin that causes diarrhea?

   (A) *Enterococcus* species
   (B) *Staphylococcus epidermidis*
   (C) *Pseudomonas aeruginosa*
   (D) *Clostridium difficile*
   (E) *Bacteroides fragilis*

6. Which one of the following microorganisms can be part of the normal vaginal flora and cause meningitis in newborns?

   (A) *Candida albicans*
   (B) *Corynebacterium* species
   (C) *Staphylococcus epidermidis*
   (D) *Ureaplasma urealyticum*
   (E) Group B streptococci

7. Dental plaque and periodontal disease can be thought of as a continuum of what type of physiological process?

   (A) Biofilm formation
   (B) Normal aging
   (C) Abnormal digestion
   (D) Exaggerated immune response
   (E) Chewing gum

8. Which one of the following microorganisms is closely associated with dental caries?

   (A) *Candida albicans*
   (B) *Streptococcus mutans*
   (C) *Prevotella melaninogenica*
   (D) *Neisseria subflava*
   (E) *Staphylococcus epidermidis*

9. Anaerobic bacteria such as *Bacteroides fragilis* occur in the sigmoid colon in a concentration of about 10^{11}/g of stool. At what concentration do facultative organisms such as *Escherichia coli* occur?

   (A) 10^{11}/g
   (B) 10^{10}/g
   (C) 10^{9}/g
   (D) 10^{8}/g
   (E) 10^{7}/g

10. *Streptococcus pneumoniae* can be part of the normal flora of 5–40% of people. At what anatomic site can it be found?

   (A) Conjunctiva
   (B) Nasopharynx
   (C) Colon
   (D) Urethra
   (E) Vagina

Answers

1. E	5. D	9. D
2. C	6. E	10. B
3. D	7. A	
4. A	8. B	

REFERENCES

Fredericks DN, Fielder TL, Marrazzo JM: Molecular identification of bacteria associated with bacterial vaginosis. N Engl J Med 2005;353:1899.

Granato PA: Pathogenic and indigenous microorganisms of humans. In: *Manual of Clinical Microbiology,* 9th ed. Murray PR et al (editors). ASM Press, 2007.

Hentges DJ: The anaerobic microflora of the human body. Clin Infect Dis 1993;16(Suppl 4):S175.

Mandell GL, Bennett JE, Dolin R (editors): *Mandell, Douglas and Bennett's Principles and Practice of Infectious Diseases,* 7th ed. Elsevier, 2010.

O'Hara AM, Shanahan F. The gut flora as a forgotten organ. EMBO reports 2006; 7:688.

The pathogenesis of periodontal diseases. J Periodontol 1999;70:457.

Thomas JG, Nakaishi LA. Managing the complexity of a dynamic biofilm. JADA 2006;137:10S.

Turnbaugh PJ et al: The human microbiome project. Nature 2008;449:804.

Spore-Forming Gram-Positive Bacilli: *Bacillus* & *Clostridium* Species

CHAPTER

11

The gram-positive spore-forming bacilli are the *Bacillus* and *Clostridium* species. These bacilli are ubiquitous, and because they form spores they can survive in the environment for many years. *Bacillus* species are aerobes, whereas clostridia are anaerobes.

Of the many species of *Bacillus* and related genera most do not cause disease and are not well characterized in medical microbiology. There are a few species, however, that cause important diseases in humans. Anthrax, a prototype disease in the history of microbiology, is caused by *Bacillus anthracis*. Anthrax remains an important disease of animals and occasionally of humans, and *B anthracis* is a major agent of bioterrorism and biologic warfare. *Bacillus cereus* causes food poisoning and occasionally eye or other localized infections.

The genus *Clostridium* is extremely heterogeneous and more than 190 species have been described. The list of pathogenic organisms, as well as novel species isolated from human feces whose pathogenic potential remain undetermined, continues to grow. Clostridia cause several important toxin-mediated diseases: *Clostridium tetani*, tetanus; *Clostridium botulinum*, botulism; *Clostridium perfringens*, gas gangrene; and *Clostridium difficile*, pseudomembranous colitis. Other clostridia are also found in mixed anaerobic infections in humans (see Chapter 21).

BACILLUS SPECIES

The genus *Bacillus* includes large aerobic, gram-positive rods occurring in chains. Most members of this genus are saprophytic organisms prevalent in soil, water, and air and on vegetation, such as *Bacillus cereus* and *Bacillus subtilis*. Some are insect pathogens. *B cereus* can grow in foods and produce an enterotoxin or an emetic toxin and cause food poisoning. Such organisms may occasionally produce disease in immunocompromised humans (eg, meningitis, endocarditis, endophthalmitis, conjunctivitis, or acute gastroenteritis). *B anthracis*, which causes **anthrax,** is the principal pathogen of the genus.

Morphology & Identification
A. Typical Organisms
The typical cells, measuring 1 × 3–4 μm, have square ends and are arranged in long chains; spores are located in the center of the nonmotile bacilli.

B. Culture
Colonies of *B anthracis* are round and have a "cut glass" appearance in transmitted light. Hemolysis is uncommon with *B anthracis* but common with *B cereus* and the saprophytic bacilli. Gelatin is liquefied, and growth in gelatin stabs resembles an inverted fir tree.

C. Growth Characteristics
The saprophytic bacilli utilize simple sources of nitrogen and carbon for energy and growth. The spores are resistant to environmental changes, withstand dry heat and certain chemical disinfectants for moderate periods, and persist for years in dry earth. Animal products contaminated with anthrax spores (eg, hides, bristles, hair, wool, bone) can be sterilized by autoclaving.

BACILLUS ANTHRACIS

Pathogenesis
Anthrax is primarily a disease of herbivores—goats, sheep, cattle, horses, etc; other animals (eg, rats) are relatively resistant to the infection. Humans become infected incidentally by contact with infected animals or their products. In animals, the portal of entry is the mouth and the gastrointestinal tract. Spores from contaminated soil find easy access when ingested with spiny or irritating vegetation. In humans, the infection is usually acquired by the entry of spores through injured skin (cutaneous anthrax) or rarely the mucous membranes (gastrointestinal anthrax), or by inhalation of spores into the lung (inhalation anthrax).

The spores germinate in the tissue at the site of entry, and growth of the vegetative organisms results in formation of a gelatinous edema and congestion. Bacilli spread via lymphatics to the bloodstream, and they multiply freely in the blood and tissues shortly before and after the animal's death.

B anthracis (see Figure 11–1) that does not produce a capsule is not virulent and does not induce anthrax in test animals. The poly-D-glutamic acid capsule is antiphagocytic. The capsule gene is on a plasmid.

Anthrax toxin is made up of three proteins: protective antigen (PA), edema factor (EF), and lethal factor (LF). PA

165

binds to specific cell receptors, and following proteolytic activation it forms a membrane channel that mediates entry of EF and LF into the cell. EF is an adenylyl cyclase; with PA it forms a toxin known as edema toxin. LF plus PA form lethal toxin, which is a major virulence factor and cause of death in infected animals and humans. When injected into laboratory animals (eg, rats) the lethal toxin can quickly kill the animals. The anthrax toxin genes are on another plasmid.

In inhalation anthrax (wool sorter disease), the spores from the dust of wool, hair, or hides are inhaled, phagocytosed in the lungs, and transported by the lymphatic drainage to the mediastinal lymph nodes, where germination occurs. This is followed by toxin production and the development of hemorrhagic mediastinitis and sepsis, which are

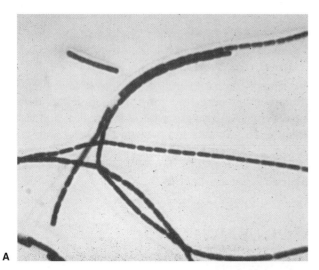

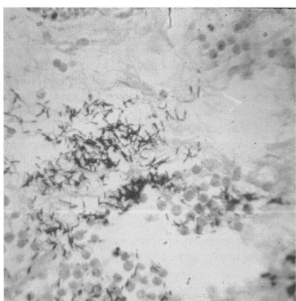

FIGURE 11–1 **A**: *Bacillus anthracis* in broth culture (original magnification ×1000). **B**: In tissue (original magnification × 400). (Courtesy of PS Brachman.)

usually rapidly fatal. In anthrax sepsis, the number of organisms in the blood exceeds 10^7/mL just prior to death. In the Sverdlovsk inhalation anthrax outbreak of 1979 and the U.S. bioterrorism inhalation cases of 2001 (see Chapter 48) the pathogenesis was the same as in inhalation anthrax from animal products.

Pathology

In susceptible animals and humans, the organisms proliferate at the site of entry. The capsules remain intact, and the organisms are surrounded by a large amount of proteinaceous fluid containing few leukocytes from which they rapidly disseminate and reach the bloodstream.

In resistant animals, the organisms proliferate for a few hours, by which time there is massive accumulation of leukocytes. The capsules gradually disintegrate and disappear. The organisms remain localized.

Clinical Findings

In humans, approximately 95% of cases are cutaneous anthrax and 5% are inhalation. Gastrointestinal anthrax is very rare; it has been reported from Africa, Asia, and the United States following occasions where people have eaten meat from infected animals.

The bioterrorism events in the fall of 2001 (see Chapter 48) resulted in 22 cases of anthrax: 11 inhalation and 11 cutaneous. Five of the patients with inhalation anthrax died. All the other patients survived.

Cutaneous anthrax generally occurs on exposed surfaces of the arms or hands, followed in frequency by the face and neck. A pruritic papule develops 1–7 days after entry of the organisms or spores through a scratch. Initially it resembles an insect bite. The papule rapidly changes into a vesicle or small ring of vesicles that coalesce, and a necrotic ulcer develops. The lesions typically are 1–3 cm in diameter and have a characteristic central black eschar. Marked edema occurs. Lymphangitis and lymphadenopathy and systemic signs and symptoms of fever, malaise, and headache may occur. After 7–10 days the eschar is fully developed. Eventually it dries, loosens, and separates; healing is by granulation and leaves a scar. It may take many weeks for the lesion to heal and the edema to subside. Antibiotic therapy does not appear to change the natural progression of the disease. In as many as 20% of patients, cutaneous anthrax can lead to sepsis, the consequences of systemic infection—including meningitis—and death.

The incubation period in inhalation anthrax may be as long as 6 weeks. The early clinical manifestations are associated with marked hemorrhagic necrosis and edema of the mediastinum. Substernal pain may be prominent, and there is pronounced mediastinal widening visible on x-ray chest films. Hemorrhagic pleural effusions follow involvement of the pleura; cough is secondary to the effects on the trachea. Sepsis occurs, and there may be hematogenous spread to the gastrointestinal tract, causing bowel ulceration, or to the meninges, causing hemorrhagic meningitis. The fatality rate in inhalation anthrax is high in the setting of known exposure; it is higher when the diagnosis is not initially suspected.

Animals acquire anthrax through ingestion of spores and spread of the organisms from the intestinal tract. This is rare in humans, and gastrointestinal anthrax is extremely uncommon. Abdominal pain, vomiting, and bloody diarrhea are clinical signs.

Diagnostic Laboratory Tests

Specimens to be examined are fluid or pus from a local lesion, blood, and sputum. Stained smears from the local lesion or of blood from dead animals often show chains of large gram-positive rods. Anthrax can be identified in dried smears by immunofluorescence staining techniques.

When grown on blood agar plates, the organisms produce nonhemolytic gray to white colonies with a rough texture and a ground-glass appearance. Comma-shaped outgrowths (Medusa head) may project from the colony. Gram stain shows large gram-positive rods. Carbohydrate fermentation is not useful. In semisolid medium, anthrax bacilli are always nonmotile, whereas related organisms (eg, *B cereus*) exhibit motility by "swarming." Virulent anthrax cultures kill mice or guinea pigs upon intraperitoneal injection. Demonstration of capsule requires growth on bicarbonate-containing medium in 5–7% carbon dioxide. Lysis by a specific anthrax γ-bacteriophage may be helpful in identifying the organism.

An enzyme-linked immunoassay (ELISA) has been developed to measure antibodies against edema and lethal toxins, but the test has not been extensively studied. Acute and convalescent sera obtained 4 weeks apart should be tested. A positive result is a fourfold change or a single titer of greater than 1:32. Some public health laboratories may also have nucleic acid amplification assays available. Clinical laboratories that recover large gram-positive rods from blood, cerebrospinal fluid, or suspicious skin lesions, which phenotypically match the description of *B anthracis* as mentioned above, should immediately contact their public health laboratory and send the organism for confirmation.

Resistance & Immunity

Immunization to prevent anthrax is based on the classic experiments of Louis Pasteur. In 1881 he proved that cultures grown in broth at 42–52°C for several months lost much of their virulence and could be injected live into sheep and cattle without causing disease; subsequently, such animals proved to be immune. Active immunity to anthrax can be induced in susceptible animals by vaccination with live attenuated bacilli, with spore suspensions, or with PAs from culture filtrates. Animals that graze in known anthrax districts should be immunized for anthrax annually.

Four countries produce vaccines for anthrax. Russia and China use attenuated spore-based vaccine administered by scarification. The United States and Great Britain use a bacteria-free filtrate of cultures adsorbed to aluminum hydroxide. The current U.S. Food and Drug Administration approved vaccine contains cell-free filtrates of a toxigenic nonencapsulated nonvirulent strain of *B anthracis*. The amount of PA present per dose is unknown and all three toxins' components (LF, EF, and PA) are present and adsorbed to aluminum hydroxide. The dose schedule is 0, 2, and 4 weeks, then 6, 12, and 18 months, followed by annual boosters. The vaccine is available only to the U.S. Department of Defense and to persons at risk for repeated exposure to *B anthracis*. Because of significant controversy in the U.S. military about the current anthrax vaccine and its use in areas where there is potential for biologic warfare, a new recombinant PA vaccine (rPA) adsorbed to aluminum hydroxide is currently in phase II clinical trials. It has been shown to be very well-tolerated and highly immunogenic (Campbell JD et al. Vaccin. 2007;3:205–11).

Treatment

Many antibiotics are effective against anthrax in humans, but treatment must be started early. Ciprofloxacin is recommended for treatment; penicillin G, along with gentamicin or streptomycin, has previously been used to treat anthrax.

In the setting of potential exposure to *B anthracis* as an agent of biologic warfare, prophylaxis with ciprofloxacin or doxycycline should be continued for 4 weeks while three doses of vaccine are being given, or for 8 weeks if no vaccine is administered.

Some other gram-positive bacilli, such as *B cereus,* are resistant to penicillin by virtue of β-lactamase production. Doxycycline, erythromycin, or ciprofloxacin may be effective alternatives to penicillin.

Epidemiology, Prevention, & Control

Soil is contaminated with anthrax spores from the carcasses of dead animals. These spores remain viable for decades. Perhaps spores can germinate in soil at pH 6.5 at proper temperature. Grazing animals infected through injured mucous membranes serve to perpetuate the chain of infection. Contact with infected animals or with their hides, hair, and bristles is the source of infection in humans. Control measures include (1) disposal of animal carcasses by burning or by deep burial in lime pits, (2) decontamination (usually by autoclaving) of animal products, (3) protective clothing and gloves for handling potentially infected materials, and (4) active immunization of domestic animals with live attenuated vaccines. Persons with high occupational risk should be immunized.

BACILLUS CEREUS

Food poisoning caused by *B cereus* has two distinct forms: the emetic type, associated with fried rice, and the diarrheal type, associated with meat dishes and sauces. *B cereus* produces toxins that cause disease that is more an intoxication than a food-borne infection. The emetic form is manifested by nausea, vomiting, abdominal cramps, and occasionally diarrhea and is self-limiting, with recovery occurring within 24 hours. It begins 1–5 hours after ingestion of rice and occasionally pasta dishes. *B cereus* is a soil organism that commonly contaminates rice. When large amounts of rice are cooked and allowed to cool slowly, the *B cereus* spores germinate, and the vegetative cells produce the toxin during log-phase growth or during sporulation. The diarrheal form has an incubation period of 1–24 hours and is manifested by profuse diarrhea with abdominal pain and cramps; fever and vomiting are uncommon. The enterotoxin may be preformed in the food or produced in the intestine. The presence of *B cereus* in a patient's stool is not sufficient to make a diagnosis of *B cereus* disease, since the bacteria may be present in normal stool specimens; a concentration of 10^5 bacteria or more per gram of food is considered diagnostic.

B cereus is an important cause of eye infections, severe keratitis, endophthalmitis, and panophthalmitis. Typically, the organisms are introduced into the eye by foreign bodies associated with trauma. *B cereus* has also been associated with localized infections and with systemic infections, including endocarditis, meningitis, osteomyelitis, and pneumonia; the presence of a medical device or intravenous drug use predisposes to these infections. *B cereus* is resistant to a variety of antimicrobial agents including penicillins and cephalosporins. Serious nonfood-borne infections should be treated with vancomycin or clindamycin with or without an aminoglycoside.

Other *Bacillus* species are rarely associated with human disease. It is difficult to differentiate superficial contamination with *Bacillus* from genuine disease caused by the organism. Five *Bacillus* species (*B thuringiensis, B popilliae* [now called *Paenibacillus popilliae*], *B sphaericus, B larvae,* and *B lentimorbus* [*Paenibacillus lentimorbus*]) are pathogens for insects, and some have been used as commercial insecticides. Genes from *B thuringiensis* coding for insecticidal compounds have been inserted into the genetic material of some commercial plants. This has been associated with concern on the part of environmental activists about genetically engineered plants and food products.

CLOSTRIDIUM SPECIES

The clostridia are large anaerobic, gram-positive, motile rods. Many decompose proteins or form toxins, and some do both. Their natural habitat is the soil or the intestinal tract of animals and humans, where they live as saprophytes. Among the pathogens are the organisms causing **botulism, tetanus, gas gangrene,** and **pseudomembranous colitis.**

Morphology & Identification

A. Typical Organisms

Spores of clostridia are usually wider than the diameter of the rods in which they are formed. In the various species, the spore is placed centrally, subterminally, or terminally. Most species of clostridia are motile and possess peritrichous flagella. A gram stain of a *Clostridium* species with terminal spores is shown in Figure 11–2.

B. Culture

Clostridia are anaerobes and grow under anaerobic conditions; a few species are aerotolerant and will also grow in ambient air. Anaerobic culture conditions are discussed in Chapter 21. In general, the clostridia grow well on the blood-enriched media used to grow anaerobes and on other media used to culture anaerobes as well.

C. Colony Forms

Some clostridia produce large raised colonies (eg, *C perfringens*); others produce smaller colonies (eg, *C tetani*). Some clostridia form colonies that spread on the agar surface. Many clostridia produce a zone of hemolysis on blood agar. *C perfringens* characteristically produces a double zone of hemolysis around colonies.

D. Growth Characteristics

Clostridia can ferment a variety of sugars; many can digest proteins. Milk is turned acid by some and digested by others and undergoes "stormy fermentation" (ie, clot torn by gas) with a third group (eg, *C perfringens*). Various enzymes are produced by different species (see below).

E. Antigenic Characteristics

Clostridia share some antigens but also possess specific soluble antigens that permit grouping by precipitin tests.

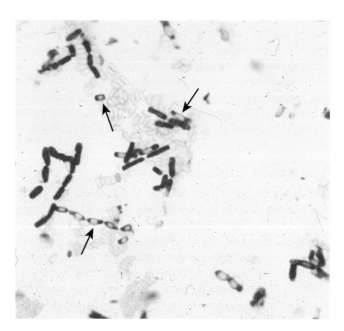

FIGURE 11–2 *Clostridium* Gram stain. Individual gram-positive bacilli are present. Many are in chains. Some of the bacilli have spores, which are the unstained or clear ovoid shapes (arrows).

CLOSTRIDIUM BOTULINUM

C botulinum, which causes **botulism,** is worldwide in distribution; it is found in soil and occasionally in animal feces.

Types of *C botulinum* are distinguished by the antigenic type of toxin they produce. Spores of the organism are highly resistant to heat, withstanding 100°C for several hours. Heat resistance is diminished at acid pH or high salt concentration.

Toxin

During the growth of *C botulinum* and during autolysis of the bacteria, toxin is liberated into the environment. Seven antigenic varieties of toxin (A–G) are known. Types A, B, and E (and occasionally F) are the principal causes of human illness. Types A and B have been associated with a variety of foods and type E predominantly with fish products. Type C produces limberneck in birds; type D causes botulism in mammals. The toxin is a 150,000-MW protein that is cleaved into 100,000-MW and 50,000-MW proteins linked by a disulfide bond. Botulinum toxin is absorbed from the gut and binds to receptors of presynaptic membranes of motor neurons of the peripheral nervous system and cranial nerves. Proteolysis—by the light chain of botulinum toxin—of the target SNARE proteins in the neurons inhibits the release of acetylcholine at the synapse, resulting in lack of muscle contraction and paralysis. The SNARE proteins are synaptobrevin, SNAP 25, and syntaxin. The toxins of *C botulinum* types A and E cleave the 25,000-MW SNAP 25. Type B toxin cleaves synaptobrevin. *C botulinum* toxins are among the most toxic substances known: The lethal dose for a human is probably about 1–2 μg/kg. The toxins are destroyed by heating for 20 minutes at 100°C. Rare strains of *C butyricum* and *C baratii* have also been shown to produce botulinum neurotoxin and cause botulism in humans. Those strains that produce toxins E and F are associated with infant botulism.

Pathogenesis

Although *C botulinum* types A and B have been implicated in cases of wound infection and botulism, most often the illness is not an infection. Rather, it is an intoxication resulting from the ingestion of food in which *C botulinum* has grown and produced toxin. The most common offenders are spiced, smoked, vacuum-packed, or canned alkaline foods that are eaten without cooking. In such foods, spores of *C botulinum* germinate; under anaerobic conditions, vegetative forms grow and produce toxin.

In infant botulism, honey is the most frequent vehicle of infection. The pathogenesis differs from the way that adults acquire infection. The infant ingests the spores of *C botulinum* (or *C butyricum* or *C baratii*), and the spores germinate within the intestinal tract. The vegetative cells produce toxin as they multiply; the neurotoxin then gets absorbed into the bloodstream.

The toxin acts by blocking release of acetylcholine at synapses and neuromuscular junctions (see above). Flaccid paralysis results. The electromyogram and edrophonium strength tests are typical.

Clinical Findings

Symptoms begin 18–24 hours after ingestion of the toxic food, with visual disturbances (incoordination of eye muscles, double vision), inability to swallow, and speech difficulty; signs of bulbar paralysis are progressive, and death occurs from respiratory paralysis or cardiac arrest. Gastrointestinal symptoms are not regularly prominent. There is no fever. The patient remains fully conscious until shortly before death. The mortality rate is high. Patients who recover do not develop antitoxin in the blood.

In the United States, infant botulism is as common as or more common than the classic form of paralytic botulism associated with the ingestion of toxin-contaminated food. The infants in the first months of life develop poor feeding, weakness, and signs of paralysis (floppy baby). Infant botulism may be one of the causes of sudden infant death syndrome. *C botulinum* and botulinum toxin are found in feces but not in serum.

Diagnostic Laboratory Tests

Toxin can often be demonstrated in serum from the patient, and toxin may be found in leftover food. Mice injected intraperitoneally die rapidly. The antigenic type of toxin is identified by neutralization with specific antitoxin in mice. *C botulinum* may be grown from food remains and tested for toxin production, but this is rarely done and is of questionable significance. In infant botulism, *C botulinum* and toxin can be demonstrated in bowel contents but not in serum. Toxin may be demonstrated by passive hemagglutination or radioimmunoassay.

Treatment

Potent antitoxins to three types of botulinum toxins have been prepared in horses. Since the type responsible for an individual case is usually not known, trivalent (A, B, E) antitoxin must be promptly administered intravenously with customary precautions. Adequate ventilation must be maintained by mechanical respirator, if necessary. These measures have reduced the mortality rate from 65% to below 25%.

Although most infants with botulism recover with supportive care alone, antitoxin therapy is recommended.

Epidemiology, Prevention, & Control

Since spores of *C botulinum* are widely distributed in soil, they often contaminate vegetables, fruits, and other materials. A large restaurant-based outbreak was associated with sautéed onions. When such foods are canned or otherwise preserved, they either must be sufficiently heated to ensure destruction of spores or must be boiled for 20 minutes before consumption. Strict regulation of commercial canning has largely overcome the danger of widespread outbreaks, but commercially prepared foods have caused deaths. A chief risk factor for botulism lies in home-canned foods, particularly string beans, corn, peppers, olives, peas, and smoked fish or vacuum-packed fresh fish in plastic bags. Toxic foods may be spoiled and rancid, and cans may "swell," or the appearance may be innocuous. The risk from home-canned foods can be reduced if the food is boiled for more than 20 minutes before consumption. Toxoids are used for active immunization of cattle in South Africa.

Botulinum toxin is considered to be a major agent for bioterrorism and biologic warfare (see Chapter 48).

CLOSTRIDIUM TETANI

C tetani, which causes **tetanus,** is worldwide in distribution in the soil and in the feces of horses and other animals. Several types of *C tetani* can be distinguished by specific flagellar antigens. All share a common O (somatic) antigen, which may be masked, and all produce the same antigenic type of neurotoxin, tetanospasmin.

Toxin

The vegetative cells of *C tetani* produce the toxin tetanospasmin (MW 150,000) that is cleaved by a bacterial protease into two peptides (MW 50,000 and 100,000) linked by a disulfide bond. The toxin initially binds to receptors on the presynaptic membranes of motor neurons. It then migrates by the retrograde axonal transport system to the cell bodies of these neurons to the spinal cord and brain stem. The toxin diffuses to terminals of inhibitory cells, including both glycinergic interneurons and aminobutyric acid-secreting neurons from the brain stem. The toxin degrades synaptobrevin, a protein required for docking of neurotransmitter vesicles on the presynaptic membrane. Release of the inhibitory glycine and γ-aminobutyric acid is blocked, and the motor neurons are not inhibited. Hyperreflexia, muscle spasms, and spastic paralysis result. Extremely small amounts of toxin can be lethal for humans.

Pathogenesis

C tetani is not an invasive organism. The infection remains strictly localized in the area of devitalized tissue (wound, burn, injury, umbilical stump, surgical suture) into which the spores have been introduced. The volume of infected tissue is small, and the disease is almost entirely a toxemia. Germination of the spore and development of vegetative organisms that produce toxin are aided by (1) necrotic tissue, (2) calcium salts, and (3) associated pyogenic infections, all of which aid establishment of low oxidation-reduction potential.

The toxin released from vegetative cells reaches the central nervous system and rapidly becomes fixed to receptors in the spinal cord and brain stem and exerts the actions described above.

Clinical Findings

The incubation period may range from 4–5 days to as many weeks. The disease is characterized by tonic contraction of voluntary muscles. Muscular spasms often involve first the area of injury and infection and then the muscles of the jaw (trismus, lockjaw), which contract so that the mouth cannot be opened. Gradually, other voluntary muscles become involved, resulting in tonic spasms. Any external stimulus may precipitate a tetanic generalized muscle spasm. The patient is fully conscious, and pain may be intense. Death usually results from interference with the mechanics of respiration. The mortality rate in generalized tetanus is very high.

Diagnosis

The diagnosis rests on the clinical picture and a history of injury, although only 50% of patients with tetanus have an injury for which they seek medical attention. The primary differential diagnosis of tetanus is strychnine poisoning. Anaerobic culture of tissues from contaminated wounds may yield *C tetani,* but neither preventive nor therapeutic use of antitoxin should ever be withheld pending such demonstration. Proof of isolation of *C tetani* must rest on production of toxin and its neutralization by specific antitoxin.

Prevention & Treatment

The results of treatment of tetanus are not satisfactory. Therefore, prevention is all-important. Prevention of tetanus depends upon (1) active immunization with toxoids; (2) proper care of wounds contaminated with soil, etc; (3) prophylactic use of antitoxin; and (4) administration of penicillin.

The intramuscular administration of 250–500 units of human antitoxin (tetanus immune globulin) gives adequate systemic protection (0.01 unit or more per milliliter of serum) for 2–4 weeks. It neutralizes the toxin that has not been fixed to nervous tissue. Active immunization with tetanus toxoid should accompany antitoxin prophylaxis.

Patients who develop symptoms of tetanus should receive muscle relaxants, sedation, and assisted ventilation. Sometimes they are given very large doses of antitoxin (3000–10,000 units of tetanus immune globulin) intravenously in an effort to neutralize toxin that has not yet been bound to nervous tissue. However, the efficacy of antitoxin for treatment is doubtful except in neonatal tetanus, where it may be lifesaving.

Surgical debridement is vitally important because it removes the necrotic tissue that is essential for proliferation of the organisms. Hyperbaric oxygen has no proved effect.

Penicillin strongly inhibits the growth of *C tetani* and stops further toxin production. Antibiotics may also control associated pyogenic infection.

When a previously immunized individual sustains a potentially dangerous wound, an additional dose of toxoid should be injected to restimulate antitoxin production. This "recall" injection of toxoid may be accompanied by a dose of antitoxin if the patient has not had current immunization or boosters or if the history of immunization is unknown.

Control

Tetanus is a totally preventable disease. Universal active immunization with tetanus toxoid should be mandatory. Tetanus toxoid is produced by detoxifying the toxin with formalin and then concentrating it. Aluminum-salt-adsorbed toxoids are employed. Three injections comprise the initial course of immunization, followed by another dose about 1 year later. Initial immunization should be carried out in all children during the first year of life. A "booster" injection of toxoid is given upon entry into school. Thereafter, "boosters" can be spaced 10 years apart to maintain serum levels of more than 0.01 unit antitoxin per milliliter. In young children, tetanus toxoid is often combined with diphtheria toxoid and acellular pertussis vaccine.

Control measures are not possible because of the wide dissemination of the organism in the soil and the long survival of its spores.

CLOSTRIDIA THAT PRODUCE INVASIVE INFECTIONS

Many different toxin-producing clostridia (*C perfringens* and related clostridia) (Figure 11–3) can produce invasive infection (including **myonecrosis** and **gas gangrene**) if introduced into damaged tissue. About 30 species of clostridia may produce such an effect, but the most common in invasive disease is *C perfringens* (90%). An enterotoxin of *C perfringens* is a common cause of food poisoning.

Toxins

The invasive clostridia produce a large variety of toxins and enzymes that result in a spreading infection. Many of these toxins have lethal, necrotizing, and hemolytic properties. In some cases, these are different properties of a single substance; in other instances, they are due to different chemical entities. The α toxin of *C perfringens* type A is a lecithinase, and its lethal action is proportionate to the rate at which it splits lecithin (an important constituent of cell membranes) to phosphorylcholine and diglyceride. The theta toxin has similar hemolytic and necrotizing effects but is not a lecithinase. DNase and hyaluronidase, a collagenase that digests collagen of subcutaneous tissue and muscle, are also produced.

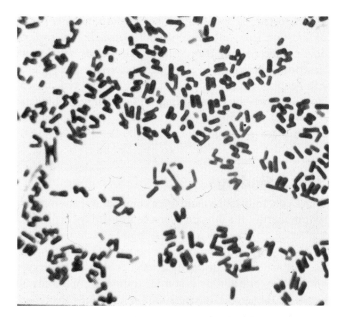

FIGURE 11–3 Gas gangrene bacilli. *Clostridium perfringens* typically does not form spores when grown on laboratory media.

Some strains of *C perfringens* produce a powerful enterotoxin, especially when grown in meat dishes. When more than 10^8 vegetative cells are ingested and sporulate in the gut, enterotoxin is formed. The enterotoxin is a protein (MW 35,000) that may be a nonessential component of the spore coat; it is distinct from other clostridial toxins. It induces intense diarrhea in 6–18 hours. The action of *C perfringens* enterotoxin involves marked hypersecretion in the jejunum and ileum, with loss of fluids and electrolytes in diarrhea. Much less frequent symptoms include nausea, vomiting, and fever. This illness is similar to that produced by *B cereus* and tends to be self-limited.

Pathogenesis

In invasive clostridial infections, spores reach tissue either by contamination of traumatized areas (soil, feces) or from the intestinal tract. The spores germinate at low oxidation-reduction potential; vegetative cells multiply, ferment carbohydrates present in tissue, and produce gas. The distention of tissue and interference with blood supply, together with the secretion of necrotizing toxin and hyaluronidase, favor the spread of infection. Tissue necrosis extends, providing an opportunity for increased bacterial growth, hemolytic anemia, and, ultimately, severe toxemia and death.

In gas gangrene (clostridial myonecrosis), a mixed infection is the rule. In addition to the toxigenic clostridia, proteolytic clostridia and various cocci and gram-negative organisms are also usually present. *C perfringens* occurs in the genital tract of 5% of women. Before legalization of abortion in the United States, clostridial uterine infections followed instrumental abortions. *Clostridium sordellii* has many of the properties of *C perfringens*. *C sordellii* has been reported to cause a toxic shock syndrome after medical abortion with mifepristone and intravaginal misoprostol. Endometrial infection with *C sordellii* is implicated. Clostridial bacteremia is a frequent occurrence in patients with neoplasms. In New Guinea, *C perfringens* type C produces a necrotizing enteritis (pigbel) that can be highly fatal in children. Immunization with type C toxoid appears to have preventive value.

Clinical Findings

From a contaminated wound (eg, a compound fracture, postpartum uterus), the infection spreads in 1–3 days to produce crepitation in the subcutaneous tissue and muscle, foul-smelling discharge, rapidly progressing necrosis, fever, hemolysis, toxemia, shock, and death. Treatment is with early surgery (amputation) and antibiotic administration. Until the advent of specific therapy, early amputation was the only treatment. At times, the infection results only in anaerobic fasciitis or cellulitis.

C perfringens food poisoning usually follows the ingestion of large numbers of clostridia that have grown in warmed meat dishes. The toxin forms when the organisms sporulate in the gut, with the onset of diarrhea—usually without vomiting or fever—in 6–18 hours. The illness lasts only 1–2 days.

Diagnostic Laboratory Tests

Specimens consist of material from wounds, pus, and tissue. The presence of large gram-positive rods in Gram-stained smears suggests gas gangrene clostridia; spores are not regularly present.

Material is inoculated into chopped meat-glucose medium and thioglycolate medium and onto blood agar plates incubated anaerobically. The growth from one of the media is transferred into milk. A clot torn by gas in 24 hours is suggestive of *C perfringens*. Once pure cultures have been obtained by selecting colonies from anaerobically incubated blood plates, they are identified by biochemical reactions (various sugars in thioglycolate, action on milk), hemolysis, and colony form. Lecithinase activity is evaluated by the precipitate formed around colonies on egg yolk media. Final identification rests on toxin production and neutralization by specific antitoxin. *C perfringens* rarely produces spores when cultured on agar in the laboratory.

Treatment

The most important aspect of treatment is prompt and extensive surgical debridement of the involved area and excision of all devitalized tissue, in which the organisms are prone to grow. Administration of antimicrobial drugs, particularly penicillin, is begun at the same time. Hyperbaric oxygen may be of help in the medical management of clostridial tissue infections. It is said to "detoxify" patients rapidly.

Antitoxins are available against the toxins of *C perfringens*, *Clostridium novyi*, *Clostridium histolyticum*, and *Clostridium septicum*, usually in the form of concentrated immune globulins. Polyvalent antitoxin (containing antibodies to several toxins) has been used. Although such antitoxin is sometimes administered to individuals with contaminated wounds containing much devitalized tissue, there is no evidence for its efficacy. Food poisoning due to *C perfringens* enterotoxin usually requires only symptomatic care.

Prevention & Control

Early and adequate cleansing of contaminated wounds and surgical debridement, together with the administration of antimicrobial drugs directed against clostridia (eg, penicillin), are the best available preventive measures. Antitoxins should not be relied on. Although toxoids for active immunization have been prepared, they have not come into practical use.

CLOSTRIDIUM DIFFICILE & DIARRHEAL DISEASE

Pseudomembranous Colitis

Pseudomembranous colitis is diagnosed by detection of one or both *C difficile* toxins in stool and by endoscopic observation of pseudomembranes or microabscesses in patients who have diarrhea and have been given antibiotics. Plaques and microabscesses may be localized to one area of the bowel. The diarrhea may be watery or bloody, and the patient frequently has associated abdominal cramps, leukocytosis, and fever. Although many antibiotics have been associated with pseudomembranous colitis, the most common are ampicillin and clindamycin and more recently, the fluoroquinolones. The disease is treated by discontinuing administration of the offending antibiotic and orally giving either metronidazole or vancomycin.

Administration of antibiotics results in proliferation of drug-resistant *C difficile* that produces two toxins. Toxin A, a potent enterotoxin that also has some cytotoxic activity, binds to the brush border membranes of the gut at receptor sites. Toxin B is a potent cytotoxin. Both toxins are found in the stools of patients with pseudomembranous colitis. Not all strains of *C difficile* produce the toxins, and the *tox* genes apparently are not carried on plasmids or phage.

Antibiotic-Associated Diarrhea

The administration of antibiotics frequently leads to a mild to moderate form of diarrhea, termed antibiotic-associated diarrhea. This disease is generally less severe than the classic form of pseudomembranous colitis. As many as 25% of cases of antibiotic-associated diarrhea may be associated with *C difficile*.

REVIEW QUESTIONS

1. A housewife who lives on a small farm is brought to the emergency room complaining of double vision and difficulty talking. Within the past 2 hours she noted a dry mouth and generalized weakness. Last night she served home-canned green beans as part of the meal. She tasted the beans before they were boiled. None of the other family members are ill. On examination, there is symmetrical descending paralysis of cranial nerves, upper extremities, and trunk. The correct diagnosis is which one of the following?

    (A) Tetanus
    (B) Strychnine poisoning
    (C) Botulism
    (D) Morphine overdose
    (E) Ricin intoxication

2. Which one of the following is an important virulence factor of *Bacillus anthracis?*

    (A) Protective antigen
    (B) Lipopolysaccharide
    (C) Pili
    (D) A toxin that inhibits peptide chain elongation factor EF-2
    (E) Lecithinase

3. A young man suffers major soft tissue injury and open fractures of his right leg after a motorcycle accident. One day later, he has a temperature of 38°C, increased heart rate, sweating, and restlessness. On examination the leg is swollen and tense, with thin, dark serous fluid draining from the wounds. The skin of the leg is cool, pale, white, and shining. Crepitus can be felt in the leg. His hematocrit is 20% (approximately 50% of normal) while his circulating hemoglobin is normal. His serum shows free hemoglobin. Which of the following microorganisms is the most likely cause of this infection?

    (A) *Clostridium tetani*
    (B) *Staphylococcus aureus*
    (C) *Escherichia coli*
    (D) *Bacillus anthracis*
    (E) *Clostridium perfringens*

4. For the patient described in Question 3, above, which of the following is likely to be responsible for the hemolysis?

    (A) EF
    (B) Tetanospasmin
    (C) Lecithinase
    (D) Streptolysin O
    (E) Toxin B

5. The reported incubation period for inhalation anthrax can be up to

    (A) 2 days
    (B) 10 days
    (C) 3 weeks
    (D) 6 weeks
    (E) 6 months

6. A food commonly associated with *Bacillus cereus* food poisoning is

    (A) Fried rice
    (B) Baked potato
    (C) Hot freshly steamed rice
    (D) Green beans
    (E) Honey

7. Tetanus toxin (tetanospasmin) diffuses to terminals of inhibitory cells in the spinal cord and brain stem and blocks which of the following?

    (A) Release of acetylcholine
    (B) Cleavage of SNARE proteins
    (C) Release of inhibitory glycine and γ-aminobutyric acid
    (D) Release of PA
    (E) Activation of acetylcholine esterase

8. A 45-year-old man who immigrated to the United States 5 years ago sustained a puncture injury to the lower part of his right leg when his rotary lawn mower threw a small stick into the leg. Six days later, he noticed spasms in the muscles of his right leg; on day 7, the spasms increased. Today—day 8—he had generalized muscle spasms, particularly noticeable in the muscles of his jaw. He was unable to open his jaw and came to the emergency room. In the emergency department you see a man who is alert and lying quietly in bed. A door slams down the hall and suddenly he has general muscle spasm with arching of his back. The correct diagnosis is which of the following?

    (A) Botulism
    (B) Anthrax
    (C) Gas gangrene
    (D) Tetanus
    (E) Toxic shock syndrome

9. Which of the following statements about tetanus and tetanus toxoid is correct?

    (A) Tetanus toxin kills neurons
    (B) Tetanus toxoid immunization has a 10% failure rate
    (C) The mortality rate of generalized tetanus is <1%
    (D) Double vision is commonly the first sign of tetanus
    (E) Tetanus toxin acts on inhibitor interneuron synapses

10. A 67-year-old man had surgery for a ruptured sigmoid colon diverticulum with an abscess. A repair was done and the abscess was drained. He was treated with intravenous gentamicin and ampicillin. Ten days later and 4 days after being discharged from the hospital, the patient developed malaise, fever, and cramping abdominal pain. He had multiple episodes of diarrhea. His stool was positive for occult blood and the presence of polymorphonuclear cells. On sigmoidoscopy the mucosa was erythematous and appeared to be inflamed, and there were many raised white to yellowish plaques 4–8 mm in diameter. Which of the following is the likely cause of the patient's problem?

    (A) *Staphylococcus aureus* enterotoxin
    (B) *Bacillus cereus* toxin
    (C) *Clostridium difficile* toxins
    (D) *Clostridium perfringens* toxin
    (E) Enterohemorrhagic *Escherichia coli*

11. Infant botulinum has been associated with all of the following *Clostridium* species *except*:

    (A) *Clostridium baratii*
    (B) *Clostridium septicum*
    (C) *Clostridium butyricum*
    (D) *Clostridium botulinum*

12. Which of the following food items is most frequently associated with infant botulism?

    (A) Corn syrup
    (B) Canned infant formula
    (C) Liquid multivitamins

(D) Honey

(E) Jarred baby food

13. All of the following are properties characteristic of *Bacillus anthracis* except:

(A) Motility on wet mount examination

(B) Medusa head colonies

(C) Poly-d-glutamic acid capsule

(D) In-vitro susceptibility to penicillin

(E) Absence of hemolysis on 5% sheep blood agar

14. Which of the following statements regarding vaccination for *Bacillus anthracis* is correct?

(A) It is routinely available for all citizens of the United States.

(B) Recombinant vaccine trials have shown good safety and efficacy

(C) The current vaccine is well-tolerated.

(D) A single dose is adequate following exposure to spores.

(E) Vaccination of animals is not useful.

15. All of the following statements regarding *Clostridium perfringens* are correct *except*:

(A) It produces an enterotoxin.

(B) It produces a double zone of β hemolysis when grown on blood agar.

(C) Some strains are aerotolerant.

(D) It is the most common cause of antibiotic-associated diarrhea.

(E) It can cause intravascular hemolysis.

Answers

1. C	5. D	9. E	13. A
2. A	6. A	10. C	14. B
3. E	7. C	11. B	15. D
4. C	8. D	12. D	

REFERENCES

Bleck TP: *Clostridium botulinum* (Botulism): In: *Mandell, Douglas and Bennett's Principles and Practice of Infectious Diseases,* 6th ed. Mandell GL, Bennett JE, Dolin R (editors). Churchill Livingstone, 2005.

Bleck TP: *Clostridium tetani* (Tetanus): In: *Mandell, Douglas and Bennett's Principles and Practice of Infectious Diseases,* 6th ed. Mandell GL, Bennett JE, Dolin R (editors). Churchill Livingstone, 2005.

Campbell JD, Clement KH, Wasserman SS, Donegan S, Chrisley L, Kotloff KL: Safety, reactogenicity, and immunogenicity of a recombinant protective antigen anthrax vaccine given to healthy adults. Human Vaccines 2007;3:205–11 [PMID: 17881903].

Dixon TC et al: Anthrax. N Engl J Med 1999;341:815. [PMID: 10477781]

Fekete F: *Bacillus* species and related genera other than *Bacillus anthracis.* In: *Mandell, Douglas and Bennett's Principles and Practice of Infectious Diseases,* 6th ed. Mandell GL, Bennett JE, Dolin R (editors). Churchill Livingstone, 2005.

Johnson EA, Summanen P, Finegold SM, Emery CL, Lyerly DM: *Clostridium.* In: *Manual of Clinical Microbiology,* 4th ed. Murray PR et al (editors). ASM Press, 2007.

Logan NA, Popovic T, Hoffmaster A: *Bacillus* and other aerobic endospore-forming bacteria. In: *Manual of Clinical Microbiology,* 9th ed. Murray PR et al (editors). ASM Press, 2007.

Lorber B: Gas gangrene and other *Clostridium*-associated diseases. In: *Mandell, Douglas and Bennett's Principles and Practice of Infectious Diseases,* 6th ed. Mandell GL, Bennett JE, Dolin R (editors). Churchill Livingstone, 2005.

Lucey D: *Bacillus anthracis* (Anthrax). In: *Mandell, Douglas and Bennett's Principles and Practice of Infectious Diseases,* 6th ed. Mandell GL, Bennett JE, Dolin R (editors). Churchill Livingstone, 2005.

Shapiro RL et al: Botulism in the United States: A clinical and epidemiologic review. Ann Intern Med 1998;129:221. [PMID: 9696731]

Thielman NM: Antibiotic-associated colitis. In: *Mandell, Douglas, and Bennett's Principles and Practice of Infectious Diseases,* 5th ed. Mandell GL, Bennett JE, Dolin R (editors). Churchill Livingstone, 2000.

Aerobic Nonspore-Forming Gram-Positive Bacilli: *Corynebacterium, Listeria, Erysipelothrix, Actinomycetes,* & Related Pathogens

The nonspore-forming gram-positive bacilli are a diverse group of aerobic and anaerobic bacteria. This chapter will focus upon the aerobic members of this group. The anaerobic, nonspore-forming gram-positive bacilli such as *Propionibacterium* species and *Actinomyces* species will be discussed in the chapter on anaerobic infections (Chapter 21). Specific genera of both groups, namely *Corynebacterium* species and *Propionibacterium* species, are members of the normal flora of skin and mucous membranes of humans and, as such, are frequently contaminants of clinical specimens submitted for diagnostic evaluation. However, among the aerobic Actinomycetes are significant pathogens such as *Corynebacterium diphtheriae*, an organism that produces a powerful exotoxin that causes diphtheria in humans, and *Mycobacterium tuberculosis*, the causative agent of tuberculosis. *Listeria monocytogenes* and *Erysipelothrix rhusiopathiae* are primarily found in animals and occasionally cause severe disease in humans. *Nocardia* species, *Gordonia* and *Tsukamurella* are emerging pathogens among immunocompromised patients.

Corynebacterium species and related bacteria tend to be clubbed or irregularly shaped; although not all isolates have the irregular shapes, the terms "coryneform or diphtheroid bacteria" are convenient ones for denoting the group. These bacteria have a high guanosine plus cytosine content and include the genera *Corynebacterium, Arcanobacterium, Brevibacterium, Mycobacterium,* and others (Table 12–1). *Actinomyces* and *Propionibacterium* are classified as anaerobes, but some isolates grow well aerobically (aerotolerant) and must be differentiated from the aerobic coryneform bacteria. Other nonspore-forming gram-positive bacilli have more regular shapes and a lower guanosine plus cytosine content. The genera include *Listeria* and *Erysipelothrix*; these bacteria are more closely related to the anaerobic *Lactobacillus* species, which sometimes grow well in air, to the spore-forming *Bacillus* and *Clostridium* species—and to the gram-positive cocci of the *Staphylococcus* and *Streptococcus* species—than they are to the coryneform bacteria. The medically important genera of gram-positive bacilli are listed in Table 12–1 and include some spore-forming and anaerobic genera. Anaerobic bacteria are discussed in Chapter 21.

There is no unifying method for identification of the gram-positive bacilli. Few laboratories are equipped to measure guanosine plus cytosine content. Growth only under anaerobic conditions implies that the isolate is an anaerobe, but many isolates of *Lactobacillus, Actinomyces,* and *Propionibacterium* species and others are aerotolerant. Most isolates of *Mycobacterium* species, *Nocardia* and *Rhodococcus* species, *Gordonia* and *Tsukamurella* are acid-fast and, therefore, readily distinguished from the coryneform bacteria. Many but not all genera of *Bacillus* and *Clostridium* produce spores, and the presence of spores readily distinguishes the isolate from the coryneform bacteria; however, *Clostridium perfringens* and other filamentous clostridia generally do not produce spores on laboratory media. Determination that an isolate is a *Lactobacillus* (or *Propionibacterium)* may require gas–liquid chromatography to measure lactic acid (or propionic acid) metabolic products, but this is generally not practical. Other tests that are used to help identify an isolate of nonspore-forming gram-positive bacilli as a member of a genus or species include catalase production, indole production, nitrate reduction, and fermentation of carbohydrates, among others. Many clinical laboratories have developed sequencing technologies targeting the 16S rRNA gene or other gene targets for identification of many of these organisms, but especially for *Mycobacterium* and *Nocardia* species recovered from clinical specimens.

CORYNEBACTERIUM DIPHTHERIAE

Morphology & Identification

Corynebacteria are 0.5–1 μm in diameter and several micrometers long. Characteristically, they possess irregular swellings at one end that give them the "club-shaped" appearance (Figure 12–1). Irregularly distributed within the rod (often near the poles) are granules staining deeply with aniline dyes (metachromatic granules) that give the rod a beaded

TABLE 12–1   Some of the More Common Gram-Positive Bacilli of Medical Importance

Aerobic Gram-Positive Bacilli with High G + C Content and Irregular Shape[a]	Aerobic Gram-Positive Bacilli with Lower G + C Content and More Regular Shape
Genera	Genera
Common	Common
Corynebacterium	*Listeria*
Uncommon	*Erysipelothrix*
Arcanobacterium	*Gardnerella*
Rothia	Aerotolerant anaerobes/strict anaerobes
Acid fast positive	*Lactobacillus*
Rhodococcus	*Clostridium* (spore-forming) (Chapter 11)
Nocardia	Aerobes
Tsukamurella	*Bacillus* (spore-forming) (Chapter 11)
Gordonia	
Many other genera of skin and environmental flora	
Aerotolerant anaerobes	
Actinomyces (Chapter 21)	
Propionibacterium (Chapter 21)	
Major pathogen: *Corynebacterium diphtheriae*	Major pathogens
Common or clinically important isolates of the genus *Corynebacterium*	*Listeria monocytogenes*
Corynebacterium amycolatum	*Erysipelothrix rhusiopathiae*
Corynebacterium minutissimum	
Corynebacterium jeikeium	
Corynebacterium pseudodiphtheriticum	
Corynebacterium striatum	
Corynebacterium urealyticum	
Corynebacterium xerosis	

G + C = guanine plus cytosine base.

[a]The medically important coryneform bacteria.

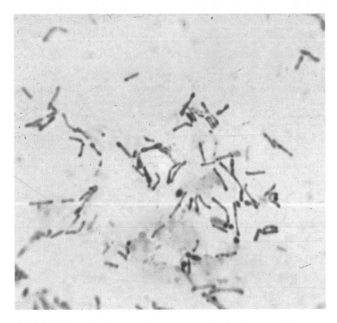

FIGURE 12–1 *Corynebacterium diphtheriae* from Pai medium stained with methylene blue. Typically they are 0.5–1 × 3–4 μm. Some of the bacteria have clubbed ends (original magnification × 1000).

appearance. Individual corynebacteria in stained smears tend to lie parallel or at acute angles to one another. True branching is rarely observed in cultures.

On blood agar, the *C diphtheriae* colonies are small, granular, and gray, with irregular edges, and may have small zones of hemolysis. On agar containing potassium tellurite, the colonies are brown to black with a brown-black halo because the tellurite is reduced intracellularly (staphylococci and streptococci can also produce black colonies). Four biotypes of *C diphtheriae* have been widely recognized: gravis, mitis, intermedius, and belfanti. These variants have been classified on the basis of growth characteristics such as colony morphology, biochemical reactions, and severity of disease produced by infection. Very few reference laboratories provide the biotype characterization; the incidence of diphtheria has greatly decreased and the association of severity of disease with biovar is not important to clinical or public health management of cases or outbreaks. If necessary in the setting of an outbreak, immunochemical and molecular methods can be used to type the *C diphtheriae* isolates.

C diphtheriae and other corynebacteria grow aerobically on most ordinary laboratory media. On Loeffler serum medium, corynebacteria grow much more readily than other respiratory organisms, and the morphology of organisms is typical in smears made from these colonies.

Corynebacteria tend to pleomorphism in microscopic and colonial morphology. When some nontoxigenic diphtheria organisms are infected with bacteriophage from certain toxigenic diphtheria bacilli, the offspring of the exposed bacteria are lysogenic and toxigenic, and this trait is subsequently hereditary. When toxigenic diphtheria bacilli are serially subcultured in specific antiserum against the temperate phage that they carry, they tend to become nontoxigenic. Thus, acquisition of phage leads to toxigenicity (lysogenic conversion). The actual production of toxin occurs perhaps only when the prophage of the lysogenic *C diphtheriae* becomes induced and lyses the cell. Whereas toxigenicity is under control of the phage gene, invasiveness is under control of bacterial genes.

Pathogenesis

The principal human pathogen of the genus *Corynebacterium* is *C diphtheriae*, the causative agent of respiratory or cutaneous diphtheria. In nature, *C diphtheriae* occurs in the respiratory tract, in wounds, or on the skin of infected persons or normal carriers. It is spread by droplets or by contact to susceptible individuals; the bacilli then grow on mucous membranes or in skin abrasions, and those that are toxigenic start producing toxin.

All toxigenic *C diphtheriae* are capable of elaborating the same disease-producing exotoxin. In vitro production of this toxin depends largely on the concentration of iron. Toxin production is optimal at 0.14 µg of iron per milliliter of medium but is virtually suppressed at 0.5 µg/mL. Other factors influencing the yield of toxin in vitro are osmotic pressure, amino acid concentration, pH, and availability of suitable carbon and nitrogen sources. The factors that control toxin production in vivo are not well understood.

Diphtheria toxin is a heat-labile polypeptide (MW 62,000) that can be lethal in a dose of 0.1 µg/kg. If disulfide bonds are broken, the molecule can be split into two fragments. Fragment B (MW=38,000), which has no independent activity, is functionally divided into a receptor domain and a translocation domain. The binding of the receptor domain to host cell membrane proteins CD-9 and heparin-binding epidermal growth factor (HBEGF)-like precursor, triggers the entry of the toxin into the cell through receptor-mediated endocytosis. Acidification of the translocation domain within a developing endosome leads to creation of a protein channel that facilitates movement of Fragment A into the host cell cytoplasm. Fragment A inhibits polypeptide chain elongation—provided nicotinamide adenine dinucleotide (NAD) is present—by inactivating the elongation factor EF-2. This factor is required for translocation of polypeptidyl-transfer RNA from the acceptor to the donor site on the eukaryotic ribosome. Toxin fragment A inactivates EF-2 by catalyzing a reaction that yields free nicotinamide plus an inactive adenosine diphosphate-ribose-EF-2 complex (ADP-ribosylation). It is assumed that the abrupt arrest of protein synthesis is responsible for the necrotizing and neurotoxic effects of diphtheria toxin. An exotoxin with a similar mode of action can be produced by strains of *Pseudomonas aeruginosa*.

Pathology

Diphtheria toxin is absorbed into the mucous membranes and causes destruction of epithelium and a superficial inflammatory response. The necrotic epithelium becomes embedded in exuding fibrin and red and white cells, so that a grayish "pseudomembrane" is formed—commonly over the tonsils, pharynx, or larynx. Any attempt to remove the pseudomembrane exposes and tears the capillaries and thus results in bleeding. The regional lymph nodes in the neck enlarge, and there may be marked edema of the entire neck. The diphtheria bacilli within the membrane continue to produce toxin actively. This is absorbed and results in distant toxic damage, particularly parenchymatous degeneration, fatty infiltration, and necrosis in heart muscle, liver, kidneys, and adrenals, sometimes accompanied by gross hemorrhage. The toxin also produces nerve damage, resulting often in paralysis of the soft palate, eye muscles, or extremities.

Wound or skin diphtheria occurs chiefly in the tropics. A membrane may form on an infected wound that fails to heal. However, absorption of toxin is usually slight and the systemic effects negligible. The small amount of toxin that is absorbed during skin infection promotes development of antitoxin antibodies. The "virulence" of diphtheria bacilli is due to their capacity for establishing infection, growing rapidly, and then quickly elaborating toxin that is effectively absorbed. *C diphtheriae* does not need to be toxigenic to establish localized infection—in the nasopharynx or skin, for example—but nontoxigenic strains do not yield the localized or systemic toxic effects. *C diphtheriae* does not actively invade deep tissues and practically never enters the bloodstream, although rare cases of endocarditis have been described.

Clinical Findings

When diphtheritic inflammation begins in the respiratory tract, sore throat and fever usually develop. Prostration and dyspnea soon follow because of the obstruction caused by the membrane. This obstruction may even cause suffocation if not promptly relieved by intubation or tracheostomy. Irregularities of cardiac rhythm indicate damage to the heart. Later, there may be difficulties with vision, speech, swallowing, or movement of the arms or legs. All of these manifestations tend to subside spontaneously.

In general, var gravis tends to produce more severe disease than var mitis, but similar illness can be produced by all types.

Diagnostic Laboratory Tests

These serve to confirm the clinical impression and are of epidemiologic significance. *Note:* Specific treatment must never be delayed for laboratory reports if the clinical picture is strongly suggestive of diphtheria. Physicians should notify the clinical laboratory before collecting or submitting samples for culture.

Dacron swabs from the nose, throat, or other suspected lesions must be obtained before antimicrobial drugs are administered. Swabs should be collected from beneath any visible membrane. The swab should then be placed in semi-solid transport media such as Amies. Smears stained with alkaline methylene blue or Gram stain show beaded rods in typical arrangement.

Specimens should be inoculated to a blood agar plate (to rule out hemolytic streptococci), a Loeffler slant, and a tellurite plate (eg, cystine-tellurite agar or modified Tinsdale's medium) and incubated at 37°C. In 12–18 hours, the Loeffler slant may yield organisms of typical "diphtheria-like" morphology. In 36–48 hours, the colonies on tellurite medium are sufficiently definite for recognition of *C diphtheriae*.

A presumptive *C diphtheriae* isolate should be subjected to testing for toxigenicity. Such tests are performed only in reference public health laboratories. There are several methods, as follows:

1. A filter paper disk containing antitoxin (10 IU/disk) is placed on an agar plate. The cultures to be tested for toxigenicity are spot innoculated 7–9 mm away from the disk. After 48 hours of incubation, the antitoxin diffusing from the paper disk has precipitated the toxin diffusing from toxigenic cultures and has resulted in precipitin bands between the disk and the bacterial growth. This is the modified Elek method described by the WHO Diphtheria Reference Unit.
2. Polymerase chain reaction-based methods have been described for detection of the diphtheria toxin gene *(tox)*. PCR assays for *tox* can also be used directly on patient specimens before culture results are available. A positive culture confirms a positive PCR assay. A negative culture following antibiotic therapy along with a positive PCR assay suggests that the patient probably has diphtheria.
3. Enzyme-linked immunosorbent assays can be used to detect diphtheria toxin from clinical *C diphtheriae* isolates.
4. An immunochromographic strip assay allows detection of diphtheria toxin in a matter of hours. This assay is highly sensitive.

The latter two assays are not widely available.

Historically, toxigenicity of a *C diphtheriae* isolate has been demonstrated by injecting two guinea pigs with the emulsified isolate. If the guinea pig protected with diphtheria antitoxin survives while the unprotected one dies, the isolate is considered to be toxigenic. This test has largely been replaced by more modern technology.

Resistance & Immunity

Since diphtheria is principally the result of the action of the toxin formed by the organism rather than invasion by the organism, resistance to the disease depends largely on the availability of specific neutralizing antitoxin in the bloodstream and tissues. It is generally true that diphtheria occurs only in persons who possess no antitoxin (or less than 0.01 Lf unit/mL). Assessment of immunity to diphtheria toxin for individual patients can best be made by review of documented diphtheria toxoid immunizations and primary or booster immunization if needed.

Treatment

The treatment of diphtheria rests largely on rapid suppression of toxin-producing bacteria by antimicrobial drugs and the early administration of specific antitoxin against the toxin formed by the organisms at their site of entry and multiplication. Diphtheria antitoxin is produced in various animals (horses, sheep, goats, and rabbits) by the repeated injection of purified and concentrated toxoid. Treatment with antitoxin is mandatory when there is strong clinical suspicion of diphtheria. From 20,000–100,000 units are injected intramuscularly or intravenously after suitable precautions have been taken (skin or conjunctival test) to rule out hypersensitivity to the animal serum. The antitoxin should be given on the day the clinical diagnosis of diphtheria is made and need not be repeated. Intramuscular injection may be used in mild cases.

Antimicrobial drugs (penicillin, erythromycin) inhibit the growth of diphtheria bacilli. Although these drugs have virtually no effect on the disease process, they arrest toxin production. They also help to eliminate coexistent streptococci and *C diphtheriae* from the respiratory tracts of patients or carriers.

Epidemiology, Prevention, & Control

Before artificial immunization, diphtheria was mainly a disease of small children. The infection occurred either clinically or subclinically at an early age and resulted in the widespread production of antitoxin in the population. An asymptomatic infection during adolescence and adult life served as a stimulus for maintenance of high antitoxin levels. Thus, most members of the population, except children, were immune.

By age 6–8 years, approximately 75% of children in developing countries where skin infections with *C diphtheriae* are common have protective serum antitoxin levels. Absorption of small amounts of diphtheria toxin from the skin infection presumably provides the antigenic stimulus for the immune response; the amount of absorbed toxin does not produce disease.

Active immunization in childhood with diphtheria toxoid yields antitoxin levels that are generally adequate until adulthood. Young adults should be given boosters of toxoid, because toxigenic diphtheria bacilli are not sufficiently prevalent in the population of many developed countries to provide the stimulus of subclinical infection with stimulation of resistance. Levels of antitoxin decline with time, and many older persons have insufficient amounts of circulating antitoxin to protect them against diphtheria.

The principal aims of prevention are to limit the distribution of toxigenic diphtheria bacilli in the population and to maintain as high a level of active immunization as possible.

To limit contact with diphtheria bacilli to a minimum, patients with diphtheria should be isolated. Without treatment, a large percentage of infected persons continue to shed diphtheria bacilli for weeks or months after recovery (convalescent carriers). This danger may be greatly reduced by active early treatment with antibiotics.

A filtrate of broth culture of a toxigenic strain is treated with 0.3% formalin and incubated at 37°C until toxicity has disappeared. This **fluid toxoid** is purified and standardized in flocculating units (Lf doses). Fluid toxoids prepared as above are adsorbed onto aluminum hydroxide or aluminum phosphate. This material remains longer in a depot after injection and is a better antigen. Such toxoids are commonly combined with tetanus toxoid (Td) and sometimes with pertussis vaccine (DPT or DaPT) as a single injection to be used in initial immunization of children. For booster injection of adults, only Td toxoids or Td toxoids combined with acellular pertussis vaccine (for a one-time injection for those individuals who received whole cell pertussis vaccine as children) are used; these combine a full dose of tetanus toxoid with a tenfold smaller dose of diphtheria toxoid in order to diminish the likelihood of adverse reactions.

All children must receive an initial course of immunizations and boosters. Regular boosters with Td are particularly important for adults who travel to developing countries, where the incidence of clinical diphtheria may be 1000-fold higher than in developed countries, where immunization is universal.

OTHER CORYNEFORM BACTERIA

Many other *Corynebacterium* species have been associated with disease in humans. The coryneform bacteria are classified as nonlipophilic or lipophilic depending upon enhancement of growth by addition of lipid to the growth medium. The lipophilic corynebacteria grow slowly on sheep blood agar, producing colonies <0.5 mm in diameter after 24 hours of incubation. Additional key reactions for the classification of the coryneform bacteria include but are not limited to the following tests: fermentative or oxidative metabolism, catalase production, motility, nitrate reduction, urease production, and esculin hydrolysis. *Corynebacterium* species are typically nonmotile and catalase-positive. The coryneform

bacteria are normal inhabitants of the mucous membranes of the skin, respiratory tract, urinary tract, and conjunctiva.

Nonlipophilic Corynebacteria

The group of nonlipophilic corynebacteria includes multiple species that can be further differentiated on the basis of fermentative or oxidative metabolism. *Corynebacterium ulcerans* and *Corynebacterium pseudotuberculosis* are closely related to *C diphtheriae* and may carry the diphtheria *tox* gene. The toxigenic *C ulcerans* can cause disease similar to clinical diphtheria, while *C pseudotuberculosis* rarely causes disease in humans. Other species in the nonlipophilic fermentative group include *Corynebacterium xerosis*, *Corynebacterium striatum*, *Corynebacterium minutissimum*, and *Corynebacterium amycolatum*. These are among the most commonly isolated coryneform bacteria. Many isolates previously identified as *C xerosis* may have been misidentified and were really *C amycolatum*. There are few well-documented cases of disease caused by *C minutissimum*, though the organism is frequently isolated from clinical specimens. Historically, *C xerosis* and *C striatum* have caused a variety of infections in humans.

Three nonfermentative corynebacteria are most frequently associated with clinical infections. *Corynebacterium auris* has been associated with ear infections in children, and *Corynebacterium pseudodiphtheriticum* has been associated with respiratory tract infections. *Corynebacterium glucuronolyticum* is often urease-positive and is a urinary tract pathogen.

Lipophilic Corynebacteria

Corynebacterium jeikeium is one of the coryneform bacterium most commonly isolated from acutely ill patients. It can cause disease in immunocompromised patients and is important because it produces infections, including bacteremia, that have a high mortality rate and because it is resistant to many commonly used antimicrobial drugs. *Corynebacterium urealyticum* is a slowly growing species that is multiply resistant to antibiotics. As its name implies, it is urease-positive. It has been associated with acute or chronic encrusted urinary tract infections manifested by alkaline urine pH and crystal formation.

Other Coryneform Genera

There are many other genera and species of coryneform bacteria. *Arcanobacterium haemolyticum* produces β-hemolysis on blood agar. It is occasionally associated with pharyngitis and can grow in media selective for streptococci. *A haemolyticum* is catalase-negative, like group A streptococci, and must be differentiated by Gram stain morphology (rods versus cocci) and biochemical characteristics. Most of the coryneform bacteria in the other genera are infrequent causes of disease and are not commonly identified in the clinical laboratory.

Rothia dentocariosa is a gram-positive rod that forms branching filaments. It has been associated with abscesses and endocarditis, presumably following entry into the blood from the mouth. The gram-positive coccus, *Stomatococcus mucilaginosis,* has been moved to the genus *Rothia.* It is a common inhabitant of the oral cavity and has been associated with bacteremia in compromised hosts and endocarditis in intravenous drug users.

LISTERIA MONOCYTOGENES

There are several species in the genus *Listeria.* Of these, *L monocytogenes* is important as a cause of a wide spectrum of disease in animals and humans. *L monocytogenes* is capable of growing and surviving over a wide range of environmental conditions. It can survive at refrigerator temperatures (4°C), under conditions of low pH and high salt conditions. Therefore, it is able to overcome food preservation and safety barriers making it an important food-borne pathogen.

Morphology & Identification

L monocytogenes is a short, gram-positive, nonspore-forming rod (Figure 12–2). It is catalase-positive and has a tumbling end-over-end motility at 22–28°C but not at 37°C; the motility test rapidly differentiates listeria from diphtheroids that are members of the normal flora of the skin.

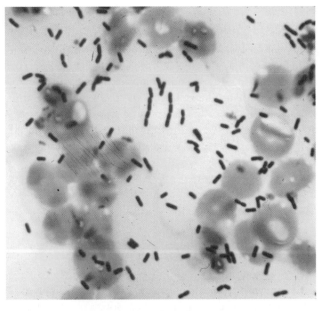

FIGURE 12–2 Gram stain of the gram-positive bacillus *Listeria monocytogenes* in a blood culture. Original magnification × 1000. Red blood cells are present in the background. Listeria isolated from clinical specimens frequently show variation in length and often in shape as well. Typically they are 0.4–0.5 μm in diameter and 0.5–2 μm long. (Courtesy of H Tran.)

Culture & Growth Characteristics

Listeria grows well on media such as 5% sheep blood agar on which it exhibits the characteristic small zone of hemolysis around and under colonies. The organism is a facultative anaerobe and is catalase-positive, esculin hydrolysis positive, and motile. Listeria produces acid but not gas from utilization of a variety of carbohydrates.

The motility at room temperature and hemolysin production are primary findings that help differentiate listeria from coryneform bacteria.

Antigenic Classification

Serologic classification is done only in reference laboratories and is primarily used for epidemiologic studies. Serotypes 1/2a, 1/2b, and 4b make up more than 95% of the isolates from humans. Serotype 4b causes most of the food-borne outbreaks.

Pathogenesis & Immunity

L monocytogenes enters the body through the gastrointestinal tract after ingestion of contaminated foods such as cheese or vegetables. The organism has several adhesin proteins (Ami, Fbp A, and flagellin proteins) that facilitate bacterial binding to the host cells and that contribute to virulence. It has a cell wall surface protein called internalin A that interacts with E-cadherin, a receptor on epithelial cells, promoting phagocytosis into the epithelial cells. After phagocytosis, the bacterium is enclosed in a phagolysosome, where the low pH activates the bacterium to produce listeriolysin O. This enzyme lyses the membrane of the phagolysosome and allows the listeriae to escape into the cytoplasm of the epithelial cell. The organisms proliferate and ActA, another listerial surface protein, induces host cell actin polymerization, which propels them to the cell membrane. Pushing against the host cell membrane, they cause formation of elongated protrusions called filopods. These filopods are ingested by adjacent epithelial cells, macrophages, and hepatocytes, the listeriae are released, and the cycle begins again. *L monocytogenes* can move from cell to cell without being exposed to antibodies, complement, or polymorphonuclear cells. *Shigella flexneri* and rickettsiae also usurp the host cells' actin and contractile system to spread their infections.

Iron is an important virulence factor. Listeriae produce siderophores and are able to obtain iron from transferrin.

Immunity to *L monocytogenes* is primarily cell-mediated, as demonstrated by the intracellular location of infection and by the marked association of infection and conditions of impaired cell-mediated immunity such as pregnancy, AIDS, lymphoma, and organ transplantation. Immunity can be transferred by sensitized lymphocytes but not by antibodies.

Clinical Findings

There are two forms of perinatal human listeriosis. Early-onset syndrome (**granulomatosis infantiseptica**) is the result of infection in utero and is a disseminated form of the disease characterized by neonatal sepsis, pustular lesions and granulomas containing *L monocytogenes* in multiple organs. Death may occur before or after delivery. The late-onset syndrome causes the development of meningitis between birth and the third week of life; it is often caused by serotype 4b and has a significant mortality rate.

Adults can develop listeria meningoencephalitis, bacteremia, and (rarely) focal infections. Meningoencephalitis and bacteremia occur most commonly in immunosuppressed patients, in whom listeria is one of the more common causes of meningitis. Clinical presentation of listeria meningitis in these patients varies from insidious to fulminant and is nonspecific. In immunocompetent individuals, illness may not occur following ingestion of contaminated food or patients may develop a symptomatic febrile gastroenteritis. This develops after an incubation period of 6–48 hours. Symptoms include fever, chills, headache, myalgias, abdominal pain, and diarrhea. Illness is usually self-limiting and most clinical laboratories do not routinely culture for *Listeria* from routine stool samples.

The diagnosis of listeriosis rests on isolation of the organism in cultures of blood and spinal fluid.

Spontaneous infection occurs in many domestic and wild animals. In ruminants (eg, sheep) listeria may cause meningoencephalitis with or without bacteremia. In smaller animals (eg, rabbits, chickens), there is septicemia with focal abscesses in the liver and heart muscle and marked monocytosis.

Many antimicrobial drugs inhibit listeria in vitro. Clinical cures have been obtained with ampicillin, with erythromycin, or with intravenous trimethoprim-sulfamethoxazole. Cephalosporins and fluoroquinolones are not active against *L monocytogenes*. Ampicillin plus gentamicin is often recommended for therapy, but gentamicin does not enter host cells and may not help treat the listeria infection.

ERYSIPELOTHRIX RHUSIOPATHIAE

Erysipelothrix rhusiopathiae is a gram-positive bacillus that produces small, transparent glistening colonies. It may be α-hemolytic on blood agar. On Gram stains it sometimes looks gram-negative because it decolorizes easily. The bacteria may appear singly, in short chains, randomly, or in long nonbranching filaments. The colony morphology and Gram stain appearance vary depending upon the growth medium, incubation temperature, and pH. Erysipelothrix is catalase-, oxidase-, and indole-negative. When *Erysipelothrix* is grown on triple sugar iron agar, hydrogen sulfide is produced, turning the TSI butt black.

E rhusiopathiae must be differentiated from *L monocytogenes*, *Arcanobacterium pyogenes*, and *A haemolyticum*, but these three species are β-hemolytic and do not produce hydrogen sulfide when grown on TSI medium. It is more difficult to differentiate *E rhusiopathiae* from aerotolerant lactobacilli; both may be α-hemolytic. They are catalase-negative and vancomycin-resistant (80% of lactobacilli). In addition, some strains of lactobacilli produce H_2S much like *E rhusiopathiae*.

E rhusiopathiae is distributed in land and sea animals worldwide, including a variety of vertebrates and invertebrates. It causes disease in domestic swine, turkeys, ducks, and sheep. The most important impact is in swine, where it causes erysipelas. In humans, erysipelas is caused by group A β-hemolytic streptococci and is much different from erysipelas of swine. People obtain *E rhusiopathiae* infection by direct inoculation from animals or animal products. Persons at greatest risk are fishermen, fish handlers, abattoir workers, butchers, and others who have contact with animal products.

The most common *E rhusiopathiae* infection in humans is called erysipeloid. It usually occurs on the fingers by direct inoculation at the site of a cut or abrasion (and has been called "seal finger" and "whale finger"). After 2–7 days' incubation, pain, which can be severe, and swelling occur. The lesion is raised, and violaceous in color. Pus is usually not present at the infection site, which helps differentiate it from staphylococcal and streptococcal skin infections. Erysipeloid can resolve after 3–4 weeks, or more rapidly with antibiotic treatment. Additional clinical forms of infection (both rare) are a diffuse cutaneous form and bacteremia with endocarditis. Erysipelothrix is highly susceptible to penicillin G, the drug of choice for severe infections. The organism is intrinsically resistant to vancomycin.

ACTINOMYCETES

The aerobic Actinomycetes are a large, diverse group of gram-positive bacilli with a tendency to form chains or filaments. They are related to the corynebacteria and include multiple genera of clinical significance such as *Mycobacteria* (discussed in Chapter 23) and saprophytic organisms such as streptomyces. As the bacilli grow, the cells remain together after division to form elongated chains of bacteria (1 μm in width) with occasional branches. The extent of this process varies in different taxa. It is rudimentary in some actinomycetes—the chains are short, break apart after formation, and resemble diphtheroids; others develop extensive substrate or aerial filaments (or both); or fragment into coccobacillary forms. Members of the aerobic Actinomycetes can be categorized on the basis of the acid-fast stain. Mycobacteria are truly positive acid-fast organisms; weakly positive genera include *Nocardia, Rhodococcus,* and a few others of clinical significance. *Streptomyces* and *Actinomadura*, two agents that cause actinomycotic mycetomas, are acid-fast stain negative.

R equi may appear to be a bacillus after a few hours of incubation in broth, but with further incubation it becomes coccoid in shape. This species of *Rhodococcus* also frequently produces pigmented colonies after 24 h of incubation that range from salmon pink to red. The organisms are generally weakly acid-fast positive when stained by the modified Kinyoun method. *R equi* occasionally causes infections such as necrotizing pneumonia in immunosuppressed patients with abnormal cell-mediated immunity (eg, AIDS patients). *R equi* is present in soil and in dung of herbivores. The organism is an occasional cause of disease in cattle, sheep, and swine and can cause severe lung infections in foals. Other species of the diverse genus *Rhodococcus* are present in the environment but rarely cause disease in humans.

NOCARDIOSIS

The genus *Nocardia* continues to undergo extensive taxonomic reclassification. New species continue to be recognized and at least 30 species have been implicated as causes of human infections.

The most common species associated with the vast majority of case of human infections are *N nova* complex, *N farcinica*, *N asteroides* type VI (*N cyriacigeorgica*), and *N brasiliensis*. Each of these is responsible for a broad range of diseases and each species/complex has unique drug susceptibility patterns. The pathogenic nocardiae, like many nonpathogenic species of nocardia, are found worldwide in soil and water. Nocardiosis is initiated by inhalation of these bacteria. The usual presentation is as a subacute to chronic pulmonary infection that may disseminate to other organs, usually the brain or skin. Nocardiae are not transmitted from person to person.

Morphology & Identification

Nocardia species are aerobic and grow on a variety of media. Over the course of several days to a week or more, they develop heaped, irregular, waxy colonies. Strains vary in their pigmentation from white to orange to red. These bacteria are gram-positive, catalase-positive, and partially acid-fast bacilli. They produce urease and can digest paraffin. Nocardiae form extensive branching substrates and aerial filaments that fragment after formation, breaking into coccobacillary cells. The cell walls contain mycolic acids that are shorter-chained than those of Mycobacteria. They are considered to be weakly acid-fast, but if they are stained with the routine acid-fast reagent (carbol-fuchsin) but decolorized with 1–4% sulfuric acid instead of the stronger acid-alcohol decolorant, most isolates will stain acid-fast. The species of nocardia are identified primarily by molecular methods such as 16S rRNA gene sequencing and RFLP analysis of amplified gene fragments such as *hsp*.

Pathogenesis & Clinical Findings

In most cases, nocardiosis is an opportunistic infection associated with several risk factors, most of which impair the cell-mediated immune responses: corticosteroid treatment, immunosuppression, organ transplantation, AIDS, tuberculosis, and alcoholism. Nocardiosis begins as chronic lobar pneumonia, and a variety of symptoms may occur, including fever, weight loss, and chest pain. The clinical manifestations are not distinctive and mimic tuberculosis and other infections. Pulmonary consolidations may develop, but granuloma formation and caseation are rare. The usual pathologic process is abscess formation. Spread from the lung often involves the central nervous system, where abscesses develop in the brain, leading to a variety of clinical presentations. Some patients have subclinical lung involvement and present with brain lesions. Dissemination may also occur to the skin, kidney, eye, or elsewhere.

Diagnostic Laboratory Tests

Specimens consist of sputum, pus, spinal fluid, and biopsy material. Gram-stained smears reveal gram-positive bacilli, coccobacillary cells, and branching filaments. With the modified acid-fast stain, most isolates will be acid-fast. *Nocardia* species grow on most laboratory media. Serologic tests are not useful.

Treatment

The treatment of choice is trimethoprim-sulfamethoxazole. If patients fail to respond, a number of other antibiotics have been used with success, such as amikacin, imipenem, minocycline, linezolide, and cefotaxime. Surgical drainage or resection may be required.

Emerging Actinomycetes: *Gordonia* & *Tsukamurella*

Members of the genera *Gordonia* and *Tsukamurella* are modified acid-fast positive bacteria that are becoming more frequently responsible for opportunistic infections among hospitalized immunocompromised patients. *Gordonia* produce orange, wrinkled colonies. On Gram stain, the organisms appear coryneform and do not branch. When these organisms are recovered from nonsterile sources like sputum, they may be disregarded as normal flora or contaminants. *Tsukamurella* sp. form whitish to orange colonies and on stain appears as long, straight, sometimes curved rods. Members of both genera are best identified by cell wall fatty acid analysis or 16S rRNA gene sequencing. These organisms have been associated with a variety of infections including postoperative wound infections, catheter associated bloodstream infections, ear drainage, and pulmonary infections. Treatment has been based upon anecdotal experiences but does require removal of catheters and drainage of abscesses. For infections caused by *Gordonia* sp. vancomycin, the carbapenems, aminoglycosides, fluoroquinolones, and linezolid have all been used successfully for treatment. In the case of *Tsukamurella* infections, in vitro susceptibility has been demonstrated with the aminoglycosides, sulfamethoxazole, fluoroquinolones, carbapenems, and clarithromycin.

ACTINOMYCETOMA

Mycetoma (Madura foot) is a localized, slowly progressive, chronic infection that begins in subcutaneous tissue and spreads to adjacent tissues. It is destructive and often painless. In many cases the cause is a soil fungus that has been implanted into the subcutaneous tissue by minor trauma. This form of mycetoma is discussed in Chapter 45. An actinomycetoma is a mycetoma caused by filamentous branching bacteria. The actinomycetoma granule is composed of tissue elements and gram-positive bacilli and bacillary chains or filaments (1 μm in diameter). The most common causes of actinomycetoma are *Nocardia asteroides*, *Nocardia brasiliensis*, *Streptomyces somaliensis*, and *Actinomadura madurae*. *N brasiliensis* may be acid-fast. These and other pathogenic actinomycetes are differentiated by biochemical tests and chromatographic analysis of cell wall components. Actinomycetomas respond well to various combinations of streptomycin, trimethoprim-sulfamethoxazole, and dapsone if therapy is begun early before extensive damage has occurred.

Oftentimes students are confused by the terms, actinomycetes and actinomycosis. The former have been described above; the latter is an infection caused by members of the anaerobic gram-positive genus, *Actinomyces*. *Actinomyces* species and the disease actinomycosis are described in more detail in Chapter 21.

REVIEW QUESTIONS

1. Three months ago, a 53-year-old woman had surgery and chemotherapy for breast cancer. Four weeks ago, she developed a cough occasionally productive of purulent sputum. About 2 weeks ago, she noted a slight but progressive weakness of her left arm and leg. On chest examination, rales were heard over the left upper back when the patient breathed deeply. Neurologic examination confirmed weakness of the left arm and leg. Chest x-ray showed a left upper lobe infiltrate. Contrast enhanced CT scan showed two lesions in the right hemisphere. Gram stain of a purulent sputum specimen showed branching gram-positive rods that were partially acid-fast. Which of the following organisms is the cause of this patient's current illness?

    (A) *Actinomyces israelii*
    (B) *Corynebacterium pseudodiphtheriticum*
    (C) *Aspergillus fumigatus*
    (D) *Nocardia asteroides*
    (E) *Erysipelothrix rhusiopathiae*

2. The drug of choice to treat this patient's infection (Question 1) is

    (A) Penicillin G
    (B) Trimethoprim-sulfamethoxazole
    (C) Gentamicin
    (D) Amphotericin B
    (E) A third-generation cephalosporin

3. It is particularly difficult to differentiate *Erysipelothrix rhusiopathiae* from

    (A) *Corynebacterium diphtheriae*
    (B) *Bacillus cereus*
    (C) *Actinomyces israelii*
    (D) *Nocardia asteroides*
    (E) *Lactobacillus* species

4. Movement of *Listeria monocytogenes* inside of host cells is caused by

    (A) Inducing host cell actin polymerization
    (B) The formation of pili (fimbriae) on the listeriae surface
    (C) Pseudopod formation
    (D) The motion of listeriae flagella
    (E) Tumbling motility

5. An 8-year-old boy develops a severe sore throat. On examination, a grayish exudate (pseudomembrane) is seen over the tonsils and pharynx. The differential diagnosis of severe pharyngitis such as this includes group A streptococcal infection, Epstein-Barr virus (EBV) infection, *Neisseria gonorrhoeae* pharyngitis, and diphtheria. The cause of the boy's pharyngitis is most likely

    (A) A gram-negative bacillus
    (B) A single-stranded positive-sense RNA virus
    (C) A catalase-positive gram-positive coccus that grows in clusters
    (D) A club-shaped gram-positive bacillus
    (E) A double stranded RNA virus

6. The primary mechanism in the pathogenesis of the boy's disease (Question 5) is

    (A) A net increase in intracellular cyclic adenosine monophosphate
    (B) Action of pyrogenic exotoxin (a superantigen)
    (C) Inactivation of acetylcholine esterase
    (D) Action of enterotoxin A
    (E) Inactivation of elongation factor 2

7. *Corynebacterium jeikeium* is

    (A) Catalase-negative
    (B) Gram-negative
    (C) Often resistant to commonly used antibiotics
    (D) Motile
    (E) Common but clinically unimportant

8. Which of the following aerobic gram-positive bacilli is modified acid-fast positive?

    (A) *Nocardia brasiliensis*
    (B) *Lactobacillus acidophilus*
    (C) *Erysipelothrix rhusiopathiae*
    (D) *Listeria monocytogenes*

9. Skin diphtheria as occurs in children in tropical areas typically

    (A) Does not occur in children who have been immunized with diphtheria toxoid
    (B) Is clinically distinct from skin infections (pyoderma, impetigo) caused by *Streptococcus pyogenes* and *Staphylococcus aureus*
    (C) Is also common in northern latitudes
    (D) Results in protective antitoxin levels in most children by the time they are 6–8 years old
    (E) Yields toxin-mediated cardiomyopathy

10. A 45-year-old fisherman imbedded a fishhook into his right forefinger. He removed it and did not seek immediate medical

therapy. Five days later, he noted fever, severe pain, and nodular type swelling of the finger. He sought medical therapy. The violaceous nodule was aspirated and after 48 hours of incubation, colonies of a gram-positive bacillus that caused greenish discoloration of the agar and formed long filaments in the broth culture were noted. The most likely cause of this infection is

(A) *Lactobacillus acidophilus*
(B) *Erysipelothrix rhusiopathiae*
(C) *Listeria monocytogenes*
(D) *Rhodococcus equi*
(E) *Nocardia brasiliensis*

11. A biochemical reaction that is useful in the identification of the causative agent of the infection in question 10 is:

(A) Catalase positivity
(B) Acid fastness using modified Kinyoun stain
(C) Esculin hydrolysis
(D) Tumbling motility
(E) Production of H$_2$S

Answers

1. D	4. A	7. C	10. B
2. B	5. D	8. A	11. E
3. E	6. E	9. D	

REFERENCES

Aerobic and facultative gram-positive bacilli. In *Koneman's Color Atlas and Textbook of Diagnostic Microbiology*, 6th ed. Winn WC Jr et al (editors). Lippincott Williams & Wilkins, 2006, pp. 765–857.

Brown-Elliott BA, Brown JM, Conville PS, Wallace RJ, Jr. Clinical and laboratory features of the *Nocardia* spp. based on current molecular taxonomy. Clin Microbiol Rev 2006; 19:259–82. [PMID 16614249]

Conville PS, Witebsky FG. *Nocardia, Rhodococcus, Gordonia, Actinomadura, Streptomyces*, and other aerobic Actinomycetes. In: *Manual of Clinical Microbiology*, 9th ed. Murray PR et al (editors). ASM Press, 2007.

Deng Q, Barbieri JT: Molecular mechanisms of the cytotoxicity of ADP-ribosylating toxins. Annu Rev Microbiol 2008; 62:271–88. [PMID 18785839]

Drevets DA, Bronze MS: *Listeria monocytogenes*: epidemiology, human disease, and mechanisms of brain invasion. FEMS Immunol Med Microbiol 2008; 53:151–65. [PMID 18462388]

Dussurget O. New insights into determinants of *Listeria monocytogenes* virulence. Int Rev Cell Mol Biol 2008;270:1–38. [PMID 19081533]

Funke G et al: Clinical microbiology of coryneform bacteria. Clin Microbiol Rev 1997;10:125. [PMID: 8993861]

Funke G, Bernard KA: Coryneform gram-positive rods. In: *Manual of Clinical Microbiology*, 9th ed. Murray PR et al (editors). ASM Press, 2007.

Reboli AC, Farrar WE: *Erysipelothrix rhusiopathiae*: an occupational pathogen. Clin Microbiol Rev 1989;2:354. [PMID: 2680056]

Sorrell TC, Mitchell DH, Iredell JR: *Nocardia* species. In: *Mandell, Douglas, and Bennett's Principles and Practice of Infectious Diseases*, 7th ed. Mandell GL, Bennett JE, Dolin R (editors). Elsevier, 2010.

CHAPTER

13

The Staphylococci

The staphylococci are gram-positive spherical cells, usually arranged in grapelike irregular clusters. They grow readily on many types of media and are active metabolically, fermenting carbohydrates and producing pigments that vary from white to deep yellow. Some are members of the normal flora of the skin and mucous membranes of humans; others cause suppuration, abscess formation, a variety of pyogenic infections, and even fatal septicemia. The pathogenic staphylococci often hemolyze blood, coagulate plasma, and produce a variety of extracellular enzymes and toxins. The most common type of food poisoning is caused by a heat-stable staphylococcal enterotoxin. Staphylococci rapidly develop resistance to many antimicrobial agents and present difficult therapeutic problems.

The genus *Staphylococcus* has at least 40 species. The three most frequently encountered species of clinical importance are *Staphylococcus aureus, Staphylococcus epidermidis,* and *Staphylococcus saprophyticus. S aureus* is **coagulase-positive**, which differentiates it from the other species. *S aureus* is a major pathogen for humans. Almost every person will have some type of *S aureus* infection during a lifetime, ranging in severity from food poisoning or minor skin infections to severe life-threatening infections. The coagulase-negative staphylococci are normal human flora and sometimes cause infection, often associated with implanted devices, such as joint prostheses, shunts, and intravascular catheters, especially in very young, old, and immunocompromised patients. Approximately 75% of these infections caused by **coagulase-negative** staphylococci are due to *S epidermidis;* infections due to *Staphylococcus lugdunensis, Staphylococcus warneri, Staphylococcus hominis,* and other species are less common. *S saprophyticus* is a relatively common cause of urinary tract infections in young women, although it rarely causes infections in hospitalized patients. Other species are important in veterinary medicine.

Morphology & Identification

A. Typical Organisms

Staphylococci are spherical cells about 1 μm in diameter arranged in irregular clusters (Figure 13–1). Single cocci, pairs, tetrads, and chains are also seen in liquid cultures. Young cocci stain strongly gram-positive; on aging, many

cells become gram-negative. Staphylococci are nonmotile and do not form spores. Under the influence of drugs like penicillin, staphylococci are lysed.

Micrococcus species often resemble staphylococci. They are found free-living in the environment and form regular packets of four or eight cocci. Their colonies can be yellow, red, or orange. Micrococci are rarely associated with disease.

B. Culture

Staphylococci grow readily on most bacteriologic media under aerobic or microaerophilic conditions. They grow most rapidly at 37°C but form pigment best at room temperature (20–25°C). Colonies on solid media are round, smooth, raised, and glistening (Figure 13–2). *S aureus* usually forms gray to deep golden yellow colonies. *S epidermidis* colonies usually are gray to white on primary isolation; many colonies develop pigment only upon prolonged incubation. No pigment is produced anaerobically or in broth. Various degrees of hemolysis are produced by *S aureus* and occasionally by other species. *Peptostreptococcus* and *Peptoniphilus* species, which are anaerobic cocci, often resemble staphylococci in morphology. The genus *Staphylococcus* contains two species, *S saccharolyticus* and *S aureus* subsp. *anaerobius,* that initially grow only under anaerobic conditions but become more aerotolerant on subcultures.

C. Growth Characteristics

The staphylococci produce catalase, which differentiates them from the streptococci. Staphylococci slowly ferment many carbohydrates, producing lactic acid but not gas. Proteolytic activity varies greatly from one strain to another. Pathogenic staphylococci produce many extracellular substances, which are discussed below.

Staphylococci are relatively resistant to drying, heat (they withstand 50°C for 30 minutes), and 9% sodium chloride but are readily inhibited by certain chemicals, eg, 3% hexachlorophene.

Staphylococci are variably sensitive to many antimicrobial drugs. Resistance falls into several classes:

1. β-Lactamase production is common, is under plasmid control, and makes the organisms resistant to many

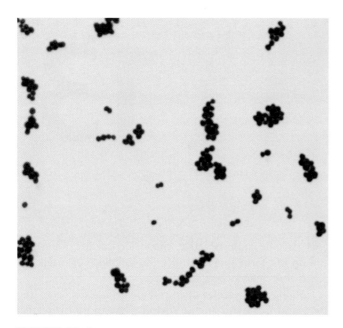

FIGURE 13–1 Gram stain of *Staphylococcus aureus* showing gram-positive cocci in pairs, tetrads, and clusters. Original magnification × 1000. (Courtesy of L Ching.)

FIGURE 13–2 Colonies of *Staphylococcus aureus* on a blood agar plate after 24 hours incubation. The yellow-gray colonies are 3-4 mm in diameter on the 10 cm plate. The colonies are surrounded by clear zones of hemolysis about 1 cm in diameter. (Courtesy of H Reyes.)

penicillins (penicillin G, ampicillin, ticarcillin, piperacillin, and similar drugs). The plasmids are transmitted by transduction and perhaps also by conjugation.

2. Resistance to nafcillin (and to methicillin and oxacillin) is independent of β-lactamase production. Resistance to nafcillin is encoded and regulated by a sequence of genes found in a region of the chromosome called the staphylococcal cassette chromosome *mec* (*SCCmec*). Specifically, the *mecA* gene on this locus encodes a low-affinity penicillin binding protein (PBP2a) that is responsible for the resistance. There are several different *SCCmec* types. Types I, II, and III are associated with hospital-acquired infections and may contain genes that encode resistance to other antimicrobials as well. *SCCmec* type IV has principally been found in community-acquired MRSA (CA-MRSA) strains that tend to be less resistant, more transmissible, and responsible for outbreaks over the last decade in the United States and some countries in Europe.

3. In the United States, *S aureus* and *S lugdunensis* are considered to be susceptible to vancomycin if the minimum inhibitory concentration (MIC) is ≤ 2 μg/mL; of intermediate susceptibility if the MIC is 4–8 μg/mL; and resistant if the MIC is ≥ 16 μg/mL. Strains of *S aureus* with intermediate susceptibility to vancomycin have been isolated in Japan, the United States, and several other countries. These are often known as vancomycin-intermediate *S aureus*, or "VISA." They generally have been isolated from patients with complex infections who have received prolonged vancomycin therapy. Often there has been vancomycin treatment failure. The mechanism of resistance is associated with increased cell wall synthesis and alterations in the cell wall and is not due to the *van*

genes found in enterococci. *S aureus* strains of intermediate susceptibility to vancomycin usually are nafcillin-resistant but generally are susceptible to oxazolidinones and to quinupristin/dalfopristin.

4. Since 2002, several isolates of vancomycin-resistant *S aureus* (VRSA) strains were isolated from patients in the United States. The isolates contained the vancomycin resistance gene *vanA* from enterococci (see Chapter 14) and the nafcillin resistance gene *mecA* (see above). Both of the initial VRSA strains were susceptible to other antibiotics. Vancomycin resistance in *S aureus* is of major concern worldwide.

5. Plasmid-mediated resistance to tetracyclines, erythromycins, aminoglycosides, and other drugs is frequent in staphylococci.

6. "Tolerance" implies that staphylococci are inhibited by a drug but not killed by it—ie, there is great difference between minimal inhibitory and minimal lethal concentrations of an antimicrobial drug. Patients with endocarditis caused by a tolerant *S aureus* may have a prolonged clinical course compared with patients who have endocarditis caused by a fully susceptible *S aureus*. Tolerance can at times be attributed to lack of activation of autolytic enzymes in the cell wall.

D. Variation

A culture of staphylococci contains some bacteria that differ from the bulk of the population in expression of colony characteristics (colony size, pigment, hemolysis), in enzyme elaboration, in drug resistance, and in pathogenicity. In vitro, the expression of such characteristics is influenced by growth conditions: When nafcillin-resistant *S aureus* is incubated at 37°C on blood agar, one in 10^7 organisms expresses nafcillin resistance; when it is incubated at 30°C on agar containing 2–5% sodium chloride, one in 10^3 organisms expresses nafcillin resistance.

Antigenic Structure

Staphylococci contain antigenic polysaccharides and proteins as well as other substances important in cell wall structure. Peptidoglycan, a polysaccharide polymer containing linked subunits, provides the rigid exoskeleton of the cell wall. Peptidoglycan is destroyed by strong acid or exposure to lysozyme. It is important in the pathogenesis of infection: It elicits production of interleukin-1 (endogenous pyrogen) and opsonic antibodies by monocytes, and it can be a chemoattractant for polymorphonuclear leukocytes, have endotoxin-like activity, and activate complement.

Teichoic acids, which are polymers of glycerol or ribitol phosphate, are linked to the peptidoglycan and can be antigenic. Antiteichoic acid antibodies detectable by gel diffusion may be found in patients with active endocarditis due to *S aureus*.

Protein A is a cell wall component of *S aureus* strains and is a bacterial surface protein that has been characterized among a group of adhesins called microbial surface components recognizing adhesive matrix molecules (MSCRAMMS). Bacterial attachment to host cells is mediated by MSCRAMMS, and these are important virulence factors. Protein A binds to the Fc portion of IgG molecules except IgG_3. The Fab portion of the IgG bound to protein A is free to combine with a specific antigen. Protein A has become an important reagent in immunology and diagnostic laboratory technology; for example, protein A with attached IgG molecules directed against a specific bacterial antigen will agglutinate bacteria that have that antigen (**"coagglutination"**).

Some *S aureus* strains have capsules, which inhibit phagocytosis by polymorphonuclear leukocytes unless specific antibodies are present. Most strains of *S aureus* have coagulase, or clumping factor, on the cell wall surface; coagulase binds nonenzymatically to fibrinogen, yielding aggregation of the bacteria.

Serologic tests have limited usefulness in identifying staphylococci.

Enzymes & Toxins

Staphylococci can produce disease both through their ability to multiply and spread widely in tissues and through their production of many extracellular substances. Some of these substances are enzymes; others are considered to be toxins, though they may function as enzymes. Many of the toxins are under the genetic control of plasmids; some may be under both chromosomal and extrachromosomal control; and for others the mechanism of genetic control is not well defined.

A. Catalase

Staphylococci produce catalase, which converts hydrogen peroxide into water and oxygen. The catalase test differentiates the staphylococci, which are positive, from the streptococci, which are negative.

B. Coagulase & Clumping Factor

S aureus produces coagulase, an enzyme-like protein that clots oxalated or citrated plasma. Coagulase binds to prothrombin; together they become enzymatically active and initiate fibrin polymerization. Coagulase may deposit fibrin on the surface of staphylococci, perhaps altering their ingestion by phagocytic cells or their destruction within such cells. Coagulase production is considered synonymous with invasive pathogenic potential.

Clumping factor is another example of an MSCRAMM that is responsible for adherence of the organisms to fibrinogen and fibrin. When mixed with plasma, *S aureus* forms clumps. Clumping factor is distinct from coagulase. Since clumping factor induces a strong immunogenic response in the host, it has been the focus of recent vaccine efforts. Humanized monoclonal antibodies that prevent the binding of clumping factor to fibrinogen are in clinical trials in combination with antibiotics for the treatment of *S aureus* bacteremia (Weems et al, 2006).

C. Other Enzymes

Other enzymes produced by staphylococci include a hyaluronidase, or spreading factor; a staphylokinase resulting in fibrinolysis but acting much more slowly than streptokinase; proteinases; lipases; and β-lactamase.

D. Exotoxins

The α-toxin is a heterogeneous protein that acts on a broad spectrum of eukaryotic cell membranes. The α-toxin is a potent hemolysin. The β-toxin degrades sphingomyelin and therefore is toxic for many kinds of cells, including human red blood cells. The δ-toxin is heterogeneous and dissociates into subunits in nonionic detergents. It disrupts biologic membranes and may have a role in *S aureus* diarrheal diseases. The γ hemolysin refers to three proteins that interact with the two proteins comprising the Panton-Valentine leukocidin (see below) to form six potential two-component toxins. All six of these protein toxins are capable of efficiently lysing white blood cells by causing pore formation in the cellular membranes that increase cation permeability. This leads to massive release of inflammatory mediators such as IL-8, leukotriene, and histamine which are responsible for necrosis and severe inflammation.

E. Panton-Valentine Leukocidin

This toxin of *S aureus* has two components. It can kill white blood cells of humans and rabbits. The two components designated as S and F act synergistically on the white blood cell membrane as described above for γ toxin. This toxin is an important virulence factor in community associated methicillin resistant *S aureus* infections.

F. Exfoliative Toxins

These epidermolytic toxins of *S aureus* are two distinct proteins of the same molecular weight. Epidermolytic toxin A is a chromosomal gene product and is heat-stable (resists boiling for 20 minutes). Epidermolytic toxin B is plasmid-mediated and heat-labile. The epidermolytic toxins yield the generalized desquamation of the staphylococcal scalded skin syndrome by dissolving the mucopolysaccharide matrix of the epidermis. The toxins are superantigens.

G. Toxic Shock Syndrome Toxin

Most *S aureus* strains isolated from patients with toxic shock syndrome produce a toxin called **toxic shock syndrome toxin-1** (TSST-1), which is the same as enterotoxin F. TSST-1 is the prototypical **superantigen** (see Chapter 8). TSST-1 binds to MHC class II molecules, yielding T cell stimulation, which promotes the protean manifestations of the toxic shock syndrome. The toxin is associated with fever, shock, and multisystem involvement, including a desquamative skin rash. The gene for TSST-1 is found in about 20% of *S aureus* isolates, including MRSA.

H. Enterotoxins

There are multiple (A–E, G–J, K–R and U, V) enterotoxins. Approximately 50% of *S aureus* strains can produce one or more of them. Like TSST-1, the enterotoxins are superantigens. The enterotoxins are heat-stable and resistant to the action of gut enzymes. Important causes of food poisoning, enterotoxins are produced when *S aureus* grows in carbohydrate and protein foods. Ingestion of 25 μg of enterotoxin B results in vomiting and diarrhea. The emetic effect of enterotoxin is probably the result of central nervous system stimulation (vomiting center) after the toxin acts on neural receptors in the gut.

The exfoliative toxins, TSST-1, and the enterotoxin genes are on a chromosomal element called a pathogenicity island. It interacts with accessory genetic elements—bacteriophages—to produce the toxins.

Pathogenesis

Staphylococci, particularly *S epidermidis,* are members of the normal flora of the human skin and respiratory and gastrointestinal tracts. Nasal carriage of *S aureus* occurs in 20–50% of humans. Staphylococci are also found regularly on clothing, bed linens, and other fomites in human environments.

The pathogenic capacity of a given strain of *S aureus* is the combined effect of extracellular factors and toxins together with the invasive properties of the strain. At one end of the disease spectrum is staphylococcal food poisoning, attributable solely to the ingestion of preformed enterotoxin; at the other end are staphylococcal bacteremia and disseminated abscesses in all organs.

Pathogenic, invasive *S aureus* produces coagulase and tends to produce a yellow pigment and to be hemolytic. Nonpathogenic, noninvasive staphylococci such as *S epidermidis* are coagulase-negative and tend to be nonhemolytic. Such organisms rarely produce suppuration but may infect orthopedic or cardiovascular prostheses or cause disease in immunosuppressed persons. They may be refractory to treatment because of the formation of biofilms. *S saprophyticus* is typically nonpigmented, novobiocin-resistant, and nonhemolytic; it causes urinary tract infections in young women.

Regulation of Virulence Determinants

The expression of staphylococcal virulence determinants is regulated by several systems that are sensitive to environmental signals. These systems consist of two proteins (two-component systems), a sensor kinase, and a response regulator. Binding of sensors to specific extracellular ligands, or to a receptor, results in a phosphorylation cascade that leads to binding of the regulator to specific DNA sequences, which ultimately leads to activation of transcription-regulating functions. There are several well-described two-component regulatory systems in *S aureus.* These include *agr,* the best described, *sae RS, srrAB, arlSR,* and *lytRS.*

The accessory gene regulator (*agr*) is essential in quorum-sensing control of gene expression. It controls the preferential expression of surface adhesins (protein A, coagulase, and fibronectin binding protein) and production of exoproteins (toxins such as TSST-1) depending upon the growth phase (and hence bacterial density).

At low cell density, the promoter P2 is off and transcriptions of transmembrane protein, AgrB, peptide precursor, AgrD, transmembrane sensor, AgrC, and transcription regulator, Agr A, are at low levels. As cell density increases during stationary growth phase, the AgrC sensor activates the regulator AgrA. AgrA is a DNA-binding protein that activates promoter P2 and promoter P3. Promoter P3 initiates transcription of δ-hemolysin and an effector called RNAIII, which down-regulates the expression of surface adhesins and activates secretion of exoproteins at both the transcriptional and translational levels. *Agr* is also positively controlled by a DNA-binding protein called SarA (encoded by *sar*) and possibly by other regulatory systems.

At least four additional two-component regulatory systems have been shown to affect virulence gene expression. These are called *sae, S aureus* exoproteins; *srrAB,* staphylococcal respiratory response; *arlS,* autolysis-related locus sensor; and *lytRS.* Sae regulates gene expression at the transcriptional level and is essential for production of α-toxin, β-hemolysins, and coagulase. Its activity is independent from that of *agr. SsrAB* is important for regulation of virulence factor expression that is influenced by environmental oxygen. The *arlSR* locus is important to the control of autolysis and also decreases the activation of the *agr* locus. The *lytRS* locus is also involved in autolysis.

Pathology

The prototype of a staphylococcal lesion is the furuncle or other localized abscess. Groups of *S aureus* established in a hair follicle lead to tissue necrosis (dermonecrotic factor). Coagulase is produced and coagulates fibrin around the lesion and within the lymphatics, resulting in formation of a wall that limits the process and is reinforced by the accumulation of inflammatory cells and, later, fibrous tissue. Within the center of the lesion, liquefaction of the necrotic tissue occurs (enhanced by delayed hypersensitivity), and the abscess "points" in the direction of least resistance. Drainage of the liquid center necrotic tissue is followed by slow filling of the cavity with granulation tissue and eventual healing.

Focal suppuration (abscess) is typical of staphylococcal infection. From any one focus, organisms may spread via the lymphatics and bloodstream to other parts of the body. Suppuration within veins, associated with thrombosis, is a common feature of such dissemination. In osteomyelitis, the primary focus of *S aureus* growth is typically in a terminal blood vessel of the metaphysis of a long bone, leading to necrosis of bone and chronic suppuration. *S aureus* may cause pneumonia, meningitis, empyema, endocarditis, or sepsis with suppuration in any organ. Staphylococci of low invasiveness are involved in many skin infections (eg, acne, pyoderma, or impetigo). Anaerobic cocci (*Peptostreptococcus*) participate in mixed anaerobic infections.

Staphylococci also cause disease through the elaboration of toxins, without apparent invasive infection. Bullous exfoliation, the scalded skin syndrome, is caused by the production of exfoliative toxins. Toxic shock syndrome is associated with TSST-1.

Clinical Findings

A localized staphylococcal infection appears as a "pimple," hair follicle infection, or abscess. There is usually an intense, localized, painful inflammatory reaction that undergoes central suppuration and heals quickly when the pus is drained. The wall of fibrin and cells around the core of the abscess tends to prevent spread of the organisms and should not be broken down by manipulation or trauma.

S aureus infection can also result from direct contamination of a wound, eg, postoperative staphylococcal wound infection or infection following trauma (chronic osteomyelitis subsequent to an open fracture, meningitis following skull fracture).

If *S aureus* disseminates and bacteremia ensues, endocarditis, acute hematogenous osteomyelitis, meningitis, or pulmonary infection can result. The clinical presentations resemble those seen with other bloodstream infections. Secondary localization within an organ or system is accompanied by the symptoms and signs of organ dysfunction and intense focal suppuration.

Food poisoning due to staphylococcal enterotoxin is characterized by a short incubation period (1–8 hours); violent nausea, vomiting, and diarrhea; and rapid convalescence. There is no fever.

Toxic shock syndrome is manifested by an abrupt onset of high fever, vomiting, diarrhea, myalgias, a scarlatiniform rash, and hypotension with cardiac and renal failure in the most severe cases. It often occurs within 5 days after the onset of menses in young women who use tampons, but it also occurs in children or in men with staphylococcal wound infections. The syndrome can recur. Toxic shock syndrome-associated *S aureus* can be found in the vagina, on tampons, in wounds or other localized infections, or in the throat but virtually never in the bloodstream.

Diagnostic Laboratory Tests

A. Specimens

Surface swab pus, blood, tracheal aspirate, or spinal fluid for culture, depending upon the localization of the process, are all appropriate specimens for testing.

B. Smears

Typical staphylococci appear as gram-positive cocci in clusters in Gram-stained smears of pus or sputum. It is not possible to distinguish saprophytic (*S epidermidis*) from pathogenic (*S aureus*) organisms on smears.

C. Culture

Specimens planted on blood agar plates give rise to typical colonies in 18 hours at 37°C, but hemolysis and pigment production may not occur until several days later and are optimal at room temperature. *S aureus* but not other staphylococci ferment mannitol. Specimens contaminated with a mixed flora can be cultured on media containing 7.5% NaCl; the salt inhibits most other normal flora but not *S aureus*. Mannitol salt agar or commercially available chromogenic media are used to screen for nasal carriers of *S aureus* and patients with cystic fibrosis.

D. Catalase Test

This test is used to detect the presence of cytochrome oxidase enzymes. A drop of 3% hydrogen peroxide solution is placed on a slide, and a small amount of the bacterial growth is placed in the solution. The formation of bubbles (the release of oxygen) indicates a positive test.

E. Coagulase Test

Citrated rabbit (or human) plasma diluted 1:5 is mixed with an equal volume of broth culture or growth from colonies on agar and incubated at 37°C. A tube of plasma mixed with sterile broth is included as a control. If clots form in 1–4 hours, the test is positive.

Coagulase-positive staphylococci are considered pathogenic for humans; however, coagulase-positive staphylococci of dogs (*Staphylococcus intermedius*) and dolphins (*Staphylococcus delphini*) rarely cause disease in humans. Infections of prosthetic devices can be caused by organisms of the coagulase-negative *S epidermidis* group.

F. Susceptibility Testing

Broth microdilution or disk diffusion susceptibility testing should be done routinely on staphylococcal isolates from clinically significant infections. Resistance to penicillin G can be predicted by a positive test for β-lactamase; approximately 90% of *S aureus* produce β-lactamase. Resistance to nafcillin (and oxacillin and methicillin) occurs in about 65% of *S aureus* and approximately 75% of *S epidermidis* isolates. Nafcillin resistance correlates with the presence of *mec*A, the gene that codes for a penicillin-binding protein (PBP 2a) not affected by these drugs. The gene can be detected using the polymerase chain reaction. Most clinical laboratories use a phenotypic method such as an oxacillin screening agar plate. Staphylococci that grow on Mueller-Hinton agar containing 4% NaCl and 6 μg/mL of oxacillin typically are *mec*A-positive and nafcillin-resistant. Alternatively, an assay for the *mec*A gene product, PBP2a, is commercially available and is much more rapid than PCR for *mec*A or than testing for resistance using growth on oxacillin-containing salt agar.

G. Serologic & Typing Tests

Serologic tests for diagnosis of *S aureus* infections have little practical value.

Antibiotic susceptibility patterns are helpful in tracing *S aureus* infections and in determining if multiple *S epidermidis* isolates from blood cultures represent bacteremia due to the same strain, seeded by a nidus of infection.

Molecular typing techniques have been used to document the spread of epidemic disease-producing clones of *S aureus*. Pulsed-field gel electrophoresis and multilocus sequence typing are highly discriminatory.

Treatment

Most persons harbor staphylococci on the skin and in the nose or throat. Even if the skin can be cleared of staphylococci (eg, in eczema), reinfection by droplets will occur almost immediately. Because pathogenic organisms are commonly spread from one lesion (eg, a furuncle) to other areas of the skin by fingers and clothing, scrupulous local antisepsis is important to control recurrent furunculosis.

Serious multiple skin infections (acne, furunculosis) occur most often in adolescents. Similar skin infections occur in patients receiving prolonged courses of corticosteroids. In acne, lipases of staphylococci and corynebacteria liberate fatty acids from lipids and thus cause tissue irritation. Tetracyclines are used for long-term treatment.

Abscesses and other closed suppurating lesions are treated by drainage, which is essential, and antimicrobial therapy. Many antimicrobial drugs have some effect against staphylococci in vitro. However, it is difficult to eradicate pathogenic staphylococci from infected persons, because the organisms rapidly develop resistance to many antimicrobial drugs and the drugs cannot act in the central necrotic part of a suppurative lesion. It is also difficult to eradicate the *S aureus* carrier state.

Acute hematogenous osteomyelitis responds well to antimicrobial drugs. In chronic and recurrent osteomyelitis, surgical drainage and removal of dead bone is accompanied by long-term administration of appropriate drugs, but eradication of the infecting staphylococci is difficult. Hyperbaric oxygen and the application of vascularized myocutaneous flaps have aided healing in chronic osteomyelitis.

Bacteremia, endocarditis, pneumonia, and other severe infections due to *S aureus* require prolonged intravenous therapy with a β-lactamase-resistant penicillin. Vancomycin is often reserved for use with nafcillin-resistant staphylococci. In recent years, an increase in MICs to vancomycin among many MRSA strains recovered from hospitalized patients has led physicians to seek alternative therapies. Alternative agents for the treatment of MRSA bacteremia and endocarditis include newer antimicrobials such as daptomycin, linezolid, and quinupristin-dalfopristin (Chapter 28). Also, these agents may be bactericidal and offer alternatives when allergies preclude the use of other compounds or the patient's infection appears to be failing clinically. However, the use of these agents should be discussed with infectious diseases physicians or pharmacists as the side effect profiles and pharmacokinetics are quite unique to each agent. If the infection is found to be due to non-β-lactamase-producing *S aureus*, penicillin G is the drug of choice, but these *S aureus* strains are rarely encountered.

S epidermidis infections are difficult to cure because they occur in prosthetic devices where the bacteria can sequester themselves in a biofilm. *S epidermidis* is more often resistant to antimicrobial drugs than is *S aureus*; approximately 75% of *S epidermidis* strains are nafcillin-resistant.

Because of the frequency of drug-resistant strains, meaningful staphylococcal isolates should be tested for antimicrobial susceptibility to help in the choice of systemic drugs. Resistance to drugs of the erythromycin group tends to emerge so rapidly that these drugs should not be used singly for treatment of chronic infection. Drug resistance (to penicillins, tetracyclines, aminoglycosides, erythromycins, etc) determined by plasmids can be transmitted among staphylococci by transduction and perhaps by conjugation.

Penicillin G-resistant *S aureus* strains from clinical infections always produce penicillinase. They constitute >95% of *S aureus* isolates in communities in the United States. They are often susceptible to β-lactamase-resistant penicillins, cephalosporins, or vancomycin. Nafcillin resistance is independent of β-lactamase production, and its clinical incidence varies greatly in different countries and at different times.

The selection pressure of β-lactamase-resistant antimicrobial drugs may not be the sole determinant for resistance to these drugs: For example, in Denmark, nafcillin-resistant *S aureus* comprised 40% of isolates in 1970 and only 10% in 1980, without notable changes in the use of nafcillin or similar drugs. In the United States, nafcillin-resistant *S aureus* accounted for only 0.1% of isolates in 1970 but in the 1990s constituted 20–30% of isolates from infections in some hospitals. In 2003, 60% of nosocomial *S aureus* among intensive care patients were resistant to nafcillin. Fortunately, *S aureus* strains of intermediate susceptibility to vancomycin have been relatively uncommon, and the isolation of vancomycin-resistant strains has been rare.

Epidemiology & Control

Staphylococci are ubiquitous human parasites. The chief sources of infection are shedding human lesions, fomites contaminated from such lesions, and the human respiratory tract and skin. Contact spread of infection has assumed added importance in hospitals, where a large proportion of the staff and patients carry antibiotic-resistant staphylococci in the nose or on the skin. Although cleanliness, hygiene, and aseptic management of lesions can control the spread of staphylococci from lesions, few methods are available to prevent the wide dissemination of staphylococci from carriers. Aerosols (eg, glycols) and ultraviolet irradiation of air have little effect.

In hospitals, the areas at highest risk for severe staphylococcal infections are the newborn nursery, intensive care units, operating rooms, and cancer chemotherapy wards. Massive introduction of "epidemic" pathogenic *S aureus* into these areas may lead to serious clinical disease. Personnel with active *S aureus* lesions and carriers may have to be excluded from these areas. In such individuals, the application of topical antiseptics such as mupirocin to nasal or perineal carriage sites may diminish shedding of dangerous organisms. Rifampin coupled with a second oral antistaphylococcal drug sometimes provides long-term suppression and possibly cure of nasal carriage; this form of therapy is usually reserved for major problems of staphylococcal carriage, because staphylococci can rapidly develop resistance to rifampin.

To diminish transmission within the hospital setting, high risk patients, such as those in intensive care units and patients transferred from chronic care facilities where prevalence is high, are frequently surveyed for anterior nares colonization. Patients who test positive by culture or PCR are placed upon contact precautions so as to minimize spread on the hands of health care workers. Health care workers should strictly adhere to infection control policies by wearing gloves and washing hands before and after patient contact.

Until relatively recently, methicillin resistant *S aureus* was confined primarily to the hospital setting. Worldwide dissemination of a few distinct clones of CA-MRSA has resulted in an increase in skin and soft tissue infections and necrotizing pneumonia, primarily in younger patients without known risk factors for MRSA acquisition. These strains appear to be more virulent. CA-MRSA isolates are characterized by the presence of the Panton-Valentine leukocidin and the presence of staphylococcal cassette chromosome *mec* type IV, which may explain the increased susceptibility to other antimicrobial agents compared to health care associated MRSA strains.

REVIEW QUESTIONS

1. A 54-year-old woman develops a right shoulder abscess with a strain of *Staphylococcus aureus* that is resistant to nafcillin. She was treated with a 2-week course of intravenous vancomycin and improved. Three weeks later (week 5), the infection recurred and she was given 2 more weeks of intravenous vancomycin and again improved. Four weeks later (week 11), the infection recurred and the patient was again started on intravenous vancomycin. The MICs for vancomycin for the *Staphylococcus aureus* isolates were as follows: initial isolate (day 1), 1 μg/mL; week 5, 2 μg/mL; and week 11, 8 μg/mL. The patient failed to improve with the third course of vancomycin, and alternative therapy was used. The mechanism that best explains the relative resistance of the patient's strain of *Staphylococcus aureus* to vancomycin is

   (A) Acquisition of the vanA gene from another microorganism
   (B) Active transport of vancomycin out of the *Staphylococcus aureus* cell
   (C) Action of β-lactamase
   (D) Increased cell wall synthesis and alterations in the cell wall structure
   (E) Phosphorylation and resultant inactivation of the vancomycin

2. An 11-year-old boy develops a mild fever and pain in his upper arm. An x-ray film of his arm shows a lytic lesion (dissolution) in the upper part of the humerus with periosteal elevation over the lesion. The patient is taken to surgery, where the lesion is debrided (dead bone and pus removed). Culture from the lesion yields gram-positive cocci. A test shows that the organism is a staphylococcus and not a streptococcus. Based on this information, you know the organism is

   (A) Susceptible to nafcillin
   (B) β-lactamase-positive
   (C) A producer of protein A
   (D) Encapsulated
   (E) Catalase-positive

3. A 36-year-old male patient has an abscess with a strain of *Staphylococcus aureus* that is β-lactamase-positive. This indicates that the organism is resistant to which of the following antibiotics?

   (A) Penicillin G, ampicillin, and piperacillin
   (B) Trimethoprim-sulfamethoxazole
   (C) Erythromycin, clarithromycin, and azithromycin
   (D) Vancomycin
   (E) Cefazolin and ceftriaxone

4. Seven days ago, a 27-year-old medical student returned from Central America, where she had spent the summer working in a clinic for indigenous people. Four days ago, she developed an erythematous sunburn-like rash. She also has had headache, muscle aches, and abdominal cramps with diarrhea. Her blood pressure is 70/40 mm Hg. Pelvic examination shows she is having her menstrual period with a tampon in place; otherwise, the pelvic examination is normal. Her kidney function tests (serum

urea nitrogen and creatinine) are abnormal, indicating mild renal failure. A blood smear for malaria is negative. Her illness is likely to be caused by which of the following?

(A) A toxin that results in greatly increased levels of intracellular cyclic adenosine monophosphate (cAMP)

(B) A toxin that degrades sphingomyelin

(C) A toxin that binds to the class II major histocompatibility complex (MHC) of an antigen-presenting cell and the Vβ region of a T cell

(D) A two-component toxin that forms pores in white blood cells and increases cation permeability

(E) A toxin that blocks elongation factor 2 (EF2)

5. Over a period of 3 weeks, a total of five newborns in the hospital nursery developed *Staphylococcus aureus* infections with *Staphylococcus aureus* bacteremia. The isolates all had the same colony morphology and hemolytic properties and identical antimicrobial susceptibility patterns, suggesting that they were the same. (Later molecular methods showed the isolates were identical.) Which of the following should be done now?

(A) Prophylactic treatment of all newborns with intravenous vancomycin

(B) Protective isolation of all newborns

(C) Closing the nursery and referring pregnant women to another hospital

(D) Hiring all new staff for the hospital nursery

(E) Culture using mannitol salt agar of the anterior nares of the physicians, nurses, and others who cared for the infected babies

6. The exfoliative toxins, TSST-1, and the enterotoxins are all superantigens. The genes for these toxins are

(A) Present in all strains of *Staphylococcus aureus*

(B) Widely distributed on the staphylococcal chromosome

(C) On both the staphylococcal chromosome (TSST-1 and exfoliative toxins) and on plasmids (enterotoxins)

(D) On the staphylococcal chromosome in a pathogenicity island

(E) On plasmids

7. A 16-year-old bone marrow transplant patient has a central venous line that has been in place for 2 weeks. He also has a urinary tract catheter, which has been in place for 2 weeks as well. He develops fever while his white blood cell count is very low and before the transplant has engrafted. Three blood cultures are done, and all grow *Staphylococcus epidermidis*. Which one of the following statements is correct?

(A) The *Staphylococcus epidermidis* organisms are likely to be susceptible to penicillin G.

(B) The *Staphylococcus epidermidis* organisms are likely to be from the surface of the urinary tract catheter.

(C) The *Staphylococcus epidermidis* organisms are likely to be resistant to vancomycin.

(D) The *Staphylococcus epidermidis* organisms are likely to be from a skin source.

(E) The *Staphylococcus epidermidis* organisms are likely to be in a biofilm on the central venous catheter surface.

8. A 65-year-old man develops an abscess on the back of his neck. Culture yields *Staphylococcus aureus*. The isolate is tested and found to be positive for the *mecA* gene, which means that

(A) The isolate is susceptible to vancomycin

(B) The isolate is resistant to vancomycin

(C) The isolate is susceptible to nafcillin

(D) The isolate is resistant to nafcillin

(E) The isolate is susceptible to clindamycin

(F) The isolate is resistant to clindamycin

9. Antimicrobial resistance has become a significant problem. Which one of the following is of major concern worldwide?

(A) Nafcillin resistance in *Staphylococcus aureus*

(B) Penicillin resistance in *Streptococcus pneumoniae*

(C) Penicillin resistance in *Neisseria gonorrhoeae*

(D) Vancomycin resistance in *Staphylococcus aureus*

(E) Tobramycin resistance in *Escherichia coli*

10. A group of six children under 8 years of age live in a semitropical country. Each of the children has several crusted weeping skin lesions of impetigo (pyoderma). The lesions are predominantly on the arms and faces. Which of the following microorganisms is a likely cause of the lesions?

(A) *Escherichia coli*

(B) *Chlamydia trachomatis*

(C) *Staphylococcus aureus*

(D) *Streptococcus pneumoniae*

(E) *Bacillus anthracis*

11. Which of the following statements regarding the role of Protein A in the pathogenesis of infections caused by *Staphylococcus aureus* is correct?

(A) It is responsible for the rash in toxic shock syndrome.

(B) It converts hydrogen peroxide into water and oxygen.

(C) It is a potent enterotoxin.

(D) It is directly responsible for lysis of neutrophils.

(E) It is a bacterial surface protein that binds to the Fc portion of IgG1.

12. Which of the following staphylococcal organisms produces coagulase and has been implicated in infections following a dog bite?

(A) *Staphylococcus intermedius*

(B) *Staphylococcus epidermidis*

(C) *Staphylococcus saprophyticus*

(D) *Staphylococcus hominis*

(E) *Staphylococcus hemolyticus*

13. All of the following statements regarding Panton-Valentine leukocidin are correct *except*:

(A) It is a two-component toxin.

(B) It is commonly produced by community associated MRSA strains.

(C) It is an important virulence factor.

(D) It is identical to one of the staphylococcal enterotoxins.

(E) It forms pores in the membranes of white blood cells.

14. Which of the following statements best describes the function of the accessory gene regulator in *Staphylococcus aureus*?

(A) It regulates production of β-hemolysins.

(B) It is influenced by environmental oxygen.

(C) It controls the preferential expression of surface adhesins.

(D) It is important in the control of autolysis.

15. All of the following are important infection control strategies in containing spread of MRSA in hospitals *except*:

(A) Aggressive hand hygiene

(B) Routine surveillance for nasal colonization among high risk individuals

(C) Contact isolation for patients who are colonized or infected with MRSA

(D) Routine antimicrobial prophylaxis for all patients hospitalized for more than 48 hours

(E) Aseptic management of skin lesions

Answers

1. D	5. E	9.D	13. D
2. E	6. D	10. C	14. C
3. A	7. E	11. E	15. D
4. C	8. D	12. A	

REFERENCES

Bronner S et al: Regulation of virulence determinants in *Staphylococcus aureus*: Complexity and applications. FEMS Microbiol Rev 2004;28:183.

Novick RP, Schlievert P, Ruzin A: Pathogenicity and resistance islands of staphylococci. Microbes Infect 2001;3:585. [PMID: 11418332]

Que, YA et al: *Staphylococcus aureus* (including staphylococcal toxic shock). In *Mandell, Douglas and Bennett's Principles and Practice of Infectious Diseases*, 7th ed. Mandell GL, Bennett JE, Dolin R (editors). Churchill Livingstone Elsevier, 2009.

Rivera J, Vannakambadi G, Hook M, Speziale P: Fibrinogen-binding proteins of Gram-positive bacteria. Thromb Haemost 2007;98:503.

Weems JJ Jr et al: Phase II, randomized, double-blind, multicenter study comparing the safety and pharmacokinetics of tefibazumab to placebo for treatment of *Staphylococcus aureus* bacteremia. Antimicrob Agents Chemother 2006;50:2751.

Winn WC et al (editors): Gram-positive cocci, Part I: Staphylococci and related gram-positive cocci. In *Koneman's Color Atlas and Textbook of Diagnostic Microbiology*, 6th ed. Winn WC Jr et al (editors). Lippincott Williams and Wilkins, 2006, p. 623.

The Streptococci

INTRODUCTION

The streptococci are gram-positive spherical bacteria that characteristically form pairs or chains during growth. They are widely distributed in nature. Some are members of the normal human flora; others are associated with important human diseases attributable in part to infection by streptococci, in part to sensitization to them. Streptococci elaborate a variety of extracellular substances and enzymes.

The streptococci are a large and heterogeneous group of bacteria and no one system suffices to classify them. Yet, understanding the classification is key to understanding their medical importance.

CLASSIFICATION OF STREPTOCOCCI

The classification of streptococci into major categories has been based on a series of observations over many years: (1) colony morphology and hemolytic reactions on blood agar; (2) serologic specificity of the cell wall group-specific substance and other cell wall or capsular antigens; (3) biochemical reactions and resistance to physical and chemical factors; and (4) ecologic features. Molecular genetics have also been used to study the streptococci. Combinations of the above methods have permitted the classification of streptococci for purposes of clinical and epidemiologic convenience, but as the knowledge evolved, new methods have been introduced with the result that several classification systems have been described. In some cases, different species names have been used to describe the same organisms; in other instances, some members of the same species have been included in another species or classified separately. The genus *Enterococcus*, eg, now includes some species previously classified as group D streptococci. The classification of streptococci described in the following paragraphs and summarized in Table 14–1 is one logical approach.

A. Hemolysis

Many streptococci are able to hemolyze red blood cells in vitro in varying degrees. Complete disruption of erythrocytes with clearing of the blood around the bacterial growth is called **β hemolysis.** Incomplete lysis of erythrocytes with reduction of hemoglobin and the formation of green pigment is called **α hemolysis.** Other streptococci are nonhemolytic (sometimes called γ (gamma) hemolysis).

The hemolysis patterns of the streptococci of medical importance to humans are shown in Table 14–1. The classification of hemolytic patterns is used primarily with the streptococci and not with other bacteria that cause disease and typically produce a variety of hemolysins.

B. Group-Specific Substance (Lancefield Classification)

This carbohydrate is contained in the cell wall of many streptococci and forms the basis of serologic grouping into **Lancefield groups A–H** and **K–U.** The serologic specificity of the group-specific carbohydrate is determined by an amino sugar. For group A streptococci, this is rhamnose-*N*-acetylglucosamine; for group B, it is rhamnose-glucosamine polysaccharide; for group C, it is rhamnose-*N*-acetylgalactosamine; for group D, it is glycerol teichoic acid containing d-alanine and glucose; and for group F, it is glucopyranosyl-*N*-acetylgalactosamine.

Extracts of group-specific antigen for grouping streptococci are prepared by a variety of methods: extraction of centrifuged culture treated with hot hydrochloric acid, nitrous acid, or formamide; by enzymatic lysis of streptococcal cells (eg, with pepsin or trypsin); or by autoclaving of cell suspensions. These extracts contain the carbohydrate group-specific substance that yield precipitin reactions specific antisera. This permits arrangement of many streptococci into groups A–H and K–U. Typing is generally done only for groups A, B, C, F, and G (Table 14–1), which cause disease in humans and for which there are reagents that allow typing using simple agglutination or color reactions.

C. Capsular Polysaccharides

The antigenic specificity of the capsular polysaccharides is used to classify *S pneumoniae* into over 90 types and to type the group B streptococci (*S agalactiae*).

D. Biochemical Reactions

Biochemical tests include sugar fermentation reactions, tests for the presence of enzymes, and tests for susceptibility

TABLE 14–1   Characteristics of Medically Important Streptococci

Name	Group-Specific Substance[a]	Hemolysis[b]	Habitat	Important Laboratory Criteria	Common and Important Diseases
Streptococcus pyogenes	A	β	Throat, skin	Large colonies (>0.5 mm), PYR[c] test positive, inhibited by bacitracin	Pharyngitis, impetigo, rheumatic fever, glomerulonephritis, toxic shock
Streptococcus agalactiae	B	β	Female genital tract, lower GI tract	Hippurate hydrolysis, CAMP-positive[d]	Neonatal sepsis and meningitis, bacteremia in adults
Streptococcus dysgalactiae subspecies *equisimilis; others*	C, G	β (human) infections), α, none	Throat	Large (>0.5 mm) colonies	Pharyngitis, pyogenic infections similar to group A streptococci
Enterococcus faecalis (and other enterococci)	D	None, α	Colon	Growth in presence of bile, hydrolyze esculin, growth in 6.5% NaCl, PYR-positive	Abdominal abscess, urinary tract infection, endocarditis
Streptococcus bovis group	D	None	Colon, biliary tree	Growth in presence of bile, hydrolyze esculin, no growth in 6.5% NaCl, degrades starch	Endocarditis, common blood isolate in colon cancer, biliary disease
Streptococcus anginosus group (*S anginosus, S intermedius, S constellatus, S milleri* group)	F (A, C, G) and untypeable	α, β, none	Throat, colon, female genital tract	Small (<0.5 mm) colony variants of β-hemolytic species. Group A are bacitracin-resistant and PYR-negative. Carbohydrate fermentation patterns	Pyogenic infections, including brain abscesses
Viridans streptococci (many species)	Usually not typed or untypeable	α, none	Mouth, throat, colon, female genital tract	Optochin-resistant. Colonies not soluble in bile. Carbohydrate fermentation patterns	Dental caries (*S mutans*), endocarditis, abscesses (with many other bacterial species), some species, such as *Streptococcus mitis*, have high-level resistance to penicillin
Streptococcus pneumoniae	None	α	Nasopharynx	Susceptible to optochin. Colonies soluble in bile, quellung reaction-positive	Pneumonia, meningitis, endocarditis, otitis media, sinusitis
Peptostreptococcus (many species) (see Chapter 21)	None	None, α	Mouth, colon, female genital tract	Obligate anaerobes	Abscesses (with multiple other bacterial species)

[a]Lancefield classification.

[b]Hemolysis observed on 5% sheep blood agar after overnight incubation.

[c]Hydrolysis of l-pyrrolidonyl-2-naphthylamide ("PYR").

[d]Christie, Atkins, Munch-Peterson test.

or resistance to certain chemical agents. Biochemical tests are most often used to classify streptococci after the colony growth and hemolytic characteristics have been observed. Biochemical tests are used for species that typically do not react with the commonly used antibody preparations for the group-specific substances, groups A, B, C, F, and G. For example, the viridans streptococci are α-hemolytic or non-hemolytic and do not react with the antibodies commonly used for the Lancefield classification. Speciation of the viridans streptococci requires a battery of biochemical tests.

Many species of streptococci, including *S pyogenes* (group A), *S agalactiae* (group B), and the enterococci (group D), are characterized by combinations of features: colony growth characteristics, hemolysis patterns on blood agar (α hemolysis, β hemolysis, or no hemolysis), antigenic composition of group-specific cell wall substances, and biochemical reactions. *S pneumoniae* (pneumococcus) types are further classified by the antigenic composition of the capsular polysaccharides. The viridans streptococci can be α-hemolytic or nonhemolytic and are generally speciated by biochemical reactions. See Table 14–1. Because biochemical reactions are labor intensive and often unreliable, laboratories with molecular capabilities, such as peptide nucleic acid hybridization and gene sequencing, are replacing phenotypic tests with these methods when identification of viridians streptococci is required.

STREPTOCOCCI OF PARTICULAR MEDICAL INTEREST

The following streptococci and enterococci are of particular medical relevance.

STREPTOCOCCUS PYOGENES

Most streptococci that contain the group A antigen are *S pyogenes*. It is a prototypical human pathogen. It is used here to illustrate general characteristics of streptococci and specific characteristics of the species. *S pyogenes* is the main human pathogen associated with local or systemic invasion and poststreptococcal immunologic disorders. *S pyogenes* typically produces large (1 cm in diameter) zones of β hemolysis around colonies greater than 0.5 mm in diameter. They are PYR-positive (hydrolysis of l-pyrrolidonyl-2-naphthylamide) and usually are susceptible to bacitracin.

Morphology & Identification

A. Typical Organisms

Individual cocci are spherical or ovoid and are arranged in chains (Figure 14–1). The cocci divide in a plane perpendicular to the long axis of the chain. The members of the chain often have a striking diplococcal appearance, and rod-like forms are occasionally seen. The lengths of the chains vary widely and are conditioned by environmental factors. Streptococci are gram-positive; however, as a culture ages and the bacteria die, they lose their gram-positivity and can appear to be gram-negative; for some streptococci, this can occur after overnight incubation.

Most group A strains (Table 14–1) produce capsules composed of hyaluronic acid. The capsules are most noticeable in very young cultures. They impede phagocytosis. The hyaluronic acid capsule likely plays a greater role in virulence than is generally appreciated and together with M protein is felt to be an important factor in the resurgence of rheumatic fever in the United States in the 1980s and 1990s. The capsule binds to hyaluronic-acid-binding protein, CD44, present on human epithelial cells. Binding induces disruption of intercellular junctions allowing microorganisms to remain extracellular as they penetrate the epithelium (Stollerman GH and Dale JB). Capsules of other streptococci (eg, *S agalactiae* and *S pneumoniae*) are different. The *S pyogenes* cell wall contains proteins (M, T, R antigens), carbohydrates (group-specific),

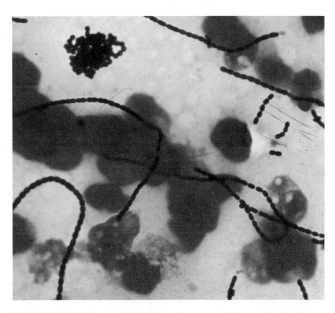

FIGURE 14–1 Streptococci grown in blood culture showing gram-positive cocci in chains. Original magnification × 1000.

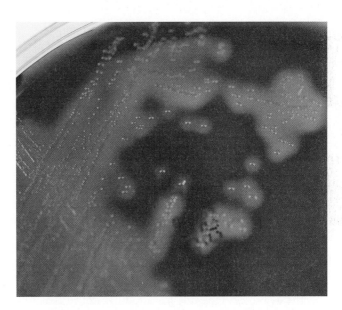

FIGURE 14–2 Group A β-hemolytic streptococci (*Streptococcus pyogenes*) after growth overnight on a 10-cm plate with 5% sheep blood agar. The small (0.5–1 mm diameter) white colonies are surrounded by diffuse zones of β-hemolysis 7–10 mm in diameter. (Courtesy of H Reyes.)

and peptidoglycans. Hair-like pili project through the capsule of group A streptococci. The pili consist partly of M protein and are covered with **lipoteichoic acid.** The latter is important in the attachment of streptococci to epithelial cells.

B. Culture

Most streptococci grow in solid media as discoid colonies, usually 1–2 mm in diameter. *S pyogenes* is β-hemolytic (Figure 14–2); other species have variable hemolytic characteristics.

C. Growth Characteristics

Energy is obtained principally from the utilization of glucose with lactic acid as the end product. Growth of streptococci tends to be poor on solid media or in broth unless enriched with blood or tissue fluids. Nutritive requirements vary widely among different species. The human pathogens are most exacting, requiring a variety of growth factors. Growth and hemolysis are aided by incubation in 10% CO_2. Most pathogenic hemolytic streptococci grow best at 37°C. Most streptococci are facultative anaerobes and grow under aerobic and anaerobic conditions. Peptostreptococci are obligate anaerobes.

D. Variation

Variants of the same streptococcus strain may show different colony forms. This is particularly marked among *S pyogenes* strains, giving rise to either matte or glossy colonies. Matte colonies consist of organisms that produce much M protein and generally are virulent. The *S pyogenes* in glossy colonies tend to produce little M protein and are often not virulent.

Antigenic Structure

A. M Protein

This substance is a major virulence factor of group A *S pyogenes*. M protein appears as hair-like projections of the streptococcal cell wall. When M protein is present, the streptococci are virulent, and in the absence of M type-specific antibodies, they are able to resist phagocytosis by polymorphonuclear leukocytes. *S pyogenes* that lack M protein are not virulent. Immunity to infection with group A streptococci is related to the presence of type-specific antibodies to M protein. Because there are many, perhaps 150, types of M protein, a person can have repeated infections with group A *S pyogenes* of different M types. Both group C and group G streptococci have genes homologous to the genes for M protein of group A, and M protein has been found on group G streptococci.

The M protein molecule has a rod-like coiled structure that separates functional domains. The structure allows for a large number of sequence changes while maintaining function, and the M protein immunodeterminants, therefore, can readily change. There are two major structural classes of M protein, classes I and II.

It appears that M protein and perhaps other streptococcal cell wall antigens have an important role in the pathogenesis of rheumatic fever. Purified streptococcal cell wall membranes induce antibodies that react with human cardiac sarcolemma; the characteristics of the cross-reactive antigens are not clear. A component of the cell wall of selected M types induces antibodies that react with cardiac muscle tissue. Conserved antigenic domains on the class IM protein cross-react with human cardiac muscle, and the class IM protein may be a virulence determinant for rheumatic fever.

B. T Substance

This antigen has no relationship to virulence of streptococci. Unlike M protein, T substance is acid-labile and heat-labile. It is obtained from streptococci by proteolytic digestion, which rapidly destroys M proteins. T substance permits differentiation of certain types of streptococci by agglutination with specific antisera, while other types share the same T substance. Yet another surface antigen has been called **R protein.**

C. Nucleoproteins

Extraction of streptococci with weak alkali yields mixtures of proteins and other substances of little serologic specificity, called **P substances,** which probably make up most of the streptococcal cell body.

Toxins & Enzymes

More than 20 extracellular products that are antigenic are elaborated by *S pyogenes,* including the following.

A. Streptokinase (Fibrinolysin)

Streptokinase is produced by many strains of group A β-hemolytic streptococci. It transforms the plasminogen of human plasma into plasmin, an active proteolytic enzyme that digests fibrin and other proteins. This process of digestion may be interfered with by nonspecific serum inhibitors and by a specific antibody, antistreptokinase. Streptokinase has been given intravenously for treatment of pulmonary emboli, coronary artery, and venous thromboses.

B. Streptodornase

Streptodornase (streptococcal deoxyribonuclease) depolymerizes DNA. The enzymatic activity can be measured by the decrease in viscosity of known DNA solutions. Purulent exudates owe their viscosity largely to deoxyribonucleoprotein. Mixtures of streptodornase and streptokinase are used in "enzymatic debridement." They help to liquefy exudates and facilitate removal of pus and necrotic tissue; antimicrobial drugs thus gain better access, and infected surfaces recover more quickly. An antibody to DNAse develops after streptococcal infections (normal limit = 100 units), especially after skin infections.

C. Hyaluronidase

Hyaluronidase splits hyaluronic acid, an important component of the ground substance of connective tissue. Thus, hyaluronidase aids in spreading infecting microorganisms (spreading factor). Hyaluronidases are antigenic and specific for each bacterial or tissue source. Following infection with hyaluronidase-producing organisms, specific antibodies are found in the serum.

D. Pyrogenic Exotoxins (Erythrogenic Toxin)

Pyrogenic exotoxins are elaborated by *S pyogenes.* There are three antigenically distinct **streptococcal pyrogenic exotoxins: A, B, and C.** Exotoxin A has been most widely studied.

It is produced by group A streptococci that carry a lysogenic phage. The streptococcal pyrogenic exotoxins have been associated with **streptococcal toxic shock syndrome** and **scarlet fever.** Most strains of group A streptococci isolated from patients with streptococcal toxic shock syndrome either produce streptococcal pyrogenic exotoxin A or have the gene that codes for it; in contrast, only about 15% of group A streptococci isolated from other patients have the gene. Streptococcal pyrogenic exotoxin C may also contribute to the syndrome, while the role for streptococcal pyrogenic exotoxin B is unclear. The group A streptococci associated with toxic shock syndrome are primarily of M protein types 1 and 3.

The pyrogenic exotoxins act as superantigens, which stimulate T cells by binding to the class II major histocompatibility complex in the V_β region of the T cell receptor. The activated T cells release cytokines that mediate shock and tissue injury. The mechanisms of action appear to be similar to those due to staphylococcal toxic syndrome toxin-1 and the staphylococcal enterotoxins.

E. Diphosphopyridine Nucleotidase

This enzyme is elaborated into the environment by some streptococci. This substance may be related to the organism's ability to kill leukocytes. Proteinases and amylase are produced by some strains.

F. Hemolysins

The β-hemolytic group A *S pyogenes* elaborates two hemolysins (streptolysins). **Streptolysin O** is a protein (MW 60,000) that is hemolytically active in the reduced state (available–SH groups) but rapidly inactivated in the presence of oxygen. Streptolysin O is responsible for some of the hemolysis seen when growth is in cuts deep into the medium in blood agar plates. It combines quantitatively with **antistreptolysin O (ASO),** an antibody that appears in humans following infection with any streptococci that produce streptolysin O. This antibody blocks hemolysis by streptolysin O. This phenomenon forms the basis of a quantitative test for the antibody. An ASO serum titer in excess of 160–200 units is considered abnormally high and suggests either recent infection with *S pyogenes* or persistently high antibody levels due to an exaggerated immune response to an earlier exposure in a hypersensitive person. **Streptolysin S** is the agent responsible for the hemolytic zones around streptococcal colonies growing on the surface of blood agar plates. It is elaborated in the presence of serum—hence the name streptolysin S. It is not antigenic, but it may be inhibited by a nonspecific inhibitor that is frequently present in the sera of humans and animals and is independent of past experience with streptococci.

Pathogenesis & Clinical Findings

A variety of distinct disease processes are associated with *S pyogenes* infections. The infections can be divided into several categories.

A. Diseases Attributable to Invasion by *S pyogenes*, β-Hemolytic Group A Streptococci

The portal of entry determines the principal clinical picture. In each case, however, there is a diffuse and rapidly spreading infection that involves the tissues and extends along lymphatic pathways with only minimal local suppuration. From the lymphatics, the infection can extend to the bloodstream.

1. Erysipelas—If the portal of entry is the skin, erysipelas results, with massive brawny edema and a rapidly advancing margin of infection.

2. Cellulitis—Streptococcal cellulitis is an acute, rapidly spreading infection of the skin and subcutaneous tissues. It follows infection associated with mild trauma, burns, wounds, or surgical incisions. Pain, tenderness, swelling, and erythema occur. Cellulitis is differentiated from erysipelas by two clinical findings: In cellulitis, the lesion is not raised, and the line between the involved and uninvolved tissue is indistinct.

3. Necrotizing fasciitis (streptococcal gangrene)—This is infection of the subcutaneous tissues and fascia. There is extensive and very rapidly spreading necrosis of the skin and subcutaneous tissues. Bacteria other than *S pyogenes* can also cause necrotizing fasciitis. The group A streptococci that cause necrotizing fasciitis have sometimes been termed "flesh-eating bacteria."

4. Puerperal fever—If the streptococci enter the uterus after delivery, puerperal fever develops, which is essentially a septicemia originating in the infected wound (endometritis).

5. Bacteremia/sepsis—Infection of traumatic or surgical wounds with streptococci results in bacteremia, which rapidly can be fatal. *S pyogenes* bacteremia can also follow skin infections, such as cellulitis and rarely pharyngitis.

B. Diseases Attributable to Local Infection with *S pyogenes* and Their By-Products

1. Streptococcal sore throat—The most common infection due to β-hemolytic *S pyogenes* is streptococcal sore throat or pharyngitis. *S pyogenes* adhere to the pharyngeal epithelium by means of lipoteichoic acid-covered surface pili and also by means of hyaluronic acid in encapsulated strains. The glycoprotein fibronectin (MW 440,000) on epithelial cells probably serves as lipoteichoic acid ligand. In infants and small children, the sore throat occurs as a subacute nasopharyngitis with a thin serous discharge and little fever but with a tendency of the infection to extend to the middle ear and the mastoid. The cervical lymph nodes are usually enlarged. The illness may persist for weeks. In older children and adults, the disease is more acute and is characterized by intense nasopharyngitis, tonsillitis, and intense redness and edema of the mucous membranes, with purulent exudate, enlarged, tender cervical lymph nodes, and (usually) a high fever. Twenty percent of infections are asymptomatic. A similar clinical picture can occur with infectious mononucleosis, diphtheria, gonococcal infection, and adenovirus infection.

S pyogenes infection of the upper respiratory tract does not usually involve the lungs. Pneumonia, when it does occur, is rapidly progressive and severe and is most commonly a sequela to viral infections, eg, influenza or measles, which seem to enhance susceptibility greatly.

2. Streptococcal pyoderma—Local infection of superficial layers of skin, especially in children, is called **impetigo.** It consists of superficial vesicles that break down and eroded areas whose denuded surface is covered with pus and later is encrusted. It spreads by continuity and is highly communicable, especially in hot, humid climates. More widespread infection occurs in eczematous or wounded skin or in burns and may progress to cellulitis. Group A streptococcal skin infections are often attributable to M types 49, 57, and 59–61 and may precede glomerulonephritis but do not often lead to rheumatic fever.

A clinically identical infection can be caused by *S aureus* and sometimes both *S pyogenes* and *S aureus* are present.

C. Invasive Group A Streptococcal Infections, Streptococcal Toxic Shock Syndrome, and Scarlet Fever

Fulminant, invasive *S pyogenes* infections with **streptococcal toxic shock syndrome** are characterized by shock, bacteremia, respiratory failure, and multiorgan failure. Death occurs in about 30% of patients. The infections tend to follow minor trauma in otherwise healthy persons with several presentations of soft tissue infection. These include necrotizing fasciitis, myositis, and infections at other soft tissue sites; bacteremia occurs frequently. In some patients, particularly those infected with group A streptococci of M types 1 or 3, the disease presents with focal soft tissue infection accompanied by fever and rapidly progressive shock with multiorgan failure. Erythema and desquamation may occur. The *S pyogenes* of the M types 1 and 3 (and types 12 and 28) that make pyrogenic exotoxin A or B are associated with the severe infections.

Pyrogenic exotoxins A–C also cause **scarlet fever** in association with *S pyogenes* pharyngitis or with skin or soft tissue infection. The pharyngitis may be severe. The rash appears on the trunk after 24 hours of illness and spreads to involve the extremities. Streptococcal toxic shock syndrome and scarlet fever are clinically overlapping diseases.

D. Poststreptococcal Diseases (Rheumatic Fever, Glomerulonephritis)

Following an acute *S pyogenes* infection, there is a latent period of 1–4 weeks, after which nephritis or rheumatic fever occasionally develops. The latent period suggests that these poststreptococcal diseases are not attributable to the direct effect of disseminated bacteria but represent instead a hypersensitivity response. Nephritis is more commonly preceded by infection of the skin; rheumatic fever is more commonly preceded by infection of the respiratory tract.

1. Acute glomerulonephritis—This sometimes develops 1–4 weeks after *S pyogenes* skin infection (pyoderma, impetigo). Some strains are particularly nephritogenic, principally with M types 2, 42, 49, 56, 57, and 60 (skin). Other nephritogenic M types associated with throat infections and glomerulonephritis are 1, 4, 12, and 25. After random streptococcal skin infections, the incidence of nephritis is less than 0.5%.

Glomerulonephritis may be initiated by antigen–antibody complexes on the glomerular basement membrane. The most important antigen is probably in the streptococcal protoplast membrane. In acute nephritis, there is blood and protein in the urine, edema, high blood pressure, and urea nitrogen retention; serum complement levels are also low. A few patients die; some develop chronic glomerulonephritis with ultimate kidney failure; and the majority recover completely.

2. Rheumatic fever—This is the most serious sequela of *S pyogenes* because it results in damage to heart muscle and valves. Certain strains of group A streptococci contain cell membrane antigens that cross-react with human heart tissue antigens. Sera from patients with rheumatic fever contain antibodies to these antigens.

The onset of rheumatic fever is often preceded by *S pyogenes* infection 1–4 weeks earlier, although the infection may be mild and may not be detected. In general, however, patients with more severe streptococcal sore throats have a greater chance of developing rheumatic fever. In the 1950s, untreated streptococcal infections were followed by rheumatic fever in up to 3% of military personnel and 0.3% of civilian children. Rheumatic fever is now relatively rare in the United States (<0.05% of streptococcal infections), but it occurs up to 100 times more frequently in tropical countries and is the most important cause of heart disease in young people in developing countries.

Typical symptoms and signs of rheumatic fever include fever, malaise, a migratory nonsuppurative polyarthritis, and evidence of inflammation of all parts of the heart (endocardium, myocardium, and pericardium). The carditis characteristically leads to thickened and deformed valves and to small perivascular granulomas in the myocardium (Aschoff bodies) that are finally replaced by scar tissue. Erythrocyte sedimentation rates, serum transaminase levels, electrocardiograms, and other tests are used to estimate rheumatic activity.

Rheumatic fever has a marked tendency to be reactivated by recurrent streptococcal infections, whereas nephritis does not. The first attack of rheumatic fever usually produces only slight cardiac damage, which, however, increases with each subsequent attack. It is therefore important to protect such patients from recurrent *S pyogenes* infections by prophylactic penicillin administration.

Diagnostic Laboratory Tests

A. Specimens

Specimens to be obtained depend upon the nature of the streptococcal infection. A throat swab, pus, or blood is obtained for culture. Serum is obtained for antibody determinations.

B. Smears

Smears from pus often show single cocci or pairs rather than definite chains. Cocci are sometimes gram-negative because the organisms are no longer viable and have lost their ability to retain the blue dye (crystal violet) and be gram-positive. If smears of pus show streptococci but cultures fail to grow, anaerobic organisms must be suspected. Smears of throat swabs are rarely contributory because viridans streptococci are always present and have the same appearance as group A streptococci on stained smears.

C. Culture

Specimens suspected of containing streptococci are cultured on blood agar plates. If anaerobes are suspected, suitable anaerobic media must also be inoculated. Incubation in 10% CO_2 often speeds hemolysis. Slicing the inoculum into the blood agar has a similar effect, because oxygen does not readily diffuse through the medium to the deeply embedded organisms, and it is oxygen that inactivates streptolysin O.

Blood cultures will grow hemolytic group A streptococci (eg, in sepsis) within hours or a few days. Certain α-hemolytic streptococci and enterococci may grow slowly, so blood cultures in cases of suspected endocarditis occasionally do not turn positive for a few days.

The degree and kind of hemolysis (and colonial appearance) may help place an organism in a definite group. *S pyogenes* can be identified by rapid tests specific for the presence of the group A-specific antigen and by the PYR test. Streptococci belonging to group A may be presumptively identified by inhibition of growth by bacitracin, but this should be used only when more definitive tests are not available.

D. Antigen Detection Tests

Several commercial kits are available for rapid detection of group A streptococcal antigen from throat swabs. These kits use enzymatic or chemical methods to extract the antigen from the swab, then use EIA or agglutination tests to demonstrate the presence of the antigen. The tests can be completed minutes to hours after the specimen is obtained. They are 60–90% sensitive, depending upon the prevalence of the disease in the population, and 98–99% specific when compared to culture methods.

E. Serologic Tests

A rise in the titer of antibodies to many group A streptococcal antigens can be estimated. Such antibodies include ASO, particularly in respiratory disease; anti-DNase and antihyaluronidase, particularly in skin infections; antistreptokinase; anti-M type-specific antibodies; and others. Of these, the anti-ASO titer is most widely used.

Immunity

Resistance against streptococcal diseases is M type-specific. Thus, a host who has recovered from infection by one group A streptococcal M type is relatively immune to reinfection by the same type but fully susceptible to infection by another M type. Anti-M type-specific antibodies can be demonstrated in a test that exploits the fact that streptococci are rapidly killed after phagocytosis. M protein interferes with phagocytosis, but in the presence of type-specific antibody to M protein, streptococci are killed by human leukocytes.

Antibody to streptolysin O develops following infection; it blocks hemolysis by streptolysin O but does not indicate immunity. High titers (>250 units) indicate recent or repeated infections and are found more often in rheumatic individuals than in those with uncomplicated streptococcal infections.

Treatment

All *S pyogenes* are susceptible to penicillin G, and most are susceptible to erythromycin. Some are resistant to tetracyclines. Antimicrobial drugs have no effect on established glomerulonephritis and rheumatic fever. In acute streptococcal infections, however, every effort must be made to rapidly eradicate streptococci from the patient, eliminate the antigenic stimulus (before day 8), and thus prevent poststreptococcal disease. Doses of penicillin or erythromycin that result in effective tissue levels for 10 days usually accomplish this. Antimicrobial drugs are also very useful in preventing reinfection with β-hemolytic group A streptococci in rheumatic fever patients.

Epidemiology, Prevention, & Control

Although humans can be asymptomatic nasopharyngeal or perineal carriers of *S pyogenes*, the organism should be considered significant if it is detected by culture or other means. The ultimate source of group A streptococci is a person harboring these organisms. The individual may have a clinical or subclinical infection or may be a carrier distributing streptococci directly to other persons via droplets from the respiratory tract or skin. The nasal discharges of a person harboring *S pyogenes* are the most dangerous source for spread of these organisms.

Many other streptococci (viridans streptococci, enterococci, etc) are members of the normal flora of the human body. They produce disease only when established in parts of the body where they do not normally occur (eg, heart valves). To prevent such accidents, particularly in the course of surgical procedures on the respiratory, gastrointestinal, and urinary tracts that result in temporary bacteremia, antimicrobial agents are often administered prophylactically to persons with known heart valve deformity and to those with prosthetic valves or joints.

Control procedures are directed mainly at the human source:

1. Detection and early antimicrobial therapy of respiratory and skin infections with group A streptococci. Prompt eradication of streptococci from early infections can effectively prevent the development of poststreptococcal disease. This requires maintenance of adequate penicillin levels in tissues for 10 days (eg, benzathine penicillin G given once intramuscularly). Erythromycin is an alternative drug, although some *S pyogenes* are resistant.

2. Antistreptococcal chemoprophylaxis in persons who have suffered an attack of rheumatic fever. This involves giving one injection of benzathine penicillin G intramuscularly, every 3–4 weeks, or daily oral penicillin or oral sulfonamide. The first attack of rheumatic fever infrequently causes major heart damage; however, such persons are particularly susceptible to reinfections with streptococci that precipitate relapses of rheumatic activity and give rise to cardiac damage. Chemoprophylaxis in such individuals, especially children, must be continued for years. Chemoprophylaxis is not used in glomerulonephritis because of the small number of nephritogenic types of streptococci. An exception may be family groups with a high rate of poststreptococcal nephritis.

3. Eradication of *S pyogenes* from carriers. This is especially important when carriers are in areas such as obstetric delivery rooms, operating rooms, classrooms, or nurseries. Unfortunately, it is often difficult to eradicate β-hemolytic streptococci from permanent carriers, and individuals may occasionally have to be shifted away from "sensitive" areas for some time.

STREPTOCOCCUS AGALACTIAE

These are the **group B streptococci**. They typically are β-hemolytic and produce zones of hemolysis that are only slightly larger than the colonies (1–2 mm in diameter). The group B streptococci hydrolyze sodium hippurate and give a positive response in the so-called CAMP test (Christie, Atkins, Munch-Peterson).

Group B streptococci are part of the normal vaginal flora and lower gastrointestinal tract in 5–25% of women. Group B streptococcal infection during the first month of life may present as fulminant sepsis, meningitis, or respiratory distress syndrome. Intravenous ampicillin given to mothers, who carry group B streptococci and are in labor, prevents colonization of their infants and group B streptococcal disease. Group B streptococcal infections are increasing among nonpregnant adults. Two expanding populations, namely the elderly and immunocompromised hosts, are most at risk for invasive disease. Predisposing factors include diabetes

mellitus, cancer, advanced age, liver cirrhosis, corticosteroid therapy, HIV, and other immune compromised states. Bacteremia, skin and soft tissue infections, respiratory infections, and genitourinary infections in descending order of frequency are the major clinical manifestations.

GROUPS C & G

These streptococci occur sometimes in the nasopharynx and may cause pharyngitis, sinusitis, bacteremia, or endocarditis. They often look like group A *S pyogenes* on blood agar medium and are β-hemolytic. They are identified by reactions with specific antisera for groups C or G. These group G streptococci have hemolysins and may have M proteins analogous to those of group A *S pyogenes*.

GROUP D *STREPTOCOCCI*

The Group D streptococci have undergone recent taxonomic changes. There are eight species in this group, many of which do not cause infections in humans. The *Streptococcus bovis* group is of most importance to human disease and is further classified into biotypes (old classification), which are important epidemiologically and more recently 4 DNA clusters. Animal species in the bovis group have been assigned to the species *S. equinus* (DNA cluster I). Biotype I (in DNA cluster II) isolates ferment mannitol and are now designated as *Streptococcus gallolyticus* subspecies *gallolyticus*. This organism causes human endocarditis and is frequently epidemiologically associated with colon carcinoma. *Streptococcus bovis* biotype II is separated into two subtypes: biotype II.1 (now called *Streptococcus infantarius* subspecies *coli*), which also causes endocarditis, and biotype II.2 (now called *S gallolyticus* subspecies *pasteurianus*), which causes not only endocarditis but also bacteremia and urinary tract infections. Biotype II bacteremias are often associated with biliary sources and less frequently with endocarditis. Finally, DNA cluster IV has one species, *S. alactolyticus*. All Group D streptococci are nonhemolytic and PYR-negative. They grow in the presence of bile and hydrolyze esculin (bile-esculin positive) but do not grow in 6.5% NaCl. They are part of the normal enteric flora.

STREPTOCOCCUS ANGINOSUS GROUP

Other species names in the *S anginosus* group are *S constellatus* and *S intermedius*. They are sometimes referred to as the *S milleri* group. These streptococci are part of the normal flora. They may be β-, α-, or nonhemolytic. *S anginosus* group includes β-hemolytic streptococci that form minute colonies (<0.5 mm in diameter) and react with groups A, C, or G antisera and all β-hemolytic group F streptococci. Those that are group A are PYR-negative. *S anginosus* are Voges-Proskauer test-positive. They may be classified as viridans streptococci.

GROUP N STREPTOCOCCI

They are rarely found in human disease states but produce normal coagulation ("souring") of milk.

GROUPS E, F, G, H, & K–U STREPTOCOCCI

These streptococci occur primarily in animals. One of the multiple species of group G streptococci, *S canis*, can cause skin infections of dogs but uncommonly infects humans; other species of group G streptococci infect humans.

VIRIDANS STREPTOCOCCI

The viridans streptococci include *S mitis, S mutans, S salivarius, S sanguis,* and others. Typically they are α-hemolytic, but they may be nonhemolytic. Their growth is not inhibited by Optochin, and colonies are not soluble in bile (deoxycholate). The viridans streptococci are the most prevalent members of the normal flora of the upper respiratory tract and are important for the healthy state of the mucous membranes there. They may reach the bloodstream as a result of trauma and are a principal cause of endocarditis on abnormal heart valves. Some viridans streptococci (eg, *S mutans*) synthesize large polysaccharides such as dextrans or levans from sucrose and contribute importantly to the genesis of dental caries.

In the course of bacteremia, viridans streptococci, pneumococci, or enterococci may settle on normal or previously deformed heart valves, producing **acute endocarditis.** Rapid destruction of the valves frequently leads to fatal cardiac failure in days or weeks unless a prosthesis can be inserted during antimicrobial therapy.

Subacute endocarditis often involves abnormal valves (congenital deformities and rheumatic or atherosclerotic lesions). Although any organism reaching the bloodstream may establish itself on thrombotic lesions that develop on endothelium injured as a result of circulatory stresses, subacute endocarditis is most frequently due to members of the normal flora of the respiratory or intestinal tract that have accidentally reached the blood. After dental extraction, at least 30% of patients have viridans streptococcal bacteremia. These streptococci, ordinarily the most prevalent members of the upper respiratory flora, are also the most frequent cause of subacute bacterial endocarditis. The group D streptococci (enterococci and *S bovis*) also are common causes of subacute endocarditis. About 5–10% of cases are due to enterococci originating in the gut or urinary tract. The lesion is slowly progressive, and a certain amount of healing accompanies the active inflammation; vegetations consist of fibrin, platelets, blood cells, and bacteria adherent to the valve leaflets. The clinical course is gradual, but the disease is invariably fatal in untreated cases. The typical clinical picture includes fever, anemia, weakness, a heart murmur, embolic phenomena, an enlarged spleen, and renal lesions.

α-Hemolytic streptococci and enterococci vary in their susceptibility to antimicrobial agents. Particularly in bacterial endocarditis, antibiotic susceptibility tests are useful to determine which drugs may be used for optimal therapy. Aminoglycosides often enhance the rate of bactericidal action of penicillin on streptococci, particularly enterococci.

NUTRITIONALLY VARIANT STREPTOCOCCI

The nutritionally variant streptococci, previously *S defectives* and *S adjacens* and additional species, are now classified in the genus *Abiotrophia* and the genus *Granulicatella*. They have also been known as "nutritionally deficient streptococci" and "pyridoxal-dependent streptococci." They require pyridoxal or cysteine for growth on blood agar or grow as satellite colonies around colonies of staphylococci and other bacteria. Routinely supplementing blood agar medium with pyridoxol allows recovery of these organisms. They are usually α-hemolytic but may be nonhemolytic. They are part of the normal flora and occasionally cause bacteremia or endocarditis and can be found in brain abscesses and other infections. Clinically, they are very much like the viridans streptococci.

PEPTOSTREPTOCOCCUS

These streptococci grow only under anaerobic or microaerophilic conditions and variably produce hemolysins. They are part of the normal flora of the mouth, upper respiratory tract, bowel, and female genital tract. They often participate with many other bacterial species in mixed anaerobic infections (Chapter 21). Such infections may occur in wounds, in the breast, in postpartum endometritis, following rupture of an abdominal viscus, the brain, or in chronic suppuration of the lung. The pus usually has a foul odor.

STREPTOCOCCUS PNEUMONIAE

The pneumococci (*S pneumoniae*) are gram-positive diplococci, often lancet-shaped or arranged in chains, possessing a capsule of polysaccharide that permits typing with specific antisera. Pneumococci are readily lysed by surface-active agents, which probably remove or inactivate the inhibitors of cell wall autolysins. Pneumococci are normal inhabitants of the upper respiratory tract of 5–40% of humans and can cause pneumonia, sinusitis, otitis, bronchitis, bacteremia, meningitis, and other infectious processes.

Morphology & Identification

A. Typical Organisms

The typical gram-positive, lancet-shaped diplococci (Figure 14–3) are often seen in specimens of young cultures. In sputum or pus, single cocci or chains are also seen. With

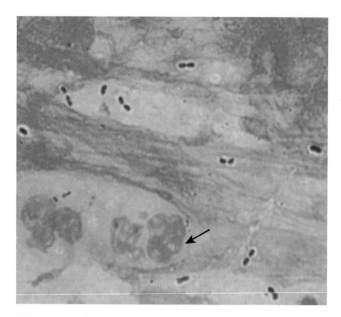

FIGURE 14–3 *Streptococcus pneumoniae* in sputum are seen as lancet-shaped gram-positive diplococci. Degenerating nuclei of polymorphonuclear cells are the large darker irregular red shapes (arrow). Mucus and amorphous debris are present in the background. Original magnification × 1000.

age, the organisms rapidly become gram-negative and tend to lyse spontaneously. Autolysis of pneumococci is greatly enhanced by surface-active agents. Lysis of pneumococci occurs in a few minutes when ox bile (10%) or sodium deoxycholate (2%) is added to a broth culture or suspension of organisms at neutral pH. Viridans streptococci do not lyse and are thus easily differentiated from pneumococci. On solid media, the growth of pneumococci is inhibited around a disk of Optochin; viridans streptococci are not inhibited by Optochin (Figure 14–4).

Other identifying points include almost uniform virulence for mice when injected intraperitoneally and the "capsule swelling test," or quellung reaction (see below).

B. Culture

Pneumococci form small round colonies, at first dome-shaped and later developing a central plateau with an elevated rim. Pneumococci are α-hemolytic on blood agar. Growth is enhanced by 5–10% CO_2.

C. Growth Characteristics

Most energy is obtained from fermentation of glucose; this is accompanied by the rapid production of lactic acid, which limits growth. Neutralization of broth cultures with alkali at intervals results in massive growth.

D. Variation

Pneumococcal isolates that produce large amounts of capsules produce large mucoid colonies. Capsule production is not essential for growth on agar medium, and capsular production is, therefore, lost after a small number of subcultures. The pneumococci will, however, again produce capsules and have enhanced virulence if injected into mice.

Antigenic Structure

A. Component Structures

The pneumococcal cell wall has peptidoglycan and teichoic acid, like other streptococci. The capsular polysaccharide is covalently bound to the peptidoglycan and to the cell wall polysaccharide. The capsular polysaccharide is immunologically distinct for each of the more than 90 types.

B. Quellung Reaction

When pneumococci of a certain type are mixed with specific antipolysaccharide serum of the same type—or with polyvalent antiserum—on a microscope slide, the capsule swells markedly, and the organisms agglutinate by cross-linking of the antibodies (Figure 14–4C). This reaction is useful for rapid identification and for typing of the organisms, either in sputum or in cultures. The polyvalent antiserum, which contains antibody to all of the types ("omniserum"), is a good reagent for rapid microscopic determination of whether or not pneumococci are present in fresh sputum.

Pathogenesis

A. Types of Pneumococci

In adults, types 1–8 are responsible for about 75% of cases of pneumococcal pneumonia and for more than half of all fatalities in pneumococcal bacteremia; in children, types 6, 14, 19, and 23 are frequent causes.

B. Production of Disease

Pneumococci produce disease through their ability to multiply in the tissues. They produce no toxins of significance. The virulence of the organism is a function of its capsule, which prevents or delays ingestion by phagocytes. A serum that contains antibodies against the type-specific polysaccharide protects against infection. If such a serum is absorbed with the type-specific polysaccharide, it loses its protective power. Animals or humans immunized with a given type of pneumococcal polysaccharide are subsequently immune to that type of pneumococcus and possess precipitating and opsonizing antibodies for that type of polysaccharide.

C. Loss of Natural Resistance

Since 40–70% of humans are at some time carriers of virulent pneumococci, the normal respiratory mucosa must possess great natural resistance to the pneumococcus. Among the

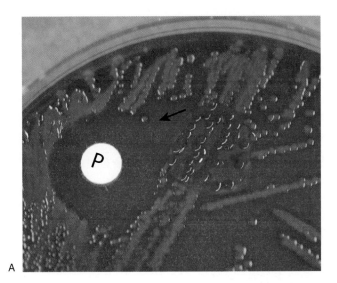

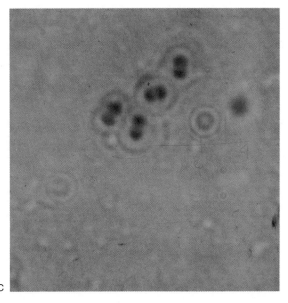

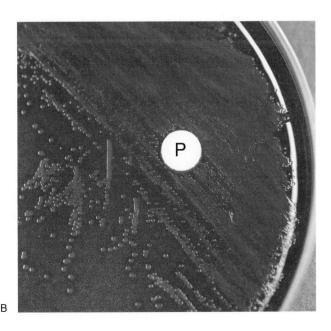

FIGURE 14–4 A: Optochin inhibition and bile solubility of *Streptococcus pneumoniae*. The *Streptococcus pneumoniae* were grown overnight on 5% sheep blood agar. The optochin (ethyl hydrocupreine HCl) or P disc was placed when the plate was inoculated. The pneumococci are α-hemolytic with greening of the agar around the colonies. The zone of inhibition around the P disc is >14 mm indicating the organisms are pneumococci rather than viridans streptococci. A drop of desoxycholate ("bile") solution was placed on the overnight growth just to the right of the P disk area (arrow); after about 20 minutes at room temperature the colonies of pneumococci were solubilized (bile soluble). **B:** The growth of viridans streptococci appears similar to the growth of pneumococci, but growth of the viridans streptococci is not inhibited by optochin. **C:** *Streptococcus pneumoniae* Quellung reaction: a small amount of growth is mixed with saline, antisera against the capsule polysaccharide, and methylene blue stain. After incubation at room temperature for 1 hour, the reaction is observed under the microscope. The organisms are outlined in light blue. A positive reaction shows clumping because of cross-linking of the antibodies and pneumococci. The halo effect around the pneumococci is apparent capsular swelling. A negative control would show no clumping or capsular swelling. (Courtesy of H Reyes.)

factors that probably lower this resistance and thus predispose to pneumococcal infection are the following:

1. Viral and other respiratory tract infections that damage surface cells; abnormal accumulations of mucus (eg, allergy), which protect pneumococci from phagocytosis; bronchial obstruction (eg, atelectasis); and respiratory tract injury due to irritants disturbing its mucociliary function.
2. Alcohol or drug intoxication, which depresses phagocytic activity, depresses the cough reflex, and facilitates aspiration of foreign material.
3. Abnormal circulatory dynamics (eg, pulmonary congestion, heart failure).
4. Other mechanisms, eg, malnutrition, general debility, sickle cell anemia, hyposplenism, nephrosis, or complement deficiency.

Pathology

Pneumococcal infection causes an outpouring of fibrinous edema fluid into the alveoli, followed by red cells and leukocytes, which results in consolidation of portions of the lung. Many pneumococci are found throughout this exudate, and they may reach the bloodstream via the lymphatic drainage of the lungs. The alveolar walls remain normally intact during the infection. Later, mononuclear cells actively phagocytose the debris, and this liquid phase is gradually reabsorbed. The pneumococci are taken up by phagocytes and digested intracellularly.

Clinical Findings

The onset of pneumococcal pneumonia is usually sudden, with fever, chills, and sharp pleural pain. The sputum is similar to the alveolar exudate, being characteristically bloody or rusty colored. Early in the disease, when the fever is high, bacteremia is present in 10–20% of cases. With antimicrobial therapy, the illness is usually terminated promptly; if drugs are given early, the development of consolidation is interrupted.

Pneumococcal pneumonia must be differentiated from pulmonary infarction, atelectasis, neoplasm, congestive heart failure, and pneumonia caused by many other bacteria. Empyema (pus in the pleural space) is a significant complication and requires aspiration and drainage.

From the respiratory tract, pneumococci may reach other sites. The sinuses and middle ear are most frequently involved. Infection sometimes extends from the mastoid to the meninges. Bacteremia from pneumonia has a triad of severe complications: meningitis, endocarditis, and septic arthritis. With the early use of chemotherapy, acute pneumococcal endocarditis and arthritis have become rare.

Diagnostic Laboratory Tests

Blood is drawn for culture; CSF and sputum are collected for demonstration of pneumococci by smear and culture. Serum antibody tests are impractical. Sputum may be examined in several ways.

A. Stained Smears

A Gram-stained film of rusty-red sputum shows typical organisms, many polymorphonuclear neutrophils, and many red cells.

B. Capsule Swelling Tests

Fresh emulsified sputum mixed with antiserum causes capsule swelling (the quellung reaction) for identification of pneumococci.

C. Culture

The culture is created by sputum cultured on blood agar and incubated in CO_2 or a candle jar. A blood culture is also taken.

Immunity

Immunity to infection with pneumococci is type-specific and depends both on antibodies to capsular polysaccharide and on intact phagocytic function. Vaccines can induce production of antibodies to capsular polysaccharides (see below).

Treatment

Since pneumococci are sensitive to many antimicrobial drugs, early treatment usually results in rapid recovery, and antibody response seems to play a much diminished role. Penicillin G is the drug of choice, but in the United States 15% of pneumococci are penicillin-resistant (MIC ≥2 μg/mL) and about 18% are moderately resistant (MIC 0.1–1 μg/mL). High-dose penicillin G with MICs of 0.1–2 μg/mL appears to be effective in treating pneumonia caused by pneumococci but would not be effective in treatment of meningitis due to the same strains. Some penicillin-resistant strains are resistant to cefotaxime. Resistance to tetracycline and erythromycin occurs also. Pneumococci remain susceptible to vancomycin.

Epidemiology, Prevention, & Control

Pneumococcal pneumonia accounts for about 60% of all bacterial pneumonias. In the development of illness, predisposing factors (see above) are more important than exposure to the infectious agent, and the healthy carrier is more important in disseminating pneumococci than the sick patient.

It is possible to immunize individuals with type-specific polysaccharides. Such vaccines can probably provide 90% protection against bacteremic pneumonia. A polysaccharide vaccine containing 23 types is licensed in the United States. This vaccine is appropriate for elderly, debilitated, or immunosuppressed individuals. A pneumococcal conjugate vaccine contains capsular polysaccharides conjugated to diphtheria CRM_{197} protein. This seven-valent vaccine is recommended for all children aged 2–23 months, to help prevent invasive infections, and for selected children aged 24–59 months.

ENTEROCOCCI

The enterococci have the group D group-specific substance and were previously classified as group D streptococci. Because the group D cell wall specific antigen is a teichoic acid, it is not an antigenically good marker; enterococci are usually identified by characteristics other than immunologic reaction with group-specific antisera. They are part of the normal enteric flora. They are usually nonhemolytic, but occasionally α-hemolytic. Enterococci are PYR-positive. They grow in the presence of bile and hydrolyze esculin (bile esculin-positive). They grow in 6.5% NaCl. They grow well at between 10°C and 45°C whereas streptococci generally grow at a much narrower temperature range. They are more resistant to penicillin G than the streptococci, and rare isolates have plasmids that encode for β-lactamase. Many isolates are vancomycin-resistant.

There are at least 12 species of enterococci. *Enterococcus faecalis* is the most common and causes 85–90% of enterococcal infections, while *Enterococcus faecium* causes 5–10%. The enterococci are among the most frequent causes of nosocomial infections, particularly in intensive care units, and are selected by therapy with cephalosporins and other antibiotics to which they are resistant. Enterococci are transmitted from one patient to another primarily on the hands of hospital personnel, some of whom may carry the enterococci in their gastrointestinal tracts. Enterococci occasionally are transmitted on medical devices. In patients, the most common sites of infection are the urinary tract, wounds, biliary tract, and blood. Enterococci may cause meningitis and bacteremia in neonates. In adults, enterococci can cause endocarditis. However, in intra-abdominal, wound, urine, and other infections, enterococci usually are cultured along with other species of bacteria, and it is difficult to define the pathogenic role of the enterococci.

Antibiotic Resistance

A major problem with the enterococci is that they can be very resistant to antibiotics. *E faecium* is usually much more antibiotic-resistant than *E faecalis*.

A. Intrinsic Resistance

Enterococci are intrinsically resistant to cephalosporins, penicillinase-resistant penicillins, and monobactams. They have intrinsic low-level resistance to many aminoglycosides, are of intermediate susceptibility or resistant to fluoroquinolones, and are less susceptible than streptococci (10- to 1000-fold) to penicillin and ampicillin. Enterococci are inhibited by β-lactams (eg, ampicillin) but generally are not killed by them.

B. Resistance to Aminoglycosides

Therapy with combinations of a cell wall-active antibiotic (a penicillin or vancomycin) plus an aminoglycoside (streptomycin or gentamicin) is essential for severe enterococcal infections, such as endocarditis. Although enterococci have intrinsic low-level resistance to aminoglycosides (MICs <500 μg/mL), they have synergistic susceptibility when treated with a cell wall-active antibiotic plus an aminoglycoside. However, some enterococci have high-level resistance to aminoglycosides (MICs >500 μg/mL) and are not susceptible to the synergism. This high-level aminoglycoside resistance is due to enterococcal aminoglycoside-modifying enzymes (Table 14–2). The genes that code for most of these enzymes are usually on conjugative plasmids or transposons. The

enzymes have differential activity against the aminoglycosides. Resistance to gentamicin predicts resistance to the other aminoglycosides except streptomycin. (Susceptibility to gentamicin does not predict susceptibility to other aminoglycosides.) Resistance to streptomycin does not predict resistance to other aminoglycosides. The result is that only streptomycin or gentamicin (or both or neither) is likely to show synergistic activity with a cell wall-active antibiotic against enterococci. Enterococci from severe infections should have susceptibility tests for high-level resistance (MICs >500 μg/mL) to gentamicin and streptomycin to predict therapeutic efficacy.

C. Vancomycin Resistance

The glycopeptide vancomycin is the primary alternative drug to a penicillin (plus an aminoglycoside) for treating enterococcal infections. In the United States, enterococci that are resistant to vancomycin have increased in frequency. These enterococci are not synergistically susceptible to vancomycin plus an aminoglycoside. Vancomycin resistance has been most common in *E faecium*, but vancomycin-resistant strains of *E faecalis* also occur.

There are multiple **vancomycin resistance phenotypes.** The VanA phenotype is manifested by inducible high-level resistance to vancomycin and teicoplanin. VanB phenotypes are inducibly resistant to vancomycin but susceptible to teicoplanin. VanC strains have intermediate to moderate resistance to vancomycin. VanC is constitutive in the less commonly isolated species, *Enterococcus gallinarum* (VanC-1) and *Enterococcus casseliflavus/Enterococcus flavescens* (VanC-2/VanC-3). The VanD phenotype is manifested by moderate resistance to vancomycin and low-level resistance or susceptibility to teicoplanin. The VanE phenotype is moderately resistant to vancomycin and susceptible to teicoplanin.

Teicoplanin is a glycopeptide with many similarities to vancomycin. It is available for patients in Europe but not in

TABLE 14–2 Enterococcal Aminoglycoside-Modifying Enzymes that Eliminate Aminoglycoside-Penicillin Synergy

Gene	Enzyme Type	Streptomycin	Gentamicin	Tobramycin	Amikacin
Aac' (6)-*le-aph* (2")-*Ia*	Bifunctional	S	R	R	R
Aph(2")-*Ib*	Phosphonotransferase	S	R	R	S
Aph(2")-*Ic*	Phosphonotransferase	S	R	R	S
Aph (2")-*Id*	Phosphonotransferase	S	R	R	S
Aac(6')-*Ii*	6'-Acetyltransferase	S	S	R	S
Aph (3')-*IIIa*	Phosphonotransferase	S	S	S	R
Ant(3")-*Ia*	Nucleotidyltransferase	R	S	S	S
Ant(4')-*Ia*	Nucleotidyltransferase	S	S	R	R
Ant(6')-*Ia*	Nucleotidyltransferase	R	S	S	S

Reprinted with permission from Chow JW. Aminoglycoside resistance in enterococci. Clin Infect Dis 2000;31:586.

the United States. It has importance in investigation of the vancomycin resistance of enterococci.

Vancomycin and teicoplanin interfere with cell wall synthesis in gram-positive bacteria by interacting with the d-alanyl-d-alanine (d-Ala-d-Ala) group of the pentapeptide chains of peptidoglycan precursors. The best-studied vancomycin resistance determinant is the VanA operon. It is a system of genes packaged in a self-transferable plasmid containing a transposon closely related to Tn*1546* (Figure 14–5). There are two open reading frames that code for transposase and resolvase; the remaining seven genes code for vancomycin resistance and accessory proteins. The *vanR* and *vanS* genes are a two-component regulatory system sensitive to the presence of vancomycin or teicoplanin in the environment. *vanH*, *vanA*, and *vanX* are required for vancomycin resistance. *vanH* and *vanA* encode for proteins that yield manufacture of the depsipeptide (d-Ala-d-lactate) rather than the normal peptide (d-Ala-d-Ala). The depsipeptide, when linked to UDP-muramyl-tripeptide, forms a pentapeptide precursor that vancomycin and teicoplanin will not bind to. *vanX* encodes a dipeptidase that depletes the environment of the normal d-Ala-d-Ala dipeptide. *vanY* and *vanZ* are not essential for vancomycin resistance. *vanY* encodes a carboxypeptidase that cleaves the terminal d-Ala from the pentapeptide, depleting the environment of any functional pentapeptide that may have been manufactured by the normal cell wall building process. *vanZ* function is unclear.

Like *vanA*, *vanB* and *vanD* code for d-Ala-d-Lac, while *vanC* and *vanE* code for d-Ala-d-Ser.

D. β-Lactamase Production and Resistance to β-Lactams

β-Lactamase-producing *E faecalis* has been isolated from patients' specimens in the United States and other countries. There is great geographic variation. The isolates from the Northeastern and Southern United States appeared to be from dissemination of a single strain, suggesting there will be spread to additional geographical areas. The gene encoding for the enterococcal β-lactamase is the same gene as found in *Staphylococcus aureus*. The gene is constitutively expressed in enterococci and inducible in staphylococci. Because enterococci may produce small amounts of the enzyme, they may appear to be susceptible to penicillin and ampicillin by routine susceptibility tests. The β-lactamase can be detected using a high inoculum and the chromogenic cephalosporin test or by other methods. High-level gentamicin resistance often accompanies the β-lactamase production. The genes coding for both of these properties reside on conjugative plasmids and can be transferred from one strain of enterococcus to another. Infections due to β-lactamase-producing enterococci can be treated with combination penicillin and β-lactamase inhibitors or vancomycin (and streptomycin), when in vitro susceptibility has been demonstrated.

E. Trimethoprim-Sulfamethoxazole Resistance

Enterococci often show susceptibility to trimethoprim-sulfamethoxazole (TMP-SMZ) by in vitro testing, but the drugs are not effective in treating infections. This discrepancy is because enterococci are able to utilize exogenous folates available in vivo and thus escape inhibition by the drugs.

OTHER CATALASE-NEGATIVE GRAM-POSITIVE COCCI

There are nonstreptococcal gram-positive cocci or coccobacilli that occasionally cause disease (Table 14–3). These organisms have many growth and morphologic characteristics like viridans streptococci. They may be α-hemolytic or nonhemolytic. Most of them are catalase-negative; others may be weakly catalase-positive. Pediococcus and leuconostoc are the genera whose members are **vancomycin-resistant.** Lactobacilli are anaerobes that can be aerotolerant and α-hemolytic, sometimes forming coccobacillary forms similar to the viridans streptococci. Most **lactobacilli** (80–90%) are vancomycin-resistant. Other organisms that occasionally cause disease and should be differentiated from streptococci and enterococci include *Lactococcus*, *Aerococcus*, and *Gemella*, genera that generally are **vancomycin-susceptible.** *Rothia mucilaginosus* was previously considered a staphylococcus, but it is catalase-negative; colonies show a distinct adherence to agar.

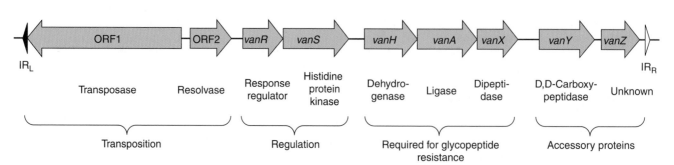

FIGURE 14–5 Schematic map of transposon Tn*1546* from *Enterococcus faecium* that codes for vancomycin resistance. IR$_L$ and IR$_R$ indicate the left and right inverted repeats of the transposon, respectively. (Adapted and reproduced with permission from Arthur M, Courvalin P: Genetics and mechanisms of glycopeptide resistance in enterococci. Antimicrob Agents Chemother 1993;37:1563.)

TABLE 14–3   Most Frequently Encountered Nonstreptococcal Catalase-Negative Gram-Positive Cocci and Coccobacilli

Genus[a]	Catalase	Gram Stain	Vancomycin Susceptibility	Comment
Abiotrophia[b] (Nutritionally variant streptococcus)	Negative	Cocci in pairs, short chains	Susceptible	Normal flora of oral cavity; isolated from cases of endocarditis
Aerococcus	Negative to weakly positive	Cocci in tetrads and clusters	Susceptible	Environmental organisms occasionally isolated from blood, urine, or sterile sites
Gemella	Negative	Cocci in pairs, tetrads, clusters, and short chains	Susceptible	Decolorize easily and may look gram-negative; grow slowly (48 hours); part of normal human flora; occasionally isolated from blood and sterile sites
Granulicatella[b] (Nutritionally variant streptococcus)	Negative	Cocci in chains, clusters	Susceptible	Normal flora of oral cavity; isolated from cases of endocarditis
Leuconostoc	Negative	Cocci in pairs and chains; coccobacilli, rods	Resistant	Environmental organisms; look like enterococci on blood agar; isolated from a wide variety of infections
Pediococcus	Negative	Cocci in pairs, tetrads, and clusters	Resistant	Present in food products and human stools; occasionally isolated from blood and abscesses
Lactobacillus	Negative	Coccobacilli, rods in pairs and chains	Resistant (90%)	Aerotolerant anaerobes generally classified as bacilli; normal vaginal flora; occasionally found in deep-seated infections

[a]Other genera where isolates from humans are rare or uncommon: Alloiococcus, Facklamia, Globicatella, Helcococcus, Lactococcus, Tetragenococcus, Vagococcus, Weissella.

[b]Require pyridoxal for growth.

REVIEW QUESTIONS

1. A 48-year-old man is admitted to hospital because of stupor. He is unkempt and homeless and lives in an encampment with other homeless people who called the authorities when he could not be easily aroused. The patient drinks a lot of fortified wine and drank excessively 2 nights previously. His temperature is 38.5°C and his blood pressure 125/80 mm Hg. He moans when attempts are made to arouse him. He has positive Kernig and Brudzinski signs, suggesting meningeal irritation. Physical examination and chest x-ray show evidence of left lower lobe lung consolidation. An endotracheal aspirate yields rust-colored sputum. Examination of a Gram-stained sputum smear shows numerous polymorphonuclear cells and numerous gram-positive lancet-shaped diplococci. On lumbar puncture, the cerebrospinal fluid is cloudy and has a white blood cell count of 570/μL with 95% polymorphonuclear cells; Gram stain shows numerous gram-positive diplococci. Based on this information, the likely diagnosis is

   (A) Pneumonia and meningitis due to Staphylococcus aureus
   (B) Pneumonia and meningitis due to Streptococcus pyogenes
   (C) Pneumonia and meningitis due to Streptococcus pneumoniae
   (D) Pneumonia and meningitis due to Enterococcus faecalis
   (E) Pneumonia and meningitis due to Neisseria meningitidis

2. The patient in Question 1 is started on antibiotic therapy to cover many possible microorganisms. Subsequently, culture of sputum and cerebrospinal fluid yields gram-positive diplococci with a minimum inhibitory concentration (MIC) to penicillin G of >2 μg/mL. The drug of choice for this patient until further susceptibility testing can be done is

   (A) Penicillin G
   (B) Nafcillin
   (C) Trimethoprim-sulfamethoxazole
   (D) Gentamicin
   (E) Vancomycin

3. This infection (Question 1) might have been prevented by

   (A) Prophylactic intramuscular benzathine penicillin every 3 weeks
   (B) A 23-valent capsular polysaccharide vaccine
   (C) A vaccine against serogroups A, C, Y, and W135 capsular polysaccharide
   (D) A vaccine of polyribosylribitol capsular polysaccharide covalently linked to a protein
   (E) Oral penicillin V daily

4. The pathogenesis of the organism causing the infection (Question 1) includes which of the following?

   (A) Invasion of cells lining the alveoli and entry into the pulmonary venule circulation
   (B) Resistance to phagocytosis mediated by M proteins
   (C) Migration to mediastinal lymph nodes where hemorrhage occurs

(D) After phagocytosis, the organism lyses the phagocytic vacuole and is released when the phagocytic cell enters the circulation

(E) Inhibition of phagocytosis by a polysaccharide capsule

5. A seven-valent capsular polysaccharide protein conjugate vaccine for the pathogen in Question 1 is recommended

(A) To age 18 and for selected adults

(B) Only on exposure to a patient with disease caused by the organism

(C) For children aged 2–23 months plus selected children through 59 months

(D) For children aged 24–72 months

(E) For all age groups above age 2 months

6. An 8-year-old boy develops a severe sore throat. On examination, a grayish-white exudate is seen on the tonsils and pharynx. The differential diagnosis includes group A streptococcal infection, Epstein-Barr virus (EBV) infection, severe adenovirus infection, and diphtheria. (*Neisseria gonorrhoeae* pharyngitis would be included also, but the patient has not been sexually abused.) The cause of the boy's pharyngitis is most likely

(A) A catalase-negative gram-positive coccus that grows in chains

(B) A single-stranded positive-sense RNA virus

(C) A catalase-positive gram-positive coccus that grows in clusters

(D) A catalase-negative gram-positive bacillus

(E) A double-stranded RNA virus

7. A primary mechanism responsible for the pathogenesis of the boy's disease (Question 6) is

(A) A net increase in intracellular cyclic adenosine monophosphate

(B) Action of M protein

(C) Action of IgA1 protease

(D) Action of enterotoxin A

(E) Inactivation of elongation factor 2

8. A 40-year-old woman develops severe headache and fever. Her neurologic examination is normal. A brain scan shows a ring-enhancing lesion of the left hemisphere. At surgery, a brain abscess is found. Culture of the abscess fluid grows an anaerobic gram-negative bacillus (*Bacteroides fragilis*) and a catalase-negative gram-positive coccus that on Gram stain is in pairs and chains. The organism is β-hemolytic and forms very small colonies (<0.5 mm in diameter). One person thought it smelled like butterscotch. It agglutinates with group F antisera. The organism most likely is

(A) *Streptococcus pyogenes* (group A)

(B) *Enterococcus faecalis* (group D)

(C) *Streptococcus agalactiae* (group B)

(D) *Streptococcus anginosus* group

(E) *Staphylococcus aureus*

9. The single most important method for classifying and speciating streptococci is

(A) Agglutination using antisera against the cell wall group-specific substance

(B) Biochemical testing

(C) Hemolytic properties (α, β, nonhemolytic)

(D) Capsular swelling (quellung) reaction

(E) None of the above

10. An 8-year-old girl develops Sydenham's chorea ("St. Vitus dance") with rapid uncoordinated facial tics and involuntary purposeless movements of her extremities, strongly suggestive of acute rheumatic fever. She has no other major manifestations of rheumatic fever (carditis, arthritis, subcutaneous nodules, skin rash). The patient's throat culture is negative for *Streptococcus pyogenes* (group A streptococci). However, she, her brother, and her mother all had sore throats 2 months ago. A test that if positive would indicate recent *Streptococcus pyogenes* infections is

(A) Antistreptolysin S antibody titer

(B) Polymerase chain reaction for antibodies against M protein

(C) ASO antibody titer

(D) Esculin hydrolysis

(E) Antihyaluronic acid antibody titer

11. All of the following statements regarding the hyaluronic acid capsule of *Streptococcus pyogenes* are correct *except*:

(A) It is responsible for the mucoid appearance of the colonies *in vitro*.

(B) It is antiphagocytic.

(C) It binds to CD44 on human epithelial cells.

(D) It is an important virulence factor.

(E) A vaccine against the capsule is currently available.

12. Enterococci can be distinguished from nonenterococcal Group D streptococci on the basis of which of the following characteristics?

(A) γ hemolysis

(B) Esculin hydrolysis

(C) Growth in 6.5% NaCl

(D) Growth in the presence of bile

(E) Gram stain morphology

13. Which of the following statements regarding the *Streptococcus bovis* group is correct?

(A) They possess Lancefield Group D antigen

(B) Some strains are vancomycin resistant.

(C) Infections caused by these organisms are benign.

(D) All subspecies are PYR positive.

(E) All subspecies are β-hemolytic.

14. Which of the following genera require pyridoxal for growth?

(A) *Aerococcus*

(B) *Granulicatella*

(C) *Enterococcus*

(D) *Leuconostoc*

(E) *Pediococcus*

15. Which of the following genera are typically resistant to vancomycin?

(A) *Aerococcus*

(B) *Gemella*

(C) *Pediococcus*

(D) *Streptococcus*

(E) *Abiotrophia*

Answers

1. C	5. C	9. E	13. A
2. E	6. A	10. C	14. B
3. B	7. B	11. E	15. C
4. E	8. D	12. C	

REFERENCES

Arias CA, Murray BE: *Enterococcus* species, *Streptococcus bovis* group, and *Leuconostoc* species. In: *Mandell, Douglas, and Bennett's Principles and Practice of Infectious Diseases,* 7th ed. Mandell GL, Bennett JE, Dolin R (editors). Elsevier, 2010.

Bisno AL: Nonsuppurative poststreptococcal sequelae: Rheumatic fever and glomerulonephritis. In: *Mandell, Douglas, and Bennett's Principles and Practice of Infectious Diseases,* 7th ed. Mandell GL, Bennett JE, Dolin R (editors). Elsevier, 2009.

Bisno AL, Stevens DL: *Streptococcus pyogenes.* In: *Mandell, Douglas, and Bennett's Principles and Practice of Infectious Diseases,* 7th ed. Mandell GL, Bennett JE, Dolin R (editors). Elsevier, 2010.

Chow JW: Aminoglycoside resistance in enterococci. Clin Infect Dis 2000;31:586.

Cunningham MW: Pathogenesis of group A streptococcal infections and their sequelae. Adv Exp Med Biol 2008;609:29.

Edwards MS, Baker CJ: *Streptococcus agalactiae* (group B streptococcus). In: *Mandell, Douglas, and Bennett's Principles and Practice of Infectious Diseases,* 7th ed. Mandell GL, Bennett JE, Dolin R (editors). Elsevier, 2010.

Facklam R, Elliott JA: Identification, classification, and clinical relevance of catalase-negative, gram-positive cocci, excluding the streptococci and enterococci. Clin Microbiol Rev 1995;8:479.

Murray BE: Vancomycin-resistant enterococcal infections. N Engl J Med 2000;342:710.

Musher DM: *Streptococcus pneumoniae.* In: *Mandell, Douglas, and Bennett's Principles and Practice of Infectious Diseases,* 7th ed. Mandell GL, Bennett JE, Dolin R (editors). Elsevier, 2010.

Petti CA, Stratton CW: *Streptococcus anginosis* group. In: *Mandell, Douglas, and Bennett's Principles and Practice of Infectious Diseases,* 7th ed. Mandell GL, Bennett JE, Dolin R (editors). Elsevier, 2010.

Ruoff KL: *Aerococcus, Abiotrophia,* and other infrequently isolated aerobic catalase-negative grampositive cocci. In: *Manual of Clinical Microbiology,* 9th ed. Murray PR et al (editors). ASM Press, 2007.

Ruoff, KL, Bisno AL: Classification of streptococci. In: *Mandell, Douglas, and Bennett's Principles and Practice of Infectious Diseases,* 7th ed. Mandell GL, Bennett JE, Dolin R (editors). Elsevier, 2010.

Ruoff KL, Whiley RA, Beighton D: *Streptococcus.* In: *Manual of Clinical Microbiology,* 9th ed. Murray PR et al (editors). ASM Press, 2007.

Schlegel L, et al: Reappraisal of the taxonomy of the *Streptococcus bovis/Streptococcus equinus* complex and related species: description of *Streptococcus gallolyticus* subsp. *gallolyticus* subsp. nov., *S. gallolyticus* subsp. *macedonicus* subsp. nov. and *S. gallolyticus* subsp. *pasteurianus* subsp. nov. International J Systematic and Evolutionary Microbiology 2003; 53:631.

Sendi P, Johansson L, Norrby-Teglund A: Invasive Group B streptococcal disease in non-pregnant adults. Infection 2008;36:100.

Sinner SW, Tunkel AR: Viridans streptococci, groups C and G streptococci, *Gemella morbillorum.* In: *Mandell, Douglas, and Bennett's Principles and Practice of Infectious Diseases,* 7th ed. Mandell GL, Bennett JE, Dolin R (editors). Elsevier, 2010.

Stollerman GH, Dale JB: The importance of the group A streptococcus capsule in the pathogenesis of human infections: a historical perspective. Clin Infect Dis 2008; 46; 1038.

Teixeira LM, Facklam RR: *Enterococcus.* In: *Manual of Clinical Microbiology,* 9th ed. Murray PR et al (editors). ASM Press, 2007.

Vaska VL, Faoagali JL: *Streptococcus bovis* bacteraemia: Identification within organism complex and association with endocarditis and colonic malignancy. Pathology 2009;41:183.

Enteric Gram-Negative Rods (Enterobacteriaceae)

The Enterobacteriaceae are a large, heterogeneous group of gram-negative rods whose natural habitat is the intestinal tract of humans and animals. The family includes many genera (*Escherichia, Shigella, Salmonella, Enterobacter, Klebsiella, Serratia, Proteus,* and others). Some enteric organisms, eg, *Escherichia coli,* are part of the normal flora and incidentally cause disease, while others, the salmonellae and shigellae, are regularly pathogenic for humans. The Enterobacteriaceae are facultative anaerobes or aerobes, ferment a wide range of carbohydrates, possess a complex antigenic structure, and produce a variety of toxins and other virulence factors. Enterobacteriaceae, enteric gram-negative rods, and enteric bacteria are the terms used in this chapter, but these bacteria may also be called coliforms.

CLASSIFICATION

The Enterobacteriaceae are the most common group of gram-negative rods cultured in the clinical laboratory and along with staphylococci and streptococci are among the most common bacteria that cause disease. The taxonomy of the Enterobacteriaceae is complex and rapidly changing since the introduction of techniques that measure evolutionary distance, such as nucleic acid hybridization and nucleic acid sequencing. According to the National Library of Medicine's Internet Taxonomy database (available at *http://www.ncbi. nlm.nih.gov//Taxonomy//Browser/wwwtax.cgl?id=543* more than 50 genera have been defined; however, the clinically significant Enterobacteriaceae comprise 20–25 species, and other species are encountered infrequently. In this chapter, taxonomic refinements will be minimized, and the names commonly employed in the medical literature will generally be used. A comprehensive approach to identification of the Enterobacteriaceae is presented in Chapters 42, 43, 44, and 45 of Murray PR et al (editors): *Manual of Clinical Microbiology,* 9th ed. ASM Press, 2007.

The family Enterobacteriaceae have the following characteristics: They are gram-negative rods, either motile with peritrichous flagella or nonmotile; they grow on peptone or meat extract media without the addition of sodium chloride or other supplements; grow well on MacConkey agar; grow aerobically and anaerobically (are facultative anaerobes); ferment rather than oxidize glucose, often with gas production; are catalase-positive, oxidase-negative, and reduce nitrate to nitrite; and have a 39–59% G + C DNA content. Examples of biochemical tests used to differentiate the species of Enterobacteriaceae are presented in Table 17–1. There are many others in addition to the ones listed. In the United States, commercially prepared kits or automated systems are used to a large extent for this purpose.

The major groups of Enterobacteriaceae are described and discussed briefly in the following paragraphs. Specific characteristics of salmonellae, shigellae, and the other medically important enteric gram-negative rods and the diseases they cause are discussed separately later in this chapter.

Morphology & Identification

A. Typical Organisms

The Enterobacteriaceae are short gram-negative rods (Figure 15-1 A). Typical morphology is seen in growth on solid media in vitro, but morphology is highly variable in clinical specimens. Capsules are large and regular in *Klebsiella,* less so in *Enterobacter,* and uncommon in the other species.

B. Culture

E coli and most of the other enteric bacteria form circular, convex, smooth colonies with distinct edges. *Enterobacter* colonies are similar but somewhat more mucoid. *Klebsiella* colonies are large and very mucoid and tend to coalesce with prolonged incubation. The salmonellae and shigellae produce colonies similar to *E coli* but do not ferment lactose. Some strains of *E coli* produce hemolysis on blood agar.

C. Growth Characteristics

Carbohydrate fermentation patterns and the activity of amino acid decarboxylases and other enzymes are used in biochemical differentiation (Table 15–1). Some tests, eg, the production of indole from tryptophan, are commonly used in rapid identification systems, while others, eg, the Voges-Proskauer reaction (production of acetylmethylcarbinol from dextrose), are used less often. Culture on "differential"

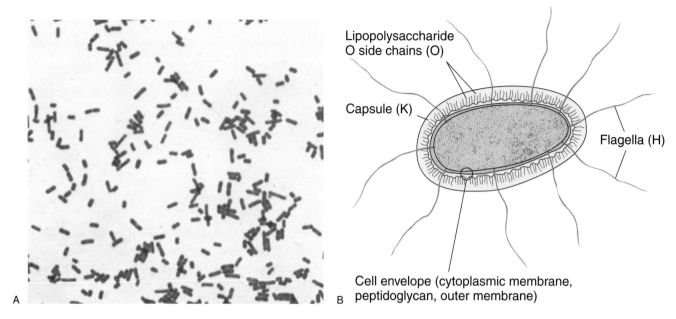

A

B

FIGURE 15–1 A: Gram stain of *Escherichia coli*. Original magnification × 1000. (Courtesy of H Reyes.) **B:** Antigenic structure of Enterobacteriaceae.

media that contain special dyes and carbohydrates (eg, eosin-methylene blue [EMB], MacConkey, or deoxycholate medium) distinguishes lactose-fermenting (colored) from nonlactose-fermenting colonies (nonpigmented) and may allow rapid presumptive identification of enteric bacteria (Table 15–2).

Many complex media have been devised to help in identification of the enteric bacteria. One such medium is triple sugar iron (TSI) agar, which is often used to help differentiate salmonellae and shigellae from other enteric gram-negative rods in stool cultures. The medium contains 0.1% glucose, 1% sucrose, 1% lactose, ferrous sulfate (for detection of H_2S production), tissue extracts (protein growth substrate), and a pH indicator (phenol red). It is poured into a test tube to produce a slant with a deep butt and is inoculated by stabbing bacterial growth into the butt. If only glucose is fermented, the slant and the butt initially turn yellow from the small amount of acid produced; as the fermentation products are subsequently oxidized to CO_2 and H_2O and released from the slant and as oxidative decarboxylation of proteins continues with formation of amines, the slant turns alkaline (red). If lactose or sucrose is fermented, so much acid is produced that the slant and butt remain yellow (acid). Salmonellae and shigellae typically yield an alkaline slant and an acid butt. Although *Proteus, Providencia,* and *Morganella* produce an alkaline slant and acid butt, they can be identified by their rapid formation of red color in Christensen's urea medium. Organisms producing acid on the slant and acid and gas (bubbles) in the butt are other enteric bacteria.

1. *Escherichia*—E coli* typically produces positive tests for indole, lysine decarboxylase, and mannitol fermentation and produces gas from glucose. An isolate from urine

can be quickly identified as *E coli* by its hemolysis on blood agar, typical colonial morphology with an iridescent "sheen" on differential media such as EMB agar, and a positive spot indole test. Over 90% of *E coli* isolates are positive for β-glucuronidase using the substrate 4-methylumbelliferyl-β-glucuronide (MUG). Isolates from anatomic sites other than urine, with characteristic properties (above plus negative oxidase tests) often can be confirmed as *E coli* with a positive MUG test.

2. *Klebsiella-Enterobacter-Serratia* group—*Klebsiella* species exhibit mucoid growth, large polysaccharide capsules, and lack of motility, and they usually give positive tests for lysine decarboxylase and citrate. Most *Enterobacter* species give positive tests for motility, citrate, and ornithine decarboxylase and produce gas from glucose. *Enterobacter aerogenes* has small capsules. *Serratia* produces DNase, lipase, and gelatinase. *Klebsiella, Enterobacter,* and *Serratia* usually give positive Voges-Proskauer reactions.

3. *Proteus-Morganella-Providencia* group—The members of this group deaminate phenylalanine, are motile, grow on potassium cyanide medium (KCN), and ferment xylose. *Proteus* species move very actively by means of peritrichous flagella, resulting in "swarming" on solid media unless the swarming is inhibited by chemicals, eg, phenylethyl alcohol or CLED (cystine-lactose-electrolyte-deficient) medium. *Proteus* species and *Morganella morganii* are urease-positive, while *Providencia* species usually are urease-negative. The *Proteus-Providencia* group ferments lactose very slowly or not at all. *Proteus mirabilis* is more susceptible to antimicrobial drugs, including penicillins, than other members of the group.

TABLE 15–1   Examples of Biochemical Reactions of Selected Enteric Gram-Negative Rods[a]

Organism	Indole Production	Methyl Red	Voges-Proskauer	Simmon Citrate	Hydrogen Sulfide	Urea Hydrolysis	Phenylalanine Deaminase	Lysine Decarboxylase	Arginine Dihydrolase	Ornithine Decarboxylase	Motility (36°C)	Gelatin Hydrolysis (22°C)	d-Glucose, Acid	d-Glucose, Gas	Lactose Fermentation	Sucrose Fermentation	d-Mannitol Fermentation	Dulcitol Fermentation	Adonitol Fermentation	d-Sorbitol Fermentation	l-Arabinose Fermentation	Raffinose Fermentation	l-Rhamnose Fermentation	d-Xylose Fermentation	Melibiose Fermentation
Citrobacter freundii	5	100	0	95	80	70	0	0	65	20	95	0	100	95	50	30	99	55	0	98	100	30	99	99	50
Enterobacter aerogenes	0	5	98	95	0	2	0	98	0	98	97	0	100	100	95	100	100	5	98	100	100	96	99	100	99
Enterobacter cloacae	0	5	100	100	0	65	0	0	97	96	95	0	100	100	93	97	100	15	25	95	100	92	100	100	90
Escherichia coli	98	99	0	1	1	1	0	90	17	65	95	0	100	95	95	50	98	60	5	94	99	50	80	95	75
Klebsiella pneumoniae	0	10	98	98	0	95	0	98	0	0	0	0	100	97	98	99	99	30	90	99	99	99	99	99	99
Klebsiella oxytoca	99	20	95	95	0	90	1	99	0	0	0	0	100	97	10	100	99	55	99	99	98	100	100	100	99
Morganella morganii	98	97	0	0	5	98	95	0	0	98	95	0	100	90	1	0	0	0	0	0	0	0	0	0	0
Proteus mirabilis	2	97	50	65	98	98	98	0	0	99	95	90	100	96	2	15	0	0	0	0	0	1	1	98	0
Salmonella choleraesuis	0	100	0	25	50	0	0	95	55	100	95	0	100	95	0	0	98	5	0	90	0	1	100	98	45
Salmonella typhi	0	100	0	0	97	0	0	98	3	0	97	0	100	0	1	0	100	0	0	99	2	0	98	82	100
Salmonella, most serotypes	1	100	0	95	95	1	0	98	70	97	95	0	100	96	1	1	100	96	0	95	99	2	95	97	95
Serratia marcescens	1	20	98	98	0	15	0	99	0	99	97	90	100	55	2	99	99	0	40	99	0	2	0	7	0
Shigella sonnei	0	100	0	0	0	0	0	0	2	98	0	0	100	0	2	1	99	0	0	2	95	3	75	2	25
Shigella dysenteriae, Shigella flexneri, Shigella boydii	50	100	0	0	0	0	0	0	5	1	0	0	100	2	0	0	99	2	0	30	60	50	5	2	50

[a]Adapted with permission from Farmer JJ III et al: Biochemical identification of new species and biogroups of Enterobacteriaceae isolated from clinical specimens. J Clin Microbiol 1984;21:46.

TABLE 15–2   Rapid, Presumptive Identification of Gram-Negative Enteric Bacteria

Lactose fermented rapidly

Escherichia coli: metallic sheen on differential media; motile; flat, nonviscous colonies

Enterobacter aerogenes: raised colonies, no metallic sheen; often motile; more viscous growth

Enterobacter cloacae: similar to *Enterobacter aerogenes*

Klebsiella pneumoniae: very viscous, mucoid growth; nonmotile

Lactose fermented slowly

Edwardsiella, Serratia, Citrobacter, Arizona, Providencia, Erwinia

Lactose not fermented

Shigella species: nonmotile; no gas from dextrose

Salmonella species: motile; acid and usually gas from dextrose

Proteus species: "swarming" on agar; urea rapidly hydrolyzed (smell of ammonia)

Pseudomonas species (see Chapter 16): soluble pigments, blue-green and fluorescing; sweetish smell

4. *Citrobacter*—These bacteria typically are citrate-positive and differ from the salmonellae in that they do not decarboxylate lysine. They ferment lactose very slowly if at all.

5. *Shigella*—Shigellae are nonmotile and usually do not ferment lactose but do ferment other carbohydrates, producing acid but not gas. They do not produce H_2S. The four *Shigella* species are closely related to *E coli*. Many share common antigens with one another and with other enteric bacteria (eg, *Hafnia alvei* and *Plesiomonas shigelloides*).

6. *Salmonella*—Salmonellae are motile rods that characteristically ferment glucose and mannose without producing gas but do not ferment lactose or sucrose. Most salmonellae produce H_2S. They are often pathogenic for humans or animals when ingested. Organisms originally described in the genus "*Arizona*" are included as subspecies in the *Salmonella* group.

7. Other Enterobacteriaceae—*Yersinia* species are discussed in Chapter 19. Other genera occasionally found in human infections include *Edwardsiella* and *Ewingella*, *Hafnia*, *Cedecea*, and *Kluyvera*.

Antigenic Structure

Enterobacteriaceae have a complex antigenic structure. They are classified by more than 150 different heat-stable somatic O (lipopolysaccharide) antigens, more than 100 heat-labile K (capsular) antigens, and more than 50 H (flagellar) antigens (Figure 15–1B). In *Salmonella typhi*, the capsular antigens are called Vi antigens.

O antigens are the most external part of the cell wall lipopolysaccharide and consist of repeating units of polysaccharide. Some O-specific polysaccharides contain unique sugars. O antigens are resistant to heat and alcohol and usually are detected by bacterial agglutination. Antibodies to O antigens are predominantly IgM.

While each genus of Enterobacteriaceae is associated with specific O groups, a single organism may carry several O antigens. Thus, most shigellae share one or more O antigens with *E coli*. *E coli* may cross-react with some *Providencia*, *Klebsiella*, and *Salmonella* species. Occasionally, O antigens may be associated with specific human diseases, eg, specific O types of *E coli* are found in diarrhea and in urinary tract infections.

K antigens are external to O antigens on some but not all Enterobacteriaceae. Some are polysaccharides, including the K antigens of *E coli*; others are proteins. K antigens may interfere with agglutination by O antisera, and they may be associated with virulence (eg, *E coli* strains producing K1 antigen are prominent in neonatal meningitis, and K antigens of *E coli* cause attachment of the bacteria to epithelial cells prior to gastrointestinal or urinary tract invasion).

Klebsiellae form large capsules consisting of polysaccharides (K antigens) covering the somatic (O or H) antigens and can be identified by capsular swelling tests with specific antisera. Human infections of the respiratory tract are caused particularly by capsular types 1 and 2; those of the urinary tract, by types 8, 9, 10, and 24.

H antigens are located on flagella and are denatured or removed by heat or alcohol. They are preserved by treating motile bacterial variants with formalin. Such H antigens agglutinate with anti-H antibodies, mainly IgG. The determinants in H antigens are a function of the amino acid sequence in flagellar protein (flagellin). Within a single serotype, flagellar antigens may be present in either or both of two forms, called phase 1 (conventionally designated by lower-case letters) and phase 2 (conventionally designated by Arabic numerals), as shown in Table 15–3. The organism tends to change from one phase to the other; this is called phase variation. H antigens on the bacterial surface may interfere with agglutination by anti-O antibody.

There are many examples of overlapping antigenic structures between Enterobacteriaceae and other bacteria. Most Enterobacteriaceae share the O14 antigen of *E coli*. The type 2 capsular polysaccharide of *Klebsiella* is very similar to the polysaccharide of type 2 pneumococci. Some K antigens cross-react with capsular polysaccharides of *Haemophilus influenzae* or *Neisseria meningitidis*. Thus, *E coli* O75:K100:H5 can induce antibodies that react with *H influenzae* type b.

The antigenic classification of Enterobacteriaceae often indicates the presence of each specific antigen. Thus, the antigenic formula of an *E coli* may be O55:K5:H21; that of *Salmonella* Schottmülleri is O1,4,5,12:Hb:1,2.

Colicins (Bacteriocins)

Many gram-negative organisms produce bacteriocins. These high-molecular-weight bactericidal proteins are produced by certain strains of bacteria active against some other

TABLE 15–3 Representative Antigenic Formulas of Salmonellae

O Group	Serotype	Antigenic Formula[a]
D	*Salmonella* Typhi	9, 12 (Vi):d:—
A	*Salmonella* Paratyphi A	1, 2, 12:a—
C₁	*Salmonella* Choleraesuis	6, 7:c:1,5
B	*Salmonella* Typhimurium	1, 4, 5, 12:i:1, 2
D	*Salmonella* Enteritidis	1, 9, 12:g, m:—

[a]O antigens: boldface numerals.

(Vi): Vi antigen if present.

Phase 1 H antigen: lower-case letter.

Phase 2 H antigen: numeral.

strains of the same or closely related species. Their production is controlled by plasmids. Colicins are produced by *E coli,* marcescens by *Serratia,* and pyocins by *Pseudomonas.* Bacteriocin-producing strains are resistant to their own bacteriocin; thus, bacteriocins can be used for "typing" of organisms.

Toxins & Enzymes

Most gram-negative bacteria possess complex lipopolysaccharides in their cell walls. These substances, cell envelope (cytoplasmic membrane, peptidoglycan, outer membrane) endotoxins, have a variety of pathophysiologic effects that are summarized in Chapter 9. Many gram-negative enteric bacteria also produce exotoxins of clinical importance. Some specific toxins are discussed in subsequent sections.

DISEASES CAUSED BY ENTEROBACTERIACEAE OTHER THAN SALMONELLA & SHIGELLA

Causative Organisms

E coli is a member of the normal intestinal flora (see Chapter 10). Other enteric bacteria (*Proteus, Enterobacter, Klebsiella, Morganella, Providencia, Citrobacter,* and *Serratia* species) are also found as members of the normal intestinal flora but are considerably less common than *E coli.* The enteric bacteria are sometimes found in small numbers as part of the normal flora of the upper respiratory and genital tracts. The enteric bacteria generally do not cause disease, and in the intestine they may even contribute to normal function and nutrition. When clinically important infections occur, they are usually caused by *E coli,* but the other enteric bacteria are causes of hospital-acquired infections and occasionally cause community-acquired infections. The bacteria become pathogenic only when they reach tissues outside of their normal intestinal or other less common normal flora sites. The most

frequent sites of clinically important infection are the urinary tract, biliary tract, and other sites in the abdominal cavity, but any anatomic site (eg, blood stream, prostate gland, lung, bone, meninges) can be the site of disease. Some of the enteric bacteria (eg, *Serratia marcescens, Enterobacter aerogenes*) are opportunistic pathogens. When normal host defenses are inadequate—particularly in infancy or old age, in the terminal stages of other diseases, after immunosuppression, or with indwelling venous or urethral catheters—localized clinically important infections can result, and the bacteria may reach the bloodstream and cause sepsis.

Pathogenesis & Clinical Findings

The clinical manifestations of infections with *E coli* and the other enteric bacteria depend on the site of the infection and cannot be differentiated by symptoms or signs from processes caused by other bacteria.

A. E coli

1. Urinary tract infection—*E coli* is the most common cause of urinary tract infection and accounts for approximately 90% of first urinary tract infections in young women (see Chapter 48). The symptoms and signs include urinary frequency, dysuria, hematuria, and pyuria. Flank pain is associated with upper tract infection. None of these symptoms or signs is specific for *E coli* infection. Urinary tract infection can result in bacteremia with clinical signs of sepsis.

Most of the urinary tract infections that involve the bladder or kidney in an otherwise healthy host are caused by a small number of O antigen types that have specifically elaborated virulence factors that facilitate colonization and subsequent clinical infections. These organisms are designated as uropathogenic *E coli.* Typically these organisms produce hemolysin, which is cytotoxic and facilitates tissue invasion. Those strains that cause pyelonephritis express K antigen and elaborate a specific type of pilus, P fimbriae, which binds to the P blood group antigen.

2. *E coli*-associated diarrheal diseases—*E coli* that cause diarrhea are extremely common worldwide. These *E coli* are classified by the characteristics of their virulence properties (see below), and each group causes disease by a different mechanism. The small or large bowel epithelial cell adherence properties are encoded by genes on plasmids. Similarly, the toxins often are plasmid- or phage-mediated. Some clinical aspects of diarrheal diseases are discussed in Chapter 48.

Enteropathogenic *E coli* (EPEC) is an important cause of diarrhea in infants, especially in developing countries. EPEC previously was associated with outbreaks of diarrhea in nurseries in developed countries. EPEC adhere to the mucosal cells of the small bowel. Chromosomally mediated factors promote tight adherence. There is loss of microvilli (effacement), formation of filamentous actin pedestals or cup-like structures, and, occasionally, entry of the EPEC into the mucosal cells. Characteristic lesions can be seen on electron micrographs of small bowel biopsy lesions. The result of EPEC infection is watery diarrhea, which is usually self-limited but can be chronic. EPEC diarrhea has been associated with multiple specific serotypes of *E coli*; strains are identified by O antigen and occasionally by H antigen typing. A two-stage infection model using HEp-2 cells also can be performed. Tests to identify EPEC are performed in reference laboratories. The duration of the EPEC diarrhea can be shortened and the chronic diarrhea cured by antibiotic treatment.

Enterotoxigenic *E coli* (ETEC) is a common cause of "traveler's diarrhea" and a very important cause of diarrhea in infants in developing countries. ETEC colonization factors specific for humans promote adherence of ETEC to epithelial cells of the small bowel. Some strains of ETEC produce a **heat-labile exotoxin** (LT) (MW 80,000) that is under the genetic control of a plasmid. Its subunit B attaches to the GM_1 ganglioside at the brush border of epithelial cells of the small intestine and facilitates the entry of subunit A (MW 26,000) into the cell, where the latter activates adenylyl cyclase. This markedly increases the local concentration of cyclic adenosine monophosphate (cAMP), which results in intense and prolonged hypersecretion of water and chlorides and inhibits the reabsorption of sodium. The gut lumen is distended with fluid, and hypermotility and diarrhea ensue, lasting for several days. LT is antigenic and cross-reacts with the enterotoxin of *Vibrio cholerae*. LT stimulates the production of neutralizing antibodies in the serum (and perhaps on the gut surface) of persons previously infected with enterotoxigenic *E coli*. Persons residing in areas where such organisms are highly prevalent (eg, in some developing countries) are likely to possess antibodies and are less prone to develop diarrhea on reexposure to the LT-producing *E coli*. Assays for LT include the following: (1) fluid accumulation in the intestine of laboratory animals; (2) typical cytologic changes in cultured Chinese hamster ovary cells or other cell lines; (3) stimulation of steroid production in cultured adrenal tumor cells; and (4) binding and immunologic assays with standardized antisera to LT. These assays are done only in reference laboratories.

Some strains of ETEC produce the **heat-stable enterotoxin** ST_a (MW 1500–4000), which is under the genetic control of a heterogeneous group of plasmids. ST_a activates guanylyl cyclase in enteric epithelial cells and stimulates fluid secretion. Many ST_a-positive strains also produce LT. The strains with both toxins produce a more severe diarrhea.

The plasmids carrying the genes for enterotoxins (LT, ST) also may carry genes for the **colonization factors** that facilitate the attachment of *E coli* strains to intestinal epithelium. Recognized colonization factors occur with particular frequency in some serotypes. Certain serotypes of ETEC occur worldwide; others have a limited recognized distribution. It is possible that virtually any *E coli* may acquire a plasmid encoding for enterotoxins. There is no definite association of ETEC with the EPEC strains causing diarrhea in children. Likewise, there is no association between enterotoxigenic strains and those able to invade intestinal epithelial cells.

Care in the selection and consumption of foods potentially contaminated with ETEC is highly recommended to help prevent traveler's diarrhea. Antimicrobial prophylaxis can be effective but may result in increased antibiotic resistance in the bacteria and probably should not be uniformly recommended. Once diarrhea develops, antibiotic treatment effectively shortens the duration of disease.

Shiga toxin producing *E coli* (STEC) are named for the cytotoxic toxins they produce. There are at least two antigenic forms of the toxin referred to as Shiga-like toxin 1 and Shiga-like toxin 2. STEC has been associated with hemorrhagic colitis, a severe form of diarrhea, and with hemolytic uremic syndrome, a disease resulting in acute renal failure, microangiopathic hemolytic anemia, and thrombocytopenia. The Shiga-like toxins have many properties that are similar to the Shiga toxin produced by some strains of *Shigella dysenteriae* type 1; however, the two toxins are antigenically and genetically distinct. Of the *E coli* serotypes that produce Shiga toxin, O157:H7 is the most common and is the one that can be identified in clinical specimens. STEC O157:H7 does not use sorbitol, unlike most other *E coli*, and is negative on sorbitol MacConkey agar (sorbitol is used instead of lactose); O157:H7 strains also are negative on MUG tests (see above). Many of the non-O157 serotypes may be sorbitol positive, when grown in culture. Specific antisera are used to identify the O157:H7 strains. Assays for Shiga toxin using commercially available enzyme immunoassays are done in many laboratories. Other sensitive test methods include cell culture cytotoxin testing using Vero cells and polymerase chain reaction for the direct detection of toxin genes directly from stool samples. Many cases of hemorrhagic colitis and its associated complications can be prevented by thoroughly cooking ground beef.

Enteroinvasive *E coli* (EIEC) produces a disease very similar to shigellosis. The disease occurs most commonly in children in developing countries and in travelers to these countries. Like *Shigella*, EIEC strains are non-lactose or late lactose fermenters and are nonmotile. EIEC produce disease by invading intestinal mucosal epithelial cells.

Enteroaggregative *E coli* (EAEC) causes acute and chronic diarrhea (>14 days in duration) in persons in developing countries. These organisms also are the cause of food-borne illnesses in industrialized countries. They are characterized by their specific patterns of adherence to human cells. EAEC produce ST-like toxin (see above) and a hemolysin.

3. Sepsis—When normal host defenses are inadequate, *E coli* may reach the bloodstream and cause sepsis. Newborns may be highly susceptible to *E coli* sepsis because they lack IgM antibodies. Sepsis may occur secondary to urinary tract infection.

4. Meningitis—*E coli* and group B streptococci are the leading causes of meningitis in infants. Approximately 75% of *E coli* from meningitis cases have the K1 antigen. This antigen cross-reacts with the group B capsular polysaccharide of

N meningitidis. The mechanism of virulence associated with the K1 antigen is not understood.

B. *Klebsiella-Enterobacter-Serratia; Proteus-Morganella-Providencia; & Citrobacter*

The pathogenesis of disease caused by these groups of enteric gram-negative rods is similar to that of the nonspecific factors in disease caused by *E coli.*

1. *Klebsiella*—*K pneumoniae* is present in the respiratory tract and feces of about 5% of normal individuals. It causes a small proportion (about 1%) of bacterial pneumonias. *K pneumoniae* can produce extensive hemorrhagic necrotizing consolidation of the lung. It produces urinary tract infection and bacteremia with focal lesions in debilitated patients. Other enterics also may produce pneumonia. *Klebsiella* sp. rank among the top ten bacterial pathogens responsible for hospital-acquired infections. Two other klebsiellae are associated with inflammatory conditions of the upper respiratory tract: *Klebsiella pneumoniae* subspecies *ozaenae* has been isolated from the nasal mucosa in ozena, a fetid, progressive atrophy of mucous membranes; and *Klebsiella pneumoniae* subspecies *rhinoscleromatis* from rhinoscleroma, a destructive granuloma of the nose and pharynx. *Klebsiella granulomatis* (formerly *Calymmatobacterium granulomatis*) causes a chronic genital ulcerative disease.

2. *Enterobacter*—Three species of *Enterobacter, E cloacae, E aerogenes,* and *E sakazakii* (recently moved to the genus *Cronobacter*), cause the majority of *Enterobacter* infections. These bacteria ferment lactose, may contain capsules that produce mucoid colonies and they are motile. These organisms cause a broad range of hospital acquired infections such as pneumonia, urinary tract infections, wound and device infections. Most strains possess a chromosomal β-lactamase called *ampC* which renders them intrinsically resistant to ampicillin and first and second generation cephalosporins. Mutants may hyperproduce β-lactamase conferring resistance to third generation cephalosporins.

3. *Serratia*—*S marcescens* is a common opportunistic pathogen in hospitalized patients. *Serratia* (usually nonpigmented) causes pneumonia, bacteremia, and endocarditis—especially in narcotics addicts and hospitalized patients. Only about 10% of isolates form the red pigment (prodigiosin) that has long characterized *Serratia marcescens.* *S marcescens* is often multiply resistant to aminoglycosides and penicillins; infections can be treated with third-generation cephalosporins.

4. *Proteus*—*Proteus* species produce infections in humans only when the bacteria leave the intestinal tract. They are found in urinary tract infections and produce bacteremia, pneumonia, and focal lesions in debilitated patients or those receiving intravenous infusions. *P mirabilis* causes urinary

tract infections and occasionally other infections. *Proteus vulgaris* and *Morganella morganii* are important nosocomial pathogens.

Proteus species produce urease, resulting in rapid hydrolysis of urea with liberation of ammonia. Thus, in urinary tract infections with *Proteus,* the urine becomes alkaline, promoting stone formation and making acidification virtually impossible. The rapid motility of *Proteus* may contribute to its invasion of the urinary tract.

Strains of *Proteus* vary greatly in antibiotic sensitivity. *P mirabilis* is often inhibited by penicillins; the most active antibiotics for other members of the group are aminoglycosides and cephalosporins.

5. *Providencia*—*Providencia* species (*Providencia rettgeri, Providencia alcalifaciens,* and *Providencia stuartii*) are members of the normal intestinal flora. All cause urinary tract infections and occasionally other infections and are often resistant to antimicrobial therapy.

6. *Citrobacter*—*Citrobacter* can cause urinary tract infections and sepsis.

Diagnostic Laboratory Tests

A. Specimens

Specimens include urine, blood, pus, spinal fluid, sputum, or other material, as indicated by the localization of the disease process.

B. Smears

The Enterobacteriaceae resemble each other morphologically. The presence of large capsules is suggestive of *Klebsiella.*

C. Culture

Specimens are plated on both blood agar and differential media. With differential media, rapid preliminary identification of gram-negative enteric bacteria is often possible (see Chapter 47).

Immunity

Specific antibodies develop in systemic infections, but it is uncertain whether significant immunity to the organisms follows.

Treatment

No single specific therapy is available. The sulfonamides, ampicillin, cephalosporins, fluoroquinolones, and aminoglycosides have marked antibacterial effects against the enterics, but variation in susceptibility is great, and laboratory tests for antibiotic susceptibility are essential. Multiple drug resistance is common and is under the control of transmissible plasmids.

Certain conditions predisposing to infection by these organisms require surgical correction, eg, relief of urinary tract obstruction, closure of a perforation in an abdominal organ, or resection of a bronchiectatic portion of lung.

Treatment of gram-negative bacteremia and impending septic shock requires rapid institution of antimicrobial therapy, restoration of fluid and electrolyte balance, and treatment of disseminated intravascular coagulation.

Various means have been proposed for the prevention of traveler's diarrhea, including daily ingestion of bismuth subsalicylate suspension (bismuth subsalicylate can inactivate *E coli* enterotoxin in vitro) and regular doses of tetracyclines or other antimicrobial drugs for limited periods. Because none of these methods are entirely successful or lacking in adverse effects, it is widely recommended that caution be observed in regard to food and drink in areas where environmental sanitation is poor and that early and brief treatment (eg, with ciprofloxacin or trimethoprim-sulfamethoxazole) be substituted for prophylaxis.

Epidemiology, Prevention, & Control

The enteric bacteria establish themselves in the normal intestinal tract within a few days after birth and from then on constitute a main portion of the normal aerobic (facultative anaerobic) microbial flora. *E coli* is the prototype. Enterics found in water or milk are accepted as proof of fecal contamination from sewage or other sources.

Control measures are not feasible as far as the normal endogenous flora is concerned. Enteropathogenic *E coli* serotypes should be controlled like salmonellae (see below). Some of the enterics constitute a major problem in hospital infection. It is particularly important to recognize that many enteric bacteria are "opportunists" that cause illness when they are introduced into debilitated patients. Within hospitals or other institutions, these bacteria commonly are transmitted by personnel, instruments, or parenteral medications. Their control depends on hand washing, rigorous asepsis, sterilization of equipment, disinfection, restraint in intravenous therapy, and strict precautions in keeping the urinary tract sterile (ie, closed drainage).

THE SHIGELLAE

The natural habitat of shigellae is limited to the intestinal tracts of humans and other primates, where they produce bacillary dysentery.

Morphology & Identification

A. Typical Organisms

Shigellae are slender gram-negative rods; coccobacillary forms occur in young cultures.

B. Culture

Shigellae are facultative anaerobes but grow best aerobically. Convex, circular, transparent colonies with intact edges reach a diameter of about 2 mm in 24 hours.

C. Growth Characteristics

All shigellae ferment glucose. With the exception of *Shigella sonnei,* they do not ferment lactose. The inability to ferment lactose distinguishes shigellae on differential media. Shigellae form acid from carbohydrates but rarely produce gas. They may also be divided into those that ferment mannitol and those that do not (Table 15–4).

Antigenic Structure

Shigellae have a complex antigenic pattern. There is great overlapping in the serologic behavior of different species, and most of them share O antigens with other enteric bacilli.

The somatic O antigens of shigellae are lipopolysaccharides. Their serologic specificity depends on the polysaccharide. There are more than 40 serotypes. The classification of shigellae relies on biochemical and antigenic characteristics. The pathogenic species are shown in Table 15–4.

Pathogenesis & Pathology

Shigella infections are almost always limited to the gastrointestinal tract; bloodstream invasion is quite rare. Shigellae are highly communicable; the infective dose is on the order of 10^3 organisms (whereas it usually is 10^5–10^8 for salmonellae and vibrios). The essential pathologic process is invasion of the mucosal epithelial cells (eg, M cells) by induced phagocytosis, escape from the phagocytic vacuole, multiplication and spread within the epithelial cell cytoplasm, and passage to adjacent cells. Microabscesses in the wall of the large intestine and terminal ileum lead to necrosis of the mucous membrane, superficial ulceration, bleeding, and formation of a "pseudomembrane" on the ulcerated area. This consists of fibrin, leukocytes, cell debris, a necrotic mucous membrane, and bacteria. As the process subsides, granulation tissue fills the ulcers and scar tissue forms.

Toxins

A. Endotoxin

Upon autolysis, all shigellae release their toxic lipopolysaccharide. This endotoxin probably contributes to the irritation of the bowel wall.

TABLE 15–4   Pathogenic Species of *Shigella*

Present Designation	Group and Type	Mannitol	Ornithine Decarboxylase
Shigella dysenteriae	A	–	–
Shigella flexneri	B	+	–
Shigella boydii	C	+	–
Shigella sonnei	D	+	+

B. *Shigella Dysenteriae* Exotoxin

S dysenteriae type 1 (Shiga bacillus) produces a heat-labile exotoxin that affects both the gut and the central nervous system. The exotoxin is a protein that is antigenic (stimulating production of antitoxin) and lethal for experimental animals. Acting as an enterotoxin, it produces diarrhea as does the *E coli* Shiga-like toxin, perhaps by the same mechanism. In humans, the exotoxin also inhibits sugar and amino acid absorption in the small intestine. Acting as a "neurotoxin," this material may contribute to the extreme severity and fatal nature of *S dysenteriae* infections and to the central nervous system reactions observed in them (ie, meningismus, coma). Patients with *Shigella flexneri* or *Shigella sonnei* infections develop antitoxin that neutralizes *S dysenteriae* exotoxin in vitro. The toxic activity is distinct from the invasive property of shigellae in dysentery. The two may act in sequence, the toxin producing an early nonbloody, voluminous diarrhea and the invasion of the large intestine resulting in later dysentery with blood and pus in stools.

Clinical Findings

After a short incubation period (1–2 days), there is a sudden onset of abdominal pain, fever, and watery diarrhea. The diarrhea has been attributed to an exotoxin acting in the small intestine (see above). A day or so later, as the infection involves the ileum and colon, the number of stools increases; they are less liquid but often contain mucus and blood. Each bowel movement is accompanied by straining and tenesmus (rectal spasms), with resulting lower abdominal pain. In more than half of adult cases, fever and diarrhea subside spontaneously in 2–5 days. However, in children and the elderly, loss of water and electrolytes may lead to dehydration, acidosis, and even death. The illness due to *S dysenteriae* may be particularly severe.

On recovery, most persons shed dysentery bacilli for only a short period, but a few remain chronic intestinal carriers and may have recurrent bouts of the disease. Upon recovery from the infection, most persons develop circulating antibodies to shigellae, but these do not protect against reinfection.

Diagnostic Laboratory Tests

A. Specimens

Specimens include fresh stool, mucus flecks, and rectal swabs for culture. Large numbers of fecal leukocytes and some red blood cells often are seen microscopically. Serum specimens, if desired, must be taken 10 days apart to demonstrate a rise in titer of agglutinating antibodies.

B. Culture

The materials are streaked on differential media (eg, MacConkey or EMB agar) and on selective media (Hektoen enteric agar or *Salmonella-Shigella* agar), which suppress other Enterobacteriaceae and gram-positive organisms. Colorless (lactose-negative) colonies are inoculated into triple sugar iron agar. Organisms that fail to produce H_2S, that produce acid but not gas in the butt and an alkaline slant in triple sugar iron agar medium, and that are nonmotile should be subjected to slide agglutination by specific *Shigella* antisera.

C. Serology

Normal persons often have agglutinins against several *Shigella* species. However, serial determinations of antibody titers may show a rise in specific antibody. Serology is not used to diagnose *Shigella* infections.

Immunity

Infection is followed by a type-specific antibody response. Injection of killed shigellae stimulates production of antibodies in serum but fails to protect humans against infection. IgA antibodies in the gut may be important in limiting reinfection; these may be stimulated by live attenuated strains given orally as experimental vaccines. Serum antibodies to somatic *Shigella* antigens are IgM.

Treatment

Ciprofloxacin, ampicillin, doxycycline, and trimethoprim-sulfamethoxazole are most commonly inhibitory for *Shigella* isolates and can suppress acute clinical attacks of dysentery and shorten the duration of symptoms. They may fail to eradicate the organisms from the intestinal tract. Multiple drug resistance can be transmitted by plasmids, and resistant infections are widespread. Many cases are self-limited. Opioids should be avoided in *Shigella* dysentery.

Epidemiology, Prevention, & Control

Shigellae are transmitted by "food, fingers, feces, and flies" from person to person. Most cases of *Shigella* infection occur in children under 10 years of age. Shigellosis has become an important problem in day care centers in the United States. *S dysenteriae* can spread widely. Mass chemoprophylaxis for limited periods of time (eg, in military personnel) has been tried, but resistant strains of shigellae tend to emerge rapidly. Since humans are the main recognized host of pathogenic shigellae, control efforts must be directed at eliminating the organisms from this reservoir by (1) sanitary control of water, food, and milk; sewage disposal; and fly control; (2) isolation of patients and disinfection of excreta; (3) detection of subclinical cases and carriers, particularly food handlers; and (4) antibiotic treatment of infected individuals.

THE *SALMONELLA-ARIZONA* GROUP

Salmonellae are often pathogenic for humans or animals when acquired by the oral route. They are transmitted from animals and animal products to humans, where they cause enteritis, systemic infection, and enteric fever.

Morphology & Identification

Salmonellae vary in length. Most isolates are motile with peritrichous flagella. Salmonellae grow readily on simple media, but they almost never ferment lactose or sucrose. They form acid and sometimes gas from glucose and mannose. They usually produce H_2S. They survive freezing in water for long periods. Salmonellae are resistant to certain chemicals (eg, brilliant green, sodium tetrathionate, sodium deoxycholate) that inhibit other enteric bacteria; such compounds are therefore useful for inclusion in media to isolate salmonellae from feces.

Classification

The classification of salmonellae is complex because the organisms are a continuum rather than a defined species. The members of the genus *Salmonella* were originally classified on the basis of epidemiology, host range, biochemical reactions, and structures of the O, H, and Vi (when present) antigens. The names (eg, *Salmonella typhi, Salmonella typhimurium*) were written as if they were genus and species; this form of the nomenclature remains in widespread but incorrect use. DNA-DNA hybridization studies have demonstrated that there are seven evolutionary groups. Currently, the genus *Salmonella* is divided into two species each with multiple subspecies and serotypes. The two species are *Salmonella enterica* and *Salmonella bongori* (formerly subspecies V). *Salmonella enterica* contains five subspecies: Subspecies *enterica* (subspecies I); subspecies *salamae* (subspecies II); subspecies *arizonae* (subspecies IIIa); subspecies *diarizonae* (subspecies IIIb); subspecies houtenae (subspecies IV); and subspecies indica (subspecies VI). Most human illness is caused by the subspecies I strains, written as *Salmonella enterica* subspecies *enterica*. Rarely human infections may be caused by subspecies IIIa and IIIb or the other subspecies frequently found in cold-blooded animals. Frequently, these infections are associated with exotic pets such as reptiles. It seems probable that the widely accepted nomenclature for classification will be as follows: *S enterica* subspecies *enterica* serotype *Typhimurium*, which can be shortened to *Salmonella* Typhimurium with the genus name in italics and the serotype name in roman type. National and international reference laboratories may use the antigenic formulas following the subspecies name because they impart more precise information about the isolates (Table 15–4).

There are more than 2500 serotypes of salmonellae, including more than 1400 in DNA hybridization group I that can infect humans. Four serotypes of salmonellae that cause enteric fever can be identified in the clinical laboratory by biochemical and serologic tests. These serotypes should be routinely identified because of their clinical significance. They are as follows: *Salmonella* Paratyphi A (serogroup A), *Salmonella* Paratyphi B (serogroup B), *Salmonella* Choleraesuis (serogroup C1), and *Salmonella* Typhi (serogroup D). *Salmonella* serotypes Enteritidis and Typhimurium are the two most common serotypes reported in the United States. The more than 1400 other salmonellae that are isolated in clinical laboratories are serogrouped by their O antigens as A, B, C_1, C_2, D, and E; some are nontypeable with this set of antisera. The isolates are then sent to reference laboratories for definitive serologic identification. This allows public health officials to monitor and assess the epidemiology of *Salmonella* infections on a statewide and nationwide basis.

Variation

Organisms may lose H antigens and become nonmotile. Loss of O antigen is associated with a change from smooth to rough colony form. Vi antigen may be lost partially or completely. Antigens may be acquired (or lost) in the process of transduction.

Pathogenesis & Clinical Findings

Salmonella Typhi, *Salmonella* Choleraesuis, and perhaps *Salmonella* Paratyphi A and *Salmonella* Paratyphi B are primarily infective for humans, and infection with these organisms implies acquisition from a human source. The vast majority of salmonellae, however, are chiefly pathogenic in animals that constitute the reservoir for human infection: poultry, pigs, rodents, cattle, pets (from turtles to parrots), and many others.

The organisms almost always enter via the oral route, usually with contaminated food or drink. The mean infective dose to produce clinical or subclinical infection in humans is 10^5–10^8 salmonellae (but perhaps as few as 10^3 *Salmonella* Typhi organisms). Among the host factors that contribute to resistance to salmonella infection are gastric acidity, normal intestinal microbial flora, and local intestinal immunity (see below).

Salmonellae produce three main types of disease in humans, but mixed forms are frequent (Table 15–5).

A. The "Enteric Fevers" (Typhoid Fever)

This syndrome is produced by only a few of the salmonellae, of which *Salmonella* Typhi (typhoid fever) is the most important. The ingested salmonellae reach the small intestine, from which they enter the lymphatics and then the bloodstream. They are carried by the blood to many organs, including the intestine. The organisms multiply in intestinal lymphoid tissue and are excreted in stools.

After an incubation period of 10–14 days, fever, malaise, headache, constipation, bradycardia, and myalgia occur. The fever rises to a high plateau, and the spleen and liver become enlarged. Rose spots, usually on the skin of the abdomen or chest, are seen briefly in rare cases. The white blood cell count is normal or low. In the preantibiotic era, the chief complications of enteric fever were intestinal hemorrhage and perforation, and the mortality rate was 10–15%.

TABLE 15-5 Clinical Diseases Induced by Salmonellae

	Enteric Fevers	Septicemias	Enterocolitis
Incubation period	7–20 days	Variable	8–48 hours
Onset	Insidious	Abrupt	Abrupt
Fever	Gradual, then high plateau, with "typhoidal" state	Rapid rise, then spiking "septic" temperature	Usually low
Duration of disease	Several weeks	Variable	2–5 days
Gastrointestinal symptoms	Often early constipation; later, bloody diarrhea	Often none	Nausea, vomiting, diarrhea at onset
Blood cultures	Positive in first to second weeks of disease	Positive during high fever	Negative
Stool cultures	Positive from second week on; negative earlier in disease	Infrequently positive	Positive soon after onset

Treatment with antibiotics has reduced the mortality rate to less than 1%.

The principal lesions are hyperplasia and necrosis of lymphoid tissue (eg, Peyer's patches), hepatitis, focal necrosis of the liver, and inflammation of the gallbladder, periosteum, lungs, and other organs.

B. Bacteremia with Focal Lesions

This is associated commonly with *S choleraesuis* but may be caused by any salmonella serotype. Following oral infection, there is early invasion of the bloodstream (with possible focal lesions in lungs, bones, meninges, etc), but intestinal manifestations are often absent. Blood cultures are positive.

C. Enterocolitis

This is the most common manifestation of salmonella infection. In the United States, *Salmonella* Typhimurium and *Salmonella* Enteritidis are prominent, but enterocolitis can be caused by any of the more than 1400 group I serotypes of salmonellae. Eight to 48 hours after ingestion of salmonellae, there is nausea, headache, vomiting, and profuse diarrhea, with few leukocytes in the stools. Low-grade fever is common, but the episode usually resolves in 2–3 days.

Inflammatory lesions of the small and large intestine are present. Bacteremia is rare (2–4%) except in immunodeficient persons. Blood cultures are usually negative, but stool cultures are positive for salmonellae and may remain positive for several weeks after clinical recovery.

Diagnostic Laboratory Tests

A. Specimens

Blood for culture must be taken repeatedly. In enteric fevers and septicemias, blood cultures are often positive in the first week of the disease. Bone marrow cultures may be useful. Urine cultures may be positive after the second week.

Stool specimens also must be taken repeatedly. In enteric fevers, the stools yield positive results from the second or third week on; in enterocolitis, during the first week.

A positive culture of duodenal drainage establishes the presence of salmonellae in the biliary tract in carriers.

B. Bacteriologic Methods for Isolation of Salmonellae

1. Differential medium cultures—EMB, MacConkey, or deoxycholate medium permits rapid detection of lactose nonfermenters (not only salmonellae and shigellae but also *Proteus*, *Serratia*, *Pseudomonas*, etc). Gram-positive organisms are somewhat inhibited. Bismuth sulfite medium permits rapid detection of salmonellae which form black colonies because of H_2S production. Many salmonellae produce H_2S.

2. Selective medium cultures—The specimen is plated on salmonella-shigella (SS) agar, Hektoen enteric agar, XLD, or deoxycholate-citrate agar, which favor growth of salmonellae and shigellae over other Enterobacteriaceae.

3. Enrichment cultures—The specimen (usually stool) also is put into selenite F or tetrathionate broth, both of which inhibit replication of normal intestinal bacteria and permit multiplication of salmonellae. After incubation for 1–2 days, this is plated on differential and selective media.

4. Final identification—Suspect colonies from solid media are identified by biochemical reaction patterns (Table 15–1) and slide agglutination tests with specific sera.

C. Serologic Methods

Serologic techniques are used to identify unknown cultures with known sera (see below) and may also be used to determine antibody titers in patients with unknown illness, although the latter is not very useful in diagnosis of *Salmonella* infections.

1. Agglutination test—In this test, known sera and unknown culture are mixed on a slide. Clumping, when it occurs, can be observed within a few minutes. This test is particularly useful for rapid preliminary identification of cultures. There are commercial kits available to agglutinate and serogroup salmonellae by their O antigens: A, B, C$_1$, C$_2$, D, and E.

2. Tube dilution agglutination test (Widal test)— Serum agglutinins rise sharply during the second and third weeks of *Salmonella* Typhi infection. The Widal test to detect these antibodies against the O and H antigens has been in use for decades. At least two serum specimens, obtained at intervals of 7–10 days, are needed to prove a rise in antibody titer. Serial dilutions of unknown sera are tested against antigens from representative salmonellae. False-positive and false-negative results occur. The interpretive criteria when single serum specimens are tested vary, but a titer against the O antigen of >1:320 and against the H antigen of >1:640 is considered positive. High titer of antibody to the Vi antigen occurs in some carriers. Alternatives to the Widal test include rapid colorimetric and enzyme immunoassay methods. There are conflicting reports in the literature regarding superiority of these methods to the Widal test. Results of serologic tests for *Salmonella* infection cannot be relied upon to establish a definitive diagnosis of typhoid fever and are most often used in resource poor areas of the world where blood cultures are not readily available.

Immunity

Infections with *Salmonella* Typhi or *Salmonella* Paratyphi usually confer a certain degree of immunity. Reinfection may occur but is often milder than the first infection. Circulating antibodies to O and Vi are related to resistance to infection and disease. However, relapses may occur in 2–3 weeks after recovery in spite of antibodies. Secretory IgA antibodies may prevent attachment of salmonellae to intestinal epithelium.

Persons with S/S hemoglobin (sickle cell disease) are exceedingly susceptible to *Salmonella* infections, particularly osteomyelitis. Persons with A/S hemoglobin (sickle cell trait) may be more susceptible than normal individuals (those with A/A hemoglobin).

Treatment

While enteric fevers and bacteremias with focal lesions require antimicrobial treatment, the vast majority of cases of enterocolitis do not. Antimicrobial treatment of *Salmonella* enteritis in neonates is important. In enterocolitis, clinical symptoms and excretion of the salmonellae may be prolonged by antimicrobial therapy. In severe diarrhea, replacement of fluids and electrolytes is essential.

Antimicrobial therapy of invasive *Salmonella* infections is with ampicillin, trimethoprim-sulfamethoxazole, or a third-generation cephalosporin. Multiple drug resistance

transmitted genetically by plasmids among enteric bacteria is a problem in *Salmonella* infections. Susceptibility testing is an important adjunct to selecting a proper antibiotic.

In most carriers, the organisms persist in the gallbladder (particularly if gallstones are present) and in the biliary tract. Some chronic carriers have been cured by ampicillin alone, but in most cases cholecystectomy must be combined with drug treatment.

Epidemiology

The feces of persons who have unsuspected subclinical disease or are carriers are a more important source of contamination than frank clinical cases that are promptly isolated, eg, when carriers working as food handlers are "shedding" organisms. Many animals, including cattle, rodents, and fowl, are naturally infected with a variety of salmonellae and have the bacteria in their tissues (meat), excreta, or eggs. The high incidence of salmonellae in commercially prepared chickens has been widely publicized. The incidence of typhoid fever has decreased, but the incidence of other *Salmonella* infections has increased markedly in the United States. The problem probably is aggravated by the widespread use of animal feeds containing antimicrobial drugs that favor the proliferation of drug-resistant salmonellae and their potential transmission to humans.

A. Carriers

After manifest or subclinical infection, some individuals continue to harbor salmonellae in their tissues for variable lengths of time (convalescent carriers or healthy permanent carriers). Three percent of survivors of typhoid become permanent carriers, harboring the organisms in the gallbladder, biliary tract, or, rarely, the intestine or urinary tract.

B. Sources of Infection

The sources of infection are food and drink that have been contaminated with salmonellae. The following sources are important:

1. Water—Contamination with feces often results in explosive epidemics.

2. Milk and other dairy products (ice cream, cheese, custard)—Contamination with feces and inadequate pasteurization or improper handling. Some outbreaks are traceable to the source of supply.

3. Shellfish—From contaminated water.

4. Dried or frozen eggs—From infected fowl or contaminated during processing.

5. Meats and meat products—From infected animals (poultry) or contamination with feces by rodents or humans.

6. "Recreational" drugs—Marijuana and other drugs.

7. Animal dyes—Dyes (eg, carmine) used in drugs, foods, and cosmetics.

8. Household pets—Turtles, dogs, cats, etc.

Prevention & Control

Sanitary measures must be taken to prevent contamination of food and water by rodents or other animals that excrete salmonellae. Infected poultry, meats, and eggs must be thoroughly cooked. Carriers must not be allowed to work as food handlers and should observe strict hygienic precautions.

Two typhoid vaccines are currently available in the United States: an oral live, attenuated vaccine and a Vi capsular polysaccharide vaccine for intramuscular use. Vaccination is recommended for travelers to endemic regions especially if the traveler visits rural areas or small villages where food choices are limited. Both vaccines have an efficacy of 50–80%. The time required for primary vaccination and age limits for each vaccine varies and individuals should consult the CDC Web site or obtain advice from a travel clinic regarding the latest vaccine information.

REVIEW QUESTIONS

1. A 20-year-old college student goes to the student health center because of dysuria, frequency, and urgency on urination for 24 hours. She has recently become sexually active. On urinalysis, many polymorphonuclear cells are seen. The most likely organism responsible for these symptoms and signs is

   (A) *Staphylococcus aureus*
   (B) *Streptococcus agalactiae*
   (C) *Gardnerella vaginalis*
   (D) *Lactobacillus* species
   (E) *Escherichia coli*

2. A 27-year-old woman is admitted to the hospital because of fever, with increasing anorexia, headache, weakness, and altered mental status of 2 days' duration. She works for an airline as a cabin attendant, flying between the Indian subcontinent and other places in Southeast Asia and the West Coast of the United States. Ten days prior to admission she had a diarrheal illness that lasted for about 36 hours. She has been constipated for the last 3 days. Her temperature is 39°C, heart rate 68/min, blood pressure 120/80 mm Hg, and respirations 18/min. She knows who she is and where she is but does not know the date. She is picking at the bedclothes. Rose spots are seen on the trunk. The remainder of the physical examination is normal. Blood cultures are done and an intravenous line is placed. The most likely cause of her illness is

   (A) Enterotoxigenic *Escherichia coli (ETEC)*
   (B) *Shigella sonnei*
   (C) *Salmonella enterica* subspecies *enterica* serotype Typhimurium (*Salmonella* Typhimurium)
   (D) *Salmonella enterica* subspecies *enterica* serotype Typhi (*Salmonella* Typhi)
   (E) Enteroinvasive *Escherichia coli (EIEC)*

3. Blood cultures from the patient in Question 2 grow a nonlactose-fermenting gram-negative bacillus. Which of the following is likely to be a constituent of this organism?

   (A) O antigen 157, H antigen 7 (O157:H7)
   (B) Vi antigen (capsule; virulence antigen)
   (C) O antigen 139 (O139)
   (D) Urease
   (E) K1 (capsular type 1)

4. A 37-year-old woman with a history of urinary tract infections comes to the emergency room with burning on urination along with frequency and urgency. She says her urine smells like ammonia. The cause of her urinary tract infection is likely to be

   (A) *Enterobacter aerogenes*
   (B) *Proteus mirabilis*
   (C) *Citrobacter freundii*
   (D) *Escherichia coli*
   (E) *Serratia marcescens*

5. An 18-year-old student has abdominal cramps and diarrhea. A plate of MacConkey agar is inoculated and grows gram-negative rods. Triple sugar iron (TSI) agar is used to screen the isolates for salmonellae and shigellae. A result suggesting one of these two pathogens would be

   (A) Production of urease
   (B) Motility in the medium
   (C) Inability to ferment lactose and sucrose
   (D) Fermentation of glucose
   (E) Production of gas in the medium

6. An uncommon serotype of *Salmonella enterica* subspecies *enterica* was found by laboratories in the health departments of adjacent states. The isolates were all from a small geographic area on either side of the border between the states, suggesting a common source for the isolates. (All the isolates were from otherwise healthy young adults who smoked marijuana; the same *Salmonella* was isolated from a specimen of the marijuana.) By what method did the laboratories determine that these isolates were the same?

   (A) Capsular (K antigen) typing
   (B) O antigen and H antigen typing
   (C) DNA sequencing
   (D) Sugar fermentation pattern determination
   (E) Decarboxylase reaction pattern determination

7. A 43-year-old diabetic man has a 4 cm nonhealing foot ulcer. Culture of the ulcer yields *Staphylococcus aureus*, *Bacteroides fragilis*, and a gram-negative bacillus that swarms across the blood agar plate covering the entire surface of the agar after 36 hours. The gram-negative bacillus is a member of the genus

   (A) *Escherichia*
   (B) *Enterobacter*
   (C) *Serratia*
   (D) *Salmonella*
   (E) *Proteus*

8. A 4-year-old boy from Kansas City who recently started attending preschool and after school daycare is brought to his pediatrician for a diarrheal illness characterized by fever to 38.2°C, severe lower abdominal pain, and initially watery diarrhea. His mother became concerned as the stools are now blood tinged 24 hours into the illness and the child appears quite ill. The mother reports that two other children who attend the same after school daycare have recently had diarrheal disease, one of whom

likewise had bloody stools. Which of the following is the most likely pathogen causing the illness in these children?

(A) An enterotoxigenic strain of *Escherichia coli*
(B) *Salmonella enterica* subspecies *enterica* serotype Typhi (*Salmonella* Typhi)
(C) *Shigella sonnei*
(D) *Edwardsiella tarda*
(E) *Klebsiella oxytoca*

9. A 5-year-old girl attended a birthday party at a local fast food restaurant. About 48 hours later she developed cramping abdominal pain, low grade fever and had five episodes of loose, bloody stools. She is taken to a local emergency room the next evening because the diarrhea has continued and she now appears pale and lethargic. On presentation, she has a temperature of 38°C, she is hypotensive and tachycardic. The abdominal exam reveals tenderness in the lower quadrants. Laboratory work is remarkable for a serum creatinine of 2.0 mg/dL, a serum hemoglobin of 8.0 mg/dL, thrombocytopenia and evidence of hemolysis. What is the most likely pathogen causing this child's illness?

(A) *Escherichia coli* O157:H7
(B) *Salmonella enterica* subspecies *enterica* serotype Typhimurium
(C) Enteropathogenic *Escherichia coli*
(D) *Edwardsiella tarda*
(E) *Plesiomonas shigelloides*

10. A 55-year-old homeless alcoholic man presents with severe multilobar pneumonia. He requires intubation and mechanical ventilation. A Gram stain of his sputum reveals numerous polymorphonuclear leukocytes and gram-negative rods that appear to have a capsule. The organism is a lactose fermenter on MacConkey agar and is very mucoid. It is nonmotile and lysine decarboxylase positive. What is the most likely organism causing this man's illness?

(A) *Serratia marcescens*
(B) *Enterobacter aerogenes*
(C) *Proteus mirabilis*
(D) *Klebsiella pneumoniae*
(E) *Morganella morganii*

11. Which of the following statements regarding O antigens is correct?

(A) All Enterobacteriaceae possess identical O antigens.
(B) They are found in the polysaccharide capsules of enteric bacteria.
(C) They are covalently linked to a polysaccharide core.

(D) They do not stimulate an immune response in the host.
(E) They are not important in the pathogenesis of infection cause by enteric bacteria.

12. Which of the following test methods is the least sensitive procedure for diagnosis of colitis caused by shiga-toxin producing *Escherichia coli*?

(A) Culture on sorbitol MacConkey agar
(B) Toxin testing using an enzyme immunoassay
(C) Cell culture cytotoxin assay using Vero cells
(D) Polymerase chain reaction for detection of the genes that encode shiga toxin

Answers

1. E	4. B	7. E	10. D
2. D	5. C	8. C	11. C
3. B	6. B	9. A	12. A

REFERENCES

Abbott S: *Klebsiella, Enterobacter, Citrobacter, Serratia, Plesiomonas,* and other Enterobacteriaceae. In: *Manual of Clinical Microbiology,* 9th ed. Murray PR et al (editors). ASM Press, 2007.

Donnenberg MS: Enterobacteriaceae. In: *Mandell, Douglas and Bennett's Principles and Practice of Infectious Diseases,* 7th ed. Mandell GL, Bennett JE, Dolin R (editors). Churchill Livingstone Elsevier, 2009.

Dupont HL: *Shigella* species (bacillary dysentery). In: *Mandell, Douglas and Bennett's Principles and Practice of Infectious Diseases,* 7th ed. Mandell GL, Bennett JE, Dolin R (editors). Churchill Livingstone Elsevier, 2009.

Farmer JJ III, Boatwright KK, Janda JM: Enterobacteriaceae: Introduction and Identification. In: *Manual of Clinical Microbiology,* 9th ed. Murray PR et al (editors). ASM Press, 2007.

Nataro JP, Bopp CA, Fields PI, Kaper JB, Strockbine NA: *Escherichia, Shigella,* and *Salmonella.* In: *Manual of Clinical Microbiology,* 9th ed. Murray PR et al (editors). ASM Press, 2007.

Pegues DA, Miller SI: *Salmonella* species, including *Salmonella typhi.* In: *Mandell, Douglas and Bennett's Principles and Practice of Infectious Diseases,* 7th ed. Mandell GL, Bennett JE, Dolin R (editors). Elsevier, 2010.

Pseudomonads, Acinetobacters, & Uncommon Gram-Negative Bacteria

The pseudomonads and acinetobacters are widely distributed in soil and water. *Pseudomonas aeruginosa* sometimes colonizes humans and is the major human pathogen of the group. *P aeruginosa* is invasive and toxigenic, produces infections in patients with abnormal host defenses, and is an important nosocomial pathogen.

Gram-negative bacteria that rarely cause disease in humans are included in this chapter. Some of these bacteria (eg, chromobacteria and chryseobacteria) are found in soil or water and are opportunistic pathogens for humans. Other gram-negative bacteria (eg, capnocytophaga, *Eikenella corrodens*, *Kingella*, and *Moraxella*) are normal flora of humans and occur in a wide variety of infections; often they are unexpected causes of disease.

THE PSEUDOMONAD GROUP

The pseudomonads are gram-negative, motile, aerobic rods some of which produce water-soluble pigments. Pseudomonads occur widely in soil, water, plants, and animals. *Pseudomonas aeruginosa* is frequently present in small numbers in the normal intestinal flora and on the skin of humans and is the major pathogen of the group. Other pseudomonads infrequently cause disease. The classification of pseudomonads is based on rRNA/DNA homology and common culture characteristics. The medically important pseudomonads are listed in Table 16–1.

PSEUDOMONAS AERUGINOSA

P aeruginosa is widely distributed in nature and is commonly present in moist environments in hospitals. It can colonize normal humans, in whom it is a saprophyte. It causes disease in humans with abnormal host defenses.

Morphology & Identification

A. Typical Organisms

P aeruginosa is motile and rod-shaped, measuring about 0.6×2 μm (Figure 16–1). It is gram-negative and occurs as single bacteria, in pairs, and occasionally in short chains.

B. Culture

P aeruginosa is an obligate aerobe that grows readily on many types of culture media, sometimes producing a sweet or grape-like or corn taco-like odor. Some strains hemolyze blood. *P aeruginosa* forms smooth round colonies with a fluorescent greenish color. It often produces the nonfluorescent bluish pigment **pyocyanin**, which diffuses into the agar. Other *Pseudomonas* species do not produce pyocyanin. Many strains of *P aeruginosa* also produce the fluorescent pigment **pyoverdin**, which gives a greenish color to the agar (Figure 16–2). Some strains produce the dark red pigment **pyorubin** or the black pigment **pyomelanin**.

P aeruginosa in a culture can produce multiple colony types (Figure 16–3). *P aeruginosa* from different colony types may also have different biochemical and enzymatic activities and different antimicrobial susceptibility patterns. Sometimes it is not clear if the colony types represent different strains of *P aeruginosa* or are variants of the same strain. Cultures from patients with cystic fibrosis (CF) often yield *P aeruginosa* organisms that form mucoid colonies as a result of overproduction of alginate, an exopolysaccharide. In CF

TABLE 16–1 Classification of Some of the Medically Important Pseudomonads[a]

rRNA Homology Group and Subgroup	Genus and Species
I Fluorescent group	*Pseudomonas aeruginosa* *Pseudomonas fluorescens* *Pseudomonas putida*
Nonfluorescent group	*Pseudomonas stutzeri* *Pseudomonas mendocina*
II	*Burkholderia pseudomallei* *Burkholderia mallei* *Burkholderia cepacia* *Ralstonia pickettii*
III	*Comamonas* species *Acidovorax* species
IV	*Brevundimonas* species
V	*Stenotrophomonas maltophilia*

[a]Many other species are occasionally encountered in clinical or environmental specimens.

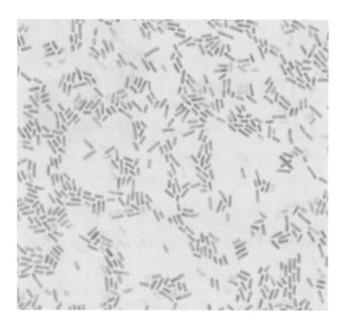

FIGURE 16–1 Gram stain of *Pseudomonas aeruginosa,* which are about 0.6 × 2 µm. Original magnification × 1000. (Courtesy of H Reyes.)

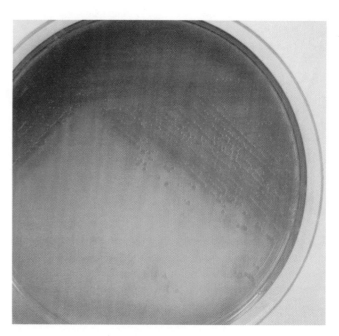

FIGURE 16–2 *Pseudomonas aeruginosa* on a 10 cm Mueller-Hinton agar plate. Individual colonies are 3–4 mm in diameter. The organism produces pyocyanin, which is blue, and pyoverdin, which is green. Together these pigments produce the blue green color that is seen in the agar around the pseudomonas growth. (Courtesy of S Lowe.)

patients, the exopolysaccharide appears to provide the matrix for the organisms to live in a biofilm (see Chapters 2 and 9).

C. Growth Characteristics

P aeruginosa grows well at 37–42°C; its growth at 42°C helps differentiate it from other *Pseudomonas* species in the fluorescent group. It is **oxidase-positive.** It does not ferment carbohydrates, but many strains oxidize glucose. Identification is usually based on colonial morphology, oxidase positivity, the presence of characteristic pigments, and growth at 42°C. Differentiation of *P aeruginosa* from other pseudomonads on the basis of biochemical activity requires testing with a large battery of substrates.

Antigenic Structure & Toxins

Pili (fimbriae) extend from the cell surface and promote attachment to host epithelial cells. The exopolysaccharide is responsible for the mucoid colonies seen in cultures from patients with CF. The lipopolysaccharide, which exists in multiple immunotypes, is responsible for many of the endotoxic properties of the organism. *P aeruginosa* can be typed by lipopolysaccharide immunotype and by pyocin (bacteriocin) susceptibility. Most *P aeruginosa* isolates from clinical infections produce extracellular enzymes, including elastases, proteases, and two hemolysins: a heat-labile phospholipase C and a heat-stable glycolipid.

Many strains of *P aeruginosa* produce exotoxin A, which causes tissue necrosis and is lethal for animals when injected in purified form. The toxin blocks protein synthesis by a mechanism of action identical to that of diphtheria toxin, though the structures of the two toxins are not identical. Antitoxins to exotoxin A are found in some human sera,

including those of patients who have recovered from serious *P aeruginosa* infections.

Pathogenesis

P aeruginosa is pathogenic only when introduced into areas devoid of normal defenses, eg, when mucous membranes and skin are disrupted by direct tissue damage; when intravenous or urinary catheters are used; or when neutropenia is present, as in cancer chemotherapy. The bacterium attaches to and colonizes the mucous membranes or skin, invades locally, and produces systemic disease. These processes are promoted by the pili, enzymes, and toxins described above. Lipopolysaccharide plays a direct role in causing fever, shock, oliguria, leukocytosis and leukopenia, disseminated intravascular coagulation, and adult respiratory distress syndrome.

P aeruginosa and other pseudomonads are resistant to many antimicrobial agents and therefore become dominant and important when more susceptible bacteria of the normal flora are suppressed.

Clinical Findings

P aeruginosa produces infection of wounds and burns, giving rise to blue-green pus; meningitis, when introduced by lumbar puncture; and urinary tract infection, when introduced by catheters and instruments or in irrigating solutions. Involvement of the respiratory tract, especially from contaminated respirators, results in necrotizing pneumonia.

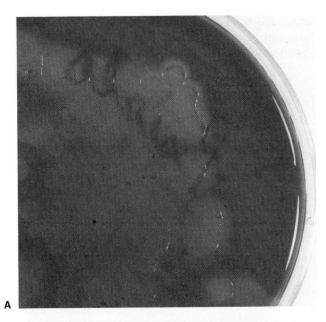

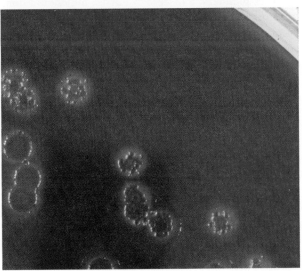

FIGURE 16–3 Variation in colony morphology of *Pseudomonas aeruginosa.* **A:** Green-gray colonies 6–8 mm in diameter on a 10 cm blood agar plate; the blood in the agar around the colonies shows hemolysis. **B:** Silver toned dry colonies on a similar blood agar plate; no hemolysis is present (the dark shadow in the lower part of the picture is from a label on the back of the petri dish). (Courtesy of H Reyes.)

The bacterium is often found in mild otitis externa in swimmers. It may cause invasive (malignant) otitis externa in diabetic patients. Infection of the eye, which may lead to rapid destruction of the eye, occurs most commonly after injury or surgical procedures. In infants or debilitated persons, *P aeruginosa* may invade the bloodstream and result in fatal sepsis; this occurs commonly in patients with leukemia or lymphoma who have received antineoplastic drugs or radiation therapy and in patients with severe burns. In most *P aeruginosa* infections, the symptoms and signs are nonspecific and are related to the organ involved. Occasionally,

verdoglobin (a breakdown product of hemoglobin) or fluorescent pigment can be detected in wounds, burns, or urine by ultraviolet fluorescence. Hemorrhagic necrosis of skin occurs often in sepsis due to *P aeruginosa;* the lesions, called **ecthyma gangrenosum,** are surrounded by erythema and often do not contain pus. *P aeruginosa* can be seen on Gram-stained specimens from ecthyma lesions, and cultures are positive. Ecthyma gangrenosum is uncommon in bacteremia due to organisms other than *P aeruginosa.*

Diagnostic Laboratory Tests

A. Specimens

Specimens from skin lesions, pus, urine, blood, spinal fluid, sputum, and other material should be obtained as indicated by the type of infection.

B. Smears

Gram-negative rods are often seen in smears. There are no specific morphologic characteristics that differentiate pseudomonads in specimens from enteric or other gram-negative rods.

C. Culture

Specimens are plated on blood agar and the differential media commonly used to grow the enteric gram-negative rods. Pseudomonads grow readily on most of these media, but they may grow more slowly than the enterics. *P aeruginosa* does not ferment lactose and is easily differentiated from the lactose-fermenting bacteria. Culture is the specific test for diagnosis of *P aeruginosa* infection.

Treatment

Clinically significant infections with *P aeruginosa* should not be treated with single-drug therapy, because the success rate is low with such therapy and because the bacteria can rapidly develop resistance when single drugs are employed. A penicillin such as piperacillin active against *P aeruginosa* is used in combination with an aminoglycoside, usually tobramycin. Other drugs active against *P aeruginosa* include aztreonam, carbapenems such as imipenem or meropenem, and the newer quinolones, including ciprofloxacin. Of the newer cephalosporins, ceftazidime and cefoperazone are active against *P aeruginosa;* ceftazidime is used with an aminoglycoside in primary therapy of *P aeruginosa* infections. The susceptibility patterns of *P aeruginosa* vary geographically, and susceptibility tests should be done as an adjunct to selection of antimicrobial therapy.

Epidemiology & Control

P aeruginosa is primarily a nosocomial pathogen, and the methods for control of infection are similar to those for other nosocomial pathogens. Since pseudomonas thrives in moist environments, special attention should be paid to sinks, water baths, showers, hot tubs, and other wet areas. For epidemiologic purposes, strains can be typed using molecular typing techniques.

BURKHOLDERIA PSEUDOMALLEI

B pseudomallei is a small, motile, oxidase-positive, aerobic gram-negative bacillus. It grows on standard bacteriologic media, forming colonies that vary from mucoid and smooth to rough and wrinkled and in color from cream to orange. It grows at 42°C and oxidizes glucose, lactose, and a variety of other carbohydrates. *B pseudomallei* causes **melioidosis** of humans, primarily in Southeast Asia and northern Australia. The organism is a natural saprophyte that has been cultured from soil, fresh water, rice paddies, and vegetable produce. Human infection probably originates from these sources by contamination of skin abrasions and possibly by ingestion or inhalation. Epizootic *B pseudomallei* infection occurs in sheep, goats, swine, horses, and other animals, though animals do not appear to be a primary reservoir for the organism.

Melioidosis may manifest itself as acute, subacute, or chronic infection. The incubation period can be as short as 2–3 days, but latent periods of months to years also occur. A localized suppurative infection can occur at the inoculation site where there is a break in the skin. This localized infection may lead to the acute septicemic form of infection with involvement of many organs. The signs and symptoms depend upon the major sites of involvement. The most common form of melioidosis is pulmonary infection, which may be a primary pneumonitis (*B pseudomallei* transmitted through the upper airway or nasopharynx) or subsequent to a localized suppurative infection and bacteremia. The patient may have fever and leukocytosis, with consolidation of the upper lobes. Subsequently, the patient may become afebrile, while upper lobe cavities develop, yielding an appearance similar to that of tuberculosis on chest films. Some patients develop chronic suppurative infection with abscesses in skin, brain, lung, myocardium, liver, bone, and other sites. Patients with chronic suppurative infections may be afebrile and have indolent disease. Latent infection is sometimes reactivated as a result of immunosuppression.

The diagnosis of melioidosis should be considered for a patient from an endemic area who has fulminant upper lobe pulmonary or unexplained systemic disease. A Gram stain of an appropriate specimen will show small gram-negative bacilli; bipolar staining (safety pin appearance) is seen with Wright's stain or methylene blue stain. A positive culture is diagnostic. A positive serologic test is diagnostically helpful and constitutes evidence of past infection.

Melioidosis has a high mortality rate if untreated. Surgical drainage of localized infection may be necessary. *B pseudomallei* are generally susceptible to ceftazidime, imipenem, meropenem, and amoxicillin-clavulanic acid (also ceftriaxone and cefotaxime). *B pseudomallei* are commonly resistant to penicillin, ampicillin, first-generation and second-generation cephalosporins, and gentamicin and tobramycin. Depending upon the clinical setting, the initial intensive therapy should be for a minimum of 10–14 days with ceftazidime, imipenem, or meropenem; sulfamethoxazole/trimethoprim can be considered in patients with severe allergy to β-lactam antimicrobials. Eradication therapy with sulfamethoxazole/trimethoprim or doxycycline should follow the intensive initial therapy and be continued for a minimum of 3 months. Recurrent disease because of failure of eradication can occur for several reasons, but the most important is noncompliance with the long-term eradication therapy.

BURKHOLDERIA MALLEI

B mallei is a small, nonmotile, nonpigmented, aerobic gram-negative rod that grows readily on most bacteriologic media. It causes **glanders,** a disease of horses, mules, and donkeys transmissible to humans. In horses, the disease has prominent pulmonary involvement, subcutaneous ulcerative lesions, and lymphatic thickening with nodules; systemic disease also occurs. Human infection, which can be fatal, usually begins as an ulcer of the skin or mucous membranes followed by lymphangitis and sepsis. Inhalation of the organisms may lead to primary pneumonia.

The diagnosis is based on rising agglutinin titers and culture of the organism from local lesions of humans or horses. Human cases can be treated effectively with the same antimicrobial regimens used to treat melioidosis.

The disease has been controlled by slaughter of infected horses and mules and at present is extremely rare. In some countries, laboratory infections are the only source of the disease.

BURKHOLDERIA CEPACIA COMPLEX & BURKHOLDERIA GLADIOLI

The prototypic species *Burkholderia cepacia* plus at least eight other species makeup the ***Burkholderia cepacia* complex**. Additional isolates cannot be assigned to these species based on molecular methods. *Burkholderia gladioli* is a closely related species. Thus, the classification of these bacteria is complex; their specific identification is difficult. These are environmental organisms able to grow in water, soil, plants, animals, and decaying vegetable materials. In hospitals, *B cepacia* has been isolated from a variety of water and environmental sources from which it can be transmitted to patients. People with CF, in particular, and those with chronic granulomatous disease are vulnerable to infection with bacteria in the *B cepacia* complex It is likely that *B cepacia* can be transmitted from one CF patient to another by close contact. They may have asymptomatic carriage, progressive deterioration over a period of months, or rapidly progressive deterioration with necrotizing pneumonia and bacteremia. Although a relative small percentage of CF patients become infected, the association with progressive disease makes *B cepacia* complex a major concern for these patients. A diagnosis of *B cepacia* infection in a CF patient may significantly alter the patient's life, because they may not be allowed association with other CF patients and they may be removed from eligibility for lung transplant.

B cepacia grows on most media used in culturing patients' specimens for gram-negative bacteria. Selective media containing colistin also can be used. *B cepacia* grows more slowly than enteric gram-negative rods, and it may take 3 days before colonies are visible. The *B cepacia* are oxidase-positive, lysine decarboxylase-positive, and produce acid from glucose, but differentiating *B cepacia* from other pseudomonads including *Stenotrophomonas maltophilia* requires a battery of biochemical tests and can be difficult. Submission of isolates to reference laboratories is recommended because of the prognostic implications of colonization in CF patients. In the United States, the CF Foundation supports a reference laboratory that uses phenotypic and genotypic methods to confirm the identity of organisms within the *B cepacia* complex. Susceptibility tests should be done on *B cepacia* complex isolates, though slow growth may make routine testing difficult. *B cepacia* complex from CF patients often are multidrug-resistant.

STENOTROPHOMONAS MALTOPHILIA

Stenotrophomonas maltophilia is a free-living gram-negative rod that is widely distributed in the environment. On blood agar, colonies have a lavender-green or gray color. The organism is oxidase-negative and lysine decarboxylase-positive.

S maltophilia is an increasingly important cause of hospital-acquired infections in patients who are receiving antimicrobial therapy and in immunocompromised patients. It has been isolated from many anatomic sites, including respiratory tract secretions, urine, skin wounds, and blood. The isolates are often part of mixed flora present in the specimens. When blood cultures are positive, it is commonly in association with use of indwelling plastic intravenous catheters.

S maltophilia is usually susceptible to trimethoprimsulfamethoxazole and ticarcillin-clavulanic acid and resistant to other commonly used antimicrobials, including cephalosporins, aminoglycosides, imipenem, and the quinolones. The widespread use of the drugs to which *S maltophilia* is resistant plays an important role in the increased frequency with which it causes disease.

OTHER PSEUDOMONADS

Some of the many genera and species of the pseudomonad group are listed in Table 16–1; occasionally these pseudomonads are opportunistic pathogens. The diagnosis of infections caused by these pseudomonads is made by culturing the bacteria and identifying them by differential reactions on a complex set of biochemical substrates. Many of the non-*P aeruginosa* species are nonfermentative and difficult to identify by routine methods; submission to a reference laboratory may be necessary to obtain definitive identification. Many of the pseudomonads have antimicrobial susceptibility patterns different from that of *P aeruginosa*.

ACINETOBACTER

Acinetobacter species are aerobic gram-negative bacteria that are widely distributed in soil and water and can occasionally be cultured from skin, mucous membranes, secretions, and the hospital environment.

Acinetobacter baumannii is the species most commonly isolated. *Acinetobacter lwoffii, Acinetobacter johnsonii, Acinetobacter haemolyticus,* and other species are isolated occasionally. Some isolates have not received species names. Acinetobacters were previously called by a number of different names, including *Mima polymorpha* and *Herellea vaginicola,* reflecting many of the organisms' characteristics.

Acinetobacters are usually coccobacillary or coccal in appearance; they resemble neisseriae on smears, because diplococcal forms predominate in body fluids and on solid media. Rod-shaped forms also occur, and occasionally the bacteria appear to be gram-positive. Acinetobacter grows well on most types of media used to culture specimens from patients. Acinetobacter recovered from meningitis and sepsis has been mistaken for *Neisseria meningitidis;* similarly, acinetobacter recovered from the female genital tract has been mistaken for *Neisseria gonorrhoeae.* However, the neisseriae produce oxidase and acinetobacter does not.

Acinetobacters often are commensals but occasionally cause nosocomial infection. *A baumannii* has been isolated from blood, sputum, skin, pleural fluid, and urine, usually in device-associated infections. *A johnsonii* is a nosocomial pathogen of low virulence and has been found in blood cultures of patients with plastic intravenous catheters. Acinetobacters encountered in nosocomial pneumonias often originate in the water of room humidifiers or vaporizers. In patients with acinetobacter bacteremias, intravenous catheters are almost always the source of infection. In patients with burns or with immune deficiencies, acinetobacters act as opportunistic pathogens and can produce sepsis. Acinetobacter strains are often resistant to antimicrobial agents, and therapy of infection can be difficult. Susceptibility testing should be done to help select the best antimicrobial drugs for therapy. Acinetobacter strains respond most commonly to gentamicin, amikacin, or tobramycin and to newer penicillins or cephalosporins.

UNCOMMON GRAM-NEGATIVE BACTERIA

ACTINOBACILLUS

Aggregatibacter actinomycetemcomitans (formerly *Actinobacillus actinomycetemcomitans*) is a small gram-negative coccobacillary organism that grows slowly. As its name implies, it is often found in actinomycosis. It also causes severe periodontal disease in adolescents, endocarditis, abscesses, osteomyelitis, and other infections. It is treatable with tetracycline or chloramphenicol and sometimes with penicillin G, ampicillin, or erythromycin.

ACHROMOBACTER & ALCALIGENES

The classification of species within the genera *Achromobacter* and *Alcaligenes* is changing and confusing. These groups include species of oxidase-positive, gram-negative rods. They have peritrichous flagella and are motile, which differentiates them from the pseudomonads. They alkalinize citrate medium and oxidation-fermentation medium containing glucose and are urease-negative. They may be part of the normal human bacterial flora and have been isolated from respirators, nebulizers, and renal dialysis systems. They are occasionally isolated from urine, blood, spinal fluid, wounds, and abscesses. *Achromobacter xylosoxidans* subspecies *xylosoxidans* has been isolated from many body sites but is uncommon as a sole cause of infection.

OCHROBACTRUM

The genus *Ochrobactrum* contains species previously classified in the genus *Achromobacter* and also has other *Ochrobactrum* species. They are similar to *Achromobacter* and *Alcaligenes*. *Ochrobactrum anthropi* is most often isolated from intravascular catheter-related bacteremia. It also may contaminate biologic products.

CAPNOCYTOPHAGA

The *Capnocytophaga* species are slow-growing capnophilic, gram-negative, fusiform or filamentous bacilli. They are fermentative and facultative anaerobes that require CO_2 for aerobic growth. They may show **gliding motility,** which can be seen as outgrowths of colonies. They produce a substance that modifies polymorphonuclear cell chemotactic activity. *Capnocytophaga ochracea, Capnocytophaga sputigena,* and *Capnocytophaga gingivalis* are members of the normal oral flora of humans. They have been associated with severe periodontal disease in juveniles. They occasionally cause bacteremia and severe systemic disease in immunocompromised patients, especially granulocytopenic patients with oral ulcerations. *Capnocytophaga canimorsus* (previously DF-2-

dysgonic fermenter 2) in the oral flora of dogs. When transmitted to humans, it occasionally causes fulminant infection in asplenic patients, alcoholics, and, rarely, healthy people. *Capnocytophaga cynodegmi* (DF-2-like) is associated with wound infections from dog or cat bites or scratches.

CARDIOBACTERIUM

Cardiobacterium hominis, another bacterium with a descriptive name, is a facultatively anaerobic, pleomorphic gram-negative rod that is part of the normal flora of the upper respiratory tract and bowel and occasionally causes endocarditis. Using modern blood culture medium, it is no longer necessary to hold blood cultures for more than the standard 5- to 7-day incubation period to grow cardiobacterium.

CHROMOBACTERIA

Chromobacterium violaceum is a gram-negative bacillus resembling pseudomonads. The organism usually produces a violet pigment. It occurs in subtropical climates in soil and water and may infect animals and humans through breaks in the skin or via the gut. This may result in abscesses, diarrhea, and sepsis, with many deaths. Chromobacteria are often susceptible to chloramphenicol, tetracyclines, and aminoglycosides.

EIKENELLA CORRODENS

E corrodens is a small, fastidious, capnophilic gram-negative rod that is part of the gingival and bowel flora of 40–70% of humans. About 50% of isolates form pits in agar during the several days of incubation required for growth. *Eikenella* is oxidase-positive and does not ferment carbohydrates. It is found in mixed flora infections associated with contamination by oral mucosal organisms; it is often present with streptococci. It occurs frequently in infections from human bites. *Eikenella* is uniformly resistant to clindamycin, which can be used to make a selective agar medium. *Eikenella* is usually susceptible to ampicillin and the newer penicillins and cephalosporins. β-Lactam-producing strains have been reported.

CHRYSEOBACTERIUM

Organisms of the *Chryseobacterium* group are long, thin, nonmotile gram-negative rods that are oxidase-positive, proteolytic, and weakly fermentative. They often form distinctive yellow colonies. Chryseobacteria are commonly found in sink drains, faucets, and on medical equipment that has been exposed to contaminated water and not sterilized. Chryseobacteria occasionally colonize the respiratory tract. *Chryseobacterium meningosepticum* rarely causes meningitis. *Chryseobacterium* species are often resistant to many antimicrobial drugs.

KINGELLA

The *Kingella* group includes three species, of which *Kingella kingae* is an oxidase-positive, nonmotile organism that is hemolytic when grown on blood agar. It is a gram-negative rod, but coccobacillary and diplococcal forms are common. It is part of the normal oral flora and occasionally causes infections of bone, joints, and tendons. The organism probably enters the circulation with minor oral trauma such as tooth brushing. It is susceptible to penicillin, ampicillin, erythromycin, and other antimicrobial drugs.

MORAXELLA

The *Moraxella* group includes six species. They are nonmotile, nonfermentative, and oxidase-positive. On staining, they appear as small gram-negative bacilli, coccobacilli, or cocci. They are members of the normal flora of the upper respiratory tract and occasionally cause bacteremia, endocarditis, conjunctivitis, meningitis, or other infections. Most of them are susceptible to penicillin and other antimicrobial drugs. *Moraxella catarrhalis* often produces β-lactamase (see Chapter 20).

REVIEW QUESTIONS

1. A sputum culture of a cystic fibrosis patient grows *Pseudomonas aeruginosa* that form very mucoid colonies. The implication of this observation is which one of the following?
   (A) The *Pseudomonas aeruginosa* are highly susceptible to the aminoglycoside antimicrobial tobramycin.
   (B) The *Pseudomonas aeruginosa* is infected with a pyocin (a bacteriocin).
   (C) The colonies are mucoid because they have polysaccharide capsule of hyaluronic acid.
   (D) The exotoxin A gene has been disabled and the *Pseudomonas aeruginosa* are no longer able to block host cell protein synthesis.
   (E) The *Pseudomonas aeruginosa* have formed a biofilm in the patient's airway.

2. An environmental gram-negative bacillus that is resistant to cephalosporins, aminoglycosides, and quinolones has become a very important nosocomial pathogen largely because it is selected by use of those antibiotics. This gram-negative bacillus can take 2–3 days to grow and must be differentiated from *Burkholderia cepacia*. It is
   (A) *Pseudomonas aeruginosa*
   (B) *Acinetobacter baumannii*
   (C) *Alcaligenes xylosoxidans*
   (D) *Klebsiella pneumoniae*
   (E) *Stenotrophomonas maltophilia*

3. This gram-negative bacillus, which is oxidase-positive and does not ferment carbohydrates, is frequently found in human bite infections
   (A) *Escherichia coli*
   (B) *Neisseria meningitidis*

   (C) *Chromobacterium violaceum*
   (D) *Eikenella corrodens*
   (E) *Proteus mirabilis*

4. A 17-year-old girl with cystic fibrosis has a slight increase in her frequent cough and production of mucoid sputum. A sputum specimen is obtained and plated on routine culture media. The predominant growths are gram-negative bacilli that form very mucoid colonies after 48 hours of incubation. These bacilli are oxidase-positive, grow at 42°C, and have a grape-like odor. These gram-negative bacilli are which of the following?
   (A) *Klebsiella pneumoniae*
   (B) *Pseudomonas aeruginosa*
   (C) *Staphylococcus aureus*
   (D) *Streptococcus pneumoniae*
   (E) *Burkholderia cepacia*

5. The sputum from the 17-year-old patient with cystic fibrosis (Question 4) also is plated on mannitol salt agar, which turns yellow (from the baseline pink) where white colonies of gram-positive cocci are growing; the cocci are catalase-positive and coagulase-positive. The microorganisms growing on the mannitol salt agar are
   (A) *Burkholderia cepacia*
   (B) *Streptococcus pneumoniae*
   (C) *Stenotrophomonas maltophilia*
   (D) *Staphylococcus aureus*
   (E) *Streptococcus pyogenes*

6. The sputum from the 17-year-old patient with cystic fibrosis (Question 4) also is plated on a colistin-containing agar. After 72 hours of incubation, the colistin-containing agar grows gram-negative bacilli that are oxidase-positive but are otherwise difficult to identify. This microorganism is of major concern. It is sent to a reference laboratory so that molecular methods can be used to identify or rule out which of the following?
   (A) *Pseudomonas aeruginosa*
   (B) *Burkholderia cepacia*
   (C) *Haemophilus influenzae*
   (D) *Pseudomonas putida*
   (E) *Burkholderia pseudomallei*

7. When a *Burkholderia cepacia* complex organism is isolated from a cystic fibrosis patient great care must be taken in identifying the organism for which one of the following reasons?
   (A) *Burkholderia cepacia* is commonly susceptible to penicillin G whereas other similar gram-negative bacilli are not.
   (B) The presence of *Burkholderia cepacia* complex in a CF patient's airway has major implications for the patient's long-term prognosis and therapeutic options.
   (C) Only *Burkholderia cepacia* produce biofilms.
   (D) *Burkholderia cepacia* complex produce an enzyme, sputolysase, which liquefies sputum making it easier for the patient to cough and clear the airway.
   (E) Selective media for *Pseudomonas aeruginosa* commonly used for CF patients' sputum cultures typically inhibit the *Burkholderia cepacia* complex organisms making it difficult to identify them.

8. Which of the following statements about *Acintobacter* species is (are) true?
   (A) They are widespread in nature and in the hospital environment.

(B) They are generally non pathogenic to healthy individuals.

(C) They may appear as gram-positive cocci.

(D) They can mimic the morphology of *Neisseria* species in gram stains of endocervical secretions to diagnose gonorrhea in women.

(E) They can be a significant cause of ventilator-associated pneumonia in intensive care unit patients.

(F) All of the above.

9. A 37-year-old firefighter suffers smoke inhalation and is hospitalized for ventilatory support. He has a severe cough and begins to expectorate purulent sputum. Gram stain of his sputum specimen shows numerous polymorphonuclear cells and numerous gram-negative rods. Sputum culture grows numerous gram-negative rods that are oxidase-positive. They grow well at 42°C. On clear agar medium they produce a green color in the agar. The agar where the green color is located fluoresces when exposed to ultraviolet light. The organism causing the patient's infection is

(A) *Pseudomonas aeruginosa*

(B) *Klebsiella pneumoniae*

(C) *Escherichia coli*

(D) *Burkholderia cepacia*

(E) *Burkholderia pseudomallei*

10. The pigment produced by the microorganism in Question 1 is

(A) Aquamarine green

(B) Aerobactin

(C) Enterochelin

(D) Pyoverdin

(E) Prodigiosin

11. *Burkholderia cepacia* is infrequently found in or on

(A) Swimming pools

(B) Soil

(C) Pond water

(D) Plants

12. Which of the following statements about *Pseudomonas aeruginosa* is correct?

(A) *Pseudomonas aeruginosa* are typically susceptible to penicillin G.

(B) *Pseudomonas aeruginosa* are readily grown in anaerobic blood cultures.

(C) *Pseudomonas aeruginosa* are able to penetrate intact normal human skin by elaborating the enzyme invasin.

(D) *Pseudomonas aeruginosa* seldom if ever cause pneumonia.

(E) *Pseudomonas aeruginosa* have fimbriae which promote attachment to epithelial cells.

13. The mechanism of action of exotoxin A of *Pseudomonas aeruginosa* is

(A) To activate acetylcholine esterase

(B) To block elongation factor 2 (EF2)

(C) To form pores in white blood cells and increase cation permeability

(D) To increase intracellular cyclic adenosine monophosphate

(E) To split lecithin into phosphorylcholine and diacylglycerol

14. The HACEK bacteria sometimes cause indolent endocarditis or other infections. This acronym represents which of the following?

(A) *Cardiobacterium hominis*

(B) *Eikenella corrodens*

(C) *Kingella kingae*

(D) *Aggregatibacter actinomycetemcomitans* (formerly *Actinobacillus actinomycetemcomitans*)

(E) *Aggregatibacter aphrophilus* (formerly *Haemophilus aphrophilus*)

(F) All of the above

Answers

1. E	5. D	9. A	13. B
2. E	6. B	10. D	14. F
3. D	7. B	11. A	
4. B	8. F	12. E	

REFERENCES

Blondell-Hill E, Henry DA, Speert DP: *Pseudomonas*. In: *Manual of Clinical Microbiology,* 9th ed. Murray PR et al (editors). ASM Press, 2007.

LiPuma JJ et al: *Burkholderia, Stenotrophomonas, Ralstonia, Cupriavidus, Pandoraea, Brevundimonas, Comamonas, Delftia,* and *Acidovorax*. In: *Manual of Clinical Microbiology,* 9th ed. Murray PR et al (editors). ASM Press, 2007.

Maschmeyer G, Göbel UB: *Stenotrophomonas maltophilia* and *Burkholderia cepacia* complex. In: *Mandell, Douglas and Bennett's Principles and Practice of Infectious Diseases,* 7th ed. Mandell GL, et al (editors). Elsevier, 2010.

Pier GB, Ramphal R: *Pseudomonas aeruginosa*. In: *Mandell, Douglas and Bennett's Principles and Practice of Infectious Diseases,* 7th ed. Mandell GL, et al (editors). Elsevier, 2010.

Schreckenberger PC et al: *Acinetobacter, Achromobacter, Chryseobacterium, Moraxella,* and other nonfermentative gram-negative rods. In: *Manual of Clinical Microbiology,* 9th ed. Murray PR et al (editors). ASM Press, 2007.

Steinberg JP, Burd EM: Other gram-negative and gram-variable bacilli. In: *Mandell, Douglas and Bennett's Principles and Practice of Infectious Diseases,* 7th ed. Mandell GL, et al (editors). Elsevier, 2010.

Vibrios, Campylobacters, Helicobacter, & Associated Bacteria

Vibrio, Aeromonas, Plesiomonas, Campylobacter, and *Helicobacter* species are gram-negative rods that are all widely distributed in nature. The vibrios are found in marine and surface waters. *Aeromonas* is found predominantly in fresh water and occasionally in cold-blooded animals. *Plesiomonas* exists in both cold-blooded and warm-blooded animals. The campylobacters are found in many species of animals, including many domesticated animals. *Vibrio cholerae* produces an enterotoxin that causes cholera, a profuse watery diarrhea that can rapidly lead to dehydration and death. *Campylobacter jejuni* is a common cause of enteritis in humans. Less commonly, aeromonas and, rarely, plesiomonas have been associated with diarrheal disease in humans. *Helicobacter pylori* has been associated with gastritis and duodenal ulcer disease.

THE VIBRIOS

Vibrios are among the most common bacteria in surface waters worldwide. They are curved aerobic rods and are motile, possessing a polar flagellum. *V cholerae* serogroups O1 and O139 cause cholera in humans, while other vibrios may cause sepsis or enteritis. The medically important vibrios are listed in Table 17–1.

TABLE 17–1   The Medically Important Vibrios

Organism BHU	Human Disease
Vibrio cholerae serogroups O1 and O139	Epidemic and pandemic cholera
Vibrio cholerae serogroups non-O1/non-O139	Cholera-like diarrhea; mild diarrhea; rarely, extraintestinal infection
Vibrio parahaemolyticus	Gastroenteritis, perhaps extraintestinal infection
Others *Vibrio mimicus, Vibrio vulnificus, Vibrio hollisae, Vibrio fluvialis, Vibrio damsela, Vibrio anginolyticus, Vibrio metschnikovii*	Ear, wound, soft tissue, and other extraintestinal infections, all uncommon

VIBRIO CHOLERAE

The epidemiology of cholera closely parallels the recognition of *V cholerae* transmission in water and the development of sanitary water systems.

Morphology & Identification

A. Typical Organisms

Upon first isolation, *V cholerae* is a comma-shaped, curved rod 2–4 μm long (Figure 17–1). It is actively motile by means of a polar flagellum. On prolonged cultivation, vibrios may become straight rods that resemble the gram-negative enteric bacteria.

B. Culture

V cholerae produces convex, smooth, round colonies that are opaque and granular in transmitted light. *V cholerae*

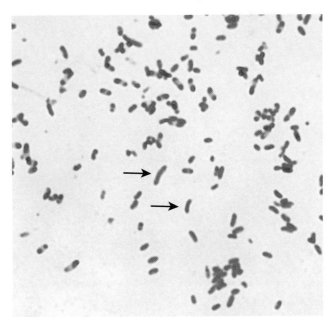

FIGURE 17–1 Gram stain of *Vibrio cholerae*. Often they are comma shaped or slightly curved (arrows) and 1 × 2–4 μm. Original magnification × 1000.

and most other vibrios grow well at 37°C on many kinds of media, including defined media containing mineral salts and asparagine as sources of carbon and nitrogen. *V cholerae* grows well on **thiosulfate-citrate-bile-sucrose (TCBS)** agar, on which it produces yellow colonies that are readily visible against the dark-green background of the agar (Figure 17–2). Vibrios are oxidase-positive, which differentiates them from enteric gram-negative bacteria. Characteristically, vibrios grow at a very high pH (8.5–9.5) and are rapidly killed by acid. Cultures containing fermentable carbohydrates therefore quickly become sterile.

In areas where cholera is endemic, direct cultures of stool on selective media, such as TCBS, and enrichment cultures in alkaline peptone water are appropriate. However, routine stool cultures on special media such as TCBS generally are not necessary or cost-effective in areas where cholera is rare.

C. Growth Characteristics

V cholerae regularly ferments sucrose and mannose but not arabinose. A positive oxidase test is a key step in the preliminary identification of *V cholerae* and other vibrios. *Vibrio* species are susceptible to the compound O/129 (2,4-diamino-6,7-diisopropylpteridine phosphate), which differentiates them from *Aeromonas* species, which are resistant to O/129. Most *Vibrio* species are halotolerant, and NaCl often stimulates their growth. Some vibrios are halophilic, requiring the presence of NaCl to grow. Another difference between vibrios and aeromonas is that vibrios grow on media containing 6% NaCl, whereas aeromonas does not.

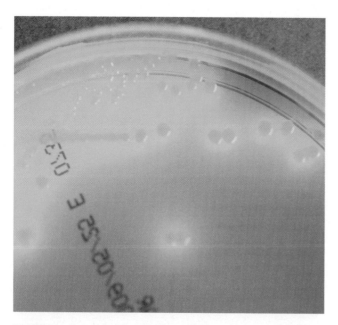

FIGURE 17–2 Colonies of *Vibrio cholerae* growing on thiosulfate, citrate, bile salts, and sucrose (TCBS) agar. The glistening yellow colonies are 2–3 mm in diameter and are surrounded by a diffuse yellowing of the indicator in the agar up to 1 cm in diameter. The plate is 10 cm in diameter.

Antigenic Structure & Biologic Classification

Many vibrios share a single heat-labile flagellar H antigen. Antibodies to the H antigen are probably not involved in the protection of susceptible hosts.

V cholerae has O lipopolysaccharides that confer serologic specificity. There are at least 139 O antigen groups. *V cholerae* strains of O group 1 and O group 139 cause classic cholera; occasionally, non-O1/non-O139 *V cholerae* causes cholera-like disease. Antibodies to the O antigens tend to protect laboratory animals against infections with *V cholerae*.

The *V cholerae* serogroup O1 antigen has determinants that make possible further typing; the serotypes are Ogawa, Inaba, and Hikojima. Two biotypes of epidemic *V cholerae* have been defined, classic and El Tor. The El Tor biotype produces a hemolysin, gives positive results on the Voges-Proskauer test, and is resistant to polymyxin B. Molecular techniques can also be used to type *V cholerae*. Typing is used for epidemiologic studies, and tests generally are done only in reference laboratories.

V cholerae O139 is very similar to *V cholerae* O1 El Tor biotype. *V cholerae* O139 does not produce the O1 lipopolysaccharide and does not have all the genes necessary to make this antigen. *V cholerae* O139 makes a polysaccharide capsule-like other non-O1 *V cholerae* strains, while *V cholerae* O1 does not make a capsule.

Vibrio cholerae Enterotoxin

V cholerae produce a heat-labile enterotoxin with a molecular weight of about 84,000, consisting of subunits A (MW 28,000) and B (see Chapter 9). Ganglioside GM_1 serves as the mucosal receptor for subunit B, which promotes entry of subunit A into the cell. Activation of subunit A_1 yields increased levels of intracellular cAMP and results in prolonged hypersecretion of water and electrolytes. There is increased sodium-dependent chloride secretion, and absorption of sodium and chloride is inhibited. Diarrhea occurs—as much as 20–30 L/day— with resulting dehydration, shock, acidosis, and death. The genes for *V cholerae* enterotoxin are on the bacterial chromosome. Cholera enterotoxin is antigenically related to LT of *Escherichia coli* and can stimulate the production of neutralizing antibodies. However, the precise role of antitoxic and antibacterial antibodies in protection against cholera is not clear.

Pathogenesis & Pathology

Under natural conditions, *V cholerae* is pathogenic only for humans. A person with normal gastric acidity may have to ingest as many as 10^{10} or more *V cholerae* to become infected when the vehicle is water, because the organisms are susceptible to acid. When the vehicle is food, as few as 10^2–10^4 organisms are necessary because of the buffering capacity of food. Any medication or condition that decreases stomach acidity makes a person more susceptible to infection with *V cholerae*.

Cholera is not an invasive infection. The organisms do not reach the bloodstream but remain within the intestinal tract. Virulent *V cholerae* organisms attach to the microvilli of the brush border of epithelial cells. There they multiply and liberate cholera toxin and perhaps mucinases and endotoxin.

Clinical Findings

About 60% of infections with classic *V cholerae* are asymptomatic, as are about 75% of infections with the El Tor biotype. The incubation period is 1–4 days for persons who develop symptoms, depending largely upon the size of the inoculum ingested. There is a sudden onset of nausea and vomiting and profuse diarrhea with abdominal cramps. Stools, which resemble "rice water," contain mucus, epithelial cells, and large numbers of vibrios. There is rapid loss of fluid and electrolytes, which leads to profound dehydration, circulatory collapse, and anuria. The mortality rate without treatment is between 25% and 50%. The diagnosis of a full-blown case of cholera presents no problem in the presence of an epidemic. However, sporadic or mild cases are not readily differentiated from other diarrheal diseases. The El Tor biotype tends to cause milder disease than the classic biotype.

Diagnostic Laboratory Tests

A. Specimens

Specimens for culture consist of mucus flecks from stools.

B. Smears

The microscopic appearance of smears made from stool samples is not distinctive. Dark-field or phase contrast microscopy may show the rapidly motile vibrios.

C. Culture

Growth is rapid in peptone agar, on blood agar with a pH near 9.0 or on TCBS agar, and typical colonies can be picked in 18 hours. For enrichment, a few drops of stool can be incubated for 6–8 hours in taurocholatepeptone broth (pH 8.0–9.0); organisms from this culture can be stained or subcultured.

D. Specific Tests

V cholerae organisms are further identified by slide agglutination tests using anti-O group 1 or group 139 antisera and by biochemical reaction patterns.

Immunity

Gastric acid provides some protection against cholera vibrios.

An attack of cholera is followed by immunity to reinfection, but the duration and degree of immunity are not known. In experimental animals, specific IgA antibodies occur in the lumen of the intestine. Similar antibodies in serum develop after infection but last only a few months. Vibriocidal antibodies in serum (titer ≥ 1:20) have been associated with protection against colonization and disease. The presence of antitoxin antibodies has not been associated with protection.

Treatment

The most important part of therapy consists of water and electrolyte replacement to correct the severe dehydration and salt depletion. Many antimicrobial agents are effective against *V cholerae*. Oral tetracycline tends to reduce stool output in cholera and shortens the period of excretion of vibrios. In some endemic areas, tetracycline resistance of *V cholerae* has emerged; the genes are carried by transmissible plasmids.

Epidemiology, Prevention, & Control

Six pandemics (worldwide epidemics) of cholera occurred between 1817 and 1923, caused most likely by *V cholerae* O1 of the classic biotype and largely originating in Asia, usually the Indian subcontinent. The seventh pandemic began in 1961 in the Celebes Islands, Indonesia, with spread to Asia, the Middle East, and Africa. This pandemic has been caused by *V cholerae* biotype El Tor. Starting in 1991, the seventh pandemic spread to Peru and then to other countries of South America and Central America. Cases also occurred in Africa. Millions of people have had cholera in this pandemic. Some consider the cholera caused by the serotype O139 strain to be the eighth pandemic that began in the Indian subcontinent in 1992–1993, with spread to Asia. The disease has been rare in North America since the mid-1800s, but an endemic focus exists on the Gulf Coast of Louisiana and Texas.

Cholera is endemic in India and Southeast Asia. From these centers, it is carried along shipping lanes, trade routes, and pilgrim migration routes. The disease is spread by contact involving individuals with mild or early illness and by water, food, and flies. In many instances, only 1–5% of exposed susceptible persons develop disease. The carrier state seldom exceeds 3–4 weeks, and the importance of carriers in transmission is unclear. Vibrios survive in water for up to 3 weeks.

V cholerae lives in aquatic environments. And such environments are the vibrios natural reservoir. *V cholerae* lives attached to algae, copepods, and crustacean shells. It can survive for years and grow, but when conditions are not suitable for growth it can become dormant.

Control rests on education and on improvement of sanitation, particularly of food and water. Patients should be isolated, their excreta disinfected, and contacts followed up. Chemoprophylaxis with antimicrobial drugs may have a place. Repeated injection of a vaccine containing either lipopolysaccharides extracted from vibrios or dense *Vibrio* suspensions can confer limited protection to heavily exposed persons (eg, family contacts) but is not effective as an epidemic control measure.

VIBRIO PARAHAEMOLYTICUS & OTHER VIBRIOS

Vibrio parahaemolyticus is a halophilic bacterium that causes acute gastroenteritis following ingestion of contaminated seafood such as raw fish or shellfish. After an incubation period of 12–24 hours, nausea and vomiting, abdominal cramps, fever, and watery to bloody diarrhea occur. Fecal leukocytes are often observed. The enteritis tends to subside spontaneously in 1–4 days with no treatment other than restoration of water and electrolyte balance. No enterotoxin has yet been isolated from this organism. The disease occurs worldwide, with highest incidence in areas where people eat raw seafood. *V parahaemolyticus* does not grow well on some of the differential media used to grow salmonellae and shigellae, but it does grow well on blood agar. It also grows well on TCBS, where it yields green colonies. *V parahaemolyticus* is usually identified by its oxidase-positive growth on blood agar.

Vibrio vulnificus can cause severe wound infections, bacteremia, and probably gastroenteritis. It is a free-living estuarine bacterium found in the United States on the Atlantic and Pacific Coasts, and especially the Gulf Coast. Infections have been reported from Korea, and the organism may be distributed worldwide. *V vulnificus* is particularly apt to be found in oysters, especially in warm months. Bacteremia with no focus of infection occurs in persons who have eaten infected oysters and who have alcoholism or liver disease. Wounds may become infected in normal or immunocompromised persons who are in contact with water where the bacterium is present. Infection often proceeds rapidly, with development of severe disease. About 50% of the patients with bacteremia die. Wound infections may be mild but often proceed rapidly (over a few hours), with development of bullous skin lesions, cellulitis, and myositis with necrosis. Several of the first deaths in Louisiana and Texas following hurricane Katrina were caused by *Vibrio vulnificus*. Because of the rapid progression of the infection, it is often necessary to treat with appropriate antibiotics before culture confirmation of the etiology can be obtained. Diagnosis is by culturing the organism on standard laboratory media; TCBS is the preferred medium for stool cultures, where most strains produce blue–green (sucrose-negative) colonies.

Tetracycline appears to be the drug of choice for *V vulnificus* infection; ciprofloxacin may be effective also based on in vitro activity.

Several other vibrios also cause disease in humans: *Vibrio mimicus* causes diarrhea after ingestion of uncooked seafood, particularly raw oysters. *Vibrio hollisae* and *Vibrio fluvialis* also cause diarrhea. *Vibrio alginolyticus* causes eye, ear, or wound infection after exposure to seawater. *Vibrio damsela* also causes wound infections. Other vibrios are very uncommon causes of disease in humans.

AEROMONAS

The taxonomy of the genus *Aeromonas* is in transition. The genus has been placed in the new family Aeromonadaceae from the family Vibrionaceae. Based on DNA hybridization groups, many genospecies have been recognized; some are renamed species, some are newly named, and some are not yet named. The following three groups are of primary clinical importance in human infections: *Aeromonas hydrophila* complex, *Aeromonas caviae* complex, and *Aeromonas veronii* biovar *sobria*.

Aeromonads are 1–4 μm long and are motile. Their colony morphology is similar to that of enteric gram-negative rods (Chapter 15), and they produce large zones of hemolysis on blood agar. *Aeromonas* species cultured from stool specimens grow readily on the differential media used to culture enteric gram-negative rods and can easily be confused with enteric bacteria. *Aeromonas* species are distinguished from the enteric gram-negative rods by finding a positive oxidase reaction in growth obtained from a blood agar plate. *Aeromonas* species are differentiated from vibrios by showing resistance to compound O/129 (see above) and lack of growth on media containing 6% NaCl.

Typically, aeromonads produce hemolysins. Some strains produce an enterotoxin. Cytotoxins and the ability to invade cells in tissue culture have been noted. However, none of these characteristics have been clearly shown to be associated with diarrheal disease in humans. Koch's postulates have not been satisfied, largely because there is no suitable animal model that reproduces human aeromonas-associated diarrhea.

Aeromonas strains are susceptible to tetracyclines, aminoglycosides, and cephalosporins.

PLESIOMONAS

Plesiomonas shigelloides is a gram-negative rod with polar flagella. Plesiomonas is most common in tropical and subtropical areas. It is a water and soil organism and has been isolated from freshwater fish and many animals. Most isolates from humans have been from stool cultures of patients with diarrhea. *Plesiomonas* grows on the differential media used to isolate *Salmonella* and *Shigella* from stool specimens (see Chapter 15). Some plesiomonas strains share antigens with *Shigella sonnei,* and cross-reactions with *Shigella* antisera occur. Plesiomonas can be distinguished from shigellae in diarrheal stools by the oxidase test: Plesiomonas is oxidase-positive and shigellae are not. Plesiomonas is positive for DNase; this and other biochemical tests distinguish it from aeromonas.

CAMPYLOBACTER

Campylobacters cause both diarrheal and systemic diseases and are among the most widespread causes of infection in the world. Campylobacter infection of domesticated animals also is widespread. The classification of bacteria within the family Campylobacteriaceae has changed frequently. Some species previously classified as campylobacters have been reclassified in the genus *Helicobacter*. The genus *Arcobacter* has been created. The organisms that cause intestinal or systemic illness are discussed in this section. *Helicobacter pylori,* which causes gastric infection, is discussed separately below. *C jejuni* is the prototype organism in the group and is a very common cause of diarrhea in humans.

CAMPYLOBACTER JEJUNI & CAMPYLOBACTER COLI

C jejuni and *Campylobacter coli* have emerged as common human pathogens, causing mainly enteritis and occasionally systemic infection. *C jejuni* and *C coli* cause infections that are clinically indistinguishable, and laboratories generally do not differentiate between the two species. Between 5% and 10% of infections reported to be caused by *C jejuni* are probably caused by *C coli.* These bacteria are at least as common as salmonellae and shigellae as a cause of diarrhea; an estimated 2 million cases occur in the United States each year.

Morphology & Identification

A. Typical Organisms

C jejuni and the other campylobacters are gram-negative rods with comma, S, or "gull wing" shapes (Figure 17–3). They are motile, with a single polar flagellum, and do not form spores.

B. Culture

The culture characteristics are most important in the isolation and identification of *C jejuni* and the other campylobacters. Selective media are needed, and incubation must

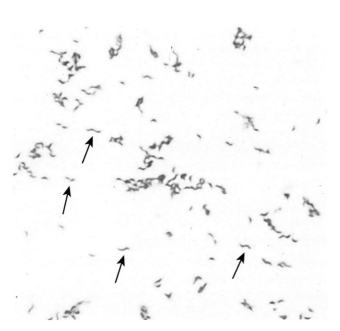

FIGURE 17–3 Gram stain of *Campylobacter jejuni* showing "comma"- or "gull wing"-shaped gram-negative bacilli (arrows). Campylobacters stain faintly and can be difficult to visualize. Original magnification × 1000.

be in an atmosphere with reduced O_2 (5% O_2) with added CO_2 (10% CO_2). A relatively simple way to produce the incubation atmosphere is to place the plates in an anaerobe incubation jar without the catalyst and to produce the gas with a commercially available gas-generating pack or by gas exchange. Incubation of primary plates for isolation of *C jejuni* should be at 42°C. Although *C jejuni* grows well at 36–37°C, incubation at 42°C prevents growth of most of the other bacteria present in feces, thus simplifying the identification of *C jejuni.* Several selective media are in widespread use. Skirrow's medium contains vancomycin, polymyxin B, and trimethoprim to inhibit growth of other bacteria. Other selective media also contain antimicrobials, including cephalothin or cefoperazone, and inhibitory compounds; because they contain a cephalosporin, they will not grow *Campylobacter fetus* and several other *Campylobacter* species. The selective media are suitable for isolation of *C jejuni* at 42°C; when media without antibiotics are incubated at 36–37°C, other campylobacters may be isolated. The colonies tend to be colorless or gray. They may be watery and spreading or round and convex, and both colony types may appear on one agar plate.

C. Growth Characteristics

Because of the selective media and incubation conditions for growth, an abbreviated set of tests is usually all that is necessary for identification. *C jejuni* and the other campylobacters pathogenic for humans are oxidase- and catalase-positive. Campylobacters do not oxidize or ferment carbohydrates. Gram-stained smears show typical morphology. Nitrate reduction, hydrogen sulfide production, hippurate tests, and antimicrobial susceptibilities can be used for further identification of species.

Antigenic Structure & Toxins

The campylobacters have lipopolysaccharides with endotoxic activity. Cytopathic extracellular toxins and enterotoxins have been found, but the significance of the toxins in human disease is not well defined.

Pathogenesis & Pathology

The infection is acquired by the oral route from food, drink, or contact with infected animals or animal products. *C jejuni* is susceptible to gastric acid, and ingestion of about 10^4 organisms is usually necessary to produce infection. This inoculum is similar to that required for *Salmonella* and *Shigella* infection but less than that for *Vibrio* infection. The organisms multiply in the small intestine, invade the epithelium, and produce inflammation that results in the appearance of red and white blood cells in the stools. Occasionally, the bloodstream is invaded and a clinical picture of enteric fever develops. Localized tissue invasion coupled with the toxic activity appears to be responsible for the enteritis.

Clinical Findings

Clinical manifestations are acute onset of crampy abdominal pain, profuse diarrhea that may be grossly bloody, headache, malaise, and fever. Usually the illness is self-limited to a period of 5–8 days, but occasionally it continues longer. *C jejuni* isolates are usually susceptible to erythromycin, and therapy shortens the duration of fecal shedding of bacteria. Most cases resolve without antimicrobial therapy.

Diagnostic Laboratory Tests

A. Specimens

Diarrheal stool is the usual specimen. Campylobacters from other types of specimens are usually incidental findings or are found in the setting of known outbreaks of disease.

B. Smears

Gram-stained smears of stool may show the typical "gull wing"-shaped rods. Dark-field or phase contrast microscopy may show the typical darting motility of the organisms.

C. Culture

Culture on the selective media described above is the definitive test to diagnose *C jejuni* enteritis. If another species of *Campylobacter* is suspected, medium without a cephalosporin should be used and incubated at 36–37°C.

Epidemiology & Control

Campylobacter enteritis resembles other acute bacterial diarrheas, particularly shigella dysentery. The source of infection may be food (eg, milk, undercooked fowl) or contact with infected animals or humans and their excreta. Outbreaks arising from a common source, eg, unpasteurized milk, may require public health control measures.

CAMPYLOBACTER FETUS

C fetus subspecies *fetus* is an opportunistic pathogen that causes systemic infections in immunocompromised patients. It may occasionally cause diarrhea. The gastrointestinal tract may be the portal of entry when *C fetus* causes bacteremia and systemic infection. *C fetus* has several surface array proteins (S protein, MW 100,000–149,000) which form a capsule-like structure on the surface of the organism (as compared with the polysaccharide capsules of pathogens such as *Neisseria meningitidis* and *Streptococcus pneumoniae*). In a mouse model of *C fetus* infection, the presence of the S protein as a surface capsule correlated with the ability of the bacteria to cause bacteremia after oral challenge and cause death in a high percentage of the animals.

OTHER CAMPYLOBACTERS

Campylobacter species other than *C jejuni* are encountered infrequently. This is partially due to the standard methods used for isolation of campylobacters from stool specimens: incubation at 42°C and use of medium containing a cephalosporin. *Campylobacter lari* is often found in seagulls and occasionally causes diarrhea in humans. *Campylobacter upsaliensis* from dogs occasionally causes diarrhea in humans. *Helicobacter fennelliae* and *Helicobacter cinaedi* can cause either diarrheal or extraintestinal disease. The *Arcobacter* species are uncommon enteric pathogens.

HELICOBACTER PYLORI

Helicobacter pylori is a spiral-shaped gram-negative rod. *H pylori* is associated with antral gastritis, duodenal (peptic) ulcer disease, gastric ulcers, and gastric carcinoma. Other *Helicobacter* species that infect the gastric mucosa exist but are rare.

Morphology & Identification

A. Typical Organisms

H pylori has many characteristics in common with campylobacters. It has multiple flagella at one pole and is actively motile.

B. Culture

Culture sensitivity can be limited by prior therapy, contamination with other mucosal bacteria, and other factors. *H pylori* grows in 3–6 days when incubated at 37°C in a microaerophilic environment, as for *C jejuni*. The media for primary isolation include Skirrow's medium with vancomycin, polymyxin B, and trimethoprim, chocolate medium, and other selective media with antibiotics (eg, vancomycin, nalidixic acid, amphotericin). The colonies are translucent and 1–2 mm in diameter.

C. Growth Characteristics

H pylori is oxidase-positive and catalase-positive, has a characteristic morphology, is motile, and is a strong producer of urease.

Pathogenesis & Pathology

H pylori grows optimally at a pH of 6.0–7.0 and would be killed or not grow at the pH within the gastric lumen. Gastric mucus is relatively impermeable to acid and has a strong buffering capacity. On the lumen side of the mucus, the pH is low (1.0–2.0) while on the epithelial side the pH is about 7.4. *H pylori* is found deep in the mucous layer near the epithelial surface where physiologic pH is present. *H pylori* also produces a protease that modifies the gastric mucus and further reduces the ability of acid to diffuse through the mucus. *H pylori* produces potent urease activity, which yields production of ammonia and further buffering of acid. *H pylori* is quite motile, even in mucus, and is able to find its way to the epithelial surface. *H pylori* overlies gastric-type but not intestinal-type epithelial cells.

In human volunteers, ingestion of *H pylori* resulted in development of gastritis and hypochlorhydria. There is a strong association between the presence of *H pylori* infection and duodenal ulceration. Antimicrobial therapy results in clearing of *H pylori* and improvement of gastritis and duodenal ulcer disease.

The mechanisms by which *H pylori* causes mucosal inflammation and damage are not well defined but probably involve both bacterial and host factors. The bacteria invade the epithelial cell surface to a limited degree. Toxins and lipopolysaccharide may damage the mucosal cells, and the ammonia produced by the urease activity may directly damage the cells also.

Histologically, gastritis is characterized by chronic and active inflammation. Polymorphonuclear and mononuclear cell infiltrates are seen within the epithelium and lamina propria. Vacuoles within cells are often pronounced. Destruction of the epithelium is common, and glandular atrophy may occur. *H pylori* thus may be a major risk factor for gastric cancer.

Clinical Findings

Acute infection can yield an upper gastrointestinal illness with nausea and pain; vomiting and fever may be present also. The acute symptoms may last for less than 1 week or as long as 2 weeks. Once colonized, the *H pylori* infection persists for years and perhaps decades or even a lifetime. About 90% of patients with duodenal ulcers and 50–80% of those with gastric ulcers have *H pylori* infection. *H pylori* also may have a role in gastric carcinoma and lymphoma.

Diagnostic Laboratory Tests

A. Specimens

Gastric biopsy specimens can be used for histologic examination or minced in saline and used for culture. Blood is collected for determination of serum antibodies.

B. Smears

The diagnosis of gastritis and *H pylori* infection can be made histologically. A gastroscopy procedure with biopsy is required. Routine stains demonstrate gastritis, and Giemsa or special silver stains can show the curved or spiraled organisms.

C. Culture

As above.

D. Antibodies

Several assays have been developed to detect serum antibodies specific for *H pylori*. The serum antibodies persist even if the *H pylori* infection is eradicated, and the role of antibody tests in diagnosing active infection or following therapy is therefore limited.

E. Special Tests

Rapid tests to detect urease activity are widely used for presumptive identification of *H pylori* in specimens. Gastric biopsy material can be placed onto a urea-containing medium with a color indicator. If *H pylori* is present, the urease rapidly splits the urea (1–2 days) and the resulting shift in pH yields a color change in the medium. In vivo tests for urease activity can be done also. ^{13}C- or ^{14}C-labeled urea is ingested by the patient. If *H pylori* is present, the urease activity generates labeled CO_2 that can be detected in the patient's exhaled breath.

Detection of *H pylori* antigen in stool specimens is appropriate as a test of cure for patients with known *H pylori* infection who have been treated.

Immunity

Patients infected with *H pylori* develop an IgM antibody response to the infection. Subsequently, IgG and IgA are produced, and these persist, both systemically and at the mucosa, in high titer in chronically infected persons. Early antimicrobial treatment of *H pylori* infection blunts the antibody response; such patients are thought to be subject to repeat infection.

Treatment

Triple therapy with metronidazole and either bismuth subsalicylate or bismuth subcitrate plus either amoxicillin or tetracycline for 14 days eradicates *H pylori* infection in 70–95% of patients. An acid-suppressing agent given for 4–6 weeks enhances ulcer healing. Proton pump inhibitors directly inhibit *H pylori* and appear to be potent urease inhibitors. Either 1 week of a proton pump inhibitor plus amoxicillin and clarithromycin or of amoxicillin plus metronidazole also is highly effective.

Epidemiology & Control

H pylori is present on the gastric mucosa of less than 20% of persons under age 30 but increases in prevalence to 40–60% of persons age 60, including persons who are asymptomatic. In developing countries, the prevalence of infection may be 80% or higher in adults. Person-to-person transmission of *H pylori* is likely because intrafamilial clustering of infection occurs. Acute epidemics of gastritis suggest a common source for *H pylori*.

REVIEW QUESTIONS

1. Long-term carriage and shedding is most likely to occur after gastrointestinal infection with which of the following species?

   (A) *Escherichia coli* O157:H7
   (B) *Shigella dysenteriae*
   (C) *Vibrio cholerae*
   (D) *Campylobacter jejuni*
   (E) *Salmonella typhi*

2. A 63-year-old man had visited his favorite oyster restaurant in a small town on the eastern shore of the Gulf Coast of Texas. He ate two dozen oysters. Two days later he was admitted to the hospital because of the abrupt onset of chills, fever, and lightheadedness when he stood up. (In the emergency room, his blood pressure was 60/40 mm Hg.) While in the emergency room, he developed erythematous skin lesions. These rapidly evolved into hemorrhagic bullae, which then formed ulcers. The man drank a six-pack of beer and one half-bottle of whisky each day. A microorganism of major concern for this patient is

   (A) *Vibrio vulnificus*
   (B) *Escherichia coli*
   (C) *Salmonella typhi*
   (D) *Clostridium perfringens*
   (E) *Streptococcus pyogenes* (group A streptococci)

3. A 10-year-old boy was playing in a slowly moving stream when he cut his foot on a sharp object. Three days later he was brought to the emergency room because of pain and swelling at the site of the wound and drainage of pus from it. The most likely cause of the infection is

   (A) *Vibrio vulnificus*
   (B) *Escherichia coli*
   (C) *Aeromonas hydrophila*
   (D) *Proteus mirabilis*
   (E) *Salmonella typhimurium*

4. A family of four persons ate a meal that included undercooked chicken. Within 3 days, three members developed an illness characterized by fever, headache, myalgia, and malaise. Two of the patients had concomitant diarrhea and abdominal pain. The third person developed diarrhea after the systemic symptoms had cleared. Stool cultures grew *Campylobacter jejuni*. Which of the following culture conditions was most likely used to isolate *Campylobacter jejuni*?

   (A) Thiosulfate-citrate-bile-sucrose medium incubated at 37°C in 5% oxygen and 10% CO_2
   (B) *Salmonella-Shigella* selective medium incubated at 37°C in ambient air
   (C) MacConkey agar and Hektoen enteric agar incubated at 42°C in 5% oxygen and 10% CO_2
   (D) 5% sheep blood agar incubated at 37°C in ambient air
   (E) A medium containing vancomycin, polymyxin B, and trimethoprim incubated at 42°C in 5% oxygen and 10% CO_2

5. Bacteremia associated with a gastrointestinal infection is most likely to occur with which of the following?

   (A) *Salmonella typhi*
   (B) *Vibrio cholerae*
   (C) *Shigella boydii*
   (D) *Vibrio parahaemolyticus*
   (E) *Campylobacter jejuni*

6. During the El Niño years in the mid- to late 1990s, the waters of Puget Sound between Washington State and British Columbia warmed considerably. During this time, many people who ate clams and oysters from these waters became ill with a disease characterized by explosive diarrhea and moderately severe abdominal cramps. The diarrhea was usually watery, but in some patients it was bloody. The diarrhea usually had onset within 24 hours after eating the shellfish. Stool cultures typically yielded a pathogenic gram-negative bacillus. The microorganism of concern in this setting is

   (A) Enterotoxigenic *Escherichia coli*
   (B) *Vibrio cholerae*
   (C) Enterohemorrhagic *Escherichia coli*
   (D) *Vibrio parahaemolyticus*
   (E) *Shigella dysenteriae*

7. A patient presents to the emergency room with non bloody diarrhea for 12 hours. The patient lives in Washington D.C. and has not recently traveled out of the area. Which one of the following is *unlikely* to be the cause of your patient's diarrhea?

   (A) *Salmonella typhimurium*
   (B) *Campylobacter jejuni*
   (C) *Shigella sonnei*
   (D) *Vibrio cholerae*

8. An 18-year-old woman in rural Bangladesh develops profuse (8 L/d) diarrhea. She has no symptoms other than the diarrhea and the manifestations of the fluid and electrolyte loss caused by the diarrhea. The most likely cause of her diarrhea is

   (A) *Campylobacter jejuni*
   (B) Enterotoxigenic *Escherichia coli*
   (C) *Salmonella typhimurium*
   (D) *Vibrio cholerae*
   (E) *Shigella dysenteriae*

9. Age and geography are major factors in the prevalence of colonization by *Helicobacter pylori*. In developing countries, the prevalence of colonization may be > 80% in adults. In the United States, the prevalence of colonization with this microorganism in adults over age 60 is

   (A) 1–2%
   (B) 5–10%
   (C) 15–20%
   (D) 40–60%
   (E) 80–95%

10. A 59-year-old man comes to the emergency room in the afternoon because of acute swelling and pain in his right leg. Earlier that morning, he had been working on a small sport fishing boat in an estuary on the Gulf Coast of Texas. While walking around the boat in shallow water, he scratched his leg, breaking the skin at the site of the current pain and swelling. He was not wearing boots. About 1 hour after the injury, the scratch became red and painful. Swelling developed. Within 3 hours, the leg below the knee had become markedly swollen. The skin was red and tender. There was serous drainage from the wound, which had ulcerated and was now much enlarged. Near the wound, bullae were forming—the largest approximately 2.5 cm in diameter. The most likely cause of this medical emergency is

    (A) *Staphylococcus aureus*
    (B) *Streptococcus pyogenes*
    (C) *Clostridium perfringens*
    (D) *Escherichia coli*
    (E) *Vibrio vulnificus*

11. The *Vibrio cholerae* factor responsible for diarrhea is a toxin that
    (A) Blocks EF-2
    (B) Yields increased intracellular levels of cAMP
    (C) Cleaves SNARE
    (D) Blocks EF-1-dependent binding of amino-acyl-tRNA to ribosomes
    (E) Cleaves VAMP

12. In September 1854, a severe epidemic of cholera occurred in the Soho/Golden Square area of London. Dr. John Snow, a father of epidemiology, studied the epidemic and helped stop it by which of the following actions
    (A) Banning the sale of apples at the local markets
    (B) Removing the handle of the Broad Street water pump
    (C) Stopping the sale of shellfish imported from Normandy
    (D) Pasteurizing milk

13. A 45-year-old man develops a gastric ulcer that can be visualized on a contrast medium-enhanced x-ray film of his stomach. A biopsy specimen is taken from the gastric mucosa at the site of the ulcer. A presumptive diagnosis can be reached most rapidly by inoculating part of the specimen on which of the following?
    (A) A medium used to detect urease incubated at 37°C
    (B) A medium containing vancomycin, polymyxin B, and trimethoprim incubated at 42°C
    (C) MacConkey agar medium incubated at 37°C
    (D) Thiosulfate-citrate-bile-sucrose medium incubated at 42°C
    (E) Blood agar medium incubated at 37°C

14. Which of the following is useful in differentiating *Vibrio* species from *Aeromonas* species?
    (A) Growth or no growth on a medium containing NaCl
    (B) The presence or absence on endotoxin in the cell wall.
    (C) Production or lack of production of a heat-labile enterotoxin.
    (D) Positivity or negativity on the enterocyte invasion test.

15. In the United States, public health officials often warn people to thoroughly cook chicken commercially obtained from supermarkets and stores. What percentage of chickens obtained from these sources are likely to be contaminated with *Campylobacter jejuni*?
    (A) 1–5%
    (B) 6–15%
    (C) 15–30%
    (D) 30–50%
    (E) > 50%

Answers

1. E	5. A	9. D	13. A
2. A	6. D	10. E	14. A
3. C	7. D	11. B	15. E
4. E	8. D	12. B	

REFERENCES

Abbott SL, Janda JM, Johnson JA Farmer JJ III: *Vibrio.* In: *Manual of Clinical Microbiology*, 9th ed. Murray PR et al (editors). ASM Press, 2007.

Allos BM, Blaser MJ: *Campylobacter jejuni* and related species. In: *Mandell, Douglas and Bennett's Principles and Practice of Infectious Diseases*, 7th ed. Mandell GL, et al (editor). Elsevier, 2010.

Blaser MJ: *Helicobacter pylori* and other gastric *Helicobacter* species. In: *Mandell, Douglas and Bennett's Principles and Practice of Infectious Diseases*, 7th ed. Mandell GL, et al (editors). Elsevier, 2010.

Fitzgerald C, Nachamkin I: *Campylobacter* and *Arcobacter.* In: *Manual of Clinical Microbiology*, 9th ed. Murray PR et al (editors). ASM Press, 2007.

Fox JG, Megraud F: *Helicobacter.* In: *Manual of Clinical Microbiology*, 9th ed. Murray PR et al (editors). ASM Press, 2007.

Horneman AJ, Ali A, Abbott SL: *Aeromonas.* In: *Manual of Clinical Microbiology*, 9th ed. Murray PR et al (editors). ASM Press, 2007.

Neill MA, Carpenter CCJ: Other pathogenic vibrios. In: *Mandell, Douglas and Bennett's Principles and Practice of Infectious Diseases*, 7th ed. Mandell GL, et al (editors). Elsevier, 2010.

Seas C, Gotuzzo E: *Vibrio cholerae.* In: *Mandell, Douglas and Bennett's Principles and Practice of Infectious Diseases*, 7th ed. Mandell GL, et al (editors). Elsevier, 2010.

Steinberg JP, Burd EM: Other gram-negative and gram-variable bacilli. In: *Mandell, Douglas and Bennett's Principles and Practice of Infectious Diseases*, 7th ed. Mandell GL, et al (editors). Elsevier, 2010.

Haemophilus, Bordetella, Brucella, & Francisella

THE *HAEMOPHILUS* SPECIES

This is a group of small, gram-negative, pleomorphic bacteria that require enriched media, usually containing blood or its derivatives, for isolation. *Haemophilus influenzae* type b is an important human pathogen; *Haemophilus ducreyi,* a sexually transmitted pathogen, causes chancroid; other *Haemophilus* species are among the normal flora of mucous membranes and only occasionally cause disease.

HAEMOPHILUS INFLUENZAE

Haemophilus influenzae is found on the mucous membranes of the upper respiratory tract in humans. It is an important cause of meningitis in children and occasionally causes respiratory tract infections in children and adults.

Morphology & Identification

A. Typical Organisms

In specimens from acute infections, the organisms are short (1.5 μm) coccoid bacilli, sometimes occurring in pairs or short chains. In cultures, the morphology depends both on age and on the medium. At 6–8 hours in rich medium, the small coccobacillary forms predominate. Later there are longer rods, lysed bacteria, and very pleomorphic forms.

Organisms in young cultures (6–18 hours) on enriched medium have a definite capsule. The capsule is the antigen used for "typing" *H influenzae* (see below).

B. Culture

On chocolate agar, flat, grayish brown colonies with diameters of 1–2 mm are present after 24 hours of incubation. IsoVitaleX in media enhances growth. *H influenzae* does not grow on sheep blood agar except around colonies of staphylococci ("satellite phenomenon"). *H haemolyticus* and *H parahaemolyticus* are hemolytic variants of *H influenzae* and *H parainfluenzae,* respectively.

C. Growth Characteristics

Identification of organisms of the *Haemophilus* group depends in part upon demonstrating the need for certain growth factors called X and V. Factor X acts physiologically as hemin; factor V can be replaced by nicotinamide adenine nucleotide (NAD) or other coenzymes. Colonies of staphylococci on sheep blood agar cause the release of NAD, yielding the satellite growth phenomenon. The requirements for X and V factors of various *Haemophilus* species are listed in Table 18–1. Carbohydrates are fermented poorly and irregularly.

D. Variation

In addition to morphologic variation, *H influenzae* has a marked tendency to lose its capsule and the associated type specificity. Nonencapsulated variant colonies lack iridescence.

E. Transformation

Under proper experimental circumstances, the DNA extracted from a given type of *H influenzae* is capable of transferring that type specificity to other cells (transformation). Resistance to ampicillin and chloramphenicol is controlled by genes on transmissible plasmids.

Antigenic Structure

Encapsulated *H influenzae* contains **capsular polysaccharides** (MW >150,000) of one of six types (a–f). The capsular antigen of type b is a polyribose-ribitol phosphate (PRP). Encapsulated *H influenzae* can be typed by slide agglutination, coagglutination with staphylococci, or agglutination of latex particles coated with type-specific antibodies. A capsule swelling test with specific antiserum is analogous to the quellung test for pneumococci. Typing can also be done by immunofluorescence. Most *H influenzae* organisms in the normal flora of the upper respiratory tract are not encapsulated.

The somatic antigens of *H influenzae* consist of outer membrane proteins. Lipooligosaccharides (endotoxins) share many structures with those of neisseriae.

Pathogenesis

H influenzae produces no exotoxin. The nonencapsulated organism is a regular member of the normal respiratory

TABLE 18–1   Characteristics and Growth Requirements of the *Haemophilus* Species Important to Humans

Species	Requires		Hemolysis
	X	V	
Haemophilus influenzae (H aegyptius)	+	+	−
Haemophilus parainfluenzae	−	+	−
Haemophilus ducreyi	+	−	−
Haemophilus haemolyticus	+	+	+
Aggregatibacter aphrophilus[a]	−	−	−
Haemophilus paraphrophaemolyticus	−	+	+
Haemophilus segnis	−	+	−

X, heme; V, nicotinamide-adenine dinucleotide.

[a] Now called *Aggregatibacter.*

microbiota of humans. The capsule is antiphagocytic in the absence of specific anticapsular antibodies. The polyribose phosphate capsule of type b *H influenzae* is the major virulence factor.

The carrier rate in the upper respiratory tract for *H influenzae* type b is 2–4%. The carrier rate for nontypeable *H influenzae* is 50–80% or higher. Type b *H influenzae* causes meningitis, pneumonia and empyema, epiglottitis, cellulitis, septic arthritis, and occasionally other forms of invasive infection. Nontypeable *H influenzae* tends to cause chronic bronchitis, otitis media, sinusitis, and conjunctivitis following breakdown of normal host defense mechanisms. The carrier rate for the encapsulated types a and c–f is low (1–2%), and these capsular types rarely cause disease. Although type b can cause chronic bronchitis, otitis media, sinusitis, and conjunctivitis, it does so much less commonly than nontypeable *H influenzae*. Similarly, nontypeable *H influenzae* only occasionally causes invasive disease (about 5% of cases).

The blood of many persons over age 3–5 years is bactericidal for *H influenzae*, and clinical infections are less frequent in such individuals. However, bactericidal antibodies have been absent from 25% of adults in the United States, and clinical infections have occurred in adults.

Clinical Findings

H influenzae type b enters by way of the respiratory tract. There may be local extension with involvement of the sinuses or the middle ear. *H influenzae* type b and pneumococci are two of the most common etiologic agents of bacterial otitis media and acute sinusitis. The organisms may reach the bloodstream and be carried to the meninges or, less frequently, may establish themselves in the joints to produce septic arthritis. Prior to the use of the conjugate vaccine, *H influenzae* was the most common cause of bacterial meningitis in children age

5 months to 5 years in the United States. Clinically, it resembles other forms of childhood meningitis, and diagnosis rests on bacteriologic demonstration of the organism.

Occasionally, a fulminating obstructive laryngotracheitis with swollen, cherry-red epiglottis develops in infants and requires prompt tracheostomy or intubation as a lifesaving procedure. Pneumonitis and epiglottitis due to *H influenzae* may follow upper respiratory tract infections in small children and old or debilitated people. Adults may have bronchitis or pneumonia due to *H influenzae*.

Diagnostic Laboratory Tests

A. Specimens

Specimens consist of nasopharyngeal swabs, pus, blood, and spinal fluid for smears and cultures.

B. Direct Identification

Commercial kits are available for immunologic detection of *H influenzae* antigens in spinal fluid. A positive test indicates that the fluid contains high concentrations of specific polysaccharide from *H influenzae* type b. These antigen detection tests generally are not more sensitive than a Gram stain and therefore are not widely used. A Gram stain of *H influenzae* in sputum is depicted in Figure 18–1.

C. Culture

Specimens are grown on IsoVitaleX-enriched chocolate agar until typical colonies appear. *H influenzae* is differentiated from related gram-negative bacilli by its requirements for

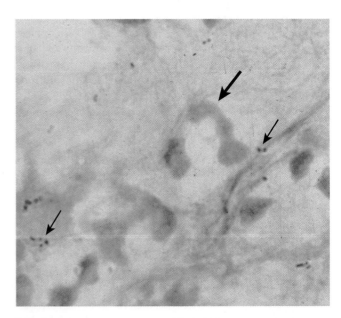

FIGURE 18–1 Gram stain of *Haemophilus influenzae* in sputum. The organisms are very small (0.3 × 1 μm) gram-negative coccobacilli (small arrows). The large irregularly-shaped objects (large arrow) are the nuclei of polymorphonuclear cells. Mucus is faintly stained pink in the background.

X and V factors and by its lack of hemolysis on blood agar (Table 18–1).

Tests for X (heme) and V (nicotinamide-adenine dinucleotide) factor requirements can be done in several ways. The *Haemophilus* species that require V factor grow around paper strips or disks containing V factor placed on the surface of agar that has been autoclaved before the blood was added (V factor is heat-labile). Alternatively, a strip containing X factor can be placed in parallel with one containing V factor on agar deficient in these nutrients. Growth of *Haemophilus* in the area between the strips indicates requirement for both factors. A better test for X factor requirement is based on the inability of *H influenzae* (and a few other *Haemophilus* species) to synthesize heme from δ-aminolevulinic acid. The inoculum is incubated with the δ-aminolevulinic acid. *Haemophilus* organisms that do not require X factor synthesize porphobilinogen, porphyrins, protoporphyrin IX, and heme. The presence of red fluorescence under ultraviolet light (approximately 360 nm) indicates the presence of porphyrins and a positive test. *Haemophilus* species that synthesize porphyrins (and thus heme) are not *H influenzae*. (See Table 18–1.)

Immunity

Infants under age 3 months may have serum antibodies transmitted from the mother. During this time *H influenzae* infection is rare, but subsequently the antibodies are lost. Children often acquire *H influenzae* infections, which are usually asymptomatic but may be in the form of respiratory disease or meningitis. *H influenzae* has been the most common cause of bacterial meningitis in children from 5 months to 5 years of age. By age 3–5 years, many unimmunized children have naturally acquired anti-PRP antibodies that promote complement-dependent bactericidal killing and phagocytosis. Immunization of children with *H influenzae* type b conjugate vaccine induces the same antibodies.

There is a correlation between the presence of bactericidal antibodies and resistance to major *H influenzae* type b infections. However, it is not known whether these antibodies alone account for immunity. Pneumonia or arthritis due to *H influenzae* can develop in adults with such antibodies.

Treatment

The mortality rate of untreated *H influenzae* meningitis may be up to 90%. Many strains of *H influenzae* type b are susceptible to ampicillin, but up to 25% produce β-lactamase under control of a transmissible plasmid and are resistant. Essentially all strains are susceptible to the third-generation cephalosporins. Cefotaxime given intravenously gives excellent results. Prompt diagnosis and antimicrobial therapy are essential to minimize late neurologic and intellectual impairment. Prominent among late complications of *H influenzae* type b meningitis is the development of a localized subdural accumulation of fluid that requires surgical drainage.

Epidemiology, Prevention, & Control

Encapsulated *H influenzae* type b is transmitted from person to person by the respiratory route. *H influenzae* type b disease can be prevented by administration of **Haemophilus b conjugate vaccine** to children. Currently three conjugate vaccines are available for use: PRPHbOC, in which the conjugate is CRM197, a nontoxic diphtheria toxin; PRP-OMPC, the outer membrane protein complex of *Neisseria meningitidis*; and PRP-T, which uses tetanus toxoid. Beginning at age 2 months, all children should be immunized with one of the conjugate vaccines. Depending upon which vaccine product is chosen, the series consists of three doses at 2, 4, and 6 months of age or two doses given at 2 and 4 months. An additional booster dose is given sometime between 12 and 15 months of age. All three conjugate vaccines can be given at the time of other vaccine administration such as DTaP. Widespread use of *H influenzae* type b vaccine has reduced the incidence of *H influenzae* type b meningitis in children by over 95%. The vaccine reduces the carrier rates for *H influenzae* type b.

Contact with patients suffering from *H influenzae* clinical infection poses little risk for adults but presents a definite risk for nonimmune siblings and other nonimmune children under age 4 years who are close contacts. Prophylaxis with rifampin is recommended for such children.

HAEMOPHILUS AEGYPTIUS

This organism was formerly called the Koch-Weeks bacillus; it is sometimes called *H influenzae* biotype III, but the current designation is *H influenzae* biogroup *aegyptius*. It resembles *H influenzae* closely and has been associated with a highly communicable form of conjunctivitis. *H aegyptius* is the cause of Brazilian purpuric fever, a disease of children characterized by fever, purpura, shock, and death.

AGGREGATIBACTER APHROPHILUS

Organisms belonging to the species *Haemophilus aphrophilus* and *Haemophilus paraphrophilus* were recently combined into the same species and the name was changed to *Aggregatibacter aphrophilus*. *Actinobacillus actinomycetemcomitans* has also been added to the genus *Aggregatibacter*. *A aphrophilus* isolates are often encountered as causes of infective endocarditis and pneumonia. These organisms are present in the oral cavity as part of the normal respiratory microbiota.

HAEMOPHILUS DUCREYI

Haemophilus ducreyi causes chancroid (soft chancre), a sexually transmitted disease. Chancroid consists of a ragged ulcer on the genitalia, with marked swelling and tenderness. The regional lymph nodes are enlarged and painful. The disease must be differentiated from syphilis, herpes simplex infection, and lymphogranuloma venereum.

The small gram-negative rods occur in strands in the lesions, usually in association with other pyogenic microorganisms. *H ducreyi* requires X factor but not V factor. It is grown best from scrapings of the ulcer base on chocolate agar containing 1% IsoVitaleX and vancomycin, 3 μg/mL, and incubated in 10% CO_2 at 33°C. There is no permanent immunity following chancroid infection. Treatment with intramuscular ceftriaxone, oral trimethoprim-sulfamethoxazole, or oral erythromycin often results in healing in 2 weeks.

OTHER *HAEMOPHILUS SPECIES*

Haemophilus haemoglobinophilus requires X factor but not V factor and has been found in dogs but not in human disease. *Haemophilus haemolyticus* is the most markedly hemolytic organism of the group in vitro; it occurs both in the normal nasopharynx and in association with rare upper respiratory tract infections of moderate severity in childhood. *Haemophilus parainfluenzae* resembles *H influenzae* and is a normal inhabitant of the human respiratory tract; it has been encountered occasionally in infective endocarditis and in urethritis. *H suis* resembles *H influenzae* bacteriologically and acts synergistically with swine influenza virus to produce the disease in hogs.

THE BORDETELLAE

There are several species of *Bordetella*. *Bordetella pertussis*, a highly communicable and important pathogen of humans, causes whooping cough (pertussis). *Bordetella parapertussis* can cause a similar disease. *Bordetella bronchiseptica* (*Bordetella bronchicanis*) causes diseases in animals such as kennel cough in dogs and snuffles in rabbits, and only occasionally causes respiratory disease and bacteremia in humans, primarily in immunocompromised hosts. *Bordetella avium* causes turkey coryza and is not known to infect humans. Newer species and their disease associations include *B hinzii* (bacteremia and respiratory illness), *B holmseii* (bacteremia among immunosuppressed patients), and *B trematum* (wounds and otitis media). *B pertussis*, *B parapertussis*, and *B bronchiseptica* are closely related, with 72–94% DNA homology and very limited differences in multilocus enzyme analysis; the three species might be considered three subspecies within a species. *B avium* is a distinct species.

BORDETELLA PERTUSSIS

Morphology & Identification

A. Typical Organisms

The organisms are minute gram-negative coccobacilli resembling *H influenzae*. With toluidine blue stain, bipolar metachromatic granules can be demonstrated. A capsule is present.

B. Culture

Primary isolation of *B pertussis* requires enriched media. Bordet-Gengou medium (potato-blood-glycerol agar) that contains penicillin G, 0.5 μg/mL, can be used; however, a charcoal-containing medium similar to that used for *Legionella pneumophila* is preferable. The plates are incubated at 35–37°C for 3–7 days in a moist environment (eg, a sealed plastic bag). The small, faintly staining gram-negative rods are identified by immunofluorescence staining. *B pertussis* is nonmotile.

C. Growth Characteristics

The organism is a strict aerobe and forms acid but not gas from glucose and lactose. It does not require X and V factors on subculture. Hemolysis of blood-containing medium is associated with virulent *B pertussis*.

D. Variation

When isolated from patients and cultured on enriched media, *B pertussis* is in the hemolytic and pertussis toxin-producing virulent phase. There are two mechanisms for *B pertussis* to shift to nonhemolytic, nontoxin-producing avirulent forms. Reversible phenotypic modulation occurs when *B pertussis* is grown under certain environmental conditions (eg, 28°C versus 37°C, the presence of $MgSO_4$, etc). Reversible phase variation follows a low-frequency mutation in the genetic locus that controls the expression of the virulence factors (see below). It is possible that these mechanisms play a role in the infectious process, but such a role has not been demonstrated clinically.

Antigenic Structure, Pathogenesis, & Pathology

B pertussis produces a number of factors that are involved in the pathogenesis of disease. One locus on the *B pertussis* chromosome acts as a central regulator of virulence genes. This locus has two *Bordetella* virulence genes, **bvgA** and **bvgS**. The products of the A and S loci are similar to those of known two-component regulatory systems. *bvgS* responds to environmental signals while *bvgA* is a transcriptional activator of the virulence genes. The **filamentous hemagglutinin** mediates adhesion to ciliated epithelial cells. **Pertussis toxin** promotes lymphocytosis, sensitization to histamine, and enhanced insulin secretion and has ADP-ribosylating activity, with an A/B structure and mechanism of action similar to that of cholera toxin. The filamentous hemagglutinin and pertussis toxin are secreted proteins and are found outside of the *B pertussis* cells. **Adenylate cyclase toxin, dermonecrotic toxin,** and **hemolysin** also are regulated by the bvg system. The **tracheal cytotoxin** inhibits DNA synthesis in ciliated cells and is not regulated by *bvg*. Pili probably play a role in adherence of the bacteria to the ciliated epithelial cells of the upper respiratory tract. The lipopolysaccharide in the cell wall may also be important in causing damage to the epithelial cells of the upper respiratory tract.

B pertussis survives for only brief periods outside the human host. There are no vectors. Transmission is largely by the respiratory route from early cases and possibly via carriers. The organism adheres to and multiplies rapidly on the epithelial surface of the trachea and bronchi and interferes with ciliary action. The blood is not invaded. The bacteria liberate the toxins and substances that irritate surface cells, causing coughing and marked lymphocytosis. Later, there may be necrosis of parts of the epithelium and polymorphonuclear infiltration, with peribronchial inflammation and interstitial pneumonia. Secondary invaders like staphylococci or *H influenzae* may give rise to bacterial pneumonia. Obstruction of the smaller bronchioles by mucous plugs results in atelectasis and diminished oxygenation of the blood. This probably contributes to the frequency of convulsions in infants with whooping cough.

Clinical Findings

After an incubation period of about 2 weeks, the "catarrhal stage" develops, with mild coughing and sneezing. During this stage, large numbers of organisms are sprayed in droplets, and the patient is highly infectious but not very ill. During the "paroxysmal" stage, the cough develops its explosive character and the characteristic "whoop" upon inhalation. This leads to rapid exhaustion and may be associated with vomiting, cyanosis, and convulsions. The "whoop" and major complications occur predominantly in infants; paroxysmal coughing predominates in older children and adults. The white blood count is high (16,000–30,000/μL), with an absolute lymphocytosis. Convalescence is slow. *B pertussis* is a common cause of prolonged (4–6 weeks) cough in adults. Rarely, whooping cough is followed by the serious and potentially fatal complication of encephalitis. Several types of adenovirus and *Chlamydia pneumoniae* can produce a clinical picture resembling that caused by *B pertussis*.

Diagnostic Laboratory Tests

A. Specimens

A saline nasal wash is the preferred specimen. Nasopharyngeal swabs or cough droplets expelled onto a "cough plate" held in front of the patient's mouth during a paroxysm are sometimes used but are not as good as the saline nasal wash.

B. Direct Fluorescent Antibody Test

The fluorescent antibody (FA) reagent can be used to examine nasopharyngeal swab specimens. However, false-positive and false-negative results may occur; the sensitivity is about 50%. The FA test is most useful in identifying *B pertussis* after culture on solid media.

C. Culture

The saline nasal wash fluid is cultured on solid medium agar (see above). The antibiotics in the media tend to inhibit other respiratory flora but permit growth of *B pertussis*. Organisms are identified by immunofluorescence staining or by slide agglutination with specific antiserum.

D. Polymerase Chain Reaction

PCR is the most sensitive method to diagnosis pertussis. Primers for both *B pertussis* and *B parapertussis* should be included. When available, the PCR test should replace the direct fluorescent antibody tests.

E. Serology

Serologic tests on patients are of little diagnostic help because a rise in agglutinating or precipitating antibodies does not occur until the third week of illness. A single serum with high titer antibodies may be helpful in diagnosing the cause of a long-term cough, one of several weeks' duration.

Immunity

Recovery from whooping cough or immunization is followed by immunity. Second infections may occur but are mild; reinfections occurring years later in adults may be severe. It is probable that the first defense against *B pertussis* infection is the antibody that prevents attachment of the bacteria to the cilia of the respiratory epithelium.

Treatment

B pertussis is susceptible to several antimicrobial drugs in vitro. Administration of erythromycin during the catarrhal stage of disease promotes elimination of the organisms and may have prophylactic value. Treatment after onset of the paroxysmal phase rarely alters the clinical course. Oxygen inhalation and sedation may prevent anoxic damage to the brain.

Prevention

Every infant should receive three injections of pertussis vaccine during the first year of life followed by a booster series for a total of five doses. There are multiple acellular pertussis vaccines licensed in the United States and elsewhere. Use of these vaccines is recommended. The acellular vaccines have at least two of the following antigens: inactivated pertussis toxin, filamentous hemagglutinin, fimbrial proteins, and pertactin.

Because different vaccines have different antigens, the same product should be used throughout an immunization series. Pertussis vaccine is usually administered in combination with toxoids of diphtheria and tetanus (DTaP). Five doses of pertussis vaccine are recommended prior to school entry. The usual schedule is administration of doses at 2, 4, 6, and

15–18 months of age and a booster dose at 4–6 years of age. In 2005, it was recommended by the Advisory Committee on Immunization Practices that all adolescents and adults receive a single booster dose of tetanus-diphtheria-acellular pertussis (Tdap) to replace the booster dose of tetanus and diphtheria toxoids alone (Td). Two acellular pertussis vaccines are available in the United States for use in adolescents and adults.

Prophylactic administration of erythromycin for 5 days may also benefit unimmunized infants or heavily exposed adults.

Epidemiology & Control

Whooping cough is endemic in most densely populated areas worldwide and also occurs intermittently in epidemics. The source of infection is usually a patient in the early catarrhal stage of the disease. Communicability is high, ranging from 30% to 90%. Most cases occur in children under age 5 years; most deaths occur in the first year of life.

Control of whooping cough rests mainly on adequate active immunization of all infants.

BORDETELLA PARAPERTUSSIS

This organism may produce a disease similar to whooping cough, but it is generally less severe. The infection is often subclinical. *Bordetella parapertussis* grows more rapidly than typical *B pertussis* and produces larger colonies. It also grows on blood agar. *B parapertussis* has a silent copy of the pertussis toxin gene.

BORDETELLA BRONCHISEPTICA

Bordetella bronchiseptica is a small gram-negative bacillus that inhabits the respiratory tracts of canines, in which it may cause "kennel cough" and pneumonitis. It causes snuffles in rabbits and atrophic rhinitis in swine. It is infrequently responsible for chronic respiratory tract infections in humans, primarily in individuals with underlying diseases. It grows on blood agar medium. *B bronchiseptica* has a silent copy of the pertussis toxin gene.

THE BRUCELLAE

The brucellae are obligate parasites of animals and humans and are characteristically located intracellularly. They are relatively inactive metabolically. *Brucella melitensis* typically infects goats; *Brucella suis*, swine; *Brucella abortus*, cattle; and *Brucella canis*, dogs. Other species are found only in animals. Although named as species, DNA relatedness studies have shown there is only one species in the genus, *Brucella melitensis*, with multiple biovars. The disease in humans, brucellosis (undulant fever, Malta fever),

is characterized by an acute bacteremic phase followed by a chronic stage that may extend over many years and may involve many tissues.

Morphology & Identification

A. Typical Organisms

The appearance in young cultures varies from cocci to rods 1.2 μm in length, with short coccobacillary forms predominating. They are gram-negative but often stain irregularly, and they are aerobic, nonmotile, and nonspore-forming.

B. Culture

Small, convex, smooth colonies appear on enriched media in 2–5 days.

C. Growth Characteristics

Brucellae are adapted to an intracellular habitat, and their nutritional requirements are complex. Some strains have been cultivated on defined media containing amino acids, vitamins, salts, and glucose. Fresh specimens from animal or human sources are usually inoculated on trypticase-soy agar or blood culture media. *B abortus* requires 5–10% CO_2 for growth, whereas the other three species grow in air.

Brucellae utilize carbohydrates but produce neither acid nor gas in amounts sufficient for classification. Catalase and oxidase are produced by the four species that infect humans. Hydrogen sulfide is produced by many strains, and nitrates are reduced to nitrites.

Brucellae are moderately sensitive to heat and acidity. They are killed in milk by pasteurization.

D. Variation

The typical virulent organism forms a smooth, transparent colony; upon culture, it tends to change to the rough form, which is avirulent.

The serum of susceptible animals contains a globulin and a lipoprotein that suppress growth of nonsmooth, avirulent types and favor the growth of virulent types. Resistant animal species lack these factors, so that rapid mutation to avirulence can occur. d-Alanine has a similar effect in vitro.

Antigenic Structure

Differentiation among *Brucella* species or biovars is made possible by their characteristic sensitivity to dyes and their production of H_2S. Few laboratories have maintained the procedures for these tests, and the brucellae are seldom placed into the traditional species. Because brucellae are hazardous in the laboratory, tests to classify them should be performed only in reference public health laboratories using appropriate biosafety precautions.

Pathogenesis & Pathology

Although each species of *Brucella* has a preferred host, all can infect a wide range of animals, including humans.

The common routes of infection in humans are the intestinal tract (ingestion of infected milk), mucous membranes (droplets), and skin (contact with infected tissues of animals). Cheese made from unpasteurized goats' milk is a particularly common vehicle. The organisms progress from the portal of entry, via lymphatic channels and regional lymph nodes, to the thoracic duct and the bloodstream, which distributes them to the parenchymatous organs. Granulomatous nodules that may develop into abscesses form in lymphatic tissue, liver, spleen, bone marrow, and other parts of the reticuloendothelial system. In such lesions, the brucellae are principally intracellular. Osteomyelitis, meningitis, or cholecystitis also occasionally occurs. The main histologic reaction in brucellosis consists of proliferation of mononuclear cells, exudation of fibrin, coagulation necrosis, and fibrosis. The granulomas consist of epithelioid and giant cells, with central necrosis and peripheral fibrosis.

The brucellae that infect humans have apparent differences in pathogenicity. *B abortus* usually causes mild disease without suppurative complications; noncaseating granulomas of the reticuloendothelial system are found. *B canis* also causes mild disease. *B suis* infection tends to be chronic with suppurative lesions; caseating granulomas may be present. *B melitensis* infection is more acute and severe.

Persons with active brucellosis react more markedly (fever, myalgia) than normal persons to injected *Brucella* endotoxin. Sensitivity to endotoxin thus may play a role in pathogenesis.

Placentas and fetal membranes of cattle, swine, sheep, and goats contain erythritol, a growth factor for brucellae. The proliferation of organisms in pregnant animals leads to placentitis and abortion in these species. There is no erythritol in human placentas, and abortion is not part of *Brucella* infection of humans.

Clinical Findings

The incubation period is 1–6 weeks. The onset is insidious, with malaise, fever, weakness, aches, and sweats. The fever usually rises in the afternoon; its fall during the night is accompanied by drenching sweat. There may be gastrointestinal and nervous symptoms. Lymph nodes enlarge, and the spleen becomes palpable. Hepatitis may be accompanied by jaundice. Deep pain and disturbances of motion, particularly in vertebral bodies, suggest osteomyelitis. These symptoms of generalized *Brucella* infection generally subside in weeks or months, although localized lesions and symptoms may continue.

Following the initial infection, a chronic stage may develop, characterized by weakness, aches and pains, low-grade fever, nervousness, and other nonspecific manifestations compatible with psychoneurotic symptoms. Brucellae

cannot be isolated from the patient at this stage, but the agglutinin titer may be high. The diagnosis of "chronic brucellosis" is difficult to establish with certainty unless local lesions are present.

Diagnostic Laboratory Tests

A. Specimens

Blood should be taken for culture, biopsy material for culture (lymph nodes, bone, etc), and serum for serologic tests.

B. Culture

Brucella agar was specifically designed to culture *Brucella* species bacteria. The medium is highly enriched and—in reduced form—is used primarily in cultures for anaerobic bacteria. In oxygenated form, the medium grows *Brucella* species bacteria very well. However, infection with *Brucella* species is often not suspected when cultures of a patient's specimens are set up, and *Brucella* agar incubated aerobically is seldom used. The *Brucella* species bacteria will grow on commonly used media, including trypticase-soy medium with or without 5% sheep blood, brain heart infusion medium, and chocolate agar. Blood culture media (see below) readily grow *Brucella* species bacteria. Liquid medium used to culture *Mycobacterium tuberculosis* also supports the growth of at least some strains. All cultures should be incubated in 8–10% CO_2 at 35–37°C and should be observed for 3 weeks before being discarded as negative; liquid media cultures should be blindly subcultured during this time.

Bone marrow and blood are the specimens from which brucellae are most often isolated. The method of choice for bone marrow is to use pediatric Isolator tubes, which do not require centrifugation, with inoculation of the entire contents of the tube onto solid media. Media used in semiautomated and automated blood culture systems readily grow brucellae, usually within 1 week; however, holding the cultures for 3 weeks is recommended. Negative cultures for *Brucella* do not exclude the disease because brucellae can be cultivated from patients only during the acute phase of the illness or during recurrence of activity.

After a few days of incubation on agar media, the brucellae form colonies in the primary streak that are <1 mm in diameter. They are nonhemolytic. The observation of tiny gram-negative coccobacilli that are catalase-positive and oxidase-positive suggests *Brucella* species. All further work on such a culture should be done in a biologic safety cabinet. A Christensen's urea slant should be inoculated and observed frequently. A positive urease test is characteristic of *Brucella* species. *B suis* and some strains of *B melitensis* can yield a positive test less than 5 minutes after inoculating the slant; other strains will take a few hours to 24 hours. Bacteria that meet these criteria should be quickly submitted to a reference public health laboratory for presumptive identification. *Brucella* species are category B select agents. Molecular methods have been developed to rapidly differentiate among the various biovars.

C. Serology

IgM antibody levels rise during the first week of acute illness, peak at 3 months, and may persist during chronic disease. Even with appropriate antibiotic therapy, high IgM levels may persist for up to 2 years in a small percentage of patients. IgG antibody levels rise about 3 weeks after onset of acute disease, peak at 6–8 weeks, and remain high during chronic disease. IgA levels parallel the IgG levels. The usual serologic tests may fail to detect infection with *B canis*.

1. Agglutination test—
To be reliable, serum agglutination tests must be performed with standardized heat-killed, phenolized, smooth *Brucella* antigens. IgG agglutinin titers above 1:80 indicate active infection. Individuals injected with cholera vaccine may develop agglutination titers to brucellae. If the serum agglutination test is negative in patients with strong clinical evidence of *Brucella* infection, tests must be made for the presence of "blocking" antibodies. These can be detected by adding antihuman globulin to the antigen-serum mixture. Brucellosis agglutinins are cross-reactive with tularemia agglutinins, and tests for both diseases should be done on positive sera; usually, the titer for one disease will be much higher than that for the other.

2. Blocking antibodies—
These are IgA antibodies that interfere with agglutination by IgG and IgM and cause a serologic test to be negative in low serum dilutions (prozone) although positive in higher dilutions. These antibodies appear during the subacute stage of infection, tend to persist for many years independently of activity of infection, and are detected by the Coombs antiglobulin method.

3. ELISA assays—
IgG, IgA, and IgM antibodies may be detected using ELISA assays, which use cytoplasmic proteins as antigens. These assays tend to be more sensitive and specific than the agglutination test.

Immunity

An antibody response occurs with infection, and it is probable that some resistance to subsequent attacks is produced. Immunogenic fractions from *Brucella* cell walls have a high phospholipid content; lysine predominates among eight amino acids; and there is no heptose (thus distinguishing the fractions from endotoxin).

Treatment

Brucellae may be susceptible to tetracyclines or ampicillin. Symptomatic relief may occur within a few days after treatment with these drugs is begun. However, because of their intracellular location, the organisms are not readily eradicated completely from the host. For best results, treatment must be prolonged. Combined treatment with a tetracycline (such as doxycycline) and either streptomycin for 2–3 weeks or rifampin for 6 weeks is recommended.

Epidemiology, Prevention, & Control

Brucellae are animal pathogens transmitted to humans by accidental contact with infected animal feces, urine, milk, and tissues. The common sources of infection for humans are unpasteurized milk, milk products, and cheese, and occupational contact (eg, farmers, veterinarians, and slaughterhouse workers) with infected animals.

Cheese made from unpasteurized goat's milk is a particularly common vehicle for transmission of brucellosis. Occasionally the airborne route may be important. Because of occupational contact, *Brucella* infection is much more frequent in men. The majority of infections remain asymptomatic (latent).

Infection rates vary greatly with different animals and in different countries. Outside the United States, infection is more prevalent. Eradication of brucellosis in cattle can be attempted by test and slaughter, active immunization of heifers with avirulent live strain 19, or combined testing, segregation, and immunization. Cattle are examined by means of agglutination tests.

Active immunization of humans against *Brucella* infection is experimental. Control rests on limitation of spread and possible eradication of animal infection, pasteurization of milk and milk products, and reduction of occupational hazards wherever possible.

FRANCISELLA TULARENSIS & TULAREMIA

Francisella species are widely found in animal reservoirs and aquatic environments. The taxonomy of this genus has undergone numerous changes over the years. There are four recognized subspecies of *Francisella tularensis*: *tularensis*, *holarctica*, *mediasiatica*, and *novicida*. Subspecies *tularensis* (type A) is the most virulent among this group and the most pathogenic for humans. It is associated with wild rabbits, ticks, and tabanid flies. Subspecies *holarctica* strains cause milder infection and are associated with hares, ticks, mosquitoes, and tabanid flies. *F tularensis* is transmitted to humans by biting arthropods and flies, direct contact with infected animal tissue, inhalation of aerosols, or ingestion of contaminated food or water. The clinical presentation depends on the route of infection: six major syndromes are described (see Pathogenesis and Clinical Findings and the updated review on tularemia provided in the reference by Nigrovic et al). An additional species, *Francisella philomiragia* also exists. Infections caused by *F philomiragia* are rare and are usually found in situations of near-drowning. This organism will not be discussed.

Morphology & Identification

A. Typical Organisms

F tularensis is a small, gram-negative, coccobacillus. It is rarely seen in smears of tissue (Figure 18–2).

B. Specimens

Blood is taken for serologic tests. The organism may be recovered in culture from lymph node aspirates, bone marrow, peripheral blood, deep tissue, and ulcer biopsies.

C. Culture

Growth requires enriched media containing cysteine. In the past, glucose-cysteine blood agar was preferred, but *F tularensis* will grow on commercially available hemin containing media such as chocolate agar, modified Thayer-Martin agar, and buffered charcoal yeast extract (BCYE) agar used to grow *Legionella* species. Media should be incubated in CO_2 at 35–37°C for 2–5 days. **Caution:** In order to avoid laboratory-acquired infections, biosafety level three (BSL III) practices are required when working with live cultures suspected of containing *F tularensis*. Clinical specimens require BSL II facilities and practice.

D. Serology

All isolates are serologically identical, possessing a polysaccharide antigen and one or more protein antigens that cross-react with brucellae. However, there are two major biogroups of strains, called Jellison type A and type B.

Type A occurs only in North America, is lethal for rabbits, produces severe illness in humans, ferments glycerol, and contains citrulline ureidase. Type B lacks these biochemical features, is not lethal for rabbits, produces milder disease in humans, and is isolated often from rodents or from water in Europe, Asia, and North America. Other biogroups are of low pathogenicity.

The usual antibody response consists of agglutinins developing 7–10 days after onset of illness.

Pathogenesis & Clinical Findings

F tularensis is highly infectious: Penetration of the skin or mucous membranes or inhalation of 50 organisms can result in infection. Most commonly, organisms enter through skin abrasions. In 2–6 days, an inflammatory, ulcerating papule develops. Regional lymph nodes enlarge and may become necrotic, sometimes draining for weeks (ulceroglandular tularemia). Inhalation of an infective aerosol results in peribronchial inflammation and localized pneumonitis (pneumonic tularemia). Oculoglandular tularemia can develop when an infected finger or droplet touches the conjunctiva. Yellowish granulomatous lesions on the lids may be accompanied by preauricular adenopathy. The other forms of the disease are glandular tularemia (lymphadenopathy but no ulcers), oropharyngeal tularemia, and typhoidal tularemia (septicemia). In all cases there is fever, malaise, headache, and pain in the involved region and regional lymph nodes.

Because of the highly infectious nature of *F tularensis*, this organism is a potential agent of bioterrorism and is currently classified on the select agent list as a category A agent. Laboratories that recover a suspected *F tularensis* should notify public health officials and should send the isolate to a reference laboratory capable of performing definitive identification.

Diagnostic Laboratory Tests

Although *F tularensis* may be recovered from the clinical specimens listed above, the diagnosis rests on serologic studies. Paired serum samples collected 2 weeks apart can show a rise in agglutination titer. A single serum titer of 1:160 is highly suggestive if the history and physical findings are compatible with the diagnosis. Because antibodies reactive in the agglutination test for tularemia also react in the test for brucellosis, both tests should be done for positive sera; the titer for the disease affecting the patient is usually fourfold greater than that for the other disease.

Treatment

Streptomycin or gentamicin therapy for 10 days almost always produces rapid improvement. Tetracycline may be equally effective, but relapses occur more frequently. *F tularensis* is resistant to all β-lactam antibiotics.

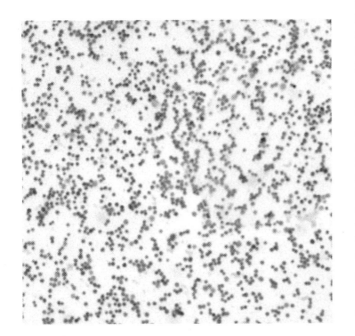

FIGURE 18–2 Gram stain of *Francisella tularensis*. These bacteria are tiny gram-negative coccobacilli approximately 0.2 × 0.7 μm. Original magnification × 1000. (Courtesy of CDC Public Health Image Library.)

Prevention & Control

Humans acquire tularemia from handling infected rabbits or muskrats or from bites by an infected tick or deer-fly. Less often, the source is contaminated water or food or contact with a dog or cat that has caught an infected wild animal. Avoidance is the key to prevention. The infection in wild animals cannot be controlled.

Persons at exceedingly high risk—particularly research laboratory personnel—may be immunized by the administration of a live attenuated strain of *F tularensis,* available only through a cooperative research agreement from the US Army Medical Research Institute of Infectious Diseases, Fort Detrick, Frederick, MD 21701. The vaccine is administered by multiple punctures through the skin and provides only partial immunity. A similar live vaccine has been administered in Russia on a large scale.

REVIEW QUESTIONS

1. A 68-year-old woman was seen in the clinic because she had felt feverish and had been experiencing increasing pain and swelling in her left knee during the past 3 weeks. Four years earlier, a prosthetic joint had been placed in her left knee. On examination the knee was swollen, and fluid could be detected. An aspirate of the fluid was obtained. There were 15,000 polymorphonuclear cells per milliliter in the fluid. No organisms were seen on Gram stain. A routine culture was done. On the fourth day of incubation, colorless colonies <1 mm in diameter were seen on the blood and chocolate agar plates. The organism was a tiny gram-negative coccobacillus that was catalase-positive and oxidase-positive. A urea slant was inoculated and was positive for urease activity after overnight incubation. The patient was probably infected with which of the following microorganisms?

   (A) *Haemophilus influenzae*
   (B) *Haemophilus ducreyi*
   (C) *Francisella tularensis*
   (D) *Brucella species*
   (E) *Staphylococcus aureus*

2. After the culture (Question 1) turned positive, additional history was obtained. Approximately 4 weeks prior to the onset of her knee pain, the patient had visited relatives in Israel and traveled to other countries in the Mediterranean area. She had a particular fondness for one food product that was the probable vehicle for her infection. The product most likely was

   (A) Bananas
   (B) Unpasteurized goat's cheese
   (C) Rare hamburger
   (D) Fresh orange juice
   (E) Green tea

3. A 55-year-old game warden in Vermont found a dead muskrat on the bank of a stream. He picked up the animal, thinking it might have been illegally trapped or shot; it was not, and the game warden buried it. Four days later he developed a 1.5 cm painful ulcer on the index finger of his right hand, a 1-cm ulcer on his right forehead, and pain in his right axilla. Physical examination also revealed right axillary lymphadenopathy. This patient is most likely infected with

   (A) *Brucella* species
   (B) *Rickettsia rickettsii*
   (C) *Salmonella* Typhi
   (D) *Haemophilus ducreyi*
   (E) *Francisella tularensis*

4. An 18-month-old boy has been playing with a child who develops *Haemophilus influenzae* meningitis. The boy's parents consult his pediatrician, who says she is comfortable that the child will be fine because he has been fully immunized with the polyribose phosphate (PRP)-protein conjugate vaccine. For what reason is it necessary to immunize infants of 2 months to 2 years of age with polysaccharide-protein conjugate vaccines?

   (A) The conjugate protein is diphtheria toxoid, and the goal is for the infant to develop simultaneous immunity to diphtheria
   (B) Infants 2 months to 2 years of age do not immunologically respond to polysaccharide vaccines that are not conjugated to a protein
   (C) The conjugate vaccine is designed for older children and adults as well as infants
   (D) Maternal (transplacental) antibodies against *Haemophilus influenzae* are gone from the infant's circulation by 2 months of age
   (E) None of the above

5. An 11-year-old boy from Peru was referred to the Brain Tumor Institute. Three months earlier he had developed headache and then slowly progressive right-sided weakness. A brain scan showed a mass lesion in the left hemisphere. He was thought to have a brain tumor. A lumbar puncture was not done because of concern about increased intracranial pressure and brain herniation through the tentorium cerebelli. At surgery, a mass lesion in the left hemisphere was found. Frozen sections of the tissue were done while the patient was in the operating room. Microscopy of the sections showed a granulomatous inflammatory reaction. No tumor was seen. Tissue was submitted for culture for *Mycobacterium tuberculosis*. Middlebrook 7H9 broth medium was used. Six days after the culture was set up, the automated machine detected that the culture was positive. An acid-fast stain and a Gram stain were both negative. Subcultures were done. Two days later, very small colonies were seen on the sheep blood agar plate. The organism was a tiny gram-negative coccobacillus that was catalase-positive and oxidase-positive. It showed urease activity after 2 hours of incubation on urea-containing medium. This child had infection with

   (A) *Brucella* species
   (B) *Mycobacterium tuberculosis*
   (C) *Francisella tularensis*
   (D) *Haemophilus influenzae*
   (E) *Moraxella catarrhalis*

6. A 3-year-old child develops *Haemophilus influenzae* meningitis. Therapy is begun with cefotaxime. Why is this third-generation cephalosporin used rather than ampicillin?

   (A) About 80% of *Haemophilus influenzae* organisms have modified penicillin-binding proteins that confer resistance to ampicillin
   (B) The drug of choice, trimethoprim-sulfamethoxazole, cannot be used because the child is allergic to sulfonamides
   (C) It is easier to administer intravenous cefotaxime than intravenous ampicillin

(D) There is concern that the child will rapidly develop penicillin (ampicillin) allergy

(E) About 20% of *Haemophilus influenzae* organisms have a plasmid that encodes for beta-lactamase

7. A 55-year-old man with severe dental caries presented with 1 month of fever, malaise, and back pain and now presents with moderately severe shortness of breath. The examination reveals a febrile man who appears pale and dyspneic. Other physical findings include conjunctival petechiae, a Grade III/VI systolic murmur and an enlarged spleen. Blood cultures grow a pleomorphic gram-negative rod that is not hemolytic and that when tested is X and V factor negative. The most likely causative pathogen is:

(A) *Haemophilus influenzae*

(B) *Haemophilus ducreyi*

(C) *Aggregatibacter aphrophilus*

(D) *Actinobacillus hominis*

(E) *Haemophilus parainfluenzae*

8. All of the following statements regarding acellular pertussis vaccines are correct *except*:

(A) All formulations of the vaccine contain at least two antigens.

(B) The acellular vaccine has replaced the whole cell vaccine in the childhood vaccine series.

(C) All children should receive five doses of the vaccine prior to school entry.

(D) The vaccine is approved only for young children and adolescents.

(E) The vaccine is safer than and as immunogenic as whole cell vaccines.

9. Which of the following subspecies of *Francisella tularensis* is the most virulent for humans?

(A) *tularensis*

(B) *holarctica*

(C) *mediasiatica*

(D) *novicida*

10. All of the following statements regarding the etiologic agent of chancroid are correct *except*:

(A) The organism is a small gram-negative rod.

(B) The organism requires X factor but not V factor.

(C) The organism grows well on standard chocolate agar.

(D) On Gram stain of lesions the organism occurs in strands.

(E) The organism is susceptible to erythromycin.

Answers

1. D	4. B	7. C	10. C
2. B	5. A	8. D	
3. E	6. E	9. A	

REFERENCES

Broder KR, et al Preventing tetanus, diphtheria, and pertussis among adolescents: Use of tetanus toxoid, reduced diphtheria toxoid and acellular pertussis vaccines Recommendations of the Advisory Committee on Immunization Practices (ACIP). MMWR Recomm Rep 2006;55(RR-3), 1.

Killian M: *Haemophilus.* In: *Manual of Clinical Microbiology,* 9th ed. Murray PR et al (editors). ASM Press, 2007.

Miscellaneous fastidious gram-negative bacilli. In: *Koneman's Color Atlas and Textbook of Diagnostic Microbiology,* 6th ed. Winn W Jr et al (editors). Lippincott Williams & Wilkins, 2006.

Murphy TF: *Haemophilus* species (including *H. influenzae* and chancroid). In: *Mandell, Douglas, and Bennett's Principles and Practice of Infectious Diseases,* 7th ed. Mandell GL, Bennett JE, Dolin R (editors). Elsevier, 2010.

Nigrovic LE, Wingerter SL. Tularemia. Infect Dis Clin N Am 2008;22:489.

Pappas G et al: Brucellosis. N Engl J Med 2005;352:2325.

Penn RL: *Francisella tularensis* (tularemia). In: *Mandell, Douglas, and Bennett's Principles and Practice of Infectious Diseases,* 7th ed. Mandell GL, Bennett JE, Dolin R (editors). Elsevier, 2010.

Waters V, Halperin S: *Bordetella* pertussis. In: *Mandell, Douglas, and Bennett's Principles and Practice of Infectious Diseases,* 7th ed. Mandell GL, Bennett JE, Dolin R (editors). Elsevier, 2010.

Young EJ: *Brucella* species. In: *Mandell, Douglas, and Bennett's Principles and Practice of Infectious Diseases,* 76th ed. Mandell GL, Bennett JE, Dolin R (editors). Elsevier, 2010.

Yersinia & Pasteurella

The organisms discussed in this chapter are short, pleomorphic gram-negative rods that can exhibit bipolar staining. They are catalase-positive, oxidase-negative, and microaerophilic or facultatively anaerobic. Most have animals as their natural hosts, but they can produce serious disease in humans.

The genus *Yersinia* includes *Yersinia pestis,* the cause of plague; *Yersinia pseudotuberculosis* and *Yersinia enterocolitica,* important causes of human diarrheal diseases; and several others considered nonpathogenic for humans. Several species of *Pasteurella* are primarily animal pathogens but can also produce human disease.

YERSINIA PESTIS & PLAGUE

Plague is an infection of wild rodents, transmitted from one rodent to another and occasionally from rodents to humans by the bites of fleas. Serious infection often results, which in previous centuries produced pandemics of "black death" with millions of fatalities. The ability of this organism to be transmitted by aerosol, and the severity and high mortality associated with pneumonic plague, make *Y pestis* a potential biological weapon.

Morphology & Identification

Y pestis is a gram-negative rod that exhibits striking bipolar staining with special stains (Figure 19–1). It is nonmotile. It grows as a facultative anaerobe on many bacteriologic media. Growth is more rapid in media containing blood or tissue fluids and fastest at 30°C. In cultures on blood agar at 37°C, colonies may be very small at 24 hours. A virulent inoculum, derived from infected tissue, produces gray and viscous colonies, but after passage in the laboratory the colonies become irregular and rough. The organism has little biochemical activity, and this is somewhat variable.

Antigenic Structure

All yersiniae possess lipopolysaccharides that have endotoxic activity when released. The three pathogenic species produce antigens and toxins that act as virulence factors. They have type III secretion systems that consist of a membrane-spanning complex that allows the bacteria to inject proteins directly into cytoplasm of the host cells. The virulent yersiniae produce V and W antigens, which are encoded by genes on a plasmid of approximately 70 kb. This is essential for virulence; the V and W antigens yield the requirement for calcium for growth at 37°C. Compared to the other pathogenic yersiniae, *Y pestis* has gained additional plasmids. pPst is a 9.5 kb plasmid that contains genes that yield plasminogen-activating protease that has temperature-dependent coagulase activity (20–28°C, the temperature of the flea) and fibrinolytic activity (35–37°C, the temperature of the host). This factor is involved in dissemination of the organism from the flea bite injection site. The pFra/pMT plasmid (80–101 kb) encodes the capsular protein (fraction F1)

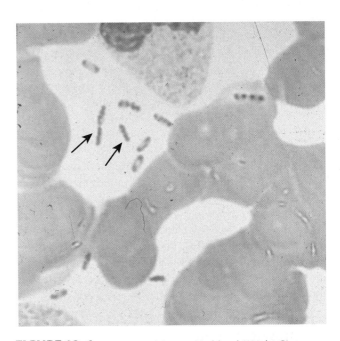

FIGURE 19–1 *Yersinia pestis* (arrows) in blood, Wright-Giemsa stain. Some of the *Yersinia pestis* have bipolar staining, which gives them a hairpin-like appearance. Original magnification × 1000. (Courtesy of K Gage, Plague Section, Centers for Disease Control and Prevention, Ft. Collins, CO.)

that is produced mainly at 37°C and confers antiphagocytic properties. In addition, this plasmid contains genes that encode phospholipase D, which is required for organism survival in the flea midgut.

The three pathogenic yersiniae have a pathogenecity island (PAI) that encodes for an iron-scavenging siderophore (see Chapter 9), yersiniabactin.

Among several exotoxins produced, one is lethal for mice in amounts of 1 μg. This homogeneous protein (MW 74,000) produces β-adrenergic blockade and is cardiotoxic in animals. Its role in human infection is unknown.

Pathogenesis & Pathology

When a flea feeds on a rodent infected with *Y pestis*, the ingested organisms multiply in the gut of the flea and, helped by the coagulase, block its proventriculus so that no food can pass through. Subsequently, the "blocked" and hungry flea bites ferociously and the aspirated blood, contaminated with *Y pestis* from the flea, is regurgitated into the bite wound. The inoculated organisms may be phagocytosed by polymorphonuclear cells and macrophages. The *Y pestis* organisms are killed by the polymorphonuclear cells but multiply in the macrophages; because the bacteria are multiplying at 37°C, they produce the antiphagocytic protein and subsequently are able to resist phagocytosis. The pathogens rapidly reach the lymphatics and an intense hemorrhagic inflammation develops in the enlarged lymph nodes, which may undergo necrosis and become fluctuant. While the invasion may stop there, *Y pestis* organisms often reach the bloodstream and become widely disseminated. Hemorrhagic and necrotic lesions may develop in all organs; meningitis, pneumonia, and serosanguineous pleuropericarditis are prominent features.

Primary pneumonic plague results from inhalation of infective droplets (usually from a coughing patient), with hemorrhagic consolidation, sepsis, and death.

Clinical Findings

After an incubation period of 2–7 days, there is high fever and painful lymphadenopathy, commonly with greatly enlarged, tender nodes (buboes) in the groin or axillae. Vomiting and diarrhea may develop with early sepsis. Later, disseminated intravascular coagulation leads to hypotension, altered mental status, and renal and cardiac failure. Terminally, signs of pneumonia and meningitis can appear, and *Y pestis* multiplies intravascularly and can be seen in blood smears.

Diagnostic Laboratory Tests

Plague should be suspected in febrile patients who have been exposed to rodents in known endemic areas. Rapid recognition and laboratory confirmation of the disease are essential in order to institute lifesaving therapy.

A. Specimens

Blood is taken for culture and aspirates of enlarged lymph nodes for smear and culture. Acute and convalescent sera may be examined for antibody levels. In pneumonia, sputum is cultured; in possible meningitis, cerebrospinal fluid is taken for smear and culture.

B. Smears

Y pestis are small gram-negative bacilli that appear as single cells or as pairs or short chains in clinical material. Wright's, Giemsa's, or Wayson's stains may be more useful when staining material from a suspected buboe or a positive blood culture because of the striking bipolar appearance (safety pin shape) of the organism using these stains that is not evident on a direct Gram stain. More specific direct staining methods (possibly available through reference laboratories) include the use of fluorescent antibody stains targeting the capsular F1 antigen.

C. Culture

All materials are cultured on blood agar, chocolate, and MacConkey agar plates and in brain-heart infusion broth. Growth on solid media may be slow, requiring more than 48 hours, but blood cultures are often positive in 24 hours. Cultures can be tentatively identified by biochemical reactions. *Y pestis* produces nonlactose-fermenting colonies on MacConkey agar, and it grows better at 28°C than at 37°C. The organism is catalase positive; indole, oxidase, urease negative; and nonmotile. The last two reactions are useful in differentiating *Y pestis* from other pathogenic yersiniae. An organism with the above characteristics should be referred to a public health laboratory for more confirmatory testing. Definite identification of cultures is best done by immunofluorescence or by lysis by a specific *Y pestis* bacteriophage (confirmation available through state health department laboratories and by consultation with the Centers for Disease Control and Prevention, Plague Branch, Fort Collins, CO).

All cultures are highly infectious and must be handled with extreme caution inside a biological safety cabinet.

D. Serology

In patients who have not been previously vaccinated, a convalescent serum antibody titer of 1:16 or greater is presumptive evidence of *Y pestis* infection. A titer rise in two sequential specimens confirms the serologic diagnosis.

Treatment

Unless promptly treated, plague may have a mortality rate of nearly 50%; pneumonic plague, nearly 100%. The drug of choice is streptomycin, but the more readily available aminoglycoside gentamicin has been shown to be as effective. Doxycycline is an alternative drug and is sometimes given in combination with streptomycin. Drug resistance has been noted in *Y pestis*.

Epidemiology & Control

Plague is an infection of wild rodents (field mice, gerbils, moles, skunks, and other animals) that occurs in many parts of the world. The chief enzootic areas are India, Southeast Asia (especially Vietnam), Africa, and North and South America. The western states of the United States and Mexico always contain reservoirs of infection. Epizootics with high mortality rates occur intermittently; at such times, the infection can spread to domestic rodents (eg, rats) and other animals (eg, cats), and humans can be infected by flea bites or by contact. The commonest vector of plague is the rat flea (*Xenopsylla cheopis*), but other fleas may also transmit the infection.

The control of plague requires surveys of infected animals, vectors, and human contacts—in the United States this is done by county and state agencies with support from the Plague Branch of the Centers for Disease Control and Prevention—and by destruction of plague-infected animals. If a human case is diagnosed, health authorities must be notified promptly. All patients with suspected plague should be isolated, particularly if pulmonary involvement has not been ruled out. All specimens must be treated with extreme caution. Contacts of patients with suspected plague pneumonia should receive doxycycline, as chemoprophylaxis.

Killed whole cell vaccines are no longer available. Because of concern for bioterrorism, numerous vaccines are currently under development.

YERSINIA ENTEROCOLITICA & *YERSINIA PSEUDOTUBERCULOSIS*

These are nonlactose-fermenting gram-negative rods that are urease-positive and oxidase-negative. They grow best at 25°C and are motile at 25°C but nonmotile at 37°C. They are found in the intestinal tract of a variety of animals, in which they may cause disease, and are transmissible to humans, in whom they can produce a variety of clinical syndromes.

Y enterocolitica exists in more than 50 serotypes; most isolates from human disease belong to serotypes O:3, O:8, and O:9. There are striking geographic differences in the distribution of *Y enterocolitica* serotypes. *Y pseudotuberculosis* exists in at least six serotypes, but serotype O:1 accounts for most human infections. *Y enterocolitica* can produce a heat-stable enterotoxin, but the role of this toxin in diarrhea associated with infection is not well defined.

Y enterocolitica has been isolated from rodents and domestic animals (eg, sheep, cattle, swine, dogs, and cats) and waters contaminated by them. Transmission to humans probably occurs by contamination of food, drink, or fomites. *Y pseudotuberculosis* occurs in domestic and farm animals and birds, which excrete the organisms in feces. Human infection probably results from ingestion of materials contaminated with animal feces. Person-to-person transmission with either of these organisms is probably rare.

Pathogenesis & Clinical Findings

An inoculum of 10^8–10^9 yersiniae must enter the alimentary tract to produce infection. During the incubation period of 4–7 days, yersiniae multiply in the gut mucosa, particularly the ileum. This leads to inflammation and ulceration, and leukocytes appear in feces. The process may extend to mesenteric lymph nodes and, rarely, to bacteremia.

Early symptoms include fever, abdominal pain, and diarrhea. Diarrhea ranges from watery to bloody and may be due to an enterotoxin or to invasion of the mucosa. At times, the abdominal pain is severe and located in the right lower quadrant, suggesting appendicitis. One to 2 weeks after onset some patients with histocompatibility antigen HLA-B 27 develop arthralgia, arthritis, and erythema nodosum, suggesting an immunologic reaction to the infection. Very rarely, *Yersinia* infection produces pneumonia, meningitis, or sepsis; in most cases, it is self-limited.

Diagnostic Laboratory Tests

A. Specimens

Specimens may be stool, blood, or material obtained at surgical exploration. Stained smears are not contributory.

B. Culture

The number of yersiniae in stool may be small and can be increased by "cold enrichment": a small amount of feces or a rectal swab is placed in buffered saline, pH 7.6, and kept at 4°C for 2–4 weeks; many fecal organisms do not survive, but *Y enterocolitica* will multiply. Subcultures made at intervals on MacConkey agar may yield yersiniae.

C. Serology

In paired serum specimens taken 2 or more weeks apart, a rise in agglutinating antibodies can be shown; however, cross reactions between yersiniae and other organisms (vibrios, salmonellae, and brucellae) may confuse the results.

Treatment

Most *Yersinia* infections with diarrhea are self-limited, and the possible benefits of antimicrobial therapy are unknown. *Y enterocolitica* is generally susceptible to aminoglycosides, chloramphenicol, tetracycline, trimethoprimsulfamethoxazole, piperacillin, third-generation cephalosporins, and fluoroquinolones; it is typically resistant to ampicillin and to first-generation cephalosporins. Proved *Yersinia* sepsis or meningitis has a high mortality rate, but deaths occur mainly in immunocompromised patients. *Yersinia* sepsis can be successfully treated with third-generation cephalosporins (possibly in combination with an aminoglycoside) or a fluoroquinolone (possibly in combination with another antimicrobial). In cases where clinical manifestations strongly point to either appendicitis or mesenteric adenitis, surgical exploration has been the rule unless several simultaneous cases indicate that *Yersinia* infection is likely.

Prevention & Control

Contact with farm and domestic animals, their feces, or materials contaminated by them probably accounts for most human infections. Meat and dairy products have occasionally been indicated as sources of infection, and group outbreaks have been traced to contaminated food or drink. Conventional sanitary precautions are probably helpful. There are no specific preventive measures.

PASTEURELLA

Pasteurella species are primarily animal pathogens, but they can produce a range of human diseases. The generic term pasteurellae formerly included all yersiniae and francisellae as well as the pasteurellae discussed below.

Pasteurellae are nonmotile gram-negative coccobacilli with a bipolar appearance on stained smears. They are aerobes or facultative anaerobes that grow readily on ordinary bacteriologic media at 37°C. They are all oxidase-positive and catalase-positive but diverge in other biochemical reactions.

Pasteurella multocida occurs worldwide in the respiratory and gastrointestinal tracts of many domestic and wild animals. It is perhaps the most common organism in human wounds inflicted by bites from cats and dogs. It is one of the common causes of hemorrhagic septicemia in a variety of animals, including rabbits, rats, horses, sheep, fowl, cats, and swine. It can also produce human infections in many systems and may at times be part of normal human flora.

Pasteurella bettyae has been recovered from infections of the human genital tract and of newborns. Its habitat is uncertain.

Pasteurella pneumotropica is a normal inhabitant of the respiratory tract and gut of mice and rats and can cause pneumonia or sepsis when the host–parasite balance is disturbed. A few human infections have followed animal bites.

Pasteurella ureae has rarely been found in animals but occurs as part of a mixed flora in human chronic respiratory disease or other suppurative infections.

Clinical Findings

The most common presentation is a history of animal bite followed within hours by an acute onset of redness, swelling, and pain. Regional lymphadenopathy is variable, and fever is often low grade. *Pasteurella* infections sometimes present as bacteremia or chronic respiratory infection without an evident connection with animals.

P multocida is susceptible to most antibiotics. Penicillin G is considered the drug of choice for *P multocida* infections resulting from animal bites. Tetracyclines and fluoroquinolones are alternative drugs.

REVIEW QUESTIONS

1. An 18-year-old male resident of Arizona came to the emergency department complaining of fever, pain in his left groin, and diarrhea for the past 2 days. On examination, he was afebrile, had a pulse rate of 126/min, respiratory rate of 20/min, and a blood pressure of 130/80 mm Hg. Left groin swelling and tenderness were noted. A groin muscle strain was diagnosed, attributed to a fall 2 days earlier. He was treated with nonsteroidal anti-inflammatory agents and released. The next day the patient reported feeling weak, had difficulty breathing, and collapsed while taking a shower. He was transported to a hospital emergency department and pronounced dead shortly after arrival. Cultures of blood samples obtained in the emergency department were positive for *Yersinia pestis*. An epidemiologic investigation indicated that the patient most likely became infected as a result of bites by *Yersinia pestis* infected fleas while walking through a prairie dog colony. (See Chapter 48.) Which of the following statements about the pathogenesis of plague is correct?
    (A) *Yersinia pestis* produces a coagulase when incubated at 28°C
    (B) There is no risk for pneumonia caused by person-to-person transmission of *Yersinia pestis*
    (C) *Yersinia pestis* organisms multiply in polymorphonuclear cells
    (D) Following the bite of an infected flea, *Yersinia pestis* infection seldom if ever disseminates beyond the site of the flea bite and the regional lymph nodes
    (E) *Yersinia pestis* is transmitted to animals (and humans) in flea feces excreted when the flea is feeding

2. The drug of choice to treat the patient in Question 1 would have been
    (A) Ampicillin
    (B) Cefotaxime
    (C) Levofloxacin
    (D) Erythromycin
    (E) Streptomycin

3. *Yersinia pestis* entered North America through San Francisco in the 1890s, carried by rats on ships that had sailed from Hong Kong, where a plague epidemic occurred. The current reservoir for *Yersinia pestis* in the United States is
    (A) Urban feral cats
    (B) Urban rats
    (C) Domestic cows
    (D) Coyotes
    (E) Rural wild rodents

4. Which of the following is generally not considered a potential agent of bioterrorism and biologic warfare?
    (A) *Yersinia pestis*
    (B) Botulinum toxin
    (C) *Streptococcus pyogenes*
    (D) *Brucella* species
    (E) *Bacillus anthracis*

5. An 8-year-old boy was bitten by a stray cat. Two days later, the wound was red and swollen and drained purulent fluid. *Pasteurella multocida* was cultured from the wound. The drug of choice to treat this infection is
    (A) Amikacin
    (B) Erythromycin
    (C) Gentamicin
    (D) Penicillin G
    (E) Clindamycin

6. Intimate contacts of patients with suspected plague pneumonia should receive which of the following agents as chemoprophylaxis?

    (A) Gentamicin
    (B) Cefazolin
    (C) Rifampin
    (D) Penicillin
    (E) Doxycycline

7. In a patient who has the bubonic form of plague, all of the following specimens are acceptable for diagnosis *except*:

    (A) Stool culture on hektoen enteric agar
    (B) Blood culture using routine laboratory media
    (C) Culture of a lymph node aspirate on blood and MacConkey agars
    (D) Acute and convalescent serology
    (E) Immunohistochemical staining of lymph node tissue

8. All of the following statements regarding the pFra/pMT plasmid of *Yersinia pestis* are true *except*:

    (A) It encodes the capsular protein (fraction FI) that confers antiphagocytic properties.
    (B) It contains genes that yield plasminogen-activating protease that has temperature-dependent coagulase activity
    (C) It contains genes that encode phospholipase D which is required for organism survival in the flea midgut.
    (D) It is unique to *Yersinia pestis.*
    (E) It encodes factors that are important for survival in both the flea and the human.

9. All of the following statements regarding the epidemiology of infections caused by *Yersinia enterocolitica* are correct *except*:

    (A) Most human infections are caused by serotype O:1.
    (B) Humans acquire the infection from ingestion of food or drinks contaminated by animals or animal products.
    (C) Person-to-person spread is quite common.
    (D) A large inoculum is required to cause infection.

    (E) Infection is more prevalent in persons with histocompatibility antigen HLA-B27.

10. Which of the following *Pasteurella* species has been associated with infections of the female genital tract and of newborns?

    (A) *Pasteurella multocida*
    (B) *Pasteurella pneumotropica*
    (C) *Pasteurella ureae*
    (D) *Pasteurella bettyae*

Answers

1. A	4. C	7. A	10. D
2. E	5. D	8. B	
3. E	6. E	9. C	

REFERENCES

Dennis DT, Mead PS: *Yersinia* species, including plague. In: *Mandell, Douglas, and Bennett's Principles and Practice of Infectious Diseases,* 7th ed. Mandell GL, Bennett JE, Dolin R (editors). Elsevier, 2010.

Prentice MB, Rahalson L: Plague. Lancet 2007;369:1196–1207.

von Graevenitz A, Zbinden R, Mutters R: *Actinobacillus, Capnocytophaga, Eikenella, Kingella, Pasteurella,* and other fastidious or rarely encountered gram-negative rods. In: *Manual of Clinical Microbiology,* 9th ed. Murray PR et al (editors). ASM Press, 2007.

Wanger A: *Yersinia.* In: *Manual of Clinical Microbiology,* 9th ed. Murray PR et al (editors). ASM Press, 2007.

Zurlo JJ: *Pasteurella* species. In: *Mandell, Douglas, and Bennett's Principles and Practice of Infectious Diseases,* 7th ed. Mandell GL, Bennett JE, Dolin R (editors). Elsevier, 2010.

The Neisseriae

The family *Neisseriaceae* includes the genera *Neisseria, Kingella, Eikenella, Simonsiella,* and *Alysiella* (see Chapter 16). The neisseriae are gram-negative cocci that usually occur in pairs. *Neisseria gonorrhoeae* (gonococci) and *Neisseria meningitidis* (meningococci) are pathogenic for humans and typically are found associated with or inside polymorphonuclear cells. Some neisseriae are normal inhabitants of the human respiratory tract, rarely if ever cause disease, and occur extracellularly. Members of the group are listed in Table 20–1.

Gonococci and meningococci are closely related, with 70% DNA homology, and are differentiated by a few laboratory tests and specific characteristics: Meningococci have polysaccharide capsules, whereas gonococci do not, and meningococci rarely have plasmids whereas most gonococci do. Most importantly, the two species are differentiated by the usual clinical presentations of the diseases they cause: Meningococci typically are found in the upper respiratory tract and cause meningitis, while gonococci cause genital infections. The clinical spectra of the diseases caused by gonococci and meningococci overlap, however.

Morphology & Identification

A. Typical Organisms

The typical neisseria is a gram-negative, nonmotile diplococcus, approximately 0.8 μm in diameter (Figures 20–1 and 20–2). Individual cocci are kidney-shaped; when the organisms occur in pairs, the flat or concave sides are adjacent.

B. Culture

In 48 hours on enriched media (eg, modified-Thayer Martin, Martin-Lewis, GC-Lect, and New York City), gonococci and meningococci form convex, glistening, elevated, mucoid colonies 1–5 mm in diameter. Colonies are transparent or opaque, nonpigmented, and nonhemolytic. *Neisseria flavescens, Neisseria cinerea, Neisseria subflava,* and *Neisseria lactamica* may have yellow pigmentation. *Neisseria sicca* produces opaque, brittle, wrinkled colonies. *M catarrhalis* produces nonpigmented or pinkish gray opaque colonies.

C. Growth Characteristics

The neisseriae grow best under aerobic conditions, but some will grow in an anaerobic environment. They have complex growth requirements. Most neisseriae oxidize carbohydrates, producing acid but not gas, and their carbohydrate patterns are a means of distinguishing them (Table 20–1). The neisseriae produce oxidase and give positive oxidase reactions; the oxidase test is a key test for identifying them. When bacteria are spotted on a filter paper soaked with tetramethylparaphenylenediamine hydrochloride (oxidase), the neisseriae rapidly turn dark purple.

Meningococci and gonococci grow best on media containing complex organic substances such as heated blood, hemin, and animal proteins and in an atmosphere containing 5% CO_2 (eg, candle jar). Growth is inhibited by some toxic constituents of the medium, eg, fatty acids or salts. The organisms are rapidly killed by drying, sunlight, moist heat, and many disinfectants. They produce autolytic enzymes that result in rapid swelling and lysis in vitro at 25°C and at an alkaline pH.

NEISSERIA GONORRHOEAE

Gonococci oxidize only glucose and differ antigenically from the other neisseriae. Gonococci usually produce smaller colonies than those of the other neisseriae. Gonococci that require arginine, hypoxanthine, and uracil (Arg⁻, Hyx⁻, and Ura⁻ auxotype) tend to grow most slowly on primary culture. Gonococci isolated from clinical specimens or maintained by selective subculture have typical small colonies containing piliated bacteria. On nonselective subculture, larger colonies containing nonpiliated gonococci are also formed. Opaque and transparent variants of both the small and large colony types also occur; the opaque colonies are associated with the presence of a surface-exposed protein, Opa.

Antigenic Structure

N gonorrhoeae is antigenically heterogeneous and capable of changing its surface structures in vitro—and presumably in vivo—to avoid host defenses. Surface structures include the following.

TABLE 20–1   Biochemical Reactions of the Neisseriae and *Moraxella catarrhalis*

| | Growth on MTM, ML, or NYC Medium[a] | Acid Formed from | | | | DNAse |
		Glucose	Maltose	Lactose	Sucrose or Fructose	
Neisseria gonorrhoeae	+	+	−	−	−	−
Neisseria meningitidis	+	+	+	−	−	−
Neisseria lactamica	+	+	+	+	−	−
Neisseria sicca	−	+	+	−	+	−
Neisseria subflava	−	+	+	−	±	−
Neisseria mucosa	−	+	+	−	+	−
Neisseria flavescens	−	−	−	−	−	−
Neisseria cinerea	±	−	−	−	−	−
Neisseria polysaccharea	±	+	+	−	−	−
Neisseria elongata	−	−/w	−	−	−	−
Moraxella catarrhalis	−	−	−	−	−	+

[a]MTM, modified Thayer-Martin medium; ML, Martin-Lewis medium; NYC, New York City medium.

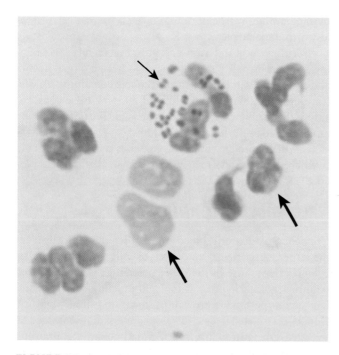

FIGURE 20–1 Gram stain of a urethral exudate of a patient with gonorrhea. Nuclei of many polymorphonuclear cells are seen (large arrows). Intracellular gram-negative diplococci (*Neisseria gonorrhoeae*) in one polymorphonuclear cell are marked by the small arrow.

A. Pili (Fimbriae)

Pili are the hair-like appendages that extend up to several micrometers from the gonococcal surface. They enhance attachment to host cells and resistance to phagocytosis. They are made up of stacked pilin proteins (MW 17,000–21,000). The amino terminal of the pilin molecule, which contains a

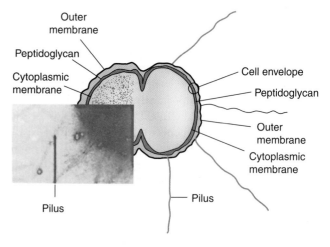

FIGURE 20–2 Collage and drawing of *Neisseria gonorrhoeae* showing pili and the three layers of the cell envelop.

high percentage of hydrophobic amino acids, is conserved. The amino acid sequence near the mid portion of the molecule also is conserved; this portion of the molecule serves in attachment to host cells and is less prominent in the immune response. The amino acid sequence near the carboxyl terminal is highly variable; this portion of the molecule is most prominent in the immune response. The pilins of almost all strains of *N gonorrhoeae* are antigenically different, and a single strain can make many antigenically distinct forms of pilin.

B. Por

Por protein extends through the gonococcal cell membrane. It occurs in trimers to form pores in the surface through which some nutrients enter the cell. Por proteins may impact intracellular killing of gonococci within neutrophils by preventing phagosome-lysosome fusion. In addition, variable resistance of gonococci to killing by normal human serum depends upon whether Por protein selectively binds to complement components C3b and C4b. The molecular weight of Por varies from 32 to 36 kDa. Each strain of gonococcus expresses only one of two types of Por, but the Por of different strains is antigenically different. Serologic typing of Por by agglutination reactions with monoclonal antibodies has distinguished 18 serovars of PorA and 28 serovars of PorB. (Serotyping is done only in reference laboratories.)

C. Opa Proteins

These proteins function in adhesion of gonococci within colonies and in attachment of gonococci to host cell receptors such as heparin-related compounds and CD66 or carcinoembryonic antigen-related cell adhesion molecules. One portion of the Opa molecule is in the gonococcal outer membrane, and the rest is exposed on the surface. The molecular weight of Opa ranges from 20 to 28 kDa. A strain of gonococcus can express no, one, two, or occasionally three types of Opa, though each strain has eleven or twelve genes for different Opas.

D. Rmp (Protein III)

This protein (MW 30–31 kDa) is antigenically conserved in all gonococci. It is a reduction-modifiable protein (Rmp) and changes its apparent molecular weight when in a reduced state. It associates with Por in the formation of pores in the cell surface.

E. Lipooligosaccharide

In contrast to the enteric gram-negative rods (see Chapters 2 and 15), gonococcal lipopolysaccharide (LPS) does not have long O-antigen side chains and is called a lipooligosaccharide. Its molecular weight is 3–7 kDa. Gonococci can express more than one antigenically different lipooligosaccharide (LOS) chain simultaneously. Toxicity in gonococcal infections is largely due to the endotoxic effects of LOS. Specifically, in the fallopian tube explant model, LOS causes ciliary loss and mucosal cell death.

In a form of molecular mimicry, gonococci make LOS molecules that structurally resemble human cell membrane glycosphingolipids. A structure is depicted in Figure 20–3. The gonococcal LOS and the human glycosphingolipid of the same structural class react with the same monoclonal antibody, indicating the molecular mimicry. The presence on the gonococcal surface of the same surface structures as human cells helps gonococci evade immune recognition.

The terminal galactose of human glycosphingolipids is often conjugated with sialic acid. Sialic acid is a nine-carbon, 5-N-acetylated ketulosonic acid also called N-acetylneuraminic acid (NANA). Gonococci do not make sialic acid but do make a sialyltransferase that functions to take NANA from the human nucleotide sugar cytidine 5′-monophospho-N-acetyl neuraminic acid (CMPNANA) and place the NANA on the terminal galactose of a gonococcal acceptor LOS. This sialylation affects the pathogenesis of gonococcal infection. It makes the gonococci resistant to killing by the human antibody-complement system and interferes with gonococcal binding to receptors on phagocytic cells.

Neisseria meningitidis and *Haemophilus influenzae* make many but not all of the same LOS structures as *N gonorrhoeae*. The biology of the LOS for the three species and for some of the nonpathogenic *Neisseria* species is similar. Four of the various serogroups of *N meningitidis* make different sialic acid capsules (see below), indicating that they also have biosynthetic pathways different from those of gonococci. These four serogroups sialylate their LOS using sialic acid from their endogenous pools.

F. Other Proteins

Several antigenically constant proteins of gonococci have poorly defined roles in pathogenesis. **Lip (H8)** is a surface-exposed protein that is heat-modifiable like Opa. The **Fbp (ferric-binding protein),** similar in molecular weight to Por, is expressed when the available iron supply is limited, eg, in human infection. Gonococci elaborate an **IgA1 protease** that splits and inactivates IgA1, a major mucosal immunoglobulin of humans. Meningococci, *Haemophilus influenzae,* and *Streptococcus pneumoniae* elaborate similar IgA1 proteases.

Genetics & Antigenic Heterogeneity

Gonococci have evolved mechanisms for frequently switching from one antigenic form (pilin, Opa, or LPS) to another antigenic form of the same molecule. This switching takes place in one in every $10^{2.5}$–10^3 gonococci, an extremely rapid rate of change for bacteria. Since pilin, Opa, and LPS are surface-exposed antigens on gonococci, they are important in the immune response to infection. The molecules' rapid switching from one antigenic form to another helps the gonococci elude the host immune system.

The switching mechanism for pilin, which has been the most thoroughly studied, is different from the mechanism for Opa.

Gonococci have multiple genes that code for pilin, but only one gene is inserted into the expression site. Gonococci can remove all or part of this pilin gene and replace it with all or part of another pilin gene. This mechanism allows gonococci to express many antigenically different pilin molecules over time.

The switching mechanism of Opa involves, at least in part, the addition or removal from the DNA of one or more of the pentameric coding repeats preceding the sequence that codes for the structural Opa gene. The switching mechanism of LPS is unknown.

The antigens and heterogeneity of types are shown in Table 20–2.

Gonococci contain several plasmids; 95% of strains have a small, "cryptic" plasmid (MW 2.4×10^6) of unknown function. Two other plasmids (MW 3.4×10^6 and 4.7×10^6)

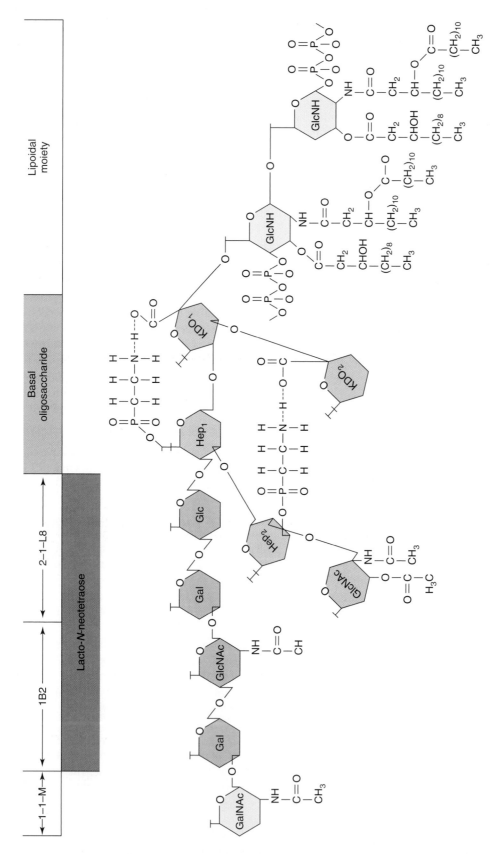

FIGURE 20–3 Structure of gonococcal lipooligosaccharide which has lacto-*N*-neotetraose and a terminal galactosamine in a structure similar to the human ganglioside glycosphingolipid series. The basal oligosaccharide is in light red and the lacto-*N*-neotetraose is in dark red. (Courtesy of JM Griffiss.)

TABLE 20–2 Antigenic Heterogeneity of *Neisseria gonorrhoeae*

Antigen	Number of Types
Pilin	Hundreds
Por (protein) (US System)	PorA with 18 subtypes PorB with 28 subtypes
Opa (protein II)	Many (perhaps hundreds)
Rmp (protein III)	One
Lipooligosaccharide	Eight or more
Fbp (iron-binding protein)	One
Lip (H8)	One
IgA1 protease	Two

contain genes that code for TEM-1 type (penicillinases) β-lactamases, which causes resistance to penicillin. These plasmids are transmissible by conjugation among gonococci; they are similar to a plasmid found in penicillinase-producing *Haemophilus* and may have been acquired from *Haemophilus* or other gram-negative organisms. Five to 20% of gonococci contain a plasmid (MW 24.5×10^6) with the genes that code for conjugation; the incidence is highest in geographic areas where penicillinase-producing gonococci are most common. High-level tetracycline resistance (minimal inhibitory concentrations [MIC] of ≥ 16 mg/L) has developed in gonococci by the insertion of a streptococcal gene *tetM* coding for tetracycline resistance into the conjugative plasmid.

Pathogenesis, Pathology, & Clinical Findings

Gonococci exhibit several morphologic types of colonies (see above), but only piliated bacteria appear to be virulent. Opa protein expression varies depending on the type of infection. Gonococci that form opaque colonies are isolated from men with symptomatic urethritis and from uterine cervical cultures at mid cycle. Gonococci that form transparent colonies are frequently isolated from men with asymptomatic urethral infection, from menstruating women, and from invasive forms of gonorrhea, including salpingitis and disseminated infection. Antigenic variation of surface proteins during infection allows the organism to circumvent host immune response.

Gonococci attack mucous membranes of the genitourinary tract, eye, rectum, and throat, producing acute suppuration that may lead to tissue invasion; this is followed by chronic inflammation and fibrosis. In males, there is usually urethritis, with yellow, creamy pus and painful urination. The process may extend to the epididymis. As suppuration subsides in untreated infection, fibrosis occurs, sometimes leading to urethral strictures. Urethral infection in men can be asymptomatic. In females, the primary infection is in the endocervix and extends to the urethra and vagina, giving rise to mucopurulent discharge. It may then progress to the uterine tubes, causing salpingitis, fibrosis, and obliteration of the tubes. Infertility occurs in 20% of women with gonococcal salpingitis. Chronic gonococcal cervicitis or proctitis is often asymptomatic.

Gonococcal bacteremia leads to skin lesions (especially hemorrhagic papules and pustules) on the hands, forearms, feet, and legs and to tenosynovitis and suppurative arthritis, usually of the knees, ankles, and wrists. Gonococci can be cultured from blood or joint fluid of only 30% of patients with gonococcal arthritis. Gonococcal endocarditis is an uncommon but severe infection. Gonococci sometimes cause meningitis and eye infections in adults; these have manifestations similar to those due to meningococci. Complement deficiency is frequently found in patients with gonoccal bacteremia. Patients with bacteremia, especially if recurrent, should be tested for total hemolytic complement activity.

Gonococcal ophthalmia neonatorum, an infection of the eye of the newborn, is acquired during passage through an infected birth canal. The initial conjunctivitis rapidly progresses and, if untreated, results in blindness. To prevent gonococcal ophthalmia neonatorum, instillation of tetracycline, erythromycin, or silver nitrate into the conjunctival sac of the newborn is compulsory in the United States.

Gonococci that produce localized infection are often serum-sensitive (killed by antibody and complement).

Diagnostic Laboratory Tests

A. Specimens

Pus and secretions are taken from the urethra, cervix, rectum, conjunctiva, throat, or synovial fluid for culture and smear. Blood culture is necessary in systemic illness, but a special culture system is helpful, since gonococci (and meningococci) may be susceptible to the polyanethol sulfonate present in standard blood culture media.

B. Smears

Gram-stained smears of urethral or endocervical exudate reveal many diplococci within pus cells. These give a presumptive diagnosis. Stained smears of the urethral exudate from men have a sensitivity of about 90% and a specificity of 99%. Stained smears of endocervical exudates have a sensitivity of about 50% and a specificity of about 95% when examined by an experienced microscopist. Additional diagnostic testing of urethral exudates from men is not necessary when the stain is positive, but nucleic acid amplification tests (NAAT) or cultures should be done for women. Stained smears of conjunctival exudates can also be diagnostic, but those of specimens from the throat or rectum are generally not helpful.

C. Culture

Immediately after collection, pus or mucus is streaked on enriched selective medium (eg, modified Thayer-Martin medium) and incubated in an atmosphere containing 5% CO_2 (candle extinction jar) at 37°C. To avoid over-growth by contaminants, the selective medium contains antimicrobial drugs (eg, vancomycin, 3 µg/mL; colistin, 7.5 µg/mL; amphotericin B, 1 µg/mL; and trimethoprim, 3 µg/mL). If immediate incubation is not possible, the specimen should be placed in a CO_2-containing transport-culture system. Forty-eight hours after culture, the organisms can be quickly identified by their appearance on a Gram-stained smear, by oxidase positivity, and by coagglutination, immunofluorescence staining, or other laboratory tests. The species of subcultured bacteria may be determined by oxidation of specific carbohydrates (Table 20–1). The gonococcal isolates from anatomic sites other than the genital tract or from children should be identified as to species using two different confirmatory tests because of the legal and social implications of a positive culture.

D. Nucleic Acid Amplification Tests

Several Food and Drug Administration-cleared nucleic acid amplification assays are available for direct detection of *N gonorrhoeae* in genitourinary specimens and these are the preferred tests from these sources. In general, these assays have excellent sensitivity and specificity in symptomatic, high-prevalence populations. Advantages include better detection, more rapid results, and the ability to use urine as a specimen source. Disadvantages include poor specificity of some assays due to cross reactivity with nongonococcal *Neisseria* species. These assays are not recommended for use for the diagnosis of extragenital gonococcal infections or for infection in children. NAATs are not recommended as tests of cure since nucleic acid may persist in patient specimens for up to 3 weeks after successful treatment.

E. Serology

Serum and genital fluid contain IgG and IgA antibodies against gonococcal pili, outer membrane proteins, and LPS. Some IgM of human sera is bactericidal for gonococci in vitro.

In infected individuals, antibodies to gonococcal pili and outer membrane proteins can be detected by immunoblotting, radioimmunoassay, and ELISA (enzyme-linked immunosorbent assay) tests. However, these tests are not useful as diagnostic aids for several reasons: gonococcal antigenic heterogeneity; the delay in development of antibodies in acute infection; and a high background level of antibodies in the sexually active population.

Immunity

Repeated gonococcal infections are common. Protective immunity to reinfection does not appear to develop as part of the disease process, because of the antigenic variety of gonococci. While antibodies can be demonstrated, including the IgA and IgG on mucosal surfaces, they either are highly strain-specific or have little protective ability.

Treatment

Since the development and widespread use of penicillin, gonococcal resistance to penicillin has gradually risen, owing to the selection of chromosomal mutants, so that many strains now require high concentrations of penicillin G for inhibition (MIC ≥ 2 µg/mL). Penicillinase-producing *N gonorrhoeae* (PPNG) also have increased in prevalence (see above). Chromosomally mediated resistance to tetracycline (MIC ≥ 2 µg/mL) is common. High-level resistance to tetracycline (MIC ≥ 32 µg/mL) also occurs. Spectinomycin resistance as well as resistance to fluoroquinolones has been noted. Single-dose fluoroquinolone treatment was recommended for treatment of gonococcal infections from 1993 until 2006. Since 2006, rates of quinolone resistance among gonococcal isolates have exceeded 5% in men who have sex with men and also heterosexual men. Because of the problems with antimicrobial resistance in *N gonorrhoeae*, the Centers for Disease Control and Prevention (CDC) recommend that uncomplicated genital or rectal infections be treated with ceftriaxone (125 mg) given intramuscularly as a single dose. Additional therapy with azithromycin 1 g orally in a single dose or with doxycycline 100 mg orally twice a day for 7 days is recommended for the possible concomitant chlamydial infection. Azithromycin has been found to be safe and effective in pregnant women, but doxycycline is contraindicated. Modifications of these therapies are recommended for other types of *N gonorrhoeae* infection. See CDC Web site http://www.cdc.gov/std/treatment/2006/updated-regimens.htm.

Since other sexually transmitted diseases may have been acquired at the same time as gonorrhea, steps must also be taken to diagnose and treat these diseases (see discussions of chlamydiae, syphilis, etc).

Epidemiology, Prevention, & Control

Gonorrhea is worldwide in distribution. In the United States its incidence rose steadily from 1955 until the late 1970s, when the incidence was between 400 and 500 cases per 100,000 population. Between 1975 and 1997, there was a 74% decline in the rate of reported gonococcal infections. Thereafter, the rates plateaued until 2005 and 2006 when they increased again slightly to 120.9 cases per 100,000 population (5.5% increase). Gonorrhea is exclusively transmitted by sexual contact, often by women and men with asymptomatic infections. The infectivity of the organism is such that the chance of acquiring infection from a single exposure to an infected sexual partner is 20–30% for men and even greater for women. The infection rate can be reduced by avoiding multiple sexual partners, rapidly eradicating gonococci from infected individuals by means of early diagnosis and treatment, and finding cases and contacts through education and screening of populations at high risk. Mechanical prophylaxis (condoms) provides partial protection. Chemoprophylaxis is of limited value because of the rise in antibiotic resistance of the gonococcus.

Gonococcal ophthalmia neonatorum is prevented by local application of 0.5% erythromycin ophthalmic ointment or 1% tetracycline ointment to the conjunctiva of newborns. Although instillation of silver nitrate solution is also effective and is the classic method for preventing ophthalmia neonatorum, silver nitrate is difficult to store and causes conjunctival irritation; its use has largely been replaced by use of erythromycin or tetracycline ointment.

NEISSERIA MENINGITIDIS

Antigenic Structure

At least 13 serogroups of meningococci have been identified by immunologic specificity of capsular polysaccharides. The most important serogroups associated with disease in humans are A, B, C, X, Y, and W-135. The group A polysaccharide is a polymer of *N*-acetylmannosamine phosphate, and that of group C is a polymer of *N*-acetyl-*O*-acetylneuraminic acid. Meningococcal antigens are found in blood and cerebrospinal fluid of patients with active disease. Outbreaks and sporadic cases in the Western Hemisphere in the last decade have been caused mainly by groups B, C, W-135, and Y; outbreaks in southern Finland and São Paulo, Brazil, were due to groups A and C; outbreaks in New Zealand have been due to a particular B strain; those in Africa were due mainly to group A. Group C and, especially, group A are associated with epidemic disease.

The outer membrane proteins of meningococci have been divided into classes on the basis of molecular weight. All strains have either class 1, class 2, or class 3 proteins; these are analogous to the Por proteins of gonococci and are responsible for the serotype specificity of meningococci. They help form pores in the meningococcal cell wall. As many as 20 serotypes have been defined; serotypes 2 and 15 have been associated with epidemic disease. The Opa (class 5) protein is comparable to Opa of the gonococci. Meningococci are piliated, but unlike gonococci, they do not form distinctive colony types indicating piliated bacteria. Meningococcal LPS is responsible for many of the toxic effects found in meningococcal disease. The highest levels of endotoxin measured in sepsis have been found in patients with meningococcemia (50- to 100-fold greater than with other gram-negative infections).

Pathogenesis, Pathology, & Clinical Findings

Humans are the only natural hosts for whom meningococci are pathogenic. The nasopharynx is the portal of entry. There, the organisms attach to epithelial cells with the aid of pili; they may form part of the transient flora without producing symptoms. From the nasopharynx, organisms may reach the bloodstream, producing bacteremia; the symptoms may be like those of an upper respiratory tract infection. Fulminant meningococcemia is more severe, with high fever and hemorrhagic rash; there may be disseminated intravascular coagulation and circulatory collapse (Waterhouse-Friderichsen syndrome).

Meningitis is the most common complication of meningococcemia. It usually begins suddenly, with intense headache, vomiting, and stiff neck, and progresses to coma within a few hours.

During meningococcemia, there is thrombosis of many small blood vessels in many organs, with perivascular infiltration and petechial hemorrhages. There may be interstitial myocarditis, arthritis, and skin lesions. In meningitis, the meninges are acutely inflamed, with thrombosis of blood vessels and exudation of polymorphonuclear leukocytes, so that the surface of the brain is covered with a thick purulent exudate.

It is not known what transforms an asymptomatic infection of the nasopharynx into meningococcemia and meningitis, but this can be prevented by specific bactericidal serum antibodies against the infecting serotype. *Neisseria* bacteremia is favored by the absence of bactericidal antibody (IgM and IgG), inhibition of serum bactericidal action by a blocking IgA antibody, or a complement component deficiency (C5, C6, C7, or C8). Meningococci are readily phagocytosed in the presence of a specific opsonin.

Diagnostic Laboratory Tests

A. Specimens

Specimens of blood are taken for culture, and specimens of spinal fluid are taken for smear, culture, and chemical determinations. Nasopharyngeal swab cultures are suitable for carrier surveys. Puncture material from petechiae may be taken for smear and culture.

B. Smears

Gram-stained smears of the sediment of centrifuged spinal fluid or of petechial aspirate often show typical neisseriae within polymorphonuclear leukocytes or extracellularly.

C. Culture

Culture media without sodium polyanethol sulfonate are helpful in culturing blood specimens. Cerebrospinal fluid specimens are plated on "chocolate" agar and incubated at 37°C in an atmosphere of 5% CO_2 (candle jar). Freshly drawn spinal fluid can be directly incubated at 37°C if agar culture media are not immediately available. A modified Thayer-Martin medium with antibiotics (vancomycin, colistin, amphotericin) favors the growth of neisseriae, inhibits many other bacteria, and is used for nasopharyngeal cultures. Presumptive colonies of neisseriae on solid media, particularly in mixed culture, can be identified by Gram's stain and the oxidase test. Spinal fluid and blood generally yield pure cultures that can be further identified by carbohydrate oxidative reactions (Table 20–1) and agglutination with type-specific or polyvalent serum.

D. Serology

Antibodies to meningococcal polysaccharides can be measured by latex agglutination or hemagglutination tests or by their bactericidal activity. These tests are done only in reference laboratories.

Immunity

Immunity to meningococcal infection is associated with the presence of specific, complement-dependent, bactericidal antibodies in the serum. These antibodies develop after subclinical infections with different strains or injection of antigens and are group-specific, type-specific, or both. The immunizing antigens for groups A, C, Y, and W-135 are the capsular polysaccharides. For group B, a specific antigen suitable for use as a vaccine has not been defined; however, group B vaccines with mixtures of antigens have been used in many parts of the world. Currently there are two vaccine types against serogroups A, C, Y, and W-135 available in the United States. A polysaccharide tetravalent vaccine (Menomune® Sanofi Pasteur, Inc.) in which each dose consists of four purified bacterial capsular polysaccharides is poorly immunogenic in children under 18 months, does not confer long-lasting immunity, and does not cause a sustainable reduction in nasopharyngeal carriage. A tetravalent conjugate vaccine approved in 2005 (Menactra™, Sanofi Pasteur, Inc.) is licensed for use in persons 11–55 years of age. It contains capsular polysaccharide conjugated to diphtheria toxoid. The advantage of this vaccine is that a T-cell–dependent response to vaccine is induced. This enhances primary response among infants and substantially reduces asymptomatic carriage. Routine vaccination of young adolescents (ages 11–12 years) before high school using the conjugated vaccine is now recommended. Vaccination is also recommended for persons 11–55 years of age who are among the following at-risk groups: persons with functional or surgical asplenia; persons with complement deficiencies; travelers to highly endemic areas (eg, sub-Saharan Africa); "closed populations" such as college freshman living in dorms and the military; populations experiencing a community outbreak; and for clinical laboratory workers (microbiologists).

Treatment

Penicillin G is the drug of choice for treating meningococcal disease. Either chloramphenicol or a third-generation cephalosporin such as cefotaxime or ceftriaxone is used in persons allergic to penicillins.

Epidemiology, Prevention, & Control

Meningococcal meningitis occurs in epidemic waves (eg, in military encampments, in religious pilgrims, and in sub-Saharan Africa; in Brazil, there were more than 15,000 cases in 1974) and a smaller number of sporadic interepidemic cases. Five to 30% of the normal population may harbor meningococci (often nontypeable isolates) in the nasopharynx during interepidemic periods. During epidemics, the carrier rate goes up to 70–80%. A rise in the number of cases is preceded by an increased number of respiratory carriers. Treatment with oral penicillin does not eradicate the carrier state. Rifampin, 600 mg orally twice daily for 2 days (or ciprofloxacin in adults, 500 mg as a single dose), can often eradicate the carrier state and serve as chemoprophylaxis for household and other close contacts. Since the appearance of many sulfonamide-resistant meningococci, chemoprophylaxis with sulfonamides is no longer reliable.

Clinical cases of meningitis present only a negligible source of infection, and isolation therefore has only limited usefulness. More important is the reduction of personal contacts in a population with a high carrier rate. This is accomplished by avoidance of crowding or administration of vaccines as discussed above. As mentioned, such vaccines are currently used in selected populations (eg, the military and in civilian epidemics).

OTHER NEISSERIAE

Neisseria lactamica very rarely causes disease but is important because it grows in the selective media (eg, modified Thayer-Martin medium) used for cultures of gonococci and meningococci from clinical specimens. *N lactamica* can be cultured from the nasopharynx of 3–40% of persons and most often is found in children. Unlike the other neisseriae, it ferments lactose.

Neisseria sicca, Neisseria subflava, Neisseria cinerea, Neisseria mucosa, and *Neisseria flavescens* are also members of the normal flora of the respiratory tract, particularly the nasopharynx, and very rarely produce disease. *N cinerea* sometimes resembles *N gonorrhoeae* because of its morphology and positive hydroxyprolyl aminopeptidase reaction.

Moraxella catarrhalis was previously named *Branhamella catarrhalis* and before that *Neisseria catarrhalis*. It is a member of the normal flora in 40–50% of healthy school children. *M catarrhalis* causes bronchitis, pneumonia, sinusitis, otitis media, and conjunctivitis. It is also of concern as a cause of infection in immunocompromised patients. Most strains of *M catarrhalis* from clinically significant infections produce β-lactamase. *M catarrhalis* can be differentiated from the neisseriae by its lack of carbohydrate fermentation and by its production of DNase. It produces butyrate esterase, which forms the basis for rapid fluorometric tests for identification.

REVIEW QUESTIONS

1. The inhabitants of a group of small villages in rural sub-Saharan Africa suffered an epidemic of meningitis. Ten percent of the

people died, most of them under the age of 15 years. The micro-organism that most likely caused this epidemic was

(A) *Streptococcus agalactiae* (group B)
(B) *Escherichia coli* K1 (capsular type 1)
(C) *Haemophilus influenzae* serotype b
(D) *Neisseria meningitidis* serogroup A
(E) West Nile virus

2. A 19-year-old man presented to the clinic with a urethral discharge for the past 24 hours. *Neisseria gonorrhoeae* was cultured from the specimen and found to be β-lactamase-positive and resistant to high levels (≥32 μg/mL) of tetracycline. Which of the following statements about these antimicrobial resistance factors is correct?

(A) β-lactamase production and high-level resistance to tetracycline are both mediated by genes on plasmids
(B) β-lactamase production is mediated by a gene on the bacterial chromosome while high-level tetracycline resistance is mediated by a gene on a plasmid
(C) β-lactamase production is mediated by a gene on a plasmid while high-level tetracycline resistance is mediated by a gene on the bacterial chromosome
(D) β-lactamase production and high-level resistance to tetracycline are both mediated by genes on the bacterial chromosome

3. A 6-year-old boy develops fever and head ache. He is taken to the emergency room where he is noted to have a stiff neck, suggesting meningeal irritation. A lumbar punc-ture is done and culture of the cerebrospinal fluid grows *Neisseria meningitidis* serogroup B Which of the following should be considered for his family (household) members?

(A) No prophylaxis or other steps are necessary
(B) They should be given *Neisseria meningitides* pilin vaccine
(C) They should be given *Neisseria meningitides* serogroup B polysaccharide capsule vaccine
(D) They should be given rifampin prophylaxis
(E) They should be given sulfonamide Prophylaxis

4. An 18-year-old woman who reports unprotected sex with a new partner 2 weeks previously develops fever and left lower quadrant abdominal pain with onset in association with her menstrual period. On pelvic examination in the emergency room there is bilateral tenderness when the uterus is palpated. A mass 2–3 cm in diameter is felt on the left, suggestive of tubo-ovarian abscess. Subsequently, *Neisseria gonorrhoeae* is cultured from her endocervix. The diagnosis is gonococcal pelvic inflammatory disease. A common sequela of this infection is

(A) Cancer of the cervix
(B) Urethral stricture
(C) Uterine fibroid tumors
(D) Infertility
(E) Vaginal-rectal fistula

5. A 38-year-old vice squad police officer comes to the emergency room with a chief complaint expressed as follows: "I have disseminated gonococcal infection again." He is correct. Cultures of his urethra and knee fluid yield *Neisseria gonorrhoeae*. He has previously had five episodes of disseminated gonococcal infection. The patient should be evaluated for

(A) Selective IgA deficiency
(B) A polymorphonuclear cell chemotactic Defect

(C) Deficiency of a late-acting complement component C5, C6, C7, or C8
(D) Absent lymphocyte adenosine deaminase activity
(E) Myeloperoxidase deficiency

6. Which of the following individuals should routinely receive vaccination with the conjugate meningococcal vaccine?

(A) A healthy young adolescent entering high school
(B) A healthy child entering kindergarten
(C) A 60-year-old man with insulin dependent diabetes
(D) A healthy 40-year-old technician who works in a cancer research lab
(E) A 65-year-old woman with coronary artery disease

7. A 25-year-old sexually active woman presents with purulent vaginal discharge and dysuria 7 days after having unprotected sexual intercourse with a new partner. Of the choices below what is the most sensitive diagnostic method for determining the likely etiologic agent?

(A) Gram's stain
(B) An enzyme immunoassay
(C) Bacterial culture on selective media
(D) A nucleic acid amplification test
(E) Serology

8. What is the currently recommended treatment for gonococcal urethritis in men who have sex with men in the United States?

(A) Single dose of an oral fluoroquinolone
(B) Seven days of oral doxycycline
(C) Ceftriaxone given intramuscularly as a single dose
(D) Spectinomycin given intramuscularly as a single dose
(E) Seven days of oral amoxicillin

9. Which of the following cell components produced by *Neisseria gonorrheae* is responsible for attachment to host cells?

(A) Lipooligosaccharide
(B) Pili (fimbriae)
(C) IgA1 protease
(D) Outer membrane porin protein
(E) Iron-binding protein

10. A 60-year-old man with severe chronic lung disease presents with fever, cough productive of purulent sputum and worsening hypoxemia. A sputum sample is collected and the specimen is sent promptly to the laboratory. Microscopic examination of a Gram's stain reveals numerous polymorphonuclear leukocytes and predominately gram-negative diplococci that are both intracellular and extracellular. The organism grows well on 5% SBA and chocolate agar and is positive for butyrate esterase. What is the most likely organism causing this man's illness?

(A) *Neisseria gonorrheae*
(B) *Neisseria lactamica*
(C) *Moraxella catarrhalis*
(D) *Haemophilus influenzae*
(E) *Neisseria meningitidis*

Answers

1. D	4. D	7. D	10. C
2. A	5. C	8. C	
3. D	6. A	9. B	

REFERENCES

Apicella MA: *Neisseria meningitidis.* In: *Mandell, Douglas, and Bennett's Principles and Practice of Infectious Diseases,* 7th ed. Mandell GL, Bennett JE, Dolin R (editors). Churchill Livingstone Elsevier, 2010.

Janda WM, Gaydos CA: Neisseria. In: *Manual of Clinical Microbiology,* 9th ed. Murray PR, Baron EJ, Jorgensen JH, Landry ML, Pfaller MA (editors). ASM Press, 2007.

Marrazzo JM: *Neisseria gonorrhoeae.* In: *Mandell, Douglas, and Bennett's Principles and Practice of Infectious Diseases,* 7th ed. Mandell GL, Bennett JE, Dolin R (editors). Churchill Livingstone Elsevier, 2010.

Infections Caused by Anaerobic Bacteria

Medically important infections due to anaerobic bacteria are common. The infections are often polymicrobial—ie, the anaerobic bacteria are found in mixed infections with other anaerobes, facultative anaerobes, and aerobes (see the glossary of definitions). Anaerobic bacteria are found throughout the human body—on the skin, on mucosal surfaces, and in high concentrations in the mouth and gastrointestinal tract—as part of the normal microbiota (see Chapter 10). Infection results when anaerobes and other bacteria of the normal flora contaminate normally sterile body sites.

Several important diseases are caused by anaerobic *Clostridium* species from the environment or from normal flora: botulism, tetanus, gas gangrene, food poisoning, and pseudomembranous colitis. These diseases are discussed in Chapters 9 and 11 and briefly later in this chapter.

GLOSSARY

Aerobic bacteria: Those that require oxygen as a terminal electron acceptor and will not grow under anaerobic conditions (ie, in the absence of O_2). Some *Micrococcus* species and *Nocardia asteroides* are obligate aerobes (ie, they must have oxygen to survive).

Anaerobic bacteria: Those that do not use oxygen for growth and metabolism but obtain their energy from fermentation reactions. A functional definition of anaerobes is that they require reduced oxygen tension for growth and fail to grow on the surface of solid medium in 10% CO_2 in ambient air. *Bacteroides* and *Clostridium* species are examples of anaerobes.

Capnophilic bacteria: Those that require carbon dioxide for growth.

Facultative anaerobes: Bacteria that can grow either oxidatively, using oxygen as a terminal electron acceptor, or anaerobically, using fermentation reactions to obtain energy. Such bacteria are common pathogens. *Streptococcus* species and the Enterobacteriaceae (eg, *Escherichia coli*) are among the many facultative anaerobes that cause disease. Often, bacteria that are facultative anaerobes are called "aerobes."

PHYSIOLOGY & GROWTH CONDITIONS FOR ANAEROBES

Anaerobic bacteria will not grow in the presence of oxygen and are killed by oxygen or toxic oxygen radicals (see below). pH and oxidation-reduction potential (E_h) are also important in establishing conditions that favor growth of anaerobes. Anaerobes grow at a low or negative E_h.

Aerobes and facultative anaerobes often have the metabolic systems listed below, whereas anaerobic bacteria frequently do not.

(1) Cytochrome systems for the metabolism of O_2.

(2) Superoxide dismutase (SOD), which catalyzes the following reaction:

$$O_2^- + O_2^- + 2H^+ \rightarrow H_2O_2 + O_2$$

(3) Catalase, which catalyzes the following reaction:

$$2H_2O_2 \rightarrow 2H_2O + O_2 \,(\text{gas bubbles})$$

Anaerobic bacteria do not have cytochrome systems for oxygen metabolism. Less fastidious anaerobes may have low levels of SOD and may or may not have catalase. Most bacteria of the *Bacteroides fragilis* group have small amounts of both catalase and SOD. There appear to be multiple mechanisms for oxygen toxicity. Presumably, when anaerobes have SOD or catalase (or both), they are able to negate the toxic effects of oxygen radicals and hydrogen peroxide and thus tolerate oxygen. **Obligate anaerobes** usually lack superoxide dismutase and catalase and are susceptible to the lethal effects of oxygen; such strict obligate anaerobes are infrequently isolated from human infections, and most anaerobic infections of humans are caused by "moderately obligate anaerobes."

The ability of anaerobes to tolerate oxygen or grow in its presence varies from species to species. Similarly, there is strain-to-strain variation within a given species (eg, one strain of *Prevotella melaninogenica* can grow at an O_2 concentration of 0.1% but not of 1%; another can grow at a concentration of 2% but not of 4%). Also, in the absence of oxygen some anaerobic bacteria will grow at a more positive E_h.

Facultative anaerobes grow as well or better under anaerobic conditions than they do under aerobic conditions. Bacteria that are facultative anaerobes are often termed "aerobes." When a facultative anaerobe such as *E coli* is present at the site of an infection (eg, abdominal abscess), it can rapidly consume all available oxygen and change to anaerobic metabolism, producing an anaerobic environment and low E_h and thus allow the anaerobic bacteria that are present to grow and produce disease.

ANAEROBIC BACTERIA FOUND IN HUMAN INFECTIONS

Since the 1990s, the taxonomic classification of the anaerobic bacteria has changed significantly due to the application of molecular sequencing and DNA–DNA hybridization technologies. The nomenclature used in this chapter refers to genera of anaerobes frequently found in human infections and to certain species recognized as important pathogens of humans. Anaerobes commonly found in human infections are listed in Table 21–1.

Gram-Negative Anaerobes

A. Gram-Negative Bacilli

1. *Bacteroides*—The *Bacteroides* species are very important anaerobes that cause human infection. They are a large group of bile-resistant, nonspore forming, slender gram-negative rods that may appear as coccobacilli. Many species previously included in the genus *Bacteroides* have been reclassified into the genus *Prevotella* or the genus *Porphyromonas*. Those species retained in the *Bacteroides* genus are members of the *B fragilis* group (~20 species).

Bacteroides species are normal inhabitants of the bowel and other sites. Normal stools contain 10^{11} *B fragilis* organisms per gram (compared with 10^8/g for facultative anaerobes). Other commonly isolated members of the *B fragilis* group include *B ovatus*, *B distasonis*, *B vulgatus*, and *B thetaiotaomicron*. *Bacteroides* species are most often implicated in intra-abdominal infections, usually under circumstances of disruption of the intestinal wall as occurs in perforations related to surgery or trauma, acute appendicitis, and diverticulitis. These infections are often polymicrobial—anaerobic cocci, *Clostridium* species and *Eubacterium* may also be found. Both *B fragilis* and *B thetaiotaomicron* are implicated in serious intrapelvic infections such as pelvic inflammatory disease and ovarian abscesses. *B fragilis* group species are the most common species recovered in some series of anaerobic bacteremia, and these organisms are associated with very high mortality. As discussed later in the chapter, *B fragilis* is capable of elaborating numerous virulence factors which contribute to its pathogenicity and mortality in the host.

2. *Prevotella*—The *Prevotella* species are gram-negative bacilli and may appear as slender rods or coccobacilli. Most commonly isolated are *P melaninogenica*, *P bivia*, and *P disiens*. *P melaninogenica* and similar species are found in infections associated with the upper respiratory tract. *P bivia* and *P disiens* occur in the female genital tract. *Prevotella* species are found in brain and lung abscesses, in empyema, and in pelvic inflammatory disease and tubo-ovarian abscesses.

In these infections the prevotellae are often associated with other anaerobic organisms that are part of the normal flora—particularly peptostreptococci, anaerobic gram-positive rods, and *Fusobacterium* species—as well as gram-positive and gram-negative facultative anaerobes that are part of the normal flora.

3. *Porphyromonas*—The *Porphyromonas* species also are gram-negative bacilli that are part of the normal oral flora and occur at other anatomic sites as well. *Porphyromonas* species can be cultured from gingival and periapical tooth infections and, more commonly, breast, axillary, perianal, and male genital infections.

4. Fusobacteria—There are approximately 13 *Fusobacterium* species, but most human infections are caused by *Fusobacterium necrophorum* and *Fusobacterium nucleatum*. Both species differ in morphology and habitat as well as the range of associated infections. *F necrophorum* is a very pleomorphic, long rod with round ends and tends to make bizarre forms. It is not a component of the healthy oral cavity. *F necrophorum* is quite virulent causing severe infections of

TABLE 21–1   Anaerobic Bacteria of Clinical Importance

Genera	Anatomic Site
Bacilli (rods)	
Gram-negative	
Bacteroides fragilis group	Colon
	Mouth
Prevotella melaninogenica	Mouth, colon, genitourinary tract
Fusobacterium	
Gram-positive	Mouth
Actinomyces	Vagina
Lactobacillus	Skin
Propionibacterium	Mouth, colon
Eubacterium, Bifidobacterium, and Arachnia	Colon[a]
Clostridium	
Cocci (spheres)	
Gram-positive	
Peptoniphilus	Colon, mouth, skin, genitourinary tract
Peptostreptococcus	Colon, mouth, skin, genitourinary tract
Peptococcus	
Finegoldia	
Gram-negative	
Veillonella	Mouth, colon

[a]Also found in soil.

the head and neck that can progress to a complicated infection called Lemierre's disease. The latter is characterized by acute jugular vein septic thrombophlebitis that progresses to sepsis with metastatic abscesses of the lungs, mediastinum, pleural space, and liver. Lemierre's disease is most common among older children and young adults and often occurs in association with infectious mononucleosis. *F necrophorum* is also seen in polymicrobial, intra-abdominal infections. *F nucleatum* is a thin rod with tapered ends (needle-shaped morphology) and is a significant component of the gingival microbiota as well as the genital, gastrointestinal, and upper respiratory tracts. As such, it is frequently encountered in a variety of clinical infections such as pleuropulmonary infections, obstetric infections, significantly chorioamnionitis, and occasionally brain abscesses complicating periodontal disease. Rarely does it cause bacteremia in neutropenic patients.

B. Gram-Negative Cocci

Veillonella species are a group of small, anaerobic, gram-negative cocci that are part of the normal flora of the mouth, the nasopharynx, and probably the intestine. Previously known by various names, they are now collectively known as the veillonellae. Though occasionally isolated in polymicrobial anaerobic infections, they are rarely the sole cause of an infection.

Gram-Positive Anaerobes

A. Gram-Positive Bacilli

1. *Actinomyces*—The *Actinomyces* group includes several species that cause actinomycosis, of which *Actinomyces israelii* and *Actinomyces gerencseriae* are the ones most commonly encountered. Several new, recently described species that are not associated with actinomycosis have been associated with infections of the groin, urogenital area, breast, and axilla and postoperative infections of the mandible, eye, and head and neck. Some species have also been implicated in cases of endocarditis particularly among substance abusers. These newly described species are aerotolerant and form small, nondescript colonies that are probably frequently overlooked as contaminants. On Gram stain, they vary considerably in length: they may be short and club-shaped or long, thin, beaded filaments. They may be branched or unbranched. Because they often grow slowly, prolonged incubation of the culture may be necessary before laboratory confirmation of the clinical diagnosis of actinomycosis can be made. Some strains produce colonies on agar that resemble molar teeth. Some *Actinomyces* species are oxygen-tolerant (aerotolerant) and grow in the presence of air; these strains may be confused with *Corynebacterium* species (diphtheroids; see Chapter 12). Actinomycosis is a chronic suppurative and granulomatous infection that produces pyogenic lesions with interconnecting sinus tracts that contain granules composed of microcolonies of the bacteria embedded in tissue elements (Figure 21–1A–C). Infection is initiated by

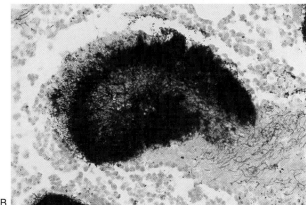

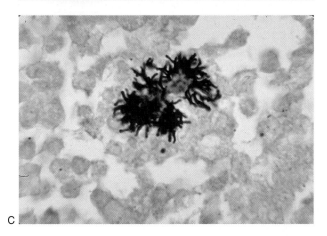

FIGURE 21–1 *Actinomyces* species. **A:** Colony of *Actinomyces* species after 72 hours growth on brain-heart infusion agar, which usually yields colonies about 2 mm in diameter; they are often termed "molar tooth" colonies. (Courtesy CDC Public Health Image Library, L Georg.) **B:** Granule of *Actinomyces* species in tissue with Brown and Breen stain. Original magnification × 400. Filaments of the branching bacilli are visible at the periphery of the granule. Such granules are commonly called "sulfur granules" because of their unstained gross yellow color. (Courtesy CDC Public Health Image Library.) **C:** *Actinomyces naeslundii* in a brain abscess stained with methylamine silver stain. Branching bacilli are visible. Original magnification × 1000. (Courtesy of CDC Public Health Image Library, L Georg.)

trauma that introduces these endogenous bacteria into the mucosa. The organisms grow in an anaerobic niche, induce a mixed inflammatory response, and spread with the formation of sinuses, which contain the granules and may drain to the surface. The infection causes swelling and may spread to neighboring organs including the bones.

Based on the site of involvement, the three common forms are cervicofacial, thoracic, and abdominal actinomycosis. Cervicofacial disease presents as a swollen, erythematosus process in the jaw area (known as "lumpy jaw"). With progression the mass becomes fluctuant, producing draining fistulas. The disease will extend to contiguous tissue, bone, and lymph nodes of the head and neck. The symptoms of thoracic actinomycosis resemble those of a subacute pulmonary infection: mild fever, cough, and purulent sputum. Eventually lung tissue is destroyed, sinus tracts may erupt through to the chest wall, and invasion of the ribs may occur. Abdominal actinomycosis often follows a ruptured appendix or an ulcer. In the peritoneal cavity, the pathology is the same, but any of several organs may be involved. Genital actinomycosis is a rare occurrence in women that results from colonization of an intrauterine device with subsequent invasion.

Diagnosis can be made by examining pus from draining sinuses, sputum, or specimens of tissue for the presence of sulfur granules. The granules are hard, lobulated, and composed of tissue and bacterial filaments, which are club-shaped at the periphery. Specimens should be cultured anaerobically on appropriate media. Treatment requires prolonged administration of a penicillin (6–12 months). Clindamycin or erythromycin is effective in penicillin-allergic patients. Surgical excision and drainage may be required.

2. Lactobacillus—*Lactobacillus* species are major members of the normal flora of the vagina. The lactic acid product of their metabolism helps maintain the low pH of the normal adult female genital tract. They rarely cause disease.

3. Propionibacterium—*Propionibacterium* species are members of the normal flora of the skin, oral cavity, large intestine, conjunctiva, and external ear canal. Their metabolic products include propionic acid, from which the genus name derives. On Gram stain, they are highly pleomorphic, showing curved, clubbed, or pointed ends, long forms with beaded uneven staining, and occasionally coccoid or spherical forms. *Propionibacterium acnes,* often considered an opportunistic pathogen, causes the disease acne vulgaris and is associated with a variety of inflammatory conditions. It causes acne by producing lipases that split free fatty acids off from skin lipids. These fatty acids can produce tissue inflammation that contributes to acne formation. In addition, *P acnes* is frequently a cause of postsurgical wound infections particularly those that involve insertion of devices, such as prosthetic joint infections, particularly of the shoulder, central nervous system shunt infections, osteomyelitis, endocarditis, and endophthalmitis. Because it is part of the normal skin flora, *P acnes* sometimes contaminates blood or

cerebrospinal fluid cultures that are obtained by penetrating the skin. It is therefore important (but often difficult) to differentiate a contaminated culture from one that is positive and indicates infection.

4. Eubacterium, Eggerthella, Bifidobacterium, & Arachnia—These four genera are made up of anaerobic, pleomorphic, gram-positive rods. There are several species. They are found in mixed infections associated with oropharyngeal or bowel flora.

5. Clostridium—Clostridia are gram-positive, spore-forming bacilli (see Chapter 11). There are more than 200 species. The major diseases associated with these bacteria are caused by exotoxins (see Chapter 9).

Spores of *Clostridium tetani*, which causes tetanus, are present throughout the environment. They germinate in devitalized tissue at an E_h of +10 mV (that of normal tissue is +120 mV). Once they are growing, the organisms elaborate the toxin tetanospasmin, a potent neurotoxin. Localized infection is often clinically insignificant. The toxin spreads along nerves to the central nervous system, where it binds to gangliosides, suppresses the release of inhibitory neurotransmitters, and yields prolonged muscle spasms. While respiratory compromise occurs as a result of upper airway obstruction or involvement of the diaphragm, autonomic dysfunction has emerged as the major cause of death. Obviously, severe trauma may predispose to development of tetanus; however, more than 50% of tetanus cases follow minor injuries. Tetanus is totally preventable: active immunity is induced with tetanus toxoid (formalinized tetanus toxin). Tetanus toxoid is part of routine childhood DTaP (diphtheria, tetanus, acellular pertussis) immunizations; adults should be given boosters every 10 years.

Clostridium botulinum causes botulism (see Chapters 9 and 11). *C botulinum* is distributed throughout the environment. The spores find their way into preserved or canned foods with low oxygen levels, low E_h, and nutrients that support growth. The organisms germinate and elaborate the toxins as growth and lysis occur. Botulinum neurotoxins are the most potent toxins known but can be neutralized by specific antibodies. The toxins are heat-labile, so properly heated food does not transmit botulism. Preformed botulinum toxin is ingested and absorbed. The toxin acts on the peripheral nervous system by inhibiting the release of acetylcholine at cholinergic synapses, causing paralysis. Once the toxin is bound, the process is irreversible. The symptoms are associated with the anticholinergic action and include dysphagia, dry mouth, diplopia, and weakness or inability to breathe. Botulism should be treated with antitoxin. Infant botulism follows the ingestion of spores, germination of the spores, and toxin production in vivo; honey is a common vehicle for spread of the spores in infants.

Clostridium perfringens causes gas gangrene. There are at least 12 different soluble antigens, many of which are toxins. All types of *C perfringens* produce the alpha toxin, a necrotizing, hemolytic exotoxin that is a lecithinase. The other toxins have varying activities, including tissue necrosis and hemolysis. *C perfringens* is present throughout the environment. **Gas gangrene** occurs when a soft tissue wound is contaminated by *C perfringens,* as occurs in trauma, septic abortion, and war wounds. Bacteremia associated with *C perfringens* can be rapidly fatal. Milder forms of disease may also occur. Once infection is initiated, the organisms elaborate necrotizing toxins; CO_2 and H_2 accumulate in tissue and are clinically detectable as gas (eg, gas gangrene). Other infectious processes yield gas in tissues and must be differentiated from clostridial gas gangrene. These infections include anaerobic streptococcal myonecrosis, synergistic nonclostridial anaerobic myonecrosis, infected vascular gangrene, and *Aeromonas hydrophila* myonecrosis. Edema occurs and the circulation is impaired, promoting spread of the anaerobic infection. Therapy involves surgical removal of the infection and administration of penicillin G.

C perfringens is a common cause of food poisoning (but less so than *Staphylococcus aureus*). The disease is caused by an enterotoxin produced and released during sporulation. The incubation period for the abdominal pain, nausea, and acute diarrhea is 8–24 hours.

Clostridium difficile causes pseudomembranous colitis. It is part of the normal gastrointestinal flora in 2–10% of humans. The organisms are relatively resistant to most commonly used antibiotics. Associated with or following antibiotic use, the normal gastrointestinal flora is suppressed and *C difficile* proliferates, producing cytopathic toxin and enterotoxin. Symptoms of the disease vary from diarrhea alone to marked diarrhea and necrosis of mucosa with accumulation of inflammatory cells and fibrin, which forms the pseudomembrane. The diagnosis is made by demonstrating neutralizable cytotoxin in the stool through its cytopathic effect in cell culture or by detecting enterotoxin by immunoassay. Many laboratories that have molecular diagnostic capabilities have developed, or are using, commercial real-time nucleic acid amplification methods for *C difficile* toxin gene detection in place of insensitive or labor-intensive phenotypic toxin assays.

Other *Clostridium* species are occasionally found in polymicrobial infections, particularly those associated with contamination of normal tissue by contents of the colon.

B. Gram-Positive Cocci

The group of anaerobic gram-positive cocci has undergone significant taxonomic expansion. Many species within the genus *Peptostreptococcus* have been reassigned to new genera such as *Anaerococcus, Finegoldia,* and *Peptoniphilus*. The species contained within these genera, as well as *Peptococcus niger,* are important members or the normal microbiota of the skin, oral cavity, upper respiratory tract, gastrointestinal tract, and female genitourinary system. The members of this group are opportunistic pathogens and are most frequently found in mixed infections particularly from specimens that have not been carefully procured. However, these organisms have been associated with serious infections such as brain abscesses, pleuropulmonary infections, necrotizing fasciitis, and other deep skin and soft tissue infections, intra-abdominal infections, and infections of the female genital tract.

PATHOGENESIS OF ANAEROBIC INFECTIONS

Infections caused by anaerobes commonly are due to combinations of bacteria that function in synergistic pathogenicity. Although studies of the pathogenesis of anaerobic infections have often focused on a single species, it is important to recognize that the anaerobic infections most often are due to several species of anaerobes acting together to cause infection.

B fragilis is a very important pathogen among the anaerobes that are part of the normal flora. The pathogenesis of anaerobic infection has been most extensively studied with *B fragilis* using a rat model of intra-abdominal infection, which in many ways mimics human disease. A characteristic sequence occurs after colon contents (including *B fragilis* and a facultative anaerobe such as *E coli*) are placed via needle, gelatin capsule, or other means into the abdomen of rats. A high percentage of the study animals die of sepsis caused by the facultative anaerobe. However, if the animals are first treated with gentamicin, a drug effective against the facultative anaerobe but not *Bacteroides,* few of the animals die, and after a few days, the surviving animals develop intra-abdominal abscesses from the *Bacteroides* infection. Treatment of the animals with both gentamicin and clindamycin, a drug effective against *Bacteroides,* prevents both the initial sepsis and the later development of abdominal abscesses.

The capsular polysaccharides of *Bacteroides* are important virulence factors. A unique feature of infections with *B fragilis* is the ability of the organism to induce abscess formation as the sole infecting organism. When injected into the rat abdomen, purified capsular polysaccharides from *B fragilis* cause abscess formation, whereas those from other bacteria (eg, *Streptococcus pneumoniae* and *E coli*) do not. The mechanism by which the *B fragilis* capsule induces abscess formation is not well understood.

Bacteroides species have lipopolysaccharides (endotoxins; see Chapter 9) but lack the lipopolysaccharide structures with endotoxic activity (including β-hydroxymyristic acid). The lipopolysaccharides of *B fragilis* are much less toxic than those of other gram-negative bacteria. Thus, infection caused by *Bacteroides* does not directly produce the clinical signs of sepsis (eg, fever and shock) so important in infections due to other gram-negative bacteria. When these clinical signs appear in *Bacteroides* infection, they are a result of the inflammatory immune response to the infection.

B fragilis elaborates a number of enzymes important in disease. In addition to proteases and neuraminidases, there is production of two cytolysins that act together to cause hemolysis of erythrocytes. An enterotoxin capable of causing diarrhea and whose gene is contained on a pathogenicity island is found in the majority of isolates that are recovered from blood cultures.

B fragilis produces a superoxide dismutase and can survive in the presence of oxygen for days. When a facultative anaerobe such as *E coli* is present at the site of infection, it can consume all available oxygen and thereby produce an environment in which *Bacteroides* and other anaerobes can grow (see above).

F necrophorum likewise possesses important virulence factors that enable it to cause Lemierre's syndrome and other seriously invasive infections. One of these factors is a leukotoxin likely responsible for the necrosis seen with these infections. Other factors include a hemagglutinin, a hemolysin, and lipopolysaccharide (endotoxin). In addition, *F necrophorum* is capable of causing platelet aggregation. The exact pathogenic interplay, if any, among these factors in the pathogenesis of human infections remains to be elucidated.

Many anaerobic bacteria produce heparinase, collagenase, and other enzymes that damage or destroy tissue. It is likely that enzymes play a part in the pathogenesis of mixed anaerobic infections, although laboratory experiments have not been able to define specific roles.

IMMUNITY IN ANAEROBIC INFECTIONS

Relatively little is known about immunity in anaerobic infections. The most complete information has been obtained from studies of animal models of *B fragilis* infections.

Many anaerobes (including *Bacteroides*, *Propionibacterium*, and *Fusobacterium* species) produce serum-independent chemotactic factors that attract polymorphonuclear cells. The capsule of *B fragilis* is both antiphagocytic and inhibitory to complement-mediated bactericidal action. *Bacteroides* species are optimally phagocytosed by polymorphonuclear cells when the organisms are opsonized by both antibody and complement. Both animals and humans produce antibodies against *Bacteroides* antigens, including the capsular material. Passive transfer of antibodies from an immune animal to a nonimmune animal is protective against *Bacteroides* bacteremia but does not prevent abdominal abscess formation; in the rat model of infection, it is a T cell-dependent immune response that prevents abscess formation. Passive transfer of immune spleen cells or a low-molecular-weight cell-free factor prevents abdominal abscess formation in the rat model.

THE POLYMICROBIAL NATURE OF ANAEROBIC INFECTIONS

Most anaerobic infections are associated with contamination of tissue by normal flora of the mucosa of the mouth, pharynx, gastrointestinal tract, or genital tract. Typically, multiple species (five or six species or more when standard culture conditions are used) are found, including both anaerobes and facultative anaerobes. Oropharyngeal, pleuropulmonary, abdominal, and female pelvic infections associated with contamination by normal mucosal flora have a relatively equal distribution of anaerobes and facultative anaerobes as causative agents: about 25% have anaerobes alone; about 25% have facultative anaerobes alone; and about 50% have both anaerobes and facultative anaerobes. Aerobic bacteria may also be present, but obligate aerobes are much less common than anaerobes and facultative anaerobes. Anaerobic bacteria and associated representative infections are listed in Table 21–2.

DIAGNOSIS OF ANAEROBIC INFECTIONS

Clinical signs suggesting possible infection with anaerobes include the following:

1. Foul-smelling discharge (due to short-chain fatty-acid products of anaerobic metabolism).
2. Infection in proximity to a mucosal surface (anaerobes are part of the normal flora).
3. Gas in tissues (production of CO_2 and H_2).
4. Negative aerobic cultures.

TABLE 21–2   Anaerobic Bacteria and Associated Representative Infections

Brain abscesses
Peptostreptococci, *Fusobacterium nucleatum* and others
Oropharyngeal infections
Oropharyngeal anaerobes; *Actinomyces, Prevotella melaninogenica, Fusobacterium* species
Pleuropulmonary infections
Peptostreptococci; *Fusobacterium* species; *Prevotella melaninogenica, Bacteroides fragilis* in 20–25%; others
Intra-abdominal infections
Liver abscess: Mixed anaerobes in 40–90%; facultative organisms Abdominal abscesses: *Bacteroides fragilis;* other gastrointestinal flora
Female genital tract infections
Vulvar abscesses: Peptostreptococci and others Tubo-ovarian and pelvic abscesses: *Prevotella bivia* and *Prevotella disiens;* peptostreptococci; others
Skin, soft tissue, and bone infections
Mixed anaerobic flora; *Propionibacterium acnes*
Bacteremia
Bacteroides fragilis; peptostreptococci; propionibacteria; Fusobacteria; *Clostridium;* others
Endocarditis
Bacteroides fragilis; Actinomyces

Diagnosis of anaerobic infection is made by anaerobic culture of properly obtained and transported specimens (see Chapter 47). Anaerobes grow most readily on complex media such as trypticase soy agar base, Schaedler blood agar, brucella agar, brain-heart infusion agar, and others—each highly supplemented (eg, with hemin, vitamin K_1, blood). A selective complex medium containing kanamycin is used in parallel. Kanamycin (like all aminoglycosides) does not inhibit the growth of obligate anaerobes; thus, it permits them to proliferate without being overshadowed by rapidly growing facultative anaerobes. Cultures are incubated at 35–37°C in an anaerobic atmosphere containing CO_2.

Colony morphology, pigmentation, and fluorescence are helpful in identifying anaerobes. Biochemical activities and production of short-chain fatty acids as measured by gas-liquid chromatography are used for laboratory confirmation.

TREATMENT OF ANAEROBIC INFECTIONS

Treatment of mixed anaerobic infections is by surgical drainage (under most circumstances) plus antimicrobial therapy.

The *B fragilis* group of organisms found in abdominal and other infections universally produces β-lactamase, as do many of the *P bivia* and *P disiens* strains found in genital tract infections in women. Fortunately these β-lactamases are inhibited by β-lactam-β-lactamase inhibitor combinations such as ampicillin/sulbactam. Therapy with antimicrobials (other than penicillin G) is necessary to treat infections with these organisms. At least two-thirds of the *P melaninogenica* strains from pulmonary and oropharyngeal infections also produce β-lactamase.

The most active drugs for treatment of anaerobic infections are clindamycin and metronidazole although clindamycin resistance among the *B fragilis* group has increased in the last decade. Clindamycin is preferred for infections above the diaphragm. Relatively few anaerobes are resistant to clindamycin and few, if any, are resistant to metronidazole. Alternative drugs include cefoxitin, cefotetan, some of the other newer cephalosporins, and piperacillin, but these drugs are not as active as clindamycin and metronidazole. The carbapenem antibiotics, ertapenem, imipenem, meropenem, and doripenem, have good activity against many anaerobes and resistance is still uncommon. Tigecycline, an agent that has FDA approval for the treatment of skin and soft tissue, and intra-abdominal infections, has good *in vitro* activity against a variety of anaerobe species including the *B fragilis* group. Penicillin G remains the drug of choice for treatment of anaerobic infections that do not involve β-lactamase-producing *Bacteroides* and *Prevotella* species.

REVIEW QUESTIONS

1. A 55-year-old man visits his physician complaining of a severe cough and production of purulent sputum. His breath has a very unpleasant fetid odor. Chest x-ray shows a large amount of fluid in the left pleural space and a 5-cm lung cavity with an air-fluid level. A needle is inserted through the chest wall and some of the fluid in the pleural space is removed; it is thick, yellow–gray in color, and malodorous. Which of the following organisms or sets of organisms are most likely to be cultured from the pleural fluid?

   (A) *Bacteroides fragilis, Escherichia coli,* and enterococci
   (B) *Prevotella bivia,* peptostreptococci, and *Staphylococcus epidermidis*
   (C) *Prevotella melaninogenica, Fusobacterium* species, and viridans streptococci
   (D) *Propionibacterium species,* peptostreptococci, and *Staphylococcus aureus*
   (E) *Streptococcus pneumoniae*

2. A 23-year-old man develops a perirectal abscess. This is drained surgically. A specimen is cultured and grows anaerobic bacteria. Clues that suggest infection with anaerobic bacteria include

   (A) Negative aerobic culture
   (B) Gas in tissues
   (C) Proximity to mucosal surface
   (D) Foul-smelling discharge
   (E) All of the above

3. A 63-year-old man with diabetes routinely injects insulin into the muscles of his left thigh. He has recently developed severe pain with swelling in his left thigh. On examination his thigh is swollen and red. Crepitus is noted on palpation, indicating gas in the tissue. Gas also is visible in the fascial planes on x-ray of the leg. Gas gangrene due to *Clostridium perfringens* is considered a likely diagnosis. What other infections must be considered?

   (A) Anaerobic streptococcal myonecrosis
   (B) Synergistic nonclostridial anaerobic myonecrosis
   (C) Infected vascular gangrene
   (D) *Aeromonas hydrophila* myonecrosis
   (E) All of the above

4. An 18-year-old man develops fever with pain in the right lower quadrant of his abdomen. After initial evaluation he is taken to the operating room. At surgery, a ruptured appendix with an abscess is found. *Bacteroides fragilis* is cultured from the abscess fluid. Which of the following factors promote abscess formation by *Bacteroides fragilis*?

   (A) Lipopolysaccharide
   (B) Capsule
   (C) Superoxide dismutase
   (D) Pili
   (E) Leukocidin toxin

5. Infections caused by *Bacteroides* species can be treated with all of the following antibiotics *except*:

   (A) Ampicillin/sulbactam
   (B) Clindamycin
   (C) Metronidazole
   (D) Penicillin
   (E) Cefoxitin

6. A 17-year-old high school senior develops infectious mononucleosis. About 2 weeks later, he develops significantly higher fever, worsening sore throat, inability to swallow, and severe neck and chest pain. Upon admission he has signs of sepsis and respiratory distress. What is the most likely organism causing this complication?

   (A) *Fusobacterium necrophorum*
   (B) *Bacteroides ovatus*

(C) *Prevotella melaninogenica*
(D) *Clostridium tetani*
(E) *Actinomyces israelii*

7. Which of the following statements regarding *Lactobacilli* is correct?

   (A) They are anaerobic gram-positive cocci.
   (B) They are most commonly found in the oral cavity.
   (C) The major product of metabolism is propionic acid.
   (D) They rarely cause disease in humans.
   (E) They form endospores.

8. Which of the following statements best describes the pathogenesis of *Clostridium botulinum*?

   (A) It elaborates a toxin that inhibits the release of acetylcholine at cholinergic synapses.
   (B) It elaborates an exotoxin that is a lecithinase that causes tissue necrosis.
   (C) It produces a polysaccharide capsule that inhibits phagocytosis and contributes to invasion of the central nervous system.
   (D) It elaborates a toxin that suppresses the release of inhibitory neurotransmitters.
   (E) It produces a leukotoxin that leads to abscess formation.

9. The drug of choice for treatment of infections caused by *Actinomyces* species is:

   (A) Tigecycline
   (B) Cefoxitin
   (C) Metronidazole
   (D) Imipenem
   (E) Penicillin

10. Infections commonly caused by *Clostridium perfringens* include all of the following *except*:

    (A) Gas gangrene
    (B) Lumpy jaw
    (C) Food poisoning
    (D) Bacteremia

Answers

1. C	4. B	7. D	10. B
2. E	5. D	8. A	
3. E	6. A	9. E	

REFERENCES

Citron DM, Poxton IR, Baron EJ: *Bacteroides, Porphyromonas, Prevotella, Fusobacterium,* and other anaerobic gram-negative rods. In: *Manual of Clinical Microbiology,* 9th ed. Murray PR et al (editors). ASM Press, 2007.

Cohen-Poradosu R, Kasper DL: Anaerobicinfections: general concepts. In *Mandell, Douglas, and Bennett's Principles and Practice of Infectious Diseases,* 7th ed. Mandell GL, Bennett JE, Dolin R (editors). Elsevier, 2010.

Finegold SM, Song Y: Anaerobic cocci. In *Mandell, Douglas, and Bennett's Principles and Practice of Infectious Diseases,* 7th ed. Mandell GL, Bennett JE, Dolin R (editors). Elsevier, 2010.

Garrett WS, Onderdonk AB: *Bacteroides, Prevotella, Porphyromonas,* and *Fusobacterium* species (and other medically important gram-negative bacilli). In *Mandell, Douglas, and Bennett's Principles and Practice of Infectious Diseases,* 7th ed. Mandell GL, Bennett JE, Dolin R (editors). Elsevier, 2010.

Hall V: *Actinomyces*—Gathering evidence of human colonization and infection. Anaerobe 2008;14:1.

Johnson EA, Summanen P, Finegold SM: *Clostridium.* In *Manual of Clinical Microbiology,* 9th ed. Murray PR et al (editors). ASM Press, 2007.

Kononen E : Anaerobic gram-positive nonsporulating bacilli. In *Mandell, Douglas, and Bennett's Principles and Practice of Infectious Diseases,* 7th ed. Mandell GL, Bennett JE, Dolin R (editors). Elsevier, 2010.

Onderdonk AB, Garrett WS: Gas gangrene and other *Clostridium*-associated diseases. In *Mandell, Douglas, and Bennett's Principles and Practice of Infectious Diseases,* 7th ed. Mandell GL, Bennett JE, Dolin R (editors). Elsevier, 2010.

Reddy P, Bleck TP: *Clostridium botulinum* (botulism). In *Mandell, Douglas, and Bennett's Principles and Practice of Infectious Diseases,* 7th ed. Mandell GL, Bennett JE, Dolin R (editors). Elsevier, 2010.

Reddy P, Bleck TP: *Clostridium tetani* (tetanus). In *Mandell, Douglas, and Bennett's Principles and Practice of Infectious Diseases,* 7th ed. Mandell GL, Bennett JE, Dolin R (editors). Elsevier, 2010.

Riordan T: Human infection with *Fusobacterium necrophorum* (Necrobacillosis) with a focus on Lemierre's syndrome. Clin Microbiol Rev 2007;20:622.

Song Y, Finegold SM: *Peptostreptococcus, Finegoldia, Anaerococcus, Peptoniphilus, Veillonella,* and other anaerobic cocci. In *Manual of Clinical Microbiology,* 9th ed. Murray PR et al (editors). ASM Press, 2007.

Wexler HM. *Bacteroides*: the good, the bad and the nitty-gritty. Clin Microbiol Rev 2007;20:593.

Legionellae, Bartonella, & Unusual Bacterial Pathogens

LEGIONELLA PNEUMOPHILA & OTHER LEGIONELLAE

A widely publicized outbreak of pneumonia in persons attending an American Legion convention in Philadelphia prompted investigations that defined *Legionella pneumophila* and the legionellae. Other outbreaks of respiratory illness caused by related organisms since 1947 have been diagnosed retrospectively. Several dozen species of *Legionella* exist, some with multiple serotypes. *L pneumophila* is the major cause of disease in humans; *Legionella micdadei* and a few other species sometimes cause pneumonia. The other legionellae are rarely isolated from patients or have been isolated only from the environment.

Morphology & Identification

L pneumophila is the prototype bacterium of the group. Legionellae of primary medical importance are listed in Table 22–1.

A. Typical Organisms

Legionellae are fastidious, aerobic gram-negative bacteria that are 0.5–1 µm wide and 2–50 µm long (Figure 22–1). They often stain poorly by Gram's method and are not seen in stains of clinical specimens. Gram-stained smears should be made for suspect *Legionella* growth on agar media. Basic fuchsin (0.1%) should be used as the counterstain, because safranin stains the bacteria very poorly.

B. Culture

Legionellae can be grown on complex media such as buffered charcoal yeast extract agar with α-ketoglutarate and iron (BCYE, at pH 6.9, temperature 35°C, and 90% humidity. Antibiotics can be added to make the medium selective for *Legionella*. The charcoal acts as a detoxifying agent. A biphasic BCYE medium can be used for blood cultures.

Legionellae grow slowly; visible colonies are usually present after 3 days of incubation. Colonies that appear after overnight incubation are not *Legionella*. Colonies are round or flat with entire edges. They vary in color from colorless to iridescent pink or blue and are translucent or speckled. Variation in colony morphology is common, and the colonies may rapidly lose their color and speckles. Many other genera of bacteria grow on BCYE medium and must be differentiated from *Legionella* by Gram staining and other tests.

Legionellae in blood cultures usually require 2 weeks or more to grow. Colonies can be seen on the agar surface of the biphasic medium.

C. Growth Characteristics

The legionellae are catalase-positive. *L pneumophila* is oxidase-positive; the other legionellae are variable in oxidase activity. *L pneumophila* hydrolyzes hippurate; the other legionellae do not. Most legionellae produce gelatinase and β-lactamase; *L micdadei* produces neither gelatinase nor β-lactamase.

Antigens & Cell Products

Antigenic specificity of *L pneumophila* is thought to be due to complex antigenic structures. There are at least 16 serogroups of *L pneumophila*; serogroup 1 was the cause

TABLE 22–1 The *Legionella* Species of Primary Medical Importance

Species	Pneumonia	Pontiac Fever
Legionella pneumophila	+	Serogroups 1 and 6
Legionella micdadei	+	
Legionella gormanii	+	
Legionella dumoffii	+	
Legionella bozemanii	+	
Legionella longbeachae	+	
Legionella wadsworthii	+	
Legionella jordanis	+	
Legionella feeleii	+	+
Legionella oakridgensis	+	

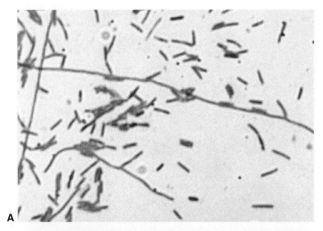

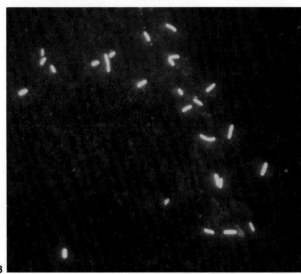

FIGURE 22–1 A: Gram stain of a *Legionella pneumophila;* the legionellae stain faintly with basic fuchsin and poorly with safranin. Original magnification × 1000. (Courtesy of CDC Public Health Image Library.) **B:** Direct fluorescent antibody stain of *Legionella* of mixed species using antibodies against legionellae genus antigens conjugated with fluorescein. Original magnification × 1000. (Courtesy of R Nadarajah.)

of the 1976 outbreak of Legionnaires' disease and remains the most common serogroup isolated from humans. *Legionella* species cannot be identified by serogrouping alone, because there is cross-reactive antigenicity among different species. Occasionally, bacteroides, bordetella, and some pseudomonads also cross-react with *L pneumophila* antisera.

The legionellae produce distinctive 14- to 17-carbon branched-chain fatty acids. Gas-liquid chromatography can be used to help characterize and determine the species of legionellae.

The legionellae make proteases, phosphatase, lipase, DNase, and RNase. A major secretory protein, a metalloprotease, has hemolytic and cytotoxic activity; however, this protein has not been shown to be a required virulence factor.

Pathology & Pathogenesis

Legionellae are ubiquitous in warm moist environments. They are found in lakes, streams, and other bodies of water. They can multiply in free-living amebas and can coexist with them in biofilms (see Epidemiology & Control section, below). Infection of debilitated or immunocompromised humans commonly follows inhalation of the bacteria from aerosols generated from contaminated air-conditioning systems, shower heads, and similar sources. *L pneumophila* usually produces a lobar, segmental, or patchy pulmonary infiltration. Histologically, the appearance is similar to that produced by many other bacterial pathogens. Acute purulent pneumonia involving the alveoli is present with a dense intraalveolar exudate of macrophages, polymorphonuclear leukocytes, red blood cells, and proteinaceous material. Most of the legionellae in the lesions are within phagocytic cells. There is little interstitial infiltration and little or no inflammation of the bronchioles and upper airways.

Knowledge of the pathogenesis of *L pneumophila* infection comes from study of isolated cells from humans and from study of susceptible animals such as guinea pigs.

L pneumophila readily enters and grows within human alveolar macrophages and monocytes and is not effectively killed by polymorphonuclear leukocytes. In vitro, when serum is present but there is no immune antibody, complement component C3 is deposited on the bacterial surface and the bacteria attach to complement receptors CR1 and CR3 on the phagocytic cell surface. Entry into the cell is by a phagocytic process involving coiling of a single pseudopod around the bacterium. When immune antibody is present, entry of the bacteria occurs by the more typical Fc-mediated phagocytosis. Once inside the cell, the individual bacteria are within phagosomal vacuoles, but the defense mechanisms of the macrophage cells stop at that point. Instead, the phagosomal vacuoles fail to fuse with lysosomal granules. The phagocyte oxidative metabolic burst is reduced. Phagosomes containing *L pneumophila* do not acidify as much as phagosomes containing other ingested particles. Ribosomes, mitochondria, and small vesicles accumulate around vacuoles containing the *L pneumophila*. The bacteria multiply within the vacuoles until they are numerous, the cells are destroyed, the bacteria are released, and infection of other macrophages then occurs. The presence of iron (transferrin-iron) is essential for the process of intracellular growth of the bacteria, but other factors important to the processes of growth, cell destruction, and tissue damage are not well understood.

Clinical Findings

Asymptomatic infection is common in all age groups, as shown by elevated titers of specific antibodies. The incidence of clinically significant disease is highest in men over age 55 years. Factors associated with high risk include smoking, chronic bronchitis and emphysema, steroid and other immunosuppressive treatment (as in renal transplantation), cancer chemotherapy, and diabetes mellitus. When pneumonia occurs in patients with these risk factors, *Legionella* should be investigated as the cause.

Infection may result in nondescript febrile illness of short duration or in a severe, rapidly progressive illness with high fever, chills, malaise, nonproductive cough, hypoxia, diarrhea, and delirium. Chest x-rays reveal patchy, often multilobar consolidation. There may be leukocytosis, hyponatremia, hematuria (and even renal failure), or abnormal liver function. During some outbreaks, the mortality rate has reached 10%. The diagnosis is based on the clinical picture and exclusion of other causes of pneumonia by laboratory tests. Demonstration of *Legionella* in clinical specimens can rapidly yield a specific diagnosis. The diagnosis can also be made by culture for *Legionella* or by serologic tests, but results of these tests are often delayed beyond the time when specific therapy must be started.

L pneumophila also produces a disease called "Pontiac fever," after the clinical syndrome that occurred in an outbreak in Michigan. The syndrome is characterized by fever and chills, myalgia, malaise, and headache that develop over 6–12 hours. Dizziness, photophobia, neck stiffness, and confusion also occur. Respiratory symptoms are much less prominent in Pontiac fever than in Legionnaires' disease and include mild cough and sore throat.

Diagnostic Laboratory Tests

A. Specimens

In human infections, the organisms can be recovered from bronchial washings, pleural fluid, lung biopsy specimens, or blood. Isolation of Legionella from sputum is more difficult because of the predominance of bacteria of the normal flora. *Legionella* is rarely recovered from other anatomic sites.

B. Smears

Legionellae are not demonstrable in Gram-stained smears of clinical specimens. Direct fluorescent antibody tests of specimens can be diagnostic, but the test has low sensitivity compared with culture. Silver stains are sometimes used on tissue specimens.

C. Culture

Specimens are cultured on BCYE agar (see above). Cultured organisms can be rapidly identified by immunofluorescence staining. BCYE agar containing antibiotics can be used to culture contaminated specimens.

D. Specific Tests

Sometimes Legionella antigens can be demonstrated in the patient's urine by immunologic methods. The urine antigen test is specific for *L pneumophila* serotype 1. Thus, the legionella urine antigen test is not useful to diagnose 20–70% of *Legionella* species infections, depending upon geographic location, and should not be relied upon as the sole test to rule out Legionella infections.

E. Serologic Tests

Levels of antibodies to legionellae rise slowly during the illness. Serologic tests have a sensitivity of 60–80% and a specificity of 95–99%. Serologic tests are most useful in obtaining a retrospective diagnosis in outbreaks of *Legionella* infections.

Immunity

Infected patients make antibodies against *Legionella*, but the peak antibody response may not occur until 4–8 weeks after infection. The roles of antibodies and cell-mediated responses in protective immunity in humans have not been defined. Animals challenged with sublethal doses of virulent *L pneumophila*, avirulent *L pneumophila*, or a major secretory protein vaccine are immune to subsequent lethal doses of *L pneumophila*. Both humoral and cell-mediated immune responses occur. The cell-mediated response is important in protective immunity because of the intracellular infection and growth of Legionella.

Treatment

L pneumophila are intracellular parasites of macrophages, other phagocytic cells, and probably of other human cells as well. Other *Legionella* species also may show significant growth within human macrophages. Thus, antimicrobials useable to treat Legionella infections must enter the phagocytes and have biological activity there. Macrolides (erythromycin, azithromycin, telithromycin, and clarithromycin), quinolones (ciprofloxacin and levofloxacin), and tetracyclines (doxycycline) are effective. β-Lactams, monobactams, and aminoglycosides are not effective; in addition, many legionellae make β-lactamase. Prolonged therapy, 3 weeks, may be required depending upon the clinical situation.

Epidemiology & Control

The natural habitats for legionellae are lakes, streams, rivers, and especially thermally heated bodies of water and soil. Legionellae grow best in warm water in the presence of amebas and water bacteria. They proliferate in amebas much as they do in pulmonary macrophages in the lung. When harsh environmental conditions occur and the amebas encyst, the amebas and legionellae both survive until better growth conditions occur, allowing excystment. The legionellae, amebas, and other microorganisms exist in biofilms; the legionellae go into a sessile state. The legionellae survive water treatment processes, and small numbers enter the water distribution systems where they proliferate.

Cooling towers and evaporative condensers can be heavily contaminated with *L pneumophila*. Presumably, aerosols exiting such towers or condensers spread the organisms to susceptible persons. Similarly, there are links between contamination of residential water systems and community-acquired Legionnaires' disease and between contamination of hospital water systems and nosocomial *L pneumophila* infection. Hyperchlorination and superheating of water can help control the multiplication of legionellae in water and in air-conditioning systems.

BARTONELLA

The three medically important species in the genus *Bartonella* are *Bartonella bacilliformis*, the cause of Oroya fever and verruga peruana; *Bartonella quintana*, the cause of trench fever of World War I and some cases of bacillary angiomatosis; and *Bartonella henselae*, which causes cat-scratch disease and has also been associated with bacillary angiomatosis. These diseases have many common characteristics. There is an additional small set of *Bartonella* species and subspecies that have rarely been associated with human disease, and there is a larger set associated with animals that is not likely to be transmitted to humans.

The *Bartonella* species are gram-negative rods that are pleomorphic, slow-growing, and difficult to isolate in the laboratory. They can be seen in infected tissues stained with the Warthin-Starry silver impregnation stain.

Bartonella bacilliformis

There are two stages of *B bacilliformis* infection: the initial stage is **Oroya fever**, a serious infectious anemia; and the eruptive stage, **verruga peruana**, which commonly begins 2–8 weeks later, though verrugae may also occur in the absence of Oroya fever.

Oroya fever is characterized by the rapid development of severe anemia due to red blood cell destruction, enlargement of the spleen and liver, and hemorrhage into the lymph nodes. Masses of bartonellae fill the cytoplasm of cells lining the blood vessels, and endothelial swelling may lead to vascular occlusion and thrombosis. The mortality rate of untreated Oroya fever is about 40%. The diagnosis is made by examining stained blood smears and blood cultures in semisolid medium.

Verruga peruana consists of vascular skin lesions that occur in successive crops; it lasts for about 1 year and produces little systemic reaction and no fatalities. Bartonellae can be seen in the granulomas; blood cultures are often positive, but there is no anemia.

B bacilliformis produces a protein that promotes deformity (indentation) of red blood cell membranes, and flagella provide the organisms with the mechanical force to invade red blood cells. *B bacilliformis* also invades endothelial cells and other types of human cells in vitro.

Bartonellosis is limited to the mountainous areas of the American Andes in tropical Peru, Colombia, and Ecuador and is transmitted by sandflies of the genus *Lutzomyia*.

B bacilliformis grows in semisolid nutrient agar containing 10% rabbit serum and 0.5% hemoglobin. After 10 days or more of incubation at 28°C, turbidity develops in the medium, and rod-shaped and granular organisms can be seen in Giemsa-stained smears.

Ciprofloxacin, doxycycline, ampicillin, or sulfamethoxazole-trimethoprim should be given for at least 1 week. Parenteral therapy can be used if the patient is unable to absorb oral medication. Chloramphenicol has been used to effectively treat *B bacilliformis* infections particularly in South America; however, chloramphenicol may not be available in the United States and should not be considered for therapy. Coupled with blood transfusions, when indicated, antimicrobial therapy greatly reduces the mortality rate. Control of the disease depends upon elimination of the sandfly vectors: insecticides, insect repellents, and elimination of sandfly breeding areas are of value. Prevention with antibiotics may be useful.

Bartonella henselae & Bartonella quintana

A. Cat-Scratch Disease

Cat-scratch disease is usually a benign, self-limited illness manifested by fever and lymphadenopathy that develop about 2 weeks after contact with a cat (usually a scratch, lick, bite, or perhaps a flea bite). A primary skin lesion (papule or pustule) develops at the site 3–10 days after the contact. The patient usually appears well but may have low-grade fever and occasionally headache, sore throat, or conjunctivitis. The regional lymph nodes are markedly enlarged and sometimes tender, and they may not subside for several weeks or even months. They may suppurate and discharge pus. More than 20,000 cases a year are thought to occur in the United States.

The diagnosis of cat-scratch disease is based on (1) a suggestive history and physical findings; (2) aspiration of pus from lymph nodes that contain no bacteria culturable by routine methods; and (3) characteristic histopathologic findings with granulomatous lesions, which may include bacteria seen on silver-impregnated stains. A positive skin test has also been included as a criterion, but is of historical interest only. A titer of 1:64 or greater in a single serum in the indirect fluorescent antibody test strongly supports the diagnosis, but development of a diagnostic titer may be delayed; enzyme immunoassays can be useful also.

Cat-scratch disease is caused by *B henselae*, a small, pleomorphic, gram-negative rod present mainly in the walls of capillaries near follicular hyperplasia or within microabscesses. The organisms are seen best in tissue sections stained with Warthin-Starry silver impregnation stain; they may also be detected by immunofluorescent stains. Culture of *B henselae* is generally not recommended for this relatively benign disease.

The reservoir for *B henselae* is the domestic cat, and one-third of cats or more (and possibly their fleas) may be infected. Contact with infected cats through skin lesions is thought to communicate the infection.

Cat-scratch disease occurs commonly in immunocompetent people and is usually self-limited. Treatment is mainly supportive, with reassurance, hot moist soaks, and analgesics. Aspiration of pus or surgical removal of an excessively large lymph node may ameliorate symptoms. Tetracycline or erythromycin therapy may be helpful.

B. Bacillary Angiomatosis

Bacillary angiomatosis is a disease predominantly of immunosuppressed individuals, particularly AIDS patients. Rare cases occur in immunocompetent persons. Bacillary angiomatosis is characterized histopathologically as circumscribed lesions with lobular capillary proliferation and round, open vessels with cuboidal endothelial cells protruding into the vascular lumen. A prominent finding is epithelioid histiocytes surrounded by a loose fibromyxoid matrix. The pleomorphic bacilli can be seen in the subendothelial tissue when stained with the Warthin-Starry silver impregnation stain. The lesions may be infiltrated by polymorphonuclear leukocytes.

In its common form, bacillary angiomatosis presents as an enlarging red (cranberry-like) papule, often with surrounding scale and erythema. The lesions enlarge and may become several centimeters in diameter and ulcerate. There may be single or many lesions. The clinical appearance is often similar to that of Kaposi sarcoma in AIDS patients, but the two diseases are different histologically. Bacillary angiomatosis occurs in virtually every organ. Involvement of the liver (and spleen) is characterized by a proliferation of cystic blood-filled spaces surrounded by a fibromyxoid matrix containing the bacteria; this form of the disease is called **peliosis hepatis** and is usually accompanied by fever, weight loss, and abdominal pain. A bacteremic form of infection with the nonspecific signs of malaise, fever, and weight loss also occurs.

The diagnosis is confirmed by the characteristic histopathologic findings and demonstration of the pleomorphic bacilli on silver-stained sections. *B henselae* and *B quintana* can be isolated by direct culture of biopsies of involved tissue carefully obtained so that no contaminating skin bacteria are present. The biopsy specimens are homogenized in supplemented tissue culture medium and inoculated onto fresh chocolate agar and heart infusion agar with 5% rabbit blood. Cultures of blood obtained by the lysis-centrifugation method can be inoculated onto the same media. The cultures should be incubated in 5% CO_2 at 36°C for a minimum of 3 weeks. Specimens can also be cultured on eukaryotic tissue culture monolayers. Biochemically, *B henselae* and *B quintana* are relatively inert, including negative catalase and oxidase reactions and negative carbohydrate utilization tests. Enzyme activity can be seen with amino acid substrates by methods to test for preformed enzymes. Definitive identification is obtained by sequencing all or part of the 16S ribosomal RNA gene amplified by the polymerase chain reaction.

Bacillary angiomatosis is treated with oral erythromycin or doxycycline (plus gentamicin for very ill patients) for a minimum of 2 months. Relapses are common but can be treated by the same drugs used initially.

The reservoir for *B henselae* usually is the domestic cat, and patients with this organism as the etiology of bacillary angiomatosis often have contact with cats or histories of cat flea bites. The only known reservoirs for *B quintana* are humans and the body louse.

BACTERIA THAT CAUSE VAGINOSIS

Bacterial vaginosis is a common vaginal condition of women of reproductive age. It is associated with premature rupture of membranes and preterm labor and birth. Bacterial vaginosis has a complex microbiology; two organisms, *Gardnerella vaginalis* and *Mobiluncus* species have been most specifically associated with the disease process.

Gardnerella vaginalis

G vaginalis is a serologically distinct organism isolated from the normal female genitourinary tract and also associated with vaginosis, so named because inflammatory cells are not present. In wet smears, this "nonspecific" vaginitis, or **bacterial vaginosis**, yields "clue cells," which are vaginal epithelial cells covered with many gram-variable bacilli, and there is an absence of other common causes of vaginitis such as trichomonas or yeasts. Vaginal discharge often has a distinct "fishy" odor and contains many anaerobes in addition to *G vaginalis*. The pH of the vaginal secretions is over 4.5 (normal pH is <4.5). The vaginosis attributed to this organism is suppressed by metronidazole, suggesting an association with anaerobes. Oral metronidazole is generally curative.

Mobiluncus

This genus comprises motile, curved, gram-variable or gram-negative, anaerobic rods isolated from "**bacterial vaginosis**," which may be a clinical variant of the vaginosis associated with *G vaginalis*. It is possible that mobiluncus may be part of the normal vaginal anaerobic flora, and it is likely that it is part of the anaerobic flora in bacterial vaginosis. The organisms are most commonly detected in Gram-stained smears of vaginal secretions, but they grow with difficulty in anaerobic cultures.

STREPTOBACILLUS MONILIFORMIS

S moniliformis is an aerobic, gram-negative, highly pleomorphic organism that forms irregular chains of bacilli interspersed with fusiform enlargements and large round bodies. It grows best at 37°C in media containing serum protein, egg yolk, or starch but ceases to grow at 22°C. L forms can easily be demonstrated in most cultures of the organism. Subculture of pure colonies of L forms in liquid media often yields the streptobacilli again. All strains of streptobacilli appear to be antigenically identical.

S moniliformis is a normal inhabitant of the throats of rats, and humans can be infected by rat bites. The human disease (**rat-bite fever**) is characterized by septic fever, blotchy and petechial rashes, and very painful polyarthritis. Diagnosis rests on cultures of blood, joint fluid, or pus; on mouse inoculation; and on serum agglutination tests.

This organism can also produce infection after being ingested in milk. The disease is called Haverhill fever and has occurred in epidemics.

Penicillin and perhaps other antibiotics are therapeutically effective.

Rat-bite fever of somewhat different clinical appearance (sodoku) is caused by *Spirillum minor* (see Chapter 24).

CALYMMATOBACTERIUM (DONOVANIA) GRANULOMATIS

C granulomatis, related to the klebsiellae, causes **granuloma inguinale**, an uncommon sexually transmitted disease characterized by genital ulcers. The organism grows with difficulty on media containing egg yolk. Ampicillin or tetracycline is effective treatment.

WHIPPLE DISEASE

Whipple disease is characterized by fever, abdominal pain, diarrhea, weight loss, and migratory polyarthralgia. The primary involvement is of the small intestine and mesenteric lymph nodes, but any organ can be affected. Histologically, there is a prominent macrophage infiltration and fat deposition. Characteristic vacuoles within the macrophage that stain with periodic acid-Schiff (PAS) stain are pathognomonic of the disease. The intracellular and extracellular PAS-positive material are bacilli. Historically, routine cultures of clinical specimens have been negative, but more recently the organism has been cultured in association with eukaryotic cells (human fibroblasts, deactivated peripheral blood monocytes). Prior to successfully culturing the organism polymerase chain reaction (PCR) amplification of bacterial 16S ribosomal RNA allowed identification of a unique sequence from the bacteria in the lesions. Phylogenetic analysis has shown the organism is a gram-positive actinomycete not closely related to any known genus. The organism has been named *Tropheryma whipplei*. The diagnosis of Whipple disease is by PCR amplification of an appropriate specimen (bowel biopsy, brain biopsy, etc) for *T whipplei*.

REVIEW QUESTIONS

1. Humans become infected with *Legionella pneumophila* by:
   (A) Drinking water contaminated with *Acanthamoeba castellani* containing *Legionella pneumophila*
   (B) Kissing a person who is a legionella carrier
   (C) Breathing aerosols from environmental water sources
   (D) Receiving a mosquito bite
   (E) Consuming undercooked pork

2. An 11-year-old girl developed acute onset of fever, chills, headache, vomiting, and severe migratory arthralgias (joint pain) and myalgias (muscle pain). Two days later, she developed a maculopapular rash over her palms, soles, and extremities. At the same time, her left knee became extremely painful and swollen. On examination, fluid was demonstrated in the knee. Further history disclosed that the patient had a pet rat which she frequently played with. Culture of the fluid from her knee on 5% sheep blood agar showed 2 mm colonies after 3 days of incubation. Broth culture showed small puffball-like growth. Gram staining showed a gram-negative bacillus 0.5 μm wide and 1–4 μm long. Some extremely long forms (up to 150 μm) with bead-like chains, fusiform swellings, and large round bodies were seen. The microbiologist who observed the Gram-stained smear immediately knew the cause of the girl's infection to be

   (A) *Treponema pallidum*
   (B) *Streptococcus moniliformis*
   (C) *Francisella tularensis*
   (D) *Bartonella bacilliformis*
   (E) *Yersinia pestis*

3. A 70-year-old man presents with bilateral pneumonia. His legionella urinary antigen test result is positive. Which of the following is the likely cause of his pneumonia?

   (A) *Legionella pneumophila* serotype 1
   (B) *Legionella micdadei* serotype 4
   (C) *Legionella bozemanii* serotype 2
   (D) *Legionella longbeachae* serotype 2
   (E) All of the above because the urinary antigen test is genus-specific and not species- or serotype-specific

4. A 28-year-old woman came to the clinic because of a whitish-gray vaginal discharge with a bad odor, first noted 6 days previously. She had been sexually active in the past month with a single partner who was new to her. Physical examination showed a thin, homogeneous, whitish-gray discharge that was adherent to the vaginal wall. There was no discharge from the cervical os. The bimanual pelvic examination was normal, as was the remainder of the physical examination. The pH of the vaginal fluid was 5.5 (normal, < 4.5). When KOH was added to vaginal fluid on a slide, an amine-like ("fishy") odor was perceived. A wet mount of the fluid showed many epithelial cells with adherent bacteria (clue cells). No polymorphonuclear cells were seen. The diagnosis was

   (A) Gonorrhea
   (B) *Trichomonas vaginalis* vaginitis
   (C) Syphilis
   (D) Yeast vaginitis
   (E) Bacterial vaginosis

5. A 70-year-old man comes to the emergency room feeling feverish and "really tired." He has a chronic cigarette cough, but this has dramatically increased in the past week and has been producing whitish sputum. The previous day he had a temperature of 38°C and watery diarrhea. Physical examination reveals inspiratory and expiratory wheezes and râles over the right lower lung field. Chest x-ray shows a patchy right lower lobe infiltrate. The differential diagnosis of this patients' disease is

   (A) *Streptococcus pneumoniae* pneumonia
   (B) *Legionella pneumophila* pneumonia
   (C) *Haemophilus influenzae* pneumonia
   (D) *Mycoplasma pneumoniae* pneumonia
   (E) All of the above

6. Routine sputum cultures for the patient in Question 5 grow normal flora. Treatment with ampicillin for 2 days yields no improvement. A diagnosis of Legionnaires' disease is considered and bronchoscopy is done to obtain bronchial alveolar lavage fluid and

deep airway specimens. Which of the following would suggest a diagnosis of disease due to *Legionella pneumophila* serotype 1?

(A) Legionella urinary antigen assay

(B) Direct fluorescent antibody on the bronchial alveolar lavage fluid

(C) Culture of the bronchial alveolar lavage on charcoal yeast extract medium with antibiotics

(D) Antibody assay on paired (acute phase and convalescent phase) sera

(E) All of the above

7. Charcoal is present in buffered charcoal yeast extract agar (BCYE) used to isolate *Legionella pneumophila* to:

(A) Provide the growth factors ordinarily provided by free-living amebas present in environmental water

(B) Serve as a carbon source for the growth of *Legionella pneumophila*

(C) Prevent hemolysis of the red blood cells in the medium

(D) Provide a dark background

(E) Act as a detoxifying agent

8. A 23-year-old woman presents with a 3-day history of low-grade fever and head ache. Examination reveals enlarged and slightly tender lymph nodes near her left elbow and in the left axilla. Approximately 2 weeks earlier she had visited a friend whose cat had scratched her on the left arm; the site later developed a reddish papule. Which of the following statements about cat-scratch disease is most correct?

(A) The diagnosis is based on a suggestive history and physical examination

(B) The diagnosis is based on negative routine bacterial cultures of pus aspirated from involved lymph nodes

(C) The disease is usually self-limited in immunocompetent people

(D) The etiologic agent is *Bartonella henselae*

(E) All of the above

9. Which of the following statements about bacillary angiomatosis is most correct?

(A) It is caused by *Bartonella bacilliformis*

(B) It is typically confined to the skin

(C) The major differential diagnosis is Kaposi sarcoma

(D) The etiologic agent can be grown in 1–2 days in routine culture on sheep blood agar

(E) Dogs are the reservoir for the etiologic agent

10. An important factor in the pathogenesis of Legionnaires' disease is that

(A) *Legionella pneumophila* kills polymorphonuclear cells

(B) Alveolar macrophages phagocytose *Legionella pneumophila* using a coiled pseudopods

(C) *Legionella pneumophila* invades pulmonary capillaries, leading to dissemination and systemic illness

(D) *Legionella pneumophila* induces alveolar macrophage phagosomes to fuse with lysosomes

(E) *Legionella pneumophila* outer surface protein A (OspA) is important for invasion of alveolar macrophages

Answers

1. C	4. E	7. E	10. B
2. B	5. E	8. E	
3. A	6. E	9. C	

REFERENCES

Chomel BB, Rolain JM: *Bartonella*. In: *Manual of Clinical Microbiology*, 9th ed. Murray PR et al (editors). ASM Press, 2007.

Edelstein PH: *Legionella*. In: *Manual of Clinical Microbiology*, 9th ed. Murray PR et al (editors). ASM Press, 2007.

Edelstein PH, Cianciotto NP: *Legionella*. In: *Mandell, Douglas, and Bennett's Principles and Practice of Infectious Diseases*, 7th ed. Mandell GL, et al (editors). Elsevier, 2010.

Hart G: Donovanosis. Clin Infect Dis 1997;25:24.

Muder RR: Other *Legionella* species. In: *Mandell, Douglas, and Bennett's Principles and Practice of Infectious Diseases*, 7th ed. Mandell GL, et al (editors). Elsevier, 2010.

Slater LN, Welch DF: *Bartonella*, including cat-scratch disease. In: *Mandell, Douglas, and Bennett's Principles and Practice of Infectious Diseases*, 7th ed. Mandell GL, et al (editors). Elsevier, 2010.

Mycobacteria

The mycobacteria are rod-shaped, aerobic bacteria that do not form spores. Although they do not stain readily, once stained they resist decolorization by acid or alcohol and are therefore called "acid-fast" bacilli. *Mycobacterium tuberculosis* causes tuberculosis and is a very important pathogen of humans. *Mycobacterium leprae* causes leprosy. *Mycobacterium avium-intracellulare* (*M avium* complex, or MAC) and other nontuberculous mycobacteria frequently infect patients with AIDS, are opportunistic pathogens in other immunocompromised persons, and occasionally cause disease in patients with normal immune systems. There are more than 125 *Mycobacterium* species, including many that are saprophytes. The mycobacteria that infect humans are listed in Table 23–1.

MYCOBACTERIUM TUBERCULOSIS

Morphology & Identification

A. Typical Organisms

In tissue, tubercle bacilli are thin straight rods measuring about $0.4 \times 3 \mu m$ (Figure 23–1). On artificial media, coccoid and filamentous forms are seen with variable morphology from one species to another. Mycobacteria cannot be classified as either gram-positive or gram-negative. Once stained by basic dyes they cannot be decolorized by alcohol, regardless of treatment with iodine. True tubercle bacilli are characterized by "acid-fastness"—ie, 95% ethyl alcohol containing 3% hydrochloric acid (acid-alcohol) quickly decolorizes all bacteria except the mycobacteria. Acid-fastness depends on the integrity of the waxy envelope. The **Ziehl-Neelsen technique** of staining is employed for identification of acid-fast bacteria. The method is detailed in Chapter 47. In smears of sputum or sections of tissue, mycobacteria can be demonstrated by yellow-orange fluorescence after staining with fluorochrome stains (eg, auramine, rhodamine). The ease with which AFB can be visualized with flurochrome stains, make them the preferred stains for clinical specimens (Figure 23–1B).

B. Culture

The media for primary culture of mycobacteria should include a nonselective medium and a selective medium.

Selective media contain antibiotics to prevent the overgrowth of contaminating bacteria and fungi. There are three general formulations that can be used for both the nonselective and selective media.

1. Semisynthetic agar media—These media (eg, Middlebrook 7H10 and 7H11) contain defined salts, vitamins, cofactors, oleic acid, albumin, catalase, and glycerol; the 7H11 medium contains casein hydrolysate also. The albumin neutralizes the toxic and inhibitory effects of fatty acids in the specimen or medium. Large inocula yield growth on these media in several weeks. Because large inocula may be necessary these media may be less sensitive than other media for primary isolation of mycobacteria.

The semisynthetic agar media are used for observing colony morphology, for susceptibility testing, and, with added antibiotics and malachite green, as selective media.

2. Inspissated egg media—These media (eg, Löwenstein-Jensen) contain defined salts, glycerol, and complex organic substances (eg, fresh eggs or egg yolks, potato flour, and other ingredients in various combinations). Malachite green is included to inhibit other bacteria. Small inocula in specimens from patients will grow on these media in 3–6 weeks.

These media with added antibiotics are used as selective media.

3. Broth media—Broth media (eg, Middlebrook 7H9 and 7H12) support the proliferation of small inocula. Ordinarily, mycobacteria grow in clumps or masses because of the hydrophobic character of the cell surface. If tweens (water-soluble esters of fatty acids) are added, they wet the surface and thus permit dispersed growth in liquid media. Growth is often more rapid than on complex media.

C. Growth Characteristics

Mycobacteria are obligate aerobes and derive energy from the oxidation of many simple carbon compounds. Increased CO_2 tension enhances growth. Biochemical activities are not characteristic, and the growth rate is much slower than that of most bacteria. The doubling time of tubercle bacilli is about 18 hours. Saprophytic forms tend to grow more rapidly, to

TABLE 23–1   Mycobacteria that Infect Humans

Species	Reservoir	Common Clinical Manifestations; Comment
SPECIES ALWAYS CONSIDERED PATHOGENS		
Mycobacterium tuberculosis	Humans	Pulmonary and disseminated tuberculosis; millions of cases annually in the world
Mycobacterium leprae	Humans	Leprosy
Mycobacterium bovis	Humans, cattle	Tuberculosis-like disease; rare in North America; *Mycobacterium bovis* is closely related to *Mycobacterium tuberculosis*
SPECIES POTENTIALLY PATHOGENIC IN HUMANS		
Moderately common causes of disease		
Mycobacterium avium complex	Soil, water, birds, fowl, swine, cattle, environment	Disseminated, pulmonary; very common in AIDS patients; occurs in other immunosuppressed patients; uncommon in patients with normal immune systems
Mycobacterium kansasii	Water, cattle	Pulmonary, other sites
Uncommon to very rare causes of disease		
Mycobacterium africanum	Humans, monkeys	Pulmonary cultures; resembles *Mycobacterium tuberculosis*; rare
Mycobacterium genavense	Humans, pet birds	Blood in AIDS patients; grows in liquid medium (BACTEC) and on solid medium supplemented with mycobactin j; grows in 2–8 weeks
Mycobacterium haemophilum	Unknown	Subcutaneous nodules and ulcers primarily in AIDS patients; requires hemoglobin or hemin; grows at 28–32°C; rare
Mycobacterium malmoense	Unknown, environment	Pulmonary, tuberculosis-like (adults), lymph nodes (children); most reported cases are from Sweden, but organism may be much more widespread; *Mycobacterium malmoense* is closely related to *Mycobacterium avium-intracellulare*; takes 8–12 weeks to grow
Mycobacterium marinum	Fish, water	Subcutaneous nodules and abscesses, skin ulcers
Mycobacterium scrofulaceum	Soil, water, moist foods	Cervical lymphadenitis; usually cured by incision, drainage, and removal of involved lymph nodes
Mycobacterium nonchromogenicum	Environment	The primary pathogen in the *Mycobacterium terrae* complex. Causes tenosynovitis of the hand.
Mycobacterium simiae	Monkeys, water	Pulmonary, disseminated in AIDS patients; rare
Mycobacterium szulgai	Unknown	Pulmonary, tuberculosis-like; rare
Mycobacterium ulcerans	Humans, environment	Subcutaneous nodules and ulcers; may be severe; *Mycobacterium ulcerans* is closely related to *Mycobacterium marinum*; takes 6–12 weeks to grow; optimal growth at 33°C suggests environmental source; rare
Mycobacterium xenopi	Water, birds	Pulmonary, tuberculosis-like with preexisting lung disease; rare
Rapid growers		
Mycobacterium abscessus	Soil, water, animals	Most frequently isolated rapid grower from pulmonary infections; skin and soft tissue infections; frequently multidrug-resistant
Mycobacterium chelonae	Soil, water, animals, marine life	Cutaneous lesions most common, subcutaneous abscesses, disseminated infections in immunocompromised patients
Mycobacterium fortuitum	Soil, water, animals	Consists of a complex of organisms that can only be differentiated by molecular methods. Associated with nail salon furunculosis; pulmonary infections similar to *Mycobacterium abscessus*
Mycobacterium immunogenum	Environment	Associated with pseudo-outbreaks linked to contaminated equipment in hospitals; isolates have been associated with joint disease, skin ulcers, catheter infections and some pulmonary disease. Closely related to *Mycobacterium chelonae-abscessus*
Mycobacterium mucogenicum	Unknown	Central venous catheter associated infections are the most important infections associated with this organism. Name reflects its mucoid appearance in culture.
SAPROPHYTIC SPECIES THAT VERY RARELY CAUSE DISEASE IN HUMANS		
Mycobacterium gordonae	Water	These saprophytic *Mycobacterium* species are very uncommon causes of disease in humans. Positive cultures for these mycobacteria usually represent environmental contamination of specimens and not disease. Many of the saprophytic mycobacteria grow best at temperatures ≤33°C. There are many other saprophytic *Mycobacterium* species not listed here that seldom if ever appear in cultures of patient's specimens.
Mycobacterium flavescens	Soil, water	
Mycobacterium fallax	Soil, water	
Mycobacterium gastri	Gastric washings	
Mycobacterium smegmatis	Soil, water	
Mycobacterium terrae complex	Soil, water	

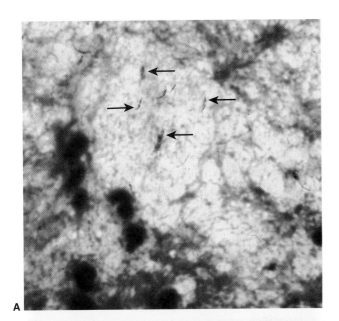

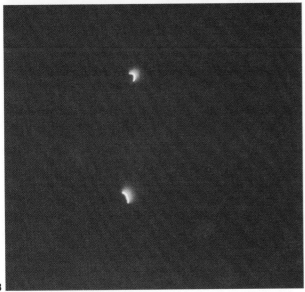

FIGURE 23–1 **A:** *Mycobacterium tuberculosis* (arrows) in a processed sputum specimen stained by Ziehl Neelsen stain. The *Mycobacterium tuberculosis* is red against a blue background. **B:** The fluorescent dye Auramine O was used to stain a sputum sample. It shows two fluorescent *Mycobacterium tuberculosis*. Original magnification ×1000. (Courtesy of G Cunningham.)

proliferate well at 22–33°C, to produce more pigment, and to be less acid-fast than pathogenic forms.

D. Reaction to Physical & Chemical Agents

Mycobacteria tend to be more resistant to chemical agents than other bacteria because of the hydrophobic nature of the cell surface and their clumped growth. Dyes (eg, malachite green) or antibacterial agents (eg, penicillin) that are bacteriostatic to other bacteria can be incorporated into media without inhibiting the growth of tubercle bacilli. Acids and

alkalies permit the survival of some exposed tubercle bacilli and are used to help eliminate contaminating organisms and for "concentration" of clinical specimens. Tubercle bacilli are resistant to drying and survive for long periods in dried sputum.

E. Variation

Variation can occur in colony appearance, pigmentation, virulence, optimal growth temperature, and many other cellular or growth characteristics.

F. Pathogenicity of Mycobacteria

There are marked differences in the ability of different mycobacteria to cause lesions in various host species. Humans and guinea pigs are highly susceptible to *M tuberculosis* infection, whereas fowl and cattle are resistant. *M tuberculosis* and *Mycobacterium bovis* are equally pathogenic for humans. The route of infection (respiratory versus intestinal) determines the pattern of lesions. In developed countries, *M bovis* has become very rare. Some "atypical" mycobacteria, now designated as nontuberculous (eg, *Mycobacterium kansasii*) produce human disease indistinguishable from tuberculosis; others (eg, *Mycobacterium fortuitum*) cause only surface lesions or act as opportunists.

Constituents of Tubercle Bacilli

The constituents listed below are found mainly in cell walls. Mycobacterial cell walls can induce delayed hypersensitivity and some resistance to infection and can replace whole mycobacterial cells in Freund's adjuvant. Mycobacterial cell contents only elicit delayed hypersensitivity reactions in previously sensitized animals.

A. Lipids

Mycobacteria are rich in lipids. These include mycolic acids (long-chain fatty acids C78–C90), waxes, and phosphatides. In the cell, the lipids are largely bound to proteins and polysaccharides. Muramyl dipeptide (from peptidoglycan) complexed with mycolic acids can cause granuloma formation; phospholipids induce caseous necrosis. Lipids are to some extent responsible for acid-fastness. Their removal with hot acid destroys acid-fastness, which depends on both the integrity of the cell wall and the presence of certain lipids. Acid-fastness is also lost after sonication of mycobacterial cells. Analysis of lipids by gas chromatography reveals patterns that aid in classification of different species.

Virulent strains of tubercle bacilli form microscopic "serpentine cords" in which acid-fast bacilli are arranged in parallel chains. Cord formation is correlated with virulence. A "cord factor" (trehalose-6,6'-dimycolate) has been extracted from virulent bacilli with petroleum ether. It inhibits migration of leukocytes, causes chronic granulomas, and can serve as an immunologic "adjuvant."

B. Proteins

Each type of mycobacterium contains several proteins that elicit the tuberculin reaction. Proteins bound to a wax fraction can, upon injection, induce tuberculin sensitivity. They can also elicit the formation of a variety of antibodies.

C. Polysaccharides

Mycobacteria contain a variety of polysaccharides. Their role in the pathogenesis of disease is uncertain. They can induce the immediate type of hypersensitivity and can serve as antigens in reactions with sera of infected persons.

Pathogenesis

Mycobacteria are emitted in droplets <25 μm in diameter when infected persons cough, sneeze, or speak. The droplets evaporate leaving organisms which are small enough, when inhaled, to be deposited in alveoli. Once inside the alveoli, the host's immune system responds by release of cytokines and lymphokines that stimulate monocytes and macrophages. Mycobacteria begin to multiply within macrophages. Some of the macrophages develop an enhanced ability to kill the organism while others may be killed by the bacilli. After 1–2 months following exposure, pathogenic lesions associated with infection, appear in the lung. Two types of lesions as described on the next page under "Pathology" may develop. Resistance and hypersensitivity of the host greatly influence development of disease and the type of lesions that are seen.

Pathology

The production and development of lesions and their healing or progression are determined chiefly by (1) the number of mycobacteria in the inoculum and their subsequent multiplication, and (2) the type of host.

A. Two Principal Lesions

1. Exudative type—This consists of an acute inflammatory reaction, with edema fluid, polymorphonuclear leukocytes, and, later, monocytes around the tubercle bacilli. This type is seen particularly in lung tissue, where it resembles bacterial pneumonia. It may heal by resolution, so that the entire exudate becomes absorbed; it may lead to massive necrosis of tissue; or it may develop into the second (productive) type of lesion. During the exudative phase, the tuberculin test becomes positive.

2. Productive type—When fully developed, this lesion, a chronic granuloma, consists of three zones: (1) a central area of large, multinucleated giant cells containing tubercle bacilli; (2) a mid zone of pale epithelioid cells, often arranged radially; and (3) a peripheral zone of fibroblasts, lymphocytes, and monocytes. Later, peripheral fibrous tissue develops, and the central area undergoes caseation necrosis. Such a lesion is called a tubercle. A caseous tubercle may break into a bronchus, empty its contents there, and form a cavity. It may subsequently heal by fibrosis or calcification.

B. Spread of Organisms in the Host

Tubercle bacilli spread in the host by direct extension, through the lymphatic channels and bloodstream, and via the bronchi and gastrointestinal tract.

In the first infection, tubercle bacilli always spread from the initial site via the lymphatics to the regional lymph nodes. The bacilli may spread farther and reach the bloodstream, which in turn distributes bacilli to all organs (miliary distribution). The bloodstream can be invaded also by erosion of a vein by a caseating tubercle or lymph node. If a caseating lesion discharges its contents into a bronchus, they are aspirated and distributed to other parts of the lungs or are swallowed and passed into the stomach and intestines.

C. Intracellular Site of Growth

Once mycobacteria establish themselves in tissue, they reside principally intracellularly in monocytes, reticuloendothelial cells, and giant cells. The intracellular location is one of the features that makes chemotherapy difficult and favors microbial persistence. Within the cells of immune animals, multiplication of tubercle bacilli is greatly inhibited.

Primary Infection & Reactivation Types of Tuberculosis

When a host has first contact with tubercle bacilli, the following features are usually observed: (1) An acute exudative lesion develops and rapidly spreads to the lymphatics and regional lymph nodes. The exudative lesion in tissue often heals rapidly. (2) The lymph node undergoes massive caseation, which usually calcifies (Ghon lesion). (3) The tuberculin test becomes positive.

This primary infection type occurred in the past, usually in childhood, but now frequently in adults who have remained free from infection and therefore tuberculin-negative in early life. In primary infections, the involvement may be in any part of the lung but is most often at the base.

The reactivation type is usually caused by tubercle bacilli that have survived in the primary lesion. Reactivation tuberculosis is characterized by chronic tissue lesions, the formation of tubercles, caseation, and fibrosis. Regional lymph nodes are only slightly involved, and they do not caseate. The reactivation type almost always begins at the apex of the lung, where the oxygen tension (PO_2) is highest.

These differences between primary infection and reinfection or reactivation are attributed to (1) resistance and (2) hypersensitivity induced by the first infection. It is not clear to what extent each of these components participates in the modified response in reactivation tuberculosis.

Immunity & Hypersensitivity

During the first infection with tubercle bacilli, a certain resistance is acquired and there is an increased capacity to localize tubercle bacilli, retard their multiplication, limit their spread, and reduce lymphatic dissemination. This can be attributed to the development of cellular immunity, with evident ability of mononuclear phagocytes to limit the multiplication of ingested organisms and even to destroy them.

In the course of primary infection, the host also acquires hypersensitivity to the tubercle bacilli. This is made evident by the development of a positive tuberculin reaction (see below). Tuberculin sensitivity can be induced by whole tubercle bacilli or by tuberculoprotein in combination with the chloroform-soluble wax D of the tubercle bacillus, but not by tuberculoprotein alone. Hypersensitivity and resistance appear to be distinct aspects of related cell-mediated reactions.

Tuberculin Test

A. Material

Old tuberculin is a concentrated filtrate of broth in which tubercle bacilli have grown for 6 weeks. In addition to the reactive tuberculoproteins, this material contains a variety of other constituents of tubercle bacilli and of growth medium. A purified protein derivative (PPD) is obtained by chemical fractionation of old tuberculin. PPD is standardized in terms of its biologic reactivity as "tuberculin units" (TU). By international agreement, the TU is defined as the activity contained in a specified weight of Seibert's PPD Lot No. 49608 in a specified buffer. This is PPD-S, the standard for tuberculin against which the potency of all products must be established by biologic assay—ie, by reaction size in humans. First-strength tuberculin has 1 TU; intermediate-strength has 5 TU; and second-strength has 250 TU. Bioequivalency of PPD products is not based on weight of the material but on comparative activity.

B. Dose of Tuberculin

A large amount of tuberculin injected into a hypersensitive host may give rise to severe local reactions and a flare-up of inflammation and necrosis at the main sites of infection (focal reactions). For this reason, tuberculin tests in surveys employ 5 TU; in persons suspected of extreme hypersensitivity, skin testing is begun with 1 TU. More concentrated material (250 TU) is administered only if the reaction to 5 TU is negative. The volume is usually 0.1 mL injected intracutaneously. The PPD preparation must be stabilized with polysorbate 80 to prevent adsorption to glass.

C. Reactions to Tuberculin

In an individual who has not had contact with mycobacteria, there is no reaction to PPD-S. An individual who has had a primary infection with tubercle bacilli develops induration, edema, erythema in 24–48 hours, and, with very intense reactions, even central necrosis. The skin test should be read in 48 or 72 hours. It is considered positive if the injection of 5 TU is followed by induration 10 mm or more in diameter. Positive tests tend to persist for several days. Weak reactions may disappear more rapidly.

The tuberculin test becomes positive within 4–6 weeks after infection (or injection of avirulent bacilli). It may be negative in the presence of tuberculous infection when "anergy" develops due to overwhelming tuberculosis, measles, Hodgkin disease, sarcoidosis, AIDS, or immunosuppression. A positive tuberculin test may occasionally revert to negative upon isoniazid treatment of a recent converter. After BCG vaccination, people convert to a positive test, but this may last for only 3–7 years. Only the elimination of viable tubercle bacilli results in reversion of the tuberculin test to negative. However, persons who were PPD-positive years ago and are healthy may fail to give a positive skin test. When such persons are retested 2 weeks later, their PPD skin test—"boosted" by the recent antigen injection—will give a positive size of induration again.

D. Interpretation of Tuberculin Test

A positive tuberculin test indicates that an individual has been infected in the past. It does not imply that active disease or immunity to disease is present. Tuberculin-positive persons are at risk of developing disease from reactivation of the primary infection, whereas tuberculin-negative persons who have never been infected are not subject to that risk, though they may become infected from an external source.

E. Gamma Interferon Release Assays for Detection of Tuberculosis

Sometimes the results of the tuberculin skin test are equivocal, particularly in persons who have been vaccinated with BCG or who live in areas where nontuberculous mycobacteria are highly prevalent in the environment. In an effort to improve diagnostic accuracy, whole-blood gamma interferon release assays have been commercially developed. These assays are based on the host's immune responses to specific *M tuberculosis* antigens ESAT-6 and CFP-10, which are absent from most nontuberculous mycobacteria and BCG. The tests detect interferon gamma that is released by sensitized CD4 T cells in response to these antigens. Currently, there are two commercial assays available in the United States. The Quantiferon-Gold (Cellestis, Valencia CA) is an ELISA assay that detects interferon gamma in whole-blood. The T-SPOT-TB (Oxford Immunotech, Oxford, UK) is an ELISA immunospot assay that uses purified peripheral blood mononuclear cells. Results for both tests are reported as positive, negative or indeterminate. These assays are still undergoing extensive evaluation. They are susceptible to the biological variation in the immune response. However, multiple studies have shown that these assays are comparable to the tuberculin skin test in evaluating latent infection particularly in persons who have received BCG. However, they should not be used in severely immunocompromised hosts or in very young children.

Clinical Findings

Since the tubercle bacillus can involve every organ system, its clinical manifestations are protean. Fatigue, weakness, weight loss, fever, and night sweats may be signs of tuberculous disease. Pulmonary involvement giving rise to chronic cough and spitting of blood usually is associated with far-advanced lesions. Meningitis or urinary tract involvement can occur in the absence of other signs of tuberculosis. Bloodstream dissemination leads to miliary tuberculosis with lesions in many organs and a high mortality rate.

Diagnostic Laboratory Tests

A positive tuberculin test does not prove the presence of active disease due to tubercle bacilli. Isolation of tubercle bacilli provides such proof.

A. Specimens

Specimens consist of fresh sputum, gastric washings, urine, pleural fluid, cerebrospinal fluid, joint fluid, biopsy material, blood, or other suspected material.

B. Decontamination & Concentration of Specimens

Specimens from sputum and other nonsterile sites should be liquefied with *N*-acetyl-L-cysteine, decontaminated with NaOH (kills many other bacteria and fungi), neutralized with buffer, and concentrated by centrifugation. Specimens processed in this way can be used for acid-fast stains and for culture. Specimens from sterile sites, such as cerebrospinal fluid, do not need the decontamination procedure but can be directly centrifuged, examined, and cultured.

C. Smears

Sputum, exudates, or other material is examined for acid-fast bacilli by staining. Stains of gastric washings and urine generally are not recommended, because saprophytic mycobacteria may be present and yield a positive stain. Fluorescence microscopy with auramine-rhodamine stain is more sensitive than traditional acid-fast stains, such as Ziehl-Neelsen and are the preferred stains for clinical material. If acid-fast organisms are found in an appropriate specimen, this is presumptive evidence of mycobacterial infection.

D. Culture, Identification, & Susceptibility Testing

Processed specimens from nonsterile sites and centrifuged specimens from sterile sites can be cultured directly onto selective and nonselective media (see above). The selective broth culture often is the most sensitive method and provides results most rapidly. A selective agar media (eg, Löwenstein-Jensen or Middlebrook 7H10/7H11 biplate with antibiotics) should be inoculated in parallel with broth media cultures. Incubation is at 35–37°C in 5–10% CO_2 for up to 8 weeks.

If cultures are negative in the setting of a positive acid-fast stain or slowly growing nontuberculous mycobacteria (see below) are suspected, then a set of inoculated media should be incubated at a lower temperature (eg, 24–33°C) and both sets incubated for 12 weeks.

Blood for culture of mycobacteria (usually *M avium* complex) should be anticoagulated and processed by one of three methods: (1) commercially available lysis centrifugation system; (2) inoculation into commercially available broth media specifically designed for blood cultures; or (3) centrifugation of the blood and inoculation of the white blood cell buffy coat layer, with or without deoxycholate lysis of the cells, into broth culture. Solid media can be used in parallel.

It is medically important to characterize and separate *M tuberculosis* from all the other species of mycobacteria. Isolated mycobacteria should be identified as to species. Conventional methods for identification of mycobacteria include observation of rate of growth, colony morphology, pigmentation, and biochemical profiles. The conventional methods often require 6–8 weeks for identification. Growth rate separates the rapid growers (growth in ≤7 days) from other mycobacteria (Table 23–1). **Photochromogens** produce pigment in light but not in darkness; **scotochromogens** develop pigment when growing in the dark; **nonchromogens** (nonphotochromogens) are nonpigmented or have light tan or buff-colored colonies. Individual species or complexes are defined by additional biochemical characteristics (eg, positive niacin test as with *M tuberculosis*, reduction of nitrate, production of urease or catalase, arylsulfatase test, and many others). The traditional classification based on the conventional methods of identification is set forth in Table 23–2. Molecular probe methods are available for four species (see below) and are much faster than the conventional methods. In the United States, the four species make up 95% or more of clinical isolates of mycobacteria, and the conventional methods are used to identify only a small percentage of the clinical isolates. The conventional methods for classifying mycobacteria are rapidly becoming of historical interest because molecular probe methods are much faster and easier. In addition, many laboratories with molecular capabilities have implemented 16S rRNA gene sequencing to rapidly identify probe-negative species or send such organisms to a reference laboratory with sequencing capability.

Molecular probes provide a rapid, sensitive, and specific method to identify mycobacteria. The probes can be used on mycobacterial growth from solid media or from broth cultures. DNA probes specific for rRNA sequences of the test organism are used in a hybridization procedure. There are approximately 10,000 copies of the rRNA per mycobacteria cell, providing a natural amplification system, enhancing detection. Double-stranded hybrids are separated from unhybridized single-stranded probes. The DNA probes are linked with chemicals that are activated in the hybrids and detected by chemiluminescence. Probes for the *M tuberculosis* complex (*M tuberculosis, M bovis,* and *Mycobacterium africanum*), *M avium* complex (*M avium, M intracellulare,* and closely

TABLE 23–2 Traditional Runyon Classification of Mycobacteria

Classification	Organism
TB complex	Mycobacterium tuberculosis Mycobacterium africanum Mycobacterium bovis
Photochromogens	Mycobacterium asiaticum Mycobacterium kansasii Mycobacterium marinum Mycobacterium simiae
Scotochromogens	Mycobacterium flavescens Mycobacterium gordonae Mycobacterium scrofulaceum Mycobacterium szulgai
Nonchromogens	Mycobacterium avium complex Mycobacterium celatum Mycobacterium haemophilum Mycobacterium gastri Mycobacterium genavense Mycobacterium malmoense Mycobacterium nonchromogenicum Mycobacterium shimoidei Mycobacterium terrae Mycobacterium trivale Mycobacterium ulcerans Mycobacterium xenopi
Rapid growers	Mycobacterium abscessus Mycobacterium fortuitum group Mycobacterium chelonae group Mycobacterium immunogenum Mycobacterium mucogenicum Mycobacterium phlei Mycobacterium smegmatis Mycobacterium vaccae

related mycobacteria), *M kansasii*, and *Mycobacterium gordonae* are in use. The use of these probes has shortened the time to identification of clinically important mycobacteria from several weeks to as little as 1 day.

High-performance liquid chromatography (HPLC) has been applied to speciation of mycobacteria. The method is based on development of profiles of mycolic acids, which vary from one species to another. HPLC to speciate mycobacteria is available in reference laboratories.

Susceptibility testing of mycobacteria is an important adjunct in selecting drugs for effective therapy. A standardized radiometric broth culture technique can be used to test for susceptibility to first-line drugs. The complex and more arduous conventional agar-based technique usually is performed in reference laboratories; first- and second-line drugs can be tested by this method.

E. DNA Detection

The polymerase chain reaction holds great promise for the rapid and direct detection of *M tuberculosis* in clinical specimens. The overall sensitivity is 55–90% with a specificity of about 99%. The test has the highest sensitivity when applied to specimens that have smears positive for acid-fast bacilli. At least two commercial molecular assays within the United States have FDA approval for detection of *M tuberculosis* in smear positive sputum samples and one of the assays is also approved for smear negative respiratory specimens.

The characterization of specific strains of *M tuberculosis* can be important for clinical and epidemiologic purposes. It allows such things as tracing transmission from one person to another, analysis of outbreaks of tuberculosis, and demonstration of reactivation versus reinfection in individual patients. DNA fingerprinting is done using a standardized protocol based on restriction fragment length polymorphism. Many copies of the insertion sequence *6110* (IS*6110*) are present in the chromosome of most strains of *M tuberculosis*, and these are located at variable positions. DNA fragments are generated by restriction endonuclease digestion and separated by electrophoresis. A probe against IS*6110* is used to determine the genotypes. This testing is done at the Centers for Disease Control and Prevention, at some state health department laboratories, and in research laboratories.

Treatment

The primary treatment for mycobacterial infection is specific chemotherapy. The drugs for treatment of mycobacterial infection are discussed in Chapter 28. Two cases of tuberculosis are presented in Chapter 48.

Between one in 10^6 and one in 10^8 tubercle bacilli are spontaneous mutants resistant to first-line antituberculosis drugs. When the drugs are used singly, the resistant tubercle bacilli emerge rapidly and multiply. Therefore, treatment regimens use drugs in combination to yield cure rates of >95%.

The two major drugs used to treat tuberculosis are **isoniazid** and **rifampin**. The other first-line drugs are **pyrazinamide**, **ethambutol**, and **streptomycin**. Second-line drugs are more toxic or less effective (or both), and they should be used in therapy only under extenuating circumstances (eg, treatment failure, multiple drug resistance). Second-line drugs include kanamycin, capreomycin, ethionamide, cycloserine, ofloxacin, and ciprofloxacin.

A four-drug regimen of isoniazid, rifampin, pyrazinamide, and ethambutol is recommended for persons in the United States who have a slight to moderate risk for being infected with drug-resistant tubercle bacilli. The risk factors include recent emigration from Latin America or Asia; persons with HIV infections or who are at risk for HIV infection and live in an area with a low prevalence of multidrug-resistant tubercle bacilli; and persons who were previously treated with a regimen that did not include rifampin.

Standard 9-month regimens are based on isoniazid and rifampin given daily; pyrazinamide, ethambutol, or streptomycin is given concomitantly until susceptibility test results are known. The isoniazid and rifampin can be administered daily for 1–2 months and twice weekly for the remainder of the 9 months, but this regimen should not be used when there is any likelihood of drug resistance. There also are several 6-month regimens for the initial treatment of tuberculosis that generally employ three or four drug regimens for 2 months followed by isoniazid and rifampin twice weekly for the total of 6 months. In noncompliant patients, directly observed therapy is important as well.

Drug resistance in *M tuberculosis* is a worldwide problem. Mechanisms explaining the resistance phenomenon for many but not all of the resistant strains have been defined. Isoniazid resistance has been associated with deletions or mutations in the catalase-peroxidase gene (*katG*); these isolates become catalase-negative or have decreased catalase activity. Isoniazid resistance has also been associated with alterations in the *inhA* gene, which encodes an enzyme that functions in mycolic acid synthesis. Streptomycin resistance has been associated with mutations in genes encoding the ribosomal S12 protein and 16S rRNA, *rpsL* and *rrs,* respectively. Rifampin resistance has been associated with alterations in the b subunit of RNA polymerase, the *rpoB* gene. Mutations in the DNA gyrase gene *gyrA* have been associated with resistance to fluoroquinolones. The possibility that drug resistance is present in a patient's *M tuberculosis* isolate must be taken into account when selecting therapy.

Multidrug-resistant *M tuberculosis* (resistant to both isoniazid and rifampin) is a major problem in tuberculosis treatment and control. Such strains are prevalent in certain geographic areas and certain populations (hospitals and prisons). There have been many outbreaks of tuberculosis with multidrug-resistant strains. They are particularly important in persons with HIV infections. Persons infected with multidrug-resistant organisms or who are at high risk for such infections, including exposure to another person with such an infection, should be treated according to susceptibility test results for the infecting strain. If susceptibility results are not available, the drugs should be selected according to the known pattern of susceptibility in the community and modified when the susceptibility test results are available. Therapy should include a minimum of three and preferably more than three drugs to which the organisms have demonstrated susceptibility.

Extensively drug resistant (XDR) strains are now globally recognized. These are defined by WHO as isolates of *M tuberculosis* with resistance to isoniazid and rifampin, any fluoroquinolone and at least three injectable second-line drugs such as amikacin, capreomycin, or kanamycin. The true prevalence of XDR tuberculosis is underestimated in resource limited countries due to the lack of available diagnostic and susceptibility tests. Factors that have contributed to the global epidemic include ineffective tuberculosis treatment, lack of proper diagnostic testing and most importantly, poor infection control practices. Persons infected with XDR tuberculosis have a poorer clinical outcome and are 64% more likely to die during treatment than persons infected with susceptible strains. In 2006, the WHO Global Task Force on XDR-TB issued multifaceted and comprehensive recommendations to address the XDR-TB epidemic (available at http://www.who.int/tb/features_archive/global_taskforce_report/en/)

Epidemiology

The most frequent source of infection is the human who excretes, particularly from the respiratory tract, large numbers of tubercle bacilli. Close contact (eg, in the family) and massive exposure (eg, in medical personnel) make transmission by droplet nuclei most likely.

Susceptibility to tuberculosis is a function of the risk of acquiring the infection and the risk of clinical disease after infection has occurred. For the tuberculin-negative person, the risk of acquiring tubercle bacilli depends on exposure to sources of infectious bacilli—principally sputum-positive patients. This risk is proportionate to the rate of active infection in the population, crowding, socioeconomic disadvantage, and inadequacy of medical care.

The development of clinical disease after infection may have a genetic component (proved in animals and suggested in humans by a higher incidence of disease in those with HLA-Bw15 histocompatibility antigen). It is influenced by age (high risk in infancy and in the elderly), by undernutrition, and by immunologic status, coexisting diseases (eg, silicosis, diabetes), and other individual host resistance factors.

Infection occurs at an earlier age in urban than in rural populations. Disease occurs only in a small proportion of infected individuals. In the United States at present, active disease has several epidemiologic patterns where individuals are at increased risk: minorities, predominantly African Americans and Hispanics; immigrants from countries of high endemicity; HIV-infected patients; homeless persons; and the very young and very old. The incidence of tuberculosis is especially high in minority persons with HIV infections. Primary infection can occur in any person exposed to an infectious source. Patients who have had tuberculosis can be infected exogenously a second time. Endogenous reactivation tuberculosis occurs most commonly among persons with AIDS immunosuppression, and elderly malnourished, or alcoholic destitute men.

Prevention & Control

1. Prompt and effective treatment of patients with active tuberculosis and careful follow-up of their contacts with tuberculin tests, x-rays, and appropriate treatment are the mainstays of public health tuberculosis control.

2. Drug treatment of asymptomatic tuberculin-positive persons in the age groups most prone to develop complications (eg, children) and in tuberculin-positive persons who must receive immunosuppressive drugs greatly reduces reactivation of infection.

3. Individual host resistance: Nonspecific factors may reduce host resistance, thus favoring the conversion of asymptomatic infection into disease. Such factors include starvation, gastrectomy, and suppression of cellular immunity by drugs (eg, corticosteroids) or infection. HIV infection is a major risk factor for tuberculosis.

4. Immunization: Various living avirulent tubercle bacilli, particularly BCG (bacillus Calmette-Guérin, an attenuated bovine organism), have been used to induce a certain amount of resistance in those heavily exposed to infection. Vaccination with these organisms is a substitute for primary infection with virulent tubercle bacilli without the danger inherent in the latter. The available vaccines are inadequate from many technical and biologic standpoints. Nevertheless, BCG is given to children in many countries. Statistical evidence indicates that an increased resistance for a limited period follows BCG vaccination.

5. The eradication of tuberculosis in cattle and the pasteurization of milk have greatly reduced *M bovis* infections.

OTHER MYCOBACTERIA

In addition to tubercle bacilli (*M tuberculosis, M bovis*), other mycobacteria of varying degrees of pathogenicity have been grown from human sources in past decades. These "atypical" mycobacteria were initially grouped according to speed of growth at various temperatures and production of pigments (see above). Several are now identified using DNA probes or by DNA sequencing. Most of them occur in the environment, are not readily transmitted from person to person, and are opportunistic pathogens (Table 23–1).

Species or complexes that are significant causes of disease are outlined below.

Mycobacterium avium Complex

The *Mycobacterium avium* complex is often called the MAC or MAI (*M avium intracellulare*) complex. These organisms grow optimally at 41°C and produce smooth, soft, nonpigmented colonies. They are ubiquitous in the environment and have been cultured from water, soil, food, and animals, including birds.

MAC organisms infrequently cause disease in immunocompetent humans. However, in the United States, disseminated MAC infection is one of the most common opportunistic infections of bacterial origin in AIDS patients. The risk of developing disseminated MAC infection in HIV-infected persons is greatly increased when the CD4-positive lymphocyte count declines to <100/μL. (See Case 17 in Chapter 48.) Gender, race, ethnic group, and individual risk factors for HIV infection do not influence the development of disseminated MAC infection, but prior *Pneumocystis jiroveci* infection, severe anemia, and interruption of antiretroviral therapy may increase the risk.

During the first 15 years of the AIDS epidemic, approximately 25% and perhaps as high as 50% of HIV-infected patients developed MAC bacteremia and disseminated infection during the course of AIDS. Subsequently, the use of highly active antiretroviral therapy (HAART) and the use of azithromycin or clarithromycin prophylaxis has greatly decreased the incidence of disseminated MAC infection in AIDS patients.

Environmental exposure can led to MAC colonization of either the respiratory or gastrointestinal tract. Transient bacteremia occurs followed by invasion of tissues. Persistent bacteremia and extensive infiltration of tissues resulting in organ dysfunction result. Any organ can be involved. In the lung, nodules, diffuse infiltrates, cavities, and endobronchial lesions are common. Other manifestations include pericarditis, soft tissue abscesses, skin lesions, lymph node involvement, bone infection, and central nervous system lesions. The patients often present with nonspecific symptoms of fever, night sweats, abdominal pain, diarrhea, and weight loss. The diagnosis is made by culturing MAC organisms from blood or tissue.

MAC organisms routinely are resistant to first-line antituberculosis drugs. Treatment with either clarithromycin or azithromycin plus ethambutol is a preferred initial therapy. Other drugs that may be useful are rifabutin (Ansamycin), clofazimine, fluoroquinolones, and amikacin. Multiple drugs often are used in combination. Therapy should be continued for life. Therapy results in decreasing counts of MAC organisms in blood and amelioration of clinical symptoms.

Mycobacterium kansasii

M kansasii is a photochromogen that requires complex media for growth at 37°C. It can produce pulmonary and systemic disease indistinguishable from tuberculosis, especially in patients with impaired immune responses. Sensitive to rifampin, it is often treated with the combination of rifampin, ethambutol, and isoniazid with good clinical response. The source of infection is uncertain, and communicability is low or absent.

Mycobacterium scrofulaceum

This is a scotochromogen occasionally found in water and as a saprophyte in adults with chronic lung disease. It causes chronic cervical lymphadenitis in children and, rarely, other granulomatous disease. Surgical excision of involved cervical lymph nodes may be curative, and resistance to antituberculosis drugs is common. (*Mycobacterium szulgai* and *Mycobacterium xenopi* are similar.)

Mycobacterium marinum & *Mycobacterium ulcerans*

These organisms occur in water, grow best at low temperature (31°C), may infect fish, and can produce superficial skin lesions (ulcers, "swimming pool granulomas") in humans. Surgical excision, tetracyclines, rifampin, and ethambutol are sometimes effective.

Mycobacterium fortuitum Complex

These are saprophytes found in soil and water that grow rapidly (3–6 days) in culture and form no pigment. They can produce superficial and systemic disease in humans on rare occasions. The organisms are often resistant to antimycobacterial drugs but may respond to amikacin, doxycycline, cefoxitin, erythromycin, or rifampin.

Mycobacterium chelonae-abscessus

These rapid growers should be differentiated because the types and severity of disease are different and because therapy for M chelonae is easier since it is more susceptible to antimicrobial agents. Both species are capable of causing skin, soft tissue and bone infections following trauma or surgery which can disseminate in immunocompromised patients. M abscessus is also frequently recovered from patients with respiratory disease in the United States especially in southeastern regions. The individuals most commonly infected are elderly, white, female nonsmokers. Patients with cystic fibrosis are also at risk and may succumb to a fulminant, rapidly progressive form of the disease. M chelonae is typically susceptible to tobramycin, clarithromycin, linezolid, and imipenem. Clarithromycin, amikacin and cefoxitin are usually used for treatment of M abscessus although drug resistance is a major problem with this organism.

Other Mycobacterium Species

The high risk for mycobacterial infection in AIDS patients has resulted in increased awareness of mycobacterial infections in general. Species previously considered to be curiosities and extremely uncommon have been more widely recognized (Table 23–1). Mycobacterium malmoense has been reported mostly from Northern Europe. It causes a pulmonary tuberculosis-like disease in adults and lymphadenitis in children. Mycobacterium haemophilum and Mycobacterium genavense cause disease in AIDS patients. The importance of these two species is not fully understood.

Saprophytic Mycobacteria Not Associated with Human Illness

Mycobacterium phlei is frequently found on plants, in soil, or in water. Mycobacterium gordonae is similar. Mycobacterium smegmatis occurs regularly in human sebaceous secretions and it might be confused with pathogenic acid-fast organisms. Mycobacterium paratuberculosis produces a chronic enteritis in cattle. There is renewed interest in this organism as a potential cause of inflammatory bowel disease.

MYCOBACTERIUM LEPRAE

Although this organism was described by Hansen in 1873 (9 years before Koch's discovery of the tubercle bacillus), it has not been cultivated on nonliving bacteriologic media. It causes leprosy. There are more than 10 million cases of leprosy, mainly in Asia.

Typical acid-fast bacilli—singly, in parallel bundles, or in globular masses—are regularly found in scrapings from skin or mucous membranes (particularly the nasal septum) in lepromatous leprosy. The bacilli are often found within the endothelial cells of blood vessels or in mononuclear cells. When bacilli from human leprosy (ground tissue nasal scrapings) are inoculated into footpads of mice, local granulomatous lesions develop with limited multiplication of bacilli. Inoculated armadillos develop extensive lepromatous leprosy, and armadillos naturally infected with leprosy have been found in Texas and Mexico. M leprae from armadillo or human tissue contains a unique o-diphenoloxidase, perhaps an enzyme characteristic of leprosy bacilli.

Clinical Findings

The onset of leprosy is insidious. The lesions involve the cooler tissue of the body: skin, superficial nerves, nose, pharynx, larynx, eyes, and testicles. The skin lesions may occur as pale, anesthetic macular lesions 1–10 cm in diameter; diffuse or discrete erythematous, infiltrated nodules 1–5 cm in diameter; or a diffuse skin infiltration. Neurologic disturbances are manifested by nerve infiltration and thickening, with resultant anesthesia, neuritis, paresthesia, trophic ulcers, and bone resorption and shortening of digits. The disfigurement due to the skin infiltration and nerve involvement in untreated cases may be extreme.

The disease is divided into two major types, lepromatous and tuberculoid, with several intermediate stages. In the lepromatous type, the course is progressive and malign, with nodular skin lesions; slow symmetric nerve involvement; abundant acid-fast bacilli in the skin lesions; continuous bacteremia; and a negative lepromin (extract of lepromatous tissue) skin test. In lepromatous leprosy, cell-mediated immunity is markedly deficient and the skin is infiltrated with suppressor T cells. In the tuberculoid type, the course is benign and nonprogressive, with macular skin lesions, severe asymmetric nerve involvement of sudden onset with few bacilli present in the lesions, and a positive lepromin skin test. In tuberculoid leprosy, cell-mediated immunity is intact and the skin is infiltrated with helper T cells.

Systemic manifestations of anemia and lymphadenopathy may also occur. Eye involvement is common. Amyloidosis may develop.

Diagnosis

Scrapings with a scalpel blade from skin or nasal mucosa or from a biopsy of earlobe skin are smeared on a slide and stained by the Ziehl-Neelsen technique. Biopsy of skin or of a thickened nerve gives a typical histologic picture. No serologic tests are of value. Nontreponemal serologic tests for syphilis frequently yield false-positive results in leprosy.

Treatment

Sulfones such as dapsone (see Chapter 28) are first-line therapy for both tuberculoid and lepromatous leprosy. Rifampin or clofazimine generally is included in the initial treatment regimens. Other drugs active against *M leprae* include minocycline, clarithromycin, and some fluoroquinolones. Regimens recommended by the World Health Organization are practical. Several years of therapy may be necessary to adequately treat leprosy.

Epidemiology

Transmission of leprosy is most likely to occur when small children are exposed for prolonged periods to heavy shedders of bacilli. Nasal secretions are the most likely infectious material for family contacts. The incubation period is probably 2–10 years. Without prophylaxis, about 10% of exposed children may acquire the disease. Treatment tends to reduce and abolish the infectivity of patients. The naturally infected armadillos found in Texas and Mexico probably play no role in transmission of leprosy to humans.

Prevention & Control

In the United States, the current recommendations for prevention of leprosy include a thorough examination of household contacts and close relatives. This should include a complete skin examination and an examination of the peripheral nervous system. The US Public Health Service National Hansen's Disease Program does not recommend routine dapsone prophylaxis. A therapeutic trial may be indicated for patients whose signs and symptoms are suggestive of leprosy but who do not have a definitive diagnosis.

REVIEW QUESTIONS

1. A 60-year-old-man has a 5-month history of progressive weakness and a weight loss of 13 kg along with intermittent fever, chills, and a chronic cough productive of yellow sputum, occasionally streaked with blood. A sputum specimen is obtained and numerous acid-fast bacteria are seen on the smear. Culture of the sputum is positive for *Mycobacterium tuberculosis*. Which treatment regimen is most appropriate for initial therapy?

    (A) Isoniazid and rifampin
    (B) Sulfamethoxazole/trimethoprim and streptomycin
    (C) Isoniazid, rifampin, pyrazinamide, and ethambutol
    (D) Isoniazid, cycloserine, and ciprofloxacin
    (E) Rifampin and streptomycin

2. If the patient's *Mycobacterium tuberculosis* isolate (Question 1) proves to be resistant to isoniazid, the likely mechanism for resistance is

    (A) β-lactamase
    (B) Mutations in the catalase-peroxidase gene
    (C) Alterations in the β subunit of RNA polymerase
    (D) Mutations in the DNA gyrase gene
    (E) Mutations in the genes encoding the S12 protein and 16S rRNA

3. A 47-year-old woman presents with a 3-month history of progressive cough, weight loss, and fever. Chest x-ray shows bilateral cavitary disease suggestive of tuberculosis. Sputum culture grows an acid-fast bacillus that is a photochromogen (makes an orange pigment when exposed to light). The organism most likely is

    (A) *Mycobacterium tuberculosis*
    (B) *Mycobacterium kansasii*
    (C) *Mycobacterium gordonae*
    (D) *Mycobacterium avium* complex
    (E) *Mycobacterium fortuitum*

4. A 31-year-old Asian woman is admitted to the hospital with a 7-week history of increasing malaise, myalgia, nonproductive cough, and shortness of breath. She has daily fevers of 38–39°C and a recent 5-kg weight loss. She had a negative chest radiograph when she entered the United States 7 years ago. The patient's grandmother died of tuberculosis when the patient was an infant. A current chest radiograph is normal; results of other tests show a decreased hematocrit and liver function test abnormalities. Liver and bone marrow biopsies show granulomas with giant cells and acid-fast bacilli. She is probably infected with

    (A) *Mycobacterium leprae*
    (B) *Mycobacterium fortuitum*
    (C) *Mycobacterium ulcerans*
    (D) *Mycobacterium gordonae*
    (E) *Mycobacterium tuberculosis*

5. It is very important that the patient (Question 4) also be evaluated for

    (A) HIV/AIDS
    (B) Typhoid fever
    (C) Liver abscess
    (D) Lymphoma
    (E) Malaria

6. Of concern regarding the patient in Question 4 (above) is that she could be infected with a mycobacterium that is

    (A) Susceptible only to isoniazid
    (B) Resistant to streptomycin
    (C) Resistant to clarithromycin
    (D) Susceptible only to ciprofloxacin
    (E) Resistant to isoniazid and rifampin

7. You observe a 40-year-old man begging on a street in a town in India. He has clawing of the fourth and fifth digits with loss of distal parts of the digits of both hands, strongly suggesting leprosy. The causative agent of this disease

    (A) Is susceptible to isoniazid and rifampin
    (B) Grows in parts of the body that are cooler than 37°C
    (C) Can be cultured in the laboratory using Middlebrook 7H11 medium
    (D) Is seen in high numbers in biopsies of tuberculoid leprosy lesions
    (E) Commonly infects people in Texas, because armadillos are hosts of *Mycobacterium leprae*

8. Which of the following statements about the purified protein derivative (PPD) and the tuberculin skin test is most correct?

    (A) It is strongly recommended that medical and other health science students have PPD skin tests every 5 years
    (B) Persons immunized with BCG rarely if ever convert to positive PPD skin tests
    (C) The intradermal skin test is usually read 4 hours after being applied

    (D) A positive tuberculin test indicates that an individual has been infected with *Mycobacterium tuberculosis* in the past and may continue to carry viable mycobacteria

    (E) A positive PPD skin test implies that a person is immune to active tuberculosis

9. A 72-year-old woman has an artificial hip joint placed because of degenerative joint disease. One week after the procedure, she has fever and joint pain. The hip is aspirated and the fluid is submitted for routine culture and for culture for acid-fast organisms. After 2 days of incubation, there is no growth on any of the media. After 4 days, however, bacilli are seen growing on the sheep blood agar plate and similar-appearing acid-fast bacilli are growing on the culture for acid-fast bacteria. The patient is most likely infected with

    (A) *Mycobacterium tuberculosis*

    (B) *Mycobacterium fortuitum/chelonae*

    (C) *Mycobacterium leprae*

    (D) *Mycobacterium kansasii*

    (E) *Mycobacterium avium* complex

10. A 10-year-old child has a primary pulmonary *Mycobacterium tuberculosis* infection. Which of the following features of tuberculosis is most correct?

    (A) In primary tuberculosis, an active exudative lesion develops and rapidly spreads to lymphatics and regional lymph nodes

    (B) The exudative lesion of primary tuberculosis often heals slowly

    (C) If tuberculosis develops years later, it is a result of another exposure to *Mycobacterium tuberculosis*

    (D) In primary tuberculosis, all of the infecting *Mycobacterium tuberculosis* organisms are killed by the patient's immune response

    (E) In primary tuberculosis, the immune system is primed but the PPD skin test remains negative until there is a second exposure to *Mycobacterium tuberculosis*

11. Which of the following statements regarding interferon gamma release assays is correct?

    (A) They are useful for evaluating immunocompromised patients for active tuberculosis.

    (B) They detect antigens present in all *Mycobacterium* species.

    (C) They are not available yet for testing in the United States.

    (D) They are performed using molecular probes that detect organism DNA.

    (E) They are used as alternatives to the tuberculin skin test to evaluate for latent tuberculosis.

12. *Mycobacterium abscessus* most often causes pulmonary disease among which group of individuals?

    (A) Young children exposed to dirt

    (B) African American smokers

    (C) Elderly, nonsmoking white females

    (D) Hispanic males who work outdoors

    (E) Persons living in the Northwestern United States

13. A newly characterized rapidly growing mycobacterium that has emerged as an important cause of central venous catheter associated infections is:

    (A) *Mycobacterium phlei*

    (B) *Mycobacterium mucogenicum*

    (C) *Mycobacterium xenopi*

    (D) *Mycobacterium smegmatis*

    (E) *Mycobacterium terrae*

14. The definition of extensively drug resistant (XDR) tuberculosis includes:

    (A) Resistance to isoniazid

    (B) Resistance to a fluoroquinolone

    (C) Resistance to capreomycin, amikacin and kanamycin

    (D) Resistance to rifampin

    (E) Resistance to all of the above

15. All of the following organisms are rapidly growing mycobacteria *except*:

    (A) *Mycobacterium fortuitum*

    (B) *Mycobacterium abscessus*

    (C) *Mycobacterium mucogenicum*

    (D) *Mycobacterium nonchromogenicum*

    (E) *Mycobacterium chelonae*

Answers

1. C	5. A	9. B	13. B
2. B	6. E	10. A	14. E
3. B	7. B	11. E	15. D
4. E	8. D	12. C	

REFERENCES

Brown-Elliott BA, Wallace RJ Jr: Infections due to nontuberculous mycobacteria other than *Mycobacterium avium-intracellulare.* In: *Mandell, Douglas, and Bennett's Principles and Practice of Infectious Diseases,* 7th ed. Mandell GL, Bennett JE, Dolin R (editors). Elsevier, 2010.

Fitzgerald D, Sterling TR, Haas DW: *Mycobacterium tuberculosis.* In: *Mandell, Douglas, and Bennett's Principles and Practice of Infectious Diseases,* 7th ed. Mandell GL, Bennett JE, Dolin R (editors). Elsevier, 2010.

Gordin FM, Horsburgh CR Jr: *Mycobacterium avium* complex. In: *Mandell, Douglas, and Bennett's Principles and Practice of Infectious Diseases,* 7th ed. Mandell GL, Bennett JE, Dolin R (editors). Elsevier, 2010.

Griffith DE et al: An Official ATS/IDSA Statement: Diagnosis, treatment, and prevention of nontuberculous mycobacterial diseases. Am J Respir Crit Care Med 2007; 175: 367.

Jassal M, Bishai WR: Extensively drug-resistant tuberculosis. Lancet Infect Dis 2009;9:19–30.

Pfyffer GE: *Mycobacterium*: General characteristics, laboratory detection, isolation, and staining procedures. In: *Manual of Clinical Microbiology,* 9th ed. Murray PR et al (editors). ASM Press, 2007.

Renault CA, Ernst JD: *Mycobacterium leprae.* In: *Mandell, Douglas, and Bennett's Principles and Practice of Infectious Diseases,* 7th ed. Mandell GL, Bennett JE, Dolin R (editors). Elsevier, 2010.

Vincent V, Gutierrez MC: *Mycobacterium*:Laboratory characteristics of slowly growing mycobacteria. In: *Manual of Clinical Microbiology,* 9th ed. Murray PR et al (editors). ASM Press, 2007.

Spirochetes & Other Spiral Microorganisms

The spirochetes are a large, heterogeneous group of spiral, motile bacteria. One family (Spirochaetaceae) of the order Spirochaetales consists of three genera of free-living, large spiral organisms. The other family (Treponemataceae) includes three genera whose members are human pathogens: (1) treponema, (2) borrelia, and (3) leptospira.

The spirochetes have many structural characteristics in common, as typified by *Treponema pallidum* (Figure 24–1). They are long, slender, helically coiled, spiral or corkscrews-haped, gram-negative bacilli. *T pallidum* has an **outer sheath** or glycosaminoglycan coating. Inside the sheath is the outer membrane, which contains peptidoglycan and maintains the structural integrity of the organisms. **Endoflagella** (axial filaments) are the flagella-like organelles in the periplasmic space encased by the outer membrane. The endoflagella begin at each end of the organism and wind around it, extending to and overlapping at the midpoint. Inside the endoflagella is the inner membrane (cytoplasmic membrane) that provides osmotic stability and covers the protoplasmic cylinder. A series of cytoplasmic tubules (body fibrils) are inside the cell near the inner membrane. Treponemes reproduce by transverse fission.

TREPONEMA

The genus *Treponema* includes *T pallidum* subspecies *pallidum*, which causes syphilis; *T pallidum* subspecies *pertenue*, which causes yaws; *T pallidum* subspecies *endemicum*, which causes endemic syphilis (also called bejel); and *Treponema carateum*, which causes pinta.

TREPONEMA PALLIDUM & SYPHILIS

Morphology & Identification

A. Typical Organisms

Slender spirals measuring about 0.2 μm in width and 5–15 μm in length. The spiral coils are regularly spaced at a distance of 1 μm from one another. The organisms are actively motile, rotating steadily around their endoflagella even after attaching to cells by their tapered ends. The long axis of the spiral is ordinarily straight but may sometimes bend, so that the organism forms a complete circle for moments at a time, returning then to its normal straight position.

The spirals are so thin that they are not readily seen unless immunofluorescent stain or dark-field illumination is employed. They do not stain well with aniline dyes, but they can be seen in tissues when stained by a silver impregnation method.

B. Culture

Pathogenic *T pallidum* has never been cultured continuously on artificial media, in fertile eggs, or in tissue culture. Nonpathogenic treponemes (eg, Reiter strain) can be cultured anaerobically in vitro. They are saprophytes antigenically related to *T pallidum*.

C. Growth Characteristics

T pallidum is a microaerophilic organism; it survives best in 1–4% oxygen. The saprophytic Reiter strain grows on a

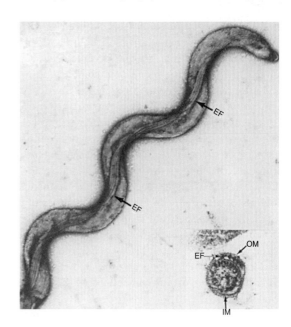

FIGURE 24–1 Electron micrograph of whole-mounted *Treponema pallidum* subspecies *pallidum*. The endoflagella are clearly visible. **Inset:** Electron micrograph of thin-sectioned *Treponema pallidum*. Note the position of the endoflagella (EF) in the periplasmic space between the inner membrane (IM) and the outer membrane (OM). (Courtesy of EM Walker.)

defined medium of 11 amino acids, vitamins, salts, minerals, and serum albumin.

In proper suspending fluids and in the presence of reducing substances, *T pallidum* may remain motile for 3–6 days at 25°C. In whole blood or plasma stored at 4°C, organisms remain viable for at least 24 hours, which is of potential importance in blood transfusions.

D. Reactions to Physical & Chemical Agents

Drying kills the spirochete rapidly, as does elevation of the temperature to 42°C. Treponemes are rapidly immobilized and killed by trivalent arsenical, mercury, and bismuth (contained in drugs of historical interest in the treatment of syphilis). Penicillin is treponemicidal in minute concentrations, but the rate of killing is slow, presumably because of the metabolic inactivity and slow multiplication rate of *T pallidum* (estimated division time is 30 hours). Resistance to penicillin has not been demonstrated in syphilis.

E. Genome

The *T pallidum* genome is a circular chromosome of approximately 1,138,000 base pairs, which is small for bacteria. Most pathogenic bacteria have transposable elements, but *T pallidum* does not, which suggests that the genome is highly conserved and may explain its continued susceptibility to penicillin. There are few genes involved in energy production and synthesis of nutrients, indicating that *T pallidum* obtains these from the host.

Antigenic Structure

The fact that *T pallidum* cannot be cultured in vitro has markedly limited the characterization of its antigens. The outer membrane surrounds the periplasmic space and the peptidoglycan-cytoplasmic membrane complex. Membrane proteins are present that contain covalently bound lipids at their amino terminals. The lipids appear to anchor the proteins to the cytoplasmic or outer membranes and keep the proteins inaccessible to antibodies. The endoflagella are in the periplasmic space. *T pallidum* subspecies *pallidum* has hyaluronidase that breaks down the hyaluronic acid in the ground substance of tissue and presumably enhances the invasiveness of the organism. The protein profiles of *T pallidum* (all the subspecies) are indistinguishable; more than 100 protein antigens have been noted. The endoflagella are composed of three core proteins that are homologous to other bacterial flagellin proteins, plus an unrelated sheath protein. Cardiolipin is an important component of the treponemal antigens.

Humans with syphilis develop antibodies capable of staining *T pallidum* by indirect immunofluorescence, immobilizing and killing live motile *T pallidum* and fixing complement in the presence of a suspension of *T pallidum* or related spirochetes. The spirochetes also cause the development of a distinct antibody-like substance, reagin, which gives positive

CF and flocculation tests with aqueous suspensions of cardiolipin extracted from normal mammalian tissues. Both reagin and antitreponemal antibody can be used for the serologic diagnosis of syphilis.

Pathogenesis, Pathology, & Clinical Findings

A. Acquired Syphilis

Natural infection with *T pallidum* is limited to the human host. Human infection is usually transmitted by sexual contact, and the infectious lesion is on the skin or mucous membranes of genitalia. In 10–20% of cases, however, the primary lesion is intrarectal, perianal, or oral. It may be anywhere on the body. *T pallidum* can probably penetrate intact mucous membranes, or they may enter through a break in the epidermis.

Spirochetes multiply locally at the site of entry, and some spread to nearby lymph nodes and then reach the bloodstream. In 2–10 weeks after infection, a papule develops at the site of infection and breaks down to form an ulcer with a clean, hard base ("hard chancre"). The inflammation is characterized by a predominance of lymphocytes and plasma cells. This "primary lesion" always heals spontaneously, but 2–10 weeks later the "secondary" lesions appear. These consist of a red maculopapular rash anywhere on the body, including the hands and feet, and moist, pale papules (condylomas) in the anogenital region, axillas, and mouth. There may also be syphilitic meningitis, chorioretinitis, hepatitis, nephritis (immune complex type), or periostitis. The secondary lesions also subside spontaneously. Both primary and secondary lesions are rich in spirochetes and highly infectious. Contagious lesions may recur within 3–5 years after infection, but thereafter the individual is not infectious. Syphilitic infection may remain subclinical, and the patient may pass through the primary or secondary stage (or both) without symptoms or signs yet develop tertiary lesions.

In about 30% of cases, early syphilitic infection progresses spontaneously to complete cure without treatment. In another 30%, the untreated infection remains latent (principally evident by positive serologic tests). In the remainder, the disease progresses to the "tertiary stage," characterized by the development of granulomatous lesions (gummas) in skin, bones, and liver; degenerative changes in the central nervous system (meningovascular syphilis, paresis, tabes); or cardiovascular lesions (aortitis, aortic aneurysm, aortic valve insufficiency). In all tertiary lesions, treponemes are very rare, and the exaggerated tissue response must be attributed to hypersensitivity to the organisms. However, treponemes can occasionally be found in the eye or central nervous system in late syphilis.

B. Congenital Syphilis

A pregnant syphilitic woman can transmit *T pallidum* to the fetus through the placenta beginning in the 10th to 15th weeks of gestation. Some of the infected fetuses die, and miscarriages result; others are stillborn at term. Others are born live but develop the signs of congenital syphilis in childhood: interstitial keratitis, Hutchinson's teeth, saddlenose, periostitis, and a variety of central nervous system anomalies. Adequate treatment of the mother during pregnancy prevents congenital syphilis. The reagin titer in the blood of the child rises with active infection but falls with time if antibody was passively transmitted from the mother. In congenital infection, the child makes IgM antitreponemal antibody.

C. Experimental Disease

Rabbits can be experimentally infected in the skin, testis, and eye with human *T pallidum*. The animal develops a chancre rich in spirochetes, and organisms persist in lymph nodes, spleen, and bone marrow for the entire life of the animal, although there is no progressive disease.

Diagnostic Laboratory Tests

A. Specimens

Specimens include tissue fluid expressed from early surface lesions for demonstration of spirochetes; blood serum for serologic tests.

B. Dark-Field Examination

A drop of tissue fluid or exudate is placed on a slide and a coverslip pressed over it to make a thin layer. The preparation is then examined under oil immersion with dark-field illumination for typical motile spirochetes.

Treponemes disappear from lesions within a few hours after the beginning of antibiotic treatment.

C. Immunofluorescence

Tissue fluid or exudate is spread on a glass slide, air dried, and sent to the laboratory. It is fixed, stained with a fluorescein-labeled antitreponeme serum, and examined by means of immunofluorescence microscopy for typical fluorescent spirochetes.

D. Serologic Tests for Syphilis

These tests use either nontreponemal or treponemal antigens.

1. Nontreponemal tests—The nontreponemal tests are universally used as screening tests for syphilis. The tests are widely available, lend themselves to automation with ease of performance in large numbers, and have low cost. In addition to their function as screening tests they can be used to follow the efficacy of therapy. The drawbacks to the nontreponemal tests are that they are not very sensitive in early syphilis and may not turn positive until a few weeks after initial infection; false-positive results can occur with many other diseases; and there may be a prozone phenomenon particularly in secondary syphilis (antibody excess produces a negative result at low serum dilutions, but positive results at higher dilutions). The

antigens in these tests contain measured amounts of cardiolipin, cholesterol, and purified lecithin in quantities sufficient to yield a standardized amount of reactivity. Historically, the cardiolipin was extracted from beef heart or liver with added lecithin and cholesterol to enhance reaction with syphilitic "reagin" antibodies. Reagin is a mixture of IgM and IgG antibodies reactive with the cardiolipin-cholesterol-lecithin complex. All of the tests are based on the fact that the particles of the lipid antigen remain dispersed in normal serum but flocculate when combining with reagin. The **Venereal Disease Research Laboratory (VDRL)** and **unheated serum reagin (USR)** tests require microscopic examination to detect flocculation. The **rapid plasma reagin (RPR)** test and **toluidine red unheated serum test (TRUST)** have colored particles that become caught in the mesh of the antigen–antibody complex allowing the tests to be read without microscopic magnification. Results develop within a few minutes, particularly if the suspension is agitated.

The nontreponemal tests can give quantitative results using serial twofold dilutions. An estimate of the amount of reagin present in serum can be expressed as the titer or as the highest dilution giving a positive result. Quantitative results are valuable in establishing a diagnosis and in evaluating the effect of treatment. Positive nontreponemal tests develop after 2–3 weeks of untreated syphilis and are positive in high titer in secondary syphilis. Positive nontreponemal tests typically revert to negative, often in 6–18 months and generally by 3 years after effective treatment of syphilis. A positive nontreponemal test late after treatment for syphilis suggests ineffective treatment or reinfection.

The VDRL test is standardized for use on cerebrospinal fluid (CSF) and becomes positive in neurosyphilis. Reagin antibodies generally do not reach the CSF from the bloodstream but are probably formed in the central nervous system in response to syphilitic infection. The serologic diagnosis of **neurosyphilis** is complex.

2. Treponemal antibody tests—The treponemal tests measure antibodies against *T pallidum* antigens. The tests are used to determine if a positive result from a nontreponemal test is truly positive or falsely positive. A positive result of a treponemal test on a serum specimen that is also positive on a nontreponemal test is a strong indication of *T pallidum* infection. The treponemal tests are less useful as screening tests because once positive following initial syphilitic infection the tests remain positive for life independent of therapy for syphilis. Serial dilutions of serum are not done in the treponemal tests, and results are reported as reactive or nonreactive (or occasionally inconclusive). The treponemal antibody tests tend to be more costly than the nontreponemal test, which is important when large groups of people (eg, blood donors) are being screened.

The ***T pallidum*-particle agglutination (TP-PA)** test is perhaps the most widely used treponemal test in the United States. Gelatin particles sensitized with *T pallidum* subspecies *pallidum* antigens are added to a standard dilution of serum. When anti-*T pallidum* antibodies react with the sensitized particles, a mat of agglutinated particles forms in the well of the microdilution tray. Gelatin particles that are not sensitized are tested with diluted serum to exclude nonspecific agglutination.

The **T pallidum hemagglutination (TPHA)** and the **microhemagglutination T pallidum (MHA-TP)** are based on the same principles as the TP-PA but use sheep erythrocytes rather than gelatin particles and may be more prone to nonspecific agglutination.

Multiple relatively similar treponemal antibody tests using the **EIA for T pallidum** (enzyme immunoassay format) are available. These tests use antigens obtained by sonication of *T pallidum*, or recombinant antigens. An aliquot of serum at a standard dilution is added to a sensitized well of a microdilution plate. Following washing, addition of an enzyme-labeled conjugate, and further washing, a precursor substrate is added. A color change indicates a reactive serum.

The **fluorescent treponemal antibody absorbed (FTA-ABS)** test is the treponemal antibody test employed for many years. Because it is difficult to perform the test is used only in selected circumstances. The test uses indirect immunofluorescence to detect reactive antibodies: killed *T pallidum*, patient's serum absorbed with sonicated saprophytic Reiter spirochetes, plus antihuman γ-globulin labeled with a fluorescent compound. The presence of IgM FTA in the blood of newborns is a good evidence of in utero infection (congenital syphilis). A negative FTA-ABS on CSF tends to exclude neurosyphilis, but a positive FTA-ABS on CSF can occur by transfer of antibodies from serum and is not helpful in the diagnosis of neurosyphilis.

Immunity

A person with active or latent syphilis or yaws appears to be resistant to superinfection with *T pallidum*. However, if early syphilis or yaws is treated adequately and the infection is eradicated, the individual again becomes fully susceptible. The various immune responses usually fail to eradicate the infection or arrest its progression.

Treatment

Penicillin in concentrations of 0.003 U/mL has definite treponemicidal activity, and penicillin is the treatment of choice. Syphilis of less than 1 year's duration is treated by a single injection of benzathine penicillin G intramuscularly. In older or latent syphilis, benzathine penicillin G intramuscularly is given three times at weekly intervals. In neurosyphilis, the same therapy is acceptable, but larger amounts of intravenous penicillin are sometimes recommended. Other antibiotics, eg, tetracyclines or erythromycin, can occasionally be substituted. Treatment of gonorrhea is thought to cure incubating syphilis. Prolonged follow-up is essential. In neurosyphilis, treponemes occasionally survive such treatment. Severe neurologic relapses of treated syphilis have occurred in

patients with acquired immunodeficiency syndrome (AIDS) who are infected with both HIV and *T pallidum*. A typical Jarisch-Herxheimer reaction may occur within hours after treatment is begun. It is due to the release of toxic products from dying or killed spirochetes.

Epidemiology, Prevention, & Control

With the exceptions of congenital syphilis and the rare occupational exposure of medical personnel, syphilis is acquired through sexual exposure. Reinfection in treated persons is common. An infected person may remain contagious for 3–5 years during "early" syphilis. "Late" syphilis, of more than 5 years' duration, is usually not contagious. Consequently, control measures depend on (1) prompt and adequate treatment of all discovered cases; (2) follow-up on sources of infection and contacts so they can be treated; and (3) safe sex with condoms. Several sexually transmitted diseases can be transmitted simultaneously. Therefore, it is important to consider the possibility of syphilis when any one sexually transmitted disease has been found.

DISEASES RELATED TO SYPHILIS

These diseases are all caused by treponemes closely related to *T pallidum*. All give positive treponemal and nontreponemal serologic tests for syphilis (STS), and some cross-immunity can be demonstrated in experimental animals and perhaps in humans. None are sexually transmitted diseases; all are commonly transmitted by direct contact. None of the causative organisms have been cultured on artificial media.

Bejel

Bejel (due to *T pallidum* subspecies *endemicum*) occurs chiefly in Africa but also in the Middle East, in Southeast Asia, and elsewhere, particularly among children, and produces highly infectious skin lesions; late visceral complications are rare. Penicillin is the drug of choice.

Yaws

Yaws is endemic, particularly among children, in many humid, hot tropical countries. It is caused by *T pallidum* subspecies *pertenue*. The primary lesion, an ulcerating papule, occurs usually on the arms or legs. Transmission is by person-to-person contact in children under age 15. Transplacental, congenital infection does not occur. Scar formation of skin lesions and bone destruction are common, but visceral or nervous system complications are very rare. It has been debated whether yaws represents a variant of syphilis adapted to transmission by nonsexual means in hot climates. There appears to be cross-immunity between yaws and syphilis. Diagnostic procedures and therapy are similar to those for syphilis. The response to penicillin treatment is dramatic.

Pinta

Pinta is caused by *T carateum* and occurs endemically in all age groups in Mexico, Central and South America, the Philippines, and some areas of the Pacific. The disease appears to be restricted to dark-skinned races. The primary lesion, a nonulcerating papule, occurs on exposed areas. Some months later, flat, hyperpigmented lesions appear on the skin; depigmentation and hyperkeratosis take place years afterward. Late cardiovascular and nervous system involvement occurs very rarely. Pinta is transmitted by nonsexual means, either by direct contact or through the agency of flies or gnats. Diagnosis and treatment are the same as for syphilis.

BORRELIA

BORRELIA SPECIES & RELAPSING FEVER

Relapsing fever in epidemic form is caused by *Borrelia recurrentis*, transmitted by the human body louse; it does not occur in the United States. Endemic relapsing fever is caused by borreliae transmitted by ticks of the genus *Ornithodoros*. The species name of the *Borrelia* genus is often the same as that of the tick. *Borrelia hermsii*, eg, the cause of relapsing fever in the western United States, is transmitted by *Ornithodoros hermsii*.

Morphology & Identification

A. Typical Organisms

The borreliae form irregular spirals 10–30 μm long and 0.3 μm wide. The distance between turns varies from 2 μm to 4 μm. The organisms are highly flexible and move both by rotation and by twisting. Borreliae stain readily with bacteriologic dyes as well as with blood stains such as Giemsa's stain or Wright's stain (Figure 24–2).

B. Culture

The organism can be cultured in fluid media containing blood, serum, or tissue, but it rapidly loses its pathogenicity for animals when transferred repeatedly in vitro. Multiplication is rapid in chick embryos when blood from patients is inoculated onto the chorioallantoic membrane.

C. Growth Characteristics

Little is known of the metabolic requirements or activity of borreliae. At 4°C, the organisms survive for several months in infected blood or in culture. In some ticks (but not in lice), spirochetes are passed from generation to generation.

D. Variation

The only significant variation of *Borrelia* is with respect to its antigenic structure.

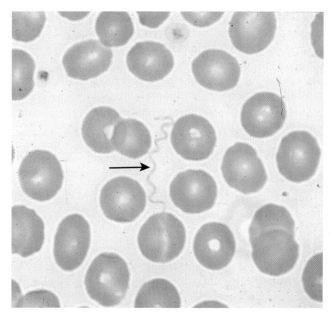

FIGURE 24–2 *Borrelia* (arrow) in a peripheral blood smear of a patient with relapsing fever. Original magnification × 1000.

Antigenic Structure

Antibodies develop in high titer after infection with borreliae. The antigenic structure of the organisms changes in the course of a single infection. The antibodies produced initially act as a selective factor that permits the survival only of antigenically distinct variants. The relapsing course of the disease appears to be due to the multiplication of such antigenic variants, against which the host must then develop new antibodies. Ultimate recovery (after 3 to 10 relapses) is associated with the presence of antibodies against several antigenic variants.

Pathology

Fatal cases show spirochetes in great numbers in the spleen and liver, necrotic foci in other parenchymatous organs, and hemorrhagic lesions in the kidneys and the gastrointestinal tract. Spirochetes have occasionally been demonstrated in the spinal fluid and brain of persons who have had meningitis. In experimental animals (guinea pigs, rats), the brain may serve as a reservoir of borreliae after they have disappeared from the blood.

Pathogenesis & Clinical Findings

The incubation period is 3–10 days. The onset is sudden, with chills and an abrupt rise of temperature. During this time, spirochetes abound in the blood. The fever persists for 3–5 days and then declines, leaving the patient weak but not ill. The afebrile period lasts 4–10 days and is followed by a second attack of chills, fever, intense headache, and malaise. There are from three to 10 such recurrences, generally of diminishing severity. During the febrile stages (especially when the temperature is rising), organisms are present in the blood; during the afebrile periods, they are absent.

Antibodies against the spirochetes appear during the febrile stage, and the attack is probably terminated by their agglutinating and lytic effects. These antibodies may select out antigenically distinct variants that multiply and cause a relapse. Several distinct antigenic varieties of borreliae may be isolated from a single patient's sequential relapses, even following experimental inoculation with a single organism.

Diagnostic Laboratory Tests

A. Specimens

Blood specimens are obtained during the rise in fever, for smears and animal inoculation.

B. Smears

Thin or thick blood smears stained with Wright's or Giemsa's stain reveal large, loosely coiled spirochetes among the red cells.

C. Animal Inoculation

White mice or young rats are inoculated intraperitoneally with blood. Stained films of tail blood are examined for spirochetes 2–4 days later.

D. Serology

Spirochetes grown in culture can serve as antigens for CF tests, but the preparation of satisfactory antigens is difficult. Patients suffering from epidemic (louse-borne) relapsing fever may develop a positive VDRL.

Immunity

Immunity following infection is usually of short duration.

Treatment

The great variability of the spontaneous remissions of relapsing fever makes evaluation of chemotherapeutic effectiveness difficult. Tetracyclines, erythromycin, and penicillin are all believed to be effective. Treatment for a single day may be sufficient to terminate an individual attack.

Epidemiology, Prevention, & Control

Relapsing fever is endemic in many parts of the world. Its main reservoir is the rodent population, which serves as a source of infection for ticks of the genus *Ornithodoros*. The distribution of endemic foci and the seasonal incidence of the disease are largely determined by the ecology of the ticks in different areas. In the United States, infected ticks are found throughout the West, especially in mountainous areas, but clinical cases are rare. In the tick, *Borrelia* may be transmitted transovarially from generation to generation.

Spirochetes are present in all tissues of the tick and may be transmitted by the bite or by crushing the tick. The tick-borne disease is not epidemic. However, when an infected individual harbors lice, the lice become infected by sucking blood; 4–5 days later, they may serve as a source of infection for other individuals. The infection of the lice is not transmitted to the next generation, and the disease is the result of rubbing crushed lice into bite wounds. Severe epidemics may occur in louse-infected populations, and transmission is favored by crowding, malnutrition, and cold climate.

In endemic areas, human infection may occasionally result from contact with the blood and tissues of infected rodents. The mortality rate of the endemic disease is low, but in epidemics it may reach 30%.

Prevention is based on avoidance of exposure to ticks and lice and on delousing (cleanliness, insecticides).

BORRELIA BURGDORFERI & LYME DISEASE

Lyme disease is named after the town of Lyme, Connecticut, where clusters of cases in children were identified. Lyme disease is caused by the spirochete *B burgdorferi* and is transmitted to humans by the bite of a small ixodes tick. The disease has early manifestations with a characteristic skin lesion, **erythema migrans**, along with flu-like symptoms, and late manifestations often with arthralgia and arthritis.

Morphology & Identification

A. Typical Organisms

B burgdorferi is a spiral organism 20–30 μm long and 0.2–0.3 μm wide. The distance between turns varies from 2 to 4 μm. The organisms have variable numbers (7–11) of endoflagella and are highly motile. *B burgdorferi* stains readily with acid and aniline dyes and by silver impregnation techniques.

B. Culture & Growth Characteristics

B burgdorferi grows most readily in a complex liquid medium, Barbour-Stoenner-Kelly medium (BSK II). Rifampin, fosfomycin (phosphonomycin), and amphotericin B can be added to BSK II to reduce the rate of culture contamination by other bacteria and fungi. *B burgdorferi* has been most easily isolated from erythema migrans skin lesions; isolation of the organism from other sites has been difficult. The organism can also be cultured from ticks. Because culture of the organism is a complex and specialized procedure with a low diagnostic yield, it is seldom used.

Antigenic Structure & Variation

B burgdorferi has a morphologic appearance similar to that of other spirochetes. The entire genome of *B burgdorferi* has been sequenced, allowing prediction of many antigenic structures. There is an unusual linear chromosome of about 950 kb and multiple circular and linear plasmids. There are a large number of sequences for lipoproteins, including outer surface proteins OspA–F. Differential expression of these proteins is thought to help *B burgdorferi* live in the very different tick and mammalian hosts. OspA and OspB along with lipoprotein 6.6 are expressed primarily in the tick. Other outer surface proteins are upregulated during tick feeding, when the organisms migrate from the tick's midgut to the salivary gland. This may explain why the tick must feed for 24–48 hours before transmission of *B burgdorferi* occurs.

Pathogenesis & Clinical Findings

The transmission of *B burgdorferi* to humans is by injection of the organism in tick saliva or by regurgitation of the tick's midgut contents. The organism adheres to proteoglycans on host cells; this is mediated by a borrelial glycosaminoglycan receptor. After injection by the tick, the organism migrates out from the site, producing the characteristic skin lesion. Dissemination occurs by lymphatics or blood to other skin and musculoskeletal sites and to many other organs.

Lyme disease, like other spirochetal diseases, occurs in stages with early and late manifestations. A unique skin lesion that begins 3 days to 4 weeks after a tick bite often marks Stage 1. The lesion, erythema migrans, begins as a flat reddened area near the tick bite and slowly expands, with central clearing. With the skin lesion there is often a flu-like illness with fever, chills, myalgia, and headache. Stage 2 occurs weeks to months later and includes arthralgia and arthritis; neurologic manifestations with meningitis, facial nerve palsy, and painful radiculopathy; and cardiac disease with conduction defects and myopericarditis. Stage 3 begins months to years later with chronic skin, nervous system, or joint involvement. Spirochetes have been isolated from all of these sites, and it is likely that some of the late manifestations are caused by deposition of antigen–antibody complexes.

Diagnostic Laboratory Tests

In some symptomatic patients, the diagnosis of early Lyme disease can be established clinically by observing the unique skin lesion. When this skin lesion is not present and at later stages of the disease, which must be differentiated from many other diseases, it is necessary to perform diagnostic laboratory tests. There is, however, no one test that is both sensitive and specific.

A. Specimens

Blood is obtained for serologic tests. CSF or joint fluid can be obtained, but culture usually is not recommended. These specimens and others can be used to detect *B burgdorferi* DNA by the polymerase chain reaction.

B. Smears

B burgdorferi has been found in sections of biopsy specimens, but examination of stained smears is an insensitive method for diagnosis of Lyme disease. *B burgdorferi* in tissue sections can sometimes be identified using antibodies and immunohistochemical methods.

C. Culture

Culture is generally not performed because it takes 6–8 weeks to complete and lacks sensitivity.

D. Molecular Probes

The polymerase chain reaction assay has been applied to detection of *B burgdorferi* DNA in many body fluids. It is rapid, sensitive, and specific, but it does not differentiate between DNA from live *B burgdorferi* in active disease and between DNA from dead *B burgdorferi* in treated or inactive disease. It has about 85% sensitivity when applied to synovial fluid samples, but the sensitivity is much lower when it is applied to CSF samples from patients with neuroborreliosis.

E. Serology

Serology has been the mainstay for the diagnosis of Lyme disease, but 3–5% of normal people and persons with other diseases (eg, rheumatoid arthritis, many infectious diseases) may be seropositive by initial EIA or indirect fluorescent antibody (IFA) assay. When the prevalence of Lyme disease is low as it is in many geographic areas, there is a much greater likelihood that a positive test is from a person who does not have Lyme disease than from a person who does have the disease (a positive predictive value of <10%). Thus, serology for Lyme disease should only be done when there are highly suggestive clinical findings. A diagnosis of Lyme disease should not be based on a positive EIA or IFA test in the absence of suggestive clinical findings. A two-stage approach to the serodiagnosis is strongly recommended; EIA or IFA followed by an immunoblot assay for reactivity with specific *B burgdorferi* antigens.

The EIA and IFA are the most widely used initial tests for Lyme disease. Multiple variations of these assays using different antigen preparations, techniques, and end points have been marketed. Results of the initial tests are generally reported as positive, negative, or indeterminate.

The immunoblot assay is generally performed to confirm results obtained by the EIA tests. Recombinant *B burgdorferi* antigens or antigens from whole cell lysates are electrophoretically separated, transferred to a nitrocellulose membrane, and reacted with a patient's serum. Interpretation of the immunoblot is based on the number and molecular size of antibody reactions with the *B burgdorferi* proteins. Blots can be analyzed for IgG or IgM. The antigen–antibody band patterns on the immunoblots should be interpreted with knowledge of known results from patients at various stages of Lyme borreliosis, and caution should be used to avoid over interpretation of minimally reactive blots.

Immunity

The immunologic response to *B burgdorferi* develops slowly. Sera obtained in Stage 1 are positive in 20–50% of patients.

Sera obtained during Stage 2 are positive in 70–90% with reactive IgG and IgM; IgG predominates in long-standing infection. In Stage 3 nearly 100% of patients have IgG reactive with *B burgdorferi*. The antibody response can expand from months to years and appears to be directed sequentially against a series of *B burgdorferi* proteins. Early antimicrobial treatment decreases the antibody response. Antibody titers fall slowly after treatment, but most patients with later manifestation of Lyme disease remain seropositive for years.

Treatment

Early infection, either local or disseminated, should be treated with doxycycline or amoxicillin—or an alternative—for 14–21 days. Treatment relieves early symptoms and promotes resolution of skin lesions. Doxycycline may be more effective than amoxicillin in preventing late manifestations. Established arthritis may respond to prolonged therapy with doxycycline or amoxicillin orally or penicillin G or ceftriaxone intravenously. In refractory cases, ceftriaxone has been effective. Nearly 50% of patients treated with doxycycline or amoxicillin early in the course of Lyme disease develop minor late complications (headache, joint pains, etc).

Epidemiology, Prevention, & Control

B burgdorferi is transmitted by a small tick of the genus *Ixodes*. The vector is *Ixodes scapularis* (also called *Ixodes dammini*) in the Northeast and Midwest and *Ixodes pacificus* on the West Coast of the United States. In Europe, the vector is *Ixodes ricinus*, and other tick vectors appear to be important in other areas of the world. The *Ixodes* ticks are quite small and often are not noticed when feeding on the skin. The larvae are about 1 mm; the nymphs about the size of a poppy seed or piece of cracked pepper (about 2 mm); and the adult female 3–4 mm. All stages are smaller by one-half, or more, than comparable stages of the dog tick *Dermacenter variabillis*. Depending upon the developmental stage and the *Ixodes* species, the ticks must feed for 2–4 days to obtain a blood meal. Transmission of *B burgdorferi* occurs late in the feeding process. Mice and deer constitute the main animal reservoirs of *B burgdorferi*, but other rodents and birds may also be infected. In the eastern part of the United States, 10–50% of ticks are infected, while in the western states the infection rate in ticks is much lower, about 2%.

Most exposures are in May through July, when the nymphal stage of the ticks is most active; however, the larval stage (August and September) and adult stage (spring and fall) also feed on humans and can transmit *B burgdorferi*.

Prevention is based on avoidance of exposure to ticks. Long sleeves and long pants tucked into socks are recommended. Careful examination of the skin for ticks after being outdoors can locate ticks for removal before they transmit *B burgdorferi*.

Environmental control of ticks using application of insecticides has provided modest success in reducing the number of nymphal ticks for a season.

LEPTOSPIRA & LEPTOSPIROSIS

Leptospirosis is a zoonosis of worldwide distribution. It is caused by spirochetes of the genus *Leptospira*. The traditional classification system is based on biochemical and serologic specificity to differentiate between the pathogenic species, *Leptospira interrogans*, and the free-living nonpathogenic species, *Leptospira biflexa*. The species are further broken down to more than 200 serovars of *L interrogans* and more than 60 serovars of *L biflexa*. The serovars are further organized into serogroups of *L interrogans* and serogroups of *L biflexa*. The serogroups are based on shared antigenicity and are primarily for laboratory use.

A second classification system is based on DNA-DNA hybridization studies, which have demonstrated a high degree of heterogeneity within the two species of the traditional classification. The molecular classification of leptospires is done by 16S rRNA sequencing. Phylogenetic analysis of these sequences indicates that there are three clades of leptospires, pathogens, saprophytes, and some of uncertain pathogenicity. The species do not correspond to the species in the traditional serologic classification. Indeed some serovars in the traditional classification occur in multiple species in the molecular classification and the serologic classification cannot be used to predict the molecular classification.

The discussion below uses the traditional serologic classification.

Morphology & Identification

A. Typical Organisms

Leptospirae are tightly coiled, thin, flexible spirochetes 5–15 μm long, with very fine spirals 0.1–0.2 μm wide; one end is often bent, forming a hook. They are actively motile, which is best seen using a dark-field microscope. Electron micrographs show a thin axial filament and a delicate membrane. The spirochete is so delicate that in the dark-field view it may appear only as a chain of minute cocci. It does not stain readily but can be impregnated with silver.

B. Culture

Leptospires grow best under aerobic conditions at 28–30°C in semisolid medium (EMJH, others) in 10 mL test tubes with 0.1% agar and 5-fluorouracil. See also Diagnostic Laboratory Tests (below). After 1–2 weeks, the leptospires produce a diffuse zone of growth near the top of the tube and later a ring of growth at a level in the tube corresponding to the level of the optimal oxygen tension for the organisms.

C. Growth Requirements

Leptospirae derive energy from oxidation of long-chain fatty acids and cannot use amino acids or carbohydrates as major energy sources. Ammonium salts are a main source of nitrogen. Leptospirae can survive for weeks in water, particularly at alkaline pH.

Antigenic Structure

The main strains ("serovars") of *L interrogans* isolated from humans or animals in different parts of the world (Table 24–1) are all serologically related and exhibit cross-reactivity in serologic tests. This indicates considerable overlapping in antigenic structure, and quantitative tests and antibody absorption studies are necessary for a specific serologic diagnosis. The outer envelope contains large amounts of lipopolysaccharide of antigenic structure that is variable from one strain to another. This variation forms the basis for the serologic classification of the *Leptospira* species. It also determines the specificity of the human immune response to leptospirae.

Pathogenesis & Clinical Findings

Human infection usually results from leptospires, often in bodies of water, entering the body through breaks in the skin (cuts and abrasions) and mucus membranes (mouth, nose, conjunctivae). Ingestion is considered to be less important. After an incubation period of 1–2 weeks, there is a variable febrile onset during which spirochetes are present in the bloodstream. They then establish themselves in the parenchymatous organs (particularly liver and kidneys), producing hemorrhage and necrosis of tissue and resulting in dysfunction of those organs (jaundice, hemorrhage, nitrogen retention). The illness is often biphasic. After initial improvement, the second phase develops when the IgM antibody titer rises. It manifests itself often as "aseptic meningitis" with intense headache, stiff neck, and pleocytosis of the CSF. Nephritis and hepatitis may also recur, and there may be skin, muscle, and eye lesions. The degree and distribution of organ involvement vary in the different diseases produced by different leptospirae in various parts of the world (Table 24–1). Many infections are mild or subclinical. Hepatitis is frequent in patients with leptospirosis.

Kidney involvement in many animal species is chronic and results in the shedding of large numbers of leptospirae in the urine; this is probably the main source of environmental contamination resulting in infection of humans. Human urine also may contain spirochetes in the second and third weeks of disease.

Agglutinating, complement-fixing, and lytic antibodies develop during the infection. Serum from convalescent patients protects experimental animals against an otherwise fatal infection. The immunity resulting from infection in humans and animals appears to be serovar-specific.

Diagnostic Laboratory Tests

A. Specimens

Specimens consist of aseptically collected blood in a heparin tube, CSF, or tissues for microscopic examination and

TABLE 24–1   Principal Leptospiral Diseases

Leptospira interrogans Serovar[a]	Source of Infection	Disease in Humans	Clinical Findings	Distribution
Autumnalis	?	Pretibial fever or Ft. Bragg fever	Fever, rash over tibia	USA, Japan
Ballum	Mice	—	Fever, rash, jaundice	USA, Europe, Israel
Bovis	Cattle, voles	—	Fever, prostration	USA, Israel, Australia
Canicola	Dog urine	Infectious jaundice	Influenza-like illness, aseptic meningitis	Worldwide
Grippotyphosa	Rodents, water	Marsh fever	Fever, prostration, aseptic meningitis	Europe, USA, Africa
Hebdomadis	Rats, mice	7-day fever	Fever, jaundice	Japan, Europe
Icterohaemorrhagiae	Rat urine, water	Weil disease	Jaundice, hemorrhages, aseptic meningitis	Worldwide
Mitis	Swine	Swineherd's disease	Aseptic meningitis	Australia
Pomona	Swine, cattle	Swineherd's disease	Fever, prostration, aseptic meningitis	Europe, USA, Australia

[a]Formerly called species.

culture. Urine should be collected using great care to avoid contamination. Serum is collected for agglutination tests.

B. Microscopic Examination

Dark-field examination or thick smears stained by the Giemsa technique occasionally show leptospirae in fresh blood from early infections. Dark-field examination of centrifuged urine may also be positive. Fluorescein-conjugated antibodies or other immunohistochemical techniques can be used also.

C. Culture

Whole fresh blood or urine can be cultured in a semisolid medium. Because of inhibitory substances in blood, only 1 or 2 drops should be placed in each of five tubes containing 5 or 10 mL of medium. Up to 0.5 mL of CSF can be used. One drop of undiluted urine can be used followed by one drop each of 10-fold serially diluted urine—for a total of four tubes. Tissue approximately 5 mm in diameter should be crushed and used as the inoculum. Growth is slow, and cultures should be kept for at least 8 weeks.

D. Serology

The diagnosis of leptospirosis in most cases is confirmed serologically. Agglutinating antibodies first appear 5–7 days after infection and develop slowly reaching a peak at 5–8 weeks. Very high titers may be attained (>1:10,000). The reference laboratory standard for detection of leptospiral antibody uses microscopic agglutination of live organisms, which can be hazardous. The test is highly sensitive, but it is difficult to standardize; the end point is 50% agglutination, which is difficult to determine. Agglutination of the live suspensions is most specific for the serovar of the infecting leptospires. Agglutination tests are generally performed only in reference laboratories. Paired sera that show a significant change in titer or a single serum with high titer agglutinins plus a compatible clinical illness can be diagnostic. Because of the difficulty in performing the definitive agglutination tests a variety of other tests have been developed for use primarily as screening tests.

Immunity

Serovar-specific immunity follows infection, but reinfection with different serovars may occur.

Treatment

Treatment of mild leptospirosis should be with oral doxycycline, ampicillin, or amoxicillin. Treatment of moderate or severe disease should be with intravenous penicillin or ampicillin.

Epidemiology, Prevention, & Control

The leptospiroses are essentially animal infections; human infection is only accidental, following contact with water or other materials contaminated with the excreta of animal hosts. Rats, mice, wild rodents, dogs, swine, and cattle are the principal sources of human infection. They excrete leptospirae in urine both during the active illness and during the asymptomatic carrier state. Leptospirae remain viable in stagnant water for several weeks; drinking, swimming, bathing, or food contamination may lead to human infection. Persons most likely to come in contact with water contaminated by rats (eg, miners, sewer workers, farmers, and fishermen) run the greatest risk of infection. Children acquire the infection from dogs more frequently than do adults. Control consists of preventing exposure to potentially contaminated water and reducing contamination by rodent control. Doxycycline, 200 mg orally once weekly during heavy exposure, is effective prophylaxis. Dogs can receive distemper-hepatitis-leptospirosis vaccinations.

OTHER SPIROCHETAL DISEASES

SPIRILLUM MINOR (SPIRILLUM MORSUS MURIS)

S minor causes one form of rat-bite fever (sodoku). This very small (3–5 μm) and rigid spiral organism is carried by rats all over the world. The organism is inoculated into humans through the bite of a rat and results in a local lesion, regional gland swelling, skin rashes, and fever of the relapsing type. The frequency of this illness depends upon the degree of contact between humans and rats. Spirillum can be isolated by inoculation of guinea pigs or mice with material from enlarged lymph nodes or blood but has not been grown in bacteriologic media. In the United States and Europe, this disease has been recognized only infrequently. Several other motile gram-negative spiral aerobic organisms can produce spirillum fever.

SPIROCHETES OF THE NORMAL MOUTH & MUCOUS MEMBRANES

A number of spirochetes occur in every normal mouth. Some of them have been named (eg, *Borrelia buccalis*), but neither their morphology nor their physiologic activity permits definitive classification. On normal genitalia, a spirochete called *Borrelia refringens* is occasionally found that may be confused with *T pallidum*. These organisms are harmless saprophytes under ordinary conditions. Most of them are strict anaerobes that can be grown in petrolatum-sealed meat infusion broth tubes with tissue added.

REVIEW QUESTIONS

1. A 28-year-old woman who is 10 weeks pregnant presents to the obstetrics clinic for prenatal care. She has a history of treatment for syphilis 7 years previously. The results of serologic tests for syphilis are as follows: nontreponemal test, RPR, nonreactive; treponemal test (TP-PA), reactive. Which of the following statements is most correct?

   (A) The mother's previous treatment for syphilis was effective
   (B) The baby is at high risk for congenital syphilis
   (C) The mother needs to be treated again for syphilis
   (D) The mother needs a lumbar puncture and a VDRL test of her CSF for neurosyphilis

2. Infections with which of the following agents can result in a false-positive nontreponemal (VDRL or RPR) test for syphilis?

   (A) Lupus erythematosis
   (B) Measles
   (C) Leprosy
   (D) Pregnancy
   (E) Blood transfusions
   (F) Malaria
   (G) All of the above

3. A 20-year-old woman presents with a 2-cm ulcer on her labia majora. The lesion has a raised border and is relatively painless. The differential diagnosis of this lesion includes

   (A) Adenovirus infection
   (B) Papilloma virus infection
   (C) *Neisseria gonorrhoeae* infection
   (D) *Chlamydia trachomatis* cervicitis
   (E) *Treponema pallidum* infection

4. A 42-year-old woman went camping in the Sierra Nevada Mountains, where she slept for two nights in an abandoned log cabin. After the second night, a tick was found on her shoulder. Six days later, she developed fever to 38°C, which lasted for 4 days. Ten days later, she had another similar episode of fever. Examination of a blood smear stained with Wright's stain showed spirochetes suggestive of *Borrelia*. Which of the following statements about relapsing fever and *Borrelia hermsii* is correct?

   (A) Each relapse is associated with an antigenically distinct variant
   (B) Blood smears should be made when the patient is afebrile
   (C) Borreliae do not pass transovarially from one generation to the next in ticks
   (D) The main reservoir for the *Borrelia* is deer
   (E) *Borrelia hermsii* is resistant to penicillin and tetracycline

5. A 23-year-old man presented with a maculopapular rash over much of his trunk but not in his mouth or on his palms. Because secondary syphilis was considered in the differential diagnosis, a RPR test was done and was positive at a 1:2 dilution. However, (TP-PA) test was negative. Which of the following diseases can be ruled out?

   (A) Secondary syphilis
   (B) Atypical measles
   (C) Coxsackie virus infection
   (D) Acute HIV 1 infection
   (E) Allergic drug reaction

6. Which of the following animals is the source of *Leptospira interrogans*?

   (A) Cattle
   (B) Dogs
   (C) Mice
   (D) Rats
   (E) Swine
   (F) All of the above

7. A 27-year-old medical resident was admitted to the hospital because of sudden onset of fever to 39°C and headache. Two weeks previously he had vacationed in rural Oregon, where he had frequently gone swimming in an irrigation canal that bordered land where cows were pastured. Blood tests done shortly after admission indicated renal function abnormality and elevated bilirubin and other liver function tests. Routine blood, urine, and CSF cultures were negative. Leptospirosis is suspected. Which of the following would be most likely to confirm this diagnosis?

   (A) Testing acute and convalescent phase sera using the RPR test
   (B) Culture of urine on human diploid fibroblast cells
   (C) Testing serum by dark-field examination for the presence of leptospires
   (D) Testing acute and convalescent phase sera for antileptospiral antibodies
   (E) Culture of CSF on blood and chocolate agar

8. A 47-year-old man presents with slowly progressive arthritis in his knees. He enjoys hiking in the coastal areas of Northern California where the prevalence of *Borrelia burgdorferi* in the *Ixodes* ticks is known to be 1–3% (considered low). The patient is concerned about Lyme disease. He never noticed a tick on his body and did not see an expanding red rash. An EIA for Lyme borreliosis is positive. What should be done now?

   (A) A biopsy specimen of the synovium of a knee joint should be examined for *Borrelia burgdorferi*
   (B) The patient should be given an antibiotic to treat Lyme disease
   (C) PCR on the patient's plasma should be done to detect *Borrelia burgdorferi*
   (D) A serum specimen should be submitted for immunoblot assay to detect antibodies reactive with *Borrelia burgdorferi* antigens

9. Infections with which of the following agents can result in a false-positive nontreponemal (VDRL or RPR) test for syphilis?

   (A) Varicella-zoster virus
   (B) *Borrelia burgdorferi*
   (C) *Streptococcus pyogenes*
   (D) Epstein-Barr virus
   (E) Hepatitis B virus
   (F) All of the above

10. Which of the following organisms principally infects the liver and kidneys?

    (A) *Streptobacillus moniliformis*
    (B) *Leptospira interrogans*
    (C) *Staphylococcus aureus*
    (D) *Escherichia coli*
    (E) *Enterococcus faecalis*
    (F) *Treponema pallidum*

Answers

1. A	4. A	7. D	10. B
2. G	5. A	8. D	
3. E	6. F	9. F	

REFERENCES

Hook EW III: Endemic treponematoses. In: *Mandell, Douglas, and Bennett's Principles and Practice of Infectious Diseases*, 7th ed. Mandell GL, et al (editors). Elsevier, 2010.

Levett PN: *Leptospira*. In: *Manual of Clinical Microbiology*, 9th ed. Murray PR et al (editors). ASM Press, 2007.

Levett PN, Hakke DA: *Leptospira* spicies (Leptospirosis). In: *Mandell, Douglas, and Bennett's Principles and Practice of Infectious Diseases*, 7th ed. Mandell GL, et al (editors). Elsevier, 2010.

Pope V, Norris SJ, Johnson RE: *Treponema* and other human host-associated spirochetes. In: *Manual of Clinical Microbiology*, 9th ed. Murray PR et al (editors). ASM Press, 2007.

Rhee KY, Johnson WD Jr: *Borrelia* species (relapsing fever). In: *Mandell, Douglas, and Bennett's Principles and Practice of Infectious Diseases*, 7th ed. Mandell GL, et al (editors). Elsevier, 2010.

Steere AC: *Borrelia burgdorferi* (Lyme disease, Lyme borreliosis). In: *Mandell, Douglas, and Bennett's Principles and Practice of Infectious Diseases*, 7th ed. Mandell GL, et al (editors). Elsevier, 2010.

Tramont EC: *Treponema pallidum* (syphilis). In: *Mandell, Douglas, and Bennett's Principles and Practice of Infectious Diseases*, 7th ed. Mandell GL, et al (editors). Elsevier, 2010.

Wilske B, Johnson BJB, Schriefer ME: *Borrelia*. In: *Manual of Clinical Microbiology*, 9th ed. Murray PR et al (editors). ASM Press, 2007.

Mycoplasmas & Cell Wall-Defective Bacteria

MYCOPLASMAS

There are over 150 species in the class of cell wall-free bacteria. At least 15 of these species are thought to be of human origin, while others have been isolated from animals and plants. In humans, four species are of primary importance: *Mycoplasma pneumoniae* causes pneumonia and has been associated with joint and other infections. *Mycoplasma hominis* sometimes causes postpartum fever and has been found with other bacteria in uterine tube infections. *Ureaplasma urealyticum* is a cause of nongonococcal urethritis in men and is associated with lung disease in premature infants of low birth weight. *Mycoplasma genitalium* is closely related to *M pneumoniae* and has been associated with urethral and other infections. Other members of the genus *Mycoplasma* are pathogens of the respiratory and urogenital tracts and joints of animals.

The smallest genome of mycoplasmas, *M genitalium*, is little more than twice the genome size of certain large viruses. Mycoplasmas are the smallest organisms that can be free-living in nature and self replicating on laboratory media. They have the following characteristics: (1) the smallest mycoplasmas are 125–250 nm in size; (2) they are highly pleomorphic because they lack a rigid cell wall and instead are bounded by a triple-layered "unit membrane" that contains a sterol (mycoplasmas require the addition of serum or cholesterol to the medium to produce sterols for growth); (3) mycoplasmas are completely resistant to penicillin because they lack the cell wall structures at which penicillin acts, but they are inhibited by tetracycline or erythromycin; (4) mycoplasmas can reproduce in cell-free media; on agar, the center of the whole colony is characteristically embedded beneath the surface; (5) growth of mycoplasmas is inhibited by specific antibody; and (6) mycoplasmas have an affinity for mammalian cell membranes.

Morphology & Identification

A. Typical Organisms

Mycoplasmas cannot be studied by the usual bacteriologic methods because of the small size of their colonies and the plasticity and delicacy of their individual cells. Growth in fluid media gives rise to many different forms. Growth on solid media consists principally of protoplasmic masses of indefinite shape that are easily distorted. These structures vary greatly in size, ranging from 50 to 300 nm in diameter. The morphology appears different according to the method of examination (eg, darkfield, immunofluorescence, Giemsa-stained films from solid or liquid media, and agar fixation).

B. Culture

Culture of mycoplasmas that cause disease in humans requires media with serum or ascites fluid, growth factors such as yeast extract, and a metabolic substrate such as glucose or urea. There is no one medium that is optimal for all the species because of different properties and substrate requirements. Following incubation at 37°C for 48–96 hours, there may be no turbidity in broth cultures; however, Giemsa's stains of the centrifuged sediment show the characteristic pleomorphic structures, and subculture on appropriate solid media yields minute colonies.

After 2–6 days on biphasic (broth over agar) and agar medium incubated in a Petri dish that has been sealed to prevent evaporation, isolated colonies measuring 20–500 μm can be detected with a hand lens. These colonies are round, with a granular surface and a dark center typically buried in the agar. They can be subcultured by cutting out a small square of agar containing one or more colonies and streaking this material on a fresh plate or dropping it into liquid medium. The organisms can be stained for microscopic study by placing a similar square on a slide and covering the colony with a cover glass onto which an alcoholic solution of methylene blue and azure has been poured and then evaporated (agar fixation). Such slides can also be stained with specific fluorescent antibody.

C. Growth Characteristics

Mycoplasmas are unique in microbiology because of (1) their extremely small size and (2) their growth on complex but cell-free media.

Mycoplasmas pass through filters with 450-nm pore size and thus are comparable to chlamydiae or large viruses. However, parasitic mycoplasmas grow on cell-free media that contain lipoprotein and sterol. This sterol requirement for growth and membrane synthesis is unique.

Many mycoplasmas use glucose as a source of energy; ureaplasmas require urea.

Some human mycoplasmas produce peroxides and hemolyze red blood cells. In cell cultures and in vivo, mycoplasmas develop predominantly at cell surfaces. Many established animal and human cell culture lines carry mycoplasmas as contaminants; often the mycoplasmas are intracellular.

D. Variation

The extreme pleomorphism of mycoplasmas is one of their principal characteristics.

Antigenic Structure

Many antigenically distinct species of mycoplasmas have been isolated from animals (eg, mice, chickens, and turkeys). In humans, at least 14 species can be identified, including *M hominis, Mycoplasma salivarium, Mycoplasma orale, Mycoplasma fermentans, M pneumoniae, M genitalium, U urealyticum*, and others.

The species are classified by biochemical and serologic features. The complement fixation (CF) antigens of mycoplasmas are glycolipids. Antigens for ELISA tests are proteins. Some species have more than one serotype.

Pathogenesis

Many pathogenic mycoplasmas have flask-like or filamentous shapes and have specialized polar tip structures that mediate adherence to host cells. These structures are a complex group of interactive proteins, adhesins, and adherence-accessory proteins. The proteins are proline-rich, which influences the protein folding and binding and is important in the adherence to cells. The mycoplasmas attach to the surfaces of ciliated and nonciliated cells, probably through the mucosal cell sialoglyco-conjugates and sulfated glycolipids. Some mycoplasmas lack the distinct tip structures but use adhesin proteins or have alternative mechanisms to adhere to host cells. The subsequent events in infection are less well understood but may include several factors as follows: direct cytotoxicity through generation of hydrogen peroxide and superoxide radicals; cytolysis mediated by antigen–antibody reactions or by chemotaxis and action of mononuclear cells; and competition for and depletion of nutrients.

Mycoplasmal Infection

The mycoplasmas appear to be host-specific, being communicable and potentially pathogenic only within a single host species. In animals, mycoplasmas appear to be intracellular parasites with a predilection for mesothelial cells (pleura, peritoneum, synovia of joints). Several extracellular products can be elaborated (eg, hemolysins).

A. Infection of Humans

Mycoplasmas have been cultivated from human mucous membranes and tissues, particularly from the genital,

urinary, and respiratory tracts. Mycoplasmas are part of the normal flora of the mouth and can be grown from normal saliva, oral mucous membranes, sputum, or tonsillar tissue. *M salivarium, M orale*, and other mycoplasmas can be recovered from the oral cavities of many healthy adults, but an association with clinical disease is uncertain. *M hominis* is found in the oropharynx of less than 5% of adults. *M pneumoniae* in the oropharynx is generally associated with disease (see below).

Some mycoplasmas are inhabitants of the genitourinary tract, particularly in females. In both men and women, genital carriage of mycoplasmas is directly related to the number of lifetime sex partners. *M hominis* can be cultured from 1% to 5% of asymptomatic men and 30% to 70% of asymptomatic women; the rates increase to 20% and over 90% positive for men and women, respectively, in sexually transmitted disease clinics. *U urealyticum* is found in the genital tracts of 5–20% of sexually active men and 40–80% of sexually active women. Approximately 20% of women attending sexually transmitted disease clinics have *M genitalium* in their lower genital tracts. Other mycoplasmas also occur in the lower genital tract.

B. Infection of Animals

Bovine pleuropneumonia is a contagious, occasionally lethal disease of cattle associated with pneumonia and pleural effusion. The disease probably has an airborne spread. Mycoplasmas are found in inflammatory exudates.

Agalactia of sheep and goats in the Mediterranean area is a generalized infection with local lesions in the skin, eyes, joints, udder, and scrotum; it leads to atrophy of lactating glands in females. Mycoplasmas are present in blood early and in milk and exudates later.

In poultry, several economically important respiratory diseases are caused by mycoplasmas. The organisms can be transmitted from hen to egg to chick. Swine, dogs, rats, mice, and other species harbor mycoplasmas that can produce infection involving particularly the pleura, peritoneum, joints, respiratory tract, and eye. In mice, a *Mycoplasma* of spiral shape (spiroplasma) can induce cataracts.

C. Infection of Plants

Aster yellows, corn stunt, and other plant diseases appear to be caused by mycoplasmas. They are transmitted by insects and can be suppressed by tetracyclines.

Diagnostic Laboratory Tests
A. Specimens

Specimens consist of throat swabs, sputum, inflammatory exudates, and respiratory, urethral, or genital secretions.

B. Microscopic Examination

Direct examination of a specimen for mycoplasmas is useless. Cultures are examined as described above.

C. Cultures

The material is inoculated onto special solid media and incubated for 3–10 days at 37°C with 5% CO_2 (under microaerophilic conditions), or into special broth and incubated aerobically. One or two transfers of media may be necessary before growth appears that is suitable for microscopic examination by staining or immunofluorescence. Colonies may have a "fried egg" appearance on agar.

D. Serology

Antibodies develop in humans infected with mycoplasmas and can be demonstrated by several methods. CF tests can be performed with glycolipid antigens extracted with chloroform–methanol from cultured mycoplasmas. HI tests can be applied to tanned red cells with adsorbed *Mycoplasma* antigens. Indirect immunofluorescence may be used. The test that measures growth inhibition by antibody is quite specific. With all these serologic techniques, there is adequate specificity for different human *Mycoplasma* species, but a rising antibody titer is required for diagnostic significance because of the high incidence of positive serologic tests in normal individuals. *M pneumoniae* and *M genitalium* are serologically cross-reactive.

Treatment

Many strains of mycoplasmas are inhibited by a variety of antimicrobial drugs, but most strains are resistant to penicillins, cephalosporins, and vancomycin. Tetracyclines and erythromycins are effective both in vitro and in vivo and are, at present, the drugs of choice in mycoplasmal pneumonia. Some ureaplasmas are resistant to tetracycline.

Epidemiology, Prevention, & Control

Isolation of infected livestock will control the highly contagious pleuropneumonia and agalactia. No vaccines are available. Mycoplasmal pneumonia behaves like a communicable viral respiratory disease (see below).

MYCOPLASMA PNEUMONIAE & ATYPICAL PNEUMONIAS

M pneumoniae is a prominent cause of pneumonia, especially in persons 5–20 years of age.

Pathogenesis

M pneumoniae is transmitted from person to person by means of infected respiratory secretions. Infection is initiated by attachment of the organism's tip to a receptor on the surface of respiratory epithelial cells (Figure 25–1). Attachment is mediated by a specific adhesin protein on the differentiated terminal structure of the organism. During infection, the organisms remain extracellular.

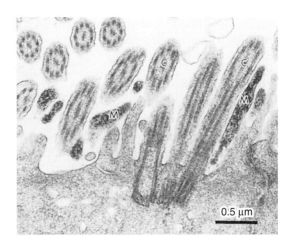

FIGURE 25–1 Electron micrograph of *Mycoplasma pneumoniae* attached to ciliated respiratory epithelial cells in a sputum sample from a patient with culture-proved *Mycoplasma pneumoniae* pneumonia. The organisms (M) are seen on the luminal border attached between cilia (C). (Courtesy of AM Collier, Department of Pediatrics, University of North Carolina.)

Clinical Findings

Mycoplasmal pneumonia is generally a mild disease. The clinical spectrum of *M pneumoniae* infection ranges from asymptomatic infection to serious pneumonitis, with occasional neurologic and hematologic (ie, hemolytic anemia) involvement and a variety of possible skin lesions. Bullous myringitis occurs in spontaneous cases and in experimentally inoculated volunteers.

The incubation period varies from 1 to 3 weeks. The onset is usually insidious, with lassitude, fever, headache, sore throat, and cough. Initially, the cough is nonproductive, but it is occasionally paroxysmal. Later there may be blood-streaked sputum and chest pain. Early in the course, the patient appears only moderately ill, and physical signs of pulmonary consolidation are often negligible compared to the striking consolidation seen on x-rays. Later, when the infiltration is at a peak, the illness may be severe. Resolution of pulmonary infiltration and clinical improvement occur slowly over 1–4 weeks. Although the course of the illness is exceedingly variable, death is very rare and is usually attributable to cardiac failure. Complications are uncommon, but hemolytic anemia may occur. The most common pathologic findings are interstitial and peribronchial pneumonitis and necrotizing bronchiolitis. Other diseases possibly related to *M pneumoniae* include erythema multiforme; central nervous system involvement, including meningitis, meningoencephalitis, and mono- and polyneuritis; myocarditis; pericarditis; arthritis; and pancreatitis.

Common causes of community-acquired bacterial pneumonia, in addition to *M pneumoniae*, include *Streptococcus pneumoniae*, *Legionella pneumophila*, *Chlamydia pneumoniae*, and *Haemophilus influenzae*. The clinical presentations of these infections can be very similar, and recognition of the subtleties of signs and symptoms is important. The causative organisms must be determined by sputum examination and culture, blood culture, and other tests.

Laboratory Tests

The diagnosis of *M pneumoniae* pneumonia is largely made by the clinical recognition of the syndrome. Laboratory tests are of secondary value. The white cell count may be slightly elevated. A sputum Gram stain is of value in not suggesting some other bacterial pathogen (eg, *S pneumoniae*). The causative mycoplasmas can be recovered by culture from the pharynx and from sputum, but culture is a highly specialized test and is almost never done to diagnose *M pneumoniae* infection. Cold hemagglutinins for group O human erythrocytes appear in about 50% of untreated patients, in rising titer, with the maximum reached in the third or fourth week after onset. A titer of 1:64 or more supports the diagnosis of *M pneumoniae* infection. There is a rise in specific antibodies to *M pneumoniae* that is demonstrable by CF tests; acute and convalescent phase sera are necessary to demonstrate a fourfold rise in the CF antibodies. EIA to detect IgM and IgG antibodies can be highly sensitive and specific, but may not be readily available. Polymerase chain reaction (PCR) assay of specimens from throat swabs or other clinical material can be diagnostic, but is generally performed only in reference laboratories.

Treatment

Tetracyclines or erythromycins can produce clinical improvement but do not eradicate the mycoplasmas.

Epidemiology, Prevention, & Control

M pneumoniae infections are endemic all over the world. In populations of children and young adults, where close contact prevails, and in families, the infection rate may be high (50–90%), but the incidence of pneumonitis is variable (3–30%). For every case of frank pneumonitis, there exist several cases of milder respiratory illness. *M pneumoniae* is apparently transmitted mainly by direct contact involving respiratory secretions. Second attacks are infrequent. The presence of antibodies to *M pneumoniae* has been associated with resistance to infection but may not be responsible for it. Cell-mediated immune reactions occur. The pneumonic process may be attributed in part to an immunologic response rather than only to infection by mycoplasmas.

MYCOPLASMA HOMINIS

M hominis has been associated with a variety of diseases but is a demonstrated cause in only a few of them. The evidence for a causal relationship in disease is from culture and serologic studies. *M hominis* can be cultured from the upper urinary tract in about 10% of patients with pyelonephritis. *M hominis* is strongly associated with infection of the uterine tubes (salpingitis) and tubo-ovarian abscesses; the organism can

be isolated from the uterine tubes of about 10% of patients with salpingitis but not from women with no signs of disease. Women with salpingitis more commonly have antibodies against *M hominis* than women with no disease. *M hominis* has been isolated from the blood of about 10% of women who have postabortal or postpartum fever and occasionally from joint fluid cultures of patients with arthritis.

UREAPLASMA UREALYTICUM

U urealyticum, like *M hominis*, has been associated with a variety of diseases but is a demonstrated cause in only a few of them. *U urealyticum*, which requires 10% urea for growth, probably causes nongonococcal urethritis in some men, but a majority of cases of nongonococcal urethritis are caused by *Chlamydia trachomatis* (Chapter 27). *U urealyticum* is common in the female genital tract, where the association with disease is weak. *U urealyticum* has been associated with lung disease in premature low-birth-weight infants who acquired the organism during birth, but a causal effect has not been clearly demonstrated. The evidence that *U urealyticum* is associated with involuntary infertility is at best marginal.

MYCOPLASMA GENITALIUM

M genitalium was originally isolated from urethral cultures of two men with nongonococcal urethritis, but culture of *M genitalium* is difficult, and subsequent observations have been based on data obtained by using the PCR, molecular probes, and serologic tests. The data suggest that *M genitalium* in men is associated with some cases of acute as well as chronic nongonococcal urethritis. In women, *M genitalium* has been associated with a variety of infections such as cervicitis, endometritis, salpingitis, and infertility.

CELL WALL-DEFECTIVE BACTERIA

L phase variants (L forms) are wall-defective microbial forms that can replicate serially as nonrigid cells and produce colonies on solid media. Some L phase variants are stable; others are unstable and revert to bacterial parental forms. Wall-defective forms are not genetically related to mycoplasmas. They can result from spontaneous mutation or from the effects of chemicals. Treatment of eubacteria with cell wall-inhibiting drugs or lysozyme can produce cell wall-defective microbial forms. **Protoplasts** are such forms usually derived from gram-positive organisms; they are osmotically fragile, with external surfaces free of cell wall constituents. **Spheroplasts** are cell wall-defective forms usually derived from gram-negative bacteria; they retain some outer membrane material.

Cell wall-defective forms continue to synthesize some antigens that are normally located in the cell wall of the parent bacteria (eg, streptococcal L forms produce M protein and capsular polysaccharide). Reversion of L forms to the parental bacterial form is enhanced by growth in the presence of

15–30% gelatin or 2.5% agar. Reversion is inhibited by inhibitors of protein synthesis.

It is uncertain whether cell wall-defective microbial forms cause tissue reactions resulting in disease. They may be important for the persistence of microorganisms in tissue and recurrence of infection after antimicrobial treatment, as in rare cases of endocarditis.

REVIEW QUESTIONS

1. *Ureplasma urealyticum* is so named because
   (A) It thrives in the upper urinary tract
   (B) It requires urea as a growth substrate
   (C) It is a frequent cause of symptomatic urinary bladder infections in young women
   (D) It causes chronic urinary tract infections in premature babies born to mothers with ureaplasmas as part of the genital flora

2. An 18-year-old sexually active woman develops left lower quadrant pain and fever. On pelvic examination there is tenderness in the left adnexa, and a mass suggestive of a uterine tube abscess is palpated. The patient is diagnosed with pelvic inflammatory disease. Which of the following bacteria is considered to be a common cause of pelvic inflammatory disease?
   (A) *Bacillus cereus*
   (B) *Haemophilus influenzae*
   (C) *Neisseria subflava*
   (D) *Mycoplasma pneumoniae*
   (E) *Chlamydia trachomatis*

3. Which of the following is important in the pathogenesis of mycoplasmal infections?
   (A) The peptidoglycan in the mycoplasmal cell wall
   (B) The presence of lacto-N-neotetraose with a terminal galactosamine as the host cell receptor
   (C) The structures and the interactive proteins that mediate adhesion to host cells
   (D) The absence of cilia on the surface of the host cells
   (E) Growth in an anatomic site where anaerobic organisms thrive

4. A 25-year-old woman is referred to the sexually transmitted diseases clinic because of contact with a male partner with gonorrhea. The woman has had 15 male sex partners since becoming sexually active. The likelihood that she also has genital *Mycoplasma hominis* infection is
   (A) 1%
   (B) 5%
   (C) 15%
   (D) 40%
   (E) 90%

5. A 25-year-old medical student has contact with a patient who has pneumonia with fever and cough. Four days later, the medical student develops fever and cough, and chest x-ray shows consolidation of the right lower lobe. Routine bacterial sputum cultures are negative. Pneumonia caused by *Mycoplasma pneumoniae* is considered. Which of the following is an appropriate way to confirm this diagnosis?
   (A) PCR amplification of *Mycoplasma pneumoniae* DNA in sputum
   (B) Culture of sputum for *Mycoplasma pneumoniae*
   (C) Gram stain of sputum smear
   (D) Culture of a lung aspirate for *Mycoplasma pneumoniae*
   (E) Complement fixation test of acute and convalescent sera

6. Each of the following is associated with genital tract infections *except*:
   (A) *Mycoplasma hominis*
   (B) *Neisseria gonorrhoeae*
   (C) *Mycoplasma pneumoniae*
   (D) *Chlamydia trachomatis*
   (E) *Mycoplasma genitalium*

7. Mycoplasmas have all of the following characteristics *except*:
   (A) Possession of both DNA and RNA
   (B) Capability for cell-free growth
   (C) Susceptibility to penicillin G
   (D) Extracellular parasitism in vivo

8. Which type of test is most readily used to obtain laboratory confirmation of *Mycoplasma pneumoniae* infection?
   (A) Culture in broth containing serum, glucose, and a penicillin (to inhibit other flora)
   (B) PCR
   (C) Electron microscopy
   (D) Complement fixation test on acute and convalescent phase sera

9. A 13-year-old boy develops infection with *Mycoplasma pneumoniae*. What is the risk for infection in other members of his household?
   (A) None; it is sexually transmitted
   (B) 1–3%
   (C) 10–15%
   (D) 20–40%
   (E) 50–90%

10. A 19-year-old man develops cough and fever. A chest film shows consolidation of the left lower lobe. A diagnosis of pneumonia is made. Which of the following bacteria is a frequent cause of community-acquired pneumonia?
    (A) *Legionella pneumophila*
    (B) *Chlamydia pneumoniae*
    (C) *Streptococcus pneumoniae*
    (D) *Mycoplasma pneumoniae*
    (E) All of the above

Answers

1. B	4. E	7. C	10. E
2. E	5. E	8. D	
3. C	6. C	9. E	

REFERENCES

Mycoplasma diseases. Vol 2, Part III, Section D. In: *Mandell, Douglas, and Bennett's Principles and Practice of Infectious Diseases*, 7th ed. Mandell GL, et al (editors) Elsevier, 2010.

Razin S, Yogev D, Naot Y: Molecular biology and pathogenicity of mycoplasmas. Microbiol Mol Biol Rev 1998;62:1094.

Waites KB, Taylor-Robinson D: *Mycoplasma* and *Ureaplasma*. In: *Manual of Clinical Microbiology*, 9th ed. Murray PR et al (editors). ASM Press, 2007.

Rickettsia & Ehrlichia

GENERAL

The human pathogens in the family Rickettsiaceae are small bacteria of the genera *Rickettsia, Orientia, Coxiella,* and *Ehrlichia.* They are obligate intracellular parasites and, except for Q fever, are transmitted to humans by arthropods. Many rickettsiae are transmitted transovarially in the arthropod, which serves as both vector and reservoir. Rickettsial infections—except Q fever and the ehrlichioses—typically are manifested by fever, rashes, and vasculitis. They are grouped on the basis of their clinical features, epidemiologic aspects, and immunologic characteristics (Table 26–1).

Properties of Rickettsiae

Rickettsiae are pleomorphic coccobacilli, appearing either as short rods (0.3 × 1–2 μm) or as cocci (0.3 μm in diameter). They do not stain well with Gram stain but are readily visible under the light microscope when stained with Giemsa's stain, Gimenez's stain, acridine orange, or other stains.

Rickettsiae grow readily in yolk sacs of embryonated eggs. Pure preparations of rickettsiae for use in laboratory testing can be obtained by differential centrifugation of yolk sac suspensions. Many strains of rickettsiae also grow in cell culture, where the generation time is 8–10 hours at 34°C. For reasons of biosafety, isolation of rickettsiae should be done only in reference laboratories.

Rickettsiae have gram-negative cell wall structures that include peptidoglycan-containing muramic acid and diaminopimelic acid. The typhus and spotted fever groups contain lipopolysaccharide. The cell wall proteins include the surface proteins OmpA and OmpB, which are important in the humoral immune response and provide the basis for serotyping.

Rickettsiae grow in different parts of the cell. Those of the typhus group are usually found in the cytoplasm; those of the spotted fever group, in the nucleus. Coxiellae grow only in cytoplasmic vacuoles.

Rickettsial growth is enhanced in the presence of sulfonamides, and rickettsial diseases are made more severe by these drugs. Tetracyclines and chloramphenicol inhibit the growth of rickettsiae and can be therapeutically effective.

Most rickettsiae survive only for short times outside of the vector or host. Rickettsiae are quickly destroyed by heat, drying, and bactericidal chemicals. Dried feces of infected lice may contain infectious *Rickettsia prowazekii* for months at room temperature. *Coxiella burnetii,* which causes Q fever, is the rickettsial agent most resistant to drying. This organism may survive pasteurization at 60°C for 30 minutes and can survive for months in dried feces or milk. This may be due to the formation of endospore-like structures by *C burnetii.*

Rickettsial Antigens & Serology

The direct immunofluorescent antibody test can be used to detect rickettsiae in ticks and sections of tissues. The test has been most useful to detect *R rickettsii* in skin biopsy specimens to aid in the diagnosis of Rocky Mountain spotted fever; however, the test is performed in only a few reference laboratories.

Serologic evidence of infection occurs no earlier than the second week of illness for any of the rickettsial diseases. Thus, serologic tests are useful only to confirm the diagnosis, which is based on clinical findings (eg, fever, headache, rash) and epidemiologic information (eg, tick bite). Therapy for potentially severe diseases, such as Rocky Mountain spotted fever and typhus, should be instituted before seroconversion occurs.

A variety of serologic tests have been used to diagnose rickettsial diseases. Most of these tests are performed only in reference laboratories. Antigens for the **complement fixation test** to diagnose Q fever and for the **indirect immunofluorescence**, latex agglutination, and enzyme immunoassay for Rocky Mountain spotted fever are commercially available. Reagents for other tests are prepared only in public health or other reference laboratories. The indirect fluorescent antibody technique may be the most widely used method, because of the availability of reagents and the ease with which it can be performed. The test is relatively sensitive, requires little antigen, and can be used to detect IgM and IgG. Rickettsiae partially purified from infected yolk sac material are tested with dilutions of a patient's serum. Reactive antibody is detected with a fluorescein-labeled antihuman globulin. The results indicate the presence of partly species-specific antibodies, but cross-reactions are observed.

TABLE 26–1   Rickettsial and Ehrlichial Diseases

Group	Organism	Disease	Geographic Distribution	Vector	Mammalian Reservoir	Clinical Features	Diagnostic Tests[a]
Typhus group	*Rickettsia prowazekii*	Epidemic typhus (louse-borne typhus), Brill-Zinsser disease	Worldwide: South America, Africa, Asia, North America	Louse	Humans	Fever, chills, myalgia, headache, rash (no eschar); severe illness if untreated	Serology
	Rickettsia typhi	Murine typhus, endemic typhus, flea-borne typhus	Worldwide (small foci)	Flea	Rodents	Fever, headache, myalgia, rash (no eschar); milder illness than epidemic typhus	Serology
Scrub Typhus group	*Orientia tsutsugamushi*	Scrub typhus	Asia, South Pacific, northern Australia	Mite	Rodents	Fever, headache, rash (50% have eschar), lymphadenopathy, atypical lymphocytes	Serology
Spotted fever group[b]	*Rickettsia rickettsii*	Rocky Mountain spotted fever	Western Hemisphere (United States, South America)	Tick[c]	Rodents, dogs	Fever, headache, rash (no eschar); many systemic manifestations	Direct FA of rickettsiae in tissue; serology
	Rickettsia akari	Rickettsial pox	United States, Korea, Russia, South Africa	Mite[c]	Mice	Mild illness, fever, headache, vesicular rash (eschar)	Serology
	Rickettsia australis	Queensland tick typhus	Australia	Tick[c]	Rodents, marsupials	Fever, rash of trunk and limbs (eschar)	Serology
	Rickettsia conorii	Fièvre boutonneuse, Mediterranean spotted fever, Israeli spotted fever, South African tick fever, African (Kenya) tick typhus, Indian tick typhus	Mediterranean countries, Africa, Middle East, India	Tick[c]	Rodents, dogs	Fever, headache, rash, "tache noire" (eschar)	Direct FA of rickettsiae in tissue not sensitive; serology
	Rickettsia sibirica	Siberian tick typhus (North Asian tick typhus)	Siberia, Mongolia	Tick[c]	Rodents	Fever, rash (eschar)	Serology
Q fever	*Coxiella burnetii*	Q fever	Worldwide	Airborne fomites, tick	Sheep, cattle, goats, others	Headache, fever, fatigue, pneumonia (no rash); can have major complications	Positive CF to phase I, II antigens
Ehrlichiae	*Erlichia chaffeensis*	Human monocyte ehrlichiosis	South-central, southeastern, and western United States	Tick	Deer, dogs, humans	Fever, headache, atypical white blood cells	Inclusions in circulating monocytes; indirect FA for antibodies
	Neorickettsia sennetsu	Human monocyte ehrlichiosis	Japan, Malaysia	Trematode-infected fish?	Mammals	Fever, headache, atypical white blood cells	
	Anaplasma phagocytophilum	Human granulocyte anaplasmosis	Upper midwestern, northwestern, and West Coast United States and Europe	Tick	Mice, other mammals	Fever, headache, myalgia	Inclusions in granulocytes; indirect FA for antibodies
	Ehrlichia ewingii	Human granulocyte ehrlichiosis	Midwestern United States	Tick	Dogs	Fever, headache, myalgia	Inclusions in granulocytes; indirect FA for antibodies

[a]CF, complement fixation test; FA, fluorescent antibody test.

[b]Other rickettsia species in the spotted fever group that infect humans include *R africae*, *R japonica*, *R honei*, and *R slovaca*.

[c]Also serves as arthropod reservoir, by maintaining the rickettsiae through transovarian transmission.

Pathology

Rickettsiae, except for *C burnetii,* multiply in endothelial cells of small blood vessels and produce vasculitis. The cells become swollen and necrotic; there is thrombosis of the vessel, leading to rupture and necrosis. Vascular lesions are prominent in the skin, but vasculitis occurs in many organs and appears to be the basis of hemostatic disturbances. Disseminated intravascular coagulation and vascular occlusion may develop. In the brain, aggregations of lymphocytes, polymorphonuclear leukocytes, and macrophages are associated with the blood vessels of the gray matter; these are called typhus nodules. The heart shows similar lesions of the small blood vessels. Other organs may also be involved.

Immunity

In cell cultures of macrophages, rickettsiae are phagocytosed and replicate intracellularly even in the presence of antibody. The addition of lymphocytes from immune animals stops this multiplication in vitro. Infection in humans is followed by partial immunity to reinfection from external sources, but relapses occur (see Brill-Zinsser disease, below).

Clinical Findings

Except for Q fever, in which there is no skin lesion, rickettsial infections are characterized by fever, headache, malaise, prostration, skin rash, and enlargement of the spleen and liver.

A. Typhus Group

1. Epidemic typhus (*Rickettsia prowazekii*)—In epidemic typhus, systemic infection and prostration are severe, and fever lasts for about 2 weeks. The disease is more severe and more often fatal in patients over 40 years of age. During epidemics, the case fatality rate has been 6–30%.

2. Endemic typhus (*Rickettsia typhi*)—The clinical picture of endemic typhus has many features in common with that of epidemic typhus, but the disease is milder and is rarely fatal except in elderly patients.

3. Scrub typhus (*Orientia tsutsugamushi*)—This disease resembles epidemic typhus clinically. One feature is the eschar, the punched-out ulcer covered with a blackened scab that indicates the location of the mite bite. Generalized lymphadenopathy and lymphocytosis are common. Cardiac and cerebral involvement may be severe.

B. Spotted Fever Group

The spotted fever group resembles typhus clinically; however, unlike the rash in other rickettsial diseases, the rash of the spotted fever group usually appears first on the extremities, moves centripetally, and involves the palms and soles. Some, like Brazilian spotted fever, may produce severe infections; others, like Mediterranean fever, are mild. The case fatality rate varies greatly. In untreated Rocky Mountain spotted fever, it is usually much greater in elderly persons (up to 50%) than in young adults or children.

Rickettsialpox is a mild disease with a rash resembling that of varicella. About a week before onset of fever, a firm red papule appears at the site of the mite bite and develops into a deep-seated vesicle that in turn forms a black eschar (see below).

C. Q Fever

This disease resembles influenza, nonbacterial pneumonia, hepatitis, or encephalopathy rather than typhus. There is a rise in the titer of specific antibodies to *C burnetii,* phase 2. Transmission results from inhalation of dust contaminated with rickettsiae from placenta, dried feces, urine, or milk or from aerosols in slaughterhouses.

Infective endocarditis occasionally develops in chronic Q fever. Blood cultures for bacteria are negative, and there is a high titer of antibodies to *C burnetii,* phase 1. Virtually all patients have preexisting valve abnormalities. Continuous treatment with tetracycline for many months, occasionally with valve replacement, can provide prolonged survival.

Laboratory Findings

Isolation of rickettsiae is technically difficult and is of only limited usefulness in diagnosis. It is also hazardous. Whole blood (or emulsified blood clot) is inoculated into guinea pigs, mice, or eggs. Rickettsiae are recovered most frequently from blood drawn soon after onset of illness.

If the guinea pigs fail to show disease (fever, scrotal swellings, hemorrhagic necrosis, and death), serum is collected for antibody tests to determine if the animal has had an inapparent infection.

Some rickettsiae can infect mice, and rickettsiae are seen in smears of peritoneal exudate. In Rocky Mountain spotted fever, skin biopsies taken from patients between the fourth and eighth days of illness may reveal rickettsiae by immunofluorescence stain.

The most widely used serologic tests are indirect immunofluorescence and complement fixation (see above). An antibody rise should be demonstrated during the course of the illness. In Rocky Mountain spotted fever, the antibody response may not occur until after the second week of illness.

The polymerase chain reaction has been used to help diagnose Rocky Mountain spotted fever, other diseases of the spotted fever group, murine typhus, scrub typhus, and Q fever. The sensitivity of the method for Rocky Mountain spotted fever is about 70%, comparable to that of skin biopsy with immunocytology.

Treatment

Tetracyclines are effective provided treatment is started early. Tetracycline is given daily orally and continued for 3–4 days after defervescence. In severely ill patients, the initial doses can be given intravenously. Chloramphenicol also can be effective.

Sulfonamides enhance the disease and are contraindicated.

The antibiotics do not free the body of rickettsiae, but they do suppress their growth. Recovery depends in part upon the immune mechanisms of the patient.

Epidemiology

A variety of arthropods, especially ticks and mites, harbor rickettsia-like organisms in the cells that line the alimentary tract. Many such organisms are not evidently pathogenic for humans.

The life cycles of different rickettsiae vary. *R prowazekii* has a life cycle in humans and the human louse (*Pediculus humanus corporis* and *Pediculus humanus capitis*). The louse obtains the organism by biting infected human beings and transmits the agent by fecal excretion on the surface of the skin of another person. Whenever a louse bites, it defecates at the same time. Scratching the area of the bite allows the rickettsiae excreted in the feces to penetrate the skin. As a result of the infection, the louse dies, but the organisms remain viable for some time in its dried feces. Rickettsiae are not transmitted from one generation of lice to another. Delousing large proportions of the population with insecticides has controlled typhus epidemics.

Brill-Zinsser disease is a recrudescence of an old typhus infection. The rickettsiae can persist for many years in the lymph nodes of an individual without any symptoms being manifest. The rickettsiae isolated from such cases behave like classic *R prowazekii;* this suggests that humans themselves are the reservoir of the rickettsiae of epidemic typhus. Typhus epidemics have been associated with war and the lowering of standards of personal hygiene, which in turn have increased the opportunities for human lice to flourish. If this occurs at the time of recrudescence of an old typhus infection, an epidemic may be set off. Brill-Zinsser disease occurs in local populations of typhus areas as well as in persons who migrate from such areas to places where the disease does not exist. Serologic characteristics readily distinguish Brill disease from primary epidemic typhus. Antibodies arise earlier and are IgG rather than the IgM detected after primary infection. They reach a maximum by the tenth day of disease. This early IgG antibody response and the mild course of the disease suggest that partial immunity is still present from the primary infection.

In the United States, *R prowazekii* has an extrahuman reservoir in the southern flying squirrel, *Glaucomys volans.* In areas where southern flying squirrels are indigenous (southern Maine to Florida to the center of the United States), human infections have occurred after bites by ectoparasites of this rodent.

R typhi has its reservoir in the rat, in which the infection is inapparent and long-lasting. Rat fleas carry the rickettsiae from rat to rat and sometimes from rats to humans, who develop endemic typhus. Cat fleas can serve as vectors. In endemic typhus, the flea cannot transmit the rickettsiae transovarially.

O tsutsugamushi has its true reservoir in the mites that infest rodents. Rickettsiae can persist in rats for over a year after infection. Mites transmit the infection transovarially. Occasionally, infected mites or rat fleas bite humans, and scrub typhus results. The rickettsiae persist in the mite-rat-mite cycle in the scrub or secondary jungle vegetation that has replaced virgin jungle in areas of partial cultivation. Such areas may become infested with rats and trombiculid mites.

R rickettsii may be found in healthy wood ticks (*Dermacentor andersoni*) and is passed transovarially. Infected ticks in the western United States occasionally bite vertebrates such as rodents, deer, and humans. In order to be infectious, the tick carrying the rickettsiae must be engorged with blood, for this increases the number of rickettsiae in the tick. Thus, there is a delay of 45–90 minutes between the time of the attachment of the tick and its becoming infective. In the eastern United States, the dog tick *Dermacentor variabilis* transmits Rocky Mountain spotted fever. Dogs are hosts to these ticks and may serve as a reservoir for tick infection. Small rodents are another reservoir. Most cases of Rocky Mountain spotted fever in the United States now occur in the eastern and southeastern regions.

R akari has its vector in bloodsucking mites of the species *Allodermanyssus sanguineus.* These mites may be found on the mice (*Mus musculus*) trapped in apartment houses in the United States where rickettsialpox has occurred. Transovarial transmission of the rickettsiae occurs in the mite. Thus, the mite may act as a true reservoir as well as a vector. *R akari* has also been isolated in Korea.

C burnetii is found in ticks, which transmit the agent to sheep, goats, and cattle. Workers in slaughterhouses and in plants that process wool and cattle hides have contracted the disease as a result of handling infected animal tissues. *C burnetii* is transmitted by the respiratory pathway rather than through the skin. There may be a chronic infection of the udder of the cow. In such cases, the rickettsiae are excreted in the milk and rarely may be transmitted to humans by ingestion of unpasteurized milk.

Infected sheep may excrete *C burnetii* in the feces and urine and heavily contaminate their skin and woolen coat. The placentas of infected cows, sheep, goats, and cats contain the rickettsiae, and parturition creates infectious aerosols. The soil may be heavily contaminated from one of the above sources, and the inhalation of infected dust leads to infection of humans and livestock. It has been proposed that endospores formed by *C burnetii* contribute to its persistence and dissemination. *Coxiella* infection is now widespread among sheep and cattle in the United States. *Coxiella* can cause endocarditis (with a rise in the titer of antibodies to *C burnetii,* phase 1) in addition to pneumonitis and hepatitis.

Geographic Occurrence

A. Epidemic Typhus

This potentially worldwide infection has disappeared from the United States, Britain, and Scandinavia. It is still present in the Balkans, Asia, Africa, Mexico, and the Andes mountains of South America. In view of its long duration in humans as a latent infection (Brill-Zinsser disease), it can emerge and flourish quickly under proper environmental conditions, as it did in Europe during World War II, because of the deterioration of community hygiene.

B. Endemic Murine Typhus

Disease exists worldwide, especially in areas of high rat infestation. It may exist in the same areas as—and may be confused with—epidemic typhus or scrub typhus.

C. Scrub Typhus

Infection is seen in the Far East, especially Myanmar (Burma), India, Sri Lanka, New Guinea, Japan, and Taiwan. The larval stage (chigger) of various trombiculid mites serves both as a reservoir, through transovarian transmission, and as a vector for infecting humans and rodents.

D. Spotted Fever Group

These infections occur around the globe, exhibiting as a rule some epidemiologic and immunologic differences in different areas. Transmission by a tick of the Ixodidae family is common to the group. The diseases that are grouped together include Rocky Mountain spotted fever and Colombian, Brazilian, and Mexican spotted fevers; Mediterranean (boutonneuse), South African tick, and Kenya fevers; North Queensland tick typhus; and North Asian tick-borne rickettsioses.

E. Rickettsialpox

The human disease has been found among inhabitants of apartment houses in the northern United States. However, the infection also occurs in Russia, Africa, and Korea.

F. Q Fever

This disease is recognized around the world and occurs mainly in persons associated with goats, sheep, dairy cattle, or parturient cats. It has attracted attention because of outbreaks in veterinary and medical centers where large numbers of people were exposed to animals shedding *Coxiella*.

Seasonal Occurrence

Epidemic typhus is more common in cool climates, reaching its peak in winter and waning in the spring. This is probably a reflection of crowding, lack of fuel, and low standards of personal hygiene, which favor louse infestation.

Rickettsial infections that must be transmitted to the human host by vector reach their peak incidence at the time the vector is most prevalent—the summer and fall months.

Control

Control must rely on breaking the infection chain, treating patients with antibiotics, and immunizing when possible. Patients with rickettsial disease who are free from ectoparasites are not contagious and do not transmit the infection.

A. Prevention of Transmission by Breaking the Chain of Infection

1. Epidemic typhus—Delousing with insecticide.

2. Murine typhus—Rat-proofing buildings and using rat poisons.

3. Scrub typhus—Clearing from campsites the secondary jungle vegetation in which rats and mites live.

4. Spotted fever—Similar measures for the spotted fevers may be used; clearing of infested land; personal prophylaxis in the form of protective clothing such as high boots, socks worn over trousers; tick repellents; and frequent removal of attached ticks.

5. Rickettsialpox—Elimination of rodents and their parasites from human domiciles.

B. Prevention of Transmission of Q Fever by Adequate Pasteurization of Milk

The presently recommended conditions of "high-temperature, short-time" pasteurization at 71.5°C for 15 seconds are adequate to destroy viable *Coxiella*.

C. Prevention by Vaccination

There is no vaccine for Rocky Mountain spotted fever, for the other diseases of the spotted fever group, or for the diseases in the typhus group. For *C burnetii* there is an investigational vaccine made from infected egg yolk sacs. This vaccine has been used for laboratory workers who handle live *C burnetii*.

EHRLICHIOSIS

The ehrlichiae that cause disease in humans have been classified in a limited number of species, based in large part on sequence analysis of rRNA genes. The pathogens are as follows: *Ehrlichia chaffeensis*, which causes human monocyte ehrlichiosis; *Ehrlichia ewingii*, which causes human granulocyte ehrlichiosis; *Anaplasma phagocytophilum*, which causes human granulocyte anaplasmosis; and *Neorickettsia sennetsu*, which causes human monocyte ehrlichiosis. The same genera contain additional species that infect animals but apparently not humans. The human pathogens in the group have animal reservoirs and can cause disease in animals as well.

The ehrlichia group organisms are obligate intracellular bacteria that are taxonomically grouped with the rickettsiae.

They have tick vectors, although *N sennetsu* may be transmitted by ingestion of trematode-infected fish. See Table 26–1.

Properties of Ehrlichiae

Ehrlichiae are small (0.5 µm) gram-negative bacteria. They infect circulating leukocytes where they multiply within phagocytic vacuoles, forming clusters with inclusion-like appearance. These clusters of ehrlichiae are called **morulae**, which is derived from the Latin word for mulberry. The ehrlichiae and chlamydiae (Chapter 27) resemble each other in that both are found in intracellular vacuoles. The ehrlichiae, however, are like the rickettsiae in that they are able to synthesize ATP; the chlamydiae are not able to synthesize ATP.

Clinical Findings

The clinical manifestations of ehrlichiosis in humans are nonspecific: fever, chills, headache, myalgia, nausea or vomiting, anorexia, and weight loss. These manifestations are very similar to those of Rocky Mountain spotted fever without the rash. *E chaffeensis* frequently and *A phagocytophilum* less often cause severe or fatal illness. Seroprevalence studies suggest that subclinical ehrlichiosis occurs frequently.

Laboratory Findings

The diagnosis is confirmed by observing typical morulae in white blood cells. The indirect fluorescent antibody test can also be used to confirm the diagnosis. Antibodies are measured against *E chaffeensis* and *A phagocytophilum. E chaffeensis* is also used as the substrate for *E ewingii,* because the two species share antigens. Seroconversion from <1:64 to ≥1:128 or a fourfold or greater rise in titer makes a confirmed serologic diagnosis of human monocytotropic ehrlichiosis in a patient with a clinically compatible illness.

Multiple methods have been described for PCR detection of ehrlichiae in EDTA-anticoagulated blood. Culture using a variety of tissue culture cell lines also can be used. PCR and culture are performed in reference laboratories and in a small number of commercial laboratories.

Treatment

Tetracycline, commonly in the form of doxycycline, is cidal for ehrlichiae and is the treatment of choice. Rifamycins also are ehrlichiacidal.

Epidemiology & Prevention

The incidence of human ehrlichioses is not well defined. In Oklahoma, which has the highest incidence of Rocky Mountain spotted fever, human monocytotropic ehrlichiosis is at least as common. Human granulocytotropic ehrlichiosis is thought to occur at a rate of about 15 cases per 100,000 population in upper midwestern Oklahoma and at higher rates in selected counties.

More than 90% of cases occur between mid April and October, and more than 80% of cases are in men. Most patients give histories of tick exposure in the month before onset of illness. Cases of human monocytotropic ehrlichiosis have occurred in over 30 states, primarily in the south-central and southeastern United States. This area corresponds to the area of distribution of the Lone Star tick, *Amblyomma americanum.* Cases of human monocytotropic ehrlichiosis in the western United States and in Europe and Africa suggest other tick vectors such as *D variabilis.* Cases of human granulocytotropic ehrlichiosis occur in the upper Midwest and East Coast states and in West Coast states. These areas correspond to the distribution of the tick vectors *Ixodes scapularis* and *Ixodes pacificus,* respectively.

REVIEW QUESTIONS

1. Morulae (intracellular inclusions in leukocytes) are characteristic of which of the following diseases?
   - (A) Malaria due to *Plasmodium falciparum* infection but not *Plasmodium malariae* infection
   - (B) Dengue
   - (C) Babesia infection
   - (D) Ehrlichiae infection
   - (E) Loa loa

2. Which of the following statements about epidemic typhus (*Rickettsia prowazekii* disease) is most correct?
   - (A) The disease occurs primarily in sub-Saharan Africa
   - (B) It is transmitted by ticks
   - (C) Mice are the reservoir
   - (D) Historically, the disease occurs in times of prosperity
   - (E) Recrudescence can occur many years after the initial infection

3. The most useful drug to treat ehrlichiosis is
   - (A) Doxycycline
   - (B) Penicillin G
   - (C) Trimethoprim-Sulfamethoxazole
   - (D) Gentamicin
   - (E) Nitrofurantoin

4. A disease characterized by malaise, headache, rigors, and fever developed in members of several families living in an unheated war-damaged house in an Eastern European country. Erythematous 2–6 mm macular red rashes appeared on the peoples' trunks and later on their extremities. Some of the people had coughs. One elderly person, although sick, was much less sick than other adults. The people huddled together to keep warm; body lice were common. Which of the following statements is most correct?
   - (A) The disease that these people had is common in the Rocky Mountain States
   - (B) The elderly person may have had acute epidemic typhus many years ago and recrudescent typhus now
   - (C) Fleas from rodents in the house were spreading *Rickettsia typhi*
   - (D) The primary host of the body louse infecting the people is the rat
   - (E) Epidemic typhus can be prevented by a vaccine

5. Which of the following statements about Ehrlichiae and ehrlichiosis is most correct?

   (A) Dogs and mice are reservoirs
   (B) Mosquitoes are the vectors
   (C) Ampicillin is the treatment of choice
   (D) Culture is a good method to confirm the diagnosis
   (E) Ehrlichiae are typically found in lymphocytes

6. A group of urban teenagers visited a sheep ranch in a large western state for a 2-week experience. While they were there, many of the pregnant ewes delivered lambs to the delight of the closely observing teenagers. About 10 days later, three of the teenagers developed flu-like illnesses characterized by malaise, cough, and fever. One had an infiltrate on chest x-ray, indicating pneumonia. The three teenagers had different doctors, but the physicians each drew a blood specimen and submitted it to the city health department for serologic testing. All three specimens were positive for Q fever. Public health investigators determined that all the teenagers had been to the sheep ranch. When the investigators contacted the ranch, they were told that there was no Q fever there and that no one who lived at the ranch had been sick. The most likely explanation for the teenagers' illnesses and the lack of illness at the ranch is

   (A) There was no Q fever at the ranch, and it was acquired elsewhere
   (B) The people at the ranch had been previously immunized against Q fever
   (C) The teenagers acquired Q fever at the ranch, and the people who lived there had all previously had Q fever and were now immune to it
   (D) The teenagers had other illnesses, and the positive Q fever serology was unrelated
   (E) The public health laboratory had errors in the Q fever serologic tests

7. A middle-aged sportsman, resident of Oklahoma, took a hike through a rural wooded and brushy area near his home. The following morning he noticed and removed a large (>1 cm) tick from his upper arm. About 1 week later, he experienced a gradual onset of fever and malaise. He now seeks medical attention because he is concerned about a possible infection transmitted by the tick. Which of the following diseases is most likely to be acquired from a tick?

   (A) Dengue
   (B) Rocky Mountain spotted fever
   (C) Typhus
   (D) Yellow fever
   (E) Malaria

8. Which of the following drugs should *not* be used to treat Rocky Mountain spotted fever (*Rickettsia rickettsii* infection)?

   (A) Trimethoprim-Sulfamethoxazole
   (B) Chloramphenicol
   (C) Doxycycline

9. Which of the following should be used to prevent Rocky Mountain spotted fever (*Rickettsia rickettsii* infection)?

   (A) Attenuated *Rickettsia rickettsii* vaccine
   (B) Prophylactic doxycycline
   (C) Preventing tick bites by wearing protective clothing
   (D) Delousing with insecticide

10. One week after deer hunting in a wooded area, a 33-year-old man developed fever to 39°C with headache and malaise. Over the subsequent 24 hours he developed nausea, vomiting, abdominal pain, and diarrhea. On day 4 he developed a rash, initially around the wrists and ankles, which then progressively evolved, involving the arms, trunk, palms, and soles. Initially the rash was macular, but it quickly evolved into maculopapules, some with central petechiae. Rocky Mountain spotted fever caused by *Rickettsia rickettsii* was diagnosed. Which of the following statements about Rocky Mountain spotted fever is correct?

    (A) The vectors of *Rickettsia rickettsii* are ticks of the genus *Ixodes*
    (B) A rash consistently appears by day 4 of illness
    (C) *Rickettsia rickettsii* forms inclusions in monocytes
    (D) The patient's antibody response may not occur until after the second week of illness
    (E) The highest incidence of this disease is in the Rocky Mountain states

Answers

1. D	4. B	7. B	10. D
2. E	5. A	8. A	
3. A	6. C	9. C	

REFERENCES

Olano JP, Aguero-Rosenfeld ME: *Ehrlichia, Anaplasma,* and related intracellular bacteria. In: *Manual of Clinical Microbiology,* 9th ed. Murray PR et al (editors). ASM Press, 2007.

Rickettsioses, ehrlichioses and AnaplasmosisVol 2, Part III, Section E. In: *Mandell, Douglas, and Bennett's Principles and Practice of Infectious Diseases,* 7th ed. Mandell GL, et al (editors). Elsevier, 2010.

Chlamydiae

Chlamydiae that infect humans are divided into three species, *Chlamydia trachomatis*, *Chlamydia (Chlamydophila) pneumoniae*, and *Chlamydia (Chlamydophila) psittaci*, on the basis of antigenic composition, intracellular inclusions, sulfonamide susceptibility, and disease production. The separation of the genus *Chlamydia* into the genera *Chlamydia* and *Chlamydophila* is controversial; in this chapter the three chlamydiae that are pathogens of humans are considered to be in the genus *Chlamydia*. Other chlamydiae infect animals but rarely if ever infect humans. All chlamydiae exhibit similar morphologic features, share a common group antigen, and multiply in the cytoplasm of their host cells by a distinctive developmental cycle. The chlamydiae can be viewed as gram-negative bacteria that lack mechanisms for the production of metabolic energy and cannot synthesize ATP. This restricts them to an intracellular existence, where the host cell furnishes energy-rich intermediates. Thus, chlamydiae are **obligate intracellular parasites**.

Developmental Cycle

All chlamydiae have a common reproductive cycle. The environmentally stable infectious particle is a small cell called the **elementary body** or **EB**. These are about 0.3 μm in diameter (Figure 27–1) with an electron-dense nucleoid. The EB membrane proteins have highly cross-linked membrane proteins. The EBs have a high affinity for host epithelial cells and rapidly enter them. There appear to be multiple adhesins, receptors, and mechanisms of entry. Heparan sulfate-like proteoglycans on the surface of *C trachomatis* are likely possibilities for mediating at least the initial interaction between EBs and host cells. Other potential adhesins include the major outer membrane protein (**MOMP**), glycosylated MOMP, and other surface proteins. The mechanisms thought to mediate entry into the host cell also varied. EBs are usually seen attached near the base of microvilli, where they are subsequently engulfed by the host cell. More than one mechanism appears to be functional: receptor-mediated endocytosis into clathrin- coated pits and pinocytosis via noncoated pits. Lysosomal fusion is inhibited creating a protected membrane-bound environment around the chlamydiae. Shortly after entry into the host cell, the disulfide

bonds of the EB membrane proteins are no longer cross-linked and the EB is reorganized into a larger structure called a **reticulate body** or **RB** measuring about 0.5–1 μm (Figure 27–1) and devoid of an electron-dense nucleoid. Within the membrane-bound vacuole, the RB grows in size and divides repeatedly by binary fission. Eventually, the entire vacuole becomes filled with elementary bodies derived from the reticulate bodies to form a cytoplasmic **inclusion**. The newly formed EBs may be liberated from the host cell to infect new cells. The developmental cycle takes 24–48 hours.

Structure & Chemical Composition

In chlamydiae, the outer **cell wall** resembles the cell wall of gram-negative bacteria. It has a relatively high lipid content. It is rigid but does not contain a typical bacterial peptidoglycan; however, the chlamydial genome contains the genes needed for peptidoglycan synthesis. Penicillin-binding proteins occur in chlamydiae, and chlamydial cell wall formation is inhibited by penicillins and other drugs that inhibit transpeptidation of bacterial peptidoglycan. Lysozyme has no effect on chlamydial cell walls. *N*-acetylmuramic acid appears to be absent from chlamydial cell walls. Both DNA and RNA are present in elementary and reticulate bodies. The reticulate bodies contain about four times as much RNA as DNA, whereas the elementary bodies contain about equal amounts of RNA and DNA. In elementary bodies, most DNA is concentrated in the electron-dense central nucleoid. Most RNA exists in ribosomes. The circular genome of chlamydiae (MW 7×10^8) is similar to that of bacterial chromosomes.

Multiple chlamydial genomes have been sequenced providing insight into the basic biology of the organisms. For example, chlamydiae have a type III secretion system, which may allow them to inject effector proteins into host cells as part of the infectious process.

Staining Properties

Chlamydiae have distinctive staining properties (similar to those of rickettsiae). Elementary bodies stain purple with

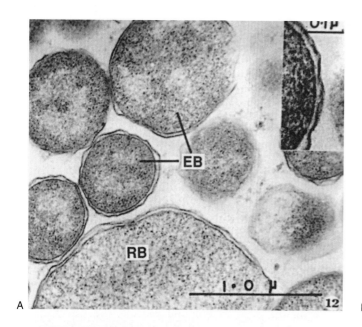

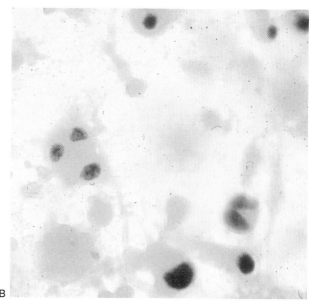

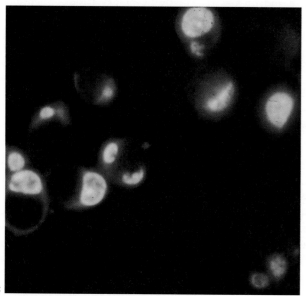

FIGURE 27–1 Chlamydiae. **A**: Thin section electron micrograph of Chlamydiae in various stages of development. (EB, elementary body particles with cell walls (inset); RB, reticulate body.) **B**: Chlamydia trachomatis grown in McCoy cells and stained with iodine. The McCoy cells stain a faint yellow in the background. The glycogen-rich intracytoplasmic inclusions of *Chlamydia trachomatis* stain a dark brown. **C**: Similar growth of *Chlamydia trachomatis* in McCoy cells stained with a fluorescein-labeled antibody against a *Chlamydia trachomatis* species antigen. The intracytoplasmic inclusions of *Chlamydia trachomatis* stain bright yellow-green. Faint outlines of the McCoy cells are visible. (Courtesy of J Schachter.)

Giemsa's stain—in contrast to the blue of host cell cytoplasm. The larger, noninfective reticulate bodies stain blue with Giemsa's stain. The Gram reaction of chlamydiae is negative or variable and is not useful in identification of the agents. Chlamydial particles and inclusions stain brightly by immunofluorescence, with group-specific, species-specific, or serovar-specific antibodies.

Fully formed, mature intracellular inclusions of *C trachomatis* are compact masses near the nucleus which are dark purple when stained with Giemsa's stain because of the densely packed mature particles. If stained with dilute Lugol's iodine solution, some of the inclusions of *C trachomatis* (but not *C pneumoniae* or *C psittaci*) appear brown because of the glycogen matrix that surrounds the particles (Figure 27–1). Inclusions of *C psittaci* are diffuse intracytoplasmic aggregates.

Antigens

Chlamydiae possess **shared group (genus)-specific antigens**. These are heat-stable lipopolysaccharides with 2-keto-3-deoxyoctanoic acid as an immunodominant component. Antibody to these genus-specific antigens can be detected by **complement fixation** (CF) and immunofluorescence. **Species-specific** or **serovar-specific** antigens are mainly outer membrane proteins. Specific antigens can best be detected by **immunofluorescence**, particularly using monoclonal antibodies. Specific antigens are shared by only a limited number of chlamydiae, but a given organism may contain several specific antigens. There are at least 18 **serovars** of *C trachomatis*; these include A, B, Ba, C–K, and L1–L3. Several serovars of *C psittaci* can be demonstrated by **CF** and **microimmunofluorescence** tests. Only one serovar of *C pneumoniae* has been described.

Growth & Metabolism

Chlamydiae require an intracellular habitat, because they are unable to synthesize ATP and depend on the host cell for energy requirements. Chlamydiae grow in cultures of a variety of eukaryotic cells lines. McCoy cells treated with cycloheximide commonly are used to isolate chlamydiae; *C pneumoniae* grows better in HL or HEp-2 cells. All types of chlamydiae proliferate in embryonated eggs, particularly in the yolk sac.

Some chlamydiae have an endogenous metabolism like other bacteria. They can liberate CO_2 from glucose, pyruvate, and glutamate; they also contain dehydrogenases. Nevertheless, they require energy-rich intermediates from the host cell to carry out their biosynthetic activities.

The replication of chlamydiae can be inhibited by many antibacterial drugs. Cell wall inhibitors such as penicillins and cephalosporins result in the production of morphologically defective forms but are not effective in clinical diseases. Inhibitors of protein synthesis (tetracyclines, erythromycins) are effective in most clinical infections. *C trachomatis* strains synthesize folates and are susceptible to inhibition by sulfonamides. Aminoglycosides are noninhibitory.

Characteristics of Host–Parasite Relationship

The outstanding biologic feature of infection by chlamydiae is the balance that is often reached between host and parasite, resulting in prolonged persistence of infection. Subclinical infection is the rule—and overt disease the exception—in the natural hosts of these agents. Spread from one species to another (eg, birds to humans, as in psittacosis) more frequently leads to disease. Antibodies to several antigens of chlamydiae are regularly produced by the infected host. These antibodies have little protective effect against reinfection. The infectious agent commonly persists in the presence of high antibody titers. Treatment with effective antimicrobial drugs (eg, tetracyclines) for prolonged periods may eliminate the chlamydiae from the infected host. Very early, intensive treatment may suppress antibody formation. Late treatment with antimicrobial drugs in moderate doses may suppress disease but permit persistence of the infecting agent in tissues.

The immunization of humans has been singularly unsuccessful in protecting against reinfection. Prior infection or immunization at most tends to result in milder disease upon reinfection, but at times the accompanying hypersensitization aggravates inflammation and scarring (eg, in trachoma).

Classification

Chlamydiae are arranged according to their pathogenic potential, host range, antigenic differences, and other methods. Three species that infect humans have been characterized (Table 27–1).

A. *Chlamydia trachomatis*

This species produces compact intracytoplasmic inclusions that contain glycogen; it is usually inhibited by sulfonamides. It includes agents of human disorders such as trachoma, inclusion conjunctivitis, nongonococcal urethritis, salpingitis, cervicitis, pneumonitis of infants, and lymphogranuloma venereum. A *C trachomatis* variant also causes mouse pneumonitis.

B. *Chlamydophila pneumoniae*

This species produces intracytoplasmic inclusions that lack glycogen; it is usually resistant to sulfonamides. It causes respiratory tract infections in humans.

TABLE 27–1 Characteristics of the Chlamydiae

	Chlamydia trachomatis	*Chlamydophila pneumoniae*	*Chlamydia psittaci*
Inclusion morphology	Round, vacuolar	Round, dense	Large, variable shape, dense
Glycogen in inclusions	Yes	No	No
Elementary body morphology	Round	Pear-shaped, round	Round
Susceptible to sulfonamides	Yes	No	No
Plasmid	Yes	No	Yes
Serovars	15	1	≥4
Natural host	Humans	Humans	Birds
Mode of transmission	Person to person, mother to infant	Airborne person to person	Airborne bird excreta to humans
Major diseases	Trachoma, STDs, infant pneumonia, lymphogranuloma venereum	Pneumonia, bronchitis, pharyngitis, sinusitis	Psittacosis, pneumonia, fever of unexplained origin

C. *Chlamydia psittaci*

This species produces diffuse intracytoplasmic inclusions that lack glycogen; it is usually resistant to sulfonamides. It includes agents of psittacosis in humans, ornithosis in birds, meningopneumonitis, feline pneumonitis, and other animal diseases.

CHLAMYDIA TRACHOMATIS OCULAR, GENITAL, & RESPIRATORY INFECTIONS

Humans are the natural host for *C trachomatis*. Monkeys and chimpanzees can be infected in the eye and genital tract. *C trachomatis* also replicates in cells in tissue culture. *C trachomatis* of different serovars replicates differently. Isolates from trachoma do not grow as well as those from lymphogranuloma venereum or genital infections. Intracytoplasmic replication results in the formation of compact inclusions with a glycogen matrix in which elementary bodies are embedded.

Immunotype-specific antisera permit typing of isolates that gives results analogous to those achieved by typing by microimmunofluorescence. The serovars specifically associated with endemic trachoma are A, B, Ba, and C; those associated with sexually transmitted disease are D–K; and those that cause lymphogranuloma venereum are L1, L2, and L3.

TRACHOMA

Trachoma is an ancient eye disease, well described in the Ebers Papyrus, which was written in Egypt 3800 years ago. It is a chronic keratoconjunctivitis that begins with acute inflammatory changes in the conjunctiva and cornea and progresses to scarring and blindness. The *C trachomatis* serovars A, B, Ba, and C are associated with clinical trachoma.

Clinical Findings

In experimental human infections, the incubation period for chlamydial conjunctival infection is 3–10 days. In endemic areas, initial infection occurs in early childhood, and the onset of the long-term consequence, trachoma, is insidious. Chlamydial infection is often mixed with bacterial conjunctivitis in endemic areas, and the two together produce the clinical picture. The earliest symptoms of trachoma are lacrimation, mucopurulent discharge, conjunctival hyperemia, and follicular hypertrophy. Microscopic examination of the cornea reveals epithelial keratitis, subepithelial infiltrates, and extension of limbal vessels into the cornea (pannus). As the pannus extends downward across the cornea, there is scarring of the conjunctiva, eyelid deformities (entropion, trichiasis), and added insult caused by eyelashes sweeping across the cornea. With secondary bacterial infection, loss of vision progresses over a period of years. There are, however, no systemic symptoms or signs of infection.

Laboratory Diagnosis

The laboratory diagnosis of chlamydial infections is discussed also in Chapter 47.

A. Culture

Typical cytoplasmic inclusions are found in epithelial cells of conjunctival scrapings stained with fluorescent antibody or by the Giemsa's method. These occur most frequently in the early stages of the disease and on the upper tarsal conjunctiva.

Inoculation of conjunctival scrapings into cycloheximide-treated McCoy cell cultures permits growth of *C trachomatis* if the number of viable infectious particles is sufficiently large. Centrifugation of the inoculum into the cells increases the sensitivity of the method. The diagnosis can sometimes be made in the first passage after 2–3 days of incubation by looking for inclusions by immunofluorescence or staining with iodine or Giemsa's stain.

B. Serology

Infected individuals often develop both group antibodies and serovar-specific antibodies in serum and in eye secretions. Immunofluorescence is the most sensitive method for their detection. Neither ocular nor serum antibodies confer significant resistance to reinfection.

C. Molecular Methods

Developing countries, where trachoma is endemic, generally do not have the resources to apply polymerase chain reaction (PCR) or other molecular methods to the diagnosis of *C trachomatis* infections of the eye. Developed countries have relatively little trachoma and little need for such tests. Thus, the molecular methods have been developed for the diagnosis of genital infections. Only research projects have used PCR in studies of trachoma.

Treatment

Clinical trials, in villages with endemic trachoma, using mass azithromycin treatment show that infection and clinical disease are greatly decreased at 6 and 12 months post therapy; this is true even with single dose therapy. Thus, azithromycin has replaced erythromycin and doxycycline in the mass treatment of endemic trachoma. Topical therapy is of little value.

Epidemiology & Control

It is believed that more than 400 million people throughout the world have trachoma and that 20 million are blinded by it. The disease is most prevalent in Africa, Asia, and the Mediterranean basin, where hygienic conditions are poor and water is scarce. In such hyperendemic areas, childhood infection may be universal, and severe blinding disease (resulting from frequent bacterial superinfection) is common. In the

United States, trachoma occurs sporadically in some areas, and endemic foci persist.

The WHO has initiated the S-A-F-E program to eliminate blinding trachoma and at least markedly reduce clinically active disease. The S-A-F-E program is as follows: Surgery for deformed eyelids; periodic Azithromycin therapy; Face washing and hygiene; and, Environmental improvement such as building latrines and decreasing the number of flies that feed on conjunctival exudates. It is clear that improved socio-economic conditions enhance the disappearance of endemic trachoma.

CHLAMYDIA TRACHOMATIS GENITAL INFECTIONS & INCLUSION CONJUNCTIVITIS

C trachomatis serovars D–K cause sexually transmitted diseases—especially in developed countries—and may also produce infection of the eye (inclusion conjunctivitis). In sexually active men, C trachomatis causes **nongonococcal urethritis** and, occasionally, **epididymitis**. In women, C trachomatis causes **urethritis**, **cervicitis**, and **pelvic inflammatory disease**, which can lead to **sterility** and predispose to **ectopic pregnancy**. Proctitis and proctocolitis may occur in men and women although these infections appear to be most common in men who have sex with men. Any of these anatomic sites of infection may give rise to symptoms and signs, or the infection may remain asymptomatic but communicable to sex partners. Up to 50% of nongonococcal urethritis (men) or the urethral syndrome (women) is attributed to chlamydiae and produces dysuria, nonpurulent discharge, and frequency of urination. Genital secretions of infected adults can be self-inoculated into the conjunctiva, resulting in inclusion conjunctivitis, an ocular infection that closely resembles acute trachoma.

The newborn acquires the infection during passage through an infected birth canal. Probably 20–60% of infants of infected mothers acquire the infection, with 15–20% of infected infants manifesting eye symptoms and 10–40% manifesting respiratory tract involvement. **Inclusion conjunctivitis of the newborn** begins as a mucopurulent conjunctivitis 7–12 days after delivery. It tends to subside with erythromycin or tetracycline treatment, or spontaneously after weeks or months. Occasionally, inclusion conjunctivitis persists as a chronic chlamydial infection with a clinical picture indistinguishable from that of subacute or chronic childhood trachoma in nonendemic areas and usually not associated with bacterial conjunctivitis.

Laboratory Diagnosis

A. Specimen Collection

Proper specimen collection is the key to the laboratory diagnosis of chlamydia infection. Because the chlamydiae are obligate intracellular bacteria, it is important that the specimens contain infected human cells as well as the extracellular material where they might also be present. Collect endocervical specimens following removal of discharge and secretions from the cervix. A swab or cytology brush is used to scrape epithelial cells from 1 to 2 cm deep into the endocervix. Dacron, cotton, rayon, or calcium alginate on a plastic shaft should be used to collect the specimen; some other swab materials and wooden shafts are toxic to chlamydiae. A similar method is used to collect specimens from the vagina, urethra, or conjunctiva. The commercial diagnostic nonculture tests for chlamydia do not require viable organisms. In general, these proprietary tests include the specimen collection swabs and transport tubes that have been demonstrated to be suitable for the specific tests. For culture, the swab specimens should be placed in a chlamydiae transport medium and kept at refrigerator temperature before transport to the laboratory.

Urine can be tested for the presence of chlamydial nucleic acid. Only the first 20 mL of the void should be collected because a larger volume of bladder urine would dilute the initial urine that passed through the urethra; this could result in a negative test because of the dilution.

B. Nucleic Acid Detection

Nucleic acid amplification tests (NAATs) are the tests of choice to diagnose genital C trachomatis infections. One is based on the PCR; another is based on strand displacement amplification. These tests have become widely used. Although they are highly sensitive and specific, they are not perfect. New assays to diagnose chlamydiae infection can be compared to combined results from two NAATs as the reference standard.

In one nucleic acid hybridization test, a DNA probe hybridizes to a specific sequence of chlamydiae 16S rRNA; chlamydiae have up to 10^4 copies of the 16S rRNA. Once the hybrids are formed they are absorbed onto beads, and the amount of hybrid is detected by chemiluminescence. Another hybridization assay used RNA probes to detect chlamydiae DNA sequences. The overall sensitivity and specificity of these tests are good but may not be quite as good as the NAATs. The hybridization assays, however, may not require as many costly resources as the NAATs.

Some of the nucleic acid detection tests have been adapted to simultaneously detect Neisseria gonorrhoeae. However, care must be taken when applying positive results from such screening tests for N gonorrhoeae; when the test to detect N gonorrhoeae is 99% sensitive and 99% specific and the prevalence of infection is 0.5–1%, the predictive value of a positive test is <50%. In this sort of setting, a positive NAAT for N gonorrhoeae should be confirmed by culture or by using a second (different) nucleic acid detection test.

C. Direct Cytologic Examination (Direct Fluorescent Antibody) & Enzyme-Linked Immunoassay

Commercially available direct fluorescent antibody (DFA) and enzyme-linked immunoassay (EIA) assays to detect

C trachomatis can be used in laboratories that lack the expertise or facilities to perform nucleic acid detection tests. The DFA uses monoclonal antibodies directed against a species-specific antigen on the chlamydial MOMP. The EIA detects the presence of genus-specific antigens extracted from elementary bodies in the specimen. The sensitivity of these tests is less than culture and considerably less than the nucleic acid detection tests.

D. Culture

Culture of *C trachomatis* has historically been used to diagnose chlamydia infections. Culture, however, is costly and arduous. Results are delayed compared to the timeliness of nucleic acid detection or other tests. Culture is generally much less sensitive than the nucleic acid detection assays; the degree of lower sensitivity is largely dependent upon the culture method used. Culture is now done in a limited number of reference laboratories. McCoy cells are grown in monolayers on coverslips in dram or shell vials. Some laboratories use flat-bottomed microdilution trays, but cultures by this method are not as sensitive as those achieved with the shell vial method. The McCoy cells are treated with cycloheximide to inhibit metabolism and increase the sensitivity of isolation of the chlamydiae. The inoculum from the swab specimen is centrifuged onto the monolayer and incubated at 35–37°C for 48–72 hours. A second monolayer can be inoculated, and after incubation, it can be sonicated and passaged to another monolayer to enhance sensitivity. The monolayers are examined by direct immunofluorescence to visualize the cytoplasmic inclusions. Chlamydial cultures by this method are about 80% sensitive but 100% specific.

E. Serology

Because of the relatively great antigenic mass of chlamydiae in genital tract infections, serum antibodies occur much more commonly than in trachoma and are of higher titer. A titer rise occurs during and after acute chlamydial infection. Because of the high prevalence of chlamydial genital tract infections in some societies, there is a high background of antichlamydial antibodies in the population; serologic tests to diagnose genital tract chlamydial infections generally are not useful.

In genital secretions (eg, cervical), antibody can be detected during active infection and is directed against the infecting immunotype (serovar).

Treatment

It is essential that chlamydial infections be treated simultaneously in both sex partners and in offspring to prevent reinfection. Tetracyclines (eg, doxycycline) are commonly used in nongonococcal urethritis and in nonpregnant infected females. Azithromycin is effective and can be given to pregnant women. Topical tetracycline or erythromycin is used for neonatal *N gonorrhoeae* infections but may not effectively prevent neonatal *C trachomatis* infection. Systemic therapy should be used for inclusion conjunctivitis as topical therapy may not cure the eye infections or prevent respiratory disease.

Epidemiology & Control

Genital chlamydial infection and inclusion conjunctivitis are sexually transmitted diseases that are spread by contact with infected sex partners. Neonatal inclusion conjunctivitis originates in the mother's infected genital tract. Prevention of neonatal eye disease depends upon diagnosis and treatment of the pregnant woman and her sex partner. As in all sexually transmitted diseases, the presence of multiple etiologic agents (gonococci, treponemes, trichomonads, herpes, etc) must be considered. Instillation of erythromycin or tetracycline into the newborn's eyes does not prevent development of chlamydial conjunctivitis. The ultimate control of this—and all—sexually transmitted disease depends on safe sex practices and on early diagnosis and treatment of infected persons.

CHLAMYDIA TRACHOMATIS AND NEONATAL PNEUMONIA

Of newborns infected by the mother, 10–20% may develop respiratory tract involvement 2–12 weeks after birth, culminating in pneumonia. *C trachomatis* may be the most common cause of neonatal pneumonia. There is striking tachypnea, characteristic paroxysmal staccato cough, absence of fever, and eosinophilia. Consolidation of lungs and hyperinflation can be seen by x-ray. The diagnosis should be suspected if pneumonitis develops in a newborn who has inclusion conjunctivitis and can be established by isolation of *C trachomatis* from respiratory secretions. In such neonatal pneumonia, an IgM antibody titer to *C trachomatis* of 1:32 or more is considered diagnostic. Systemic erythromycin is effective treatment in severe cases.

LYMPHOGRANULOMA VENEREUM

Lymphogranuloma venereum is a sexually transmitted disease caused by *C trachomatis* and characterized by suppurative inguinal adenitis; it is most common in tropical climates.

Properties of the Agent

The particles contain CF heat-stable chlamydial group antigens that are shared with all other chlamydiae. They also contain one of three serovar antigens (L1–L3), which can be defined by immunofluorescence.

Clinical Findings

Several days to several weeks after exposure, a small, evanescent papule or vesicle develops on any part of the external genitalia, anus, rectum, or elsewhere. The lesion may ulcerate, but usually it remains unnoticed and heals in a few days. Soon thereafter, the regional lymph nodes enlarge and tend to become matted and painful. In males, inguinal nodes are most commonly involved both above and below Poupart's ligament, and the overlying skin often turns purplish as the nodes suppurate and eventually discharge pus through multiple sinus tracts. In females and in homosexual males, the perirectal nodes are prominently involved, with proctitis and a bloody mucopurulent anal discharge. Lymphadenitis may be most marked in the cervical chains.

During the stage of active lymphadenitis, there are often marked systemic symptoms including fever, headaches, meningismus, conjunctivitis, skin rashes, nausea and vomiting, and arthralgias. Meningitis, arthritis, and pericarditis occur rarely. Unless effective antimicrobial drug treatment is given at that stage, the chronic inflammatory process progresses to fibrosis, lymphatic obstruction, and rectal strictures. The lymphatic obstruction may lead to elephantiasis of the penis, scrotum, or vulva. The chronic proctitis of women or homosexual males may lead to progressive rectal strictures, rectosigmoid obstruction, and fistula formation.

Laboratory Diagnosis

A. Smears

Pus, buboes, or biopsy material may be stained, but particles are rarely recognized.

B. Culture

Suspected material is inoculated into McCoy cell cultures. The inoculum can be treated with an aminoglycoside (but not with penicillin) to lessen bacterial contamination. The agent is identified by morphology and serologic tests.

C. Serology

Antibodies are commonly demonstrated by the CF reaction. The test becomes positive 2–4 weeks after onset of illness, at which time skin hypersensitivity can sometimes also be demonstrated. In a clinically compatible case, a rising antibody level or a single titer of more than 1:64 is good evidence of active infection. If treatment has eradicated the lymphogranuloma venereum infection, the CF titer falls. Serologic diagnosis of lymphogranuloma venereum can employ immunofluorescence, but the antibody is broadly reactive with many chlamydial antigens.

Immunity

Untreated infections tend to be chronic, with persistence of the agent for many years. Little is known about active immunity. The coexistence of latent infection, antibodies, and cell-mediated reactions is typical of many chlamydial infections.

Treatment

The sulfonamides and tetracyclines have been used with good results, especially in the early stages. In some drug-treated persons there is a marked decline in complement-fixing antibodies, which may indicate that the infective agent has been eliminated from the body. Late stages require surgery.

Epidemiology & Control

Although the highest incidence of lymphogranuloma venereum has been reported from subtropical and tropical areas, the infection occurs all over the world. The disease is most often spread by sexual contact, but not exclusively so. The portal of entry may sometimes be the eye (conjunctivitis with an oculoglandular syndrome). The genital tracts and rectums of chronically infected (but at times asymptomatic) persons serve as reservoirs of infection. Laboratory personnel exposed to aerosols of *C trachomatis* serovars L1–L3 can develop a chlamydial pneumonitis with mediastinal and hilar adenopathy. If the infection is recognized, treatment with tetracycline or erythromycin is effective.

The measures used for the control of other sexually transmitted diseases apply also to the control of lymphogranuloma venereum. Case-finding and early treatment and control of infected persons are essential.

CHLAMYDOPHILA PNEUMONIAE & RESPIRATORY INFECTIONS

The first *C pneumoniae* (TWAR) strain was obtained in the 1960s in chick embryo yolk sac culture. Following the development of cell culture methods, this initial strain was thought to be a member of the species *C psittaci*. Subsequently, *C pneumoniae* has been firmly established as a new species that causes respiratory disease. Humans are the only known host.

Properties of the Agent

C pneumoniae produces round, dense, glycogen-negative inclusions that are sulfonamide-resistant, much like *C psittaci* (Table 27–1). The elementary bodies sometimes have a pear-shaped appearance. The genetic relatedness of *C pneumoniae* isolates is >95%. Only one serovar has been demonstrated.

Clinical Findings

Most infections with *C pneumoniae* are asymptomatic or associated with mild illness, but severe disease has been reported. There are no signs or symptoms that specifically differentiate *C pneumoniae* infections from those caused by many other agents. Both upper and lower airway diseases occur. Pharyngitis is common. Sinusitis and otitis media may occur and be accompanied by lower airway disease. An atypical pneumonia similar to that caused by *Mycoplasma pneumoniae* is the primary recognized illness. Five to 20% of community-acquired pneumonia in young persons is thought to be caused by *C pneumoniae*.

Laboratory Diagnosis

A. Smears

Direct detection of elementary bodies in clinical specimens using fluorescent antibody techniques is insensitive. Other stains do not effectively demonstrate the organism.

B. Culture

Swab specimens of the pharynx should be put into a chlamydiae transport medium and placed at 4°C; *C pneumoniae* is rapidly inactivated at room temperature. It grows poorly in cell culture, forming inclusions smaller than those formed by the other chlamydiae. *C pneumoniae* grows better in HL and HEp-2 cells than in HeLa 229 or McCoy cells; the McCoy cells are widely used to culture *C trachomatis*. The sensitivity of the culture is increased by incorporation of cycloheximide into the cell culture medium to inhibit the eukaryotic cell metabolism and by centrifugation of the inoculum onto the cell layer. Growth is better at 35°C than 37°C. After 3 days' incubation, the cells are fixed and inclusions detected by fluorescent antibody staining with genus- or species-specific antibody or, preferably, with a *C pneumoniae*-specific monoclonal antibody conjugated with fluorescein. Giemsa staining is insensitive, and the glycogen-negative inclusions do not stain with iodine. It is moderately difficult to grow *C pneumoniae*—as evidenced by the number of isolates described compared with the incidence of infection.

C. Serology

Serology using the microimmunofluorescence test is the most sensitive method for diagnosis of *C pneumoniae* infection. The test is species-specific and can detect IgG or IgM antibodies by using the appropriate reagents. Primary infection yields IgM antibody after about 3 weeks followed by IgG antibody at 6–8 weeks. In reinfection, the IgM response may be absent or minimal and the IgG response occurs in 1–2 weeks. The following criteria have been suggested for the serologic diagnosis of *C pneumoniae* infection: a single IgM titer of ≥1:16; a single IgG titer of ≥1:512; and a fourfold rise in either the IgM or IgG titers.

The CF test can be used, but it is group-reacting, does not differentiate *C pneumoniae* infection from psittacosis or lymphogranuloma venereum, and is less sensitive than the microimmunofluorescence test.

Immunity

Little is known about active or potentially protective immunity. Prolonged infections can occur with *C pneumoniae*, and asymptomatic carriage may be common.

Treatment

C pneumoniae is susceptible to the macrolides and tetracyclines and to some fluoroquinolones. Treatment with doxycycline, azithromycin, or clarithromycin appears to significantly benefit patients with *C pneumoniae* infection, but there are only limited data on the efficacy of antibiotic treatment. Reports indicate that the symptoms may continue or recur after routine courses of therapy with erythromycin, doxycycline, or tetracycline, and these drugs should be given for 10- to 14-day courses.

Epidemiology

Infection with *C pneumoniae* is common. Worldwide, 30–50% of people have antibody to *C pneumoniae*. Few young children have antibody, but after the age of 6–8 years, the prevalence of antibody increases through young adulthood. Infection is both endemic and epidemic, with multiple outbreaks attributed to *C pneumoniae*. There is no known animal reservoir, and transmission is presumed to be from person to person, predominantly by the airborne route.

Lines of evidence suggesting that *C pneumoniae* is associated with atherosclerotic coronary artery and cerebrovascular disease consist of seroepidemiologic studies, detection of *C pneumoniae* in atherosclerotic tissues, cell culture studies, animal models, and trials of prevention using antibiotic agents. However, other studies have shown no association. The possible link between *C pneumoniae* infection and coronary artery disease remains controversial.

CHLAMYDIA PSITTACI & PSITTACOSIS

The term "psittacosis" is applied to the human *C psittaci* disease acquired from contact with birds and also the infection of psittacine birds (parrots, parakeets, cockatoos, etc). The term "ornithosis" is applied to infection with similar agents in all types of domestic birds (pigeons, chickens, ducks, geese, turkeys, etc) and free-living birds (gulls, egrets, petrels, etc). In humans, *C psittaci* produces a spectrum of clinical manifestations ranging from severe pneumonia and sepsis with a high mortality rate to a mild inapparent infection.

Properties of the Agent

C psittaci can be propagated in embryonated eggs, in mice and other animals, and in some cell cultures. The heat-stable group-reactive CF antigen resists proteolytic enzymes and appears to be a lipopolysaccharide. Treatment of *C psittaci* infection with deoxycholate and trypsin yields extracts that contain group-reactive CF antigens, whereas the cell walls retain the species-specific antigen. Antibodies to the species-specific antigen are able to neutralize toxicity and infectivity. Specific serovars characteristic for certain mammalian and avian species may be demonstrated by immunofluorescence typing. Neutralization of infectivity of the agent by specific antibody or cross-protection of immunized animals can also be used for serotyping, and the results parallel those of immunofluorescence typing.

Pathogenesis & Pathology

The agent enters through the respiratory tract, is found in the blood during the first 2 weeks of the disease, and may be found in the sputum at the time the lung is involved.

Psittacosis causes a patchy inflammation of the lungs in which consolidated areas are sharply demarcated. The exudates are predominantly mononuclear. Only minor changes occur in the large bronchioles and bronchi. The lesions are similar to those found in pneumonitis caused by some viruses and mycoplasmas. Liver, spleen, heart, and kidney are often enlarged and congested.

Clinical Findings

A sudden onset of illness taking the form of influenza or nonbacterial pneumonia in a person exposed to birds is suggestive of psittacosis. The incubation period averages 10 days. The onset is usually sudden, with malaise, fever, anorexia, sore throat, photophobia, and severe headache. The disease may progress no further, and the patient may improve in a few days. In severe cases, the signs and symptoms of bronchial pneumonia appear at the end of the first week of the disease. The clinical picture often resembles that of influenza, nonbacterial pneumonia, or typhoid fever. The mortality rate may be as high as 20% in untreated cases, especially in the elderly.

Laboratory Diagnosis

A. Culture

Culture of *C psittaci* can be dangerous, and detection of the organism using immunoassays or PCR is preferred. If necessary, *C psittaci* can be cultured from blood or sputum or from lung tissue by culture in tissue culture cells, embryonated eggs, or mice. Isolation of *C psittaci* is confirmed by the serial transmission, its microscopic demonstration, and serologic identification.

B. Detection of *Chlamydia psittaci*

Antigen detection by direct fluorescent antibody staining or by immunoassay or molecular diagnosis by PCR is done in reference or research laboratories.

C. Serology

A diagnosis of psittacosis is usually confirmed by demonstrating complement-fixing or microimmunofluorescent antibodies in serum specimens. A confirmed case is one with a positive culture or associated with a compatible clinical illness plus a fourfold or greater change in antibody titer to at least 1:32 or a microimmunofluorescence IgM titer of at least 1:16. A probable case is one associated with a compatible illness linked epidemiologically with a confirmed case or a titer of at least 1:32 in a single specimen. The CF test is cross-reactive with *C trachomatis* and *C pneumoniae*. The microimmunofluorescence test (MIF) is more sensitive and specific than the CF test, but cross-reactions do occur. MIF allows detection of IgM and IgG. Although antibodies usually develop within 10 days, the use of antibiotics may delay their development for 20–40 days or suppress it altogether.

In live birds, infection is suggested by a positive CF test and an enlarged spleen or liver. This can be confirmed by demonstration of particles in smears or sections of organs and by passage of the agent in mice and eggs.

D. Molecular Methods

Multiple PCR assays have been developed to detect *C psittaci* in respiratory tract specimens, vascular tissues, serum, and mononuclear cells from peripheral blood. These tests are done in reference or research laboratories.

Immunity

Immunity in animals and humans is incomplete. A carrier state in humans can persist for 10 years after recovery. During this period, the agent may continue to be excreted in the sputum.

Live or inactivated vaccines induce only partial resistance in animals. They have not been used in humans.

Treatment

Because of the difficulty in obtaining laboratory confirmation of *C psittaci* infection, most infections are treated based only on the clinical diagnosis. Information on therapeutic efficacy comes from several clinical trials. Azithromycin, clarithromycin, and erythromycin (and doxycycline in adults) clear most, but not all, respiratory *C psittaci* infections. All the patients improve clinically, even those with persistent infection.

Epidemiology & Control

Outbreaks of human disease can occur whenever there is close and continued contact between humans and infected birds that excrete or shed large amounts of infectious agent. Birds often acquire infection as fledglings in the nest, may develop diarrheal illness or no illness, and often carry the infectious agent for their normal life span. When subjected to stress (eg, malnutrition, shipping), birds may become sick and die. The agent is present in tissues (eg, spleen) and is often excreted in feces by healthy birds. The inhalation of infected dried bird feces is a common method of human infection. Another source of infection is the handling of infected tissues (eg, in poultry rendering plants) and inhalation of an infected aerosol.

Birds kept as pets have been an important source of human infection. Foremost among these were the many imported psittacine birds. Latent infections often flared up in these birds during transport and crowding, and sick birds excreted exceedingly large quantities of infectious agent. Control of bird shipment, quarantine, testing of imported birds for psittacosis infection, and prophylactic tetracyclines in bird feed have helped to control this source. Pigeons kept for racing or as pets or raised for squab meat have been important sources of infection. Pigeons populating buildings and thoroughfares in many cities, if infected, shed relatively small quantities of agent.

REVIEW QUESTIONS

1. Which of the following statements about chlamydial antigens is correct?

   (A) Chlamydiae have shared group or genus-specific antigens
   (B) There is no cross-reaction between *Chlamydia trachomatis* and *Chlamydophila pneumoniae* antigens
   (C) All five serovars of *Chlamydophila pneumoniae* cross react with *Chlamydia psittaci*
   (D) One serovar of *Chlamydia trachomatis* causes eye infections and the second serovar causes genital infections

2. The following are part of the control of *Chlamydia psittaci* and psittacosis in birds *except*

   (A) Quarantine of psittacine birds imported into the United States
   (B) Only allowing sale of psittacine birds hatched in the United States
   (C) Testing of birds for *Chlamydia psittaci* infection
   (D) Controlling the shipment of psittacine birds
   (E) Putting tetracycline in the feed of psittacine birds

3. All of the following statements about perinatal *Chlamydia trachomatis* infections are correct *except*

   (A) Between 15 and 40% of infants born to infected women develop inclusion conjunctivitis
   (B) Between 10 and 20% of infants born to infected women develop infant pneumonia
   (C) The incubation period for *Chlamydia trachomatis* inclusion conjunctivitis is 1–2 days
   (D) The incubation period for infant pneumonia is typically 2–12 weeks
   (E) Ocular prophylaxis with erythromycin or tetracycline for neonatal *Neisseria gonorrhoeae* infection is generally not effective against neonatal *Chlamydia trachomatis* infection
   (F) Infant pneumonia due to *Chlamydia trachomatis* often presents with a staccato cough

4. An adolescent female came to the clinic because of a new and unusual vaginal discharge. She had recently become sexually active and had two new partners during the previous month. On pelvic examination a purulent discharge was seen at the opening of her endocervical canal. Which of the following statements about this case is most correct?

   (A) A serologic test for syphilis is not indicated because her symptoms are not those of syphilis
   (B) A Gram stain of her endocervical specimen would show *Chlamydia trachomatis* inside polymorphonuclear cells
   (C) The differential diagnosis includes infection with *Neisseria gonorrhoeae* or *Chlamydia trachomatis*
   (D) The endocervical specimen should be analyzed for herpes simplex
   (E) Initial treatment is with ampicillin

5. The following statements about trachoma are correct *except*

   (A) It follows chronic or recurrent eye infection with *Chlamydia trachomatis*
   (B) Millions of people worldwide have trachoma
   (C) Trachoma is readily prevented by a chlamydial vaccine
   (D) Progression of trachoma can be slowed by intermittent treatment with azithromycin
   (E) Trachoma involves scarring of the conjunctiva, eyelid deformities, and eyelash injury to the cornea

6. Elimination of blinding trachoma involves all of the following *except*

   (A) Periodic administration of azithromycin
   (B) Face washing and hygiene
   (C) Periodic culture screening of conjunctiva swab specimens for *Chlamydia trachomatis*
   (D) Environment improvements to sewage systems to decrease the number of flies
   (E) Surgery on deformed eyelids

7. Which one of the following statements about *Chlamydophila pneumoniae* is most correct?

   (A) Transmission from person to person is by the airborne route
   (B) It makes glycogen-rich inclusions that stain with iodine
   (C) There are multiple serovars including three that cause a systemic illness
   (D) They are resistant to macrolides
   (E) The reservoir is house cats

8. The serovars of *Chlamydia trachomatis* generally can be divided into groups representing their clinical infections/anatomic site infected. Which of the following statements about the *Chlamydia trachomatis* serovars is most correct?

   (A) There is no immunologic cross-reaction between *Chlamydia trachomatis* serovars A, B, Ba, and D and the *Chlamydophila pneumoniae* serovar.
   (B) Serovars L1, L2, and L3 are associated with lymphogranuloma venereum
   (C) The same *Chlamydia trachomatis* serovars are associated with blinding trachoma and sexually transmitted infections
   (D) The antibody titer rise seen beginning about around 6–8 years follows infections with *Chlamydia trachomatis* serovars D–K

9. In the United States, it has long been known that the positive seroprevalence for *Chlamydia trachomatis* infection increases greatly during the primary school years (ages 6–10). A likely reason for this is

   (A) Frequent adenovirus infections
   (B) Increased incidence of infections with *Chlamydia trachomatis*

(C) Cross-reactive antibodies with M protein of group A streptococci (*Streptococcus pyogenes*)

(D) Children often have psittacosis

(E) Frequent infections with *Chlamydophila pneumoniae*

10. All of the following statements about lymphogranuloma venereum (LGV) are correct *except*

   (A) Chronic LGV proctitis can lead to rectal strictures and fistula formation

   (B) The disease is more common in northern latitudes

   (C) There may be marked systemic symptoms including fever, nausea, vomiting, headache, and meningismus

   (D) Chronic inflammation with LGV can lead to lymphatic obstruction

   (E) Inguinal lymph nodes may become enlarged and matted, draining pus through the skin

   (F) A few days or weeks after exposure the disease manifests itself as a genital papule or vesicle

Answers

1. A	4. C	7. A	10. B
2. B	5. C	8. B	
3. C	6. C	9. E	

REFERENCES

Chlamydial diseases. Vol 2 Part III, Section C. In *Mandell, Douglas, and Bennett's Principles and Practice of Infectious Diseases*, 7th ed. Mandell GL, et al (editors). Elsevier, 2010.

Essig A: *Chlamydia* and *Chlamydophila*. In *Manual of Clinical Microbiology*, 9th ed. Murray PR et al (editors). ASM Press, 2007.

Antimicrobial Chemotherapy

28

Drugs have been used for the treatment of infectious diseases since the 17th century (eg, quinine for malaria, emetine for amebiasis); however, chemotherapy as a science began in the first decade of the 20th century with understanding of the principles of selective toxicity, the specific chemical relationships between microbial pathogens and drugs, the development of drug resistance, and the role of combined therapy. Experiments led to the arsphenamines for syphilis, the first planned chemotherapeutic regimen.

The current era of antimicrobial chemotherapy began in 1935 with the discovery of the sulfonamides. In 1940, it was demonstrated that penicillin, discovered in 1929, could be an effective therapeutic substance. During the next 25 years, research on chemotherapeutic agents centered largely around substances of microbial origin called antibiotics. The isolation, concentration, purification, and mass production of penicillin were followed by the development of streptomycin, tetracyclines, chloramphenicol, and many other agents. These substances were originally isolated from filtrates of media in which their respective molds had grown. Synthetic modification of previously described drugs has been prominent in the development of new antimicrobial agents.

Antimicrobial agents commonly employed in treatment of bacterial infections are presented in this chapter. The chemotherapy of viruses, fungi, and parasites is discussed in Chapters 30, 45, and 46, respectively. Additional comments on antimicrobial susceptibility testing for bacteria are to be found in Chapter 47.

MECHANISMS OF ACTION OF ANTIMICROBIAL DRUGS

Antimicrobial drugs act in one of several ways: by selective toxicity, by inhibition of cell membrane synthesis and function, by inhibition of protein synthesis, or by inhibition of nucleic acid synthesis.

SELECTIVE TOXICITY

An ideal antimicrobial agent exhibits selective toxicity, which means that the drug is harmful to a pathogen without being harmful to the host. Often, selective toxicity is relative rather than absolute; this implies that a drug in a concentration tolerated by the host may damage an infecting microorganism.

Selective toxicity may be a function of a specific receptor required for drug attachment, or it may depend on the inhibition of biochemical events essential to the pathogen but not to the host. The mechanisms of action of antimicrobial drugs can be discussed under four headings:

1. Inhibition of cell wall synthesis.
2. Inhibition of cell membrane function.
3. Inhibition of protein synthesis (ie, inhibition of translation and transcription of genetic material).
4. Inhibition of nucleic acid synthesis.

INHIBITION OF CELL WALL SYNTHESIS

Bacteria have a rigid outer layer, the cell wall. The cell wall maintains the shape and size of the microorganism, which has a high internal osmotic pressure. Injury to the cell wall (eg, by lysozyme) or inhibition of its formation may lead to lysis of the cell. In a hypertonic environment (eg, 20% sucrose), damaged cell wall formation leads to formation of spherical bacterial "protoplasts" from gram-positive organisms or "spheroplasts" from gram-negative organisms; these forms are limited by the fragile cytoplasmic membrane. If such **protoplasts** or **spheroplasts** are placed in an environment of ordinary tonicity, they take up fluid rapidly, swell, and may explode. Specimens from patients being treated with cell wall-active antibiotics often show swollen or misshapen bacteria.

The cell wall contains a chemically distinct complex polymer "mucopeptide" ("peptidoglycan") consisting of polysaccharides and a highly cross-linked polypeptide. The polysaccharides regularly contain the amino sugars N-acetylglucosamine and acetylmuramic acid. The latter is found only in bacteria. To the amino sugars are attached short peptide chains. The final rigidity of the cell wall is imparted by cross-linking of the peptide chains (eg, through pentaglycine bonds) as a result of transpeptidation reactions carried out by several enzymes. The peptidoglycan layer is much thicker in the cell wall of gram-positive than of gram-negative bacteria.

All β-lactam drugs are selective inhibitors of bacterial cell wall synthesis and therefore active against growing bacteria. This inhibition is only one of several different activities of these drugs, but it is the best understood. The initial step in drug action consists of binding of the drug to cell receptors (**penicillin-binding proteins; PBPs**). There are three to six PBPs (MW $4–12 \times 10^5$), some of which are transpeptidation enzymes. Different receptors have different affinities for a drug, and each may mediate a different effect. For example, attachment of penicillin to one PBP may result chiefly in abnormal elongation of the cell, whereas attachment to another PBP may lead to a defect in the periphery of the cell wall, with resulting cell lysis. PBPs are under chromosomal control, and mutations may alter their number or their affinity for β-lactam drugs.

After a β-lactam drug has attached to one or more receptors, the transpeptidation reaction is inhibited and peptidoglycan synthesis is blocked. The next step probably involves removal or inactivation of an inhibitor of autolytic enzymes in the cell wall. This activates the lytic enzyme and results in lysis if the environment is isotonic. In a markedly hypertonic environment, the microbes change to protoplasts or spheroplasts, covered only by the fragile cell membrane. In such cells, synthesis of proteins and nucleic acids may continue for some time.

The inhibition of the transpeptidation enzymes by penicillins and cephalosporins may be due to a structural similarity of these drugs to acyl-d-alanyl-d-alanine. The transpeptidation reaction involves loss of a d-alanine from the pentapeptide.

The remarkable lack of toxicity of β-lactam drugs to mammalian cells must be attributed to the absence, in animal cells, of a bacterial type cell wall, with its peptidoglycan. The difference in susceptibility of gram-positive and gram-negative bacteria to various penicillins or cephalosporins probably depends on structural differences in their cell walls (eg, amount of peptidoglycan, presence of receptors and lipids, nature of cross-linking, activity of autolytic enzymes) that determine penetration, binding, and activity of the drugs.

Resistance to penicillins may be determined by the organism's production of penicillin-destroying enzymes (β-lactamases). **β-Lactamases** open the β-lactam ring of penicillins and cephalosporins and abolish their antimicrobial activity. β-Lactamases have been described for many species of gram-positive and gram-negative bacteria. Some β-lactamases are plasmid-mediated (eg, penicillinase of *S aureus*), while others are chromosomally mediated (eg, many species of gram-negative bacteria). All of the more than 30 plasmid-mediated β-lactamases are produced constitutively and have a high propensity to move from one species of bacteria to another (eg, β-lactamase-producing *Neisseria gonorrhoeae, Haemophilus influenzae,* and enterococci). Chromosomally mediated β-lactamases may be constitutively produced (eg, *Bacteroides,*

Acinetobacter), or they may be inducible (eg, *Enterobacter, Citrobacter, Pseudomonas*).

There is one group of β-lactamases that is occasionally found in certain species of gram-negative bacilli, usually *Klebsiella pneumoniae* and *Escherichia coli.* These enzymes are termed **extended-spectrum β-lactamases (ESBLs)** because they confer upon the bacteria the additional ability to hydrolyze the β-lactam rings of cefotaxime, ceftazidime, or aztreonam.

The classification of β-lactamases is complex, based upon the genetics, biochemical properties, and substrate affinity for a β-lactamase inhibitor (clavulanic acid). Clavulanic acid, sulbactam, and tazobactam are β-lactamase inhibitors that have a high affinity for and irreversibly bind some β-lactamases (eg, penicillinase of *S aureus*) but are not hydrolyzed by the β-lactamase. These inhibitors protect simultaneously present hydrolyzable penicillins (eg, ampicillin, amoxicillin, and ticarcillin) from destruction. Certain penicillins (eg, cloxacillin) also have a high affinity for β-lactamases.

Shortly after their first description almost three decades ago, the most common extended spectrum β-lactamases were of the Class A TEM and SHV plasmid mediated types (see Table 28–1). Currently throughout much of the world the CTX-M enzymes have become more prevalent. Of most concern is the emergence of *Klebsilla pneumoniae* carbapenemases (KPC) which are ESBL-type enzymes that confer resistance to third and fourth generation cephalosporins and carbapenems. This resistance mechanism is plasmid mediated and has spread nosocomially among many hospitals throughout the United States and other countries.

There are two other types of resistance mechanisms. One is due to the absence of some penicillin receptors (PBPs) and occurs as a result of chromosomal mutation; the other results from failure of the β-lactam drug to activate the autolytic enzymes in the cell wall. As a result, the organism is inhibited but not killed. Such **tolerance** has been observed especially with staphylococci and certain streptococci.

Examples of agents acting by inhibition of cell wall synthesis are penicillins, the cephalosporins, vancomycin, and cycloserine. Several other drugs, including bacitracin, teicoplanin, vancomycin, ristocetin, and novobiocin, inhibit early steps in the biosynthesis of the peptidoglycan. Since the early stages of synthesis take place inside the cytoplasmic membrane, these drugs must penetrate the membrane to be effective.

INHIBITION OF CELL MEMBRANE FUNCTION

The cytoplasm of all living cells is bounded by the cytoplasmic membrane, which serves as a selective permeability barrier, carries out active transport functions, and thus controls the internal composition of the cell. If the functional integrity

TABLE 28–1 Classification of β-Lactamases

Bush-Jacoby Medeiros System Group	Enzyme Type	Inhibition by Clavulanate	Ambler System	Main Attributes
1	Cephalosporinase	NO	C	Chromosomal; resistant to all β-lactams except carbapenems
2a	Penicillinase	YES	A (serine)	Staphylococcal penicillinase
2b	Broad-spectrum	YES	A	TEM-1, TEM-2, SHV-1
2be	Extended-spectrum	YES	A	TEM-3-160,SHV2-101
2br	Inhibitor-resistant	Diminished	A	Inhibitor resistant TEM
2c	Carbenicillinase	YES	A	Carbenicillin-hydrolyzing
2d	Cloxacillinase	YES	D or A	Cloxacillin-hydrolyzing (OXA)
2e	Cephalosporinase	YES	A	Cephalosporinases
2f	Carbapenemase	YES	A	Carbapenemases inhibited by clavulanate (eg, IMP-1)
3a, 3b, 3c	Metalloenzymes	NO	B	Zinc-dependent carbapenemases
4	Penicillinase	NO	A	Misc enzymes

Modified with permission from Perez F, Endimiani A, Hujer KM, Bonomo RA. Curr Opin Pharmacol 2007;7:459, andOpal SM and Pop-Vicas A.Molecular mechanisms of antibiotic resistance in bacteria. In: *Mandell, Douglas and Bennett's Principles and Practice of Infectious Diseases,* 7th ed. Mandell GL, Bennett JE, Dolin R (editors). p 282–283, 2010.

of the cytoplasmic membrane is disrupted, macromolecules and ions escape from the cell, and cell damage or death ensues. The cytoplasmic membrane of bacteria and fungi has a structure different from that of animal cells and can be more readily disrupted by certain agents. Consequently, selective chemotherapy is possible.

Detergents, which contain lipophilic and hydrophilic groups, disrupt cytoplasmic membranes and kill the cell (Chapter 4). One class of antibiotics, the polymyxins, consists of detergent-like cyclic peptides that selectively damage membranes containing phosphatidylethanolamine, a major component of bacterial membranes. A number of antibiotics specifically interfere with biosynthetic functions of the cytoplasmic membranes—eg, nalidixic acid and novobiocin inhibit DNA synthesis, and novobiocin also inhibits teichoic acid synthesis.

A third class of membrane-active agents is the ionophores, compounds that permit rapid diffusion of specific cations through the membrane. Valinomycin, for example, specifically mediates the passage of potassium ions. Some ionophores act by forming hydrophilic pores in the membrane; others act as lipid-soluble ion carriers that behave as though they shuttle back and forth within the membrane. Ionophores can kill cells by discharging the membrane potential, which is essential for oxidative phosphorylation, as well as for other membrane-mediated processes; they are not selective for bacteria but act on the membranes of all cells.

Daptomycin is a new lipopeptide antibiotic that is rapidly bactericidal by binding to the cell membrane in a calcium-dependent manner causing depolarization of bacterial membrane potential. This leads to intracellular potassium release. Currently this agent is approved for use in the treatment of

Staphylococcus aureus blood stream infections and skin and soft-tissue infections caused by gram-positive bacteria, particularly those organisms that are highly resistant to β-lactam agents and vancomycin.

Other examples of agents acting by inhibition of cell membrane function are amphotericin B, colistin, and the imidazoles and triazoles.

INHIBITION OF PROTEIN SYNTHESIS

It is established that erythromycins, lincomycins, tetracyclines, aminoglycosides, and chloramphenicol can inhibit protein synthesis in bacteria. The precise mechanisms of action are not fully established for these drugs.

Bacteria have 70S ribosomes, whereas mammalian cells have 80S ribosomes. The subunits of each type of ribosome, their chemical composition, and their functional specificities are sufficiently different to explain why antimicrobial drugs can inhibit protein synthesis in bacterial ribosomes without having a major effect on mammalian ribosomes.

In normal microbial protein synthesis, the mRNA message is simultaneously "read" by several ribosomes that are strung out along the mRNA strand. These are called **polysomes**.

Examples of drugs acting by inhibition of protein synthesis are the erythromycins, lincomycins, tetracyclines, glycylcyclines, aminoglycosides, and chloramphenicol.

Aminoglycosides

The mode of action of streptomycin has been studied far more intensively than that of other aminoglycosides, but all probably act similarly. The first step is the attachment of the aminoglycoside to a specific receptor protein (P 12 in the case of streptomycin) on the 30S subunit of the microbial ribosome. Second, the aminoglycoside blocks the normal activity of the "initiation complex" of peptide formation (mRNA + formyl methionine + tRNA). Third, the mRNA message is misread on the "recognition region" of the ribosome; consequently, the wrong amino acid is inserted into the peptide, resulting in a nonfunctional protein. Fourth, aminoglycoside attachment results in the breakup of polysomes and their separation into **monosomes** incapable of protein synthesis. These activities occur more or less simultaneously, and the overall effect is usually an irreversible event—killing of the bacterium.

Chromosomal resistance of microbes to aminoglycosides principally depends on the lack of a specific protein receptor on the 30S subunit of the ribosome. Plasmid-dependent resistance to aminoglycosides depends on the production by the microorganism of adenylylating, phosphorylating, or acetylating enzymes that destroy the drugs. A third type of resistance consists of a "permeability defect," an outer membrane change that reduces active transport of the aminoglycoside into the cell so that the drug cannot reach the ribosome. Often this is plasmid-mediated.

Macrolides, Azalides, & Ketolides

These drugs (erythromycins, azithromycin, clarithromycin, and roxithromycin and the ketolide, telithromycin) bind to the 50S subunit of the ribosome, and the binding site is a 23S rRNA. They may interfere with formation of initiation complexes for peptide chain synthesis or may interfere with aminoacyl translocation reactions. Some macrolide-resistant bacteria lack the proper receptor on the ribosome (through methylation of the rRNA). This may be under plasmid or chromosomal control.

Lincomycins

Clindamycin binds to the 50S subunit of the microbial ribosome and resembles macrolides in binding site, antibacterial activity, and mode of action. Chromosomal mutants are resistant because they lack the proper binding site on the 50S subunit.

Tetracyclines

Tetracyclines bind to the 30S subunit of microbial ribosomes. They inhibit protein synthesis by blocking the attachment of charged aminoacyl-tRNA. Thus, they prevent introduction of new amino acids to the nascent peptide chain. The action is usually inhibitory and reversible upon withdrawal of the drug. Resistance to tetracyclines occurs by three mechanisms—efflux, ribosomal protection, and chemical modification. The first two are the most important and occur as follows: Efflux pumps, located in the bacterial

cell cytoplasmic membrane, are responsible for pumping the drug out of the cell. *Tet* gene products are responsible for protecting the ribosome, likely through mechanisms that induce conformational changes. These conformational changes either prevent binding of the tetracyclines or cause their dissociation from the ribosome. This is often plasmid-controlled. Mammalian cells do not actively concentrate tetracyclines.

Glycylcyclines

The glycylcyclines are synthetic analogues of the tetracyclines. The agent that is available for use in the United States and Europe is tigecycline, a derivative of minocycline. The glycylcyclines inhibit protein synthesis in a manner similar to the tetracyclines; however, they are bactericidal, likely due to their more avid binding to the ribosome. Tigecycline is active against a broad range of gram-positive and gram-negative bacteria, including strains resistant to the typical tetracyclines. The clinical activity of this agent is still undergoing investigation, but currently its major use appears to be in the treatment of skin and skin structure infections and in intra-abdominal infections, particularly caused by bacterial pathogens resistant to a variety of other antimicrobial agents.

Chloramphenicol

Chloramphenicol binds to the 50S subunit of the ribosome. It interferes with the binding of new amino acids to the nascent peptide chain, largely because chloramphenicol inhibits peptidyl transferase. Chloramphenicol is mainly bacteriostatic, and growth of microorganisms resumes when the drug is withdrawn. Microorganisms resistant to chloramphenicol produce the enzyme chloramphenicol acetyltransferase, which destroys drug activity. The production of this enzyme is usually under control of a plasmid.

Streptogramins

Quinupristin/dalfopristin is a combination of two pristinamycin derivatives. These two agents act synergistically to achieve bactericidal activity against gram-positive bacteria not seen with either agent alone. The mechanism of action appears to be irreversible binding to different sites on the 50S ribosome.

Oxazolidinones

The oxazolidinones are a relatively new class of antimicrobial agents that possess a unique mechanism of inhibition of protein synthesis primarily in gram-positive bacteria. These compounds interfere with translation by inhibiting the formation of N-formylmethionyl-tRNA, the initiation complex at the 30S ribosome. Linezolid is the agent that is currently commercially available.

INHIBITION OF NUCLEIC ACID SYNTHESIS

Examples of drugs acting by inhibition of nucleic acid synthesis are the quinolones, pyrimethamine, rifampin, sulfonamides, trimethoprim, and trimetrexate. Rifampin inhibits bacterial growth by binding strongly to the DNA-dependent RNA polymerase of bacteria. Thus, it inhibits bacterial RNA synthesis. Rifampin resistance results from a change in RNA polymerase due to a chromosomal mutation that occurs with high frequency. The mechanism of rifampin action on viruses is different. It blocks a late stage in the assembly of poxviruses.

All quinolones and fluoroquinolones inhibit microbial DNA synthesis by blocking DNA gyrase.

For many microorganisms, p-aminobenzoic acid (PABA) is an essential metabolite. The specific mode of action of PABA involves an adenosine triphosphate (ATP)-dependent condensation of a pteridine with PABA to yield dihydropteroic acid, which is subsequently converted to folic acid. PABA is involved in the synthesis of folic acid, an important precursor to the synthesis of nucleic acids. Sulfonamides are structural analogs of PABA and inhibit dihydropteroate synthetase.

p-Aminobenzoic acid (PABA) | **Basic ring structure of sulfonamides**

Sulfonamides can enter into the reaction in place of PABA and compete for the active center of the enzyme. As a result, nonfunctional analogs of folic acid are formed, preventing further growth of the bacterial cell. The inhibiting action of sulfonamides on bacterial growth can be counteracted by an excess of PABA in the environment (competitive inhibition). Animal cells cannot synthesize folic acid and must depend upon exogenous sources. Some bacteria, like animal cells, are not inhibited by sulfonamides. Many other bacteria, however, synthesize folic acid as mentioned above and consequently are susceptible to action by sulfonamides.

Trimethoprim (3,4,5-trimethoxybenzylpyrimidine) inhibits dihydrofolic acid reductase 50,000 times more efficiently in bacteria than in mammalian cells. This enzyme reduces dihydrofolic to tetrahydrofolic acid, a stage in the sequence leading to the synthesis of purines and ultimately of DNA. Sulfonamides and trimethoprim each can be used alone to inhibit bacterial growth. If used together, they produce sequential blocking, resulting in a marked enhancement (synergism) of activity. Such mixtures of sulfonamide (five parts) plus trimethoprim (one part) have been used in the treatment of pneumocystis pneumonia, malaria, shigella enteritis, systemic salmonella infections, urinary tract infections, and many others.

Pyrimethamine also inhibits dihydrofolate reductase, but it is more active against the enzyme in mammalian cells and therefore is more toxic than trimethoprim. Pyrimethamine plus sulfonamide or clindamycin is the current treatment of choice in toxoplasmosis and some other protozoal infections.

RESISTANCE TO ANTIMICROBIAL DRUGS

There are many different mechanisms by which microorganisms might exhibit resistance to drugs.

(1) Microorganisms produce enzymes that destroy the active drug. *Examples:* Staphylococci resistant to penicillin G produce a β-lactamase that destroys the drug. Other β-lactamases are produced by gram-negative rods. Gram-negative bacteria resistant to aminoglycosides (by virtue of a plasmid) produce adenylylating, phosphorylating, or acetylating enzymes that destroy the drug.

(2) Microorganisms change their permeability to the drug. *Examples:* Tetracyclines accumulate in susceptible bacteria but not in resistant bacteria. Resistance to polymyxins is also associated with a change in permeability to the drugs. Streptococci have a natural permeability barrier to aminoglycosides. This can be partly overcome by the simultaneous presence of a cell wall-active drug, eg, a penicillin. Resistance to amikacin and to some other aminoglycosides may depend on a lack of permeability to the drugs, apparently due to an outer membrane change that impairs active transport into the cell.

(3) Microorganisms develop an altered structural target for the drug (see also [5], below). *Examples:* Erythromycin-resistant organisms have an altered receptor on the 50S subunit of the ribosome, resulting from methylation of a 23S ribosomal RNA. Resistance to some penicillins and cephalosporins may be a function of the loss or alteration of PBPs. Penicillin resistance in *Streptococcus pneumoniae* and enterococci is due to altered PBPs.

(4) Microorganisms develop an altered metabolic pathway that bypasses the reaction inhibited by the drug. *Example:* Some sulfonamide-resistant bacteria do not require extracellular PABA but, like mammalian cells, can utilize preformed folic acid.

(5) Microorganisms develop an altered enzyme that can still perform its metabolic function but is much less affected by the drug. *Example:* In trimethoprim-resistant bacteria, the dihydrofolic acid reductase is inhibited far less efficiently than in trimethoprim-susceptible bacteria.

ORIGIN OF DRUG RESISTANCE

Nongenetic Origin of Drug Resistance

Active replication of bacteria is required for most antibacterial drug actions. Consequently, microorganisms that are metabolically inactive (nonmultiplying) may be phenotypically resistant to drugs. However, their offspring are fully susceptible. ***Example:*** Mycobacteria often survive in tissues for many years after infection yet are restrained by the host's defenses and do not multiply. Such "persisting" organisms are resistant to treatment and cannot be eradicated by drugs. Yet if they start to multiply (eg, following suppression of cellular immunity in the patient), they are fully susceptible to the same drugs.

Microorganisms may lose the specific target structure for a drug for several generations and thus be resistant. ***Example:*** Penicillin-susceptible organisms may change to cell wall-deficient L forms during penicillin administration. Lacking cell walls, they are resistant to cell wall-inhibitor drugs (penicillins, cephalosporins) and may remain so for several generations. When these organisms revert to their bacterial parent forms by resuming cell wall production, they are again susceptible to penicillin.

Microorganisms may infect the host at sites where antimicrobials are excluded or are not active. ***Examples:*** Aminoglycosides such as gentamicin are not effective in treating salmonella enteric fevers because the salmonellae are intracellular and the aminoglycosides do not enter the cells. Similarly, only drugs that enter cells are effective in treating Legionnaires' disease because of the intracellular location of *Legionella pneumophila*.

Genetic Origin of Drug Resistance

Most drug-resistant microbes emerge as a result of genetic change and subsequent selection processes by antimicrobial drugs.

Chromosomal Resistance

This develops as a result of spontaneous mutation in a locus that controls susceptibility to a given antimicrobial drug. The presence of the antimicrobial drug serves as a selecting mechanism to suppress susceptible organisms and favor the growth of drug-resistant mutants. Spontaneous mutation occurs with a frequency of 10^{-12} to 10^{-7} and thus is an infrequent cause of the emergence of clinical drug resistance in a given patient. However, chromosomal mutants resistant to rifampin occur with high frequency (about 10^{-7} to 10^{5}). Consequently, treatment of bacterial infections with rifampin as the sole drug often fails. Chromosomal mutants are most commonly resistant by virtue of a change in a structural receptor for a drug. Thus, the P 12 protein on the 30S subunit of the bacterial ribosome serves as a receptor for streptomycin attachment. Mutation in the gene controlling that structural protein results in streptomycin resistance. Mutation can also result in the loss of PBPs, making such mutants resistant to β-lactam drugs.

Extrachromosomal Resistance

Bacteria often contain extrachromosomal genetic elements called plasmids. Their features are described in Chapter 7.

Some plasmids carry genes for resistance to one—and often several—antimicrobial drugs. Plasmid genes for antimicrobial resistance often control the formation of enzymes capable of destroying the antimicrobial drugs. Thus, plasmids determine resistance to penicillins and cephalosporins by carrying genes for the formation of β-lactamases. Plasmids code for enzymes that acetylate, adenylate, or phosphorylate various aminoglycosides; for enzymes that determine the active transport of tetracyclines across the cell membrane; and for others.

Genetic material and plasmids can be transferred by transduction, transformation, and conjugation. These processes are discussed in Chapter 7.

CROSS-RESISTANCE

Microorganisms resistant to a certain drug may also be resistant to other drugs that share a mechanism of action. Such relationships exist mainly between agents that are closely related chemically (eg, different aminoglycosides) or that have a similar mode of binding or action (eg, macrolides-lincomycins). In certain classes of drugs, the active nucleus of the chemical is so similar among many congeners (eg, tetracyclines) that extensive cross-resistance is to be expected.

LIMITATION OF DRUG RESISTANCE

Emergence of drug resistance in infections may be minimized in the following ways: (1) by maintaining sufficiently high levels of the drug in the tissues to inhibit both the original population and first-step mutants; (2) by simultaneously administering two drugs that do not give cross-resistance, each of which delays the emergence of mutants resistant to the other drug (eg, rifampin and isoniazid in the treatment of tuberculosis); and (3) by avoiding exposure of microorganisms to a particularly valuable drug by limiting its use, especially in hospitals.

CLINICAL IMPLICATIONS OF DRUG RESISTANCE

A few examples will illustrate the impact of the emergence of drug-resistant organisms and their selection by the widespread use of antimicrobial drugs.

Gonococci

When sulfonamides were first employed in the late 1930s for the treatment of gonorrhea, virtually all isolates of gonococci were susceptible and most infections were cured. A few years later, most strains had become resistant to sulfonamides, and gonorrhea was rarely curable by these drugs. Most gonococci were still highly susceptible to penicillin. Over the next decades, there was a gradual increase in resistance to penicillin, but large doses of that drug were still curative. In the 1970s, β-lactamase-producing gonococci appeared, first in the Philippines and in West Africa, and then spread to form endemic foci worldwide. Such infections could not be treated effectively by penicillin but were treated with spectinomycin. Resistance to spectinomycin has appeared. Third-generation cephalosporins or quinolones are recommended to treat gonorrhea. However, the emergence of quinolone resistance in some geographic locations has subsequently limited their use.

Meningococci

Until 1962, meningococci were uniformly susceptible to sulfonamides, and these drugs were effective for both prophylaxis and therapy. Subsequently, sulfonamide-resistant meningococci spread widely, and the sulfonamides have now lost their usefulness against meningococcal infections. Penicillins remain effective for therapy, and rifampin is employed for prophylaxis. However, rifampin-resistant meningococci have emerged (as high as 27% of isolates) which may then cause invasive infections.

Staphylococci

In 1944, most staphylococci were susceptible to penicillin G, though a few resistant strains had been observed. After massive use of penicillin, 65–85% of staphylococci isolated from hospitals in 1948 were β-lactamase producers and thus resistant to penicillin G. The advent of β-lactamase-resistant penicillins (eg, nafcillin) provided a temporary respite, but infections due to nafcillin-resistant staphylococci are common. Presently, penicillin-resistant staphylococci include not only those acquired in hospitals but also 80–90% of those isolated in the community. These organisms also tend to be resistant to other drugs, eg, tetracyclines. Nafcillin-resistant staphylococci are common in tertiary hospitals. Vancomycin has been the major drug used for treatment of nafcillin-resistant *S aureus* infections, but recovery of isolates with intermediate resistance and the reports of several cases of high-level resistance to vancomycin have spurred the search for newer agents.

Pneumococci

S pneumoniae was uniformly susceptible to penicillin G until 1963, when relatively penicillin-resistant strains were found in New Guinea. Penicillin-resistant pneumococci subsequently were found in South Africa, Japan, Spain, and later worldwide. In the United States, approximately 15% of pneumococci are resistant to penicillin G (MICs of >2 μg/mL) and approximately 18% are intermediate (MICs of 0.1–1 μg/mL). The penicillin resistance is due to altered penicillin-binding proteins. Penicillin resistance in pneumococci tends to be clonal. Pneumococci also are frequently resistant to trimethoprim-sulfamethoxazole, erythromycin and tetracycline. Isolated quinolone resistance is also beginning to emerge.

Enterococci

The enterococci have intrinsic resistance to multiple antimicrobials: penicillin G and ampicillin with high MICs; cephalosporins with very high MICs; low-level resistance to aminoglycosides; and resistance to trimethoprim-sulfamethoxazole in vivo. The enterococci also have shown acquired resistance to almost all if not all other antimicrobials as follows: altered PBPs and resistance to β-lactams; high-level resistance to aminoglycosides; and resistance to fluoroquinolones, macrolides, azalides, and tetracyclines. Some enterococci have acquired a plasmid that encodes for β-lactamase and are fully resistant to penicillin and ampicillin. Of greatest importance is the development of resistance to vancomycin, which has become common in Europe and North America though there is geographic variation in the percentages of enterococci that are vancomycin-resistant. *Enterococcus faecium* is the species that is most commonly vancomycin-resistant. In outbreaks of infections due to vancomycin-resistant enterococci, the isolates may be clonal or genetically diverse. Resistance to the streptogramins (quinupristin-dalfopristin) also occurs in enterococci.

Gram-Negative Enteric Bacteria

Most drug resistance in enteric bacteria is attributable to the widespread transmission of resistance plasmids among different genera. About half the strains of *Shigella* species in many parts of the world are now resistant to multiple drugs.

Salmonellae carried by animals have developed resistance also, particularly to drugs (especially tetracyclines) incorporated into animal feeds. The practice of incorporating drugs into animal feeds causes farm animals to grow more rapidly but is associated with an increase in drug-resistant enteric organisms in the fecal flora of farm workers. A concomitant rise in drug-resistant salmonella infections in Britain led to a restriction on antibiotic supplements in animal feeds. Continued use of tetracycline supplements in animal feeds in the United States may contribute to the spread of resistance plasmids and of drug-resistant salmonellae.

Plasmids carrying drug resistance genes occur in many gram-negative bacteria of the normal gut flora. The abundant use of antimicrobial drugs—particularly in hospitalized patients—leads to the suppression of drug-susceptible organisms in the gut flora and favors the persistence and growth of drug-resistant bacteria, including *Enterobacter, Klebsiella, Proteus, Pseudomonas,* and *Serratia*—and fungi. Such organisms present particularly difficult problems in granulocytopenic and immunocompromised patients. The closed environments of hospitals favor transmission of such resistant organisms through personnel and fomites as well as by direct contact.

Mycobacterium tuberculosis

Primary drug resistance in *M tuberculosis* occurs in about 10% of isolates and most commonly is to isoniazid or streptomycin. Resistance to rifampin or ethambutol is less common. Isoniazid and rifampin are the primary drugs used in most standard treatment regimens; other first-line drugs are pyrazinamide, ethambutol, and streptomycin. Resistance to isoniazid and rifampin is considered multiple drug resistance. In the United States, multiple drug resistance of *M tuberculosis* has significantly decreased. Worldwide, the highest rates of multidrug-resistant tuberculosis have been reported from Eastern European countries, particularly among countries of the former Soviet Union. Poor compliance with drug treatment is a major factor in the development of drug resistance during therapy. Control of multidrug-resistant tuberculosis is a significant worldwide problem.

ANTIMICROBIAL ACTIVITY IN VITRO

Antimicrobial activity is measured in vitro in order to determine (1) the potency of an antibacterial agent in solution, (2) its concentration in body fluids or tissues, and (3) the susceptibility of a given microorganism to known concentrations of the drug.

FACTORS AFFECTING ANTIMICROBIAL ACTIVITY

Among the many factors that affect antimicrobial activity in vitro, the following must be considered, because they significantly influence the results of tests.

pH of Environment

Some drugs are more active at acid pH (eg, nitrofurantoin); others, at alkaline pH (eg, aminoglycosides, sulfonamides).

Components of Medium

Sodium polyanetholsulfonate (in blood culture media) and other anionic detergents inhibit aminoglycosides. PABA in tissue extracts antagonizes sulfonamides. Serum proteins bind penicillins in varying degrees, ranging from 40% for methicillin to 98% for dicloxacillin. Addition of NaCl to the medium enhances the detection of methicillin resistance in *S aureus*.

Stability of Drug

At incubator temperature, several antimicrobial agents lose their activity. Penicillins are inactivated slowly, whereas aminoglycosides and ciprofloxacin are quite stable for long periods.

Size of Inoculum

In general, the larger the bacterial inoculum, the lower the apparent "susceptibility" of the organism. Large bacterial populations are less promptly and completely inhibited than small ones. In addition, a resistant mutant is much more likely to emerge in large populations.

Length of Incubation

In many instances, microorganisms are not killed but only inhibited upon short exposure to antimicrobial agents. The longer incubation continues, the greater the chance for resistant mutants to emerge or for the least susceptible members of the antimicrobial population to begin multiplying as the drug deteriorates.

Metabolic Activity of Microorganisms

In general, actively and rapidly growing organisms are more susceptible to drug action than those in the resting phase. Metabolically inactive organisms that survive long exposure to a drug may have offspring that are fully susceptible to the same drug.

MEASUREMENT OF ANTIMICROBIAL ACTIVITY

Determination of the susceptibility of a bacterial pathogen to antimicrobial drugs can be done by one of two principal methods: dilution or diffusion. It is important to use a standardized method that controls for all the factors that affect antimicrobial activity; in the United States, the tests are performed according to the methods of the Clinical and Laboratory Standards Institute (CLSI) (formerly the National Committee for Clinical Laboratory Standards [NCCLS]). These tests also are discussed in Chapter 47.

Using an appropriate standard test organism and a known sample of drug for comparison, these methods can be employed to estimate either the potency of antibiotic in the sample or the susceptibility of the microorganism.

Dilution Method

Graded amounts of antimicrobial substances are incorporated into liquid or solid bacteriologic media. Commonly, twofold ($\log_2$) dilutions of the antimicrobial substances are used. The media are subsequently inoculated with test bacteria and incubated. The end point is taken as that amount of antimicrobial substance required to inhibit the growth of—or to kill—the test bacteria. Agar dilution susceptibility tests are time consuming, and their use is limited to special circumstances. Broth dilution tests were cumbersome and little used when dilutions had to be made in test tubes; however, the advent of prepared broth dilution series for many different drugs in microdilution plates has greatly enhanced and simplified the method. The advantage of microbroth dilution tests is that they permit a quantitative result to be reported, indicating the amount of a given drug necessary to inhibit (or kill) the microorganisms tested.

Diffusion Method

The most widely used method is the disk diffusion test. A filter paper disk containing a measured quantity of a drug is placed on the surface of a solid medium that has been inoculated on the surface with the test organism. After incubation, the diameter of the clear zone of inhibition surrounding the disk is taken as a measure of the inhibitory power of the drug against the particular test organism. This method is subject to many physical and chemical factors in addition to the simple interaction of drug and organisms (eg, the nature of the medium and diffusibility, molecular size, and the stability of the drug). Nevertheless, standardization of conditions permits determination of the susceptibility of the organism.

Interpretation of the results of diffusion tests must be based on comparisons between dilution and diffusion methods. Such comparisons have led to the establishment of reference standards. Linear regression lines can express the relationship between log of minimum inhibitory concentration in dilution tests and diameter of inhibition zones in diffusion tests.

Use of a single disk for each antibiotic with careful standardization of the test conditions permits the report of susceptible or resistant for a microorganism by comparing the size of the inhibition zone against a standard of the same drug.

Inhibition around a disk containing a certain amount of antimicrobial drug does not imply susceptibility to that same concentration of drug per milliliter of medium, blood, or urine.

ANTIMICROBIAL ACTIVITY IN VIVO

Analysis of the activity of antimicrobial agents in vivo is much more complex than the circumstances in vitro. The activity involves not only the drug and organism but also a third factor, the host. Drug–pathogen and host–pathogen relationships are discussed in the following paragraphs. Host–drug relationships (absorption, excretion, distribution, metabolism, and toxicity) are dealt with mainly in pharmacology texts.

DRUG–PATHOGEN RELATIONSHIPS

Several important interactions between drug and pathogen have been discussed in the preceding pages. The following are additional important in vivo factors.

Environment

In the host, varying environmental influences affect microorganisms located in different tissues and in different parts of the body—in contrast to the test tube or Petri dish, where the environment is constant for all members of a microbial population.

Therefore, the response of the microbial population is much less uniform within the host than in the test tube.

A. State of Metabolic Activity

In the body, the state of metabolic activity is diverse—undoubtedly, many organisms exist at a low level of biosynthetic activity and are thus relatively insusceptible to drug action. These "dormant" microorganisms often survive exposure to high concentrations of drugs and subsequently may produce a clinical relapse of the infection.

B. Distribution of Drug

In the body, the antimicrobial agent is unequally distributed in tissues and fluids. Many drugs do not reach the central nervous system effectively. The concentration in urine is often much greater than the concentration in blood or other tissue. The tissue response induced by the microorganism may protect it from the drug. Necrotic tissue or pus may adsorb the drug and thus prevent its contact with bacteria.

C. Location of Organisms

In the body, microorganisms often are located within tissue cells. Drugs enter tissue cells at different rates. Some (eg, tetracyclines) reach about the same concentration inside monocytes as in the extracellular fluid. With others (eg, gentamicin), the drug probably does not enter host cells at all. This is in contrast to the test tube, where microorganisms come into direct contact with the drug.

D. Interfering Substances

The biochemical environment of microorganisms in the body is very complex and results in significant interference with drug action. The drug may be bound by blood and tissue proteins or phospholipids; it may also react with nucleic acids in pus and may be physically adsorbed onto exudates, cells, and necrotic debris. In necrotic tissue, the pH may be highly acid and thus unfavorable for drug action (eg, aminoglycosides).

Concentration

In the body, microorganisms are not exposed to a constant concentration of drug; in the test tube they are.

A. Absorption

The absorption of drugs from the intestinal tract (if taken by mouth) or from tissues (if injected) is irregular. There is also a continuous excretion as well as inactivation of the drug. Consequently, the levels of drug in body compartments fluctuate continually, and the microorganisms are exposed to varying concentrations of the antimicrobial agent.

B. Distribution

The distribution of drugs varies greatly with different tissues. Some drugs penetrate certain tissues poorly (eg, central nervous system, prostate). Drug concentrations following systemic administration may therefore be inadequate for effective treatment. On surface wounds or mucous membranes such as the conjunctivae, local (topical) application of poorly absorbed drugs permits highly effective local concentrations without toxic side effects. Alternatively, some drugs applied topically on surface wounds are well absorbed. Drug concentrations in urine are often much higher than in blood.

C. Variability of Concentration

It is critical to maintain an effective concentration of a drug where the infecting microorganisms proliferate. This concentration must be maintained for a sufficient length of time to eradicate the microorganisms. Because the drug is administered intermittently and is absorbed and excreted irregularly, the levels constantly fluctuate at the site of infection. In order to maintain sufficient drug concentrations for a sufficient time, the time–dose relationship must be considered. The larger each individual drug dose, the longer the permissible interval between doses. The smaller the individual dose, the shorter the interval that will ensure adequate drug levels.

D. Postantibiotic Effect

The postantibiotic effect is the delayed regrowth of bacteria after exposure to antimicrobial agents. It is a property of most antimicrobials, except that most β-lactams do not show the postantibiotic effect with gram-negative bacilli. The carbapenems do have a postantibiotic effect with the gram-negative bacilli. Aminoglycosides and fluoroquinolones have prolonged (up to several hours) in vitro postantibiotic effects against gram-negative bacilli.

HOST–PATHOGEN RELATIONSHIPS

Host–pathogen relationships may be altered by antimicrobial drugs in several ways.

Alteration of Tissue Response

The inflammatory response of the tissue to infections may be altered if the drug suppresses the multiplication of microorganisms but does not eliminate them from the body. An acute process may in this way be transformed into a chronic one. Conversely, the suppression of inflammatory reactions in tissues by impairment of cell-mediated immunity in recipients of tissue transplants or antineoplastic therapy or by immunocompromise as a result of disease (eg, AIDS) causes enhanced susceptibility to infection and impaired responsiveness to antimicrobial drugs.

Alteration of Immune Response

If an infection is modified by an antimicrobial drug, the immune response of the host may also be altered. One example illustrates this phenomenon: Pharyngeal infection with β-hemolytic group A streptococci is followed frequently by the development of antistreptococcal antibodies, and if there is a hyperimmune response the infection may be followed by rheumatic fever. If the infective process can be interrupted early and completely with antimicrobial drugs, the development of an immune response and of rheumatic fever can be prevented (presumably by rapid elimination of the antigen). Drugs and dosages that rapidly eradicate the infecting streptococci (eg, penicillin) are more effective in preventing rheumatic fever than those which merely suppress the microorganisms temporarily (eg, tetracycline).

Alteration of Microbial Flora

Antimicrobial drugs affect not only the microorganisms causing disease but also susceptible members of the normal microbial flora. An imbalance is thus created that in itself may lead to disease. A few examples are of interest.

1. In hospitalized patients who receive antimicrobials, the normal microbial flora is suppressed. This creates a partial void that is filled by the organisms most prevalent in the environment, particularly drug-resistant gram-negative aerobic bacteria (eg, pseudomonads, staphylococci). Such superinfecting organisms subsequently may produce serious drug-resistant infections.
2. In women taking antibiotics by mouth, the normal vaginal flora may be suppressed, permitting marked overgrowth of candida. This leads to unpleasant local inflammation (vulvovaginitis) and itching that are difficult to control.
3. In the presence of urinary tract obstruction, the tendency to bladder infection is great. When such urinary tract infection due to a sensitive microorganism (eg, *E coli*) is treated with an appropriate drug, the organism may be eradicated. However, it often happens that reinfection due to another drug-resistant gram-negative bacillus occurs after the drug-sensitive microorganisms are eliminated. A similar process accounts for respiratory tract superinfections in patients given antimicrobials for chronic bronchitis.
4. In persons receiving antimicrobial drugs for several days, parts of the normal intestinal flora may be suppressed. Drug-resistant organisms may establish themselves in the bowel in great numbers and may precipitate serious enterocolitis (*Clostridium difficile*, etc).

CLINICAL USE OF ANTIBIOTICS

SELECTION OF ANTIBIOTICS

The rational selection of antimicrobial drugs depends upon the following considerations.

Diagnosis

A specific etiologic diagnosis must be formulated. This can often be done on the basis of a clinical impression. Thus, in typical lobar pneumonia or acute urinary tract infection, the relationship between clinical picture and causative agent is sufficiently constant to permit selection of the antibiotic of choice on the basis of clinical impression alone. Even in these cases, however, as a safeguard against diagnostic error, it is preferable to obtain a representative specimen for bacteriologic study before giving antimicrobial drugs.

In most infections, the relationship between causative agent and clinical picture is not constant. It is therefore important to obtain proper specimens for bacteriologic identification of the causative agent. As soon as such specimens have been secured, chemotherapy can be started on the basis of the "best guess." Once the causative agent has been identified by laboratory procedures, the initial regimen can be modified as necessary.

The "best guess" of a causative organism is based on the following considerations, among others: (1) the site of infection (eg, pneumonia, urinary tract infection); (2) the age of the patient (eg, meningitis: neonatal, young child, adult); (3) the place where the infection was acquired (hospital versus community); (4) mechanical predisposing factors (indwelling vascular catheter, urinary catheter, ventilator, exposure to vector); and (5) predisposing host factors (immunodeficiency, corticosteroids, transplant, cancer chemotherapy, etc).

When the causative agent of a clinical infection is known, the drug of choice can often be selected on the basis of current clinical experience. At other times, laboratory tests for antibiotic susceptibility (see below) are necessary to determine the drug of choice.

Susceptibility Tests

Laboratory tests for antibiotic susceptibility are indicated in the following circumstances: (1) when the microorganism recovered is of a type that is often resistant to antimicrobial drugs (eg, gram-negative enteric bacteria); (2) when an infectious process is likely to be fatal unless treated specifically (eg, meningitis, septicemia); and (3) in certain infections where eradication of the infectious organisms requires the use of drugs that are rapidly bactericidal, not merely bacteriostatic (eg, infective endocarditis). The basic principles of antimicrobial susceptibility testing are presented earlier in this chapter. Additional laboratory aspects of antimicrobial susceptibility testing are discussed in Chapter 47.

DANGERS OF INDISCRIMINATE USE

The indications for administration of antibiotics must sometimes be qualified by the following concerns:

1. Widespread sensitization of the population, with resulting hypersensitivity, anaphylaxis, rashes, fever, blood disorders, cholestatic hepatitis, and perhaps collagen-vascular diseases.
2. Changes in the normal flora of the body, with disease resulting from "superinfection" due to overgrowth of drug-resistant organisms.
3. Masking serious infection without eradicating it. For example, the clinical manifestations of an abscess may be suppressed while the infectious process continues.
4. Direct drug toxicity (eg, granulocytopenia or thrombocytopenia with cephalosporins and penicillins and renal damage or auditory nerve damage due to aminoglycosides).
5. Development of drug resistance in microbial populations, chiefly through the elimination of drug-sensitive microorganisms from antibiotic-saturated environments (eg, hospitals) and their replacement by drug-resistant microorganisms.

ANTIMICROBIAL DRUGS USED IN COMBINATION

Indications

Possible reasons for employing two or more antimicrobials simultaneously instead of a single drug are as follows:

1. To give prompt treatment in desperately ill patients suspected of having a serious microbial infection. A good guess, usually based on available antibiogram data, about the most probable two or three pathogens, is made, and drugs are aimed at those organisms. Before such treatment is started, it is essential that adequate specimens be obtained for identifying the etiologic agent in the laboratory. Suspected gram-negative or staphylococcal sepsis in immunocompromised patients and bacterial meningitis in children are foremost indications in this category.
2. To delay the emergence of microbial mutants resistant to one drug in chronic infections by the use of a second or third noncross-reacting drug. The most prominent example is active tuberculosis.
3. To treat mixed infections, particularly those following massive trauma or those involving vascular structures. Each drug is aimed at an important pathogenic microorganism.
4. To achieve bactericidal synergism or to provide bactericidal action (see below). In a few infections, eg, enterococcal sepsis, a combination of drugs is more likely to eradicate the infection than either drug used alone. Such synergism is only partially predictable, and a given drug pair may be synergistic for only a single microbial strain. Occasionally, simultaneous use of two drugs permits significant reduction in dose and thus avoids toxicity but still provides satisfactory antimicrobial action.

Disadvantages

The following disadvantages of using antimicrobial drugs in combinations must always be considered:

1. The physician may feel that since several drugs are already being given, everything possible has been done for the patient, leading to relaxation of the effort to establish a specific diagnosis. It may also give a false sense of security.
2. The more drugs that are administered, the greater the chance for drug reactions to occur or for the patient to become sensitized to drugs.
3. The cost is unnecessarily high.
4. Antimicrobial combinations usually accomplish no more than an effective single drug.
5. Very rarely, one drug may antagonize a second drug given simultaneously (see below).

Mechanisms

When two antimicrobial agents act simultaneously on a homogeneous microbial population, the effect may be one of the following: (1) indifference, ie, the combined action is no greater than that of the more effective agent when used alone; (2) addition, ie, the combined action is equivalent to the sum of the actions of each drug when used alone; (3) synergism, ie, the combined action is significantly greater than the sum of both effects; or (4) antagonism, ie, the combined action is less than that of the more effective agent when used alone. All these effects may be observed in vitro (particularly in terms of bactericidal rate) and in vivo.

The effects that can be achieved with combinations of antimicrobial drugs vary with different combinations and are specific for each strain of microorganism. Thus, no combination is uniformly synergistic.

Combined therapy should not be used indiscriminately; every effort should be made to employ the single antibiotic of choice. In resistant infections, detailed laboratory study can at times define synergistic drug combinations that may be essential to eradicate the microorganisms.

Antimicrobial synergism can occur in several types of situations. Synergistic drug combinations must be selected by complex laboratory procedures.

1. Two drugs may sequentially block a microbial metabolic pathway. Sulfonamides inhibit the use of extracellular *p*-aminobenzoic acid by some microbes for the synthesis of folic acid. Trimethoprim or pyrimethamine inhibits the next metabolic step, the reduction of dihydro- to tetrahydrofolic acid. The simultaneous use of a sulfonamide plus trimethoprim is effective in some bacterial (shigellosis, salmonellosis, serratia) and some other infections (pneumocystosis, malaria). Pyrimethamine plus a sulfonamide or clindamycin is used in toxoplasmosis.

2. A drug such as a cell wall-inhibitor (a penicillin or cephalosporin) may enhance the entry of an aminoglycoside into bacteria and thus produce synergistic effects. Penicillins enhance the uptake of gentamicin or streptomycin by enterococci. Thus, ampicillin plus gentamicin may be essential for the eradication of *Enterococcus faecalis,* particularly in endocarditis. Similarly, piperacillin plus tobramycin may be synergistic against some strains of *Pseudomonas.*

3. One drug may affect the cell membrane and facilitate the entry of the second drug. The combined effect may then be greater than the sum of its parts. For example, amphotericin has been synergistic with flucytosine against certain fungi (eg, *Cryptococcus, Candida*).

4. One drug may prevent the inactivation of a second drug by microbial enzymes. Thus, inhibitors of β-lactamase (eg, clavulanic acid, sulbactam, tazobactam) can protect amoxicillin, ticarcillin, or piperacillin from inactivation by β-lactamases. In such circumstances, a form of synergism takes place.

Antimicrobial antagonism is sharply limited by time–dose relationships and is therefore a rare event in clinical antimicrobial therapy. Antagonism resulting in higher morbidity and mortality rates has been most clearly demonstrated in bacterial meningitis. It occurred when a bacteriostatic drug (which inhibited protein synthesis in bacteria) such as chloramphenicol or tetracycline was given with a bactericidal drug such as a penicillin or an aminoglycoside. Antagonism occurred mainly if the bacteriostatic drug reached the site of infection before the bactericidal drug; if the killing of bacteria was essential for cure; and if only minimal effective doses of either drug in the pair were present. Another example is combining β-lactam drugs in treatment of *Pseudomonas aeruginosa* infections (eg, imipenem and piperacillin, where imipenem is a potent β-lactamase inducer and the β-lactamase breaks down the less stable piperacillin).

ANTIMICROBIAL CHEMOPROPHYLAXIS

Anti-infective chemoprophylaxis implies the administration of antimicrobial drugs to prevent infection. In a broader sense, it also includes the use of antimicrobial drugs soon after the acquisition of pathogenic microorganisms (eg, after compound fracture) but before the development of signs of infection.

Useful chemoprophylaxis is limited to the action of a specific drug on a specific organism. An effort to prevent all types of microorganisms in the environment from establishing themselves only selects the most drug-resistant organisms as the cause of a subsequent infection. In all proposed uses of prophylactic antimicrobials, the risk of the patient's acquiring an infection must be weighed against the toxicity, cost, inconvenience, and enhanced risk of superinfection resulting from the prophylactic drug.

Prophylaxis in Persons of Normal Susceptibility Exposed to a Specific Pathogen

In this category, a specific drug is administered to prevent one specific infection. Outstanding examples are the injection of benzathine penicillin G intramuscularly once every 3–4 weeks to prevent reinfection with group A hemolytic streptococci in rheumatic patients; prevention of meningitis by eradicating the meningococcal carrier state with rifampin; prevention of syphilis by the injection of benzathine penicillin G; prevention of plague pneumonia by oral administration of tetracycline in persons exposed to infectious droplets; prevention of clinical rickettsial disease (but not of infection) by the daily ingestion of tetracycline during exposure; and prevention of leptospirosis with oral administration of doxycycline in a hyperendemic environment.

Early treatment of an asymptomatic infection is sometimes called prophylaxis. Thus, administration of isoniazid, 6–10 mg/kg/day (maximum, 300 mg/day) orally for 6 months, to an asymptomatic person who converts from a negative to a positive tuberculin skin test may prevent later clinically active tuberculosis.

Prophylaxis in Persons of Increased Susceptibility

Certain anatomic or functional abnormalities predispose to serious infections. It may be feasible to prevent or abort such infections by giving a specific drug for short periods. Some important examples are listed below.

A. Heart Disease

Persons with heart valve abnormalities or with prosthetic heart valves are unusually susceptible to implantation of microorganisms circulating in the bloodstream. This infective endocarditis can sometimes be prevented if the proper drug can be used during periods of bacteremia. Large numbers of viridans streptococci are pushed into the circulation during dental procedures and operations on the mouth or throat. At such times, the increased risk warrants the use of a prophylactic antimicrobial drug aimed at viridans streptococci. For example, amoxicillin taken orally before the procedure and 2 hours later can be effective. Persons allergic to penicillin can take erythromycin orally. Other oral and parenteral dosage schedules can be effective.

Enterococci cause 5–15% of cases of infective endocarditis. They reach the bloodstream from the urinary, gastrointestinal, or female genital tract. During procedures in these areas, if the patient is infected or colonized with enterococci, then ampicillin combined with an aminoglycoside (eg, gentamicin) should be given intramuscularly or intravenously 30 minutes before the procedure, to the person with significant heart valve abnormalities and prostheses.

During and after cardiac catheterization, blood cultures may be positive in 10–20% of patients. Many of these persons also have fever, but very few acquire endocarditis. Prophylactic antimicrobials do not appear to influence these events.

B. Respiratory Tract Disease

Trimethoprim-sulfamethoxazole orally or pentamidine by aerosol is used for prophylaxis for pneumocystis pneumonia in AIDS patients.

C. Recurrent Urinary Tract Infection

For certain women who are subject to frequently recurring urinary tract infections, the oral intake either daily or three times weekly of nitrofurantoin or trimethoprim-sulfamethoxazole can markedly reduce the frequency of symptomatic recurrences over long periods.

Certain women tend to develop symptoms of cystitis after sexual intercourse. The ingestion of a single dose of antimicrobial drug (nitrofurantoin, trimethoprim-sulfamethoxazole, etc) can prevent postcoital cystitis by early inhibition of growth of bacteria moved from the introitus into the proximal urethra or bladder during intercourse.

D. Opportunistic Infections in Severe Granulocytopenia

Immunocompromised patients receiving organ transplants or antineoplastic chemotherapy often develop profound leukopenia. When the neutrophil count falls below 1000/μL, they become unusually susceptible to opportunistic infections, most often gram-negative sepsis. Such persons are sometimes given a fluoroquinolone or cephalosporin or a drug combination (eg, vancomycin, gentamicin, cephalosporin) directed at the most prevalent opportunists at the earliest sign—or even without clinical evidence—of infection. This is continued for several days until the granulocyte count rises again. Several studies suggest that there is benefit from empiric therapy. Two clinical cases—liver and bone marrow transplants—presented in Chapter 48 illustrate the infections that occur in these patients and the antimicrobials used for prophylaxis and treatment.

Prophylaxis in Surgery

A major portion of all antimicrobial drugs used in hospitals is employed on surgical services with the stated intent of prophylaxis.

Several general features of surgical prophylaxis merit consideration:

1. In clean elective surgical procedures (ie, procedures during which no tissue bearing normal flora is traversed other than the prepared skin), the disadvantages of "routine" antibiotic prophylaxis (allergy, toxicity, superinfection) may outweigh the possible benefits except when hardware (eg, artificial hip joint) is being placed. However, even in

"clean" herniorrhaphy, a single preoperative dose of a cephalosporin resulted in measurable benefit.

2. Prophylactic administration of antibiotics should generally be considered only if the expected rate of infectious complications is 3–5%. An exception to this rule is the elective insertion of prostheses (cardiovascular, orthopedic), where a possible infection would have a catastrophic effect.

3. The initial dose of systemic prophylactic antibiotic should be given at the time of induction of anesthesia. An exception is elective colonic surgery, in which case oral antibiotics should be given hours before the procedure.

4. Prolonged administration of antimicrobial drugs tends to alter the normal flora of organ systems, suppressing the susceptible microorganisms and favoring the implantation of drug-resistant ones. Thus, antimicrobial prophylaxis should usually continue for no more than 1 day after the procedure and ideally should be given only intraoperatively.

5. Systemic levels of antimicrobial drugs usually do not prevent wound infection, pneumonia, or urinary tract infection if physiologic abnormalities or foreign bodies are present.

Topical antimicrobials for prophylaxis (intravenous catheter site, closed urinary drainage, within a surgical wound, acrylic bone cement, etc) have limited usefulness.

Recent studies have demonstrated increased morbidity and mortality with *S aureus* postsurgical wound infections, particularly if the infection is caused by methicillin-resistant *S aureus* (MRSA). Many hospitals perform presurgical nares surveillance screening for MRSA using either culture or nucleic acid detection. Patients who are found to be colonized are treated with mupirocin ointment to the nares for 3–5 days along with chlorhexidine for bathing in an attempt to eliminate colonization prior to the procedure.

Disinfectants

Disinfectants and antiseptics differ from systemically active antimicrobials in that they possess little selective toxicity: They are toxic not only for microbial pathogens but for host cells as well. Therefore, they can be used only to inactivate microorganisms in the inanimate environment or, to a limited extent, on skin surfaces. They cannot be administered systemically.

The antimicrobial action of disinfectants is determined by concentration, time, and temperature, and the evaluation of their effect may be complex. A few examples of disinfectants that are used in medicine or public health are listed in Table 28–2.

ANTIMICROBIAL DRUGS FOR SYSTEMIC ADMINISTRATION

Refer to Table 28–3 for a list of infecting organisms and their respective primary and alternative drug choices.

PENICILLINS

The penicillins are derived from molds of the genus *Penicillium* (eg, *Penicillium notatum*) and obtained by extraction of submerged cultures grown in special media. The most widely used natural penicillin is penicillin G. From fermentation brews of penicillium, 6-aminopenicillanic acid has been isolated on a large scale. This makes it possible to synthesize an almost unlimited variety of penicillin compounds by coupling the free amino group of the penicillanic acid to free carboxyl groups of different radicals.

All penicillins share the same basic structure (see 6-aminopenicillanic acid in Figure 28–1). A thiazolidine ring is attached to a β-lactam ring that carries a free amino group. The acidic radicals attached to the amino group can be split off by bacterial and other amidases. The structural integrity of the 6-aminopenicillanic acid nucleus is essential to the biologic activity of the compounds. If the β-lactam ring is enzymatically cleaved by β-lactamases (penicillinases), the resulting product, penicilloic acid, is devoid of antibacterial activity. However, it carries an antigenic determinant of the penicillins and acts as a sensitizing hapten when attached to carrier proteins.

The different radicals (R) attached to the aminopenicillanic acid determine the essential pharmacologic properties of the resulting drugs. The clinically important penicillins fall into four principal groups: (1) highest activity against gram-positive organisms, spirochetes, and some others but susceptible to hydrolysis by β-lactamases and acid-labile (eg, penicillin G); (2) relative resistance to β-lactamases but lower activity against gram-positive organisms and inactivity against gram-negatives (eg, nafcillin); (3) relatively high activity against both gram-positive and gram-negative organisms but destroyed by β-lactamases (eg, ampicillin, piperacillin); and (4) relative stability to gastric acid and suitable for oral administration (eg, penicillin V, cloxacillin, amoxicillin). Some representatives are shown in Figure 28–1. Most penicillins are dispensed as sodium or potassium salts of the free acid. Potassium penicillin G contains about 1.7 meq of K^+ per million units (2.8 meq/g). Procaine salts and benzathine salts of penicillin provide repository forms for intramuscular injection. In dry form, penicillins are stable, but solutions rapidly lose their activity and must be prepared fresh for administration.

Antimicrobial Activity

The initial step in penicillin action is binding of the drug to cell receptors. These receptors are PBPs, at least some of which are enzymes involved in transpeptidation reactions. From three to six (or more) PBPs per cell can be present. After penicillin molecules have attached to the receptors, peptidoglycan synthesis is inhibited as final transpeptidation is blocked. A final bactericidal event is the removal or inactivation of an inhibitor of autolytic enzymes in the cell wall. This activates the autolytic enzymes and results in cell lysis. Organisms

TABLE 28–2 Chemical Disinfectants, Antiseptics, and Topical Antimicrobial Agents

Disinfection of the inanimate environment	
Tabletops, instruments	Lysol or other phenolic compound
	Formaldehyde
	Aqueous glutaraldehyde
	Quaternary ammonium compounds
Excreta, bandages, bedpans	Sodium hypochlorite
	Lysol or other phenolic compound
Air	Propylene glycol mist or aerosol
	Formaldehyde vapor
Heat-sensitive instruments	Ethylene oxide gas (alkylates nucleic acids; residual gas must be removed by aeration)
Disinfection of skin or wounds	Washing with soap and water
	Soaps or detergents containing hexachlorophene or trichlorocarbanilide or chlorhexidine
	Tincture of iodine
	Ethyl alcohol; isopropyl alcohol
	Povidone-iodine (water-soluble)
	Peracids (hydrogen peroxide, peracetic acid)
	Nitrofurazone jelly or solution
Topical drugs to skin or mucous membranes	
In candidiasis	Nystatin cream
	Candicidin ointment
	Miconazole creams
In burns	Mafenide acetate cream
	Silver sulfadiazine
In dermatophytosis	Undecylenic acid powder or cream
	Tolnaftate cream
	Azole cream
In pyoderma	Bacitracin-neomycin-polymyxin ointment
	Potassium permanganate
In pediculosis	Malathion or permethrin lotion
In nasal decolonization	Mupirocin
Topical application of drugs to eyes	
For gonorrhea prophylaxis	Erythromycin or tetracycline ointment
For bacterial conjunctivitis	Sulfacetamide ointment
	Gentamicin or tobramycin ointment
	Ciprofloxacin ointment
	Moxifloxacin ophthalmic solution
	Gatifloxacin solution
	Levofloxacin solution

with defective autolysin function are inhibited but not killed by β-lactam drugs, and they are said to be "tolerant."

Since active cell wall synthesis is required for penicillin action, metabolically inactive microorganisms are not susceptible.

Penicillin G and penicillin V are often measured in units (1 million units = 0.6 g), but the semisynthetic penicillins are measured in grams. Whereas 0.002–1 µg/mL of penicillin G is lethal for a majority of susceptible gram-positive organisms, 10–100 times more is required to kill gram-negative bacteria (except neisseriae).

Resistance

Resistance to penicillins falls into several categories: (1) Production of β-lactamases by staphylococci, gram-negative bacteria, haemophili, gonococci, and others. More than 50 different β-lactamases are known, most of them produced under the control of bacterial plasmids. Some β-lactamases are inducible by the newer cephalosporins. (2) Lack of penicillin receptors (PBPs) or altered PBPs (eg, pneumococci, enterococci) or inaccessibility of receptors because of permeability barriers of bacterial outer membranes. These are often under chromosomal control. (3) Failure of activation of

TABLE 28–3   Drugs of Choice for Suspected or Proved Microbial Pathogens[a]

Suspected or Proved Etiologic Agent	Drug(s) of First Choice	Alternative Drug(s)
Gram-negative cocci		
Moraxella catarrhalis	Cefuroxime, a fluoroquinolone[c]	TMP-SMZ,[b] cefotaxime, ceftizoxime, cefpodoxime, an erythromycin,[d] a tetracycline,[e] azithromycin, amoxicillin-clavulanic acid, clarithromycin
Neisseria gonorrhoeae (gonococcus)	Ceftriaxone	Cefixime, cefotaxmine, penicillin G
Neisseria meningitidis (meningococcus)	Penicillin G[f]	Cefotaxime, ceftizoxime, ceftriaxone, ampicillin, chloramphenicol ,a fluroquinolone
Gram-positive cocci		
Streptococcus pneumoniae (pneumococcus)[h]	Penicillin G[f] or V; amoxicillin	An erythromycin,[d] a cephalosporin,[g] vancomycin, TMP-SMZ,[b] clindamycin, azithromycin, clarithromycin, a tetracycline,[e] imipenem, meropenem or ertapenem, quinupristin-dalfopristin, certain fluoroquinolones,[c] linezolid
Streptococcus, hemolytic, groups A, B, C, G	Penicillin G[f] or V; ampicillin	An erythromycin,[d] a cephalosporin,[g] vancomycin, clindamycin, azithromycin, clarithromycin, linezolid, daptomycin
Viridans streptococci	Penicillin G[f] ± gentamicin	A cephalosporin,[g] vancomycin
Staphylococcus, methicillin-resistant	Vancomycin ± gentamicin ± rifampin	TMP-SMZ,[b] doxycycline, a fluoroquinolone,[c] linezolid, quinupristin-dalfopristin, daptomycin, tigecycline
Staphylococcus, non-penicillinase-producing	Penicillin[f]	A cephalosporin,[h] vancomycin, imipenem, meropenem, a fluoroquinolone,[c] clindamycin
Staphylococcus, penicillinase producing	Penicillinase-resistant penicillin[i]	Vancomycin, a cephalosporin,[g] clindamycin, amoxicillin-clavulanic acid, ampicillin-sulbactam, piperacillin-tazobactam, imipenem, meropenem, a fluoroquinolone,[c] TMP-SMZ,[b] daptomycin, linezolid
Enterococcus faecalis	Ampicillin + gentamicin[j]	Vancomycin + gentamicin
Enterococcus faecium	Vancomycin + gentamicin[j]	Quinupristin-dalfopristin, linezolid; daptomycin
Gram-negative rods		
Acinetobacter	Imipenem or meropenem	Doxycycline, TMP-SMZ,[b] doxycycline, aminoglycosides,[k] ceftazidime, a fluoroquinolone,[c] piperacillin-tazobactam, sulbactam, colistin
Prevotella, oropharyngeal strains	Clindamycin	Penicillin,[f] metronidazole, cefoxitin, cefotetan
Bacteroides	Metronidazole	Chloramphenicol, imipenem, meropenem, ertapenem, ticarcillin-clavulanic acid, ampicillin-sulbactam, piperacillin-tazobactam; amoxicillin/clavulanate
Brucella	Tetracycline + rifampin[e]	TMP-SMZ[b] ± gentamicin; chloramphenicol ± gentamicin; doxycycline + gentamicin; ciprofloxacin + rifampin
Campylobacter jejuni	Erythromycin[d] or azithromycin	Tetracycline,[e] a fluoroquinolone,[c] gentamicin
Enterobacter	Imipenem, meropenem or cefepime	Aminoglycoside, ciprofloxacin, piperacillin/tazobactam, TMP-SMZ,[b] aztreonam, third-generation cephalosporin, tigecycline, aztreonam
Escherichia coli (sepsis)	Cefotaxime, ceftriaxone, ceftazidime, cefepime	Imipenem, meropenem or ertapenem, aminoglycosides,[k] a fluoroquinolone[c]
Escherichia coli (uncomplicated urinary infection)	Fluoroquinolones,[c] nitrofurantoin	TMP-SMZ,[b] oral cephalosporin, fosfomycin
Haemophilus (meningitis and other serious infections)	Cefotaxime, ceftriaxone	Chloramphenicol, meropenem

(Continued)

TABLE 28–3 **Drugs of Choice for Suspected or Proved Microbial Pathogens[a] (Continued)**

Suspected or Proved Etiologic Agent	Drug(s) of First Choice	Alternative Drug(s)
Gram-negative rods (continued)		
Haemophilus (respiratory infections, otitis)	TMP-SMZ[b]	Ampicillin, amoxicillin, doxycycline, azithromycin, clarithromycin, cefotaxime, ceftizoxime, ceftriaxone, cefuroxime, cefuroxime axetil, a fluoroquinolone, a tetracycline, amoxicillin/clavulanate
Helicobacter pylori	Proton pump inhibitor + clarithromycin + either amoxicillin or metronidazole	Bismuth subsalicylate + metronidazole + tetracycline HCl + proton pump inhibitor or H$_2$-blocker
Klebsiella pneumoniae	Cefotaxime, ceftriaxone, cefepime, or ceftazidime	TMP-SMZ,[b] aminoglycoside,[k] imipenem, meropenem or ertapenem, a fluoroquinolone,[c] piperacillin/tazobactam, aztreonam, ticarcillin/clavulanate, tigecycline
Legionella species (pneumonia)	Azithromycin, or fluoroquinolones[c] ± rifampin	TMP-SMZ,[b] doxycycline ± rifampin, erythromycin
Proteus mirabilis	Ampicillin	An aminoglycoside,[k] TMP-SMZ,[b] a fluoroquinolone,[e] a cephalosporin,[g] imipenem, meropenem or ertapenem, ticarcillin/clavulanate, piperacillin/tazobactam
Proteus vulgaris and other species (*Morganella, Providencia*)	Cefotaxime, ceftriaxone, ceftazidime, cefepime	Aminoglycoside,[k] TMP-SMZ,[b] a fluoroquinolone,[c] imipenem, meropenem, or ertapenem, aztreonam, ticarcillin/clavulanate, piperacillin/tazobactam, ampicillin/sulbactam, amoxicillin/clavulanate.
Pseudomonas aeruginosa	Aminoglycoside[k] + antipseudomonal penicillin[l]	Ceftazidime ± aminoglycoside; imipenem or meropenem ± aminoglycoside; aztreonam ± aminoglycoside; ciprofloxacin; cefepime
Burkholderia pseudomallei (melioidosis)	Ceftazidime, imipenem	Chloramphenicol + tetracycline,[e] TMP-SMZ,[b] amoxicillin-clavulanic acid, meropenem
Burkholderia mallei (glanders)	Streptomycin + tetracycline[e]	Chloramphenicol + streptomycin; imipenem
Salmonella (bacteremia)	Cefotaxime, ceftriaxone, or a fluoroquinolone[c]	TMP-SMZ,[b] ampicillin, chloramphenicol
Serratia	Imipenem or meropenem	TMP-SMZ,[b] aminoglycosides,[k] a fluoroquinolone,[c] ceftriaxone, cefotaxime, ceftizoxime, ceftazidime, cefepime
Shigella	A fluoroquinolone[c]	Ampicillin, TMP-SMZ,[b] ceftriaxone, azithromycin
Vibrio (cholera, sepsis)	Tetracycline[e]	TMP-SMZ,[b] a fluoroquinolone[c]
Yersinia pestis (plague)	Streptomycin ± a tetracycline[e]	Chloramphenicol, TMP-SMZ,[b] ciprofloxacin, gentamicin
Gram-positive rods		
Actinomyces	Penicillin[f]	Doxycycline,[e] clindamycin, erythromycin
Bacillus (including anthrax)	Penicillin[f] (ciprofloxacin or doxycycline for anthrax)	Erythromycin,[d] tetracycline,[e] a fluoroquinolone[c]
Bacillus anthracis	Ciprofloxacin, a tetracycline	Penicillin G, amoxicillin, erythromycin, imipenem, clindamycin, levofloxacin
Bacillus cereus (subtilis)	Vancomycin	Imipenem or meropenem, clindamycin
Clostridium (eg, gas gangrene, tetanus)	Penicillin G[f]; clindamycin	Metronidazole, chloramphenicol, imipenem, meropenem, or ertapenem
Corynebacterium diphtheriae	Erythromycin[d]	Penicillin G[f]
Corynebacterium jeikeium	Vancomycin	Penicillin G + gentamicin, erythromycin
Listeria monocytogenes	Ampicillin ± aminoglycoside[k]	TMP-SMZ[b]

(Continued)

TABLE 28–3   Drugs of Choice for Suspected or Proved Microbial Pathogens[a] (Continued)

Suspected or Proved Etiologic Agent	Drug(s) of First Choice	Alternative Drug(s)
Acid-fast rods		
Mycobacterium tuberculosis[m]	INH + rifampin + pyrazinamide ± ethambutol or streptomycin	A fluoroquinolone; cycloserine; capreomycin or kanamycin or amikacin; ethionamide; PAS
Mycobacterium leprae	Dapsone + rifampin ± clofazimine	Minocycline; ofloxacin; clarithromycin
Mycobacterium kansasii	INH + rifampin ± ethambutol or streptomycin	Ethionamide; cycloserine; clarithromycin, or azithromycin
Mycobacterium avium complex	Clarithromycin or azithromycin + one or more of the following: ethambutol ± rifabutin	Amikacin, ciprofloxacin
Mycobacterium fortuitum-chelonae	Amikacin + clarithromycin	Cefoxitin, sulfonamide, doxycycline, linezolid, rifampin, ethambutol
Nocardia	TMP-SMZ[b]	Imipenem or meropenem, sulfisoxazole, linezolid, a tetracycline, amikacin; ceftriaxone; cycloserine
Spirochetes		
Borrelia burgdorferi (Lyme disease)	Doxycycline, amoxicillin, cefuroxime axetil	Ceftriaxone, cetotaxime, penicillin G, azithromycin, clarithromycin
Borrelia recurrentis (relapsing fever)	Doxycycline[e]	Penicillin[f] G; erythromycin
Leptospira	Penicillin G[f]	Doxycycline,[e] ceftriaxone
Treponema pallidum (syphilis)	Penicillin G[f]	Doxycycline, ceftriaxone
Treponema pertenue (yaws)	Penicillin G[f]	Doxycycline[e]
Mycoplasmas	Erythromycin[d] or doxycycline; clarithromycin; azithromycin	A fluoroquinolone[c]
Chlamydiae		
Chlamydia psittaci	A tetracycline	Chloramphenicol
Chlamydia trachomatis (urethritis or pelvic inflammatory disease)	Doxycycline or azithromycin	Ofloxacin; erythromycin; amoxicillin
Chlamydia pneumoniae	A tetracycline, erythromycin,[d] clarithromycin, azithromycin	A fluoroquinolone[c,n]
Rickettsiae	Doxycycline	Chloramphenicol, a fluoroquinolone[c]

[a]Data from Med Lett Drugs Ther 5(Issue 57), May 2007.

[b]TMP-SMZ is a mixture of 1 part trimethoprim and 5 parts sulfamethoxazole.

[c]Fluoroquinolones include ciprofloxacin, ofloxacin, levofloxacin, moxifloxacin, gatifloxacin, and others (see text). Gatifloxacin, levofloxacin, and moxifloxacin have the best activity against gram-positive organisms, including penicillin-resistant *S pneumoniae* and methicillin-sensitive *S aureus*. Activity against enterococci and *S epidermidis* is variable. Ciprofloxacin has the best activity against *P aeruginosa*.

[d]Erythromycin estolate is best absorbed orally but carries the highest risk of hepatitis; erythromycin stearate and erythromycin ethylsuccinate are also available.

[e]All tetracyclines have similar activity against most microorganisms. Minocycline and doxycycline have increased activity against *S aureus*. Dosage is determined by rates of absorption and excretion of various preparations.

[f]Penicillin G is preferred for parenteral injection; penicillin V for oral administration—to be used only in treating infections due to highly sensitive organisms.

[g]Most intravenous cephalosporins (with the exception of ceftazidime) have good activity against gram-positive cocci.

[h]Intermediate and high-level resistance to penicillin has been described. Infections caused by strains with intermediate resistance may respond to high doses of penicillin, cefotaxime, or ceftriaxone. Infections caused by highly resistant strains should be treated with vancomycin ± rifampin. Many strains of penicillin-resistant pneumococci are resistant to erythromycin, macrolides, TMP-SMZ, and chloramphenicol.

[i]Parenteral nafcillin or oxacillin; oral dicloxacillin, cloxacillin, or oxacillin.

[j]Addition of gentamicin indicated only for severe enterococcal infections (eg, endocarditis, meningitis).

[k]Aminoglycosides—gentamicin, tobramycin, amikacin, netilmicin—should be chosen on the basis of local patterns of susceptibility.

[l]Antipseudomonal penicillins: ticarcillin, piperacillin.

[m]Resistance may be a problem, and susceptibility testing should be done.

[n]Ciprofloxacin has inferior antichlamydial activity compared with newer fluoroquinolones.

Site of amidase action

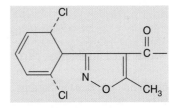

6-Aminopenicillanic acid

Site of penicillinase action
(break in β-lactam ring)

The following structures can each be substituted at the R to produce a new penicillin.

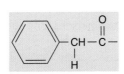

Penicillin G (benzylpenicillin):
High activity against gram-positive bacteria. Low activity against gram-negative bacteria. Acid-labile. Destroyed by β-lactamase. 60% protein-bound.

Oxacillin (no Cl atoms); cloxacillin (one Cl in structure); dicloxacillin (2 Cls in structure); flucloxacillin (one Cl and one F in structure) (isoxazolyl penicillins):
Similar to methicillin in β-lactamase resistance, but acid-stable. Can be taken orally. Highly protein-bound (95–98%).

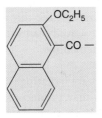

Nafcillin (ethoxynaphthamidopenicillin):
Similar to isoxazolyl penicillins. Less strongly protein-bound (90%). Can be given by mouth or by vein. Resistant to staphylococcal β-lactamase.

Ampicillin (alpha-aminobenzylpenicillin):
Similar to penicillin G (destroyed by β-lactamase) but acid-stable and more active against gram-negative bacteria. Carbenicillin has —COONa instead of —NH₂ group.

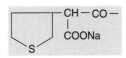

Ticarcillin:
Similar to carbenicillin but gives higher blood levels. Piperacillin, azlocillin, and mezlocillin resemble ticarcillin in action against gram-negative aerobes.

Amoxicillin:
Similar to ampicillin but better absorbed, gives higher blood levels.

FIGURE 28–1 Structures of some penicillins.

autolytic enzymes in cell wall, which can result in inhibition without killing bacteria (eg, tolerance of some staphylococci). (4) Failure to synthesize peptidoglycans, eg, in mycoplasmas, L forms, or metabolically inactive bacteria.

Absorption, Distribution, & Excretion

After intramuscular or intravenous administration, absorption of most penicillins is rapid and complete. After oral administration, only 5–30% of the dose of most penicillins is absorbed, depending on acid stability, binding to foods, presence of

buffers, etc. Amoxicillin is well absorbed. After absorption, penicillins are widely distributed in tissues and body fluids.

Special dosage forms have been designed for delayed absorption to yield drug levels for long periods. After a single intramuscular dose of benzathine penicillin, 1.5 g (2.4 million units), serum levels of 0.03 unit/mL are maintained for 10 days and levels of 0.005 unit/mL for 3 weeks. Procaine penicillin given intramuscularly yields therapeutic levels for 24 hours.

In many tissues, penicillin concentrations are similar to those in serum. Lower levels occur in the eyes, the prostate, and the central nervous system. However, in meningitis, penetration is enhanced, and levels of 0.5–5 μg/mL occur in the cerebrospinal fluid with a daily parenteral dose of 12 g.

Most of the penicillins are rapidly excreted by the kidneys. About 10% of renal excretion is by glomerular filtration and 90% by tubular secretion. The latter can be partially blocked by probenecid to achieve higher systemic and cerebrospinal fluid levels. In the newborn and in persons with renal failure, penicillin excretion is reduced and systemic levels remain elevated longer. Some penicillins (eg, nafcillin) are eliminated mainly by nonrenal mechanisms.

Clinical Uses

Penicillins are the most widely used antibiotics, particularly in the following areas.

Penicillin G is the drug of choice in most infections caused by streptococci, pneumococci, meningococci, spirochetes, clostridia, aerobic gram-positive rods, nonpenicillinase-producing staphylococci and gonococci, and actinomycetes.

Penicillin G is inhibitory for enterococci (*E faecalis*), but for bactericidal effects (eg, in enterococcal endocarditis) an aminoglycoside must be added. Penicillin G in ordinary doses is excreted into the urine in sufficiently high concentrations to inhibit some gram-negative organisms unless they produce a large amount of β-lactamases.

Benzathine penicillin G is a salt of very low solubility given intramuscularly for low but prolonged drug levels. A single injection of 1.2 million units (0.7 g) is satisfactory treatment for group A streptococcal pharyngitis and primary syphilis. The same injection once every 3–4 weeks is satisfactory prophylaxis against group A streptococcal reinfection in rheumatic fever patients.

Infection with β-lactamase-producing staphylococci is the only indication for the use of penicillinase-resistant penicillins, eg, nafcillin or oxacillin. Cloxacillin or dicloxacillin by mouth can be given for milder staphylococcal infections. Staphylococci resistant to oxacillin and nafcillin have the *mecA* gene and make a low-affinity penicillin-binding protein.

Oral amoxicillin is better absorbed than ampicillin and yields higher levels. Amoxicillin given together with clavulanic acid is active against β-lactamase-producing *H influenzae*. Ticarcillin resembles ampicillin but is more active against gram-negative rods. It is usually given in gram-negative sepsis in conjunction with an aminoglycoside (eg, gentamicin). Piperacillin is more effective against aerobic gram-negative rods, especially pseudomonads. Piperacillin combined with the β-lactamase inhibitor tazobactam has increased activity against some β-lactamase-producing gram-negative rods. The piperacillin-tazobactam combination however, is no more active against *P aeruginosa* than piperacillin alone.

Side Effects

Penicillins possess less direct toxicity than most of the other antimicrobial drugs. Most serious side effects are due to hypersensitivity.

All penicillins are cross-sensitizing and cross-reacting. Any material (including milk, cosmetics) containing penicillin may induce sensitization. The responsible antigens are degradation products (eg, penicilloic acid) bound to host protein. Skin tests with penicilloyl-polylysine, with alkaline hydrolysis products, and with undegraded penicillin identify many hypersensitive persons. Among positive reactors to skin tests, the incidence of major immediate allergic reactions is high. Such reactions are associated with cell-bound IgE antibodies. IgG antibodies to penicillin are common and are not associated with allergic reactions other than rare cases of hemolytic anemia. A history of a penicillin reaction in the past is not reliable, but the drug must be administered with caution to such persons, or a substitute drug should be chosen.

Allergic reactions may occur as typical anaphylactic shock, typical serum sickness type reactions (urticaria, joint swelling, angioneurotic edema, pruritus, respiratory embarrassment within 7–12 days of penicillin dosage), and a variety of skin rashes, fever, nephritis, eosinophilia, vasculitis, etc. The incidence of hypersensitivity to penicillin is negligible in children but may be 1–5% among adults in the United States. Acute anaphylactic life-threatening reactions are very rare (0.5%). Corticosteroids can sometimes suppress allergic manifestations to penicillins.

Very high doses may produce central nervous system concentrations that are irritating. In patients with renal failure, smaller doses may produce encephalopathy, delirium, and convulsions. With such doses, direct cation toxicity (K⁺) may also occur. Nafcillin occasionally causes granulocytopenia. Oral penicillins can cause diarrhea. High doses of penicillins may cause a bleeding tendency. Some penicillins have become obsolete because of their enhanced toxicities. Methicillin too frequently causes interstitial nephritis. Carbenicillin too frequently decreases normal platelet aggregation, which can lead to clinically significant bleeding.

CEPHALOSPORINS

Some cephalosporium fungi yield antimicrobial substances called cephalosporins. These are β-lactam compounds with a nucleus of 7-aminocephalosporanic acid (Figure 28–2) instead of the penicillins' 6-aminopenicillanic acid. Natural cephalosporins have low antibacterial activity, but the attachment of various R side-groups has resulted in the proliferation of an enormous array of drugs with varying pharmacologic properties and antimicrobial spectra and activity. Cephamycins are similar to cephalosporins but are derived from actinomycetes.

The mechanism of action of cephalosporins is analogous to that of penicillins: (1) binding to specific PBPs that serve as drug receptors on bacteria; (2) inhibiting cell wall synthesis by blocking the transpeptidation of peptidoglycan; and (3) activating autolytic enzymes in the cell wall that can produce lesions resulting in bacterial death. Resistance to cephalosporins can

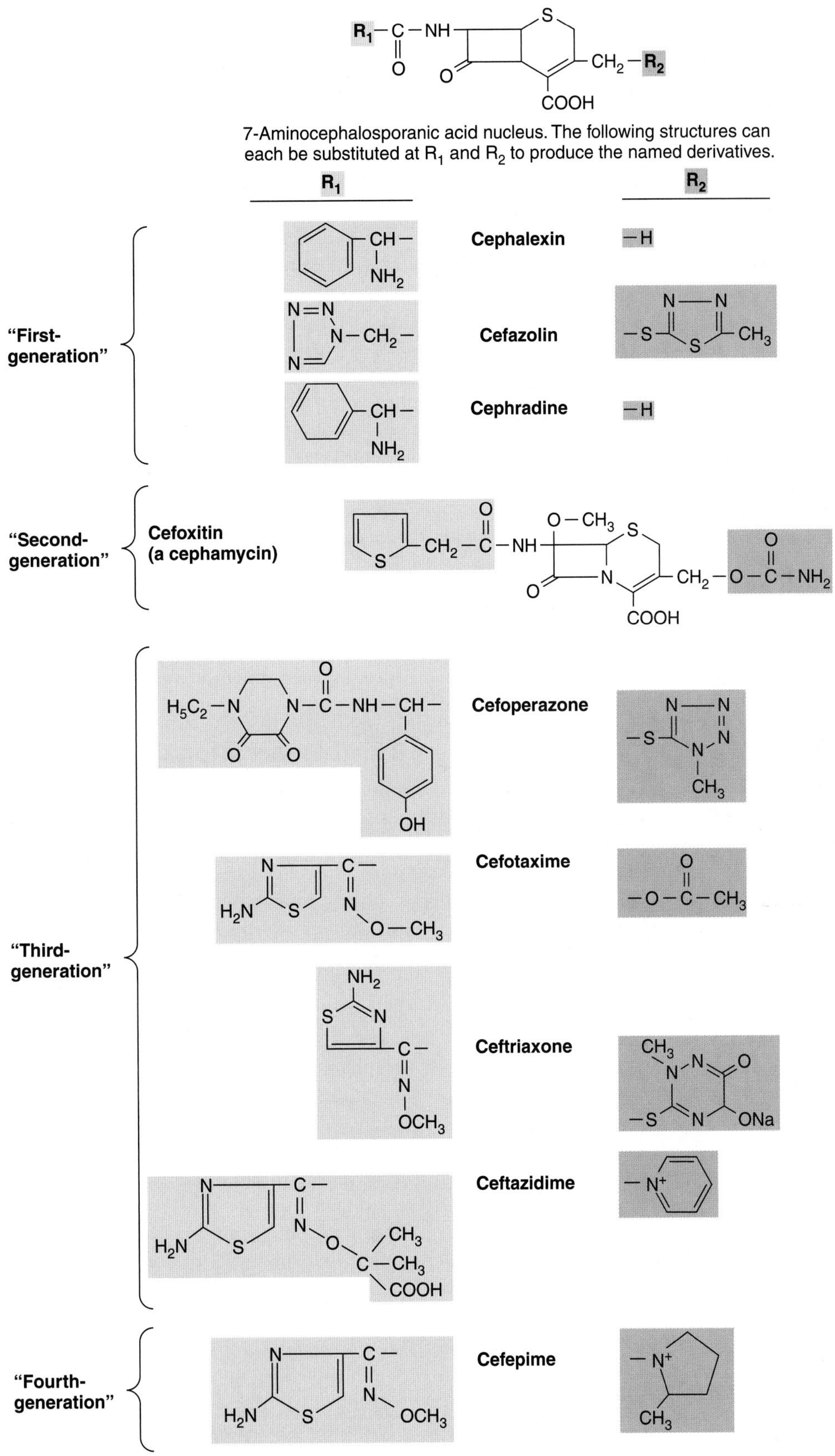

FIGURE 28–2 Structures of some cephalosporins.

be attributed to (1) poor permeation of bacteria by the drug; (2) lack of PBP for a specific drug; and (3) degradation of drug by β-lactamases, many of which exist. Certain second- and third-generation cephalosporins can induce special β-lactamases in gram-negative bacteria. In general, however, cephalosporins tend to be resistant to the β-lactamases produced by staphylococci and common gram-negative bacteria that hydrolyze and inactivate many penicillins.

For ease of reference, cephalosporins have been arranged into major groups, or "generations," discussed below (Table 28–4). Many cephalosporins are excreted mainly by the kidney and may accumulate and induce toxicity in renal insufficiency.

First-Generation Cephalosporins

First-generation cephalosporins are very active against gram-positive cocci—except enterococci and methicillin-resistant staphylococci (MRSA)—and moderately active against some gram-negative rods—primarily *E coli*, *Proteus*, and *Klebsiella*. Anaerobic cocci are often sensitive, but *Bacteroides fragilis* is not.

TABLE 28–4   Major Groups of Cephalosporins

First-generation
Cephalothin
Cephapirin
Cefazolin
Cephalexin[a]
Cephradine[a]
Cefadroxil
Second-generation
Cefamandole
Cefuroxime
Cefonicid
Ceforanide
Cefaclor[a]
Cefoxitin
Cefotetan
Cefprozil[a]
Cefuroxime axetil[a]
Cefmetazole
Third-generation
Cefotaxime
Ceftizoxime
Ceftriaxone
Ceftazidime
Cefoperazone
Cefixime[a]
Cefpodoxime proxetil[a]
Ceftibuten[a]
Cefdinir[a]
Fourth-generation
Cefepime

[a]Oral agents.

Cephalexin, cephradine, and cefadroxil are absorbed from the gut to a variable extent and can be used to treat urinary and respiratory tract infections. Other first-generation cephalosporins must be injected to give adequate levels in blood and tissues. Cefazolin is a choice for surgical prophylaxis because it gives the highest (90–120 μg/mL) levels with every 8-hour dosing. Cephalothin and cephapirin in the same dose give lower levels. None of the first-generation drugs penetrate the central nervous system, and they are not drugs of first choice for any infection.

Second-Generation Cephalosporins

The second-generation cephalosporins are a heterogeneous group. All are active against organisms covered by first-generation drugs but have extended coverage against gram-negative rods—including *Klebsiella* and *Proteus* but not *P aeruginosa*.

Some (not all) oral second-generation cephalosporins can be used to treat sinusitis and otitis caused by *H influenzae*, including β-lactamase-producing strains.

Cefoxitin and cefotetan are active against *B fragilis* and thus are used in mixed anaerobic infections, including peritonitis or pelvic inflammatory disease. However, resistance to these agents among the *B fragilis* group is increasing.

Third-Generation Cephalosporins

Third-generation cephalosporins have decreased activity against gram-positive cocci, except for *S pneumoniae*; enterococci are intrinsically resistant to cephalosporins and often produce superinfections during their use. Most third-generation cephalosporins are active against staphylococci, but ceftazidime is only weakly active. A major advantage of third-generation drugs is their enhanced activity against gram-negative rods. Where second-generation drugs tend to fail against *P aeruginosa*, ceftazidime or cefoperazone may succeed. Thus, third-generation drugs are very useful in the management of hospital-acquired gram-negative bacteremia. In immunocompromised patients, these drugs are often combined with an aminoglycoside. Ceftazidime may also be lifesaving in severe melioidosis (*Burkholderia pseudomallei* infection).

Another important distinguishing feature of several third-generation drugs—except cefoperazone—is the ability to reach the central nervous system and to appear in the spinal fluid in sufficient concentrations to treat meningitis caused by gram-negative rods. Cefotaxime, ceftriaxone, or ceftizoxime given intravenously may be used for management of gram-negative bacterial sepsis and meningitis.

Several new agents have recently been approved or are about to be so in the United States. Cefditoren is an oral third-generation cephalosporin with excellent activity against many gram-positive and gram-negative species. This agent has bactericidal activity and stability against many β-lactamase enzymes. Cefditoren is the most potent orally administered cephalosporin against *S pneumoniae*. Two agents claim activity against MRSA. These include ceftaroline and ceftobiprole. Ceftaroline has enhanced anti–gram-positive activity including MRSA and penicillin nonsusceptible pneumococci. Ceftobiprole has a spectrum of activity similar to that of other cephalosporins but, in addition, is active against MRSA, *E faecalis*, and penicillin-resistant *S pneumoniae*. All three agents will likely play a role in the treatment of skin and soft-tissue infections and community-acquired pneumonia.

Fourth-Generation Cephalosporins

Cefepime is the only fourth-generation cephalosporin now in clinical use in the United States. It has enhanced activity against *Enterobacter* and *Citrobacter* species that are resistant to third-generation cephalosporins. Cefepime has activity comparable to that of ceftazidime against *P aeruginosa*. The activity against streptococci and methicillin-susceptible staphylococci is greater than that of ceftazidime and comparable to that of the other third-generation compounds. Cefpirome is a fourth-generation cephalosporin available outside of the United States.

Adverse Effects of Cephalosporins

Cephalosporins are sensitizing and can elicit a variety of hypersensitivity reactions, including anaphylaxis, fever, skin rashes, nephritis, granulocytopenia, and hemolytic anemia. The frequency of cross-allergy between cephalosporins and penicillins is approximately 5%. Patients with minor penicillin allergy can often tolerate cephalosporins, but those with a history of anaphylaxis cannot.

Thrombophlebitis can occur after intravenous injection. Hypoprothrombinemia is frequent with cephalosporins that have a methylthiotetrazole group (eg, cefamandole, cefmetazole, cefotetan, cefoperazone). Oral administration of vitamin K (10 mg) twice weekly can prevent this complication. These same drugs can also cause severe disulfiram reactions, and use of alcohol must be avoided.

Since many second-, third-, and fourth-generation cephalosporins have little activity against gram-positive organisms, particularly enterococci, superinfection with these organisms and with fungi may occur.

OTHER β-LACTAM DRUGS

Monobactams

Monobactams have a monocyclic β-lactam ring and are resistant to β-lactamases. They are active against gram-negative rods but not against gram-positive bacteria or anaerobes. The first such drug to become available was aztreonam, which resembles aminoglycosides in activity and is given intravenously or intramuscularly every 8 or 12 hours. Patients with IgE-mediated penicillin allergy can tolerate it without reaction, and—apart from skin rashes and minor

aminotransferase disturbances—no major toxicity has been reported. Superinfections with staphylococci and enterococci can occur.

Carbapenems

These drugs are structurally related to β-lactam antibiotics. Imipenem, the first drug of this type, has good activity against many gram-negative rods, gram-positive organisms, and anaerobes. It is resistant to β-lactamases but is inactivated by dihydropeptidases in renal tubules. Consequently, it is administered together with a peptidase inhibitor, cilastatin.

Imipenem penetrates body tissues and fluids well, including cerebrospinal fluid. The drug is given intravenously every 6–8 hours and in reduced dosage in renal insufficiency. Imipenem may be indicated for infections due to organisms resistant to other drugs. *Pseudomonas* species rapidly develop resistance, and the concomitant use of an aminoglycoside is therefore required; however, this does not delay the development of resistance. Such a combination may be effective treatment for febrile neutropenic patients.

Adverse effects of imipenem include vomiting, diarrhea, skin rashes, and reactions at infusion sites. Excessive levels in patients with renal failure may lead to seizures. Patients allergic to penicillins may be allergic to imipenem as well.

Meropenem is similar to imipenem in pharmacology and antimicrobial spectrum of activity. However, it is not inactivated by dipeptidases and is less likely to cause seizures than imipenem.

Ertapenem has a long half-life suitable for once-daily administration. It is useful for treatment of complicated infections not involving hospital pathogens. It has poor activity against *Enterococcus* spp. and *P aeruginosa* and other glucose nonfermenting gram-negative rods.

Doripenem is the most recent carbapenem to be approved for use in the United States. The sulfamoylamimoethyl-pyrrolidinylthio group in its side chain at position 2 enhances its activity against glucose nonfermenting gram-negative rods. This drug has been reported to have strong affinity for PBP that are species specific. For example, doripenem has affinity for PBP3 in *P aeruginosa*. It is reported that doripenem is more active against *P aeruginosa* than imipenem but equal in active to meropenem. None of the carbapenems have activity against *Stenotrophomonas maltophilia*.

TETRACYCLINES

The tetracyclines are a group of drugs that differ in physical and pharmacologic characteristics but have virtually identical antimicrobial properties and give complete cross-resistance. All tetracyclines are readily absorbed from the intestinal tract and distributed widely in tissues but penetrate into the cerebrospinal fluid poorly. Some can also be administered intramuscularly or intravenously. They are excreted in stool and into bile and urine at varying rates. With doses of tetracycline hydrochloride, 2 g/day orally, blood levels reach 8 µg/mL. Minocycline and doxycycline are excreted more slowly and therefore are administered at longer intervals.

The tetracyclines have the basic structure shown below. The following radicals occur in the different forms:

	R	R₁	R₂	Renal clearance (mL/min)
Tetracycline	—H	—CH₃	—H	65
Doxycycline	—H	—CH₃	—OH	16
Minocycline	—N(CH₃)₂	—H	—H	< 10

Antimicrobial Activity

Tetracyclines are concentrated by susceptible bacteria and inhibit protein synthesis by inhibiting the binding of aminoacyl-tRNA to the 30S unit of bacterial ribosomes. Resistant bacteria fail to concentrate the drug. This resistance is under the control of transmissible plasmids.

The tetracyclines are principally bacteriostatic agents. They inhibit the growth of susceptible gram-positive and gram-negative bacteria (inhibited by 0.1–10 µg/mL) and are drugs of choice in infections caused by rickettsiae, chlamydiae, and *Mycoplasma pneumoniae*. Tetracyclines are used in cholera to shorten excretion of vibrios. Tetracycline hydrochloride or doxycycline orally for 7 days is effective against chlamydial genital infection. Tetracyclines are sometimes employed in combination with streptomycin to treat *Brucella*, *Yersinia*, and *Francisella* infections. Minocycline is often active against *Nocardia* and can eradicate the meningococcal carrier state. Low doses of tetracycline for many months are given for acne to suppress both skin bacteria and their lipases, which promote inflammatory changes.

Tetracyclines do not inhibit fungi. They temporarily suppress parts of the normal bowel flora, but superinfections may occur, particularly with tetracycline-resistant pseudomonads, protei, staphylococci, and yeasts.

Side Effects

The tetracyclines produce varying degrees of gastrointestinal upset (nausea, vomiting, diarrhea), skin rashes, mucous membrane lesions, and fever in many patients, particularly when administration is prolonged and dosage high. Replacement of bacterial flora (see above) occurs commonly. Overgrowth of yeasts on anal and vaginal mucous membranes during tetracycline administration leads to inflammation and pruritus. Overgrowth of organisms in the intestine may lead to enterocolitis.

Tetracyclines are deposited in bony structures and teeth, particularly in the fetus and during the first 6 years of life. Discoloration and fluorescence of the teeth occur in newborns if tetracyclines are taken for prolonged periods by pregnant women. Hepatic damage may occur. Minocycline can cause marked vestibular disturbances.

Bacteriologic Examination

Organisms that are susceptible to tetracycline are also considered susceptible to doxycycline and minocycline. However, resistance to tetracycline cannot be used to predict resistance to the other agents.

GLYCYLCYCLINES

Glycylcyclines are synthetic analogs of the tetracyclines. Only one agent is currently available for use—tigecycline. Tigecycline is the 9-*tert*-butyl-glycylamido derivative of minocycline. Tigecycline shares the same binding site on the ribosome as the tetracyclines. It binds more avidly to the ribosome, and this stronger binding is likely responsible for the enhanced activity against tetracycline-resistant organisms. Tigecycline is active against a broad spectrum of gram-positive and gram-negative pathogens. Compared to the tetracyclines, it is more active against methicillin-resistant *S aureus* and *S epidermidis*, drug-susceptible and drug-resistant *S pneumoniae*, and enterococci. In terms of the gram-negative aerobes, in addition to the spectrum of the other tetracyclines, tigecycline has enhanced activity against several *Enterobacteriaceae*, including *Salmonella* and *Shigella* species, and *Acinetobacter* species. It does not have good activity against *P aeruginosa*, *S maltophilia*, or *Burkholderia cepacia*. Tigecycline also has good activity against many anaerobic bacteria, including *B fragilis*.

Tigecycline is currently available only as a parenteral agent because of poor bioavailability. The drug has extensive and rapid distribution in tissues. Protein binding ranges from 73% to 79%. Tigecycline is not metabolized to pharmacologically active metabolites. The half-life is long, approximately 40 hours. The major route of elimination is via the biliary tract and through the feces; renal clearance is a secondary route of elimination. Currently, tigecycline is approved in the United States for treatment of complicated skin and soft tissue infections, as well as complicated intraabdominal infections.

CHLORAMPHENICOL

Chloramphenicol is a substance produced originally from cultures of *Streptomyces venezuelae* but now manufactured synthetically.

Crystalline chloramphenicol is a stable compound that is rapidly absorbed from the gastrointestinal tract and widely distributed into tissues and body fluids, including the central nervous system and cerebrospinal fluid; it penetrates cells

O
||
C — CHCl₂
|
O_2N —⟨benzene ring⟩— C — C — CH₂OH
| |
OH N

Chloramphenicol

well. Most of the drug is inactivated in the liver by conjugation with glucuronic acid or by reduction to inactive arylamines. Excretion is mainly in the urine, 90% in inactive form. Although chloramphenicol is usually administered orally, the succinate can be injected intravenously in similar dosage.

Chloramphenicol is a potent inhibitor of protein synthesis in microorganisms. It blocks the attachment of amino acids to the nascent peptide chain on the 50S unit of ribosomes by interfering with the action of peptidyl transferase. Chloramphenicol is principally bacteriostatic, and its spectrum, dosage, and blood levels are similar to those of the tetracyclines. Chloramphenicol has been used to treat many types of infection (eg, due to salmonellae, meningococci, *H influenzae*), but it is no longer the drug of choice for any infection.

Chloramphenicol resistance is due to destruction of the drug by an enzyme (chloramphenicol acetyltransferase) that is under plasmid control.

Chloramphenicol infrequently causes gastrointestinal upsets. However, administration of more than 3 g/day regularly induces disturbances in red cell maturation, elevation of serum iron, and anemia. These changes are reversible upon discontinuance of the drug. Very rarely, individuals exhibit an apparent idiosyncrasy to chloramphenicol and develop severe or fatal aplastic anemia that is distinct from the dose-related reversible effect described above. For these reasons, the use of chloramphenicol is generally restricted to those infections where it is clearly the most effective drug by laboratory test or experience.

In premature and newborn infants, chloramphenicol can induce collapse ("gray syndrome") because the normal mechanism of detoxification (glucuronide conjugation in the liver) is not yet developed.

ERYTHROMYCINS

Erythromycin is obtained from *Streptomyces erythreus* and has the chemical formula $C_{37}H_{67}NO_{13}$. Drugs related to erythromycin are clarithromycin, azithromycin, and others. Erythromycins attach to a receptor (a 23S rRNA) on the 50S subunit of the bacterial ribosome. They inhibit protein synthesis by interfering with translocation reactions and the formation of initiation complexes. Resistance to erythromycins results from an alteration (methylation) of the rRNA receptor. This is under control of a transmissible plasmid. The activity of erythromycins is greatly enhanced at alkaline pH.

Erythromycins in concentrations of 0.1–2 μg/mL are active against gram-positive bacteria, including pneumococci,

streptococci, and corynebacteria. *M pneumoniae, Chlamydia trachomatis, L pneumophila,* and *Campylobacter jejuni* are also susceptible. Resistant variants occur in susceptible microbial populations and tend to emerge during treatment, especially in staphylococcal infections.

Erythromycins may be drugs of choice in infections caused by the organisms listed above and are substitutes for penicillins in persons hypersensitive to the latter. Erythromycin stearate, succinate, or estolate orally four times a day yields serum levels of 0.5–2 μg/mL. Other forms are given intravenously.

Undesirable side effects are drug fever, mild gastrointestinal upsets, and cholestatic hepatitis as a hypersensitivity reaction, especially to the estolate. Hepatotoxicity may be increased during pregnancy. Erythromycin tends to increase levels of simultaneously administered anticoagulants, cyclosporine, and other drugs by depressing microsomal enzymes.

Dirithromycin is a macrolide with a spectrum of antimicrobial activity similar to that of erythromycin. Dirithromycin has a long serum half-life and is conveniently administered once a day.

Clarithromycin and azithromycin are azalides chemically related to erythromycin. Like erythromycin, both clarithromycin and azithromycin are active against staphylococci and streptococci. Clarithromycin has enhanced activity against *L pneumophila, Helicobacter pylori, Moraxella catarrhalis, C trachomatis,* and *Borrelia burgdorferi.* Azithromycin has enhanced activity against *C jejuni, H influenzae, M pneumoniae, M catarrhalis, N gonorrhoeae,* and *B burgdorferi.* Both drugs are active against *Mycobacterium avium* complex, and both drugs inhibit most strains of *Mycobacterium chelonei* and *Mycobacterium fortuitum.* Bacteria resistant to erythromycin are also resistant to clarithromycin and azithromycin. The chemical modifications prevent the metabolism of clarithromycin and azithromycin to inactive forms, and the drugs are given twice daily (clarithromycin) or once daily (azithromycin). Both drugs are associated with a much lower incidence of gastrointestinal side effects than erythromycin.

The ketolides are semisynthetic derivatives of erythromycin. They are more active than the macrolides, particularly against some macrolide-resistant bacteria, and have improved pharmacokinetics. Telithromycin is the agent currently approved for use in the United States. It is administered orally for the treatment of acute upper and lower respiratory

Erythromycin (R = CH₃)

tract infections. Its mechanism of action and side effect profile are similar to the macrolides.

CLINDAMYCIN & LINCOMYCIN

Lincomycin (derived from *Streptomyces lincolnensis*) and clindamycin (a chlorine-substituted derivative) resemble erythromycins in mode of action, antibacterial spectrum, and ribosomal receptor site but are chemically distinct. Clindamycin is active against *Bacteroides* and other anaerobes.

The drugs are acid-stable and can be given by mouth or intravenously. They are widely distributed in tissues, except the central nervous system. Excretion is mainly through the liver, bile, and urine.

Probably the most important indication for intravenous clindamycin is the treatment of severe anaerobic infections, including those caused by *B fragilis*. Successful treatment of staphylococcal infections of bone with lincomycins has been recorded. Clindamycin has been used extensively more recently in the treatment of skin and skin structure infections caused by community-associated MRSA. Lincomycins should not be used in meningitis. Clindamycin has been prominent in antibiotic-associated colitis caused by *C difficile*; however, most antimicrobials have been associated with *C difficile* colitis.

GLYCOPEPTIDES

Vancomycin

Vancomycin (MW 1450) is produced by *Streptomyces orientalis*. It is poorly absorbed from the intestine.

Vancomycin is markedly bactericidal for staphylococci, some clostridia, and some bacilli. The drug inhibits early stages in cell wall peptidoglycan synthesis. Drug-resistant strains do not emerge rapidly. Vancomycin is given intravenously for serious systemic staphylococcal infections, including endocarditis, especially if resistant to nafcillin. For enterococcal sepsis or endocarditis, vancomycin can be effective if combined with an aminoglycoside. Oral vancomycin is indicated in antibiotic-associated pseudomembranous colitis (see Clindamycin & Lincomycin).

The development of vancomycin resistance in enterococci has had a major impact on the treatment of severe multidrug-resistant enterococcal infections. See the section Clinical Implications of Drug Resistance earlier in this chapter and Chapter 15.

S aureus of intermediate susceptibility to vancomycin in vitro has been isolated from patients in several countries, including the United States. These patients have tended to have complex illnesses that included long-term therapy with vancomycin. In some cases, the infections appeared to have failed vancomycin therapy.

High-level vancomycin resistance in *S aureus* is of major international concern. The mechanism is the same as or similar to the transposon-mediated vancomycin resistance in enterococci (acquisition of vanA genes [see Chapter 15]). Such isolates have been cultured from several patients and may occur in more patients in the future.

Undesirable side effects are thrombophlebitis, skin rashes, nerve deafness, leukopenia, and perhaps kidney damage when used in combination with an aminoglycoside.

Teicoplanin

Teicoplanin has a structure similar to that of vancomycin. It is active against staphylococci (including methicillin-resistant strains), streptococci, enterococci, and many other gram-positive bacteria. Enterococci with VanA resistance to vancomycin are also resistant to teicoplanin, but enterococci with VanB vancomycin resistance are susceptible to teicoplanin. The drug has a long half-life and is administered once a day. Adverse effects include localized irritation at injection sites, hypersensitivity, and the potential for ototoxicity and nephrotoxicity. Teicoplanin is available in Europe but not in the United States.

DAPTOMYCIN

Daptomycin is a naturally occurring cyclic lipopeptide produced by *Streptomyces roseoporus*. Structurally, it has a 10-member amino acid ring, a 10-carbon decanoic acid attached to a terminal L-tryptophan. It is bactericidal by causing depolarization of the bacterial membrane in a calcium-dependent manner. It is available in a parenteral form administered once daily. It is highly protein bound and excreted in the kidney as parent drug. Dosage adjustment is required in patients with creatinine clearance <30 mL/min.

A major adverse effect of daptomycin is reversible myopathy. Weekly monitoring of creatine phosphokinase is recommended, and the drug should be discontinued when levels reach five times normal. Currently daptomycin is approved for use in the United States for treatment of skin and soft tissue infections caused by susceptible and resistant gram-positive cocci and for *S aureus* bacteremia. Synergy is seen when daptomycin is combined with gentamicin.

STREPTOGRAMINS

Quinupristin-dalfopristin is an injectable streptogramin antibiotic consisting of a 30:70 mixture of two semisynthetic derivatives of pristinamycin (a group B streptogramin) and dalfopristin (a group A streptogramin). The two components act synergistically to inhibit a wide spectrum of gram-positive bacteria including methicillin-resistant staphylococci, vancomycin-resistant enterococci, and penicillin-resistant pneumococci. Quinupristin-dalfopristin is active against some anaerobes and certain gram-negative bacteria (eg, *N gonorrhoeae, H influenzae*) but not against Enterobacteriaceae, *P aeruginosa,* or acinetobacters. Vancomycin-resistant enterococci that are resistant also to quinupristin-dalfopristin occur but are uncommon.

OXAZOLIDINONES

Oxazolidinones are in a new class of synthetic antimicrobials discovered in 1987. Linezolid is the only commercially available agent. The antimicrobial spectrum is similar to that of the glycopeptides. The mechanism of action of linezolid is seen early in protein synthesis—interference with translation by inhibiting the formation of *N*-formylmethionyl-tRNA, the initiation complex at the 30S ribosome. Linezolid is 100% bioavailable and is superior to vancomycin in that it has excellent penetration into respiratory secretions. It also diffuses well into bone, fat, and urine. Linezolid is most frequently used to treat pneumonia, bacteremia, and skin and soft-tissue infections caused by glycopeptide-resistant staphylococci and enterococci. Its major side effect is reversible thrombocytopenia.

BACITRACIN

Bacitracin is a polypeptide obtained from a strain (Tracy strain) of *Bacillus subtilis*. It is stable and poorly absorbed from the intestinal tract. Its only use is for topical application to skin, wounds, or mucous membranes.

Bacitracin is mainly bactericidal for gram-positive bacteria, including penicillin-resistant staphylococci. For topical use, concentrations of 500–2000 units per milliliter of solution or gram of ointment are used. In combination with polymyxin B or neomycin, bacitracin is useful for the suppression of mixed bacterial flora in surface lesions.

Bacitracin is toxic for the kidney, causing proteinuria, hematuria, and nitrogen retention. For this reason, it has no place in systemic therapy. Bacitracin is said not to induce hypersensitivity readily.

POLYMYXINS

Polymyxins are basic cationic polypeptides that are nephrotoxic and neurotoxic. Polymyxins can be bactericidal for many gram-negative aerobic rods—including pseudomonads and serratiae—by binding to cell membranes rich in phosphatidylethanolamine and destroying membrane functions of active transport and permeability barrier. Until recently, because of their toxicity and poor distribution to tissues, polymyxins were used primarily topically and rarely for systemic infections. Polymyxin E (colistin), available parenterally as colistimethate sodium, has undergone renewed interest and increasing utilization as an alternative agent for treatment of multidrug resistant *Acinetobacter baumannii*

and *P aeruginosa* and as salvage therapy for carbapenemase-resistant *Klebsiella* infections. Colistin is bactericidal against these gram-negative organisms. When used wisely, observed toxicity has been less than previously described.

AMINOGLYCOSIDES

Aminoglycosides are a group of drugs sharing chemical, antimicrobial, pharmacologic, and toxic characteristics. At present, the group includes streptomycin, neomycin, kanamycin, amikacin, gentamicin, tobramycin, sisomicin, netilmicin, and others. All inhibit protein synthesis of bacteria by attaching to and inhibiting the function of the 30S subunit of the bacterial ribosome. Resistance is based on (1) a deficiency of the ribosomal receptor (chromosomal mutant), (2) enzymatic destruction of the drug (plasmid-mediated transmissible resistance of clinical importance), or (3) lack of permeability to the drug molecule and lack of active transport into the cell. The last can be chromosomal (eg, streptococci are relatively impermeable to aminoglycosides), or it can be plasmid-mediated (eg, in gram-negative enteric bacteria). Anaerobic bacteria are often resistant to aminoglycosides because transport through the cell membrane is an energy-requiring process that is oxygen-dependent.

All aminoglycosides are more active at alkaline pH than at acid pH. All are potentially ototoxic and nephrotoxic, though to different degrees. All can accumulate in renal failure; therefore, marked dosage adjustments must be made when nitrogen retention occurs. Aminoglycosides are used most widely against gram-negative enteric bacteria or when there is suspicion of sepsis. In the treatment of bacteremia or endocarditis caused by fecal streptococci or some gram-negative bacteria, the aminoglycoside is given together with a penicillin that facilitates the entry of the aminoglycoside. Aminoglycosides are selected according to recent susceptibility patterns in a given area or hospital until susceptibility tests become available on a specific isolate. The clinical usefulness of aminoglycosides has declined with the advent of cephalosporins and quinolones, but they continue to be used in combinations (eg, with cephalosporins for multidrug-resistant gram-negative bacteremias). All positively charged aminoglycosides are inhibited in blood cultures by sodium polyanetholsulfonate and other polyanionic detergents. Some aminoglycosides (especially streptomycin) are useful as antimycobacterial drugs.

Neomycin & Kanamycin

Kanamycin is a close relative of neomycin, with similar activity and complete cross-resistance. Paromomycin is also closely related and is used in amebiasis. These drugs are stable and poorly absorbed from the intestinal tract and other surfaces. Neither drug is used systemically because of ototoxicity and neurotoxicity. Oral doses of both neomycin and kanamycin are used for reduction of intestinal flora before large bowel surgery, often in combination with erythromycin. Otherwise, these drugs are mainly limited to topical application on infected surfaces (skin and wounds).

Amikacin

Amikacin is a semisynthetic derivative of kanamycin. It is relatively resistant to several of the enzymes that inactivate gentamicin and tobramycin and therefore can be employed against some microorganisms resistant to the latter drugs. However, bacterial resistance due to impermeability to amikacin is slowly increasing. Many gram-negative enteric bacteria are inhibited by amikacin in concentrations obtained after injection. Central nervous system infections require intrathecal or intraventricular injection.

Like all aminoglycosides, amikacin is nephrotoxic and ototoxic (particularly for the auditory portion of the eighth nerve). Its level should be monitored in patients with renal failure.

Gentamicin

In concentrations of 0.5–5 µg/mL, gentamicin is bactericidal for many gram-positive and gram-negative bacteria, including many strains of *Proteus, Serratia,* and *Pseudomonas.* Gentamicin is ineffective against streptococci and *Bacteroides.*

Gentamicin has been used in serious infections caused by gram-negative bacteria resistant to other drugs. Penicillins may precipitate gentamicin in vitro (and thus must not be mixed), but in vivo they may facilitate the aminoglycoside entrance into streptococci and gram-negative rods and result in bactericidal synergism, beneficial in sepsis and endocarditis.

Gentamicin is toxic, particularly in the presence of impaired renal function. Gentamicin sulfate, 0.1%, has been used topically in creams or solutions for infected burns or skin lesions. Such creams tend to select gentamicin-resistant bacteria, and patients receiving them should remain in strict isolation.

Tobramycin

This aminoglycoside closely resembles gentamicin, and there is some cross-resistance between them. Separate susceptibility tests are desirable. Tobramycin has slightly enhanced activity against *P aeruginosa* when compared with gentamicin.

The pharmacologic properties of tobramycin are virtually identical to those of gentamicin. Most of the drug is excreted by glomerular filtration. In renal failure, the drug dosage must be reduced, and monitoring of blood levels is desirable.

Like other aminoglycosides, tobramycin is ototoxic but perhaps less nephrotoxic than gentamicin. It should not be used concurrently with other drugs having similar adverse effects or with diuretics, which tend to enhance aminoglycoside tissue concentrations.

Netilmicin

Netilmicin shares many characteristics with gentamicin and tobramycin, but it is not inactivated by some bacteria that are resistant to the other drugs.

The principal indication for netilmicin may be iatrogenic infections in immunocompromised and severely ill patients at very high risk for gram-negative bacterial sepsis in the hospital setting.

Netilmicin may be somewhat less ototoxic and nephrotoxic than the other aminoglycosides.

Streptomycin

Streptomycin was the first aminoglycoside—it was discovered in the 1940s as a product of *Streptomyces griseus.* It was studied in great detail and became the prototype of this class of drugs. For this reason, its properties are listed here, though widespread resistance among microorganisms has greatly reduced its clinical usefulness.

After intramuscular injection, streptomycin is rapidly absorbed and widely distributed in tissues except the central nervous system. Only 5% of the extracellular concentration of streptomycin reaches the interior of the cell. Absorbed streptomycin is excreted by glomerular filtration into the urine. After oral administration, it is poorly absorbed from the gut; most of it is excreted in feces.

Streptomycin may be bactericidal for enterococci (eg, in endocarditis) when combined with a penicillin. In tularemia and plague, it may be given with a tetracycline. In tuberculosis, it is used in combination with other antituberculous drugs (isoniazid, rifampin). Streptomycin should not be used alone to treat any infection.

The therapeutic effectiveness of streptomycin is limited by the rapid emergence of resistant mutants. All microbial strains produce streptomycin-resistant chromosomal mutants with relatively high frequency. Chromosomal mutants have an alteration in the P 12 receptor on the 30S ribosomal subunit. Plasmid-mediated resistance results in enzymatic destruction of the drug. Enterococci resistant to high levels of streptomycin (2000 µg/mL) or gentamicin (500 µg/mL) are resistant to the synergistic actions of these drugs with penicillin.

Fever, skin rashes, and other allergic manifestations may result from hypersensitivity to streptomycin. This occurs most frequently upon prolonged contact with the drug, in patients receiving a protracted course of treatment (eg, for tuberculosis), or in personnel preparing and handling the drug. (Those preparing solutions should wear gloves.)

Streptomycin is markedly toxic for the vestibular portion of the eighth cranial nerve, causing tinnitus, vertigo, and ataxia, which are often irreversible. It is moderately nephrotoxic.

Spectinomycin

Spectinomycin is an aminocyclitol antibiotic (related to aminoglycosides) for intramuscular administration. Its sole application is in the single-dose treatment of gonorrhea caused by β-lactamase-producing gonococci or occurring in individuals hypersensitive to penicillin. About 5–10% of gonococci are probably resistant. There is usually pain at the injection site, and there may be nausea and fever.

QUINOLONES

Quinolones are synthetic analogs of nalidixic acid. The currently available quinolones are listed in Table 28–5. The mode of action of all quinolones involves inhibition of bacterial DNA synthesis by blocking of the DNA gyrase.

The earlier quinolones (nalidixic acid, oxolinic acid, and cinoxacin) did not achieve systemic antibacterial levels after oral intake and thus were useful only as urinary antiseptics (see below). The fluorinated derivatives (eg, ciprofloxacin, norfloxacin, and others; see Figure 28–3 for structures of some of them) have greater antibacterial activity and low toxicity and achieve clinically useful levels in blood and tissues.

Antimicrobial Activity

The fluoroquinolones inhibit many types of bacteria, though the spectrum of activity varies from one drug to another (Tables 28–5 and 28–6). The drugs are highly active against Enterobacteriaceae, including those resistant to third-generation cephalosporins, *Haemophilus* species, neisseriae, chlamydiae, and others. *P aeruginosa* and legionellae are inhibited by somewhat larger amounts of these drugs. The quinolones vary in their activity against gram-positive pathogens. Some are active against multidrug-resistant *S pneumoniae* (see Table 28–6). They may be active against methicillin-resistant staphylococci and *E faecalis*. Vancomycin-resistant enterococci are usually resistant to the quinolones. Newer fluoroquinolones have increased activity against anaerobic bacteria, allowing them to be used as monotherapy in the treatment of mixed aerobic and anaerobic infections.

Fluoroquinolones may also have activity against *M tuberculosis, M fortuitum, M kansasii,* and sometimes *M chelonei.*

During fluoroquinolone therapy, the emergence of resistance of pseudomonads, staphylococci, and other pathogens has been observed. Chromosomal resistance develops by mutation and involves one of two mechanisms—either an alteration in the A subunit of the target enzyme, DNA gyrase; or a change in outer membrane permeability, resulting in decreased drug accumulation in the bacterium.

Absorption & Excretion

After oral administration, representative fluoroquinolones are well absorbed and widely distributed in body fluids and tissues to varying degrees, but they do not reach the central nervous system to a significant extent. The serum half-life is variable (3–8 hours) and can be prolonged in renal failure depending upon the specific drug used.

The fluoroquinolones are mainly excreted into the urine via the kidney, but some of the dose may be metabolized in the liver.

Clinical Uses

Fluoroquinolones are generally effective in urinary tract infections, and several of them benefit prostatitis. Some fluoroquinolones (eg, ofloxacin) are valuable in the treatment of sexually transmitted diseases caused by *N gonorrhoeae* and *C trachomatis* but have no effect on *T pallidum*. These drugs can control lower respiratory infections due to *H influenzae* (but may not be drugs of choice) and enteritis caused by salmonellae, shigellae, or campylobacters. Fluoroquinolones

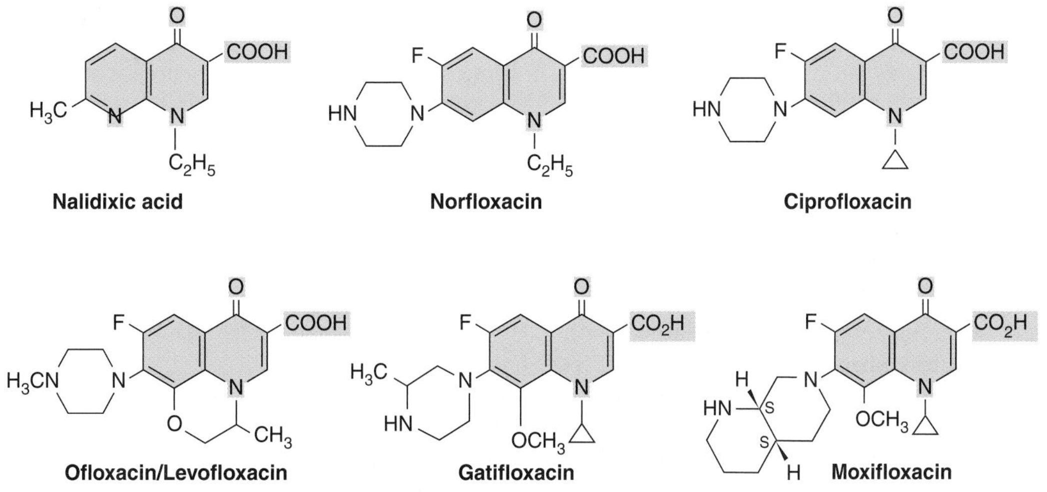

FIGURE 28–3 Structures of some fluoroquinolones.

TABLE 28–5   The Quinolones[a]

First-generation
Nalidixic acid
Cinoxacin
Oxolinic acid
Second-generation
Ciprofloxacin
Enoxacin
Lomefloxacin
Ofloxacin
Third- and fourth-generation
Clinafloxacin
Gatifloxacin
Gemifloxacin
Levofloxacin
Moxifloxacin
Pefloxacin
Sparfloxacin
Garenoxacin

[a]Courtesy of B Joseph Guglielmo, PharmD.

TABLE 28–6   Relative Spectrum of Antibacterial Activity of the Quinolones[a]

Strong	Moderate	Weak
Gram-positive activity		
Clinafloxacin	Ofloxacin	Lomefloxacin
Gatifloxacin	Ciprofloxacin	Norfloxacin
Gemifloxacin	Enoxacin	Pefloxacin
Levofloxacin		
Moxifloxacin		
Garenoxacin		
Gram-negative activity		
Clinafloxacin	Enoxacin	Norfloxacin
Ciprofloxacin	Gatifloxacin	
Pefloxacin	Gemifloxacin	
	Levofloxacin	
	Lomefloxacin	
	Moxifloxacin	
	Ofloxacin	
	Sparfloxacin	
	Garenoxacin	
Anaerobe activity		
Clinafloxacin	Sparfloxacin	Ciprofloxacin
Gatifloxacin	Levofloxacin	Lomefloxacin
Gemifloxacin	Ofloxacin	Enoxacin
Moxifloxacin		Pefloxacin
Garenoxacin		

[a]Courtesy of B Joseph Guglielmo, PharmD.

may be suitable for the treatment of major gynecologic and soft tissue bacterial infections and for osteomyelitis of gram-negative origin. While they can benefit some exacerbations of cystic fibrosis caused by pseudomonads, about one-third of such mucoid organisms are drug resistant.

Side Effects

The most prominent adverse effects are nausea, insomnia, headache, and dizziness. Occasionally, there are other gastrointestinal disturbances, impaired liver function, skin rashes, and superinfections, particularly with enterococci and staphylococci. In puppies, prolonged administration of fluoroquinolones produces joint damage, and for that reason fluoroquinolones have been seldom prescribed for children but are used as needed in cystic fibrosis patients. Disturbances of blood glucose leading to significant hypoglycemia have been reported with newer agents, but have been seen most often in patients treated with gatifloxacin.

SULFONAMIDES & TRIMETHOPRIM

The sulfonamides are a group of compounds with the basic formula shown earlier in this chapter. By substituting various R-radicals, a series of compounds is obtained with somewhat varying physical, pharmacologic, and antibacterial properties. The basic mechanism of action of all of these compounds is the competitive inhibition of PABA utilization. The simultaneous use of sulfonamides with trimethoprim results in the inhibition of sequential metabolic steps and possible antibacterial synergism.

The sulfonamides are bacteriostatic for some gram-negative and gram-positive bacteria, chlamydiae, nocardiae, and protozoa.

The "soluble" sulfonamides (eg, trisulfapyrimidines, sulfisoxazole) are readily absorbed from the intestinal tract after oral administration and are distributed in all tissues and body fluids. Most sulfonamides are excreted rapidly in the urine. Some (eg, sulfamethoxypyridazine) are excreted very slowly and thus tend to be toxic. At present, sulfonamides are particularly useful in the treatment of nocardiosis and first attacks of urinary tract infections due to coliform bacteria. By contrast, many meningococci, shigellae, group A streptococci, and organisms causing recurrent urinary tract infections are now resistant. A mixture of five parts sulfamethoxazole plus one part trimethoprim is widely used in urinary tract infections, shigellosis, and salmonellosis and infections with other gram-negative bacterial infections and in pneumocystis pneumonia.

Trimethoprim alone can be effective treatment for uncomplicated urinary tract infections.

Resistance

Microorganisms that do not use extracellular PABA but, like mammalian cells, can use preformed folic acid are resistant to sulfonamides. In some sulfonamide-resistant mutants, the tetrahydropteroic acid synthetase has a much higher affinity for PABA than for sulfonamides. The opposite is true for sulfonamide-susceptible organisms.

Side Effects

The soluble sulfonamides may produce side effects that fall into two categories, allergy and toxicity. Many individuals develop hypersensitivity to sulfonamides after initial contact with these drugs and, on reexposure, may develop fever, hives, skin rashes, and chronic vascular diseases such as polyarteritis nodosa. Toxic effects are manifested by fever, skin rashes, gastrointestinal disturbances, depression of the bone marrow leading to anemia or agranulocytosis, hemolytic anemia, and liver and kidney function abnormalities. Toxicity is especially frequent in patients with AIDS.

Bacteriologic Examination

When culturing specimens from patients receiving sulfonamides, the incorporation of PABA (5 mg/dL) into the medium overcomes sulfonamide inhibition.

OTHER DRUGS WITH SPECIALIZED USES

Trimetrexate

Trimetrexate is a folinic acid analog whose mechanism of action is inhibition of dihydrofolate reductase. The primary use of trimetrexate is in the treatment of *P jiroveci* infections in AIDS patients who are intolerant of or refractory to trimethoprim-sulfamethoxazole and pentamidine isethionate. Because trimetrexate is lipophilic, it passively diffuses across host cell membranes with associated toxicity, primarily bone marrow suppression. Therefore, it must be coadministered with leucovorin calcium, a reduced folate coenzyme, which is transported into and protects the host cells but not *P jiroveci.*

Dapsone

Dapsone is a sulfone closely related to the sulfonamides. Combined therapy with dapsone and rifampin is often given in the initial therapy of leprosy. Dapsone may also be used to treat pneumocystis pneumonia in AIDS patients. Dapsone is well absorbed from the gastrointestinal tract and is widely distributed in tissues. Side effects are common, including hemolytic anemia, gastrointestinal intolerance, fever, itching, and rashes.

Dapsone

Metronidazole

Metronidazole is an antiprotozoal drug used in treating trichomonas, giardia, and amebic infections. It also has striking effects in anaerobic bacterial infections, eg, those due to *Bacteroides* species, and in bacterial vaginosis. It appears to be effective for the preoperative preparation of the colon and in antibiotic-associated diarrhea caused by toxigenic *C difficile.* Adverse effects include stomatitis, diarrhea, and nausea.

Urinary Antiseptics

These are drugs with antibacterial effects limited to the urine. They fail to produce significant levels in tissues and thus have no effect on systemic infections. However, they effectively lower bacteria counts in the urine and thus greatly diminish the symptoms of lower urinary tract infection. They are used only in the management of urinary tract infections.

The following are commonly used urinary antiseptics: nitrofurantoin, fosfomycin, nalidixic acid, methenamine mandelate, and methenamine hippurate. Nitrofurantoin is active against many bacteria but may cause gastrointestinal distress. Fosfomycin is a derivative of phosphonic acid and is used primarily in the United States as single-dose therapy for urinary tract infections caused by *E coli* and other Enterobacteriaceae and enterococci. Nalidixic acid, a quinolone, is effective only in urine, but resistant bacteria may rapidly emerge in the urine. Both methenamine mandelate and methenamine hippurate acidify the urine and liberate formaldehyde there. Other substances that acidify urine (eg, methionine, cranberry juice) may result in bacteriostasis in urine.

Systemically absorbed oral drugs that are excreted in high concentrations in urine are usually preferred in acute urinary tract infections. These include ampicillin, amoxicillin, sulfonamides, quinolones, and others.

DRUGS USED PRIMARILY TO TREAT MYCOBACTERIAL INFECTIONS

Isoniazid

Isoniazid has little effect on most bacteria but is strikingly active against mycobacteria, especially *M tuberculosis.* Most tubercle bacilli are inhibited and killed in vitro by isoniazid, 0.1–1 μg/mL, but large populations of tubercle bacilli usually contain some isoniazid-resistant organisms. For this reason, the drug is employed in combination with other antimycobacterial agents (especially ethambutol or rifampin) to reduce the emergence of resistant tubercle bacilli. Isoniazid acts on mycobacteria by inhibiting the synthesis of mycolic acids. Isoniazid and pyridoxine are structural analogs. Patients receiving isoniazid excrete pyridoxine in excessive amounts, which results in peripheral neuritis. This can be prevented by the administration of pyridoxine, which does not interfere with the antituberculous action of isoniazid.

O
‖
C—NH—NH₂
|
C
HC⫶   ⫶CH
| ‖
HC   CH
　⫶   ‖
　　N

Isoniazid

CH₂OH
|
C
HO—C⫶   ⫶C—CH₂OH
| ‖
H₃C—C   CH
　　⫶ ‖
　　　N

Pyridoxine

Isoniazid is rapidly and completely absorbed from the gastrointestinal tract and is in part acetylated and in part excreted in the urine. With usual doses, toxic manifestations, eg, hepatitis, are infrequent. Isoniazid freely diffuses into tissue fluids, including the cerebrospinal fluid.

In converters from negative to positive tuberculin skin tests who have no evidence of disease, isoniazid may be used as prophylaxis.

Ethambutol

Ethambutol is a synthetic water-soluble, heat-stable D-isomer of the structure shown below.

　　CH₂OH　　　　　　　　C₂H₅
　　　|　　　　　　　　　　　|
H—C—NH—(CH₂)₂—HN—C—H
　　　|　　　　　　　　　　　|
　　C₂H₅　　　　　　　　CH₂OH

Ethambutol

Many strains of *M tuberculosis* and of "atypical" mycobacteria are inhibited in vitro by ethambutol, 1–5 μg/mL.

Ethambutol is well absorbed from the gut. About 20% of the drug is excreted in feces and 50% in urine in unchanged form. Excretion is delayed in renal failure. In meningitis, ethambutol appears in the cerebrospinal fluid.

Resistance to ethambutol emerges fairly rapidly among mycobacteria when the drug is used alone. Therefore, ethambutol is always given in combination with other antituberculous drugs.

Ethambutol is usually given as a single oral daily dose. Hypersensitivity to ethambutol occurs infrequently. The most common side effects are visual disturbances, but these are rare at standard dosages: Reduction in visual acuity, optic neuritis, and perhaps retinal damage occur in some patients given high doses for several months. Most of these changes apparently regress when ethambutol is discontinued. However, periodic visual acuity testing is mandatory during treatment. With low doses, visual disturbances are very rare.

Rifampin

Rifampin is a semisynthetic derivative of rifamycin, an antibiotic produced by *Streptomyces mediterranei*. It is active in vitro against some gram-positive and gram-negative cocci, some enteric bacteria, mycobacteria, chlamydiae, and poxviruses. Although many meningococci and mycobacteria are inhibited by less than 1 μg/mL, highly resistant mutants occur in all microbial populations in a frequency of 10^{-6} to 10^{-5}. The prolonged administration of rifampin as a single drug permits the emergence of these highly resistant mutants. There is no cross-resistance to other antimicrobial drugs.

Rifampin binds strongly to DNA-dependent RNA polymerase and thus inhibits RNA synthesis in bacteria. It blocks a late stage in the assembly of poxviruses. Rifampin penetrates phagocytic cells well and can kill intracellular organisms. Rifampin-resistant mutants exhibit an altered RNA polymerase.

Rifampin is well absorbed after oral administration, widely distributed in tissues, and excreted mainly through the liver and to a lesser extent into the urine.

In tuberculosis, a single oral dose is administered together with ethambutol, isoniazid, or another antituberculous drug in order to delay the emergence of rifampin-resistant mycobacteria. A similar regimen may apply to atypical mycobacteria. In short-term treatment schedules for tuberculosis, rifampin is given orally, first daily (together with isoniazid) and then two or three times weekly for 6–9 months. However, no less than two doses weekly should be given to avoid a "flu syndrome" and anemia. Rifampin used in conjunction with a sulfone is effective in leprosy.

Oral rifampin can eliminate a majority of meningococci from carriers. Unfortunately, some highly resistant meningococcal strains are selected out by this procedure. Close contacts of children with *H influenzae* infections (eg, in the family or in day care centers) can receive rifampin as prophylaxis. In urinary tract infections and in chronic bronchitis, rifampin is not useful because resistance emerges promptly.

Rifampin imparts a harmless orange color to urine, sweat, and contact lenses. Occasional adverse effects include rashes, thrombocytopenia, light chain proteinuria, and impairment of liver function. Rifampin induces microsomal enzymes (eg, cytochrome P450).

Rifabutin is a related antimycobacterial drug, active in the prevention of infection due to *M avium* complex.

Rifaximin is a derivative of rifampin that possesses an additional pyridoimidazole ring. It is a nonabsorbed oral agent useful in the treatment of traveler's diarrhea and as salvage therapy for recurrent *C difficile* disease.

Pyrazinamide

Pyrazinamide is related to nicotinamide. It is readily absorbed from the gastrointestinal tract and widely distributed in tissues. *M tuberculosis* readily develops resistance to pyrazinamide, but there is no cross-resistance with isoniazid or other antituberculous drugs. The major adverse effects of pyrazinamide are hepatotoxicity (1–5%), nausea, vomiting, hypersensitivity, and hyperuricemia.

Pyrazinamide (PZA)

REVIEW QUESTIONS

1. The antimicrobial agent whose structure is presented above is considered the drug of choice to treat infections caused by which one of the following microorganisms?

   (A) *Bacteroides fragilis*
   (B) *Pseudomonas aeruginosa*
   (C) Herpes simplex virus
   (D) *Streptococcus pyogenes* (group A streptococci)
   (E) *Mycobacterium tuberculosis*

2. Resistance of *Staphylococcus aureus* to the drug shown above is due to

   (A) The action of acetyltransferase
   (B) The action of β-lactamase
   (C) Substitution of the D-Ala-D-Ala dipeptide with the D-Ala-D-Lac dipeptide in the cell wall peptidoglycan
   (D) Decreased permeability of the bacterial cell wall to the drug.
   (E) *Staphylococcus aureus* being an intracellular pathogen

3. *Streptococcus pneumoniae* resistance to the drug shown above is due to

   (A) The action of acetyltransferase
   (B) The action of β-lactamase
   (C) Substitution of the D-Ala-D-Ala dipeptide with D-Ala-D-Lac dipeptide in the cell wall peptidoglycan
   (D) Decreased permeability of the bacterial cell wall
   (E) Genetically modified binding proteins in the bacterial cell wall

4. All of the following statements about antimicrobial resistance of enterococci are correct *except*:

   (A) Enterococci are resistant to sulfamethoxazole-trimethoprim in vivo.
   (B) Cephalosporins are not active against enterococci.
   (C) Resistance to the streptogramins (quinupristin-dalfopristin) has emerged.
   (D) Vancomycin-resistant enterococci are rare in Europe and the United States.
   (E) Vancomycin-resistant enterococci once consistently clonal are now heterogeneous.

5. A 20-year-old Asian woman, a recent immigrant to the United States, develops fever and a cough productive of blood-streaked sputum. She has lost 6 kg body weight in the last 6 weeks. Her chest film shows bilateral upper lobe infiltrates with cavities. Given the history and chest x-ray findings, which of the following drug regimens would be the best appropriate initial therapy while awaiting culture results?

   (A) Isoniazid, rifampin, pyrazinamide, and ethambutol
   (B) Penicillin G and rifampin
   (C) Cefotaxime, clindamycin, and trimethoprim-sulfamethoxazole
   (D) Ampicillin/sulbactam
   (E) Vancomycin, gentamicin, and clindamycin

6. Aminoglycoside antibiotics typically cause which of the following adverse events?

   (A) They cause aplastic anemia
   (B) They cause nonspecific stimulation of B cells
   (C) They cause ototoxicity and nephrotoxicity
   (D) They cause photosensitivity

7. Which one of the following groups of antimicrobial agents acts on microorganisms by inhibiting protein synthesis?

   (A) Fluoroquinolones
   (B) Aminoglycosides
   (C) Penicillins
   (D) Glycopeptides (eg, vancomycin)
   (E) Polymyxins

8. There are many bacterial-antimicrobial resistance combinations. Which one of the following is of major international concern?

   (A) Sulfonamide resistance in *Neisseria meningitidis*
   (B) Penicillin G resistance in *Neisseria gonorrhoeae*
   (C) Ampicillin resistance in *Haemophilus influenzae*
   (D) Erythromycin resistance in *Streptococcus pyogenes* (group A streptococci)
   (E) Vancomycin resistance in *Staphylococcus aureus*

9. Which of the following factors is not generally considered when selecting initial antimicrobial therapy for an infection?

   (A) Age of the patient
   (B) Anatomic site of the infection (eg, meningitis or urinary tract infection)
   (C) Whether or not the patient is immunocompromised
   (D) Whether or not the patient has implanted devices in place (eg, artificial hip joint, artificial heart valve, urinary catheter)
   (E) Waiting for culture and susceptibility test results

10. All of the following agents have good activity against gram-positive organisms *except*:

    (A) Daptomycin
    (B) Vancomycin
    (C) Aztreonam
    (D) Quinupristin-dalfopristin
    (E) Tigecycline

11. Tigecycline, a new glycylcycline antibiotic with good activity against a variety of pathogens, is best used for treatment of which of the following infections?

    (A) Meningitis
    (B) Intra-abdominal infections caused by mixed aerobic and anaerobic bacteria
    (C) Neonatal sepsis

(D) Urethritis caused by *Chlamydia trachomatis*

(E) As monotherapy for bacteremia caused by *Acinetobacter baumannii*

12. Which of the following carbapenem antibiotics has no activity against *Pseudomonas aeruginosa*?

(A) imipenem

(B) meropenem

(C) doripenem

(D) ertapenem

13. Which of the following agents would not be expected to demonstrate postantibiotic affect against gram-negative bacilli?

(A) imipenem

(B) ciprofloxacin

(C) gentamicin

(D) ampicillin

14. All of the following are common mechanisms of resistance to the penicillins *except*:

(A) Production of β-lactamases

(B) Alterations in target receptors (PBPs)

(C) Inability to activate autolytic enzymes

(D) Failure to synthesize peptidoglycans

(E) Methylation of ribosomal RNA

15. The drug of first choice for the treatment of serious anaerobic infections caused by *Bacteroides fragilis* is:

(A) Clindamycin

(B) Ampicillin

(C) Cefoxitin

(D) Metronidazole

(E) Amoxicillin-clavulanate

Answers

1. D	5. A	9. E	13. D
2. B	6. C	10. C	14. E
3. E	7. B	11. B	15. D
4. D	8. E	12. D	

REFERENCES

Lundstrom TS, Sobel JD: Antibiotics for gram-positive bacterial infections: vancomycin, quinupristin-dalfopristin, linezolid, and daptomycin. Infect Dis Clin North Am 2004;18:651.

Meagher AK, et al: Pharmacokinetic/pharmacodynamic profile for tigecycline—A new glycylcycline antimicrobial agent. Diagn Microbiol Infect Dis 2005;52:165.

Moellering RC et al: Anti-infective therapy. Volume I Part I Section E, In: *Mandell, Douglas, and Bennett's Principles and Practice of Infectious Diseases,* 5th ed. Mandell GL, Bennett JE, Dolin R (editors). Churchill Livingstone, 2000.

O'Donnell JA, Gelone SP: The newer fluoroquinolones. Infect Dis Clin North Am 2004;18:691.

Opal SM and Pop-Vicas A: Molecular mechanisms of antibiotic resistance in bacteria. In *Mandell, Douglas and Bennett's Principles and Practice of Infectious Diseases,* 7th ed. Mandell GL, Bennett JE, Dolin R (editors). Churchill Livingstone Elsevier, 2010.

Perez F, Endimiani A, Hujer KM, Bonomo RA. The continuing challenge of ESBLs. Curr Opin Pharmacol 2007;7:459.

Richter SS, Heilmann KP, Dohrn CL, Riahi F, Beekman SE, Doem GV: Changing epidemiology of antimicrobial resistant *Streptococcus pneumoniae* in the United States, 2004–2005. Clin Infect Dis 2009;48:e23.

Yu VL, Merrigan TC Jr, Barriere SL (editors): *Antimicrobial Therapy and Vaccines.* Williams & Wilkins, 1999.

General Properties of Viruses

Viruses are the smallest infectious agents (ranging from about 20 nm to about 300 nm in diameter) and contain only one kind of nucleic acid (RNA or DNA) as their genome. The nucleic acid is encased in a protein shell, which may be surrounded by a lipid-containing membrane. The entire infectious unit is termed a virion. Viruses are inert in the extracellular environment; they replicate only in living cells, being parasites at the genetic level. The viral nucleic acid contains information necessary for programming the infected host cell to synthesize virus-specific macromolecules required for the production of viral progeny. During the replicative cycle, numerous copies of viral nucleic acid and coat proteins are produced. The coat proteins assemble together to form the capsid, which encases and stabilizes the viral nucleic acid against the extracellular environment and facilitates the attachment and penetration by the virus upon contact with new susceptible cells. The virus infection may have little or no effect on the host cell or may result in cell damage or death.

The universe of viruses is rich in diversity. Viruses vary greatly in structure, genome organization and expression, and strategies of replication and transmission. The host range for a given virus may be broad or extremely limited. Viruses are known to infect unicellular organisms such as mycoplasmas, bacteria, and algae and all higher plants and animals. General effects of viral infection on the host are considered in Chapter 30.

Much information on virus–host relationships has been obtained from studies on bacteriophages, the viruses that attack bacteria. This subject is discussed in Chapter 7. Properties of individual viruses are discussed in Chapters 31–44.

TERMS & DEFINITIONS IN VIROLOGY

Schematic diagrams of viruses with icosahedral and helical symmetry are shown in Figure 29–1. Indicated viral components are described below.

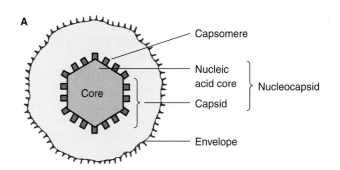

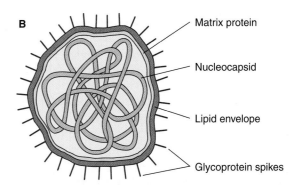

FIGURE 29–1 Schematic diagram illustrating the components of the complete virus particle (the virion). **A:** Enveloped virus with icosahedral symmetry. **B:** Virus with helical symmetry.

Capsid: The protein shell, or coat, that encloses the nucleic acid genome.

Capsomeres: Morphologic units seen in the electron microscope on the surface of icosahedral virus particles. Capsomeres represent clusters of polypeptides, but the morphologic units do not necessarily correspond to the chemically defined structural units.

Defective virus: A virus particle that is functionally deficient in some aspect of replication.

Envelope: A lipid-containing membrane that surrounds some virus particles. It is acquired during viral maturation by a budding process through a cellular membrane. Virus-encoded glycoproteins are exposed on the surface of the envelope. These projections are called **peplomers**.

Nucleocapsid: The protein–nucleic acid complex representing the packaged form of the viral genome. The term is commonly used in cases where the nucleocapsid is a substructure of a more complex virus particle.

Structural units: The basic protein building blocks of the coat. They are usually a collection of more than one nonidentical protein subunit. The structural unit is often referred to as a **protomer**.

Subunit: A single folded viral polypeptide chain.

Virion: The complete virus particle. In some instances (eg, papillomaviruses, picornaviruses), the virion is identical with the nucleocapsid. In more complex virions (herpesviruses, orthomyxoviruses), this includes the nucleocapsid plus a surrounding envelope. This structure, the virion, serves to transfer the viral nucleic acid from one cell to another.

EVOLUTIONARY ORIGIN OF VIRUSES

The origin of viruses is not known. There are profound differences among the DNA viruses, the RNA viruses, and viruses that utilize both DNA and RNA as their genetic material during different stages of their life cycle. It is possible that different types of agents are of different origins. Two theories of viral origin can be summarized as follows:

1. Viruses may be derived from DNA or RNA nucleic acid components of host cells that became able to replicate autonomously and evolve independently. They resemble genes that have acquired the capacity to exist independently of the cell. Some viral sequences are related to portions of cellular genes encoding protein functional domains. It seems likely that at least some viruses evolved in this fashion.
2. Viruses may be degenerate forms of intracellular parasites. There is no evidence that viruses evolved from bacteria, though other obligately intracellular organisms, eg, rickettsiae and chlamydiae, presumably did so. However, poxviruses are so large and complex that they might represent evolutionary products of some cellular ancestor.

CLASSIFICATION OF VIRUSES

Basis of Classification

The following properties have been used as a basis for the classification of viruses. The amount of information available in each category is not the same for all viruses. The way in which viruses are characterized is changing rapidly. Genome sequencing is now often performed early in virus identification, and comparisons with databases obviate the need to obtain more classic data (virion buoyant density, etc). Genomic sequence data are advancing taxonomic criteria (eg, gene order) and may provide the basis for the identification of new virus families.

1. Virion morphology, including size, shape, type of symmetry, presence or absence of peplomers, and presence or absence of membranes.
2. Virus genome properties, including type of nucleic acid (DNA or RNA), size of genome in kilobases (kb) or kilobase pairs (kbp), strandedness (single or double), whether linear or circular, sense (positive, negative, ambisense), segments (number, size), nucleotide sequence, G + C content, and presence of special features (repetitive elements, isomerization, 5′-terminal cap, 5′-terminal covalently linked protein, 3′-terminal poly(A) tract).
3. Physicochemical properties of the virion, including molecular mass, buoyant density, pH stability, thermal stability, and susceptibility to physical and chemical agents, especially ether and detergents.
4. Virus protein properties, including number, size, and functional activities of structural and nonstructural proteins, amino acid sequence, modifications (glycosylation, phosphorylation, myristylation), and special functional activities (transcriptase, reverse transcriptase, neuraminidase, fusion activities).
5. Genome organization and replication, including gene order, number and position of open reading frames, strategy of replication (patterns of transcription, translation), and cellular sites (accumulation of proteins, virion assembly, virion release).
6. Antigenic properties.
7. Biologic properties, including natural host range, mode of transmission, vector relationships, pathogenicity, tissue tropisms, and pathology.

Universal System of Virus Taxonomy

A system has been established in which viruses are separated into major groupings—called families—on the basis of virion morphology, genome structure, and strategies of replication. Virus family names have the suffix **-viridae**. Table 29–1 sets forth a convenient scheme used for classification. Diagrams of animal virus families are shown in a later figure (see Figure 29–3).

Within each family, subdivisions called genera are usually based on physicochemical or serologic differences. Criteria

TABLE 29–1 Families of Animal Viruses that Contain Members Able to Infect Humans

Nucleic Acid Core	Capsid Symmetry	Virion: Enveloped or Naked	Ether Sensitivity	Number of Capsomeres	Virus Particle Size (nm)[a]	Size of Nucleic Acid in Virion (kb/kbp)	Physical Type of Nucleic Acid[b]	Virus Family
DNA	Icosahedral	Naked	Resistant	32	18–26	5.6	ss	Parvoviridae
				72	45	5	ds circular	Polyomaviridae
				72	55	8	ds circular	Papillomaviridae
				252	70–90	26–45	ds	Adenoviridae
		Enveloped	Sensitive	180	40–48	3.2	ds circular[c]	Hepadnaviridae
				162	150–200	125–240	ds	Herpesviridae
	Complex	Complex coats	Resistant[d]		230 × 400	130–375	ds	Poxviridae
RNA	Icosahedral	Naked	Resistant	32	28–30	7.2–8.4	ss	Picornaviridae
					28–30	6.4–7.4	ss	Astroviridae
				32	27–40	7.4–8.3	ss	Caliciviridae
					27–34	7.2	ss	Hepeviridae
					60–80	16–27	ds segmented	Reoviridae
		Enveloped	Sensitive	42	50–70	9.7–11.8	ss	Togaviridae
	Unknown or complex	Enveloped	Sensitive		40–60	9.5–12.5	ss	Flaviviridae
					50–300	10–14	ss segmented	Arenaviridae
					120–160	27–32	ss	Coronaviridae
					80–110	7–11[e]	ss diploid	Retroviridae
	Helical	Enveloped	Sensitive		80–120	10–13.6	ss segmented	Orthomyxoviridae
					80–120	11–21	ss segmented	Bunyaviridae
					80–125	8.5–10.5	ss	Bornaviridae
					75×180	13–16	ss	Rhabdoviridae
					150–300	16–20	ss	Paramyxoviridae
					80×1000[f]	19.1	ss	Filoviridae

[a]Diameter, or diameter × length.

[b]ds, double-stranded; ss, single-stranded.

[c]The negative-sense strand has a constant length of 3.2 kb; the other varies in length, leaving a large single-stranded gap.

[d]The genus *Orthopoxvirus*, which includes the better-studied poxviruses (eg, vaccinia), is ether-resistant; some of the poxviruses belonging to other genera are ether-sensitive.

[e]Size of monomer.

[f]Filamentous forms vary greatly in length.

used to define genera vary from family to family. Genus names carry the suffix *-virus*. In four families (Poxviridae, Herpesviridae, Parvoviridae, Paramyxoviridae), a larger grouping called subfamilies has been defined, reflecting the complexity of relationships among member viruses. Virus orders may be used to group virus families that share common characteristics. For example, order Mononegavirales encompasses the Bornaviridae, Filoviridae, Paramyxoviridae, and Rhabdoviridae families.

By 2000, the International Committee on Taxonomy of Viruses had organized more than 4000 animal and plant viruses into 56 families and 233 genera, with hundreds of viruses still unassigned. Of these, 24 families contained viruses that infect humans and animals.

Properties of the major families of animal viruses that contain members important in human disease are summarized in Table 29–1. They are discussed briefly below, in the order shown in Table 29–1, and are considered in greater detail in the chapters that follow.

Survey of DNA-Containing Viruses

A. Parvoviruses

Parvoviruses are very small viruses with a particle size of about 18–26 nm. The particles have cubic symmetry, with 32 capsomeres, but they have no envelope. The genome is linear, single-stranded DNA, 5.6 kb in size. Replication occurs only in actively dividing cells; capsid assembly takes place in the nucleus of the infected cell. Many parvoviruses replicate autonomously, but the adeno-associated satellite viruses are defective, requiring the presence of an adenovirus or herpesvirus as "helper." Human parvovirus B19 replicates in immature erythroid cells and causes several adverse consequences, including aplastic crisis, fifth disease, and fetal death. (See Chapter 31.)

B. Polyomaviruses

Small (45 nm), nonenveloped, heat-stable, ether-resistant viruses exhibiting cubic symmetry, with 72 capsomeres. The genome is circular, double-stranded DNA, 5 kbp in size. These agents have a slow growth cycle, stimulate cell DNA synthesis, and replicate within the nucleus. The most well-known human polyomaviruses are JC virus, the causative agent of progressive multifocal leukoencephalopathy, and BK virus, associated with nephropathy in transplant recipients. SV40 also infects humans and has been recovered from human tumors. Most animal species harbor one or more polyomaviruses. They produce chronic infections in their natural hosts, and all can induce tumors in some animal species. Polyomaviruses were formerly a part of the Papovaviridae family before it was split into two families. (See Chapter 43.)

C. Papillomaviruses

Also a former member of the Papovaviridae family. Similar to polyomaviruses in some respects, but with a larger genome (8 kbp) and particle size (55 nm). There are many genotypes of human papillomaviruses, also known as "wart" viruses; certain types are causative agents of genital cancers in humans. Papillomaviruses are very host- and tissue-specific. Many animal species carry papillomaviruses. (See Chapter 43.)

D. Adenoviruses

Medium-sized (70–90 nm), nonenveloped viruses exhibiting cubic symmetry, with 252 capsomeres. Fibers protrude from the vertex capsomeres. The genome is linear, double-stranded DNA, 26–45 kbp in size. Replication occurs in the nucleus. Complex splicing patterns produce mRNAs. At least 51 types infect humans, especially in mucous membranes, and some types can persist in lymphoid tissue. Some adenoviruses cause acute respiratory diseases, conjunctivitis, and gastroenteritis. Some human adenoviruses can induce tumors in newborn hamsters. There are many serotypes that infect animals. (See Chapters 32 and 43.)

E. Hepadnaviruses

Small (40–48 nm) viruses containing circular double-stranded DNA molecules that are 3.2 kbp in size. The viral DNA in the particles contains a large single-stranded gap. Replication involves repair of the single-stranded gap in the DNA, transcription of RNA, and reverse transcription of the RNA to make genomic DNA. The virus consists of a 27-nm icosahedral nucleocapsid core within a closely adherent envelope that contains lipid and the viral surface antigen. The surface protein is characteristically overproduced during replication of the virus, which takes place in the liver, and is shed into the bloodstream. Hepadnaviruses cause acute and chronic hepatitis; persistent infections are associated with a high risk of developing liver cancer. Viral types are known that infect mammals and ducks. (See Chapter 35.)

F. Herpesviruses

A large family of viruses 150–200 nm in diameter. The nucleocapsid is 100 nm in diameter, with cubic symmetry and 162 capsomeres, surrounded by a lipid-containing envelope. The genome is linear, double-stranded DNA, 125–240 kbp in size. The presence of terminal and internal reiterated sequences results in several isomeric forms of genomic DNA. Virions contain over 30 proteins. Latent infections may last for the life span of the host, usually in ganglial or lymphoblastoid cells. Human herpesviruses include herpes simplex types 1 and 2 (oral and genital lesions), varicella-zoster virus (chickenpox and shingles), cytomegalovirus, Epstein-Barr virus (infectious mononucleosis and association with human neoplasms), human herpesviruses 6 and 7 (T lymphotropic), and human herpesvirus 8 (associated with Kaposi sarcoma). Other herpesviruses occur in many animals. (See Chapters 33 and 43.)

G. Poxviruses

Large brick-shaped or ovoid viruses 220–450 nm long × 140–260 nm wide × 140–260 nm thick. Particle structure is complex, with a lipid-containing envelope. The genome is linear, covalently closed, double-stranded DNA, 130–375 kbp in size. Poxvirus particles contain about 100 proteins, including many with enzymatic activities, such as a DNA-dependent RNA polymerase. Replication occurs entirely within the cell cytoplasm. All poxviruses tend to produce skin lesions. Some are pathogenic for humans (smallpox, vaccinia, molluscum contagiosum); others that are pathogenic for animals can infect humans (cowpox, monkeypox). (See Chapter 34.)

Survey of RNA-Containing Viruses

A. Picornaviruses

Small (28–30 nm), ether-resistant viruses exhibiting cubic symmetry. The RNA genome is single-stranded and positive-sense (ie, it can serve as an mRNA) and is 7.2–8.4 kb in size. The groups infecting humans are enteroviruses (polioviruses, coxsackieviruses, and echoviruses), rhinoviruses (more than 100 serotypes causing common colds), and hepatovirus (hepatitis A). Rhinoviruses are acid-labile and have a high density; enteroviruses are acid-stable and have a lower density. Picornaviruses infecting animals include foot-and-mouth disease of cattle and encephalomyocarditis of rodents. (See Chapter 36.)

B. Astroviruses

Similar in size to picornaviruses (28–30 nm), but particles display a distinctive star-shaped outline on their surface. The genome is linear, positive-sense, single-stranded RNA, 6.4–7.4 kb in size. These agents may be associated with gastroenteritis in humans and animals. (See Chapter 37.)

C. Caliciviruses

Similar to picornaviruses but slightly larger (27–40 nm). Particles appear to have cup-shaped depressions on the surface. The genome is single-stranded, positive-sense RNA, 7.4–8.3 kb in size; the virion has no envelope. An important human pathogen is Norwalk virus, the cause of epidemic acute gastroenteritis. Other agents infect cats and sea lions as well as primates. (See Chapter 37.)

D. Hepeviruses

Similar to caliciviruses, particles are small (27–34 nm) and ether-resistant. The genome is single-stranded, positive-sense RNA, 7.2 kb in size. Human hepatitis E virus belongs in this group (see Chapter 35).

E. Reoviruses

Medium-sized (60–80 nm), ether-resistant, nonenveloped viruses having icosahedral symmetry. Particles have two or three protein shells with channels extending from the surface to the core; short spikes extend from the virion surface. The genome is linear, double-stranded, segmented RNA (10–12 segments), totaling 16–27 kbp in size. Individual RNA segments range in size from 680 to 3900 bp. Replication occurs in the cytoplasm; genome segment reassortment occurs readily. Reoviruses of humans include rotaviruses, which have a distinctive wheel-shaped appearance and cause gastroenteritis. Antigenically similar reoviruses infect many animals. The genus *Coltivirus* includes Colorado tick fever virus of humans. (See Chapter 37.)

F. Arboviruses

An ecologic grouping (not a virus family) of viruses with diverse physical and chemical properties. All of these viruses (there are over 350 of them) have a complex cycle involving arthropods as vectors that transmit the viruses to vertebrate hosts by their bite. Viral replication does not seem to harm the infected arthropod. Arboviruses infect humans, mammals, birds, and snakes and use mosquitoes and ticks as vectors. Human pathogens include dengue, yellow fever, encephalitis viruses, and others. Arboviruses belong to several virus families, including togavirus, flavivirus, bunyavirus, rhabdovirus, arenavirus, and reovirus. (See Chapter 38.)

G. Togaviruses

Many arboviruses that are major human pathogens, called alphaviruses—as well as rubella virus—belong in this group. They have a lipid-containing envelope and are ether-sensitive, and their genome is single-stranded, positive-sense RNA, 9.7–11.8 kb in size. The enveloped virion measures 70 nm. The virus particles mature by budding from host cell membranes. An example is eastern equine encephalitis virus. Rubella virus has no arthropod vector. (See Chapters 38 and 40.)

H. Flaviviruses

Enveloped viruses, 40–60 nm in diameter, containing single-stranded, positive-sense RNA. Genome sizes vary from 9.5 kb (hepatitis C) to 11 kb (flaviviruses) to 12.5 kb (pestiviruses). Mature virions accumulate within cisternae of the endoplasmic reticulum. This group of arboviruses includes yellow fever virus and dengue viruses. Most members are transmitted by blood-sucking arthropods. Hepatitis C virus has no known vector. (See Chapters 35 and 38.)

I. Arenaviruses

Pleomorphic, enveloped viruses ranging in size from 50 to 300 nm (mean, 110–130 nm). The genome is segmented, circular, single-stranded RNA that is negative-sense and ambisense, 10–14 kb in total size. Replication occurs in the cytoplasm with assembly via budding on the plasma membrane. The virions incorporate host cell ribosomes during maturation, which gives the particles a "sandy" appearance. Most members of this family are unique to tropical America (ie, the Tacaribe complex). All arenaviruses pathogenic for humans cause chronic infections in rodents. Lassa fever virus of Africa is one example. These viruses require maximum containment conditions in the laboratory. (See Chapter 38.)

J. Coronaviruses

Enveloped 120- to 160-nm particles containing an unsegmented genome of positive-sense, single-stranded RNA, 27–32 kb in size; the nucleocapsid is helical, 9–11 nm in diameter. Coronaviruses resemble orthomyxoviruses but have petal-shaped surface projections arranged in a fringe, like a solar corona. Coronavirus nucleocapsids develop in the cytoplasm and mature by budding into cytoplasmic vesicles. These viruses have narrow host ranges. Most human coronaviruses cause mild acute upper respiratory tract illnesses—"colds"—but a new coronavirus identified in 2003 causes a severe acute respiratory syndrome (SARS). Toroviruses, which cause gastroenteritis, form a distinct genus. Coronaviruses of animals readily establish persistent infections and include mouse hepatitis virus and avian infectious bronchitis virus. (See Chapter 41.)

K. Retroviruses

Spherical, enveloped viruses (80–110 nm in diameter) whose genome contains two copies of linear, positive-sense, single-stranded RNA of the same polarity as viral mRNA. Each monomer RNA is 7–11 kb in size. Particles contain a helical nucleocapsid within an icosahedral capsid. Replication is unique; the virion contains a reverse transcriptase enzyme that produces a DNA copy of the RNA genome. This DNA becomes circularized and integrated into host chromosomal DNA. The virus is then replicated from the integrated "provirus" DNA copy. Virion assembly occurs by budding on plasma membranes. Hosts remain chronically infected. Retroviruses are widely distributed; there are also endogenous proviruses resulting from ancient infections of germ cells transmitted as inherited genes in most species. Leukemia and sarcoma viruses of animals and humans (see Chapter 43), foamy viruses of primates, and lentiviruses (human immunodeficiency viruses; visna of sheep) (see Chapters 42 and 44) are included in this group. Retroviruses cause acquired immunodeficiency syndrome (AIDS) (see Chapter 44) and made possible the identification of cellular oncogenes (see Chapter 43).

L. Orthomyxoviruses

Medium-sized, 80- to 120-nm enveloped viruses exhibiting helical symmetry. Particles are either round or filamentous, with surface projections that contain hemagglutinin or neuraminidase activity. The genome is linear, segmented, negative-sense, single-stranded RNA, totaling 10–13.6 kb in size. Segments range from 900 to 2350 nucleotides each. The internal nucleoprotein helix measures 9–15 nm. During replication, the nucleocapsid is assembled in the nucleus, whereas the hemagglutinin and neuraminidase accumulate in the cytoplasm. The virus matures by budding at the cell membrane. All orthomyxoviruses are influenza viruses that infect humans or animals. The segmented nature of the viral genome permits ready genetic reassortment when two influenza viruses infect the same cell, presumably fostering the high rate of natural variation among influenza viruses. Transmission from other species is thought to explain the emergence of new human pandemic strains of influenza A viruses. (See Chapter 39.)

M. Bunyaviruses

Spherical or pleomorphic, 80- to 120-nm enveloped particles. The genome is made up of a triple-segmented, circular, single-stranded, negative-sense or ambisense RNA, 11–19 kb in overall size. Virion particles contain three circular, helically symmetric nucleocapsids about 2.5 nm in diameter and 200–3000 nm in length. Replication occurs in the cytoplasm, and an envelope is acquired by budding into the Golgi. The majority of these viruses are transmitted to vertebrates by arthropods (arboviruses). Hantaviruses are transmitted not by arthropods but by persistently infected rodents, via aerosols of contaminated excreta. They cause hemorrhagic fevers and nephropathy as well as a severe pulmonary syndrome. (See Chapter 38.)

N. Bornaviruses

Enveloped, spherical (80–125 nm) viruses. The genome is linear, single-stranded, nonsegmented, negative-sense RNA, 8.5–10.5 kb in size. Unique among nonsegmented, negative-sense RNA viruses, replication and transcription of the viral genome occur in the nucleus. Borna disease virus is neurotropic in animals; postulated association with neuropsychiatric disorders of humans is unproved (see Chapter 42).

O. Rhabdoviruses

Enveloped virions resembling a bullet, flat at one end and round at the other, measuring about 75 × 180 nm. The envelope has 10-nm spikes. The genome is linear, single-stranded, nonsegmented, negative-sense RNA, 13–16 kb in size. Particles are formed by budding from the cell membrane. Viruses have broad host ranges. Rabies virus is a member of this group. (See Chapter 42.)

P. Paramyxoviruses

Similar to but larger (150–300 nm) than orthomyxoviruses. Particles are pleomorphic. The internal nucleocapsid measures 13–18 nm, and the linear, single-stranded, nonsegmented, negative-sense RNA is 16–20 kb in size. Both the nucleocapsid and the hemagglutinin are formed in the cytoplasm. Those infecting humans include mumps, measles, parainfluenza, and respiratory syncytial viruses. These viruses have narrow host ranges. In contrast to influenza viruses, paramyxoviruses are genetically stable. (See Chapter 40.)

Q. Filoviruses

Enveloped, pleomorphic viruses that may appear very long and thread-like. They typically are 80 nm wide and about 1000 nm long. The envelope contains large peplomers. The genome is linear, negative-sense, single-stranded RNA, 19 kb in size. Marburg and Ebola viruses cause severe hemorrhagic fever in Africa. These viruses require maximum containment conditions (Biosafety Level 4) for handling. (See Chapter 38.)

R. Other Viruses

Insufficient information to permit classification. This applies to some viruses of gastroenteritis (see Chapter 37).

S. Viroids

Small infectious agents that cause diseases of plants. Viroids are agents that do not fit the definition of classic viruses. They are nucleic acid molecules (MW 70,000–120,000) without a protein coat. Plant viroids are single-stranded, covalently closed circular RNA molecules consisting of about 360 nucleotides and with a highly base-paired rod-like structure. Viroids replicate by an entirely novel mechanism. Viroid RNA does not encode any protein products; the devastating plant diseases induced by viroids occur by an unknown mechanism. To date, viroids have been detected only in plants; none have been demonstrated to exist in animals or humans.

T. Prions

Infectious particles composed solely of protein with no detectable nucleic acid. Highly resistant to inactivation by heat, formaldehyde, and ultraviolet light that inactivate viruses. The prion protein is encoded by a single cellular gene. Prion diseases, called "transmissible spongiform encephalopathies," include scrapie in sheep, mad cow disease in cattle, and kuru and Creutzfeldt-Jakob disease in humans. Prions do not appear to be viruses. (See Chapter 42.)

PRINCIPLES OF VIRUS STRUCTURE

Viruses come in many shapes and sizes. Structural information is necessary for virus classification and for establishing structure–function relationships of viral proteins. The particular structural features of each virus family are determined by the functions of the virion: morphogenesis and release from infected cells; transmission to new hosts; and attachment, penetration, and uncoating in newly infected cells. Knowledge of virus structure furthers our understanding of the mechanisms of certain processes such as the interaction of virus particles with cell surface receptors and neutralizing antibodies. It may lead also to the rational design of antiviral drugs capable of blocking viral attachment, uncoating, or assembly in susceptible cells.

Types of Symmetry of Virus Particles

Electron microscopy, cryoelectron microscopy, and x-ray diffraction techniques have made it possible to resolve fine differences in the basic morphology of viruses. The study of viral symmetry by standard electron microscopy requires the use of heavy metal stains (eg, potassium phosphotungstate) to emphasize surface structure. The heavy metal permeates the virus particle like a cloud and brings out the surface structure of viruses by virtue of "negative staining." The typical level of resolution is 3–4 nm. (The size of a DNA double helix is 2 nm.) However, conventional methods of sample preparation often cause distortions and changes in particle morphology. Cryoelectron microscopy uses virus samples quick-frozen in vitreous ice; fine structural features are preserved, and the use of negative stains is avoided. Three-dimensional structural information can be obtained by the use of computer image processing procedures. Examples of image reconstructions of virus particles are shown in following chapters. (See Chapters 32 and 37.)

X-ray crystallography can provide atomic resolution information, generally at a level of 0.2–0.3 nm. The specimen must be crystalline, and this has only been achieved with small, nonenveloped viruses. However, it is possible to obtain high-resolution structural data on well-defined substructures prepared from the more complex viruses.

Genetic economy requires that a viral structure be made from many identical molecules of one or a few proteins. Viral architecture can be grouped into three types based on the arrangement of morphologic subunits: (1) cubic symmetry, eg, adenoviruses; (2) helical symmetry, eg, orthomyxoviruses; and (3) complex structures, eg, poxviruses.

A. Cubic Symmetry

All cubic symmetry observed with animal viruses is of the icosahedral pattern, the most efficient arrangement for subunits in a closed shell. The icosahedron has 20 faces (each an equilateral triangle), 12 vertices, and fivefold, threefold, and twofold axes of rotational symmetry. The vertex units have five neighbors (pentavalent), and all others have six (hexavalent).

There are exactly 60 identical subunits on the surface of an icosahedron. In order to build a particle size adequate to encapsidate viral genomes, viral shells are composed of multiples of 60 structural units. The use of larger numbers of chemically identical protein subunits, while maintaining the rules of icosahedral symmetry, is accomplished by subtriangulation of each face of an icosahedron. An overview of the packing of subunits in picornaviruses is presented in Figure 29–2.

Most viruses that have icosahedral symmetry do not have an icosahedral shape—rather, the physical appearance of the particle is spherical.

The viral nucleic acid is condensed within the isometric particles; virus-encoded core proteins—or, in the case of

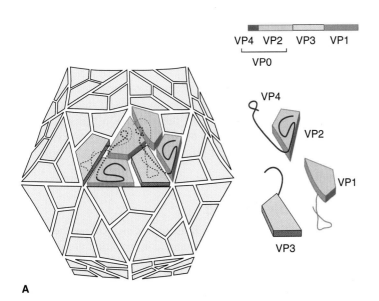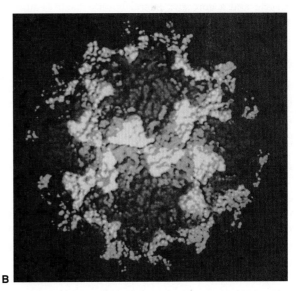

FIGURE 29–2 **A:** Overview of the packing of subunits in picornaviruses. The proteins are cleaved from a precursor, as shown. VP1–VP3 are represented by wedge-shaped blocks (the viral capsid β-barrel domains) with amino and carboxyl terminal extensions. The amino terminal extensions interdigitate to form an internal framework. VP4 is, in effect, part of the amino terminal extension of VP0. In poliovirus and rhinovirus, the prominent GH of the VP1 loop lies across VP2 and VP3 as shown. An "exploded" view of one protomer is shown at the right. **B:** Surface view of poliovirus in the same orientation as in **(A)**. (Reproduced with permission from Harrison SC, Skehel JJ, Wiley DC: Virus structure. In: *Fields Virology*, 3rd ed. Fields BN et al [editors]. Lippincott-Raven, 1996.)

polyomaviruses and papillomaviruses, cellular histones—are involved in condensation of the nucleic acid into a form suitable for packaging. "Packaging sequences" on viral nucleic acid are involved in assembly into virus particles. There are size constraints on the nucleic acid molecules that can be packaged into a given icosahedral capsid. Icosahedral capsids are formed independently of nucleic acid. Most preparations of isometric viruses will contain some "empty" particles devoid of viral nucleic acid. Expression of capsid proteins from cloned genes often results in self-assembly and formation of empty "virus-like particles." Both DNA and RNA viral groups exhibit examples of cubic symmetry.

B. Helical Symmetry

In cases of helical symmetry, protein subunits are bound in a periodic way to the viral nucleic acid, winding it into a helix. The filamentous viral nucleic acid–protein complex (nucleocapsid) is then coiled inside a lipid-containing envelope. Thus, as is not the case with icosahedral structures, there is a regular, periodic interaction between capsid protein and nucleic acid in viruses with helical symmetry. It is not possible for "empty" helical particles to form.

All known examples of animal viruses with helical symmetry contain RNA genomes and, with the exception of rhabdoviruses, have flexible nucleocapsids that are wound into a ball inside envelopes (Figures 29–1B, 29–3, and 42–1).

C. Complex Structures

Some virus particles do not exhibit simple cubic or helical symmetry but are more complicated in structure. For example, poxviruses are brick-shaped, with ridges on the external surface and a core and lateral bodies inside (Figures 29–3 and 34–1).

Measuring the Sizes of Viruses

Small size and ability to pass through filters that hold back bacteria are classic attributes of viruses. However, because some bacteria may be smaller than the largest viruses, filterability is not regarded as a unique feature of viruses.

Direct observation in the electron microscope is the most widely used method for estimating particle size. Viruses can be visualized in preparations from tissue extracts and in ultrathin sections of infected cells. Another method that can be used is sedimentation in the ultracentrifuge. In an ultracentrifuge, forces of more than 100,000 times gravity may be used to drive the particles to the bottom of the tube. The relationship between the size and shape of a particle and its rate of sedimentation permits determination of particle size.

A. Comparative Measurements

See Table 29–1. For purposes of reference, the following data should be recalled: (1) *Staphylococcus* has a diameter of about 1000 nm. (2) Bacterial viruses (bacteriophages) vary in size

(10–100 nm). Some are spherical or hexagonal and have short or long tails. (3) Representative protein molecules range in diameter from serum albumin (5 nm) and globulin (7 nm) to certain hemocyanins (23 nm).

The relative sizes and morphology of various virus families are shown in Figure 29–3. Particles with a twofold difference in diameter have an eightfold difference in volume. Thus, the mass of a poxvirus is about 1000 times greater than that of the poliovirus particle, and the mass of a small bacterium is 50,000 times greater.

CHEMICAL COMPOSITION OF VIRUSES

Viral Protein

The structural proteins of viruses have several important functions. Their major purpose is to facilitate transfer of the viral nucleic acid from one host cell to another. They serve to protect the viral genome against inactivation by nucleases, participate in the attachment of the virus particle to a susceptible cell, and provide the structural symmetry of the virus particle.

The proteins determine the antigenic characteristics of the virus. The host's protective immune response is directed against antigenic determinants of proteins or glycoproteins exposed on the surface of the virus particle. Some surface proteins may also exhibit specific activities, eg, influenza virus hemagglutinin agglutinates red blood cells.

Some viruses carry enzymes (which are proteins) inside the virions. The enzymes are present in very small amounts and are probably not important in the structure of the virus particles; however, they are essential for the initiation of the viral replicative cycle when the virion enters a host cell. Examples include an RNA polymerase carried by viruses with negative-sense RNA genomes (eg, orthomyxoviruses, rhabdoviruses) that is needed to copy the first mRNAs, and reverse transcriptase, an enzyme in retroviruses that makes a DNA copy of the viral RNA, an essential step in replication and transformation. At the extreme in this respect are the poxviruses, the cores of which contain a transcriptional system; many different enzymes are packaged in poxvirus particles.

Viral Nucleic Acid

Viruses contain a single kind of nucleic acid—either DNA or RNA—that encodes the genetic information necessary for replication of the virus. The genome may be single-stranded or double-stranded, circular or linear, and segmented or nonsegmented. The type of nucleic acid, its strandedness, and its size are major characteristics used for classifying viruses into families (Table 29–1).

The size of the viral DNA genome ranges from 3.2 kbp (hepadnaviruses) to 375 kbp (poxviruses). The size of the viral RNA genome ranges from about 7 kb (some picornaviruses and astroviruses) to 30 kb (coronaviruses).

DNA viruses

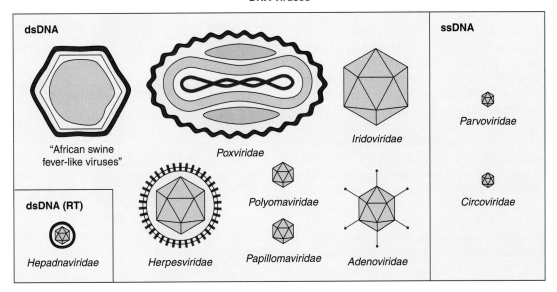

RNA viruses

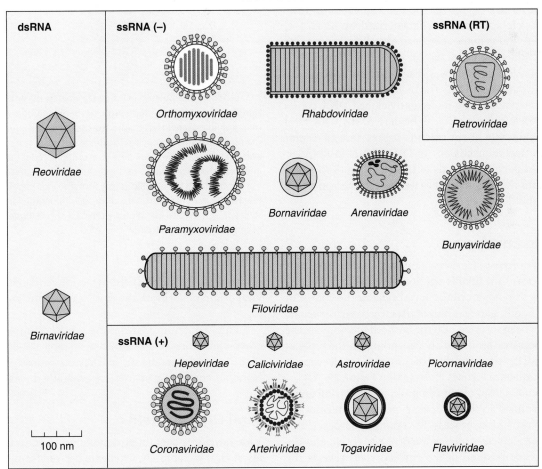

FIGURE 29–3 Shapes and relative sizes of animal viruses of families that infect vertebrates. In some diagrams, certain internal structures of the particles are represented. Only those families that include human pathogens are listed in Table 29–1 and described in the text. (Reproduced with permission from van Regenmortel MHV et al [editors]: Virus taxonomy: Classification and nomenclature of viruses. Seventh report of the International Committee on Taxonomy of Viruses. Academic Press, 2000.)

All major DNA viral groups in Table 29–1 have genomes that are single molecules of DNA and have a linear or circular configuration.

Viral RNAs exist in several forms. The RNA may be a single linear molecule (eg, picornaviruses). For other viruses (eg, orthomyxoviruses), the genome consists of several segments of RNA that may be loosely associated within the virion. The isolated RNA of viruses with positive-sense genomes (ie, picornaviruses, togaviruses) is infectious, and the molecule functions as an mRNA within the infected cell. The isolated RNA of the negative-sense RNA viruses, such as rhabdoviruses and orthomyxoviruses, is not infectious. For these viral families, the virions carry an RNA polymerase that in the cell transcribes the genomic RNA molecules into several complementary RNA molecules, each of which may serve as an mRNA.

The sequence and composition of nucleotides of each viral nucleic acid are distinctive. Many viral genomes have been sequenced. The sequences can reveal genetic relationships among isolates, including unexpected relationships between viruses not thought to be closely related. The number of genes in a virus can be estimated from the open reading frames deduced from the nucleic acid sequence.

Viral nucleic acid may be characterized by its G + C content. DNA viral genomes can be analyzed and compared using restriction endonucleases, enzymes that cleave DNA at specific nucleotide sequences. Each genome will yield a characteristic pattern of DNA fragments after cleavage with a particular enzyme. Using molecularly cloned DNA copies of RNA, restriction maps also can be derived for RNA viral genomes. Polymerase chain reaction assays and molecular hybridization techniques (DNA to DNA, DNA to RNA, or RNA to RNA) permit the study of transcription of the viral genome within the infected cell as well as comparison of the relatedness of different viruses.

Viral Lipid Envelopes

A number of different viruses contain lipid envelopes as part of their structure. The lipid is acquired when the viral nucleocapsid buds through a cellular membrane in the course of maturation. Budding occurs only at sites where virus-specific proteins have been inserted into the host cell membrane. The budding process varies markedly depending on the replication strategy of the virus and the structure of the nucleocapsid. Budding by influenza virus is illustrated in Figure 29–4.

The specific phospholipid composition of a virion envelope is determined by the specific type of cell membrane involved in the budding process. For example, herpesviruses bud through the nuclear membrane of the host cell, and the phospholipid composition of the purified virus reflects the lipids of the nuclear membrane. The acquisition of a lipid-containing membrane is an integral step in virion morphogenesis in some viral groups (see Replication of Viruses, below).

There are always viral glycosylated proteins protruding from the envelope and exposed on the external surface

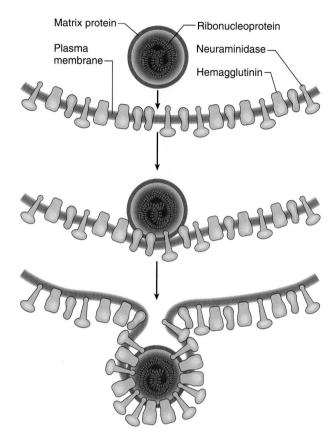

FIGURE 29–4 Release of influenza virus by plasma membrane budding. First, viral envelope proteins (hemagglutinin and neuraminidase) are inserted into the host plasma membrane. Then the nucleocapsid approaches the inner surface of the membrane and binds to it. At the same time viral proteins collect at the site, and host membrane proteins are excluded. Finally, the plasma membrane buds to simultaneously form the viral envelope and release the mature virion. (Reproduced with permission from Willey JM, Sherwood LM, Woolverton CJ: *Prescott, Harley, and Klein's Microbiology*, 7th ed. McGraw-Hill, 2008.)

of the virus particle. There are unglycosylated proteins of viral origin underneath the envelope that anchor the particle together.

Lipid-containing viruses are sensitive to treatment with ether and other organic solvents (Table 29–1), indicating that disruption or loss of lipid results in loss of infectivity. Nonlipid-containing viruses are generally resistant to ether.

Viral Glycoproteins

Viral envelopes contain glycoproteins. In contrast to the lipids in viral membranes, which are derived from the host cell, the envelope glycoproteins are virus-encoded. However, the sugars added to viral glycoproteins often reflect the host cell in which the virus is grown.

It is the surface glycoproteins of an enveloped virus that attach the virus particle to a target cell by interacting with a cellular receptor. They are also often involved in the membrane fusion step of infection. The glycoproteins are also important viral antigens. As a result of their position at the outer surface of the virion, they are frequently involved in the interaction of the virus particle with neutralizing antibody. Extensive glycosylation of viral surface proteins may prevent effective neutralization of a virus particle by specific antibody. The three-dimensional structures of the externally exposed regions of both of the influenza virus membrane glycoproteins (hemagglutinin, neuraminidase) have been determined by x-ray crystallography (see Figure 39–2). Such studies provide insights into the antigenic structure and functional activities of viral glycoproteins.

CULTIVATION & ASSAY OF VIRUSES

Cultivation of Viruses

Many viruses can be grown in cell cultures or in fertile eggs under strictly controlled conditions. Growth of virus in animals is still used for the primary isolation of certain viruses and for studies of the pathogenesis of viral diseases and of viral oncogenesis. Diagnostic laboratories attempt to recover viruses from clinical samples to establish disease causes (see Chapter 47). Research laboratories cultivate viruses as the basis for detailed analyses of viral expression and replication.

Cells grown in vitro are central to the cultivation and characterization of viruses. There are three basic types of cell cultures. Primary cultures are made by dispersing cells (usually with trypsin) from freshly removed host tissues. In general, they are unable to grow for more than a few passages in culture. Diploid cell lines are secondary cultures which have undergone a change that allows their limited culture (up to 50 passages) but which retain their normal chromosome pattern. Continuous cell lines are cultures capable of more prolonged—perhaps indefinite—growth that have been derived from diploid cell lines or from malignant tissues. They invariably have altered and irregular numbers of chromosomes. The type of cell culture used for viral cultivation depends on the sensitivity of the cells to a particular virus.

A. Detection of Virus-Infected Cells

Multiplication of a virus can be monitored in a variety of ways:

1. Development of cytopathic effects, ie, morphologic changes in the cells. Types of virus-induced cytopathic effects include cell lysis or necrosis, inclusion formation, giant cell formation, and cytoplasmic vacuolization (Figure 29–5A, B, and C). Most viruses produce some obvious cytopathic effect in infected cells.
2. Appearance of a virus-encoded protein, such as the hemagglutinin of influenza virus. Specific antisera can be used to detect the synthesis of viral proteins in infected cells.
3. Detection of virus-specific nucleic acid. Molecular-based assays such as polymerase chain reaction provide rapid, sensitive, and specific methods of detection.

4. Adsorption of erythrocytes to infected cells, called hemadsorption, due to the presence of virus-encoded hemagglutinin (parainfluenza, influenza) in cellular membranes. This reaction becomes positive before cytopathic changes are visible and in some cases occurs in the absence of cytopathic effects (Figure 29–5D).
5. Viral growth in an embryonated chick egg may result in death of the embryo (eg, encephalitis viruses), production of pocks or plaques on the chorioallantoic membrane (eg, herpes, smallpox, vaccinia), or development of hemagglutinins in the embryonic fluids or tissues (eg, influenza).

B. Inclusion Body Formation

In the course of viral multiplication within cells, virus-specific structures called inclusion bodies may be produced. They become far larger than the individual virus particle and often have an affinity for acid dyes (eg, eosin). They may be situated

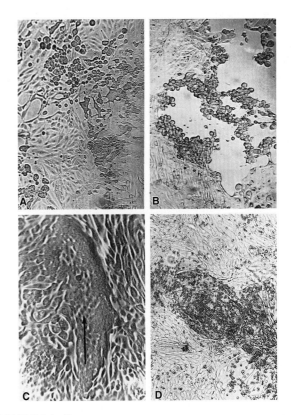

FIGURE 29–5 Cytopathic effects produced in monolayers of cultured cells by different viruses. The cultures are shown as they would normally be viewed in the laboratory, unfixed and unstained (60×). **A:** Enterovirus—rapid rounding of cells progressing to complete cell destruction. **B:** Herpesvirus—focal areas of swollen rounded cells. **C:** Paramyxovirus—focal areas of fused cells (syncytia). **D:** Hemadsorption. Erythrocytes adhere to those cells in the monolayer that are infected by a virus that causes a hemagglutinin to be incorporated into the plasma membrane. Many enveloped viruses that mature by budding from cytoplasmic membranes produce hemadsorption. (Used with permission from I Jack.)

in the nucleus (herpesvirus; see Figure 33–3), in the cytoplasm (poxvirus), or in both (measles virus; see Figure 40–5). In many viral infections, the inclusion bodies are the site of development of the virions (the viral factories). Variations in the appearance of inclusion material depend largely upon the tissue fixative used.

The presence of inclusion bodies may be of considerable diagnostic aid. The intracytoplasmic inclusion in nerve cells—the Negri body—is pathognomonic for rabies.

Quantitation of Viruses

A. Physical Methods

Quantitative nucleic acid-based assays such as the polymerase chain reaction can determine the number of viral genome copies in a sample. Both infectious and noninfectious genomes are detected. Virus sequence variation may reduce virus detection and quantitation by this method.

A variety of serologic tests such as radioimmunoassays (RIA) and enzyme-linked immunosorbent assays (ELISA; see Chapter 47) can be standardized to quantitate the amount of virus in a sample. These tests do not distinguish infectious from noninfectious particles and sometimes detect viral proteins not assembled into particles.

Certain viruses contain a protein (hemagglutinin) that has the ability to agglutinate red blood cells of humans or some animal. Hemagglutination assays are an easy and rapid method of quantitating these types of viruses (see Chapter 47). Both infective and noninfective particles give this reaction; thus, hemagglutination measures the total quantity of virus present.

Virus particles can be counted directly in the electron microscope by comparison with a standard suspension of latex particles of similar small size. However, a relatively concentrated preparation of virus is necessary for this procedure, and infectious virus particles cannot be distinguished from noninfectious ones.

B. Biologic Methods

End point biologic assays depend on the measurement of animal death, animal infection, or cytopathic effects in tissue culture at a series of dilutions of the virus being tested. The titer is expressed as the 50% infectious dose (ID_{50}), which is the reciprocal of the dilution of virus that produces the effect in 50% of the cells or animals inoculated. Precise assays require the use of a large number of test subjects.

The most widely used assay for infectious virus is the plaque assay. Monolayers of host cells are inoculated with suitable dilutions of virus and after adsorption are overlaid with medium containing agar or carboxymethylcellulose to prevent virus spreading throughout the culture. After several days, the cells initially infected have produced virus that spreads only to surrounding cells. Multiple cycles of replication and cell killing produce a small area of infection, or plaque. The length of time from infection to when plaques can be visualized for counting

depends on the replication cycle of the virus and can range from a few days (eg, poliovirus) to 2 weeks or more (eg, SV40). Under controlled conditions, a single plaque can arise from a single infectious virus particle, termed a plaque-forming unit (PFU). The cytopathic effect of infected cells within the plaque can be distinguished from uninfected cells of the monolayer with or without suitable staining, and plaques can usually be counted macroscopically. The ratio of the number of infectious particles to the total number of virus particles varies widely, from near unity to less than one per 1000, but often is one per several hundred.

Certain viruses, eg, herpes and vaccinia, form pocks when inoculated onto the chorioallantoic membrane of an embryonated egg. Such viruses can be quantitated by relating the number of pocks counted to the viral dilution inoculated.

PURIFICATION & IDENTIFICATION OF VIRUSES

Purification of Virus Particles

Pure virus must be available in order for certain types of studies on the properties and molecular biology of the agent to be carried out. For purification studies, the starting material is usually large volumes of tissue culture medium, body fluids, or infected cells. The first step frequently involves concentration of the virus particles by precipitation with ammonium sulfate, ethanol, or polyethylene glycol or by ultrafiltration. Hemagglutination and elution can be used to concentrate orthomyxoviruses (see Chapter 39). Once concentrated, virus can then be separated from host materials by differential centrifugation, density gradient centrifugation, column chromatography, and electrophoresis.

More than one step is usually necessary to achieve adequate purification. A preliminary purification will remove most nonviral material. This first step may include centrifugation; the final purification step almost always involves density gradient centrifugation. In rate-zonal centrifugation, a sample of concentrated virus is layered onto a preformed linear density gradient of sucrose or glycerol, and during centrifugation the virus sediments as a band at a rate determined primarily by the size and weight of the virus particle.

Viruses can also be purified by high-speed centrifugation in density gradients of cesium chloride, potassium tartrate, potassium citrate, or sucrose. The gradient material of choice is the one that is least toxic to the virus. Virus particles migrate to an equilibrium position where the density of the solution is equal to their buoyant density and form a visible band.

Additional methods for purification are based on the chemical properties of the viral surface. In column chromatography, virus is bound to a substance such as diethylaminoethyl or phosphocellulose and then eluted by changes in pH or salt concentration. Zone electrophoresis permits the separation of virus particles from contaminants on the basis of charge. Specific antisera also can be used to remove virus particles from host materials.

Icosahedral viruses are easier to purify than enveloped viruses. Because the latter usually contain variable amounts of envelope per particle, the viral population is heterogeneous in both size and density.

It is very difficult to achieve complete purity of viruses. Small amounts of cellular material tend to adsorb to particles and copurify. The minimal criteria for purity are a homogeneous appearance in electron micrographs and the failure of additional purification procedures to remove "contaminants" without reducing infectivity.

Identification of a Particle as a Virus

When a characteristic physical particle has been obtained, it should fulfill the following criteria before it is identified as a virus particle:

1. The particle can be obtained only from infected cells or tissues.
2. Particles obtained from various sources are identical regardless of the cellular origin in which the virus is grown.
3. Particles contain nucleic acid (DNA or RNA), the sequence of which is not the same as the species of host cells from which the particles were obtained.
4. The degree of infective activity of the preparation varies directly with the number of particles present.
5. Destruction of the physical particle by chemical or physical means is associated with a loss of viral activity.
6. Certain properties of the particles and infectivity must be shown to be identical, eg, their sedimentation behavior in the ultracentrifuge and their pH stability curves.
7. Antisera prepared against the infectious virus should react with the characteristic particle and vice versa. Direct observation of an unknown virus can be accomplished by electron microscopic examination of aggregate formation in a mixture of antisera and crude viral suspension.
8. The particles should be able to induce the characteristic disease in vivo (if such experiments are feasible).
9. Passage of the particles in tissue culture should result in the production of progeny with biologic and antigenic properties of the virus.

LABORATORY SAFETY

Many viruses are human pathogens, and laboratory-acquired infections can occur. Laboratory procedures are often potentially hazardous if proper technique is not followed. Among the common hazards that might expose laboratory personnel to the risk of infection are the following: (1) aerosols—generated by homogenization of infected tissues, centrifugation, ultrasonic vibration, broken glassware; (2) ingestion—from mouth pipetting, eating or smoking in the laboratory, inadequate washing of hands; (3) skin penetration—from needle sticks, broken glassware, hand contamination by leaking containers, handling of infected tissues, animal bites; and (4) splashes into the eye.

Good biosafety practices include the following: (1) training in and use of aseptic techniques; (2) interdiction of mouth pipetting; (3) no eating, drinking, or smoking in the laboratory; (4) use of personal protective equipment (coats, gloves, masks, etc) not to be worn outside the laboratory; (5) sterilization of experimental wastes; (6) use of biosafety hoods; and (7) immunization if relevant vaccines are available. Additional precautions and special containment facilities (Biosafety Level 4) are necessary when personnel are working with high-risk agents such as the filoviruses (see Chapter 38) and rabies virus.

REACTION TO PHYSICAL & CHEMICAL AGENTS

Heat & Cold

There is great variability in the heat stability of different viruses. Icosahedral viruses tend to be stable, losing little infectivity after several hours at 37°C. Enveloped viruses are much more heat-labile, rapidly dropping in titer at 37°C. Viral infectivity is generally destroyed by heating at 50–60°C for 30 minutes, though there are some notable exceptions (eg, hepatitis B virus, polyomaviruses).

Viruses can be preserved by storage at subfreezing temperatures, and some may withstand lyophilization and can thus be preserved in the dry state at 4°C or even at room temperature. Viruses that withstand lyophilization are more heat-resistant when heated in the dry state. Enveloped viruses tend to lose infectivity after prolonged storage even at –90°C and are particularly sensitive to repeated freezing and thawing.

Stabilization of Viruses by Salts

Many viruses can be stabilized by salts in concentrations of 1 mol/L, ie, the viruses are not inactivated even by heating at 50°C for 1 hour. The mechanism by which the salts stabilize viral preparations is not known. Viruses are preferentially stabilized by certain salts. $MgCl_2$, 1 mol/L, stabilizes picornaviruses and reoviruses; $MgSO_4$, 1 mol/L, stabilizes orthomyxoviruses and paramyxoviruses; and Na_2SO_4, 1 mol/L, stabilizes herpesviruses.

The stability of viruses is important in the preparation of vaccines. The ordinary nonstabilized oral polio vaccine must be stored at freezing temperatures to preserve its potency. However, with the addition of salts for stabilization of the virus, potency can be maintained for weeks at ambient temperatures even in the high temperatures of the tropics.

pH

Viruses are usually stable between pH values of 5.0 and 9.0. Some viruses (eg, enteroviruses) are resistant to acidic conditions. All viruses are destroyed by alkaline conditions. In hemagglutination reactions, variations of less than one pH unit may influence the result.

Radiation

Ultraviolet, x-ray, and high-energy particles inactivate viruses. The dose varies for different viruses. Infectivity is the most radiosensitive property because replication requires expression of the entire genetic contents. Irradiated particles that are unable to replicate may still be able to express some specific functions in host cells.

Ether Susceptibility

Ether susceptibility can be used to distinguish viruses that possess an envelope from those that do not. Ether sensitivity of different virus groups is shown in Table 29–1.

Detergents

Nonionic detergents—eg, Nonidet P40 and Triton X-100—solubilize lipid constituents of viral membranes. The viral proteins in the envelope are released (undenatured). Anionic detergents, eg, sodium dodecyl sulfate, also solubilize viral envelopes; in addition, they disrupt capsids into separated polypeptides.

Formaldehyde

Formaldehyde destroys viral infectivity by reacting with nucleic acid. Viruses with single-stranded genomes are inactivated much more readily than those with double-stranded genomes. Formaldehyde has minimal adverse effects on the antigenicity of proteins and therefore has been used frequently in the production of inactivated viral vaccines.

Photodynamic Inactivation

Viruses are penetrable to a varying degree by vital dyes such as toluidine blue, neutral red, and proflavine. These dyes bind to the viral nucleic acid, and the virus then becomes susceptible to inactivation by visible light. Neutral red is commonly used to stain plaque assays so that plaques are more readily seen. The assay plates must be protected from bright light once the neutral red has been added; otherwise, there is the risk that progeny virus will be inactivated and plaque development will cease.

Antibiotics & Other Antibacterial Agents

Antibacterial antibiotics and sulfonamides have no effect on viruses. Some antiviral drugs are available, however (see Chapter 30).

Quaternary ammonium compounds, in general, are not effective against viruses. Organic iodine compounds are also ineffective. Larger concentrations of chlorine are required to destroy viruses than to kill bacteria, especially in the presence of extraneous proteins. For example, the chlorine treatment of stools adequate to inactivate typhoid bacilli is inadequate to destroy poliomyelitis virus present in feces. Alcohols, such as isopropanol and ethanol, are relatively ineffective against certain viruses, especially picornaviruses.

Common Methods of Inactivating Viruses for Various Purposes

Viruses may be inactivated for various reasons: to sterilize laboratory supplies and equipment, disinfect surfaces or skin, make drinking water safe, and produce inactivated virus vaccines. Different methods and chemicals are used for these purposes.

Sterilization may be accomplished by steam under pressure, dry heat, ethylene oxide, and γ-irradiation. Surface disinfectants include sodium hypochlorite, glutaraldehyde, formaldehyde, and peracetic acid. Skin disinfectants include chlorhexidine, 70% ethanol, and iodophores. Vaccine production may involve the use of formaldehyde, β-propiolactone, psoralen + ultraviolet irradiation, or detergents (subunit vaccines) to inactivate the vaccine virus.

REPLICATION OF VIRUSES: AN OVERVIEW

Viruses multiply only in living cells. The host cell must provide the energy and synthetic machinery and the low-molecular-weight precursors for the synthesis of viral proteins and nucleic acids. The viral nucleic acid carries the genetic specificity to code for all the virus-specific macromolecules in a highly organized fashion.

In order for a virus to replicate, viral proteins must be synthesized by the host cell protein-synthesizing machinery. Therefore, the virus genome must be able to produce a usable mRNA. Various mechanisms have been identified which allow viral RNAs to compete successfully with cellular mRNAs to produce adequate amounts of viral proteins.

The unique feature of viral multiplication is that, soon after interaction with a host cell, the infecting virion is disrupted and its measurable infectivity is lost. This phase of the growth cycle is called the **eclipse period;** its duration varies depending on both the particular virus and the host cell, and it is followed by an interval of rapid accumulation of infectious progeny virus particles. The eclipse period is actually one of intense synthetic activity as the cell is redirected toward fulfilling the needs of the viral "pirate." In some cases, as soon as the viral nucleic acid enters the host cell, the cellular metabolism is redirected exclusively toward the synthesis of new virus particles, and the cell will be destroyed. In other cases, the metabolic processes of the host cell are not altered significantly, although the cell synthesizes viral proteins and nucleic acids, and the cell is not killed.

After the synthesis of viral nucleic acid and viral proteins, the components assemble to form new infectious virions. The yield of infectious virus per cell ranges widely, from modest numbers to more than 100,000 particles. The duration of the virus replication cycle also varies widely, from 6 to 8 hours (picornaviruses) to more than 40 hours (some herpesviruses).

Not all infections lead to new progeny virus. **Productive** infections occur in **permissive** cells and result in the production of infectious virus. **Abortive** infections fail to produce infectious progeny, either because the cell may be **nonpermissive** and unable to support the expression of all viral genes or because the infecting virus may be **defective**, lacking some functional viral gene. A **latent** infection may ensue, with the persistence of viral genomes, the expression of no or a few viral genes, and the survival of the infected cell. The pattern of replication may vary for a given virus, depending on the type of host cell infected.

General Steps in Viral Replication Cycles

A variety of different viral strategies have evolved for accomplishing multiplication in parasitized host cells. Although the details vary from group to group, the general outline of the replication cycles is similar. The growth cycles of a double-stranded DNA virus and a positive-sense, single-stranded RNA virus are diagrammed in Figure 29–6. Details are included in the following chapters devoted to specific virus groups.

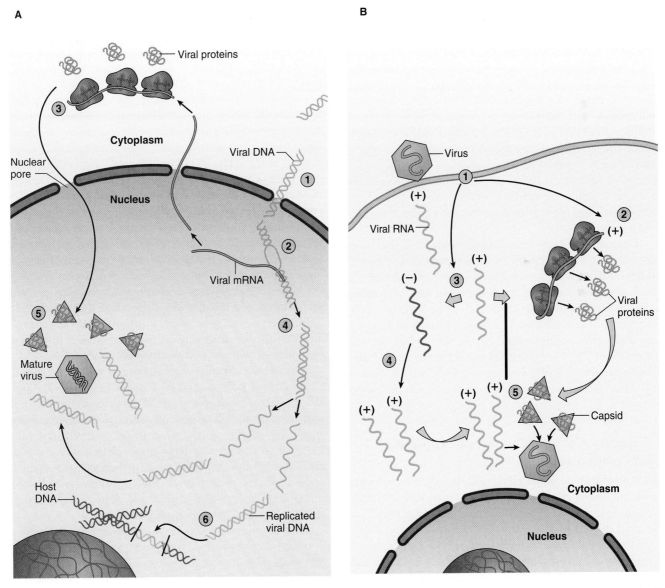

FIGURE 29–6 Example of viral growth cycles. **A:** The growth cycle of a nonenveloped, double-stranded DNA virus. **B:** The growth cycle of a positive-sense, single-stranded RNA virus. (Reproduced with permission from Talaro KP: *Foundations in Microbiology: Basic Principles,* 6th ed. McGraw-Hill, 2008.)

A. Attachment, Penetration, & Uncoating

The first step in viral infection is **attachment**, interaction of a virion with a specific receptor site on the surface of a cell. Receptor molecules differ for different viruses but are generally glycoproteins. In some cases the virus binds protein sequences (eg, picornaviruses) and in others oligosaccharides (eg, orthomyxoviruses and paramyxoviruses). Receptor binding is believed to reflect fortuitous configurational homologies between a virion surface structure and a cell surface component. For example, human immunodeficiency virus binds to the CD4 receptor on cells of the immune system, rhinoviruses bind ICAM-1, and Epstein-Barr virus recognizes the CD21 receptor on B cells. The presence or absence of receptors plays an important determining role in cell tropism and viral pathogenesis. Not all cells in a susceptible host will express the necessary receptors; for example, poliovirus is able to attach only to cells in the central nervous system and intestinal tract of primates. Each susceptible cell may contain up to 100,000 receptor sites for a given virus. The attachment step may initiate irreversible structural changes in the virion.

After binding, the virus particle is taken up inside the cell. This step is referred to as **penetration** or engulfment. In some systems, this is accomplished by receptor-mediated endocytosis, with uptake of the ingested virus particles within endosomes. There are also examples of direct penetration of virus particles across the plasma membrane. In other cases, there is fusion of the virion envelope with the plasma membrane of the cell. Those systems involve the interaction of a viral fusion protein with a second cellular receptor or "coreceptor" (eg, chemokine receptors for human immunodeficiency virus).

Uncoating occurs concomitantly with or shortly after penetration. Uncoating is the physical separation of the viral nucleic acid from the outer structural components of the virion so that it can function. The genome may be released as free nucleic acid (picornaviruses) or as a nucleocapsid (reoviruses). The nucleocapsids usually contain polymerases. Uncoating may require acidic pH in the endosome. The infectivity of the parental virus is lost at the uncoating stage. Viruses are the only infectious agents for which dissolution of the infecting agent is an obligatory step in the replicative pathway.

B. Expression of Viral Genomes & Synthesis of Viral Components

The synthetic phase of the viral replicative cycle ensues after uncoating of the viral genome. The essential theme in viral replication is that specific mRNAs must be transcribed from the viral nucleic acid for successful expression and duplication of genetic information. Once this is accomplished, viruses use cell components to translate the mRNA. Various classes of viruses use different pathways to synthesize the mRNAs depending upon the structure of the viral nucleic acid. Table 29–2 summarizes the various pathways of transcription (but not necessarily those of replication) of the nucleic acids of different classes of viruses. Some viruses (eg, rhabdoviruses) carry RNA polymerases to synthesize mRNAs. RNA viruses of this type are called negative-strand (negative-sense) viruses, as their single-strand RNA genome is complementary to mRNA, which is conventionally designated positive-strand (positive-sense). The negative-strand viruses must supply their own RNA polymerase, as eukaryotic cells lack enzymes able to synthesize mRNA off an RNA template.

In the course of viral replication, all the virus-specified macromolecules are synthesized in a highly organized sequence. In some viral infections, notably those involving double-stranded DNA-containing viruses, early viral proteins are synthesized soon after infection and late proteins are made only late in infection, after viral DNA synthesis. Early genes may or may not be shut off when late products are made. In contrast, most if not all of the genetic information of RNA-containing viruses is expressed at the same time. In addition to these temporal controls, quantitative controls also exist, since not all viral proteins are made in the same amounts. Virus-specific proteins may regulate the extent of transcription of the genome or the translation of viral mRNA.

Small animal viruses and bacteriophages are good models for studies of gene expression. The total nucleotide sequences of many viruses have been elucidated. This led to the discovery of overlapping genes in which some sequences in DNA are utilized in the synthesis of two different polypeptides, either by the use of two different reading frames or by two mRNA molecules using the same reading frame but different starting points. A viral system (adenovirus) first revealed the mRNA processing phenomenon called "splicing," whereby the mRNA sequences that code for a given protein are generated from separated sequences in the template, with noncoding intervening sequences spliced out of the transcript. Recently, several DNA viruses (herpesviruses, adenovirus, polyomavirus) were found to encode microRNAs; these small (~22 nucleotide) RNAs function at a new level of posttranscriptional gene regulation, either by mediating degradation of target mRNAs or by inducing inhibition of translation of those mRNAs.

The widest variation in strategies of gene expression is found among RNA-containing viruses (Table 29–3). Some virions carry polymerases (orthomyxoviruses, reoviruses); some systems utilize subgenomic messages, sometimes generated by splicing (orthomyxoviruses, retroviruses); and some viruses synthesize large polyprotein precursors that are processed and cleaved to generate the final gene products (picornaviruses, retroviruses). The viral protease of human immunodeficiency virus is what is inhibited by the class of antiviral drugs called protease inhibitors.

The extent to which virus-specific enzymes are involved in these processes varies from group to group. DNA viruses that replicate in the nucleus generally use host cell DNA and RNA polymerases and processing enzymes. The larger viruses (herpesviruses, poxviruses) are more independent of cellular functions than are the smaller viruses. This is one reason the larger viruses are more susceptible to antiviral chemotherapy

TABLE 29–2   **Pathways of Nucleic Acid Transcription for Various Virus Classes**

Type of Viral Nucleic Acid	Intermediates	Type of mRNA	Example	Comments
± ds DNA	None	+ mRNA	Most DNA viruses (eg, herpesvirus, adenovirus)	
+ ss DNA	± ds DNA	+ mRNA	Parvoviruses	
± ds RNA	None	+ mRNA	Reoviruses	Virion contains RNA polymerase that transcribes each segment to mRNA
+ ss RNA	± ds RNA	+ mRNA	Picornaviruses, togaviruses, flaviviruses	Viral nucleic acid is infectious and serves as mRNA. For togaviruses, smaller + mRNA is also formed for certain proteins
− ss RNA	None	+ mRNA	Rhabdoviruses, paramyxoviruses, orthomyxoviruses	Viral nucleic acid is not infectious; virion contains RNA polymerase which forms + mRNAs smaller than the genome. For orthomyxoviruses, + mRNAs are transcribed from each segment
+ ss RNA	− DNA, ± DNA	+ mRNA	Retroviruses	Virion contains reverse transcriptase; viral RNA is not infectious, but complementary DNA from transformed cell is

− indicates negative strand; + indicates positive strand; ± indicates a helix containing a positive and a negative strand; ds, double-stranded; ss, single-stranded.

TABLE 29–3   **Comparison of Replication Strategies of Several Important RNA Virus Families**

	Grouping Based on Genomic RNA[a]					
	Positive-Strand Viruses			**Negative-Strand Viruses**		**Double-Stranded Viruses**
Characteristic	Picornaviridae	Togaviridae	Retroviridae	Orthomyxoviridae	Paramyxoviridae and Rhabdoviridae	Reoviridae
Structure of genomic RNA	ss	ss	ss	ss	ss	ds
Sense of genomic RNA	Positive	Positive	Positive	Negative	Negative	
Segmented genome	0	0	0[b]	+	0	+
Genomic RNA infectious	+	+	0	0	0	0
Genomic RNA acts as messenger	+	+	+	0	0	0
Virion-associated polymerase	0	0	+[c]	+	+	+
Subgenomic messages	0	+	+	+	+	+
Polyprotein precursors	+	+	+	0	0	0

[a] +, indicated property applies to that virus family; 0, indicated property does not apply to that virus family; ds, double-stranded; negative, complementary to mRNA; positive, same sense as mRNA; ss, single-stranded.

[b] Retroviruses contain a diploid genome (two copies of nonsegmented genomic RNA).

[c] Retroviruses contain a reverse transcriptase (RNA-dependent DNA polymerase).

(see Chapter 30)—because more virus-specific processes are available as targets for drug action.

The intracellular sites where the different events in viral replication take place vary from group to group (Table 29–4). A few generalizations are possible. Viral protein is synthesized in the cytoplasm on polyribosomes composed of virus-specific mRNA and host cell ribosomes. Many viral proteins undergo modifications (glycosylation, acylation, cleavages, etc). Viral DNA is usually replicated in the nucleus. Viral genomic RNA is generally duplicated in the cell cytoplasm, though there are exceptions.

C. Morphogenesis & Release

Newly synthesized viral genomes and capsid polypeptides assemble together to form progeny viruses. Icosahedral capsids can condense in the absence of nucleic acid, whereas nucleocapsids of viruses with helical symmetry cannot form without viral RNA. In general, nonenveloped viruses accumulate in infected cells, and the cells eventually lyse and release the virus particles.

Enveloped viruses mature by a budding process. Virus-specific envelope glycoproteins are inserted into cellular membranes; viral nucleocapsids then bud through the membrane at these modified sites and in so doing acquire an envelope. Budding frequently occurs at the plasma membrane but may involve other membranes in the cell. Enveloped viruses are not infectious until they have acquired their envelopes. Therefore, infectious progeny virions typically do not accumulate within the infected cell.

Viral maturation is sometimes an inefficient process. Excess amounts of viral components may accumulate and be involved in the formation of inclusion bodies in the cell. As a result of the profound deleterious effects of viral replication, cellular cytopathic effects eventually develop and the cell dies. However, there are instances in which the cell is not damaged by the virus and long-term, persistent infections evolve (see Chapter 30). Virus-induced mechanisms may regulate apoptosis, a genetically programmed event that makes cells undergo self-destruction. Some virus infections delay early apoptosis, which allows time for the production of high yields of progeny virus. Additionally, some viruses actively

TABLE 29–4   Summary of Replication Cycles of Major Virus Families

Virus Family	Presence of Virion Envelope	Intracellular Location[a]			Multiplication Cycle (Hours)[c]
		Replication of Genome	Formation of Nucleocapsid[b]	Virion Maturation	
DNA viruses					
Parvoviridae	0	N	N	N	
Polyomaviridae	0	N	N	N	48
Adenoviridae	0	N	N	N	25
Hepadnaviridae	+	N	C	M–E	
Herpesviridae	+	N	N	M	15–72
Poxviridae	0	C	C	C	20
RNA viruses					
Picornaviridae	0	C	C	C	6–8
Reoviridae	0	C	C	C	15
Togaviridae	+	C	C	M–P	10–24
Flaviviridae	+	C	C	M–E	
Retroviridae	+	N	C	M–P	
Bunyaviridae	+	C	C	M–G	24
Orthomyxoviridae	+	N	N	M–P	15–30
Paramyxoviridae	+	C	C	M–P	10–48
Rhabdoviridae	+	C	C	M–P	6–10

[a]C, cytoplasm; M, membranes; M–E, endoplasmic reticulum membranes; M–G, Golgi membranes; M–P, plasma membranes; N, nucleus.

[b]The synthesis of viral proteins always occurs in the cytoplasm.

[c]The values shown for duration of the multiplication cycle are approximate; ranges indicate that various members within a given family replicate with different kinetics. Different host cell types also influence the kinetics of viral replication.

induce apoptosis at late stages which would facilitate spread of progeny virus to new cells.

GENETICS OF ANIMAL VIRUSES

Genetic analysis is a powerful approach toward understanding the structure and function of the viral genome, its gene products, and their roles in infection and disease. Variation in viral properties is of great importance for human medicine. Viruses that have stable antigens on their surfaces (poliovirus, measles virus) can be controlled by vaccination. Other viruses that exist as many antigenic types (rhinoviruses) or change frequently (influenza virus A) are difficult to control by vaccination; viral genetics may help develop more effective vaccines. Some types of viral infections recur repetitively (parainfluenza viruses) or persist (retroviruses) in the presence of antibody and may be better controlled by antiviral drugs. Genetic analysis will help identify virus-specific processes that may be appropriate targets for the development of antiviral therapy.

The following terms are basic to a discussion of genetics: **Genotype** refers to the genetic constitution of an organism. **Phenotype** refers to the observable properties of an organism, which are produced by the genotype in cooperation with the environment. A **mutation** is a heritable change in the genotype. The **genome** is the sum of the genes of an organism. **Wild-type virus** denotes the original virus from which mutants are derived and with which the mutants are compared; the term may not accurately characterize the virus as it is isolated in nature. Fresh virus isolates from the natural host are referred to as **field isolates** or **primary isolates**.

Mapping of Viral Genomes

The rapid and precise techniques of molecular biology have facilitated the identification of viral gene products and the mapping of these on the viral genome. Biochemical and physical mapping can be done much more rapidly than genetic mapping using classic genetic techniques.

For isolates that can be cloned, sequence analysis and comparison with known viruses is often used in place of the approaches described below for mapping viral genomes.

The technique of reassortment mapping has been used with influenza A viruses, which have a genome of eight segments of RNA, each coding for one viral protein. Under suitable conditions, the RNA genome segments and the polypeptides of different influenza A viruses migrate at different rates in polyacrylamide gels, so that strains can be distinguished. By analyzing the recombinants (reassortants) formed between different influenza viruses, the RNA segment coding for each protein was determined.

Restriction endonucleases can be used for identification of specific strains of DNA viruses. Viral DNA is isolated and incubated with a specific endonuclease until DNA sequences susceptible to the nuclease are cleaved. The fragments are then resolved on the basis of size by gel electrophoresis. The large fragments are most retarded by the sieving effect of the gel, so that an inverse relationship between size and migration is observed. The position of the DNA fragments can be determined by radioautography or by specialized staining techniques. Such physical mapping techniques were useful in distinguishing viral types in systems in which the viruses cannot be cultured (eg, papillomaviruses).

Physical maps can be correlated with genetic maps. This allows viral gene products to be mapped to individual regions of the genome defined by the restriction enzyme fragments. Transcription of mRNAs throughout the replication cycle can be assigned to specific DNA fragments. Using mutagenesis, mutations can be introduced into defined sites of the genome. Viral genome fragments generated by polymerase chain reaction can be used in place of restriction enzyme fragments in mapping and mutagenesis studies.

Types of Virus Mutants

Classic genetic studies with animal viruses require a sensitive and accurate quantitative assay method, such as a plaque assay for viral infectivity, and good mutants (resulting from single mutations) that are easily scored and reasonably stable. Some markers commonly used include plaque morphology, antibody escape or resistance to neutralizing antisera, loss of a virus protein, drug resistance, host range, and inability to grow at low or high temperatures. Mutants with such markers are obtained after treatment with a mutagen, after engineering in a mutation by molecular techniques, or after spontaneous mutation.

Cloned viral sequences are commonly used for molecular genetic analysis. RNA virus genomes are cloned as cDNA copies. This permits genetic analysis of viruses that cannot be cultured and of RNA viruses. Different types of mutations can be introduced into precise sites in cloned viral DNAs for functional analysis of coding sequences and viral cis-acting elements.

Conditional-lethal mutants are mutants that are lethal (in that no infectious virus is produced) under one set of conditions—termed nonpermissive conditions—but that yield normal infectious progeny under other conditions—termed permissive conditions. Temperature-sensitive mutants grow at low (permissive) temperatures but not at high (nonpermissive) temperatures. Host-range mutants are able to grow in one kind of cell (permissive cell), whereas abortive infection occurs in another type (nonpermissive cell). Mixed infection studies with pairs of mutants under permissive and nonpermissive conditions can yield information concerning gene functions and mechanisms of viral replication at the molecular level.

Defective Viruses

A defective virus is one that lacks one or more functional genes required for viral replication. Defective viruses require helper activity from another virus for some step in replication or maturation.

One type of defective virus lacks a portion of its genome (ie, deletion mutant). The extent of loss by deletion may vary from a short base sequence to a large amount of the genome. Spontaneous deletion mutants may interfere with the replication of homologous virus and are called defective interfering virus particles. Defective interfering particles have lost essential segments of genome but contain normal capsid proteins; they require infectious homologous virus as helper for replication, and they interfere with the multiplication of that homologous virus.

Another category of defective virus requires an unrelated replication-competent virus as helper. Examples include the adeno-associated satellite viruses and hepatitis D virus (delta agent), which replicate only in the presence of coinfecting human adenovirus or hepatitis B virus, respectively. No nondefective isolates of this type of defective virus have been recovered. The essential helper function supplied by the helper virus varies, depending on the system.

Pseudovirions, a different type of defective particle, contain host cell DNA rather than the viral genome. During viral replication, the capsid sometimes encloses random pieces of host nucleic acid rather than viral nucleic acid. Such particles look like ordinary virus particles when observed by electron microscopy, but they do not replicate. Pseudovirions theoretically might be able to transduce cellular nucleic acid from one cell to another.

The transforming retroviruses are usually defective. A portion of the viral genome has been deleted and replaced with a piece of DNA of cellular origin that encodes a transforming protein. These viruses allowed the identification of cellular oncogenes (see Chapter 43). Another retrovirus is required as helper in order for the transforming virus to replicate.

Interactions Among Viruses

When two or more virus particles infect the same host cell, they may interact in a variety of ways. They must be sufficiently closely related, usually within the same viral family, for most types of interactions to occur. Genetic interaction results in some progeny that are heritably (genetically) different from either parent. Progeny produced as a consequence of nongenetic interaction are similar to the parental viruses. In genetic interactions the actual nucleic acid molecules interact, whereas the products of the genes are involved in nongenetic interactions.

A. Recombination

Recombination results in the production of progeny virus (recombinant) that carries traits not found together in either parent. The classic mechanism is that the nucleic acid strands break, and part of the genome of one parent is joined to part of the genome of the second parent. The recombinant virus is genetically stable, yielding progeny like itself upon replication. Viruses vary widely in the frequency with which they undergo recombination. In the case of viruses with segmented genomes, eg, influenza virus, the formation of recombinants is due to reassortment of individual genome fragments rather than to an actual crossover event, and it occurs with ease (see Chapter 39).

B. Complementation

This refers to the interaction of viral gene products in cells infected with two viruses, one or both of which may be defective. It results in the replication of one or both under conditions in which replication would not ordinarily occur. The basis for complementation is that one virus provides a gene product in which the second is defective, allowing the second virus to grow. The genotypes of the two viruses remain unchanged. If both mutants are defective in the same gene product, they will not be able to complement each other's growth.

C. Phenotypic Mixing

A special case of complementation is phenotypic mixing, or the association of a genotype with a heterologous phenotype. This occurs when the genome of one virus becomes randomly incorporated within capsid proteins specified by a different virus or a capsid consisting of components of both viruses. If the genome is encased in a completely heterologous protein coat, this extreme example of phenotypic mixing may be called "phenotypic masking" or "transcapsidation." Such mixing is not a stable genetic change because, upon replication, the phenotypically mixed parent will yield progeny encased in capsids homologous to the genotype.

Phenotypic mixing usually occurs between different members of the same virus family; the intermixed capsid proteins must be able to interact correctly to form a structurally intact capsid. However, phenotypic mixing also can occur between enveloped viruses, and in this case the viruses do not have to be closely related. The nucleocapsid of one virus becomes encased within an envelope specified by another, a phenomenon designated "pseudotype formation." There are many examples of pseudotype formation among the RNA tumor viruses (see Chapter 43). The nucleocapsid of vesicular stomatitis virus, a rhabdovirus, has an unusual propensity for being involved in pseudotype formation with unrelated envelope material.

D. Interference

Infection of either cell cultures or whole animals with two viruses often leads to an inhibition of multiplication of one of the viruses, an effect called interference. Interference in animals is distinct from specific immunity. Furthermore, interference does not occur with all viral combinations; two viruses may infect and multiply within the same cell as efficiently as in single infections.

Several mechanisms have been elucidated as causes of interference: (1) One virus may inhibit the ability of the second to adsorb to the cell, either by blocking its receptors (retroviruses, enteroviruses) or by destroying its receptors (orthomyxoviruses). (2) One virus may compete with the second for components of the replication apparatus (eg, polymerase, translation initiation factor). (3) The first virus may cause the infected cell to produce an inhibitor (interferon; see Chapter 30) that prevents replication of the second virus.

Viral Vectors

Recombinant DNA technology has revolutionized the production of biologic materials, hormones, vaccines, interferon, and other gene products. Viral genomes have been engineered to serve as replication and expression vectors for both viral and cellular genes. Almost any virus can be converted to a vector if enough is known about its replication functions, transcription controls, and packaging signals. Viral vector technology is based on both DNA viruses (eg, SV40, parvovirus, bovine papillomavirus, adenovirus, herpesviruses, vaccinia virus) and RNA viruses (eg, poliovirus, Sindbis virus, retroviruses). Each system has distinct advantages and disadvantages.

Typical eukaryotic expression vectors contain viral regulatory elements (promoters or enhancers) that control transcription of the desired cloned gene placed adjacent, signals for efficient termination and polyadenylation of transcripts, and an intronic sequence bounded by splice donor and acceptor sites. There may be sequences that enhance translation or affect expression in a particular cell type. The principles of recombinant DNA technology are described and illustrated in Chapter 7. This approach offers the possibility of producing large amounts of a pure antigen for structural studies or for vaccine purposes.

NATURAL HISTORY (ECOLOGY) & MODES OF TRANSMISSION OF VIRUSES

Ecology is the study of interactions between living organisms and their environment. Different viruses have evolved ingenious and often complicated mechanisms for survival in nature and transmission from one host to the next. The mode of transmission utilized by a given virus depends on the nature of the interaction between the virus and the host.

Viruses may be transmitted in the following ways: (1) Direct transmission from person to person by contact. The major means of transmission include droplet or aerosol infection (eg, influenza, measles, smallpox); by sexual contact (eg, papillomavirus, hepatitis B, herpes simplex type 2, human immunodeficiency virus); by hand–mouth, hand–eye, or mouth–mouth contact (eg, herpes simplex, rhinovirus, Epstein-Barr virus); or by exchange of contaminated blood (eg, hepatitis B, human immunodeficiency virus). (2) Indirect transmission by the fecal–oral route (eg, enteroviruses, rotaviruses, infectious hepatitis A) or by fomites (eg, Norwalk

virus, rhinovirus). (3) Transmission from animal to animal, with humans an accidental host. Spread may be by bite (rabies) or by droplet or aerosol infection from rodent-contaminated quarters (eg, arenaviruses, hantaviruses). (4) Transmission by means of an arthropod vector (eg, arboviruses, now classified primarily as togaviruses, flaviviruses, and bunyaviruses).

At least three different transmission patterns have been recognized among the arthropod-borne viruses:

1. **Human–arthropod cycle:** *Examples:* Urban yellow fever, dengue.

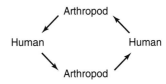

2. **Lower vertebrate–arthropod cycle with tangential infection of humans:** *Examples:* Jungle yellow fever, St. Louis encephalitis. The infected human is a "dead end" host. This is a more common transmission mechanism.

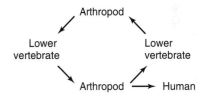

3. **Arthropod–arthropod cycle with occasional infection of humans and lower vertebrates:** *Examples:* Colorado tick fever, LaCrosse encephalitis.

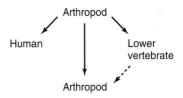

In this cycle, the virus may be transmitted from the adult arthropod to its offspring through the egg (transovarian passage); thus, the cycle may continue with or without intervention of a viremic vertebrate host.

In vertebrates, the invasion of most viruses evokes a violent reaction, usually of short duration. The result is decisive. Either the host succumbs or it lives through the production of antibodies that neutralize the virus. Regardless of the outcome, the sojourn of the active virus is usually short, although persistent or latent infections that last for months to years may occur (hepatitis B, herpes simplex, cytomegalovirus, retroviruses). In arthropod vectors of the virus, the relationship is usually quite different. The viruses produce little or no ill effect and remain active in the arthropod throughout the latter's natural life. Thus arthropods, in contrast to vertebrates, act as permanent hosts and reservoirs.

Emerging Viral Diseases

Owing to wide-reaching changes in social attitudes, technology, and the environment—plus the decreased effectiveness of previous approaches to disease control—the spectrum of infectious diseases is expanding today. New agents appear, and diseases once thought to be under control are increasing in incidence as pathogens evolve and spread. The term "emerging infectious diseases" denotes these phenomena.

Viral diseases emerge following one of three general patterns: recognition of a new agent, abrupt increase in illnesses caused by an endemic agent, and invasion of a new host population.

Combinations of factors contribute to disease emergence. Some factors increase human exposure to once-obscure pathogens; others provide for dissemination of once-localized infections; and still others force changes in viral properties or host responses to infection. Factors include (1) environmental changes (deforestation, damming or other changes in water ecosystems, flood or drought, famine); (2) human behavior (sexual behavior, drug use, outdoor recreation); (3) socioeconomic and demographic phenomena (war, poverty, population growth and migration, urban decay); (4) travel and commerce (highways, international air travel); (5) food production (globalization of food supplies, changes in methods of food processing and packaging); (6) healthcare (new medical devices, blood transfusions, organ and tissue transplantation, drugs causing immunosuppression, widespread use of antibiotics); (7) microbial adaptation (changes in virulence, development of drug resistance, cofactors in chronic diseases); and (8) public health measures (inadequate sanitation and vector control measures, curtailment of prevention programs, lack of trained personnel in sufficient numbers).

Examples of emerging viral infections in different regions of the world include Ebola virus, Nipah virus, hantavirus pulmonary disease, human immunodeficiency virus infection, dengue hemorrhagic fever, West Nile virus, Rift Valley fever, and bovine spongiform encephalopathy (the latter a prion disease).

Of potential concern also is the possible use of animal organs as xenografts in humans. Because the numbers of available human donor organs cannot meet the needs of all waiting patients, xenotransplantation of nonhuman primate and porcine organs is considered an alternative. Concerns exist about the potential accidental introduction of new viral pathogens from the donor species into humans.

Bioterrorism Agents

Bioterrorism agents are microorganisms (or toxins) that could be used to produce death and disease in humans, animals, or plants for terrorist purposes. Such microorganisms could be genetically modified to increase their virulence, make them resistant to drugs or vaccines, or enhance their ability to be disseminated in the environment.

Potential bioterrorism agents are classified into risk categories based on the ease of dissemination or transmission from person to person, mortality rates, ability to cause public panic, and requirement for public health preparedness. Viral agents in the highest risk category are smallpox and the viral hemorrhagic fevers; highest risk bacteria include the agents of anthrax, botulism, plague, and tularemia.

REVIEW QUESTIONS

1. Some viruses are characterized by helical symmetry of the viral nucleocapsid. Which of the following statements about viruses with helical symmetry is most accurate?

    (A) All enveloped viruses with helical symmetry are classified into the same virus family
    (B) Helical nucleocapsids are found primarily in DNA-containing viruses
    (C) All human viruses with helical nucleocapsids possess an envelope
    (D) Excess empty helical particles containing no nucleic acid are commonly produced in infected cells

2. Virus-infected cells often develop morphologic changes referred to as cytopathic effects. Which of the following statements about virus-induced cytopathic changes is most accurate?

    (A) Are pathognomonic for an infecting virus
    (B) Are rarely associated with cell death
    (C) May include giant cell formation
    (D) Can only be seen with an electron microscope

3. Viruses usually initiate infection by first interacting with receptors on the surface of cells. Which of the following statements is most accurate about cellular receptors for viruses?

    (A) Cellular receptors for viruses have no known cellular function
    (B) All viruses within a given family use the same cellular receptor
    (C) All cells in a susceptible host will express the viral receptor
    (D) Successful infection of a cell by a virus may involve interaction with more than one type of receptor

4. Which of the following can be used to quantitate the infectious titer of viruses?

    (A) Plaque assay
    (B) Electron microscopy
    (C) Hemagglutination
    (D) Polymerase chain reaction
    (E) Enzyme immunoassay

5. Which one of the following states a principle regarding viral nucleic acid?

    (A) Viruses contain both RNA and DNA
    (B) Some viruses contain a segmented genome
    (C) Purified viral nucleic acid from any virus is usually infectious
    (D) Viral genome sizes are similar among known human viruses

6. Two mutants of poliovirus have been isolated, one (MutX) with a mutation in gene X and the second (MutY) with a mutation in gene Y. If cells are infected with each mutant alone, no virus

is produced. If a cell is coinfected with both MutX and MutY, which one of the following is most likely to occur?

(A) Reassortment of genome segments may occur and give rise to a viable wild-type virus

(B) The genomes may be reverse transcribed to DNA and both MutX and MutY viruses produced

(C) Complementation between the mutant gene products may occur and both MutX and MutY viruses produced

(D) The cells will transform at high frequency, as they will not be killed by the poliovirus mutants

7. Which one of the following viruses possesses an RNA genome that is infectious when purified?

(A) Influenza virus
(B) Poliovirus
(C) Papillomavirus
(D) Measles virus
(E) Rotavirus

8. Viruses belonging to which of the following groups are likely to establish latent infections?

(A) Poxviruses
(B) Filoviruses
(C) Herpesviruses
(D) Influenza viruses
(E) Caliciviruses

9. Some viruses encode for a viral RNA-dependent RNA polymerase. Which of the following states a principle about viral RNA polymerases?

(A) All RNA viruses carry RNA polymerase molecules inside virus particles as they are needed to initiate the next infectious cycle

(B) Antibodies against the viral RNA polymerase neutralize virus infectivity

(C) Negative-strand RNA viruses supply their own RNA-dependent RNA polymerase as eukaryotic cells lack such enzymes

(D) The viral RNA polymerase protein also serves as a major core structural protein in the virus particle

10. Which of the following statements regarding virus morphology is true?

(A) All RNA viruses are spherical in shape
(B) Some viruses contain flagella
(C) Some viruses with DNA genomes contain a primitive nucleus
(D) Viruses are smaller than bacteria
(E) Viruses are larger than mitochondria

11. Many viruses can be grown in the laboratory. Which of the following statements about virus propagation is not true?

(A) Some viruses can be propagated in cell-free media
(B) Some mammalian viruses can be cultivated in hen's eggs

(C) Some viruses with broad host ranges can multiply in many types of cells

(D) Some human viruses can be grown in mice

(E) Most virus preparations have particle-to-infectious unit ratios greater than one

12. Laboratory infections can be acquired when working with viruses unless good laboratory safety practices are followed. Which of the following is not a good biosafety practice?

(A) Use of aseptic techniques
(B) Use of personal protective equipment
(C) No pipetting by mouth
(D) Flushing experimental waste down laboratory sink
(E) No eating or drinking in the laboratory

Answers

1. C	4. A	7. B	10. D
2. C	5. B	8. C	11. A
3. D	6. C	9. C	12. D

REFERENCES

Chiu W, Burnett RM, Garcea RL (editors): *Structural Biology of Viruses.* Oxford University Press, 1997.

Espy MJ et al: Real-time PCR in clinical microbiology: Applications for routine laboratory testing. Clin Microbiol Rev 2006;19:165. [PMID: 16418529]

Fauquet CM et al (editors): Virus taxonomy: Classification and nomenclature of viruses. Eighth report of the International Committee on Taxonomy of Viruses. Academic Press, 2005.

Girones R: Tracking viruses that contaminate environments. Microbe 2006;1:19.

Guideline for hand hygiene in health-care settings. Recommendations of the Healthcare Infection Control Practices Advisory Committee and the HICPAC/SHEA/APIC/IDSA Hand Hygiene Task Force. Society for Healthcare Epidemiology of America/Association for Professionals in Infection Control/Infectious Diseases Society of America. MMWR Recomm Rep 2002;51(RR-16):1.

Knipe DM et al (editors): *Fields Virology,* 5th ed. Lippincott Williams & Wilkins, 2007.

Preventing emerging infectious diseases: A strategy for the 21st century. Overview of the updated CDC plan. MMWR Morb Mortal Wkly Rep 1998;47(RR-15):1.

Woolhouse MEJ: Where do emerging pathogens come from? Microbe 2006;1:511.

Pathogenesis & Control of Viral Diseases

PRINCIPLES OF VIRAL DISEASES

The fundamental process of viral infection is the viral replicative cycle. The cellular response to that infection may range from no apparent effect to cytopathology with accompanying cell death to hyperplasia or cancer.

Viral disease is some harmful abnormality that results from viral infection of the host organism. Clinical disease in a host consists of overt signs and symptoms. A syndrome is a specific group of signs and symptoms. Viral infections that fail to produce any symptoms in the host are said to be inapparent (subclinical). In fact, most viral infections do not result in the production of disease (Figure 30–1).

Important principles that pertain to viral disease include the following: (1) many viral infections are subclinical; (2) the same disease may be produced by a variety of viruses;

(3) the same virus may produce a variety of diseases; (4) the disease produced bears no relationship to viral morphology; and (5) the outcome in any particular case is determined by both viral and host factors and is influenced by the genetics of each.

Viral pathogenesis is the process that occurs when a virus infects a host. **Disease pathogenesis** is a subset of events during an infection that results in disease manifestation in the host. A virus is **pathogenic** for a particular host if it can infect and cause signs of disease in that host. A strain of a certain virus is more **virulent** than another strain if it commonly produces more severe disease in a susceptible host. Viral virulence in intact animals should not be confused with cytopathogenicity for cultured cells; viruses highly cytocidal in vitro may be harmless in vivo, and, conversely, noncytocidal viruses may cause severe disease.

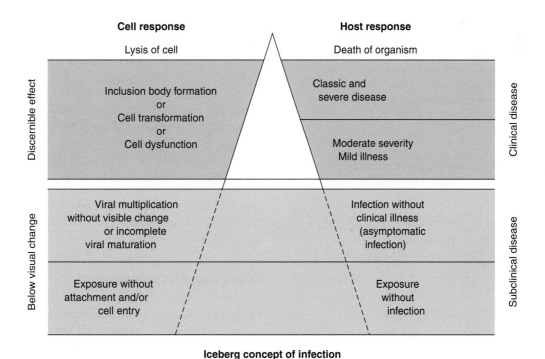

FIGURE 30–1 Types of host and cellular responses to virus infection. (Modified with permission from Evans AS: Epidemiological concepts. In: Evans AS, Brachman PS [editors]: *Bacterial Infections of Humans,* 2nd ed. Plenum, 1991.)

Important features of two general categories of acute viral diseases (local, systemic) are compared in Table 30–1.

PATHOGENESIS OF VIRAL DISEASES

To produce disease, viruses must enter a host, come in contact with susceptible cells, replicate, and produce cell injury. Understanding mechanisms of viral pathogenesis at the molecular level is necessary to design effective and specific antiviral strategies. Much of our knowledge of viral pathogenesis is based on animal models, because such systems can be readily manipulated and studied.

Steps in Viral Pathogenesis

Specific steps involved in viral pathogenesis are the following: viral entry into the host, primary viral replication, viral spread, cellular injury, host immune response, viral clearance or establishment of persistent infection, and viral shedding.

A. Entry & Primary Replication

In order for host infection to occur, a virus must first attach to and enter cells of one of the body surfaces—skin, respiratory tract, gastrointestinal tract, urogenital tract, or conjunctiva. Most viruses enter their hosts through the mucosa of the respiratory or gastrointestinal tract (Table 30–2). Major exceptions are those viruses that are introduced directly into the bloodstream by needles (hepatitis B, human immunodeficiency virus [HIV]), by blood transfusions, or by insect vectors (arboviruses).

Viruses usually replicate at the primary site of entry. Some, such as influenza viruses (respiratory infections) and noroviruses (gastrointestinal infections), produce disease at the portal of entry and likely have no necessity for further systemic spread. They spread locally over the epithelial surfaces, but there is no spread to distant sites.

TABLE 30–1   Important Features of Acute Viral Diseases

	Local Infections	Systemic Infections
Specific disease example	Respiratory (rhinovirus)	Measles
Site of pathology	Portal of entry	Distant site
Incubation period	Relatively short	Relatively long
Viremia	Absent	Present
Duration of immunity	Variable—may be short	Usually lifelong
Role of secretory antibody (IgA) in resistance	Usually important	Usually not important

B. Viral Spread & Cell Tropism

Many viruses produce disease at sites distant from their point of entry (eg, enteroviruses, which enter through the gastrointestinal tract but may produce central nervous system disease). After primary replication at the site of entry, these viruses then spread within the host (Figure 30–2). Mechanisms of viral spread vary, but the most common route is via the bloodstream or lymphatics. The presence of virus in the blood is called **viremia**. Virions may be free in the plasma (eg, enteroviruses, togaviruses) or associated with particular cell types (eg, measles virus) (Table 30–3). Some viruses even multiply within those cells. The viremic phase is short in many viral infections. In some instances, neuronal spread is involved; this is apparently how rabies virus reaches the brain to cause disease and how herpes simplex virus moves to the ganglia to initiate latent infections.

Viruses tend to exhibit organ and cell specificities. Thus, tropism determines the pattern of systemic illness produced during a viral infection. As an example, hepatitis B virus has a tropism for liver hepatocytes, and hepatitis is the primary disease caused by the virus.

Tissue and cell tropism by a given virus usually reflect the presence of specific cell surface receptors for that virus. Receptors are components of the cell surface with which a region of the viral surface (capsid or envelope) can specifically interact and initiate infection. Receptors are cell constituents that function in normal cellular metabolism but also happen to have an affinity for a particular virus. The identity of the specific cellular receptor is known for some viruses but is unknown in many cases.

Factors affecting viral gene expression are important determinants of cell tropism. Enhancer regions that show some cell-type specificity may regulate transcription of viral genes. For example, the JC polyomavirus enhancer is much more active in glial cells than in other cell types.

Another mechanism dictating tissue tropism involves proteolytic enzymes. Certain paramyxoviruses are not infectious until an envelope glycoprotein undergoes proteolytic cleavage. Multiple rounds of viral replication will not occur in tissues that do not express the appropriate activating enzymes.

Viral spread may be determined in part by specific viral genes. Studies with reovirus have demonstrated that the extent of spread from the gastrointestinal tract is determined by one of the outer capsid proteins.

C. Cell Injury & Clinical Illness

Destruction of virus-infected cells in the target tissues and physiologic alterations produced in the host by the tissue injury are partly responsible for the development of disease. Some tissues, such as intestinal epithelium, can rapidly regenerate and withstand extensive damage better than others, such as the brain. Some physiologic effects may result from nonlethal impairment of specialized functions

TABLE 30–2 **Common Routes of Viral Infection in Humans**

Route of Entry	Virus Group	Produce Local Symptoms at Portal of Entry	Produce Generalized Infection Plus Specific Organ Disease
Respiratory tract	Parvovirus		B19
	Adenovirus	Most types	
	Herpesvirus	Epstein-Barr virus, herpes simplex virus	Varicella virus
	Poxvirus		Smallpox virus
	Picornavirus	Rhinoviruses	Some enteroviruses
	Togavirus		Rubella virus
	Coronavirus	Most types	
	Orthomyxovirus	Influenza virus	
	Paramyxovirus	Parainfluenza viruses, respiratory syncytial virus	Mumps virus, measles virus
Mouth, intestinal tract	Adenovirus	Some types	
	Herpesvirus	Epstein-Barr virus, herpes simplex virus	Cytomegalovirus
	Picornavirus		Some enteroviruses, including poliovirus, and hepatitis A virus
	Reovirus	Rotaviruses	
Skin			
Mild trauma	Papillomavirus	Most types	
	Herpesvirus	Herpes simplex virus	
	Poxvirus	Molluscum contagiosum virus, orf virus	
Injection	Hepadnavirus		Hepatitis B
	Herpesvirus		Epstein-Barr virus, cytomegalovirus
	Retrovirus		Human immunodeficiency virus
Bites	Togavirus		Many species, including eastern equine encephalitis virus
	Flavivirus		Many species, including yellow fever virus
	Rhabdovirus		Rabies virus

of cells, such as loss of hormone production. Clinical illness from viral infection is the result of a complex series of events, and many of the factors that determine degree of illness are unknown. General symptoms associated with many viral infections, such as malaise and anorexia, may result from host response functions such as cytokine production. Clinical illness is an insensitive indicator of viral infection; inapparent infections by viruses are very common.

D. Recovery from Infection

The host either succumbs or recovers from viral infection. Recovery mechanisms include both innate and adaptive immune responses. Interferon (IFN) and other cytokines, humoral and cell-mediated immunity, and possibly other host defense factors are involved. The relative importance of each component differs with the virus and the disease.

The importance of host factors in influencing the outcome of viral infections is illustrated by an incident in the 1940s in which 45,000 military personnel were inoculated with hepatitis B virus-contaminated yellow fever virus

vaccine. Although the personnel were presumably subjected to comparable exposures, clinical hepatitis occurred in only 2% (914 cases), and of those only 4% developed serious disease. The genetic basis of **host susceptibility** remains to be determined for most infections.

In acute infections, recovery is associated with viral clearance. However, there are times when the host remains persistently infected with the virus. Such long-term infections are described below.

E. Virus Shedding

The last stage in pathogenesis is the shedding of infectious virus into the environment. This is a necessary step to maintain a viral infection in populations of hosts. Shedding usually occurs from the body surfaces involved in viral entry (Figure 30–2). Shedding occurs at different stages of disease depending on the particular agent involved. It represents the time at which an infected individual is infectious to contacts. In some viral infections, such as rabies, humans represent dead-end infections, and shedding does not occur.

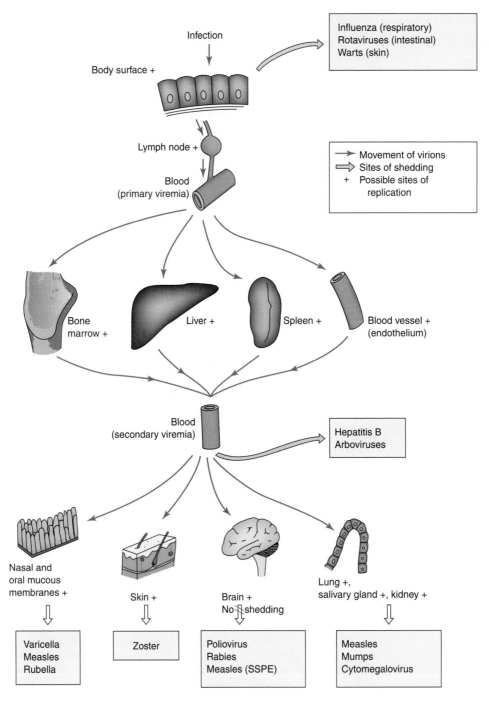

FIGURE 30–2 Mechanisms of spread of virus through the body in human viral infections. + indicates possible sites of viral replication; large arrows indicate sites of shedding of virus, with illustrative examples of diseases in which that route of excretion is important. Transfer from blood is by transfusion with hepatitis B and by mosquito bite in certain arboviral infections. SSPE, subacute sclerosing panencephalitis. (Modified and reproduced with permission from Mims CA, White DO: *Viral Pathogenesis and Immunology.* Blackwell, 1984.)

Host Immune Response

The outcome of viral infections reflects the interplay between viral and host factors. Nonspecific host defense mechanisms are usually elicited very soon after viral infection. The most prominent among the innate immune responses is the induction of IFNs (see below). These responses help inhibit viral growth during the time it takes to induce specific humoral and cell-mediated immunity.

Both humoral and cellular components of the immune response are involved in control of viral infections. Viruses elicit a tissue response different from the response to pathogenic bacteria. Whereas polymorphonuclear leukocytes form the principal cellular response to the acute inflammation

TABLE 30–3 Viruses Spread Via the Bloodstream[a]

Cell Type Associated	Examples	
	DNA Viruses	**RNA Viruses**
Lymphocytes	Epstein-Barr virus, cytomegalovirus, hepatitis B virus, JC virus, BK virus	Mumps, measles, rubella, human immunodeficiency virus
Monocytes-macrophages	Cytomegalovirus	Poliovirus, human immunodeficiency virus, measles virus
Neutrophils		Influenza virus
Red blood cells	Parvovirus B19	Colorado tick fever virus
None (free in plasma)		Togavirus, picornavirus

[a]Modified with permission from Tyler KL, Fields BN: Pathogenesis of viral infections. In: *Fields Virology*, 3rd ed. Fields BN et al (editors). Lippincott-Raven, 1996.

caused by pyogenic bacteria, infiltration with mononuclear cells and lymphocytes characterizes the inflammatory reaction of uncomplicated viral lesions.

Virus-encoded proteins serve as targets for the immune response. Virus-infected cells may be lysed by cytotoxic T lymphocytes as a result of recognition of viral polypeptides on the cell surface. Humoral immunity protects the host against reinfection by the same virus. Neutralizing antibody directed against capsid proteins blocks the initiation of viral infection, presumably at the stage of attachment, entry, or uncoating. Secretory IgA antibody is important in protecting against infection by viruses through the respiratory or gastrointestinal tracts.

Special characteristics of certain viruses may have profound effects on the host's immune response. Some viruses infect and damage cells of the immune system. The most dramatic example is the human retrovirus associated with acquired immunodeficiency syndrome (AIDS) that infects T lymphocytes and destroys their ability to function (see Chapter 44).

Host susceptibility and response to infection are genetically determined; these differences are often in immune response genes.

Viruses have evolved a variety of ways that serve to suppress or evade the host immune response and thus avoid being eradicated. Oftentimes, the viral proteins involved in modulating the host response are not essential for growth of the virus in tissue culture, and their properties are realized only in pathogenesis experiments in animals. In addition to infecting cells of the immune system and abrogating their function (HIV), they may infect neurons that express little or no class I MHC (herpesvirus), or they may encode immunomodulatory proteins that inhibit MHC function (adenovirus, herpesvirus) or inhibit cytokine activity (poxvirus, measles virus). Viruses may mutate and change antigenic sites on virion proteins (influenza virus, HIV) or may

downregulate the level of expression of viral cell surface proteins (herpesvirus). Most viruses have anti-IFN strategies (see below).

A type of immunopathologic disorder was observed in humans immunized with vaccines containing killed measles or respiratory syncytial virus (no longer in use). A few persons developed unusual immune responses that gave rise to serious consequences when they later were exposed to the naturally occurring infective virus. Dengue hemorrhagic fever with shock syndrome, which develops in persons who already have had at least one prior infection with another dengue serotype, may be a naturally occurring manifestation of the same type of immunopathology.

Another potential adverse effect of the immune response is the development of autoantibodies. If a viral antigen were to elicit antibodies that fortuitously recognized an antigenic determinant on a cellular protein in normal tissues, cellular injury or loss of function unrelated to viral infection might result. The magnitude of this potential problem in human disease is currently unknown.

Comparison of Pathogenesis of a Viral Disease of the Skin & of the Central Nervous System

The pathogenesis of mousepox, a disease of the skin, and of human poliomyelitis, a disease of the central nervous system, are outlined in Figure 30–3. Both viruses multiply at the primary site of entry prior to systemic spread to target organs.

In mousepox, the virus enters the body through minute abrasions of the skin and multiplies in the epidermal cells. At the same time, it is carried by the lymphatics to the regional lymph nodes, where multiplication also occurs. The few virus particles entering the blood by way of the efferent lymphatics are taken up by the macrophages of the liver and spleen. The virus multiplies rapidly in both organs. Following release of virus from the liver and spleen, it moves by way of the bloodstream and localizes in the basal epidermal layers of the skin, in the conjunctival cells, and near the lymph follicles in the intestine. The virus may occasionally also localize in the epithelial cells of the kidney, lung, submaxillary gland, and pancreas. A primary lesion occurs at the site of entry of the virus. It appears as a localized swelling that rapidly increases in size, becomes edematous, ulcerates, and goes on to scar formation. A generalized rash follows that is responsible for the release of large quantities of virus into the environment.

In poliomyelitis, virus enters by way of the alimentary tract, multiplies locally at the initial sites of viral implantation (tonsils, Peyer's patches) or the lymph nodes that drain these tissues, and begins to appear in the throat and in the feces. Secondary viral spread occurs by way of the bloodstream to other susceptible tissues—specifically, other lymph nodes and the central nervous system. Within the central nervous system, the virus spreads along nerve fibers. If a high level of multiplication occurs as the virus spreads through the central

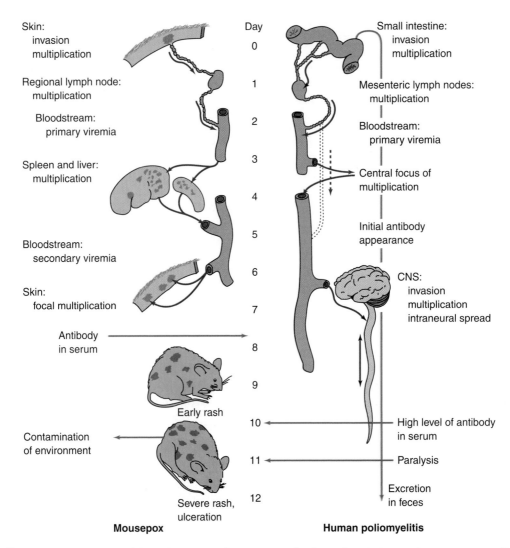

FIGURE 30–3 Schematic illustrations of the pathogenesis of mousepox and poliomyelitis. CNS, central nervous system. (Courtesy of F Fenner.)

nervous system, motor neurons are destroyed and paralysis occurs. The shedding of virus into the environment does not depend on secondary viral spread to the central nervous system. Spread to the central nervous system is readily prevented by the presence of antibodies induced by prior infection or vaccination.

Viral Persistence: Chronic & Latent Virus Infections

Infections are **acute** when a virus first infects a susceptible host. Viral infections are usually self-limiting. Sometimes, however, the virus persists for long periods of time in the host. Long-term virus–host interaction may take several forms. **Chronic infections** (also called **persistent infections**) are those in which replicating virus can be continuously detected, often at low levels; mild or no clinical symptoms may be evident. **Latent infections** are those in which the virus persists in an occult (hidden or cryptic) form most of the time when no new virus is produced. There will be intermittent flare-ups

of clinical disease; infectious virus can be recovered during flare-ups. Viral sequences may be detectable by molecular techniques in tissues harboring latent infections. **Inapparent or subclinical infections** are those that give no overt sign of their presence.

Chronic infections occur with a number of animal viruses, and the persistence in certain instances depends upon the age of the host when infected. In humans, for example, rubella virus and cytomegalovirus infections acquired in utero characteristically result in viral persistence that is of limited duration, probably because of development of the immunologic capacity to react to the infection as the infant matures. Infants infected with hepatitis B virus frequently become persistently infected (chronic carriers); most carriers are asymptomatic (see Chapter 35). In chronic infections with RNA viruses, the viral population often undergoes many genetic and antigenic changes.

Herpesviruses typically produce latent infections. Herpes simplex viruses enter the sensory ganglia and persist in a noninfectious state (Figure 30–4). There may be periodic

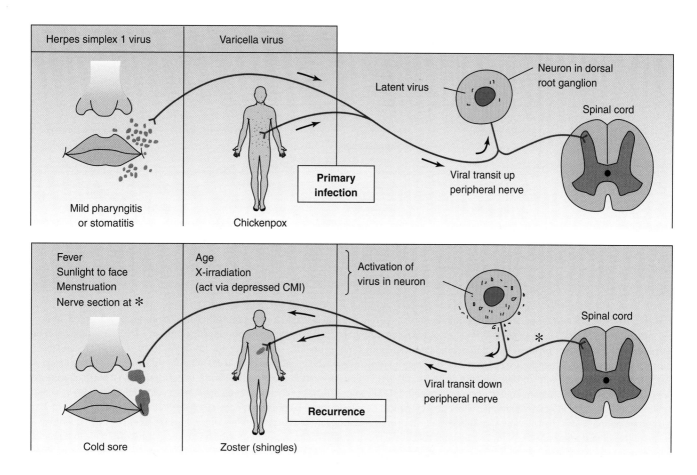

FIGURE 30–4 Latent infections by herpesviruses. Examples are shown for both herpes simplex and varicella-zoster viruses. Primary infections occur in childhood or adolescence, followed by establishment of latent virus in cerebral or spinal ganglia. Later activation causes recurrent herpes simplex or zoster. Recurrences are rare for zoster. CMI, cell-mediated immunity. (Reproduced with permission from Mims CA, White DO: *Viral Pathogenesis and Immunology.* Blackwell, 1984.)

reactivations during which lesions containing infectious virus appear at peripheral sites (eg, fever blisters). Chickenpox virus (varicella-zoster) also becomes latent in sensory ganglia. Recurrences are rare and occur years later, usually following the distribution of a peripheral nerve (shingles). Other members of the herpesvirus family also establish latent infections, including cytomegalovirus and Epstein-Barr virus. All may be reactivated by immunosuppression. Consequently, reactivated herpesvirus infections may be a serious complication for persons receiving immunosuppressant therapy.

Persistent viral infections may play a far-reaching role in human disease. Persistent viral infections are associated with certain types of cancers in humans (see Chapter 43) as well as with progressive degenerative diseases of the central nervous system of humans (see Chapter 42). Examples of different types of persistent viral infections are presented in Figure 30–5.

Spongiform encephalopathies are a group of chronic, progressive, fatal infections of the central nervous system caused by unconventional, transmissible agents called prions (see Chapter 42). Prions are thought not to be viruses. The best examples of this type of "slow" infection are scrapie in

sheep and bovine spongiform encephalopathy in cattle; kuru and Creutzfeldt-Jakob disease occur in humans.

Overview of Acute Viral Respiratory Infections

Many types of viruses gain access to the human body via the respiratory tract, primarily in the form of aerosolized droplets or saliva. This is the most frequent means of viral entry into the host. Successful infection occurs despite normal host protective mechanisms, including the mucus covering most surfaces, ciliary action, collections of lymphoid cells, alveolar macrophages, and secretory IgA. Many infections remain localized in the respiratory tract, although some viruses produce their characteristic disease symptoms following systemic spread (eg, chickenpox, measles, rubella; Table 30–2, Figure 30–2).

Disease symptoms exhibited by the host depend on whether the infection is concentrated in the upper or lower respiratory tract (Table 30–4). Although definitive diagnosis requires isolation of the virus, identification of viral gene sequences, or demonstration of a rise in antibody titer, the

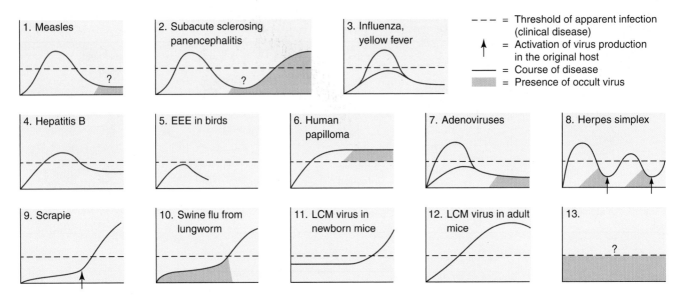

FIGURE 30–5 Different types of virus–host interactions: apparent (clinical disease), inapparent (subclinical), chronic, latent, occult, and slow infections. **(1)** Measles runs an acute, almost always clinically apparent course resulting in long-lasting immunity. **(2)** Measles may also be associated with persistence of latent infection in subacute sclerosing panencephalitis (see Chapter 40). **(3)** Yellow fever and influenza follow a pattern similar to that of measles except that infection may be more often subclinical than clinical. **(4)** In hepatitis B, recovery from clinical disease may be associated with chronic infection in which fully active virus persists in the blood. **(5)** Some infections are, in a particular species, always subclinical, such as eastern equine encephalomyelitis (EEE) in some species of birds that then act as reservoirs of the virus. **(6)** In human papilloma, the course of infection is chronic; when cervical cancer develops, the virus present is occult (not replicating). **(7)** Infection of humans with certain adenoviruses may be clinical or subclinical. There may be a long latent infection during which virus is present in small quantity; virus may also persist after the illness. **(8)** The periodic reactivation of latent herpes simplex virus, which may recur throughout life in humans, often follows an initial acute episode of stomatitis in childhood. **(9)** Infection may be unrecognized for long periods of time before it becomes apparent. Examples of such "slow" infections characterized by long incubation periods are scrapie in sheep and kuru in humans (thought to be caused by prions, not viruses). **(10)** In pigs that have eaten virus-bearing lungworms, swine "flu" is occult until the appropriate stimulus induces viral production and, in turn, clinical disease. **(11)** Lymphocytic choriomeningitis (LCM) virus may be established in mice by in utero infection. A form of immunologic tolerance develops in which virus-specific T cells are not activated. Antibody is produced against viral proteins; this antibody and circulating LCM virus form antigen–antibody complexes that produce immune complex disease in the host. The presence of LCM virus in this chronic infection (circulating virus with little or no apparent disease) may be revealed by transmission to an indicator host, eg, adult mice from a virus-free stock. **(12)** All adult mice develop classic acute symptoms of lymphocytic choriomeningitis and frequently die. **(13)** The possibility is shown of infection with an occult virus that is not detectably replicating. Proof of the presence of such a virus remains a difficult task which, however, is attracting the attention of cancer investigators (see Chapter 43).

TABLE 30–4   Viral Infections of the Respiratory Tract

| Syndromes | Main Symptoms | Most Common Viral Causes | | |
		Infants	Children	Adults
Common cold	Nasal obstruction, nasal discharge	Rhino Adeno	Rhino Adeno	Rhino Corona
Pharyngitis	Sore throat	Adeno Herpes simplex	Adeno Coxsackie	Adeno Coxsackie
Laryngitis/croup	Hoarseness, "barking" cough	Parainfluenza Influenza	Parainfluenza Influenza	Parainfluenza Influenza
Tracheobronchitis	Cough	Parainfluenza Influenza	Parainfluenza Influenza	Influenza Adeno
Bronchiolitis	Cough, dyspnea	Respiratory syncytial Parainfluenza	Rare	Rare
Pneumonia	Cough, chest pain	Respiratory syncytial Influenza	Influenza Parainfluenza	Influenza Adeno

specific viral disease can frequently be deduced by considering the major symptoms, the patient's age, the time of year, and any pattern of illness in the community.

The severity of respiratory infection can range from inapparent to overwhelming. The most severe illness is usually seen in infants infected with certain paramyxoviruses and in elderly or chronically ill adults infected with influenza virus.

Overview of Viral Infections of the Gastrointestinal Tract

Many viruses initiate infection via the alimentary tract. A few agents, such as herpes simplex virus and Epstein-Barr virus, probably infect cells in the mouth. Viruses are exposed in the intestinal tract to harsh elements involved in the digestion of food—acid, bile salts (detergents), and proteolytic enzymes. Consequently, viruses able to initiate infection by this route are all acid- and bile salts-resistant. There may also be virus-specific secretory IgA and nonspecific inhibitors of viral replication to overcome.

Acute gastroenteritis is the designation for short-term gastrointestinal disease with symptoms ranging from mild, watery diarrhea to severe febrile illness characterized by vomiting, diarrhea, and prostration. Rotaviruses, Norwalk viruses, and caliciviruses are major causes of gastroenteritis. Infants and children are affected most often.

Some viruses that produce enteric infections utilize host proteases to facilitate infection. In general, proteolytic digestion alters the viral capsid by partial cleavage of a viral surface protein that then facilitates a specific event such as virus attachment or membrane fusion.

Enteroviruses, coronaviruses, and adenoviruses also infect the gastrointestinal tract, but those infections are often asymptomatic. Some enteroviruses, notably polioviruses, and hepatitis A virus are important causes of systemic disease but do not produce intestinal symptoms.

Overview of Viral Skin Infections

The skin is a tough and impermeable barrier to the entry of viruses. However, a few viruses are able to breach this barrier and initiate infection of the host (Table 30–2). Some obtain entry through small abrasions of the skin (poxviruses, papillomaviruses, herpes simplex viruses), others are introduced by the bite of arthropod vectors (arboviruses) or infected vertebrate hosts (rabies virus, herpes B virus), and still others are injected during blood transfusions or other manipulations involving contaminated needles, such as acupuncture and tattooing (hepatitis B virus, HIV).

A few agents remain localized and produce lesions at the site of entry (papillomaviruses and molluscum contagiosum); most spread to other sites. The epidermal layer is devoid of blood vessels and nerve fibers, so viruses that infect epidermal cells tend to stay localized. Viruses that are introduced deeper into the dermis have access to blood vessels, lymphatics, dendritic cells, and macrophages and usually spread and cause systemic infections.

Many of the generalized skin rashes associated with viral infections develop because virus spreads to the skin via the bloodstream following replication at some other site. Such infections originate by another route (eg, measles virus infections occur via the respiratory tract), and the skin becomes infected from below.

Lesions in skin rashes are designated as macules, papules, vesicles, or pustules. Macules, which are caused by local dilation of dermal blood vessels, progress to papules if edema and cellular infiltration are present in the area. Vesicles occur if the epidermis is involved, and they become pustules if an inflammatory reaction delivers polymorphonuclear leukocytes to the lesion. Ulceration and scabbing follow. Hemorrhagic and petechial rashes occur when there is more severe involvement of the dermal vessels.

Skin lesions frequently play no role in viral transmission. Infectious virus is not shed from the maculopapular rash of measles or from rashes associated with arbovirus infections. In contrast, skin lesions are important in the spread of poxviruses and herpes simplex viruses. Infectious virus particles are present in high titers in the fluid of these vesiculopustular rashes, and they are able to initiate infection by direct contact with other hosts. However, even in these instances, it is believed that virions in oropharyngeal secretions may be more important to disease transmission than the skin lesions.

Overview of Viral Infections of the Central Nervous System

Invasion of the central nervous system by viruses is always a serious matter. Viruses can gain access to the brain by two routes: by the bloodstream (hematogenous spread) and by peripheral nerve fibers (neuronal spread). Access from the blood may occur by growth through the endothelium of small cerebral vessels, by passive transport across the vascular endothelium, by passage through the choroid plexus to the cerebrospinal fluid, or by transport within infected monocytes, leukocytes, or lymphocytes. Once the blood–brain barrier is breached, more extensive spread throughout the brain and spinal cord is possible. There tends to be a correlation between the level of viremia achieved by a blood-borne neurotropic virus and its neuroinvasiveness.

The other pathway to the central nervous system is via peripheral nerves. Virions can be taken up at sensory nerve or motor endings and be moved within axons, through endoneural spaces, or by Schwann cell infections. Herpesviruses travel in axons to be delivered to dorsal root ganglia neurons.

The routes of spread are not mutually exclusive, and a virus may utilize more than one method. Many viruses, including herpes-, toga-, flavi-, entero-, rhabdo-, paramyxo-, and bunyaviruses, can infect the central nervous system and cause meningitis, encephalitis, or both. Encephalitis caused by herpes simplex virus is the most common cause of sporadic encephalitis in humans.

Pathologic reactions to cytocidal viral infections of the central nervous system include necrosis, inflammation, and phagocytosis by glial cells. The cause of symptoms in some other central nervous system infections, such as rabies, is unclear. The postinfectious encephalitis that occurs after measles infections (about one per 1000 cases) and more rarely after rubella infections is characterized by demyelination without neuronal degeneration and is probably an autoimmune disease.

There are several rare neurodegenerative disorders, called slow virus infections, that are uniformly fatal. Features of these infections include a long incubation period (months to years) followed by the onset of clinical illness and progressive deterioration, resulting in death in weeks to months; usually only the central nervous system is involved. Some slow virus infections, such as progressive multifocal leukoencephalopathy (JC polyomavirus) and subacute sclerosing panencephalitis (measles virus), are caused by typical viruses. In contrast, the subacute spongiform encephalopathies, typified by scrapie, are prion diseases caused by unconventional agents called prions. In those infections, characteristic neuropathologic changes occur, but no inflammatory or immune response is elicited.

Overview of Congenital Viral Infections

Few viruses produce disease in the human fetus. Most maternal viral infections do not result in viremia and fetal involvement. However, if the virus crosses the placenta and infection occurs in utero, serious damage may be done to the fetus.

Three principles involved in the production of congenital defects are (1) the ability of the virus to infect the pregnant woman and be transmitted to the fetus; (2) the stage of gestation at which infection occurs; and (3) the ability of the virus to cause damage to the fetus directly, by infection of the fetus, or indirectly, by infection of the mother resulting in an altered fetal environment (eg, fever). The sequence of events that may occur prior to and following viral invasion of the fetus is shown in Figure 30–6.

Rubella virus and cytomegalovirus are presently the primary agents responsible for congenital defects in humans (see Chapters 33 and 40). Congenital infections can also occur with herpes simplex, varicella-zoster, hepatitis B, measles, and mumps virus and with HIV, parvovirus, and some enteroviruses (Table 30–5).

In utero infections may result in fetal death, premature birth, intrauterine growth retardation, or persistent postnatal infection. Developmental malformations, including congenital heart defects, cataracts, deafness, microcephaly, and

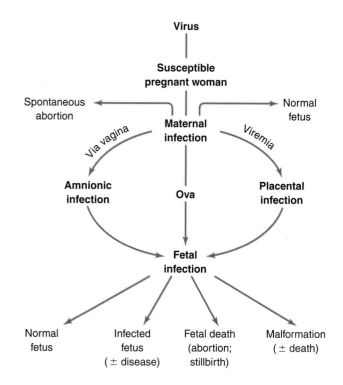

FIGURE 30–6 Viral infection of the fetus. (Courtesy of L Catalano and J Sever.)

TABLE 30–5   Acquisition of Significant Perinatal Viral Infections

| Virus | Frequency of Time of Infection | | | Neonatal Incidence (Per 1000 Live Births) |
	Prenatal (In Utero)	Natal (during Delivery)	Postnatal (after Delivery)	
Rubella	+	−	Rare	0.1–0.7
Cytomegalovirus	+	+ +	+	5–25
Herpes simplex	+	+ +	+	0.03–0.5
Varicella-zoster	+	Rare	Rare	Rare
Hepatitis B	+	+ +	+	0–7
Enterovirus	+	+ +	+	Uncommon
HIV	+	+ +	Rare	Variable
Parvovirus B19	+	−	Rare	Rare

limb hypoplasia, may result. Fetal tissue is rapidly proliferating. Viral infection and multiplication may destroy cells or alter cell function. Lytic viruses, such as herpes simplex, may result in fetal death. Less cytolytic viruses, such as rubella, may slow the rate of cell division. If this occurs during a critical phase in organ development, structural defects and congenital anomalies may result.

Many of the same viruses can produce serious disease in newborns (Table 30–5). Such infections may be contracted from the mother during delivery (natal) from contaminated genital secretions, stool, or blood. Less commonly, infections may be acquired during the first few weeks after birth (postnatal) from maternal sources, family members, hospital personnel, or blood transfusions.

Effect of Host Age

Host age is a factor in viral pathogenicity. More severe disease is often produced in newborns. In addition to maturation of the immune response with age, there seem to be age-related changes in the susceptibility of certain cell types to viral infection. Viral infections usually can occur in all age groups but may have their major impact at different times of life. Examples include rubella, which is most serious during gestation; rotavirus, which is most serious for infants; and St. Louis encephalitis, which is most serious in the elderly.

PREVENTION & TREATMENT OF VIRAL INFECTIONS

Antiviral Chemotherapy

Unlike viruses, bacteria and protozoans do not rely on host cellular machinery for replication, so processes specific to these organisms provide ready targets for the development of antibacterial and antiprotozoal drugs. Because viruses are obligate intracellular parasites, antiviral agents must be capable of selectively inhibiting viral functions without damaging the host, making the development of such drugs very difficult. Another limitation is that many rounds of virus replication occur during the incubation period and the virus has spread before symptoms appear, making a drug relatively ineffective.

There is a need for antiviral drugs active against viruses for which vaccines are not available or not highly effective—the latter perhaps because of a multiplicity of serotypes (eg, rhinoviruses) or because of a constantly changing virus (eg, influenza, HIV). Antivirals can be used to treat established infections when vaccines would not be effective. Antivirals are needed to reduce morbidity and economic loss due to viral infections and to treat increasing numbers of immunosuppressed patients who are at increased risk of infection.

Molecular virology studies are succeeding in identifying virus-specific functions that can serve as targets for antiviral therapy. The most amenable stages to target in viral infections include attachment of virus to host cells; uncoating of the viral genome; viral nucleic acid synthesis; translation of viral proteins; and assembly and release of progeny virus particles. In reality, it has been very difficult to develop antivirals that can distinguish viral from host replicative processes.

However, in the last decade a number of compounds have been developed that are of value in treatment of some viral diseases, particularly against herpesviruses and HIV infections (Table 30–6). Examples are summarized below. The mechanisms of action vary among antivirals. Oftentimes the drug must be activated by enzymes in the cell before it can act as an inhibitor of viral replication; the most selective drugs are activated by a virus-encoded enzyme in the infected cell.

Future work is necessary to learn how to minimize the emergence of drug-resistant variant viruses and to design more specific antivirals based on molecular insights into the structure and replication of different classes of agents.

A. Nucleoside Analogs

The majority of available antiviral agents are nucleoside analogs. They inhibit nucleic acid replication by inhibition of polymerases for nucleic acid replication. In addition, some analogs can be incorporated into the nucleic acid and block further synthesis or alter its function.

Analogs can inhibit cellular enzymes as well as virus-encoded enzymes. The most effective analogs are those able to specifically inhibit virus-encoded enzymes, with minimal inhibition of analogous host cell enzymes. Virus variants resistant to the drug usually arise over time, sometimes quite rapidly. The use of combinations of antiviral drugs can delay the emergence of resistant variant (eg, "triple drug" therapy used to treat HIV infections).

Examples of nucleoside analogs include acyclovir (acycloguanosine), lamivudine (3TC), ribavirin, vidarabine (adenine arabinoside), and zidovudine (azidothymidine; AZT).

B. Nucleotide Analogs

Nucleotide analogs differ from nucleoside analogs in having an attached phosphate group. Their ability to persist in cells for long periods of time increases their potency. Cidofovir (HPMPC) is an example.

C. Nonnucleoside Reverse Transcriptase Inhibitors

Nevirapine was the first member of the class of nonnucleoside reverse transcriptase inhibitors. It does not require phosphorylation for activity and does not compete with nucleoside triphosphates. It acts by binding directly to reverse transcriptase and disrupting the enzyme's catalytic site. Resistant mutants emerge rapidly.

D. Protease Inhibitors

Saquinavir was the first protease inhibitor to be approved for treatment of HIV infection. It is a peptidomimetic agent designed by computer modeling as a molecule that fits into

TABLE 30-6   **Examples of Antiviral Compounds Used for Treatment of Viral Infections**

Drug	Nucleoside Analog	Mechanism of Action	Viral Spectrum[a]
Acyclovir	Yes	Viral polymerase inhibitor	Herpes simplex, varicella-zoster
Amantadine	No	Blocks viral uncoating	Influenza A
Cidofovir	No	Viral polymerase inhibitor	Cytomegalovirus, herpes simplex, polyomavirus
Didanosine (ddl)	Yes	Reverse transcriptase inhibitor	HIV-1, HIV-2
Entecavir	Yes	Reverse transcriptase inhibitor	HBV
Foscarnet	No	Viral polymerase inhibitor	Herpesviruses, HIV-1, HBV
Fuzeon	No	HIV fusion inhibitor (blocks viral entry)	HIV-1
Ganciclovir	Yes	Viral polymerase inhibitor	Cytomegalovirus
Indinavir	No	HIV protease inhibitor	HIV-1, HIV-2
Lamivudine (3TC)	Yes	Reverse transcriptase inhibitor	HIV-1, HIV-2, HBV
Lopinavir	No	HIV protease inhibitor	HIV-1
Maraviroc	No	Entry inhibitor (blocks binding to CCR5)	HIV-1
Nevirapine	No	Reverse transcriptase inhibitor	HIV-1
Oseltamivir	No	Viral neuraminidase inhibitor	Influenza A and B
Raltegravir	No	Integrase inhibitor	HIV-1
Ribavirin	Yes	Perhaps blocks capping of viral mRNA	Respiratory syncytial virus, influenza A and B, Lassa fever, hepatitis C, others
Ritonavir	No	HIV protease inhibitor	HIV-1, HIV-2
Saquinavir	No	HIV protease inhibitor	HIV-1, HIV-2
Stavudine (d4T)	Yes	Reverse transcriptase inhibitor	HIV-1, HIV-2
Trifluridine	Yes	Viral polymerase inhibitor	Herpes simplex, cytomegalovirus, vaccinia
Valacyclovir	Yes	Viral polymerase inhibitor	Herpesviruses
Vidarabine	Yes	Viral polymerase inhibitor	Herpesviruses, vaccinia, HBV
Zalcitabine (ddC)	Yes	Reverse transcriptase inhibitor	HIV-1, HIV-2, HBV
Zidovudine (AZT)	Yes	Reverse transcriptase inhibitor	HIV-1, HIV-2, HTLV-1

[a] HBV, hepatitis B virus; HIV-1, HIV-2, human immunodeficiency virus types 1 and 2; HTLV-1, human T cell leukemia virus type 1.

the active site of the HIV protease enzyme. Such drugs inhibit the viral protease that is required at the late stage of the replicative cycle to cleave the viral *gag* and *gag-pol* polypeptide precursors to form the mature virion core and activate the reverse transcriptase that will be used in the next round of infection. Inhibition of the protease yields noninfectious virus particles. Protease inhibitors include indinavir and ritonavir and others not listed here.

E. Fusion Inhibitor

Fuzeon is a large peptide that blocks the virus and cellular membrane fusion step involved in entry of HIV-1 into cells.

F. Other Types of Antiviral Agents

A number of other types of compounds have been shown to possess some antiviral activity under certain conditions.

1. Amantadine & rimantadine—These synthetic amines specifically inhibit influenza A viruses by blocking viral uncoating. They must be administered prophylactically to have a significant protective effect.

2. Foscarnet (phosphonoformic acid)—Foscarnet, an organic analog of inorganic pyrophosphate, selectively inhibits viral DNA polymerases and reverse transcriptases at the pyrophosphate-binding site.

3. Methisazone—Methisazone is of historical interest as an inhibitor of poxviruses. It was the first antiviral agent to be described and contributed to the campaign to eradicate smallpox. It blocked a late stage in viral replication, resulting in the formation of immature, noninfectious virus particles.

Interferons

IFNs are host-coded proteins that are members of the large cytokine family and which inhibit viral replication. They are produced very quickly (within hours) in response to viral infection or other inducers and are one of the body's first responders in the defense against viral infection. IFN was the first cytokine to be recognized. IFNs are central to the innate antiviral immune response. They also modulate humoral and cellular immunity and have broad cell growth regulatory activities, but the focus here will be on their antiviral effects.

A. Properties of IFNs

There are multiple species of IFNs that fall into three general groups, designated IFN-α, IFN-β, and IFN-γ (Table 30–7). Both IFN-α and IFN-β are considered type I or viral IFNs, whereas IFN-γ is type II or immune IFN. The IFN-α family is large, being coded by at least 20 genes in the human genome; the IFN-β and IFN-γ families are coded by one gene each. The three gene families have diverged so that the coding sequences now are not closely related.

The different IFNs are similar in size, but the three classes are antigenically distinct. IFN-α and IFN-β are resistant to low pH. IFN-β and IFN-γ are glycosylated, but the sugars are not necessary for biologic activity, so cloned IFNs produced in bacteria are biologically active. Dendritic cells are potent IFN producers; under the same virus challenge conditions, dendritic cells can secrete up to 1000 times more IFN than fibroblasts.

B. Synthesis of IFNs

IFNs are produced by all vertebrate species. Normal cells do not generally synthesize IFN until they are induced to do so. Infection with viruses is a potent insult leading to induction; RNA viruses are stronger inducers of IFN than DNA viruses. IFNs also can be induced by double-stranded RNA, bacterial endotoxin, and small molecules such as tilorone. IFN-γ is not produced in response to most viruses but is induced by mitogen stimulation.

The different classes of IFN are produced by different cell types. IFN-α and IFN-β are synthesized by many cell types, but IFN-γ is produced mainly by lymphocytes, especially T cells and natural killer (NK) cells. Dendritic cells are potent IFN producers; under the same virus challenge conditions, dendritic cells can secrete up to 1000 times more IFN than fibroblasts.

C. Antiviral Activity & Other Biologic Effects

IFNs were first recognized by their ability to interfere with viral infection in cultured cells. IFNs are detectable soon after viral infection in intact animals, and viral production then decreases (Figure 30–7). Antibody does not appear in the blood of the animal until several days after viral production has abated. This temporal relationship suggests that IFN plays a primary role in the nonspecific defense of the host

TABLE 30–7   **Properties of Human Interferons**

Property	Type		
	Alpha	**Beta**	**Gamma**
Current nomenclature	IFN-α	IFN-β	IFN-γ
Former designation	Leukocyte	Fibroblast	Immune interferon
Type designation	Type I	Type I	Type II
Number of genes that code for family	≥20	1	1
Principal cell source	Most cell types	Most cell types	Lymphocytes
Inducing agent	Viruses; dsRNA	Viruses; dsRNA	Mitogens
Stability at pH 2.0	Stable	Stable	Labile
Glycosylated	No	Yes	Yes
Introns in genes	No	No	Yes
Homology with IFN-α	80–95%	30%	<10%
Chromosomal location of genes	9	9	12
Size of secreted protein (number of amino acids)	165	166	143
IFN receptor	IFNAR	IFNAR	IFNGR
Chromosomal location of IFN receptor genes	21	21	6

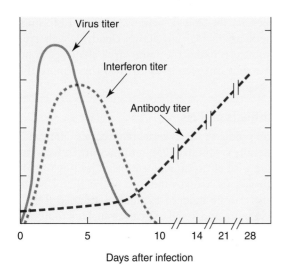

FIGURE 30–7 Illustration of kinetics of interferon and antibody synthesis after respiratory viral infection. The temporal relationships suggest that interferons are involved in the host's early defense system against viral infections.

against viral infections. This conclusion is also supported by observations that agammaglobulinemic individuals usually recover from primary viral infections about as well as normal people.

IFN does not protect the virus-infected cell that produces it, and IFN itself is not the antiviral agent. Rather, IFN moves to other cells where it induces an antiviral state by prompting the synthesis of other proteins that actually inhibit viral replication. IFN molecules bind to specific cell surface receptors on target cells. IFN-α and IFN-β have the same receptor, whereas IFN-γ recognizes a different receptor. Receptor binding triggers tyrosine phosphorylation and activation of transcription factors (STAT proteins) in the cytoplasm, which then translocate into the nucleus and mediate transcription of IFN-inducible genes (which occurs within minutes after IFN binding). This results in the synthesis of several enzymes believed to be instrumental in the development of the antiviral state. Several pathways appear to be involved, including the following: (1) a dsRNA-dependent protein kinase, PKR, which phosphorylates and inactivates cellular initiation factor eIF-2 and thus prevents formation of the initiation complex needed for viral protein synthesis; (2) an oligonucleotide synthetase, 2-5A synthetase, which activates a cellular endonuclease, RNase L, which in turn degrades mRNA; (3) a phosphodiesterase, which inhibits peptide chain elongation; and (4) nitric oxide synthetase, which is induced by IFN-γ in macrophages. These explanations, however, fail to reveal why the antiviral state acts selectively against viral mRNAs and not cellular mRNAs. Other steps in viral replication may also be inhibited by IFN.

IFNs are almost always host species-specific in function but are not specific for a given virus. The replication of a wide variety of DNA and RNA viruses can be inhibited. When IFN is added to cells prior to infection, there is marked inhibition of viral replication but nearly normal cell function. IFNs are extremely potent, so that very small amounts are required for function. It has been estimated that fewer than 50 molecules of IFN per cell are sufficient to induce the antiviral state.

D. Virus Mechanisms to Counteract IFN

Viruses display different mechanisms that block the inhibitory activities of IFNs on virus replication, processes necessary to surmount this line of host defense. Examples: specific viral proteins may block induction of expression of IFN (herpesvirus, papillomavirus, filovirus, hepatitis C virus, rotavirus); may block the activation of the key PKR protein kinase (adenovirus, herpesviruses); may activate a cellular inhibitor of PKR (influenza, poliovirus); may block IFN-induced signal transduction (adenovirus, herpesviruses, hepatitis B virus); or may neutralize IFN-γ by acting as a soluble IFN receptor (myxoma virus).

E. Clinical Studies

It was originally hoped that IFNs might be the answer to prevention of many viral diseases, such as respiratory infections in which many different viruses may be involved. However, their use turns out to be impractical because for them to be effective, high doses must be given prior to virus exposure or early in infection before the appearance of clinical signs of disease. Recombinant IFN-α is beneficial in controlling hepatitis B and hepatitis C viral infections of the liver (Chapter 35), though relapse after cessation of treatment is common. Topical IFN in the eye may suppress herpetic keratitis and accelerate healing.

Several IFN preparations are approved for clinical use. IFNs cause many side effects, most commonly systemic and hematologic.

Viral Vaccines

The purpose of viral vaccines is to utilize the immune response of the host to prevent viral disease. Several vaccines have proved to be remarkably effective at reducing the annual incidence of viral disease (Figure 30–8). Vaccination is the most cost-effective method of prevention of serious viral infections.

A. General Principles

Immunity to viral infection is based on the development of an immune response to specific antigens located on the surface of virus particles or virus-infected cells. For enveloped viruses, the important antigens are the surface glycoproteins. Although infected animals may develop antibodies against virion core proteins or nonstructural proteins involved in viral replication, that immune response is believed to play little or no role in the development of resistance to infection.

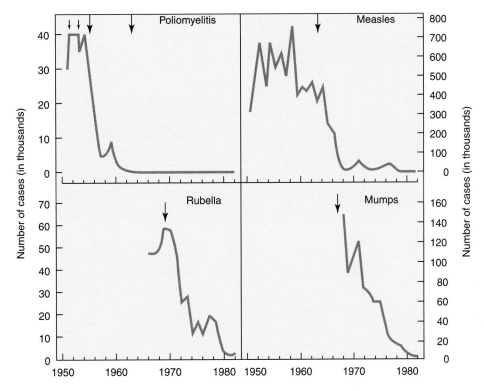

FIGURE 30–8 Annual incidence of various viral diseases in the United States. Date of introduction of vaccine indicated by arrow. (Data from the Centers for Disease Control and Prevention.)

Vaccines are available for the prevention of several significant human diseases. Currently available vaccines (Table 30–8) are described in detail in the chapters dealing with specific virus families and diseases.

The pathogenesis of a particular viral infection influences the objectives of immunoprophylaxis. Mucosal immunity (local IgA) is important in resistance to infection by viruses that replicate in mucosal membranes (rhinoviruses, influenza viruses, rotaviruses) or invade through the mucosa (papillomavirus). Viruses that have a viremic mode of spread (polio, hepatitis A and B, yellow fever, varicella, mumps, measles) are controlled by serum antibodies. Cell-mediated immunity also is involved in protection against systemic infections (measles, herpes).

Certain characteristics of a virus or of a viral disease may complicate the generation of an effective vaccine. The existence of many serotypes, as with rhinoviruses, and of large numbers of animal reservoirs, as with influenza virus, makes vaccine production difficult. Other hurdles include the integration of viral DNA into host chromosomal DNA (retroviruses) and infection of cells of the host's immune system (HIV).

B. Killed-Virus Vaccines

Inactivated (killed-virus) vaccines are made by purifying viral preparations to a certain extent and then inactivating viral infectivity in a way that does minimal damage to the viral structural proteins; mild formalin treatment is frequently used (Table 30–9). For some diseases, killed-virus vaccines are currently the only ones available.

Killed-virus vaccines prepared from whole virions generally stimulate the development of circulating antibody against the coat proteins of the virus, conferring some degree of resistance.

Advantages of inactivated vaccines are that there is no reversion to virulence by the vaccine virus and that vaccines can be made when no acceptable attenuated virus is available.

The following disadvantages apply to killed-virus vaccines:

1. Extreme care is required in their manufacture to make certain that no residual live virulent virus is present in the vaccine.

2. The immunity conferred is often brief and must be boosted, which not only involves the logistic problem of repeatedly reaching the persons in need of immunization but also has caused concern about the possible effects (hypersensitivity reactions) of repeated administration of foreign proteins.

3. Parenteral administration of killed-virus vaccine, even when it stimulates circulating antibody (IgM, IgG) to satisfactory levels, has sometimes given limited protection because local resistance (IgA) is not induced adequately at the natural portal of entry or primary site of multiplication of the wild virus infection—eg, nasopharynx for respiratory viruses, alimentary tract for poliovirus (see Figure 30–9 and Chapters 36 and 39).

TABLE 30–8   **Virus Vaccines Approved in the United States (2008)**

Use	Vaccine	Type	Cell Substrate
Common	Hepatitis A	Killed	Human diploid fibroblasts (MRC-5)
	Hepatitis B	Subunit (HBsAg)	Yeast (recombinant DNA)
	Influenza A and B	Killed	Embryonated chicken eggs
	Influenza A and B	Live (intranasal)	Embryonated chicken eggs
	Measles	Live	Chicken embryo fibroblasts
	Mumps	Live	Embryonated chicken eggs and chicken embryo fibroblasts
	Papilloma	Subunit (L1)	Yeast (recombinant DNA)
	Poliovirus (IPV)	Killed	Monkey kidney cells (Vero)
	Poliovirus (OPV)	Live	Monkey kidney cells
	Rabies	Killed	Human diploid fibroblasts (MRC-5) or rhesus fetal lung diploid cells or chicken fibroblasts
	Rotavirus[a]	Live	Monkey kidney cells (Vero)
	Rubella	Live	Human diploid fibroblasts (WI-38)
	Varicella	Live	Human diploid fibroblasts (MRC-5)
	Zoster	Live	Human diploid fibroblasts (MRC-5)
Special situations	Adenovirus[b]	Live	Human diploid fibroblasts (WI-38)
	Japanese encephalitis[c]	Killed	Mouse brain
	Smallpox	Live	Calf lymph
	Yellow fever[c]	Live	Embryonated chicken eggs

[a]A live rotavirus vaccine was withdrawn from the market in 1999 because of an association with intussusception of infants. The vaccine approved in 2006 is different and has not been associated with intussusception.

[b]Used by U.S. military; no longer available.

[c]Used when traveling in endemic areas.

4. The cell-mediated response to inactivated vaccines is generally poor.
5. Some killed-virus vaccines have induced hypersensitivity to subsequent infection, perhaps owing to an unbalanced immune response to viral surface antigens that fails to mimic infection with natural virus.

C. Attenuated Live-Virus Vaccines

Live-virus vaccines utilize virus mutants that antigenically overlap with wild-type virus but are restricted in some step in the pathogenesis of disease (Table 30–9).

The genetic basis for the attenuation of most viral vaccines is not known, as they were selected empirically by serial passages in animals or cell cultures (usually from a species different from the natural host). As more is learned about viral genes involved in disease pathogenesis, attenuated candidate vaccine viruses can be engineered in the laboratory.

Attenuated live-virus vaccines have the advantage of acting like the natural infection with regard to their effect on immunity. They multiply in the host and tend to stimulate longer-lasting antibody production, to induce a good cell-mediated response, and to induce antibody production and resistance at the portal of entry (Figure 30–9).

The disadvantages of attenuated live-virus vaccines include the following:

1. The risk of reversion to greater virulence during multiplication within the vaccinee. Although reversion has not proved to be a problem in practice, its potential exists.
2. Unrecognized adventitious agents latently infecting the culture substrate (eggs, primary cell cultures) may enter the vaccine stocks. Viruses found in vaccines have included avian leukosis virus, simian polyomavirus SV40, and simian cytomegalovirus. The problem of adventitious contaminants may be circumvented through the use of normal cells serially propagated in culture (eg, human diploid cell lines) as substrates for cultivation of vaccine viruses.
3. The storage and limited shelf life of attenuated vaccines present problems, but this can be overcome in some cases by the use of viral stabilizers (eg, $MgCl_2$ for polio vaccine).
4. Interference by coinfection with a naturally occurring, wild-type virus may inhibit replication of the vaccine virus and decrease its effectiveness. This has been noted with the vaccine strains of poliovirus, which can be inhibited by concurrent infections by various enteroviruses.

D. Proper Use of Present Vaccines

One fact cannot be overemphasized: An effective vaccine does not protect against disease until it is administered in the proper dosage to susceptible individuals. Failure to reach all sectors of the population with complete courses of immunization is reflected in the continued occurrence of measles in unvaccinated persons. Preschool children in

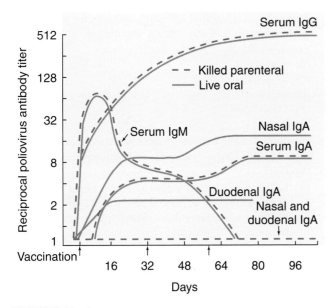

FIGURE 30-9 Serum and secretory antibody response to orally administered, live attenuated polio vaccine and to intramuscular inoculation of killed polio vaccine. (Reproduced with permission from Ogra PL et al: Rev Infect Dis 1980;2:352.)

TABLE 30-9 **Comparison of Characteristics of Killed and Live Viral Vaccines**

Characteristic	Killed Vaccine	Live Vaccine
Number of doses	Multiple	Single
Need for adjuvant	Yes	No
Duration of immunity	Shorter	Longer
Effectiveness of protection (more closely mimics natural infection)	Lower	Greater
Immunoglobulins produced	IgG	IgA and IgG
Mucosal immunity produced	Poor	Yes
Cell-mediated immunity produced	Poor	Yes
Residual virulent virus in vaccine	Possible	No
Reversion to virulence	No	Possible
Excretion of vaccine virus and transmission to nonimmune contacts	No	Possible
Interference by other viruses in host	No	Possible
Stability at room temperature	High	Low

poverty areas are the least adequately vaccinated group in the United States.

Certain viral vaccines are recommended for use by the general public. Other vaccines are recommended only for use by persons at special risk due to occupation, travel, or lifestyle. In general, live-virus vaccines are contraindicated for pregnant women.

There was a theoretical possibility that antibody response might be diminished or that interference might occur if two or more live-virus vaccines were given at the same time. In practice, however, simultaneous administration of live-virus vaccines can be safe and effective. Trivalent live oral polio vaccine or a combined live measles, mumps, and rubella vaccine is effective. Antibody response to each component of these combination vaccines is comparable with antibody response to the individual vaccines given separately.

E. Future Prospects

Molecular biology and modern technologies are combining to devise novel approaches to vaccine development. Many of these approaches avoid the incorporation of viral nucleic acid in the final product, improving vaccine safety. Examples of what is ongoing in this field can be listed as follows. The ultimate success of these new approaches remains to be determined.

1. Use of recombinant DNA techniques to insert the gene coding for the protein of interest into the genome of an avirulent virus that can be administered as the vaccine (such as vaccinia virus).

2. Including in the vaccine only those subviral components needed to stimulate protective antibody, thus minimizing the occurrence of adverse reactions to the vaccine.

3. Use of purified proteins isolated from purified virus or synthesized from cloned genes (a recombinant hepatitis B virus vaccine contains viral proteins synthesized in yeast cells). Expression of cloned gene(s) sometimes results in formation of empty virus-like particles.

4. Use of synthetic peptides that correspond to antigenic determinants on a viral protein, thus avoiding any possibility of reversion to virulence since no viral nucleic acid would be present—although the immune response induced by synthetic peptides is considerably weaker than that induced by intact protein.

5. Development of edible vaccines whereby transgenic plants synthesizing antigens from pathogenic viruses may provide new cost-effective ways of delivering vaccines.

6. Use of naked DNA vaccines—potentially simple, cheap, and safe—in which recombinant plasmids carrying the gene for the protein of interest are injected into hosts and the DNA produces the immunizing protein.

7. Administration of vaccine locally to stimulate antibody at the portal of entry (such as aerosol vaccines for respiratory disease viruses).

REVIEW QUESTIONS

1. Interferons are an important part of the host defense against viral infections. What is interferon's principal mode of action?

    (A) It is present in the serum of healthy individuals, providing a viral surveillance role

(B) It coats viral particles and blocks their attachment to cells

(C) It induces synthesis of one or more cellular proteins that inhibit translation or transcription

(D) It protects the virus-infected cell that produced it from cell death

2. A 9-month-old girl is taken to the emergency room because of fever and persistent cough. Rales are heard in her left chest on physical examination. An infiltrate in her left lung is seen on the chest x-ray film. Pneumonia is diagnosed. Which of the following is the most likely cause?

(A) Rotavirus

(B) Rhinovirus

(C) Adenovirus

(D) Respiratory syncytial virus

(E) Coxsackievirus

3. Which one of the following is a fundamental principle of viral disease causation?

(A) One virus type induces a single disease syndrome

(B) Many viral infections are subclinical and do not produce clinical disease

(C) The type of disease produced by a virus can be predicted by the morphology of that virus

(D) A particular disease syndrome has a single viral cause

4. The skin is an impenetrable barrier to virus entry, but a few viruses are able to breach this barrier and initiate infection of the host. Which of the following is an example of a virus that enters through skin abrasions?

(A) Adenovirus

(B) Rotavirus

(C) Rhinovirus

(D) Papillomavirus

(E) Influenza virus

5. A 40-year-old man has HIV/AIDS characterized by a low CD4 count and a high viral load. Highly active antiretroviral therapy (HAART) will be initiated. One of the drugs under consideration is a nucleoside analog that inhibits viral reverse transcriptase and is active against both HIV and HBV. That drug is

(A) Acyclovir

(B) Amantadine

(C) Ribavirin

(D) Saquinavir

(E) Lamivudine

(F) Fuzeon

6. Regarding the HIV/AIDS patient in Question 5, a peptidomimetic agent that blocks virus-mediated cleavage of viral structural protein precursors is chosen as a second drug. That drug is

(A) Acyclovir

(B) Amantadine

(C) Ribavirin

(D) Saquinavir

(E) Lamivudine

(F) Fuzeon

7. A 63-year-old woman is hospitalized for treatment of leukemia. One day after admission she develops chills, fever, cough, headache, and myalgia. She states that her husband had a similar illness a few days earlier. There is major concern about a respiratory virus outbreak in the staff of the chemotherapy ward and in the patients on that ward. A synthetic amine that inhibits influenza

A virus by blocking viral uncoating is chosen for prophylactic treatment of the staff and patients. That drug is

(A) Acyclovir

(B) Amantadine

(C) Ribavirin

(D) Saquinavir

(E) Lamivudine

(F) Fuzeon

8. Which one of the following statements describes an advantage of killed-virus vaccines over attenuated live-virus vaccines?

(A) Killed-virus vaccines induce a broader range of immune responses than do attenuated live-virus vaccines

(B) Killed-virus vaccines more closely mimic natural infections than do attenuated live-virus vaccines

(C) Killed-virus vaccines pose no risk that vaccine virus might be transmitted to susceptible contacts

(D) Killed-virus vaccines are efficacious against respiratory virus infections because they induce good mucosal immunity

9. What type of hepatitis B vaccine is currently in use in the United States?

(A) Synthetic peptide vaccine

(B) Killed-virus vaccine

(C) Attenuated live-virus vaccine

(D) Subunit vaccine produced using recombinant DNA

10. Which one of the following phrases accurately describes viral neutralizing antibodies?

(A) Directed against viral protein determinants on the outside of the virus particle

(B) Appear in the host sooner after viral infection than interferon

(C) Directed against viral nucleic acid sequences

(D) Induced only by disease-causing viruses

(E) Of little importance to immunity to viral infection

11. Many viruses use the respiratory tract as the route of entry to initiate infections. Which of the following virus groups does not?

(A) Adenovirus

(B) Coronavirus

(C) Hepadnavirus

(D) Paramyxovirus

(E) Poxvirus

12. Which of the following licensed virus vaccines is a subunit vaccine prepared using recombinant DNA technology?

(A) Measles–mumps–rubella

(B) Varicella

(C) Hepatitis A

(D) Papilloma

(E) Rotavirus

(F) Rabies

13. Which of the following viruses is the most common cause of neonatal infections in the United States?

(A) Rubella

(B) Parvovirus B19

(C) Hepatitis B

(D) Cytomegalovirus

(E) Varicella

(F) HIV

Answers

1. C	5. E	9. D	13. D
2. D	6. D	10. A	
3. B	7. B	11. C	
4. D	8. C	12. D	

REFERENCES

Ada G: Vaccines and vaccination. N Engl J Med 2001;345:1042. [PMID: 11586958]

Alcami A, Koszinowski UH: Viral mechanisms of immune evasion. Trends Microbiol 2000;8:410. [PMID: 10989308]

Bonjardim CA: Interferons (IFNs) are key cytokines in both innate and adaptive antiviral immune responses—and viruses counteract IFN action. Microbes and Infection 2005;7:569. [PMID: 15792636]

Bonthius DJ, Perlman S: Congenital viral infections of the brain: Lessons learned from lymphocytic choriomeningitis virus in the neonatal rat. PLoS Pathogens 2007;3:e149. [PMID: 18052527]

Centers for Disease Control and Prevention. General recommendations on immunization. Recommendations of the Advisory Committee on Immunization Practices (ACIP). MMWR Morb Mortal Wkly Rep 2006;55(No. RR-15).

Centers for Disease Control and Prevention. Recommended childhood and adolescent immunization schedule—United States, 2006. MMWR Morb Mortal Wkly Rep 2006;54(Nos. 51 & 52).

Espy MJ et al: Real-time PCR in clinical microbiology: Applications for routine laboratory testing. Clin Microbiol Rev 2006;19:165. [PMID: 16418529]

Hawley RJ, Eitzen EM Jr: Biological weapons—a primer for microbiologists. Annu Rev Microbiol 2001;55:235. [PMID: 11544355]

Immunization of health-care workers: Recommendations of the Advisory Committee on Immunization Practices (ACIP) and the Hospital Infection Control Practices Advisory Committee (HICPAC). MMWR Morb Mortal Wkly Rep 1997;46 (RR-18):1.

Pereira L, Maidji E, McDonagh S, Tabata T: Insights into viral transmission at the uterine-placental interface. Trends Microbiol 2005;13:164. [PMID: 15817386]

Plotkin SA: Correlates of vaccine-induced immunity. Clin Infect Dis 2008;47:401. [PMID: 18558875]

Preventing emerging infectious diseases: A strategy for the 21st century. Overview of the updated CDC plan. MMWR Morb Mortal Wkly Rep 1998;47 (RR-15):1.

Randall RE, Goodbourn S: Interferons and viruses: An interplay between induction, signalling, antiviral responses and virus countermeasures. J Gen Virol 2008;89:1. [PMID: 18089727]

Virgin S: Pathogenesis of viral infection. In: *Fields Virology,* 5th ed. Knipe DM et al (editors). Lippincott Williams & Wilkins, 2007.

Parvoviruses

Parvoviruses are the simplest DNA animal viruses. Because of the small coding capacity of their genome, viral replication is dependent on functions supplied by replicating host cells or by coinfecting helper viruses. Parvovirus B19 is pathogenic for humans and has a tropism for erythroid progenitor cells. It is the cause of erythema infectiosum ("fifth disease"), a common childhood exanthem; of a polyarthralgia-arthritis syndrome in normal adults; of aplastic crisis in patients with hemolytic disorders; of chronic anemia in immunocompromised individuals; and of fetal death. A newly discovered parvovirus, human bocavirus, has been detected in respiratory specimens from children with acute respiratory disease, but a role in the disease is unproven.

PROPERTIES OF PARVOVIRUSES

Important properties of parvoviruses are listed in Table 31–1. It is noteworthy that there are both autonomously replicating and defective parvoviruses.

Structure & Composition

The icosahedral, nonenveloped particles are 18–26 nm in diameter (Figure 31–1). The particles have a molecular weight of $5.5–6.2 \times 10^6$, a heavy buoyant density of 1.39–1.42 g/cm^3, and an $S_{20,w}$ of 110–122. Virions are extremely resistant to inactivation. They are stable between pH 3 and 9 and withstand heating at 56°C for 60 minutes, but they can be inactivated by formalin, β-propiolactone, and oxidizing agents.

Virions contain two coat proteins that are encoded by an overlapping, in-frame DNA sequence, so that VP2 is identical in sequence to the carboxy portion of VP1. The major capsid protein, VP2, represents about 90% of virion protein. The genome is about 5 kb, linear, single-stranded DNA. An autonomous virus, B19, contains 5596 nucleotides, whereas a defective parvovirus, AAV-2, contains 4680 bases. Autonomous parvoviruses usually encapsidate primarily DNA strands complementary to viral mRNA; defective viruses tend to encapsidate DNA strands of both polarities with equal frequency into separate virions.

Classification

There are two subfamilies of Parvoviridae: the **Parvovirinae**, which infect vertebrates, and the **Densovirinae**, which infect insects. The Parvovirinae comprise several genera. Human parvovirus B19 is the most common member of the *Erythrovirus* genus. Two new human genotypes (types 2 and 3) were identified in humans recently (strains K71 and V9, respectively), each of which differs about 10% in nucleotide sequence from B19 (type 1). The human bocavirus is in the *Bocavirus* genus. Feline panleukopenia virus and canine parvovirus, both serious causes of veterinary diseases, are classified as members of the *Parvovirus* genus, as are isolates from many other animals. The genus *Dependovirus* contains members that are defective and depend on a helper virus (an adenovirus or herpesvirus) for replication. Human "adeno-associated viruses" have not been linked with any disease.

Parvovirus Replication

The replication cycle of human B19 parvovirus is summarized in Figure 31–2. Nonstructural viral protein NS1 is essential. The virus is highly tropic for human erythroid cells. The cellular receptor for B19 is blood group P antigen (globoside). P antigen is expressed on mature erythrocytes, erythroid progenitors, megakaryocytes, endothelial cells, placenta, and

TABLE 31–1   Important Properties of Parvoviruses

Virion: Icosahedral, 18–26 nm in diameter, 32 capsomeres
Composition: DNA (20%), protein (80%)
Genome: Single-stranded DNA, linear, 5.6 kb, MW 1.5–2.0 million
Proteins: One major (VP2) and one minor (VP1)
Envelope: None
Replication: Nucleus, dependent on functions of dividing host cells

Outstanding characteristics:
    Very simple viruses
    Human pathogen, B19, has tropism for red blood cell progenitors
    One genus is replication-defective and requires a helper virus

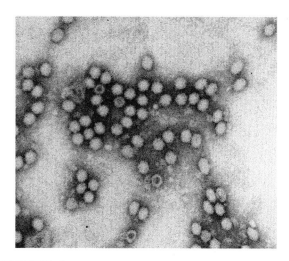

FIGURE 31–1 Electron micrograph of parvovirus particles. (Courtesy of FA Murphy and EL Palmer.)

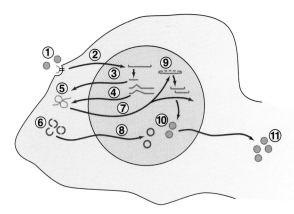

FIGURE 31–2 Life cycle of B19 parvovirus. **1** Binding to erythrocyte P antigen and entry; **2** translocation of viral DNA to the nucleus; **3** transcription of nonstructural RNA and **4** later capsid protein RNA, followed by **5** protein translation. Not separable temporally are **6** capsid self-assembly, **7** nonstructural protein action on viral DNA, **8** capsid translocation to nucleus, **9** DNA replication, **10** insertion of DNA into intact capsids, and **11** virus release and cell lysis. (Reproduced with permission from Young NS: Parvoviruses. In: *Fields Virology*, 3rd ed. Fields BN et al [editors]. Lippincott-Raven, 1996.)

fetal liver and heart, which helps explain the narrow tissue tropism of B19 virus.

The parvoviruses are highly dependent on cellular functions for replication. Viral DNA replication occurs in the nucleus. It is necessary for the host cell to go through S phase, but the parvoviruses do not have the ability to stimulate resting cells to initiate DNA synthesis. One or more cellular DNA polymerases are involved; terminal sequences on the linear parvovirus DNA are used as primers to initiate DNA synthesis. There are two capsid proteins. The nonstructural protein, which is required for virus replication, may be important in the pathogenesis of some B19-associated diseases by modulating host cell genes. Viral replication results in cell death.

PARVOVIRUS INFECTIONS IN HUMANS

Pathogenesis & Pathology

A typical course of human parvovirus B19 infection in adults is illustrated in Figure 31–3. B19 has been implicated as the causative agent of several diseases (Table 31–2). Immature cells in the erythroid lineage are principal targets for human B19 parvovirus. Hence, the major sites of virus replication in patients are assumed to be the adult marrow, some blood cells, and the fetal liver. Viral replication causes cell death, interrupting red cell production. In immunocompromised patients, persistent B19 infections occur, resulting in chronic anemia. In cases of fetal death, chronic infections may have caused severe anemia in the fetus.

As nondefective parvoviruses require dividing host cells in order to replicate, known parvovirus diseases reflect that target specificity (Figure 31–4).

Both virus-specific IgM and IgG antibodies are made following B19 infections. Persistent parvovirus infections occur in patients with immune deficiencies who fail to make virus-neutralizing antibodies, resulting in anemia. Persistence of low levels of B19 DNA, and to a lesser extent virus type

2 DNA, has also been detected in blood, skin, tonsil, liver, and synovial tissues of immunocompetent volunteers. The rash associated with erythema infectiosum is at least partly immune complex-mediated.

B19 can be found in blood and respiratory secretions of infected patients. Transmission is presumably by the respiratory route. There is no evidence of virus excretion in feces or urine. The virus can be transmitted parenterally by blood transfusions or by infected blood products (clotting and immunoglobulin concentrates) and vertically from mother to fetus. Because B19 is resistant to harsh treatments that inactivate enveloped viruses, some clotting factor concentrates end up contaminated. The prevalence of antibodies to B19 is higher among hemophiliacs than the general population; however, the minimal level of virus in blood products able to cause infections is not known.

TABLE 31–2   Human Diseases Associated with B19 Parvovirus[a]

Syndrome	Host or Condition	Clinical Features
Erythema infectiosum	Children (fifth disease) Adults	Cutaneous rash Arthralgia-arthritis
Transient aplastic crisis	Underlying hemolysis	Severe acute anemia
Pure red cell aplasia	Immunodeficiencies	Chronic anemia
Hydrops fetalis	Fetus	Fatal anemia

[a]Modified with permission from Young NS: Parvoviruses. In: *Fields Virology*, 3rd ed. Fields BN et al (editors). Lippincott-Raven, 1996.

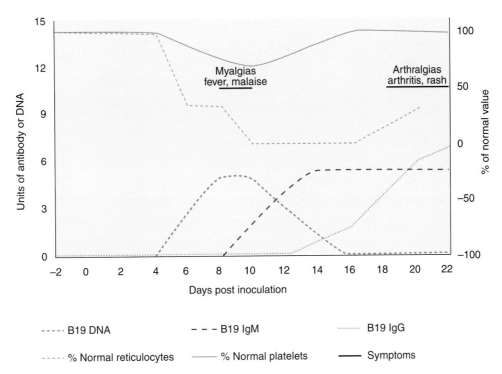

FIGURE 31–3 Clinical and laboratory findings during the course of human parvovirus B19 infection in adult volunteers. The first phase of illness with flu-like symptoms coincides with viremia (days 6–12); the second phase of illness with rash appears on about day 18. (Reproduced with permission from Anderson LJ: Human parvovirus B19. In: *Clinical Virology*, 2nd ed. Richman DD, Whitley RJ, Hayden FG [editors]. ASM Press, 2002; data taken from Anderson MJ et al: Experimental parvoviral infection in humans. J Infect Dis 1985;152:257.)

The pathogenesis of human bocavirus infection is not yet known. As it has been found in respiratory specimens, it is presumed to infect the respiratory tract and be transmitted by the respiratory route. It has also been detected in stool and serum samples.

Several pathogenic parvoviruses of animals replicate in intestinal mucosal cells and cause enteritis.

Clinical Findings

A. Erythema Infectiosum (Fifth Disease)

The most common manifestation of human parvovirus B19 infection is erythema infectiosum, or fifth disease. This erythematous illness is most common in children of early school age and occasionally affects adults. Mild constitutional symptoms may accompany the rash, which has a typical "slapped cheek" appearance. Both sporadic cases and epidemics have been described. Joint involvement is a prominent feature in adult cases; joints in the hands and the knees are most frequently affected. The symptoms mimic rheumatoid arthritis, and the arthropathy may persist for weeks, months, or years.

The incubation period is usually 1–2 weeks but may extend to 3 weeks. Viremia occurs 1 week after infection and persists for about 5 days. During the period of viremia, virus is present in nasal washes and gargle specimens, identifying the upper respiratory tract—most probably the pharynx—as the site of viral shedding. The first phase of illness occurs at the end of the first week; symptoms are flu-like, including fever, malaise, myalgia, chills, and itching. The first episode of illness coincides in time with viremia and reticulocytopenia and with detection of circulating IgM–parvovirus immune complexes. After an incubation period of about 17 days, a second phase of illness begins. The appearance of an erythematous facial rash and a lacelike rash on the limbs or trunk may be accompanied by joint symptoms, especially in adults. The illness is short-lived, with the rash fading after 2–4 days, although the joint symptoms may persist longer. Specific IgG antibodies appear about 15 days postinfection.

B. Transient Aplastic Crisis

Parvovirus B19 is the cause of transient aplastic crisis that may complicate chronic hemolytic anemia, eg, in patients with sickle cell disease, thalassemias, and acquired hemolytic anemias in adults. Transient aplastic crisis may also occur after bone marrow transplantation. The syndrome is an abrupt cessation of red blood cell synthesis in the bone marrow and is reflected in the absence of erythroid precursors in the marrow, accompanied by a rapid worsening of anemia. The infection lowers production of erythrocytes, causing a reduction in the hemoglobin level of peripheral blood. The temporary arrest of production of red blood cells becomes apparent only in patients with chronic hemolytic anemia because of the short life span of their erythrocytes; a 7-day interruption in erythropoiesis would not be expected to cause

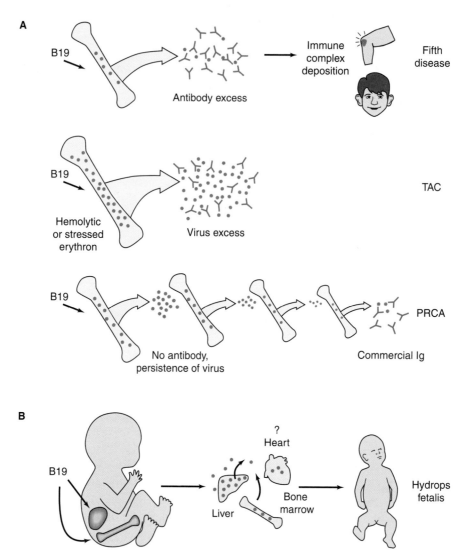

FIGURE 31–4 Pathogenesis of diseases caused by B19 parvovirus. **A:** In children and adults. (TAC, transient aplastic crisis; PRCA, pure red cell aplasia.) **B:** In fetal infections. (Modified with permission from Young NS: Parvoviruses. In: *Fields Virology*, 3rd ed. Fields BN et al [editors]. Lippincott-Raven, 1996.)

detectable anemia in a normal person. Few anemia patients have a rash. Symptoms of transient aplastic crisis occur during the viremic phase of infection.

C. B19 Infection in Immunodeficient Patients

B19 may establish persistent infections and cause chronic suppression of bone marrow and chronic anemia in immunocompromised patients. The disease is called pure red cell aplasia. The anemia is severe, and patients are dependent on blood transfusions. It has been observed in patient populations with congenital immunodeficiency, malignancies, AIDS, and organ transplants.

D. B19 Infection during Pregnancy

Maternal infection with B19 virus may pose a serious risk to the fetus, resulting in hydrops fetalis and fetal death due to severe anemia. The overall risk of human parvovirus infection during pregnancy is low; fetal loss occurs in less than 10% of primary maternal infections. Fetal death occurs most commonly before the 20th week of pregnancy. Although there is frequent intrauterine transmission of human parvovirus (with estimates of vertical transmission rates of 30% or higher), there is no evidence that B19 infection causes physical abnormalities. Maternal–fetal transmission may occur most commonly in pregnant women with high plasma viral loads.

E. Human Bocavirus Respiratory Infection

Human bocavirus has been detected in 1.5–11.3% of respiratory tract samples from young children with respiratory infections. It is prevalent among children with acute wheezing. However, bocavirus is often found in mixed infections with other viruses, so it remains unclear if bocavirus is the cause of acute respiratory disease in children.

Laboratory Diagnosis

The most sensitive tests detect viral DNA. Available tests are polymerase chain reaction, probe hybridization of serum or tissue extracts, and in situ hybridization of fixed tissue. Polymerase chain reaction is the most sensitive assay. B19 DNA has been detected in serum, blood cells, tissue samples, and respiratory secretions. During acute infections, viral loads in the blood can reach approximately 10^{11} genome copies/mL. Polymerase chain reaction assays based on B19 may miss non-B19 strains due to sequence differences. The only assay currently available for human bocavirus is polymerase chain reaction. Bocavirus DNA has been found in serum, stool samples, and respiratory specimens.

Serologic assays based on recombinant parvovirus B19 antigens produced in vitro using bacterial or baculovirus expression systems are used to measure antibodies. VP2 virus-like particles appear to be optimal as antigen for antibody detection. Detection of B19 IgM antibody is indicative of recent infection; it is present for 2–3 months after infection. B19 IgG antibody against conformational epitopes on VP1 and VP2 persists for years, although antibody responses against linear epitopes decline within months postinfection. Antibody may not be found in immunodeficient patients with chronic B19 infections. In those patients, chronic infection is diagnosed by detecting viral DNA.

Antigen detection assays can identify high-titered B19 virus in clinical samples. Immunohistochemistry has been used to detect B19 antigens in fetal tissues and in bone marrow.

Human B19 and human bocaviruses are difficult to grow. Virus isolation is not used to detect infection.

Epidemiology

The B19 virus is widespread. Infections can occur throughout the year, in all age groups, and as outbreaks or as sporadic cases. Infections are most commonly seen as outbreaks in schools. Parvovirus infection is common in childhood; antibody most often develops between the ages of 5 and 19 years. Up to 60% of all adults and 90% of elderly people are seropositive.

B19 infection seems to be transmitted via the respiratory tract. The viruses are stable in the environment, and contaminated surfaces may also be involved in transmission. Transfer among siblings and children in schools and daycare centers is the main path of transmission. The source of maternal infection during pregnancy is often the mother's older child. Many infections are subclinical. Estimates of attack rates in susceptible contacts range from 20% to 50%.

Transmission of B19 from patients with aplastic crisis to members of the hospital staff has been documented. Patients with aplastic crisis are likely to be infectious during the course of their illness, whereas patients with fifth disease are probably no longer infectious by the time of onset of rash.

The epidemiology of human bocavirus is not known. It has been found in young children and appears to be global in distribution.

Treatment

Fifth disease and transient aplastic crisis are treated symptomatically. The latter may require transfusion therapy.

Commercial immunoglobulin preparations contain neutralizing antibodies to human parvovirus. They can sometimes ameliorate persistent B19 infections in immunocompromised patients and in those with anemia.

There is no treatment for human bocavirus infections.

Prevention & Control

There is no vaccine against human parvovirus, though prospects are good that a vaccine can be developed. There are effective vaccines against animal parvoviruses for use in cats, dogs, and pigs. There is no antiviral drug therapy.

Good hygienic practices, such as hand washing and not sharing drinks, should help prevent the spread of B19 through respiratory secretions, aerosols, and fomites. Standard infection control practices should be followed to prevent transmission of B19 to health care workers from patients with aplastic crisis and from immunodeficient patients with chronic B19 infection.

REVIEW QUESTIONS

1. Which one of the following best describes a physicochemical property of parvoviruses?

    (A) Enveloped virus particle
    (B) Single-stranded DNA genome
    (C) Infectivity is inactivated by ether treatment
    (D) Virion exhibits helical symmetry
    (E) Virion is about the same size as herpesviruses

2. An 8-year-old child recently had erythema infectiosum. Her 33-year-old mother subsequently developed arthralgia followed by painful arthritis with swelling in the small joints of both hands. In addition to the apparent tropism for joints, human parvovirus B19 is highly tropic for which cell type?

    (A) CD4 T lymphocytes
    (B) Renal tubule cells
    (C) Erythroid cells
    (D) Glial cells
    (E) Peyer's patches

3. The 8-year-old child in Question 2 had an illness with more than one phase. Which symptoms coincide with the second phase of the illness?

    (A) Sore throat
    (B) Skin rash
    (C) Headache
    (D) Diarrhea
    (E) Cough

4. A 42-year-old man with HIV/AIDS presented with aplastic anemia. Using the polymerase chain reaction, parvovirus B19 was detected in his serum. The patient presumably acquired his parvovirus B19 infection from another person. The most likely route of transmission is

    (A) By contact with respiratory secretions or droplets
    (B) By contact with a skin rash
    (C) Through sexual activity
    (D) Through a recent blood transfusion

5. Which one of the following is a disease in which the role of parvovirus B19 has not been established?

   (A) Erythema infectiosum (fifth disease)
   (B) Transient aplastic crisis
   (C) Hydrops fetalis
   (D) Fulminant hepatitis

6. Which one of the following best describes the replication of human parvovirus B19?

   (A) Stimulates resting cells to proliferate
   (B) Uses blood group antigen P as cellular receptor
   (C) Readily establishes persistent infections
   (D) Entire replication cycle occurs in cytoplasm
   (E) Production of infectious progeny requires the presence of a helper virus

7. Which one of the following statements is most accurate concerning human infections by parvovirus B19?

   (A) Parvovirus B19 is transmitted readily by sexual intercourse
   (B) Patients with disseminated disease caused by parvovirus B19 should be treated with acyclovir
   (C) Parvovirus B19 does not cause any human disease
   (D) There is no vaccine for human parvovirus

8. Human bocavirus is a newly discovered parvovirus. It has been detected most frequently in which type of sample?

   (A) Urine
   (B) Cord blood
   (C) Respiratory secretions
   (D) Fetal liver
   (E) Bone marrow

9. Which of the following is available as a treatment or preventative for parvovirus B19 infections?

   (A) Commercial immunoglobulin
   (B) Vaccine containing recombinant VP2 viral antigen
   (C) Bone marrow transplantation
   (D) Antiviral drug that blocks virus–receptor interaction

Answers

1. B	4. A	7. D
2. C	5. D	8. C
3. B	6. B	9. A

REFERENCES

Allander T et al: Human bocavirus and acute wheezing in children. Clin Infect Dis 2007;44:904. [PMID: 17342639]

Azzi A, Morfini M, Mannucci PM: The transfusion-associated transmission of parvovirus B19. Transfusion Med Rev 1999;13:194. [PMID: 10425692]

Corcoran A, Doyle S: Advances in the biology, diagnosis, and host-pathogen interactions of parvovirus B19. J Med Microbiol 2004;53:459. [PMID: 15150324]

Faisst S, Rommelaere J (editors): *Parvoviruses: From Molecular Biology to Pathology and Therapeutic Uses*. Karger, 2000.

Magro CM, Dawood MR, Crowson AN: The cutaneous manifestations of human parvovirus B19 infection. Hum Pathol 2000;31:488. [PMID: 10821497]

Norja P et al: Bioportfolio: Lifelong persistence of variant and prototypic erythrovirus DNA genomes in human tissue. Proc Natl Acad Sci USA 2006;103:7450. [PMID: 16651522]

Saldanha J et al: Establishment of the first World Health Organization International Standard for human parvovirus B19 DNA nucleic acid amplification techniques. Vox Sang 2002;82:24. [PMID: 11856464]

Adenoviruses

Adenoviruses can replicate and produce disease in the respiratory, gastrointestinal, and urinary tracts and in the eye. Many adenovirus infections are subclinical, and virus may persist in the host for months. About one-third of the 51 known human serotypes are responsible for most cases of human adenovirus disease. A few types serve as models for cancer induction in animals. Adenoviruses are especially valuable systems for molecular and biochemical studies of eukaryotic cell processes.

PROPERTIES OF ADENOVIRUSES

Important properties of adenoviruses are listed in Table 32–1.

Structure & Composition

Adenoviruses are 70–90 nm in diameter and display icosahedral symmetry, with capsids composed of 252 capsomeres. There is no envelope. Adenoviruses are unique among icosahedral viruses in that they have a structure called a "fiber" projecting from each of the 12 vertices, or penton bases (Figures 32–1 and 32–2). The rest of the capsid is composed of 240 hexon capsomeres. The hexons, pentons, and fibers constitute the major adenovirus antigens important in viral classification and disease diagnosis.

The DNA genome (26–45 kbp) is linear and double-stranded. The entire DNA sequences of genomes of many adenovirus types are known. The viral genome for type 2 contains 36,000 base pairs. The guanine-plus-cytosine content of the DNA is lowest (48–49%) in group A (types 12, 18, and 31) adenoviruses, the most strongly oncogenic types, and ranges as high as 61% in other types. This is one criterion used in grouping human isolates. Viral DNA contains a virus-encoded protein that is covalently linked to each 5′ end of the linear genome. The DNA can be isolated in an infectious form, and the relative infectivity of that DNA is reduced at least 100-fold if the terminal protein is removed by proteolysis. The DNA is condensed in the core of the virion; a virus-encoded protein, polypeptide VII (Figure 32–2B), is important in forming the core structure.

There are an estimated 11 virion proteins; their structural positions in the virion are shown in Figure 32–2B. Hexon and penton capsomeres are the major components on the surface of the virus particle. There are group- and type-specific epitopes on both the hexon and fiber polypeptides. All human adenoviruses display this common hexon antigenicity. Pentons occur at the 12 vertices of the capsid and have fibers protruding from them. The penton base carries a toxin-like activity that causes rapid appearance of cytopathic effects and detachment of cells from the surface on which they are growing. Another group-reactive antigen is exhibited by the penton base. The fibers contain type-specific antigens that are important in serotyping. Fibers are associated with hemagglutinating activity. Because the hemagglutinin is type-specific, HI tests are commonly used for typing isolates. It is possible, however, to recover isolates that are recombinants and give discordant reactions in Nt and HI assays.

Classification

Adenoviruses have been recovered from a wide variety of species and grouped into four genera. All the human adenoviruses are classified in the *Mastadenovirus* genus. At least 51 distinct antigenic types have been isolated from humans and many other types from various animals.

TABLE 32–1   Important Properties of Adenoviruses

Virion: Icosahedral, 70–90 nm in diameter, 252 capsomeres; fiber projects from each vertex

Composition: DNA (13%), protein (87%)

Genome: Double-stranded DNA, linear, 26–45 kbp, protein-bound to termini, infectious

Proteins: Important antigens (hexon, penton base, fiber) are associated with the major outer capsid proteins

Envelope: None

Replication: Nucleus

Outstanding characteristic: Excellent models for molecular studies of eukaryotic cell processes

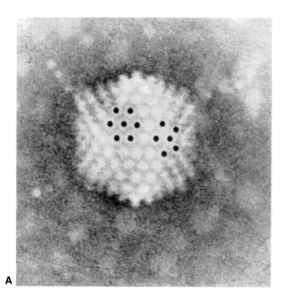
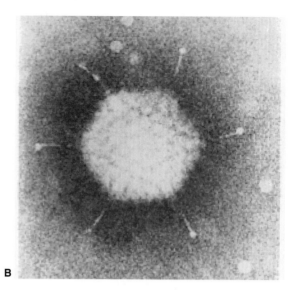

FIGURE 32–1 Electron micrographs of adenovirus. **A:** The viral particle displays cubic symmetry and is nonenveloped. A hexon capsomere (surrounded by six identical hexons) and a penton capsomere (surrounded by five hexons) are marked with dots. **B:** Note the fiber structures projecting from the vertex penton capsomeres (285,000×). (Reproduced with permission from Valentine RC, Pereira HG: Antigens and structure of the adenovirus. J Mol Biol 1965;13:13.)

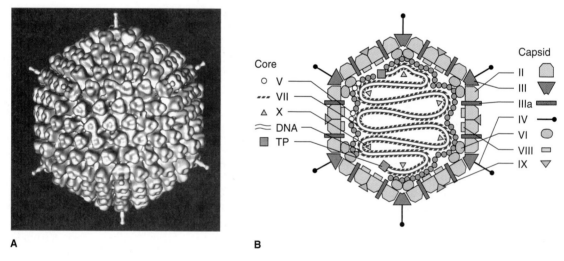

FIGURE 32–2 Models of the adenovirus virion. **A:** A three-dimensional image reconstruction of the intact adenovirus particle viewed along an icosahedral threefold axis. (Reproduced with permission from Stewart PL et al: Image reconstruction reveals the complex molecular organization of adenovirus. Cell 1991;67:145. Copyright © 1991 by Cell Press.) **B:** A stylized section of the adenovirus particle showing polypeptide components and DNA. No real section of the icosahedral virion would contain all components. Virion constituents are designated by their polypeptide numbers with the exception of the terminal protein (TP). (Reproduced with permission from Stewart PL, Burnett RM: Adenovirus structure as revealed by x-ray crystallography, electron microscopy and difference imaging. Jpn J Appl Phys 1993;32:1342.)

Human adenoviruses are divided into six groups (A–F) on the basis of their genetic, physical, chemical, and biologic properties (Table 32–2). Adenoviruses of a given group have fibers of a characteristic length, display considerable DNA homology (>85%, as compared to <20% with members of other groups), and exhibit similar capacities to agglutinate erythrocytes from either monkeys or rats. Members of a given adenovirus group resemble one another in the guanine-plus-cytosine content of their DNA and in their potential to produce tumors in newborn

rodents. Importantly, viruses within a group tend to behave similarly with respect to epidemiologic spread and disease association.

Adenovirus Replication

Adenoviruses replicate well only in cells of epithelial origin. The replicative cycle is sharply divided into early and late events. The carefully regulated expression of sequential events in the adenovirus cycle is summarized in Figure 32–3. The

TABLE 32–2 Classification Schemes for Human Adenoviruses

| Group | Serotypes | Hemagglutination | | Percentage of G + C[a] in DNA | Oncogenic Potential | |
		Group	Result		Tumorigenicity In Vivo[b]	Transformation of Cells
A	12, 18, 31	IV	None	48–49	High	+
B	3, 7, 11, 14, 16, 21, 34, 35, 50	I	Monkey (complete)	50–52	Moderate	+
C	1, 2, 5, 6	III	Rat (partial)	57–59	Low or none	+
D	8–10, 13, 15, 17, 19, 20, 22–30, 32, 33, 36–39, 42–49, 51	II	Rat (complete)	57–61	Low or none[c]	+
E	4	III	Rat (partial)	57	Low or none	+
F	40, 41	III	Rat (partial)	57–59	Low or none	+

[a]Guanine plus cytosine.

[b]Tumor induction in newborn hamsters.

[c]Adenovirus 9 can induce mammary tumors in rats.

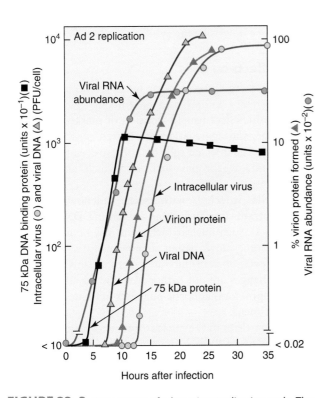

FIGURE 32–3 Time course of adenovirus replication cycle. The time between infection and the first appearance of progeny virus is the eclipse period. Note the sequential regulation of specific events in the virus replication cycle. "PFU" means "plaque-forming unit," a measure of infectious virus. (Courtesy of M Green.)

A. Virus Attachment, Penetration, & Uncoating

The virus attaches to cells via the fiber structures. The host cell receptor for some serotypes is CAR (coxsackie–adenovirus receptor), a member of the immunoglobulin gene superfamily. The interaction of the penton base with cellular integrins following attachment promotes the internalization step. Adsorption and internalization are separate steps in the adenovirus infection process, requiring the interaction of fiber and penton proteins with different cellular target proteins. Adsorbed virus is internalized into endosomes; the majority of particles (≈90%) move rapidly from endosomes into the cytosol (half-life ≈5 minutes) by a process triggered by the acidic pH of the endosome. Microtubules are probably involved in the transport of virus particles across the cytoplasm to the nucleus. Uncoating commences in the cytoplasm and is completed in the nucleus, with release of the DNA perhaps occurring at the nuclear membrane. Uncoating is an organized, sequential process that systematically breaks down the stabilizing interactions that were established during maturation of the virus particle.

B. Early Events

The steps that occur before the onset of viral DNA synthesis are defined as early events. The goals of the early events are to induce the host cell to enter the S phase of the cell cycle to create conditions conducive to viral replication, to express viral functions that protect the infected cell from host defense mechanisms, and to synthesize viral gene products needed for viral DNA replication.

The early ("E") transcripts come from seven widely separated regions of the viral genome and from both viral DNA strands. More than 20 early proteins, many of which are nonstructural and are involved in viral DNA replication, are synthesized in adenovirus-infected cells. The E1A early

distinction between early and late events is not absolute in infected cells; early genes continue to be expressed throughout the cycle; a few genes begin to be expressed at "intermediate" times; and low levels of late gene transcription may occur soon after infection.

gene is especially important; it must be expressed in order for the other early regions to be transcribed. Modulation of the cell cycle is accomplished by the E1A gene products. The E1B early region encodes proteins that block cell death (apoptosis) that occurs due to E1A functions; this is necessary to prevent premature cell death that would adversely affect virus yields. The E1A and E1B regions contain the only adenovirus genes involved in cell transformation; those gene products bind cellular proteins (eg, pRb, p300, p53) that regulate cell cycle progression. The early proteins are represented by the 75-kDa DNA-binding protein shown in Figure 32–3.

C. Replication of Viral DNA & Late Events

Viral DNA replication takes place in the nucleus. The virus-encoded, covalently linked terminal protein functions as a primer for initiation of viral DNA synthesis.

Late events begin concomitantly with the onset of viral DNA synthesis. The major late promoter controls the expression of the late ("L") genes coding for viral structural proteins. There is a single large primary transcript (≈29,000 nucleotides in length) that is processed by splicing to generate at least 18 different late mRNAs. These mRNAs are grouped (L1 to L5) based on the utilization of common poly(A) addition sites. The processed transcripts are transported to the cytoplasm, where the viral proteins are synthesized.

Although host genes continue to be transcribed in the nucleus late in the course of infection, few host genetic sequences are transported to the cytoplasm. A complex involving the E1B 55-kDa polypeptide and the E4 34-kDa polypeptide inhibits the cytoplasmic accumulation of cellular mRNAs and facilitates accumulation of viral mRNAs, perhaps by relocalizing a putative cellular factor required for mRNA transport. Very large amounts of viral structural proteins are made.

It is noteworthy that studies with adenovirus hexon mRNA led to the profound discovery that eukaryotic mRNAs are usually not colinear with their genes but are spliced products of separated coding regions in the genomic DNA.

D. Viral Assembly & Maturation

Virion morphogenesis occurs in the nucleus. Each hexon capsomere is a trimer of identical polypeptides. The penton is composed of five penton base polypeptides and three fiber polypeptides. A late L4-encoded "scaffold protein" assists in the aggregation of hexon polypeptides but is not part of the final structure.

Capsomeres self-assemble into empty-shell capsids in the nucleus. Naked DNA then enters the preformed capsid. A cis-acting DNA element near the left-hand end of the viral chromosome serves as a packaging signal, necessary for the DNA–capsid recognition event. Another viral scaffolding protein, encoded in the L1 group, facilitates DNA encapsidation. Finally, precursor core proteins are cleaved, which allows the particle to tighten its configuration, and the pentons are added. A virus-encoded cysteine proteinase functions in some cleavages of precursor proteins. The mature particle is then stable, infectious, and resistant to nucleases. The adenovirus infectious cycle takes about 24 hours. The assembly process is inefficient; about 80% of hexon capsomeres and 90% of viral DNA are not used. Nevertheless, about 100,000 virus particles are produced per cell. Structural proteins associated with mature virus particles are catalogued in Figure 32–2B.

E. Virus Effects on Host Defense Mechanisms

Adenoviruses encode several gene products that counter antiviral host defense mechanisms. The small, abundant VA RNAs afford protection from the antiviral effect of interferon by preventing activation of an interferon-inducible kinase that phosphorylates and inactivates eukaryotic initiation factor 2. Adenovirus E3 region proteins, which are nonessential for viral growth in tissue culture, inhibit cytolysis of infected cells by host responses. The E3 gp19-kDa protein blocks movement of the MHC class I antigen to the cell surface, thereby protecting the infected cell from cytotoxic T lymphocyte (CTL)-mediated lysis. Other E3-encoded proteins block induction of cytolysis by the cytokine TNF-α.

F. Virus Effects on Cells

Adenoviruses are cytopathic for human cell cultures, particularly primary kidney and continuous epithelial cells. The cytopathic effect usually consists of marked rounding, enlargement, and aggregation of affected cells into grape-like clusters. The infected cells do not lyse even though they round up and leave the glass surface on which they have been grown.

In cells infected with some adenovirus types, rounded intranuclear inclusions containing DNA are seen (Figure 32–4). Such nuclear inclusions may be mistaken for those of cytomegalovirus, but adenovirus infections do not induce syncytia or multinucleated giant cells. Although the cytologic changes are not pathognomonic for adenoviruses, they are helpful for diagnostic purposes in tissue culture and biopsy specimens.

Virus particles in the nucleus frequently exhibit crystalline arrangements. Cells infected with group B viruses also contain crystals composed of protein without nucleic acid. Virus particles remain within the cell after the cycle is complete and the cell is dead.

Human adenoviruses exhibit a narrow host range. When cells derived from species other than humans are infected, the human adenoviruses usually undergo an abortive replication cycle and no infectious progeny are produced.

Gene Therapy

Adenoviruses are being used as gene delivery vehicles for cancer therapy, gene therapy, and genetic immunization studies. Adenoviruses are attractive because recombinant, replication-defective viruses possess the advantages of high

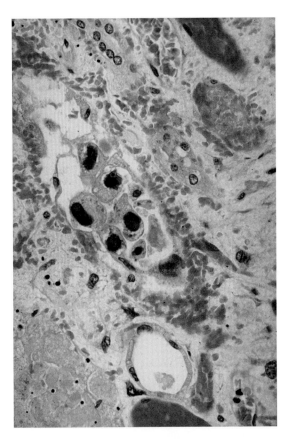

FIGURE 32–4 Adenovirus cytopathology in human tissue. Tubular epithelial cells with basophilic inclusion bodies from a patient with necrotizing tubulointerstitial nephritis (450×). (Courtesy of M Ito.)

transduction efficiencies of many cell types and high levels of short-term expression of transduced genes; however, significant limitations include their high immunogenicity and the high prevalence of preexisting immunity in humans to subgroup C adenoviruses (types 2 and 5 are widely used as vectors). Other limitations are variable receptor (CAR) expression on different cells and the failure to integrate into chromosomal DNA to facilitate long-term transgene expression. Efforts are under way with vector design and targeting techniques to surmount these limitations.

A novel anticancer therapy utilizes an attenuated replication-competent adenovirus engineered to replicate only in targeted cancer cells. This "oncolytic therapy" is aimed at directly killing tumor cells due to viral lytic replication.

Animal Susceptibility & Transformation of Cells

Most laboratory animals are not readily infected with human adenoviruses, though newborn hamsters sustain a fatal infection with type 5, and young adult animals are permissive for adenovirus 5 replication in the lung. Several serotypes, especially types 12, 18, and 31, are able to induce tumors when inoculated into newborn hamsters (Table 32–2). All adenoviruses can morphologically transform cells in culture regardless of their oncogenic potential in vivo (see Chapter 43). Only a small part (<20%) of the adenovirus genome is present in most transformed cells.

The transforming genes of human adenoviruses are located in the early region (E1A and E1B) at the left-hand end of the viral genome. An exception is type 9; with it, the E4 gene is required for mammary tumorigenesis in rats. Studies of adenovirus transforming genes have revealed cellular growth control mechanisms that are altered in many types of cancer cells.

The highly oncogenic nature of adenovirus type 12 may be related to the observation that one effect of its early region is to turn off the synthesis of class I major histocompatibility antigens (H2 or HLA) in some infected and transformed cells, thereby preventing destruction by CTLs.

Adenoviruses are not thought to be important in human cancer.

ADENOVIRUS INFECTIONS IN HUMANS

Pathogenesis

Adenoviruses infect and replicate in epithelial cells of the respiratory tract, eye, gastrointestinal tract, and urinary tract. They usually do not spread beyond the regional lymph nodes. Group C viruses persist as latent infections for years in adenoids and tonsils and are shed in the feces for many months after the initial infection. In fact, the name "adenovirus" reflects the recovery of the initial isolate from explants of human adenoids.

Most human adenoviruses replicate in intestinal epithelium after ingestion but usually produce subclinical infections rather than overt symptoms.

Clinical Findings

About one-third of the known human serotypes are commonly associated with human illness. It should be noted that a single serotype may cause different clinical diseases and, conversely, that more than one type may cause the same clinical illness. Adenoviruses 1–7 are the most common types worldwide and account for most instances of adenovirus-associated illness.

Adenoviruses are responsible for about 5% of acute respiratory disease in young children, but they account for much less in adults. Most infections are mild and self-limited. The viruses occasionally cause disease in other organs, particularly the eye and the gastrointestinal tract.

A. Respiratory Diseases

Typical symptoms include cough, nasal congestion, fever, and sore throat. This syndrome is most commonly manifested in infants and children and usually involves group C viruses.

These cases are difficult to distinguish from other mild viral respiratory infections that may exhibit similar symptoms.

Adenoviruses—particularly types 3, 7, and 21—are thought to be responsible for about 10–20% of pneumonias in childhood. Adenoviral pneumonia has been reported to have an 8–10% mortality rate in the very young.

An outbreak of severe respiratory disease, sometimes fatal, occurred in 2007 that was caused by a new variant of adenovirus 14. Patients of all ages were affected, including healthy young adults.

Adenoviruses are the cause of an acute respiratory disease syndrome among military recruits. This syndrome is characterized by fever, sore throat, nasal congestion, cough, and malaise, sometimes leading to pneumonia. It occurs in epidemic form among young military recruits under conditions of fatigue, stress, and crowding soon after induction. This disease is caused by types 4 and 7 and occasionally by type 3. Because of vaccine unavailability, the United States military stopped vaccinating against adenoviruses (types 4 and 7) in the 1990s; this was followed by large epidemics affecting thousands of trainees.

B. Eye Infections

Mild ocular involvement may be part of the respiratory-pharyngeal syndromes caused by adenoviruses. Pharyngoconjunctival fever tends to occur in outbreaks, such as at children's summer camps ("swimming pool conjunctivitis"), and is associated with types 3 and 7. Duration of conjunctivitis is 1–2 weeks, and complete recovery with no lasting sequelae is the common outcome.

A more serious disease is epidemic keratoconjunctivitis. This disease occurs mainly in adults and is highly contagious. Adenoviruses can remain viable for several weeks on sinks and hand towels, and these may be a source of transmission. The disease is characterized by acute conjunctivitis, followed by keratitis that usually resolves in 2 weeks but may leave subepithelial opacities in the cornea for up to 2 years. It is caused by types 8, 19, and 37.

A study in Japan (1990–2001) where type 37 is the major cause of epidemic keratoconjunctivitis showed that mutations in the viral genome occurred chronologically and that certain mutations were correlated with epidemics of disease.

C. Gastrointestinal Disease

Many adenoviruses replicate in intestinal cells and are present in stools, but the presence of most serotypes is not associated with gastrointestinal disease. However, two serotypes (types 40 and 41) have been etiologically associated with infantile gastroenteritis and may account for 5–15% of cases of viral gastroenteritis in young children. Adenovirus types 40 and 41 are abundantly present in diarrheal stools. The enteric adenoviruses are very difficult to cultivate.

D. Other Diseases

Immunocompromised patients may suffer from a variety of casual and severe adenovirus infections. The most common problem caused by adenovirus infection in transplant patients is respiratory disease that may progress to severe pneumonia and may be fatal (usually types 1–7). Children receiving liver transplants may develop adenovirus hepatitis in the allograft. In addition, children with heart transplants who develop myocardial adenovirus infections are at increased risk of graft loss. Pediatric recipients of hematopoietic stem cell transplants may develop infections with a wide variety of adenovirus types. Patients with acquired immunodeficiency syndrome (AIDS) may suffer adenovirus infections, especially in the gastrointestinal tract.

Types 11 and 21 may cause acute hemorrhagic cystitis in children, especially boys. Virus commonly occurs in the urine of such patients.

Immunity

In contrast to most respiratory infectious agents, the adenoviruses induce effective and long-lasting immunity against reinfection. This may reflect the fact that adenoviruses also infect the regional lymph nodes and lymphoid cells in the gastrointestinal tract. Resistance to clinical disease appears to be directly related to the presence of circulating neutralizing antibodies, which probably persist for life. Although type-specific neutralizing antibodies may protect against disease symptoms, they may not always prevent reinfection. (Infections with adenoviruses frequently occur without the production of overt illness.)

Maternal antibodies usually protect infants against severe adenovirus respiratory infections. Neutralizing antibodies against one or more types have been detected in over 50% of infants 6–11 months old. Normal, healthy adults generally have antibodies to several types.

A group-reactive antibody response, different from the type-specific neutralizing antibody, may be measured by CF, IF, or ELISA testing. Group-specific antibodies are not protective, decline with time, and do not reveal the serotypes of previous viral infections.

Laboratory Diagnosis

A. Detection, Isolation, & Identification of Virus

Samples should be collected from affected sites early in the illness to optimize virus isolation. Depending on the clinical disease, virus may be recovered from stool or urine or from a throat, conjunctival, or rectal swab. Duration of adenovirus excretion varies among different illnesses: 1–3 days, throat of adults with common cold; 3–5 days, throat, stool, and eye, for pharyngoconjunctival fever; 2 weeks, eye, for keratoconjunctivitis; 3–6 weeks, throat and stool of children with respiratory illnesses; 2–12 months, urine, throat, and stool of immunocompromised patients.

Virus isolation in a cell culture requires human cells. Primary human embryonic kidney cells are most susceptible but usually unavailable. Established human epithelial cell lines, such as HEp-2, HeLa, and KB, are sensitive but are difficult to maintain without degeneration for the length of time (28 days) required to detect some slow-growing natural isolates. Isolates can be identified as adenoviruses by immunofluorescence tests using an antihexon antibody on infected cells. HI and Nt tests measure type-specific antigens and can be used to identify specific serotypes.

Infectious adenovirus detection may be made rapidly using the shell vial technique. Viral specimens are centrifuged directly onto tissue culture cells; cultures are incubated for 1–2 days and are then tested with monoclonal antibodies directed against a group-reactive epitope on the hexon antigen. Also, nasal epithelial cells from a patient may be stained directly to detect viral antigens.

Polymerase chain reaction (PCR) assays can be used for diagnosis of adenovirus infections in tissue samples or body fluids, usually by using primers from a conserved viral sequence (eg, hexon, VA I) that can detect all serotypes. PCR assays have been described that use single primer pairs which target conserved segments that bracket a hypervariable region in the hexon gene. The assays can detect all known serotypes of human adenoviruses, and sequencing of the amplicon allows serotype identification. This method is rapid compared to the weeks required for virus isolation followed by neutralization assays. However, the sensitivity of the PCR assay may result in detection of latent adenoviruses in some patients.

Characterization of viral DNA by hybridization or by restriction endonuclease digestion patterns can identify an isolate as an adenovirus and group it. These approaches are especially useful for types that are difficult to cultivate.

The fastidious enteric adenoviruses can be detected by direct examination of fecal extracts by electron microscopy, by ELISA, or by latex agglutination tests. With difficulty, they can be isolated in a line of human embryonic kidney cells transformed with a fragment of adenovirus 5 DNA (293 cells).

Since adenoviruses can persist in the gut and in lymphoid tissue for long periods and since recrudescent viral shedding can be precipitated by other infections, the significance of a viral isolation must be interpreted with caution. Viral recovery from the eye, lung, or genital tract is diagnostic of current infection. Isolation of virus from throat secretions of a patient with respiratory illness can be considered relevant to the clinical disease. Viral isolation from fecal specimens is inconclusive unless one of the fastidious types is recovered from a patient with gastroenteritis.

B. Serology

Infection of humans with any adenovirus type stimulates a rise in complement-fixing antibodies to adenovirus group antigens shared by all types. The CF test is an easily applied method for detecting infection by any member of the adenovirus group, although the test has low sensitivity. A fourfold or greater rise in complement-fixing antibody titer between acute-phase and convalescent-phase sera indicates recent infection with an adenovirus, but it gives no clue about the specific type involved.

If specific identification of a patient's serologic response is required, Nt or HI tests can be used. The Nt test is the most sensitive. In most cases, the neutralizing antibody titer of infected persons shows a fourfold or greater rise against the adenovirus type recovered from the patient.

Epidemiology

Adenoviruses exist in all parts of the world. They are present year-round and usually do not cause community outbreaks of disease. The most common serotypes in clinical samples are the low-numbered respiratory types (1, 2, 3, 5, 7) and the gastroenteritis types (40, 41). Adenoviruses are spread by direct contact, by the fecal–oral route, by respiratory droplets, or by contaminated fomites. Most adenovirus-related diseases are not clinically pathognomonic, and many infections are subclinical.

Infections with types 1, 2, 5, and 6 occur chiefly during the first years of life; types 3 and 7 are contracted during school years; and other types (such as 4, 8, and 19) are not encountered until adulthood.

While adenoviruses cause only 2–5% of all respiratory illness in the general population, respiratory disease due to types 3, 4, and 7 is common among military recruits. Adenovirus disease can cause great morbidity among recruits. However, adenovirus disease is not a problem in seasoned troops.

An outbreak of acute respiratory disease caused by type 11 occurred in 1997 among young adults living in a job training facility—the first such outbreak recognized among civilians.

Eye infections can be transmitted in several ways, but hand-to-eye transfer is particularly important. Outbreaks of swimming pool conjunctivitis are presumably waterborne, usually occur in the summer, and are commonly caused by types 3 and 7. Epidemic keratoconjunctivitis is a highly contagious and serious disease. The disease, caused by type 8, spread in 1941 from Australia via the Hawaiian Islands to the Pacific Coast. It spread rapidly through the shipyards (hence the name "shipyard eye") and across the United States. In the United States, the incidence of neutralizing antibody to type 8 in the general population is very low (about 1%), whereas in Japan it is more than 30%. More recently, adenovirus types 19 and 37 have caused epidemics of typical epidemic keratoconjunctivitis. Outbreaks of conjunctivitis traced to ophthalmologists' offices were presumably caused by contaminated ophthalmic solutions or diagnostic equipment.

The incidence of adenovirus infection in patients undergoing bone marrow transplantation has been estimated to be from about 5% to as high as 30%. The reported incidence is higher in pediatric patients than in adults. Patients may develop fatal disseminated infections. Types 34 and 35 are found most often in bone marrow and renal transplant recipients. The most likely source of infection in transplant patients is endogenous viral reactivation, though primary infections may be a factor in the pediatric population.

Treatment

There is no specific treatment for adenovirus infections.

Prevention & Control

Careful hand washing is the easiest way to prevent infections. Environmental surfaces can be disinfected with sodium hypochlorite. In group settings, paper towels may be advisable because dirty towels can be a source of infection in outbreaks. The risk of waterborne outbreaks of conjunctivitis can be minimized by chlorination of swimming pools and waste water. Strict asepsis during eye examinations, coupled with adequate sterilization of equipment, is essential for the control of epidemic keratoconjunctivitis.

Attempts to control adenovirus infections in the military have focused on vaccines. Live adenovirus vaccine containing types 4 and 7, encased in gelatin-coated capsules and given orally, was introduced in 1971. In this way virus bypasses the respiratory tract, where it could cause disease, and is released in the intestine, where it replicates and induces neutralizing antibody. It does not spread from a vaccinated person to contacts. The vaccine proved highly effective, but after 1999 was no longer available as its manufacture had been discontinued.

REVIEW QUESTIONS

In what follows, singular may be construed as plural (or vice versa) as the sense dictates.

1. What adenovirus protein or proteins regulate early transcription of the viral genes and modulate the cell cycle?

    (A) Fiber
    (B) Hexon
    (C) Penton
    (D) Terminal protein
    (E) E1 region protein
    (F) Cysteine proteinase
    (G) E3 region protein

2. What adenovirus protein serves as primer for initiation of viral DNA synthesis?

    (A) Fiber
    (B) Hexon
    (C) Penton
    (D) Terminal protein
    (E) E1 region protein
    (F) Cysteine proteinase
    (G) E3 region protein

3. What adenovirus protein comprises the majority of capsomeres making up the virus capsid?

    (A) Fiber
    (B) Hexon
    (C) Penton
    (D) Terminal protein
    (E) E1 region protein
    (F) Cysteine proteinase
    (G) E3 region protein

4. A 3-month-old infant had watery diarrhea and fever for 10 days. Rotavirus or adenovirus types 40 and 41 are the suspected agents. What type of specimen would be most appropriate for detection of adenovirus type 40 and 41 infection in this patient?

    (A) Blood
    (B) Urine
    (C) Conjunctival swab
    (D) Stool
    (E) Throat swab
    (F) Cerebrospinal fluid

5. Which of the following human diseases has not been associated with adenoviruses?

    (A) Cancer
    (B) Common colds
    (C) Acute respiratory diseases
    (D) Keratoconjunctivitis
    (E) Gastroenteritis
    (F) Hemorrhagic cystitis

6. A 2½-year-old child attending nursery school acquires a mild respiratory infection. Other children in the nursery school have similar illnesses. Which adenovirus types are the most likely causes of the illnesses?

    (A) Types 40, 41
    (B) Types 8, 19, 37
    (C) Types 1, 2, 5, 6
    (D) Types 3, 4, 7
    (E) Types 21, 22, 34, 35

7. Which adenovirus types are frequent causes of acute respiratory disease among military recruits?

    (A) Types 40, 41
    (B) Types 8, 19, 37
    (C) Types 1, 2, 5, 6
    (D) Types 3, 4, 7
    (E) Types 21, 22, 34, 35

8. Which of the following events led to reappearance of acute respiratory disease outbreaks among U.S. military recruits in the late 1990s?

    (A) Emergence of a new virulent strain of adenovirus
    (B) Cessation of adenovirus vaccination program for recruits
    (C) Change in military housing and training conditions for recruits
    (D) Cessation of adenovirus antiviral drug therapy program for recruits

9. Your summer research project is to study the viruses that cause gastroenteritis. You recover a virus from a stool sample and notice that the growth medium on the infected cultures is highly acidic. You find that the viral genome is double-stranded DNA. Of the following, which one is the most appropriate conclusion you could draw?

    (A) There is a high likelihood the agent is a rotavirus
    (B) You need to determine the viral serotype to establish whether the virus was important in causing the disease
    (C) The patient should have been treated with the antiviral drug amantadine to shorten the duration of symptoms
    (D) The virus particle would contain a reverse transcriptase enzyme

10. Which of the following groups of individuals is at the lowest risk of adenovirus disease?
    (A) Healthy adults
    (B) Young children
    (C) Bone marrow transplant recipients
    (D) Military recruits
    (E) AIDS patients

11. Adenoviruses can cause eye infections that are highly contagious. Which of the following is least likely to be a means of transmission during an outbreak of epidemic keratoconjunctivitis?
    (A) Swimming pools
    (B) Hand towels
    (C) Mosquito bites
    (D) Hand-to-eye
    (E) Contaminated ophthalmic equipment

Answers

1. E	4. D	7. D	10. A
2. D	5. A	8. B	11. C
3. B	6. C	9. B	

REFERENCES

Berk AJ: *Adenoviridae:* The viruses and their replication. In: *Fields Virology,* 5th ed. Knipe DM et al (editors). Lippincott Williams & Wilkins, 2007.

Berk AJ: Recent lessons in gene expression, cell cycle control, and cell biology from adenovirus. Oncogene 2005;24:7673. [PMID: 16299528]

Kolavic-Gray SA et al: Large epidemic of adenovirus type 4 infection among military trainees: Epidemiological, clinical, and laboratory studies. Clin Infect Dis 2002;35:808. [PMID: 12228817]

Mahony JB: Detection of respiratory viruses by molecular methods. Clin Microbiol Rev 2008;21:716. [PMID: 18854489]

Russell WC: Adenoviruses: Update on structure and function. J Gen Virol 2009;90:1. [PMID: 19088268]

Sarantis H, Johnson G, Brown M, Petric M, Tellier R: Comprehensive detection and serotyping of human adenoviruses by PCR and sequencing. J Clin Microbiol 2004;42:3963. [PMID: 15364976]

Herpesviruses

The herpesvirus family contains several of the most important human viral pathogens. Clinically, the herpesviruses exhibit a spectrum of diseases. Some have a wide host-cell range, whereas others have a narrow host-cell range. The outstanding property of herpesviruses is their ability to establish lifelong persistent infections in their hosts and to undergo periodic reactivation. Their frequent reactivation in immunosuppressed patients causes serious health complications. Curiously, the reactivated infection may be clinically quite different from the disease caused by the primary infection. Herpesviruses possess a large number of genes, some of which have proved to be susceptible to antiviral chemotherapy.

The herpesviruses that commonly infect humans include herpes simplex virus types 1 and 2 (HSV-1, HSV-2), varicella-zoster virus, cytomegalovirus, Epstein-Barr virus (EBV), herpesviruses 6 and 7, and herpesvirus 8 (Kaposi sarcoma-associated herpesvirus [KSHV]). Herpes B virus of monkeys can also infect humans. There are nearly 100 viruses of the herpes group that infect many different animal species.

PROPERTIES OF HERPESVIRUSES

Important properties of herpesviruses are summarized in Table 33–1.

Structure & Composition

Herpesviruses are large viruses. Different members of the group share architectural details and are indistinguishable by electron microscopy. All herpesviruses have a core of double-stranded DNA, in the form of a toroid, surrounded by a protein coat that exhibits icosahedral symmetry and has 162 capsomeres. The nucleocapsid is surrounded by an envelope that is derived from the nuclear membrane of the infected cell and contains viral glycoprotein spikes about 8 nm long. An amorphous, sometimes asymmetric structure between the capsid and envelope is designated the tegument. The enveloped form measures 150–200 nm; the "naked" virion, 125 nm.

The double-stranded DNA genome (125–240 kbp) is linear. A striking feature of herpesvirus DNAs is their sequence arrangement (Figure 33–1). Herpesvirus genomes possess

terminal and internal repeated sequences. Some members, such as the HSVs, undergo genome rearrangements, giving rise to different genome "isomers." The base composition of herpesvirus DNAs varies from 31% to 75% (G + C). There is little DNA homology among different herpesviruses except for HSV-1 and HSV-2, which show 50% sequence homology, and human herpesviruses 6 and 7, which display limited (30–50%) sequence homology. Treatment with restriction endonucleases yields characteristically different cleavage patterns for herpesviruses and even for different strains of each type. This "fingerprinting" of strains allows epidemiologic tracing of a given strain.

The herpesvirus genome is large and encodes at least 100 different proteins. Of these, more than 35 polypeptides are involved in the structure of the virus particle; at least 10 are part of the viral envelope. Herpesviruses encode an array of virus-specific enzymes involved in nucleic acid metabolism, DNA synthesis, gene expression, and protein regulation (DNA polymerase, helicase-primase, thymidine kinase, transcription factors, protein kinases). Many herpesvirus genes appear to be viral homologs of cellular genes.

Classification

Classification of the numerous members of the herpesvirus family is complicated. A useful division into subfamilies

TABLE 33–1   Important Properties of Herpesviruses

Virion: Spherical, 150–200 nm in diameter (icosahedral)

Genome: Double-stranded DNA, linear, 125–240 kbp, reiterated sequences

Proteins: More than 35 proteins in virion

Envelope: Contains viral glycoproteins, Fc receptors

Replication: Nucleus, bud from nuclear membrane

Outstanding characteristics:
- Encode many enzymes
- Establish latent infections
- Persist indefinitely in infected hosts
- Frequently reactivated in immunosuppressed hosts
- Some are cancer-causing

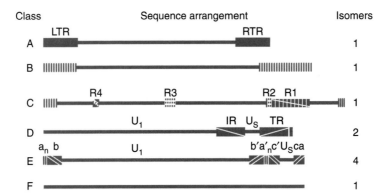

FIGURE 33–1 Schematic diagram of sequence arrangements of herpesvirus DNAs. Genome classes A, B, C, D, E, and F are exemplified by channel catfish virus, herpesvirus saimiri, EBV, varicella-zoster virus, HSVs, and tupaia herpesvirus, respectively. Horizontal lines represent unique regions. Reiterated domains are shown as rectangles: left and right terminal repeats (LTR and RTR) for class A; repeats R1 to R4 for internal repeats of class C; and internal and terminal repeats (IR and TR) of class D. In class B, terminal sequences are reiterated numerous times at both termini. The termini of class E consist of two elements. The terminal sequences (ab and ca) are inserted in an inverted orientation separating the unique sequences into long (U$_l$) and short (U$_s$) domains. Genomes of class F have no terminal reiterations. The components of the genomes in classes D and E invert. In class D (varicella-zoster virus), the short component inverts relative to the long, and the DNA forms two populations (isomers) differing in the orientation of the short component. In class E (HSV), both the short and long components can invert, and viral DNA consists of four isomers. (Reproduced with permission from Roizman B: Herpesviridae: A brief introduction. Pages 1787–1793 in: *Virology*, 2nd ed. Fields BN et al [editors]. Raven Press, 1990.)

TABLE 33–2   Classification of Human Herpesviruses

| Subfamily ("-herpesvirinae") | Biologic Properties | | | Examples | |
	Growth Cycle and Cytopathology	Latent Infections	Genus ("-virus")	Official Name ("Human -Herpesvirus")	Common Name
Alpha	Short, cytolytic	Neurons	*Simplex*	1	Herpes simplex virus type 1
				2	Herpes simplex virus type 2
			Varicello	3	Varicella-zoster virus
Beta	Long, cytomegalic	Glands, kidneys	*Cytomegalo*	5	Cytomegalovirus
	Long, lymphoproliferative	Lymphoid tissue	*Roseolo*	6	Human herpesvirus 6
				7	Human herpesvirus 7
Gamma	Variable, lymphoproliferative	Lymphoid tissue	*Lymphocrypto*	4	Epstein-Barr virus
			Rhadino	8	Kaposi sarcoma-associated herpesvirus

is based on biologic properties of the agents (Table 33–2). Alphaherpesviruses are fast-growing, cytolytic viruses that tend to establish latent infections in neurons; HSV (genus *Simplexvirus*) and varicella-zoster virus (genus *Varicellovirus*) are members. Betaherpesviruses are slow-growing and may be cytomegalic (massive enlargements of infected cells) and become latent in secretory glands and kidneys; cytomegalovirus is classified in the *Cytomegalovirus* genus. Also included here, in the genus *Roseolovirus*, are human herpesviruses 6 and 7; by biologic criteria, they are more like gammaherpesviruses because they infect lymphocytes (T lymphotropic), but molecular analyses of their genomes reveal that they are more closely related to the betaherpesviruses. Gammaherpesviruses, exemplified by EBV (genus *Lymphocryptovirus*), infect and become latent in lymphoid cells. KSHV, designated as human herpesvirus 8, is classified in the *Rhadinovirus* genus.

Many herpesviruses infect animals, the most notable being B virus (herpesvirus simiae or cercopithecine herpesvirus 1) in the *Simplexvirus* genus; herpesviruses saimiri and ateles of monkeys, both in genus *Rhadinovirus*; marmoset herpesvirus (genus *Simplexvirus*); and pseudorabies virus of pigs and infectious bovine rhinotracheitis virus of cattle, both in genus *Varicellovirus*.

There is little antigenic relatedness among members of the herpesvirus group. Only HSV-1 and HSV-2 share a significant number of common antigens. Human herpesviruses 6 and 7 exhibit a few cross-reacting epitopes.

Herpesvirus Replication

The replication cycle of HSV is summarized in Figure 33–2. The virus enters the cell by fusion with the cell membrane

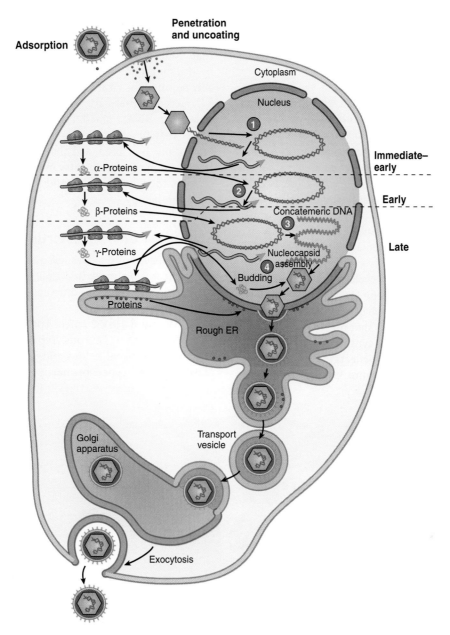

FIGURE 33–2 Replication cycle of herpes simplex virus. **1:** Virus fuses with plasma membrane and viral DNA is released from capsid at nuclear pore, followed by circularization of genome and transcription of immediate-early genes. **2:** α-proteins, products of immediate-early genes, stimulate transcription of early genes. **3:** β-proteins, products of early genes, function in DNA replication, yielding concatemeric DNA. Late genes are transcribed. **4:** γ-proteins, products of late genes and consisting primarily of viral structural proteins, participate in virion assembly. Unit-length viral DNA is cleaved from concatemers and packaged into capsids. Enveloped viral particles accumulate in the endoplasmic reticulum and are transported from the cell. (Reproduced with permission from Willey JM, Sherwood LM, Woolverton CJ: *Prescott, Harley, and Klein's Microbiology,* 7th ed. McGraw Hill, 2008.)

after binding to specific cellular receptors via envelope glycoproteins. Several herpesviruses bind to cell surface glycosaminoglycans, principally heparan sulfate. Virus attachment also involves binding to one of several coreceptors (eg, members of the immunoglobulin superfamily). After fusion, the capsid is transported through the cytoplasm to a nuclear pore; uncoating occurs; and the DNA becomes associated with the nucleus. The viral DNA forms a circle

immediately upon release from the capsid. Expression of the viral genome is tightly regulated and sequentially ordered in a cascade fashion. VP16, a tegument protein, complexes with several cellular proteins and activates initial viral gene expression. Immediate-early genes are expressed, yielding "α" proteins. These proteins permit expression of the early set of genes, which are translated into "β" proteins. Viral DNA replication begins, and late transcripts are produced that give rise to "γ" proteins. More than 50 different proteins are synthesized in herpesvirus-infected cells. Many α and β proteins are enzymes or DNA-binding proteins; most of the γ proteins are structural components.

Viral DNA is transcribed throughout the replicative cycle by cellular RNA polymerase II but with the participation of viral factors. Viral DNA is synthesized by a rolling-circle mechanism. Herpesviruses differ from other nuclear DNA viruses in that they encode a large number of enzymes involved in DNA synthesis. (These enzymes are good targets for antiviral drugs.) Newly synthesized viral DNA is packaged into preformed empty nucleocapsids in the cell nucleus.

Maturation occurs by budding of nucleocapsids through the altered inner nuclear membrane. Enveloped virus particles are then transported by vesicular movement to the surface of the cell.

The length of the replication cycle varies from about 18 hours for HSV to over 70 hours for cytomegalovirus. Cells productively infected with herpesviruses are invariably killed. Host macromolecular synthesis is shut off early in infection; normal cellular DNA and protein synthesis virtually stop as viral replication begins. Cytopathic effects induced by human herpesviruses are quite distinct (Figure 33–3).

The number of potential protein-coding open-reading frames in herpesvirus genomes ranges from about 70 to more than 200. In the case of HSV, about half the genes are not needed for growth in cultured cells. The other genes are probably required for viral survival in vivo in natural hosts.

Herpesviruses have recently been found to express multiple microRNAs, small (~22 nucleotides) single-stranded RNAs that function posttranscriptionally to regulate gene expression. It is predicted that these viral microRNAs are important in regulating entry into and/or exit from the latent phase of the virus life cycle and may be attractive targets for antiviral therapy.

Overview of Herpesvirus Diseases

A wide variety of diseases are associated with infection by herpesviruses. Primary infection and reactivated disease by a given virus may involve different cell types and present different clinical pictures.

HSV-1 and HSV-2 infect epithelial cells and establish latent infections in neurons. Type 1 is classically associated with oropharyngeal lesions and causes recurrent attacks of "fever blisters." Type 2 primarily infects the genital mucosa

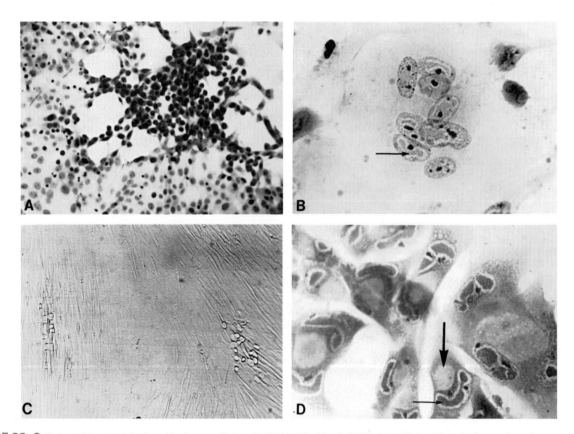

FIGURE 33–3 Cytopathic effects induced by herpesviruses. **A:** HSV in HEp-2 cells (H&E stain, 57×), with early focus of swollen, rounded cells. **B:** Varicella-zoster virus in human kidney cells (H&E stain, 228×), with multinucleated giant cell containing acidophilic intranuclear inclusions (arrow). **C:** Cytomegalovirus in human fibroblasts (unstained, 35×) with two foci of slowly developing cytopathic effect. **D:** Cytomegalovirus in human fibroblasts (H&E stain, 228×), showing giant cells with acidophilic inclusions in the nuclei (small arrow) and cytoplasm (large arrow), the latter being characteristically large and round. (Courtesy of I Jack; reproduced with permission from White DO, Fenner FJ: *Medical Virology*, 3rd ed. Academic Press, 1986.)

and is mainly responsible for genital herpes. Both viruses also cause neurologic disease. HSV-1 is the leading cause of sporadic encephalitis in the United States. Both type 1 and type 2 can cause neonatal infections which are often severe.

Varicella-zoster virus causes chickenpox (varicella) on primary infection and establishes latent infection in neurons. Upon reactivation, the virus causes zoster (shingles). Adults who are infected for the first time with varicella-zoster virus are apt to develop serious viral pneumonia.

Cytomegalovirus replicates in epithelial cells of the respiratory tract, salivary glands, and kidneys and persists in lymphocytes. It causes an infectious mononucleosis (heterophil-negative). In newborns, cytomegalic inclusion disease may occur. Cytomegalovirus is an important cause of congenital defects and mental retardation.

Human herpesvirus 6 infects T lymphocytes. It is typically acquired in early infancy and causes exanthem subitum (roseola infantum). Human herpesvirus 7, also a T-lymphotropic virus, has not yet been linked to any specific disease.

EBV replicates in epithelial cells of the oropharynx and parotid gland and establishes latent infections in lymphocytes. It causes infectious mononucleosis and is the cause of human lymphoproliferative disorders, especially in immunocompromised patients. Human herpesvirus 8 appears to be associated with the development of Kaposi sarcoma, a vascular tumor that is common in patients with AIDS.

Herpes B virus of macaque monkeys can infect humans. Such infections are rare, but those that occur usually result in severe neurologic disease and are frequently fatal.

Human herpesviruses are frequently reactivated in immunosuppressed patients (eg, transplant recipients, cancer patients) and may cause severe disease, such as pneumonia or lymphomas.

Herpesviruses have been linked with malignant diseases in humans and lower animals: EBV with Burkitt lymphoma of African children, with nasopharyngeal carcinoma, and with other lymphomas; KSHV with Kaposi sarcoma; Marek disease virus with a lymphoma of chickens; and a number of primate herpesviruses with reticulum cell sarcomas and lymphomas in monkeys.

HERPESVIRUS INFECTIONS IN HUMANS

HERPES SIMPLEX VIRUSES

HSVs are extremely widespread in the human population. They exhibit a broad host range, being able to replicate in many types of cells and to infect many different animals. They grow rapidly and are highly cytolytic. The HSVs are responsible for a spectrum of diseases, ranging from gingivostomatitis to keratoconjunctivitis, encephalitis, genital disease, and infections of newborns. The HSVs establish latent infections in nerve cells; recurrences are common.

Properties of the Viruses

There are two distinct HSV: type 1 and type 2 (HSV-1, HSV-2) (Table 33–3). Their genomes are similar in organization and exhibit substantial sequence homology. However, they can be distinguished by sequence analysis or by restriction enzyme analysis of viral DNA. The two viruses cross-react serologically, but some unique proteins exist for each type. They differ in their mode of transmission; HSV-1 is spread by contact, usually involving infected saliva, whereas HSV-2 is transmitted sexually or from a maternal genital infection to a newborn. This results in different clinical features of human infections.

The HSV growth cycle proceeds rapidly, requiring 8–16 hours for completion. The HSV genome is large (about 150 kbp) and can encode at least 70 polypeptides; the functions of many of the proteins in replication or latency are not known. At least eight viral glycoproteins are among the viral late gene products. One (gD) is the most potent inducer of neutralizing antibodies. Glycoprotein C is a complement (C3b)-binding protein, and gE is an Fc receptor, binding to the Fc portion of IgG. Glycoprotein G is type-specific and allows for antigenic discrimination between HSV-1 (gG-1) and HSV-2 (gG-2).

Pathogenesis & Pathology

A. Pathology

Because HSV causes cytolytic infections, pathologic changes are due to necrosis of infected cells together with the inflammatory response. Lesions induced in the skin and mucous membranes by HSV-1 and HSV-2 are the same and resemble those of varicella-zoster virus. Changes induced by HSV are similar for primary and recurrent infections but vary in degree, reflecting the extent of viral cytopathology.

Characteristic histopathologic changes include ballooning of infected cells, production of Cowdry type A intranuclear inclusion bodies, margination of chromatin, and formation of multinucleated giant cells. Cell fusion provides an efficient method for cell-to-cell spread of HSV, even in the presence of neutralizing antibody.

B. Primary Infection

HSV is transmitted by contact of a susceptible person with an individual excreting virus. The virus must encounter mucosal surfaces or broken skin in order for an infection to be initiated (unbroken skin is resistant). HSV-1 infections are usually limited to the oropharynx, and virus is spread by respiratory droplets or by direct contact with infected saliva. HSV-2 is usually transmitted by genital routes. Viral replication occurs first at the site of infection. Virus then invades local nerve endings and is transported by retrograde axonal flow to dorsal root ganglia, where, after further replication, latency is established. Oropharyngeal HSV-1 infections result in latent infections in the trigeminal ganglia, whereas genital HSV-2 infections lead to latently infected sacral ganglia. Viremia is more common during primary HSV-2 infections than during HSV-1 infections.

TABLE 33–3   Comparison of HSV Type 1 and Type 2[a]

Characteristics	HSV-1	HSV-2
Biochemical		
Viral DNA base composition (G + C)	67%	69%
Buoyant density of DNA (g/cm³)	1.726	1.728
Buoyant density of virions (g/cm³)	1.271	1.267
Homology between viral DNAs	~50%	~50%
Biologic		
Animal vectors or reservoirs	None	None
Site of latency	Trigeminal ganglia	Sacral ganglia
Epidemiologic		
Age of primary infection	Young children	Young adults
Transmission	Contact (often saliva)	Sexual
Clinical		
Primary infection:		
Gingivostomatitis	+	−
Pharyngotonsillitis	+	−
Keratoconjunctivitis	+	−
Neonatal infections	±	+
Recurrent infection:		
Cold sores, fever blisters	+	−
Keratitis	+	−
Primary or recurrent infection:		
Cutaneous herpes		
Skin above the waist	+	±
Skin below the waist	±	+
Hands or arms	+	+
Herpetic whitlow	+	+
Eczema herpeticum	+	−
Genital herpes	±	+
Herpes encephalitis	+	−
Herpes meningitis	±	+

[a]Modified with permission from Oxman MN: Herpes stomatitis. In: *Infectious Diseases and Medical Microbiology*, 2nd ed. Braude AI, Davis CE, Fierer J (editors). Saunders, 1986:752.

Primary HSV infections are usually mild; in fact, most are asymptomatic. Only rarely does systemic disease develop. Widespread organ involvement can result when an immunocompromised host is not able to limit viral replication and viremia ensues.

C. Latent Infection

Virus resides in latently infected ganglia in a nonreplicating state; only a very few viral genes are expressed. Viral persistence in latently infected ganglia lasts for the lifetime of the host. No virus can be recovered between recurrences at or near the usual site of recurrent lesions. Provocative stimuli can reactivate virus from the latent state, including axonal injury, fever, physical or emotional stress, and exposure to ultraviolet light. The virus follows axons back to the peripheral site, and replication proceeds at the skin or mucous membranes. Spontaneous reactivations occur in spite of HSV-specific humoral and cellular immunity in the host. However, this immunity limits local viral replication, so that recurrent infections are less extensive and less severe. Many recurrences are asymptomatic, reflected only by viral shedding in secretions. When symptomatic, episodes of recurrent HSV-1 infection are usually manifested as cold sores (fever blisters) near the lip. More than 80% of the human population harbor HSV-1 in a latent form, but only a small portion experience recurrences. It is not known why some individuals suffer reactivations and others do not.

Clinical Findings

HSV-1 and HSV-2 may cause many clinical entities, and the infections may be primary or recurrent (Table 33–3). Primary infections occur in persons without antibodies and in most individuals are clinically inapparent but result in antibody production and establishment of latent infections in sensory ganglia. Recurrent lesions are common.

A. Oropharyngeal Disease

Primary HSV-1 infections are usually asymptomatic. Symptomatic disease occurs most frequently in small children (1–5 years of age) and involves the buccal and gingival mucosa of the mouth (Figure 33–4A). The incubation period is short (about 3–5 days, with a range of 2–12 days), and clinical illness lasts 2–3 weeks. Symptoms include fever, sore throat, vesicular and ulcerative lesions, gingivostomatitis, and malaise. Gingivitis (swollen, tender gums) is the most striking and common lesion. Primary infections in adults commonly cause pharyngitis and tonsillitis. Localized lymphadenopathy may occur.

Recurrent disease is characterized by a cluster of vesicles most commonly localized at the border of the lip (Figure 33–4B). Intense pain occurs at the outset but fades over 4–5 days. Lesions progress through the pustular and crusting stages, and healing without scarring is usually complete in 8–10 days. The lesions may recur, repeatedly and at various intervals, in the same location. The frequency of recurrences varies widely among individuals. Many recurrences of oral shedding are asymptomatic and of short duration (24 hours).

B. Keratoconjunctivitis

HSV-1 infections may occur in the eye, producing severe keratoconjunctivitis. Recurrent lesions of the eye are common and appear as dendritic keratitis or corneal ulcers or as vesicles on the eyelids. With recurrent keratitis, there may be progressive involvement of the corneal stroma, with permanent opacification and blindness. HSV-1 infections are second only to trauma as a cause of corneal blindness in the United States.

C. Genital Herpes

Genital disease is usually caused by HSV-2, although HSV-1 can also cause clinical episodes of genital herpes. Primary genital herpes infections can be severe, with illness lasting about 3 weeks. Genital herpes is characterized by

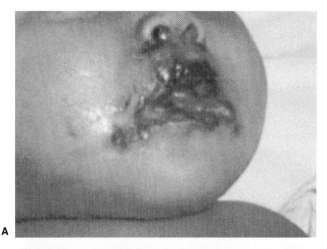

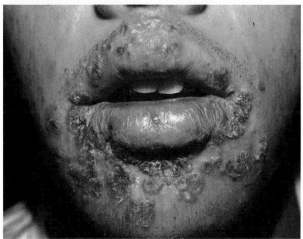

FIGURE 33–4 A: Primary herpes simplex gingivostomatitis. (Courtesy of JD Millar. Source: Centers for Disease Control and Prevention, Public Health Image Library, ID# 2902, 2008). **B:** Recurrent herpes simplex labialis. (Used with permission from Berger TG, Dept Dermatology, UCSF. Reproduced from McPhee SJ, Papadakis MA [editors]: *Current Medical Diagnosis & Treatment*, 48th ed. McGraw Hill, 2009.)

vesiculoulcerative lesions of the penis of the male or of the cervix, vulva, vagina, and perineum of the female. The lesions are very painful and may be associated with fever, malaise, dysuria, and inguinal lymphadenopathy. Complications include extragenital lesions (≈20% of cases) and aseptic meningitis (≈10% of cases). Viral excretion persists for about 3 weeks.

Because of the antigenic cross-reactivity between HSV-1 and HSV-2, preexisting immunity provides some protection against heterotypic infection. An initial HSV-2 infection in a person already immune to HSV-1 tends to be less severe.

Recurrences of genital herpetic infections are common and tend to be mild. A limited number of vesicles appear and heal in about 10 days. Virus is shed for only a few days. Some recurrences are asymptomatic with anogenital shedding lasting less than 24 hours. Whether a recurrence is symptomatic or asymptomatic, a person shedding virus can transmit the infection to sexual partners.

D. Skin Infections

Intact skin is resistant to HSV, so cutaneous HSV infections are uncommon in healthy persons. Localized lesions caused by HSV-1 or HSV-2 may occur in abrasions that become contaminated with the virus (traumatic herpes). These lesions are seen on the fingers of dentists and hospital personnel (herpetic whitlow) and on the bodies of wrestlers (herpes gladiatorum or mat herpes).

Cutaneous infections are often severe and life-threatening when they occur in individuals with disorders of the skin, such as eczema or burns, that permit extensive local viral replication and spread. Eczema herpeticum is a primary infection, usually with HSV-1, in a person with chronic eczema. In rare instances, the illness may be fatal.

E. Encephalitis

A severe form of encephalitis may be produced by herpesvirus. HSV-1 infections are considered the most common cause of sporadic, fatal encephalitis in the United States. The disease carries a high mortality rate, and those who survive often have residual neurologic defects. About half of patients with HSV encephalitis appear to have primary infections, and the rest appear to have recurrent infection.

F. Neonatal Herpes

HSV infection of the newborn may be acquired in utero, during birth, or after birth. The mother is the most common source of infection in all cases. Neonatal herpes is estimated to occur in about one in 5000 deliveries per year. The newborn infant seems to be unable to limit the replication and spread of HSV and has a propensity to develop severe disease.

The most common route of infection (≈75% of cases) is for HSV to be transmitted to the newborn during birth by contact with herpetic lesions in the birth canal. To avoid infection, delivery by cesarean section has been used in pregnant women with genital herpes lesions. However, many fewer cases of neonatal HSV infection occur than cases of recurrent genital herpes, even when virus is present at term.

Neonatal herpes can be acquired postnatally by exposure to either HSV-1 or HSV-2. Sources of infection include family members and hospital personnel who are shedding virus. About 75% of neonatal herpes infections are caused by HSV-2. There do not appear to be any differences between the nature and severity of neonatal herpes in premature or full-term infants, in infections caused by HSV-1 or HSV-2, or in disease when virus is acquired during delivery or postpartum.

Neonatal herpes infections are almost always symptomatic. The overall mortality rate of untreated disease is 50%. Babies with neonatal herpes exhibit three categories of disease: (1) lesions localized to the skin, eye, and mouth; (2) encephalitis with or without localized skin involvement; and (3) disseminated disease involving multiple organs, including the central nervous system. The worst prognosis (mortality rate about 80%) applies to infants with disseminated infection, many of whom develop encephalitis. The cause of death of babies with disseminated disease is usually viral pneumonitis or intravascular coagulopathy. Many survivors of severe infections are left with permanent neurologic impairment.

G. Infections in Immunocompromised Hosts

Immunocompromised patients are at increased risk of developing severe HSV infections. These include patients immunosuppressed by disease or therapy (especially those with deficient cellular immunity) and individuals with malnutrition. Renal, cardiac, and bone marrow transplant recipients are at particular risk for severe herpes infections. Patients with hematologic malignancies and patients with AIDS suffer more frequent and more severe HSV infections. Herpes lesions may spread and involve the respiratory tract, esophagus, and intestinal mucosa. Malnourished children are prone to fatal disseminated HSV infections. In most cases, the disease reflects reactivation of latent HSV infection.

Immunity

Many newborns acquire passively transferred maternal antibodies. These antibodies are lost during the first 6 months of life, and the period of greatest susceptibility to primary herpes infection occurs between ages 6 months and 2 years. Transplacentally acquired antibodies from the mother are not totally protective against infection of newborns, but they seem to ameliorate infection if not prevent it. HSV-1 antibodies begin to appear in the population in early childhood; by adolescence, they are present in most persons. Antibodies to HSV-2 rise during the age of adolescence and sexual activity.

During primary infections, IgM antibodies appear transiently and are followed by IgG and IgA antibodies that persist for long periods. The more severe the primary infection or the more frequent the recurrences, the greater the level of antibody response. However, the pattern of antibody response has not correlated with the frequency of disease recurrence. Cell-mediated immunity and nonspecific host factors (natural killer cells, interferon) are important in controlling both primary and recurrent HSV infections.

After recovery from a primary infection (inapparent, mild, or severe), the virus is carried in a latent state in the presence of antibodies. These antibodies do not prevent reinfection or reactivation of latent virus but may modify subsequent disease.

Laboratory Diagnosis

A. Cytopathology

A rapid cytologic method is to stain scrapings obtained from the base of a vesicle (eg, with Giemsa's stain); the presence of multinucleated giant cells indicates that herpesvirus (HSV-1,

HSV-2, or varicella-zoster) is present, distinguishing lesions from those caused by coxsackieviruses and nonviral entities.

B. Isolation & Identification of Virus

Virus isolation remains the definitive diagnostic approach. Virus may be isolated from herpetic lesions and may also be found in throat washings, cerebrospinal fluid, and stool, both during primary infection and during asymptomatic periods. Therefore, the isolation of HSV is not in itself sufficient evidence to indicate that the virus is the causative agent of a disease under investigation.

Inoculation of tissue cultures is used for viral isolation. HSV is easy to cultivate, and cytopathic effects usually occur in only 2–3 days. The agent is then identified by Nt test or immunofluorescence staining with specific antiserum. Typing of HSV isolates may be done using monoclonal antibody or by restriction endonuclease analysis of viral DNA but is only useful for epidemiologic studies.

C. Polymerase Chain Reaction

Polymerase chain reaction (PCR) assays can be used to detect virus and are sensitive and specific. PCR amplification of viral DNA from cerebrospinal fluid has replaced viral isolation from brain tissue obtained by biopsy or at postmortem examination as the standard assay for specific diagnosis of HSV infections of the central nervous system.

D. Serology

Antibodies appear in 4–7 days after infection and reach a peak in 2–4 weeks. They persist with minor fluctuations for the life of the host. Methods available include Nt, immunofluorescence, and ELISA.

The diagnostic value of serologic assays is limited by the multiple antigens shared by HSV-1 and HSV-2. There may also be some heterotypic anamnestic responses to varicella-zoster virus in persons infected with HSV and vice versa. The use of HSV type-specific antibodies, available in some research laboratories, allows more meaningful serologic tests.

Epidemiology

HSV are worldwide in distribution. No animal reservoirs or vectors are involved with the human viruses. Transmission is by contact with infected secretions. The epidemiology of HSV-1 and HSV-2 differs.

HSV-1 is probably more constantly present in humans than any other virus. Primary infection occurs early in life and is usually asymptomatic; occasionally, it produces oropharyngeal disease (gingivostomatitis in young children, pharyngitis in young adults). Antibodies develop, but the virus is not eliminated from the body; a carrier state is established that lasts throughout life and is punctuated by transient recurrent attacks of herpes.

The highest incidence of HSV-1 infection occurs among children 6 months to 3 years of age. By adulthood,

70–90% of persons have type 1 antibodies. There is a high rate of geographic variation in seroprevalence. Middle-class individuals in developed countries acquire antibodies later in life than those in lower socioeconomic populations. Presumably this reflects more crowded living conditions and poorer hygiene among the latter. The virus is spread by direct contact with infected saliva or through utensils contaminated with the saliva of a virus shedder. The source of infection for children is usually an adult with a symptomatic herpetic lesion or with asymptomatic viral shedding in saliva.

The frequency of recurrent HSV-1 infections varies widely among individuals. At any given time, 1–5% of normal adults will be excreting virus, often in the absence of clinical symptoms.

HSV-2 is usually acquired as a sexually transmitted disease, so antibodies to this virus are seldom found before puberty. It is estimated that there are about 40–60 million infected individuals in the United States. Antibody prevalence studies have been complicated by the cross-reactivity between HSV types 1 and 2. Surveys using type-specific glycoprotein antigens recently determined that 20% of adults in the United States possess HSV-2 antibodies, with seroprevalence higher among women than men and higher among blacks than whites.

Recurrent genital infections may be symptomatic or asymptomatic. Either situation provides a reservoir of virus for transmission to susceptible persons. Studies have estimated that transmission of genital herpes in over 50% of cases resulted from sexual contact in the absence of lesions or symptoms.

Maternal genital HSV infections pose risks to both mother and fetus. Rarely, pregnant women may develop disseminated disease after primary infection, with a high mortality rate. Primary infection before 20 weeks of gestation has been associated with spontaneous abortion. The fetus may acquire infection as a result of viral shedding from recurrent lesions in the mother's birth canal at the time of delivery. Estimates of the frequency of cervical shedding of virus among pregnant women vary widely.

Genital HSV infections increase acquisition of human immunodeficiency virus (HIV) type 1 infections because the ulcerative lesions are openings in the mucosal surface.

Treatment, Prevention, & Control

Several antiviral drugs have proved effective against HSV infections, including acyclovir, valacyclovir, and vidarabine (see Chapter 30). All are inhibitors of viral DNA synthesis. Acyclovir, a nucleoside analog, is monophosphorylated by the HSV thymidine kinase and is then converted to the triphosphate form by cellular kinases. The acyclovir triphosphate is efficiently incorporated into viral DNA by the HSV polymerase, where it then prevents chain elongation. The drugs may suppress clinical manifestations, shorten time to healing, and reduce recurrences of genital herpes. However,

HSV remains latent in sensory ganglia. Drug-resistant virus strains may emerge.

Newborns and persons with eczema should be protected from exposure to persons with active herpetic lesions.

Patients with genital herpes should be counseled that asymptomatic shedding is frequent and that the risk of transmission can be reduced by antiviral therapy and condom usage.

Experimental vaccines of various types are being developed. One approach is to use purified glycoprotein antigens found in the viral envelope, expressed in some recombinant systems. Such vaccines might be helpful for the prevention of primary infections. A recombinant HSV-2 glycoprotein vaccine tested in a recent multicenter trial prevented genital herpes infections in women who were seronegative for both HSV-1 and HSV-2; it was not effective in women who were seropositive for HSV-1 or in men.

VARICELLA-ZOSTER VIRUS

Varicella (chickenpox) is a mild, highly contagious disease, chiefly of children, characterized clinically by a generalized vesicular eruption of the skin and mucous membranes. The disease may be severe in adults and in immunocompromised children.

Zoster (shingles) is a sporadic, incapacitating disease of adults or immunocompromised individuals that is characterized by a rash limited in distribution to the skin innervated by a single sensory ganglion. The lesions are similar to those of varicella.

Both diseases are caused by the same virus. Varicella is the acute disease that follows primary contact with the virus, whereas zoster is the response of the partially immune host to reactivation of varicella virus present in latent form in neurons in sensory ganglia.

Properties of the Virus

Varicella-zoster virus is morphologically identical to HSV. It has no animal reservoir. The virus propagates in cultures of human embryonic tissue and produces typical intranuclear inclusion bodies (see Figure 33–3B). Cytopathic changes are more focal and spread much more slowly than those induced by HSV. Infectious virus remains strongly cell-associated, and serial propagation is more easily accomplished by passage of infected cells than of tissue culture fluids.

The same virus causes chickenpox and zoster. Viral isolates from the vesicles of chickenpox or zoster patients exhibit no significant genetic variation. Inoculation of zoster vesicle fluid into children produces chickenpox.

Pathogenesis & Pathology

A. Varicella

The route of infection is the mucosa of the upper respiratory tract or the conjunctiva (Figure 33–5). Following initial

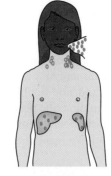

Incubation period

{ Inoculation of respiratory mucosa
{ Viral replication in regional nodes → virus-infected cells into capillaries

{ Primary viremia → replication in liver/spleen

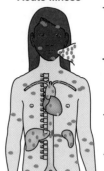

Acute illness

{ Secondary viremia: mononuclear cell transport to skin and mucous membranes

{ Virus release into respiratory secretions

{ Replication in epidermal cells
  Virus in dorsal root ganglia

{ VZV specific immunity → resolution of replication

FIGURE 33–5 The pathogenesis of primary infection with varicella-zoster virus (VZV). The incubation period with primary viremia lasts from 10 to 21 days. A secondary viremic phase results in the transport of virus to skin and respiratory mucosal sites. Replication in epidermal cells causes the characteristic rash of varicella, referred to as chickenpox. The induction of varicella-zoster virus-specific immunity is required to terminate viral replication. The virus gains access to cells of the trigeminal and dorsal root ganglia during primary infection and establishes latency. (Reproduced with permission from Arvin AM: Varicella-zoster virus. In: *Fields Virology*, 3rd ed. Fields BN et al [editors]. Lippincott-Raven, 1996.)

replication in regional lymph nodes, primary viremia spreads virus and leads to replication in liver and spleen. Secondary viremia involving infected mononuclear cells transports virus to the skin, where the typical rash develops. Swelling of epithelial cells, ballooning degeneration, and the accumulation of tissue fluids result in vesicle formation (Figure 33–6).

Varicella-zoster virus replication and spread are limited by host humoral and cellular immune responses. Interferon may also be involved.

B. Zoster

The skin lesions of zoster are histopathologically identical to those of varicella. There is also an acute inflammation of the sensory nerves and ganglia. Often only a single ganglion may be involved. As a rule, the distribution of lesions in the skin corresponds closely to the areas of innervation from an individual dorsal root ganglion.

It is not clear what triggers reactivation of latent varicella-zoster virus infections in ganglia. It is believed that

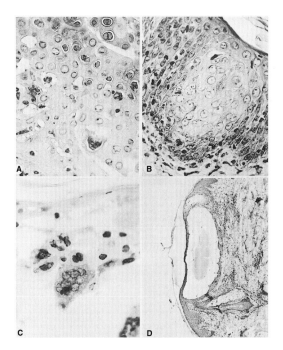

FIGURE 33-6 Characteristic histologic changes of varicella-zoster virus infection. Punch biopsies of varicella-zoster virus vesicles were fixed and stained with hematoxylin-eosin. **A:** Early infection showing "balloon degeneration" of cells with basophilic nuclei and marginated chromatin (reduced from 480×). **B:** Later infection showing eosinophilic intranuclear inclusions surrounded by wide clear zones (reduced from 480×). **C:** Multinucleated giant cell in the roof of a varicella vesicle (reduced from 480×). **D:** Low-power view of an early vesicle showing separation of the epidermis (acantholysis), dermal edema, and mononuclear cell infiltration (reduced from 40×). (Reproduced with permission from Gelb LD: Varicella-zoster virus. In: *Virology*, 2nd ed. Fields BN et al [editors]. Raven Press, 1990.)

waning immunity allows viral replication to occur in a ganglion, causing intense inflammation and pain. Virus travels down the nerve to the skin and induces vesicle formation. Cell-mediated immunity is probably the most important host defense in containment of varicella-zoster virus. Reactivations are sporadic and recur infrequently.

Clinical Findings

A. Varicella

Subclinical varicella is unusual. The incubation period of typical disease is 10–21 days. Malaise and fever are the earliest symptoms, soon followed by the rash, first on the trunk and then on the face, the limbs, and the buccal and pharyngeal mucosa in the mouth. Successive fresh vesicles appear in crops, so that all stages of macules, papules, vesicles, and crusts may be seen at one time (Figure 33–7). The rash lasts about 5 days, and most children develop several hundred skin lesions.

Complications are rare in normal children, and the mortality rate is very low. Encephalitis does occur in rare cases

and can be life-threatening. Survivors of varicella encephalitis may be left with permanent sequelae. In neonatal varicella, the infection is contracted from the mother just before or after birth but without sufficient immune response to modify the disease. Virus is often widely disseminated and may prove fatal. Cases of congenital varicella syndrome following maternal cases of chickenpox during pregnancy have been described.

Varicella pneumonia is rare in healthy children but is the most common complication in neonates, adults, and immunocompromised patients. It is responsible for many varicella-related deaths.

Immunocompromised patients are at increased risk of complications of varicella, including those with malignancies, organ transplants, or HIV infection and those receiving high doses of corticosteroids. Disseminated intravascular coagulation may occur that is rapidly fatal. Children with leukemia are especially prone to develop severe, disseminated varicella-zoster virus disease.

B. Zoster

Zoster usually occurs in persons immunocompromised as a result of disease, therapy, or aging, but it occasionally develops in healthy young adults. It usually starts with severe pain in the area of skin or mucosa supplied by one or more groups of sensory nerves and ganglia. Within a few days after onset, a crop of vesicles appears over the skin supplied by the affected nerves. The trunk, head, and neck are most commonly affected (Figure 33–8), with the ophthalmic division of the trigeminal nerve involved in 10–15% of cases. The most common complication of zoster in the elderly is postherpetic neuralgia—protracted pain that may continue for months. It is especially common after ophthalmic zoster. Visceral disease, especially pneumonia, is responsible for deaths that occur in immunosuppressed patients with zoster (<1% of patients).

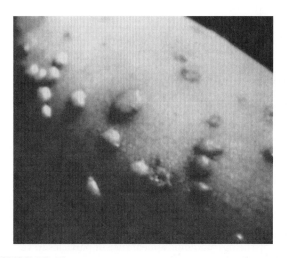

FIGURE 33-7 Multiple stages or "crops" of varicella skin lesions. (Reproduced with permission from Gelb LD: Varicella-zoster virus. In: *Virology*, 2nd ed. Fields BN et al [editors]. Raven Press, 1990.)

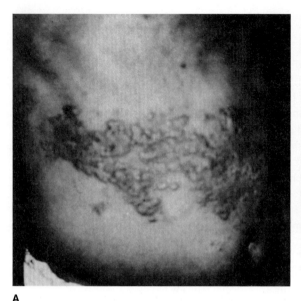 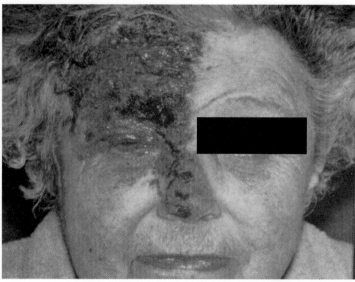

FIGURE 33–8 A: Herpes zoster in the distribution of thoracic nerves. (Courtesy of AA Gershon.) **B:** Herpes zoster ophthalmicus. (Courtesy MN Oxman, University of California, San Diego. Reproduced from MMWR Morb Mortal Wkly Rep 2008;57(RR-5):1.)

Immunity

Varicella and zoster viruses are identical, the two diseases being the result of differing host responses. Previous infection with varicella is believed to confer lifelong immunity to varicella. Antibodies induced by varicella vaccine persist for at least 20 years. Zoster occurs in the presence of neutralizing antibody to varicella.

Increases in varicella antibody titer may occur in persons with HSV infections.

The development of varicella-zoster virus-specific cell-mediated immunity is important in recovery from both varicella and zoster. Appearance of local interferon may also contribute to recovery.

Varicella-zoster virus, like other herpesviruses, encodes means of evading host immune responses. For example, it downregulates major histocompatibility complex class I and II antigen expression.

Laboratory Diagnosis

In stained smears of scrapings or swabs of the base of vesicles (Tzanck smear), multinucleated giant cells are seen (Figure 33–6). These are absent in nonherpetic vesicles. Intracellular viral antigens can be demonstrated by immunofluorescence staining of similar smears.

Rapid diagnostic procedures are clinically useful for varicella-zoster virus. Virus-specific antigens or viral DNA can be detected in vesicle fluid, in skin scrapings, or in biopsy material. Herpesviruses can be differentiated from poxviruses by the morphologic appearance of particles in vesicular fluids examined by electron microscopy (Figure 33–9).

Virus can be isolated from vesicle fluid early in the course of illness using cultures of human cells in 3–7 days. Varicella-

zoster virus in vesicle fluid is very labile, and cell cultures should be inoculated promptly.

A rise in specific antibody titer can be detected in the patient's serum by various tests, including fluorescent antibody and enzyme immunoassay. The choice of assay to use depends on the purpose of the test and the laboratory facilities available. Cell-mediated immunity is important, but difficult to demonstrate.

Epidemiology

Varicella and zoster occur worldwide. Varicella (chickenpox) is highly communicable and is a common epidemic disease of childhood (most cases occur in children under 10 years of age). Adult cases do occur. It is much more common in winter and spring than in summer in temperate climates. Zoster occurs sporadically, chiefly in adults and without seasonal prevalence. Ten to 20 percent of adults will experience at least one zoster attack during their lifetime, usually after the age of 50.

A live attenuated varicella vaccine is available. In the pre-vaccine era, varicella caused about 4 million illnesses, 11,000 hospitalizations, and 100 deaths annually in the United States. Since the vaccine was introduced in 1995, there has been a steady decline in the incidence of varicella diseases; however, varicella outbreaks continue to occur among school children, because some children are unvaccinated and the vaccine is 80–85% effective in vaccinated persons.

Varicella spreads readily by airborne droplets and by direct contact. A varicella patient is probably infectious (capable of transmitting the disease) from shortly before the appearance of rash to the first few days of rash. Contact infection is less common in zoster, perhaps because the virus

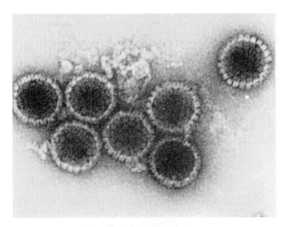

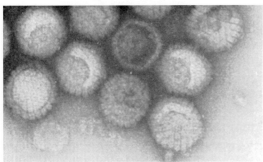

FIGURE 33–9 **Top:** Herpesvirus particles from human vesicle fluid, stained with uranyl acetate to show DNA core (140,000×). **Bottom:** Virions stained to show protein capsomeres of the virus coat (140,000×). **Note:** Different herpesviruses cannot be distinguished by electron microscopy. (Courtesy of KO Smith and JL Melnick.)

is absent from the upper respiratory tract in typical cases. Zoster patients can be the source of varicella in susceptible children. Varicella-zoster virus DNA has been detected, using a PCR amplification method, in air samples from hospital rooms of patients with active varicella (82%) and zoster (70%) infections.

Treatment

Varicella in normal children is a mild disease and requires no treatment. Neonates and immunocompromised patients with severe infections should be treated.

γ-Globulin of high varicella-zoster virus antibody titer (varicella-zoster immune globulin) can be used to prevent the development of the illness in patients exposed to varicella who are at high risk of developing severe disease. It has no therapeutic value once varicella has started. Standard immune globulin is without value because of its low titer of varicella antibodies.

The manufacturer of the only United States-licensed varicella-zoster immune globulin discontinued its production in 2004; however, in 2006, a new investigational (not licensed) product became available. It can be requested for patients at increased risk for severe disease.

Several antiviral compounds provide effective therapy for varicella, including acyclovir, valacyclovir, famciclovir, and foscarnet. Acyclovir can prevent the development of systemic disease in varicella-infected immunosuppressed patients and can halt the progression of zoster in adults. Acyclovir does not appear to prevent postherpetic neuralgia.

Prevention & Control

A live attenuated varicella vaccine was approved in 1995 for general use in the United States. A similar vaccine has been used successfully in Japan for about 30 years. The vaccine is highly effective at inducing protection from varicella in children (80–85% effective), but less so in adults (70%). The vaccine is about 95% effective in preventing severe disease. About 5% of individuals develop a mild vaccine-associated rash 1 month after immunization. Transmission of the vaccine virus is rare but can occur when the vaccinee has a rash. The duration of protective immunity induced by the vaccine is unknown, but is probably long term. Varicella infections can occur in vaccinated persons, but they are usually mild illnesses.

A zoster (shingles) vaccine was licensed in the United States in 2006. It is a more potent version of the varicella vaccine. It has been shown to be effective in older adults at reducing both the frequency of outbreaks of zoster and the severity of disease that does occur. The zoster vaccine is recommended for those with chronic medical conditions and for persons over 60 years of age.

CYTOMEGALOVIRUS

Cytomegaloviruses are ubiquitous herpesviruses that are common causes of human disease. Cytomegaloviruses are the agents of the most common congenital infection.

Cytomegalic inclusion disease is a generalized infection of infants caused by intrauterine or early postnatal infection with the cytomegaloviruses. The name for the classic cytomegalic inclusion disease derives from the propensity for massive enlargement of cytomegalovirus-infected cells. Cytomegalovirus poses an important public health problem because of its high frequency of congenital infections, which may lead to severe congenital anomalies. Inapparent infection is common during childhood and adolescence. Severe cytomegalovirus infections are frequently found in adults who are immunosuppressed.

Properties of the Virus

Cytomegalovirus has the largest genetic content of the human herpesviruses. Its DNA genome (240 kbp) is significantly larger than that of HSV. Only a few of the many proteins encoded by the virus (over 200) have been characterized. One, a cell surface glycoprotein, acts as an Fc receptor that can nonspecifically bind the Fc portion of immunoglobulins. This may help infected cells evade immune elimination by providing a protective coating of irrelevant host immunoglobulins.

The major immediate-early promoter-enhancer of cytomegalovirus is one of the strongest known enhancers, due to the concentration of binding sites for cellular transcription factors. It is used experimentally to support high-level expression of foreign genes.

Many genetically different strains of cytomegalovirus are circulating in the human population. The strains are sufficiently related antigenically, however, so that strain differences are probably not important determinants in human disease.

Cytomegaloviruses are very species-specific and cell type-specific. All attempts to infect animals with human cytomegalovirus have failed. A number of animal cytomegaloviruses exist, all of them species-specific.

Human cytomegalovirus replicates in vitro only in human fibroblasts, although the virus is often isolated from epithelial cells of the host. Cytomegalovirus replicates very slowly in cultured cells, with growth proceeding more slowly than that of HSV or varicella-zoster virus. Very little virus becomes cell-free; infection is spread primarily cell-to-cell. It may take several weeks for an entire monolayer of cultured cells to become involved.

Cytomegalovirus produces a characteristic cytopathic effect (see Figure 33–3C). Perinuclear cytoplasmic inclusions form in addition to the intranuclear inclusions typical of herpesviruses. Multinucleated cells are seen. Many affected cells become greatly enlarged. Inclusion-bearing cytomegalic cells can be found in samples from infected individuals (Figure 33–10).

Pathogenesis & Pathology

A. Normal Hosts

Cytomegalovirus may be transmitted person-to-person in several different ways, all requiring close contact with virus-bearing material. There is a 4- to 8-week incubation period in normal older children and adults after viral exposure. The virus causes a systemic infection; it has been isolated from lung, liver, esophagus, colon, kidneys, monocytes, and T and B lymphocytes. The disease is an infectious mononucleosis-like syndrome, although most cytomegalovirus infections are subclinical. Like all herpesviruses, cytomegalovirus establishes lifelong latent infections. Virus can be shed intermittently from the pharynx and in the urine for months to years after primary infection (Figure 33–11). Prolonged cytomegalovirus infection of the kidney does not seem to be deleterious in normal persons. Salivary gland involvement is common and is probably chronic.

Cell-mediated immunity is depressed with primary infections (Figure 33–11), and this may contribute to the persistence of viral infection. It may take several months for cellular responses to recover.

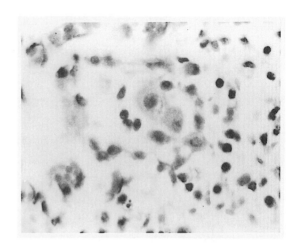

FIGURE 33–10 Massively enlarged "cytomegalic" cells typical of cytomegalovirus infection present in the lung of a premature infant who died of disseminated cytomegalovirus disease. (Courtesy of GJ Demmler.)

B. Immunosuppressed Hosts

Primary cytomegalovirus infections in immunosuppressed hosts are much more severe than in normal hosts. Individuals at greatest risk for cytomegalovirus disease are those receiving organ transplants, those with malignant tumors who are receiving chemotherapy, and those with AIDS. Viral excretion is increased and prolonged, and the infection is more apt to become disseminated. Pneumonia is the most common complication.

The host immune response presumably maintains cytomegalovirus in a latent state in seropositive individuals. Reactivated infections are associated with disease much more often in immunocompromised patients than in normal hosts. Although usually less severe, reactivated infections may be as virulent as primary infections.

C. Congenital & Perinatal Infections

Fetal and newborn infections with cytomegalovirus may be severe (Figure 33–12). About 1% of live births annually in the United States have congenital cytomegalovirus infections, and about 5–10% of those will suffer cytomegalic inclusion disease. A high percentage of babies with this disease will exhibit developmental defects and mental retardation.

The virus can be transmitted in utero with both primary and reactivated maternal infections. About one-third of pregnant women with primary infection transmit the virus. Generalized cytomegalic inclusion disease results most often from primary maternal infections. There is no evidence that gestational age at the time of maternal infection affects expression of disease in the fetus. Intrauterine transmission occurs in about 1% of seropositive women. Fetal damage seldom results from these reactivated maternal infections; the infection of the infant remains subclinical though chronic (Figure 33–11).

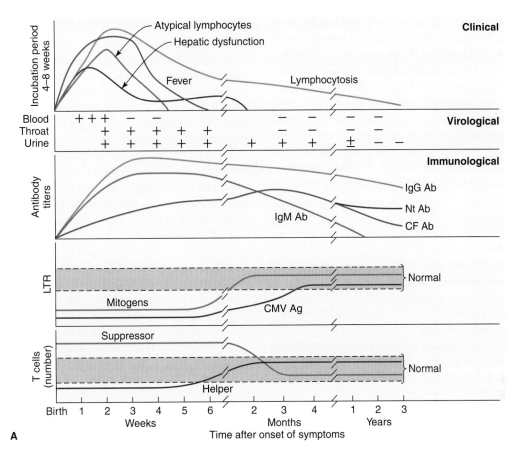

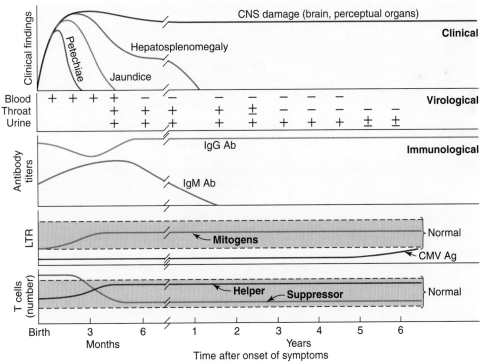

FIGURE 33–11 Clinical, virologic, and immunologic features of cytomegalovirus (CMV) infection **A:** in normal individuals and **B:** in congenitally infected infants. LTR, lymphocyte transformation response. (Reproduced with permission from Alford CA, Britt WJ: Cytomegalovirus. In: *Virology*, 2nd ed. Fields BN et al [editors]. Raven Press, 1990.)

Cytomegalovirus can also be acquired by the infant from exposure to virus in the mother's genital tract during delivery and from maternal breast milk. In these cases, the infants usually have received some maternal antibody, and the perinatally acquired cytomegalovirus infections tend to be subclinical. Transfusion-acquired cytomegalovirus infections in newborns will vary, depending on the amount of virus received and the serologic status of the blood donor. Whether cytomegalovirus is acquired in utero or perinatally, a more chronic infection results—with respect to viral excretion—than when the virus is acquired later in life (Figure 33–11).

Clinical Findings

A. Normal Hosts

Primary cytomegalovirus infection of older children and adults is usually asymptomatic but occasionally causes a spontaneous infectious mononucleosis syndrome. Cytomegalovirus is estimated to cause 20–50% of heterophil-negative (non-EBV) mononucleosis cases.

Cytomegalovirus mononucleosis is a mild disease, and complications are rare. Subclinical hepatitis is common. In younger children (under 7 years old), hepatosplenomegaly is frequently observed.

An association has been observed between the presence of cytomegalovirus and restenosis following coronary angioplasty. It is speculated that the virus may be contributing to the proliferation of smooth muscle cells, leading to restenosis.

B. Immunocompromised Hosts

Both morbidity and mortality rates are increased with primary and recurrent cytomegalovirus infections in immunocompromised individuals. Pneumonia is a frequent complication. Interstitial pneumonitis caused by cytomegalovirus occurs in 10–20% of bone marrow transplant recipients. Virus-associated leukopenia is common in solid organ transplant recipients; also seen are obliterative bronchiolitis in lung transplants, graft atherosclerosis after heart transplantation, and cytomegalovirus-related rejection of renal allografts. Cytomegalovirus often causes disseminated disease in untreated AIDS patients; gastroenteritis and chorioretinitis are common problems, the latter often leading to progressive blindness.

C. Congenital & Perinatal Infections

Congenital infection may result in death of the fetus in utero (Figure 33–12). Cytomegalic inclusion disease of newborns is characterized by involvement of the central nervous system and the reticuloendothelial system. Clinical features include intrauterine growth retardation, jaundice, hepatosplenomegaly, thrombocytopenia, microcephaly, and retinitis. Mortality rates are about 20%. The majority of survivors will develop significant central nervous system defects within 2 years; severe hearing loss, ocular abnormalities, and mental retardation are common. About 10% of infants with subclinical congenital cytomegalovirus infection will develop deafness. It has been estimated that one in every 1000 infants born in

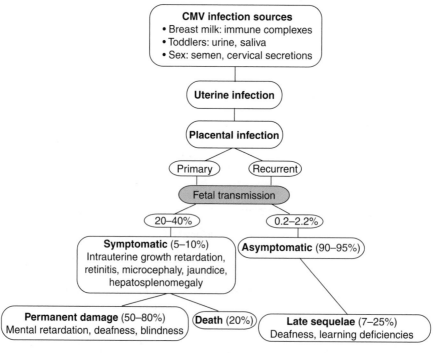

FIGURE 33–12 Congenital infections by cytomegalovirus and birth defects in symptomatic and asymptomatic children. Cytomegalovirus is the most common intrauterine infection associated with congenital defects. (Reproduced with permission from Pereira L et al: Insights into viral transmission at the uterine-placental interface. Trends Microbiol 2005;13:164.)

the United States is seriously retarded as a result of congenital cytomegalovirus infection.

Many women infected previously with cytomegalovirus show reactivation and begin to excrete the virus from the cervix during pregnancy. At the time of delivery through the infected birth canal, infants may become infected, though they possess high titers of maternal antibody acquired transplacentally. These infants begin to shed virus at about 8–12 weeks of age. They continue to excrete the virus for several years but remain healthy.

Acquired infection with cytomegalovirus is common and usually inapparent. The virus is shed in the saliva and urine of infected individuals for weeks or months. Cytomegalovirus may be a cause of isolated pneumonia in infants less than 6 months of age.

Immunity

Antibodies to cytomegalovirus in human sera in the United States increase with age, from about 40% in teenagers to more than 80% in those over 60 years old. Reactivation of latent infection occurs in the presence of humoral immunity. The presence of antibody in breast milk does not prevent transmission of infection to breast-feeding infants. Maternal antibody protects more against development of serious disease in the infant than viral transmission.

Laboratory Diagnosis

A. Polymerase Chain Reaction & Antigen Detection Assays

PCR assays have replaced virus isolation for routine detection of cytomegalovirus infections. Cell culture methods of viral isolation are too slow to be useful in guiding therapy, particularly in immunosuppressed patients. The PCR assays are designed to detect replicating virus, not latent viral genomes. Blood and urine are most commonly tested. PCR assays can provide viral load data, which appears to be important in predicting cytomegalovirus disease. Monoclonal antibodies against viral antigens can be used to detect virus-positive leukocytes from patients.

B. Isolation of Virus

Human fibroblasts are used for virus isolation attempts. The virus can be recovered most readily from throat washings and urine. In cultures, 2–3 weeks are usually needed for the appearance of cytologic changes, consisting of small foci of swollen, translucent cells with large intranuclear inclusions (see Figure 33–3C and D). The virus stays cell-associated.

C. Serology

Many types of assays can detect cytomegalovirus IgG antibodies, indicative of past infection (and the potential to undergo reactivation). Detection of viral IgM antibodies suggests a current infection. Serologic assays are not informative for immunocompromised patients. Furthermore, serologic techniques cannot distinguish strain differences among clinical isolates.

Epidemiology

Cytomegalovirus is endemic in all parts of the world; epidemics are unknown. It is present throughout the year, with no seasonal variation seen in infection rates.

The prevalence of infection varies with socioeconomic status, living conditions, and hygienic practices. Antibody prevalence may be moderate (40–70%) in adults in high socioeconomic groups in developed countries—in contrast to a prevalence of 90% in children and adults in developing nations and in low socioeconomic groups in developed countries.

New infections are almost always asymptomatic. After infection, virus is shed from multiple sites. Viral shedding may continue for years, often intermittently, as latent virus becomes reactivated. Thus, exposures to cytomegalovirus are widespread and common.

Humans are the only known host for cytomegalovirus. Transmission requires close person-to-person contact. Virus may be shed in urine, saliva, semen, breast milk, and cervical secretions and is carried in circulating white blood cells. Oral and respiratory spread are probably the dominant routes of cytomegalovirus transmission. Cytomegalovirus can be transmitted by blood transfusion. Estimated risk varies widely but is about 1–5% per unit of whole blood. Seronegative solid organ transplantation recipients are at risk, as a seropositive organ transmits the virus in 60–80% of cases.

Intrauterine infection may produce serious disease in the newborn. About 1% of infants born in the United States are infected with cytomegalovirus. The majority have subclinical but chronic infections; 5–10% have cytomegalic inclusion disease with attendant developmental defects and high mortality. Congenital infections, whether subclinical or clinically apparent, result in chronic infections, with viral shedding detectable for years. Many more infants become infected with cytomegalovirus in the first months of life, often from infected breast milk or by nursery spread. Most of these infections are subclinical but are usually chronic, with persistent viral shedding.

Transmission in utero occurs in about 40% of primary infections of mothers. Such primary maternal infections during pregnancy are responsible for most cases of cytomegalic inclusion disease. Infants and children with subclinical cytomegalovirus infections are the major source of exposure. Other congenital infections are due to reactivations of latent maternal infections. Transmission in utero from such reactivations is uncommon (~1%).

Cytomegalovirus infections are markedly increased in immunosuppressed populations; transplant recipients often develop infections, most of which are due to reactivations of their own latent virus.

Treatment & Control

Drug treatments of cytomegalovirus infections have shown some encouraging results. Ganciclovir, a nucleoside structurally related to acyclovir, has been used successfully to treat life-threatening cytomegalovirus infections in immunosuppressed patients. The severity of cytomegalovirus retinitis, esophagitis, and colitis is reduced by ganciclovir. In addition, early treatment with ganciclovir reduces the incidence of cytomegalovirus pneumonia in bone marrow allograft recipients. Ganciclovir also controls progressive hearing loss in neonates with congenital infections. Foscarnet, an analog of inorganic pyrophosphate, is recommended for treatment of cytomegalovirus retinitis. Acyclovir and valacyclovir have shown some benefits in bone marrow and renal transplant patients.

Specific control measures are not available to prevent cytomegalovirus spread. Isolation of newborns with generalized cytomegalic inclusion disease from other newborns is advisable.

Screening of transplant donors and recipients for cytomegalovirus antibody may prevent some transmissions of primary cytomegalovirus. The cytomegalovirus-seronegative transplant recipient population represents a high-risk group for cytomegalovirus infections. Administration of human IgG prepared from plasma pools obtained from healthy persons with high titers of cytomegalovirus antibodies (cytomegalovirus immune globulin) has given discordant results in tests to decrease the incidence of viral infections in transplant recipients. Cytomegalovirus immune globulin is in limited supply.

The use of blood from seronegative donors has been recommended when infants will require multiple transfusions. This approach would eliminate transfusion-acquired cytomegalovirus infections, but it is difficult to implement.

Both live and recombinant cytomegalovirus vaccines are under development.

EPSTEIN-BARR VIRUS

EBV is a ubiquitous herpesvirus that is the causative agent of acute infectious mononucleosis and is associated with nasopharyngeal carcinoma, Burkitt lymphoma, Hodgkin and non-Hodgkin lymphomas, other lymphoproliferative disorders in immunodeficient individuals, and gastric carcinoma.

Properties of the Virus

The EBV DNA genome contains about 172 kbp, has a G + C content of 59%, and encodes about 100 genes. There are two major strains of EBV (types A and B).

A. Biology of Epstein-Barr Virus

The major target cell for EBV is the B lymphocyte. When human B lymphocytes are infected with EBV, continuous cell lines can be established, indicating that cells have been immortalized by the virus. Very few of the immortalized cells produce infectious virus. Laboratory studies of EBV are hampered by the lack of a fully permissive cell system able to propagate the virus.

EBV initiates infection of B cells by binding to the viral receptor, which is the receptor for the C3d component of complement (CR2 or CD21). EBV directly enters a latent state in the lymphocyte without undergoing a period of complete viral replication. The hallmarks of latency are viral persistence, restricted virus expression, and the potential for reactivation and lytic replication.

The efficiency of B cell immortalization by EBV is quite high. When virus binds to the cell surface, cells are activated to enter the cell cycle. Subsequently, a limited repertoire of EBV genes are expressed, and the cells are able to proliferate indefinitely. The linear EBV genome forms a circle and is amplified during the cell cycle S phase; the majority of viral DNA in the immortalized cells exists as circular episomes.

EBV-immortalized B lymphocytes express differentiated functions, such as secretion of immunoglobulin. B cell activation products (eg, CD23) are also expressed. Several patterns of latent viral gene expression are recognized, based on the spectrum of proteins and transcripts expressed. These include EBV nuclear antigens (EBNA1, 2, 3A-3C, LP), latent membrane proteins (LMP1, 2), and small untranslated RNAs (EBERs).

At any given time, very few cells (<10%) in an immortalized population release virus particles. Latency can be disrupted and the EBV genome activated to replicate in a cell by a variety of stimuli, including chemical inducing agents or cross-linking cell surface immunoglobulin.

EBV can replicate in vivo in epithelial cells of the oropharynx, parotid gland, and uterine cervix; it is found in epithelial cells of some nasopharyngeal carcinomas. Although epithelial cells in vivo contain an EBV receptor, the receptor is lost from cultured cells.

EBV is associated with a number of lymphoproliferative disorders. Viral gene expression in these cells is limited and varies from only EBNA1 to the full complement of proteins found in latently infected B cells.

B. Viral Antigens

EBV antigens are divided into three classes, based on the phase of the viral life cycle in which they are expressed: (1) Latent phase antigens are synthesized by latently infected cells. These include the EBNAs and the LMPs. Their expression reveals that an EBV genome is present. Only EBNA1, needed to maintain the viral DNA episomes, is invariably expressed; expression of the other latent phase antigens may be regulated in different cells. LMP1 mimics an activated growth factor receptor. (2) Early antigens are nonstructural proteins whose synthesis is not dependent on viral DNA replication. The expression of early antigens indicates the onset of productive viral replication. (3) Late antigens are the structural components of the viral capsid (viral capsid antigen) and viral envelope (glycoproteins). They are produced abundantly in cells undergoing productive viral infection.

C. Experimental Animal Infections

EBV is highly species-specific for humans. However, cotton-top tamarins inoculated with EBV frequently develop fatal malignant lymphomas.

Pathogenesis & Pathology

A. Primary Infection

EBV is commonly transmitted by infected saliva and initiates infection in the oropharynx. Viral replication occurs in epithelial cells (or surface B lymphocytes) of the pharynx and salivary glands. Many people shed low levels of virus for weeks to months after infection. Infected B cells spread the infection from the oropharynx throughout the body. In normal individuals, most virus-infected cells are eliminated, but small numbers of latently infected lymphocytes persist for the lifetime of the host (one in 10^5–10^6 B cells).

Primary infections in children are usually subclinical, but if they occur in young adults acute infectious mononucleosis often develops. Mononucleosis is a polyclonal stimulation of lymphocytes. EBV-infected B cells synthesize immunoglobulin. Autoantibodies are typical of the disease, with heterophil antibody that reacts with antigens on sheep erythrocytes the classic autoantibody.

B. Reactivation from Latency

Reactivations of EBV latent infections can occur, as evidenced by increased levels of virus in saliva and of DNA in blood cells. These are usually clinically silent. Immunosuppression is known to reactivate infection, sometimes with serious consequences.

Clinical Findings

Most primary infections in children are asymptomatic. In adolescents and young adults, the classic syndrome associated with primary infection is infectious mononucleosis (about 50% of infections). EBV is also associated with several types of cancer.

A. Infectious Mononucleosis

After an incubation period of 30–50 days, symptoms of headache, fever, malaise, fatigue, and sore throat occur. Enlarged lymph nodes and spleen are characteristic. Some patients develop signs of hepatitis.

The typical illness is self-limited and lasts 2–4 weeks. During the disease, there is an increase in the number of circulating white blood cells, with a predominance of lymphocytes. Many of these are large, atypical T lymphocytes. Low-grade fever and malaise may persist for weeks to months after acute illness. Complications are rare in normal hosts.

B. Cancer

EBV is associated with Burkitt lymphoma, nasopharyngeal carcinoma, Hodgkin and non-Hodgkin lymphomas, and gastric carcinoma. EBV-associated posttransplant lymphoproliferative disorders are a complication for immunodeficient patients. Sera from patients with Burkitt lymphoma or nasopharyngeal carcinoma contain elevated levels of antibody to virus-specific antigens, and the tumor tissues contain EBV DNA and express a limited number of viral genes.

Burkitt lymphoma is a tumor of the jaw in African children and young adults (see Chapter 43). Most African tumors (>90%) contain EBV DNA and express EBNA1 antigen. In other parts of the world, only about 20% of Burkitt lymphomas contain EBV DNA. It is speculated that EBV may be involved at an early stage in Burkitt lymphoma by immortalizing B cells. Malaria, a recognized cofactor, may foster enlargement of the pool of EBV-infected cells. Finally, there are characteristic chromosome translocations that involve immunoglobulin genes and result in deregulation of expression of the c-*myc* proto-oncogene.

Nasopharyngeal carcinoma is a cancer of epithelial cells and is common in males of Chinese origin. EBV DNA is regularly found in nasopharyngeal carcinoma cells, and patients have high levels of antibody to EBV. EBNA1 and LMP1 are expressed. Genetic and environmental factors are believed to be important in the development of nasopharyngeal carcinoma.

Immunodeficient patients are susceptible to EBV-induced lymphoproliferative diseases that may be fatal. From 1% to 10% of transplant patients develop an EBV-associated lymphoproliferative disorder, often when experiencing a primary infection. Aggressive monoclonal B cell lymphomas may develop.

AIDS patients are susceptible to EBV-associated lymphomas and oral hairy leukoplakia, a wart-like growth that develops on the tongue; it is an epithelial focus of EBV replication. Virtually all central nervous system non-Hodgkin lymphomas are associated with EBV, whereas less than 50% of systemic lymphomas are EBV-positive. In addition, EBV is associated with classic Hodgkin disease, with the viral genome detected in the malignant Reed-Sternberg cells in up to 50% of cases.

Immunity

EBV infections elicit an intense immune response consisting of antibodies against many virus-specific proteins, a number of cell-mediated responses, and secretion of lymphokines. Cell-mediated immunity and cytotoxic T cells are important in limiting primary infections and controlling chronic infections.

Serologic testing to determine the pattern of specific antibodies to different classes of EBV antigens is the usual means of ascertaining a patient's status with regard to EBV infection.

Laboratory Diagnosis

A. Isolation & Identification of Virus

Nucleic acid hybridization is the most sensitive means of detecting EBV in patient materials. EBER RNAs are abundantly expressed in both latently infected and lytically infected cells and provide a useful diagnostic target for detection of EBV-infected cells by hybridization. Viral antigens can be demonstrated directly in lymphoid tissues and in nasopharyngeal carcinomas. During the acute phase of infection, about 1% of circulating lymphocytes will contain EBV markers; after recovery from infection, about one in 1 million B lymphocytes will carry the virus.

EBV can be isolated from saliva, peripheral blood, or lymphoid tissue by immortalization of normal human lymphocytes, usually obtained from umbilical cord blood. This assay is laborious and time-consuming (6–8 weeks), requires specialized facilities, and is seldom performed. It is also possible to culture "spontaneously transformed" B lymphocytes from EBV DNA or virus-infected patients. Any recovered immortalizing agent is confirmed as EBV by detection of EBV DNA or virus-specific antigens in the immortalized lymphocytes.

EBV is present in the saliva of many immunosuppressed patients. Up to 20% of healthy adults will also yield virus-positive throat washings.

B. Serology

Common serologic procedures for detection of EBV antibodies include ELISA tests, immunoblot assays, and indirect immunofluorescence tests using EBV-positive lymphoid cells.

The typical pattern of antibody responses to EBV-specific antigens after a primary infection is shown in Figure 33–13. Early in acute disease, a transient rise in IgM antibodies to viral capsid antigen occurs, replaced within weeks by IgG antibodies to this antigen, which persist for life. Slightly later, antibodies to the early antigen develop that persist for several months. Several weeks after acute infection, antibodies to EBNA and the membrane antigen arise and persist throughout life.

The less-specific heterophil agglutination test may be used to diagnose EBV infections. In the course of infectious mononucleosis, most patients develop transient heterophil antibodies that agglutinate sheep cells. Commercially available spot tests are convenient. Accidental antigenic relationships provide for the specificity of this heterophil reaction.

Serologic tests for EBV antibodies require some interpretation. The presence of antibody of the IgM type to the viral capsid antigen is indicative of current infection. Antibody of the IgG type to the viral capsid antigen is a marker of past infection and indicates immunity. Early antigen antibodies are generally evidence of current viral infection, though such antibodies are often found in patients with Burkitt lymphoma or nasopharyngeal carcinoma. Antibodies to the EBNA antigens reveal past infection with EBV, though detection of a rise in anti-EBNA antibody would suggest a primary infection. Not all persons develop antibody to EBNA.

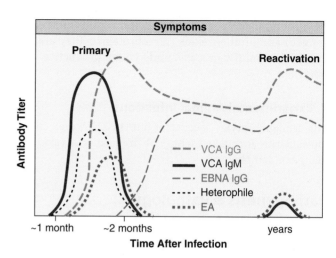

FIGURE 33–13 Typical pattern of antibody formation to EBV-specific antigens after a primary infection. Individuals with recent infection have IgM and IgG antibodies to the viral capsid antigen (VCA IgM, VCA IgG); only the IgG antibodies persist for years. Transient heterophil antibodies develop that can agglutinate sheep cells. Antibodies to early antigens (EA) develop in many patients and persist for several months. Several weeks after acute infection, antibodies to EBV nuclear-associated antigens (EBNA) and membrane antigen appear and persist for life. (Reprinted from Gulley ML, Tang W: Laboratory assays for Epstein-Barr virus-related disease. J Mol Diagnost 2008;10:279–292 with permission from the American Society for Investigative Pathology and the Association for Molecular Pathology.)

Epidemiology

EBV is common in all parts of the world, with over 90% of adults being seropositive. It is transmitted primarily by contact with oropharyngeal secretions. In developing areas, infections occur early in life; more than 90% of children are infected by age 6. These infections in early childhood usually occur without any recognizable disease. The inapparent infections result in permanent immunity to infectious mononucleosis. In industrialized nations, more than 50% of EBV infections are delayed until late adolescence and young adulthood. In almost half of cases, the infection is manifested by infectious mononucleosis. There are an estimated 100,000 cases of infectious mononucleosis annually in the United States.

Prevention, Treatment, & Control

There is no EBV vaccine available.

Acyclovir reduces EBV shedding from the oropharynx during the period of drug administration, but it does not affect the number of EBV-immortalized B cells. Acyclovir has no effect on the symptoms of mononucleosis and is of no proved benefit in the treatment of EBV-associated lymphomas in immunocompromised patients.

Adoptive transfer of EBV-reactive T cells shows promise as a treatment for EBV-related lymphoproliferative disease.

HUMAN HERPESVIRUS 6

The T-lymphotropic human herpesvirus 6 was first recognized in 1986. Initial isolations were made from cultures of peripheral blood mononuclear cells from patients with lymphoproliferative disorders.

Properties of the Virus

The viral DNA is about 160–170 kbp in size and has a mean composition of 43–44% (G + C). The genetic arrangement of the human herpesvirus 6 genome resembles that of human cytomegalovirus.

Human herpesvirus 6 appears to be unrelated antigenically to the other known human herpesviruses except for some limited cross-reactivity with human herpesvirus 7. Isolates of human herpesvirus 6 segregate into two closely related but distinct antigenic groups (designated A and B).

The virus grows well in CD4 T lymphocytes. Other cell types also support viral replication, including B cells and cells of glial, fibroblastoid, and megakaryocyte origin. Cells in the oropharynx must become infected, since virus is present in saliva. It is not known which cells in the body become latently infected. Human CD46 is the cellular receptor for the virus.

Epidemiology & Clinical Findings

Seroepidemiologic studies using immunofluorescence tests for serum antibodies or PCR assays for viral DNA in saliva or blood cells have shown that human herpesvirus 6 is widespread in the population. It is estimated that over 90% of children over age 1 and adults are virus positive.

Infections with human herpesvirus 6 typically occur in early childhood. This primary infection causes exanthem subitum (roseola infantum, or "sixth disease"), the mild common childhood disease characterized by high fever and skin rash. The 6B variant appears to be the cause of this disease. The virus is associated with febrile seizures in children.

The mode of transmission of human herpesvirus 6 is presumed to be via oral secretions. The fact that it is a ubiquitous agent suggests that it must be shed into the environment from an infected carrier.

Infections persist for life. Reactivation appears to be common in transplant patients and during pregnancy. The consequences of reactivated infection remain to be determined. Human herpesvirus 6 reactivation occurs in close to half of patients who undergo hematopoietic stem cell transplantation. Those reactivations occur soon after transplant and have been associated with delayed engraftment, central nervous system dysfunction, and increased mortality.

HUMAN HERPESVIRUS 7

A T-lymphotropic human herpesvirus, designated human herpesvirus 7, was first isolated in 1990 from activated T cells recovered from peripheral blood lymphocytes of a healthy individual.

Human herpesvirus 7 is immunologically distinct from human herpesvirus 6, though they share about 50% homology at the DNA level.

Human herpesvirus 7 appears to be a ubiquitous agent, with most infections occurring in childhood but later than the very early age of infection noted with human herpesvirus 6. Persistent infections are established in salivary glands, and the virus can be isolated from saliva of most individuals. In a longitudinal study of healthy adults, 75% of subjects excreted infectious virus in saliva one or more times during a 6-month observation period. Similar to human herpesvirus 6, primary infection with human herpesvirus 7 has been linked with roseola infantum in infants and young children. Any other disease associations of human herpesvirus 7 remain to be established.

HUMAN HERPESVIRUS 8

A new herpesvirus, designated human herpesvirus 8 and also called KSHV, was first detected in 1994 in Kaposi sarcoma specimens. KSHV is lymphotropic and is more closely related to EBV and herpesvirus saimiri than to other known herpesviruses. The KSHV genome (about 165 kbp) contains numerous genes related to cellular regulatory genes involved in cell proliferation, apoptosis, and host responses (cyclin D, cytokines, chemokine receptor) that presumably contribute to viral pathogenesis. This molecular piracy of cell regulatory genes is a striking feature of the virus. KSHV is the cause of Kaposi sarcomas, vascular tumors of mixed cellular composition, and is involved in the pathogenesis of body cavity-based lymphomas occurring in AIDS patients and of multicentric Castleman disease.

KSHV is not as ubiquitous as other herpesviruses; about 5% of the general population in the United States and northern Europe have serologic evidence of KSHV infection. Contact with oral secretions is likely the most common route of transmission. The virus can also be transmitted sexually, vertically, by blood, and through organ transplants. Viral DNA has also been detected in breast-milk samples in Africa. Infections are common in Africa (>50%) and are acquired early in life.

Viral DNA can be detected in patient specimens using PCR assays. Direct virus culture is difficult and impractical. Serologic assays are available to measure persistent antibody to KSHV, using indirect immunofluorescence, Western blot, and ELISA formats.

Foscarnet, ganciclovir, and cidofovir have activity against KSHV replication. The rate of new Kaposi sarcomas is markedly reduced in HIV-positive patients on effective antiretroviral therapy, probably reflecting reconstituted immune surveillance against KSHV-infected cells.

B VIRUS

Herpes B virus of Old World monkeys is highly pathogenic for humans. Transmissibility of virus to humans is limited, but those infections that do occur are associated with a high mortality rate (about 60%). B virus disease of humans is an acute ascending myelitis and encephalomyelitis.

Properties of the Virus

B virus is a typical herpesvirus that is indigenous in macaques, Old World monkeys in Asia. B virus is enzootic in rhesus, cynomolgus, and other macaque monkeys (genus *Macaca*). It is designated cercopithecine herpesvirus 1, replacing the older name of *Herpes simiae*. Its genome organization is similar to that of HSV, with many genes arranged colinearly. Its genome is 75% G + C, the highest among herpesviruses. As with all herpesviruses, B virus establishes latent infections in infected hosts. The virus grows well in cultures of monkey kidney, rabbit kidney, and human cells with a short growth cycle. Cytopathic effects are similar to those of HSV.

Pathogenesis & Pathology

B virus infections seldom cause disease in rhesus monkeys. Vesicular lesions of the oropharynx may occur and resemble those induced in humans by HSV. Genital lesions also occur. Many rhesus monkeys carry latent B virus infections that may be reactivated by conditions of stress.

The virus is transmissible to other monkeys, rabbits, guinea pigs, rats, and mice. Rabbits routinely develop fatal infections after B virus inoculation.

B virus infections in humans usually result from a monkey bite, though infection by the respiratory route or ocular splash exposure is possible. The striking feature of B virus infections in humans is the very strong propensity to cause neurologic disease. Many survivors are left with neurologic impairment.

Epidemiology & Clinical Findings

B virus is transmitted by direct contact with virus or virus-containing material. Transmission occurs among *Macaca* monkeys, between monkeys and humans, and rarely from humans to humans. Virus may be present in saliva, conjunctival and vesicular fluids, genital areas, and feces of monkeys. Respiratory transmission can occur. Other sources of infection include direct contact with animal cages and with infected monkey cell cultures.

Infection in the natural host is rarely associated with obvious disease. Infections with B virus are very common in colonies of rhesus monkeys. Seroprevalence in adult animals is 70% or higher. As latent infections may be reactivated, seropositive animals are reservoirs for transmission of B virus infections. The frequency of excretion of B virus by monkeys is probably no more than 3%.

Animal workers and persons handling macaque monkeys are at risk of acquiring B virus infection, including medical researchers, veterinarians, pet owners, and zoo workers.

Individuals having intimate contact with animal workers exposed to the monkeys are also at some risk.

Treatment & Control

There is no specific treatment once the clinical disease is manifest. However, treatment with acyclovir is recommended immediately after exposure. γ-Globulin has not proved to be effective treatment for human B virus infections. No vaccine is available.

The risk of B virus infections can be reduced by proper procedures in the laboratory and in the handling and management of macaque monkeys. This risk makes macaques unsuitable as pets.

REVIEW QUESTIONS

1. A previously healthy 3-year-old boy develops a classic viral childhood illness. Which of the following primary viral infections of childhood is usually symptomatic?

    (A) Cytomegalovirus
    (B) EBV
    (C) Hepatitis B virus
    (D) Varicella-zoster virus
    (E) Parvovirus B19

2. Which one of the following is a recommended therapy for HSV genital infection?

    (A) Acyclovir
    (B) Attenuated live virus vaccine
    (C) Herpes immune globulin
    (D) Interferon α
    (E) Ribavirin

3. Most herpesvirus infections are endemic worldwide. Which one of the following viruses shows marked geographic differences in seroprevalence?

    (A) Cytomegalovirus
    (B) EBV
    (C) HSV-2
    (D) KSHV
    (E) Varicella-zoster virus

4. A 19-year-old female college student has fever, sore throat, and lymphadenopathy accompanied by lymphocytosis with atypical cells and an increase in sheep cell agglutinins. The diagnosis is most likely

    (A) Infectious hepatitis
    (B) Infectious mononucleosis
    (C) Chickenpox
    (D) Herpes simplex infection
    (E) Viral meningitis

5. A Tzanck smear of a scraping obtained from a vesicle on the skin demonstrates multinucleated giant cells. Multinucleated giant cells are associated with which of the following viruses?

    (A) Varicella-zoster
    (B) Variola major
    (C) Coxsackievirus
    (D) Molluscum contagiosum

6. Which of the following statements about betaherpesviruses is not true?

   (A) They establish latent infections and persist indefinitely in infected hosts

   (B) They are reactivated in immunocompromised patients

   (C) Most infections are subclinical

   (D) They can infect lymphoid cells

   (E) They have short, cytolytic growth cycles in cultured cells

7. A 28-year-old woman has recurrent genital herpes. Which of the following statements about genital herpes infections is true?

   (A) Reactivation of latent virus during pregnancy poses no threat to the newborn

   (B) Virus cannot be transmitted in the absence of apparent lesions

   (C) Recurrent episodes due to reactivation of latent virus tend to be more severe than the primary infection

   (D) They can be caused by either HSV-1 or HSV-2

   (E) Latent HSV can be found in dendritic cells

8. Which of the following viruses causes a mononucleosis-like syndrome and is excreted in the urine?

   (A) Cytomegalovirus

   (B) EBV

   (C) Human herpesvirus 6

   (D) Varicella-zoster virus

   (E) HSV-2

9. A 53-year-old woman develops fever and focal neurologic signs. Magnetic resonance imaging shows a left temporal lobe lesion. Which of the following tests would be most appropriate to confirm a diagnosis of herpes simplex encephalitis in this patient?

   (A) Brain biopsy

   (B) Tzanck smear

   (C) PCR assay for viral DNA in cerebrospinal fluid

   (D) Serologic test for viral IgM antibody

10. Which of the following tumors is caused by a virus other than EBV?

    (A) Posttransplant lymphomas

    (B) Hodgkin disease

    (C) Kaposi sarcoma

    (D) AIDS-related central nervous system non-Hodgkin lymphomas

    (E) Burkitt lymphoma

11. An outbreak of a rash called "mat herpes" occurred among high school students who had competed in a wrestling tournament. Which of the following statements is most accurate?

    (A) The rash is not contagious among wrestlers

    (B) Causative agent is HSV-1

    (C) Causative agent is varicella-zoster

    (D) Lesions typically last one month or longer

    (E) Students should be vaccinated before participating in wrestling tournaments

12. The shingles vaccine is recommended for which of the following groups?

    (A) Healthy adolescents

    (B) Individuals over age 60

    (C) Pregnant women

    (D) Those who never had chickenpox

13. The most common congenital infection is caused by

    (A) Varicella-zoster virus

    (B) HSV-2

    (C) Human herpesvirus 8 (KSHV)

    (D) Cytomegalovirus

    (E) Parvovirus

14. Which of the following groups are at increased risk for herpes zoster?

    (A) Persons at advanced age

    (B) Patients with atopic dermatitis

    (C) Pregnant women

    (D) Persons who have been vaccinated with varicella vaccine

    (E) Infants with congenital infections

Answers

1. D	5. A	9. C	13. D
2. A	6. E	10. C	14. A
3. D	7. D	11. B	
4. B	8. A	12. B	

REFERENCES

Ashley RL, Wald A: Genital herpes: Review of the epidemic and potential use of type-specific serology. Clin Microbiol Rev 1999;12:1. [PMID: 9880471]

Espy MJ et al: Real-time PCR in clinical microbiology: Applications for routine laboratory testing. Clin Microbiol Rev 2006;19:165. [PMID: 16418529]

Gulley ML, Tang W: Laboratory assays for Epstein-Barr virus-related disease. J Mol Diagnost 2008;10:279. [PMID: 18556771]

Hassan J, Connell J: Translational mini-review series on infectious disease: Congenital cytomegalovirus infection: 50 years on. Clin Exp Immunol 2007;149:205. [PMID: 17635529]

Hengel H, Brune W, Koszinowski UH: Immune evasion by cyto-megalovirus—Survival strategies of a highly adapted opportunist. Trends Microbiol 1998;6:190. [PMID: 9614343]

Huff JL, Barry PA: B-virus (Cercopithecine herpesvirus 1) infection in humans and macaques: Potential for zoonotic disease. Emerging Infect Dis 2003;9:246. [PMID: 12603998]

Kimberlin DW, Whitley RJ: Human herpesvirus-6: Neurologic implications of a newly-described viral pathogen. J Neurovirol 1998;4:474. [PMID: 9839645]

Knipe DM et al (editors): Herpesviridae. In: Fields Virology, 5th ed. Lippincott Williams & Wilkins, 2007. [9 chapters]

Prevention of herpes zoster. Recommendations of the Advisory Committee on Immunization Practices (ACIP). MMWR Morb Mortal Wkly Rep 2008;57(RR-5):1.

Prevention of varicella. Recommendations of the Advisory Committee on Immunization Practices (ACIP). MMWR Morb Mortal Wkly Rep 1996;45(RR-11):1.

Poxviruses

Poxviruses are the largest and most complex of viruses. The family encompasses a large group of agents that are similar morphologically and share a common nucleoprotein antigen. Infections with most poxviruses are characterized by a rash, although lesions induced by some members of the family are markedly proliferative. The group includes variola virus, the etiologic agent of smallpox—the viral disease that has affected humans throughout recorded history.

Even though smallpox was declared eradicated from the world (in 1980) after an intensive campaign coordinated by the World Health Organization, there is concern that the virus could be reintroduced as a biologic weapon. There is a continuing need to be familiar with vaccinia virus (used for smallpox vaccinations) and its possible complications in humans. It is also necessary to be aware of other poxvirus diseases that may resemble smallpox and must be differentiated from it by laboratory means. Lastly, vaccinia virus is under intensive study as a vector for introducing active immunizing genes as live-virus vaccines for a variety of viral diseases of humans and domestic animals.

PROPERTIES OF POXVIRUSES

Important properties of the poxviruses are listed in Table 34–1.

Structure & Composition

Poxviruses are large enough to be seen as featureless particles by light microscopy. By electron microscopy, they appear to be brick-shaped or ellipsoid particles measuring about 300–400 × 230 nm. Their structure is complex and conforms to neither icosahedral nor helical symmetry. The external surface of particles contains ridges. There is an outer lipoprotein membrane, or envelope, that encloses a core and two structures of unknown function called lateral bodies (Figure 34–1).

The core contains the large viral genome of linear double-stranded DNA (130–375 kbp). The complete genomic sequence is known for several poxviruses, including vaccinia and variola. The vaccinia genome contains about 185 open reading frames. The DNA contains inverted terminal repeats of variable length, and the strands are connected at the ends by terminal hairpin loops. The inverted terminal repeats may include coding regions, so some genes are present at both ends of the genome. The DNA is rich in adenine and thymine bases.

The chemical composition of a poxvirus resembles that of a bacterium. Vaccinia virus is composed predominantly of protein (90%), lipid (5%), and DNA (3%). More than 100 structural polypeptides have been detected in virus particles. A number of the proteins are glycosylated or phosphorylated. The lipids are cholesterol and phospholipids.

The virion contains a multiplicity of enzymes, including a transcriptional system that can synthesize, polyadenylate, cap, and methylate viral mRNA.

Classification

Poxviruses are divided into two subfamilies based on whether they infect vertebrate or insect hosts. The vertebrate poxviruses fall into nine genera, with the members of a given genus displaying similar morphology and host range as well as some antigenic relatedness.

TABLE 34–1 Important Properties of Poxviruses

Virion: Complex structure, oval or brick-shaped, 300–400 nm in length × 230 nm in diameter; external surface shows ridges; contains core and lateral bodies

Composition: DNA (3%), protein (90%), lipid (5%)

Genome: Double-stranded DNA, linear; size 130–375 kbp; has terminal loops; has low G + C content (30–40%) except for *Parapoxvirus* (63%)

Proteins: Virions contain more than 100 polypeptides; many enzymes are present in core, including transcriptional system

Envelope: Virion assembly involves formation of multiple membranes

Replication: Cytoplasmic factories

Outstanding characteristics:
Largest and most complex viruses; very resistant to inactivation
Virus-encoded proteins help evade host immune defense system
Smallpox was the first viral disease eradicated from the world

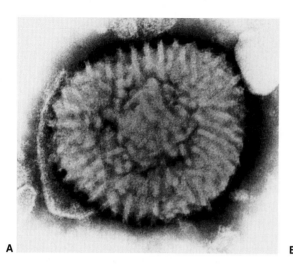
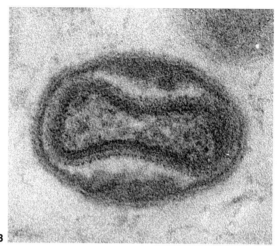

FIGURE 34–1 Electron micrographs of vaccinia (*Orthopoxvirus*) virions. **A:** Negatively stained particle showing ridges or tubular elements covering the surface (228,000×). (Reproduced with permission from Dales S: J Cell Biol 1963;18:51.) **B:** Thin section of vaccinia virion showing a central biconcave core, two lateral bodies, and an outer membrane (220,000×). (Reproduced with permission from Pogo BGT, Dales S: Proc Natl Acad Sci USA 1969;63:820.)

Most of the poxviruses that can cause disease in humans are contained in the genera *Orthopoxvirus* and *Parapoxvirus*; there are also several that are classified in the genera *Yatapoxvirus* and *Molluscipoxvirus* (Table 34–2).

The orthopoxviruses have a broad host range affecting several vertebrates. They include ectromelia (mousepox), camelpox, cowpox, monkeypox, vaccinia, and variola (smallpox) viruses. The last four are infectious for humans. Vaccinia virus differs only in minor morphologic respects from variola and cowpox viruses. It is the prototype of poxviruses in terms of structure and replication. Monkeypox can infect rodents, monkeys, and humans and may resemble smallpox clinically.

Some poxviruses have a restricted host range and infect only rabbits (fibroma and myxoma) or only birds. Others infect mainly sheep and goats (sheeppox, goatpox) or cattle (pseudocowpox, or milker's nodule).

Parapoxviruses are morphologically distinctive. Compared to the orthopoxviruses, parapoxviruses are somewhat smaller particles (260 × 160 nm), and their surfaces exhibit a crisscross pattern (Figure 34–2). Their genomes are smaller (about 135 kbp) and have a higher guanine plus cytosine content (63%) than those of the orthopoxviruses (170–250 kbp; G + C, 30–40%).

All vertebrate poxviruses share a common nucleoprotein antigen in the inner core. There is serologic cross-reactivity among viruses within a given genus but very limited reactivity across genera. Consequently, immunization with vaccinia virus affords no protection against disease induced by parapoxviruses or the unclassified poxviruses.

TABLE 34–2   Poxviruses Causing Disease in Humans

Genus	Virus	Primary Host	Disease
Orthopoxvirus	Variola	Humans	Smallpox (now eliminated)
	Vaccinia	Humans	Localized lesion; used for smallpox vaccination
	Buffalopox	Water buffalo	Human infections rare; localized lesion
	Monkeypox	Rodents, monkeys	Human infections rare; generalized disease
	Cowpox	Cows	Human infections rare; localized ulcerating lesion
Parapoxvirus	Orf	Sheep	Human infections rare; localized lesion
	Pseudocowpox	Cows	
	Bovine papular stomatitis	Cows	
Molluscipoxvirus	Molluscum contagiosum	Humans	Many benign skin nodules
Yatapoxvirus	Tanapox	Monkeys	Human infections rare; localized lesion
	Yabapox	Monkeys	Human infections very rare and accidental; localized skin tumors

Poxvirus Replication

The replication cycle of vaccinia virus is summarized in Figure 34–3. Poxviruses are unique among DNA viruses in that the entire multiplication cycle takes place in the cytoplasm of infected cells. It is possible, however, that nuclear factors may be involved in transcription and virion assembly. Poxviruses are further distinguished from all other animal viruses by the fact that the uncoating step requires a newly synthesized, virus-encoded protein.

A. Virus Attachment, Penetration, & Uncoating

Virus particles establish contact with the cell surface and fuse with the cell membrane. Some particles may appear within vacuoles. Viral cores are released into the cytoplasm. Among the several enzymes inside the poxvirus particle, there is a viral RNA polymerase that transcribes about half the viral genome into early mRNA. These mRNAs are transcribed within the viral core and are then released into the cytoplasm. Because the necessary enzymes are contained within the viral core, early transcription is not affected by inhibitors of protein synthesis. The "uncoating" protein that acts on the cores is among the more than 50 polypeptides made early after infection. The second-stage uncoating step liberates viral DNA from the cores; it requires both RNA and protein synthesis. The synthesis of host cell macromolecules is inhibited at this stage.

Poxviruses inactivated by heat can be reactivated either by viable poxviruses or by poxviruses inactivated by nitrogen mustards (which inactivate the DNA). This process is called **nongenetic reactivation** and is due to the action of the uncoating protein. Heat-inactivated virus alone cannot

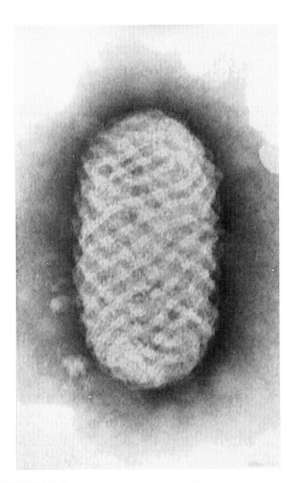

FIGURE 34–2 Electron micrograph of orf virus (*Parapoxvirus*). Note distinctive crisscross pattern of surface of virion (200,000×). (Courtesy of FA Murphy and EL Palmer.)

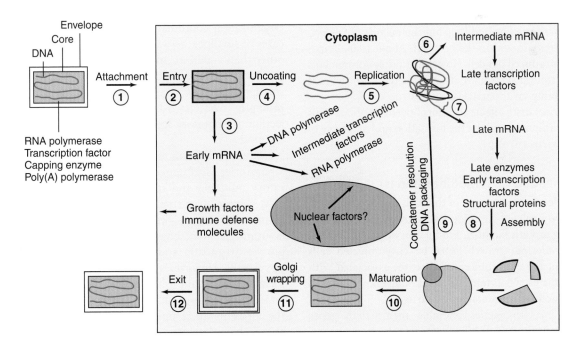

FIGURE 34–3 Outline of replication cycle of vaccinia virus. (Reproduced with permission from Moss B: Poxviridae: The viruses and their replication. In: *Fields Virology.* Fields BN et al [editors]. Lippincott-Raven, 1996.)

cause second-stage uncoating because of the heat lability of the RNA polymerase. Apparently, the heat-killed virus provides the template and the second virus provides the enzymes needed for transcription. Any vertebrate poxvirus can reactivate any other vertebrate poxvirus.

B. Replication of Viral DNA & Synthesis of Viral Proteins

Among the early proteins made after vaccinia virus infection are enzymes involved in DNA replication, including a DNA polymerase and thymidine kinase. Viral DNA replication occurs in the cytoplasm and appears to be accomplished by viral coded enzymes. Viral DNA replication starts soon after the release of viral DNA in the second stage of uncoating. It occurs 2–6 hours after infection in discrete areas of the cytoplasm, which appear as "factories" or inclusion bodies in electron micrographs. The number of inclusion bodies per cell is proportionate to the multiplicity of infection, suggesting that each infectious particle can induce a "factory." High rates of homologous recombination occur within poxvirus-infected cells. This has been exploited experimentally to construct and map mutations.

The pattern of viral gene expression changes markedly with the onset of replication of viral DNA. The synthesis of many of the early proteins is inhibited. There is a small intermediate class of genes whose expression temporally precedes the expression of the late class of genes. Late viral mRNA is translated into large amounts of structural proteins and small amounts of other viral proteins and enzymes.

C. Maturation

The assembly of the virus particle from the manufactured components is a complex process. Some of the particles are released from the cell by budding, but the majority of poxvirus particles remain within the host cell. About 10,000 virus particles are produced per cell. How the multiple components of the transcription system are incorporated within the core of the assembling virus particle is unknown.

An antiviral drug affects the morphogenesis of poxvirus particles. Rifampin can block the formation and assembly of the vaccinia virus envelope.

D. Virus-Encoded Host Modifier Genes

A polypeptide encoded by one of the early genes of vaccinia virus is closely related to epidermal growth factor and to transforming growth factor-α. Production of growth factors similar to epidermal growth factor by virus-infected cells could account for the proliferative diseases associated with members of the poxvirus family such as Shope fibroma, Yaba tumor, and molluscum contagiosum viruses.

Several poxvirus genes resemble mammalian genes for proteins that would inhibit host defense mechanisms. Examples include tumor necrosis factor receptor, γ interferon receptor, IL-1 receptor, and a complement-binding protein. These poxvirus-encoded host defense modifiers presumably counter the complement and cytokine networks important in the host immune response to viral infection, allowing enhanced virus replication and, perhaps, facilitating virus transmission.

POXVIRUS INFECTIONS IN HUMANS: VACCINIA & VARIOLA

Control & Eradication of Smallpox

Control of smallpox by deliberate infection with mild forms of the disease was practiced for centuries. This process, called variolation, was dangerous but decreased the disastrous effects of major epidemics, reducing the case-fatality rate from 25% to 1%. Edward Jenner introduced vaccination with live cowpox virus in 1798.

In 1967, the World Health Organization introduced a worldwide campaign to eradicate smallpox. Epidemiologic features of the disease (described below) made it feasible to attempt total eradication. At that time, there were 33 countries with endemic smallpox and 10–15 million cases per year. The last Asiatic case occurred in Bangladesh in 1975, and the last natural victim was diagnosed in Somalia in 1977. Smallpox was officially declared eliminated in 1980. There were three main reasons for this outstanding success: The vaccine was easily prepared, stable, and safe; it could be given simply by personnel in the field; and mass vaccination of the world population was not necessary. Cases of smallpox were traced, and contacts of the patient and those in the immediate area were vaccinated.

Even though there has been no evidence of smallpox transmission anywhere in the world, the World Health Organization coordinated the investigation of 173 possible cases of smallpox between 1979 and 1984. All were diseases other than smallpox, most commonly chickenpox or other illnesses that produce a rash. Even so, a suspected case of smallpox becomes a public health emergency and must be promptly investigated by means of clinical evaluation, collection of laboratory specimens, and preliminary laboratory diagnosis.

The presence of stocks of virulent smallpox virus in laboratories is of concern because of the danger of laboratory infection and subsequent spread into the community. Variola virus stocks supposedly were destroyed in all laboratories except two World Health Organization collaborating centers (one in Atlanta and one in Moscow) that pursue diagnostic and research work on variola-related poxviruses. However, in the 1990s it was learned that the former Soviet Union had used smallpox virus in its biologic warfare program. How many countries may possess the virus today is unknown. Smallpox virus is considered to be a dangerous potential biothreat agent. Because of the worldwide eradication of variola virus and subsequent discontinuation of vaccination programs, today's human population possesses low or nonexistent smallpox immunity and thus is highly susceptible to infection with smallpox virus.

Research scientists may obtain portions of the variola virus genome from the collaborating centers but not a complete genome. The distribution, synthesis, and handling of variola virus DNA is governed by recommendations from the World Health Organization.

Comparison of Vaccinia & Variola Viruses

Vaccinia virus, the agent used for smallpox vaccination, is a distinct species of *Orthopoxvirus*. Restriction endonuclease maps of the genome of vaccinia virus are distinctly different from those of cowpox virus, which was believed to be its ancestor. At some time after Jenner's original use of "cowpox" virus, the vaccine virus became "vaccinia virus"; the time and reasons for the change are not known. Vaccinia virus may be the product of genetic recombination, a new species derived from cowpox virus or variola virus by serial passage, or the descendant of a now extinct viral genus.

Variola has a narrow host range (only humans and monkeys), whereas vaccinia has a broad host range that includes rabbits and mice. Some strains of vaccinia can cause a severe disease in laboratory rabbits that has been called rabbitpox. Vaccinia virus has also infected cattle and water buffalo, and the disease in buffalo has persisted in India (buffalopox). Both vaccinia and variola viruses grow on the chorioallantoic membrane of the 10- to 12-day-old chick embryo, but the latter produce much smaller pocks. Both grow in several types of chick and primate cell lines.

The nucleotide sequences of variola (186 kb) and vaccinia (192 kb) are similar, with the most divergence in terminal regions of the genomes. Of 187 putative proteins, 150 were markedly similar in sequence between the two viruses; the remaining 37 diverged or were variola-specific and may represent potential virulence determinants. The sequences do not reveal variola virus origins or explain its strict human host range or its particular virulence.

Pathogenesis & Pathology of Smallpox

Although smallpox has been eradicated, the pathogenesis of the disease (described here in the past tense) is instructive for other poxvirus infections. The pathogenesis of mousepox is illustrated in Figure 30–3.

The portal of entry of variola virus was the mucous membranes of the upper respiratory tract. After viral entry, the following are believed to have taken place: (1) primary multiplication in the lymphoid tissue draining the site of entry; (2) transient viremia and infection of reticuloendothelial cells throughout the body; (3) a secondary phase of multiplication in those cells, leading to (4) a secondary, more intense viremia; and (5) the clinical disease.

In the preeruptive phase, the disease was barely infective. By the sixth to ninth day, lesions in the mouth tended to ulcerate and discharge virus. Thus, early in the disease, infectious virus originated in lesions in the mouth and upper respiratory tract. Later, pustules broke down and discharged virus into the environment of the smallpox patient.

Histopathologic examination of the skin showed proliferation of the prickle-cell layer. Those proliferated cells contained many cytoplasmic inclusions. There was infiltration with mononuclear cells, particularly around the vessels in the corium. Epithelial cells of the malpighian layer became swollen through distention of cytoplasm and underwent "ballooning degeneration." The vacuoles in the cytoplasm enlarged. The cell membrane broke down and coalesced with neighboring, similarly affected cells, resulting in the formation of vesicles. The vesicles enlarged and then became filled with white cells and tissue debris. All the layers of the skin were involved, and there was actual necrosis of the corium. Thus, scarring occurred after variola infection. Similar histopathology is seen with vaccinia, though vaccinia virus ordinarily causes localized pustular lesions only at the site of inoculation.

Clinical Findings

The incubation period of variola (smallpox) was 10–14 days. The onset was usually sudden. One to 5 days of fever and malaise preceded the appearance of the exanthems, which began as macules, then papules, then vesicles, and finally pustules. These formed crusts that fell off after about 2 weeks, leaving pink scars that faded slowly. In each affected area, the lesions were generally found in the same stage of development (in contrast to chickenpox).

A "Smallpox Recognition Card" prepared by the World Health Organization shows the typical rash (Figure 34–4). Lesions were most abundant on the face and less so on the trunk. In severe cases, the rash was hemorrhagic. The case-fatality rate varied from 5% to 40%. In mild variola, called variola minor, or in vaccinated persons, the mortality rate was under 1%.

Immunity

All viruses within the *Orthopoxvirus* genus are so closely related antigenically that they cannot be easily differentiated serologically. Infection with one induces an immune response that reacts with all other members of the group.

An attack of smallpox gave complete protection against reinfection. Vaccination with vaccinia induced immunity against variola virus for at least 5 years and sometimes longer. Antibodies alone are not sufficient for recovery from primary poxvirus infection. In the human host, neutralizing antibodies develop within a few days after onset of smallpox but do not prevent progression of lesions, and patients may die in the pustular stage with high antibody levels. Cell-mediated immunity is probably more important than circulating antibody. Patients with hypogammaglobulinemia generally react normally to vaccination and develop immunity despite the apparent absence of antibody. Patients who have defects

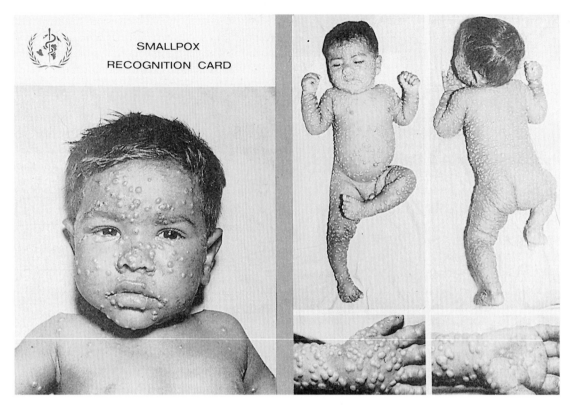

FIGURE 34–4 Smallpox rash. A "Smallpox Recognition Card" from the World Health Organization illustrates the distribution and nature of the typical rash of smallpox in an unvaccinated child. (Courtesy of F Fenner and the World Health Organization.)

in both cellular immune response and antibody response develop a progressive, usually fatal disease upon vaccination.

Production of interferon (see Chapter 30) is another possible immune mechanism. Irradiated animals without detectable antibody or delayed hypersensitivity recovered from vaccinia infection as rapidly as untreated control animals.

Laboratory Diagnosis

Several tests are available to confirm the diagnosis of smallpox. Now that the disease is presumably eradicated, it is important to diagnose any cases that resemble smallpox. The tests depend upon identification of viral DNA or antigen from the lesion, direct microscopic examination of material from skin lesions, recovery of virus from the patient, and, least importantly, demonstration of antibody in the blood.

A. Isolation & Identification of Virus

Skin lesions are the specimen of choice for viral detection and isolation. Poxviruses are stable and will remain viable in specimens for weeks even without refrigeration.

Direct examination of clinical material in the electron microscope is used for rapid identification of virus particles (in about 1 hour) and can readily differentiate a poxvirus infection from chickenpox (the latter is caused by a herpesvirus). Orthopoxviruses cannot be distinguished from one another by electron microscopy because they are similar in size and morphology. However, they can be easily differentiated from tanapoxvirus and parapoxviruses.

Polymerase chain reaction (PCR) tests that are specific for various poxviruses are available and can be used for detection and identification purposes.

Viral antigen can be detected by immunohistochemistry in tissues and in material collected from skin lesions. Many antigens are cross-reactive and identify orthopoxviruses as a group. The use of PCR or restriction enzyme cleavage of viral DNA or the analysis of polypeptides in poxvirus-infected cells can demonstrate distinct characteristics for variola, vaccinia, monkeypox, and cowpox. Smallpox-like illnesses must be identified to ascertain that variola has indeed been eradicated and has not reappeared.

Virus isolation is carried out by inoculation of vesicular fluid onto the chorioallantoic membrane of chick embryos. This test is the easiest way of distinguishing cases of smallpox from generalized vaccinia, for the lesions produced by these viruses on the membrane differ markedly. In 2–3 days, vaccinia pocks are large with necrotic centers whereas variola pocks are much smaller. Cowpox and monkeypox produce distinctive hemorrhagic lesions. The parapoxviruses, molluscum contagiosum virus, and tanapoxvirus do not grow on the membrane.

Cell cultures can also be used for virus isolation. Human and nonhuman primate cells are most susceptible. The orthopoxviruses grow well in cultured cells; parapoxviruses and tanapoxvirus grow less well, and molluscum contagiosum virus cannot be cultured.

B. Serology

Virus isolation is necessary for quick and accurate identification of poxvirus infections. However, antibody assays can be used to confirm a diagnosis. Antibodies appear after the first week of infection that can be detected by HI, Nt, ELISA, RIA, or immunofluorescence tests. None of these tests will distinguish among the orthopoxviruses.

Treatment

Vaccinia immune globulin is prepared from blood from persons vaccinated with the vaccinia virus. It is recommended for treatment of all complications except postvaccinial encephalitis. As of 2003, stocks of vaccinia immune globulin were very limited and available only from the Centers for Disease Control and Prevention.

Methisazone is a chemotherapeutic agent of some value against poxviruses. It is effective as prophylaxis but is not useful in treatment of established disease. Cidofovir, a nucleotide analog, shows activity against poxviruses in vitro and in vivo.

Studies in monkeys have revealed that antiviral treatment is more effective than postexposure smallpox vaccination at reducing mortality from lethal virus infection.

Epidemiology

Transmission of smallpox occurred by contact between cases. Smallpox was highly contagious. The virus was stable in the extracellular environment but was most commonly transmitted by respiratory spread. The dried virus in crusts from skin lesions could survive on clothes or other materials and result in infections.

Patients were most highly infectious during the first week of rash once fever had begun. Respiratory droplets were infectious earlier than skin lesions.

The following epidemiologic features made smallpox amenable to total eradication: There was no known nonhuman reservoir. There was one stable serotype. There was an effective vaccine. Subclinical infectious cases did not occur. Chronic asymptomatic carriage of the virus did not occur. Since virus in the environment of the patient derived from lesions in the mouth and throat (and later in the skin), patients with infection sufficiently severe to transmit the disease were likely to be so ill that they quickly reached the attention of medical authorities. The close contact required for effective spread of the disease generally made for ready identification of a patient's contacts so that specific control measures could be instituted to interrupt the cycle of transmission.

The World Health Organization was successful in eradicating smallpox by using a surveillance-containment program. The source of each outbreak was determined, and all susceptible contacts were identified and vaccinated.

Vaccination with Vaccinia

Vaccinia virus for vaccination is prepared from vesicular lesions ("lymph") produced in the skin of calves, or it can be grown in chick embryos. The final vaccine contains 40% glycerol to stabilize the virus and 0.4% phenol to destroy bacteria. World Health Organization standards require that smallpox vaccines have a potency of no fewer than 10^8 pock-forming units per milliliter. A new cell culture-produced vaccine is under development. The vaccinia vaccine does not contain smallpox (variola) virus.

The success of smallpox eradication has meant that routine vaccination is no longer recommended. Routine smallpox vaccination of children in the United States was stopped in 1971. Hence, all those born from 1972 on are susceptible to infection.

Vaccinia virus is used in research and has resulted in laboratory-acquired infections. Current recommendations are that laboratory workers who handle cultures or animals infected with vaccinia or other orthopoxviruses that infect humans should be vaccinated. Recent concerns about a possible terrorist attack involving smallpox have resulted in recommendations for using smallpox vaccine on a limited scale (eg, starting with health care workers).

The following summary of vaccination is given also because vaccinia virus is under consideration as a vector for introducing foreign genes for immunization purposes and as oncolytic viral therapy for cancer.

A. Time of Vaccination

Complications of vaccination (see below) occur most commonly under the age of 1 year. Therefore, vaccinating between 1 and 2 years of age is preferable to vaccinating in the first year of life. Revaccination has been done at 3-year intervals.

B. Reactions & Interpretations

1. Primary take—In the fully susceptible person, a papule surrounded by hyperemia appears on the third or fourth day. The papule increases in size until vesiculation appears (on the fifth or sixth day). The vesicle reaches its maximum size by the ninth day and then becomes pustular. Desiccation follows and is complete in about 2 weeks, leaving a depressed pink scar that ultimately turns white. The reading of the result is usually done on the seventh day. A person is considered fully protected after a vesicular or pustular response surrounding a central lesion (scab or ulcer) occurs. If this reaction is not observed, vaccination should be repeated.

2. Revaccination—A successful revaccination shows in 6–8 days a vesicular or pustular lesion or an area of palpable induration surrounding a central lesion, which may be a scab or an ulcer. Only this reaction indicates with certainty that viral multiplication has taken place. Equivocal reactions may represent immunity but may also represent merely allergic reactions to a vaccine that has become inactivated. When an equivocal reaction occurs, the revaccination should be repeated using a new lot of vaccine.

C. Adverse Reactions of Vaccination

Smallpox vaccination was associated with a definite measurable risk. In the United States, the risk of death from all complications was 1 per million for primary vaccinees and 0.6 per million for revaccinees. For children under 1 year of age, the risk of death was 5 per million primary vaccinations. Severe complications of vaccination occurred in conjunction with immunodeficiency, immunosuppression, malignancies, and pregnancy. Those conditions are contraindications for vaccinia vaccine use, as well as eczema, allergy to a vaccine component, and living in a household with someone having a vaccination contraindication.

1. Inadvertent autoinoculation—This occurs when a part of the body distant from the inoculation site becomes infected through scratching or through inanimate objects such as clothing. This is the most common complication, occurring about 25 times per million. Ocular vaccinia was the most frequent problem and sometimes resulted in residual visual defects. The most common nonocular sites were the face, nose, mouth, lips, and genitalia.

2. Contact transmission—Contact transmission of vaccinia virus occurs when virus is transferred from a vaccinee to a close contact. Virus can be shed until the scab heals. Virus can survive for several days on clothing, bedding, and other inanimate objects. Infection acquired through contact transmission can result in the same adverse reactions as from vaccination.

3. Eczema vaccinatum—This is a localized or generalized rash syndrome that can occur anywhere on the body. Persons with atopic dermatitis (ie, eczema) are at highest risk, with the disease most severe in young children (Figure 34–5). Eczema vaccinatum may occur concurrently with the development of the local vaccinial lesion in vaccinees and up to 3 weeks after exposure in close contacts of the vaccinee. This condition has a high mortality; patients benefit from treatment with vaccinia immune globulin.

4. Generalized vaccinia—This is manifested by the occurrence of crops of vaccinial lesions anywhere on the body 4 days or more after vaccination (23 cases per million). The skin lesions are thought to contain virus spread by the hematogenous route. Generalized vaccinia is usually self-limited in immunocompetent hosts, but is often more severe in persons with immunodeficiency.

5. Progressive vaccinia—This rare and severe complication results when a vaccination site fails to heal and vaccinia virus replication persists. It occurs in persons with underlying humoral or cellular immune deficiency and is often fatal. The incidence of disease in the United States was about 1 per million. Congenital or acquired immunodeficiency and immunosuppression are contraindications to vaccination.

6. Fetal vaccinia—Very rarely, a woman vaccinated any time during pregnancy may transmit vaccinia virus to the fetus, usually resulting in stillbirth. Therefore, vaccination should be withheld during pregnancy.

7. Postvaccinial central nervous system disease—This rare, serious reaction is an inflammation of the parenchyma of the central nervous system after smallpox vaccination, such as postvaccinial encephalitis. The mortality rate of this serious complication is high. The incidence in the United States was about 3 per million among primary vaccinees of all ages. It is most common among infants less than 1 year of age. The onset occurs about 12 days after vaccination. The cause is not clear.

MONKEYPOX INFECTIONS

Monkeypox virus is a species of *Orthopoxvirus*. The disease was first recognized in captive monkeys in 1958. Human infections with this virus were discovered in the early 1970s in West Africa and central Africa after the eradication of smallpox from those regions.

The disease is a rare zoonosis that has been detected in remote villages in tropical rain forests, particularly in the Congo basin countries of Africa and perhaps in West Africa. It is probably acquired by direct contact with wild animals killed for food and skins. The primary reservoir host is not known but squirrels and rodents can be infected.

The clinical features of human monkeypox have been established based on an examination of 282 infected patients in Zaire from 1980 to 1985. Patients were of all ages, but the majority (90%) were less than 15 years old. Clinical symptoms were similar to ordinary and modified forms of smallpox. "Cropping" of the rash occurred in some patients, posing a diagnostic problem with chickenpox. Pronounced lymphadenopathy occurred in most patients, a feature not seen with smallpox or chickenpox.

Complications were common and often serious. These were generally pulmonary distress and secondary bacterial infections. In unvaccinated patients, the fatality rate was about 11%. Vaccination with vaccinia either protects against monkeypox or lessens the severity of disease.

Human monkeypox infection is generally believed not to be easily transmitted from person to person. Previous estimates were that only about 15% of susceptible family contacts acquired monkeypox from patients. However, an outbreak in Zaire in 1996 and 1997 suggested a higher potential for person-to-person transmission.

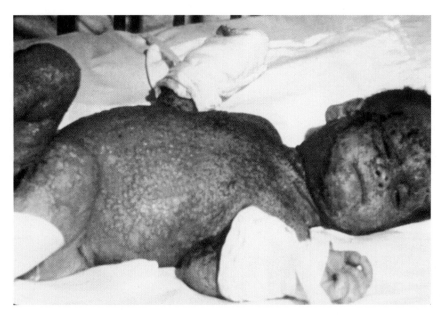

FIGURE 34–5 Eczema vaccinatum in a young child with eczema. The disease developed following exposure to a newly vaccinated family member. (Courtesy of AE Kaye; Centers for Disease Control and Prevention Public Health Image Library.)

The first outbreak of monkeypox in the western hemisphere occurred in the United States in 2003. Over 80 human cases (no deaths) were diagnosed, mostly in midwestern states. The source was traced to an exotic pet store where apparently an imported African rat spread the virus to pet prairie dogs and they transmitted it to humans. It is likely that the isolate of monkeypox virus introduced was a naturally attenuated virus from West Africa that was less pathogenic in humans than isolates from central Africa.

COWPOX INFECTIONS

Cowpox virus is another species of *Orthopoxvirus*. This disease of cattle is milder than the pox diseases of other animals, the lesions being confined to the teats and udders (Figure 34–6A). Infection of humans occurs by direct contact during milking, and the lesion in milkers is usually confined to the hands (Figure 34–6D). The disease is more severe in unvaccinated persons than in those vaccinated with vaccinia virus.

Cowpox virus is similar to vaccinia virus immunologically and in host range. It is also closely related immunologically to variola virus. Jenner observed that those who have had cowpox are immune to smallpox. Cowpox virus can be distinguished from vaccinia virus by the deep red hemorrhagic lesions that cowpox virus produces on the chorioallantoic membrane of the chick embryo.

The natural reservoir of cowpox seems to be a rodent, and both cattle and humans are only accidental hosts. Domestic cats also are susceptible to cowpox virus. More than 50 cases in felines have been reported from the United Kingdom, but transmission from cats to humans is believed to be uncommon. Cowpox is no longer enzootic in cattle, although bovine and associated human cases occasionally occur. Feline cowpox is sporadic, and transmission is probably from a small wild rodent, including field voles. Human cases (with hemorrhagic skin lesions, fever, and general malaise) may occur without any known animal contact and may not be diagnosed. There is no treatment.

BUFFALOPOX INFECTIONS

Buffalopox virus is a derivative of vaccinia virus that has persisted in India in water buffalo since smallpox vaccination was discontinued. The disease in buffalo—and occasionally in cattle—is indistinguishable from cowpox. Buffalopox can be transmitted to humans, and localized pox lesions develop. There is some concern that human-to-human transmission may also occur.

ORF VIRUS INFECTIONS

The virus of orf is a species of *Parapoxvirus*. It causes a disease in sheep and goats that is prevalent worldwide (Figure 34–6C). The disease is also called contagious pustular dermatitis or sore mouth.

Orf is transmitted to humans by direct contact with an infected animal. It is an occupational disease of sheep and goat handlers. Recent reports from the United States emphasized the temporal association between human lesions and recent flock vaccination with live orf virus. Infection by orf virus is facilitated by skin trauma. Infection of humans occurs usually as a single lesion on a finger, hand, or forearm (Figure 34–6F)

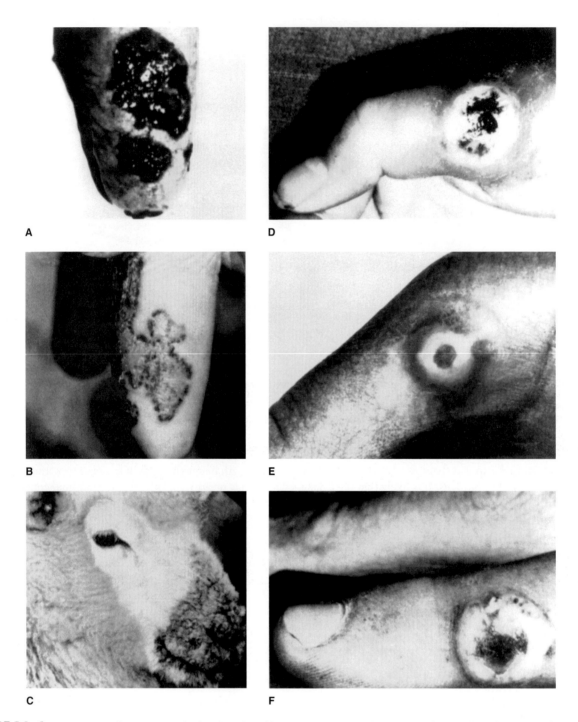

FIGURE 34–6 Cowpox, pseudocowpox, and orf in animals and humans. **A:** Cowpox ulcer on teat of cow 7 days after onset of signs. **B:** Pseudocowpox (milker's nodule virus) on teat of cow. **C:** Scabby mouth in a lamb, caused by orf virus. **D, E, F:** Hand lesions caused by these viruses. **D:** Cowpox. **E:** Milker's nodule (pseudocowpox). **F:** Orf. (**A** and **B** courtesy of EPJ Gibbs; **C** courtesy of A Robinson; **D** courtesy of AD McNae; **E** and **F** courtesy of J Nagington.)

but may appear on the face or neck. Lesions are large nodules, rather painful, with surrounding inflamed skin. The infection is seldom generalized. Healing takes several weeks.

Electron microscopy can confirm a parapoxvirus infection, but only PCR can definitively identify a parapoxvirus as orf virus.

MOLLUSCUM CONTAGIOSUM

Molluscum contagiosum is a benign epidermal tumor that occurs only in humans (though there is evidence of a closely related virus in horses). The causative agent is classified as the sole member of the *Molluscipoxvirus* genus.

The virus has not been transmitted to animals and has not been grown in tissue culture. It has been studied in the human lesion by electron microscopy. The purified virus is oval or brick-shaped and measures about 230 nm by 330 nm; it resembles vaccinia. Antibodies to the virus do not cross-react with any other poxviruses.

The viral DNA resembles that of vaccinia virus with respect to terminal cross-linking and inverted terminal repeats. It has an overall G + C content of about 60%. The sequence of the entire genome of molluscum contagiosum virus (≈190 kbp) is known. It contains at least 163 genes, about two-thirds of which resemble genes of smallpox and cowpox viruses. The large number of dissimilar genes must account for the different human illnesses produced by molluscum contagiosum and the smallpox virus.

The lesions of this disease are small, pink, wart-like tumors on the face, arms, back, and buttocks (Figure 34–7). They are rarely found on the palms, soles, or mucous membranes. The disease occurs throughout the world in both sporadic and epidemic forms and is more frequent in children than in adults. It is spread by direct and indirect contact (eg, by barbers, common use of towels, swimming pools).

The incidence of molluscum contagiosum as a sexually transmitted disease in young adults is increasing. It is seen also in some patients with AIDS. The skin of late-stage AIDS patients may be covered with many papules. Although the typical lesion is an umbilicated papule, lesions in moist genital areas may become inflamed or ulcerated and may be confused with those produced by herpes simplex virus (HSV). Specimens from such lesions are often submitted to viral diagnostic laboratories for isolation of HSV (see below).

The incubation period may extend for up to 6 months. Lesions may itch, leading to autoinoculation. The lesions may persist for up to 2 years but will eventually regress spontaneously. The virus is a poor immunogen; about one-third of patients never produce antibodies against it. Second attacks are common.

Although molluscum contagiosum virus has not been serially propagated in cell culture, it can infect human and primate cells and undergo an abortive infection. Uncoating occurs to produce cores, followed by a transient characteristic cytopathic effect. The cellular changes can be mistaken for those produced by HSV; thus, isolates from specimens suspected to contain HSV should be specifically identified. In a 1985 study of 137 specimens cultured for HSV with the use of human fibroblast cells, 49 contained HSV; six others produced cytopathic effects but were negative for HSV antigens. Electron microscopy confirmed the presence of molluscum contagiosum virus in those HSV-negative, cytopathic-effect-positive samples.

The diagnosis of molluscum contagiosum can usually be made clinically. However, a semisolid caseous material can be expressed from the lesions and used for laboratory diagnosis. PCR can detect viral DNA sequences, and electron microscopy can detect poxvirus particles.

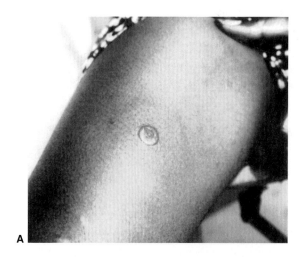

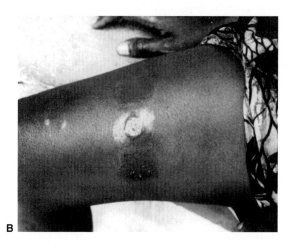

FIGURE 34–8 Lesions produced by tanapox virus. **A:** Ten days after first appearance of the lesion. **B:** Thirty-one days after appearance of the lesion. (Courtesy of Z Jezek.)

FIGURE 34–7 Lesions of molluscum contagiosum in humans. (Courtesy of D Lowy.)

TANAPOX & YABA MONKEY TUMOR POXVIRUS INFECTIONS

Tanapox is a fairly common skin infection in parts of Africa, mainly in Kenya and the Democratic Republic of Congo. Its natural host is probably monkeys, though it is possible that there is another reservoir and that monkeys are only incidental hosts. The mode of transmission is not known.

Tanapox and Yaba monkey tumor viruses are serologically related to each other but are distinct from all other poxviruses. They are classified in the *Yatapoxvirus* genus. They are morphologically similar to orthopoxviruses. The tanapox virus genome is 160 kbp in size, whereas that of Yaba monkey tumor poxvirus is smaller (145 kbp; 32.5% G + C). The viruses grow only in cultures of monkey and human cells, with cytopathic effects. They do not grow on the chorioallantoic membrane of embryonated eggs.

Tanapox begins with a febrile period of 3–4 days and can include severe headache and prostration. There are usually only one or two skin lesions; pustulation never occurs (Figure 34–8). Healing may take 4–7 weeks.

Yaba monkey tumor poxvirus causes benign histiocytomas 5–20 days after subcutaneous or intramuscular administration to monkeys. The tumors regress after about 5 weeks. Intravenous administration of the virus causes the appearance of multiple histiocytomas in the lungs, heart, and skeletal muscles. True neoplastic changes do not occur. The virus is easily isolated from tumor tissue, and characteristic inclusions are found in the tumor cells. Monkeys of various species and humans are susceptible to the cellular proliferative effects of the virus, but other laboratory animals are insusceptible. Although animal handlers have become infected, Yaba virus infections of humans have not been observed naturally in Africa.

REVIEW QUESTIONS

1. A 40-year-old emergency services worker is considering smallpox vaccination because of the potential for bioterrorism. He inquires about the vaccine and finds that the most common adverse reaction or complication after vaccinia (smallpox) vaccination is

    (A) Generalized vaccinia
    (B) Eczema vaccinatum
    (C) Progressive vaccinia
    (D) Severe allergic reaction
    (E) Inadvertent autoinoculation

2. The emergency services worker in Question 1 also learns about contraindications to smallpox vaccination. Which one of the following conditions is not a contraindication for the use of vaccinia (smallpox) vaccine under routine nonemergency conditions?

    (A) Immunosuppression
    (B) Severe allergy to a component of the vaccine
    (C) Household contact with a person with eczema
    (D) Pregnancy
    (E) Previous smallpox vaccination

3. Which of the following poxviruses infects only humans?

    (A) Monkeypox
    (B) Molluscum contagiosum
    (C) Tanapox
    (D) Cowpox
    (E) Yaba tumor virus

4. A 7-year-old boy has pox-like lesions on his left hand and arm. He has a pet rodent imported from West Africa. Monkeypox is diagnosed in the boy and the rodent. Which of the following statements about monkeypox virus is most correct?

    (A) Clinical disease resembles smallpox
    (B) Human infections are never fatal
    (C) Smallpox vaccination is not protective
    (D) Infections are readily transmitted among family members
    (E) Virus particles can be distinguished from smallpox virus by electron microscopy

5. Which of the following best describes the currently licensed vaccinia (smallpox) vaccine?

    (A) Live attenuated smallpox virus
    (B) Inactivated smallpox virus
    (C) Live vaccinia virus
    (D) Inactivated vaccinia virus
    (E) Reassortant vaccine containing both vaccinia and smallpox viruses

6. Which of the following does not apply to vaccinia virus replication in cultured cells?

    (A) Viral replication cycle takes place in the cytoplasm of infected cells
    (B) The uncoating step leading to release of the viral genome requires a newly synthesized viral protein
    (C) Early transcription of more than 50 viral genes occurs within viral cores and precedes viral DNA replication
    (D) Newly formed virus particles mature by budding through the nuclear membrane

7. A 37-year-old emergency room nurse is vaccinated for smallpox because of the bioterrorism threat. Twelve days later she has a major complication from the vaccine. Vaccinia immune globulin therapy is considered. For which of the following conditions is treatment with vaccinia immunoglobulin of no benefit?

    (A) Severe generalized vaccinia
    (B) Progressive vaccinia
    (C) Postvaccinial encephalitis
    (D) Eczema vaccinatum
    (E) Ocular vaccinia

8. Another nurse from the same emergency room as the patient in Question 7 also is vaccinated for smallpox. At what point is she considered to be fully protected from smallpox?

    (A) Ten days after the first dose of vaccine, regardless of the response at the site of administration
    (B) Ten days after the second dose of vaccine, regardless of the response at the site of administration
    (C) After the appearance of any reaction at the site of administration
    (D) After the appearance of a vesicular or pustular lesion at the site of administration
    (E) After the appearance of a generalized rash in the vaccinated person

9. Which of the following does not fulfill the criteria for exposure to vaccinia?

(A) Smallpox vaccination
(B) Close contact with a recent smallpox vaccinee
(C) Intrauterine exposure
(D) Injection of vaccinia immune globulin

10. A researcher wishes to obtain a full-length genome of variola virus for vaccine studies. Which of the following is the appropriate source of the viral DNA?

(A) The Centers for Disease Control and Prevention
(B) A World Health Organization collaborating center
(C) The American Type Culture Collection
(D) A colleague with a variola virus clone
(E) Distribution of a full-length viral genome is prohibited

11. Laboratory scientists who work with vaccinia virus-infected cultures or animals are at risk of accidental exposure to the virus. Which of the following procedures by the laboratory worker is of least benefit in protecting against inadvertent infection with vaccinia virus?

(A) Proper use of personal protective equipment such as gloves and goggles
(B) Cleaning of laboratory work space before experimentation
(C) Smallpox vaccination
(D) Safe needle-handling practices
(E) Use of biosafety hoods

Answers

1. E	4. A	7. C	10. E
2. E	5. C	8. D	11. B
3. B	6. D	9. D	

REFERENCES

Haig DM: Poxvirus interference with the host cytokine response. Vet Immunol Immunopathol 1998;63:149. [PMID: 9656450]

Li Y et al: On the origin of smallpox: Correlating variola phylogenics with historical smallpox records. Proc Natl Acad Sci USA 2007;104:15787. [PMID: 17901212]

McFadden G: Poxvirus tropism. Nature Rev Microbiol 2005;3:201. [PMID: 15738948]

Mercer A et al: Molecular genetic analyses of parapoxviruses pathogenic for humans. Arch Virol 1997;13 (Suppl):25.

Moss B: Genetically engineered poxviruses for recombinant gene expression, vaccination, and safety. Proc Natl Acad Sci USA 1996;93:11341. [PMID: 8876137]

Surveillance guidelines for smallpox vaccine (vaccinia) adverse reactions. MMWR Morb Mortal Wkly Rep 2006;55(RR-1):1.

Vaccinia (smallpox) vaccine: Recommendations of the Advisory Committee on Immunization Practices (ACIP), 2001. MMWR Recomm Rep 2001;50(RR-10):1.

Wharton M et al: Recommendations for using smallpox vaccine in a pre-event vaccination program: Supplemental recommendations of the Advisory Committee on Immunization Practices (ACIP) and the Healthcare Infection Control Practices Advisory Committee (HICPAC). MMWR Recomm Rep 2003;52(RR-7):1.

WHO recommendations concerning the distribution, handling and synthesis of variola virus DNA, May 2008. World Health Org Wkly Epidemiol Rec 2008;83:393.

Hepatitis Viruses

Viral hepatitis is a systemic disease primarily involving the liver. Most cases of acute viral hepatitis in children and adults are caused by one of the following agents: hepatitis A virus (HAV), the etiologic agent of viral hepatitis type A (infectious hepatitis); hepatitis B virus (HBV), which is associated with viral hepatitis B (serum hepatitis); hepatitis C virus (HCV), the agent of hepatitis C (common cause of posttransfusion hepatitis); or hepatitis E virus (HEV), the agent of enterically transmitted hepatitis. Additional well-characterized viruses that can cause sporadic hepatitis, such as yellow fever virus, cytomegalovirus, Epstein-Barr virus, herpes simplex virus, rubella virus, and the enteroviruses, are discussed in other chapters. Hepatitis viruses produce acute inflammation of the liver, resulting in a clinical illness characterized by fever, gastrointestinal symptoms such as nausea and vomiting, and jaundice. Regardless of the virus type, identical histopathologic lesions are observed in the liver during acute disease.

PROPERTIES OF HEPATITIS VIRUSES

The characteristics of the five known hepatitis viruses are shown in Table 35–1. Nomenclature of the hepatitis viruses, antigens, and antibodies is presented in Table 35–2.

Hepatitis Type A

HAV is a distinct member of the picornavirus family (see Chapter 36). HAV is a 27- to 32-nm spherical particle with cubic symmetry, containing a linear single-stranded RNA genome with a size of 7.5 kb. Although it was first provisionally classified as enterovirus 72, the nucleotide and amino acid sequences of HAV are sufficiently distinct to assign it to a new picornavirus genus, *Hepatovirus*. Only one serotype is known. There is no antigenic cross-reactivity with HBV or with the other hepatitis viruses. Genomic sequence analysis of a variable region involving the junction of the 1D and 2A genes divided HAV isolates into seven genotypes. Important properties of the family Picornaviridae are listed in Table 36–1.

HAV is stable to treatment with 20% ether, acid (pH 1.0 for 2 hours), and heat (60°C for 1 hour), and its infectivity can be preserved for at least 1 month after being dried and stored at 25°C and 42% relative humidity or for years at –20°C. The virus is destroyed by autoclaving (121°C for 20 minutes), by boiling in water for 5 minutes, by dry heat (180°C for 1 hour), by ultraviolet irradiation (1 minute at 1.1 watts), by treatment with formalin (1:4000 for 3 days at 37°C), or by treatment with chlorine (10–15 ppm for 30 minutes). Heating food to >85°C (185°F) for 1 minute and disinfecting surfaces with sodium hypochlorite (1:100 dilution of chlorine bleach) are necessary to inactivate HAV. The relative resistance of HAV to disinfection procedures emphasizes the need for extra precautions in dealing with hepatitis patients and their products.

HAV initially was identified in stool and liver preparations by employing immune electron microscopy as the detection system (Figure 35–1). Sensitive serologic assays and polymerase chain reaction (PCR) methods have made it possible to detect HAV in stools and other samples and to measure specific antibody in serum.

Various primate cell lines will support growth of HAV, though fresh isolates of virus are difficult to adapt and grow. Usually, no cytopathic effects are apparent. Mutations in the viral genome are selected during adaptation to tissue culture.

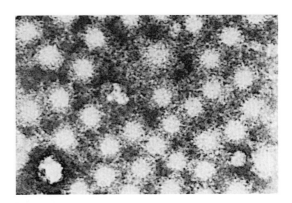

FIGURE 35–1 Electron micrograph of 27-nm hepatitis A virus aggregated with antibody (222,000×). Note the presence of an antibody "halo" around each particle. (Courtesy of DW Bradley, CL Hornbeck, and JE Maynard.)

TABLE 35–1   Characteristics of Hepatitis Viruses

Virus	Hepatitis A	Hepatitis B	Hepatitis C	Hepatitis D	Hepatitis E
Family	Picornaviridae	Hepadnaviridae	Flaviviridae	Unclassified	Hepeviridae
Genus	*Hepatovirus*	*Orthohepadnavirus*	*Hepacivirus*	*Deltavirus*	*Hepevirus*
Virion	27 nm, icosahedral	42 nm, spherical	60 nm, spherical	35 nm, spherical	30–32 nm, icosahedral
Envelope	No	Yes (HBsAg)	Yes	Yes (HBsAg)	No
Genome	ssRNA	dsDNA	ssRNA	ssRNA	ssRNA
Genome size (kb)	7.5	3.2	9.4	1.7	7.2
Stability	Heat- and acid-stable	Acid-sensitive	Ether-sensitive, acid-sensitive	Acid-sensitive	Heat-stable
Transmission	Fecal–oral	Parenteral	Parenteral	Parenteral	Fecal–oral
Prevalence	High	High	Moderate	Low, regional	Regional
Fulminant disease	Rare	Rare	Rare	Frequent	In pregnancy
Chronic disease	Never	Often	Often	Often	Never
Oncogenic	No	Yes	Yes	?	No

TABLE 35–2   Nomenclature and Definitions of Hepatitis Viruses, Antigens, and Antibodies

Disease	Component of System	Definition
Hepatitis A	HAV	Hepatitis A virus. Etiologic agent of infectious hepatitis. A picornavirus, the prototype of genus *Hepatovirus*
	Anti-HAV	Antibody to HAV. Detectable at onset of symptoms; lifetime persistence
	IgM anti-HAV	IgM class antibody to HAV. Indicates recent infection with hepatitis A; positive up to 4–6 months after infection
Hepatitis B	HBV	Hepatitis B virus. Etiologic agent of serum hepatitis. A hepadnavirus
	HBsAg	Hepatitis B surface antigen. Surface antigen(s) of HBV detectable in large quantity in serum; several subtypes identified
	HBeAg	Hepatitis B e antigen. Associated with HBV nucleocapsid; indicates viral replication; circulates as soluble antigen in serum
	HBcAg	Hepatitis B core antigen
	Anti-HBs	Antibody to HBsAg. Indicates past infection with and immunity to HBV, presence of passive antibody from HBIG, or immune response from HBV vaccine
	Anti-HBe	Antibody to HBeAg. Presence in serum of HBsAg carrier suggests lower titer of HBV
	Anti-HBc	Antibody to HBcAg. Indicates infection with HBV at some undefined time in the past
	IgM anti-HBc	IgM class antibody to HBcAg. Indicates recent infection with HBV; positive for 4–6 months after infection
Hepatitis C	HCV	Hepatitis C virus, a common etiologic agent of posttransfusion hepatitis. A flavivirus, genus *Hepacivirus*
	Anti-HCV	Antibody to HCV
Hepatitis D	HDV	Hepatitis D virus. Etiologic agent of delta hepatitis; causes infection only in presence of HBV
	HDAg	Delta antigen (delta-Ag). Detectable in early acute HDV infection
	Anti-HDV	Antibody to delta-Ag (anti-delta). Indicates past or present infection with HDV
Hepatitis E	HEV	Hepatitis E virus. Enterically transmitted hepatitis virus. Causes large epidemics in Asia, North and West Africa, and Mexico; fecal–oral or waterborne transmission. Unclassified
Immune globulins	IG	Immune globulin USP. Contains antibodies to HAV; no antibodies to HBsAg, HCV, or HIV
	HBIG	Hepatitis B immune globulin. Contains high titers of antibodies to HBV

Hepatitis Type B

HBV is classified as a hepadnavirus (Table 35–3). HBV establishes chronic infections, especially in those infected as infants; it is a major factor in the eventual development of liver disease and hepatocellular carcinoma in those individuals.

A. Structure & Composition

Electron microscopy of HBsAg-positive serum reveals three morphologic forms (Figures 35–2 and 35–3A). The most numerous are spherical particles measuring 22 nm in diameter (Figure 35–3B). These small particles are made up exclusively of HBsAg—as are tubular or filamentous forms, which have the same diameter but may be over 200 nm long—and result from overproduction of HBsAg. Larger, 42-nm spherical virions (originally referred to as Dane particles) are less frequently observed (Figure 35–2). The outer surface, or envelope, contains HBsAg and surrounds a 27-nm inner nucleocapsid core that contains HBcAg (Figure 35–3C). The variable length of a single-stranded region of the circular DNA genome results in genetically heterogeneous particles with a wide range of buoyant densities.

The viral genome (Figure 35–4) consists of partially double-stranded circular DNA, 3200 bp in length. Different HBV isolates share 90–98% nucleotide sequence homology. The full-length DNA minus strand (L or long strand) is complementary to all HBV mRNAs; the positive strand (S or short strand) is variable and between 50% and 80% of unit length.

There are four open reading frames that encode seven polypeptides. These include structural proteins of the virion surface and core, a small transcriptional transactivator (X), and a large polymerase (P) protein that includes DNA polymerase, reverse transcriptase, and RNase H activities. The S gene has three in-frame initiation codons and encodes the major HBsAg, as well as polypeptides containing in addition pre-S2 or pre-S1 and pre-S2 sequences. The C gene has two in-frame initiation codons and encodes HBcAg plus the HBe protein, which is processed to produce soluble HBeAg.

The particles containing HBsAg are antigenically complex. Each contains a group-specific antigen, *a*, in addition

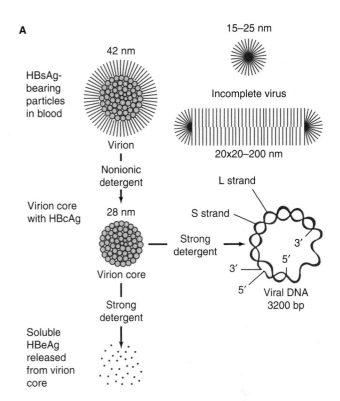

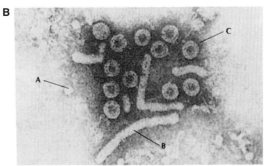

FIGURE 35–2 Hepatitis B viral and subviral forms. **A:** Schematic representation of three HBsAg-containing forms that can be identified in serum from HBV carriers. The 42-nm spherical Dane particle can be disrupted by nonionic detergents to release the 28-nm core that contains the partially double-stranded viral DNA genome. A soluble antigen, termed HBeAg, may be released from core particles by treatment with strong detergent. **B:** Electron micrograph showing three distinct HBsAg-bearing forms: 20-nm pleomorphic spherical particles (A), filamentous forms (B), and 42-nm spherical Dane particles, the infectious form of HBV (C). (Courtesy of FB Hollinger.)

TABLE 35–3 Important Properties of Hepadnaviruses[a]

Virion: About 42 nm in diameter overall (nucleocapsids, 18 nm)
Genome: One molecule of double-stranded DNA, circular, 3.2 kbp. In virion, negative DNA strand is full-length and positive DNA strand is partially complete. The gap must be completed at beginning of replication cycle
Proteins: Two major polypeptides (one glycosylated) are present in HBsAg; one polypeptide is present in HBcAg
Envelope: Contains HBsAg and lipid
Replication: By means of an intermediate RNA copy of the DNA genome (HBcAg in nucleus; HBsAg in cytoplasm). Both mature virus and 22-nm spherical particles consist of HBsAg secreted from the cell surface
Outstanding characteristics:
Family is made up of many types that infect humans and lower animals (eg, woodchucks, squirrels, ducks)
Cause acute and chronic hepatitis, often progressing to permanent carrier states and hepatocellular carcinoma

[a]For HAV, see properties of picornaviruses (Table 36–1); for HCV, see description of flaviviruses (Table 38–1).

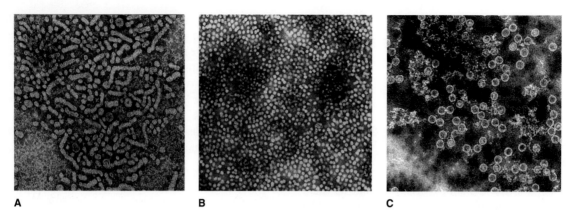

FIGURE 35–3 A: Unfractionated HBsAg-positive human plasma. Filaments, 22-nm spherical particles, and a few 42-nm virions are shown (77,000×). **B:** Purified HBsAg (55,000×). (Courtesy of RM McCombs and JP Brunschwig.) **C:** HBcAg purified from infected liver nuclei (122,400×). The diameter of the core particles is 27 nm. (Courtesy of HA Fields, GR Dreesman, and G Cabral.)

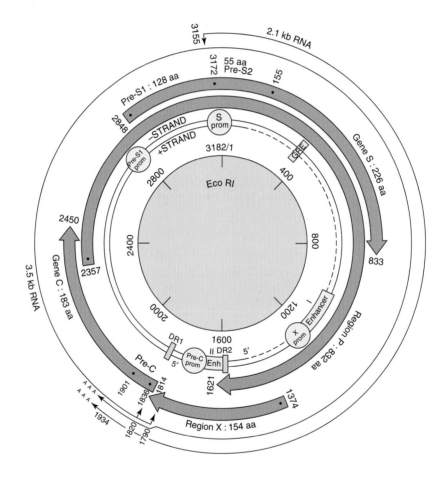

FIGURE 35–4 Genetic organization of the HBV genome. Four open reading frames encoding seven peptides are indicated by large arrows. Regulatory sequences (promoters [prom], enhancers [Enh], and glucocorticoid responsive element [GRE]) are marked. Only the two major transcripts (core/pre-genome and S mRNAs) are represented. DR1 and DR2 are two directly repeated sequences of 11 bp at the 5′ extremities of the minus- and plus-strand DNA. (Reproduced with permission from Buendia MA: Hepatitis B viruses and hepatocellular carcinoma. Adv Cancer Res 1992;59:167. Academic Press, Inc., 1992.)

to two pairs of mutually exclusive subdeterminants, *d/y* and *w/r*. Thus, four phenotypes of HBsAg have been observed: *adw, ayw, adr,* and *ayr.* In the United States, *adw* is the predominant subtype. These virus-specific markers are useful

in epidemiologic investigations, as secondary cases have the same subtype as the index case.

The stability of HBsAg does not always coincide with that of the infectious agent. However, both are stable at –20°C for over 20 years and stable to repeated freezing and thawing. The virus also is stable at 37°C for 60 minutes and remains viable after being dried and stored at 25°C for at least 1 week. HBV (but not HBsAg) is sensitive to higher temperatures (100°C for 1 minute) or to longer incubation periods (60°C for 10 hours). HBsAg is stable at pH 2.4 for up to 6 hours, but HBV infectivity is lost. Sodium hypochlorite, 0.5% (eg, 1:10 chlorine bleach), destroys antigenicity within 3 minutes at low protein concentrations, but undiluted serum specimens require higher concentrations (5%). HBsAg is not destroyed by ultraviolet irradiation of plasma or other blood products, and viral infectivity may also resist such treatment.

B. Replication of Hepatitis B Virus

The infectious virion attaches to cells and becomes uncoated (Figure 35–5). In the nucleus, the partially double-stranded viral genome is converted to covalently closed circular double-stranded DNA (cccDNA). The cccDNA serves as template for all viral transcripts, including a 3.5-kb pregenome RNA. The pregenome RNA becomes encapsidated with newly synthesized HBcAg. Within the cores, the viral polymerase synthesizes by reverse transcription a negative-strand DNA copy. The polymerase starts to synthesize the positive DNA strand, but the process is not completed. Cores bud from the pre-Golgi membranes, acquiring HBsAg-containing envelopes, and may exit the cell. Alternatively, cores may be reimported into the nucleus and initiate another round of replication in the same cell.

Hepatitis Type C

Clinical and epidemiologic studies and cross-challenge experiments in chimpanzees in the past had suggested that there were several non-A, non-B (NANB) hepatitis agents which, based on serologic tests, were not related to HAV or HBV. The major agent was identified as hepatitis C virus (HCV). HCV is a positive-stranded RNA virus, classified as family Flaviviridae, genus *Hepacivirus*. Various viruses can be differentiated by RNA sequence analysis into at least six major genotypes (clades) and more than 100 subtypes. Clades differ from each other by 25–35% at the nucleotide level; subtypes differ from each other by 15–25%. The genome is 9.4 kb in size and encodes a core protein, two envelope glycoproteins, and several nonstructural proteins (Figure 35–6). The expression of cDNA clones of HCV in yeast led to the development of serologic tests for antibodies to HCV. Most cases of posttransfusion NANB hepatitis were caused by HCV.

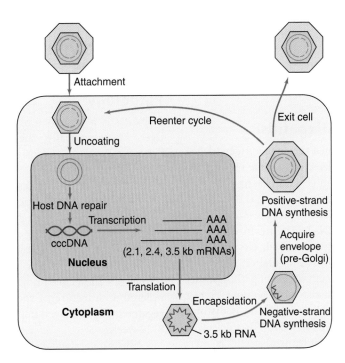

FIGURE 35–5 HBV replication cycle. HBV attachment to a receptor on the surface of hepatocytes occurs via a portion of the pre-S region of HBsAg. After uncoating of the virus, unidentified cellular enzymes convert the partially double-stranded DNA to covalent closed circular (ccc) DNA that can be detected in the nucleus. The cccDNA serves as the template for the production of HBV mRNAs and the 3.5-kb RNA pregenome. The pre-genome is encapsidated by a packaging signal located near the 5′ end of the RNA into newly synthesized core particles, where it serves as template for the HBV reverse transcriptase encoded within the polymerase gene. An RNase H activity of the polymerase removes the RNA template as the negative-strand DNA is being synthesized. Positive-strand DNA synthesis does not proceed to completion within the core, resulting in replicative intermediates consisting of full-length minus-strand DNA plus variable-length (20–80%) positive-strand DNA. Core particles containing these DNA replicative intermediates bud from pre-Golgi membranes (acquiring HBsAg in the process) and may either exit the cell or reenter the intracellular infection cycle. (Reproduced with permission from Butel JS, Lee TH, Slagle BL: Is the DNA repair system involved in hepatitis-B-virus-mediated hepatocellular carcinogenesis? Trends Microbiol 1996;4:119.)

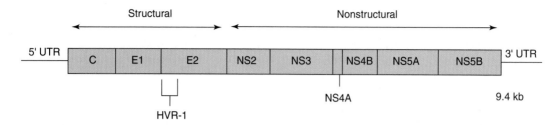

FIGURE 35–6 Genetic organization of the HCV genome. The single open reading frame is expressed as a polyprotein that gets processed; the positions of structural and nonstructural domains are shown. HVR-1 represents the highly variable region of an envelope glycoprotein. (Redrawn with permission from Chung RT, Liang TJ: Hepatitis C virus and hepatocellular carcinoma. In: *Microbes and Malignancy: Infection as a Cause of Human Cancers.* Parsonnet J [editor]. Oxford University Press, 1999.)

Most new infections with HCV are subclinical. The majority (70–90%) of HCV patients develop chronic hepatitis, and many are at risk of progressing to chronic active hepatitis and cirrhosis (10–20%). In some countries, as in Japan, HCV infection often leads to hepatocellular carcinoma. About 25,000 individuals die annually of chronic liver disease and cirrhosis in the United States; HCV appears to be a major contributor to this burden (approximately 40%).

HCV displays genomic diversity, with different genotypes (clades) predominating in different parts of the world. The virus undergoes sequence variation during chronic infections. This complex viral population in a host is referred to as "quasi-species." This genetic diversity is not correlated with differences in clinical disease, although differences do exist in response to antiviral therapy according to viral genotype.

Hepatitis Type D (Delta Hepatitis)

An antigen-antibody system termed the delta antigen (delta-Ag) and antibody (anti-delta) is detected in some HBV infections. The antigen is found within certain HBsAg particles. In blood, HDV (delta agent) contains delta-Ag (HDAg) surrounded by an HBsAg envelope. It has a particle size of 35–37 nm and a buoyant density of 1.24–1.25 g/mL in CsCl. The genome of HDV consists of single-stranded, circular, negative-sense RNA, 1.7 kb in size. It is the smallest of known human pathogens and resembles subviral plant pathogens, ie, viroids. No homology exists with the HBV genome. HDAg is the only protein coded for by HDV RNA and is distinct from the antigenic determinants of HBV. HDV is a defective virus that acquires an HBsAg coat for transmission. It is often associated with the most severe forms of hepatitis in HBsAg-positive patients. It is classified in the *Deltavirus* genus, which is not assigned to any virus family.

Hepatitis Type E

HEV is transmitted enterically and occurs in epidemic form in developing countries, where water or food supplies are sometimes fecally contaminated. It was first documented in samples collected during the New Delhi outbreak of 1955,

when 29,000 cases of icteric hepatitis occurred after sewage contamination of the city's drinking water supply. Pregnant women may have a high (20%) mortality rate if fulminant hepatitis develops. The viral genome has been cloned and is a positive-sense, single-stranded RNA 7.2 kb in size. The virus resembles caliciviruses but has been placed in a new virus family, Hepeviridae, in the genus *Hepevirus*. Animal strains of HEV are common throughout the world. There is evidence of HEV or HEV-like infections in rodents, pigs, sheep, and cattle in the United States. There is the possibility of spread of virus from animals to humans.

HEPATITIS VIRUS INFECTIONS IN HUMANS

Pathology

Hepatitis is a general term meaning inflammation of the liver. Microscopically, there is spotty parenchymal cell degeneration, with necrosis of hepatocytes, a diffuse lobular inflammatory reaction, and disruption of liver cell cords. These parenchymal changes are accompanied by reticuloendothelial (Kupffer) cell hyperplasia, periportal infiltration by mononuclear cells, and cell degeneration. Localized areas of necrosis are frequently observed. Later in the course of the disease, there is an accumulation of macrophages near degenerating hepatocytes. Preservation of the reticulum framework allows hepatocyte regeneration so that the highly ordered architecture of the liver lobule can be ultimately regained. The damaged hepatic tissue is usually restored in 8–12 weeks.

Chronic carriers of HBsAg may or may not have demonstrable evidence of liver disease. Persistent (unresolved) viral hepatitis, a mild benign disease that may follow acute hepatitis B in 8–10% of adult patients, is characterized by sporadically abnormal aminotransferase values and hepatomegaly. Histologically, the lobular architecture is preserved, with portal inflammation, swollen and pale hepatocytes (cobblestone arrangement), and slight to absent fibrosis. This lesion is frequently observed in asymptomatic carriers, usually does not progress toward cirrhosis, and has a favorable prognosis.

Chronic active hepatitis features a spectrum of histologic changes from inflammation and necrosis to collapse of the normal reticulum framework with bridging between the portal triads or terminal hepatic veins. HBV is detected in 10–50% of these patients.

Occasionally during acute viral hepatitis, more extensive damage may occur that prevents orderly liver cell regeneration. Such fulminant or massive hepatocellular necrosis is seen in 1–2% of jaundiced patients with hepatitis B. It is 10 times more common in those coinfected with HDV than in the absence of HDV.

None of the hepatitis viruses are typically cytopathogenic, and it is believed that the cellular damage seen in hepatitis is immune-mediated.

Both HBV and HCV have significant roles in the development of hepatocellular carcinoma that may appear many (15–60) years after establishment of chronic infection.

Clinical Findings

The clinical features of infections by HAV, HBV, and HCV are summarized in Table 35–4. In individual cases, it is not possible to make a reliable clinical distinction among cases caused by the hepatitis viruses.

Other viral diseases that may present as hepatitis are infectious mononucleosis, yellow fever, cytomegalovirus infection, herpes simplex, rubella, and some enterovirus infections. Hepatitis may occasionally occur as a complication of leptospirosis, syphilis, tuberculosis, toxoplasmosis, and amebiasis, all of which are susceptible to specific drug therapy. Noninfectious causes include biliary obstruction, primary biliary cirrhosis, Wilson disease, drug toxicity, and drug hypersensitivity reactions.

In viral hepatitis, onset of jaundice is often preceded by gastrointestinal symptoms such as nausea, vomiting,

TABLE 35–4   Epidemiologic and Clinical Features of Viral Hepatitis Types A, B, and C

Feature	Viral Hepatitis Type A	Viral Hepatitis Type B	Viral Hepatitis Type C
Incubation period	10–50 days (avg., 25–30)	50–180 days (avg., 60–90)	15–160 days (avg., 50)
Principal age distribution	Children,[a] young adults	15–29 years,[b] babies	Adults[b]
Seasonal incidence	Throughout the year but tends to peak in autumn	Throughout the year	Throughout the year
Route of infection	Predominantly fecal–oral	Predominantly parenteral	Predominantly parenteral
Occurrence of virus			
Blood	2 weeks before to ≤1 week after jaundice	Months to years	Months to years
Stool	2 weeks before to 2 weeks after jaundice	Absent	Probably absent
Urine	Rare	Absent	Probably absent
Saliva, semen	Rare (saliva)	Frequently present	Present (saliva)
Clinical and laboratory features			
Onset	Abrupt	Insidious	Insidious
Fever >38°C (100.4°F)	Common	Less common	Less common
Duration of aminotransferase elevation	1–3 weeks	1–6+ months	1–6+ months
Immunoglobulins (IgM levels)	Elevated	Normal to slightly elevated	Normal to slightly elevated
Complications	Uncommon, no chronicity	Chronicity in 5–10% (95% of neonates)	Chronicity in 70–90%
Mortality rate (icteric cases)	<0.5%	<1–2%	0.5–1%
HBsAg	Absent	Present	Absent
Immunity			
Homologous	Yes	Yes	Probably no
Heterologous	No	No	No
Duration	Probably lifetime	Probably lifetime	?
Immune globulin intramuscular (IG, γ-globulin, ISG)	Regularly prevents jaundice	Prevents jaundice only if immune globulin is of sufficient potency against HBV	?

[a]Nonicteric hepatitis is common in children.

[b]Among the age group 15–29 years, hepatitis B and C are often associated with drug abuse or promiscuous sexual behavior. Patients with transfusion-associated HBV or HCV are generally over age 29.

anorexia, and mild fever. Jaundice may appear within a few days of the prodromal period, but anicteric hepatitis is more common.

Extrahepatic manifestations of viral hepatitis (primarily type B) include a transient serum sickness-like prodrome consisting of fever, skin rash, and polyarthritis; necrotizing vasculitis (polyarteritis nodosa); and glomerulonephritis. Circulating immune complexes have been suggested as the cause of these syndromes. Diseases associated with chronic HCV infections include mixed cryoglobulinemia and glomerulonephritis. Extrahepatic manifestations are unusual with HAV infections.

Uncomplicated viral hepatitis rarely continues for more than 10 weeks without improvement. Relapses occur in 5–20% of cases and are manifested by abnormalities in liver function with or without the recurrence of clinical symptoms.

The median incubation period is different for each type of viral hepatitis (Table 35–4). However, there is considerable overlap in timing, and the patient may not know when exposure occurred, so the incubation period is not very useful in determining the specific viral cause.

The onset of disease tends to occur abruptly with HAV (within 24 hours), in contrast to a more insidious onset with HBV and HCV. Complete recovery occurs in most hepatitis A cases; chronicity has not been observed (Table 35–5). The disease is more severe in adults than in children, in whom it often goes unnoticed. Relapses of HAV infection can occur 1–4 months after initial symptoms have resolved.

The outcome after infection with HBV varies, ranging from complete recovery to progression to chronic hepatitis and, rarely, death due to fulminant disease. In adults, 65–80% of infections are inapparent, with 90–95% of all patients recovering completely. In contrast, 80–95% of infants and young children infected with HBV become chronic carriers (Table 35–6), and their serum remains positive for HBsAg. The vast majority of individuals with chronic HBV remain asymptomatic for many years; there may or may not be biochemical and histologic evidence of liver disease. Chronic carriers are at high risk of developing hepatocellular carcinoma.

Fulminant hepatitis occasionally develops during acute viral hepatitis, defined as hepatic encephalopathy within the

TABLE 35–5   Outcomes of Infection with Hepatitis A Virus[a]

Outcome	Children	Adults
Inapparent (subclinical) infection	80–95%	10–25%
Icteric disease	5–20%	75–90%
Complete recovery	>98%	>98%
Chronic disease	None	None
Mortality rate	0.1%	0.3–2.1%

[a]Adapted with permission from Hollinger FB, Ticehurst JR: Hepatitis A virus. In: *Fields Virology*, 3rd ed. Fields BN et al (editors). Lippincott-Raven, 1996.

first 8 weeks of disease in patients without preexisting liver disease. It is fatal in 70–90% of cases, with survival uncommon over the age of 40 years. Fulminant HBV disease is associated with superinfection by other agents, including HDV. In most patients who survive, complete restoration of the hepatic parenchyma and normal liver function is the rule. Fulminant disease rarely occurs with HAV or HCV infections.

Hepatitis C is usually clinically mild, with only minimal to moderate elevation of liver enzymes. Hospitalization is unusual, and jaundice occurs in less than 25% of patients. Despite the mild nature of the disease, 70–90% of cases progress to chronic liver disease. Most patients are asymptomatic, but histologic evaluation often reveals evidence of chronic active hepatitis, especially in those whose disease is acquired following transfusion. Many patients (20–50%) develop cirrhosis and are at high risk for hepatocellular carcinoma (5–25%) decades later. About 40% of chronic liver disease is HCV-related, resulting in an estimated 8000–10,000 deaths annually in the United States. End-stage liver disease associated with HCV is the most frequent indication for adult liver transplants.

Laboratory Features

Liver biopsy permits a tissue diagnosis of hepatitis. Tests for abnormal liver function, such as serum alanine aminotransferase (ALT) and bilirubin, supplement the clinical, pathologic, and epidemiologic findings.

TABLE 35–6   Transmission of Hepatitis B Virus and Spectrum of Outcomes of Infection

	Transmission[a]		
	Vertical (Asia)	**Contact (Africa)**	**Parenteral, Sexual**
Age at infection	Newborns, infants	Young children	Teenagers, adults
Recovery from acute infection (%)	5	20	90–95
Progression to chronic infection (%)	95	80	5–10
Chronic carriers[b] (% of total population)	10–20	10–20	0.5

[a]Vertical and contact-associated transmission occurs in endemic regions; parenteral and sexual transmission are the main modes of transmission in nonendemic regions.
[b]At high risk of developing hepatocellular carcinoma.

A. Hepatitis A

The clinical, virologic, and serologic events following exposure to HAV are shown in Figure 35–7. Virus particles have been detected by immune electron microscopy in fecal extracts of hepatitis A patients (Figure 35–1). Virus appears early in the disease and disappears within 2 weeks following the onset of jaundice.

HAV can be detected in the liver, stool, bile, and blood of naturally infected humans and experimentally infected non-human primates by immunoassays, nucleic acid hybridization assays, or PCR. HAV is detected in the stool from about 2 weeks prior to the onset of jaundice up to 2 weeks after.

Anti-HAV appears in the IgM fraction during the acute phase, peaking about 2 weeks after elevation of liver enzymes (Table 35–7). Anti-HAV IgM usually declines to nondetectable levels within 3–6 months. Anti-HAV IgG appears soon after the onset of disease and persists for decades. Thus, detection of IgM-specific anti-HAV in the blood of an acutely infected patient confirms the diagnosis of hepatitis A. ELISA is the method of choice for measuring HAV antibodies.

B. Hepatitis B

Clinical and serologic events following exposure to HBV are depicted in Figure 35–8 and summarized in Table 35–8. DNA polymerase activity, HBV DNA, and HBeAg, which are representative of the viremic stage of hepatitis B, occur early in the incubation period, concurrently or shortly after the first appearance of HBsAg. High concentrations of HBV particles may be present in the blood (up to 10^{10} particles/mL) during the initial phase of infection; communicability is

TABLE 35–7 Interpretation of HAV, HCV, and HDV Serologic Markers in Patients with Hepatitis

Assay Results	Interpretation
Anti-HAV IgM-positive	Acute infection with HAV
Anti-HAV IgG-positive	Past infection with HAV
Anti-HCV-positive	Current or past infection with HCV
Anti-HDV-positive, HBsAg-positive	Infection with HDV
Anti-HDV-positive, anti-HBc IgM-positive	Coinfection with HDV and HBV
Anti-HDV-positive, anti-HBc IgM-negative	Superinfection of chronic HBV infection with HDV

highest at this time. HBsAg is usually detectable 2–6 weeks in advance of clinical and biochemical evidence of hepatitis and persists throughout the clinical course of the disease but typically disappears by the sixth month after exposure.

High levels of IgM-specific anti-HBc are frequently detected at the onset of clinical illness. Because this antibody is directed against the 27-nm internal core component of HBV, its appearance in the serum is indicative of viral replication. Antibody to HBsAg is first detected at a variable period after the disappearance of HBsAg. It is present in low concentrations. Before HBsAg disappears, HBeAg is replaced by anti-HBe, signaling the start of resolution of the disease. Anti-HBe levels often are no longer detectable after 6 months.

By definition, HBV chronic carriers are those in whom HBsAg persists for more than 6 months in the presence of HBeAg or anti-HBe. HBsAg may persist for years after loss of HBeAg. In contrast to the high titers of IgM-specific anti-HBc observed in acute disease, low titers of IgM anti-HBc are found in the sera of most chronic HBsAg carriers. Small amounts of HBV DNA are usually detectable in the serum as long as HBsAg is present.

The most useful detection methods are ELISA for HBV antigens and antibodies and PCR for viral DNA.

C. Hepatitis C

Clinical and serologic events associated with HCV infections are shown in Figure 35–9. Most primary infections are asymptomatic or clinically mild (20–30% have jaundice, 10–20% have only nonspecific symptoms such as anorexia, malaise, and abdominal pain). Serologic assays are available for diagnosis of HCV infection. Enzyme immunoassays (EIA) detect antibodies to HCV but do not distinguish between acute, chronic, or resolved infection (Table 35–7). Anti-HCV antibodies can be detected in 50–70% of patients at onset of symptoms, whereas in others antibody appearance is delayed 3–6 weeks. Antibodies are directed against core, envelope, and NS3 and NS4 proteins and tend to be relatively low in titer. Nucleic acid-based assays (eg, reverse transcription-polymerase chain reaction) detect the presence of circulating HCV RNA and are

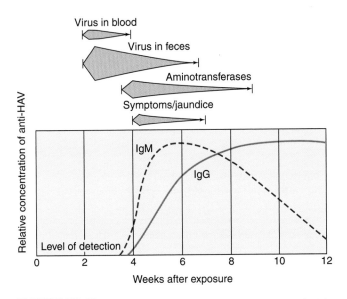

FIGURE 35–7 Immunologic and biologic events associated with human infection with hepatitis A virus. (Reproduced from Hollinger FB, Ticehurst JR: Hepatitis A virus. In: *Fields Virology*, 3rd ed. Fields BN et al [editors]. Lippincott-Raven, 1996. Modified with permission from Hollinger FB, Dienstag JL: Hepatitis viruses. In: *Manual of Clinical Microbiology*, 4th ed. American Society for Microbiology, 1985.)

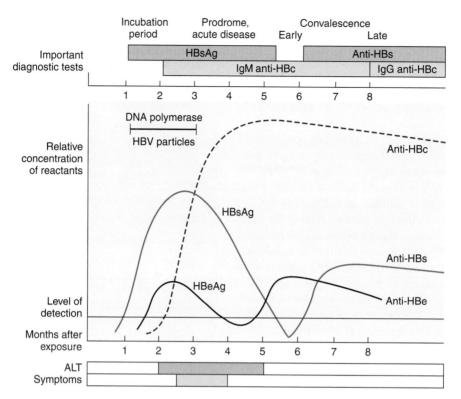

FIGURE 35–8 Clinical and serologic events occurring in a patient with acute hepatitis B virus infection. The common diagnostic tests and their interpretation are presented in Table 35–8. (Reproduced with permission from Hollinger FB, Dienstag JL, Murray PR: Hepatitis B and D viruses. In: *Manual of Clinical Microbiology,* 6th ed. American Society for Microbiology, 1995.)

TABLE 35–8   Interpretation of HBV Serologic Markers in Patients with Hepatitis[a]

	Assay Results		
HBsAg	**Anti-HBs**	**Anti-HBc**	**Interpretation**
Positive	Negative	Negative	Early acute HBV infection. Confirmation is required to exclude nonspecific reactivity
Positive	(±)	Positive	HBV infection, either acute or chronic. Differentiate with IgM anti-HBc. Determine level of replicative activity (infectivity) with HBeAg or HBV DNA
Negative	Positive	Positive	Indicates previous HBV infection and immunity to hepatitis B
Negative	Negative	Positive	Possibilities include: HBV infection in remote past; "low-level" HBV carrier; "window" between disappearance of HBsAg and appearance of anti-HBs; or false-positive or nonspecific reaction. Investigate with IgM anti-HBc. When present, anti-HBe helps validate the anti-HBc reactivity
Negative	Negative	Negative	Never infected with HBV. Possibilities include another infectious agent, toxic injury to liver, disorder of immunity, hereditary disease of the liver, or disease of the biliary tract
Negative	Positive	Negative	Vaccine-type response

[a]Modified and reproduced with permission from Hollinger FB: Hepatitis B virus. In: *Fields Virology,* 3rd ed. Fields BN et al (editors). Lippincott-Raven, 1996.

useful for monitoring patients on antiviral therapy. Nucleic acid assays also are used to genotype HCV isolates.

Occult HBV infections occur frequently (about 33%) in patients with chronic HCV liver disease. Occult infections are those in which the patients lack detectable HBsAg but HBV DNA can be identified in liver or serum samples. These unrecognized HBV coinfections may be clinically significant.

D. Hepatitis D

Serologic patterns following HDV infection are shown in Figure 35–10 and listed in Table 35–7. Because HDV is dependent on a coexistent HBV infection, acute type D infection will occur either as a simultaneous infection (coinfection) with HBV or as a superinfection of a person chronically infected with HBV. In the coinfection pattern, antibody to

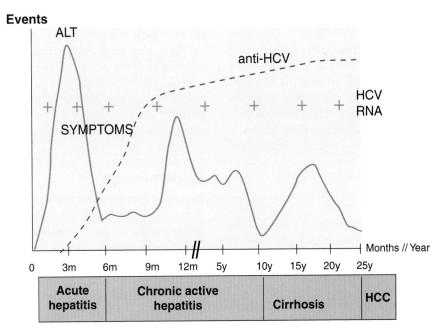

FIGURE 35–9 Clinical and serologic events associated with hepatitis C virus infection. ALT, alanine aminotransferase; HCC, hepatocellular carcinoma. (Reproduced with permission from Garnier L, Inchauspé G, Trépo C: Hepatitis C virus. In: *Clinical Virology,* 2nd ed. Richman DD, Whitley RJ, Hayden FG [editors]. ASM Press, 2002.)

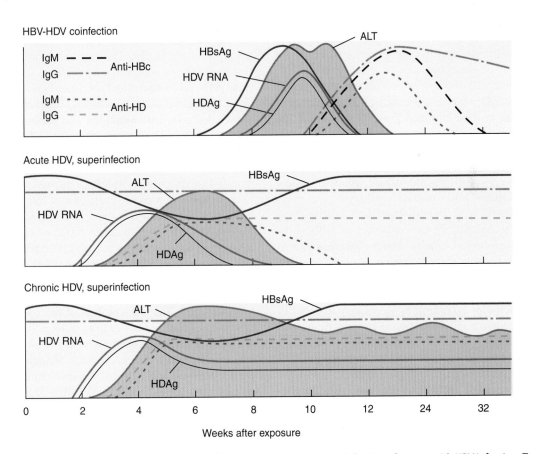

FIGURE 35–10 Serologic patterns of type D hepatitis following coinfection or superinfection of person with HBV infection. **Top:** Coexistent acute hepatitis B and hepatitis D. **Middle:** Acute hepatitis D superimposed on a chronic hepatitis B virus infection. **Bottom:** Acute hepatitis D progressing to chronic hepatitis, superimposed on a chronic hepatitis B virus infection. (Reproduced with permission from Purcell RH et al: Hepatitis. In: *Diagnostic Procedures for Viral, Rickettsial and Chlamydial Infections,* 6th ed. Schmidt NJ, Emmons RW [editors]. American Public Health Association, 1989.)

HDAg develops late in the acute phase of infection and may be of low titer. Assays for HDAg or HDV RNA in the serum or for IgM-specific anti-HDV are preferable. All markers of HDV replication disappear during convalescence; even the HDV antibodies may disappear within months to years. However, superinfection by HDV usually results in persistent HDV infection (over 70% of cases). High levels of both IgM and IgG anti-HD persist, as do levels of HDV RNA and HDAg. HDV superinfections may be associated with fulminant hepatitis.

Virus–Host Interactions

Currently there is evidence for five hepatitis viruses—types A, B, C, D, and E. A single infection with any is believed to confer homologous but not heterologous protection against reinfection. A possible exception may be HCV, where reinfection may occur.

Most cases of hepatitis type A presumably occur without jaundice during childhood, and by late adulthood there is a widespread resistance to reinfection. However, serologic studies in the United States and several Asian countries indicate that the incidence of infection may be declining as a result of improvements in sanitation commensurate with a rise in the standard of living, coupled with expanded use of the vaccine in some countries. It has been estimated that as many as 60–90% of young middle- to upper-income adults in the United States may be susceptible to type A hepatitis.

Infection with HBV of a specific subtype, eg, HBsAg/*adw*, appears to confer immunity to other HBsAg subtypes, probably because of their common group *a* specificity. The immunopathogenetic mechanisms that result in viral persistence and hepatocellular injury in type B hepatitis remain to be elucidated. As the virus is not cytopathic, it is believed that hepatocellular injury during acute disease represents a host immune attack against HBV-infected hepatocytes.

Host responses, both immunologic and genetic, have been proposed to account for the frequency of HBV chronicity in those infected as infants. About 95% of newborns infected at birth become chronic carriers of virus, often for life (Table 35–6). This risk decreases steadily with time, so that the risk of infected adults becoming carriers decreases to 10%. Hepatocellular carcinoma is most likely to occur in adults who experienced HBV infection at a very early age and became carriers. Therefore, for vaccination to be maximally effective against the carrier state, cirrhosis, and hepatoma, it must be carried out during the first week of life.

HCV genotypes 1–4 are the predominant types circulating in Western countries and display some differential characteristics. Genotype 1 is predominant in North America, Japan, and Western Europe. It shows the poorest response to interferon therapy and may have a more deleterious effect on the progression of human immunodeficiency virus (HIV) type 1 disease than other HCV genotypes. In contrast, HCV genotype 2 responds the best to interferon-based therapies. Genotype 3 shows the highest rate of spontaneous clearance, whereas genotype 4 seems to have the highest frequency leading to chronic infection after acute infection.

Less is known about host immune responses to HCV. The majority of acute infections are asymptomatic or mild, and chronic infections usually progress slowly and insidiously. It appears that the immune response is slow to develop and relatively weak, reflecting the fact that HCV has particularly effective immune evasion mechanisms.

Epidemiology

The global distributions of hepatitis A, B, and C infections are shown in Figure 35–11. There are marked differences in the epidemiologic features of these infections (Table 35–4).

The risk of these viruses being transmitted by transfusion today in the United States is markedly reduced as a result of improved screening tests and the establishment of volunteer donor populations. It was calculated in 1996 that the risk of transmission of HBV by blood transfusion was 1:63,000 and for HCV 1:103,000.

A. Hepatitis A

HAV is widespread throughout the world. Outbreaks of type A hepatitis are common in families and institutions, summer camps, day care centers, neonatal intensive care units, and among military troops. The most likely mode of transmission under these conditions is by the fecal–oral route through close personal contact. Stool specimens may be infectious for up to 2 weeks before to 2 weeks after onset of jaundice.

Under crowded conditions and poor sanitation, HAV infections occur at an early age; most children in such circumstances become immune by age 10. Clinical illness is uncommon in infants and children; disease is most often manifest in children and adolescents, with the highest rates in those between 5 and 14 years of age. The ratio of anicteric to icteric cases in adults is about 1:3; in children, it may be as high as 12:1. However, fecal excretion of HAV antigen and RNA persists longer in the young than in adults.

Recurrent epidemics are a prominent feature. Sudden, explosive epidemics of type A hepatitis usually result from fecal contamination of a single source (eg, drinking water, food, or milk). The consumption of raw oysters or improperly steamed clams obtained from water polluted with sewage has also resulted in several outbreaks of hepatitis. The largest outbreak of this type occurred in Shanghai in 1988, when over 300,000 cases of hepatitis A were attributed to uncooked clams from polluted water. A multistate foodborne outbreak that was traced to frozen strawberries occurred in the United States in 1997.

Other identified sources of potential infection are nonhuman primates. There have been more than 35 outbreaks in which primates, usually chimpanzees, have infected humans in close personal contact with them.

HAV is seldom transmitted by the use of contaminated needles and syringes or through the administration of blood.

Countries/areas with moderate to high risk of infection   Source: WHO, 2001

A

Countries/areas with moderate to high risk of infection   Source: WHO, 2001

B

FIGURE 35–11 Global distribution of hepatitis viruses causing human disease in 2001. **A:** Hepatitis A virus. **B:** Hepatitis B virus. **C:** Hepatitis C virus. (Source, World Health Organization, 2001.) (Continued)

Transfusion-associated hepatitis A is rare because the viremic stage of infection occurs during the prodromal phase and is of short duration, the titer of virus in the blood is low, and there is no carrier state. However, a 1996 report documented the transmission of HAV to hemophiliacs through clotting factor concentrates. There is little evidence for HAV transmission by exposure to urine or nasopharyngeal secretions of infected patients. Hemodialysis plays no role in the spread of hepatitis A infections to either patients or staff.

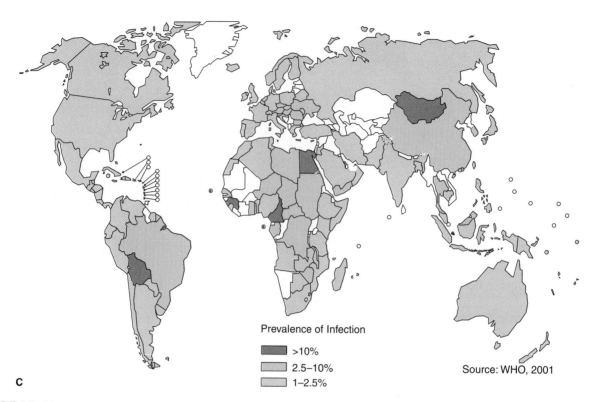

Prevalence of Infection

■ >10%
■ 2.5–10%   Source: WHO, 2001
■ 1–2.5%

C

FIGURE 35–11, Cont. Global distribution of hepatitis viruses causing human disease in 2001. **A:** Hepatitis A virus. **B:** Hepatitis B virus. **C:** Hepatitis C virus. (Source, World Health Organization, 2001.)

In the United States in the prevaccine era, there were an estimated 271,000 infections per year. Since the advent of hepatitis A vaccines, infection rates have declined sharply to an estimated 3500 cases in 2006.

Groups who are at increased risk of acquiring hepatitis A are travelers to developing countries from developed countries, men who have sex with men, users of injection and noninjection drugs, persons with clotting factor disorders, and persons working with nonhuman primates. Individuals with chronic liver disease are at increased risk for fulminant hepatitis if a hepatitis A infection occurs. These groups should be vaccinated.

B. Hepatitis B

HBV is worldwide in distribution. Transmission modes and response to infection vary, depending on the age at time of infection (Table 35–6). Most individuals infected as infants develop chronic infections. As adults they are subject to liver disease and are at high risk of developing hepatocellular carcinoma. There are more than 250 million carriers, of whom about 1 million live in the United States; 25% of carriers develop chronic active hepatitis. Worldwide, 1 million deaths a year are attributed to HBV-related liver disease and hepatocellular carcinoma.

The major modes of HBV transmission during infancy are from an infected mother to her newborn during delivery and from an infected household contact to an infant.

There is no seasonal trend for HBV infection and no high predilection for any age group, although there are definite high-risk groups such as parenteral drug abusers, institutionalized persons, health care personnel, multiply transfused patients, organ transplant patients, hemodialysis patients and staff, highly promiscuous persons, and newborn infants born to mothers with hepatitis B. Since mandatory screening of blood donors for HBsAg was instituted, the number of cases of transfusion-associated hepatitis has been dramatically reduced. People have been infected by improperly sterilized syringes, needles, or scalpels and even by tattooing or ear piercing. The estimated ratio of anicteric to icteric infections is reported to be as high as 4:1.

Other modes of transmission of hepatitis B exist. HBsAg can be detected in saliva, nasopharyngeal washings, semen, menstrual fluid, and vaginal secretions as well as in blood. Transmission from carriers to close contacts by the oral route or by sexual or other intimate exposure occurs. There is strong evidence of transmission from persons with subclinical cases and carriers of HBsAg to homosexual and heterosexual long-term partners. Transmission by the fecal–oral route has not been documented. Recalling that there may be more than 1 billion virions per milliliter of blood from an HBeAg-positive carrier and that the virus is resistant to drying, it should be assumed that all bodily fluids from HBV-infected patients may be infectious. Subclinical infections are common, and these unrecognized infections represent the principal hazard to hospital personnel.

Health care personnel (medical and dental surgeons, pathologists, other physicians, nurses, laboratory technicians, and blood bank personnel) have a higher incidence of hepatitis and prevalence of detectable HBsAg or anti-HBs than those who have no occupational exposure to patients or blood products. The risk that these apparently healthy HBsAg carriers (especially medical and dental surgeons) represent to the patients under their care remains to be determined but is probably small.

Hepatitis B infections are common among patients and staff of hemodialysis units. As many as 50% of the renal dialysis patients who contract hepatitis B may become chronic carriers of HBsAg compared with 2% of the staff group, emphasizing differences in the host immune response. Family contacts are also at increased risk.

The incubation period of hepatitis B is 50–180 days, with a mean between 60 and 90 days. It appears to vary with the dose of HBV administered and the route of administration, being prolonged in patients who receive a low dose of virus or who are infected by a nonpercutaneous route.

C. Hepatitis C

Infections by HCV are extensive throughout the world. The World Health Organization estimated in 1997 that about 3% of the world population has been infected, with population subgroups in Africa having prevalence rates as high as 10%. Other high-prevalence areas are found in South America and Asia. It is estimated that there are more than 170 million chronic carriers worldwide who are at risk of developing liver cirrhosis, liver cancer, or both—and that over 3 million of them are in the United States.

HCV is transmitted primarily through direct percutaneous exposures to blood, though in 10–50% of cases the source of HCV cannot be identified. In roughly decreasing order of prevalence of infection are injecting drug users (about 80%), hemophiliacs treated with clotting factor products before 1987, recipients of transfusions from HCV-positive donors, chronic hemodialysis patients (10%), persons who engage in high-risk sexual practices, and health care workers (1%). The virus can be transmitted from mother to infant, though not as frequently as for HBV. Estimates of mother-to-child vertical transmission vary from 3% to 10%. Mothers with higher HCV viral loads or coinfection with HIV more frequently transmit HCV. No risk of transmission has been associated with breast feeding.

HCV was found in saliva from more than a third of patients with HCV and HIV coinfections. HCV has been transmitted by commercial intravenous immune globulin preparations, including an outbreak in the United States in 1994. The population of Egypt has a high prevalence of HCV (about 20%). Transmission of HCV has been linked to an attempt (from the 1950s to 1980s) to treat the parasitic disease schistosomiasis by therapy that involved multiple injections, often with improperly sterilized or reused needles. HCV infection has been associated with tattooing and, in some countries, with folk medicine practices.

The average incubation period for HCV is 6–7 weeks. The average time from exposure to seroconversion is 8–9 weeks, and about 90% of patients are anti-HCV-positive within 5 months.

D. Hepatitis D (Delta Agent)

HDV is found throughout the world but with a nonuniform distribution. Its highest prevalence has been reported in Italy, the Middle East, central Asia, West Africa, and South America. HDV infects all age groups. Persons who have received multiple transfusions, intravenous drug abusers, and their close contacts are at high risk.

The primary routes of transmission are believed to be similar to those of HBV, though HDV does not appear to be a sexually transmitted disease. Infection is dependent on HBV replication, as HBV provides an HBsAg envelope for HDV. The incubation period varies from 2 to 12 weeks, being shorter in HBV carriers who are superinfected with the agent than in susceptible persons who are simultaneously infected with both HBV and HDV. HDV has been transmitted perinatally, but fortunately it is not prevalent in regions of the world (such as Asia) where perinatal transmission of HBV occurs frequently.

Two epidemiologic patterns of delta infection have been identified. In Mediterranean countries, delta infection is endemic among persons with hepatitis B, and most infections are thought to be transmitted by intimate contact. In nonendemic areas, such as the United States and northern Europe, delta infection is confined to persons exposed frequently to blood and blood products, primarily drug addicts and hemophiliacs.

Delta hepatitis may occur in explosive outbreaks and affect entire localized pockets of hepatitis B carriers. Outbreaks of severe, often fulminant and chronic delta hepatitis have occurred for decades in isolated populations in the Orinoco and Amazon basins of South America. In the United States, HDV has been found to participate in 20–30% of cases of chronic hepatitis B, acute exacerbations of chronic hepatitis B, and fulminant hepatitis B, and 3–12% of blood donors with serum HBsAg have antibodies to HDV. Delta hepatitis is not a new disease, because globulin lots prepared from plasma collected in the United States more than 40 years ago contain antibodies to HDV.

Treatment

Treatment of patients with hepatitis is supportive and directed at allowing hepatocellular damage to resolve and repair itself. Only HBV and HCV have specific treatments, and those are only partially effective.

Recombinant interferon-α and pegylated interferon-α are currently the therapy of proven benefit in the treatment of patients chronically infected with HBV or HCV. Many who responded clinically and biochemically relapsed after cessation of treatment. Only about 35% of patients with chronic

HBV infections have long-lasting remissions, and about 25% of those with chronic HCV infection have a sustained response. Interferon-based therapy is associated with many side effects.

Several antiviral drugs are available for use against chronic hepatitis infections. With nucleoside and nucleotide analogs, such as lamivudine (Table 30–6), HBV DNA levels are reduced, but the virus is rarely eliminated and viral replication resumes in the majority of patients when treatment is stopped. The emergence of drug-resistant virus mutants in long-term therapy is a major problem. Combination therapy of interferon-α and ribavirin against chronic hepatitis C gives a sustained response rate of up to 50%, though that therapy is less successful in patients with genotype 1.

Orthotopic liver transplantation is a treatment for chronic hepatitis B and C end-stage liver damage. However, the risk of reinfection on the graft is at least 80% with HBV and 50% with HCV, presumably from extrahepatic reservoirs in the body.

Prevention & Control

Viral vaccines and protective immune globulin preparations are available against HAV and HBV. Neither type of reagent is currently available to prevent HCV infections.

A. Standard Precautions

Simple environmental procedures can limit the risk of infection to health care workers, laboratory personnel, and others. With this approach, all blood and body fluids and materials contaminated with them are treated as if they are infectious for HIV, HBV, HCV, and other blood-borne pathogens. Exposures that might place workers at risk of infection include percutaneous injury (eg, needlestick) or contact of mucous membrane or nonintact skin (eg, chapped, cuts, dermatitis) with blood, tissue, or other body fluids that are potentially infectious. Methods are devised to prevent contact with such samples. Examples of specific precautions include the following: Gloves should be used when handling all potentially infectious materials; protective garments should be worn and removed before leaving the work area; masks and eye protection should be worn whenever splashes or droplets from infectious material pose a risk; only disposable needles should be used; needles should be discarded directly into special containers without resheathing; work surfaces should be decontaminated using a bleach solution; and laboratory personnel should refrain from mouth-pipetting, eating, drinking, and smoking in the work area. Metal objects and instruments can be disinfected by autoclaving or by exposure to ethylene oxide gas.

B. Hepatitis A

Formalin-inactivated HAV vaccines made from cell culture-adapted virus were licensed in the United States in 1995. The vaccines are safe, effective, and recommended for use in persons over 1 year of age.

Routine vaccination of all children is now recommended, as is vaccination of persons at increased risk, including international travelers, men who have sex with men, and drug users.

Until all susceptible at-risk groups are immunized, prevention and control of hepatitis A still must emphasize interrupting the chain of transmission and using passive immunization.

The appearance of hepatitis in camps or institutions is often an indication of poor sanitation and poor personal hygiene. Control measures are directed toward the prevention of fecal contamination of food, water, or other sources by the individual. Reasonable hygiene—such as hand washing, the use of disposable plates and eating utensils, and the use of 0.5% sodium hypochlorite (eg, 1:10 dilution of chlorine bleach) as a disinfectant—is essential in preventing the spread of HAV during the acute phase of the illness.

Immune (γ) globulin (IG) is prepared from large pools of normal adult plasma and confers passive protection in about 90% of those exposed when given within 1–2 weeks after exposure to hepatitis A. Its prophylactic value decreases with time, and its administration more than 2 weeks after exposure or after onset of clinical symptoms is not indicated. In the doses generally prescribed, IG does not prevent infection but rather makes the infection mild or subclinical and permits active immunity to develop. HAV vaccine produces a more enduring immunity and should replace the use of IG.

C. Hepatitis B

A vaccine for hepatitis B has been available since 1982. The initial vaccine was prepared by purifying HBsAg associated with the 22-nm particles from healthy HBsAg-positive carriers and treating the particles with virus-inactivating agents (formalin, urea, heat). Preparations containing intact 22-nm particles have been highly effective in reducing HBV infection. Although plasma-derived vaccines are still in use in certain countries, they have been replaced in the United States by recombinant DNA-derived vaccines. These vaccines consist of HBsAg produced by a recombinant DNA in yeast cells or in continuous mammalian cell lines. The HBsAg expressed in yeast forms particles 15–30 nm in diameter, with the morphologic characteristics of free surface antigen in plasma though the polypeptide antigen produced by recombinant yeast is not glycosylated. The vaccine formulated using this purified material has a potency similar to that of vaccine made from plasma-derived antigen.

Preexposure prophylaxis with a commercially available hepatitis B vaccine currently is recommended by the World Health Organization, the Centers for Disease Control and Prevention, and the Advisory Committee on Immunization Practices for all susceptible, at-risk groups. In the United States, HBV vaccine is recommended for all children as part of their regular immunization schedule.

Hepatitis B vaccination is the most effective measure to prevent HBV and its consequences. A comprehensive public health strategy exists to eliminate HBV transmission in the United States. It involves universal vaccination of infants, routine screening of all pregnant women for HBsAg, postexposure immunoprophylaxis of infants born to HBsAg-positive mothers, vaccination of children and adolescents not previously vaccinated, and vaccination of unvaccinated adults at increased risk for infection.

Immunosuppressed groups, such as hemodialysis patients or those receiving cancer chemotherapy or infected with HIV, respond to vaccination less well than healthy individuals.

Studies on passive immunization using specific hepatitis B immune globulin (HBIG) have shown a protective effect if it is given soon after exposure. HBIG is not recommended for preexposure prophylaxis because the HBV vaccine is available and effective. Persons exposed to HBV percutaneously or by contamination of mucosal surfaces should immediately receive both HBIG and HBsAg vaccine administered simultaneously at different sites to provide immediate protection with passively acquired antibody followed by active immunity generated by the vaccine.

Immune globulin isolated from plasma by the cold ethanol fractionation method has not been documented to transmit HBV, HAV, HCV, or HIV in the United States. Immune globulins prepared outside the United States by other methods have been implicated in outbreaks of hepatitis B and C.

Women who are HBV carriers or who acquire type B hepatitis while pregnant can transmit the disease to their infants. The effectiveness of hepatitis vaccine and HBIG in preventing hepatitis B in infants born to HBV-positive mothers has been substantiated. Reduction in the cost of vaccine for public health programs has made vaccination of newborns feasible in areas of high endemicity. The high cost of HBIG precludes its use in most countries.

Patients with acute type B hepatitis generally need not be isolated so long as blood and instrument precautions are stringently observed, both in the general patient care areas and in the laboratories. Because spouses and intimate contacts of persons with acute type B hepatitis are at risk of acquiring clinical type B hepatitis, they need to be informed about practices that might increase the risk of infection or transmission. There is no evidence that asymptomatic HBsAg-positive food handlers pose a health risk to the general public.

D. Hepatitis C

There is no vaccine for hepatitis C although several candidate vaccines are undergoing tests. Control measures focus on prevention activities that reduce risks for contracting HCV. These include screening and testing blood, plasma, organ, tissue, and semen donors; virus inactivation of plasma-derived products; counseling of persons with high-risk drug or sexual practices; implementation of infection control practices in health care and other settings; and professional and public education.

E. Hepatitis D

Delta hepatitis can be prevented by vaccinating HBV-susceptible persons with hepatitis B vaccine. However, vaccination does not protect hepatitis B carriers from superinfection by HDV.

REVIEW QUESTIONS

1. A 24-year-old woman in New York City is admitted to the hospital because of jaundice. On workup she is found to have HCV infection. The major risk factor for HCV infection in the United States is

   (A) Tattoos
   (B) Injecting drug use
   (C) Blood transfusion
   (D) Sexual activity
   (E) Working in health care occupations

2. Which of the following exposures poses a risk for hepatitis infection?

   (A) A nurse sustains a needlestick while drawing up insulin to administer to an HBV-infected patient with diabetes
   (B) While cleaning the bathroom, a housekeeper's intact skin has contact with feces
   (C) An operating room technician with chapped and abraded hands notices blood under his gloves after assisting in an operation on a patient with HCV infection
   (D) A child drinks out of the same cup as her mother, who has an HAV infection
   (E) A shopper eats a sandwich prepared by a worker with an asymptomatic HBV infection

3. An epidemic of jaundice caused by HEV occurred in New Delhi. HEV is

   (A) Found in rodents and pigs
   (B) A major cause of blood-borne hepatitis
   (C) The cause of a disease that resembles hepatitis C
   (D) Capable of establishing chronic infections
   (E) Associated with an increased risk of liver cancer

4. HDV (delta agent) is found only in patients who have either acute or chronic infection with HBV. HDV

   (A) Is a defective mutant of HBV
   (B) Depends on HBV surface antigen for virion formation
   (C) Induces an immune response indistinguishable from that induced by HBV
   (D) Is related to HCV
   (E) Contains a circular DNA genome

5. A 23-year-old woman is planning a 1-year trip through Europe, Egypt, and the Indian subcontinent and receives a vaccine for hepatitis A. The current hepatitis A vaccine is

   (A) A live attenuated virus vaccine
   (B) A recombinant DNA vaccine
   (C) A formalin-inactivated virus vaccine
   (D) An envelope glycoprotein subunit vaccine
   (E) A chimeric poliovirus that expresses HAV neutralizing epitopes

6. The following statements about HCV infection and associated chronic liver disease in the United States are correct except that

   (A) HCV is responsible for 40% of chronic liver disease
   (B) Chronic infection develops in most (70–90%) HCV-infected persons

(C) HCV-associated liver disease is the major cause for liver transplantation

(D) HCV viremia occurs transiently during early stages of infection

(E) HCV-infected patients are at high risk (5–20%) for liver cancer

7. A middle-aged man complained of acute onset of fever, nausea, and pain in the right upper abdominal quadrant. There was jaundice, and dark urine had been observed several days earlier. A laboratory test was positive for HAV IgM antibody. The physician can tell the patient that

(A) He probably acquired the infection from a recent blood transfusion

(B) He will probably develop chronic hepatitis

(C) He will be at high risk of developing hepatocellular carcinoma

(D) He will be resistant to infection with hepatitis E

(E) He may transmit the infection to family members by person-to-person spread for up to 2 weeks

8. Several different viruses can cause hepatitis. One of the following statements applies to all four viruses: HAV, HCV, HDV, and HEV.

(A) Contains a single-stranded RNA genome

(B) Is transmitted primarily by the parenteral route

(C) Is transmitted primarily by the fecal–oral route

(D) Is associated with fulminant hepatitis

(E) Undergoes sequence variation during chronic infection

9. A 30-year-old student goes to the emergency room because of fever and anorexia for the past 3 days. She appears jaundiced. Her liver is enlarged and tender. A laboratory test shows elevated aminotransferases. She reports a history of having received hepatitis B vaccine 2 years ago but has not had hepatitis A vaccine. The results of her hepatitis serologic tests are as follows: HAV IgM-negative, HAV IgG-positive, HBsAg-negative, HBsAb-positive, HBcAb-negative, HCV Ab-positive. The most accurate conclusion is that she probably

(A) Has hepatitis A now, has not been infected with HBV, and had hepatitis C in the past

(B) Has hepatitis A now and has been infected with both HBV and HCV in the past

(C) Has been infected with HAV and HCV in the past and has hepatitis B now

(D) Has been infected with HAV in the past, has not been infected with HBV, and has hepatitis C now

(E) Has been infected with HAV and HCV in the past, has not been infected with HBV, and has hepatitis E now

10. A 36-year-old nurse is found to be both HBsAg-positive and HBeAg-positive. The nurse most likely

(A) Has acute hepatitis and is infectious

(B) Has both HBV and HEV infections

(C) Has a chronic HBV infection

(D) Has cleared a past HBV infection

(E) Was previously immunized with HBV vaccine prepared from healthy HBsAg-positive carriers

11. The following persons are at increased risk for HAV infection and should be routinely vaccinated except for which group?

(A) Persons traveling to or working in countries that have high levels of HAV infection

(B) Men who have sex with men

(C) Users of illegal drugs (both injecting and noninjecting)

(D) Persons who have an occupational risk for infection

(E) Persons who have a clotting factor disorder

(F) Susceptible persons who have chronic liver disease

(G) Teachers in elementary schools

12. There is global variation in the prevalence of HBV infection. Which of the following geographic areas has low endemicity (HBsAg prevalence of <2%)?

(A) Southeast Asia

(B) The Pacific Islands

(C) Eastern Europe

(D) Australia

(E) Sub-Saharan Africa

13. Which of the following persons are not recommended to receive hepatitis B vaccine because they have a risk factor for HBV infection?

(A) Sexually active persons who are not in a long-term, mutually monogamous relationship

(B) Injection drug users

(C) Pregnant women

(D) Persons who live in a household with a person who is HBsAg positive

(E) Persons seeking treatment for a sexually transmitted disease

14. Which of the following statements regarding HBIG is not true?

(A) HBIG provides temporary protection when administered in standard doses

(B) HBIG typically is used instead of hepatitis B vaccine for postexposure immunoprophylaxis to prevent HBV infection

(C) No evidence exists that HBV, HCV, or HIV have ever been transmitted by HBIG in the United States

(D) HBIG is not used as protection against HCV infection

Answers

1. B	5. C	9. D	13. C
2. C	6. D	10. A	14. B
3. A	7. E	11. G	
4. B	8. A	12. D	

REFERENCES

A comprehensive immunization strategy to eliminate transmission of hepatitis B virus infection in the United States. Recommendations of the Advisory Committee on Immunization Practices (ACIP). Part II: Immunization of adults. MMWR Morb Mortal Wkly Rep 2006;55(RR-16).

Advisory Committee on Immunization Practices: A comprehensive immunization strategy to eliminate transmission of hepatitis B virus infection in the United States. MMWR Morb Mortal Wkly Rep 2005;54(RR-16).

Advisory Committee on Immunization Practices: Prevention of hepatitis A through active or passive immunization. MMWR Morb Mortal Wkly Rep 2006;55(RR-7).

Emerson SU, Purcell RH: Hepatitis E virus. In: *Fields Virology,* 5th ed. Knipe DM et al (editors). Lippincott, Williams & Wilkins, 2007.

Hollinger FB, Emerson SU: Hepatitis A virus. In: *Fields Virology,* 5th ed. Knipe DM et al (editors). Lippincott Williams & Wilkins, 2007.

Lemon SM, Walker C, Alter MJ, Yi M: Hepatitis C virus. In: *Fields Virology,* 5th ed. Knipe DM et al (editors). Lippincott, Williams & Wilkins, 2007.

Recommendations for prevention and control of hepatitis C virus (HCV) infection and HCV-related chronic disease. MMWR Morb Mortal Wkly Rep 1998;47(RR-19).

Seeger C, Zoulim F, Mason WS: Hepadnaviruses. In: *Fields Virology,* 5th ed. Knipe DM et al (editors). Lippincott, Williams & Wilkins, 2007.

Taylor JM, Farci P, Purcell RH: Hepatitis D (delta) virus. In: *Fields Virology,* 5th ed. Knipe DM et al (editors). Lippincott, Williams & Wilkins, 2007.

Updated U.S. Public Health Service guidelines for the management of occupational exposures to HBV, HCV, and HIV and recommendations for postexposure prophylaxis. MMWR Morb Mortal Wkly Rep 2001;50(RR-11).

Weinbaum C, Lyerla R, Margolis HS: Prevention and control of infections with hepatitis viruses in correctional settings. MMWR Recomm Rep 2003;52(RR-1):1.

Weiss U: Insight: Hepatitis C. Nature 2005;436:929–978.

Picornaviruses (Enterovirus & Rhinovirus Groups)

Picornaviruses represent a very large virus family with respect to the number of members but one of the smallest in terms of virion size and genetic complexity. They include two major groups of human pathogens: **enteroviruses** and **rhinoviruses**. Enteroviruses are transient inhabitants of the human alimentary tract and may be isolated from the throat or lower intestine. Rhinoviruses are isolated chiefly from the nose and throat.

Many picornaviruses cause diseases in humans ranging from severe paralysis to aseptic meningitis, pleurodynia, myocarditis, vesicular and exanthematous skin lesions, mucocutaneous lesions, respiratory illnesses, undifferentiated febrile illness, conjunctivitis, and severe generalized disease of infants. However, subclinical infection is far more common than clinically manifest disease. Etiology is difficult to establish, as different viruses may produce the same syndrome; the same picornavirus may cause more than a single syndrome; and some clinical symptoms cannot be distinguished from those caused by other types of viruses. The most serious disease caused by any enterovirus is poliomyelitis.

A worldwide effort is under way with the goal of total eradication of poliomyelitis.

PROPERTIES OF PICORNAVIRUSES

Important properties of picornaviruses are shown in Table 36–1.

Structure & Composition

The virion of enteroviruses and rhinoviruses consists of a capsid shell of 60 subunits, each of four proteins (VP1–VP4) arranged with icosahedral symmetry around a genome made up of a single strand of positive-sense RNA (Figure 36–1). Parechoviruses are similar except that their capsids contain only three proteins, as VP0 does not get cleaved into VP2 and VP4.

By means of x-ray diffraction studies, the molecular structures of poliovirus and rhinovirus have been determined. The three largest viral proteins, VP1–VP3, have a very similar core structure, in which the peptide backbone of the protein loops back on itself to form a barrel of eight strands held together by hydrogen bonds (the β barrel). The amino acid chain between the β barrel and the amino and carboxyl terminal portions of the protein contains a series of loops. These loops include the main antigenic sites that are found on the surface of the virion and are involved in the neutralization of viral infection.

There is a prominent cleft or canyon around each pentameric vertex on the surface of the virus particle. The receptor binding site used to attach the virion to a host cell is thought to be located near the floor of the canyon. This location would presumably protect the crucial cell attachment site from structural variation influenced by antibody selection in hosts, as the canyon is too narrow to permit deep penetration of antibody molecules (Figure 36–1).

The genome RNA ranges in size from 7.2 kb (human rhinovirus) to 7.4 kb (poliovirus, hepatitis A virus) to 8.4 kb (aphthovirus). The organization of the genome is similar for all (Figure 36–2). The genome is polyadenylated at the 3′ end and has a small viral coded protein (VPg) covalently bound to the 5′ end. The positive-sense genomic RNA is infectious.

TABLE 36–1 Important Properties of Picornaviruses

Virion: Icosahedral, 28–30 nm in diameter, contains 60 subunits

Composition: RNA (30%), protein (70%)

Genome: Single-stranded RNA, linear, positive-sense, 7.2–8.4 kb in size, MW 2.5 million, infectious, contains genome-linked protein (VPg)

Proteins: Four major polypeptides cleaved from a large precursor polyprotein. Surface capsid proteins VP1 and VP3 are major antibody-binding sites. VP4 is an internal protein.

Envelope: None

Replication: Cytoplasm

Outstanding characteristic: Family is made up of many enterovirus and rhinovirus types that infect humans and lower animals, causing various illnesses ranging from poliomyelitis to aseptic meningitis to the common cold.

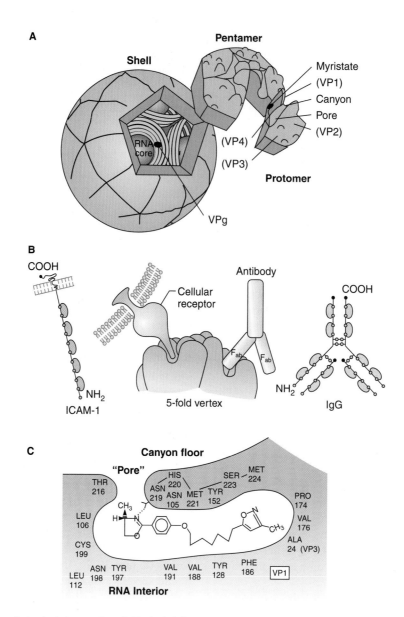

FIGURE 36–1 Structure of a typical picornavirus. **A:** Exploded diagram showing internal location of the RNA genome surrounded by capsid composed of pentamers of proteins VP1, VP2, VP3, and VP4. Note the "canyon" depression surrounding the vertex of the pentamer. **B:** Binding of cellular receptor to the floor of the canyon. The major rhinovirus receptor (ICAM-1 molecule) has a diameter roughly half that of an IgG antibody molecule. **C:** Location of a drug binding site in VP1 of a rhinovirus. The antiviral drug shown, WIN 52084, prevents viral attachment by deforming part of the canyon floor. (Reproduced with permission from Rueckert RR: Picornaviridae: The viruses and their replication. In: *Fields Virology,* 3rd ed. Fields BN et al [editors]. Lippincott-Raven, 1996.)

Enteroviruses are stable at acid pH (3.0–5.0) for 1–3 hours, whereas rhinoviruses are acid-labile. Enteroviruses and some rhinoviruses are stabilized by magnesium chloride against thermal inactivation. Enteroviruses have a buoyant density in cesium chloride of about 1.34 g/mL; human rhinoviruses, about 1.4 g/mL.

Classification

The **Picornaviridae** family contains nine genera, including *Enterovirus* (enteroviruses), *Rhinovirus* (rhinoviruses), *Hepatovirus* (hepatitis A virus), *Parechovirus* (parechoviruses),

Aphthovirus (foot-and-mouth disease viruses), and *Cardiovirus* (cardioviruses). The first four groups contain important human pathogens.

Enteroviruses of human origin are subdivided into six species based mainly upon sequence analyses. The former taxonomy for these viruses included the following: (1) polioviruses, types 1–3; (2) coxsackieviruses of group A, types 1–24 (there is no type 23); (3) coxsackieviruses of group B, types 1–6; (4) echoviruses, types 1–33 (no types 8, 10, 22, 23, 28, or 34); and (5) enteroviruses, types 68–78 (no type 72) (Table 36–2). Additional enteroviruses types 79–101 have been tentatively identified. Since 1969, new enterovirus types

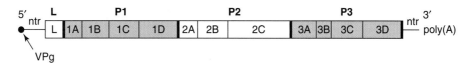

FIGURE 36-2 Structure of picornavirus RNA and genetic organization of its polyprotein (open bar). The RNA is organized 5'-VPg-ntr-polyprotein-ntr-poly(A). ntr refers to nontranslated regions flanking the polyprotein. L specifies a leader protein found in cardioviruses and aphthoviruses but not in enteroviruses, human rhinoviruses, or human hepatitis virus A. P1, P2, and P3 refer to precursor proteins cleaved by virus-coded proteinases into four, three, and four end products, respectively. (Reproduced with permission from Rueckert RR: Picornaviridae: The viruses and their replication. In: *Fields Virology*, 3rd ed. Fields BN et al [editors]. Lippincott-Raven, 1996.)

TABLE 36-2   Characteristics of Human Picornaviruses

Property		Enteroviruses				Parechoviruses[c]	Rhinoviruses[d]
	Polio	Coxsackie		Echo[a]	Entero[b]		
		A[a]	B				
Serotypes	1–3	1–24	1–6	1–33	68–78	1–3	>100
Acid pH (pH 3.0)	Stable	Stable	Stable	Stable	Stable	Stable	Labile
Density (g/mL)	1.34	1.34	1.34	1.34	1.34		1.4
Optimal temperature for growth	37°C	37°C	37°C	37°C	37°C	37°C	33°C
Common sites of isolation from humans							
Nose	0	0	0	0	0	0	+
Throat	+	+	+	+	+		+
Lower intestine	+	+	+	+	+	+	0
Infect newborn mice[e]	0	+	+	0			0

[a]Because of reclassifications there is no coxsackievirus A23, echovirus type 8, 10, 22, 23, 28, or 34, or enterovirus type 72.

[b]Since 1969, new enteroviruses have been assigned a number rather than being subclassified as coxsackieviruses or echoviruses. Enteroviruses 79–101 await inclusion in the International Committee on Taxonomy of Viruses classification.

[c]Parechoviruses 1 and 2 were previously classified as echovirus types 22 and 23.

[d]Rhinovirus 87 is considered to be the same as enterovirus 68.

[e]Some variability exists in this property.

have been assigned enterovirus type numbers rather than being subclassified as coxsackieviruses or echoviruses. The vernacular names of the previously identified enteroviruses have been retained. The coxsackie A viruses fall primarily into human enterovirus species (HEV)-A and HEV-C, and coxsackie B viruses and echoviruses into HEV-B.

Enteroviruses also exist in many animals, including cattle, pigs, monkeys, and mice.

Human rhinoviruses include more than 100 antigenic types. Rhinoviruses of other host species include those of horses and cattle.

Hepatitis A virus was originally classified as enterovirus type 72 but is now assigned to a separate genus. It is described in Chapter 35.

Parechoviruses, previously classified as echoviruses 22 and 23, were found to differ significantly from the enteroviruses in both biologic properties and molecular characteristics and were placed into a new genus (*Parechovirus*).

Other picornaviruses are foot-and-mouth disease virus of cattle (*Aphthovirus*) and encephalomyocarditis virus of rodents (*Cardiovirus*).

The host range of picornaviruses varies greatly from one type to the next and even among strains of the same type. Many enteroviruses (polioviruses, echoviruses, some coxsackieviruses) can be grown at 37°C in human and monkey cells; most rhinovirus strains can be recovered only in human cells at 33°C. Coxsackieviruses are pathogenic for newborn mice.

Picornavirus Replication

The picornavirus replication cycle occurs in the cytoplasm of cells (Figure 36–3). First, the virion attaches to a specific receptor in the plasma membrane. The receptors for poliovirus and human rhinovirus are members of the immunoglobulin gene superfamily, which includes antibodies and some

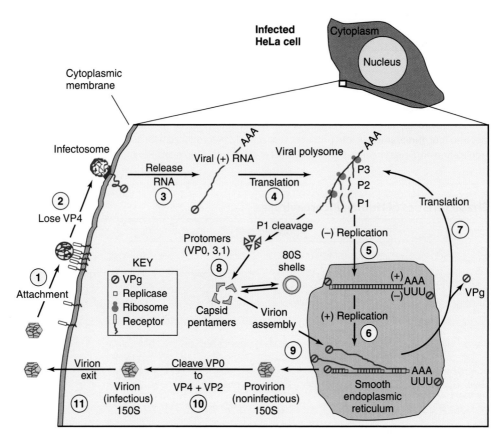

FIGURE 36–3 Overview of the picornavirus infection cycle. (Reproduced with permission from Rueckert RR: Picornaviridae: The viruses and their replication. In: *Fields Virology*, 3rd ed. Fields BN et al [editors]. Lippincott-Raven, 1996.)

cell surface adhesion molecules. In contrast, echoviruses recognize a member of the integrin adhesion superfamily. Not all rhinoviruses or echoviruses use the same cellular receptor. Receptor binding triggers a conformational change in the virion which results in release of the viral RNA into the cell cytosol. VPg is removed from the viral RNA as it associates with ribosomes. Translation occurs via a cap-independent mechanism, using the internal ribosome entry site (IRES) downstream from the 5′ end of the viral genome. This bypasses the need for intact cellular initiation factor complex (eIF4F), required by many capped cellular mRNAs. eIF4 is often cleaved by a viral protease, leading to shut-off of host protein synthesis and preferential translation of viral RNAs.

The infecting viral RNA is translated into a polyprotein that contains both coat proteins and essential replication proteins. This polyprotein is rapidly cleaved into fragments by proteinases encoded in the polyprotein (Figure 36–4). Synthesis of new viral RNA cannot begin until the virus-coded replication proteins, including an RNA-dependent RNA polymerase, are produced. The infecting viral RNA strand is copied, and that complementary strand serves as template for the synthesis of new plus strands. Many plus strands are generated from each minus-strand template. Some new plus strands are recycled

as templates to amplify the pool of progeny RNA; many plus strands get packaged into virions.

Maturation involves several cleavage events. Coat precursor protein P1 (Figure 36–4) is cleaved to form aggregates of VP0, VP3, and VP1. When an adequate concentration is reached, these "protomers" assemble into pentamers that package plus-stranded VPg-RNA to form "provirions." The provirions are not infectious until a final cleavage changes VP0 to VP4 and VP2. The mature virus particles are released when the host cell disintegrates. The multiplication cycle for most picornaviruses takes 5–10 hours.

ENTEROVIRUS GROUP

POLIOVIRUSES

Poliomyelitis is an acute infectious disease that in its serious form affects the central nervous system. The destruction of motor neurons in the spinal cord results in flaccid paralysis. However, most poliovirus infections are subclinical.

Poliovirus has served as a model picornavirus in many laboratory studies of the molecular biology of picornavirus replication.

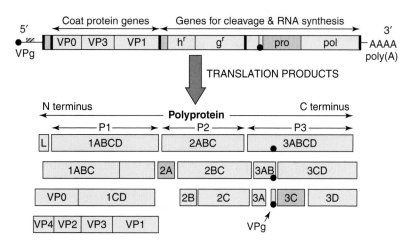

FIGURE 36–4 Organization and expression of the picornavirus genome. **TOP:** The genome-linked VPg protein is at the 5′ end. The striped bar over the 5′-nontranslated region indicates the presence of a polycytidylic acid tract found only in cardioviruses and aphthoviruses. Synthesis of the protein is from left (N terminus) to right (C terminus). Proteins needed for RNA synthesis and proteinases required to cleave the polyprotein are encoded downstream from the capsid protein. **BOTTOM:** Cleavage of the polyprotein is accomplished by virus-coded proteinases 2A and 3C. The maturation cleavage (VP0 → VP4 + VP2) occurs only after the RNA has been packaged in the protein shell. Protein 2A performs early cleavages of the polyprotein, and all other cleavages are performed by proteinase 3C or a precursor form, 3CD. (Reproduced with permission from Rueckert RR: Picornaviridae: The viruses and their replication. In: *Fields Virology*, 3rd ed. Fields BN et al [editors]. Lippincott-Raven, 1996.)

Properties of the Virus

A. General Properties

Poliovirus particles are typical enteroviruses (see above). They are inactivated when heated at 55°C for 30 minutes, but Mg^{2+}, 1 mol/L, prevents this inactivation. Whereas purified poliovirus is inactivated by a chlorine concentration of 0.1 ppm, much higher concentrations of chlorine are required to disinfect sewage containing virus in fecal suspensions and in the presence of other organic matter. Polioviruses are not affected by ether or sodium deoxycholate.

B. Animal Susceptibility & Growth of Virus

Polioviruses have a very restricted host range. Most strains will infect monkeys when inoculated directly into the brain or spinal cord. Chimpanzees and cynomolgus monkeys can also be infected by the oral route; in chimpanzees, the infection is usually asymptomatic and the animals become intestinal carriers of the virus.

Most strains can be grown in primary or continuous cell line cultures derived from a variety of human tissues or from monkey kidney, testis, or muscle but not from tissues of lower animals.

Poliovirus requires a primate-specific membrane receptor for infection, and the absence of this receptor on the surface of nonprimate cells makes them virus-resistant. This restriction can be overcome by transfection of infectious poliovirus RNA into resistant cells. Introduction of the viral receptor gene converts resistant cells to susceptible cells. Transgenic mice harboring the primate receptor gene have been developed; they are susceptible to human polioviruses.

C. Antigenic Properties

There are three antigenic types of polioviruses.

Pathogenesis & Pathology

The mouth is the portal of entry of the virus, and primary multiplication takes place in the oropharynx or intestine. The virus is regularly present in the throat and in the stools before the onset of illness. One week after infection there is little virus in the throat, but virus continues to be excreted in the stools for several weeks even though high antibody levels are present in the blood.

The virus may be found in the blood of patients with nonparalytic poliomyelitis. Antibodies to the virus appear early in the disease, usually before paralysis occurs.

It is believed that the virus first multiplies in the tonsils, the lymph nodes of the neck, Peyer's patches, and the small intestine. The central nervous system may then be invaded by way of the circulating blood.

Poliovirus can spread along axons of peripheral nerves to the central nervous system, where it continues to progress along the fibers of the lower motor neurons to increasingly involve the spinal cord or the brain. Poliovirus invades certain types of nerve cells, and in the process of its intracellular multiplication it may damage or completely destroy these cells.

Poliovirus does not multiply in muscle in vivo. The changes that occur in peripheral nerves and voluntary muscles are secondary to the destruction of nerve cells. Some cells that lose their function may recover completely. Inflammation occurs secondary to the attack on the nerve cells.

In addition to pathologic changes in the nervous system, there may be myocarditis, lymphatic hyperplasia, and ulceration of Peyer's patches.

Clinical Findings

When an individual susceptible to infection is exposed to the virus, the response ranges from inapparent infection without symptoms, to a mild febrile illness, to severe and permanent paralysis. Most infections are subclinical; only about 1% of infections result in clinical illness.

The incubation period is usually 7–14 days, but it may range from 3 to 35 days.

A. Mild Disease

This is the most common form of disease. The patient has only a minor illness, characterized by fever, malaise, drowsiness, headache, nausea, vomiting, constipation, and sore throat in various combinations. Recovery occurs in a few days.

B. Nonparalytic Poliomyelitis (Aseptic Meningitis)

In addition to the symptoms and signs listed in the preceding paragraph, the patient with the nonparalytic form has stiffness and pain in the back and neck. The disease lasts 2–10 days, and recovery is rapid and complete. Poliovirus is only one of many viruses that produce aseptic meningitis. In a small percentage of cases, the disease advances to paralysis.

C. Paralytic Poliomyelitis

The predominating complaint is flaccid paralysis resulting from lower motor neuron damage. However, incoordination secondary to brain stem invasion and painful spasms of nonparalyzed muscles may also occur. The amount of damage varies greatly. Maximal recovery usually occurs within 6 months, with residual paralysis lasting much longer.

D. Progressive Postpoliomyelitis Muscle Atrophy

A recrudescence of paralysis and muscle wasting has been observed in individuals decades after their experience with paralytic poliomyelitis. Although progressive postpoliomyelitis muscle atrophy is rare, it is a specific syndrome. It does not appear to be a consequence of persistent infection but rather a result of physiologic and aging changes in paralytic patients already burdened by loss of neuromuscular functions.

Laboratory Diagnosis

The virus may be recovered from throat swabs taken soon after onset of illness and from rectal swabs or stool samples collected over long periods. No permanent carriers have been identified among immunocompetent individuals, but long-term excretion of poliovirus has been observed in some immunodeficient persons. Poliovirus is uncommonly recovered from the cerebrospinal fluid—unlike some coxsackieviruses and echoviruses.

Specimens should be kept frozen during transit to the laboratory. Cultures of human or monkey cells are inoculated, incubated, and observed. Cytopathogenic effects appear in 3–6 days. An isolated virus is identified and typed by neutralization with specific antiserum. Virus can also be identified by polymerase chain reaction (PCR) assays.

Paired serum specimens are required to show a rise in antibody titer during the course of the disease. Only first infection with poliovirus produces strictly type-specific responses. Subsequent infections with heterotypic polioviruses induce antibodies against a group antigen shared by all three types.

Immunity

Immunity is permanent to the virus type causing the infection and is predominantly antibody mediated. There may be a low degree of heterotypic resistance induced by infection, especially between type 1 and type 2 polioviruses.

Passive immunity is transferred from mother to offspring. The maternal antibodies gradually disappear during the first 6 months of life. Passively administered antibody lasts only 3–5 weeks.

Virus-neutralizing antibody forms soon after exposure to the virus, often before the onset of illness, and apparently persists for life. Its formation early in the disease reflects the fact that viral multiplication occurs in the body before the invasion of the nervous system. As the virus in the brain and spinal cord is not influenced by high titers of antibodies in the blood, immunization is of value only if it precedes the onset of symptoms referable to the nervous system.

The VP1 surface protein of poliovirus contains several virus-neutralizing epitopes, each of which may contain fewer than ten amino acids. Each epitope is capable of inducing virus-neutralizing antibodies.

Global Eradication

A major campaign was launched by the World Health Organization in 1988 to eradicate poliovirus from the world as was done for smallpox virus. There were an estimated 350,000 cases of polio worldwide in 1988. The Americas were certified as free from wild poliovirus in 1994, the Western Pacific Region in 2000, and Europe in 2002. Progress is being made globally; fewer than 2000 cases of polio still occur each year, principally in Africa and the Indian subcontinent.

In 2009, only four countries remained polio-endemic; however, as occurred in 2004–2005, wild poliovirus may spread into polio-free countries if vaccination programs have been discontinued.

Epidemiology

Poliomyelitis has had three epidemiologic phases: endemic, epidemic, and the vaccine era. The first two reflect prevaccine patterns. The generally accepted explanation is that improved systems of hygiene and sanitation in cooler climates promoted the transition from endemic to epidemic paralytic disease in those societies.

Before global eradication efforts began, poliomyelitis occurred worldwide—year-round in the tropics and during summer and fall in the temperate zones. Winter outbreaks are rare.

The disease occurs in all age groups, but children are usually more susceptible than adults because of the acquired immunity of the adult population. In developing areas, where living conditions favor the wide dissemination of virus, poliomyelitis is a disease of infancy and early childhood ("infantile paralysis"). In developed countries, before the advent of vaccination, the age distribution shifted so that most patients were over age 5 and 25% were over age 15 years. The case fatality rate is variable. It is highest in the oldest patients and may reach 5–10%.

Humans are the only known reservoir of infection. Under crowded conditions of poor hygiene and sanitation in warm areas, where almost all children become immune early in life, polioviruses maintain themselves by continuously infecting a small part of the population. In temperate zones with high levels of hygiene, epidemics have been followed by periods of little spread of virus until sufficient numbers of susceptible children have grown up to provide a pool for transmission in the area. Virus can be recovered from pharynx and intestine of patients and healthy carriers. The prevalence of infection is highest among household contacts.

In temperate climates, infection with enteroviruses, including poliovirus, occurs mainly during the summer. Virus is present in sewage during periods of high prevalence and can serve as a source of contamination of water used for drinking, bathing, or irrigation. There is a direct correlation between poor hygiene, sanitation, and crowding and the acquisition of infection and antibodies at an early age.

Prevention & Control

Both live-virus and killed-virus vaccines are available. Formalinized vaccine (Salk) is prepared from virus grown in monkey kidney cultures. Killed-virus vaccine induces humoral antibodies but does not induce local intestinal immunity so that virus is still able to multiply in the gut. Oral vaccines contain live attenuated virus grown in primary monkey or human diploid cell cultures. The vaccine can be stabilized by magnesium chloride so that it can be kept without losing potency for a year at 4°C and for weeks at moderate room temperature (about 25°C). Nonstabilized vaccine must be kept frozen until used.

The live polio vaccine infects, multiplies, and thus immunizes. In the process, infectious progeny of the vaccine virus are disseminated in the community. The vaccine produces not only IgM and IgG antibodies in the blood but also secretory IgA antibodies in the intestine, which then becomes resistant to reinfection (see Figure 30–9).

Both killed-virus and live-virus vaccines induce antibodies and protect the central nervous system from subsequent invasion by wild virus. However, the gut develops a far greater degree of resistance after administration of live-virus vaccine.

A potential limiting factor for oral vaccine is interference. If the alimentary tract of a child is infected with another enterovirus at the time the vaccine is given, the establishment of polio infection and immunity may be blocked. This may be an important problem in areas—particularly in tropical regions—where enterovirus infections are common.

The vaccine viruses—particularly types 2 and 3—may mutate in the course of their multiplication in vaccinated children. However, only extremely rare cases of paralytic poliomyelitis have occurred in recipients of oral polio vaccine or their close contacts (no more than one vaccine-associated case for every two million persons vaccinated).

Trivalent oral polio vaccine was generally used in the United States. However, in 2000 the Advisory Committee on Immunization Practices recommended a switch to the use of only inactivated polio vaccine (four doses) for children in the United States. The change was made because of the reduced risk for wild virus-associated disease resulting from continuing progress in global eradication of poliovirus. This schedule will reduce the incidence of vaccine-associated disease while maintaining individual and population immunity against polioviruses.

Once global eradication is achieved, the use of oral polio vaccine will cease. Continuation of its use could lead to the reemergence of polio due to mutation and increased transmissibility and neurovirulence of vaccine virus.

Pregnancy is neither an indication for nor a contraindication to required immunization. Live-virus vaccine should not be administered to immunodeficient or immunosuppressed individuals or their household contacts. Only killed-virus (Salk) vaccine is to be used in those cases.

There are no antiviral drugs for treatment of poliovirus infection. Quarantine of patients or intimate contacts is ineffective in controlling the spread of the disease. This is understandable in view of the large number of inapparent infections that occur. Immune globulin can provide protection for a few weeks against the paralytic disease but does not prevent subclinical infection. Immune globulin is effective only if given shortly before infection; it is of no value after clinical symptoms develop.

Before the beginning of vaccination campaigns in the United States, there were about 21,000 cases of paralytic poliomyelitis per year.

COXSACKIEVIRUSES

Coxsackieviruses, a large subgroup of the enteroviruses, are divided into two groups, A and B, having different pathogenic potentials for mice. They produce a variety of illnesses in humans, including aseptic meningitis and respiratory and undifferentiated febrile illnesses. Herpangina (vesicular pharyngitis), hand-foot-and-mouth disease, and acute hemorrhagic conjunctivitis are caused by certain coxsackievirus group A serotypes; pleurodynia (epidemic myalgia), myocarditis, pericarditis, and severe generalized disease of infants are caused by some group B coxsackieviruses. In addition to these, a number of group A and B serotypes can give rise to meningoencephalitis and paralysis. Generally, paralysis produced by nonpolio enteroviruses is incomplete and reversible. Coxsackie B viruses are the most commonly identified causative agents of viral heart disease in humans (Table 36–3). The coxsackieviruses tend to be more pathogenic than the echoviruses. Some of the more recent isolates of enteroviruses exhibit properties similar to the coxsackieviruses.

Properties of the Virus

Coxsackieviruses are highly infective for newborn mice, in contrast to most other human enteroviruses. Certain strains (B1–6, A7, 9, 16, and 24) also grow in monkey kidney cell culture. Some group A strains grow in human amnion and

TABLE 36–3   Human Enteroviruses and Commonly Associated Clinical Syndromes[a]

| Syndrome | Poliovirus Types 1–3 | Coxsackievirus | | Echovirus Types 1–33 | Enterovirus Types 68–71 | Parechovirus Types 1–3 |
		Group A Types 1–24	Group B Types 1–6			
Neurologic						
Aseptic meningitis	1–3	Many	1–6	Many	71	1
Paralysis	1–3	7, 9	2–5	2, 4, 6, 9, 11, 30	70, 71	3
Encephalitis		2, 5–7, 9	1–5	2, 6, 9, 19	70, 71	
Skin and mucosa						
Herpangina		2–6, 8, 10			71	
Hand-foot-and-mouth disease		5, 10, 16			71	
Exanthems		Many	5	2, 4, 6, 9, 11, 16, 18		
Cardiac and muscular						
Pleurodynia (epidemic myalgia)			1–5	1, 6, 9		
Myocarditis, pericarditis			1–5	1, 6, 9, 19		1
Ocular						
Acute hemorrhagic conjunctivitis		24			70	
Respiratory						
Colds		21, 24	1, 3, 4, 5	4, 9, 11, 20, 25		1
Pneumonia			4, 5		68	1
Pneumonitis of infants		9, 16				
Pulmonary edema					71	
Gastrointestinal						
Diarrhea		18, 20–22, 24[b]		Many[b]		1
Hepatitis		4, 9	5	4, 9		
Other						
Undifferentiated febrile illness	1–3		1–6			
Generalized disease of infants			1–5	11		
Diabetes mellitus			3, 4			

[a]Examples are not all-inclusive. Other enterovirus types may be associated with a given disease.

[b]Causality not established.

human embryonic lung fibroblast cells. Type A14 produces poliomyelitis-like lesions in adult mice and in monkeys, but only myositis in suckling mice. Type A7 strains produce paralysis and severe central nervous system lesions in monkeys. Group A viruses produce widespread myositis in the skeletal muscles of newborn mice, resulting in flaccid paralysis without other observable lesions. The genetic makeup of inbred strains of mice determines their susceptibility to coxsackie B viruses.

Pathogenesis & Pathology

Virus has been recovered from the blood in the early stages of natural infection in humans. Virus is also found in the throat for a few days early in the infection and in the stools for up to 5–6 weeks. Virus distribution is similar to that of the other enteroviruses.

Clinical Findings

The incubation period of coxsackievirus infection ranges from 2 to 9 days. The clinical manifestations of infection with various coxsackieviruses are diverse and may present as distinct disease entities (Table 36–3). They range from mild febrile illness to central nervous system, skin, cardiac, and respiratory diseases. The examples shown are not all-inclusive; different serotypes may be associated with a particular outbreak.

Aseptic meningitis is caused by all types of group B coxsackieviruses and by many group A coxsackieviruses, most commonly A7 and A9. Fever, malaise, headache, nausea, and abdominal pain are common early symptoms. The disease sometimes progresses to mild muscle weakness suggestive of paralytic poliomyelitis. Patients almost always recover completely from nonpoliovirus paresis.

Herpangina is a severe febrile pharyngitis that is caused by certain group A viruses. Despite its name, it has nothing to do with herpesviruses. There is an abrupt onset of fever and sore throat with discrete vesicles on the posterior half of the palate, pharynx, tonsils, or tongue. The illness is self-limited and most frequent in small children.

Hand-foot-and-mouth disease is characterized by oral and pharyngeal ulcerations and a vesicular rash of the palms and soles that may spread to the arms and legs. Vesicles heal without crusting, which clinically differentiates them from the vesicles of herpesviruses and poxviruses. This disease has been associated particularly with coxsackievirus A16. Virus may be recovered not only from the stool and pharyngeal secretions but also from vesicular fluid. It is not to be confused with foot-and-mouth disease of cattle, caused by an unrelated picornavirus that does not infect humans.

Pleurodynia (also known as epidemic myalgia) is caused by group B viruses. Fever and stabbing chest pain are usually abrupt in onset but are sometimes preceded by malaise, headache, and anorexia. The chest pain may last from 2 days to 2 weeks. Abdominal pain occurs in approximately half of cases, and in children this may be the chief complaint. The illness is self-limited and recovery is complete, though relapses are common.

Myocarditis is a serious disease. It is an acute inflammation of the heart or its covering membranes (pericarditis). Coxsackievirus B infections are a cause of primary myocardial disease in adults as well as children. About 5% of all symptomatic coxsackievirus infections induce heart disease. Infections may be fatal in neonates or may cause permanent heart damage at any age. Persistent viral infections of heart muscle may occur, sustaining chronic inflammation.

Enteroviruses are estimated to cause 15–20% of respiratory tract infections, especially in the summer and fall. A number of coxsackieviruses have been associated with **common colds** and with **undifferentiated febrile illnesses**.

Generalized disease of infants is an extremely serious disease in which the infant is overwhelmed by simultaneous viral infections of multiple organs, including heart, liver, and brain. The clinical course may be rapidly fatal, or the patient may recover completely. The disease is caused by group B coxsackieviruses. In severe cases, myocarditis or pericarditis can occur within the first 8 days of life; it may be preceded by a brief episode of diarrhea and anorexia. The disease may sometimes be acquired transplacentally.

Although the gastrointestinal tract is the primary site of replication for enteroviruses, they do not cause marked disease there. Certain group A coxsackieviruses have been associated with **diarrhea** in children, but causality is unproved.

Laboratory Diagnosis

A. Recovery of Virus

Virus can be isolated from throat washings during the first few days of illness and from stools during the first few weeks. In coxsackievirus A21 infections, the largest amount of virus is found in nasal secretions. In cases of aseptic meningitis, strains have been recovered from the cerebrospinal fluid as well as from the alimentary tract. In hemorrhagic conjunctivitis cases, A24 virus is isolated from conjunctival swabs, throat swabs, and feces.

Specimens are inoculated into tissue cultures and also into suckling mice. In tissue culture, a cytopathic effect appears within 5–14 days. In suckling mice, signs of illness appear usually within 1 week with group A strains and 2 weeks with group B strains. The virus is identified by the pathologic lesions it produces and by immunologic means. Because of the difficulty of the technique, virus isolation in suckling mice is rarely attempted.

B. Nucleic Acid Detection

Methods for the direct detection of enteroviruses provide rapid and sensitive assays useful for clinical samples. Reverse transcription-PCR tests can be broadly reactive (detect many serotypes) or more specific. Such assays have advantages over cell culture methods, as many enterovirus clinical isolates have poor growth characteristics. Real-time PCR assays are comparable in sensitivity to conventional PCR assays, but are less labor intensive to perform.

C. Serology

Neutralizing antibodies appear early during the course of infection, tend to be specific for the infecting virus, and persist for years. Serum antibodies can also be detected and titrated by the immunofluorescence technique, using infected cell cultures on coverslips as antigens. Serologic tests are difficult to evaluate (because of the multiplicity of types) unless the antigen used in the test has been isolated from a specific patient or during an epidemic outbreak.

Immunity

In humans, neutralizing antibodies are transferred passively from mother to fetus. Adults have antibodies against more types of coxsackieviruses than do children, indicating that multiple experiences with these viruses are common and increasingly so with age.

Epidemiology

Viruses of the coxsackie group have been encountered around the globe. Isolations have been made mainly from human feces, pharyngeal swabbings, sewage, and flies. Antibodies to various coxsackieviruses are found in serum collected from persons all over the world and in pooled immune globulin.

The most frequent types of coxsackieviruses recovered worldwide over an 8-year period (1967–1974) were types A9 and B2–B5. In the United States, from 1970 to 2005, the most common coxsackievirus detections were types A9, B2, and B4 in endemic patterns and type B5 in an epidemic pattern. However, in any given year or area, another type may predominate. An epidemic pattern is characterized by fluctuations in circulation levels, whereas an endemic pattern shows stable, low-level circulation with few peaks.

Coxsackieviruses are recovered much more frequently in summer and early fall. Children develop antibodies in summer, indicating infection by coxsackieviruses during this period. Such children have much higher incidence rates for acute, febrile minor illnesses during the summer than children who fail to develop coxsackievirus antibodies.

Familial exposure is important in the acquisition of infections with coxsackieviruses. Once the virus is introduced into a household, all susceptible persons usually become infected, although all do not develop clinically apparent disease.

The coxsackieviruses share many properties with other enteroviruses, including the echoviruses and polioviruses. Because of their epidemiologic similarities, various enteroviruses may occur together in nature, even in the same human host or the same specimens of sewage.

Control

There are no vaccines or antiviral drugs currently available for prevention or treatment of diseases caused by coxsackieviruses.

OTHER ENTEROVIRUSES

Echoviruses (*e*nteric *c*ytopathogenic *h*uman *o*rphan viruses), based on historical terminology, were grouped together because they infect the human enteric tract and because they can be recovered from humans only by inoculation of certain tissue cultures. More than 30 serotypes are known, but not all have been associated with human illness. More recent isolates are designated as numbered enteroviruses. Aseptic meningitis, encephalitis, febrile illnesses with or without rash, common colds, and ocular disease are among the diseases caused by echoviruses and other enteroviruses.

Clinical Findings

To establish etiologic association of an enterovirus with disease, the following criteria are used: (1) There is a much higher rate of recovery of virus from patients with the disease than from healthy individuals of the same age and socioeconomic level living in the same area at the same time. (2) Antibodies against the virus develop during the course of the disease. If the clinical syndrome can be caused by other known agents, virologic or serologic evidence must be negative for concurrent infection with such agents. (3) The virus is isolated from body fluids or tissues manifesting lesions, eg, from the cerebrospinal fluid in cases of aseptic meningitis.

Many echoviruses have been associated with aseptic meningitis. Rashes are most common in young children. Infantile diarrhea may be associated with some types, but causality has not been established. For many echoviruses, no disease entities have been defined.

Enterovirus 70 is the chief cause of acute hemorrhagic conjunctivitis. It was isolated from the conjunctiva of patients with this striking eye disease, which occurred in pandemic form in 1969–1971 in Africa and Southeast Asia. Acute hemorrhagic conjunctivitis has a sudden onset of subconjunctival hemorrhage. The disease is most common in adults, with an incubation period of 1 day and a duration of 8–10 days. Complete recovery is the rule. The virus is highly communicable and spreads rapidly under crowded or unhygienic conditions.

Enterovirus 71 has been isolated from patients with meningitis, encephalitis, and paralysis resembling poliomyelitis. It is one of the main causes of central nervous system disease, sometimes fatal, around the world. An outbreak of hand-foot-and-mouth disease caused by enterovirus 71 occurred in China in 2008 and involved about 4500 cases and 22 deaths in infants and young children.

With the virtual elimination of poliomyelitis in developed countries, the central nervous system syndromes associated with coxsackieviruses, echoviruses, and other enteroviruses have assumed greater prominence. The latter in children under age 1 year may lead to neurologic sequelae and mental impairment. Enteroviruses recovered from fecal samples of patients with acute flaccid paralysis in Australia between 1996 and 2004 included coxsackieviruses A24 and B5; echoviruses 9, 11, and 18; and enteroviruses 71 and 75. Enterovirus 71 was most common.

Laboratory Diagnosis

It is impossible in an individual case to diagnose an echovirus infection on clinical grounds. However, in the following epidemic situations, echoviruses must be considered: (1) summer outbreaks of aseptic meningitis and (2) summer epidemics, especially in young children, of a febrile illness with rash.

The diagnosis is dependent upon laboratory tests. Nucleic acid detection assays, such as PCR, are more rapid than virus isolation for diagnosis. Although the specific virus may not be identified by PCR, it is often not necessary to determine the specific serotype of infecting enterovirus associated with a disease.

Virus isolation may be accomplished from throat swabs, stools, rectal swabs, and, in aseptic meningitis, cerebrospinal fluid. Serologic tests are impractical—because of the many different viral types—except when a virus has been isolated from a patient or during an outbreak of typical clinical illness. Neutralizing and hemagglutination-inhibiting antibodies are type-specific and may persist for years.

If an agent is isolated in tissue culture, it can be tested against different pools of antisera against enteroviruses. Determination of the type of virus present is by either immunofluorescence or Nt tests. Infection with two or more enteroviruses may occur simultaneously.

Epidemiology

The epidemiology of echoviruses is similar to that of other enteroviruses. They occur in all parts of the globe and are more apt to be found in the young than in the old. In the temperate zone, infections occur chiefly in summer and autumn and are about five times more prevalent in children of lower-income families than in those living in more favorable circumstances.

The most commonly recovered echoviruses worldwide in the period 1967–1974 were types 4, 6, 9, 11, and 30. In the United States, from 1970 to 2005, the most commonly detected echoviruses were types 6, 9, 11, 13, and 30, along with coxsackieviruses A9, B2, B4, and B5, and enterovirus 71, and the diseases most often seen in those patients were aseptic meningitis and encephalitis. However, as with all enteroviruses, dissemination of different serotypes may occur in waves and spread widely.

There appears to be a core group of consistently circulating enteroviruses that determine the bulk of disease burden. Fifteen serotypes accounted for 83% of reports in the United States from 1970 to 2005. Children under 1 year of age accounted for 44% of reports of disease.

Studies of families into which enteroviruses were introduced demonstrated the ease with which these agents spread and the high frequency of infection in persons who had formed no antibodies from earlier exposures. This is true for all enteroviruses.

Control

Avoidance of contact with patients exhibiting acute febrile illness is advisable for very young children. There are no antivirals or vaccines (other than polio vaccines) available for the treatment or prevention of any enterovirus diseases.

ENTEROVIRUSES IN THE ENVIRONMENT

Humans are the only known reservoir for members of the human enterovirus group. These viruses are generally shed for longer periods of time in stools than in secretions from the upper alimentary tract. Thus, fecal contamination (hands, utensils, food, water) is the usual avenue of virus spread. Enteroviruses are present in variable amounts in sewage. This may serve as a source of contamination of water supplies used for drinking, bathing, irrigation, or recreation (Figure 36–5). Enteroviruses survive exposure to the sewage treatments and chlorination in common practice, and human wastes in much of the world are discharged into natural waters with little or no treatment. Waterborne outbreaks due to enteroviruses are difficult to recognize, and it has been shown that the viruses can travel long distances from the source of contamination and remain infectious. Adsorption to organics and sediment material protects viruses from inactivation and helps in transport. Filter-feeding shellfish (oysters, clams, mussels) have been found to concentrate viruses from water and, if inadequately cooked, may transmit disease. Bacteriologic standards using fecal coliform indices as a monitor of water quality probably are not an adequate reflection of a potential for transmission of viral disease.

PARECHOVIRUS GROUP

This genus was defined in the 1990s and contains six types, of which types 1 and 2 were originally classified as echoviruses 22 and 23. Parechoviruses are highly divergent from enteroviruses, with no protein sequence having greater than 30% identity with the corresponding protein of other picornaviruses. The capsid contains three proteins, as the VP0 precursor protein does not get cleaved.

Parechovirus infections are often acquired in early childhood. The viruses replicate in the respiratory and gastrointestinal tract. They have been reported to cause diseases similar to other enteroviruses, such as gastrointestinal and respiratory illness, meningoencephalitis, otitis media, and neonatal diseases.

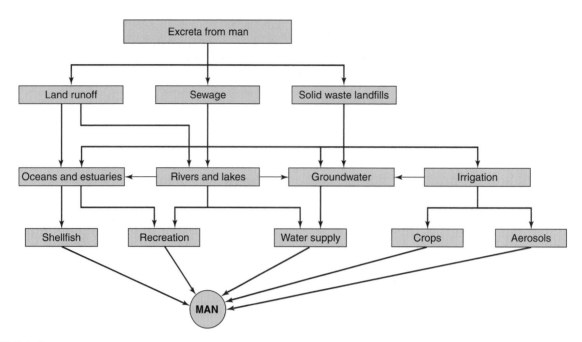

FIGURE 36–5 Routes of potential enteric virus transmission in the environment. (Reproduced with permission from Melnick JL, Gerba CP, Wallis C: Viruses in water. Bull World Health Org 1978;56:499.)

RHINOVIRUS GROUP

Rhinoviruses are the common cold viruses. They are the most commonly recovered agents from people with mild upper respiratory illnesses. They are usually isolated from nasal secretions but may also be found in throat and oral secretions. These viruses—as well as coronaviruses, adenoviruses, enteroviruses, parainfluenza viruses, and influenza viruses—cause upper respiratory tract infections, including the common cold syndrome. Rhinoviruses are also responsible for about one-half of asthma exacerbations.

Classification

Human rhinovirus isolates are numbered sequentially. More than 100 species are known. Isolates within a species share more than 70% sequence identity within certain protein-coding regions.

Human rhinoviruses can be divided into major and minor receptor groups. Viruses of the major group use intercellular adhesion molecule-1 (ICAM-1) as receptor and those of the minor group bind members of the low-density lipoprotein receptor (LDLR) family.

Properties of the Virus

A. General Properties

Rhinoviruses are picornaviruses similar to enteroviruses but differ from them in having a buoyant density in cesium chloride of 1.40 g/mL and in being acid-labile. Virions are unstable below pH 5–6 and complete inactivation occurs at pH 3.0.

Rhinoviruses are more thermostable than enteroviruses and may survive for hours on environmental surfaces.

Nucleotide sequence identity over the entire genome is more than 50% among all rhinoviruses and between enteroviruses and rhinoviruses. There is greater or less identity for particular genomic regions.

In 2009, the genomes of all known strains of rhinovirus were sequenced, defining conserved and divergent regions. This information will facilitate new understanding of pathogenic potential and the design of antiviral drugs and vaccines.

B. Animal Susceptibility & Growth of Virus

These viruses are infectious only for humans, gibbons, and chimpanzees. They can be grown in a number of human cell lines, including the WI-38 and MRC-5 lines. Organ cultures of ferret and human tracheal epithelium may be necessary for some fastidious strains. Most grow better at 33°C, which is similar to the temperature of the nasopharynx in humans, than at 37°C.

C. Antigenic Properties

More than 100 serotypes are known. New serotypes are based on the absence of cross-reactivity in Nt tests using polyclonal antisera. Human rhinovirus 87 is now considered the same serotype as human enterovirus 68.

Pathogenesis & Pathology

The virus enters via the upper respiratory tract. High titers of virus in nasal secretions—which can be found as early as 2–4 days after exposure—are associated with maximal illness.

Thereafter, viral titers fall, although illness persists. In some instances, virus may remain detectable for 3 weeks. There is a direct correlation between the amount of virus in secretions and the severity of illness.

Replication is limited to the surface epithelium of the nasal mucosa. Biopsies have shown that histopathologic changes are limited to the submucosa and surface epithelium. These include edema and mild cellular infiltration. Nasal secretion increases in quantity and in protein concentration.

Rhinoviruses rarely cause lower respiratory tract disease in healthy individuals, although they are associated with the majority of acute asthma exacerbations. Experiments under controlled conditions have shown that chilling, including the wearing of wet clothes, does not produce a cold or increase susceptibility to the virus. Chilliness is an early symptom of the common cold.

Clinical Findings

The incubation period is brief—from 2 to 4 days—and the acute illness usually lasts for 7 days although a nonproductive cough may persist for 2–3 weeks. The average adult has 1–2 attacks each year. Usual symptoms in adults include sneezing, nasal obstruction, nasal discharge, and sore throat; other symptoms may include headache, mild cough, malaise, and a chilly sensation. There is little or no fever. The nasal and nasopharyngeal mucosa become red and swollen. There are no distinctive clinical findings that permit an etiologic diagnosis of colds caused by rhinoviruses versus colds caused by other viruses. Secondary bacterial infection may produce acute otitis media, sinusitis, bronchitis, or pneumonitis, especially in children.

Immunity

Neutralizing antibody to the infecting virus develops in serum and secretions of most persons. Depending on the test used, estimates of the frequency of response have ranged from 37% to over 90%.

Antibody develops 7–21 days after infection; the time of appearance of neutralizing antibody in nasal secretions parallels that of serum antibodies. Because recovery from illness usually precedes appearance of antibodies, it seems that recovery is not dependent on antibody. However, antibody may accomplish final clearance of infection. Serum antibody persists for years but decreases in titer.

Epidemiology

The disease occurs throughout the world. In the temperate zones, the attack rates are highest in early fall and late spring. Prevalence rates are lowest in summer. Members of isolated communities form highly susceptible groups.

The virus is believed to be transmitted through close contact, by means of virus-contaminated respiratory secretions. The fingers of a person with a cold are usually contaminated, and transmission to susceptible persons then occurs by hand to hand, hand to eye, or hand to object (eg, doorknob) to hand contamination. Rhinoviruses can survive for hours on contaminated environmental surfaces. Self-inoculation after hand contamination may be a more important mode of spread than that by airborne particles.

Infection rates are highest among infants and children and decrease with increasing age. The family unit is a major site of spread of rhinoviruses. Introduction of virus is generally attributable to preschool-aged and school-aged children. Secondary attack rates in the family vary from 30% to 70%. Infections in young children are symptomatic whereas infections in adults are often asymptomatic.

In a single community, multiple rhinovirus serotypes cause outbreaks of disease in a single season, and different serotypes predominate during different respiratory disease seasons. There are usually a limited number of serotypes causing disease at any given time.

Treatment & Control

No specific prevention method or treatment is available. The development of a potent rhinovirus vaccine is unlikely because of the difficulty in growing rhinoviruses to high titer in culture, the fleeting immunity, and the multiplicity of serotypes causing colds.

Antiviral drugs are thought to be a more likely control measure for rhinoviruses because of the problems with vaccine development. Many compounds effective in vitro have failed to be effective clinically.

A 5-day course of high doses of intranasal interferon α has been shown to be effective in preventing the spread of rhinoviruses from an index case within a family. It was not effective as therapy of established infections.

FOOT-AND-MOUTH DISEASE (APHTHOVIRUS OF CATTLE)

This highly infectious disease of cloven-hoofed animals such as cattle, sheep, pigs, and goats is rare in the United States but endemic in other countries. It may be transmitted to humans by contact or ingestion. In humans, the disease is characterized by fever, salivation, and vesiculation of the mucous membranes of the oropharynx and of the skin of the palms, soles, fingers, and toes.

The virus is a typical picornavirus and is acid-labile (particles are unstable below pH 6.8). It has a buoyant density in cesium chloride of 1.43 g/mL. There are at least seven types with more than 50 subtypes.

The disease in animals is highly contagious in the early stages of infection when viremia is present and when vesicles in the mouth and on the feet rupture and liberate large amounts of virus. Excreted material remains infectious for long periods. The mortality rate in animals is usually low but may reach 70%. Infected animals become poor producers of milk and meat. Many cattle serve as foci of infection for up to 8 months. Immunity after infection is of short duration.

A variety of animals are susceptible to infection, and the virus has been recovered from at least 70 species of mammals. The typical disease can be reproduced by inoculating the virus into the pads of the foot. Formalin-treated vaccines have been prepared from virus grown in tissue cultures, but such vaccines do not produce long-lasting immunity. New vaccines are being developed based on recombinant DNA techniques.

The methods of control of the disease are dictated by its high degree of contagiousness and the resistance of the virus to inactivation. Should a focus of infection occur in the United States, all exposed animals are slaughtered and their carcasses destroyed. Strict quarantine is established, and the area is not presumed to be safe until susceptible animals fail to develop symptoms within 30 days. Another method is to quarantine the herd and vaccinate all unaffected animals. Other countries have successfully employed systematic vaccination schedules. Some nations (eg, the United States and Australia) forbid the importation of potentially infective materials such as fresh meat, and the disease has been eliminated in these areas.

REVIEW QUESTIONS

1. Which of the following statements about rhinoviruses is correct?
   (A) There are three antigenic types
   (B) Amantadine protects against infection
   (C) They do not survive on environmental surfaces
   (D) They are the most frequent causative agent of the common cold
   (E) They share physicochemical similarities with coronaviruses

2. A 26-year-old man develops myopericarditis with mild congestive heart failure that increases over several weeks. Coxsackievirus B5 infection is diagnosed. Which of the following clinical syndromes is not associated with coxsackievirus infections?
   (A) Herpangina
   (B) Myocarditis/pericarditis
   (C) Aseptic meningitis
   (D) Acute hemorrhagic conjunctivitis
   (E) Progressive postpolio muscle atrophy

3. A 3-month-old child develops fever, restlessness, and unusual crying. These are followed by apparent lethargy. Physical examination shows a normal-appearing infant who is minimally responsive to stimuli. A lumbar puncture yields cerebrospinal fluid with 200 white blood cells per microliter, predominantly lymphocytes. Acute aseptic meningitis is diagnosed, probably caused by an enterovirus. Enteroviruses are characterized by
   (A) Latency in sensory ganglia and reactivation primarily in immunocompromised patients
   (B) Transmission primarily by the fecal–oral route
   (C) The presence of a DNA polymerase enzyme
   (D) The entry of cells following binding to the ICAM-1 receptor
   (E) Undergoing antigenic shift and drift

4. Picornavirus vaccines have been used for several decades in the prevention of human disease. Which of the following statements is correct?
   (A) The live, attenuated poliovirus vaccine produces gastrointestinal tract resistance
   (B) There is an effective killed vaccine against the three major types of rhinoviruses
   (C) The live, attenuated poliovirus vaccine induces protective immunity against the closely related coxsackie B viruses
   (D) None of the available echovirus vaccines should be given to immunocompromised patients
   (E) Only the live attenuated poliovirus vaccine is currently recommended for use in the United States

5. One month after school has been let out for the summer, a 16-year-old girl develops fever, myalgia, and headache. An outbreak of an illness with similar symptoms caused by an echovirus is known to be occurring in the community. The primary anatomic site of echovirus multiplication in the human host is
   (A) The muscular system
   (B) The central nervous system
   (C) The alimentary tract
   (D) The blood and lymph system
   (E) The respiratory system

6. Which of the following properties of enteroviruses is not shared by rhinoviruses?
   (A) Single-stranded RNA genome
   (B) Production by cleavage of viral proteins from a polyprotein precursor
   (C) Resistance to lipid solvents
   (D) Stability at acid pH (pH 3.0)
   (E) Icosahedral symmetry

7. A person with asthma suffers an acute exacerbation with increased lower respiratory illness. A virus is recovered. The isolate is most likely to be which of the following virus types?
   (A) Parainfluenza
   (B) Parechovirus
   (C) Rhinovirus
   (D) Respiratory syncytial virus
   (E) Echovirus

8. The use of live oral polio vaccine has been replaced by inactivated polio vaccine in many countries. Which of the following is the primary reason?
   (A) It is more cost effective to use the inactivated vaccine
   (B) There is a greater risk of vaccine-induced disease than wild virus-induced disease in areas where poliovirus has been eradicated
   (C) Only a single dose of inactivated vaccine is necessary compared to multiple doses of the oral vaccine
   (D) Circulating poliovirus strains have changed and the live vaccine is no longer effective in many countries

9. Outbreaks of hand-foot-and-mouth disease, characterized by oral ulcerations and vesicular rashes, occur and may result in infant deaths. The disease is caused by
   (A) Foot-and-mouth disease virus
   (B) Chickenpox virus
   (C) Nonpolio enteroviruses
   (D) Rhinoviruses
   (E) Rubella virus

10. Epidemiological studies indicate that a core group of enteroviruses is consistently circulating in the United States. Which of the following statements is most accurate?

    (A) Members of the core group all display an epidemic pattern of outbreaks of disease

    (B) The group includes about one-half of known enteroviruses

    (C) Disease occurs predominantly in adolescents and adults

    (D) Members of the group are all classified as coxsackie A and B viruses

    (E) This core group determines the majority of enterovirus disease

Answers

1. D	4. A	7. C	10. E
2. E	5. C	8. B	
3. B	6. D	9. C	

REFERENCES

Chumakov K, Ehrenfeld E: New generation of inactivated poliovirus vaccines for universal immunization after eradication of poliomyelitis. Clin Infect Dis 2008;47:1587. [PMID: 18990066]

Enterovirus surveillance—United States, 1970–2005. MMWR Morb Mortal Wkly Rep 2006;55(SS-8).

Harvala H, Simmonds P: Human parechoviruses: Biology, epidemiology and clinical significance. J Clin Virol 2009;45:1. [PMID: 19372062]

Mahony JB: Detection of respiratory viruses by molecular methods. Clin Microbiol Rev 2008;21:716. [PMID: 18854489]

Pallansch M, Roos R: Enteroviruses: Polioviruses, coxsackieviruses, echoviruses, and newer enteroviruses. In: *Fields Virology,* 5th ed. Knipe DM et al (editors). Lippincott Williams & Wilkins, 2007.

Rotbart HA: Antiviral therapy for enteroviruses and rhinoviruses. Antiviral Chem Chemother 2000;11:261. [PMID: 10950388]

Technical Consultative Group to the World Health Organization on the Global Eradication of Poliomyelitis: "Endgame" issues for the global polio eradication initiative. Clin Infect Dis 2002;34:72. [PMID: 11731948]

Turner RB, Couch RB: Rhinoviruses. In: *Fields Virology,* 5th ed. Knipe DM et al (editors). Lippincott Williams & Wilkins, 2007.

Whitton JL, Cornell CT, Feuer R: Host and virus determinants of picornavirus pathogenesis and tropism. Nature Rev Microbiol 2005;3:765. [PMID: 16205710]

Reoviruses, Rotaviruses, & Caliciviruses

37

Reoviruses are medium-sized viruses with a double-stranded, segmented RNA genome. The family includes human rotaviruses, the most important cause of infantile gastroenteritis around the world (Figure 37–1). Acute gastroenteritis is a very common disease with significant public health impact. In developing countries it is estimated to cause as many as 1.5 million deaths of preschool children annually, of which rotavirus is responsible for about 600,000 deaths. In the United States, acute gastroenteritis is second only to acute respiratory infections as a cause of disease in families.

Caliciviruses are small viruses with a single-stranded RNA genome. The family contains Norwalk virus, the major cause of nonbacterial epidemic gastroenteritis worldwide. Astroviruses also cause gastroenteritis.

REOVIRUSES & ROTAVIRUSES

Important properties of reoviruses are summarized in Table 37–1.

Structure & Composition

The virions measure 60–80 nm in diameter and possess two concentric capsid shells, each of which is icosahedral.

(Rotaviruses have a triple-layered structure.) There is no envelope. Single-shelled virus particles that lack the outer capsid are 50–60 nm in diameter. The inner core of the particles is 33–40 nm in diameter (Figure 37–2). The double-shelled particle is the complete infectious form of the virus.

The reovirus genome consists of double-stranded RNA in 10–12 discrete segments with a total genome size of 16–27 kbp, depending on the genus. Rotaviruses contain 11 genome segments, whereas orthoreoviruses and orbiviruses each possess ten segments and coltiviruses have 12 segments. The individual RNA segments vary in size from 680 bp (rotavirus) to 3900 bp (orthoreovirus). The virion core contains several enzymes needed for transcription and capping of viral RNAs.

Rotaviruses are stable to heat at 50°C, to a 3.0–9.0 range of pH, and to lipid solvents, such as ether and chloroform, but they are inactivated by 95% ethanol, phenol, and chlorine. Limited treatment with proteolytic enzymes increases infectivity.

Classification

The family **Reoviridae** is divided into 15 genera. Four of the genera are able to infect humans and animals: *Orthoreovirus, Rotavirus, Coltivirus,* and *Orbivirus.* The genera can be divided into two groups; one group contains viruses with large spikes at the 12 vertices on the particle

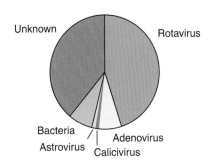

A   **Developed countries**

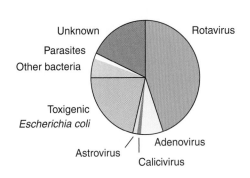

B   **Developing countries**

FIGURE 37–1 An estimate of the role of etiologic agents in severe diarrheal illnesses requiring hospitalization of infants and young children. **A:** In developed countries. **B:** In developing countries. (Reproduced with permission from Kapikian AZ: Viral gastroenteritis. JAMA 1993;269:627.)

TABLE 37–1   Important Properties of Reoviruses

Virion: Icosahedral, 60–80 nm in diameter, double capsid shell

Composition: RNA (15%), protein (85%)

Genome: Double-stranded RNA, linear, segmented (10–12 segments); total genome size 16–27 kbp

Proteins: Nine structural proteins; core contains several enzymes

Envelope: None (transient pseudoenvelope is present during rotavirus particle morphogenesis)

Replication: Cytoplasm; virions not completely uncoated

Outstanding characteristics:
Genetic reassortment occurs readily
Rotaviruses are the major cause of infantile diarrhea
Reoviruses are good models for molecular studies of viral pathogenesis

(eg, *Orthoreovirus*) whereas members of the second group appear more smooth, lacking the large surface projections (eg, *Rotavirus*).

There are at least five species or groups of rotaviruses (A–E), plus two tentative species (F and G), of which three species (A, B, C) infect humans. Strains of human and animal origin may fall in the same serotype. Other rotavirus groups and serotypes are found only in animals. Three different serotypes of reovirus are recognized, along with about 100 different orbivirus serotypes and two coltivirus serotypes.

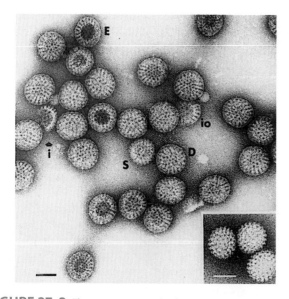

FIGURE 37–2 Electron micrograph of a negatively stained preparation of human rotavirus. (D, double-shelled particles; E, empty capsids; i, fragment of inner shell; io, fragments of a combination of inner and outer shell; S, single-shelled particles.) **Inset:** Single-shelled particles obtained by treatment of the viral preparation with sodium dodecyl sulfate. Bars, 50 nm. (Courtesy of J Esparza and F Gil.)

Reovirus Replication

Viral particles attach to specific receptors on the cell surface (Figure 37–3). The cell attachment protein for reoviruses is the viral hemagglutinin (σ1 protein), a minor component of the outer capsid.

After attachment and penetration, uncoating of virus particles occurs in lysosomes in the cell cytoplasm. Only the outer shell of the virus is removed, and a core-associated RNA transcriptase is activated. This transcriptase transcribes mRNA molecules from the minus strand of each genome double-stranded RNA segment contained in the intact core. There are short terminal sequences at both ends of the RNA segments that are conserved among all isolates of a given subgroup. These conserved sequences may be recognition signals for the viral transcriptase. The functional mRNA molecules correspond in size to the genome segments. Most RNA segments encode a single protein, although a few (depending on the virus) encode two proteins. Reovirus cores contain all enzymes necessary for transcribing, capping, and extruding the mRNAs from the core, leaving the double-stranded RNA genome segments inside.

Once extruded from the core, the mRNAs are translated into primary gene products. Some of the full-length transcripts are encapsidated to form immature virus particles. A viral replicase is responsible for synthesizing negative-sense strands to form the double-stranded genome segments. This replication to form progeny double-stranded RNA occurs in partially completed core structures. The mechanisms that ensure assembly of the correct complement of genome segments into a developing viral core are unknown. However, genome reassortment occurs readily in cells coinfected with different viruses of the same subgroup, giving rise to virus particles containing RNA segments from the different parental strains. Viral polypeptides probably self-assemble to form the inner and outer capsid shells.

Reoviruses produce inclusion bodies in the cytoplasm in which virus particles are found. These viral factories are closely associated with tubular structures (microtubules and intermediate filaments). Rotavirus morphogenesis involves budding of single-shelled particles into the rough endoplasmic reticulum. The "pseudoenvelopes" so acquired are then removed and the outer capsids are added (Figure 37–3). This unusual pathway is utilized because the major outer capsid protein of rotaviruses is glycosylated.

Cell lysis results in release of progeny virions.

ROTAVIRUSES

Rotaviruses are a major cause of diarrheal illness in human infants and young animals, including calves and piglets. Infections in adult humans and animals are also common. Among rotaviruses are the agents of human infantile diarrhea, Nebraska calf diarrhea, epizootic diarrhea of infant mice, and SA11 virus of monkeys.

Rotaviruses resemble reoviruses in terms of morphology and strategy of replication.

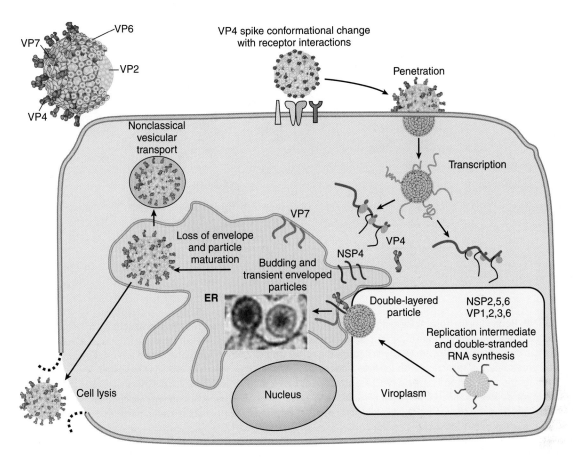

FIGURE 37–3 Overview of the rotavirus replication cycle. (Courtesy of MK Estes.)

Classification & Antigenic Properties

Rotaviruses have been classified into five species (A–E), plus two tentative species (F and G), based on antigenic epitopes on the internal structural protein VP6. These can be detected by immunofluorescence, ELISA, and immune electron microscopy (IEM). Group A rotaviruses are the most frequent human pathogens. Outer capsid proteins VP4 and VP7 carry epitopes important in neutralizing activity, with VP7 glycoprotein being the predominant antigen. These type-specific antigens differentiate among rotaviruses and are demonstrable by Nt tests. Five serotypes are responsible for the majority of human disease. Serotype distributions differ geographically. Multiple serotypes have been identified among human and animal rotaviruses. Some animal and human rotaviruses share serotype specificity. For example, monkey virus SA11 is antigenically very similar to human serotype 3. The gene-coding assignments responsible for the structural and antigenic specificities of rotavirus proteins are shown in Figure 37–4.

Molecular epidemiologic studies have analyzed isolates based on differences in the migration of the 11 genome segments following electrophoresis of the RNA in polyacrylamide gels (Figure 37–5). These differences in electropherotypes can be used to differentiate group A viruses from other groups, but they cannot be used to predict serotypes.

Animal Susceptibility

Rotaviruses have a wide host range. Most isolates have been recovered from newborn animals with diarrhea. Cross-species infections can occur in experimental inoculations, but it is not clear if they occur in nature. Swine rotavirus infects both newborn and weanling piglets. Newborns often exhibit subclinical infection due perhaps to the presence of maternal antibody, whereas overt disease is more common in weanling animals.

Propagation in Cell Culture

Rotaviruses are fastidious agents to culture. Most group A human rotaviruses can be cultivated if pretreated with the proteolytic enzyme trypsin and if low levels of trypsin are included in the tissue culture medium. This cleaves an outer capsid protein and facilitates uncoating. Very few non-group A rotavirus strains have been cultivated.

Pathogenesis

Rotaviruses infect cells in the villi of the small intestine (gastric and colonic mucosa are spared). They multiply in the cytoplasm of enterocytes and damage their transport mechanisms. One of the rotavirus-encoded proteins, NSP4,

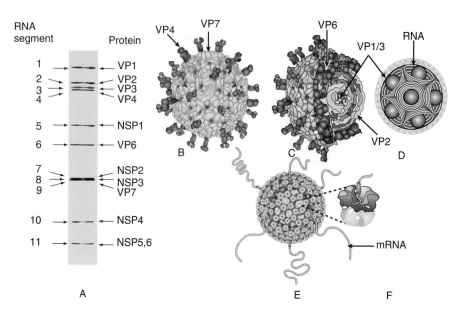

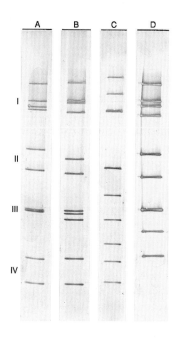

FIGURE 37–4 Rotavirus structure. **A:** Gel diagram showing the 11 segments of the genome. The structural (VP) and nonstructural (NSP) proteins encoded by these segments are indicated. **B:** Surface representation of the rotavirus structure from cryo-electron microscopic analysis. The two outer layer proteins are VP4, which forms the spikes, and VP7, which forms the capsid layer. **C:** Cut-away view showing the triple-layered organization of the virion, with the intermediate VP6 layer and the innermost VP2 layer indicated. The enzymes required for endogenous transcription (VP1) and capping (VP3) are attached as heterodimeric complexes to the inner surface of the VP2 layer. **D:** Proposed organization of the double-stranded RNA genome inside the VP2 layer along with transcription enzyme complexes (VP1/3) depicted as balls. **E:** Exit of transcripts from the channels at the fivefold vertices of actively transcribing double-layered particles. **F:** Close-up view of one of the exit channels. (Courtesy of BVV Prasad.)

FIGURE 37–5 Electrophoretic profiles of rotavirus RNA segments. Viral RNAs were electrophoresed in 10% polyacrylamide gels and visualized by silver stain. Different rotavirus groups and RNA patterns are illustrated: a group A monkey virus (SA11; lane A), a group A human rotavirus (lane B), a group B human adult diarrhea virus (lane C), and a group A rabbit virus that exhibits a "short" RNA pattern (lane D). Rotaviruses contain 11 genome RNA segments, but sometimes two or three segments migrate closely together and are difficult to separate. (Photograph provided by T Tanaka and MK Estes.)

is a viral enterotoxin and induces secretion by triggering a signal transduction pathway. Damaged cells may slough into the lumen of the intestine and release large quantities of virus, which appear in the stool (up to 10^{12} particles per gram of feces). Viral excretion usually lasts 2–12 days in otherwise healthy patients but may be prolonged in those with poor nutrition. Diarrhea caused by rotaviruses may be due to impaired sodium and glucose absorption as damaged cells on villi are replaced by nonabsorbing immature crypt cells. It may take 3–8 weeks for normal function to be restored.

Clinical Findings & Laboratory Diagnosis

Rotaviruses cause the major portion of diarrheal illness in infants and children worldwide but not in adults (Table 37–2). There is an incubation period of 1–3 days. Typical symptoms include watery diarrhea, fever, abdominal pain, and vomiting, leading to dehydration.

In infants and children, severe loss of electrolytes and fluids may be fatal unless treated. Patients with milder cases have symptoms for 3–8 days and then recover completely. However, viral excretion in the stool may persist up to 50 days after onset of diarrhea. Asymptomatic infections, with seroconversion, occur. In children with immunodeficiencies, rotavirus can cause severe and prolonged disease.

Adult contacts may be infected, as evidenced by seroconversion, but they rarely exhibit symptoms, and virus is infrequently detected in their stools. A common source of

TABLE 37–2   Viruses Associated with Acute Gastroenteritis in Humans[a]

Virus	Size (nm)	Epidemiology	Important as a Cause of Hospitalization
Rotaviruses			
Group A	60–80	Single most important cause (viral or bacterial) of endemic severe diarrheal illness in infants and young children worldwide (in cooler months in temperate climates)	Yes
Group B	60–80	Outbreaks of diarrheal illness in adults and children in China	No
Group C	60–80	Sporadic cases and occasional outbreaks of diarrheal illness in children	No
Enteric adenovirus	70–90	Second most important viral agent of endemic diarrheal illness of infants and young children worldwide	Yes
Caliciviruses			
Norwalk	27–40	Important cause of outbreaks of vomiting and diarrheal illness in older children and adults in families, communities, and institutions; frequently associated with ingestion of food	No
Sapporo	27–40	Sporadic cases and occasional outbreaks of diarrheal illness in infants, young children, and the elderly	No
Astroviruses	28–30	Sporadic cases and occasional outbreaks of diarrheal illness in infants, young children, and the elderly	No

Source: Kapikian AZ: Viral gastroenteritis. JAMA 1993;269:627.

infection is contact with pediatric cases. However, epidemics of severe disease have occurred in adults, especially in closed populations, as in a geriatric ward. Group B rotaviruses have been implicated in large outbreaks of severe gastroenteritis in adults in China (Table 37–2).

Laboratory diagnosis rests on demonstration of virus in stool collected early in the illness and on a rise in antibody titer. Virus in stool is demonstrated by enzyme immunoassays (EIAs) or IEM. The EIA test is more sensitive than the IEM. Genotyping of rotavirus nucleic acid from stool specimens by the polymerase chain reaction is the most sensitive detection method. Serologic tests can be used to detect an antibody titer rise, particularly ELISA.

Epidemiology & Immunity

Rotaviruses are the single most important worldwide cause of gastroenteritis in young children. Estimates range from 3 billion to 5 billion for annual diarrheal episodes in children under 5 years of age in Africa, Asia, and Latin America, resulting in as many as 1 million deaths. Developed countries have a high morbidity rate but a low mortality rate. Typically, up to 50% of cases of acute gastroenteritis of hospitalized children throughout the world are caused by rotaviruses.

Rotavirus infections usually predominate during the winter season. Symptomatic infections are most common in children between ages 6 months and 2 years, and transmission appears to be by the fecal–oral route. Nosocomial infections are frequent.

Rotaviruses are ubiquitous. By age 3 years, 90% of children have serum antibodies to one or more types. This high prevalence of rotavirus antibodies is maintained in adults,

suggesting subclinical reinfections by the virus. Rotavirus reinfections are common; it has been shown that young children can suffer up to five reinfections by 2 years of age. Asymptomatic infections are more common with successive reinfections. Local immune factors, such as secretory IgA or interferon, may be important in protection against rotavirus infection. Asymptomatic infections are common in infants before age 6 months, the time during which protective maternal antibody acquired passively by newborns should be present. Such neonatal infection does not prevent reinfection, but it does protect against the development of severe disease during reinfection.

Treatment & Control

Treatment of gastroenteritis is supportive, to correct the loss of water and electrolytes that may lead to dehydration, acidosis, shock, and death. Management consists of replacement of fluids and restoration of electrolyte balance either intravenously or orally, as feasible. The infrequent mortality from infantile diarrhea in developed countries is due to routine use of effective replacement therapy.

In view of the fecal–oral route of transmission, wastewater treatment and sanitation are significant control measures.

An oral live attenuated rhesus-based rotavirus vaccine was licensed in the United States in 1998 for vaccination of infants. It was withdrawn a year later because of reports of intussusception (bowel blockages) as an uncommon but serious side effect associated with the vaccine. In 2006, an oral pentavalent bovine-based rotavirus vaccine was licensed in the United States, followed by licensing of an oral monovalent human-based rotavirus vaccine in 2008. Neither vaccine appears to be associated with intussusception. A safe and effective vaccine remains the best hope for reducing the worldwide burden of rotavirus disease.

REOVIRUSES

The viruses of this genus, which have been studied most thoroughly by molecular biologists, are not known to cause human disease.

Classification & Antigenic Properties

Reoviruses are ubiquitous, with a very wide range of mammalian, avian, and reptilian hosts. Three distinct but related types of reovirus have been recovered from many species and are demonstrable by Nt and HI tests. Reoviruses contain a hemagglutinin for human O or bovine erythrocytes.

Epidemiology

Reoviruses cause many inapparent infections, because most people have serum antibodies by early adulthood. Antibodies are also present in other species. All three types have been recovered from healthy children, from young children during outbreaks of minor febrile illness, from children with enteritis or mild respiratory disease, and from chimpanzees with epidemic rhinitis.

Human volunteer studies have failed to demonstrate a clear cause-and-effect relationship of reoviruses to human illness. In inoculated volunteers, reovirus is recovered far more readily from feces than from the nose or throat.

Pathogenesis

Reoviruses have become important model systems for the study of the pathogenesis of viral infection at the molecular level. Defined recombinants from two reoviruses with differing pathogenic phenotypes are used to infect mice. Segregation analysis is then used to associate particular features of pathogenesis with specific viral genes and gene products. The pathogenic properties of reoviruses are primarily determined by the protein species found on the outer capsid of the virion.

ORBIVIRUSES & COLTIVIRUSES

Orbiviruses are a genus within the reovirus family. They commonly infect insects, and many are transmitted by insects to vertebrates. About 100 serotypes are known. None of these viruses cause serious clinical disease in humans, but they may cause mild fevers. Serious animal pathogens include bluetongue virus of sheep and African horse sickness virus. Antibodies to orbiviruses are found in many vertebrates, including humans.

The genome consists of ten segments of double-stranded RNA, with a total genome size of 18 kbp. The replicative cycle is similar to that of reoviruses. Orbiviruses are sensitive to low pH, in contrast with the general stability of other reoviruses.

Coltiviruses form another species within the **Reoviridae**. Their genome consists of 12 segments of double-stranded RNA, totaling about 29 kbp. Colorado tick fever virus, transmitted by ticks, is able to infect humans.

CALICIVIRUSES

In addition to rotaviruses and noncultivable adenoviruses, members of the family **Caliciviridae** are important agents of viral gastroenteritis in humans. The most significant member is Norwalk virus. Properties of caliciviruses are summarized in Table 37–3.

Classification & Antigenic Properties

Caliciviruses are similar to picornaviruses but are slightly larger (27–40 nm) and contain a single major structural protein (Figure 37–6). They exhibit a distinctive morphology in the electron microscope (Figure 37–7). The family Caliciviridae is divided into four genera: *Norovirus*, which includes the Norwalk viruses; *Sapovirus*, which includes the Sapporo-like viruses; *Lagovirus*, the rabbit hemorrhagic disease virus; and *Vesivirus*, which includes vesicular exanthem virus of swine, feline calicivirus, and marine viruses found in pinnipeds, whales, and fish. The first two genera contain human viruses that cannot be cultured; the latter two genera contain only animal strains that can be grown in vitro. Rabbit hemorrhagic disease virus was introduced in 1995 in Australia as a biologic control agent to reduce that country's population of wild rabbits.

Human calicivirus serotypes are not defined. Multiple genotypes of noroviruses have been detected. Three genogroups are associated with human gastroenteritis.

TABLE 37–3   Important Properties of Caliciviruses

Virion: Icosahedral, 27–40 nm in diameter, cup-like depressions on capsid surface
Genome: Single-stranded RNA, linear, positive-sense, nonsegmented; 7.4–8.3 kb in size; contains genome-linked protein (VPg)
Proteins: Polypeptides cleaved from a precursor polyprotein; capsid is composed of a single protein
Envelope: None
Replication: Cytoplasm
Outstanding characteristics: Noroviruses are major cause of nonbacterial epidemic gastroenteritis Human viruses are noncultivable

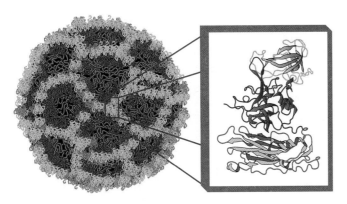

FIGURE 37–6 X-ray structure of the Norwalk virus capsid (*left*). The capsid subunit structure is illustrated (*right panel*). S, P1, and P2 domains are shaded in dark gray, medium gray, and light gray, respectively. (Courtesy of B V V Prasad.)

Clinical Findings & Laboratory Diagnosis

Noroviruses (Norwalk virus) are the most important cause of epidemic viral gastroenteritis in adults (Table 37–2). Epidemic nonbacterial gastroenteritis is characterized by (1) absence of bacterial pathogens; (2) gastroenteritis with rapid onset and recovery and relatively mild systemic signs; and (3) an epidemiologic pattern of a highly communicable disease that spreads rapidly with no particular predilection in terms of age or geography. Various descriptive terms have been used in reports of different outbreaks (eg, epidemic viral gastroenteritis, viral diarrhea, winter vomiting disease) depending on the predominant clinical feature.

Norwalk viral gastroenteritis has an incubation period of 24–48 hours. Onset is rapid, and the clinical course is brief, lasting 12–60 hours; symptoms include diarrhea, nausea, vomiting, low-grade fever, abdominal cramps, headache, and malaise. The illness can be incapacitating during the symptomatic phase, but hospitalization is rarely required. Norovirus infections are more likely to induce vomiting than those with Sapporo-like viruses. Dehydration is the most common complication in the young and elderly. Viral shedding may persist for as long as 1 month. No sequelae have been reported.

Volunteer experiments have clearly shown that the appearance of Norwalk virus coincides with clinical illness. Antibody develops during the illness and is usually protective on a short-term basis against reinfection with the same agent. Long-term immunity does not correspond well to the presence of serum antibodies. Some volunteers can be reinfected with the same virus after about 2 years.

Reverse transcriptase-polymerase chain reaction is the most widely used technique for detection of human caliciviruses in clinical specimens (feces, vomitus) and environmental samples (contaminated food, water). Because of the genetic diversity among circulating strains, the choice of polymerase chain reaction primer pairs is very important.

Electron microscopy is frequently used to detect virus particles in stool samples. However, norovirus particles are usually present in low concentration and are difficult to recognize; they should be identified by IEM. ELISA immunoassays based on recombinant virus-like particles can detect antibody responses, with a fourfold or greater rise in IgG antibody titer in acute and convalescent-phase sera indicative of a recent infection. However, the necessary reagents are not widely available, and the antigens are not able to detect responses to all antigenic types of noroviruses.

Epidemiology & Immunity

Human caliciviruses have worldwide distribution. Noroviruses are the most common cause of nonbacterial gastroenteritis in the United States, causing an estimated 23 million cases annually.

The viruses are most often associated with epidemic outbreaks of waterborne, foodborne, and shellfish-associated gastroenteritis. Community outbreaks can occur in any season. All age groups can be affected. Outbreaks occur throughout the year, with a seasonal peak during cooler months. Most outbreaks involve foodborne or person-to-person transmission via fomites or aerosolization of contaminated body fluids (vomitus, fecal material).

Characteristics of norovirus include a low infectious dose (as few as 10 virus particles), relative stability in the environment, and multiple modes of transmission. It survives 10 ppm chlorine and heating to 60°C; it can be maintained in steamed oysters.

Fecal–oral spread is probably the primary means of transmission of Norwalk virus. During a 5-year period in the United States (1996–2000), food was implicated in 39% of outbreaks of Norwalk gastroenteritis, person-to-person contact in 12%, and water in 3%, with the source in 18% unknown. Among all foodborne disease outbreaks in the United States (1998–2002), norovirus caused 30%. Ill food-service workers are often involved in norovirus outbreaks.

Viruses, predominantly norovirus, were involved in 10% of waterborne disease outbreaks associated with recreational water in the United States (2003–2004).

Outbreaks of Norwalk gastroenteritis occur in multiple settings. From 1996 to 2000, 39% occurred in restaurants, 29% in nursing homes and hospitals, 12% in schools and daycare centers, 10% in vacation settings, including cruise ships, and 9% in other settings. In 2006, following Hurricane Katrina, a norovirus outbreak occurred in a crowded evacuee setting in Texas.

No in vitro neutralization assay is available to study immunity. Volunteer challenge studies have shown that about 50% of adults are susceptible to illness. Norwalk virus antibody is acquired later in life than rotavirus antibody, which develops early in childhood. In developing countries, most children have developed norovirus antibodies by 4 years of age.

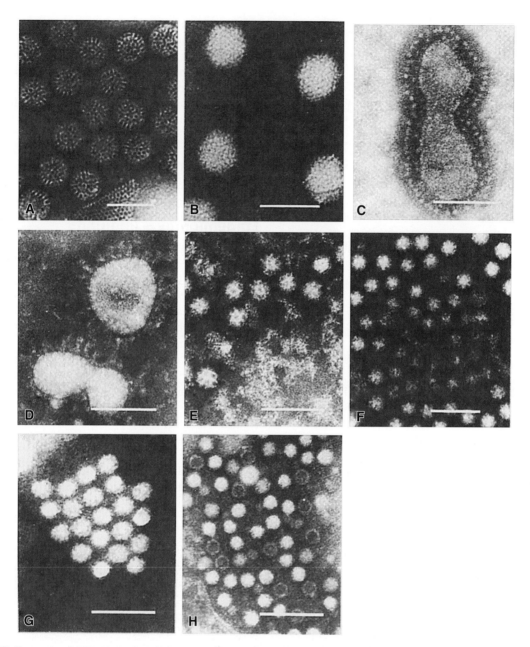

FIGURE 37–7 Electron micrographs of virus particles found in stools of patients with gastroenteritis. These viruses were visualized following negative staining. Specific viruses and the original magnifications of the micrographs are as follows. **A:** Rotavirus (185,000×). **B:** Enteric adenovirus (234,000×). **C:** Coronavirus (249,000×). **D:** Torovirus (coronavirus) (249,000×). **E:** Calicivirus (250,000×). **F:** Astrovirus (196,000×). **G:** Norwalk virus (calicivirus) (249,000×). **H:** Parvovirus (249,000×). The electron micrographs in panels C–H were originally provided by T Flewett; panel E was originally obtained from CR Madeley. Bars, 100 nm. (Reproduced with permission from Graham DY, Estes MK: Viral infections of the intestine. In: *Principles and Practice of Gastroenterology and Hepatology*. Gitnick G et al [editors]. Elsevier Science Co., 1988;566.)

Treatment & Control

Treatment is symptomatic. The low infectious dose permits efficient transmission of the virus. Because of the infectious nature of the stools, care should be taken in their disposal. Effective hand washing can decrease transmission in family or institutional settings. Containment and disinfection of soiled areas and bedding can help decrease viral spread. Careful processing of food and education of food handlers are important, as many foodborne outbreaks occur. Purification

of drinking water and swimming pool water should decrease norovirus outbreaks. There is no vaccine.

ASTROVIRUSES

Astroviruses are about 28–30 nm in diameter and exhibit a distinctive star-like morphology in the electron microscope (Figure 37–7). They contain single-stranded, positive-sense

RNA, 6.4–7.4 kb in size. The family Astroviridae contains two genera; all human viruses are classified in the *Mamastrovirus* genus. At least eight serotypes of human viruses are recognized by IEM and neutralization.

Astroviruses cause diarrheal illness and may be shed in extraordinarily large quantities in feces. The viruses are transmitted by the fecal–oral route through contaminated food or water, person-to-person contact, or contaminated surfaces. They are recognized as pathogens for infants and children, elderly institutionalized patients, and immunocompromised persons (Table 37–2). They may be shed for prolonged periods by immunocompromised hosts.

Animal astroviruses are found in a variety of mammals and birds and have recently been identified in several species of bats.

REVIEW QUESTIONS

1. A 36-year-old man enjoyed a meal of raw oysters. Twenty-four hours later he became ill, with sudden onset of vomiting, diarrhea, and headache. The most likely cause of his gastroenteritis is

   (A) Astrovirus
   (B) Hepatitis A virus
   (C) Norwalk virus
   (D) Rotavirus, group A
   (E) Echovirus

2. This virus is the most important cause of gastroenteritis in infants and young children. It causes infections that are often severe and may be life-threatening, especially in infants.

   (A) Echovirus
   (B) Norwalk virus
   (C) Rotavirus, group A
   (D) Orbivirus
   (E) Parvovirus

3. An outbreak of epidemic gastroenteritis occurred at a wooded summer camp 24 hours after a party for visiting families. Some of the visiting parents became ill also. Samples taken 2 weeks later from the well that was the source of drinking water at the camp were negative for fecal coliforms. The most likely source of the outbreak was

   (A) Mosquitoes or ticks, present in high numbers in the area
   (B) Contaminated food served at the party
   (C) A nearby stream used for fishing
   (D) A visiting parent who was developing pneumonia
   (E) The swimming pool

4. This viral gastroenteritis agent has a segmented, double-stranded RNA genome and a double-shelled capsid. It is a member of which virus family?

   (A) Adenoviridae
   (B) Astroviridae
   (C) Caliciviridae
   (D) Reoviridae
   (E) Coronaviridae

5. Rotavirus and Norwalk virus are distinctly different viruses. However, they share which one of the following characteristics?

   (A) Fecal–oral mode of transmission
   (B) They mainly cause disease in infants and young children
   (C) They induce generally mild disease in young children
   (D) Infection patterns show no seasonal variation
   (E) A double-stranded RNA genome

6. Because rotavirus infections can be serious, a vaccine would be beneficial. Which of the following is most correct regarding a rotavirus vaccine?

   (A) A killed human rotavirus group A vaccine is licensed for use in the United States (2008)
   (B) A live attenuated vaccine was withdrawn from use because of reports of intussusception (1998)
   (C) Vaccine development is complicated by rapid antigenic variation by the virus
   (D) Available antiviral drugs make a vaccine unnecessary
   (E) Vaccine development is complicated because the virus cannot be grown in cell culture

7. Rotaviruses and astroviruses share a number of characteristics. Which of the following is not shared?

   (A) Multiple serotypes exist
   (B) Can cause gastroenteritis in infants and children
   (C) Can cause gastroenteritis in elderly institutionalized patients
   (D) Live vaccine available
   (E) Fecal–oral route of transmission

Answers

1. C	3. B	5. A	7. D
2. C	4. D	6. B	

REFERENCES

Bresee JS, Nelson EA, Glass RI (guest editors): Rotavirus in Asia: Epidemiology, burden of disease, and current status of vaccines. J Infect Dis 2005;192(Suppl 1). [Entire issue.]

Dennehy PH: Rotavirus vaccines: An overview. Clin Microbiol Rev 2008;21:198. [PMID: 18202442]

Estes MK, Kapikian AZ: Rotaviruses. In: *Fields Virology*, 5th ed. Knipe DM et al (editors). Lippincott Williams & Wilkins, 2007.

Green KY: *Caliciviridae*: The noroviruses. In: *Fields Virology*, 5th ed. Knipe DM et al (editors). Lippincott Williams & Wilkins, 2007.

Monroe SS, Ando T, Glass RI (guest editors): International Workshop on Human Caliciviruses. J Infect Dis 2000;181(Suppl 12). [Entire issue.]

Prevention of rotavirus among infants and children. Recommendations of the Advisory Committee on Immunization Practices. MMWR Morb Mortal Wkly Rep 2006;55(RR-12):1.

Prevention of rotavirus gastroenteritis among infants and children. Recommendations of the Advisory Committee on Immunization Practices (ACIP). MMWR Morb Mortal Wkly Rep 2009;58(RR-2).

WHO position paper: Rotavirus vaccines. World Health Org Wkly Epidemiol Record 2007;82:285.

Arthropod-Borne & Rodent-Borne Viral Diseases

The **arthropod-borne viruses** (arboviruses) and **rodent-borne viruses** represent ecologic groupings of viruses with complex transmission cycles involving arthropods or rodents. These viruses have diverse physical and chemical properties and are classified in several virus families.

Arboviruses and rodent-borne viruses are classified among the **Arenaviridae, Bunyaviridae, Flaviviridae, Reoviridae,** and **Togaviridae** families. The African hemorrhagic fever viruses are classified in the **Filoviridae** family (Table 38–1, Figure 38–1). A number of the diseases described here are considered emerging infectious diseases (see Chapter 29).

The arboviruses are transmitted by bloodsucking arthropods from one vertebrate host to another. The vector acquires a lifelong infection through the ingestion of blood from a viremic vertebrate. The viruses multiply in the tissues of the arthropod without evidence of disease or damage. Some arboviruses are maintained in nature by transovarian transmission in arthropods.

The major arbovirus diseases worldwide are yellow fever, dengue, Japanese B encephalitis, St. Louis encephalitis, western equine encephalitis, eastern equine encephalitis, Russian spring-summer encephalitis, West Nile fever, and sandfly fever. In the United States the most important arboviral infections are La Crosse encephalitis, West Nile fever, St. Louis encephalitis, eastern equine encephalitis, and western equine encephalitis.

Rodent-borne viral diseases are maintained in nature by direct intraspecies or interspecies transmission from rodent to rodent without participation of arthropod vectors. Viral infection is usually persistent. Transmission occurs by contact with body fluids or excretions.

Major rodent-borne viral diseases include hantavirus infections, Lassa fever, and South American hemorrhagic fevers. In the United States, the most important rodent-borne diseases are hantavirus pulmonary syndrome and Colorado tick fever. Also considered here are the African hemorrhagic fevers—Marburg and Ebola. Their reservoir hosts are unknown, but are suspected to be rodents or bats.

HUMAN ARBOVIRUS INFECTIONS

There are several hundred arboviruses, of which about 100 are known human pathogens. Those infecting humans are all believed to be zoonotic, with humans the accidental hosts who play no important role in the maintenance or transmission cycle of the virus. Exceptions are urban yellow fever and dengue. Some of the natural cycles are simple and involve infection of a nonhuman vertebrate host (mammal or bird) transmitted by a species of mosquito or tick (eg, jungle yellow fever, Colorado tick fever). Others, however, are more complex. For example, tick-borne encephalitis can occur following ingestion of raw milk from goats and cows infected by grazing in tick-infested pastures where a tick–rodent cycle is occurring.

Individual viruses were sometimes named after a disease (dengue, yellow fever) or after the geographic area where the virus was first isolated (St. Louis encephalitis, West Nile fever). Arboviruses are found in all temperate and tropical zones, but they are most prevalent in the tropics with its abundance of animals and arthropods.

Diseases produced by arboviruses may be divided into three clinical syndromes: (1) fevers of an undifferentiated type with or without a maculopapular rash and usually benign; (2) encephalitis (inflammation of the brain), often with a high case-fatality rate; and (3) hemorrhagic fevers, also frequently severe and fatal. These categories are somewhat arbitrary, and some arboviruses may be associated with more than one syndrome, eg, dengue.

The degree of viral multiplication and its predominant site of localization in tissues determine the clinical syndrome. Thus, individual arboviruses can produce a minor febrile illness in some patients and encephalitis or a hemorrhagic diathesis in others.

Arbovirus infections occur in distinct geographic distributions and vector patterns (Figure 38–2). Each continent tends to have its own arbovirus pattern, and names are usually suggestive, eg, Venezuelan equine encephalitis, Japanese

TABLE 38–1   Classification & Properties of Some Arthropod-Borne & Rodent-Borne Viruses

Taxonomic Classification	Important Arbovirus & Rodent-Borne Virus Members	Virus Properties
Arenaviridae		
Genus *Arenavirus*	New World: Guanarito, Junin, Machupo, ,Sabia, and Whitewater Arroyo viruses. Old World: Lassa and lymphocytic choriomeningitis viruses. Rodent-borne	Spherical, 50–300 nm in diameter (mean, 110–130 nm). Genome: double-segmented, negative-sense and ambisense, single-stranded RNA, 10–14 kb in overall size. Virion contains a transcriptase. Four major polypeptides. Envelope. Replication: cytoplasm. Assembly: incorporate ribosomes and bud from plasma membrane
Bunyaviridae		
Genus *Orthobunyavirus*	Anopheles A and B, Bunyamwera, California encephalitis, Guama, La Crosse, Oropouche, and Turlock viruses. Arthropod-borne (mosquitoes)	Spherical, 80–120 nm in diameter. Genome: triple-segmented, negative-sense or ambisense, single-stranded RNA, 11–19 kb in total size. Virion contains a transcriptase. Four major polypeptides. Envelope. Replication: cytoplasm. Assembly: budding into the Golgi
Genus *Hantavirus*	Hantaan virus (Korean hemorrhagic fever), Seoul virus (hemorrhagic fever with renal syndrome), Sin Nombre virus (hantavirus pulmonary syndrome). Rodent-borne	
Genus *Nairovirus*	Crimean-Congo hemorrhagic fever, Nairobi sheep disease, and Sakhalin viruses. Arthropod-borne (ticks)	
Genus *Phlebovirus*	Rift Valley fever, sandfly (*Phlebotomus*) fever, and Uukuniemi viruses. Arthropod-borne (mosquitoes, sandflies, ticks)	
Filoviridae		
Genus *Marburgvirus*	Marburg viruses	Long filaments, 80 nm in diameter × varying length (>10,000 nm), though most average about 1000 nm. Genome: negative-sense, nonsegmented, single-stranded RNA, 19 kb in size. Seven polypeptides. Envelope. Replication: cytoplasm. Assembly: budding from plasma membrane
Genus *Ebolavirus*	Ebola viruses	
Flaviviridae		
Genus *Flavivirus*	Brazilian encephalitis (Rocio virus), dengue, Japanese B encephalitis, Kyasanur Forest disease, louping ill, Murray Valley encephalitis, Omsk hemorrhagic fever, Russian spring-summer encephalitis, St. Louis encephalitis, tick-borne encephalitis, West Nile fever, and yellow fever viruses. Arthropod-borne (mosquitoes, ticks)	Spherical, 40–60 nm in diameter. Genome: positive-sense, single-stranded RNA, 11 kb in size. Genome RNA infectious. Envelope. Three structural polypeptides, two glycosylated. Replication: cytoplasm. Assembly: within endoplasmic reticulum. All viruses serologically related
Reoviridae		
Genus *Coltivirus*	Colorado tick fever virus. Arthropod-borne (ticks, mosquitoes)	Spherical, 60–80 nm in diameter. Genome: 10–12 segments of linear, double-stranded RNA, 16–27 kbp total size. No envelope. Ten to 12 structural polypeptides. Replication and assembly: cytoplasm (see Chapter 37)
Genus *Orbivirus*	African horse sickness and bluetongue viruses. Arthropod-borne (mosquitoes)	
Togaviridae		
Genus *Alphavirus*	Chikungunya, eastern, western, and Venezuelan equine encephalitis viruses, Mayaro, O'Nyong-nyong, Ross River, Semliki Forest, and Sindbis viruses. Arthropod-borne (mosquitoes)	Spherical, 70 nm in diameter, nucleocapsid has 42 capsomeres. Genome: positive-sense, single-stranded RNA, 11–12 kb in size. Envelope. Three or four major structural polypeptides, two glycosylated. Replication: cytoplasm. Assembly: budding through host cell membranes. All viruses serologically related

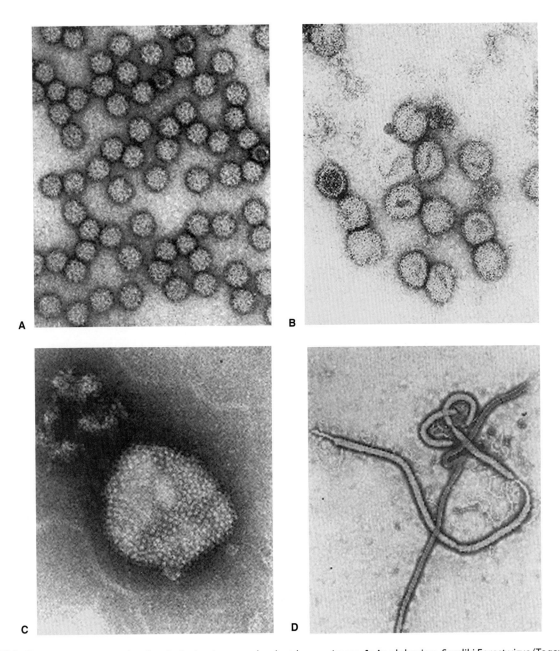

FIGURE 38–1 Electron micrographs of typical arboviruses and rodent-borne viruses. **A:** An alphavirus, Semliki Forest virus (Togaviridae). **B:** A representative member of Bunyaviridae, Uukuniemi virus. **C:** An arenavirus, Tacaribe virus (Arenaviridae). **D:** Ebola virus (Filoviridae). (Courtesy of FA Murphy and EL Palmer.)

B encephalitis, Murray Valley (Australia) encephalitis. Many encephalitides are alphavirus and flavivirus infections spread by mosquitoes, although the group of California encephalitis diseases is caused by bunyaviruses. On a given continent there may be a shifting distribution depending on viral hosts and vectors in a given year.

Several arboviruses cause significant human infections in the United States (Table 38–2). The numbers of cases vary widely from year to year.

TOGAVIRUS & FLAVIVIRUS ENCEPHALITIS

Classification & Properties of Togaviruses & Flaviviruses

In the Togaviridae family, the *Alphavirus* genus consists of about 30 viruses 70 nm in diameter that possess a single-stranded, positive-sense RNA genome (Table 38–1). The envelope surrounding the particle contains two glycoproteins.

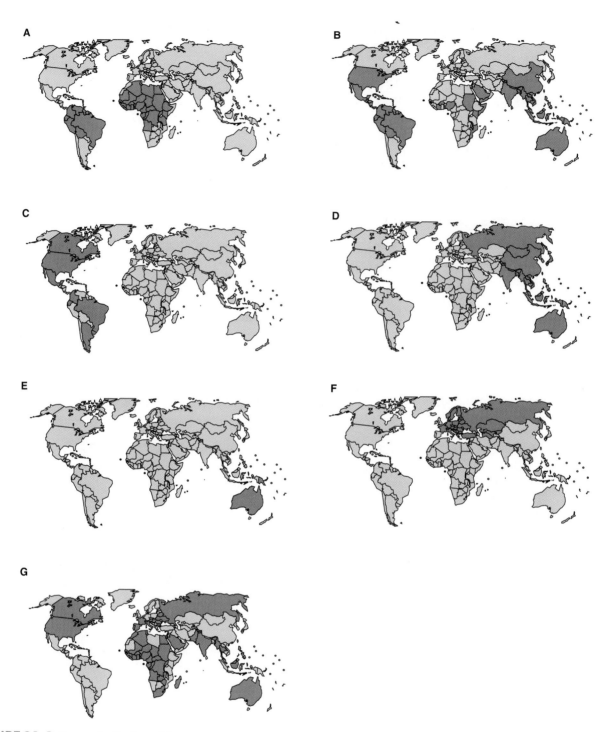

FIGURE 38–2 Known distributions of flaviviruses causing human disease. **A:** Yellow fever virus. **B:** Dengue virus. **C:** St. Louis encephalitis virus. **D:** Japanese B encephalitis virus. **E:** Murray Valley encephalitis virus. **F:** Tick-borne encephalitis virus. **G:** West Nile virus. (Reproduced with permission from Monath TP, Tsai TF: Flaviviruses. In: *Clinical Virology,* 2nd ed. Richman DD, Whitley RJ, Hayden FG [editors]. ASM Press, 2002.)

Alphaviruses often establish persistent infections in mosquitoes and are transmitted between vertebrates by mosquitoes or other blood-feeding arthropods. They have a worldwide distribution. All alphaviruses are antigenically related. The viruses are inactivated by acid pH, heat, lipid solvents, detergents, bleach, phenol, 70% alcohol, and formaldehyde. Most possess hemagglutinating ability (Figure 38–1). Rubella

virus, classified in a separate genus in the Togaviridae family, has no arthropod vector and is not an arbovirus (see Chapter 40).

Arboviruses are in the *Flavivirus* genus in the Flaviviridae family. Initially, the flaviviruses were included in the togavirus family as "group B arboviruses" but were moved to a separate family because of differences in viral

TABLE 38–2   Summary of Major Human Arbovirus & Rodent-Borne Virus Infections that Occur in the United States

Diseases[a]	Exposure	Distribution	Major Vectors	Infection Case Ratio (Age Incidence)	Sequelae[b]	Mortality Rate
Eastern equine encephalitis (*Alphavirus*)	Rural	Atlantic, southern coastal	*Aedes, Culex*	10:1 (infants) 50:1 (middle-aged) 20:1 (elderly)	+	30–70%
Western equine encephalitis (*Alphavirus*)	Rural	Pacific, Mountain, Southwest	*Culex tarsalis, Aedes*	50:1 (under 5) 1000:1 (over 15)	+	3–7%
Venezuelan equine encephalitis (*Alphavirus*)	Rural	South (also South and Central America)	*Aedes, Psorophora, Culex*	25:1 (under 15) 1000:1 (over 15)	±	Fatalities rare
St. Louis encephalitis (*Flavivirus*)	Urban–rural	Widespread	*Culex*	800:1 (under 9) 400:1 (9–59) 85:1 (over 60)	±	3–10% (under 65) 30% (over 65)
West Nile fever (*Flavivirus*)	Urban–rural	Widespread	*Culex, Aedes, Anopheles*	150:1	Unknown	3–15%
California encephalitis (La Crosse) (*Orthobunyavirus*)	Rural	North central, Atlantic, South	*Aedes triseriatus*	Unknown ratio (most cases under 20)	Rare	About 1%
Hantavirus pulmonary syndrome (*Hantavirus*)	Rural	Southwest, West	*Peromyscus maniculatus*[c]	Unknown	Unknown	30%
Colorado tick fever (*Coltivirus*)	Rural	Pacific, Mountain	*Dermacentor andersoni*	Unknown ratio (all ages affected)	Rare	Fatalities rare

[a]Shown in parentheses under the name of the disease is the genus in which the causative virus(es) is(are) classified. Virus families are indicated and described in Table 38–1.

[b]Sequelae: +, common; ±, occasional.

[c]Rodent reservoir; no vector.

genome organization. The Flaviviridae family consists of about 70 viruses 40–60 nm in diameter that have a single-stranded, positive-sense RNA genome. The viral envelope contains two glycoproteins. Some flaviviruses are transmitted between vertebrates by mosquitoes and ticks, whereas others are transmitted among rodents or bats without any known insect vectors. Many have worldwide distribution. All flaviviruses are antigenically related. Flaviviruses are inactivated similarly to alphaviruses, and many also exhibit hemagglutinating ability. Hepatitis C virus, classified in a separate genus in the Flaviviridae family, has no arthropod vector and is not an arbovirus (see Chapter 35).

Replication of Togaviruses & Flaviviruses

The alphavirus RNA genome is positive-sense (Figure 38–3). Genomic length and subgenomic (26S) mRNAs are produced during transcription. The genomic-length transcript produces a precursor polyprotein encoding the nonstructural proteins (ie, replicase, transcriptase) needed for viral RNA replication. The subgenomic mRNA encodes structural proteins. The proteins are elaborated by posttranslational cleavage. Alphaviruses replicate in the cytoplasm and mature by budding nucleocapsids through the plasma membrane. Sequence data indicate that western equine encephalitis virus is a genetic recombinant of eastern equine encephalitis and Sindbis viruses.

The flavivirus RNA genome also is positive-sense. A large precursor protein is produced from genome-length mRNAs during viral replication; it is cleaved by viral and host proteases to yield all the viral proteins, both structural and nonstructural. Flaviviruses replicate in the cytoplasm, and particle assembly occurs in intracellular vesicles (Figure 38–4). Proliferation of intracellular membranes is a characteristic of flavivirus-infected cells.

Antigenic Properties of Togaviruses & Flaviviruses

All alphaviruses are antigenically related. Because of common antigenic determinants, the viruses show cross-reactions in immunodiagnostic techniques. HI, ELISA, and IF tests define eight antigenic complexes or serogroups of alphaviruses, four of which are typified by western equine encephalitis, eastern

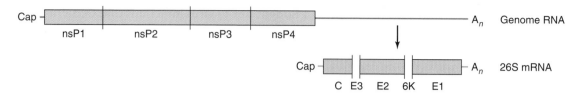

FIGURE 38–3 Genomic organization of alphaviruses. The nonstructural proteins (nsP) are translated from the genomic RNA as a polyprotein that is processed into four nonstructural proteins by a viral protease present in nsP2. The structural proteins are translated from a subgenomic 26S mRNA as a polyprotein that is processed by a combination of viral and cellular proteases into a capsid protein (C), three envelope glycoproteins (E3, E2, and E1), and a membrane-associated protein named 6K. C, E2, and E1 are major components of virions and are shaded in the figure. (Reproduced with permission from Strauss JH, Strauss EG, Kuhn RJ: Budding of alphaviruses. Trends Microbiol 1995;3:346.)

equine encephalitis, Venezuelan equine encephalitis, and Semliki Forest virus. Identification of a specific virus can be accomplished using Nt tests. Similarly, all flaviviruses share antigenic sites. At least eight antigenic complexes have been identified based on Nt tests. The envelope (E) protein is the viral hemagglutinin and contains the group-, serocomplex-, and type-specific determinants. Sequence comparisons of the E glycoprotein gene show that viruses within a serocomplex share over 70% amino acid sequences, whereas amino acid homology across serocomplexes is <50%.

Pathogenesis & Pathology

In susceptible vertebrate hosts, primary viral multiplication occurs either in myeloid and lymphoid cells or in vascular

endothelium. Multiplication in the central nervous system depends on the ability of the virus to pass the blood–brain barrier and to infect nerve cells. In natural infection of birds and mammals, an inapparent infection is usual. For several days there is viremia, and arthropod vectors acquire the virus by sucking blood during this period—the first step in its dissemination to other hosts.

The disease in experimental animals provides insights into human disease. Mice have been used to study the pathogenesis of encephalitis. After subcutaneous inoculation, virus replication occurs in local tissues and regional lymph nodes. Virus then enters the bloodstream and is disseminated. Depending on the specific agent, different tissues support further virus replication, including monocyte-macrophages, endothelial cells, lung, liver, and muscles. Virus crosses the

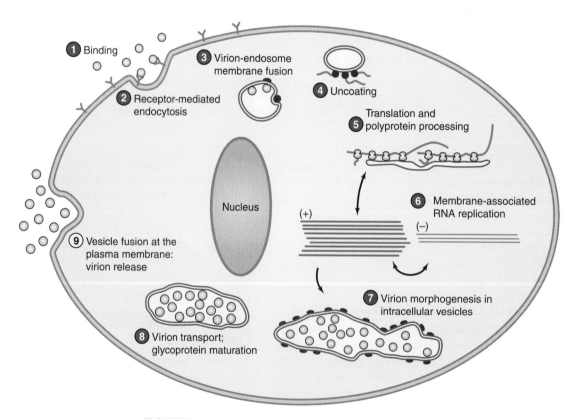

FIGURE 38–4 The flavivirus life cycle. (Courtesy of CM Rice.)

blood–brain barrier by unknown mechanisms, perhaps involving olfactory neurons or cerebral vascular cells, and spreads. Widespread neuronal degeneration occurs in all arbovirus-induced encephalitides.

In the vast majority of infections, the virus is controlled before neuroinvasion occurs. Invasion depends on many factors, including the level of viremia, the genetic background of the host, the host innate and adaptive immune responses, and the virulence of the virus strain. Humans show an age-dependent susceptibility to central nervous system infections, with infants and the elderly being most susceptible.

The equine encephalitides in horses are diphasic. In the first phase (minor illness), the virus multiplies in nonneural tissue and is present in the blood several days before the first signs of involvement of the central nervous system. In the second phase (major illness), the virus multiplies in the brain, cells are injured and destroyed, and encephalitis becomes clinically apparent. High concentrations of virus in brain tissue are necessary before the clinical disease becomes manifest.

Clinical Findings

Incubation periods of the encephalitides are between 4 and 21 days. Inapparent infections are common. Some infected persons develop mild flu-like illness, whereas others develop encephalitis. There is a sudden onset with severe headache, chills and fever, nausea and vomiting, generalized pains, and malaise. Within 24–48 hours, marked drowsiness develops and the patient may become stuporous. Mental confusion, tremors, convulsions, and coma develop in severe cases. Fever lasts 4–10 days. The mortality rate in encephalitides varies (Table 38–2). With Japanese B encephalitis, the mortality rate in older age groups may be as high as 80%. Sequelae may be mild to severe and include mental deterioration, personality changes, paralysis, aphasia, and cerebellar signs.

Laboratory Diagnosis

A. Recovery of Virus & Direct Detection

Virus isolation attempts require appropriate biosafety precautions to prevent laboratory infections. Virus occurs in the blood only early in the infection, usually before the onset of symptoms. Virus can also be found in cerebrospinal fluid and tissue specimens, depending on the agent. Alphaviruses and flaviviruses are usually able to grow in common cell lines, such as Vero, BHK, HeLa, and MRC-5. Mosquito cell lines are useful. Intracerebral inoculation of suckling mice or hamsters may also be used for virus isolation.

Antigen detection and polymerase chain reaction assays are available for direct detection of viral RNA or proteins in clinical specimens for some arboviruses. The use of virus-specific monoclonal antibodies in immunofluorescence assays has facilitated rapid virus identification in clinical samples.

B. Serology

Neutralizing and hemagglutination-inhibiting antibodies are detectable within a few days after the onset of illness. The neutralizing and the hemagglutination-inhibiting antibodies endure for years. The HI test is the simplest diagnostic test, but it identifies the group rather than the specific causative virus. The most sensitive serologic assays detect virus-specific IgM in serum or cerebrospinal fluid by ELISA.

It is necessary to establish a fourfold or greater rise in specific antibodies during infection to confirm a diagnosis. The first sample of serum should be taken as soon after the onset as possible and the second sample 2–3 weeks later. The cross-reactivity within the alphavirus or flavivirus group must be considered in making the diagnosis. Following a single infection by one member of the group, antibodies to other members may also appear. Serologic diagnosis becomes difficult when an epidemic caused by one member of the serologic group occurs in an area where another group member is endemic.

Immunity

Immunity is believed to be permanent after a single infection. Both humoral antibody and cellular immune responses are thought to be important in protection and recovery from infection. In endemic areas, the population may build up immunity as a result of inapparent infections; the proportion of persons with antibodies to the local arthropod-borne virus increases with age.

Because of common antigens, the response to immunization or to infection with one of the viruses of a group may be modified by prior exposure to another member of the same group. This mechanism may be important in conferring protection on a community against an epidemic of another related agent (eg, no Japanese B encephalitis in areas endemic for West Nile fever).

Epidemiology

In highly endemic areas, almost the entire human population may become infected with an arbovirus, and most infections are asymptomatic. High infection-to-case ratios exist among specified age groups for many arbovirus infections (Table 38–2). Most cases occur in the summer months in the northern hemisphere when arthropods are most active.

A. Eastern & Western Equine Encephalitis

Eastern equine encephalitis is the most severe of the arboviral encephalitides, with the highest case-fatality rate. Infections are rare and sporadic in the United States, averaging five confirmed cases per year. Inapparent infections are unusual. In the case of western equine encephalitis, transmission occurs at a low level in the rural West, where birds and *Culex tarsalis* mosquitoes are involved in the maintenance cycle of the virus. Infections of humans average about 15 confirmed cases annually. However, there have been instances in the past (most recently in 1987) when humans and equines became infected at epidemic and epizootic levels. Outbreaks have affected wide areas of the western United States and Canada.

B. St. Louis Encephalitis

St. Louis encephalitis virus is the most important cause of epidemic encephalitis of humans in North America (Figure 38–2), having caused about 10,000 cases and 1000 deaths since it was first recognized in 1933. Seroprevalence rates are generally low, and the incidence of St. Louis encephalitis varies each year in the United States. There are currently an average of 130 confirmed cases annually. Less than 1% of viral infections are clinically apparent. The presence of infected mosquitoes is required before human infections can occur, although socioeconomic and cultural factors (air conditioning, screens, mosquito control) affect the degree of exposure of the population to these virus-carrying vectors.

C. West Nile Fever

West Nile fever is caused by a member of the Japanese B encephalitis antigenic complex of flaviviruses. It occurs in Europe, the Middle East, Africa, the former Soviet Union, Southwest Asia, and, more recently, the United States. It appeared unexpectedly in the New York City area in 1999, resulting in seven deaths and extensive mortality in a range of domestic and exotic birds. Sequence analysis of virus isolates showed that it originated in the Middle East; it probably crossed the Atlantic in an infected bird, mosquito, or human traveler.

Within 3 years West Nile virus had completed transcontinental movement across the United States and was established as a permanent presence in temperate North America. It is now the leading cause of arboviral encephalitis in the United States. It is estimated that about 80% of West Nile infections are asymptomatic, with about 20% causing West Nile fever and <1% causing neuroinvasive disease (meningitis, encephalitis, or acute flaccid paralysis). Fatal encephalitis is more common in older people. A genetic deficiency resulting in a nonfunctional variant of chemokine receptor CCR5 has been identified as a risk factor for symptomatic West Nile infections.

An epidemic in 2002 in the United States was the largest arbovirus meningoencephalitis epidemic documented in the western hemisphere. There were 3389 reported cases of human West Nile virus illness, 69% with meningoencephalitis and 21% with West Nile fever. Over 9000 cases occurred in horses. In 2006, a total of 1491 cases of West Nile virus neuroinvasive disease in humans was reported in the United States. It is estimated there would have been 41,750 cases of West Nile fever and a total of over 208,000 human infections in 2006. West Nile virus was detected in all 48 contiguous states.

The 2002 West Nile virus epidemic included the first documented cases of person-to-person transmission through organ transplantation, blood transfusion, in utero, and perhaps breast feeding. Screening of blood donations for West Nile virus was implemented in the United States in 2003.

West Nile virus produces viremia and an acute, mild febrile disease with lymphadenopathy and rash. Transitory meningeal involvement may occur during the acute stage. Only one antigenic type of virus exists, and immunity is presumably permanent.

A West Nile vaccine for horses became available in 2003. There is no human vaccine. Prevention of West Nile virus disease depends on mosquito control and protection against mosquito bites.

D. Japanese B Encephalitis

Japanese B encephalitis is the leading cause of viral encephalitis in Asia (Figure 38–2). About 50,000 cases occur annually in China, Japan, Korea, and the Indian subcontinent, with 10,000 deaths, mostly among children and the elderly. Mortality can exceed 30%. A high percentage of survivors (up to 30%) are left with neurologic and psychiatric sequelae. Infections during the first and second trimesters of pregnancy have reportedly led to fetal death.

Seroprevalence studies indicate nearly universal exposure to Japanese B encephalitis virus by adulthood. The estimated ratio of asymptomatic to symptomatic infections is 300:1.

Arbovirus Host–Vector Transmission Cycles

Infection of humans by mosquito-borne encephalitis viruses occurs when a mosquito or another arthropod bites first an infected animal and later a human.

The equine encephalitides—eastern, western, and Venezuelan—are transmitted by culicine mosquitoes to horses or humans from a mosquito-bird-mosquito cycle (Figure 38–5). Equines, like humans, are unessential hosts for the maintenance of the virus. Both eastern and Venezuelan equine encephalitis in horses are severe, with up to 90% of affected animals dying. Epizootic western equine encephalitis is less frequently fatal for horses. In addition, eastern equine encephalitis produces severe epizootics in certain domestic game birds. A mosquito-bird-mosquito cycle also occurs in St. Louis encephalitis, West Nile virus, and Japanese B encephalitis. Swine are an important host of Japanese B encephalitis. Mosquitoes remain infected for life (several weeks to months). Only the female feeds on blood and can feed and transmit the virus more than once. The cells of the mosquito's midgut are the site of primary viral multiplication. This is followed by viremia and invasion of organs—chiefly salivary glands and nerve tissue, where secondary viral multiplication occurs. The arthropod remains healthy.

Infection of insectivorous bats with arboviruses produces a viremia that lasts 6–12 days without any illness or pathologic changes in the bat. While the viral concentration

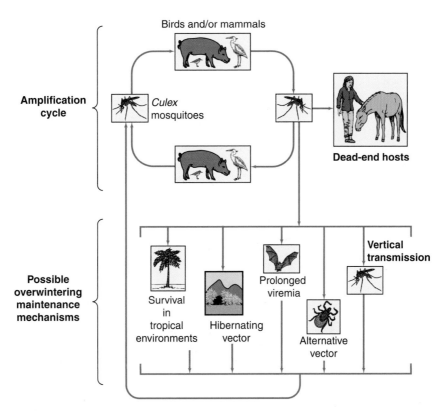

FIGURE 38-5 Generalized transmission cycle of mosquito-borne flaviviruses causing encephalitis. Summertime amplification and possible overwintering mechanisms are shown. Humans are dead-end hosts and do not contribute to perpetuation of virus transmission. Wild birds are the most common viremic hosts, but pigs play an important role in the case of Japanese encephalitis virus. The pattern shown applies to many (but not all) flaviviruses. (Adapted with permission from Monath TP, Heinz FX: Flaviviruses. In: *Fields Virology*, 3rd ed. Fields BN et al [editors]. Lippincott-Raven, 1996.)

is high, the infected bat may infect mosquitoes that are then able to transmit the infection to wild birds and domestic fowl as well as to other bats.

There are also tick-borne flavivirus encephalitides, such as Russian spring-summer encephalitis. This disease occurs chiefly in the early summer, particularly in humans exposed to the ticks *Ixodes persulcatus* and *Ixodes ricinus* in uncleared forest areas. Ticks can become infected at any stage in their metamorphosis, and virus can be transmitted transovarially (Figure 38-6). Virus is secreted in the milk of infected goats for long periods, and infection may be transmitted to those who drink unpasteurized milk. Powassan encephalitis virus was the first member of the Russian spring-summer complex isolated in North America. The original fatal case was reported from Canada in 1959. Human infection is rare.

Overwintering of Arboviruses

The epidemiology of the arthropod-borne encephalitides must account for the maintenance and dissemination of the viruses in nature in the absence of humans. Viruses have been isolated from mosquitoes and ticks, which serve as reservoirs of infection. In ticks, the viruses may pass from generation to generation by the transovarian route, and in such instances the tick acts as a true reservoir of the virus as

well as its vector (Figure 38-6). In tropical climates, where mosquito populations are present throughout the year, arboviruses cycle continually between mosquitoes and reservoir animals.

In temperate climates, the virus may be reintroduced each year from the outside (eg, by birds migrating from tropical areas) or it may survive the winter in the local area. Possible but unproved overwintering mechanisms include the following (Figures 38-5 and 38-6): (1) Hibernating mosquitoes at the time of their emergence may reinfect birds; (2) the virus may remain latent in winter within birds, mammals, or arthropods; and (3) cold-blooded vertebrates (snakes, turtles, lizards, alligators, frogs) may act as winter reservoirs. Mosquitoes can be infected by feeding on emerged snakes and then transmit the virus. Virus has been found in the blood of wild snakes.

Mosquitoes are closely associated with bats, both in summer and during the winter (in hibernation sites). The mosquito-bat-mosquito cycle may be a possible overwintering mechanism for some arboviruses.

Treatment & Control

There is no specific treatment. Biologic control of the natural vertebrate host is generally impractical, especially when the

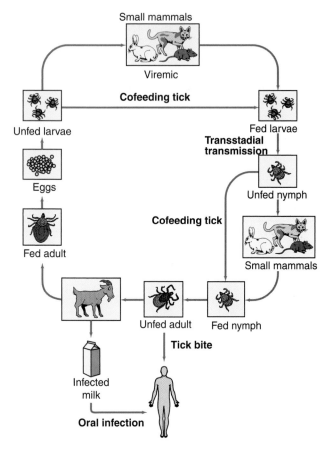

FIGURE 38–6 Generalized transmission cycle of tick-borne flaviviruses, showing hosts for larval, nymphal, and adult ticks. Virus is passed to succeeding tick stages during moulting (transstadial transmission), as well as transovarially to progeny of adult ticks. Both male and female ticks are involved in transmission. Tick-borne encephalitis virus may be transmitted to uninfected ticks cofeeding on a vertebrate host without the requirement for active viremic infection of the host. (Adapted with permission from Monath TP, Heinz FX: Flaviviruses. In: *Fields Virology*, 3rd ed. Fields BN et al [editors]. Lippincott-Raven, 1996.)

hosts are wild birds. The most effective method is arthropod control, such as spraying of insecticides to kill mosquitoes. Personal measures include avoiding mosquitoes by using repellents and wearing protective clothing. Houses should have adequate window screens.

Effective killed-virus vaccines have been developed to protect horses against eastern, western, and Venezuelan equine encephalitis. An attenuated live-virus vaccine for Venezuelan equine encephalitis is available for curtailing epidemics among horses. These vaccines are not for human use. Experimental inactivated human vaccines against eastern, western, and Venezuelan equine encephalitis viruses are available on an investigational basis to protect laboratory workers. Both killed-virus and attenuated live-virus Japanese B encephalitis vaccines for humans are in use in several Asian countries. Vaccine is available in the United States for individuals traveling to endemic countries.

YELLOW FEVER

Yellow fever virus is the prototype member of the Flaviviridae family. It causes yellow fever, an acute, febrile, mosquito-borne illness that occurs in the tropics and subtropics of Africa and South America (Figure 38–2). Severe cases are characterized by liver and renal dysfunction and hemorrhage, with high mortality.

Based on sequence analysis, at least seven genotypes of yellow fever virus have been identified, five in Africa and two in South America. There is a single serotype.

Yellow fever virus multiplies in many different types of animals and in mosquitoes and grows in embryonated eggs, chick embryo cell cultures, and cell lines, including those of monkey, human, hamster, and mosquito origin.

Pathogenesis & Pathology

The virus is introduced by a mosquito through the skin where it multiplies. It spreads to the local lymph nodes, liver, spleen, kidney, bone marrow, and myocardium, where it may persist for days. It is present in the blood early during infection.

The lesions of yellow fever are due to the localization and propagation of the virus in a particular organ. Infections may result in necrotic lesions in the liver and kidney. Degenerative changes also occur in the spleen, lymph nodes, and heart. Serious disease is characterized by hemorrhage and circulatory collapse. Virus injury to the myocardium may contribute to shock.

Clinical Findings

The incubation period is 3–6 days. At the abrupt onset, the patient has fever, chills, headache, dizziness, myalgia, and backache—followed by nausea, vomiting, and bradycardia. During this initial period, which lasts several days, the patient is viremic and a source of infection for mosquitoes. Most patients recover at this point, but in about 15% of cases the disease progresses to a more severe form, with fever, jaundice, renal failure, and hemorrhagic manifestations. The vomitus may be black with altered blood. When the disease progresses to the severe stage (hepato-renal failure), the mortality rate is high (20% or higher), especially among young children and the elderly. Death occurs on day 7–10 of illness. Encephalitis is rare.

On the other hand, the infection may be so mild as to go unrecognized. Regardless of severity, there are no sequelae; patients either die or recover completely.

Laboratory Diagnosis

A. Virus Detection or Isolation

Virus antigen or nucleic acid can be identified in tissue specimens using immunohistochemistry, ELISA antigen capture, or polymerase chain reaction tests. The virus may be recovered from the blood the first 4 days after onset or from postmortem tissue by intracerebral inoculation of mice or by use of cell lines.

B. Serology

IgM antibodies appear during the first week of illness. The detection of IgM antibody by ELISA capture in a single sample provides a presumptive diagnosis, with confirmation by a fourfold or greater rise in titer of neutralizing antibody between acute phase and convalescent phase serum samples. Older serologic methods, such as HI, have largely been replaced by ELISA. Specific hemagglutination-inhibiting antibodies appear first, followed rapidly by antibodies to other flaviviruses.

Immunity

Neutralizing antibodies develop about a week into the illness and are responsible for viral clearance. Neutralizing antibodies endure for life and provide complete protection from disease. Demonstration of neutralizing antibodies is the only useful test for immunity to yellow fever.

Epidemiology

Two major epidemiologic cycles of transmission of yellow fever are recognized: (1) urban yellow fever and (2) jungle yellow fever (Figure 38–7). Urban yellow fever involves person-to-person transmission by domestic *Aedes* mosquitoes. In the western hemisphere and West Africa, this species is primarily *Aedes aegypti,* which breeds in the accumulations of water that accompany human settlement. In areas where *A aegypti* has been eliminated or suppressed, urban yellow fever has disappeared.

Jungle yellow fever is primarily a disease of monkeys. In South America and Africa, it is transmitted from monkey to monkey by arboreal mosquitoes (ie, *Haemagogus, Aedes*) that inhabit the moist forest canopy. The infection in animals may be severe or inapparent. The virus multiplies in mosquitoes, which remain infectious for life. Persons involved in forest clearing activities come in contact with these mosquitoes in the forest and become infected.

Yellow fever has not invaded Asia, even though the vector, *A aegypti,* is widely distributed there.

Yellow fever continues to infect and kill thousands of persons worldwide because they have failed to be immunized.

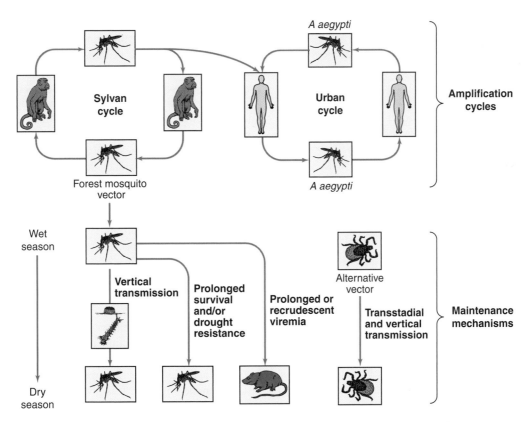

FIGURE 38–7 Transmission cycles of yellow fever and dengue viruses. These viruses have enzootic maintenance cycles involving *Aedes* vectors and nonhuman primates. Dengue viruses are transmitted principally between humans and *Aedes aegypti* that breed in domestic water containers. In the case of yellow fever, sylvatic (jungle) transmission is widespread throughout the geographic distribution of the virus. In tropical America, human yellow fever cases derive from contact with forest mosquito vectors, and there have been no cases of urban (*Aedes aegypti*-borne) yellow fever for over 50 years. In Africa, sylvatic vectors are responsible for monkey-monkey and interhuman virus transmission, and there is frequent involvement of *Aedes aegypti* in urban and dry savanna regions. (Adapted with permission from Monath TP, Heinz FX: Flaviviruses. In: *Fields Virology,* 3rd ed. Fields BN et al [editors]. Lippincott-Raven, 1996.)

It is estimated that annually, yellow fever strikes 200,000 persons, of whom about 30,000 die. The majority of outbreaks (~90%) occur in Africa. Epidemics usually occur in a typical emergence zone for yellow fever: humid and semihumid savanna adjoining a rain forest where the sylvatic cycle is maintained in a large monkey population. During epidemics in Africa, the infection:case ratio ranges from 20:1 to 2:1. All age groups are susceptible.

Yellow fever in the Americas presents epidemiologic features typical of its jungle cycle: Most cases are in males aged 15–45 years and engaged in agricultural or forestry activities.

Treatment, Prevention, & Control

There is no antiviral drug therapy.

Vigorous mosquito abatement programs have virtually eliminated urban yellow fever throughout much of South America; however, vector control is impractical in many parts of Africa. The last reported outbreak of yellow fever in the United States occurred in 1905. However, with the speed of modern air travel, the threat of a yellow fever outbreak exists wherever *A aegypti* is present. Most countries insist upon proper mosquito control on airplanes and vaccination of all persons at least 10 days before arrival in or from an endemic zone.

The 17D strain of yellow fever virus is an excellent attenuated live-virus vaccine. During the serial passage of a pantropic strain of yellow fever virus through tissue cultures, the relatively avirulent 17D strain was recovered. This strain lost its capacity to induce viscerotropic or neurotropic disease and has been used as a vaccine for over 70 years.

The virulent Asibi strain of yellow fever virus has been sequenced and its sequence compared with that of the 17D vaccine strain, which was derived from it. These two strains are separated by more than 240 passages. The two RNA genomes (10,862 nucleotides long) differ at 68 nucleotide positions, resulting in a total of 32 amino acid differences.

Vaccine is prepared in eggs and dispensed as a dried powder. It is a live virus and must be kept cold. A single dose produces a good antibody response in over 95% of vaccinated persons that persists for at least 30 years. After vaccination, the virus multiplies and may be isolated from the blood before antibodies develop.

Vaccination is contraindicated for infants under 9 months of age, during pregnancy, and in persons with egg allergies or altered immune systems (eg, human immunodeficiency virus infection, malignancy, organ transplantation).

The 17D vaccine is safe. More than 400 million doses of yellow fever vaccine have been administered and severe adverse reactions are extremely rare. There have been about two dozen cases worldwide of vaccine-associated neurotropic disease (postvaccinal encephalitis), most of which occurred in infants. In 2000, a serious syndrome called yellow fever vaccine-associated viscerotropic disease was described. Less than 20 cases of multiple organ system failure in vaccine recipients have been reported worldwide.

Vaccination is the most effective preventive measure against yellow fever, a potentially severe infection with a high death rate for which there is no specific treatment.

DENGUE

Dengue (**breakbone fever**) is a mosquito-borne infection caused by a flavivirus that is characterized by fever, severe headache, muscle and joint pain, nausea and vomiting, eye pain, and rash. A severe form of the disease, dengue hemorrhagic fever/dengue shock syndrome, principally affects children. Dengue is endemic in more than 100 countries.

Clinical Findings

Clinical disease begins 4–7 days (range of 3–14 days) after an infective mosquito bite. The onset of fever may be sudden or there may be prodromal symptoms of malaise, chills, and headache. Pains soon develop, especially in the back, joints, muscles, and eyeballs. Fever lasts from 2 to 7 days, corresponding to peak viral load. The temperature may subside on about the third day and rise again about 5–8 days after onset ("saddleback" form). Myalgia and deep bone pain (breakbone fever) are characteristic. A rash may appear on the third or fourth day and last for 1–5 days. Lymph nodes are frequently enlarged. Classic dengue fever is a self-limited disease. Convalescence may take weeks, although complications and death are rare. Especially in young children, dengue may be a mild febrile illness lasting a short time.

A severe syndrome—**dengue hemorrhagic fever/dengue shock syndrome**—may occur in individuals (usually children) with passively acquired (as maternal antibody) or preexisting nonneutralizing heterologous dengue antibody due to a previous infection with a different serotype of virus. Although initial symptoms simulate normal dengue, the patient's condition worsens. The key pathological feature of dengue hemorrhagic fever is increased vascular permeability with plasma leakage into the interstitial spaces associated with increased levels of vasoactive cytokines. This can lead to life-threatening shock in some patients. Circumstantial evidence suggests that secondary infection with dengue type 2 following a type 1 infection is a particular risk factor for severe disease.

The pathogenesis of the severe syndrome involves preexisting dengue antibody. It is postulated that virus–antibody complexes are formed within a few days of the second dengue infection and that the nonneutralizing enhancing antibodies promote infection of higher numbers of mononuclear cells, followed by the release of cytokines, vasoactive mediators, and procoagulants, leading to the disseminated intravascular coagulation seen in the hemorrhagic fever syndrome. Cross-reactive cellular immune responses to dengue virus may also be involved.

Laboratory Diagnosis

Reverse transcriptase-polymerase chain reaction-based methods are available for rapid identification and serotyping of dengue virus in acute-phase serum, roughly during the period of fever. Isolation of the virus is difficult. The current favored approach is inoculation of a mosquito cell line with patient serum, coupled with nucleic acid assays to identify a recovered virus.

Serological diagnosis is complicated by cross-reactivity of IgG antibodies to heterologous flavivirus antigens. A variety of methods are available; the most commonly used methods are E/M viral protein-specific capture IgM or IgG ELISA and the hemagglutination inhibition test. IgM antibodies develop within a few days of illness. Neutralizing and hemagglutination-inhibiting antibodies appear within a week after onset of dengue fever. Analysis of paired acute and convalescent sera to show a significant rise in antibody titer is the most reliable evidence of an active dengue infection.

Immunity

Four serotypes of the virus exist that can be distinguished by molecular-based assays and by Nt tests. Infection confers lifelong protection against that serotype, but cross-protection between serotypes is of short duration. Reinfection with a virus of a different serotype after the primary attack is more apt to result in severe disease (dengue hemorrhagic fever).

Epidemiology

Dengue viruses are distributed worldwide in tropical regions (Figure 38–2). Most subtropical and tropical regions around the world where *Aedes* vectors exist are endemic areas. In the last 20 years, epidemic dengue has emerged as a problem in the Americas. In 1995, more than 200,000 cases of dengue and over 5500 cases of dengue hemorrhagic fever occurred in Central and South America. The changing disease patterns are probably related to rapid urban population growth, overcrowding, and lax mosquito control efforts.

Dengue in 2008 was the most important mosquito-borne viral disease affecting humans. There are an estimated 50 million or more cases of dengue annually worldwide, with 400,000 cases of dengue hemorrhagic fever. The latter is a leading cause of childhood death in several Asian countries.

The risk of the hemorrhagic fever syndrome is about 0.2% during the first dengue infection but is at least tenfold higher during infection with a second dengue virus serotype. The fatality rate with dengue hemorrhagic fever can reach 15%, but can be reduced to less than 1% with proper treatment.

The ratio of inapparent to apparent infections is variable but may be about 15:1 for primary infections; the ratio is lower in secondary infections.

In urban communities, dengue epidemics are explosive and involve appreciable portions of the population. They often start during the rainy season, when the vector mosquito, *A aegypti,* is abundant (Figure 38–7). The mosquito breeds in tropical or semitropical climates in water-holding receptacles or in plants close to human dwellings.

A aegypti is the primary vector mosquito for dengue in the western hemisphere. The female acquires the virus by feeding upon a viremic human. After a period of 8–14 days, mosquitoes are infective and probably remain so for life (1–3 months). In the tropics, mosquito breeding throughout the year maintains the disease.

World War II was responsible for the spread of dengue from Southeast Asia throughout the Pacific region. Only dengue type 2 was present in the Americas for years. Then, in 1977, a dengue type 1 virus was detected; this was the first time type 1 virus had been isolated in the western hemisphere. In 1981, dengue type 4 was first recognized in the western hemisphere, followed in 1994 by dengue type 3. The viruses are now spread throughout Central and South America, and dengue hemorrhagic fever is endemic in many countries.

Endemic dengue in the Caribbean and Mexico is a constant threat to the United States, where *A aegypti* mosquitoes are prevalent in the summer months. Concurrent with the increased epidemic activity of dengue in the tropics, there has been an increase in the number of cases imported into the United States. The first locally acquired case of dengue hemorrhagic fever in the United States occurred in south Texas in 2005.

A albopictus, a mosquito of Asian origin, was discovered in Texas in 1985; by 1989 it had spread throughout the southeastern United States, where *A aegypti,* the principal vector of dengue virus, is prevalent. In contrast to *A aegypti,* which cannot overwinter in northern states, *A albopictus* can overwinter farther north, increasing the risk of epidemic dengue in the United States.

Treatment & Control

There is no antiviral drug therapy. Dengue hemorrhagic fever can be treated by fluid replacement therapy. There is no vaccine, but candidate vaccines are under development. Vaccine development is difficult because a vaccine must provide protection against all four serotypes of virus.

Control depends upon antimosquito measures, eg, elimination of breeding places and the use of insecticides. Screened windows and doors can reduce exposure to the vectors.

BUNYAVIRUS ENCEPHALITIS

The Bunyaviridae family contains more than 300 viruses, mostly arthropod-transmitted. Spherical particles measuring 80–120 nm contain a single-stranded, negative-sense or ambisense, triple-segmented RNA genome 11–19 kb in total size. The envelope has two glycoproteins. Several member viruses produce mosquito-borne encephalitides of humans and animals; others cause hemorrhagic fevers.

Transovarial transmission occurs in some mosquitoes. Some are transmitted by sandflies. Hantavirus pulmonary syndrome is caused by a virus transmitted by rodents. Bunyaviruses are sensitive to inactivation by heat, detergents, formaldehyde, and low pH; some are hemagglutinating (Figure 38–1).

The California encephalitis virus complex comprises 14 antigenically related viruses in the *Orthobunyavirus* genus of the family. This includes La Crosse virus, a significant human pathogen in the United States (Table 38–2). La Crosse virus is a major cause of encephalitis and aseptic meningitis in children, particularly in the upper Midwest. Most cases occur between July and September in children under the age of 16. There are about 70 cases of La Crosse encephalitis reported per year.

The viruses are transmitted by various woodland mosquitoes, primarily *Aedes triseriatus*. Principal vertebrate hosts are small mammals such as squirrels, chipmunks, and rabbits. Human infection is tangential. Overwintering can occur in eggs of the mosquito vector. The virus is transmitted transovarially, and adult mosquitoes that develop from infected eggs can transmit the virus by bite.

The onset of California encephalitis viral infection is abrupt, typically with severe headache, fever, and in some cases vomiting and convulsions. About half the patients develop seizures, and the case-fatality rate is about 1%. Less frequently, there is only aseptic meningitis. The illness lasts 10–14 days, although convalescence may be prolonged. Neurologic sequelae are rare. There are many infections for every case of encephalitis. Serologic confirmation by HI, ELISA, or Nt tests is done on acute and convalescent specimens.

SANDFLY FEVER

Sandfly fever is a mild, insect-borne disease that occurs commonly in countries bordering the Mediterranean Sea and in Russia, Iran, Pakistan, India, Panama, Brazil, and Trinidad. Sandfly fever (also called *Phlebotomus* fever) is caused by a bunyavirus in the *Phlebovirus* genus (Table 38–1).

The disease is transmitted by the female sandfly, *Phlebotomus papatasii*, a midge only a few millimeters in size. In the tropics, the sandfly is prevalent all year; in cooler climates, only during the warm seasons. Transovarial transmission occurs.

In endemic areas, infection is common in childhood. When nonimmune adults (eg, troops) arrive, large outbreaks can occur among the new arrivals and are occasionally mistaken for malaria.

In humans, the bite of the sandfly results in small itching papules on the skin that persist for up to 5 days. The disease begins abruptly after an incubation period of 3–6 days. The virus is found in the blood briefly near the time of onset of symptoms. Clinical features consist of headache, malaise, nausea, fever, photophobia, stiffness of the neck and back, abdominal pain, and leukopenia. All patients recover. There is no specific treatment.

Sandflies are most common just above the ground. Because of their small size, they can pass through ordinary screens and mosquito nets. The insect feeds primarily at night. Prevention of disease in endemic areas relies on use of insect repellents during the night and residual insecticides around living quarters.

RIFT VALLEY FEVER

The agent of this disease, a bunyavirus of the *Phlebovirus* genus, is a mosquito-borne zoonotic virus pathogenic primarily for domestic livestock. Humans are secondarily infected during the course of epizootics in domesticated animals. Infection among laboratory workers is common.

Epizootics occur periodically following heavy rains that allow hatches of the primary vector and reservoir (*Aedes* species mosquitoes). Viremia in animals leads to infection of other vectors with collateral transmission to humans. Transmission to humans is primarily by contact with infected animal blood and body fluids and mosquito bites.

Disease in humans is usually a mild febrile illness that is short-lived, and recovery almost always is complete. Complications include retinitis, encephalitis, and hemorrhagic fever. Permanent loss of vision may occur (1–10% of cases with retinitis). About 1% of infected patients die.

Rift Valley fever exists in most countries of sub-Saharan Africa. It spread in 1977 to Egypt, where it caused enormous losses of sheep and cattle and thousands of human cases, with 600 deaths. A large outbreak occurred in West Africa in 1987 and in East Africa in 1997. The first documented spread of Rift Valley fever virus outside of Africa occurred in 2000 in Yemen and Saudi Arabia.

COLORADO TICK FEVER

A few arboviruses are members of the family Reoviridae (see Chapter 37). Colorado tick fever is classified in the genus *Coltivirus*. African horse sickness and bluetongue viruses are in the genus *Orbivirus*. Rotaviruses and orthoreoviruses have no arthropod vectors.

Colorado tick fever, also called mountain fever or tick fever, is transmitted by a tick (Table 38–1). The virus appears to be antigenically distinct from other known viruses, and only one antigenic type is recognized.

Colorado tick fever is a mild febrile disease, without rash. The incubation period is 4–6 days. The disease has a sudden onset with fever and myalgia. Symptoms include headache, muscle and joint pains, lethargy, and nausea and vomiting. The temperature is usually diphasic. After the first bout of 2 days, the patient may feel well, but symptoms reappear and last 3–4 more days. The disease in humans is self-limited (Table 38–2).

The virus may be isolated from whole blood by inoculation of cell cultures. Viremia may persist for 4 weeks or longer. Reverse transcriptase-polymerase chain reaction assays can detect viral RNA in red blood cells and in plasma. Specific neutralizing antibodies appear in the second week of illness that can be detected by plaque reduction tests. Other serologic assays include ELISA and fluorescent antibody tests. A single infection is believed to produce lasting immunity.

There are several hundred reported cases of Colorado tick fever annually, but that is believed to be only a fraction of total cases. The disease is limited to areas where the wood tick *Dermacentor andersoni* is distributed, primarily high altitudes in the western United States and southwestern Canada. Patients have been in a tick-infested area before onset of symptoms. Cases occur chiefly in young men, the group with greatest exposure to ticks. *D andersoni* collected in nature can carry the virus. This tick is a true reservoir, and the virus is transmitted transovarially by the adult female. Natural infection occurs in rodents, which act as hosts for immature stages of the tick.

There is no specific therapy. The disease can be prevented by avoiding tick-infested areas and by using protective clothing or repellent chemicals.

RODENT-BORNE HEMORRHAGIC FEVERS

The zoonotic rodent-borne hemorrhagic fevers include Asian (eg, Hantaan and Seoul viruses), South American (eg, Junin and Machupo viruses), and African (Lassa virus) fevers. Hantaviruses also cause a hantavirus pulmonary syndrome in the Americas (eg, Sin Nombre virus). The natural reservoirs of Marburg and Ebola viruses (African hemorrhagic fever) are not known but are suspected to be rodents or bats. Causative agents are classified as bunyaviruses, arenaviruses, and filoviruses (Table 38–1).

BUNYAVIRUS DISEASES

Hantaviruses are classified in the *Hantavirus* genus of the Bunyaviridae family. The viruses are found worldwide and cause two serious and often fatal human diseases: hemorrhagic fever with renal syndrome (HFRS) and hantavirus pulmonary syndrome (HPS). It is estimated there are 100,000–200,000 cases of hantavirus infection annually worldwide. There are several distinct hantaviruses, each associated with a specific rodent host. The virus infections in rodents are lifelong and without deleterious effects. Transmission among rodents seems to occur horizontally, and transmission to humans occurs by inhaling aerosols of rodent excreta (urine, feces, saliva). The presence of hantavirus-associated diseases is determined by the geographic distribution of the rodent reservoirs.

Hemorrhagic Fever with Renal Syndrome

Hemorrhagic fever with renal syndrome (HFRS) is an acute viral infection that causes an interstitial nephritis that can lead to acute renal insufficiency and renal failure in severe forms of the disease. Hantaan and Dobrava viruses cause the severe disease that occurs in Asia, particularly in China, Russia, and Korea, and in Europe, primarily in the Balkans. Generalized hemorrhage and shock may occur, with a case-fatality rate of 5–15%. A moderate form of HFRS caused by Seoul virus occurs throughout Eurasia. In a mild clinical form, called nephropathia epidemica, which is caused by Puumala virus and is prevalent in Scandinavia, the nephritis generally resolves without hemorrhagic complications, and fatalities are rare (<1%).

More than 2000 cases of HFRS occurred among United Nations troops during the Korean war, but Hantaan virus was not isolated until 1976 in Korea from a rodent, *Apodemus agrarius.*

Urban rats are known to be persistently infected with hantaviruses, and it has been suggested that rats on trading ships may have dispersed hantaviruses worldwide. Serosurveys indicated that brown Norway rats in the United States are infected with Seoul virus. Infected laboratory rats were proved to be sources of Hantaan outbreaks in scientific institutes in Europe and Asia, but such infections have not been detected in laboratory rats raised in the United States. Hantavirus infections have occurred in persons whose occupations place them in contact with rats (eg, longshoremen).

HFRS is treated using supportive therapy. Prevention depends on rodent control and protection from exposure to rodent droppings and contaminated material.

Hantavirus Pulmonary Syndrome

In 1993 an outbreak of severe respiratory illness occurred in the United States, now designated the hantavirus pulmonary syndrome (HPS). It was found to be caused by a novel hantavirus (Sin Nombre virus). This agent was the first hantavirus recognized to cause disease in North America and the first to cause primarily an adult respiratory distress syndrome. Since that time, numerous hantaviruses have been detected in rodents in North, Central, and South America (Table 38–2).

The deer mouse (*Peromyscus maniculatus*) is the primary rodent reservoir for Sin Nombre virus. Deer mice are widespread and about 10% of those tested show evidence of infection with Sin Nombre virus. Other hantaviruses known to cause HPS in the United States include New York virus, Black Creek Canal virus, and Bayou virus, each having a different rodent host. HPS is more common in South America than in the United States. Andes virus is one causative hantavirus and is found in Argentina and Chile. Choclo virus has been identified in Panama.

Infections with hantaviruses are not common, and subclinical infections appear to be unusual, particularly with Sin Nombre virus. HPS is generally severe, with reported mortality rates of 30% or greater. This case-fatality rate is substantially higher than that of other hantavirus infections. The disease begins with fever, headache, and myalgia, followed by rapidly progressive pulmonary edema, often leading to severe respiratory compromise. There are no signs of hemorrhage. Hantaviral antigens are detected in endothelial cells and macrophages in lung, heart, spleen, and lymph nodes. Pathogenesis of HPS involves the functional impairment of vascular endothelium. Person-to-person transmission of hantaviruses seldom occurs, though it has been observed during outbreaks of HPS caused by Andes virus.

Laboratory diagnosis depends on detection of viral nucleic acid by reverse transcriptase-polymerase chain reaction, detection of viral antigens in fixed tissues by immunohistochemistry, or detection of specific antibodies using recombinant proteins. An ELISA test to detect IgM antibodies may be used to diagnose acute infections. A fourfold rise in IgG antibody titer between acute and convalescent sera is diagnostic. IgG antibodies are long lasting. Isolation of hantaviruses is difficult and requires the use of containment facilities.

Current therapy for HPS consists of maintenance of adequate oxygenation and support of hemodynamic functioning. The antiviral drug ribavirin is of some benefit as therapy in HPS. Preventive measures are based on rodent control and avoidance of contact with rodents and rodent droppings. Care must be taken to avoid inhaling aerosolized dried excreta when cleaning rodent-infested structures.

ARENAVIRUS DISEASES

Arenaviruses are typified by pleomorphic particles that contain a segmented RNA genome; are surrounded by an envelope with large, club-shaped peplomers; and measure 50–300 nm in diameter (mean, 110–130 nm) (Figure 38–1). The arenavirus genome consists of two single-stranded RNA molecules with unusual ambisense genetic organization.

Based on sequence data, arenaviruses are divided into Old World viruses (eg, Lassa virus) and New World viruses. The latter division is divided into three groups, with Group A including Pichinde virus and Group B containing the human pathogenic viruses, such as Machupo virus. Some isolates, such as Whitewater Arroyo virus, appear to be recombinants between New World lineages A and B.

Arenaviruses establish chronic infections in rodents. Each virus is generally associated with a single rodent species. The geographic distribution of a given arenavirus is determined in part by the range of its rodent host. Humans are infected when they come in contact with rodent excreta. Some viruses cause severe hemorrhagic fever. Several arenaviruses are known to infect the fetus and may cause fetal death in humans.

Multiple arenaviruses cause human disease, including Lassa, Junin, Machupo, Guanarito, Sabia, Whitewater Arroyo, and lymphocytic choriomeningitis (LCM) (Table 38–1). Because these arenaviruses are infectious by aerosols, great care must be taken when processing rodent and human specimens. High-level containment conditions are required in the laboratory. Transmission of arenaviruses in the natural rodent hosts may occur by vertical and horizontal routes. Milk, saliva, and urine may be involved in transmission. Arthropod vectors are believed not to be involved.

A generalized replication cycle is shown in Figure 38–8. Host ribosomes are encapsidated during the morphogenesis of virus particles. Arenaviruses typically do not cause cytopathic effects when replicating in cultured cells.

Lassa Fever

The first recognized cases of Lassa fever occurred in 1969 among Americans stationed in the Nigerian village of Lassa. Lassa virus is highly virulent—the mortality rate is about 15% for patients hospitalized with Lassa fever. Overall, about 1% of Lassa virus infections are fatal. In western Africa, estimates are that the annual toll may reach several hundred thousand infections and 5000 deaths. Lassa virus is active in all western African countries situated between Senegal and Republic of Congo. Occasional cases identified outside the endemic area usually are imported, often by persons returning from West Africa.

The incubation period for Lassa fever is 1–3 weeks from time of exposure. The disease can involve many organ systems, although symptoms may vary in the individual patient. Onset is gradual, with fever, vomiting, and back and chest pain. The disease is characterized by very high fever, mouth ulcers, severe muscle aches, skin rash with hemorrhages, pneumonia, and heart and kidney damage. Deafness is a common complication, affecting about 25% of cases during recovery; hearing loss is often permanent.

Lassa virus infections cause fetal death in more than 75% of pregnant women. During the third trimester, maternal mortality is increased (30%) and fetal mortality is very high (>90%). Benign febrile cases do occur.

Diagnosis usually involves detection of IgM and IgG antibodies by ELISA. Immunohistochemistry can be used to detect viral antigens in postmortem tissue specimens. Viral sequences can be detected using reverse transcriptase-polymerase chain reaction assays in research laboratories.

A house rat (Mastomys natalensis) is the principal rodent reservoir of Lassa virus. Rodent control measures are one way to minimize virus spread but are often impractical in endemic areas. The virus can be transmitted by human-to-human contact. When the virus spreads within a hospital, human contact is the mode of transmission. Meticulous barrier nursing procedures and standard precautions to avoid contact with virus-contaminated blood and body fluids can prevent transmission to hospital personnel.

The antiviral drug ribavirin is the drug of choice for Lassa fever and is most effective if given early in the disease process. No vaccine exists, although a vaccinia virus recombinant that expresses the glycoprotein gene of Lassa virus is able to induce protective immunity both in guinea pigs and in monkeys.

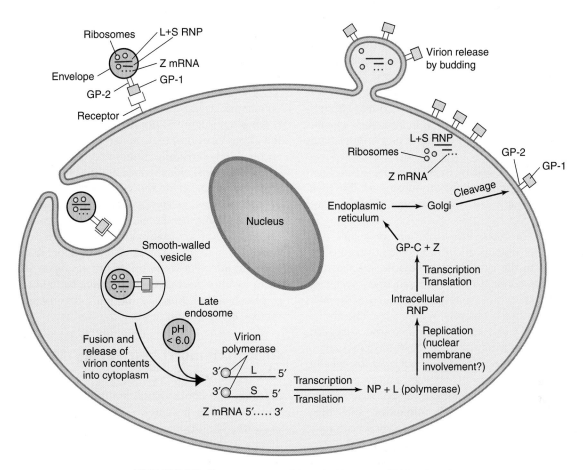

FIGURE 38–8 The arenavirus life cycle. (Courtesy of PJ Southern.)

South American Hemorrhagic Fevers

Based on both serologic and phylogenetic studies of viral RNA, the South American arenaviruses are all considered to be members of the Tacaribe complex. Most have cricetid rodent reservoirs. The viruses tend to be prevalent in a particular area, limited in their distribution. Numerous viruses have been discovered; serious human pathogens are the closely related Junin, Machupo, Guanarito, and Sabia viruses. Bleeding is more common in Argentine (Junin) and other South American hemorrhagic fevers than in Lassa fever.

Junin hemorrhagic fever (Argentine hemorrhagic fever) is a major public health problem in certain agricultural areas of Argentina; over 18,000 cases were reported between 1958 and 1980, with a mortality rate of 10–15% in untreated patients. Many cases continue to occur each year. The disease has a marked seasonal variation, and the infection occurs almost exclusively among workers in maize and wheat fields who are exposed to the reservoir rodent, *Calomys musculinus.*

Junin virus produces both humoral and cell-mediated immunodepression; deaths due to Junin hemorrhagic fever may be related to an inability to initiate a cell-mediated immune response. Administration of convalescent human plasma to patients during the first week of illness reduced the mortality rate from 15–30% to 1%. Some of these patients develop a self-limited neurologic syndrome 3–6 weeks later. An effective live-attenuated Junin virus vaccine is used to vaccinate high-risk individuals in South America.

The first outbreak of **Machupo hemorrhagic fever** (Bolivian hemorrhagic fever) was identified in Bolivia in 1962. It is estimated that 2000–3000 persons were affected by the disease, with a case-fatality rate of 20%. An effective rodent control program directed against infected *Calomys callosus,* the host of Machupo virus, was undertaken in Bolivia and has greatly reduced the number of cases of Machupo hemorrhagic fever.

Guanarito virus (the agent of **Venezuelan hemorrhagic fever**) was identified in 1990; it has a mortality rate of about 33%. Its emergence was tied to clearance of forest land for small farm use. **Sabia virus** was isolated in 1990 from a fatal case of hemorrhagic fever in Brazil. Both Guanarito virus and Sabia virus induce a clinical disease resembling that of Argentine hemorrhagic fever and probably have similar mortality rates.

Lymphocytic Choriomeningitis

LCM virus was discovered in 1933 and is widespread in Europe and the Americas. Its natural vector is the wild house mouse, *Mus musculus*. It is endemic in mice but can also infect other rodents. About 5% of mice throughout the United States carry the virus. It may chronically infect mouse or hamster colonies and may infect pet rodents.

LCM virus is occasionally transmitted to humans, presumably via mouse droppings. There is no evidence of horizontal person-to-person spread. LCM in humans is an acute disease manifested by aseptic meningitis or a mild systemic influenza-like illness. Rarely is there a severe encephalomyelitis or a fatal systemic disease in healthy people (mortality is less than 1%). Many infections are subclinical. The incubation period is usually 1–2 weeks, and the illness lasts 1–3 weeks.

LCM virus infections can be serious in people with impaired immune systems. In 2005, four solid-organ transplant recipients in the United States became infected from a common organ donor. Three of the four organ recipients died 23–27 days after transplantation. The source of the virus was determined to be a pet hamster recently purchased by the organ donor. The LCM virus also can be transmitted vertically from mother to fetus and infection of the fetus early in pregnancy can lead to serious defects, such as hydrocephalus, blindness, and fetal death.

Infections are usually diagnosed retrospectively by serology using ELISA for IgM and IgG antibodies. Other diagnostic approaches include immunohistochemical staining of tissues for viral antigens, reverse transcriptase-polymerase chain reaction for viral nucleic acid, and viral culture using Vero cells. Serological studies in urban areas have shown infection rates in humans ranging from 2% to 5%.

Experimental studies have shown that the immune response may be protective or deleterious in LCM virus-infected mice. T cells are required to control the infection but may also induce immune-mediated disease. The result depends on the age, immune status, and genetic background of the mouse and the route of inoculation of the virus. Mice infected as adults may develop a rapidly fatal disease due to a T cell-mediated inflammatory response in the brain. Congenitally or neonatally infected mice do not become acutely ill but carry a lifelong persistent infection. They fail to clear the infection because they were infected before the cellular immune system matured. They make a strong antibody response that may lead to circulating viral antigen-antibody complexes and immune complex disease.

FILOVIRUS DISEASES

Classification & Properties of Filoviruses

Filoviruses are pleomorphic particles, appearing as long filamentous threads or as odd-shaped forms 80 nm in diameter (Figure 38–1). Unit-length particles are from 665 nm (Marburg) to 805 nm (Ebola). The two known filoviruses (Marburg virus and Ebola virus) are antigenically distinct and are classified in separate genera (Table 38–1). The four subtypes of Ebola virus (Zaire, Sudan, Reston, Ivory Coast) differ from one another by up to 40% at the nucleotide level but share some common epitopes. The subtypes appear to be stable over time.

The large filovirus genome is single-stranded, nonsegmented, negative-sense RNA 19 kb in size and contains seven genes (Figure 38–9). An unusual coding strategy with the Ebola viruses is that the envelope glycoprotein (GP) is encoded in two reading frames and requires transcriptional editing or translational frame-shifting to be expressed. The glycoprotein makes up the viral surface spikes in the form of trimers 10 nm in length. Virions are released via budding from the plasma membrane.

Filoviruses are highly virulent and require maximum containment facilities (Biosafety Level 4) for laboratory work. Filovirus infectivity is destroyed by heating for 30 minutes at 60°C, by ultraviolet and γ-irradiation, by lipid solvents, and by bleach and phenolic disinfectants. The natural hosts and vectors, if any, are unknown.

African Hemorrhagic Fevers (Marburg & Ebola Viruses)

Marburg and Ebola viruses are highly virulent in humans and nonhuman primates, with infections usually ending in death. The incubation period is 3–9 days for Marburg disease and 2–21 days for Ebola. They cause similar acute diseases characterized by fever, headache, sore throat, and muscle pain, followed by abdominal pain, vomiting, diarrhea, and rash, with both internal and external bleeding, often leading to shock and death. Filoviruses have a tropism for cells of the macrophage system, dendritic cells, interstitial fibroblasts, and endothelial cells. Very high titers of virus are present in many tissues, including liver, spleen, lungs, and kidneys, and in blood and other fluids. These viruses have the highest mortality rates (25–90%) of all the viral hemorrhagic fevers.

Marburg virus disease was recognized in 1967 among laboratory workers exposed to tissues of African green monkeys (*Cercopithecus aethiops*) imported into Germany and Yugoslavia. Transmission from patients to medical personnel occurred, with high mortality rates. Antibody surveys have indicated that the virus is present in East Africa and causes infection in monkeys and humans. Recorded cases of the disease are rare, but outbreaks have been documented in Kenya, South Africa, Democratic Republic of the Congo, and, in 2005, in Angola. Marburg virus can infect guinea pigs, mice, hamsters, monkeys, and various cell culture systems.

Ebola virus was discovered in 1976 when two severe epidemics of hemorrhagic fever occurred in Sudan and Zaire (now the Democratic Republic of the Congo). The outbreaks involved more than 500 cases and at least 400 deaths due to clinical hemorrhagic fever. In each outbreak, hospital staff became infected through close and prolonged contact with patients, their blood, or their excreta. These subtypes of Ebola

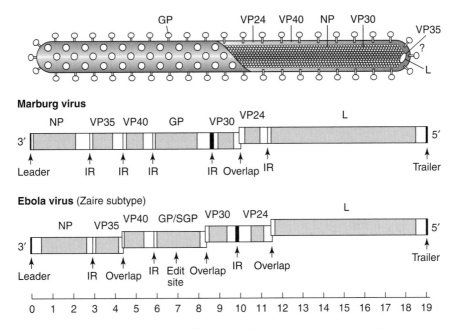

FIGURE 38–9 Virion structure and genome organization of filoviruses. The genome organization of Marburg virus and the Zaire subtype of Ebola virus are shown. The diagram of the virion shows the single-strand, negative-sense RNA encased in the nucleocapsid and enveloped in a lipid bilayer membrane. Structural proteins associated with the nucleocapsid are the nucleoprotein (NP), VP30, VP35, and the polymerase (L) protein. Membrane-associated proteins are the matrix protein (VP40), VP24, and the GP (peplomer glycoprotein). The genes encoding the structural proteins are identified and drawn to scale in the genome structures. Shaded areas denote the coding regions and white areas the noncoding sequences. Genes begin with a conserved transcriptional start site and end with a transcriptional stop (polyadenylation) site; adjoining genes are either separated from one another by an intergenic region (IR) or overlap one another. The site at which the additional A is added within the GP gene during transcriptional editing is indicated in the diagram of Ebola. The primary gene product of the GP gene of Ebola viruses is the SGP, a nonstructural secreted glycoprotein. At the extreme 3′ and 5′ ends of the genomes are the complementary leader and trailer sequences, respectively. (Adapted with permission from Peters CJ et al: Filoviridae: Marburg and Ebola viruses. In: *Fields Virology*, 3rd ed. Fields BN et al [editors]. Lippincott-Raven, 1996.)

virus (Zaire, Sudan) are highly virulent. The mean time to death from the onset of symptoms is 7–8 days.

Subsequent outbreaks of Ebola hemorrhagic fever have occurred in Uganda (2000), the Republic of the Congo (1995, 2001, 2002, 2003), Gabon (1994, 1996, 1997, 2002), South Africa (1996), and Sudan (2004). Epidemics are often stopped by the institution of barrier nursing methods and training of hospital personnel.

Since Ebola virus was discovered, approximately 1850 cases had been recognized by 2004, with more than 1200 deaths. The outbreak in 2003 was first recognized by a large number of dead gorillas and chimpanzees.

In 1989, infections caused by a filovirus closely related to Ebola virus were detected in cynomolgus monkeys (*Macaca fascicularis*) imported into the United States from the Philippines and held in a quarantine facility in Virginia. The infection spread to only a few of the 149 persons who came in contact with the infected monkeys or their tissues, but none of the workers became sick, indicating that the virus (Reston strain) possesses low pathogenicity for humans.

High mortality among pigs in the Philippines in 2008 led to the discovery of Ebola Reston virus in animals other than primates. Five individuals who had contact with sick pigs

developed antibodies to Ebola Reston, but remained healthy, confirming that this virus strain can infect humans without causing disease.

Filovirus infections appear to be immunosuppressive. Fatal cases often show impaired humoral immune responses. However, filovirus antibodies appear as patients recover that are detectable by ELISA. Viral antigens in serum can be detected by ELISA, providing a rapid screening test of human samples. Reverse transcriptase-polymerase chain reaction can also be used on clinical specimens. One hazard to performing tests for filoviruses is that patient sera and other specimens may contain virulent virus. Tests can only be conducted under maximum biologic containment conditions. Fresh virus isolates can be cultured in cell lines such as Vero and MA-104 monkey cell lines.

It is probable that Marburg and Ebola viruses have a reservoir host, perhaps a rodent or a bat, and become transmitted to humans only accidentally. Monkeys are not considered to be reservoir hosts as most infected animals die too rapidly to sustain virus survival. Human infections are highly communicable to human contacts, generally by direct contact with blood or body fluids. Typically, outbreaks of Ebola virus infection are associated with the introduction of virus into the community by one infected person, followed by dissemination by person-to-person spread, often within medical facilities.

Because the natural reservoirs of Marburg and Ebola viruses are still unknown, no control activities can be organized. The use of isolation facilities in hospital settings remains the most effective means of controlling Ebola disease outbreaks. Strict barrier nursing techniques should be implemented. Extreme care must be taken with infected blood, secretions, tissues, and wastes. Personnel involved in the transportation and care of nonhuman primates should be instructed about the potential hazards of handling such animals.

There are no specific antiviral therapies available. Treatment is directed at maintaining renal function and electrolyte balance and combating hemorrhage and shock. There is no vaccine, but candidate vaccines are under development.

REVIEW QUESTIONS

1. A 74-year-old man develops fever, malaise, and a sore throat, followed shortly thereafter by nausea, vomiting, and then stupor. Eastern equine encephalitis is diagnosed. Control of this disease in humans could be accomplished by eradication of which of the following?

    (A) Horses
    (B) Birds
    (C) Sandflies
    (D) Mosquitoes
    (E) Ticks

2. An arbovirus common in the Middle East, Africa, and Southwest Asia first appeared in New York in 1999. By 2002 the virus had spread throughout the continental United States. This arbovirus, a member of the Japanese B encephalitis antigenic complex, is which of the following?

    (A) Japanese B encephalitis virus
    (B) Tick-borne encephalitis virus
    (C) West Nile virus
    (D) Dengue virus
    (E) Rift Valley fever virus

3. Which of the following descriptions of or statements about Lassa fever is correct?

    (A) It is found in eastern Africa
    (B) Human-to-human transmission does not occur
    (C) It seldom causes death or complications
    (D) It occurs from contact with the house rat *Mastomys*
    (E) There is no drug that is effective in treating Lassa fever

4. Arboviruses are transmitted by bloodsucking arthropods from one vertebrate host to another. Arboviruses are found in the following virus families except which of the following?

    (A) Togaviridae
    (B) Flaviviridae
    (C) Bunyaviridae
    (D) Reoviridae
    (E) Arenaviridae

5. A 27-year-old man develops fever, chills, headache, and backache. Four days later he develops high fever and jaundice. Yellow fever is diagnosed. Which of the following statements concerning yellow fever is correct?

    (A) The virus is transmitted by culicine mosquitoes in the urban form of disease
    (B) Monkeys in the jungle are a major reservoir of yellow fever virus
    (C) Yellow fever often has long-term complications
    (D) All infections lead to apparent disease
    (E) Ribavirin is specific therapy

6. Regarding the case in Question 5, yellow fever occurs in which region or regions of the world?

    (A) Asia
    (B) Africa and South America
    (C) North America
    (D) Africa and Middle East
    (E) Throughout the world

7. African hemorrhagic fevers, Marburg and Ebola, are severe diseases often ending in death. Which of the following is most accurate about Ebola virus?

    (A) Is spread by contact with blood or other body fluids
    (B) Is transmitted by mosquitoes
    (C) Is a flavivirus
    (D) Causes infections but no disease in nonhuman primates
    (E) Is antigenically related to Lassa fever virus

8. Which of the following groups can be vaccinated routinely with yellow fever vaccine without special safety considerations?

    (A) Children less than 9 months old
    (B) Pregnant women
    (C) Persons with compromised immune systems
    (D) All of the above
    (E) None of the above

9. Hantaviruses, which are emerging pathogens in the United States, can be described by which of the following?

    (A) They are arenaviruses
    (B) They are readily transmitted human-to-human
    (C) They cause influenza-like symptoms followed rapidly by acute respiratory failure
    (D) They are acquired by inhalation of aerosols of deer urine
    (E) They show a high frequency of antigenic variation

10. A microbiologist was performing a necropsy in a laminar flow biosafety cabinet on a blue-jay submitted as part of a state's arbovirus surveillance program. He lacerated his thumb while using a scalpel to remove the bird's brain. Four days later, he developed headache, myalgia, and malaise followed by chills, sweats, and lymph node swelling. Two days later, a rash began on his face and spread to the trunk, arms, and legs, persisting for about 3 days. He sought medical care and reported a history of dengue fever and vaccinations with yellow fever and Japanese B encephalitis vaccines. A serum sample taken the day of the injury contained anti-flavivirus IgG antibody by ELISA. A serum sample collected 13 days after onset of illness showed an increased titer of anti-flavivirus IgG antibody and the presence of West Nile virus IgM antibody. The physician could conclude that the most likely cause of the microbiologist's illness was which virus?

    (A) Dengue virus
    (B) Yellow fever virus
    (C) West Nile virus
    (D) St. Louis encephalitis

(E) Not identifiable until neutralizing antibody titers from paired sera could be assessed against a panel of arboviruses

11. Which of the following statements about dengue virus is not true?

(A) It is the most important mosquito-borne viral disease affecting humans

(B) It is distributed worldwide in tropical regions

(C) It can cause a severe hemorrhagic fever

(D) There is a single antigenic type

(E) One form of disease is characterized by increased vascular permeability

12. Which of the following diseases occurring in the United States lacks a known vector?

(A) Hantavirus pulmonary syndrome

(B) West Nile fever

(C) La Crosse encephalitis

(D) Colorado tick fever

(E) St. Louis encephalitis

Answers

1. D	4. E	7. A	10. C
2. C	5. B	8. E	11. D
3. D	6. B	9. C	12. A

REFERENCES

Brinton MA: The molecular biology of West Nile Virus: A new invader of the western hemisphere. Annu Rev Microbiol 2002;56:371. [PMID: 12142476]

Calisher CH, Childs JE, Field HE, Holmes KV, Schountz T: Bats: Important reservoir hosts of emerging viruses. Clin Microbiol Rev 2006;19:531. [PMID: 16847084]

Feldmann H, Geisbert T, Kawaoka Y (guest editors): Filoviruses: Recent advances and future challenges. J Infect Dis 2007;196(Suppl 2). [Entire issue.]

Griffin DE: Alphaviruses. In: *Fields Virology,* 5th ed. Knipe DM et al (editors). Lippincott Williams & Wilkins, 2007.

Gubler DJ, Kuno G, Markoff L: Flaviviruses. In: *Fields Virology,* 5th ed. Knipe DM et al (editors). Lippincott Williams & Wilkins, 2007.

Mills JN et al: Hantavirus pulmonary syndrome—United States: Updated recommendations for risk reduction. Centers for Disease Control and Prevention. MMWR Recomm Rep 2002;51(RR-9):1.

Shu PY, Huang JH: Current advances in dengue diagnosis. Clin Diagnostic Lab Immunol 2004;11:642. [PMID: 15242935]

Soldan SS, González-Scarano F: Emerging infectious diseases: The *Bunyaviridae.* J Neurovirol 2005;11:412. [PMID: 16287682]

Süss J: Epidemiology and ecology of TBE relevant to the production of effective vaccines. Vaccine 2003;21(Suppl 1):S19. [PMID: 12628811]

WHO position paper: Japanese encephalitis vaccines. World Health Org Wkly Epidemiol Rec 2006;81:331.

WHO position paper: Yellow fever vaccine. World Health Org Wkly Epidemiol Rec 2003;78:349.

Orthomyxoviruses (Influenza Viruses)

Respiratory illnesses are responsible for more than half of all acute illnesses each year in the United States. The **Orthomyxoviridae** (influenza viruses) are a major determinant of morbidity and mortality caused by respiratory disease, and outbreaks of infection sometimes occur in worldwide epidemics. Influenza has been responsible for millions of deaths worldwide. Mutability and high frequency of genetic reassortment and resultant antigenic changes in the viral surface glycoproteins make influenza viruses formidable challenges for control efforts. Influenza type A is antigenically highly variable and is responsible for most cases of epidemic influenza. Influenza type B may exhibit antigenic changes and sometimes causes epidemics. Influenza type C is antigenically stable and causes only mild illness in immunocompetent individuals.

PROPERTIES OF ORTHOMYXOVIRUSES

Three immunologic types of influenza viruses are known, designated A, B, and C. Antigenic changes continually occur within the type A group of influenza viruses and to a lesser degree in the type B group, whereas type C appears to be antigenically stable. Influenza A strains are also known for aquatic birds, chickens, ducks, pigs, horses, and seals. Some of the strains isolated from animals are antigenically similar to strains circulating in the human population.

The following descriptions are based on influenza virus type A, the best-characterized type (Table 39–1).

Structure & Composition

Influenza virus particles are usually spherical and about 100 nm in diameter (80–120 nm), although virions may display great variation in size (Figure 39–1).

The single-stranded, negative-sense RNA genomes of influenza A and B viruses occur as eight separate segments; influenza C viruses contain seven segments of RNA, lacking a neuraminidase gene. Sizes and protein-coding assignments are known for all the segments (Table 39–2). Most of the segments code for a single protein. The complete nucleotide sequence is known for many influenza viruses. The first

12–13 nucleotides at each end of each genomic segment are conserved among all eight RNA segments; these sequences are important in viral transcription.

Influenza virus particles contain nine different structural proteins. The nucleoprotein (NP) associates with the viral RNA to form a ribonucleoprotein (RNP) structure 9 nm in diameter that assumes a helical configuration and forms the viral nucleocapsid. Three large proteins (PB1, PB2, and PA) are bound to the viral RNP and are responsible for RNA transcription and replication. The matrix (M_1) protein, which forms a shell underneath the viral lipid envelope, is important in particle morphogenesis and is a major component of the virion (about 40% of viral protein).

A lipid envelope derived from the cell surrounds the virus particle. Two virus-encoded glycoproteins, the hemagglutinin (HA) and the neuraminidase (NA), are inserted into the envelope and are exposed as spikes about 10 nm long on the surface of the particle. These two surface glycoproteins are the important antigens that determine antigenic variation of influenza viruses and host immunity. The HA represents

TABLE 39–1   Important Properties of Orthomyxoviruses[a]

Virion: Spherical, pleomorphic, 80–120 nm in diameter (helical nucleocapsid, 9 nm)

Composition: RNA (1%), protein (73%), lipid (20%), carbohydrate (6%)

Genome: Single-stranded RNA, segmented (eight molecules), negative-sense, 13.6 kb overall size

Proteins: Nine structural proteins, one nonstructural

Envelope: Contains viral hemagglutinin (HA) and neuraminidase (NA) proteins

Replication: Nuclear transcription; capped 5′ termini of cellular RNA scavenged as primers; particles mature by budding from plasma membrane

Outstanding characteristics:
   Genetic reassortment common among members of the same genus
   Influenza viruses cause worldwide epidemics

[a]Description for influenza A virus, genus *Influenzavirus A*.

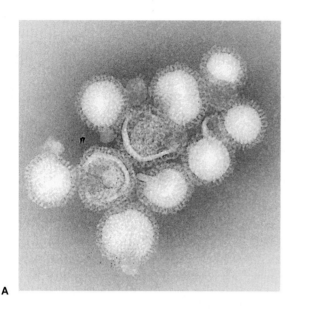

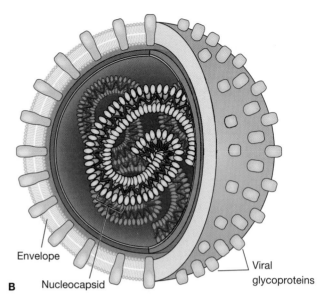

A

B   Envelope

Nucleocapsid

Viral glycoproteins

FIGURE 39–1 Influenza virus. **A:** Electron micrograph of influenza virus A/Hong Kong/1/68(H3N2). Note pleomorphic shapes and glycoprotein projections covering particle surfaces (315,000×). (Courtesy of FA Murphy and EL Palmer.) **B:** Schematic view of influenza. Virus particles have segmented genomes consisting of 7–8 different RNA molecules, each coated by capsid proteins and forming helical nucleocapsids. Viral glycoproteins (hemagglutinin and neuraminidase) protrude as spikes through the lipid envelope. (Reproduced with permission from Willey JM, Sherwood LM, Woolverton CJ: *Prescott, Harley, and Klein's Microbiology,* 7th ed. McGraw Hill, 2008.)

TABLE 39–2   Coding Assignments of Influenza Virus A RNA Segments[a]

Genome Segment		Encoded Polypeptide			
Number[b]	Size (Number of Nucleotides)	Designation	Predicted Molecular Weight[c]	Approximate Number of Molecules per Virion	Function
1	2341	PB2	85,700	30–60	RNA transcriptase components
2	2341	PB1	86,500		
3	2233	PA	84,200		
4	1778	HA	61,500	500	Hemagglutinin; trimer; envelope glycoprotein; mediates virus attachment to cells; activated by cleavage; fusion activity at acid pH
5	1565	NP	56,100	1000	Associated with RNA and polymerase proteins; helical structure; nucleocapsid
6	1413	NA	50,000	100	Neuraminidase; tetramer; envelope glycoprotein; enzyme
7	1027	M_1	27,800	3000	Matrix protein; major component of virion; lines inside of envelope; involved in assembly; interacts with viral RNPs and NS_2
		M_2	11,000	20–60	Integral membrane protein; ion channel; essential for virus uncoating; from spliced mRNA
8	890	NS_1	26,800	0	Nonstructural; high abundance; inhibits pre-mRNA splicing; reduces interferon response
		NS_2	14,200	130–200	Minor component of virions; nuclear export of viral RNPs; from spliced mRNA

[a]Adapted with permission from Lamb RA, Krug RM: Orthomyxoviridae: The viruses and their replication. In: *Fields Virology,* 3rd ed. Fields BN et al (editors). Lippincott-Raven, 1996.

[b]RNA segments are numbered in order of decreasing size.

[c]The molecular weights of the two glycoproteins, HA and NA, appear larger (about 76,000 and 56,000, respectively) because of the added carbohydrate.

about 25% of viral protein, and the NA about 5%. The M_2 ion channel protein and the NS_2 protein are also present in the envelope but at only a few copies per particle.

Because of the segmented nature of the genome, when a cell is coinfected by two different viruses of a given type, mixtures of parental gene segments may be assembled into progeny virions. This phenomenon, called **genetic reassortment,** may result in sudden changes in viral surface antigens—a property that explains the epidemiologic features of influenza and poses significant problems for vaccine development.

Influenza viruses are relatively hardy in vitro and may be stored at 0–4°C for weeks without loss of viability. Lipid solvents, protein denaturants, formaldehyde, and irradiation destroy infectivity. Both infectivity and hemagglutination are more resistant to inactivation at alkaline pH than at acid pH.

Classification & Nomenclature

Genus *Influenzavirus A* contains human and animal strains of influenza type A; *Influenzavirus B* contains human strains of type B; and *Influenzavirus C* contains influenza type C viruses of humans and swine.

Antigenic differences exhibited by two of the internal structural proteins, the nucleocapsid (NP) and matrix (M) proteins, are used to divide influenza viruses into types A, B, and C. These proteins possess no cross-reactivity among the three types. Antigenic variations in the surface glycoproteins, HA and NA, are used to subtype the viruses. Only type A has designated subtypes.

The standard nomenclature system for influenza virus isolates includes the following information: type, host of origin, geographic origin, strain number, and year of isolation. Antigenic descriptions of the HA and the NA are given in parentheses for type A. The host of origin is not indicated for human isolates, eg, A/Hong Kong/03/68(H3N2), but it is indicated for others, eg, A/swine/Iowa/15/30(H1N1).

So far, 15 subtypes of HA (H1–H15) and nine subtypes of NA (N1–N9), in many different combinations, have been recovered from birds, animals, or humans. Four HA (H1–H3, H5) and two NA (N1, N2) subtypes have been recovered from humans.

Structure & Function of Hemagglutinin

The HA protein of influenza virus binds virus particles to susceptible cells and is the major antigen against which neutralizing (protective) antibodies are directed. Variability in HA is primarily responsible for the continual evolution of new strains and subsequent influenza epidemics. Hemagglutinin derives its name from its ability to agglutinate erythrocytes under certain conditions.

The amino acid sequence for HA can be calculated from the sequence of the HA gene, and the three-dimensional structure of the protein has been revealed by x-ray crystallography, so it is possible to correlate functions of the HA molecule with its structure.

The primary sequence of HA contains 566 amino acids (Figure 39–2A). A short signal sequence at the amino terminal inserts the polypeptide into the endoplasmic reticulum; the signal is then removed. The HA protein is cleaved into two subunits, HA1 and HA2, that remain tightly associated by a disulfide bridge. A hydrophobic stretch near the carboxyl terminal of HA2 anchors the HA molecule in the membrane, with a short hydrophilic tail extending into the cytoplasm. Oligosaccharide residues are added at several sites.

The HA molecule is folded into a complex structure (Figure 39–2B). Each linked HA1 and HA2 dimer forms an elongated stalk capped by a large globule. The base of the stalk anchors it in the membrane. Five antigenic sites on the HA molecule exhibit extensive mutations. These sites occur at regions exposed on the surface of the structure, are apparently not essential to the molecule's stability, and are involved in viral neutralization. Other regions of the HA molecule are conserved in all isolates, presumably because they are necessary for the molecule to retain its structure and function.

The HA spike on the virus particle is a trimer, composed of three intertwined HA1 and HA2 dimers (Figure 39–2C). The trimerization imparts greater stability to the spike than could be achieved by a monomer. The cellular receptor binding site (viral attachment site) is a pocket located at the top of each large globule. The pocket is inaccessible to antibody.

The cleavage that separates HA1 and HA2 is necessary for the virus particle to be infectious and is mediated by cellular proteases. Influenza viruses normally remain confined to the respiratory tract because the protease enzymes that cleave HA are common only at those sites. Examples have been noted of more virulent viruses that have adapted to use a more ubiquitous enzyme, such as plasmin, to cleave HA and promote widespread infection of cells. The amino terminal of HA2, generated by the cleavage event, is necessary for the viral envelope to fuse with the cell membrane, an essential step in the process of viral infection. Low pH triggers a conformational change that activates the fusion activity.

Structure & Function of Neuraminidase

The antigenicity of NA, the other glycoprotein on the surface of influenza virus particles, is also important in determining the subtype of influenza virus isolates.

The spike on the virus particle is a tetramer, composed of four identical monomers (Figure 39–2D). A slender stalk is topped with a box-shaped head. There is a catalytic site for NA on the top of each head, so that each NA spike contains four active sites.

The NA functions at the end of the viral replication cycle. It is a sialidase enzyme that removes sialic acid from glycoconjugates. It facilitates release of virus particles from infected cell surfaces during the budding process and helps prevent self-aggregation of virions by removing sialic acid residues from viral glycoproteins. It is possible that NA helps the virus negotiate through the mucin layer in the respiratory tract to reach the target epithelial cells.

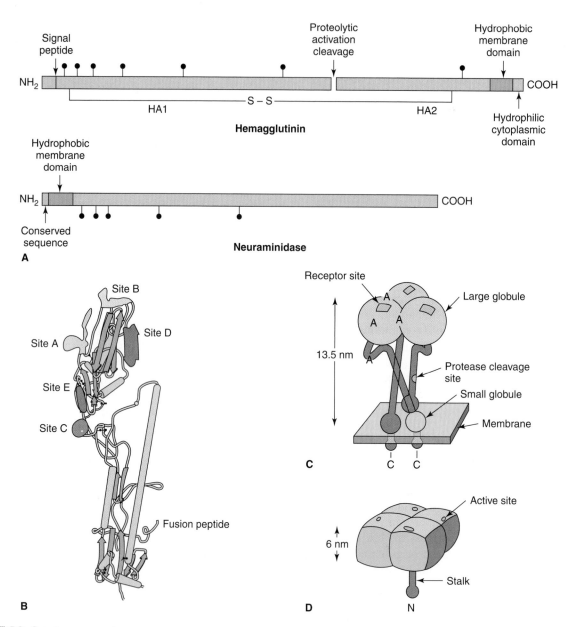

FIGURE 39–2 Influenza virus hemagglutinin and neuraminidase surface glycoproteins. **A:** Primary structures of HA and NA polypeptides. The cleavage of HA into HA1 and HA2 is necessary for virus to be infectious. HA1 and HA2 remain linked by a disulfide bond (S–S). No posttranslational cleavage occurs with NA. Carbohydrate attachment sites (⬍) are shown. The hydrophobic amino acids that anchor the proteins in the viral membrane are located near the carboxyl terminal of HA and the amino terminal of NA. **B:** Folding of the HA1 and HA2 polypeptides in an HA monomer. Five major antigenic sites (sites A–E) that undergo change are shown as shaded areas. The amino terminal of HA2 provides fusion activity (fusion peptide). The fusion particle is buried in the molecule until it is exposed by a conformational change induced by low pH. **C:** Structure of the HA trimer as it occurs on a virus particle or the surface of infected cells. Some of the sites involved in antigenic variation are shown (A). Carboxyl terminal residues (C) protrude through the membrane. **D:** Structure of the NA tetramer. Each NA molecule has an active site on its upper surface. The amino terminal region (N) of the polypeptides anchors the complex in the membrane. (Redrawn with permission from **[A, B]** Murphy BR, Webster RG: Influenza viruses, page 1179, and **[C, D]** Kingsbury DW: Orthomyxo- and paramyxoviruses and their replication, page 1157. In: *Virology.* Fields BN et al [editors]. Raven Press, 1985.)

Antigenic Drift & Antigenic Shift

Influenza viruses are remarkable because of the frequent antigenic changes that occur in HA and NA. Antigenic variants of influenza virus have a selective advantage over the parental virus in the presence of antibody directed against the original strain. This phenomenon is responsible for the unique epidemiologic features of influenza. Other respiratory tract agents do not display significant antigenic variation.

The two surface antigens of influenza undergo antigenic variation independent of each other. Minor antigenic changes are termed **antigenic drift;** major antigenic changes in HA or NA, called **antigenic shift,** result in the appearance of a new subtype (Figure 39–3). Antigenic shift is most likely to result in an epidemic.

Antigenic drift is due to the accumulation of point mutations in the gene, resulting in amino acid changes in the protein. Sequence changes can alter antigenic sites on the molecule such that a virion can escape recognition by the host's immune system. The immune system does not cause the antigenic variation, but rather functions as a selection force that allows new antigenic variants to expand. A variant must sustain two or more mutations before a new, epidemiologically significant strain emerges.

Antigenic shift reflects drastic changes in the sequence of a viral surface protein, changes too extreme to be explained by mutation. The segmented genomes of influenza viruses reassort readily in doubly infected cells. The mechanism for shift is genetic reassortment between human and avian influenza viruses. Influenza B and C viruses do not exhibit antigenic shift because few related viruses exist in animals.

Influenza Virus Replication

The replication cycle of influenza virus is summarized in Figure 39–4. Influenza is unusual among nononcogenic RNA viruses because all of its RNA transcription and replication occur in the nucleus of infected cells. The viral multiplication cycle proceeds rapidly. There is the shut-off of host cell protein synthesis by about 3 hours postinfection, permitting selective translation of viral mRNAs. New progeny viruses are produced within 8–10 hours.

A. Viral Attachment, Penetration, & Uncoating

The virus attaches to cell-surface sialic acid via the receptor site located on the top of the large globule of the HA. Virus

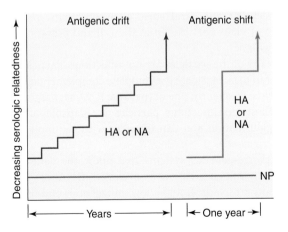

FIGURE 39–3 Antigenic drift and antigenic shift account for antigenic changes in the two surface glycoproteins (HA and NA) of influenza virus. Antigenic drift is a gradual change in antigenicity due to point mutations that affect major antigenic sites on the glycoprotein. Antigenic shift is an abrupt change due to genetic reassortment with an unrelated strain. Changes in HA and NA occur independently. Internal proteins of the virus, such as the nucleoprotein (NP), do not undergo antigenic changes.

particles are then internalized within endosomes by a process called receptor-mediated endocytosis. The next step involves fusion between the viral envelope and cell membrane, triggering uncoating. The low pH within the endosome is required for virus-mediated membrane fusion that releases viral RNPs into the cytosol. Acid pH causes a conformational change in the HA structure to bring the HA2 "fusion peptide" in correct contact with the membrane. The M_2 ion channel protein present in the virion permits the entry of ions from the endosome into the virus particle, triggering the conformational change in HA. Viral nucleocapsids are then released into the cell cytoplasm.

B. Transcription & Translation

Transcription mechanisms used by orthomyxoviruses differ markedly from those of other RNA viruses in that cellular functions are more intimately involved. Viral transcription occurs in the nucleus. The mRNAs are produced from viral nucleocapsids. The virus-encoded polymerase, consisting of a complex of the three P proteins, is primarily responsible for transcription. Its action must be primed by scavenged capped and methylated 5′ terminals from cellular transcripts that are newly synthesized by cellular RNA polymerase II. This explains why influenza virus replication is inhibited by dactinomycin and α-amanitin, which block cellular transcription, whereas other RNA viruses are not affected because they do not use cellular transcripts in viral RNA synthesis.

Six of the genome segments yield monocistronic mRNAs that are translated in the cytoplasm into six viral proteins. The other two transcripts undergo splicing, each yielding two mRNAs that are translated in different reading frames. At early times after infection, the NS_1 and NP proteins are preferentially synthesized. At later times, the structural proteins are synthesized at high rates. The two glycoproteins, HA and NA, are modified using the secretory pathway.

The influenza virus nonstructural protein NS_1 has a posttranscriptional role in regulating viral and cellular gene expression. The NS_1 protein binds to poly(A) sequences, inhibits pre-mRNA splicing, and inhibits the nuclear export of spliced mRNAs, ensuring a pool of donor cellular molecules to provide the capped primers needed for viral mRNA synthesis. The NS_2 protein interacts with M_1 protein and is involved in nuclear export of viral RNPs.

C. Viral RNA Replication

Viral genome replication is accomplished by the same virus-encoded polymerase proteins involved in transcription. The mechanisms that regulate the alternative transcription and replication roles of the same proteins are related to the abundance of one or more of the viral nucleocapsid proteins.

As with all other negative-strand viruses, templates for viral RNA synthesis remain coated with nucleoproteins. The only completely free RNAs are mRNAs. The first step in genome replication is production of positive-strand copies of each segment. These antigenome copies differ from mRNAs at

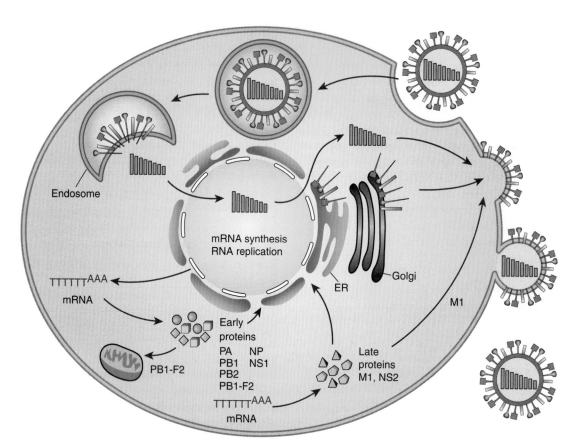

FIGURE 39–4 Schematic diagram of the life cycle of influenza virus. After receptor-mediated endocytosis, the viral ribonucleoprotein complexes are released into the cytoplasm and transported to the nucleus, where replication and transcription take place. Messenger RNAs are exported to the cytoplasm for translation. Early viral proteins required for replication and transcription are transported back to the nucleus. The assembly and budding of progeny virions occurs at the plasma membrane. (Reproduced with permission from Neumann G, Noda T, Kawaoka Y: Emergence and pandemic potential of swine-origin H1N1 influenza virus. Nature 2009;459:931.)

both terminals; the 5′ ends are not capped, and the 3′ ends are neither truncated nor polyadenylated. These copies serve as templates for synthesis of faithful copies of genomic RNAs.

Because there are common sequences at both ends of all viral RNA segments, they can be recognized efficiently by the RNA-synthesizing machinery. Intermingling of genome segments derived from different parents in coinfected cells is presumably responsible for the high frequency of genetic reassortment typical of influenza viruses within a genus. Frequencies of reassortment as high as 40% have been observed.

D. Maturation

The virus matures by budding from the surface of the cell. Individual viral components arrive at the budding site by different routes. Nucleocapsids are assembled in the nucleus and move out to the cell surface. The glycoproteins, HA and NA, are synthesized in the endoplasmic reticulum; are modified and assembled into trimers and tetramers, respectively; and are inserted into the plasma membrane. The M_1 protein serves as a bridge, linking the nucleocapsid to the cytoplasmic ends of the glycoproteins. Progeny virions bud off the cell. During this sequence of events, the HA is cleaved

into HA1 and HA2 if the host cell possesses the appropriate proteolytic enzyme. The NA removes terminal sialic acids from cellular and viral surface glycoproteins, facilitating release of virus particles from the cell and preventing their aggregation.

Many of the particles are not infectious. Particles sometimes fail to encapsidate the complete complement of genome segments; frequently, one of the large RNA segments is missing. These noninfectious particles are capable of causing hemagglutination and can interfere with the replication of intact virus.

Reverse-genetics systems that allow the generation of infectious influenza viruses from cloned cDNAs of viral RNA segments are available and allow mutagenesis and functional studies.

INFLUENZA VIRUS INFECTIONS IN HUMANS

A comparison of influenza A virus with other viruses that infect the human respiratory tract is shown in Table 39–3. Influenza virus is considered here.

TABLE 39–3   Comparison of Viruses That Infect the Human Respiratory Tract

Virus	Disease	Number of Serotypes	Lifelong Immunity to Disease	Vaccine Available	Viral Latency
RNA viruses					
Influenza A virus	Influenza	Many	No	+	−
Parainfluenza virus	Croup	Many	No	−	−
Respiratory syncytial virus	Bronchiolitis	One	No	−	−
Rubella virus	Rubella	One	Yes	+	−
Measles virus	Measles	One	Yes	+	−
Mumps virus	Parotitis, meningitis	One	Yes	+	−
Rhinovirus	Common cold	Many	No	−	−
Coronavirus	Common cold	Many	No	−	−
Coxsackievirus	Herpangina, pleurodynia	Many	No	−	−
DNA viruses					
Herpes simplex virus type 1	Gingivostomatitis	One	No	−	+
Epstein-Barr virus	Infectious mononucleosis	One	Yes	−	+
Varicella-zoster virus	Chickenpox, shingles	One	Yes[a]	+	+
Adenovirus	Pharyngitis, pneumonia	Many	No	−	+

[a]Lifelong immunity to reinfections with varicella (chickenpox) but not to reactivation of zoster (shingles).

Pathogenesis & Pathology

Influenza virus spreads from person to person by airborne droplets or by contact with contaminated hands or surfaces. A few cells of respiratory epithelium are infected if deposited virus particles avoid removal by the cough reflex and escape neutralization by preexisting specific IgA antibodies or inactivation by nonspecific inhibitors in the mucous secretions. Progeny virions are soon produced and spread to adjacent cells, where the replicative cycle is repeated. Viral NA lowers the viscosity of the mucous film in the respiratory tract, laying bare the cellular surface receptors and promoting the spread of virus-containing fluid to lower portions of the tract. Within a short time, many cells in the respiratory tract are infected and eventually killed.

The incubation period from exposure to virus and the onset of illness varies from 1 day to 4 days, depending upon the size of the viral dose and the immune status of the host. Viral shedding starts the day preceding onset of symptoms, peaks within 24 hours, remains elevated for 1–2 days, and then declines over the next 5 days. Infectious virus is very rarely recovered from blood.

Interferon is detectable in respiratory secretions about 1 day after viral shedding begins. Influenza viruses are sensitive to the antiviral effects of interferon, and it is believed that the interferon response contributes to host recovery from infection. Specific antibody and cell-mediated responses cannot be detected for another 1–2 weeks.

Influenza infections cause cellular destruction and desquamation of superficial mucosa of the respiratory tract but do not affect the basal layer of epithelium. Complete reparation of cellular damage probably takes up to 1 month. Viral damage to the respiratory tract epithelium lowers its resistance to secondary bacterial invaders, especially staphylococci, streptococci, and *Haemophilus influenzae.*

Edema and mononuclear infiltrations in response to cell death and desquamation due to viral replication probably account for local symptoms. The prominent systemic symptoms associated with influenza probably reflect the production of cytokines.

Clinical Findings

Influenza attacks mainly the upper respiratory tract. It poses a serious risk for the elderly, the very young, and people with underlying medical conditions such as lung, kidney, or heart problems, diabetes, or cancer.

A. Uncomplicated Influenza

Symptoms of classic influenza usually appear abruptly and include chills, headache, and dry cough, followed closely by high fever, generalized muscular aches, malaise, and anorexia. The fever usually lasts 3–5 days, as do the systemic symptoms. Respiratory symptoms typically last another 3–4 days. The cough and weakness may persist for 2–4 weeks after major symptoms subside. Mild or asymptomatic infections may occur. These symptoms may be induced by any strain of influenza A or B. In contrast, influenza C rarely causes the influenza syndrome, causing instead a common cold illness. Coryza and cough may last for several weeks.

Clinical symptoms of influenza in children are similar to those in adults, although children may have higher fever and a higher incidence of gastrointestinal manifestations such as vomiting. Febrile convulsions can occur. Influenza A viruses are an important cause of croup in children under

1 year of age, which may be severe. Finally, otitis media may develop.

When influenza appears in epidemic form, clinical findings are consistent enough that the disease can be diagnosed. Sporadic cases cannot be diagnosed on clinical grounds, as disease manifestations cannot be distinguished from those caused by other respiratory tract pathogens. However, those other agents rarely cause severe viral pneumonia, which is a complication of influenza A virus infection.

B. Pneumonia

Serious complications usually occur only in the elderly and debilitated, especially those with underlying chronic disease. Pregnancy appears to be a risk factor for lethal pulmonary complications in some epidemics. The lethal impact of an influenza epidemic is reflected in the excess deaths due to pneumonia and cardiopulmonary diseases.

Pneumonia complicating influenza infections can be viral, secondary bacterial, or a combination of the two. Increased mucous secretion helps carry agents into the lower respiratory tract. Influenza infection enhances susceptibility of patients to bacterial superinfection. This is attributed to loss of ciliary clearance, dysfunction of phagocytic cells, and provision of a rich bacterial growth medium by the alveolar exudate. Bacterial pathogens are most often *Staphylococcus aureus, Streptococcus pneumoniae,* and *H influenzae.*

Combined viral–bacterial pneumonia is approximately three times more common than primary influenza pneumonia. *S aureus* coinfection has been reported to have a fatality rate of up to 42%. A molecular basis for a synergistic effect between virus and bacteria may be that some *S aureus* strains secrete a protease able to cleave the influenza HA, thereby allowing production of much higher titers of infectious virus in the lungs.

C. Reye's Syndrome

Reye's syndrome is an acute encephalopathy of children and adolescents, usually between 2 and 16 years of age. The mortality rate is high (10–40%). The cause of Reye's syndrome is unknown, but it is a recognized rare complication of influenza B, influenza A, and herpesvirus varicella-zoster infections. There is a possible relationship between salicylate use and subsequent development of Reye's syndrome. The incidence of the syndrome has decreased with the reduced use of salicylates in children with flu-like symptoms.

Immunity

Immunity to influenza is long-lived and subtype-specific. Antibodies against HA and NA are important in immunity to influenza, whereas antibodies against the other virus-encoded proteins are not protective. Resistance to initiation of infection is related to antibody against the HA, whereas decreased severity of disease and decreased ability

to transmit virus to contacts are related to antibody directed against the NA. Antibodies against the ribonucleoprotein are type-specific and are useful in typing viral isolates (as influenza A or B).

Protection correlates with both serum antibodies and secretory IgA antibodies in nasal secretions. The local secretory antibody is probably important in preventing infection. Serum antibodies persist for many months to years, whereas secretory antibodies are of shorter duration (usually only several months). Antibody also modifies the course of illness. A person with low titers of antibody may be infected but will experience a mild form of disease. Immunity can be incomplete, as reinfection with the same virus can occur.

The three types of influenza viruses are antigenically unrelated and therefore induce no cross-protection. When a viral type undergoes antigenic drift, a person with preexisting antibody to the original strain may suffer only mild infection with the new strain. Subsequent infections or immunizations reinforce the antibody response to the first subtype of influenza experienced years earlier, a phenomenon called "original antigenic sin."

The primary role of cell-mediated immune responses in influenza is believed to be clearance of an established infection; cytotoxic T cells lyse infected cells. The cytotoxic T lymphocyte response is cross-reactive (able to lyse cells infected with any subtype of virus) and appears to be directed against both internal proteins (NP, M) and the surface glycoproteins.

Laboratory Diagnosis

Clinical characteristics of viral respiratory infections can be produced by many different viruses. Consequently, diagnosis of influenza relies on identification of viral antigens or viral nucleic acid in specimens, isolation of the virus, or demonstration of a specific immunologic response by the patient.

Nasal washings, gargles, and throat swabs are the best specimens for diagnostic testing and should be obtained within 3 days after the onset of symptoms.

A. Polymerase Chain Reaction

Rapid tests based on detection of influenza RNA in clinical specimens using reverse-transcription polymerase chain reaction (RT-PCR) are preferred for diagnosis of influenza. RT-PCR is rapid (<1 day), sensitive, and specific. However, it is not currently available in all settings.

B. Isolation & Identification of Virus

The sample to be tested for virus isolation should be held at 4°C until inoculation into cell culture, as freezing and thawing reduce the ability to recover virus. However, if storage time will exceed 5 days, the sample should be frozen at −70°C.

Viral culture procedures take 3–10 days. Classically, embryonated eggs and primary monkey kidney cells have been the isolation methods of choice for influenza viruses, although some continuous cell lines may be used. Inoculated cell cultures are incubated in the absence of serum, which may contain nonspecific viral inhibitory factors, and in the presence of trypsin, which cleaves and activates the HA so that replicating virus will spread throughout the culture.

Cell cultures can be tested for the presence of virus by hemadsorption 3–5 days after inoculation, or the culture fluid can be examined for virus after 5–7 days by hemagglutination. If the results are negative, a passage is made into fresh cultures. This passage may be necessary, because primary viral isolates are often fastidious and grow slowly.

Viral isolates can be identified by hemagglutination inhibition, a procedure that permits rapid determination of the influenza type and subtype. To do this, reference sera to currently prevalent strains must be used. Hemagglutination by the new isolate will be inhibited by antiserum to the homologous subtype.

For rapid diagnosis, cell cultures on coverslips in shell vials may be inoculated and stained 1–4 days later with monoclonal antibodies to respiratory agents. Rapid viral cultures can also be tested by RT-PCR to identify a cultured agent.

It is possible to identify viral antigen directly in exfoliated cells in nasal aspirates using fluorescent antibodies. This test is rapid (taking only a few hours) but is not as sensitive as viral isolation, does not provide full details about the viral strain, and does not yield an isolate that can be characterized. Rapid influenza antigen detection tests are commercially available that take less than 15 minutes. However, these tests vary in sensitivity and specificity.

C. Serology

Antibodies to several viral proteins (hemagglutinin, neuraminidase, nucleoprotein, and matrix) are produced during infection with influenza virus. The immune response against the HA glycoprotein is associated with resistance to infection.

Routine serodiagnostic tests in use are based on hemagglutination inhibition (HI) and ELISA. Paired acute and convalescent sera are necessary, because normal individuals usually have influenza antibodies. A fourfold or greater increase in titer must occur to indicate influenza infection. Human sera often contain nonspecific mucoprotein inhibitors that must be destroyed before testing by HI.

The HI test reveals the strain of virus responsible for infection only if the correct antigen is available for use. Neutralization tests are the most specific and the best predictor of susceptibility to infection but are more unwieldy and more time-consuming to perform than the other tests. The ELISA test is more sensitive than other assays.

Complications may be encountered in attempting to identify the strain of infecting influenza virus by the patient's

antibody response because anamnestic responses frequently occur.

Epidemiology

Influenza viruses occur worldwide and cause annual outbreaks of variable intensity. It is estimated that annual epidemics of seasonal influenza cause 3–5 million cases of severe illness and 250,000–500,000 deaths worldwide. The economic impact of influenza A outbreaks is significant because of the morbidity associated with infections. Economic costs have been estimated at $10–60 million per million population in industrialized countries, depending on the size of the epidemic.

The three types of influenza vary markedly in their epidemiologic patterns. Influenza C is least significant; it causes mild, sporadic respiratory disease but not epidemic influenza. Influenza B sometimes causes epidemics, but influenza type A can sweep across continents and around the world in massive epidemics called pandemics.

The incidence of influenza peaks during the winter. In the United States, influenza epidemics usually occur from January through April (and from May to August in the Southern Hemisphere). A continuous person-to-person chain of transmission must exist for maintenance of the agent between epidemics. Some viral activity can be detected in large population centers throughout each year, indicating that the virus remains endemic in the population and causes a few subclinical or minor infections.

A. Antigenic Change

Periodic outbreaks appear because of antigenic changes in one or both surface glycoproteins of the virus. When the number of susceptible persons in a population reaches a sufficient level, the new strain of virus causes an epidemic. The change may be gradual (hence the term "antigenic drift"), due to point mutations reflected in alterations at major antigenic sites on the glycoprotein (see Figure 39–3), or drastic and abrupt (hence the term "antigenic shift"), owing to genetic reassortment during coinfection with an unrelated strain.

All three types of influenza virus exhibit antigenic drift. However, only influenza A undergoes antigenic shift, presumably because types B and C are restricted to humans, whereas related influenza A viruses circulate in animal and bird populations. These animal strains account for antigenic shift by genetic reassortment of the glycoprotein genes. Influenza A viruses have been recovered from many aquatic birds, especially ducks; from domestic poultry, such as turkeys, chickens, geese, and ducks; from pigs and horses; and even from seals and whales.

Influenza outbreaks occur in waves, although there is no regular periodicity in the occurrence of epidemics. The experience in any given year will reflect the interplay between extent of antigenic drift of the predominant virus and waning immunity in the population. The period between epidemic waves of influenza A tends to be 2–3 years; the interepidemic period for type B is longer (3–6 years). Every 10–40 years, when a new subtype of influenza A appears, a pandemic results. This happened in 1918 (H1N1), 1957 (H2N2), and 1968 (H3N2). The H1N1 subtype reemerged in 1977, although no epidemic materialized. Since 1977, influenza A (H1N1) and (H3N2) viruses and influenza B viruses have been in global circulation.

A novel swine-origin H1N1 virus appeared in early 2009 and reached pandemic spread by mid-year. It was a quadruple reassortant, containing genes from both North American and Eurasian swine viruses, as well as from avian and human influenza viruses. The virus was readily transmissible among humans and the severity of illness, at least initially, was comparable to that of seasonal flu.

Surveillance for influenza outbreaks is necessary to identify the early appearance of new strains, with the aim of preparing vaccines against them before an epidemic occurs. That surveillance may extend into animal populations, especially birds, pigs, and horses. Isolation of a virus with an altered hemagglutinin in the late spring during a mini-epidemic signals a possible epidemic the following winter. This warning sign, termed a "herald wave," has been observed to precede influenza A and B epidemics.

B. Avian Influenza

Sequence analyses of influenza A viruses isolated from many hosts in different regions of the world support the theory that all mammalian influenza viruses derive from the avian influenza reservoir. Of the 15 HA subtypes found in birds, only a few have been transferred to mammals (H1, H2, H3, and H5 in humans; H1 and H3 in swine; and H3 and H7 in horses). The same pattern holds for NA; nine NA subtypes are known for birds, only two of which are found in humans (N1, N2). The influenza viruses do not appear to undergo antigenic change in the birds, perhaps because of their brief life span. This means the genes that caused previous influenza pandemics in humans still exist unchanged in the aquatic bird reservoir.

Avian influenza ranges from inapparent infections to highly lethal infections in chickens and turkeys. Most influenza infections in ducks are avirulent. Influenza viruses of ducks multiply in cells lining the intestinal tract and are shed in high concentrations in fecal material into water, where they remain viable for days or weeks, especially at low temperatures. It is likely that avian influenza is a waterborne infection, moving from wild to domestic birds and pigs.

To date, all human pandemic strains have been reassortants between avian and human influenza viruses. Evidence supports the model that pigs serve as mixing vessels for reassortants as their cells contain receptors recognized by both human and avian viruses (Figure 39–5). The pandemic strain of 2009 was a novel reassortant that contained swine-origin viral genes as well as those from avian and human influenza

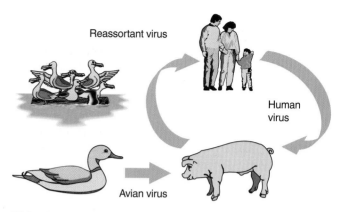

FIGURE 39–5 The pig may act as an intermediate host for the generation of human–avian reassortant influenza viruses with pandemic potential. (Reproduced with permission from Claas ECJ, Osterhaus ADME: New clues to the emergence of flu pandemics. Nat Med 1998;4:1122.)

viruses. School-age children are the predominant vectors of influenza transmission. Crowding in schools favors the aerosol transmission of virus, and children take the virus home to the family.

In 1997, in Hong Kong, the first documented infection of humans by avian influenza A virus (H5N1) occurred. The source was domestic poultry. By 2006, the geographic presence of this highly pathogenic H5N1 avian influenza virus in both wild and domestic birds had expanded to include many countries in Asia, Africa, Europe, and the Middle East. Outbreaks were the largest and most severe on record. Of about 425 laboratory-confirmed human cases, by May 2009 more than half have been fatal. So far, isolates from human cases have contained all RNA gene segments from avian viruses, indicating that, in those infections, the avian virus had jumped directly from bird to human. All evidence to date indicates that close contact with diseased birds has been the source of human H5N1 infection. The concern is that, given enough opportunities, the highly pathogenic H5N1 avian influenza virus will acquire the ability to spread efficiently and be sustained among humans, either by reassortment or by adaptive mutation. This would result in a devastating influenza pandemic.

C. Seroarcheology

Human influenza virus was first isolated in 1933 using ferrets. The subtypes that circulated prior to that time have been deduced using retrospective seroepidemiology. This technique is based on screening HI titers against numerous HA subtypes of virus with sera from many individuals in different age groups.

The range of the influenza antibody spectrum is narrow in early life, but it becomes progressively broader in later years. Antibodies acquired from initial infections in childhood reflect dominant antigens of the prevailing strains. Later exposures to viruses result in an antibody spectrum broadening toward a larger number of common antigens of influenza viruses. Exposures later in life to antigenically related strains result in progressive reinforcement of the primary antibody. The highest antibody levels in a particular age group therefore reflect dominant antigens of the virus responsible for childhood infections of the group. Thus, a serologic recapitulation of past infection with influenza viruses of different antigenic makeup can be obtained by studying age distribution of influenza antibodies in normal populations.

This approach suggests that the epidemic of 1890 was probably caused by an H2N8 subtype and the epidemic of 1900 by an H3N8 virus. The catastrophic pandemic of 1918–1919 (Spanish flu) was caused by the abrupt appearance of the H1N1 subtype, the swine-like influenza. (More than 20 million people died during this pandemic, mainly from complicating bacterial pneumonias.) Subsequent antigenic shifts have been documented by viral isolations; H2N2 (Asian flu) appeared in 1957 and was replaced in 1968 by the H3N2 subtype (Hong Kong flu). The H1N1 strain reappeared in 1977 (Russian flu).

D. Reconstruction of 1918 Influenza Virus

PCR technology has yielded gene fragments of influenza virus from archival lung tissue specimens from victims of the 1918 Spanish flu epidemic. The complete coding sequences of all eight viral RNA segments have been determined, and the sequences document that it was an H1N1 influenza A virus. It appears that the 1918 virus was not a reassortant, but was derived entirely from an avian source that adapted to humans. Using reverse genetics, an infectious virus containing all the gene segments from the 1918 pandemic virus was constructed. In contrast to ordinary influenza viruses, the 1918 virus was highly pathogenic, including being able to kill mice rapidly. The 1918 HA and polymerase genes appeared to be responsible for the high virulence.

Prevention & Treatment by Drugs

Amantadine hydrochloride and an analog, rimantadine, are M_2 ion channel inhibitors for systemic use in the treatment and prophylaxis of influenza A. The NA inhibitors zanamivir and oseltamivir were approved in 1999 for treatment of both influenza A and influenza B. To be maximally effective, the drugs must be administered very early in the disease. Resistant viruses emerge more frequently during therapy with M_2 inhibitors than with NA inhibitors and more frequently in children than adults. During the influenza season in the United States in late 2005, 92% of influenza A (H3N2) virus isolates were resistant to M_2 inhibitors.

Prevention & Control by Vaccines

Inactivated viral vaccines are the primary means of prevention of influenza in the United States. However, certain characteristics of influenza viruses make prevention and control

of the disease by immunization especially difficult. Existing vaccines are continually being rendered obsolete as the viruses undergo antigenic drift and shift. Surveillance programs by government agencies and the World Health Organization constantly monitor subtypes of influenza circulating around the world to promptly detect the appearance and spread of new strains.

Several other problems are worthy of mention. Although protection can reach 70–100% in healthy adults, frequency of protection is lower (30–60%) among the elderly and among young children. Inactivated viral vaccines usually do not generate good local IgA or cell-mediated immune responses. The immune response is influenced by whether the person is "primed" by having had prior antigenic experience with an influenza A virus of the same subtype.

A. Preparation of Inactivated Viral Vaccines

Inactivated influenza A and B virus vaccines are licensed for parenteral use in humans. Federal bodies and the World Health Organization make recommendations each year about which strains should be included in the vaccine. The vaccine is usually a cocktail containing one or two type A viruses and a type B virus of the strains isolated in the previous winter's outbreaks.

Selected seed strains are grown in embryonated eggs, the substrate used for vaccine production. Sometimes the natural isolates grow too poorly in eggs to permit vaccine production, in which case a reassortant virus is made in the laboratory. The reassortant virus, which carries the genes for the surface antigens of the desired vaccine with the replication genes from an egg-adapted laboratory virus, is then used for vaccine production.

Virus is harvested from the egg allantoic fluid, purified, concentrated by zonal centrifugation, and inactivated with formalin or β-propiolactone. The quantity of HA is standardized in each vaccine dose (approximately 15 μg of antigen), but the quantity of NA is not standardized, as it is more labile under purification and storage conditions. Each dose of vaccine contains the equivalent of about 10 billion virus particles.

Vaccines are either whole virus (WV), subvirion (SV), or surface antigen preparations. The WV vaccine contains intact, inactivated virus; the SV vaccine contains purified virus disrupted with detergents; and the surface antigen vaccines contain purified HA and NA glycoproteins. All are efficacious.

B. Live-Virus Vaccines

A live-virus vaccine must be attenuated so as not to induce the disease it is designed to prevent. In view of the constantly changing face of influenza viruses in nature and the extensive laboratory efforts required to attenuate a virulent virus, the only feasible strategy is to devise a way to transfer defined attenuating genes from an attenuated master donor virus to each new epidemic or pandemic isolate.

A cold-adapted donor virus, able to grow at 25°C but not at 37°C—the temperature of the lower respiratory tract—should replicate in the nasopharynx, which has a cooler temperature (33°C). A live attenuated, cold-adapted, temperature-sensitive, trivalent influenza virus vaccine administered by nasal spray was licensed in the United States in 2003. It was the first live-virus influenza vaccine approved in the United States, as well as the first nasally administered vaccine in the United States.

C. Use of Influenza Vaccines

The only contraindication to vaccination is a history of allergy to egg protein. Since vaccine strains are grown in eggs, some egg protein antigens are present in the vaccine.

Annual influenza vaccination is recommended for all children aged 6 months to 18 years and for high-risk groups. These include individuals at increased risk of complications associated with influenza infection (those with either chronic heart or lung disease, including children with asthma, or metabolic or renal disorders; residents of nursing homes; persons infected with the human immunodeficiency virus [HIV]; and those 65 years of age and older) and persons who might transmit influenza to high-risk groups (medical personnel, employees in chronic care facilities, household members). The live-virus intranasal vaccine is not currently recommended for individuals in the high-risk groups.

Prevention by Hand Hygiene

Although transmission of influenza virus occurs primarily by aerosol spread, hand transmission is potentially important also. Studies have shown that hand washing with soap and water or the use of alcohol-based hand rubs is highly effective at reducing the amount of virus on human hands.

REVIEW QUESTIONS

1. Which of the following statements regarding the prevention and treatment of influenza is correct?
   (A) Booster doses of vaccine are not recommended
   (B) Drugs that inhibit neuraminidase are active only against influenza A
   (C) As with some other live vaccines, the influenza vaccine should not be given to pregnant women
   (D) The influenza vaccine contains several serotypes of virus
   (E) The virus strains in the influenza vaccine do not vary from year to year

2. Which of the following statements about the neuraminidase of influenza virus is not correct?
   (A) Is embedded in the outer surface of the viral envelope
   (B) Forms a spike structure composed of four identical monomers, each with enzyme activity
   (C) Facilitates release of virus particles from infected cells
   (D) Lowers the viscosity of the mucous film in the respiratory tract
   (E) Is antigenically similar among all mammalian influenza viruses

3. Which of the following statements reflects the pathogenesis of influenza?
   (A) The virus enters the host in airborne droplets
   (B) Viremia is common
   (C) The virus frequently establishes persistent infections in the lung
   (D) Pneumonia is not associated with secondary bacterial infections
   (E) Viral infection does not kill cells in the respiratory tract

4. Which of the following symptoms is not typical of influenza?
   (A) Fever
   (B) Muscular aches
   (C) Malaise
   (D) Dry cough
   (E) Rash

5. The type-specific antigen (A, B, or C) of influenza viruses is found on which viral constituent?
   (A) Hemagglutinin
   (B) Neuraminidase
   (C) Nucleocapsid
   (D) Polymerase complex
   (E) Major nonstructural protein
   (F) Lipid in the viral envelope

6. A 70-year-old nursing home patient refused the influenza vaccine and subsequently developed influenza. She died of acute pneumonia 1 week after contracting the flu. The most common cause of acute postinfluenza pneumonia is which of the following?
   (A) Legionella
   (B) *Staphylococcus aureus*
   (C) Measles
   (D) Cytomegalovirus
   (E) Listeria

7. Which of the following statements concerning antigenic drift in influenza viruses is correct?
   (A) It results in major antigenic changes
   (B) It is exhibited only by influenza A viruses
   (C) It is due to frameshift mutations in viral genes
   (D) It results in new subtypes over time
   (E) It affects predominantly the matrix protein

8. A 32-year-old male physician developed a "flu-like" syndrome with fever, sore throat, headache, and myalgia. In order to provide laboratory confirmation of influenza, a culture for the virus was ordered. Which of the following would be the best specimen for isolating the virus responsible for this infection?
   (A) Stool
   (B) Nasopharyngeal washing
   (C) Vesicle fluid
   (D) Blood
   (E) Saliva

9. Which of the following statements about isolation of influenza viruses is correct?
   (A) Diagnosis of an influenza virus infection can only be made by isolating the virus
   (B) Isolation of influenza virus is done using newborn mice
   (C) Isolation of virus can help determine the epidemiology of the disease
   (D) Primary influenza virus isolates grow readily in cell culture

10. The principal reservoir for the antigenic shift variants of influenza virus appears to be which of the following?

    (A) Chronic human carriers of the virus
    (B) Sewage
    (C) Pigs, horses, and fowl
    (D) Mosquitoes
    (E) Rodents

11. Highly pathogenic H5N1 avian influenza (HPAI) can infect humans with a high mortality rate, but it has not yet resulted in a pandemic. The following are characteristics of HPAI, except for one. Which one is not?

    (A) Efficient human-to-human transmission
    (B) Presence of avian influenza genes
    (C) Efficient infection of domestic poultry
    (D) Contains segmented RNA genome

12. Which of the following statements about diagnostic testing for influenza is true?

    (A) Clinical symptoms reliably distinguish influenza from other respiratory illnesses
    (B) Viral culture is the "gold standard" diagnostic test because it is the most rapid assay
    (C) Patient antibody responses are highly specific for the strain of infecting influenza virus
    (D) Reverse-transcription PCR is preferred for its speed, sensitivity, and specificity

Answers

1. D	4. E	7. D	10. C
2. E	5. C	8. B	11. A
3. A	6. B	9. C	12. D

REFERENCES

Avian influenza fact sheet. World Health Org Wkly Epidemiol Rec 2006;81:129.

Couch RB: Prevention and treatment of influenza. N Engl J Med 2000;343:1778. [PMID: 11114318]

Gambotto A et al: Human infection with highly pathogenic H5N1 influenza virus. Lancet 2008;371:1464. [PMID: 18440429]

Horimoto T, Kawaoka Y: Influenza: Lessons from past pandemics, warnings from current incidents. Nature Rev Microbiol 2005;3:591. [PMID: 16064053]

Influenza vaccination of health-care personnel. Recommendations of the Healthcare Infection Control Practices Advisory Committee and the Advisory Committee on Immunization Practices. MMWR Morb Mortal Wkly Rep 2006;55(RR-2):1.

Neumann G, Noda T, Kawaoka Y: Emergence and pandemic potential of swine-origin H1N1 influenza virus. Nature 2009;459:931. [PMID: 19525932]

Olsen B et al: Global patterns of influenza A virus in wild birds. Science 2006;312:384. [PMID: 16627734]

Palese P, Shaw ML: Orthomyxoviridae: The viruses and their replication. In: *Fields Virology,* 5th ed. Knipe DM et al (editors). Lippincott Williams & Wilkins, 2007.

Prevention and control of influenza. Recommendations of the Advisory Committee on Immunization Practices (ACIP), 2008. MMWR Morb Mortal Wkly Rep 2008;57(RR-7):1.

Seasonal and pandemic influenza: At the crossroads, a global opportunity. J Infect Dis 2006;194(Suppl 2). [Entire issue.]

Special section: Novel 2009 influenza A H1N1 (swine variant). J Clin Virol 2009;45:169. [10 articles.]

Taubenberger JK, Morens DM (guest editors): Influenza. Emerging Infect Dis 2006;12:1. [Entire issue.]

Wright PF, Neumann G, Kawaoka Y: Orthomyxoviruses. In: *Fields Virology,* 5th ed. Knipe DM et al (editors). Lippincott Williams & Wilkins, 2007.

Paramyxoviruses & Rubella Virus

The paramyxoviruses include the most important agents of respiratory infections of infants and young children (respiratory syncytial virus and the parainfluenza viruses) as well as the causative agents of two of the most common contagious diseases of childhood (mumps and measles). The World Health Organization estimates that acute respiratory infections and pneumonia are responsible every year worldwide for the deaths of 4 million children under 5 years of age. Paramyxoviruses are the major respiratory pathogens in this age group.

All members of the **Paramyxoviridae** family initiate infection via the respiratory tract. Replication of the respiratory pathogens is limited to the respiratory epithelia, whereas measles and mumps become disseminated throughout the body and produce generalized disease.

Rubella virus, though classified as a togavirus because of its chemical and physical properties (see Chapter 29), can be considered with the paramyxoviruses on an epidemiologic basis.

PROPERTIES OF PARAMYXOVIRUSES

Major properties of paramyxoviruses are listed in Table 40–1.

Structure & Composition

The morphology of **Paramyxoviridae** is pleomorphic, with particles 150 nm or more in diameter, occasionally ranging up to 700 nm. A typical particle is shown in Figure 40–1. The envelope of paramyxoviruses seems to be fragile, making virus particles labile to storage conditions and prone to distortion in electron micrographs.

The viral genome is linear, negative-sense, single-stranded, nonsegmented RNA, about 15 kb in size (Figure 40–2). As the genome is not segmented, this negates any opportunity for frequent genetic reassortment, resulting in the fact that all members of the paramyxovirus group are antigenically stable.

Most paramyxoviruses contain six structural proteins. Three proteins are complexed with the viral RNA—the nucleoprotein (N) that forms the helical nucleocapsid (13 or 18 nm in diameter) and represents the major internal protein and two other large proteins (designated P and L), which are

involved in the viral polymerase activity that functions in transcription and RNA replication.

Three proteins participate in the formation of the viral envelope. A matrix (M) protein underlies the viral envelope; it has an affinity for both the N and the viral surface glycoproteins and is important in virion assembly. The nucleocapsid is surrounded by a lipid envelope that is studded with 8- to 12-nm spikes of two different transmembrane glycoproteins. The activities of these surface glycoproteins help differentiate the various genera of the Paramyxoviridae family (Table 40–2). The larger glycoprotein (HN or G) may or may not possess hemagglutination and neuraminidase activities and is responsible for attachment to the host cell. It is assembled as a tetramer in the mature virion. The other glycoprotein (F) mediates membrane fusion and hemolysin activities. The pneumoviruses and metapneumoviruses contain two additional small envelope proteins (M2-1 and SH).

A diagram of a paramyxovirus particle is shown in Figure 40–3.

Classification

The Paramyxoviridae family is divided into two subfamilies and seven genera, six of which contain human pathogens

TABLE 40–1 Important Properties of Paramyxoviruses

Virion: Spherical, pleomorphic, 150 nm or more in diameter (helical nucleocapsid, 13–18 nm)

Composition: RNA (1%), protein (73%), lipid (20%), carbohydrate (6%)

Genome: Single-stranded RNA, linear, nonsegmented, negative-sense, noninfectious, about 15 kb

Proteins: Six to eight structural proteins

Envelope: Contains viral glycoprotein (G, H, or HN) (which sometimes carries hemagglutinin or neuraminidase activity) and fusion (F) glycoprotein; very fragile

Replication: Cytoplasm; particles bud from plasma membrane

Outstanding characteristics:
Antigenically stable
Particles are labile yet highly infectious

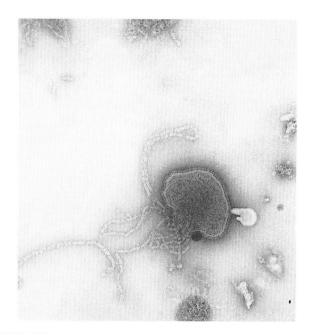

FIGURE 40–1 Ultrastructure of parainfluenza virus type 1. The virion is partially disrupted, showing the nucleocapsid. Surface projections are visible along the edge of the particle. (Courtesy of FA Murphy and EL Palmer.)

(Table 40–2). Most of the members are monotypic (ie, they consist of a single serotype); all are antigenically stable.

The genus *Respirovirus* contains two serotypes of human parainfluenza viruses, and the genus *Rubulavirus* contains two other parainfluenza viruses as well as mumps virus. Some animal viruses are related to the human strains. Sendai virus of mice, which was the first parainfluenza virus isolated and is now recognized as a common infection in mouse colonies, is a subtype of human type 1 virus. Simian parainfluenza virus 5 (SV5 or PIV5), a common contaminant of primary monkey cells, is the same as canine parainfluenza virus type 2, whereas shipping fever virus of cattle and sheep is a subtype of type 3. Newcastle disease virus, the prototype avian parainfluenza virus of genus *Avulavirus*, is also related to the human viruses.

Members within a genus share common antigenic determinants. Although the viruses can be distinguished antigenically using well-defined reagents, hyperimmunization stimulates cross-reactive antibodies that react with all four parainfluenza viruses, mumps virus, and Newcastle disease virus. Such heterotypic antibody responses, which include antibodies directed against both internal and surface proteins of the virus, are commonly observed in older people. This

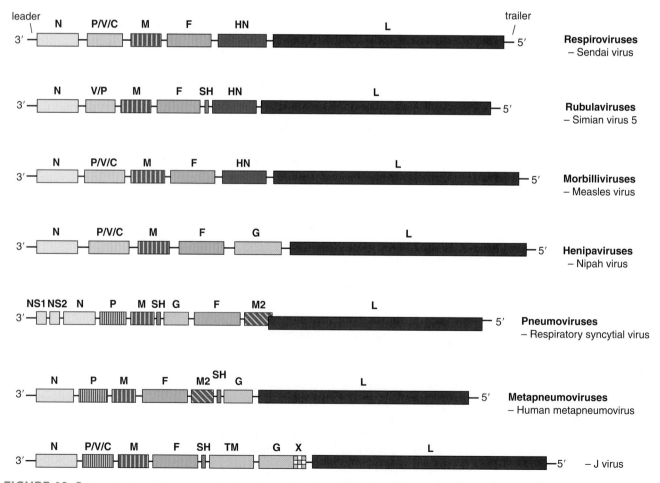

FIGURE 40–2 Genetic maps of representative members of the genera of the family Paramyxoviridae. Gene sizes (boxes) are drawn approximately to scale. (Copyright GD Parks and RA Lamb, 2006.)

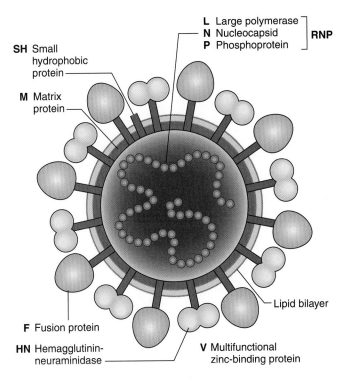

FIGURE 40–3 Schematic diagram of a paramyxovirus showing major components (not drawn to scale). The viral matrix protein (M) underlies the lipid bilayer. Inserted through the viral membrane are the hemagglutinin-neuraminidase (HN) attachment glycoprotein and the fusion (F) glycoprotein. Only some paramyxoviruses contain the SH protein. Inside the virus is the negative-strand virion RNA, which is encased in the nucleocapsid protein (N). Associated with the nucleocapsid are the L and P proteins, and together this complex has RNA-dependent RNA transcriptase activity. The V protein is only found in rubulavirus virions. (Copyright GD Parks and RA Lamb, 2006.)

phenomenon makes it difficult to determine by serodiagnosis the most likely infecting type. All members of the genera *Respirovirus* and *Rubulavirus* possess hemagglutinating and neuraminidase activities, both carried by the HN glycoprotein, as well as membrane fusion and hemolysin properties, both functions of the F protein.

The *Morbillivirus* genus contains measles virus (rubeola) of humans as well as canine distemper virus, rinderpest virus of cattle, and aquatic morbilliviruses that infect marine mammals. These viruses are antigenically related to each other but not to members of the other genera. The F protein is highly conserved among the morbilliviruses, whereas the HN/G proteins display more variability. Measles virus has a hemagglutinin but lacks neuraminidase activity. Measles virus induces formation of intranuclear inclusions, whereas other paramyxoviruses do not.

The *Henipavirus* genus contains zoonotic paramyxoviruses that are able to infect and cause disease in humans. Hendra and Nipah viruses, both indigenous to fruit bats, are members of the genus. These viruses lack neuraminidase activity.

Respiratory syncytial viruses of humans and cattle and pneumonia virus of mice constitute the genus *Pneumovirus*.

There are two antigenically distinct strains of respiratory syncytial virus of humans, subgroups A and B. The larger surface glycoprotein of pneumoviruses lacks hemagglutinating and neuraminidase activities characteristic of respiroviruses and rubulaviruses, so it is designated the G protein. The F protein of respiratory syncytial virus exhibits membrane fusion activity but no hemolysin activity. Newly recognized respiratory pathogens of humans are classified in the genus *Metapneumovirus*.

Paramyxovirus Replication

The typical paramyxovirus replication cycle is illustrated in Figure 40–4.

A. Virus Attachment, Penetration, & Uncoating

Paramyxoviruses attach to host cells via the hemagglutinin glycoprotein (HN, H, or G protein). In the case of measles virus, the receptor is the membrane CD46 or the CD150 molecule. Next, the virion envelope fuses with the cell membrane by the action of the fusion glycoprotein F_1 cleavage product. The F_1 protein undergoes complex refolding during the process of viral and cellular membrane fusion. If the F_0 precursor is not cleaved, it has no fusion activity; virion penetration does not occur; and the virus particle is unable to initiate infection. Fusion by F_1 occurs at the neutral pH of the extracellular environment, allowing release of the viral nucleocapsid directly into the cell. Thus, paramyxoviruses are able to bypass internalization through endosomes.

B. Transcription, Translation, & RNA Replication

Paramyxoviruses contain a nonsegmented, negative-strand RNA genome. Messenger RNA transcripts are made in the cell cytoplasm by the viral RNA polymerase. There is no need for exogenous primers and therefore no dependence on cell nuclear functions. The mRNAs are much smaller than genomic size; each represents a single gene. Transcriptional regulatory sequences at gene boundaries signal transcriptional start and termination. The position of a gene relative to the 3′ end of the genome correlates with transcription efficiency. The most abundant class of transcripts produced by an infected cell is from the N gene, located nearest the 3′ end of the genome, whereas the least abundant is from the L gene, located at the 5′ end (Figure 40–2).

Viral proteins are synthesized in the cytoplasm, and the quantity of each gene product corresponds to the level of mRNA transcripts from that gene. Viral glycoproteins are synthesized and glycosylated in the secretory pathway.

The viral polymerase protein complex (P and L proteins) is also responsible for viral genome replication. For successful synthesis of a positive-strand antigenome intermediate template, the polymerase complex must disregard the termination signals interspersed at gene boundaries. Full-length progeny genomes are then copied from the antigenome template.

TABLE 40–2   Characteristics of Genera in the Subfamilies of the Family Paramyxoviridae

Property	Paramyxovirinae				Pneumovirinae	
	Respirovirus	*Rubulavirus*	*Morbillivirus*	*Henipavirus*[a]	*Pneumovirus*	*Metapneumovirus*
Human viruses	Parainfluenza 1, 3	Mumps, parainfluenza 2, 4a, 4b	Measles	Hendra, Nipah	Respiratory syncytial virus	Human metapneumovirus
Serotypes	1 each	1 each	1	?	2	1
Diameter of nucleocapsid (nm)	18	18	18	18	13	13
Membrane fusion (F protein)	+	+	+	+	+	+
Hemolysin[b]	+	+	+	?	0	0
Hemagglutinin[c]	+	+	+	0	0	0
Hemadsorption	+	+	+	0	0	0
Neuraminidase[c]	+	+	0	0	0	0
Inclusions[d]	C	C	N,C	C	C	?

[a]Zoonotic paramyxoviruses.

[b]Hemolysin activity carried by F glycoprotein.

[c]Hemagglutination and neuraminidase activities carried by HN glycoprotein of respiroviruses and rubulaviruses; H glycoprotein of morbilliviruses lacks neuraminidase activity; G glycoprotein of other paramyxoviruses lacks both activities.

[d]C, cytoplasm; N, nucleus.

The nonsegmented genome of paramyxoviruses negates the possibility of gene segment reshuffling (ie, genetic reassortment) so important to the natural history of influenza viruses. The HN/H/G and F surface proteins of paramyxoviruses exhibit minimal antigenic variation over long periods of time. It is surprising that they do not undergo antigenic drift as a result of mutations introduced during replication, as RNA polymerases tend to be error-prone. One possible explanation is that nearly all the amino acids in the primary structures of paramyxovirus glycoproteins may be involved in structural or functional roles, leaving little opportunity for substitutions that would not markedly diminish the viability of the virus.

C. Maturation

The virus matures by budding from the cell surface. Progeny nucleocapsids form in the cytoplasm and migrate to the cell surface. They are attracted to sites on the plasma membrane that are studded with viral HN/H/G and F_0 glycoprotein spikes. The M protein is essential for particle formation, serving to link the viral envelope to the nucleocapsid. During budding, most host proteins are excluded from the membrane.

The neuraminidase activity of the HN protein of parainfluenza viruses and mumps virus presumably functions to prevent self-aggregation of virus particles. Other paramyxoviruses do not possess neuraminidase activity (Table 40–2).

If appropriate host cell proteases are present, F_0 proteins in the plasma membrane will be activated by cleavage. Activated fusion protein will then cause fusion of adjacent cell membranes, resulting in formation of large syncytia (Figure 40–5). Syncytium formation is a common response to paramyxovirus infection. Acidophilic cytoplasmic inclusions are regularly formed (Figure 40–5). Inclusions are believed to reflect sites of viral synthesis and have been found to contain recognizable nucleocapsids and viral proteins. Measles virus also produces intranuclear inclusions (Figure 40–5).

PARAINFLUENZA VIRUS INFECTIONS

Parainfluenza viruses are ubiquitous and cause common respiratory illnesses in persons of all ages. They are major pathogens of severe respiratory tract disease in infants and young children. Only respiratory syncytial virus, and perhaps human metapneumovirus, causes more cases of serious respiratory disease in children. Reinfections with parainfluenza viruses are common.

Pathogenesis & Pathology

Parainfluenza virus replication in the immunocompetent host appears to be limited to respiratory epithelia. Viremia, if it occurs at all, is uncommon. The infection may involve only the nose and throat, resulting in a harmless "common cold" syndrome. However, infection may be more extensive and, especially with types 1 and 2, may involve the larynx and upper trachea, resulting in croup (laryngotracheobronchitis). Croup is characterized by respiratory obstruction due to swelling of the larynx and related structures. The infection

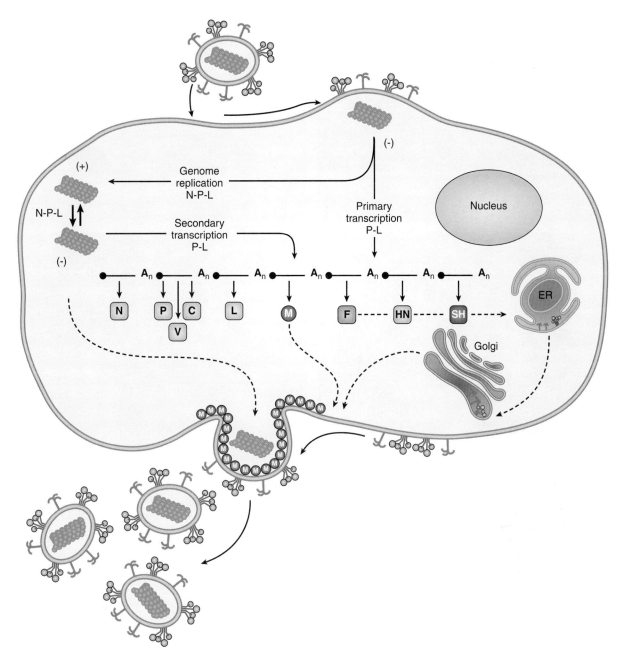

FIGURE 40–4 Paramyxovirus life cycle. The infecting virus particle fuses with the plasma membrane and releases the viral nucleocapsid into the cytoplasm. Solid lines represent transcription and genome replication. Dotted lines indicate transport of newly synthesized viral proteins to plasma membrane. Progeny virions are released from the cell by a budding process. The entire paramyxovirus replication cycle takes place in the cell cytoplasm. (Copyright GD Parks and RA Lamb, 2006.)

may spread deeper to the lower trachea and bronchi, culminating in pneumonia or bronchiolitis, especially with type 3, but at a much lower frequency than that observed with respiratory syncytial virus.

Duration of parainfluenza virus shedding is about 1 week after onset of illness; some children may excrete virus several days prior to illness. Type 3 may be excreted for up to 4 weeks after onset of primary illness. This persistent shedding from young children facilitates spread of infection. Prolonged viral

shedding may occur in children with compromised immune function and in adults with chronic lung disease.

Factors that determine the severity of parainfluenza virus disease are unclear but include both viral and host properties, such as susceptibility of the protein to cleavage by different proteases, production of an appropriate protease by host cells, immune status of the patient, and airway hyperreactivity.

The production of virus-specific IgE antibodies during primary infections has been associated with disease severity.

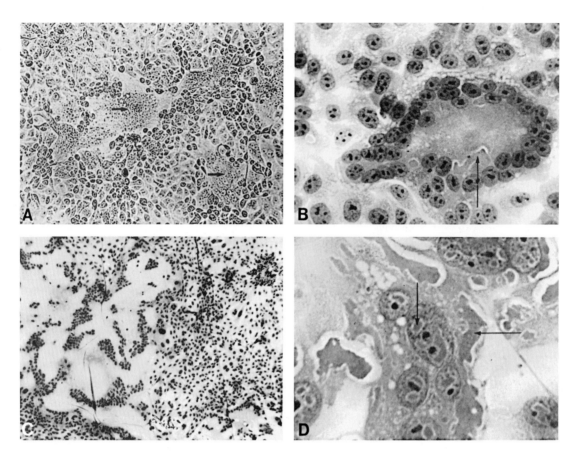

FIGURE 40–5 Syncytial formation induced by paramyxoviruses. **A:** Respiratory syncytial virus in MA104 cells (unstained, 100×). Syncytia (arrows) result from fusion of plasma membranes; nuclei are accumulated in the center. **B:** Respiratory syncytial virus in HEp-2 cells (H&E stain, 400×). Syncytium contains many nuclei and acidophilic cytoplasmic inclusions (arrow). **C:** Measles virus in human kidney cells (H&E stain, 30×). Huge syncytium contains hundreds of nuclei. **D:** Measles virus in human kidney cells (H&E stain, 400×). Multinucleated giant cell contains acidophilic nuclear inclusions (vertical arrow) and cytoplasmic inclusions (horizontal arrow). (Used with permission from I Jack.)

The mechanism may involve release of mediators of inflammation which alter airway function.

Clinical Findings

The relative importance of parainfluenza viruses as a cause of respiratory diseases in different age groups is indicated in Table 30–4. Their presence in lower respiratory tract infections in young children is shown in Figure 40–6.

Primary infections in young children usually result in rhinitis and pharyngitis, often with fever and some bronchitis. However, children with primary infections caused by parainfluenza virus type 1, 2, or 3 may have serious illness, ranging from laryngotracheitis and croup (particularly with types 1 and 2) to bronchiolitis and pneumonia (particularly with type 3). The severe illness associated with type 3 occurs mainly in infants under the age of 6 months; croup or laryngotracheobronchitis is more likely to occur in older children, between ages 6 months and 18 months. More than one-half of initial infections with parainfluenza virus types 1–3 result in febrile illness. It is estimated that only 2–3% develop into

croup. Parainfluenza virus type 4 does not cause serious disease, even on first infection.

The most common complication of parainfluenza virus infection is otitis media.

Immunocompromised children and adults are susceptible to severe infections. Mortality rates following parainfluenza infection in bone marrow transplant recipients range from 10% to 20%.

Newcastle disease virus is an avian paramyxovirus that produces pneumoencephalitis in young chickens and "influenza" in older birds. In humans, it may produce inflammation of the conjunctiva. Recovery is complete in 10–14 days. Infection in humans is an occupational disease limited to workers handling infected birds.

Immunity

Parainfluenza virus types 1–3 are distinct serotypes that lack significant cross-neutralization (Table 40–2). Virtually all infants have maternal antibodies to the viruses in serum, yet these antibodies do not prevent infection or disease.

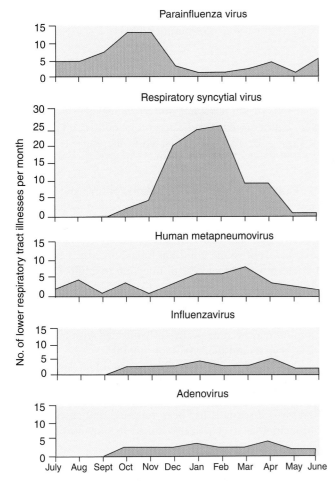

FIGURE 40–6 Patterns of lower respiratory tract infections in infants and young children with paramyxoviruses and other viruses. Data from 25 years of surveillance (1976–2001), involving 2009 children from birth to age 5 years. (Reproduced with permission from Williams JV et al: Human metapneumovirus and lower respiratory tract disease in otherwise healthy infants and children. N Engl J Med 2004;350:443.)

Reinfection of older children and adults also occurs in the presence of antibodies elicited by an earlier infection. However, those antibodies modify the disease, as such reinfections usually present simply as nonfebrile upper respiratory infections (colds).

Natural infection stimulates appearance of IgA antibody in nasal secretions and concomitant resistance to reinfection. The secretory IgA antibodies are most important for providing protection against reinfection but disappear within a few months. Reinfections are thus common even in adults.

Serum antibodies are made to both HN and F viral surface proteins, but their relative roles in determining resistance are unknown. As successive reinfections occur, the antibody response becomes less specific because of shared antigenic determinants among parainfluenza viruses and mumps virus. This makes it difficult to diagnose the specific paramyxovirus associated with a given infection using serologic assays.

Laboratory Diagnosis

The immune response to the initial parainfluenza virus infection in life is type-specific. However, with repeated infections the response becomes less specific, and cross-reactions extend even to mumps virus. Antigen detection methods are useful for rapid diagnosis. Definitive diagnosis relies on viral isolation from appropriate specimens or detection of viral RNA by reverse transcription-polymerase chain reaction (RT-PCR).

A. Antigen Detection

Direct identification of viral antigens in specimens is commonly done. Antigens may be detected in exfoliated nasopharyngeal cells by direct or indirect immunofluorescence tests. These methods are rapid but less sensitive than viral isolation and must be carefully controlled. Highly specific immune reagents are essential if specific serotype identification is desired.

B. Isolation & Identification of Virus

Nasal washes are good specimens for viral isolation. Bronchoalveolar lavage fluid and lung tissue have also been used. A continuous monkey kidney cell line, LLC-MK$_2$, is suitable for isolation of parainfluenza viruses. Prompt inoculation of samples into cell cultures is important for successful viral isolation, as viral infectivity drops rapidly if clinical specimens are stored.

For rapid diagnosis, samples are inoculated onto cells growing on coverslips in shell vials and centrifuged (30 minutes at 700 × g), and the cultures are incubated. Twenty-four to 72 hours later, the cells are fixed and tested by immunofluorescence using monoclonal antibodies. If desired, pools of antibodies to multiple respiratory viruses can be used, followed by specific typing of positive samples with individual antibodies.

Parainfluenza viruses grow slowly and produce very little cytopathic effect. Another way to detect the presence of virus is to perform hemadsorption using guinea pig erythrocytes. Depending on the amount of virus, 10 days or more of incubation may be necessary before the cultures become hemadsorption-positive.

C. Nucleic Acid Detection

Polymerase chain reaction assays can be used to detect viral RNA in nasal washes or nose and throat swabs. RT-PCR assays are as sensitive as cell culture methods. Sequence analyses are useful in molecular epidemiology studies of parainfluenza virus infections.

D. Serology

Serodiagnosis should be based on paired sera. Antibody responses can be measured using Nt, HI, or ELISA tests. A fourfold rise in titer is indicative of infection with a parainfluenza virus, as is the appearance of specific IgM antibody.

However, because of the problem of shared antigens, it is impossible to be confident of the specific virus type involved.

Epidemiology

Parainfluenza viruses are a major cause of lower respiratory tract disease in young children (Figure 40–6). Parainfluenza viruses are widely distributed geographically. Type 3 is most prevalent, with about two-thirds of infants infected during the first year of life; virtually all have antibodies to type 3 by age 2 years. Infections with types 1 and 2 occur at a lower rate, reaching prevalences of about 75% and 60%, respectively, by 5 years of age.

Type 3 is endemic, with some increase during the spring, whereas types 1 and 2 tend to cause epidemics during the fall or winter, frequently on a 2-year cycle.

Reinfections are common throughout childhood and in adults and result in mild upper respiratory tract illnesses. Reportedly, 67% of children are reinfected with parainfluenza type 3 during the second year of life. Reinfections may necessitate hospitalization of adults with chronic lung diseases (eg, asthma).

Parainfluenza viruses are transmitted by direct person-to-person contact or by large-droplet aerosols. Type 1 has been recovered from air samples collected in the vicinity of infected patients. Infections can occur through both the nose and the eyes.

Parainfluenza viruses are usually introduced into a group by preschool children and then spread readily from person to person. The incubation period appears to be 5–6 days. Type 3 virus especially will generally infect all susceptible individuals in a semiclosed population, such as a family or a nursery, within a short time. Parainfluenza viruses are troublesome causes of nosocomial infection in pediatric wards in hospitals. Other high-risk situations include day care centers and schools.

Treatment & Prevention

Contact isolation precautions are necessary to manage nosocomial outbreaks of parainfluenza virus. These include restriction of visitors, isolation of infected patients, and gowning and hand washing by medical personnel.

The antiviral drug ribavirin has been used with some benefit in treatment of immunocompromised patients with lower respiratory tract disease.

No vaccine is available.

RESPIRATORY SYNCYTIAL VIRUS INFECTIONS

Respiratory syncytial virus is the most important cause of lower respiratory tract illness in infants and young children, usually outranking all other microbial pathogens as the cause of bronchiolitis and pneumonia in infants under 1 year of age.

It is estimated to account for approximately 25% of pediatric hospitalizations due to respiratory disease.

Pathogenesis & Pathology

Respiratory syncytial virus replication occurs initially in epithelial cells of the nasopharynx. Virus may spread into the lower respiratory tract and cause bronchiolitis and pneumonia. Viral antigens can be detected in the upper respiratory tract and in shed epithelial cells. Viremia occurs rarely if at all.

The incubation period between exposure and onset of illness is 3–5 days. Viral shedding may persist for 1–3 weeks from infants and young children, whereas adults shed for only 1–2 days. High viral titers are present in respiratory tract secretions from young children. Inoculum size is an important determinant of successful infection in adults (and possibly in children as well).

An intact immune system seems to be important in resolving an infection, as patients with impaired cell-mediated immunity may become persistently infected with respiratory syncytial virus and shed virus for months.

Although the airways of very young infants are narrow and more readily obstructed by inflammation and edema, only a subset of young babies develops severe respiratory syncytial virus disease. It has been reported that susceptibility to bronchiolitis is genetically linked to polymorphisms in innate immune genes.

Clinical Findings

The spectrum of respiratory illness caused by respiratory syncytial virus ranges from inapparent infection or the common cold through pneumonia in infants to bronchiolitis in very young babies. Bronchiolitis is the distinct clinical syndrome associated with this virus. About one-third of primary respiratory syncytial virus infections involve the lower respiratory tract severely enough to require medical attention. The child may wheeze. Almost 2% of infected babies require hospitalization, resulting in an estimated 51,000–82,000 hospitalizations annually in the United States, with the peak occurrence at 2 months of age.

Progression of symptoms may be very rapid, culminating in death. With availability of modern pediatric intensive care, the mortality rate in normal infants is low (about 1% of hospitalized patients). But if a respiratory syncytial virus infection is superimposed on preexisting disease, such as congenital heart disease, the mortality rate may be high.

Reinfection is common in both children and adults. Although reinfections tend to be symptomatic, the illness is usually limited to the upper respiratory tract, resembling a cold, in healthy individuals.

Respiratory syncytial virus infections account for about one-third of respiratory infections in bone marrow transplant patients. Pneumonia develops in about one-half of infected immunocompromised children and adults, especially if infection occurs in the early posttransplant period. Reported mortality rates range from 20% to 80%.

Infections in the elderly may cause symptoms similar to influenza virus disease. Pneumonia may develop. Estimates of respiratory syncytial virus prevalence in long-term care facilities include infection rates of 5–10%, pneumonia in 10–20% of those infected, and mortality rates of 2–5%.

Children who suffered from respiratory syncytial virus bronchiolitis and pneumonia as infants often exhibit recurrent episodes of wheezing illness for many years. However, no causal relationship has been shown between respiratory syncytial virus infections and long-term abnormalities. It may be that certain individuals have underlying physiologic traits that predispose them both to severe respiratory syncytial virus infections and to reactive airway disease.

Respiratory syncytial virus is an important cause of otitis media. It is estimated that 30–50% of wintertime episodes in infants may be due to respiratory syncytial virus infection.

Immunity

High levels of neutralizing antibody that is maternally transmitted and present during the first several months of life are believed to be critical in protective immunity against lower respiratory tract illness. Severe respiratory syncytial disease begins to occur in infants at 2–4 months of age, when maternal antibody levels are falling. However, primary infection and reinfection can occur in the presence of viral antibodies. Serum neutralizing antibody appears to be strongly correlated with immunity against disease of the lower respiratory tract but not of the upper respiratory tract.

Respiratory syncytial virus is not an effective inducer of interferon—in contrast to influenza and parainfluenza virus infections, in which interferon levels are high and correlate with disappearance of virus.

Both serum and secretory antibodies are made in response to respiratory syncytial virus infection. Primary infection with one subgroup induces cross-reactive antibodies to virus of the other subgroup (Table 40–2). Younger infants have lower IgG and IgA secretory antibody responses to respiratory syncytial virus than do older infants. It is not clear if secretory IgA in nasal secretions is involved in protection against reinfection. Cellular immunity is important in recovery from infection.

An association has been noted between virus-specific IgE antibody and severity of disease. Viral secretory IgE antibodies have been correlated with occurrence of bronchiolitis.

It is apparent that immunity is only partially effective and is often overcome under natural conditions; reinfections are common, but the severity of ensuing disease is lessened.

Laboratory Diagnosis

Isolation of virus and detection of viral RNA or viral antigen in respiratory secretions are the procedures of choice to diagnose respiratory syncytial virus infection. Respiratory syncytial virus differs from other paramyxoviruses in that it does not have a hemagglutinin; therefore, diagnostic methods cannot use hemagglutination or hemadsorption assays.

A. Antigen Detection

Direct identification of viral antigens in clinical samples is rapid, requiring only a few hours. Immunofluorescence on exfoliated cells or ELISA on nasopharyngeal secretions is commonly used. A nasal wash or nasal aspirate is a good source of virus. Large amounts of virus are present in nasal washes from young children (10^3–10^8 plaque-forming units [PFU] per milliliter), but much less is present in specimens from adults (<100 PFU/mL). Antigen detection is not a sensitive test in many adults. ELISA kits are useful for rapid diagnosis, which is desirable because antiviral therapy is available.

B. Isolation & Identification of Virus

Respiratory syncytial virus can be isolated from nasal secretions. It is extremely labile. Samples should be inoculated into cell cultures immediately; freezing of clinical specimens may result in complete loss of infectivity.

Human heteroploid cell lines HeLa and HEp-2 are the most sensitive for viral isolation. The presence of respiratory syncytial virus can usually be recognized by development of giant cells and syncytia in inoculated cultures (Figure 40–5). It may take as long as 10 days for cytopathic effects to appear. Definitive diagnosis can be established by detecting viral antigen in infected cells using a defined antiserum and the immunofluorescence test. More rapid isolation of respiratory syncytial virus can be achieved by spin-amplified inoculation of vials containing tissue cultures growing on coverslips. Cells can be tested 24–48 hours later by immunofluorescence or RT-PCR.

Detection of respiratory syncytial virus is strong evidence that the virus is involved in a current illness because it is almost never found in healthy people.

C. Nucleic Acid Detection

RT-PCR assays are available for the detection of respiratory syncytial virus in respiratory secretions. The sensitivity of these assays equals or exceeds that of cell culture. The test takes about 1 day. RT-PCR testing is especially useful for adult specimens in which only small amounts of virus are often present. Such assays also are useful for subtyping respiratory syncytial virus isolates and for the analysis of genetic variation in outbreaks.

D. Serology

Serum antibodies can be assayed in a variety of ways—immunofluorescence, ELISA, and Nt tests are all used. Measurements of serum antibody are important for epidemiologic studies but play only a small role in clinical decision making.

Epidemiology

Respiratory syncytial virus is distributed worldwide and is recognized as the major pediatric respiratory tract pathogen (Figure 40–6). About 70% of infants are infected by age 1 and almost all by age 2 years. Serious bronchiolitis or pneumonia is most apt to occur in infants between the ages of 6 weeks and 6 months, with peak incidence at 2 months. The virus can be isolated from most infants under age 6 months suffering from bronchiolitis, but it is almost never isolated from healthy infants. Subgroup A infections appear to cause more severe illness than subgroup B infections. Respiratory syncytial virus is the most common cause of viral pneumonia in children under age 5 years but may also cause pneumonia in the elderly or in immunocompromised persons. Respiratory syncytial virus infection of older infants and children results in milder respiratory tract infection than in those under age 6 months.

Respiratory syncytial virus is spread by large droplets and direct contact. Although the virus is very labile, it can survive on environmental surfaces for up to 6 hours. The main portal of entry into the host is through the nose and eyes.

Reinfection occurs frequently (in spite of the presence of specific antibodies), but resulting symptoms are those of a mild upper respiratory infection (a cold). In families with an identified case of respiratory syncytial infection, virus spread to siblings and adults is common.

Respiratory syncytial virus spreads extensively in children every year during the winter season. Although the virus persists throughout the summer months, outbreaks tend to peak in February or March in the Northern Hemisphere. In tropical areas, respiratory syncytial virus epidemics may coincide with rainy seasons.

Respiratory syncytial virus causes nosocomial infections in nurseries and on pediatric hospital wards. Transmission occurs primarily via hospital staff members.

Respiratory syncytial virus can also cause symptomatic disease in healthy young adults in crowded conditions (military recruits in basic training). In a study in 2000, respiratory syncytial virus was identified in 11% of recruits with respiratory symptoms. This compared to identification of adenoviruses (48%), influenza viruses (11%), and parainfluenza virus 3 (3%) in the symptomatic recruits.

Treatment & Prevention

Treatment of serious respiratory syncytial virus infections depends primarily on supportive care (eg, removal of secretions, administration of oxygen). The antiviral drug ribavirin is approved for treatment of lower respiratory tract disease due to respiratory syncytial virus, especially in infants at high risk for severe disease. The drug is administered in an aerosol for 3–6 days. Oral ribavirin is not useful.

Immune globulin with high-titer antibodies against respiratory syncytial virus is of marginal benefit. Humanized antiviral monoclonal antibodies are available.

Much research effort has been expended in an attempt to develop a respiratory syncytial virus vaccine. In the late 1960s, an experimental formalin-inactivated respiratory syncytial virus vaccine was tested. Recipients developed high titers of nonneutralizing serum antibodies, but when immunized children encountered a subsequent infection with wild-type respiratory syncytial virus, they suffered significantly more severe lower respiratory tract illness than did children from the control group. It has been suggested that the formalin treatment destroyed protective epitopes on the virus and/or that due to lack of stimulation of Toll-like receptors, the vaccine induced only low-avidity antibodies which were not protective. No vaccine is available today.

Respiratory syncytial virus poses special problems for vaccine development. The target group, newborns, would have to be immunized soon after birth to afford protection at the time of greatest risk of serious respiratory syncytial virus infection, and eliciting a protective immune response at this early age is difficult in the presence of maternal antibody. A strategy being tested is maternal immunization with a vaccine. The aim is to ensure transfer of protective levels of virus-specific neutralizing antibody to infants that would persist for 3–5 months, the period of greatest vulnerability of newborns to severe respiratory syncytial virus disease.

Control measures necessary when nosocomial outbreaks occur are the same as described above for parainfluenza viruses (contact isolation, hand washing, and restriction of visitors).

HUMAN METAPNEUMOVIRUS INFECTIONS

Human metapneumovirus is a respiratory pathogen first described in 2001 (Table 40–2). It was detected using a molecular (polymerase chain reaction) approach on clinical samples from children with respiratory illnesses but with negative viral test results. It appears to be widespread, with 100% seroprevalence in young adults and older persons. Human metapneumovirus is able to cause a wide range of respiratory illnesses from mild upper respiratory symptoms to severe lower respiratory tract disease. In general, symptoms are similar to those caused by respiratory syncytial virus.

It appears that infections with human metapneumovirus in young children are less common than with respiratory syncytial virus but more common than with the parainfluenza viruses (Figure 40–6). In a 25-year surveillance study in the United States (1976–2001) involving over 2000 children from birth to 5 years of age, 20% of stored nasal wash specimens from children with lower respiratory tract infections and previously negative for viruses by culture were found to be positive for human metapneumovirus by polymerase chain reaction. The majority of infections occurred between December and April.

Human metapneumovirus also causes respiratory disease in adults with hematologic malignancies and in institutionalized elderly persons.

MUMPS VIRUS INFECTIONS

Mumps is an acute contagious disease characterized by nonsuppurative enlargement of one or both salivary glands. Mumps virus mostly causes a mild childhood disease, but in adults complications including meningitis and orchitis are fairly common. More than one-third of all mumps infections are asymptomatic.

Pathogenesis & Pathology

Humans are the only natural hosts for mumps virus. Primary replication occurs in nasal or upper respiratory tract epithelial cells. Viremia then disseminates the virus to the salivary glands and other major organ systems. Involvement of the parotid gland is not an obligatory step in the infectious process.

The incubation period may range from 2 to 4 weeks but is typically about 14–18 days. Virus is shed in the saliva from about 3 days before to 9 days after the onset of salivary gland swelling. About one-third of infected individuals do not exhibit obvious symptoms (inapparent infections) but are equally capable of transmitting infection. It is difficult to control transmission of mumps because of the variable incubation periods, the presence of virus in saliva before clinical symptoms develop, and the large number of asymptomatic but infectious cases.

Mumps is a systemic viral disease with a propensity to replicate in epithelial cells in various visceral organs. Virus frequently infects the kidneys and can be detected in the urine of most patients. Viruria may persist for up to 14 days after the onset of clinical symptoms. The central nervous system is also commonly infected and may be involved in the absence of parotitis.

Clinical Findings

The clinical features of mumps reflect the pathogenesis of the infection. At least one-third of all mumps infections are subclinical, including the majority of infections in children under 2 years of age. The most characteristic feature of symptomatic cases is swelling of the salivary glands, which occurs in about 50% of patients.

A prodromal period of malaise and anorexia is followed by rapid enlargement of parotid glands as well as other salivary glands. Swelling may be confined to one parotid gland, or one gland may enlarge several days before the other. Gland enlargement is associated with pain.

Central nervous system involvement is common (10–30% of cases). Mumps causes aseptic meningitis and is more common among males than females. Meningoencephalitis usually occurs 5–7 days after inflammation of the salivary glands, but up to half of patients will not have clinical evidence of parotitis. Meningitis is reported in up to 15% of cases and encephalitis in less than 0.3%. Cases of mumps meningitis and meningoencephalitis usually resolve without sequelae, although unilateral deafness occurs in about 5:100,000 cases. The mortality rate from mumps encephalitis is about 1%.

The testes and ovaries may be affected, especially after puberty. Twenty to fifty percent of men who are infected with mumps virus develop orchitis (often unilateral). Because of the lack of elasticity of the tunica albuginea, which does not allow the inflamed testis to swell, the complication is extremely painful. Atrophy of the testis may occur as a result of pressure necrosis, but only rarely does sterility result. Mumps oophoritis occurs in about 5% of women. Pancreatitis is reported in about 4% of cases.

Immunity

Immunity is permanent after a single infection. There is only one antigenic type of mumps virus, and it does not exhibit significant antigenic variation (Table 40–2).

Antibodies to the HN glycoprotein (V antigen), the F glycoprotein, and the internal NP nucleocapsid protein (S antigen) develop in serum following natural infection. Antibodies to S antigen appear earliest (3–7 days after onset of clinical symptoms) but are transient and are usually gone within 6 months. Antibodies to V antigen develop more slowly (about 4 weeks after onset) but persist for years.

Antibodies against the HN antigen correlate well with immunity. Even subclinical infections are thought to generate lifelong immunity. A cell-mediated immune response also develops. Interferon is induced early in mumps infection. In immune individuals, IgA antibodies secreted in the nasopharynx exhibit neutralizing activity.

Passive immunity is transferred from mother to offspring; thus, it is rare to see mumps in infants under 6 months of age.

Laboratory Diagnosis

The diagnosis of typical cases usually can be made on the basis of clinical findings. However, other infectious agents, drugs, and conditions can cause similar symptoms. In cases without parotitis, the laboratory can be helpful in establishing the diagnosis. Tests include isolation of infectious virus, detection of viral nucleic acid by RT-PCR, and serology.

A. Isolation & Identification of Virus

The most appropriate clinical samples for viral isolation are saliva, cerebrospinal fluid, and urine collected within a few days after onset of illness. Virus can be recovered from the urine for up to 2 weeks.

Monkey kidney cells are preferred for viral isolation. Samples should be inoculated shortly after collection, as mumps virus is thermolabile. For rapid diagnosis, immunofluorescence using mumps-specific antiserum can detect mumps virus antigens as early as 2–3 days after the inoculation of cell cultures in shell vials.

In traditional culture systems, cytopathic effects typical of mumps virus consist of cell rounding and giant cell formation. As not all primary isolates show characteristic syncytial formation, the hemadsorption test may be used to demonstrate the presence of a hemadsorbing agent 1 and 2 weeks after inoculation. An isolate can be confirmed as mumps virus by hemadsorption inhibition using mumps-specific antiserum.

B. Nucleic Acid Detection

RT-PCR is a very sensitive method that can detect mumps genome sequences in clinical samples. It can detect the virus in many clinical samples that are negative in virus isolation attempts. RT-PCR assays can identify virus strains and provide useful information in epidemiological studies.

C. Serology

Simple detection of mumps antibody is not adequate to diagnose an infection. Rather, an antibody rise can be demonstrated using paired sera: a fourfold or greater rise in antibody titer is evidence of mumps infection. The ELISA or HI test is commonly used. Antibodies against the HN protein are neutralizing.

ELISA can be designed to detect either mumps-specific IgM antibody or mumps-specific IgG antibody. Mumps IgM is uniformly present early in the illness and seldom persists longer than 60 days. Therefore, demonstration of mumps-specific IgM in serum drawn early in illness strongly suggests recent infection. Heterotypic antibodies induced by parainfluenza virus infections do not cross-react in the mumps IgM ELISA.

Epidemiology

Mumps occurs endemically worldwide. Cases appear throughout the year in hot climates and peak in winter and spring in temperate climates. Outbreaks occur where crowding favors dissemination of the virus. Mumps is primarily an infection of children. The disease reaches its highest incidence in children aged 5–9 years, but epidemics may occur in army camps. In children under 5 years of age, mumps may commonly cause upper respiratory tract infection without parotitis.

Mumps is quite contagious; most susceptible individuals in a household will acquire infection from an infected member. The virus is transmitted by direct contact, airborne droplets, or fomites contaminated with saliva or urine. Closer contact is necessary for transmission of mumps than for transmission of measles or varicella.

About one-third of infections with mumps virus are inapparent. During the course of inapparent infection, the patient can transmit the virus to others. Individuals with subclinical mumps acquire immunity.

The overall mortality rate for mumps is low (1 death per 10,000 cases in the United States), mostly due to encephalitis.

The incidence of mumps and associated complications has declined markedly since introduction of the live-virus vaccine. In 1967, the year mumps vaccine was licensed, there were about 200,000 mumps cases (and 900 patients with encephalitis) in the United States. In 2001–2003, there were fewer than 300 mumps cases each year.

In 2006, there was an outbreak of mumps in the United States that resulted in over 5700 cases. Six states in the Midwest reported 84% of the cases. The outbreak started on a college campus among young adults and spread to all age groups. The outbreak probably spread because of close living conditions on college campuses and the accumulation of susceptible persons who were not successfully immunized.

Treatment, Prevention, & Control

There is no specific therapy.

Immunization with attenuated live mumps virus vaccine is the best approach to reducing mumps-associated morbidity and mortality rates. Attempts to minimize viral spread during an outbreak by using isolation procedures are futile because of the high incidence of asymptomatic cases and the degree of viral shedding before clinical symptoms appear; however, students and health care workers who acquire mumps illness should be excluded from school and work until 5 days after the onset of parotitis.

An effective attenuated live-virus vaccine made in chick embryo cell culture was licensed in the United States in 1967. It produces a subclinical, noncommunicable infection. Mumps vaccine is available in combination with measles and rubella (MMR) live-virus vaccines. Combination live-virus vaccines produce antibodies to each of the viruses in about 78–95% of vaccinees. There is no increased risk of aseptic meningitis after MMR vaccination. Other live attenuated mumps virus vaccines have been developed in Japan, Russia, and Switzerland.

Two doses of MMR vaccine are recommended for school entry. Because of the 2006 outbreak of mumps, updated vaccination recommendations for prevention of mumps transmission in settings with high risk for spread of infection were released. Two doses of vaccine should be given to health care workers born before 1957 without evidence of mumps immunity, and a second dose of vaccine should be considered for those who had received only a single dose.

MEASLES (RUBEOLA) VIRUS INFECTIONS

Measles is an acute, highly infectious disease characterized by fever, respiratory symptoms, and a maculopapular rash. Complications are common and may be quite serious.

The introduction of an effective live-virus vaccine has dramatically reduced the incidence of this disease in the United States, but measles is still a leading cause of death of young children in many developing countries.

Pathogenesis & Pathology

Humans are the only natural hosts for measles virus, although numerous other species, including monkeys, dogs, and mice, can be experimentally infected. The natural history of measles infection is shown in Figure 40–7.

The virus gains access to the human body via the respiratory tract, where it multiplies locally; the infection then spreads to the regional lymphoid tissue, where further multiplication occurs. Primary viremia disseminates the virus, which then replicates in the reticuloendothelial system. Finally, a secondary viremia seeds the epithelial surfaces of the body, including the skin, respiratory tract, and conjunctiva, where focal replication occurs. Measles can replicate in certain lymphocytes, which aids in dissemination throughout the body. Multinucleated giant cells with intranuclear inclusions are seen in lymphoid tissues throughout the body (lymph nodes, tonsils, appendix). The described events occur during the incubation period, which typically lasts 8–12 days but may last up to 3 weeks in adults.

During the prodromal phase (2–4 days) and the first 2–5 days of rash, virus is present in tears, nasal and throat secretions, urine, and blood. The characteristic maculopapular rash appears about day 14 just as circulating antibodies become detectable, the viremia disappears, and the fever falls. The rash develops as a result of interaction of immune T cells with virus-infected cells in the small blood vessels and lasts about 1 week. (In patients with defective cell-mediated immunity, no rash develops.)

Involvement of the central nervous system is common in measles (Figure 40–8). Symptomatic encephalitis develops in about 1:1000 cases. Because infectious virus is rarely recovered from the brain, it has been suggested that an autoimmune reaction is the mechanism responsible for this complication. In contrast, progressive measles inclusion body encephalitis may develop in patients with defective cell-mediated immunity. Actively replicating virus is present in the brain in this usually fatal form of disease.

A rare late complication of measles is subacute sclerosing panencephalitis (SSPE). This fatal disease develops years after the initial measles infection and is caused by virus that remains in the body after acute measles infection. Large amounts of measles antigens are present within inclusion bodies in infected brain cells, but only a few virus particles mature. Viral replication is defective owing to lack of production of one or more viral gene products, often the matrix protein.

Clinical Findings

Infections in nonimmune hosts are almost always symptomatic. After an incubation period of 8–12 days, measles is typically a 7- to 11-day illness (with a prodromal phase of 2–4 days followed by an eruptive phase of 5–8 days).

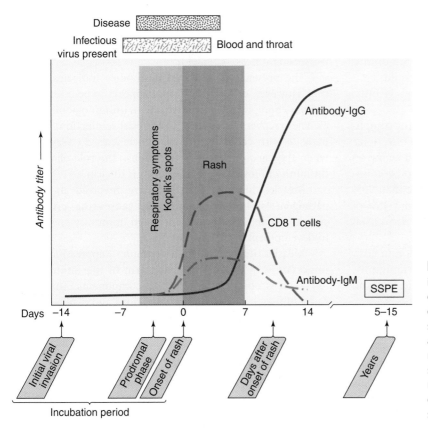

FIGURE 40–7 Natural history of measles infection. Viral replication begins in the respiratory epithelium and spreads to monocyte-macrophages, endothelial cells, and epithelial cells in the blood, spleen, lymph nodes, lung, thymus, liver, and skin and to the mucosal surfaces of the gastrointestinal, respiratory, and genitourinary tracts. The virus-specific immune response is detectable when the rash appears. Clearance of virus is approximately coincident with fading of the rash. (SSPE, subacute sclerosing panencephalitis.)

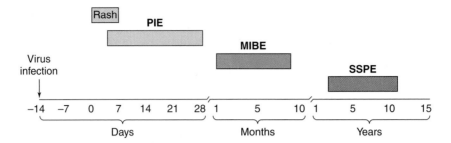

FIGURE 40–8 Timing of neurologic complications of measles. PIE, postinfectious encephalomyelitis (also called acute disseminated encephalomyelitis); MIBE, measles inclusion body encephalitis; SSPE, subacute sclerosing panencephalitis. Encephalitis occurs in about one out of every 1000 cases of measles, whereas SSPE is a rare late complication that develops in about one out of 300,000 cases. (Adapted with permission from Griffin DE, Bellini WJ: Measles virus. In: *Fields Virology,* 3rd ed. Fields BN et al [editors]. Lippincott-Raven, 1996.)

The prodromal phase is characterized by fever, sneezing, coughing, running nose, redness of the eyes, Koplik spots, and lymphopenia. The cough and coryza reflect an intense inflammatory reaction involving the mucosa of the respiratory tract. The conjunctivitis is commonly associated with photophobia. Koplik spots—pathognomonic for measles—are small, bluish-white ulcerations on the buccal mucosa opposite the lower molars. These spots contain giant cells and viral antigens and appear about 2 days before the rash. The fever and cough persist until the rash appears and then subside within 1–2 days. The rash, which starts on the head and then spreads progressively to the chest, the trunk, and down the limbs, appears as light pink, discrete maculopapules that coalesce to form blotches, becoming brownish in 5–10 days. The fading rash resolves with desquamation. Symptoms are most marked when the rash is at its peak but subside rapidly thereafter.

Modified measles occurs in partially immune persons, such as infants with residual maternal antibody. The incubation period is prolonged, prodromal symptoms are diminished, Koplik spots are usually absent, and rash is mild.

The most common complication of measles is otitis media (5–9% of cases).

Pneumonia is the most common life-threatening complication of measles, caused by secondary bacterial infections. This occurs in less than 10% of cases in developed countries but is much more frequent (20–80%) in developing countries. Pulmonary complications account for more than 90% of measles-related deaths. Pneumonia develops in 3–15% of adults with measles, but most cases are due to the virus itself rather than bacteria. Fatalities are rare.

Giant cell pneumonia is a serious complication in children and adults with deficiencies in cell-mediated immunity. It is believed to be due to unrestrained viral replication and has a high fatality rate.

Complications involving the central nervous system are the most serious. About 50% of children with regular measles register electroencephalographic changes. Acute encephalitis occurs in about 1:1000 cases. There is no apparent correlation between the severity of the measles and the appearance of neurologic complications. Postinfectious encephalomyelitis (acute disseminated encephalomyelitis) is an autoimmune disease associated with an immune response to myelin basic protein. The mortality rate in encephalitis associated with measles is about 10–20%. The majority of survivors have neurologic sequelae.

SSPE, the rare late complication of measles infection, occurs with an incidence of about 1:300,000 cases. The disease begins insidiously 5–15 years after a case of measles; it is characterized by progressive mental deterioration, involuntary movements, muscular rigidity, and coma. It is usually fatal within 1–3 years after onset. Patients with SSPE exhibit high titers of measles antibody in cerebrospinal fluid and serum and defective measles virus in brain cells. With the widespread use of measles vaccine, SSPE has become less common.

Immunity

There is only one antigenic type of measles virus (Table 40–2). Infection confers lifelong immunity. Most so-called second attacks represent errors in diagnosis of either the initial or the second illness.

The presence of humoral antibodies indicates immunity. However, cellular immunity appears to be essential for recovery and protection: Patients with immunoglobulin deficiencies recover from measles and resist reinfection, whereas patients with cellular immune deficiencies do very poorly when they acquire measles infections. The role of mucosal immunity in resistance to infections is unclear.

Measles immune responses are involved in disease pathogenesis. Local inflammation causes the prodromal symptoms, and specific cell-mediated immunity plays a role in development of the rash.

Measles infection causes immune suppression—most importantly in the cell-mediated arm of the immune system—but is observed to affect all components. This is the cause of the serious secondary infections and may persist for months after measles infection.

Laboratory Diagnosis

Typical measles is reliably diagnosed on clinical grounds; laboratory diagnosis may be necessary in cases of modified or atypical measles.

A. Antigen & Nucleic Acid Detection

Measles antigens can be detected directly in epithelial cells from respiratory secretions, the nasopharynx, conjunctiva, and urine. Antibodies to the nucleoprotein are useful because that is the most abundant viral protein in infected cells.

Detection of viral RNA by RT-PCR is a sensitive method that can be applied to a variety of clinical samples for measles diagnosis.

B. Isolation & Identification of Virus

Nasopharyngeal and conjunctival swabs, blood samples, respiratory secretions, and urine collected from a patient during the febrile period are appropriate sources for viral isolation. Monkey or human kidney cells or a lymphoblastoid cell line (B95-a) are optimal for isolation attempts. Measles virus grows slowly; typical cytopathic effects (multinucleated giant cells containing both intranuclear and intracytoplasmic inclusion bodies) take 7–10 days to develop (Figure 40–5). Shell vial culture tests can be completed in 2–3 days using fluorescent antibody staining to detect measles antigens in the inoculated cultures. However, virus isolation is technically difficult.

C. Serology

Serologic confirmation of measles infection depends on a fourfold rise in antibody titer between acute-phase and convalescent-phase sera or on demonstration of measles-specific IgM antibody in a single serum specimen drawn between 1 and 2 weeks after the onset of rash. ELISA, HI, and Nt tests all may be used to measure measles antibodies, though ELISA is the most practical method.

Dried blood spots and oral fluids appear to be useful alternatives to serum for detection of measles antibody in areas where serum samples are difficult to collect and handle.

The major part of the immune response is directed against the viral nucleoprotein. Patients with SSPE display an exaggerated antibody response, with titers 10- to 100-fold higher than those seen in typical convalescent sera.

Epidemiology

The key epidemiologic features of measles are as follows: The virus is highly contagious, there is a single serotype, there is no animal reservoir, inapparent infections are rare, and infection confers lifelong immunity. Prevalence and age incidence of measles are related to population density, economic and environmental factors, and the use of an effective live-virus vaccine.

Transmission occurs predominantly via the respiratory route (by inhalation of large droplets of infected secretions). Fomites do not appear to play a significant role in transmission. Hematogenous transplacental transmission can occur when measles occurs during pregnancy.

A continuous supply of susceptible individuals is required for the virus to persist in a community. A population size approaching 500,000 is necessary to sustain measles as an endemic disease; in smaller communities, the virus disappears until it is reintroduced from the outside after a critical number of nonimmune persons accumulates.

Measles is endemic throughout the world. In general, epidemics recur regularly every 2–3 years. A population's state of immunity is the determining factor; the disease will flare up when there is an accumulation of susceptible children. The severity of an epidemic is a function of the number of susceptible individuals. When the disease is introduced into isolated communities where it has not been endemic, an epidemic builds rapidly and attack rates are almost 100%. All age groups develop clinical measles, and the mortality rate may be as high as 25%.

In industrialized countries, measles occurs in 5- to 10-year-old children, whereas in developing countries it commonly infects children under 5 years of age. Measles rarely causes death in healthy people in developed countries. However, in malnourished children in developing countries where adequate medical care is unavailable, measles is a leading cause of infant mortality. The World Health Organization estimated in 2005 that there were 30–40 million measles cases and 530,000 deaths annually worldwide. Measles is the fifth leading global cause of mortality among children under 5 years of age, and measles deaths occur disproportionately in Africa and Southeast Asia.

The World Health Organization and the United Nations International Children's Emergency Fund (UNICEF) established a plan in 2005 to reduce measles mortality through immunization activities and better clinical care of cases. Between 2000 and 2007 the number of measles cases and of measles deaths were estimated to be reduced by more than two-thirds.

Measles cases occur throughout the year in temperate climates. Epidemics tend to occur in late winter and early spring.

There were 540 measles cases in the United States from 1997 to 2001, 67% of which were linked to imports (persons infected outside the United States). Over an 8-year period (1996–2004), 117 passengers with imported measles cases were considered infectious while traveling by aircraft. Despite the highly infectious nature of the virus, only four secondary-spread cases were identified.

In 2008, 131 cases of measles occurred in the United States. Only 17 were importations; the other cases occurred largely among unvaccinated school-age children. To sustain elimination of measles transmission, vaccine coverage rates need to exceed 90%.

Treatment, Prevention, & Control

Vitamin A treatment in developing countries has decreased mortality and morbidity. Measles virus is susceptible in vitro to inhibition by ribavirin, but clinical benefits have not been proved.

A highly effective and safe attenuated live measles virus vaccine has been available since 1963. Measles vaccine is available in monovalent form and in combination with live attenuated rubella vaccine (MR), live attenuated rubella and mumps vaccines (MMR), and live attenuated varicella vaccine (MMRV). However, because of failure to vaccinate children and because of infrequent cases of vaccine failure, measles has not been eliminated. The vaccine has reduced indigenous measles in the United States from prevaccine levels of more than 500,000 cases annually to only 37 cases in 2004.

Mild clinical reactions (fever or mild rash) will occur in 2–5% of vaccinees, but there is little or no virus excretion and no transmission. Immunosuppression occurs as with measles, but it is transient and clinically insignificant. Antibody titers tend to be lower than after natural infection, but studies have shown that vaccine-induced antibodies persist for up to 33 years, indicating that immunity is probably lifelong.

It is recommended that all children, health care workers, and international travelers be vaccinated. Contraindications to vaccination include pregnancy, allergy to eggs or neomycin, immune compromise (except that due to infection with human immunodeficiency virus), and recent administration of immunoglobulin.

The use of killed measles virus vaccine was discontinued by 1970, as certain vaccinees became sensitized and developed severe atypical measles when infected with wild virus.

Quarantine is not effective as a control measure because transmission of measles occurs during the prodromal phase.

HENDRA VIRUS & NIPAH VIRUS INFECTIONS

Two zoonotic paramyxoviruses that represent a new genus (*Henipavirus*) were recognized in the late 1990s in disease outbreaks in Australasia (Table 40–2). An outbreak of severe encephalitis in Malaysia in 1998 and 1999 was caused by Nipah virus. There was a high mortality rate (>35%) among more than 250 cases; a few survivors had persistent neurologic deficits. It appeared that the infections were caused by direct viral transmission from pigs to humans. Some patients (<10%) may develop late onset encephalitis months to several years after the initial Nipah virus infection. Hendra virus—an equine virus—has caused many horse fatalities and a few human fatalities in Australia.

Fruit bats (flying foxes) are the natural host for both Nipah and Hendra viruses. Ecologic changes, including land use and animal husbandry practices, are probably the reason for the emergence of these two infectious diseases.

Both viruses are of public health concern because of their high mortality, wide host range, and ability to jump species barriers. They are classified as Biosafety Level 4 pathogens. No vaccines are available.

RUBELLA (GERMAN MEASLES) VIRUS INFECTIONS

Rubella (German measles; 3-day measles) is an acute febrile illness characterized by a rash and lymphadenopathy that affects children and young adults. It is the mildest of common viral exanthems. However, infection during early pregnancy may result in serious abnormalities of the fetus, including congenital malformations and mental retardation. The consequences of rubella in utero are referred to as the congenital rubella syndrome.

Classification

Rubella virus, a member of the **Togaviridae** family, is the sole member of the genus *Rubivirus*. Although its morphologic features and physicochemical properties place it in the togavirus group, rubella is not transmitted by arthropods. Togavirus structure and replication are described in Chapter 38.

There is significant sequence diversity among rubella virus isolates. They are currently classified into two distantly related groups (clades) and nine genotypes.

For clarity in presentation, postnatal rubella and congenital rubella infections are described separately.

POSTNATAL RUBELLA

Pathogenesis & Pathology

Neonatal, childhood, and adult infections occur through the mucosa of the upper respiratory tract. Rubella has an incubation period of about 12 days or longer. Initial viral replication probably occurs in the respiratory tract, followed by multiplication in the cervical lymph nodes. Viremia develops after 7–9 days and lasts until the appearance of antibody on about day 13–15. The development of antibody coincides with the appearance of the rash, suggesting an immunologic basis for the rash. After the rash appears, the virus remains detectable only in the nasopharynx, where it may persist for several weeks (Figure 40–9). In 20–50% of cases, primary infection is subclinical.

Clinical Findings

Rubella usually begins with malaise, low-grade fever, and a morbilliform rash appearing on the same day. The rash starts on the face, extends over the trunk and extremities, and rarely lasts more than 3 days. No feature of the rash is pathognomonic for rubella. Unless an epidemic occurs, the disease is difficult to diagnose clinically, as the rash caused by other viruses (eg, enteroviruses) is similar.

Transient arthralgia and arthritis are commonly seen in adults, especially women. Despite certain similarities, rubella arthritis is not etiologically related to rheumatoid arthritis. Rare complications include thrombocytopenic purpura and encephalitis.

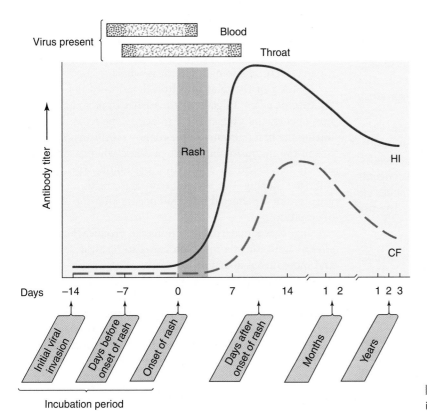

FIGURE 40-9 Natural history of primary rubella infection: virus production and antibody responses.

Immunity

Rubella antibodies appear in the serum of patients as the rash fades and the antibody titer rises rapidly over the next 1–3 weeks. Much of the initial antibody consists of IgM antibodies, which generally do not persist beyond 6 weeks after the illness. IgM rubella antibodies found in a single serum sample obtained 2 weeks after the rash give evidence of recent rubella infection. IgG rubella antibodies usually persist for life.

One attack of the disease confers lifelong immunity, as only one antigenic type of the virus exists. Because of the nondescript nature of the rash, a history of "rubella" is not a reliable index of immunity. Immune mothers transfer antibodies to their offspring, who are then protected for 4–6 months.

Laboratory Diagnosis

Clinical diagnosis of rubella is unreliable because many viral infections produce symptoms similar to those of rubella. Certain diagnosis rests on specific laboratory studies (isolation of virus, detection of viral RNA, or evidence of seroconversion).

A. Isolation & Identification of Virus

Nasopharyngeal or throat swabs taken 6 days before and after onset of rash are a good source of rubella virus. Various cell lines of monkey or rabbit origin may be used. Rubella produces a rather inconspicuous cytopathic effect in most of the cell lines. Using cells cultured in shell vials, viral antigens can be detected by immunofluorescence 3–4 days postinoculation.

B. Nucleic Acid Detection

RT-PCR can be used to detect rubella virus nucleic acid directly in clinical samples or in cell cultures used for virus isolation. Molecular typing can identify virus subtypes and genotypes and is useful in surveillance studies. Throat swabs are appropriate samples for molecular typing.

C. Serology

The HI test is a standard serologic test for rubella. However, serum must be pretreated to remove nonspecific inhibitors before testing. ELISA tests are preferred because serum pretreatment is not required and they can be adapted to detect specific IgM.

Detection of IgG is evidence of immunity, as there is only one serotype of rubella virus. To accurately confirm a recent rubella infection (critically important in the case of a pregnant woman), either a rise in antibody titer must be demonstrated between two serum samples taken at least 10 days apart or rubella-specific IgM must be detected in a single specimen.

Accurate serologic testing for rubella antibodies is so important that various diagnostic kits in a variety of formats are commercially available. Most individuals are unable to assess their rubella immunity status reliably, because subclinical infections are common and rashes induced by other viruses may be mistaken for rubella.

Epidemiology

Rubella is worldwide in distribution. Infection occurs throughout the year with a peak incidence in the spring. Epidemics occur every 6–10 years, with explosive pandemics every 20–25 years. Infection is transmitted by the respiratory route, but rubella is not as contagious as measles.

A worldwide rubella epidemic occurred in 1962–1965. There were over 12 million cases in the United States, resulting in 2000 cases of encephalitis, over 11,000 fetal deaths, 2000 neonatal deaths, and 20,000 infants born with congenital rubella syndrome. The economic impact of this epidemic in the United States was estimated at $1.5 billion. The use of rubella vaccine eliminated both epidemic and endemic rubella in the United States by 2005. A program is underway to also eliminate rubella and rubella congenital syndrome from Central and South America.

Treatment, Prevention, & Control

Rubella is a mild, self-limited illness, and no specific treatment is indicated.

Laboratory-proved rubella in the first 3–4 months of pregnancy is almost uniformly associated with fetal infection. Immune globulin intravenous (IGIV) injected into the mother does not protect the fetus against rubella infection because it is usually not given early enough to prevent viremia.

Attenuated live rubella vaccines have been available since 1969. The vaccine is available as a single antigen or combined with measles and mumps vaccine. The primary purpose of rubella vaccination is to prevent congenital rubella infections. The vaccine virus multiplies in the body and is shed in small amounts, but it does not spread to contacts. Vaccinated children pose no threat to mothers who are susceptible and pregnant. In contrast, nonimmunized children can bring home wild virus and spread it to susceptible family contacts. The vaccine induces lifelong immunity in at least 95% of recipients.

The vaccine is safe and causes few side effects in children. In adults, the only significant side effects are transient arthralgia and arthritis in about one-fourth of vaccinated women.

Vaccination in the United States decreased the incidence of rubella from about 70,000 cases in 1969 to less than 10 in 2004, cases that occurred predominantly among persons born outside the United States. The virus was subsequently declared eliminated from the United States. Cost-benefit studies in both developed and developing countries have shown that the benefits of rubella vaccination outweigh the costs.

CONGENITAL RUBELLA SYNDROME

Pathogenesis & Pathology

Maternal viremia associated with rubella infection during pregnancy may result in infection of the placenta and fetus. Only a limited number of fetal cells become infected. The growth rate of infected cells is reduced, resulting in fewer numbers of cells in affected organs at birth. The infection may lead to deranged and hypoplastic organ development, resulting in structural anomalies in the newborn.

Timing of the fetal infection determines the extent of teratogenic effect. In general, the earlier in pregnancy infection occurs, the greater the damage to the fetus. Infection during the first trimester of pregnancy results in abnormalities in the infant in about 85% of cases, whereas detectable defects are found in about 16% of infants who acquired infection during the second trimester. Birth defects are uncommon if maternal infection occurs after the 20th week of gestation.

Inapparent maternal infections can produce these anomalies as well. Rubella infection can also result in fetal death and spontaneous abortion.

Intrauterine infection with rubella is associated with chronic persistence of the virus in the newborn. At birth, virus is easily detectable in pharyngeal secretions, multiple organs, cerebrospinal fluid, urine, and rectal swabs. Viral excretion may last for 12–18 months after birth, but the level of shedding gradually decreases with age.

Clinical Findings

Rubella virus has been isolated from many different organs and cell types from infants infected in utero, and rubella-induced damage is similarly widespread.

Clinical features of congenital rubella syndrome may be grouped into three broad categories: (1) transient effects in infants, (2) permanent manifestations that may be apparent at birth or become recognized during the first year, and (3) developmental abnormalities that appear and progress during childhood and adolescence.

The classic triad of congenital rubella consists of cataracts, cardiac abnormalities, and deafness. Infants may also display transient symptoms of growth retardation, rash, hepatosplenomegaly, jaundice, and meningoencephalitis.

Central nervous system involvement is more global. The most common developmental manifestation of congenital rubella is moderate to profound mental retardation. Problems with balance and motor skills develop in preschool children. Severely affected infants may require institutionalization.

Progressive rubella panencephalitis, a rare complication that develops in the second decade of life in children with congenital rubella, is a severe neurologic deterioration that inevitably progresses to death.

Immunity

Normally, maternal rubella antibody in the form of IgG is transferred to infants and is gradually lost over a period of 6 months. Demonstration of rubella antibodies of the IgM class in infants is diagnostic of congenital rubella. As IgM antibodies do not cross the placenta, their presence indicates that they must have been synthesized by the infant in utero. Children with congenital rubella exhibit impaired cell-mediated immunity specific for rubella virus.

Treatment, Prevention, & Control

There is no specific treatment for congenital rubella. It can be prevented by childhood immunization with rubella vaccine to ensure that women of childbearing age are immune.

REVIEW QUESTIONS

1. A 4-year-old boy develops an acute febrile illness. His pediatrician diagnoses mumps. The organ most commonly exhibiting signs of mumps is
   - (A) Lungs
   - (B) Ovary
   - (C) Parotid glands
   - (D) Skin
   - (E) Testes

2. The paramyxoviruses include the most important causes of respiratory infections in infants and young children. Which of the following is not characteristic of paramyxoviruses?
   - (A) Genome is negative-sense RNA
   - (B) Envelope contains a glycoprotein with fusion activity
   - (C) Paramyxoviruses do not undergo genetic reassortment
   - (D) Replication cycle occurs in cytoplasm of susceptible cells
   - (E) Genome is segmented

3. A 3-month-old infant developed a respiratory illness that the pediatrician diagnosed as bronchiolitis. The most likely cause of the disease is
   - (A) Parainfluenza virus type 4
   - (B) Respiratory syncytial virus
   - (C) Influenza virus
   - (D) Adenovirus type 7
   - (E) Measles virus

4. Several paramyxoviruses can cause pneumonia in infants or children. For which of the following paramyxoviruses is there an effective vaccine available that would prevent pneumonia?
   - (A) Parainfluenza virus type 1
   - (B) Measles virus
   - (C) Respiratory syncytial virus
   - (D) Mumps virus
   - (E) Metapneumovirus

5. A 27-year-old woman who is 2 months pregnant develops fever, malaise, and arthralgia. A fine maculopapular rash appears on her face, trunk, and extremities. Rubella is diagnosed, and there is concern that the fetus will be infected, resulting in the congenital rubella syndrome. Which of the following statements about this syndrome is correct?
   - (A) The disease can be prevented by vaccination of school-age children with measles vaccine
   - (B) Congenital abnormalities occur when a nonimmune pregnant woman is infected at any time during pregnancy
   - (C) Deafness is a common defect associated with congenital rubella syndrome
   - (D) Only rare strains of rubella virus are teratogenic
   - (E) None of the above

6. A 5-year-old child develops a low-grade fever, coryza, conjunctivitis, and Koplik spots. The physician can conclude that
   - (A) The child has probably not been successfully vaccinated with the MMR vaccine
   - (B) The child's pregnant mother is at risk of becoming infected and her unborn child developing congenital abnormalities, including mental retardation
   - (C) A rash will soon develop on the child's face and will last only 2–3 days
   - (D) Treatment of the child with the antiviral drug ribavirin should be initiated immediately to minimize the chance of development of acute encephalitis

7. Parainfluenza viruses are ubiquitous and cause respiratory illnesses in people of all ages. However, reinfections with parainfluenza viruses are common because
   - (A) Many antigenic types of parainfluenza viruses exist and exposure to new strains results in new infections
   - (B) Infections in the respiratory tract do not elicit a systemic immune response
   - (C) Limited virus replication occurs which fails to stimulate antibody production
   - (D) Secretory IgA antibody in the nose is short-lived, disappearing a few months after infection

8. A 20-month-old boy had an illness characterized by fever, irritability, conjunctivitis, and a brick-red rash initially on the face but spreading downward and outward. At age 9 years the boy had the gradual onset of severe, generalized neurologic deterioration. SSPE was diagnosed. Which of the following statements about SSPE is correct?
   - (A) Defective varicella-zoster virus is present in brain cells
   - (B) High titers of measles antibody are found in cerebrospinal fluid
   - (C) The incidence of the disease is rising since the introduction of MMR vaccine
   - (D) Rapidly progressive deterioration of brain function occurs
   - (E) The disease is a rare, late complication of rubella infection

9. Which of the following paramyxoviruses has an HN surface glycoprotein lacking hemagglutinin activity?
   - (A) Measles virus
   - (B) Mumps virus
   - (C) Parainfluenza virus type 1
   - (D) Respiratory syncytial virus
   - (E) Rubella virus

10. A 3-year-old girl develops an acute respiratory virus infection that requires hospitalization. Ribavirin therapy is considered. Ribavirin is approved for treatment of which of the following situations?
    - (A) Lower respiratory tract disease due to respiratory syncytial virus in infants
    - (B) Congenital rubella syndrome
    - (C) Aseptic meningitis due to mumps infection
    - (D) Pneumonia caused by measles virus in adults
    - (E) Encephalitis related to Nipah virus
    - (F) All of the above

11. RT-PCR assays are useful in diagnosis of paramyxovirus infections. Which of the following statements about RT-PCR is not accurate?
    - (A) More sensitive assay than virus isolation
    - (B) Can identify virus strains

(C) More rapid assay than antigen detection

(D) Can provide data about genetic variation for molecular epidemiology studies

(E) More specific assay for parainfluenza viruses than serology

Answers

1. C	4. B	7. D	10. A
2. E	5. C	8. B	11. C
3. B	6. A	9. D	

REFERENCES

Calisher CH et al: Bats prove to be rich reservoirs for emerging viruses. Microbe 2008;3:521.

Delgado MF et al: Lack of antibody affinity maturation due to poor Toll-like receptor stimulation leads to enhanced respiratory syncytial virus disease. Nature Med 2009;15:34. [PMID: 19079256]

Eaton BT, Broder CC, Middleton D, Wang LF: Hendra and Nipah viruses: Different and dangerous. Nature Rev Microbiol 2006;4:23. [PMID: 16357858]

Falsey AR et al: Human metapneumovirus infections in young and elderly adults. J Infect Dis 2003;187:785. [PMID: 12599052]

Hall CB: Respiratory syncytial virus and parainfluenza virus. N Engl J Med 2001;344:1917. [PMID: 11419430]

Henrickson KJ: Parainfluenza viruses. Clin Microbiol Rev 2003;16:242. [PMID: 12692097]

Kahn JS: Epidemiology of human metapneumovirus. Clin Microbiol Rev 2006;19:546. [PMID: 16847085]

Lamb RA, Parks GD: Paramyxoviridae: The viruses and their replication. In: *Fields Virology*, 5th ed. Knipe DM et al (editors). Lippincott Williams & Wilkins, 2007.

Mahony JB: Detection of respiratory viruses by molecular methods. Clin Microbiol Rev 2008;21:716. [PMID: 18854489]

Mumps virus vaccines. WHO position paper. Wkly Epidemiol Rec 2007;82:51.

Coronaviruses

41

Coronaviruses are large, enveloped RNA viruses. The human coronaviruses cause common colds and have been implicated in gastroenteritis in infants. A novel coronavirus was identified as the cause of a worldwide outbreak of a severe acute respiratory syndrome (SARS) in 2003. Animal coronaviruses cause diseases of economic importance in domestic animals. Coronaviruses of lower animals establish persistent infections in their natural hosts. The human viruses are difficult to culture and therefore are more poorly characterized.

PROPERTIES OF CORONAVIRUSES

Important properties of the coronaviruses are listed in Table 41–1.

Structure & Composition

Coronaviruses are enveloped, 120- to 160-nm particles that contain an unsegmented genome of single-stranded positive-sense RNA (27–32 kb), the largest genome among RNA viruses. The genomes are polyadenylated at the 3′ end. Isolated genomic RNA is infectious. The helical nucleocapsid is 9–11 nm in diameter. There are 20-nm-long club- or petal-shaped projections that are widely spaced on the outer surface of the envelope, suggestive of a solar corona (Figure 41–1). The viral structural proteins include a 50–60 kDa phosphorylated nucleocapsid (N) protein, a 20–35 kDa membrane (M) glycoprotein that serves as a matrix protein embedded in the envelope lipid bilayer and interacting with the nucleocapsid, and the spike (S; 180–220 kDa) glycoprotein that makes up the petal-shaped peplomers. Some viruses, including human coronavirus OC43, contain a third glycoprotein (HE; 65 kDa) that causes hemagglutination and has acetylesterase activity.

The genome organizations of representative coronaviruses are shown in Figure 41–2. The gene order for the proteins encoded by all coronaviruses is Pol-S-E-M-N-3′. Several open reading frames encoding nonstructural proteins and the HE protein differ in number and gene order among coronaviruses. The SARS virus contains a comparatively large number of interspersed genes for nonstructural proteins at the 3′ end of the genome.

Classification

The Coronaviridae is one of two families, along with Arteriviridae, within the order Nidovirales. Characteristics used to classify Coronaviridae include particle morphology, unique RNA replication strategy, genome organization, and nucleotide sequence homology. There are two genera in the Coronaviridae family: *Coronavirus* and *Torovirus*. The toroviruses are widespread in ungulates and appear to be associated with diarrheas.

There seem to be two serogroups of human coronaviruses, represented by strains 229E and OC43. The novel coronavirus recovered in 2003 from patients with SARS is in the same group (group 2) as OC43. Coronaviruses of domestic animals and rodents are included in these two groups. There is a third distinct antigenic group which contains the avian infectious bronchitis virus of chickens. There appears to be significant antigenic heterogeneity among viral strains within a major antigenic group (ie, 229E-like). Cross-reactions occur between some human and some animal strains. Some strains have hemagglutinins.

TABLE 41–1 Important Properties of Coronaviruses

Virion: Spherical, 120–160 nm in diameter, helical nucleocapsid

Genome: Single-stranded RNA, linear, nonsegmented, positive-sense, 27–32 kb, capped and polyadenylated, infectious

Proteins: Two glycoproteins and one phosphoprotein. Some viruses contain a third glycoprotein (hemagglutinin esterase)

Envelope: Contains large, widely spaced, club- or petal-shaped spikes

Replication: Cytoplasm; particles mature by budding into endoplasmic reticulum and Golgi

Outstanding characteristics:

Cause colds and SARS

Display high frequency of recombination

Difficult to grow in cell culture

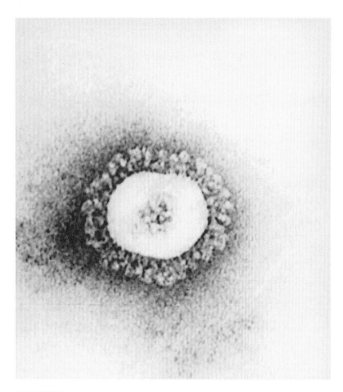

FIGURE 41–1 Human coronavirus OC43. Note the characteristic large, widely spaced spikes that form a "corona" around the virion (297,000×). (Courtesy of FA Murphy and EL Palmer.)

Viruses can also be placed into the same groups based on genome sequence analysis.

Coronavirus Replication

Because human coronaviruses do not grow well in cell culture, details of viral replication have come from studies with mouse hepatitis virus, which is closely related to human strain OC43 (Figure 41–3). The replication cycle takes place in the cytoplasm of cells.

The virus attaches to receptors on target cells by the glycoprotein spikes on the viral envelope (either by S or HE). The receptor for human coronavirus 229E is aminopeptidase N, whereas a functional receptor for the SARS virus is angiotensin-converting enzyme 2. Multiple isoforms of the carcinoembryonic antigen-related glycoprotein family serve as receptors for mouse coronavirus. The particle is then internalized, probably by absorptive endocytosis. The S glycoprotein may cause fusion of the viral envelope with the cell membrane.

The first event after uncoating is translation of the viral genomic RNA to produce a virus-specific RNA-dependent RNA polymerase. The viral polymerase transcribes a full-length complementary (minus-strand) RNA that serves as the template for a nested set of five to seven subgenomic mRNAs. Only the 5′ terminal gene sequence of each mRNA is translated. Full-length genomic RNA copies are also transcribed off the complementary RNA. As each subgenomic mRNA

is translated into a single polypeptide, polyprotein precursors are not common in coronavirus infections, although the genomic RNA encodes a large polyprotein that gets processed to yield the viral RNA polymerase.

Newly synthesized genomic RNA molecules interact in the cytoplasm with the nucleocapsid protein to form helical nucleocapsids. There is a preferred binding site for N protein within the leader RNA. The nucleocapsids bud through membranes of the rough endoplasmic reticulum and the Golgi apparatus in areas that contain the viral glycoproteins. Mature virions may then be transported in vesicles to the cell periphery for exit or may wait until the cell dies to be released.

Virions are apparently not formed by budding at the plasma membrane. Large numbers of particles may be seen on the exterior of infected cells and are presumably adsorbed to it after virion release. Certain coronaviruses induce cell fusion; this is mediated by the S glycoprotein and requires pH 6.5 or higher. Some coronaviruses establish persistent infections of cells rather than being cytocidal.

Coronaviruses exhibit a high frequency of mutation during each round of replication, including the generation of a high incidence of deletion mutations. Coronaviruses undergo a high frequency of recombination during replication; this is unusual for an RNA virus with a nonsegmented genome and may contribute to the evolution of new virus strains.

CORONAVIRUS INFECTIONS IN HUMANS

Pathogenesis

Coronaviruses tend to be highly species-specific. Little is known about the pathogenesis of coronavirus disease in humans. Most of the known animal coronaviruses display a tropism for epithelial cells of the respiratory or gastrointestinal tract. Coronavirus infections in vivo may be disseminated, such as with mouse hepatitis virus, or localized. Coronavirus infections in humans usually remain limited to the upper respiratory tract.

In contrast, the outbreak of SARS in 2003 was characterized by serious respiratory illness, including pneumonia and progressive respiratory failure. Virus can also be detected in other organs, including kidney, liver, and small intestine, and in stool. The SARS virus probably originated in a nonhuman host, most likely bats, was amplified in palm civets, and was transmitted to humans in live animal markets. Chinese horseshoe bats are natural reservoirs of SARS-like coronaviruses. In rural regions of southern China, where the outbreak began, people, pigs, and domestic fowl live close together, and there is widespread use of wild species for food and traditional medicine—conditions that promote the emergence of new viral strains.

Coronaviruses are suspected of causing some gastroenteritis in humans. There are several animal models for enteric coronaviruses, including porcine transmissible gastroenteritis virus (TGEV). Disease occurs in young animals and is

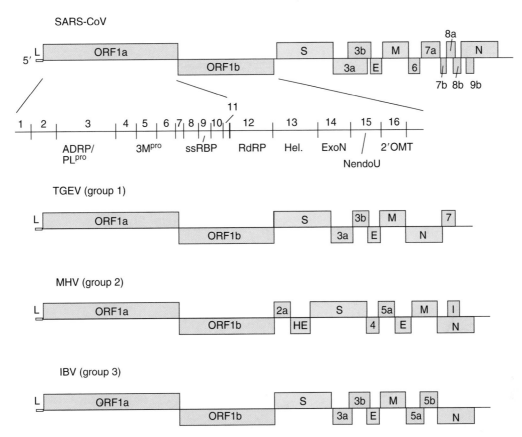

FIGURE 41–2 Genomic organization of coronaviruses. The SARS coronavirus (SARS-CoV) genome is about 29.7 kb. The genomes for porcine TGEV, mouse hepatitis virus (MHV), and avian infectious bronchitis virus (IBV) are approximately 28.5, 31.2, and 27.6 kb. Boxes shaded in yellow represent open reading frames (ORFs) encoding structural proteins; other boxes encode nonstructural proteins. The separate ORFs within each gene are translated from a single mRNA species. The ORF1 cleavage products are designated nsp1–16. S, spike; E, envelope; M, trans-membrane; N, nucleocapsid; ADRP, adenosine diphosphate-ribose 1′-phosphatase; PL^pro, papainlike cysteine proteinase; M^pro, the main cysteine proteinase (also called 3CL^pro); ssRBP, single-strand RNA-binding protein; RdRP, RNA-dependent RNA polymerase; Hel., superfamily 1 helicase; ExoN, 3′ → 5′ exonuclease; NendoU, poly(U)-specific endoribonuclease; 2′OMT, S-adenosylmethionine-dependent ribose 2′-O-methyltransferase. (Reproduced with permission from Lai MMC, Perlman S, Anderson LJ: Coronaviridae. In: *Fields Virology*, 5th ed. Knipe DM et al [editors]. Lippincott Williams & Wilkins, 2007.)

marked by epithelial cell destruction and loss of absorptive capacity. It is of interest that a novel porcine respiratory coronavirus (PRCV) appeared in Europe in the 1980s and caused widespread epizootics in pigs. Sequence analysis showed that PRCV was derived from TGEV by a large deletion in the S1 glycoprotein.

Clinical Findings

The human coronaviruses produce "colds," usually afebrile, in adults. The symptoms are similar to those produced by rhinoviruses, typified by nasal discharge and malaise. The incubation period is from 2 days to 5 days, and symptoms usually last about 1 week. The lower respiratory tract is seldom involved, although pneumonia in military recruits has been attributed to coronavirus infection. Asthmatic children may suffer wheezing attacks, and chronic pulmonary disease in adults may exacerbate respiratory symptoms.

The SARS coronavirus causes severe respiratory disease. The incubation period averages about 6 days. Common early symptoms include fever, malaise, chills, headache, dizziness, cough, and sore throat, followed a few days later by shortness of breath. Many patients have abnormal chest radiographs. Some cases progress rapidly to acute respiratory distress, requiring ventilatory support. Death from progressive respiratory failure occurs in almost 10% of cases, with the death rate highest among the elderly.

Clinical features of coronavirus-associated enteritis have not been clearly described. They appear to be similar to those of rotavirus infections.

Immunity

As with other respiratory viruses, immunity develops but is not absolute. Immunity against the surface projection antigen is probably most important for protection. Resistance to reinfection may last several years, but reinfections with similar strains are common.

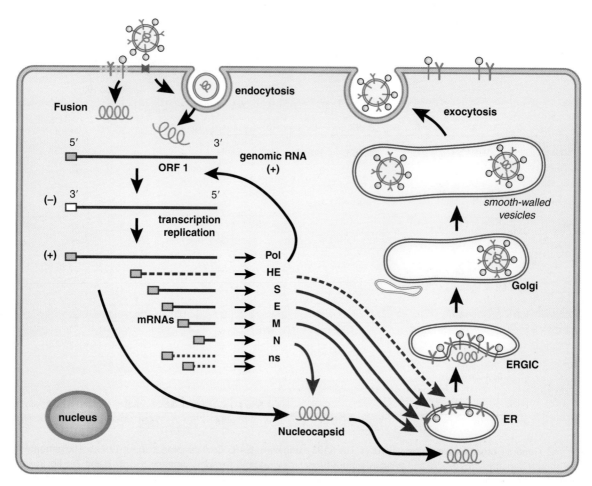

FIGURE 41–3 Coronavirus replication cycle. Virions bind to specific receptor glycoproteins or glycans via the spike protein. Penetration and uncoating occur by S protein-mediated fusion of the viral envelope with the plasma membrane or endosomal membranes. Gene 1 of viral genomic RNA is translated into a polyprotein, which is processed to yield the transcriptase–replicase complex. Genomic RNA is used as a template to synthesize negative-stranded RNAs, which are used to synthesize full-length genomic RNA and subgenomic mRNAs. Each mRNA is translated to yield only the protein encoded by the 5′ end of the mRNA, including nonstructural proteins. The N protein and newly synthesized genomic RNA assemble to form helical nucleocapsids. Membrane glycoprotein M is inserted in the endoplasmic reticulum (ER) and anchored in the Golgi apparatus. Nucleocapsid (N plus genomic RNA) binds to M protein at the budding compartment (ERGIC). E and M proteins interact to trigger the budding of virions, enclosing the nucleocapsid. S and HE glycoproteins are glycosylated and trimerized, associate with M protein, and are incorporated into the maturing virus particles. Virions are released by exocytosis-like fusion of vesicles with the plasma membrane. Virions may remain adsorbed to the plasma membranes of infected cells. The entire cycle of coronavirus replication occurs in the cytoplasm. (Reproduced with permission from Lai MMC, Perlman S, Anderson LJ: Coronaviridae. In: *Fields Virology*, 5th ed. Knipe DM et al [editors]. Lippincott Williams & Wilkins, 2007.)

Most patients (>95%) with SARS developed an antibody response to viral antigens detectable by a fluorescent antibody test or ELISA. It was important that the convalescent serum be collected more than 28 days after symptom onset.

Laboratory Diagnosis

A. Antigen & Nucleic Acid Detection

Coronavirus antigens in cells in respiratory secretions may be detected using the ELISA test if a high-quality antiserum is available. Enteric coronaviruses can be detected by examination of stool samples by electron microscopy. Polymerase chain reaction (PCR) assays are useful to detect coronavirus nucleic acid in respiratory secretions and in stool samples. SARS virus RNA was detectable in plasma by PCR, with viremia most readily detectable between days 4 and 8 of infection.

B. Isolation & Identification of Virus

Isolation of human coronaviruses in cell culture has been difficult. However, the SARS virus was recovered from oropharyngeal specimens using Vero monkey kidney cells.

C. Serology

Because of the difficulty of virus isolation, serodiagnosis using acute and convalescent sera is the practical means of confirming coronavirus infections. ELISA and hemagglutination tests may be used. Serologic diagnosis of infections with strain 229E is possible using a passive hemagglutination test in which red cells coated with coronavirus antigen are agglutinated by antibody-containing sera.

Epidemiology

Coronaviruses are distributed worldwide. They are a major cause of respiratory illness in adults during some winter months when the incidence of colds is high, but the isolation of rhinoviruses or other respiratory viruses is low. They tend to be associated with well-defined outbreaks.

It is estimated that coronaviruses cause 15–30% of all colds. The incidence of coronavirus infections varies markedly from year to year, ranging in one 3-year study from 1% to 35%.

Antibodies to respiratory coronaviruses appear in childhood, increase in prevalence with age, and are found in more than 90% of adults. It appears that reinfection with symptoms can occur after a period of 1 year.

Coronaviruses are commonly associated with acute respiratory disease in the elderly, along with rhinoviruses, influenza virus, and respiratory syncytial virus. The frequency of coronavirus infection is estimated to be about half that of rhinoviruses and equivalent to those of the latter two viruses.

It has been shown that SARS coronavirus can be airborne in a health care setting, suggesting that airborne transmission might occur. SARS contamination of frequently touched surfaces, such as a bed table (fomites), has been observed also.

The outbreak of SARS erupted in southern China in late 2002 and, by the time it waned in mid-2003, had resulted in over 8000 cases in 29 countries, with over 800 deaths (case fatality rate of 9.6%). In almost all cases there was a history of close contact with a SARS patient or of recent travel to an area where SARS was reported. International air travel allowed SARS to spread around the world with unprecedented speed. The experience with SARS illustrated that in a globalized world, an infectious disease outbreak anywhere places every country at risk.

Interestingly, a few persons with SARS were identified as "super spreaders"; each appeared to have infected more than ten contacts. Super spreaders have been described for other diseases such as rubella, Ebola, and tuberculosis and presumably reflect a certain constellation of host, viral, and environmental factors.

Very little is known about the epidemiology of enteric coronavirus infections.

Treatment, Prevention, & Control

There is no proven treatment for coronavirus infections and no vaccine.

Control measures that were effective in stopping the spread of SARS included isolation of patients, quarantine of those who had been exposed, and travel restrictions, as well as the use of gloves, gowns, goggles, and respirators by health care workers.

REVIEW QUESTIONS

1. A 63-year-old woman develops fever, headache, malaise, myalgia, and cough. It is early in the winter respiratory virus season, and the patient's physician does not know what viruses are present in the community. Which of the following viruses is not a cause of acute respiratory disease?
   (A) Influenza virus
   (B) Adenovirus
   (C) Respiratory syncytial virus
   (D) Coronavirus
   (E) Rotavirus

2. Based on sequence analysis and serologic assays, the most likely origin of the SARS coronavirus is which of the following?
   (A) Recombination between a human and an animal coronavirus that created a new virus
   (B) Jump of an animal coronavirus into humans
   (C) Mutation of a human coronavirus that resulted in increased virulence
   (D) Acquisition of human cellular genes by a human coronavirus via recombination that allowed viral evasion of the host immune response

3. The coronavirus SARS epidemic of 2002–2003 resulted in many cases and deaths. What is the primary route of transmission of human coronaviruses?
   (A) Fecal–oral
   (B) Respiratory
   (C) Blood
   (D) Perinatal mother-to-infant
   (E) Sexual activity

4. Coronavirus infections in humans usually cause a common cold syndrome. However, a recent outbreak of SARS was characterized by pneumonia and progressive respiratory failure. The prevention or treatment of these diseases can be accomplished by
   (A) A subunit vaccine
   (B) A cold-adapted live-attenuated vaccine
   (C) The antiviral drug amantadine
   (D) Infection control measures, including isolation and wearing of protective gear
   (E) The antiviral drug acyclovir

5. An epidemic of acute respiratory virus infections occurred among the elderly residents of a nursing home. Influenza viruses and coronaviruses, which can cause serious respiratory disease in the elderly, are suspected. Which of the following characteristics is shared by these viruses?
   (A) Segmented genome
   (B) Infectious RNA genome
   (C) High frequency of recombination during replication
   (D) Single serotype infects humans
   (E) Negative-sense genome

6. The following are common characteristics of coronaviruses, *except* for one. Which is not correct?
   (A) Possess cross-reactive antigens with influenza viruses
   (B) Contain the largest genomes among RNA viruses
   (C) Can cause gastroenteritis
   (D) Are distributed worldwide

Answers

1. E	3. B	5. C
2. B	4. D	6. A

REFERENCES

Baric RS, Hu Z (editors): SARS-CoV pathogenesis and replication. Virus Res 2008;133(Issue 1). [Entire issue.]

Booth TF et al: Detection of airborne severe acute respiratory syndrome (SARS) coronavirus and environmental contamination in SARS outbreak units. J Infect Dis 2005;191:1472. [PMID: 15809906]

Cheng VCC et al: Severe acute respiratory syndrome coronavirus as an agent of emerging and reemerging infection. Clin Microbiol Rev 2007;20:660. [PMID: 17934078]

Lee N et al: A major outbreak of severe acute respiratory syndrome in Hong Kong. N Engl J Med 2003;348:1986. [PMID: 12682352]

Mahony JB: Detection of respiratory viruses by molecular methods. Clin Microbiol Rev 2008;21:716. [PMID: 18854489]

Perlman S, Dandekar AA: Immunopathogenesis of coronavirus infections: Implications for SARS. Nature Rev Immunol 2005;5:917. [PMID: 16322745]

Rota PA et al: Characterization of a novel coronavirus associated with severe acute respiratory syndrome. Science 2003;30:1394. [PMID: 12730500]

Rabies, Slow Virus Infections, & Prion Diseases

42

Many different viruses can invade the central nervous system and cause disease. This chapter discusses rabies, a viral encephalitis feared since antiquity that is still an incurable disease; slow virus infections; and transmissible spongiform encephalopathies—rare neurodegenerative disorders that are caused by unconventional agents called "prions."

RABIES

Rabies is an acute infection of the central nervous system that is almost always fatal. The virus is usually transmitted to humans from the bite of a rabid animal. Although the number of human cases is small, rabies is a major public health problem because it is widespread among animal reservoirs.

Properties of the Virus

A. Structure

Rabies virus is a rhabdovirus with morphologic and biochemical properties in common with vesicular stomatitis virus of cattle and several animal, plant, and insect viruses (Table 42–1). The rhabdoviruses are rod- or bullet-shaped particles measuring 75×180 nm (Figure 42–1). The particles are surrounded by a membranous envelope with protruding spikes, 10 nm long. The peplomers (spikes) are composed of trimers of the viral glycoprotein. Inside the envelope is a ribonucleocapsid. The genome is single-stranded, negative-sense RNA (12 kb; MW 4.6×10^6). Virions contain an RNA-dependent RNA polymerase. The particles have a buoyant density in CsCl of about 1.19 g/cm^3 and a molecular weight of $300–1000 \times 10^6$.

B. Classification

The viruses are classified in the family **Rhabdoviridae**. Rabies viruses belong to the genus *Lyssavirus*, whereas the vesicular stomatitis-like viruses are members of the genus *Vesiculovirus*. The rhabdoviruses are very widely distributed in nature, infecting vertebrates, invertebrates, and plants. Rabies is the only medically important rhabdovirus. Many of the animal rhabdoviruses infect insects, but rabies virus does not.

C. Reactions to Physical & Chemical Agents

Rabies virus survives storage at 4°C for weeks and at –70°C for years. It is inactivated by CO_2, so on dry ice it must be stored in glass-sealed vials. Rabies virus is killed rapidly by exposure to ultraviolet radiation or sunlight, by heat (1 hour at 50°C), by lipid solvents (ether, 0.1% sodium deoxycholate), by trypsin, by detergents, and by extremes of pH.

D. Virus Replication

The rhabdovirus replication cycle is shown in Figure 42–2. Rabies virus attaches to cells via its glycoprotein spikes; the nicotinic acetylcholine receptor may serve as a cellular receptor for rabies virus. The single-stranded RNA genome is transcribed by the virion-associated RNA polymerase to five mRNA species. The template for transcription is the genome RNA in the form of ribonucleoprotein (RNP) (encased in N protein and containing the viral transcriptase). The monocistronic mRNAs code for the five virion proteins: nucleocapsid (N), polymerase proteins (L, P), matrix (M), and glycoprotein (G). The genome RNP is a template for complementary positive-sense RNA, which is responsible for the generation of negative-sense progeny RNA. The same viral proteins serve as polymerase for viral RNA replication as well as for transcription. Ongoing translation is required for replication, particularly of viral N and P proteins. The newly replicated

TABLE 42–1   Important Properties of Rhabdoviruses

Virion: Bullet-shaped, 75 nm in diameter × 180 nm in length
Composition: RNA (4%), protein (67%), lipid (26%), carbohydrate (3%)
Genome: Single-stranded RNA, linear, nonsegmented, negative-sense, MW 4.6 million, 12 kb
Proteins: Five major proteins; one is the envelope glycoprotein
Envelope: Present
Replication: Cytoplasm; virions bud from plasma membrane
Outstanding characteristics: Wide array of viruses with broad host range Group includes the deadly rabies virus

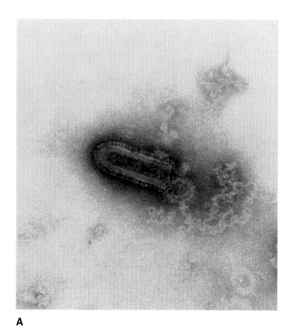

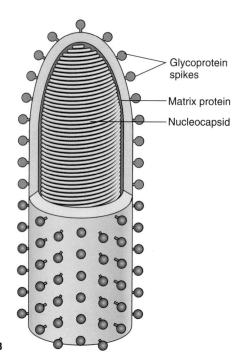

FIGURE 42–1 Structure of rhabdoviruses. **A**: Electron micrograph of bullet-shaped particle typical of the rhabdovirus family (100,000×). Shown here is vesicular stomatitis virus negatively stained with potassium phosphotungstate. (Courtesy of RM McCombs, M Benyesh-Melnick, and JP Brunschwig.) **B**: Schematic model of rabies virus showing the surface glycoprotein spikes extending from the lipid envelope that surrounds the internal nucleocapsid and the matrix protein lining the envelope. The nucleocapsid comprises the single RNA genome plus nucleoprotein and the polymerase proteins. (Reproduced with permission from Cowan MK, Talaro KP: Microbiology. A Systems Approach, 2nd ed. McGraw-Hill, 2009.)

genomic RNA associates with the viral transcriptase and nucleoprotein to form RNP cores in the cytoplasm. The particles acquire an envelope by budding through the plasma membrane. The viral matrix protein forms a layer on the inner side of the envelope, whereas the viral glycoprotein is on the outer layer and forms the spikes.

E. Animal Susceptibility & Growth of Virus

Rabies virus has a wide host range. All warm-blooded animals, including humans, can be infected. Susceptibility varies among mammalian species, ranging from very high (foxes, coyotes, wolves) to low (opossums); those with intermediate susceptibility include skunks, raccoons, and bats (Table 42–2). The virus is widely distributed in infected animals, especially in the nervous system, saliva, urine, lymph, milk, and blood. Recovery from infection is rare except in certain bats, where the virus has become peculiarly adapted to the salivary glands. Vampire bats may transmit the virus for months without themselves ever showing any signs of disease.

When freshly isolated in the laboratory, the strains are referred to as street virus. Such strains show long and variable incubation periods (usually 21–60 days in dogs) and regularly produce intracytoplasmic inclusion bodies. Serial brain-to-brain passage in rabbits yields a "fixed" virus that no longer multiplies in extraneural tissues. This fixed (or mutant) virus multiplies rapidly, and the incubation period is shortened to 4–6 days. Inclusion bodies are found only with difficulty.

F. Antigenic Properties

There is a single serotype of rabies virus. However, there are strain differences among viruses isolated from different species (raccoons, foxes, skunks, canines, bats) in different geographic areas. These viral strains can be distinguished by epitopes in the nucleoprotein and glycoprotein recognized by monoclonal antibodies as well as by specific nucleotide sequences. There are at least seven antigenic variants found in terrestrial animals and bats.

The G glycoprotein is a major factor in rabies virus neuroinvasiveness and pathogenicity. Avirulent mutants of rabies virus have been selected using certain monoclonal antibodies against the viral glycoprotein. A substitution at amino acid position 333 of the glycoprotein results in loss of virulence, indicating some essential role for that site of the protein in disease pathogenesis.

Purified spikes containing the viral glycoprotein elicit neutralizing antibody in animals. Antiserum prepared against the purified nucleocapsid is used in diagnostic immunofluorescence for rabies.

Pathogenesis & Pathology

Rabies virus multiplies in muscle or connective tissue at the site of inoculation and then enters peripheral nerves at neuromuscular junctions and spreads up the nerves to the central nervous system. However, it is also possible for rabies virus to enter the nervous system directly without local replication.

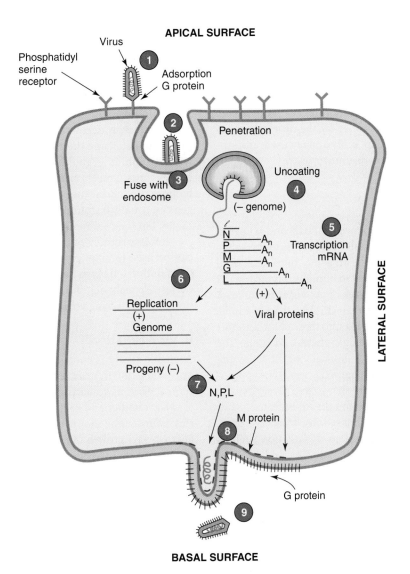

FIGURE 42–2 Steps in the replication of a rhabdovirus: **(1)** virus attachment; **(2)** penetration within an endosome; **(3)** fusion of virus with endosomal membrane, releasing core into cytoplasm; **(4)** uncoating of nucleocapsid; **(5)** viral negative-sense genomic RNA transcribed into positive-sense RNA; **(6)** positive-sense RNA serves as template for synthesis of viral genome, plus mRNA that gives rise to viral proteins; **(7)** negative-sense RNA becomes incorporated into nucleocapsids (N); **(8)** nucleocapsids join matrix protein (M) at cell surface; **(9)** budding of virus from cell surface. (Reproduced with permission from Levy JA, Fraenkel-Conrat H, Owens RA: Virology, 3rd ed. Prentice Hall, 1994.)

It multiplies in the central nervous system and progressive encephalitis develops. The virus then spreads through peripheral nerves to the salivary glands and other tissues. The organ with the highest titers of virus is the submaxillary salivary gland. Other organs where rabies virus has been found include pancreas, kidney, heart, retina, and cornea. Rabies virus has not been isolated from the blood of infected persons.

Susceptibility to infection and the incubation period may depend on the host's age, genetic background, and immune status, the viral strain involved, the amount of inoculum, the severity of lacerations, and the distance the virus has to travel from its point of entry to the central nervous system. There is a higher attack rate and shorter incubation period in persons bitten on the face or head; the lowest mortality occurs in those bitten on the legs.

TABLE 42–2   Animal Susceptibility to Rabies

Very High	High	Moderate	Low
Foxes	Hamsters	Dogs	Opossums
Coyotes	Skunks	Sheep	
Jackals	Raccoons	Goats	
Wolves	Cats	Horses	
Cotton rats	Bats	Nonhuman primates	
	Rabbits		
	Cattle		

Modified from Baer GM, Bellini WJ, Fishbein DB: Rhabdoviruses. In: *Fields Virology*. Fields BN et al (editors). Raven Press, 1990.

Rabies virus produces a specific eosinophilic cytoplasmic inclusion, the Negri body, in infected nerve cells. Negri bodies are filled with viral nucleocapsids. The presence of such inclusions is pathognomonic of rabies but is not observed in at least 20% of cases. Therefore, the absence of Negri bodies does not rule out rabies as a diagnosis. The importance of Negri bodies in rabies diagnosis has been lessened by the development of the more sensitive fluorescent antibody and reverse transcription-polymerase chain reaction diagnostic tests.

Clinical Findings

Rabies is primarily a disease of lower animals and is spread to humans by bites of rabid animals or by contact with saliva from rabid animals. The disease is an acute, fulminant, fatal encephalitis. The incubation period in humans is typically 1–2 months but may be as short as 1 week or as long as many years (up to 19 years). It is usually shorter in children than in adults. The clinical spectrum can be divided into three phases: a short prodromal phase, an acute neurologic phase, and coma. The prodrome, lasting 2–10 days, may show any of the following nonspecific symptoms: malaise, anorexia, headache, photophobia, nausea and vomiting, sore throat, and fever. Usually there is an abnormal sensation around the wound site.

During the acute neurologic phase, which lasts 2–7 days, patients show signs of nervous system dysfunction such as nervousness, apprehension, hallucinations, and bizarre behavior. General sympathetic overactivity is observed, including lacrimation, pupillary dilatation, and increased salivation and perspiration. A large fraction of patients will exhibit hydrophobia (fear of water) or aerophobia (fear when feeling a breeze). The act of swallowing precipitates a painful spasm of the throat muscles. This phase is followed by convulsive seizures or coma and death. The major cause of death is respiratory paralysis. Paralytic rabies occurs in about 20% of patients, most frequently in those infected with bat rabies virus. The disease course is slower, with some patients surviving 30 days. Recovery and survival are extremely rare.

Rabies should be considered in any case of encephalitis or myelitis of unknown cause even in the absence of an exposure history, and particularly in a person who has lived or traveled outside the United States. Most cases of rabies in the United States are in individuals with no known exposure. Because of the long incubation period, people may forget a possible exposure incident. People who contract bat rabies often have no recollection of being bitten by a bat.

The usual incubation period in dogs ranges from 3 to 8 weeks, but it may be as short as 10 days. Clinically, the disease in dogs is divided into the same three phases as human rabies.

Laboratory Diagnosis

A. Rabies Antigens or Nucleic Acids

Tissues infected with rabies virus are currently identified most rapidly and accurately by means of immunofluorescence or immunoperoxidase staining using antirabies monoclonal antibodies. A biopsy specimen is usually taken from the skin of the neck at the hairline. Impression preparations of brain or cornea tissue may be used.

A definitive pathologic diagnosis of rabies can be based on the finding of Negri bodies in the brain or the spinal cord. They are sharply demarcated, more or less spherical, and 2–10 µm in diameter, and they have a distinctive internal structure with basophilic granules in an eosinophilic matrix. Negri bodies contain rabies virus antigens and can be demonstrated by immunofluorescence. Both Negri bodies and rabies antigen can usually be found in animals or humans infected with rabies, but they are rarely found in bats.

Reverse transcription-polymerase chain reaction testing can be used to amplify parts of a rabies virus genome from fixed or unfixed brain tissue. Although unusual as a diagnostic test, sequencing of amplified products allows identification of the infecting virus strain.

B. Viral Isolation

Available tissue is inoculated intracerebrally into suckling mice. Infection in mice results in encephalitis and death. The central nervous system of the inoculated animal is examined for Negri bodies and rabies antigen. In specialized laboratories, hamster and mouse cell lines can be inoculated for rapid (2- to 4-day) growth of rabies virus; this is much faster than virus isolation in mice. An isolated virus is identified by fluorescent antibody tests with specific antiserum. Virus isolation takes too long to be useful in making a decision about whether to give vaccine.

C. Serology

Serum antibodies to rabies can be detected by immunofluorescence or Nt tests. Such antibodies develop slowly in infected persons or animals during progression of the disease but promptly after vaccination with cell-derived vaccines. Antibodies in cerebrospinal fluid are produced in rabies-infected individuals but not in response to vaccination.

D. Animal Observation

All animals considered "rabid or suspected rabid" (Table 42–3) should be sacrificed immediately for laboratory examination of neural tissues. Other animals should be held for observation for 10 days. If they show any signs of encephalitis, rabies, or unusual behavior, they should be killed humanely and the tissues examined in the laboratory. If they appear normal after 10 days, decisions must be made on an individual basis in consultation with public health officials.

Immunity & Prevention

Only one antigenic type of rabies virus is known. More than 99% of infections in humans and other mammals who develop symptoms end fatally. Survival after the onset of rabies symptoms is extremely rare. It is therefore essential

TABLE 42–3   Rabies Postexposure Prophylaxis Guide—United States, 2008

The following recommendations are only a guide. In applying them, take into account the animal species involved, the circumstances of the bite or other exposure, the vaccination status of the animal, and the presence of rabies in the region. Note: Local or state public health officials should be consulted if questions arise about the need for rabies prophylaxis.

Animal Type	Evaluation of Animal	Treatment of Exposed Person[a]
Domestic		
Dogs, cats, and ferrets	Healthy and available for 10 days of observation	None, unless animal develops symptoms of rabies
	Rabid or suspected rabid	Immediately begin prophylaxis
	Unknown (escaped)	Consult public health officials
Wild		
Skunks, raccoons, bats, foxes, coyotes, and other carnivores	Regard as rabid unless animal is proved negative by laboratory tests	Consider immediate prophylaxis
Other		
Livestock, rodents, and lagomorphs (rabbits and hares)	Consider individually. Local and state public health officials should be consulted about the need for rabies prophylaxis. Bites of squirrels, hamsters, guinea pigs, gerbils, chipmunks, rats, mice, other rodents, rabbits, and hares almost never require rabies prophylaxis.	

Source: MMWR Morb Mortal Wkly Rep 2008;57(RR-3):1.

[a]Prophylaxis consists of immediate and thorough cleansing of bites and wounds with soap and water, administration of rabies immune globulin, and vaccination.

that individuals at high risk receive preventive immunization, that the nature and risk of any exposure be evaluated, and that individuals be given postexposure prophylaxis if their exposure is believed to have been dangerous (Table 42–3). Because treatment is of no benefit after the onset of clinical disease, it is essential that postexposure treatment be initiated promptly. Postexposure rabies prophylaxis consists of the immediate and thorough cleansing of all wounds with soap and water, administration of rabies immune globulin, and a vaccination regimen.

A. Pathophysiology of Rabies Prevention by Vaccine

Presumably the virus must be amplified in muscle near the site of inoculation until the concentration of virus is sufficient to accomplish infection of the central nervous system. If immunogenic vaccine or specific antibody can be administered promptly, virus replication can be depressed and virus can be prevented from invading the central nervous system. The action of passively administered antibody is to neutralize some of the inoculated virus and lower the concentration of virus in the body, providing additional time for a vaccine to stimulate active antibody production to prevent entry into the central nervous system.

B. Types of Vaccines

All vaccines for human use contain only inactivated rabies virus. Two vaccines are available in the United States, though a number of others are in use in other countries. Both rabies vaccines available in the United States are equally safe and efficacious.

1. Human diploid cell vaccine (HDCV)—To obtain a rabies virus suspension free from nervous system and foreign proteins, rabies virus is grown in the MRC-5 human diploid cell line. The rabies virus preparation is concentrated by ultrafiltration and inactivated with β-propiolactone. No serious anaphylactic or encephalitic reactions have been reported. This vaccine has been used in the United States since 1980.

2. Rabies vaccine, adsorbed (RVA)—A vaccine made in a diploid cell line derived from fetal rhesus monkey lung cells was licensed in the United States in 1988. This vaccine virus is inactivated with β-propiolactone and concentrated by adsorption to aluminum phosphate. This vaccine is no longer available in the United States (2009).

3. Purified chick embryo cell vaccine (PCEC)—This vaccine is prepared from the fixed rabies virus strain Flury LEP grown in chicken fibroblasts. It is inactivated with β-propiolactone and further purified by zonal centrifugation. It became available in the United States in 1997.

4. Nerve tissue vaccine—This is made from infected sheep, goat, or mouse brains and is used in many parts of the world including Asia, Africa, and South America. It has a low potency per dose, and a complete treatment involves up to 23 painful injections. It causes sensitization to nerve tissue and results in postvaccinal encephalitis (an allergic disease) with substantial frequency (0.05%). Estimates of its efficacy in persons bitten by rabid animals vary from 5% to 50%.

5. Duck embryo vaccine—Duck embryo vaccine was developed to minimize the problem of postvaccinal encephalitis. The rabies virus is grown in embryonated duck eggs. Anaphylactic reactions are infrequent but the antigenicity of the vaccine is low, so that many (16–25) doses have to be given to obtain a satisfactory postexposure antibody response. It is no longer manufactured.

6. Live attenuated viruses

—Live attenuated viruses adapted to growth in chick embryos (eg, Flury strain) are used for animals but *not* for humans. Occasionally, such vaccines can cause death from rabies in injected cats or dogs. Rabies viruses grown in various animal cell cultures have also been used as vaccines for domestic animals.

A recombinant viral vaccine consisting of vaccinia virus carrying the rabies surface glycoprotein gene has successfully immunized animals following oral administration. This vaccine may prove valuable in the immunization of both wildlife reservoir species and domestic animals.

C. Types of Rabies Antibody

1. Rabies immune globulin, human (HRIG)

—HRIG is a gamma globulin prepared by cold ethanol fractionation from the plasma of hyperimmunized humans. There are fewer adverse reactions to human rabies immune globulin than to equine antirabies serum.

2. Antirabies serum, equine

—This is concentrated serum from horses hyperimmunized with rabies virus. It has been used in countries where HRIG is not available.

D. Preexposure Prophylaxis

This is indicated for persons at high risk of contact with rabies virus (research and diagnostic laboratory workers, spelunkers) or with rabid animals (veterinarians, animal control and wildlife workers). The goal is to attain an antibody level presumed to be protective by means of vaccine administration prior to any exposure. It is recommended that antibody titers of vaccinated individuals be monitored periodically and that boosters be given when required.

E. Postexposure Prophylaxis

Although few (0–5) cases of human rabies occur in the United States per year, more than 20,000 persons receive some treatment every year for possible bite wound exposure. The decision to administer rabies antibody, rabies vaccine—or both—depends on several factors: (1) the nature of the biting animal (species, state of health, domestic or wild) and its vaccination status; (2) the availability of the animal for laboratory examination (*all* bites by wild animals and bats require rabies immune globulin and vaccine); (3) the existence of rabies in the area; (4) the manner of attack (provoked or unprovoked); (5) the severity of the bite and contamination by saliva of the animal; and (6) advice from local public health officials (Table 42–3). Schedules for postexposure prophylaxis involving the administration of rabies immune globulin and vaccine are available from the Centers for Disease Control and Prevention and state public health offices.

Epidemiology

Rabies is believed to be the tenth most common cause of death in humans due to infections.

Rabies is enzootic in both wild and domestic animals. Worldwide, at least 50,000 cases of human rabies occur each year; however, rabies is grossly underreported in many countries. Almost all rabies deaths (>99%) occur in developing countries, with Asia accounting for over 90% of all rabies fatalities. In these countries, where canine rabies is still endemic, most human cases develop from bites of rabid dogs. Children aged 5–15 years are at particular risk. An estimated 10 million persons are given postexposure prophylaxis annually.

In the United States, Canada, and western Europe, where canine rabies has been controlled, dogs are responsible for very few cases. Rather, human rabies develops from bites of wild animals (especially bats, raccoons, skunks, and foxes) or occurs in travelers bitten by dogs elsewhere in the world. The most serious problem in livestock appears to be vampire bat-transmitted rabies in Latin America. The increase in wildlife rabies in the United States and some other developed countries presents a far greater risk to humans than do dogs or cats.

Primarily as a result of the successful control of rabies in domestic dogs, the incidence of human rabies in the United States declined to fewer than three persons per year during the 1990s.

Antigenic analysis with monoclonal antibodies and genotyping by nucleotide sequence analysis can distinguish rabies virus isolates from different animal reservoirs. From 1990 to 2003, there were 35 diagnosed human rabies cases in the United States, 26 (74%) of which were proved to be due to bat-associated virus. Five patients with imported rabies had dog-associated strains.

Raccoons are an important reservoir for rabies in the United States and account for over half of all reported cases of animal rabies. It is believed that raccoon rabies was introduced into the mid-Atlantic region in the 1970s, when infected raccoons were transported there from the southeastern United States to replenish hunting stocks. The raccoon rabies epizootic has spread and now covers the eastern United States into Canada.

Bats present a special problem because they may carry rabies virus while they appear to be healthy, excrete it in saliva, and transmit it to other animals and to humans. Among human rabies cases in the United States attributed to bat-associated variants, about 70% were caused by the silver-haired bat and eastern pipistrelle bat variants. However, only two cases were associated with a history of bat bite, as most bat exposures go undetected. Bat caves may contain aerosols of rabies virus and present a risk to spelunkers. Migrating fruit-eating bats exist in many countries and are a source of infection for many animals and humans. Bat rabies may be important in the initiation of terrestrial enzootics in new regions. Australia, long considered to be a rabies-free continent, was found in 1996 to harbor rabies virus in fruit bats. All persons bitten by bats must receive postexposure rabies prophylaxis.

Human-to-human rabies infection is very rare. The only documented cases involve rabies transmitted by corneal and organ transplants. One example involves corneal transplants—the corneas came from donors who died with undiagnosed central nervous system diseases, and the recipients died from rabies 50–80 days later. The first documented case involving solid organ transplants occurred in the United States in 2004. The liver and kidneys from a single donor were transplanted into three recipients, all of whom died of confirmed rabies 5–7 weeks later. Transmission likely occurred via neuronal tissue in the transplanted organs, as rabies virus is not spread in the blood. Theoretically, rabies could originate from the saliva of a patient who has rabies and exposes attending personnel, but such transmission has never been documented.

Treatment & Control

There is no successful treatment for clinical rabies. Interferons, ribavirin, and other drugs have shown no beneficial effects. Symptomatic treatment may prolong life, but the outcome is almost always fatal.

Historically, several key events have contributed to the control of human rabies: the development of a human rabies vaccine (1885), the discovery of the diagnostic Negri body (1903), the use of rabies vaccines for dogs (1940s), the addition of rabies immune globulin to human postexposure vaccination treatments (1954), the growth of rabies virus in cultured cells (1958), and the development of diagnostic fluorescent antibody tests (1959).

Preexposure vaccination is desirable for all persons who are at high risk of contact with rabid animals, such as veterinarians, animal care personnel, certain laboratory workers, and spelunkers. Persons traveling to developing countries where rabies control programs for domestic animals are not optimal should be offered preexposure prophylaxis if they plan to stay for more than 30 days. However, preexposure prophylaxis does not eliminate the need for prompt postexposure prophylaxis if an exposure to rabies occurs.

Isolated countries (eg, Great Britain) that have no indigenous rabies in wild animals can establish quarantine procedures for dogs and other pets to be imported. In countries where dog rabies exists, stray animals should be destroyed and vaccination of pet dogs and cats should be mandatory. In countries where wildlife rabies exists and where contact between domestic animals, pets, and wildlife is inevitable, all domestic animals and pets should be vaccinated.

An oral vaccinia–rabies glycoprotein recombinant virus vaccine (V-RG) proved effective at controlling rabies in foxes in Europe. Added to baits, the oral vaccine is being used to curtail rabies epizootics in wildlife in the United States.

BORNA DISEASE

Borna disease, a central nervous system disease primarily of horses and sheep in certain areas of Germany, is manifested by behavioral abnormalities usually ending in death. Inflammatory cell infiltrates are present in the brain. The disorder is immune-mediated.

Borna disease virus (BDV) is an enveloped, nonsegmented, negative-stranded RNA virus in the family **Bornaviridae** (Table 42–4). Although similar to the rhabdoviruses and paramyxoviruses, BDV is novel in that it transcribes and replicates its genome in the nucleus and uses RNA splicing for regulation of gene expression. BDV is noncytolytic and highly neurotropic; it establishes persistent infections. There is a single recognized serotype of BDV. Titers of neutralizing antibodies produced in host species are usually very low.

Many species can be infected by bornaviruses, including humans. Serologic data suggest that BDV may be associated with neuropsychiatric disorders in humans, although it remains to be established whether BDV is etiologically involved in the pathophysiology of certain human mental disorders.

SLOW VIRUS INFECTIONS & PRION DISEASES

Some chronic degenerative diseases of the central nervous system in humans are caused by "slow" or chronic, persistent infections by classic viruses. Among these are subacute sclerosing panencephalitis and progressive multifocal leukoencephalopathy. Other diseases known as transmissible spongiform encephalopathies—eg, Creutzfeldt-Jakob disease (CJD)—are caused by unconventional transmissible agents termed "prions" (Table 42–5). The progressive neurologic diseases produced by these agents may have incubation periods of years before clinical manifestations of the infections become evident (Table 42–5).

TABLE 42–4   Important Properties of Bornaviruses

Virion: Spherical, 90 nm in diameter
Genome: Single-stranded RNA, linear, nonsegmented, negative-sense, 8.9 kb, MW 3 million
Proteins: Six structural proteins
Envelope: Present
Replication: Nucleus; site of maturation not identified
Outstanding characteristics: 　Broad host range 　Neurotropic 　Cause neurobehavioral abnormalities

TABLE 42-5   Slow Virus and Prion Diseases

Disease	Agent	Hosts	Incubation Period	Nature of Disease
Diseases of humans				
Subacute sclerosing panencephalitis	Measles virus variant	Humans	2–20 years	Chronic sclerosing panencephalitis
Progressive multifocal leukoencephalopathy	Polyomavirus JCV	Humans	Years	Central nervous system demyelination
CJD	Prion	Humans, chimpanzees, monkeys	Months to years	Spongiform encephalopathy
Variant CJD[a]	Prion	Humans, cattle	Months to years	Spongiform encephalopathy
Kuru	Prion	Humans, chimpanzees, monkeys	Months to years	Spongiform encephalopathy
Diseases of animals				
Visna	Retrovirus	Sheep	Months to years	Central nervous system demyelination
Scrapie	Prion	Sheep, goats, mice, hamsters	Months to years	Spongiform encephalopathy
Bovine spongiform encephalopathy	Prion	Cattle	Months to years	Spongiform encephalopathy
Transmissible mink encephalopathy	Prion	Mink, other animals	Months	Spongiform encephalopathy
Chronic wasting disease	Prion	Mule deer, elk	Months to years	Spongiform encephalopathy

CJD, Creutzfeldt-Jakob disease.

[a] Associated with exposure to bovine spongiform encephalopathy-contaminated material.

Slow Virus Infections

A. Visna

Visna and **progressive pneumonia (maedi) viruses** are closely related agents that cause slowly developing infections in sheep. These viruses are classified as retroviruses (genus *Lentivirus*; see Chapter 44).

Visna virus infects all the organs of the body of the infected sheep; however, pathologic changes are confined primarily to the brain, lungs, and reticuloendothelial system. Inflammatory lesions develop in the central nervous system soon after infection, but there is usually a long incubation period (months to years) before observable neurologic symptoms appear. Disease progression can be either rapid (weeks) or slow (years).

Virus can be recovered for the life of the animal, but viral expression is restricted in vivo so that only minimal amounts of infectious virus are present. Antigenic variation occurs during the long-term persistent infections. Many mutations occur in the structural gene that codes for viral envelope glycoproteins. Infected animals develop antibodies to the virus.

B. Subacute Sclerosing Panencephalitis

This is a rare disease of young adults caused by measles virus, with slowly progressive demyelination in the central nervous system ending in death (see Chapter 40). Large numbers of viral nucleocapsid structures are produced in neurons and glial cells. There is restricted expression of the viral genes that encode envelope proteins, so the virus in persistently infected neural cells lacks proteins needed for the production of infectious particles. Patients with subacute sclerosing panencephalitis have high titers of antimeasles antibody except that antibody to the M protein is frequently lacking. Reduced efficiency of measles virus transcription in differentiated brain cells is important in maintaining the persistent infection that leads to subacute sclerosing panencephalitis.

C. Progressive Multifocal Leukoencephalopathy

JC virus (JCV), a member of the family **Polyomaviridae** (see Chapter 43), is the etiologic agent of progressive multifocal leukoencephalopathy, a central nervous system complication that occurs in some immunosuppressed individuals. Once exceedingly rare, the disease occurs in a significant proportion (about 5%) of patients with AIDS; however, as antiviral drugs slow the progression of human immunodeficiency virus infections, fewer patients develop this disease. Demyelination in the central nervous system of patients with progressive multifocal leukoencephalopathy results from oligodendrocyte infection by polyomaviruses.

Transmissible Spongiform Encephalopathies (Prion Diseases)

Degenerative central nervous system diseases—kuru, CJD, Gerstmann-Sträussler-Scheinker syndrome, fatal familial insomnia of humans, scrapie of sheep, transmissible encephalopathy of mink, bovine spongiform encephalopathy of cattle, and chronic wasting disease of deer—have similar pathologic features. These diseases are described as transmissible spongiform encephalopathies. The causative agents are not conventional viruses; infectivity is associated with proteinaceous material devoid of detectable amounts of nucleic acid. The term **"prion"** is used to designate this novel class of agents.

The different types of prions appear to have common mechanisms of pathogenesis. Species barriers exist for all transmissible spongiform encephalopathies, but some prions have crossed such barriers.

These agents are unusually resistant to standard means of inactivation. They are resistant to treatment with formaldehyde (3.7%), urea (8 M), dry heat, boiling, ethanol (50%), proteases, deoxycholate (5%), and ionizing radiation. However, they are sensitive to phenol (90%), household bleach, ether, NaOH (2 N), strong detergents (10% sodium dodecyl sulfate), and autoclaving (1 hour, 121°C). Guanidine thiocyanate is highly effective in decontaminating medical supplies and instruments.

There are several distinguishing hallmarks of diseases caused by these unconventional agents. Although the etiologic agent may be recoverable from other organs, the diseases are confined to the nervous system. The basic features are neurodegeneration and spongiform changes. Amyloid plaques may be present. Long incubation periods (months to decades) precede the onset of clinical illness and are followed by chronic progressive disease (weeks to years). The diseases are always fatal, with no known cases of remission or recovery. The host shows no inflammatory response and no immune response (the agents do not appear to be antigenic); no production of interferon is elicited; and there is no effect on host B cell or T cell function. Immunosuppression of the host has no effect on pathogenesis; however, chronic inflammation induced by other factors (viruses, bacteria, autoimmunity) may affect prion pathogenesis. It has been observed that prions accumulate in organs with chronic lymphocytic inflammation. When coincident with nephritis, prions were excreted in urine.

A. Scrapie

Scrapie shows marked differences in susceptibility of different breeds of animal. Susceptibility to experimentally transmitted scrapie ranges from zero to over 80% in various breeds of sheep, whereas goats are almost 100% susceptible. The transmission of scrapie to mice and hamsters, in which the incubation period is greatly reduced, has facilitated study of the disease.

Infectivity can be recovered from lymphoid tissues early in infection, and high titers of the agent are found in the brain, spinal cord, and eye (the only places where pathologic changes are observed). Prion infectivity has also been detected in milk from sheep incubating natural scrapie. Maximum titers of infectivity are reached in the brain long before neurologic symptoms appear. The disease is characterized by the development of amyloid plaques in the central nervous system of infected animals. These areas represent extracellular accumulations of protein; they stain with Congo red.

A protease-resistant protein of molecular mass 27–30 kDa can be purified from scrapie-infected brain and is designated prion protein PrP. Preparations containing only PrP and no detectable nucleic acid are infectious. PrP is derived from a larger host-encoded protein, PrP^{Sc}, that is an altered version of a normal cellular protein (PrP^C). The protein is a glycolipid-anchored membrane protein. The level of PrP^{Sc} is increased in infected brains because the protein becomes resistant to degradation. Genetic susceptibility to scrapie infection is associated with point mutations in the PrP^C gene, and mice genetically altered to be devoid of PrP^C are resistant to scrapie. A conformational model for prion replication proposes that PrP^{Sc} forms a heterodimer with PrP^C and refolds it so that it becomes like PrP^{Sc}. "Strains" of prions are speculated to reflect different conformations of PrP^{Sc}. It is still uncertain whether this protein represents the essential structural element of the infectious agent or a pathologic product that accumulates as a result of the disease; however, mouse prion protein produced in bacteria caused disease when inoculated in vivo, suggesting that prions are infectious proteins.

B. Bovine Spongiform Encephalopathy & New Variant CJD

A disease similar to scrapie, designated bovine spongiform encephalopathy (BSE), or "mad cow disease," emerged in cattle in Great Britain in 1986. This outbreak was traced to the use of cattle feed that contained contaminated bone meal from scrapie-infected sheep and BSE-infected cattle carcasses. The use of such cattle feed was prohibited in 1988. The epidemic of "mad cow disease" peaked in Great Britain in 1993. It is estimated that over 1 million cattle were infected. BSE has also been found in other European countries. In 1996, a new variant form of CJD was recognized in the United Kingdom that occurred in younger people and had distinctive pathologic characteristics similar to those of BSE. It is now accepted that the new variant forms of CJD and BSE are caused by a common agent, indicating that the BSE agent had infected humans. Through 2006, over 150 people had been diagnosed with new variant CJD in England, and most had died. A particular polymorphism in the amino acid sequence of the human prion protein seems to influence susceptibility to disease.

C. Kuru & Classic CJD

Two human spongiform encephalopathies are kuru and the classic form of CJD. Brain homogenates from patients have transmitted both diseases to nonhuman primates. Kuru occurred only in the eastern highlands of New Guinea and was spread by customs surrounding ritual cannibalism of dead relatives. Since the practice has ceased, the disease has disappeared. CJD in humans develops gradually, with progressive dementia, ataxia, and myoclonus, and leads to death in 5–12 months. Sporadic CJD occurs with a frequency of approximately one case per million population per year in the United States and Europe and involves patients over 50 years of age. The estimated incidence is less than one case per 200 million for persons under 30 years of age. However, the new variant form of CJD linked to BSE (above) has mainly affected people under the age of 30.

Two familial forms of CJD are Gerstmann-Sträussler-Scheinker syndrome and fatal familial insomnia. These diseases are rare (10–15% of CJD cases) and are due to inheritance of mutations in the PrP gene.

Iatrogenic CJD has been transmitted accidentally by contaminated growth hormone preparations from human cadaver pituitary glands, by corneal transplant, by contaminated surgical instruments, and by cadaveric human dura mater grafts used for surgical repair of head injury. It appears that recipients of contaminated dura mater grafts remain at risk of developing CJD for more than 20 years following receipt of grafts. There is currently no suggestion of CJD transmission by blood or blood products, although the potential is there.

A protein very similar to scrapie PrP^{Sc} is present in brain tissue infected with classic CJD. It has been speculated that the agent of CJD was derived originally from scrapie-infected sheep and transmitted to humans by ingestion of poorly cooked sheep brains.

D. Chronic Wasting Disease

A scrapie-like disease, designated chronic wasting disease, is found in mule deer and elk in the United States. It is laterally transmitted with high efficiency, but there is no evidence that it has been transmitted to humans. Infectivity has been detected in saliva from deer infected with chronic wasting disease.

E. Alzheimer Disease

There are some neuropathologic similarities between CJD and Alzheimer disease, including the appearance of amyloid plaques. However, the disease has not been transmitted experimentally to primates or rodents, and the amyloid material in the brains of Alzheimer patients does not contain PrP^{Sc} protein.

REVIEW QUESTIONS

1. Rabies virus is rapidly destroyed by
   (A) Ultraviolet radiation
   (B) Heating at 56°C for 1 hour
   (C) Ether treatment
   (D) Trypsin treatment
   (E) All of the above

2. Prions are readily destroyed by
   (A) Ionizing radiation
   (B) Formaldehyde
   (C) Boiling
   (D) Proteases
   (E) None of the above

3. The presence in neurons of eosinophilic cytoplasmic inclusion bodies, called Negri bodies, is characteristic of which of the following central nervous system infections?
   (A) Borna disease
   (B) Rabies
   (C) Subacute sclerosing panencephalitis
   (D) New variant Creutzfeldt-Jakob disease
   (E) Postvaccinal encephalitis

4. Which of the following statements about rabies vaccines for human use is true?
   (A) Contain live, attenuated rabies virus
   (B) Contain multiple antigenic type of rabies virus
   (C) May cause allergic encephalitis when prepared from nerve tissue
   (D) Can be used only for preexposure prophylaxis
   (E) Duck embryo vaccine is highly antigenic and only a single dose is needed

5. A 22-year-old man is a resident of a small town near London. He likes to eat beefsteak. He develops a severe progressive neurologic disease characterized by psychiatric symptoms, cerebellar signs, and dementia. Probable bovine spongiform encephalopathy (BSE) is diagnosed. New variant Creutzfeldt-Jakob disease in humans and BSE appear to be caused by the same agent. Which of the following statements is true of both diseases?
   (A) Immunosuppression of the host is a predisposing factor
   (B) It is an immune-mediated degenerative neurologic disorder
   (C) There is a long incubation period (months to years) from time of exposure to appearance of symptoms
   (D) The agent is recoverable only from the central nervous system of an infected host
   (E) The interferon response persists throughout the incubation period
   (F) There is a high-titer antibody response toward PrP^{Sc} protein of the agent

6. Rabies virus has a wide host range and the ability to infect all warm-blooded animals, including humans. Which statement about the epidemiology of human rabies is true?
   (A) Africa accounts for the majority of rabies fatalities
   (B) Dog bites cause most cases of human rabies in England
   (C) Domestic animals are the source of most human rabies in the United States
   (D) Human-to-human rabies transmission places medical personnel at serious risk
   (E) Bat rabies caused most human rabies cases in the United States in the 1990s

7. Infectious scrapie agent can be detected in amyloid plaques in infected brains of sheep and hamsters. The genome of the infectious agent is characterized by which of the following nucleic acid types?
   (A) Negative-sense, single-stranded RNA
   (B) Small interfering RNA, smallest known infectious RNA
   (C) DNA copy of RNA genome, integrated in mitochondrial DNA
   (D) Single-stranded, circular DNA
   (E) No detectable nucleic acid

8. A 49-year-old man visited a neurologist after 2 days of increasing right arm pain and paresthesias. The neurologist diagnosed an atypical neuropathy. The symptoms increased and were accompanied by hand spasms and sweating on the right side of the face and trunk. The patient was admitted to the hospital the day after developing dysphagia, hypersalivation, agitation, and generalized muscle twitching. Vital signs and blood tests were normal, but within hours the patient became confused. The consulting neurologist suspected rabies. Rabies immune globulin, vaccine,

and acyclovir were administered. The patient was placed on mechanical ventilation the following day. Renal failure developed, and the patient died 3 days later. Rabies test results were positive. The patient's wife reported the patient had suffered no bites by dogs or wild animals. The most likely explanation for treatment failure is

(A) The rabies test results were falsely positive and the patient did not have rabies

(B) Treatment was initiated after the onset of clinical symptoms of rabies

(C) The vaccine was directed against dog rabies and the patient was infected with bat rabies

(D) The rabies immune globulin should not have been administered as it interfered with the vaccine

(E) Interferons—and not the treatment regimen administered—are the treatment of choice once rabies symptoms develop

9. Which of the following animals is most commonly reported rabid in the United States?

(A) Squirrels

(B) Raccoons

(C) Rabbits

(D) Swine

(E) Rats

10. A runner reports an "unprovoked bite" from a neighborhood dog. The dog was captured by local animal control authorities, and it appears healthy. What is the appropriate action?

(A) Confine and observe the dog for 10 days for signs suggestive of rabies

(B) Begin postexposure prophylaxis of the bitten person

(C) Immediately euthanize the dog

(D) Because canine rabies has been eliminated in the United States, dog bites are no longer an indication for postexposure prophylaxis, and no further action is needed

(E) Test the dog for rabies antibody

Answers

1. E	4. C	7. E	10. A
2. E	5. C	8. B	
3. B	6. E	9. B	

REFERENCES

Beisel CE, Morens DM: Variant Creutzfeldt-Jakob disease and the acquired and transmissible spongiform encephalopathies. Clin Infect Dis 2004;38:697. [PMID: 14986255]

Compendium of animal rabies prevention and control, 2008. National Association of State Public Health Veterinarians, Inc. MMWR Morb Mortal Wkly Rep 2008;57(RR-2):1.

De Serres G et al: Bat rabies in the United States and Canada from 1950 through 2007: Human cases with and without bat contact. Clin Infect Dis 2008;46:1329. [PMID: 18419432]

Human rabies prevention—United States, 2008. Recommendations of the Advisory Committee on Immunization Practices. MMWR Morb Mortal Wkly Rep 2008;57(RR-3):1.

Jackson AC, Fu ZF: Pathogenesis of rabies—Editorial. J Neurovirol 2005;11:74.

Lyles DS, Rupprecht CE: Rhabdoviridae. In: *Fields Virology,* 5th ed. Knipe DM et al (editors). Lippincott Williams & Wilkins, 2007.

Mabbott NA, MacPherson GG: Prions and their lethal journey to the brain. Nature Rev Microbiol 2006;4:201. [PMID: 16462753]

Priola SA: How animal prions cause disease in humans. Microbe 2008;3:568.

Rabies vaccines: WHO position paper. Weekly Epidemiol Record 2007;82:425.

Rutala WA, Weber DJ: Creutzfeldt-Jakob disease: Recommendations for disinfection and sterilization. Clin Infect Dis 2001;32:1348. [PMID: 11303271]

Staeheli P et al: Epidemiology of Borna disease virus. J Gen Virol 2000;81:2123. [PMID: 10950968]

Human Cancer Viruses

Viruses are etiologic factors in the development of several types of human tumors, including two of great significance worldwide—cervical cancer and liver cancer. At least 15–20% of all human tumors worldwide have a viral cause. The viruses that have been strongly associated with human cancers are listed in Table 43–1. They include human papillomaviruses (HPVs), Epstein-Barr virus (EBV), human herpesvirus 8, hepatitis B virus, hepatitis C virus, and two human retroviruses plus several candidate human cancer viruses. New cancer-associated viruses are being discovered by the use of molecular techniques. Many viruses can cause tumors in animals, either as a consequence of natural infection or after experimental inoculation.

Animal viruses are studied to learn how a limited amount of genetic information (one or a few viral genes) can profoundly alter the growth behavior of cells, ultimately converting a normal cell into a neoplastic one. Such studies reveal insights into growth regulation in normal cells. Tumor viruses are agents that can produce tumors when they infect appropriate animals. Many studies are done using cultured animal cells rather than intact animals, because it is possible to analyze events at cellular and subcellular levels. In such cultured cells, tumor viruses can cause "transformation." However, animal studies are essential to study many steps in carcinogenesis, including complex interactions between virus and host and host responses to tumor formation.

Studies with RNA tumor viruses revealed the involvement of cellular oncogenes in neoplasia; DNA tumor viruses established a role for cellular tumor suppressor genes. These discoveries revolutionized cancer biology and provided the conceptual framework for the molecular basis of carcinogenesis.

GENERAL FEATURES OF VIRAL CARCINOGENESIS

Tenets of viral carcinogenesis are summarized in Table 43–2.

Tumor Viruses Are of Different Types

Like other viruses, tumor viruses are classified among different virus families according to the nucleic acid of their genome and the biophysical characteristics of their virions. Most recognized tumor viruses either have a DNA genome or generate a DNA provirus after infection of cells (hepatitis C virus is an exception).

DNA tumor viruses are classified among the papilloma-, polyoma-, adeno-, herpes-, hepadna-, and poxvirus groups. DNA tumor viruses encode viral oncoproteins that are important for viral replication but also affect cellular growth control pathways.

Most RNA tumor viruses belong to the retrovirus family. Retroviruses carry an RNA-directed polymerase (reverse transcriptase) that constructs a DNA copy of the RNA genome of the virus. The DNA copy (provirus) becomes integrated into the DNA of the infected host cell, and it is from this integrated DNA copy that all proteins of the virus are translated.

RNA tumor viruses are of two general types with respect to tumor induction. The highly oncogenic (direct-transforming) viruses carry an oncogene of cellular origin. The weakly oncogenic (slowly transforming) viruses do not contain an oncogene and induce leukemias after long incubation periods by indirect mechanisms. The two known cancer-causing retroviruses in humans act indirectly. Hepatitis C virus, a flavivirus, does not generate a provirus and appears to induce cancer indirectly.

Multistep Carcinogenesis

Carcinogenesis is a multistep process, ie, multiple genetic changes must occur to convert a normal cell into a malignant one. Intermediate stages have been identified and designated by terms such as "immortalization," "hyperplasia," and "preneoplastic." Tumors usually develop slowly over a long period of time. The natural history of human and animal cancers suggests a multistep process of cellular evolution, probably involving cellular genetic instability and repeated selection of rare cells with some selective growth advantage. The number of mutations underlying this process is estimated to range from five to eight. Observations suggest that activation of multiple cellular oncogenes and inactivation of tumor suppressor genes are involved in the evolution of tumors whether or not a virus is involved.

TABLE 43–1   Association of Viruses with Human Cancers[a]

Virus Family	Virus	Human Cancer
Papillomaviridae	Human papillomaviruses	Genital tumors
		Squamous cell carcinoma
		Oropharyngeal carcinoma
Herpesviridae	Epstein-Barr virus	Nasopharyngeal carcinoma
		Burkitt lymphoma
		Hodgkin disease
		B cell lymphoma
	Human herpesvirus 8	Kaposi sarcoma
Hepadnaviridae	Hepatitis B virus	Hepatocellular carcinoma
Retroviridae	Human T cell lymphoma virus	Adult T cell leukemia
	Human immunodeficiency virus	AIDS-related malignancies
Flaviviridae	Hepatitis C virus	Hepatocellular carcinoma

[a]Candidate human tumor viruses include additional types of papillomaviruses and polyomaviruses SV40, JC, and BK.

It appears that a tumor virus usually acts as a cofactor, providing only some of the steps required to generate malignant cells. Viruses are necessary—but not sufficient—for development of tumors with a viral etiology. Viruses often act as initiators of the neoplastic process and may do so by different mechanisms.

Interactions of Tumor Viruses with Their Hosts

A. Persistent Infections

The pathogenesis of a viral infection and the response of the host are integral to understanding how cancer might arise

TABLE 43–2   Tenets of Viral Carcinogenesis

1. Viruses can cause cancer in animals and humans
2. Tumor viruses frequently establish persistent infections in natural hosts
3. Host factors are important determinants of virus-induced tumorigenesis
4. Viruses are seldom complete carcinogens
5. Virus infections are more common than virus-related tumor formation
6. Long latent periods usually elapse between initial virus infection and tumor appearance
7. Viral strains may differ in oncogenic potential
8. Viruses may be either direct- or indirect-acting carcinogenic agents
9. Oncogenic viruses modulate growth control pathways in cells
10. Animal models may reveal mechanisms of viral carcinogenesis
11. Viral markers are usually present in tumor cells
12. One virus may be associated with more than one type of tumor

Reproduced with permission from Butel JS: Viral carcinogenesis: Revelation of molecular mechanisms and etiology of human disease. Carcinogenesis 2000;21:405.

from that background. The known tumor viruses establish long-term persistent infections in humans. Because of differences in individual genetic susceptibilities and host immune responses, levels of virus replication and tissue tropisms may vary among persons. Even though very few cells in the host may be infected at any given time, the chronicity of infection presents the long-term opportunity for a rare event to occur that allows survival of a cell with growth control mechanisms that are virus-modified.

B. Host Immune Responses

Viruses that establish persistent infections must avoid detection and recognition by the immune system that would eliminate the infection. Different viral evasion strategies have been identified, including restricted expression of viral genes that makes infected cells nearly invisible to the host (EBV in B cells); infection of sites relatively inaccessible to immune responses (HPV in the epidermis); mutation of viral antigens that allows escape from antibody and T cell recognition (human immunodeficiency virus [HIV]); modulation of host major histocompatibility complex class I molecules in infected cells (adenovirus, cytomegalovirus); inhibition of antigen processing (EBV); and infection and suppression of essential immune cells (HIV).

It is believed that host immune surveillance mechanisms usually eliminate the rare neoplastic cells that may arise in normal individuals infected with cancer viruses. However, if the host is immunosuppressed, cancer cells are more likely to proliferate and escape host immune control. Immunosuppressed organ transplant recipients and HIV-infected individuals are at increased risk of EBV-associated lymphomas and of HPV-related diseases. It is possible that variations in individual immune responses may contribute to susceptibility to virus-induced tumors in normal hosts.

C. Mechanisms of Action by Human Cancer Viruses

Tumor viruses mediate changes in cell behavior by means of a limited amount of genetic information. There are two general patterns by which this is accomplished: The tumor virus introduces a new "transforming gene" into the cell (direct-acting) or the virus alters the expression of a preexisting cellular gene or genes (indirect-acting). In either case, the cell loses control of normal regulation of growth processes. DNA repair pathways are frequently affected, leading to genetic instability and a mutagenic phenotype.

Viruses usually do not behave as complete carcinogens. In addition to changes mediated by viral functions, other alterations are necessary to disable the multiple regulatory pathways and checkpoints in normal cells to allow a cell to become completely transformed. There is no single mode of transformation underlying viral carcinogenesis. At the molecular level, oncogenic mechanisms by human tumor viruses are very diverse.

Cellular transformation may be defined as a stable, heritable change in the growth control of cells in culture. No set of characteristics invariably distinguishes transformed cells from their normal counterparts. In practice, transformation is recognized by the cells' acquisition of some growth property not exhibited by the parental cell type. Transformation to a malignant phenotype is recognized by tumor formation when transformed cells are injected into appropriate test animals.

Indirect-acting tumor viruses are not able to transform cells in culture.

D. Cell Susceptibility to Viral Infections & Transformation

At the cellular level, host cells are either permissive or nonpermissive for replication of a given virus. Permissive cells support viral growth and production of progeny virus; nonpermissive cells do not. Especially with the DNA viruses, permissive cells are often killed by virus replication and are not transformed unless the viral replicative cycle that results in death of the host cell is blocked in some way; nonpermissive cells may be transformed. However, there are situations in which DNA virus replication does not lyse the host cell and such cells may be transformed. Nevertheless, transformation is a rare event. A characteristic property of RNA tumor viruses is that they are not lethal for the cells in which they replicate. Cells that are permissive for one virus may be nonpermissive for another.

Not all cells from the natural host species are susceptible to viral replication or transformation or both. Most tumor viruses exhibit marked tissue specificity, a property that probably reflects the variable presence of surface receptors for the virus, the ability of the virus to cause disseminated versus local infections, or intracellular factors necessary for viral gene expression.

Some viruses are associated with a single tumor type, whereas others are linked to multiple tumor types. These differences reflect the tissue tropisms of the viruses.

E. Retention of Tumor Virus Nucleic Acid in a Host Cell

The stable genetic change from a normal to a neoplastic cell generally requires the retention of viral genes in the cell. Oftentimes but not always, this is accomplished by the integration of certain viral genes into the host cell genome. With DNA tumor viruses, a portion of the DNA of the viral genome may become integrated into the host cell chromosome. Sometimes, episomal copies of the viral genome are maintained in tumor cells. With retroviruses, the proviral DNA copy of the viral RNA is integrated in the host cell DNA. Genome RNA copies of hepatitis C virus that are not integrated are maintained in tumor cells.

In some viral systems, virus-transformed cells may release growth factors that affect the phenotype of neighboring uninfected cells, thereby contributing to tumor formation. It is also possible that as tumor cells collect genetic mutations during tumor growth, the need for the viral genes that drove tumor initiation may become unnecessary and viral markers will be lost from some cells.

RETROVIRUSES

Retroviruses contain an RNA genome and an RNA-directed DNA polymerase (reverse transcriptase). RNA tumor viruses in this family mainly cause tumors of the reticuloendothelial and hematopoietic systems (leukemias, lymphomas) or of connective tissue (sarcomas).

Important properties of the retroviruses are listed in Table 43–3.

Structure & Composition

The retrovirus genome consists of two identical subunits of single-stranded, positive-sense RNA, each 7–11 kb in size. The

TABLE 43–3 Important Properties of Retroviruses

Virion: Spherical, 80–110 nm in diameter, helical nucleoprotein within icosahedral capsid

Composition: RNA (2%), protein (about 60%), lipid (about 35%), carbohydrate (about 3%)

Genome: Single-stranded RNA, linear, positive-sense, 7–11 kb, diploid; may be defective; may carry oncogene

Proteins: Reverse transcriptase enzyme contained inside virions

Envelope: Present

Replication: Reverse transcriptase makes DNA copy from genomic RNA; DNA (provirus) integrates into cellular chromosome; provirus is template for viral RNA

Maturation: Virions bud from plasma membrane

Outstanding characteristics:

  Infections do not kill cells

  May transduce cellular oncogenes, may activate expression of cell genes

  Proviruses remain permanently associated with cells and are frequently not expressed

  Many members are tumor viruses

reverse transcriptase contained in virus particles is essential for viral replication.

Retrovirus particles contain the helical ribonucleoprotein within an icosahedral capsid that is surrounded by an outer membrane (envelope) containing glycoprotein and lipid. Type-specific or subgroup-specific antigens are associated with the glycoproteins in the viral envelope, which are encoded by the *env* gene; group-specific antigens are associated with the virion core, which are encoded by the *gag* gene.

Three morphologic classes of extracellular retrovirus particles—as well as an intracellular form—are recognized,

based on electron microscopy. They reflect slightly different processes of morphogenesis by different retroviruses. Examples of each are shown in Figure 43–1.

Type A particles occur only intracellularly and appear to be noninfectious. Intracytoplasmic type A particles, 75 nm in diameter, are precursors of extracellular type B viruses, whereas intracisternal type A particles, 60–90 nm in diameter, are unknown entities. Type B viruses are 100–130 nm in diameter and contain an eccentric nucleoid. The prototype of this group is the mouse mammary tumor virus, which occurs in "high mammary cancer" strains of inbred mice and is found

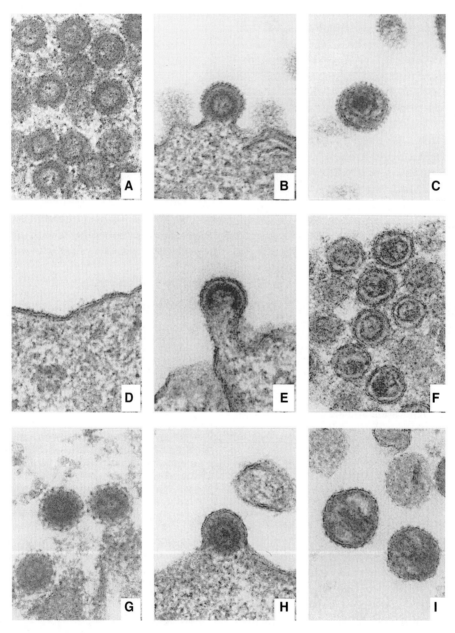

FIGURE 43–1 Comparative morphology of type A, B, C, and D retroviruses. **A:** Intracytoplasmic type A particles (representing immature precursor of budding type B virus). **B:** Budding type B virus. **C:** Mature, extracellular type B virus. **D:** Lack of morphologically recognizable intracytoplasmic form for type C virus. **E:** Budding type C virus. **F:** Mature, extracellular type C virus. **G:** Intracytoplasmic type A particle (representing immature precursor form of type D virus). **H:** Budding type D virus. **I:** Mature, extracellular type D virus. All micrographs are approximately 87,000×. Thin sections were double-stained with uranyl acetate and lead citrate. (Courtesy of D Fine and M Gonda.)

in particularly large amounts in lactating mammary tissue and milk. It is readily transferred to suckling mice, in whom the incidence of subsequent development of adenocarcinoma of the breast is high. The type C viruses represent the largest group of retroviruses. The particles are 90–110 nm in diameter, and the electron-dense nucleoids are centrally located. The type C viruses may exist as exogenous or endogenous entities (see below). The lentiviruses are also type C viruses. Finally, the type D retroviruses are poorly characterized. The particles are 100–120 nm in diameter, contain an eccentric nucleoid, and exhibit surface spikes shorter than those on type B particles.

Classification

A. Genera

The Retroviridae family is divided into seven genera: *Alpharetrovirus* (which contains avian leukosis and sarcoma viruses), *Betaretrovirus* (mouse mammary tumor virus), *Gammaretrovirus* (mammalian leukemia and sarcoma viruses), *Deltaretrovirus* (human T-lymphotropic viruses and bovine leukemia virus), *Epsilonretrovirus* (fish viruses), *Spumavirus* (which contains viruses able to cause "foamy" degeneration of inoculated cells but which are not associated with any known disease process), and *Lentivirus* (which encompasses agents able to cause chronic infections with slowly progressive neurologic impairment, including HIV; see Chapter 44).

Retroviruses can be organized in various ways depending on their morphologic, biologic, and genetic properties. Differences in genome sequences and natural host range are frequently used, but antigenic properties are not. Retroviruses may be grouped morphologically (types B, C, and D); the vast majority of isolates display type C characteristics.

B. Host of Origin

Retroviruses have been isolated from virtually all vertebrate species. Natural infections by a given virus are usually limited to a single species, though infections across species barriers may occur. Group-specific antigenic determinants on the major internal (core) protein are shared by viruses from the same host species. All mammalian viruses are more closely related to one another than to those from avian species.

The RNA tumor viruses most widely studied experimentally are the sarcoma viruses of chickens and mice and the leukemia viruses of mice, cats, chickens, and humans.

C. Exogenous or Endogenous

Exogenous retroviruses are spread horizontally and behave as typical infectious agents. They initiate infection and transformation only after contact. In contrast to endogenous viruses, which are found in all cells of all individuals of a given species, gene sequences of exogenous viruses are found only in infected cells. The pathogenic retroviruses all appear to be exogenous viruses.

Retroviruses may also be transmitted vertically through the germ line. Viral genetic information that is a constant part of the genetic constitution of an organism is designated as "endogenous." An integrated retroviral provirus behaves like a cluster of cellular genes and is subject to regulatory control by the cell. This cellular control usually results in partial or complete repression of viral gene expression. Its location in the cellular genome and the presence of appropriate cellular transcription factors determine to a great extent if (and when) viral expression will be activated. It is not uncommon for normal cells to maintain the endogenous viral infection in a quiescent form for extended periods of time.

Many vertebrates, including humans, possess multiple copies of endogenous RNA viral sequences. The endogenous viral sequences are of no apparent benefit to the animal. However, endogenous proviruses of mammary tumor virus carried by inbred strains of mice express superantigen activities that influence the T cell repertoires of the animals.

Endogenous viruses are usually not pathogenic for their host animals. They do not produce any disease and cannot transform cells in culture. (There are examples of disease caused by replication of endogenous viruses in inbred strains of mice.)

Important features of endogenous viruses are as follows: (1) DNA copies of RNA tumor virus genomes are covalently linked to cellular DNA and are present in all somatic and germ cells in the host; (2) endogenous viral genomes are transmitted genetically from parent to offspring; (3) the integrated state subjects the endogenous viral genomes to host genetic control; and (4) the endogenous virus may be induced to replicate either spontaneously or by treatment with extrinsic (chemical) factors.

D. Host Range

The presence or absence of an appropriate cell surface receptor is a major determinant of the host range of a retrovirus. Infection is initiated by an interaction between the viral envelope glycoprotein and a cell surface receptor. **Ecotropic** viruses infect and replicate only in cells from animals of the original host species. **Amphotropic** viruses exhibit a broad host range (able to infect cells not only of the natural host but of heterologous species as well) because they recognize a receptor that is widely distributed. **Xenotropic** viruses can replicate in some heterologous (foreign) cells but not in cells of the natural host. Many endogenous viruses have xenotropic host ranges.

E. Genetic Content

Retroviruses have a simple genetic content, but there is some variation in the number and type of genes contained. The genetic makeup of a virus influences its biologic properties. Genomic structure is a useful way of categorizing RNA tumor viruses (Figure 43–2).

The standard leukemia viruses (*Alpharetrovirus* and *Gammaretrovirus*) contain genes required for viral replication: *gag*, which encodes the core proteins (group-specific antigens); *pro*, which encodes a protease enzyme; *pol*, which encodes the reverse transcriptase enzyme (polymerase); and *env*, which encodes the glycoproteins that form projections

A

B

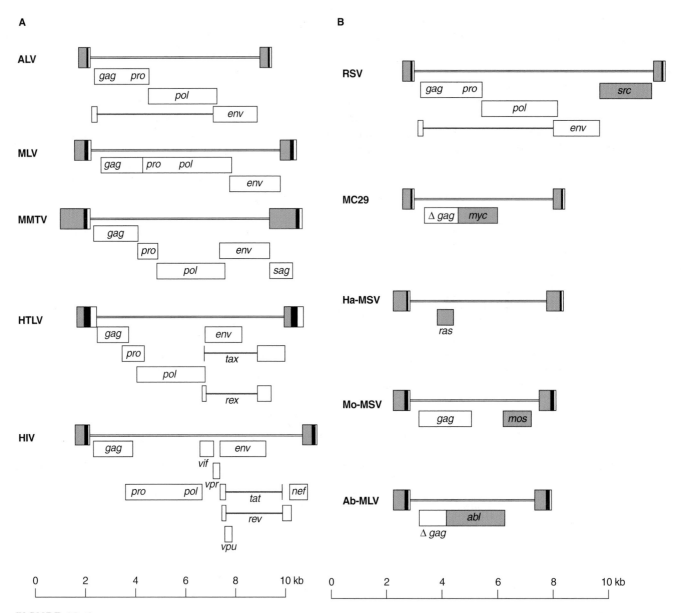

FIGURE 43–2 Genetic organization of representative retroviruses. **A:** Nondefective, replication-competent viruses. Examples of retroviruses with simple and complex genomes are shown. An open rectangle shows the open reading frame for the indicated gene. If the rectangles are offset vertically, their reading frames are different. Horizontal lines connecting two rectangles indicate that this segment is spliced out. **Simple genomes:** ALV, avian leukosis virus (*Alpharetrovirus*); MLV, murine leukemia virus (*Gammaretrovirus*); MMTV, mouse mammary tumor virus (*Betaretrovirus*). **Complex genomes:** HTLV, human T-lymphotropic virus (*Deltaretrovirus*); HIV, human immunodeficiency virus type 1 (*Lentivirus*). **B:** Viruses carrying oncogenes. Several examples are shown, with the oncogene shaded; all are defective except RSV. RSV, Rous sarcoma virus (*src* oncogene) (*Alpharetrovirus*); MC29, avian myelocytomatosis virus (*myc* oncogene) (*Alpharetrovirus*); Ha-MSV, Harvey murine sarcoma virus (*ras* oncogene) (*Gammaretrovirus*); Mo-MSV, Moloney murine sarcoma virus (*mos* oncogene) (*Gammaretrovirus*); Ab-MLV, Abelson murine leukemia virus (*abl* oncogene) (*Gammaretrovirus*). The scale for genome sizes is shown at the bottom of each panel. (Modified with permission from Vogt VM: Retroviral virions and genomes. In: *Retroviruses.* Coffin JM, Hughes SH, Varmus HE [editors]. Cold Spring Harbor Laboratory Press, 1997.)

on the envelope of the particle. The gene order in all retroviruses is 5′-*gag-pro-pol-env*-3′.

Some viruses, exemplified by the human retroviruses (*Deltaretrovirus* and *Lentivirus*), contain additional genes downstream from the *env* gene. One is a transactivating regulatory gene (*tax* or *tat*) that encodes a nonstructural protein which alters the transcription or translational efficiency

of other viral genes. The lentiviruses, including HIV, have a more complex genome and contain several additional accessory genes (see Chapter 44).

Retroviruses with either of these two genomic structures will be replication-competent (in appropriate cells). Because they lack a transforming (*onc*) gene, they cannot transform cells in tissue culture. However, they may have the ability to transform precursor cells in blood-forming tissues in vivo.

The directly transforming retroviruses carry an *onc* gene. The transforming genes carried by various RNA tumor viruses represent cellular genes that have been appropriated by those viruses at some time in the distant past and incorporated into their genomes (Figure 43–2).

Such viruses are highly oncogenic in appropriate host animals and can transform cells in culture. With very few exceptions, the addition of the cellular DNA results in the loss of portions of the viral genome. Consequently, the sarcoma viruses usually are replication-defective; progeny virus is produced only in the presence of helper viruses. The helper viruses are generally other retroviruses (leukemia viruses), which may recombine in various ways with the defective viruses. These defective transforming retroviruses have been the source of many of the recognized cellular oncogenes.

F. Oncogenic Potential

The retroviruses that contain oncogenes are highly oncogenic. They are sometimes referred to as "acute transforming" agents because they induce tumors in vivo after very short latent periods and rapidly induce morphologic transformation of cells in vitro. The viruses that do not carry an oncogene have a much lower oncogenic potential. Disease (usually of blood cells) appears after a long latent period (ie, "slow-transforming"); cultured cells are not transformed.

Briefly, neoplastic transformation by retroviruses is the result of a cellular gene that is normally expressed at low, carefully regulated levels becoming activated and expressed constitutively. In the case of the acute transforming viruses, a cellular gene has been inserted by recombination into the viral genome and is expressed as a viral gene under the control of the viral promoter. In the case of the slow-transforming leukemia viruses, the viral promoter or enhancer element is inserted adjacent to or near the cellular gene in the cellular chromosome.

Replication of Retroviruses

A schematic outline of a typical retrovirus replication cycle, represented by human T-lymphotropic virus (HTLV), is shown in Figure 43–3. The *pol* gene encodes the unique polymerase (reverse transcriptase) protein that has four enzymatic activities (protease, polymerase, RNase H, and integrase). After virus particles have adsorbed to and penetrated host cells, the viral RNA serves as the template for the synthesis of viral DNA through the action of the viral enzyme reverse transcriptase, functioning as an RNA-dependent DNA polymerase. By a complex process, sequences from both ends of the viral RNA become duplicated, forming the long terminal repeat located at each end of the viral DNA (Figure 43–4). Long terminal repeats are present only in viral DNA. The newly formed viral DNA

becomes integrated into the host cell DNA as a provirus. The structure of the provirus is constant, but its integration into host cell genomes can occur at different sites. The very precise orientation of the provirus after integration is achieved by specific sequences at the ends of both long terminal repeats.

Progeny viral genomes may then be transcribed from the provirus DNA into viral RNA. The U3 sequence in the long terminal repeat contains both a promoter and an enhancer. The enhancer may help confer tissue specificity on viral expression. The proviral DNA is transcribed by the host enzyme, RNA polymerase II. Full-length transcripts (capped, polyadenylated) serve as genomic RNA for encapsidation in progeny virions. Some transcripts are spliced, and the subgenomic mRNAs are translated to produce viral precursor proteins that are modified and cleaved to form the final protein products.

If the virus happens to contain a transforming gene, the oncogene plays no role in replication. This is in marked contrast to the DNA tumor viruses, in which the transforming genes are also essential viral replication genes.

Virus particles assemble and emerge from infected host cells by budding from plasma membranes. The viral protease then cleaves the Gag and Pol proteins from the precursor polyprotein, producing a mature infectious virion prepared for reverse transcription when the next cell is infected.

A salient feature of retroviruses is that they are not cytolytic, ie, they do not kill the cells in which they replicate. The exceptions are the lentiviruses, which may be cytolytic (see Chapter 44). The provirus remains integrated within the cellular DNA for the life of the cell. There is no known way to cure a cell of a chronic retrovirus infection.

Human Retroviruses

A. Human T-Lymphotropic Viruses

Only a few retroviruses are linked to human tumors. The HTLV group of retroviruses has probably existed in humans for thousands of years. HTLV-1 has been established as the causative agent of adult T cell leukemia-lymphomas (ATL) as well as a nervous system degenerative disorder called tropical spastic paraparesis. It does not carry an oncogene. A related human virus, HTLV-2, has been isolated but has not been conclusively associated with a specific disease. HTLV-1 and HTLV-2 share about 65% sequence homology and display significant serologic cross-reactivity.

The human lymphotropic viruses have a marked affinity for mature T cells. HTLV-1 is expressed at very low levels in infected individuals. It appears that the viral promoter-enhancer sequences in the long terminal repeat may be responsive to signals associated with the activation and proliferation of T cells. If so, the replication of the viruses may be linked to the replication of the host cells—a strategy that would ensure efficient propagation of the virus.

The human retroviruses are transregulating (Figure 43–2). They carry a gene, *tax,* whose product alters the expression of other viral genes. Transactivating regulatory genes are believed to be necessary for viral replication in vivo and may contribute to oncogenesis by also modulating cellular genes that regulate cell growth.

There are several genetic subtypes of HTLV-1, with the major ones being subtypes A, B, and C (these do not represent distinct serotypes).

The virus is distributed worldwide, with an estimated 10 to 20 million infected individuals. Clusters of HTLV-associated disease are found in certain geographic areas (southern Japan, Melanesia, the Caribbean, Central and South America, and parts of Africa) (Figure 43–5). Although fewer than 1% of people worldwide have HTLV-1 antibody, more than 10% of the population in endemic areas are seropositive, and antibody may be found in 50% of relatives of virus-positive leukemia patients.

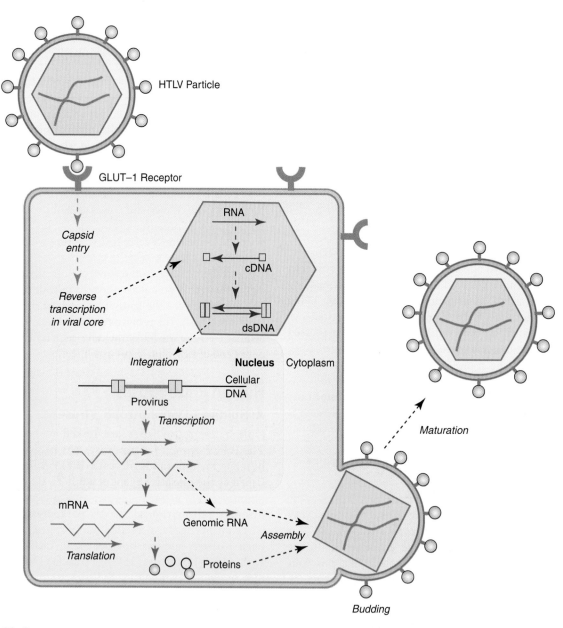

FIGURE 43–3 Overview of retrovirus HTLV replication cycle. The virus particle attaches to a cell surface receptor, and the viral capsid enters the cell. The viral reverse transcriptase enzyme produces a DNA copy of the genome RNA within the capsid in the cytoplasm. The DNA enters the nucleus and is integrated at random into cell DNA, forming the provirus. The integrated provirus serves as template for the synthesis of viral transcripts, some of which are unspliced and will be encapsidated as genomic RNAs and others, some of which are spliced, will serve as mRNAs. Viral proteins are synthesized; the proteins and genome RNAs assemble; and particles bud from the cell. Capsid proteins are proteolytically processed by the viral protease producing mature, infectious virions, shown schematically as conversion from a square to an icosahedral core. (Courtesy of SJ Marriott.)

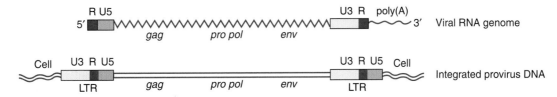

FIGURE 43–4 Comparison of structures of retrovirus RNA genome and integrated provirus DNA. A virus particle contains two identical copies of the single-stranded RNA genome. The 5′ terminal is capped, and the 3′ terminal is polyadenylated. A short sequence, R, is repeated at both ends; unique sequences are located near the 5′ (U5) and 3′ (U3) ends. U3 contains promoter and enhancer sequences. The integrated provirus DNA is flanked at each end by the long terminal repeat (LTR) structure generated during synthesis of the DNA copy by reverse transcription. Each long terminal repeat contains U3, R, and U5 sequences. The long terminal repeats and coding regions of the retrovirus genome are not drawn to scale.

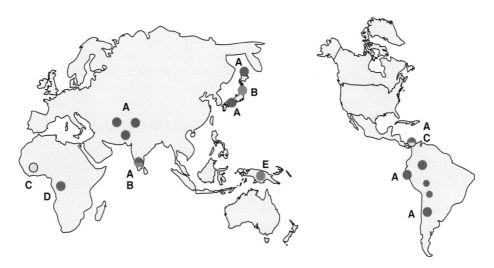

FIGURE 43–5 Subtypes of HTLV-1 are geographically distributed in endemic foci. **A:** Japan, India, the Caribbean, and the Andes; **B:** Japan and India; **C:** West Africa and the Caribbean; **D:** Central Africa; **E:** Papua New Guinea. (Courtesy of N Mueller.)

ATL is poorly responsive to therapy. The 5-year survival rate for patients with this cancer is <5%.

Transmission of HTLV-1 seems to involve cell-associated virus. Mother-to-child transmission via breast feeding is an important mode. Efficiency of transmission from infected mother to child is estimated at 15–25%. Such early-life infections are associated with the greatest risk of ATL. Blood transfusion is an effective means of transmission, as are sharing blood-contaminated needles (drug abusers) and sexual intercourse.

Seroepidemiology has linked infection with HTLV-1 to a syndrome called HTLV-1-associated myelopathy/tropical spastic paraparesis (HAM/TSP). The primary clinical feature is development of progressive weakness of the legs and lower body. The patient's mental faculties remain intact. HAM/TSP is described as being of the same magnitude and importance in the tropics as is multiple sclerosis in Western countries.

B. Human Immunodeficiency Viruses

A group of human retroviruses has been established as the cause of AIDS (see Chapter 44). The HIV are cytolytic and nontransforming and are classified as lentiviruses. However, AIDS patients are at elevated risk of several types of cancer because of the immune suppression associated with HIV infection. These cancers include lymphomas and cervical cancer.

C. Other

A new human gammaretrovirus, XMRV, was discovered in 2006 using a microarray containing genetic sequences from over 1000 viruses. It is related to, but distinct from, xenotropic murine leukemia viruses and lacks an oncogene. XMRV was detected in tumors from patients with familial prostate cancer, but any role in prostate carcinogenesis has not been established.

The simian foamy viruses from the *Spumavirus* genus are highly prevalent in captive nonhuman primates. Humans occupationally exposed to the primates can be infected with foamy viruses, but these infections have not resulted in any recognized disease.

CELLULAR ONCOGENES

"Oncogene" is the general term given to genes that are involved in cancer causation. Normal versions of these transforming genes are present in normal cells and have been designated proto-oncogenes.

The discovery of cellular oncogenes came from studies with acutely transforming retroviruses. It was found that normal cells contained highly related (but not identical) copies of various retrovirus transforming genes; cellular sequences had been captured and incorporated into the retrovirus genomes. Transduction of the cellular genes was probably an accident, as the presence of the cellular sequences is of no benefit to the viruses. Many other known cellular oncogenes that have not been segregated into retrovirus vectors have been detected using molecular methods.

Cellular oncogenes are partly responsible for the molecular basis of human cancer. They represent individual components of complicated pathways responsible for regulating cell proliferation, division, and differentiation and for maintaining the integrity of the genome. Incorrect expression of any component might interrupt that regulation, resulting in uncontrolled growth of cells (cancer). Examples exist of tyrosine-specific protein kinases (eg, *src*), growth factors (*sis* is similar to human platelet-derived growth factor, a potent mitogen for cells of connective tissue origin), mutated growth factor receptors (*erb*-B is a truncated epidermal growth factor receptor), GTP-binding proteins (Ha-*ras*), and nuclear transcription factors (*myc, jun*).

The molecular mechanisms responsible for activating a benign proto-oncogene and converting it into a cancer gene vary—but all involve genetic damage. The gene may be overexpressed, and a dosage effect of the overproduced oncogene product may be important in cellular growth changes. These mechanisms might result in constitutive activity (loss of normal regulation), so that the gene is expressed at the wrong time during the cell cycle or in inappropriate tissue types. Mutations might alter the carefully regulated interaction of a proto-oncogene protein with other proteins or nucleic acids. Insertion of a retroviral promoter adjacent to a cellular oncogene may result in enhanced expression of that gene (ie, "promoter-insertion oncogenesis"). Expression of a cellular gene also may be increased through the action of nearby viral "enhancer" sequences.

TUMOR SUPPRESSOR GENES

A second class of human cancer genes is involved in tumor development. These are the negative regulators of cell growth, tumor suppressor genes. They were identified because they form complexes with oncoproteins of certain DNA tumor viruses. The **inactivation** or functional loss of both alleles of such a gene is required for tumor formation—in contrast to the **activation** that occurs with cellular oncogenes. The prototype of this inhibitory class of genes is the retinoblastoma (*Rb*) gene. The Rb protein inhibits entry of cells into S phase by binding to key transcription factors that regulate expression of S phase genes. The function of normal Rb protein is regulated by phosphorylation. The loss of *Rb* gene function is causally related to the development of retinoblastoma—a rare ocular tumor of children—and other human tumors.

Another crucial tumor suppressor gene is the *p53* gene. It also blocks cell cycle progression; p53 acts as a transcription factor and regulates the synthesis of a protein that inhibits the function of certain cell cycle kinases. It also causes cells with DNA damage to undergo apoptosis. The loss of p53 function allows cells with damaged DNA to progress through the cell cycle, leading to the eventual accumulation of genetic mutations. The *p53* gene is mutated in over half of all human cancers.

DNA TUMOR VIRUSES

Fundamental differences exist between the oncogenes of DNA and RNA tumor viruses. The transforming genes carried by DNA tumor viruses encode functions required for viral replication and do not have normal homologs in cells. In contrast, retroviruses either carry transduced cellular oncogenes that have no role in viral replication or they act through indirect mechanisms. The DNA virus transforming proteins complex with normal cell proteins and alter their function. To understand the mechanism of action of DNA virus transforming proteins, it is important to identify the cellular targets with which they interact. Examples of such interactions are shown in Table 43–4.

POLYOMAVIRUSES

Important properties of polyomaviruses are listed in Table 43–5.

Classification

The Polyomaviridae family contains a single genus designated *Polyomavirus*, formerly part of the Papovaviridae family (which no longer exists). Polyomaviruses are small viruses (diameter 45 nm) that possess a circular genome of double-stranded DNA (5 kbp; MW 3×10^6) enclosed within a nonenveloped capsid exhibiting icosahedral symmetry (Figure 43–6). Cellular histones are used to condense viral DNA inside virus particles.

SV40 from monkeys and humans, BK, JC, KI, WU, and Merkel cell viruses from humans, and murine polyoma virus from mice are simple DNA-containing tumor viruses that

TABLE 43–4 Examples of DNA Virus Oncoproteins & Cellular Protein Interactions[a]

Virus	Viral Oncoproteins	Cellular Targets
Polyomavirus SV40	Large T antigen	p53, pRb
	Small t antigen	PP2A
Human papillomavirus	E6	p53, DLG, MAGI-1, MUPP1
	E7	pRb
Bovine papillomavirus	E5	PDGFβ receptor
Adenovirus	E1A	pRb
	E1B-55K	p53
Adenovirus 9	E4ORF1	DLG, MAGI-1, MUPP1
Herpesvirus EBV	LMP1	TRAFs

[a]p53, product of *p53* gene; pRb, retinoblastoma gene product; PP2A, protein phosphatase 2A; PDGF, platelet-derived growth factor; EBV, Epstein-Barr virus; TRAF, tumor necrosis factor receptor-associated factor. DLG, MAGI-1, and MUPP1 are members of a family of cellular proteins that contain PDZ domains.

TABLE 43–5 Important Properties of Polyomaviruses[a]

Virion: Icosahedral, 45 nm in diameter

Composition: DNA (10%), protein (90%)

Genome: Double-stranded DNA, circular, 5 kbp, MW 3 million

Proteins: Three structural proteins; cellular histones condense DNA in virion

Envelope: None

Replication: Nucleus

Outstanding characteristics:
   Stimulate cell DNA synthesis
   Viral oncoproteins interact with cellular tumor suppressor proteins
   Important model tumor viruses
   Human viruses can cause human neurologic and renal disease
   May cause human cancer

[a]Formerly classified in Papovaviridae family.

possess a limited amount of genetic information (six or seven genes). Many species of mammals and some birds have been found to carry their own species of polyomavirus.

Polyomavirus Replication

The polyomavirus genome contains "early" and "late" regions (Figure 43–7). The early region is expressed soon after infection of cells; it contains genes that code for early proteins—eg, the SV40 large tumor (T) antigen, which is necessary for the replication of viral DNA in permissive cells, and the small tumor (t) antigen. The murine polyoma virus genome encodes three early proteins (small, middle, and large T antigens). One or two of the T antigens are the only viral gene products required for transformation of cells. Usually, the transforming proteins must be continuously synthesized for cells to stay transformed. The late region consists of genes that code for the synthesis of coat protein; they have no role in transformation and usually are not expressed in transformed cells.

SV40 T antigen interacts with the cellular tumor suppressor gene products, p53 and pRb family members (Table 43–4). Interactions of T antigen with the cellular proteins are important in the replicative cycle of the virus. Complex formation functionally inactivates the growth inhibitory properties of pRb and p53, allowing cells to enter S phase so that viral DNA may be replicated. Likewise, functional inactivation of the cellular proteins by T antigen binding is central to the virus-mediated transformation process. As p53 senses DNA damage and either blocks cell cycle progression or initiates apoptosis, abolishing its function would lead to accumulation of T antigen-expressing cells with genomic mutations that might promote tumorigenic growth.

Pathogenesis & Pathology

The human polyomaviruses (BK and JC) are widely distributed in human populations, as evidenced by the presence of specific antibody in 70–80% of adult sera. Infection usually occurs during early childhood. Both viruses may persist in the kidneys and lymphoid tissues of healthy individuals after primary infection and may reactivate when the host's immune response is impaired, eg, by renal transplantation, during pregnancy, or increasing age. Viral reactivation and shedding in urine are asymptomatic in immunocompetent persons. The viruses are most commonly isolated from immunocompromised patients, in whom disease may occur. BK virus causes hemorrhagic cystitis in bone

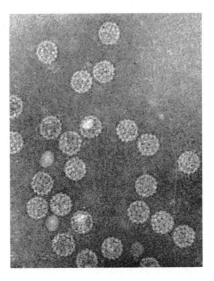

FIGURE 43–6 Polyomavirus SV40. Purified preparation negatively stained with phosphotungstate (150,000×). (Courtesy of S McGregor and H Mayor.)

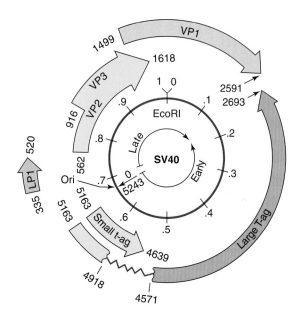

FIGURE 43–7 Genetic map of the polyomavirus SV40. The thick circle represents the circular SV40 DNA genome. The unique *Eco*RI site is shown at map unit 0/1. Nucleotide numbers begin and end at the origin (Ori) of viral DNA replication (0/5243). Boxed arrows indicate the open reading frames that encode the viral proteins. Arrowheads point in the direction of transcription; the beginning and end of each open reading frame are indicated by nucleotide numbers. Various shadings depict different reading frames used for different viral polypeptides. Note that large T antigen (T-ag) is coded by two noncontiguous segments on the genome. The genome is divided into "early" and "late" regions that are expressed before and after the onset of viral DNA replication, respectively. Only the early region is expressed in transformed cells. (Reproduced with permission from Butel JS, Jarvis DL: Biochim Biophys Acta 1986;865:171.)

marrow transplant recipients. It is the cause of polyomavirus-associated nephropathy in renal transplant recipients, a serious disease that occurs in up to 5% of recipients and which results in graft failure in up to 50% of those affected patients. JC virus is the cause of progressive multifocal leukoencephalopathy (PML), a fatal brain disease that occurs in some immunocompromised persons, especially those with depressed cell-mediated immunity resulting from immunosuppressive therapies or infection by HIV. PML affects about 5% of AIDS patients. The two viruses are antigenically distinct, but both encode a T antigen that is related to SV40 T antigen. These human viruses can transform rodent cells and induce tumors in newborn hamsters. JC virus has been associated with human brain tumors, but an etiologic role is not yet established.

KI and WU viruses were discovered in 2007 in nasopharyngeal aspirates from children with respiratory infections. Merkel cell polyomavirus was identified in 2008 in Merkel cell carcinomas, rare skin tumors of neuroendocrine origin. Seroprevalence studies suggest that KI, WU, and Merkel cell virus infections are widespread and likely occur in childhood. Because of their recent discoveries, information on disease

associations is limited, although Merkel cell virus DNA appears to be present and integrated in a large fraction of Merkel cell carcinomas.

SV40 replicates in certain types of monkey and human cells; it is highly tumorigenic in experimentally inoculated hamsters and in transgenic mice and can transform many types of cells in culture. Tumor induction in the natural host—the rhesus monkey—is rarely observed. SV40 may cause a PML-like disease in rhesus monkeys.

SV40 contaminated early lots of live and killed poliovirus vaccines that had been grown in monkey cells unknowingly infected with SV40. Millions of people worldwide received such SV40-contaminated vaccines between 1955 and 1963. SV40 is detected in humans today, including in individuals too young to have been exposed via vaccination. Evidence suggests that it (and other polyomaviruses) may be transmitted by the fecal–oral route in humans. The prevalence of SV40 infections in humans appears to be low.

SV40 DNA has been detected in selected types of human tumors, including brain tumors, mesotheliomas, bone tumors, and lymphomas. The role SV40 is playing in formation of human cancers is under investigation.

The host range for polyomaviruses is often highly restricted. Usually a single species can be infected and only certain cell types within that species. Exceptions are the primate polyomaviruses SV40 and BK virus; SV40 can infect also humans and human cells and BK virus can infect some monkeys and monkey cells. Cell types that fail to support polyomavirus replication may be transformed by a virus.

PAPILLOMAVIRUSES

Important properties of papillomaviruses are listed in Table 43–6.

TABLE 43–6   Important Properties of Papillomaviruses[a]

Virion: Icosahedral, 55 nm in diameter
Composition: DNA (10%), protein (90%)
Genome: Double-stranded DNA, circular, 8 kbp, MW 5 million
Proteins: Two structural proteins; cellular histones condense DNA in virion
Envelope: None
Replication: Nucleus
Outstanding characteristics:
Stimulate cell DNA synthesis
Restricted host range and tissue tropism
Significant cause of human cancer, especially cervical cancer
Viral oncoproteins interact with cellular tumor suppressor proteins

[a]Formerly classified in Papovaviridae family.

Classification

The Papillomaviridae family is a very large virus family currently divided into 16 genera, of which five contain members that infect humans (*Alpha-*, *Beta-*, *Gamma-*, *Mupa-*, and *Nupapapillomavirus*). The papillomaviruses are former members of the Papovaviridae family. Although papillomaviruses and polyomaviruses share similarities in morphology, nucleic acid composition, and transforming capabilities, differences in genome organization and biology led to their separation into distinct virus families. The papillomaviruses are slightly larger in diameter (55 nm) than the polyomaviruses (45 nm) and contain a larger genome (8 kbp versus 5 kbp). The organization of the papillomavirus genome is more complex (Figure 43–8). There is widespread diversity among papillomaviruses. Because neutralization tests cannot be done since there is no in vitro infectivity assay, papillomavirus isolates

are classified using molecular criteria. Virus "types" are at least 10% dissimilar in the sequence of their L1 genes. Almost 200 distinct HPV types have been recovered.

Papillomavirus Replication

Papillomaviruses are highly tropic for epithelial cells of the skin and mucous membranes. Viral nucleic acid can be found in basal stem cells, but late gene expression (capsid proteins) is restricted to the uppermost layer of differentiated keratinocytes (Figure 43–9). Stages in the viral replicative cycle are dependent on specific factors that are present in sequential differentiated states of epithelial cells. This strong dependence of viral replication on the differentiated state of the host cell is responsible for the difficulties in propagating papillomaviruses in vitro.

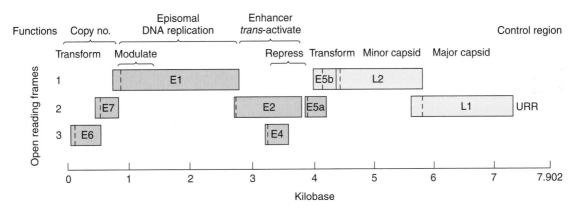

FIGURE 43–8 Map of the human papillomavirus genome (HPV-6, 7902 base pairs). The papillomavirus genome is circular but is shown linearized in the upstream regulatory region (URR). The upstream regulatory region contains the origin of replication and enhancer and promoter sequences. Early (E1–E7) and late (L1, L2) open reading frames and their functions are shown. All the open reading frames are on the same strand of viral DNA. Biologic functions are extrapolated from studies with the bovine papillomavirus. The organization of the papillomavirus genome is much more complex than that of a typical polyomavirus (compare with Figure 43–7). (Reproduced with permission from Broker TR: Structure and genetic expression of papillomaviruses. Obstet Gynecol Clin North Am 1987;14:329.)

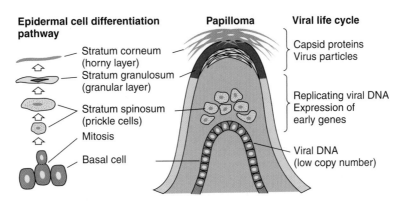

FIGURE 43–9 Schematic representation of a skin wart (papilloma). The papillomavirus life cycle is tied to epithelial cell differentiation. The terminal differentiation pathway of epidermal cells is shown on the left. Events in the virus life cycle are noted on the right. Late events in viral replication (capsid protein synthesis and virion morphogenesis) occur only in terminally differentiated cells. (Reproduced with permission from Butel JS: Papovaviruses. In: *Medical Microbiology*, 3rd ed. Baron S [editor]. Churchill Livingstone, 1991.)

Pathogenesis & Pathology

Transmission of viral infections occurs by close contact. Viral particles are released from the surface of papillomatous lesions. It is likely that microlesions allow infection of proliferating basal layer cells at other sites or within different hosts.

Papillomaviruses cause infections at cutaneous and mucosal sites, sometimes leading to the development of different kinds of warts, including skin warts, plantar warts, flat warts, anogenital warts, laryngeal papillomas, and several cancers, including those of the cervix, vulva, penis and anus, and a subset of head and neck cancers (Table 43–7). The multiple types of HPV isolates are preferentially associated with certain clinical lesions, though distribution patterns are not absolute. HPV genital infections are sexually transmitted and represent the most common sexually transmitted disease in the United States. Cervical cancer is the second most frequent cancer in women worldwide (about 500,000 new cases annually) and is a major cause of cancer deaths in developing countries.

Based on the relative occurrence of viral DNA in certain cancers, HPV types 16 and 18 are considered to be high cancer risk; about 15 other less common types are also considered high risk. Many HPV types are considered benign.

Integrated copies of viral DNA are usually present in cervical cancer cells, though HPV DNA is generally not integrated (episomal) in noncancerous cells or premalignant lesions. Skin carcinomas appear to harbor HPV genomes in an episomal state. Viral early proteins E6 and E7 are synthesized in cancer tissue. These are HPV transforming proteins, able to complex with Rb and p53 and other cellular proteins (Table 43–4).

The behavior of HPV lesions is influenced by immunologic factors. Cell-mediated immunity is important. Nearly all HPV infections are cleared and become undetectable within 2–3 years.

Cervical cancer develops slowly, sometimes taking years to decades. It is thought that multiple factors are involved in progression to malignancy; however, persistent infection with a high-risk HPV is a necessary component to the process (Figure 43–10).

Clinical Findings & Epidemiology

An estimated 660 million people worldwide have HPV genital infections, the most common viral infection of the reproductive tract. An estimated 6.2 million new infections occur annually in the United States. The peak incidence of HPV infections occurs in adolescents and young adults under 25 years of age.

HPVs are accepted as the cause of anogenital cancers. Over 99% of cervical cancer cases and over 80% of anal cancer cases are linked to genital infections with HPVs. Papillomaviruses illustrate the concept that natural viral strains may differ in oncogenic potential. Although many different HPV types cause genital infections, HPV-16 or HPV-18 is found most frequently in cervical carcinomas, though some cancers contain DNA from other types, such as HPV type 31. Epidemiologic studies indicate that HPV-16 and HPV-18 are responsible for more than 70% of all cervical cancers, with type 16 being most common. HeLa cells, a widely used tissue culture cell line derived many years ago from a cervical carcinoma, contain HPV-18 DNA.

Anal cancer is associated with high-risk HPV infection. Immunocompromised patients are especially at risk, as are men who have sex with men. Oropharyngeal cancers, a subset of head and neck squamous cell carcinomas, are also linked to HPV infections, especially by type 16.

The role of men as carriers of HPV as well as vectors for transmission of infections is well documented; however, most penile HPV infections in men are subclinical and do not result in HPV-associated disease.

Anogenital warts are usually (90%) caused by low-risk HPV types 6 and 11. Laryngeal papillomas in children, also called recurrent respiratory papillomatosis, are caused

TABLE 43–7   Examples of Association of Human Papillomaviruses with Clinical Lesions

Human Papillomavirus Type[a]	Clinical Lesion	Suspected Oncogenic Potential
1	Plantar warts	Benign
2, 4, 27, 57	Common skin warts	Benign
3, 10, 28, 49, 60, 76, 78	Cutaneous lesions	Low
5, 8, 9, 12, 17, 20, 36, 47	Epidermodysplasia verruciformis	Mostly benign, but some progress to malignancy
6, 11, 40, 42–44, 54, 61, 70, 72, 81	Anogenital condylomas; laryngeal papillomas; dysplasias and intraepithelial neoplasias (mucosal sites)	Low
7	Hand warts of butchers	Low
16, 18, 30, 31, 33, 35, 39, 45, 51–53, 56, 58, 59, 66, 68, 73, 82	High-grade dysplasias and carcinomas of genital mucosa; laryngeal and esophageal carcinomas	High correlation with genital and oral carcinomas, especially cervical cancer

[a]Not all papillomavirus types are listed.

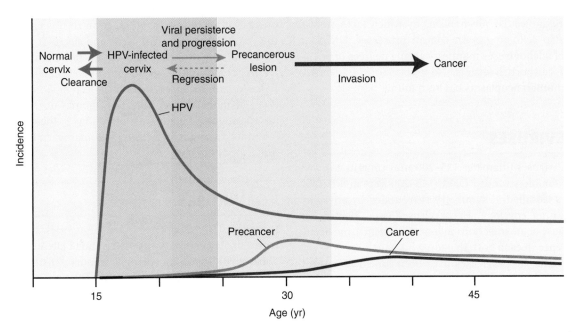

FIGURE 43–10 Relationship among cervical HPV infection, precancer, and cancer. The HPV curve shows the high incidence of infection soon after women initiate sexual activity and the subsequent decrease because many infections are self-limited. Precancer incidence curve illustrates the delay between acquisition of HPV infection and precancer development and that only a subset of infected women develop precancers. The cancer incidence curve shows the relatively long interval between precancer and progression to invasive cancer. (Reproduced from Lowy DR, Schiller JT: Prophylactic human papillomavirus vaccines. J Clin Invest 2006;116:1167. Modified with permission from Schiffman M, Castle, PE: The promise of global cervical-cancer prevention. N Engl J Med 2005;353:2101.)

by HPV-6 and HPV-11, the same viruses that cause benign genital condylomas. The infection is acquired during passage through the birth canal of a mother with genital warts. While laryngeal papillomas are rare, the growths may obstruct the larynx and must be removed repeatedly by surgical means. About 3000 cases of this disease are diagnosed annually; up to 3% of children will die.

There is a high prevalence of HPV DNA in normal skin from healthy adults. It appears that these asymptomatic HPV infections are acquired early in infancy. A great multiplicity of HPV types are detected in normal skin. Transmission is thought to occur from those in close contact with the child, with a high concordance (about 60%) between types detected in infants and their mothers.

Immunosuppressed patients experience an increased incidence of warts and cancer of the cervix. All HPV-associated cancers occur more frequently in persons with HIV/AIDS.

Prevention & Control

Vaccines against HPV are expected to be a cost-effective way to reduce anogenital HPV infections, the incidence of cervical cancer, and the HPV-associated health care burden. A quadrivalent HPV vaccine was approved in the United States in 2006 and a bivalent vaccine in 2007. Both are noninfectious recombinant vaccines containing virus-like particles composed of HPV L1 proteins. The quadrivalent vaccine contains particles derived from HPV types 6,

11, 16, and 18, whereas the bivalent vaccine contains particles from types 16 and 18. Both vaccines are effective at preventing persistent infections by the targeted HPV types and the development of HPV-related genital precancerous lesions. They are not effective against established HPV disease. Adolescent and young adult females make up the initial target population for vaccination. It is not known how long vaccine-induced immunity lasts, but it appears to extend for at least 5 years.

HPV vaccines are not recommended for pregnant females.

ADENOVIRUSES

The adenoviruses (see Chapter 32) comprise a large group of agents widely distributed in nature. They are medium-sized, nonenveloped viruses containing a linear genome of double-stranded DNA (26–45 kbp). Replication is species-specific, occurring in cells of the natural hosts. Adenoviruses commonly infect humans, causing mild acute illnesses, mainly of the respiratory and intestinal tracts.

Adenoviruses can transform rodent cells and induce the synthesis of virus-specific early antigens that localize in both the nucleus and the cytoplasm of transformed cells. The E1A early proteins complex with the cellular Rb protein as well as with several other cellular proteins. Other early proteins, E1B and E4ORF1, bind p53 and other cellular signaling proteins (Table 43–4). The adenoviruses are important models for studying the molecular mechanisms by which DNA tumor viruses usurp cellular growth control processes. Different serotypes of adenoviruses manifest varying degrees of oncogenicity in newborn hamsters. No association of adenoviruses with human neoplasms has been found.

HERPESVIRUSES

These large viruses (diameter 125–200 nm) contain a linear genome of double-stranded DNA (125–240 kbp) and have a capsid with icosahedral symmetry surrounded by an outer lipid-containing envelope. Herpesviruses (see Chapter 33) typically cause acute infections followed by latency and eventual recurrence in each host, including humans.

In humans, herpesviruses have been linked to several specific types of tumors. Herpesvirus EBV causes acute infectious mononucleosis when it infects B lymphocytes of susceptible humans. Normal human lymphocytes have a limited life span in vitro, but EBV can immortalize such lymphocytes into lymphoblast cell lines that grow indefinitely in culture.

EBV is etiologically linked to Burkitt lymphoma, a tumor most commonly found in children in central Africa; to nasopharyngeal carcinoma, more common in Cantonese Chinese and Alaskan Eskimos than other populations; to posttransplant lymphomas; and to Hodgkin disease. These tumors usually contain EBV viral DNA (both integrated and episomal forms) and viral antigens.

EBV encodes a viral oncogene protein (LMP1) that mimics an activated growth factor receptor. LMP1 is able to transform rodent fibroblasts and is essential for transformation of B lymphocytes (Table 43–4). Several EBV-encoded nuclear antigens (EBNAs) are necessary for immortalization of B cells; EBNA1 is the only viral protein consistently expressed in Burkitt lymphoma cells. EBV is very successful at avoiding immune elimination; this may be due in part to the function of EBNA1 in inhibition of antigen processing to allow infected cells to escape killing by cytotoxic T lymphocytes.

Malaria may be a cofactor of African Burkitt lymphoma. Most of those tumors also show characteristic chromosomal translocations between the c-myc gene and immunoglobulin loci, leading to the constitutive activation of myc expression. Consumption of salted or dried fish may be a dietary cofactor in EBV-related nasopharyngeal carcinoma.

Kaposi sarcoma-associated herpesvirus, also known as human herpesvirus 8 (KSHV/HHV8), is not as ubiquitous as most other human herpesviruses. It is believed to be the cause of Kaposi sarcoma, primary effusion lymphoma, and multicentric Castleman disease, a lymphoproliferative disorder. KSHV has a number of genes related to cellular regulatory genes that may stimulate cellular proliferation and modify host defense mechanisms.

Some herpesviruses are associated with tumors in lower animals. Marek disease is a highly contagious lymphoproliferative disease of chickens that can be prevented by vaccination with an attenuated strain of Marek disease virus. The prevention of cancer by vaccination in this case establishes the virus as the causative agent and suggests the possibility of a similar approach to prevention of human tumors with a virus as a causative agent. Other examples of herpesvirus-induced tumors in animals include lymphomas of certain types of monkeys and adenocarcinomas of frogs. The simian viruses cause inapparent infections in their natural hosts but induce malignant T cell lymphomas when transmitted to other species of monkeys.

POXVIRUSES

Poxviruses (see Chapter 34) are large, brick-shaped viruses with a linear genome of double-stranded DNA (130–375 kbp). Yaba virus produces benign tumors (histiocytomas) in its natural host, monkeys. Shope fibroma virus produces fibromas in some rabbits and is able to alter cells in culture. Molluscum contagiosum virus produces small benign growths in humans. Very little is known about the nature of these proliferative diseases, but the poxvirus-encoded growth factor that is related to epidermal growth factors and to transforming growth factor may be involved.

HEPATITIS B VIRUS & HEPATITIS C VIRUS

Hepatitis B virus (see Chapter 35), a member of the Hepadnaviridae family, is characterized by 42-nm spherical virions with a circular genome of double-stranded DNA (3.2 kbp). One strand of the DNA is incomplete and variable in length. Studies of the virus are hampered because it has not been grown in cell culture.

In addition to causing hepatitis, hepatitis B virus is a risk factor in the development of liver cancer in humans. Epidemiologic and laboratory studies have proved persistent infection with hepatitis B virus to be an important cause of chronic liver disease and the development of hepatocellular carcinoma. Hepatitis B virus infections occurring in adults are usually resolved, but primary infections in neonates and young children tend to become chronic in up to 90% of cases. It is these persistent hepatitis B virus infections established early in life that carry the highest risk of hepatocellular carcinoma later in life. The mechanism of oncogenesis remains obscure. Persistent viral infection leads to necrosis, inflammation, and liver regeneration which over time result in cirrhosis; hepatocellular carcinoma usually arises out of this background. The hepatitis B virus transactivator protein, X protein, is a potential viral oncoprotein. A dietary carcinogen, aflatoxin, may be a cofactor for hepatocellular carcinoma, especially in Africa and China.

The advent of an effective hepatitis B vaccine for the prevention of primary infection raised the possibility of prevention of hepatocellular carcinoma, particularly in areas of the world where infection with hepatitis B virus is hyperendemic (eg, Africa, China, Southeast Asia). Twenty years after the initiation of a universal hepatitis B vaccination program in Taiwan, chronic hepatitis B virus infection rates and liver cancer incidence rates were markedly reduced.

Woodchucks are an excellent model for hepatitis B virus infections of humans. A similar virus, woodchuck hepatitis virus, establishes chronic infections in both newborn and adult woodchucks, many of which develop hepatocellular carcinomas within a 3-year period.

Hepatitis C virus (see Chapter 35), a member of the Flaviviridae family, contains a genome of single-stranded RNA 9.4 kb in size. It appears that the majority of infections become persistent, even in adults. Chronic infection with hepatitis C virus is also considered to be a causative factor in hepatocellular carcinoma. Most probably, hepatitis C virus acts indirectly in the development of hepatocellular carcinoma.

There are currently over 250 million people worldwide persistently infected with hepatitis B virus and over 170 million chronic carriers of hepatitis C virus—a large pool of individuals at risk of developing liver cancer.

HOW TO PROVE THAT A VIRUS CAUSES HUMAN CANCER

It is clear that viruses are involved in the genesis of several types of human tumors. Proving a causal relationship between a virus and a given type of cancer is, in general, very difficult.

If a virus is the only etiologic agent of a specific cancer, the geographic distribution of viral infection should coincide with that of the tumor; the presence of viral markers should be higher in cases than in controls; and viral infection should precede the tumor. These criteria can be difficult to establish if other environmental or genetic factors cause some cases of the same type of cancer. Only if the continued expression of a viral function is necessary for maintenance of transformation will viral genes necessarily persist in every tumor cell. If the virus provides an early step in multistep carcinogenesis, the viral genome may be lost as the tumor progresses to more altered stages. Conversely, a virus may be found associated frequently with a tumor but be there simply as a passenger because of an affinity for the cell type.

Tumor viruses are usually not replicating in transformed cells, so it is necessary to use very sensitive methods to search for viral nucleic acids or proteins in cells to detect virus presence. Viral structural proteins are frequently not expressed, but virus-encoded nonstructural proteins may be expressed as markers of virus presence.

Tumor induction in laboratory animals and transformation of human cells in culture are good circumstantial lines of evidence that a virus is tumorigenic, and those systems can provide models for molecular analyses of the transformation process. However, they do not constitute proof that the virus causes a particular human cancer.

The most definitive proof of a causal relationship is decreased tumor incidence by prevention of viral infection.

Intervention methods should be effective in reducing the occurrence of the cancer even if the virus is only one of several cofactors.

REVIEW QUESTIONS

1. Viruses can cause cancer in animals and humans. A principle of viral carcinogenesis is that
   (A) Retroviruses cause most types of human cancer
   (B) Not all infections with a human cancer virus lead to tumor formation
   (C) Short latent periods elapse between time of virus infection and tumor appearance
   (D) Animal models seldom predict cellular mechanisms in human cancer
   (E) Host factors are insignificant in influencing the development of virus-induced human cancer

2. Cellular oncogenes represent activated genes involved in cancer. A second class of cancer genes is involved in cancer development only when both alleles of such a gene are inactivated. The second class of genes is called
   (A) Proto-oncogenes
   (B) T antigen genes
   (C) Tumor suppressor genes
   (D) Transduced genes
   (E) Silent genes

3. A 38-year-old woman with many lifetime sexual partners is diagnosed with cervical cancer. This cancer is common worldwide and has a sexually transmitted viral etiology. The causative agent of human cervical cancer is
   (A) Hepatitis C virus
   (B) Hepatitis B virus
   (C) Human papillomaviruses, high-risk types
   (D) Polyomaviruses
   (E) Herpesviruses

4. Retroviruses encode an enzyme called reverse transcriptase. The function of the reverse transcriptase enzyme is
   (A) DNase activity
   (B) RNA-dependent DNA polymerase activity
   (C) DNA-dependent RNA polymerase activity
   (D) RNA-dependent RNA polymerase activity
   (E) Topoisomerase activity

5. Two months after a kidney transplant, a 47-year-old man developed nephropathy. Up to 5% of renal allograft recipients develop nephropathy. A viral cause of some of the nephropathy cases has been identified as
   (A) Polyomavirus BK
   (B) Human papillomavirus, all types
   (C) Human papillomavirus, low-risk types
   (D) Hepatitis C virus
   (E) Human cytomegalovirus

6. Human papillomavirus can cause cancer in humans and is most commonly associated with
   (A) Rectal polyps
   (B) Breast cancer
   (C) Prostate cancer
   (D) Anogenital cancers
   (E) Mesotheliomas

7. A virus that causes human cancer is also associated with a nervous system disorder called tropical spastic paraparesis. That virus is

(A) Polyomavirus JC
(B) Polyomavirus SV40
(C) Herpes simplex virus
(D) Human T-lymphotropic virus
(E) Human immunodeficiency virus

8. The polyomaviruses encode oncoproteins called T antigens. These viral gene products

(A) Are not needed for virus replication
(B) Interact with cellular tumor suppressor proteins
(C) Function to integrate the viral provirus into the cellular chromosome
(D) Mutate rapidly to allow the virus to escape immune clearance by the host
(E) Are not able to transform cells in culture

9. Cancer viruses are classified in several virus families. The following virus family contains a human cancer virus with an RNA genome

(A) Adenoviridae
(B) Herpesviridae
(C) Hepadnaviridae
(D) Papillomaviridae
(E) Flaviviridae

10. Laryngeal papillomas in children are generally caused by the same viruses that cause benign genital condylomas. These viruses are

(A) Papillomaviruses, types 6 and 11
(B) Polyomavirus JC
(C) Epstein-Barr virus
(D) Molluscum contagiosum virus
(E) Papillomaviruses, types 16 and 18

11. Vaccines against the most common HPV types that cause genital infections were approved in 2006 and 2007. They are aimed for use in the following population(s)

(A) All adults, both men and women
(B) All female adults
(C) Women with precancerous cervical lesions
(D) All adolescent and young adults, both boys and girls
(E) Adolescent and young adult females

12. Which of the following best describes available HPV vaccines?

(A) Live attenuated virus
(B) Live recombinant virus
(C) Noninfectious subunit
(D) Toxoid

Answers

1. B	4. B	7. D	10. A
2. C	5. A	8. B	11. E
3. C	6. D	9. E	12. C

REFERENCES

Brechot C (guest editor): Hepatocellular carcinoma. Oncogene Rev 2006;25:3753. [Entire issue.]

Butel JS: Viral carcinogenesis: Revelation of molecular mechanisms and etiology of human disease. Carcinogenesis 2000;21:405. [PMID: 10688861]

Chang MH: Cancer prevention by vaccination against hepatitis B. Recent Results Cancer Res 2009;181:85. [PMID: 19213561]

Dalianis T, Ramqvist T, Andreasson K, Kean JM, Garcea RL: KI, WU and Merkel cell polyomaviruses: A new era for human polyomavirus research. Semin Cancer Biol 2009;19:270. [PMID: 19416753]

de Villiers EM et al: Classification of papillomaviruses. Virology 2004;324:17. [PMID: 15183049]

Goff SP: Retroviridae: The retroviruses and their replication. In: *Fields Virology*, 5th ed. Knipe DM et al (editors). Lippincott Williams & Wilkins, 2007.

Howley PM, Lowy DR: Papillomaviruses. In: *Fields Virology*, 5th ed. Knipe DM et al (editors). Lippincott Williams & Wilkins, 2007.

Human papillomavirus vaccines. WHO position paper. Wkly Epidemiol Rec 2009;84:118.

Imperiale MJ, Major EO: Polyomaviruses. In: *Fields Virology*, 5th ed. Knipe DM et al (editors). Lippincott Williams & Wilkins, 2007.

Javier RT, Butel JS: The history of tumor virology. Cancer Res 2008;68:7693. [PMID: 18829521]

Jeang KT, Yoshida M (guest editors): HTLV-1 and adult T-cell leukemia: 25 years of research on the first human retrovirus. Oncogene Rev 2005;24:5923. [Entire issue.]

Jones-Engel L et al: Diverse contexts of zoonotic transmission of simian foamy viruses in Asia. Emerg Infect Dis 2008;14:1200. [PMID: 18680642]

Kean JM, Rao S, Wang M, Garcea RL: Seroepidemiology of human polyomaviruses. PLoS Pathogens 2009;5:e1000363. [PMID: 19325891]

Khalili K, Raab-Traub N (editors): Cancer viruses. Oncogene Rev 2003;22 (No. 2). [Entire issue.]

Münger K et al: Mechanisms of human papillomavirus-induced oncogenesis. J Virol 2004;78:11451. [PMID: 15479788]

Parsonnet J (editor): *Microbes and Malignancy: Infection as a Cause of Human Cancers.* Oxford University Press, 1999.

Schiffman M, Castle PE, Jeronimo J, Rodriguez AC, Wacholder S: Human papillomavirus and cervical cancer. Lancet 2007;370:890. [PMID: 17826171]

Shroyer KR, Dunn ST (editors): Update on molecular diagnostics for the detection of human papillomavirus. J Clin Virol 2009;45 (Suppl 1):S1. [Entire issue.]

Trottier H, Franco EL: The epidemiology of genital human papillomavirus infection. Vaccine 2006;24(Suppl 1):S1. [PMID: 16406226]

AIDS & Lentiviruses

Human immunodeficiency virus (HIV) types, derived from primate lentiviruses, are the etiologic agents of AIDS. The illness was first described in 1981, and HIV-1 was isolated by the end of 1983. Since then, AIDS has become a worldwide epidemic, expanding in scope and magnitude as HIV infections have affected different populations and geographic regions. Millions are now infected worldwide; once infected, individuals remain infected for life. Within a decade, if left untreated, the vast majority of HIV-infected individuals develop fatal opportunistic infections as a result of HIV-induced deficiencies in the immune system. AIDS is one of the most important public health problems worldwide at the start of the 21st century. The development of highly active antiretroviral therapy (HAART) for chronic suppression of HIV replication and prevention of AIDS has been a major achievement in HIV medicine.

PROPERTIES OF LENTIVIRUSES

Important properties of lentiviruses, members of a genus in the **Retroviridae** family, are summarized in Table 44–1.

Structure & Composition

HIV is a retrovirus, a member of the *Lentivirus* genus, and exhibits many of the physicochemical features typical of the family (see Chapter 43). The unique morphologic characteristic of HIV is a cylindrical nucleoid in the mature virion (Figure 44–1). The diagnostic bar-shaped nucleoid is visible in electron micrographs in those extracellular particles that happen to be sectioned at the appropriate angle.

The RNA genome of lentiviruses is more complex than that of transforming retroviruses (Figure 44–2). Lentiviruses contain the four genes required for a replicating retrovirus—*gag, pro, pol,* and *env*—and follow the general pattern for retrovirus replication (see Chapter 43). Up to six additional genes regulate viral expression and are important in disease pathogenesis in vivo. Although these auxiliary genes show little sequence homology among lentiviruses, their functions are conserved. (The feline and ungulate viruses encode fewer accessory genes.) One early-phase replication protein, the

Tat protein, functions in "transactivation," whereby a viral gene product is involved in transcriptional activation of other viral genes. Transactivation by HIV is highly efficient and may contribute to the virulent nature of HIV infections. The Rev protein is required for the expression of viral structural proteins. Rev facilitates the export of unspliced viral transcripts from the nucleus; structural proteins are translated from unspliced mRNAs during the late phase of viral replication. The Nef protein increases viral infectivity, facilitates activation of resting T cells, and downregulates expression of CD4 and MHC class I. The *nef* gene is necessary for simian immunodeficiency virus (SIV) to be pathogenic in monkeys. The Vpr protein increases transport of the viral preintegration complex into the nucleus and also arrests cells in the G2 phase of the cell cycle. The Vpu protein promotes CD4 degradation.

TABLE 44–1 Important Properties of Lentiviruses (Nononcogenic Retroviruses)

Virion: Spherical, 80–100 nm in diameter, cylindric core
Genome: Single-stranded RNA, linear, positive-sense, 9–10 kb, diploid; genome more complex than that of oncogenic retroviruses, contains up to six additional replication genes
Proteins: Envelope glycoprotein undergoes antigenic variation; reverse transcriptase enzyme contained inside virions; protease required for production of infectious virus
Envelope: Present
Replication: Reverse transcriptase makes DNA copy from genomic RNA; provirus DNA is template for viral RNA. Genetic variability is common.
Maturation: Particles bud from plasma membrane
Outstanding characteristics:
Members are nononcogenic and may be cytocidal
Infect cells of the immune system
Proviruses remain permanently associated with cells
Viral expression is restricted in some cells in vivo
Cause slowly progressive, chronic diseases
Replication is usually species-specific
Group includes the causative agents of AIDS

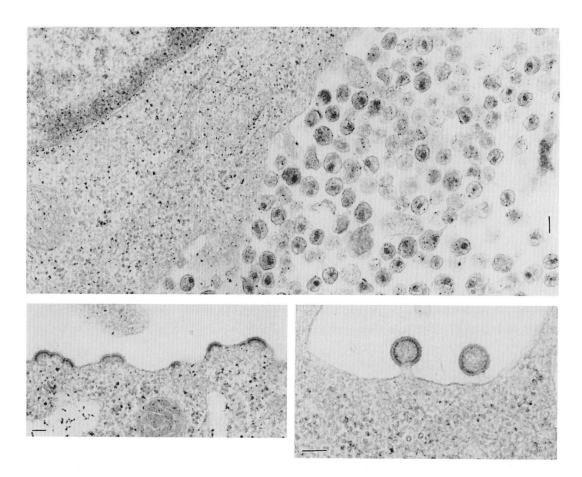

FIGURE 44–1 Electron micrographs of HIV-infected lymphocytes, showing a large accumulation of freshly produced virus at the cell surface (**top**, 46,450×, bar = 100 nm); newly formed virus budding from cytoplasmic membrane (**lower left**, 49,000×, bar = 100 nm); two virions about to be cast off from cell surface (**lower right**, 75,140×, bar = 100 nm).

Cells contain intracellular antiviral inhibitory proteins referred to as restriction factors. One type is APOBEC3G, a cytidine deaminase that inhibits HIV replication. The Vif protein promotes viral infectivity by suppressing the effects of APOBEC3G. Another inhibitory protein is TRIM5α, which binds to incoming retrovirus particles and recruits them to proteasomes before much viral DNA synthesis occurs.

The many different isolates of HIV are not identical but appear to comprise a spectrum of related viruses (see Classification). Heterogeneous populations of viral genomes are found in an infected individual. This heterogeneity reflects high rates of viral replication and the high error rate of the viral reverse transcriptase. The regions of greatest divergence among different isolates are localized to the *env* gene, which codes for the viral envelope proteins (Figure 44–3). The SU (gp120) product of the *env* gene contains binding domains responsible for virus attachment to the CD4 molecule and coreceptors, determines lymphocyte and macrophage tropisms, and carries the major antigenic determinants that elicit neutralizing antibodies. The HIV glycoprotein has five variable (V) regions that diverge among isolates, with the V3 region important in neutralization. The TM (gp41) *env*

product contains both a transmembrane domain that anchors the glycoprotein in the viral envelope and a fusion domain that facilitates viral penetration into target cells. The divergence in the envelope of HIV complicates efforts to develop an effective vaccine for AIDS.

Lentiviruses are completely exogenous viruses; in contrast to the transforming retroviruses, the lentiviral genome does not contain any conserved cellular genes (see Chapter 43). Individuals become infected by the introduction of virus from outside sources.

Classification

Lentiviruses have been isolated from many species (Table 44–2), including more than two dozen different African nonhuman primate species. There are two distinct types of human AIDS viruses: HIV-1 and HIV-2. The two types are distinguished on the basis of genome organization and phylogenetic (evolutionary) relationships with other primate lentiviruses. Sequence divergence between HIV-1 and HIV-2 exceeds 50%.

Based on *env* gene sequences, HIV-1 comprises three distinct virus groups (M, N, and O); the predominant

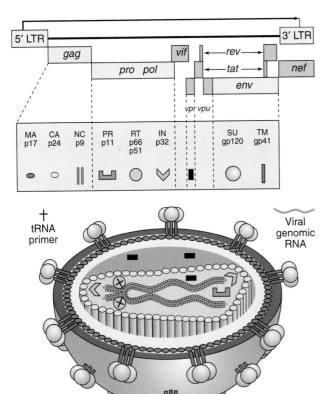

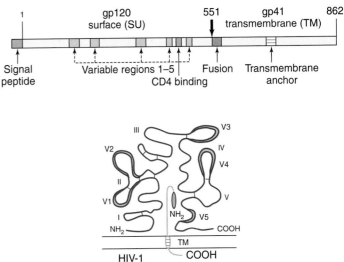

FIGURE 44–2 HIV genome and virion structure. The HIV-1 genome is shown at the top. Viral proteins are synthesized as precursor polyproteins (Gag-Pol [Pr160], Gag [Pr55], and Env [gp160]), which are enzymatically processed to yield mature virion proteins. Gag-Pol and Gag are cleaved by the viral protease PR to produce the indicated smaller proteins. Env is cleaved by a cellular PR, producing SU gp120 and TM gp41. The placements of virion proteins in the virus particle are indicated by symbols (bottom of figure). Exact positions of the proteins PR, RT, and IN in the viral core are not known. HIV-2 and SIV lack the *vpu* gene but contain a *vpx* gene. (Reproduced from Peterlin BM: Molecular biology of HIV. In: *The Retroviridae*. Levy JA [editor]. Plenum, 1995. Modified there from Luciw PA, Shacklett BL in: *HIV: Molecular Organization, Pathogenicity and Treatment*. Morrow WJW, Haigwood NL [editors]. Elsevier, 1993.)

FIGURE 44–3 HIV-1 envelope proteins. The gp160 precursor polypeptide is shown at the top. The gp120 subunit is on the outside of the cell, and gp41 is a transmembrane protein. Hypervariable domains in gp120 are designated V1 through V5; the positions of disulfide bonds are shown as connecting lines in the loops. Important regions in the gp41 subunit are the fusion domain at the amino terminal and the transmembrane domain (TM). Amino (NH₂) and carboxyl (COOH) terminals are labeled for both subunits. (Reproduced from Peterlin BM: Molecular biology of HIV. In: *The Viruses*. Vol 4: *The Retroviridae*. Levy JA [editor]. Plenum, 1995. Modified there from Myers G et al: *Human Retroviruses and AIDS 1993: A Compilation and Analysis of Nucleic Acid and Amino Acid Sequences*. Theoretical Biology and Biophysics Group T-10, Los Alamos National Library, Los Alamos, New Mexico.)

M group contains at least ten subtypes or "clades" (A–J). Recombinant forms of virus are also found in circulation in humans in different geographic regions. Similarly, five subtypes of HIV-2 (A–E) have been identified. Within each subtype there is extensive variability. The genetic clades do not seem to correspond to neutralization serotype groups, and there is currently no evidence that subtypes differ in biology or pathogenesis.

Numerous lentivirus isolates have been obtained from nonhuman primate species. The primate lentiviruses fall into six major phylogenetic lineages (Table 44–2). SIV from sooty mangabeys (a type of monkey in West Africa) and HIV-2 are considered to be variants of the same virus, as are chimpanzee isolates and HIV-1. The SIVs from African green monkeys, Sykes monkeys, mandrills, and colobus monkeys represent additional discrete lineages.

The organization of the genomes of primate lentiviruses (human and simian) is very similar. One difference is that HIV-1 and the chimpanzee virus carry a *vpu* gene, whereas HIV-2 and the SIV_sm group have a *vpx* gene. Other SIV isolates have neither *vpu* nor *vpx* genes. The sequences of the *gag* and *pol* genes are highly conserved. There is significant divergence among the envelope glycoprotein genes; the sequences of the transmembrane protein portion are more conserved than the external glycoprotein sequences (the protein component exposed on the exterior of the virus particle).

The SIVs appear to be nonpathogenic in their host species of origin (eg, chimpanzee, African green monkey, sooty mangabey), species known to be infected in their natural habitats. In contrast, rhesus monkeys are not infected naturally in the wild in Asia but are susceptible to induction of simian AIDS by various SIV isolates. The virus first recovered from captive rhesus monkeys (SIV_mac) is the sooty mangabey/HIV-2 strain.

The nonprimate lentiviruses establish persistent infections affecting various animal species. These viruses cause chronic debilitating diseases and sometimes immunodeficiency. The prototype agent, visna virus (also called maedi virus), causes neurologic symptoms or pneumonia in sheep in Iceland. Other viruses cause infectious anemia in horses and arthritis

TABLE 44–2   **Representative Members of the *Lentivirus* Genus**

Origin of Isolates	Virus	Diseases
Humans	HIV-1 (SIV$_{cpz}$)[a]	AIDS
	HIV-2 (SIV$_{sm}$)	
Nonhuman primates[b]		
Chimpanzee	SIV$_{cpz}$	Simian AIDS
Sooty mangabey	SIV$_{sm}$	
Macaques[c]	SIV$_{mac}$	
African green monkey	SIV$_{agm}$	
Sykes monkey	SIV$_{syk}$	
Mandrill	SIV$_{mnd}$	
l'Hoest monkey[c]	SIV$_{lhoest}$	
Colobus monkey	SIV$_{col}$	
Nonprimates[d]		
Cat	Feline immunodeficiency virus	Feline AIDS
Cow	Bovine immunodeficiency virus	
Sheep	Visna/maedi virus	Lung, central nervous system disease
Horse	Equine infectious anemia virus	Anemia
Goat	Caprine arthritis encephalitis virus	Arthritis, encephalitis

[a]The origins of HIV-1 and HIV-2 were cross-species transmissions of SIV$_{cpz}$ and SIV$_{sm}$, respectively.

[b]Disease not caused in host of origin by SIVs but requires transmission to a different species of monkey (rhesus are the most susceptible to disease). The Asian macaques (rhesus) show no evidence of SIV infection in the wild; SIV$_{sm}$ was probably accidentally introduced to macaques in captivity.

[c]Indention indicates that the virus is in the same phylogenetic lineage as the one above it.

[d]Nonprimate lentiviruses cause disease in species of origin.

and encephalitis in goats. Feline and bovine lentiviruses may cause an immunodeficiency. Nonprimate lentiviruses are not known to infect any primates, including humans.

Origin of AIDS

HIV in humans originated from cross-species infections by simian viruses in rural Africa, probably due to direct human contact with infected primate blood. Current evidence is that the primate counterparts of HIV-1 and HIV-2 were transmitted to humans on multiple (at least seven) different occasions. Sequence evolution analyses place the introduction of SIV$_{cpz}$ into humans that gave rise to HIV-1 group M at about 1930, although some estimates push the date back to about 1908. Presumably, such transmissions occurred repeatedly over the ages, but particular social, economic, and behavioral changes that occurred in the mid 20th century provided circumstances that allowed these virus infections to expand, become well-established in humans, and reach epidemic proportions.

Disinfection & Inactivation

HIV is completely inactivated ($\geq 10^5$ units of infectivity) by treatment for 10 minutes at room temperature with any of the following: 10% household bleach, 50% ethanol, 35% isopropanol, 1% Nonidet P40, 0.5% Lysol, 0.5% paraformaldehyde, or 0.3% hydrogen peroxide. The virus is also inactivated by extremes of pH (pH 1.0, pH 13.0). When HIV is present in clotted or unclotted blood in a needle or syringe, exposure to undiluted bleach for at least 30 seconds is necessary for inactivation.

The virus is not inactivated by 2.5% Tween 20. Although paraformaldehyde inactivates virus free in solution, it is not known if it penetrates tissues sufficiently to inactivate all virus that might be present in cultured cells or tissue specimens.

HIV is readily inactivated in liquids or 10% serum by heating at 56°C for 10 minutes, but dried proteinaceous material affords marked protection. Lyophilized blood products would need to be heated at 68°C for 72 hours to ensure inactivation of contaminating virus.

Animal Lentivirus Systems

Insights into the biologic characteristics of lentivirus infections have been gained from experimental infections, including sheep with visna virus (Table 44–2). Natural disease patterns vary among species, but certain common features are recognized.

1. Viruses are transmitted by exchange of body fluids.
2. Virus persists indefinitely in infected hosts, though it may be present at very low levels.
3. Viruses have high mutation rates, and different mutants will be selected under different conditions (host factors, immune responses, tissue types). Infected hosts contain

"swarms" of closely related viral genomes, known as quasi species.

4. Virus infection progresses slowly through specific stages. Cells in the macrophage lineage play central roles in the infection. Lentiviruses differ from other retroviruses in that they can infect nondividing, terminally differentiated cells. However, those cells must be activated before viral replication ensues and progeny virus is produced. Virus is cell-associated in monocytes and macrophages, but only about one cell per million is infected. Monocytes carry the virus around the body in a form that the immune system cannot recognize, seeding other tissues. Lymphocyte-tropic strains of virus tend to cause highly productive infections, whereas replication of macrophage-tropic virus is restricted.

5. It may take years for disease to develop. Infected hosts usually make antibodies, but they do not clear the infection, so virus persists lifelong. New antigenic variants periodically arise in infected hosts, with most mutations occurring in envelope glycoproteins. Clinical symptoms may develop at any time from 3 months to many years after infection. The exceptions to long incubation periods for lentivirus disease include AIDS in children, infectious anemia in horses, and encephalitis in young goats.

Host factors important in pathogenesis of disease include age (the young are at greater risk), stress (may trigger disease), genetics (certain breeds of animals are more susceptible), and concurrent infections (may exacerbate disease or facilitate virus transmission).

The diseases in ungulates (horses, cattle, sheep, and goats) are not complicated by opportunistic secondary infections. Equine infectious anemia virus can be spread among horses by blood-sucking horseflies, the only lentivirus known to be transmitted by an insect vector.

Simian lentiviruses share molecular and biologic characteristics with HIV and cause an AIDS-like disease in rhesus macaques. The SIV model is important for understanding disease pathogenesis and developing vaccine and treatment strategies.

Virus Receptors

All primate lentiviruses use as a receptor the CD4 molecule, which is expressed on macrophages and T lymphocytes. A second coreceptor in addition to CD4 is necessary for HIV-1 to gain entry to cells. The second receptor is required for fusion of the virus with the cell membrane. The virus first binds to CD4 and then to the coreceptor. These interactions cause conformational changes in the viral envelope, activating the gp41 fusion peptide and triggering membrane fusion. Chemokine receptors serve as HIV-1 second receptors. (Chemokines are soluble factors with chemoattractant and cytokine properties.) CCR5, the receptor for chemokines RANTES, MIP-1α, and MIP-1β, is the predominant coreceptor for macrophage-tropic strains of HIV-1, whereas CXCR4,

the receptor for chemokine SDF-1, is the coreceptor for lymphocyte-tropic strains of HIV-1. The chemokine receptors used by HIV for cell entry are found on lymphocytes, macrophages, and thymocytes as well as on neurons and cells in the colon and cervix. Individuals who possess homozygous deletions in CCR5 and produce mutant forms of the protein may be protected from infection by HIV-1; mutations in the CCR5 gene promoter appear to delay disease progression. The requirement for a coreceptor for HIV fusion with cells provided new targets for antiviral therapeutic strategies, with the first HIV entry inhibitor licensed in the United States in 2003.

Another molecule, integrin α-4 β-7, appears to function as a receptor for HIV in the gut. A dendritic cell-specific lectin, DC-SIGN, appears to bind HIV-1 but not to mediate cell entry. Rather, it may facilitate transport of HIV by dendritic cells to lymphoid organs and enhance infection of T cells.

HIV INFECTIONS IN HUMANS

Pathogenesis & Pathology

A. Overview of Course of HIV Infection

The typical course of untreated HIV infection spans about a decade (Figure 44–4). Stages include the primary infection, dissemination of virus to lymphoid organs, clinical latency, elevated HIV expression, clinical disease, and death. The duration between primary infection and progression to clinical disease averages about 10 years. In untreated cases, death usually occurs within 2 years after the onset of clinical symptoms.

Following primary infection, there is a 4- to 11-day period between mucosal infection and initial viremia; the viremia is detectable for about 8–12 weeks. Virus is widely disseminated throughout the body during this time, and the lymphoid organs become seeded. An acute mononucleosis-like syndrome develops in many patients (50–75%) 3–6 weeks after primary infection. There is a significant drop in numbers of circulating CD4 T cells at this early time. An immune response to HIV occurs 1 week to 3 months after infection, plasma viremia drops, and levels of CD4 cells rebound. However, the immune response is unable to clear the infection completely, and HIV-infected cells persist in the lymph nodes.

This period of clinical latency may last for as long as 10 years. During this time, there is a high level of ongoing viral replication. It is estimated that 10 billion HIV particles are produced and destroyed each day. The half-life of the virus in plasma is about 6 hours, and the virus life cycle (from the time of infection of a cell to the production of new progeny that infect the next cell) averages 2.6 days. CD4 T lymphocytes, major targets responsible for virus production, appear to have similar high turnover rates. Once productively infected, the half-life of a CD4 lymphocyte is about 1.6 days. Because of this rapid viral proliferation and the inherent error rate of the

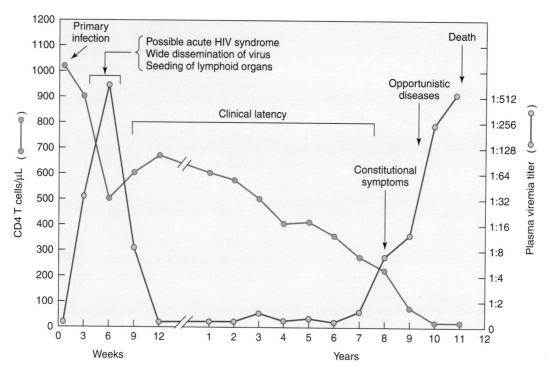

FIGURE 44–4 Typical course of untreated HIV infection. During the early period after primary infection, there is widespread dissemination of virus and a sharp decrease in the number of CD4 T cells in peripheral blood. An immune response to HIV ensues, with a decrease in detectable viremia followed by a prolonged period of clinical latency. Sensitive assays for viral RNA show that virus is present in the plasma at all times. The CD4 T cell count continues to decrease during the following years until it reaches a critical level below which there is a substantial risk of opportunistic diseases. (Reproduced with permission from Pantaleo G, Graziosi C, Fauci AS: New concepts in the immunopathogenesis of human immunodeficiency virus infection. N Engl J Med 1993;328:327.)

HIV reverse transcriptase, it is estimated that every nucleotide of the HIV genome probably mutates on a daily basis.

Eventually, the patient will develop constitutional symptoms and clinically apparent disease, such as opportunistic infections or neoplasms. Higher levels of virus are readily detectable in the plasma during the advanced stages of infection. HIV found in patients with late-stage disease is usually much more virulent and cytopathic than the strains of virus found early in infection. Often, a shift from monocyte-tropic or macrophage-tropic (M-tropic) strains of HIV-1 to lymphocyte-tropic (T-tropic) variants accompanies progression to AIDS.

B. CD4 T Lymphocytes, Memory Cells, & Latency

The cardinal feature of HIV infection is the depletion of T helper-inducer lymphocytes—the result of HIV replication in this population of lymphocytes as well as of the death of uninfected T cells by indirect mechanisms. The T cells express the CD4 phenotypic marker on their surface. The CD4 molecule is the major receptor for HIV; it has a high affinity for the viral envelope. The HIV coreceptor on lymphocytes is the CXCR4 chemokine receptor.

Early in infection, primary HIV isolates are M-tropic. However, all strains of HIV infect primary CD4 T lymphocytes (but not immortalized T cell lines in vitro). As the infection progresses, the dominant M-tropic viruses are replaced by T-tropic viruses. Laboratory adaptation of these primary isolates in immortalized T cell lines results in loss of ability to infect monocytes and macrophages.

The consequences of CD4 T cell dysfunction caused by HIV infection are devastating because the CD4 lymphocyte plays a critical role in the human immune response. It is responsible directly or indirectly for induction of a wide array of lymphoid and nonlymphoid cell functions. These effects include activation of macrophages; induction of functions of cytotoxic T cells, natural killer cells, and B cells; and secretion of a variety of soluble factors that induce growth and differentiation of lymphoid cells and affect hematopoietic cells.

At any given time, only a small fraction of CD4 T cells are productively infected. Many infected T cells are killed, but a fraction survives and reverts to a resting memory state. There is little or no virus gene expression in the memory cells, and they provide a long-term, stable latent reservoir for the virus. Less than 1 cell per million resting CD4 T cells harbor latent HIV-1 provirus in patients on successful antiretroviral therapy. Even after 10 years of treatment, patients show very little change in the size of the latent HIV reservoir because the latent reservoir of infected memory cells decays very slowly. It is unlikely that an HIV infection can be cured; if there were a million infected memory cells in the body, it would take about 70 years for them to decay. When exposed to antigen or when drug therapy is discontinued, the memory cells become activated and release infectious virus. It is possible that other drug-insensitive reservoirs may also exist among macrophages, hematopoietic stem cells, or brain cells.

C. Monocytes & Macrophages

Monocytes and macrophages play a major role in the dissemination and pathogenesis of HIV infection. Certain subsets of monocytes express the CD4 surface antigen and therefore bind to the envelope of HIV. The HIV coreceptor on monocytes and macrophages is the CCR5 chemokine receptor. In the brain, the major cell types infected with HIV appear to be the monocytes and macrophages, and this may have important consequences for the development of neuropsychiatric manifestations associated with HIV infection. Infected pulmonary alveolar macrophages may play a role in the interstitial pneumonitis seen in certain patients with AIDS.

Macrophage-tropic strains of HIV predominate early after infection, and these strains are responsible for initial infections even when the transmitting source contains both M-tropic and T-tropic viruses.

It is believed that monocytes and macrophages serve as major reservoirs for HIV in the body. Unlike the CD4 T lymphocyte, the monocyte is relatively refractory to the cytopathic effects of HIV, so that the virus not only can survive in this cell but can be transported to various organs in the body (such as the lungs and brain). Infected macrophages may continue to produce virus for a long period of time.

D. Lymphoid Organs

Lymphoid organs play a central role in HIV infection. Lymphocytes in the peripheral blood represent only about 2% of the total lymphocyte pool, the remainder being located chiefly in lymphoid organs. It is in the lymphoid organs that specific immune responses are generated. The network of follicular dendritic cells in the germinal centers of lymph nodes traps antigens and stimulates an immune response. Throughout the course of untreated infection—even during the stage of clinical latency—HIV is actively replicating in lymphoid tissues. The microenvironment of the lymph node is ideal for the establishment and spread of HIV infection. Cytokines are released, activating a large pool of CD4 T cells that are highly susceptible to HIV infection. As the late stages of HIV disease progress, the architecture of the lymph nodes becomes disrupted.

E. Neural Cells

Neurologic abnormalities are common in late stages of infection and are an AIDS-defining condition. Central nervous system disease occurs to varying degrees in 40–90% of patients. These include HIV encephalopathy, peripheral neuropathies, and—most serious—AIDS dementia complex. Both direct and indirect pathogenic mechanisms might explain the neuropsychiatric manifestations of HIV infection. The predominant cell types in the brain that are infected with HIV are monocytes and macrophages. Virus may enter the brain through infected monocytes and release cytokines that are toxic to neurons as well as chemotactic factors that lead to infiltration of the brain with inflammatory cells. HIV is present rarely, if at all, in neurons, oligodendrocytes, and astrocytes.

F. Viral Coinfections

Activation signals are required for the establishment of a productive HIV infection. In the HIV-infected individual, a wide range of in vivo antigenic stimuli seem to serve as cellular activators. For example, active infection by *Mycobacterium tuberculosis* substantially increases plasma viremia. The damaging effects of HIV on the immune system leave patients vulnerable to many types of infection. The World Health Organization reports that infection with HIV increases the risk of getting tuberculosis as much as 20-fold. Of the 9 million new tuberculosis cases worldwide in 2007, it is estimated that 15% occurred in persons infected with HIV.

Other concomitant viral infections—with Epstein-Barr virus, cytomegalovirus, herpes simplex virus, or hepatitis B virus—may serve as cofactors of AIDS. Hepatitis C virus coinfection, which occurs in 15–30% of HIV cases in the United States and often results in liver disease, is a leading cause of morbidity and mortality in HIV-infected persons. There is also a high prevalence of cytomegalovirus infection in HIV-positive individuals.

Coinfections with two different strains of HIV can occur. There are documented cases of superinfection with a second strain in an HIV-infected individual, even in the presence of a strong CD8 T cell response to the first strain. HIV superinfection is considered to be a rare event.

Clinical Findings

Symptoms of acute HIV infection are nonspecific and include fatigue, rash, headache, nausea, and night sweats. AIDS is characterized by pronounced suppression of the immune system and development of a wide variety of severe opportunistic infections or unusual neoplasms (especially Kaposi sarcoma). The more serious symptoms in adults are often preceded by a prodrome ("diarrhea and dwindling") that can include fatigue, malaise, weight loss, fever, shortness of breath, chronic diarrhea, white patches on the tongue (hairy leukoplakia, oral candidiasis), and lymphadenopathy. Disease symptoms in the gastrointestinal tract from the esophagus to the colon are a major cause of debility. With no treatment, the interval between primary infection with HIV and the first appearance of clinical disease is usually long in adults, averaging about 8–10 years. Death occurs about 2 years later.

A. Plasma Viral Load

The amount of HIV in the blood (viral load) is of significant prognostic value. There are continual rounds of viral replication and cell killing in each patient, and the steady-state level of virus in the blood (viral set point) varies from individual to individual during the asymptomatic period. This level reflects the total number of productively infected cells and their average burst size. It turns out that a single measurement of plasma viral load about 6 months after infection is able to predict the subsequent risk of development of AIDS in men several years later (Figure 44–5). High set points tend to correlate with rapid disease progression and poorer responses to treatment. However, more recent data suggest a gender difference in this parameter—in women, the viral load may be less predictive of progression to AIDS. Plasma HIV RNA levels can be determined using a variety of commercially available assays. The plasma viral load appears to be the best predictor of long-term clinical outcome, whereas CD4 lymphocyte counts are the best predictor of short-term risk of developing an opportunistic disease. Plasma viral load measurements are a critical element in assessing the effectiveness of antiretroviral drug therapy.

B. Pediatric AIDS

The responses of infected neonates are different from those observed in HIV-infected adults. Pediatric AIDS—acquired from infected mothers—usually presents with clinical symptoms by 2 years of age; death follows in another 2 years. The neonate is particularly susceptible to the devastating effects of HIV because the immune system has not developed at the time of primary infection. Clinical findings may include lymphoid interstitial pneumonitis, pneumonia, severe oral candidiasis, encephalopathy, wasting, generalized lymphadenopathy,

bacterial sepsis, hepatosplenomegaly, diarrhea, and growth retardation.

Children with perinatally acquired HIV-1 infection—*if untreated*—have a very poor prognosis. A high rate of disease progression occurs in the first few years of life. High levels of plasma HIV-1 load appear to predict infants at risk of rapid progression of disease. The pattern of viral replication in infants differs from that in adults. Viral RNA load levels are generally low at birth, suggesting infection acquired close to that time; RNA levels then rise rapidly within the first 2 months of life and are followed by a slow decline until the age of 24 months, suggesting that the immature immune system has difficulty containing the infection. A small percentage of infants (≤5%) display transient HIV infections, suggesting that some infants can clear the virus.

C. Neurologic Disease

Neurologic dysfunction occurs frequently in HIV-infected persons. Forty to ninety percent of patients have neurologic symptoms, and many are found during autopsy to have neuropathologic abnormalities.

Several distinct neurologic syndromes that occur frequently include subacute encephalitis, vacuolar myelopathy, aseptic meningitis, and peripheral neuropathy. AIDS dementia complex, the most common neurologic syndrome, occurs as a late manifestation in 25–65% of AIDS patients and is characterized by poor memory, inability to concentrate, apathy, psychomotor retardation, and behavioral changes. Other neurologic diseases associated with HIV infection include toxoplasmosis, cryptococcosis, primary lymphoma of the central nervous system, and JC virus-induced progressive multifocal leukoencephalopathy. Mean survival time from onset of severe dementia is usually less than 6 months.

Pediatric AIDS patients also display neurologic abnormalities. These include seizure disorders, progressive loss of behavioral developmental milestones, encephalopathy, attention deficit disorders, and developmental delays. HIV encephalopathy may occur in as many as 12% of children, usually accompanied by profound immune deficiency. Bacterial pathogens predominate in pediatric AIDS as the most common cause of meningitis.

As children born with HIV infection are living to adolescence due to antiretroviral therapy, many appear to be at high risk for psychiatric disorders. The most common problems are anxiety disorders.

D. Opportunistic Infections

The predominant causes of morbidity and mortality among patients with late-stage HIV infection are opportunistic infections, ie, severe infections induced by agents that rarely cause serious disease in immune-competent individuals. Opportunistic infections usually do not occur in HIV-infected patients until CD4 T cell counts have dropped from the normal level of about 1000 cells/μL to less than

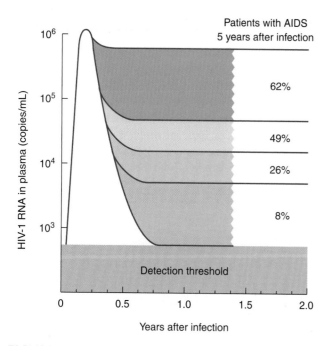

FIGURE 44–5 Prognostic value of HIV-1 RNA levels in the plasma (viral load). The virologic setpoint predicts the long-term clinical outcome. (Reproduced with permission from Ho DD: Viral counts count in HIV infection. Science 1996;272:1124.)

200 cells/μL. As treatments are developed for some common opportunistic pathogens and management of AIDS patients permits longer survivals, the spectrum of opportunistic infections changes.

The most common opportunistic infections in untreated AIDS patients include the following:

1. Protozoa: *Toxoplasma gondii*, *Isospora belli*, *Cryptosporidium* species.
2. Fungi: *Candida albicans*, *Cryptococcus neoformans*, *Coccidioides immitis*, *Histoplasma capsulatum*, *Pneumocystis jiroveci*.
3. Bacteria: *Mycobacterium avium-intracellulare*, *M tuberculosis*, *Listeria monocytogenes*, *Nocardia asteroides*, *Salmonella* species, *Streptococcus* species.
4. Viruses: Cytomegalovirus, herpes simplex virus, varicella-zoster virus, adenovirus, polyomavirus, JC virus, hepatitis B virus, hepatitis C virus.

Herpesvirus infections are common in AIDS patients, and multiple herpesviruses are frequently detected being shed in saliva. Cytomegalovirus retinitis is the most common severe ocular complication of AIDS.

E. Cancer

AIDS patients exhibit a marked predisposition to the development of cancer, another consequence of immune suppression. AIDS-associated cancers tend to be those with a viral cofactor and include non-Hodgkin lymphoma (both systemic and central nervous system types), Kaposi sarcoma, cervical cancer, and anogenital cancers. Epstein-Barr viral DNA is found in the majority of B cell malignancies classified as Burkitt lymphoma and those of the central nervous system (but is not found in most of the systemic lymphomas). Polyomavirus SV40 has been detected in some non-Hodgkin lymphomas. Burkitt lymphoma occurs 1000 times more commonly in AIDS patients than in the general population.

Kaposi sarcoma is a vascular tumor thought to be of endothelial origin that appears in skin, mucous membranes, lymph nodes, and visceral organs. Before this type of malignancy was observed in AIDS patients, it was considered to be a very rare cancer. Kaposi sarcoma is 20,000 times more common in untreated AIDS patients than in the general population. Kaposi sarcoma-associated herpesvirus, or HHV8, appears to be causally related to the cancer (see Chapter 33). Cervical cancer is caused by high-risk papillomaviruses; the anogenital cancers also arise as a result of coinfections with human papillomaviruses (see Chapter 43).

Effective antiretroviral drug therapy has resulted in a marked reduction in the occurrence of Kaposi sarcomas but has had less of an effect on the incidence of non-Hodgkin lymphomas in HIV-infected individuals.

As HIV-infected persons live longer lives due to effective antiretroviral therapy, they are developing a broad spectrum of cancers at higher frequencies than the noninfected population. These HIV-associated malignancies include head and neck cancer, lung cancer, Hodgkin lymphoma, liver cancer, melanoma, and oral cancer. There does not appear to be an increased risk of breast, colon, or prostate cancer.

Immunity

HIV-infected persons develop both humoral and cell-mediated responses against HIV-related antigens. Antibodies to a number of viral antigens develop soon after infection (Table 44–3).

Most infected individuals make neutralizing antibodies against HIV, directed against the envelope glycoprotein. However, the levels of neutralizing activity are low; many anti-envelope antibodies are nonneutralizing. It is believed that the dense glycosylation may inhibit binding of neutralizing antibody to the envelope protein. The envelope glycoprotein shows great sequence variability. This natural variation may allow the evolution of successive populations of resistant virus that escape recognition by existing neutralizing antibodies.

The neutralizing antibodies can be measured in vitro by inhibiting HIV infection of susceptible lymphocyte cell lines. Viral infection is quantified by (1) reverse transcriptase assay, which measures the enzyme activity of released HIV particles; (2) indirect immunofluorescence assay, which measures the percentage of infected cells; and (3) reverse transcriptase-polymerase chain reaction (RT-PCR) or branched-chain DNA (bDNA) amplification assays that measure HIV nucleic acids.

Cellular responses develop that are directed against HIV proteins. Cytotoxic T lymphocytes (CTLs) recognize *env*, *pol*, *gag*, and *nef* gene products; this reactivity is mediated

TABLE 44–3 Major Gene Products of HIV that Are Useful in Diagnosis of Infection

Gene Product[a]	Description
gp160[b]	Precursor of envelope glycoproteins
gp120[b]	Outer envelope glycoprotein of virion, SU[c]
p66	Reverse transcriptase and RNase H from polymerase gene product
p55	Precursor of core proteins, polyprotein from *gag* gene
p51	Reverse transcriptase, RT
gp41[b]	Transmembrane envelope glycoprotein, TM
p32	Integrase, IN
p24[b]	Nucleocapsid core protein of virion, CA
p17	Matrix core protein of virion, MA

[a]Number refers to the approximate molecular mass of the protein in kilodaltons.

[b]Antibodies to these viral proteins are the most commonly detected.

[c]Two-letter abbreviation for viral protein.

by major histocompatibility complex-restricted CD3–CD8 lymphocytes. The *env*-specific reactivity occurs in nearly all infected people and decreases with progression of disease. Natural killer (NK) cell activity has also been detected against HIV-1 gp120.

It is not clear which host responses are important in providing protection against HIV infection or development of disease. A problem confronting AIDS vaccine research is that the correlates of protective immunity are not known, including the relative importance of humoral and cell-mediated immune responses.

Laboratory Diagnosis

Evidence of infection by HIV can be detected in three ways: (1) virus isolation; (2) serologic determination of antiviral antibodies; and (3) measurement of viral nucleic acid or antigens.

A. Virus Isolation

HIV can be cultured from lymphocytes in peripheral blood (and occasionally from specimens from other sites). The numbers of circulating infected cells vary with the stage of disease (Figure 44–4). Higher titers of virus are found in the plasma and in peripheral blood cells of patients with AIDS as compared with asymptomatic individuals. The magnitude of plasma viremia appears to be a better correlate of the clinical stage of HIV infection than the presence of any antibodies (Figure 44–6). The most sensitive virus isolation technique is to cocultivate the test sample with uninfected, mitogen-stimulated peripheral blood mononuclear cells. Primary isolates of HIV grow very slowly compared with laboratory-adapted strains. Viral growth is detected by testing culture

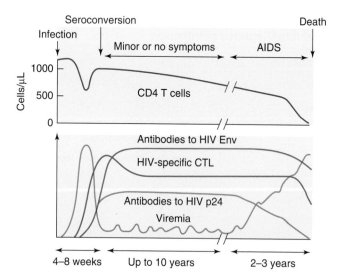

FIGURE 44–6 Pattern of HIV antibody responses related to the course of HIV infection. (PBL, peripheral blood lymphocytes; CTL, cytotoxic T lymphocytes.) (Reproduced with permission from Weiss RA: How does HIV cause AIDS? Science 1993;260:1273.)

supernatant fluids after about 7–14 days for viral reverse transcriptase activity or for virus-specific antigens (p24).

The vast majority of HIV-1 antibody-positive persons will have virus that can be cultured from their blood cells. However, virus isolation techniques are time-consuming and laborious and are limited to research studies. PCR amplification techniques are commonly used for detection of virus in clinical specimens.

B. Serology

Test kits are commercially available for measuring antibodies by enzyme-linked immunoassay (EIA). If properly performed, these tests have a sensitivity and specificity exceeding 98%. When EIA-based antibody tests are used for screening populations with a low prevalence of HIV infections (eg, blood donors), a positive test in a serum sample must be confirmed by a repeat test. If the repeat EIA test is reactive, a confirmation test is performed to rule out false-positive EIA results. The most widely used confirmation assay is the Western blot technique, in which antibodies to HIV proteins of specific molecular weights can be detected. Antibodies to viral core protein p24 or envelope glycoproteins gp41, gp120, or gp160 are most commonly detected.

The response pattern against specific viral antigens changes over time as patients progress to AIDS. Antibodies to the envelope glycoproteins (gp41, gp120, gp160) are maintained, but those directed against the Gag proteins (p17, p24, p55) decline. The decline of anti-p24 may herald the beginning of clinical signs and other immunologic markers of progression (Figure 44–6).

Simple, rapid tests for detecting HIV antibodies are available for use in laboratories ill-equipped to perform EIA tests and in settings where test results are desired with little delay. The simple tests can be performed on blood or oral fluid and are based on principles such as particle agglutination or immunodot reactions. There are rapid tests that can detect HIV antibodies in whole blood specimens that require no processing. These tests can be performed outside the traditional laboratory setting.

Home testing kits are available. The procedure involves placing drops of blood from a finger prick on a specially treated card. The card is then mailed to a licensed laboratory for testing.

The mean time to seroconversion after HIV infection is 3–4 weeks. Most individuals will have detectable antibodies within 6–12 weeks after infection, whereas virtually all will be positive within 6 months. HIV infection for longer than 6 months without a detectable antibody response is very uncommon.

C. Detection of Viral Nucleic Acid or Antigens

Amplification assays such as the RT-PCR, DNA PCR, and bDNA tests are commonly used to detect viral RNA in clinical specimens. The RT-PCR assay uses an enzymatic method to amplify HIV RNA; the bDNA assay amplifies viral RNA by sequential oligonucleotide hybridization steps. The tests can be quantitative when reference standards are used; appropriate positive and negative controls must be included with each test. These molecular-based tests are very sensitive and form the basis for plasma viral load determinations. HIV sequence heterogeneity may limit the sensitivity of these assays to detect HIV infections. The HIV RNA levels are important predictive markers of disease progression and valuable tools with which to monitor the effectiveness of antiviral therapies.

Early diagnosis of HIV infection in infants born to infected mothers can be accomplished using plasma HIV-1 RNA tests. The presence of maternal antibodies makes serologic tests uninformative.

Low levels of circulating HIV-1 p24 antigen can be detected in the plasma by EIA soon after infection. The antigen often becomes undetectable after antibodies develop (because the p24 protein is complexed with p24 antibodies) but may reappear late in the course of infection, indicating a poor prognosis.

Epidemiology

A. Worldwide Spread of AIDS

AIDS was first recognized in the United States in 1981 as a new disease entity in homosexual men. Twenty years later, AIDS had become a worldwide epidemic that continues to expand. The Joint United Nations Program on HIV/AIDS estimated that by the end of 2007, a total of 33 million people worldwide were living with HIV/AIDS, the majority having been infected by heterosexual contact (Figure 44–7). It was estimated that in that year, 2.0 million people died of AIDS and that 2.7 million new infections with HIV occurred, including

370,000 children, many of whom were babies infected perinatally. By the year 2005, the World Health Organization estimated that more than 25 million people worldwide had died of AIDS and that over 15 million children had been orphaned, 12 million of whom were living in sub-Saharan Africa.

The epidemic varies by geographic region. Based on 2007 data, sub-Saharan Africa had the highest number of HIV infections (Figure 44–7). In certain high-prevalence cities in Africa, as many as one of every three adults was infected with the virus. The epidemic here appears to have stabilized, although often at high levels. Antiretroviral therapies are being introduced in some of these countries. Infections were spreading also in southern and southeastern Asia (especially in India, China, and Russia). Because AIDS tends to strike young adults and workers in their prime, the AIDS epidemic is having devastating effects on social and economic structures in some countries.

Group M viruses are responsible for most HIV-1 infections worldwide, but subtype distributions vary. Subtype C predominates in southern Africa, subtype A in West Africa, and subtype B in the United States, Europe, and Australia. HIV-2 has remained localized primarily to West Africa.

The World Health Organization estimates that of the 2.7 million new HIV infections each year, 90% are occurring in developing countries. In those countries, AIDS is overwhelmingly a heterosexually transmitted disease, and there are about equal numbers of male and female cases.

It is hypothesized that the rapid dissemination of HIV globally in the latter part of the 20th century was fostered by massive migration of rural inhabitants to urban centers, coupled with international movement of infected individuals as a consequence of civil disturbances, tourism, and business travels.

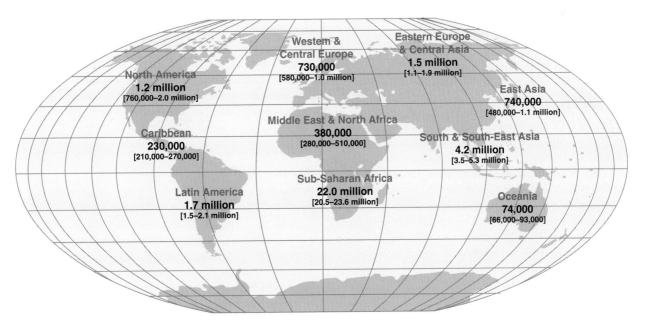

FIGURE 44–7 Adults and children estimated to be living with HIV/AIDS, by continent or region, as of December 2007. It is estimated that about 2.0 million people worldwide died of HIV/AIDS in 2007. (Data from the Joint United Nations Program on HIV/AIDS.)

B. United States

The face of the AIDS epidemic has changed in the United States since 1981. At first, most of the cases occurred in homosexual men. Then the disease was identified in injecting drug users. By 2005, racial and ethnic minority communities were disproportionately affected, accounting for about two-thirds of reported HIV/AIDS cases. Heterosexual transmission was increasingly more common, and about one-quarter of new diagnoses were in women. Most heterosexually acquired AIDS cases were attributed to sexual contact with an injecting drug user or a partner with HIV infection. Despite recommendations issued by the Centers for Disease Control and Prevention in 2006 to have HIV screening be part of routine medical care for persons aged 13–64 years, it is estimated that one-quarter of those living with HIV are unaware of their infection.

By the end of 2007, over 1.5 million HIV/AIDS cases were estimated to have occurred (of which over 500,000 had resulted in death). Over 1 million persons are living with HIV/AIDS in the United States, and an estimated 40,000 new cases occur each year. The death rate decreased for the first time in 1996, reflecting the use of antiretroviral combination therapy and prevention of secondary opportunistic infections.

Pediatric AIDS increased as the number of HIV-infected women increased. It was estimated that 1650 newborns acquired the virus in 1991 in the United States. The numbers of new infections have been reduced dramatically by the development in 1994 of zidovudine antenatal, intrapartum, and neonatal therapy (see below). From transmission rates of 25–30% with no interventions, drug treatments have reduced transmission rates in the United States to less than 2%. Mother-to-child transmission continues to occur because of undiagnosed HIV infection in the mother and lack of medical treatment.

The success in reducing perinatal HIV transmission in the United States has not been achieved in many poorer countries. Especially in sub-Saharan Africa, mother-to-child transmission rates remain high.

C. Routes of Transmission

High titers of HIV are found in two body fluids—blood and semen. HIV is transmitted during sexual contact (including genital-oral sex), through parenteral exposure to contaminated blood or blood products, and from mother to child during the perinatal period. The presence of other sexually transmitted diseases such as syphilis, gonorrhea, or herpes simplex type 2 increases the risk of sexual HIV transmission as much as a 100-fold because the inflammation and sores facilitate the transfer of HIV across mucosal barriers. Asymptomatic virus-positive individuals can transmit the virus. Since the first description of AIDS, promiscuous homosexual activity has been recognized as a major risk factor for acquisition of the disease. The risk increases in proportion to the number of sexual encounters with different partners.

Transfusion of infectious blood or blood products is an effective route for viral transmission. For example, over 90% of hemophiliac recipients of contaminated clotting factor concentrates in the United States (before HIV was detected) developed antibodies to HIV. Injection users of illicit drugs are commonly infected through the use of contaminated needles. Injection drug use accounts for a substantial proportion of new AIDS cases.

Careful testing is necessary to ensure a safe blood supply. The World Health Organization has reported that voluntary unremunerated blood donation is far safer than paid donations. It was reported in 1996 that the risk of transfusion-transmitted HIV infection in the United States was very small (about 1:500,000).

Mother-to-infant transmission rates vary from 13% to 40% in untreated women. Infants can become infected in utero, during the birth process, or, more commonly, through breastfeeding. In the absence of breastfeeding, about 30% of infections occur in utero and 70% during delivery. Data indicate that from one-third to one-half of perinatal HIV infections in Africa are due to breastfeeding. Transmission during breastfeeding usually occurs early (by 6 months). High maternal viral loads are a risk factor for transmission.

Health care workers have been infected by HIV following a needlestick with contaminated blood. The numbers of infections are relatively few in comparison with the number of needlesticks that have occurred involving contaminated blood (estimated risk of transmission is about 0.3%). The risk of transmission is even lower after a mucous membrane exposure to infected blood (about 0.09%). This contrasts with the risk of hepatitis C virus infection after a needlestick of about 1.8% and hepatitis B virus infection of 6–30%.

The routes of transmission (blood, sex, and birth) described above account for almost all HIV infections. There has been considerable concern that in rare circumstances other types of transmission may occur, such as through "casual" contact with HIV-infected persons or insect vectors, but there is no evidence of virus transmission under these casual conditions.

Prevention, Treatment, & Control
A. Antiviral Drugs

A growing number of antiviral drugs are approved for treatment of HIV infections (see Chapter 30). Classes of drugs include both nucleoside and nonnucleoside inhibitors of the viral enzyme reverse transcriptase and inhibitors of the viral protease enzyme. The protease inhibitors are potent antiviral drugs because the protease activity is absolutely essential for production of infectious virus, and the viral enzyme is distinct from human cell proteases. Newer classes of drugs include fusion inhibitors, first approved in 2003, that block virus entry into cells; entry inhibitors, first approved in 2007, that block coreceptor CCR5 binding by HIV; and integrase inhibitors, first approved in 2007, that interfere with the viral enzyme required for HIV replication.

Therapy with combinations of antiretroviral drugs, referred to as HAART, became available in 1996. It oftentimes can suppress viral replication to below limits of detection in plasma, decrease viral loads in lymphoid tissues, allow the recovery of immune responses to opportunistic pathogens, and prolong patient survival. However, HAART has failed to cure HIV-1 infections. The virus persists in reservoirs of long-lived, latently infected cells, including memory CD4 T cells. When HAART is discontinued or there is treatment failure, virus production rebounds.

Whereas monotherapy usually results in the rapid emergence of drug-resistant mutants of HIV, combination therapy, which targets multiple steps in virus replication, usually delays selection of HIV mutants. However, mutants that arise which are resistant to one protease inhibitor are often resistant to other protease inhibitors as well.

Transmission of drug-resistant variants may affect future therapy options. In 2004 and 2005, treatment-naïve patients with newly diagnosed HIV infections were found to carry virus with drug-resistant mutations in 8% and 10% of cases in the United States and Europe, respectively. Among perinatally infected infants in the United States in 2002, 19% had virus with drug-resistant mutations.

Results with combination therapy have been successful and have turned HIV infection into a chronic, treatable disease. Prolonged suppression of viral replication can be achieved, along with restoration of immune function, but treatment must be maintained for life and drug resistance can develop. In addition, current drug regimens are expensive, cannot be tolerated by all patients, and may have side effects (such as lipodystrophy). The majority of infected persons worldwide do not have access to any HIV drugs.

Zidovudine (azidothymidine; AZT) can significantly reduce the transmission of HIV from mother to infant. A regimen of AZT therapy of the mother during pregnancy and during the birth process and of the baby after birth reduced the risk of perinatal transmission by 65–75% (from about 25% to less than 2%). This treatment decreases vertical transmission at all levels of maternal viral load. A shorter course of AZT given to infected mothers or a simple nevirapine regimen has been shown to reduce transmission by 50% and to be safe for use in developing nations. However, the high rate of HIV transmission by breastfeeding can undermine the benefits of maternal perinatal drug treatment.

B. Vaccines Against HIV

A safe and effective vaccine offers the best hope of controlling the worldwide AIDS epidemic. Viral vaccines are typically preventive, ie, given to uninfected individuals to prevent either infection or disease. However, all candidate HIV vaccines tested as of 2009 proved ineffective at preventing infection.

Vaccine development is difficult because HIV mutates rapidly, is not expressed in all cells that are infected, and is not completely cleared by the host immune response after primary infection. HIV isolates show a marked variation, especially in the envelope antigens—variability that may promote the emergence of neutralization-resistant mutants. As the correlates of protective immunity are not known, it is unclear what cellular and/or humoral immune responses a vaccine should elicit.

Because of the safety concerns, vaccines based on attenuated or inactivated HIV or on simian isolates are viewed with apprehension. Recombinant viral proteins—especially those of the envelope glycoproteins—are likely candidates, whether delivered with adjuvants or with heterologous viral vectors. Many novel vaccination methods are also under investigation. Gene therapy approaches are being developed that are designed to achieve "intracellular immunization," ie, to genetically alter target cells in such a way as to make them resistant to HIV.

A large hurdle for vaccine development is the lack of an appropriate animal model for HIV. Chimpanzees are the only animals that are susceptible to HIV. Not only is the supply scarce, but chimpanzees develop only viremia and antibodies; they do not develop immunodeficiency. The SIV–macaque model of simian AIDS does develop disease and is useful for vaccine development studies.

C. Topical Microbicides

In many countries in the world, women make up at least 50% of those living with HIV/AIDS, and the majority of those became infected through heterosexual contact. Efforts are underway to develop safe and effective topical microbicides to prevent sexual transmission of HIV. To date, none of the candidate compounds tested has proved effective in clinical trials.

D. Control Measures

Without control by drugs or vaccines, the only way to avoid epidemic spread of HIV is to maintain a lifestyle that minimizes or eliminates the high-risk factors discussed above. No cases have been documented to result from such common exposures as sneezing, coughing, sharing meals, or other casual contacts.

Because HIV may be transmitted in blood, all donor blood should be tested for antibody. Properly conducted antibody tests appear to detect almost all HIV-1 and HIV-2 carriers. In settings with widespread screening of blood donors for viral exposure and the rejection of contaminated blood, transmission by blood transfusion has virtually disappeared.

Public health authorities have recommended that persons reported to have an HIV infection be given the following information and advice:

1. Almost all persons will remain infected for life and will develop the disease.
2. Although asymptomatic, such individuals may transmit HIV to others. Regular medical evaluation and follow-up are advised.

3. Infected persons should refrain from donating blood, plasma, body organs, other tissues, or sperm.

4. There is a risk of infecting others by sexual intercourse (vaginal or anal), by oral-genital contact, or by sharing of needles. The consistent and proper use of condoms can reduce transmission of the virus, though protection is not absolute.

5. Toothbrushes, razors, and other implements that could become contaminated with blood should not be shared.

6. Seropositive women or women with seropositive sexual partners are themselves at increased risk of acquiring HIV. If they become pregnant, their offspring also are at high risk of acquiring HIV.

7. After accidents that result in bleeding, contaminated surfaces should be cleaned with household bleach freshly diluted 1:10 in water.

8. Devices that have punctured the skin—eg, hypodermic and acupuncture needles—should be steam-sterilized by autoclaving before reuse or should be safely discarded. Dental instruments should be heat-sterilized between patients. Whenever possible, disposable needles and equipment should be used.

9. When seeking medical or dental care for intercurrent illness, infected persons should inform those responsible for their care that they are seropositive, so that appropriate evaluation can be undertaken and precautions taken to prevent transmission to others.

10. Testing for HIV antibody should be offered to persons who may have been infected as a result of their contact with seropositive individuals (eg, sexual partners, persons with whom needles have been shared, infants born to seropositive mothers).

11. Most persons with a positive test for HIV do not need to consider a change in employment unless their work involves significant potential for exposing others to their blood or other body fluids. There is no evidence of HIV transmission by food handling.

12. Seropositive persons in the health care professions who perform invasive procedures or have skin lesions should take precautions similar to those recommended for hepatitis B carriers to protect patients from the risk of infection.

13. Children with positive tests should be allowed to attend school, since casual person-to-person contact of schoolchildren poses no risk. However, a more restricted environment is advisable for preschool children or children who lack control of their body secretions, display biting behavior, or have oozing lesions.

E. Health Education

Without a vaccine or treatment, the prevention of cases of AIDS relies on the success of education projects involving behavioral changes. The health education messages for the general public have been summarized as follows: (1) Any sexual intercourse (outside of mutually monogamous HIV antibody-negative relationships) should be protected by a condom; (2) Do not share unsterile needles or syringes; (3) All women who have been potentially exposed should seek HIV antibody testing before becoming pregnant and, if the test is positive, should consider avoiding pregnancy; and (4) HIV-infected mothers should avoid breast feeding to reduce transmission of the virus to their children if safe alternative feeding options are available.

REVIEW QUESTIONS

1. HIV-1 is classified as a member of the *Lentivirus* genus in the Retroviridae family. Lentiviruses
   (A) Contain a DNA genome
   (B) Cause tumors in mice
   (C) Infect cells of the immune system
   (D) Have related sequences endogenous in normal cells
   (E) Cause rapidly progressive neurologic disease

2. HIV-1 encodes an envelope glycoprotein, gp120. This protein
   (A) Causes membrane fusion
   (B) Binds to the viral coreceptor on the cell surface
   (C) Is highly conserved among different isolates
   (D) Fails to elicit neutralizing antibody
   (E) Induces chemokine production

3. HIV/AIDS has become a worldwide epidemic that continues to expand. In 2007, the geographic area that had the largest number of HIV-infected people after sub-Saharan Africa was
   (A) Central and South America and the Caribbean
   (B) East Asia, including China
   (C) North America
   (D) South/Southeast Asia
   (E) Eastern Europe and Central Asia

4. The typical course of an untreated HIV infection extends over 10 or more years. There is usually a long period (clinical latency) between the time of primary HIV infection and the development of AIDS. During this period of clinical latency
   (A) HIV is not detectable in the plasma
   (B) CD4 cell counts remain unchanged
   (C) Virus cannot be transmitted to others
   (D) Virus is present in lymphoid organs
   (E) Neutralizing antibodies are not elicited

5. Viral coinfections occur in HIV-1-infected individuals and may contribute to morbidity and mortality. The most common coinfection in HIV-1-positive persons in the United States involves
   (A) Hepatitis C virus
   (B) Hepatitis D virus
   (C) HIV type 2
   (D) Human T-lymphotropic virus
   (E) Kaposi sarcoma herpesvirus

6. Which of the following individuals may be at increased risk of acquiring an HIV infection?
   (A) A grandmother living in the same household with a relative who is HIV-positive
   (B) A tourist in Botswana who has sex with a prostitute
   (C) A receptionist at an AIDS clinic in a hospital
   (D) A teacher with an HIV-positive child in her classroom
   (E) A baseball player whose teammate is HIV-positive

7. A 36-year-old nurse suffered a needlestick with blood from an HIV-positive patient. Six months later, the nurse's serum was positive in an EIA test, gave equivocal results in a repeat EIA test, and was negative by Western blot. The nurse

   (A) Is probably infected with HIV
   (B) Is in the window between acute infection with HIV and seroconversion
   (C) Is probably not infected with HIV
   (D) May be infected with a drug-resistant strain of HIV
   (E) May be a long-term nonprogressor

8. A 41-year-old HIV-infected male who had refused antiretroviral therapy is diagnosed with *Pneumocystis jiroveci* infection. This patient

   (A) Probably has a CD4 T cell count below 200 cells/μL
   (B) Is at elevated risk for lung cancer
   (C) Has a life expectancy of about 5 years
   (D) Probably has declining levels of plasma viremia
   (E) Is unlikely to develop dementia at this stage

9. A 48-year-old HIV-positive man with a CD4 count of 40 complains of memory loss to his doctor. Four months later, he becomes paralyzed and dies. An autopsy reveals demyelination of many neurons in the brain and electron microscopy shows clusters of nonenveloped viral particles in the neurons. The most likely cause of the disease is

   (A) Adenovirus type 12
   (B) Coxsackievirus B2
   (C) Parvovirus B19
   (D) Epstein-Barr virus
   (E) JC virus

10. Highly active antiretroviral combination therapy for HIV infection usually includes a protease inhibitor such as saquinavir. Such a protease inhibitor

    (A) Is effective against HIV-1 but not HIV-2
    (B) Seldom gives rise to resistant mutants of HIV
    (C) Inhibits a late step in virus replication
    (D) Degrades the CD4 receptor on cells
    (E) Interferes with virus interaction with coreceptor

11. In a person with HIV infection, potentially infectious fluids include all of the following except

    (A) Blood
    (B) Saliva visibly contaminated with blood
    (C) Urine not visibly contaminated with blood
    (D) Genital secretions
    (E) Amniotic fluid

12. Of the more than 1 million persons estimated to be living with HIV in the United States at the end of 2007, how many are thought to be unaware of their infection?

    (A) About 5%
    (B) About 10%

(C) About 20%
(D) About 25%
(E) About 30%
(F) About 50%

Answers

1. C	4. D	7. C	10. C
2. B	5. A	8. A	11. C
3. D	6. B	9. E	12. D

REFERENCES

Apetrei C, Robertson DL, Marx PA: The history of SIVs and AIDS: Epidemiology, phylogeny, and biology of isolates from naturally SIV infected non-human primates (NHP) in Africa. Front Biosci 2004;9:225. [PMID: 14766362]

Desrosiers RC: Nonhuman lentiviruses. In: *Fields Virology,* 5th ed. Knipe DM et al (editors). Lippincott Williams & Wilkins, 2007.

Freed EO, Martin MA: HIVs and their replication. In: *Fields Virology,* 5th ed. Knipe DM et al (editors). Lippincott Williams & Wilkins, 2007.

Moore JP, Doms RW: The entry of entry inhibitors: A fusion of science and medicine. Proc Natl Acad Sci USA 2003;100:10598. [PMID: 12960367]

Patel K et al: Long-term effectiveness of highly active antiretroviral therapy on the survival of children and adolescents with HIV infection: A 10-year follow-up study. Clin Infect Dis 2008;46:507. [PMID: 18199042]

Patrick MK, Johnston JB, Power C: Lentiviral neuropathogenesis: Comparative neuroinvasion, neurotropism, neurovirulence, and host neurosusceptibility. J Virol 2002;76:7923. [PMID: 12133996]

Paul ME, Schearer WT: Pediatric human immunodeficiency virus infection. In: *Pediatric Allergy: Principles and Practice.* Leung DYM, Sampson HA, Geha RS, Szefler SJ (editors). Mosby, 2003.

Revised recommendations for HIV testing of adults, adolescents, and pregnant women in health-care settings. MMWR Recomm Rep 2006;55(RR-14):1.

Special issue: 25 years of HIV. Trends Microbiol 2008;16(No. 12). [Entire issue.]

Twenty-five years of HIV/AIDS—United States, 1981–2006. MMWR Morb Mortal Wkly Rep 2006;55(No. 21):585.

Updated U.S. Public Health Service guidelines for the management of occupational exposures to HIV and recommendations for postexposure prophylaxis. MMWR Recomm Rep 2005;54(RR-9):1.

U.S. Public Health Service guidelines for testing and counseling blood and plasma donors for human immunodeficiency virus type 1 antigen. MMWR Recomm Rep 1996;45(RR-2):1.

C H A P T E R

45

Medical Mycology

Thomas G. Mitchell, PhD*

Mycology is the study of fungi. Approximately 80,000 species of fungi have been described, but fewer than 400 are medically important, and less than 50 species cause more than 90% of the fungal infections of humans and other animals. Rather, most species of fungi are beneficial to humankind. They reside in nature and are essential in breaking down and recycling organic matter. Some fungi greatly enhance our quality of life by contributing to the production of food and spirits, including cheese, bread, and beer. Other fungi have served medicine by providing useful bioactive secondary metabolites such as antibiotics (eg, penicillin) and immunosuppressive drugs (eg, cyclosporine). Fungi have been exploited by geneticists and molecular biologists as model systems for the investigation of a variety of eukaryotic processes. Fungi exert their greatest economic impact as phytopathogens; the agricultural industry sustains huge crop losses every year as a result of fungal diseases of rice, corn, grains, and other plants.

All fungi are eukaryotic organisms, and each fungal cell has at least one nucleus and nuclear membrane, endoplasmic reticulum, mitochondria, and secretory apparatus. Most fungi are obligate or facultative aerobes. They are chemotrophic, secreting enzymes that degrade a wide variety of organic substrates into soluble nutrients which are then passively absorbed or taken into the cell by active transport.

Fungal infections are **mycoses**. Most pathogenic fungi are exogenous, their natural habitats being water, soil, and organic debris. The mycoses with the highest incidence—candidiasis and dermatophytosis—are caused by fungi that are part of the normal microbial flora or highly adapted to survival on the human host. For convenience, mycoses may be classified as superficial, cutaneous, subcutaneous, systemic, and opportunistic (Table 45–1). Grouping mycoses in these categories reflects their usual portal of entry and initial site of involvement. However, there is considerable overlap, since systemic mycoses can have subcutaneous manifestations and vice versa. Most patients who develop opportunistic infections have serious underlying diseases and compromised host defenses. But primary systemic mycoses also occur in such patients, and the opportunists may also infect immunocompetent individuals. During infection, most patients develop significant cellular and humoral immune responses to the fungal antigens.

As medical advances have significantly prolonged the survival of patients with cancer, AIDS, and stem cell or organ transplants, the incidence of opportunistic mycoses has increased dramatically. Pathogenic fungi do not produce potent toxins, and the mechanisms of fungal pathogenicity are complex and polygenic. Most mycoses are difficult to treat. Because fungi are eukaryotes, they share numerous homologous genes, gene products, and pathways with their human hosts. Consequently, there are few unique targets for chemotherapy and effective antibiotics. Fortunately, there is growing interest in medically significant fungi and in the search for virulence factors and potential therapeutic targets.

* Associate Professor, Department of Molecular Genetics and Microbiology, Duke University Medical Center, Durham, North Carolina.

GLOSSARY

Conidia: Asexual reproductive structures (mitospores) produced either from the transformation of a vegetative yeast or hyphal cell or from a specialized conidiogenous cell, which may be simple or complex and elaborate. Conidia may be formed on specialized hyphae, termed **conidiophores**. **Microconidia** are small, and **macroconidia** are large or multicellular.

Arthroconidia (arthrospores): Conidia that result from the fragmentation of hyphal cells (Figure 45–1).

Blastoconidia (blastospores): Conidial formation through a budding process (eg, yeasts).

Chlamydospores (chlamydoconidia): Large, thick-walled, usually spherical conidia produced from terminal or intercalary hyphal cells.

Phialoconidia: Conidia that are produced by a "vase-shaped" conidiogenous cell termed a **phialide** (eg, *Aspergillus fumigatus*, Figure 45–6).

Dematiaceous fungi: Fungi whose cell walls contain melanin, which imparts a brown to black pigment.

Dimorphic fungi: Fungi that have two growth forms, such as a mold and a yeast, which develop under different growth conditions (eg, *Blastomyces dermatitidis* forms hyphae in vitro and yeasts in tissue).

Hyphae: Tubular, branching filaments (2–10 μm in width) of fungal cells, the mold form of growth. Most hyphal cells are separated by porous cross-walls or **septa**, but the zygomycetous hyphae are characteristically sparsely septate. Vegetative or substrate hyphae anchor the colony and absorb nutrients. Aerial hyphae project above the colony and bear the reproductive structures.

Imperfect fungi: Fungi that lack sexual reproduction; they are represented only by an **anamorph**, the mitotic or asexual reproductive state. They are identified on the basis of asexual reproductive structures (ie, mitospores).

Mold: Hyphal or mycelial colony or form of growth.

Mycelium: Mass or mat of hyphae, mold colony.

Perfect fungi: Fungi that are capable of sexual reproduction, which is the **teleomorph**.

Pseudohyphae: Chains of elongated buds or blastoconidia.

Septum: Hyphal cross-wall, typically perforated.

Sporangiospores: Asexual structures characteristic of zygomycetes; they are mitotic spores produced within an enclosed **sporangium**, often supported by one **sporangiophore** (Figure 45–2 and 45–3).

Spore: A specialized propagule with enhanced survival value, such as resistance to adverse conditions or structural features that promote dispersion. Spores may result from asexual (eg, conidia, sporangiospores) or sexual (see below) reproduction. During sexual reproduction, haploid cells of compatible strains mate through a process of plasmogamy, karyogamy, and meiosis.

Ascospores: Following meiosis, four to eight meiospores form within an **ascus**.

Basidiospores: Following meiosis, four meiospores usually form on the surface of a specialized structure, a club-shaped **basidium**.

Zygospores: Following meiosis, a large, thick-walled **zygospore** develops.

Yeasts: Unicellular, spherical to ellipsoid (3–15 μm) fungal cells that usually reproduce by budding.

TABLE 45–1   The Major Mycoses and Causative Fungi

Category	Mycosis	Causative Fungal Agents
Superficial	Pityriasis versicolor	*Malassezia* species
	Tinea nigra	*Hortaea werneckii*
	White piedra	*Trichosporon* species
	Black piedra	*Piedraia hortae*
Cutaneous	Dermatophytosis	*Microsporum* species, *Trichophyton* species, and *Epidermophyton floccosum*
	Candidiasis of skin, mucosa, or nails	*Candida albicans* and other *Candida* species
Subcutaneous	Sporotrichosis	*Sporothrix schenckii*
	Chromoblastomycosis	*Phialophora verrucosa*, *Fonsecaea pedrosoi*, and others
	Mycetoma	*Pseudallescheria boydii*, *Madurella mycetomatis*, and others
	Phaeohyphomycosis	*Exophiala*, *Bipolaris*, *Exserohilum*, and other dematiaceous molds
Endemic (primary, systemic)	Coccidioidomycosis	*Coccidioides posadasii* and *Coccidioides immitis*
	Histoplasmosis	*Histoplasma capsulatum*
	Blastomycosis	*Blastomyces dermatitidis*
	Paracoccidioidomycosis	*Paracoccidioides brasiliensis*
Opportunistic	Systemic candidiasis	*Candida albicans* and other *Candida* species
	Cryptococcosis	*Cryptococcus neoformans* and *Cryptococcus gattii*
	Aspergillosis	*Aspergillus fumigatus* and other *Aspergillus* species
	Hyalohyphomycosis	Species of *Fusarium*, *Paecilomyces*, *Trichosporon* and other hyaline molds
	Phaeohyphomycosis	*Cladophialophora bantiana*; species of *Alternaria*, *Cladosporium*, *Bipolaris*, *Exserohilum* and numerous other dematiaceous molds
	Mucormycosis (zygomycosis)	Species of *Rhizopus*, *Absidia*, *Cunninghamella*, and other zygomycetes
	Penicilliosis	*Penicillium marneffei*

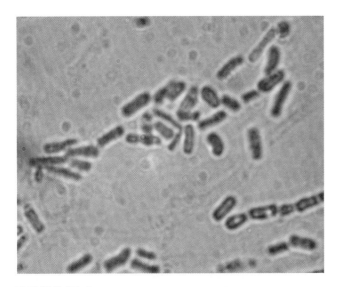

FIGURE 45–1 Arthroconidia formed by the fragmentation of hyphal cells into compact conidia. 400×.

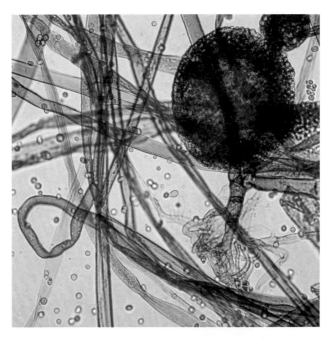

FIGURE 45–2 *Rhizopus.* The sporangium of this zygomycetous mold has released its sporangiospores but remains attached to the supporting sporangiophore, and rhizoids are apparent at the base of the sporangiophore. 200×.

GENERAL PROPERTIES & CLASSIFICATION OF FUNGI

As indicated in Chapter 1, fungi grow in two basic forms, as **yeasts** and **molds** (or **moulds**). Growth in the mold form occurs by production of multicellular filamentous colonies. These colonies consist of branching cylindric tubules called **hyphae**, varying in diameter from 2 μm to 10 μm. The mass of intertwined hyphae that accumulates during active growth is a **mycelium**. Some hyphae are divided into cells by cross-walls or **septa**, typically forming at regular intervals during hyphal growth. One group of medically important molds, the zygomycetes, produces hyphae that are rarely septated. Hyphae that penetrate the supporting medium and absorb nutrients are the vegetative or substrate hyphae. In contrast, aerial hyphae project above the surface of the mycelium and usually bear the reproductive structures of the mold. Under standardized growth conditions in the laboratory, molds produce colonies with characteristic features such as rates of growth, texture, and pigmentation. The genus—if not the species—of most clinical molds isolated can be determined by microscopic examination of the ontogeny and morphology of their asexual reproductive spores, or conidia. (See Figures 45–2 through 45–8.)

Yeasts are single cells, usually spherical to ellipsoid in shape and varying in diameter from 3 μm to 15 μm. Most yeasts reproduce by budding. Some species produce buds that characteristically fail to detach and become elongated; continuation of the budding process then produces a chain of elongated yeast cells called **pseudohyphae**. Yeast colonies are usually soft, opaque, 1–3 mm in size, and cream-colored. Because the colonies and microscopic morphology of many yeasts are quite similar, yeast species are identified on the basis of physiologic tests and a few key morphologic differences. Some species of fungi are dimorphic and capable of growth as a yeast or mold depending on environmental conditions.

All fungi have an essential rigid cell wall that determines their shape. Cell walls are composed largely of carbohydrate layers—long chains of polysaccharides—as well as glycoproteins and lipids. During infection, fungal cell walls have important pathobiologic properties. The surface components of the cell wall mediate attachment of the fungus to host cells. Cell wall polysaccharides may activate the complement cascade and provoke an inflammatory reaction; they are poorly degraded by the host and can be detected with special stains. Cell walls release immunodominant antigens that may elicit cellular immune responses and diagnostic antibodies. Some yeasts and molds have melanized cell walls, imparting a brown or black pigment. Such fungi are **dematiaceous**. In several studies, melanin has been associated with virulence.

In addition to their vegetative growth as yeasts or molds, fungi can produce spores to enhance their survival. Spores can be readily dispersed, are more resistant to adverse conditions, and can germinate when conditions for growth are favorable. Spores can derive from asexual or sexual reproduction—the anamorphic and teleomorphic states, respectively. Asexual spores are mitotic progeny (ie, mitospores) and genetically identical. The medical fungi produce two major types of asexual spores, **conidia**, and, in the zygomycetes,

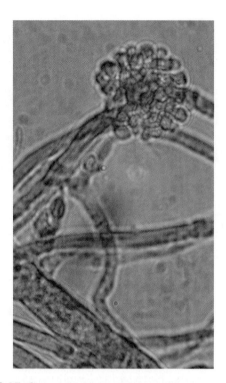

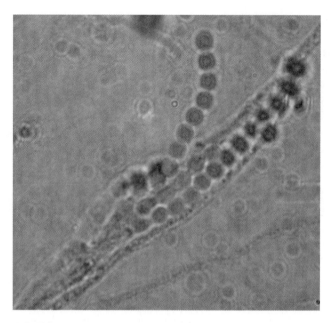

FIGURE 45–4 *Penicillium*. Chains of conidia are generated by phialides, which are supported by a branched conidiophore. The basal conidium is newest. 400×.

FIGURE 45–3 *Cunninghamella bertholletiae* is another pathogenic zygomycete. Its sporangiospores are produced within sporangiola that are attached to a vesicle and supported by a sporangiophore. 400×.

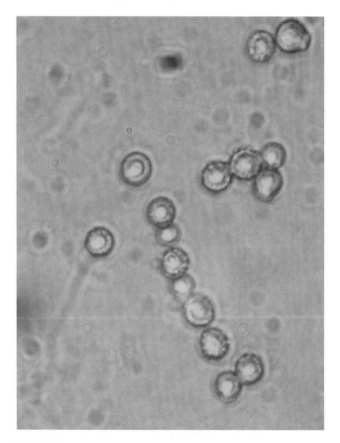

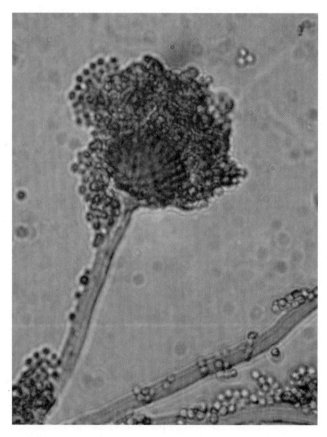

FIGURE 45–5 *Scopulariopsis*. This chain of conidia was produced by an annellide, which is another type of conidiogenous cell. 400×.

FIGURE 45–6 *Aspergillus fumigatus*. Phialides form on top of a swollen vesicle at the end of a long conidiophore. The basal conidia are the youngest. Mature conidia have rough walls. 400×.

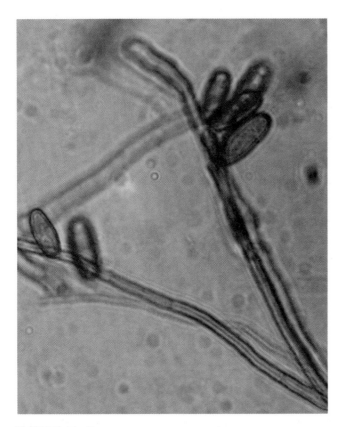

FIGURE 45–7 *Bipolaris*. Dematiaceous mold that produces characteristic thick-walled macroconidia. 400×.

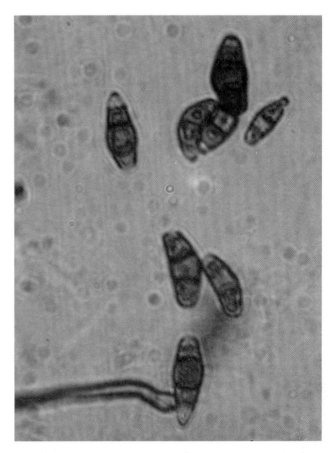

FIGURE 45–8 *Curvularia*. Dematiaceous mold that produces characteristic curved macroconidia with distinctly larger central cells. 400×.

sporangiospores. Informative features of spores include their ontogeny (some molds produce complex conidiogenic structures) as well as their morphology (size, shape, texture, color, and unicellularity or multicellularity). In some fungi, vegetative cells may transform into conidia (eg, arthroconidia, chlamydospores). In others, conidia are produced by a conidiogenous cell, such as a phialide, which itself may be attached to a specialized hypha called a conidiophore. In the zygomycetes, sporangiospores result from mitotic replication and spore production within a sac-like structure called a sporangium, which is supported by a sporangiophore.

Classification

The fungi are classified in four phyla: Chytridiomycota, Zygomycota, Ascomycota, and Basidiomycota. The largest phylum is the Ascomycota (or ascomycetes), which includes more than 60% of the known fungi and about 85% of the human pathogens. The remaining pathogenic fungi are zygomycetes or basidiomycetes. A fungal species is assigned to a phylum, as well as the appropriate Class, Order, and Family, based on its mode of sexual reproduction, phenotypic properties (eg, morphology and physiology), and phylogenetic relationships. The latter methods are used to classify anamorphic or asexual species. Sexual reproduction typically occurs when mating-compatible strains of a species are stimulated by pheromones to undergo plasmogamy, nuclear fusion, and meiosis, resulting in the exchange of genetic information. Asexual isolates and their spores reproduce clonally. Many species have been given different names that reflect their sexual (teleomorphic) and asexual (anamorphic) reproductive forms.

A. Zygomycota (Zygomycetes)

Sexual reproduction results in a zygospore; asexual reproduction occurs via sporangia. Vegetative hyphae are sparsely septate. **Examples**: *Rhizopus, Absidia, Mucor, Cunninghamella, Pilobolus.*

B. Ascomycota (Ascomycetes)

Sexual reproduction involves a sac or ascus in which karyogamy and meiosis occur, producing ascospores. Asexual reproduction is via conidia. Molds have septate hyphae. **Examples**: Most yeasts (*Saccharomyces, Candida*) and molds (*Coccidioides, Blastomyces, Trichophyton*).

C. Basidiomycota (Basidomycetes)

Sexual reproduction results in four progeny basidiospores supported by a club-shaped basidium. Hyphae have complex septa. **Examples**: Mushrooms, *Cryptococcus.*

GROWTH & ISOLATION OF FUNGI

Most fungi occur in nature and grow readily on simple sources of nitrogen and carbohydrate. The traditional mycological medium, Sabouraud's agar, which contains glucose and modified peptone (pH 7.0), has been used because it does not readily support the growth of bacteria. The morphologic characteristics of fungi used for identification have been described from growth on Sabouraud's agar. However, other media, such as inhibitory mold agar, have facilitated the recovery of fungi from clinical specimens. To culture medical fungi from nonsterile specimens, antibacterial antibiotics (eg, gentamicin, chloramphenicol) and cycloheximide are added to the media to inhibit bacteria and saprophytic molds, respectively. The specimens used for isolation of fungi and other media used to isolate them are discussed in Chapter 47.

SUPERFICIAL MYCOSES

Pityriasis Versicolor

Pityriasis versicolor is a chronic mild superficial infection of the stratum corneum caused by *Malassezia globosa, M restricta,* and other members of the *M furfur* complex. Invasion of the cornified skin and the host responses are both minimal. Discrete, serpentine, hyper- or hypopigmented maculae occur on the skin, usually on the chest, upper back, arms, or abdomen. The lesions are chronic and occur as macular patches of discolored skin that may enlarge and coalesce, but scaling, inflammation, and irritation are minimal. Indeed, this common affliction is largely a cosmetic problem.

Malassezia species are lipophilic yeasts, and most require lipid in the medium for growth. The diagnosis is confirmed by direct microscopic examination of scrapings of infected skin, treated with 10–20% KOH or stained with calcofluor white. Short unbranched hyphae and spherical cells are observed. The lesions also fluoresce under Wood's lamp. Pityriasis versicolor is treated with daily applications of selenium sulfide. Topical or oral azoles are also effective. Rarely, *Malassezia* may cause an opportunistic fungemia in patients—usually infants—receiving total parenteral nutrition, as a result of contamination of the lipid emulsion. In most cases, the fungemia is transient and corrected by replacing the fluid and intravenous catheter. Some individuals develop folliculitis due to *Malassezia*. Species of *Malassezia* are considered part of the microbial flora and can be isolated from normal skin and scalp. They have been implicated as a cause of or contributor to seborrheic dermatitis, or dandruff. This hypothesis is supported by the observation that many cases are alleviated by treatment with ketoconazole.

Tinea Nigra

Tinea nigra (or tinea nigra palmaris) is a superficial chronic and asymptomatic infection of the stratum corneum caused by the dematiaceous fungus *Hortaea (Exophiala) werneckii*. This condition is more prevalent in warm coastal regions and among young women. The lesions appear as a dark (brown to black) discoloration, often on the palm. Microscopic examination of skin scrapings from the periphery of the lesion will reveal branched, septate hyphae and budding yeast cells with melaninized cell walls. Tinea nigra will respond to treatment with keratolytic solutions, salicylic acid, or azole antifungal drugs.

Piedra

Black piedra is a nodular infection of the hair shaft caused by *Piedraia hortai* (Figure 45–9B). White piedra, due to infection with *Trichosporon* species, presents as larger, softer, yellowish nodules on the hairs (Figure 45–9A). Axillary, pubic, beard, and scalp hair may be infected. Treatment for both types consists of removal of hair and application of a topical antifungal agent. Piedra is endemic in tropical underdeveloped countries.

CUTANEOUS MYCOSES

Dermatophytosis

Cutaneous mycoses are caused by fungi that infect only the superficial keratinized tissue (skin, hair, and nails). The most important of these are the dermatophytes, a group of about 40 related fungi that belong to three genera: *Microsporum, Trichophyton,* and *Epidermophyton*. Dermatophytes are probably restricted to the nonviable skin because most are unable to grow at 37°C or in the presence of serum. Dermatophytoses are among the most prevalent infections in the world. Although they can be persistent and troublesome, they are not debilitating or life-threatening—yet millions of dollars are expended annually in their treatment. Being superficial, dermatophyte (ringworm) infections have been recognized since antiquity. In skin they are diagnosed by the presence of hyaline, septate, branching hyphae or chains of arthroconidia. In culture, the many species are closely related and often difficult to identify. They are speciated on the basis of subtle differences in the appearance of the colonies and microscopic morphology as well as a few vitamin requirements. Despite their similarities in morphology, nutritional requirements, surface antigens, and other features, many species have developed keratinases, elastases, and other enzymes that enable them to be quite host-specific. For some species of dermatophytes, a sexual reproductive state has been discovered, and all dermatophytes with a sexual form produce ascospores and belong to the teleomorphic genus *Arthroderma*.

Dermatophytes are classified as geophilic, zoophilic, or anthropophilic depending on whether their usual habitat is soil, animals, or humans. Several dermatophytes that normally reside in soil or are associated with particular animal species are still able to cause human infections. In general, as a species evolves from habitation in soil to a specific animal or human host, it loses the ability to produce asexual conidia and to reproduce sexually. Anthropophilic species, which cause the greatest number of human infections, cause relatively mild and chronic infections in humans, produce few conidia in culture, and may be difficult to eradicate. Conversely, geophilic and zoophilic dermatophytes, being less adapted to human hosts, produce more acute inflammatory infections that tend to resolve more quickly. Dermatophytes are acquired by contact with contaminated soil or with infected animals or humans.

Some anthropophilic species are geographically restricted, but others, such as *Epidermophyton floccosum*, *Trichophyton mentagrophytes* var *interdigitale*, *T rubrum*, and *T tonsurans*, are globally distributed. The most common geophilic

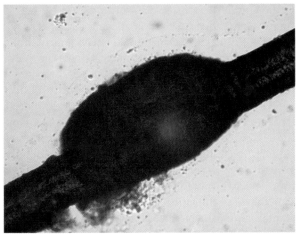

FIGURE 45–9 Piedra. **A:** White piedra hair with nodule due to growth of *Trichosporon*. 200×. **B:** Black piedra hair with a hard, black nodule, caused by growth of the dematiaceous mold, *Hortaea werneckii*. 200×.

species causing human infections is *Microsporum gypseum*. Cosmopolitan zoophilic species (and their natural hosts) include *Microsporum canis* (dogs and cats), *Microsporum gallinae* (fowl), *Microsporum nanum* (pigs), *Trichophyton equinum* (horses), and *Trichophyton verrucosum* (cattle).

Morphology & Identification

Dermatophytes are identified by their colonial appearance and microscopic morphology after growth for 2 weeks at 25°C on Sabouraud's dextrose agar. *Trichophyton* species, which may infect hair, skin, or nails, develop cylindric, smooth-walled macroconidia and characteristic microconidia (Figure 45–10A). Depending on the variety, colonies of *T mentagrophytes* may be cottony to granular; both types display abundant grape-like clusters of spherical microconidia on terminal branches. Coiled or spiral hyphae are commonly found in primary isolates. The typical colony of *T rubrum* has a white, cottony surface and a deep red, nondiffusible pigment when viewed from the reverse side of the colony. The microconidia are small and piriform (pear-shaped). *T tonsurans* produces a flat, powdery to velvety colony on the obverse surface that becomes reddish-brown on reverse; the microconidia are mostly elongate (Figure 45–10A).

Microsporum species tend to produce distinctive multicellular macroconidia with echinulate walls (Figure 45–10B). Both types of conidia are borne singly in these genera. *M canis* forms a colony with a white cottony surface and a deep yellow color on reverse; the thick-walled, 8- to 15-celled macroconidia frequently have curved or hooked tips. *M gypseum* produces a tan, powdery colony and abundant thin-walled, four- to six-celled macroconidia. *Microsporum* species infect only hair and skin.

Epidermophyton floccosum, which is the only pathogen in this genus, produces only macroconidia, which are smooth-walled, clavate, two- to four-celled, and formed in small clusters (Figure 45–10C). The colonies are usually flat and velvety with a tan to olive-green tinge. *E floccosum* infects the skin and nails but not the hair.

In addition to gross and microscopic morphology, a few nutritional or other tests, such as growth at 37°C or a test for in vitro hair perforation, are useful in differentiating certain species.

Epidemiology & Immunity

Dermatophyte infections begin in the skin after trauma and contact. There is evidence that host susceptibility may be enhanced by moisture, warmth, specific skin chemistry, composition of sebum and perspiration, youth, heavy exposure, and genetic predisposition. The incidence is higher in hot, humid climates and under crowded living conditions. Wearing shoes provides warmth and moisture, a setting for infections of the feet. The source of infection is soil or an infected animal in the case of geophilic and zoophilic dermatophytes, respectively. The conidia can remain viable for long periods. Anthropophilic species may

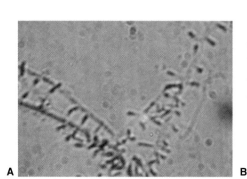
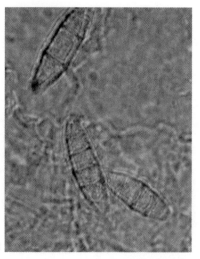
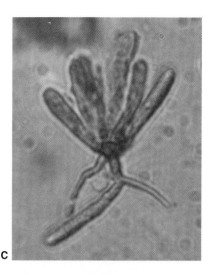

FIGURE 45–10 Examples of the three genera of dermatophytes. **A**: *Trichophyton tonsurans* is characterized by the production of elongated microcondia attached to a supporting hypha. **B**: *Microsporum gypseum* produces individual thin- and rough-walled macroconidia. **C**: *Epidermophyton floccosum* has club-shaped, thin- and smooth-walled macroconidia that typically arise in small clusters.

be transmitted by direct contact or through fomites, such as contaminated towels, clothing, shared shower stalls, and similar examples.

Trichophytin is a crude antigen preparation that can be used to detect immediate- or delayed-type hypersensitivity to dermatophytic antigens. Many patients who develop chronic, noninflammatory dermatophyte infections have poor cell-mediated immune responses to dermatophyte antigen. These patients often are atopic and have immediate-type hypersensitivity and elevated IgE concentrations. In the normal host, immunity to dermatophytosis varies in duration and degree depending on the host, the site, and the species of fungus causing the infection.

Clinical Findings

Dermatophyte infections were mistakenly termed ringworm or tinea because of the raised circular lesions. The clinical forms are based on the site of involvement. A single species is able to cause more than one type of clinical infection. Conversely, a single clinical form, such as tinea corporis, may be caused by more than one dermatophyte species. The more common agents associated with particular clinical forms are listed in Table 45–2. Very rarely, immunocompromised patients may develop systemic infection by a dermatophyte.

A. Tinea Pedis (Athlete's Foot)

Tinea pedis is the most prevalent of all dermatophytoses. It usually occurs as a chronic infection of the toe webs. Other varieties are the vesicular, ulcerative, and moccasin types, with hyperkeratosis of the sole. Initially, there is itching between the toes and the development of small vesicles that rupture and discharge a thin fluid. The skin of the toe webs becomes macerated and peels, whereupon cracks appear that are prone to develop secondary bacterial infection. When the fungal infection becomes chronic, peeling and cracking of the skin are the principal manifestations, accompanied by pain and pruritus.

B. Tinea Unguium (Onychomycosis)

Nail infection may follow prolonged tinea pedis. With hyphal invasion, the nails become yellow, brittle, thickened, and crumbly. One or more nails of the feet or hands may be involved.

C. Tinea Corporis, Tinea Cruris, & Tinea Manus

Dermatophytosis of the glabrous skin commonly gives rise to the annular lesions of ringworm, with a clearing, scaly center surrounded by a red advancing border that may be dry or vesicular. The dermatophyte grows only within dead, keratinized tissue, but fungal metabolites, enzymes, and antigens diffuse through the viable layers of the epidermis to cause erythema, vesicle formation, and pruritus. Infections with geophilic and zoophilic dermatophytes produce more irritants and are more inflammatory than anthropophilic species. As hyphae age, they often form chains of arthroconidia. The lesions expand centrifugally and active hyphal growth is at the periphery, which is the most likely region from which to obtain material for diagnosis. Penetration into the newly forming stratum corneum of the thicker plantar and palmar surfaces accounts for the persistent infections at those sites.

When the infection occurs in the groin area, it is called tinea cruris, or jock itch. Most such infections involve males and present as dry, itchy lesions that often start on the scrotum and spread to the groin. Tinea manus refers to ringworm

TABLE 45-2   Some Clinical Features of Dermatophyte Infection

Skin Disease	Location of Lesions	Clinical Features	Fungi Most Frequently Responsible
Tinea corporis (ringworm)	Nonhairy, smooth skin	Circular patches with advancing red, vesiculated border and central scaling. Pruritic	*Trichophyton rubrum, Epidermophyton floccosum*
Tinea pedis (athlete's foot)	Interdigital spaces on feet of persons wearing shoes	Acute: itching, red vesicular. Chronic: itching, scaling, fissures	*Trichophyton rubrum, Trichophyton mentagrophytes, Epidermophyton floccosum*
Tinea cruris (jock itch)	Groin	Erythematous scaling lesion in intertriginous area. Pruritic	*Trichophyton rubrum, Trichophyton mentagrophytes, Epidermophyton floccosum*
Tinea capitis	Scalp hair. Endothrix: fungus inside hair shaft. Ectothrix: fungus on surface of hair	Circular bald patches with short hair stubs or broken hair within hair follicles. Kerion rare. *Microsporum*-infected hairs fluoresce	*Trichophyton mentagrophytes, Microsporum canis*
Tinea barbae	Beard hair	Edematous, erythematous lesion	*Trichophyton mentagrophytes*
Tinea unguium (onychomycosis)	Nail	Nails thickened or crumbling distally; discolored; lusterless. Usually associated with tinea pedis	*Trichophyton rubrum, Trichophyton mentagrophytes, Epidermophyton floccosum*
Dermatophytid (id reaction)	Usually sides and flexor aspects of fingers. Palm. Any site on body	Pruritic vesicular to bullous lesions. Most commonly associated with tinea pedis	No fungi present in lesion. May become secondarily infected with bacteria

of the hands or fingers. Dry scaly lesions may involve one or both hands, single fingers, or two or more fingers.

D. Tinea Capitis & Tinea Barbae

Tinea capitis is dermatophytosis or ringworm of the scalp and hair. The infection begins with hyphal invasion of the skin of the scalp, with subsequent spread down the keratinized wall of the hair follicle. Infection of the hair takes place just above the hair root. The hyphae grow downward on the nonliving portion of the hair and at the same rate as the hair grows upward. The infection produces dull gray, circular patches of alopecia, scaling, and itching. As the hair grows out of the follicle, the hyphae of *Microsporum* species produce a chain of spores that form a sheath around the hair shaft (ectothrix). These spores impart a greenish to silvery fluorescence when the hairs are examined under Wood's light (365 nm). In contrast, *T tonsurans*, the chief cause of "black dot" tinea capitis, produces spores within the hair shaft (endothrix). These hairs do not fluoresce; they are weakened and typically break easily at the follicular opening. In prepubescent children, epidemic tinea capitis is usually self-limiting.

Zoophilic species may induce a severe combined inflammatory and hypersensitivity reaction called a **kerion**. Another manifestation of tinea capitis is favus, an acute inflammatory infection of the hair follicle caused by *T schoenleinii*, which leads to the formation of scutula (crusts) around the follicle. In favic hairs, the hyphae do not form spores but can be found within the hair shaft. Tinea barbae involves the bearded region. Especially when a zoophilic dermatophyte is involved, a highly inflammatory reaction may be elicited that closely resembles pyogenic infection.

E. Trichophytid Reaction

In the course of dermatophytosis, the individual may become hypersensitive to constituents or products of the fungus and may develop allergic manifestations—called dermatophytids (usually vesicles)—elsewhere on the body, most often on the hands. The trichophytin skin test is markedly positive in such persons.

Diagnostic Laboratory Tests

A. Specimens

Specimens consist of scrapings from both the skin and the nails plus hairs plucked from involved areas. The ectothrix spores of *Microsporum*-infected hairs fluoresce under Wood's light in a darkened room.

B. Microscopic Examination

Specimens are placed on a slide in a drop of 10–20% potassium hydroxide, with or without calcofluor white, which is a nonspecific fungal cell wall stain viewed with a fluorescent microscope. A coverslip is added, and the specimen is examined immediately and again after 20 minutes. In skin or nails, regardless of the infecting species, branching hyphae or chains of arthroconidia (arthrospores) are seen. In hairs, most *Microsporum* species form dense sheaths of spores around the hair (ectothrix). *T tonsurans* and *T violaceum* are noted for producing arthroconidia inside the hair shaft (endothrix).

C. Culture

The identification of dermatophyte species requires cultures. Specimens are inoculated onto inhibitory mold agar or Sabouraud's agar slants containing cycloheximide and chloramphenicol to suppress mold and bacterial growth, incubated for 1–3 weeks at room temperature, and further examined in slide cultures if necessary. Species are identified on the basis of colonial morphology (growth rate, surface texture, and any pigmentation), microscopic morphology (macroconidia, microconidia), and, in some cases, nutritional requirements.

Treatment

Therapy consists of thorough removal of infected and dead epithelial structures and application of a topical antifungal chemical or antibiotic. To prevent reinfection the area should be kept dry, and sources of infection, such as an infected pet or shared bathing facilities, should be avoided.

A. Tinea Capitis

Scalp infections are treated for several weeks with oral administration of griseofulvin or terbinafine. Frequent shampoos and miconazole cream or other topical antifungal agents may be effective if used for weeks. Alternatively, ketoconazole and itraconazole are quite effective.

B. Tinea Corporis, Tinea Pedis, & Related Infections

The most effective drugs are itraconazole and terbinafine. However, a number of topical preparations may be used, such as miconazole nitrate, tolnaftate, and clotrimazole. If applied for at least 2–4 weeks, the cure rates are usually 70–100%. Treatment should be continued for 1–2 weeks after clearing of the lesions. For troublesome cases, a short course of oral griseofulvin can be administered.

C. Tinea Unguium

Nail infections are the most difficult to treat, often requiring months of oral itraconazole or terbinafine as well as surgical removal of the nail. Relapses are common.

SUBCUTANEOUS MYCOSES

The fungi that cause subcutaneous mycoses normally reside in soil or on vegetation. They enter the skin or subcutaneous tissue by traumatic inoculation with contaminated material. In general, the lesions become granulomatous and expand slowly from the area of implantation. Extension via the lymphatics draining the lesion is slow except in sporotrichosis. These mycoses are usually confined to the subcutaneous tissues, but in rare cases they become systemic and produce life-threatening disease.

SPOROTRICHOSIS

Sporothrix schenckii is a thermally dimorphic fungus that lives on vegetation. It is associated with a variety of plants—grasses, trees, sphagnum moss, rose bushes, and other horticultural plants. At ambient temperatures, it grows as a mold, producing branching, septate hyphae and conidia, and in tissue or in vitro at 35–37°C as a small budding yeast. Following traumatic introduction into the skin, *S schenckii* causes **sporotrichosis**, a chronic granulomatous infection. The initial episode is typically followed by secondary spread with involvement of the draining lymphatics and lymph nodes.

Morphology & Identification

S schenckii grows well on routine agar media, and at room temperature the young colonies are blackish and shiny, becoming wrinkled and fuzzy with age. Strains vary in pigmentation from shades of black and gray to whitish. The organism produces branching, septate hyphae and distinctive small (3–5 μm) conidia, delicately clustered at the ends of tapering conidiophores. Isolates may also form larger conidia directly from the hyphae. *S schenckii* is thermally dimorphic, and at 35°C on a rich medium it converts to growth as small, often multiply budding yeast cells that are variable in shape but often fusiform (about $1–3 \times 3–10$ μm), as shown in Figure 45–11.

Antigenic Structure

Heat-killed saline suspensions of cultures or carbohydrate fractions (sporotrichin) will elicit positive delayed skin tests in infected humans or animals. A variety of serologic tests have been developed, and most patients, as well as some normal individuals, have specific or cross-reactive antibodies.

Pathogenesis & Clinical Findings

The conidia or hyphal fragments of *S schenckii* are introduced into the skin by trauma. Patients frequently recall a history of trauma associated with outdoor activities and plants. The initial lesion is usually located on the extremities but can be found anywhere (children often present with facial lesions). About 75% of cases are lymphocutaneous; ie, the initial lesion develops as a granulomatous nodule that may progress to form a necrotic or ulcerative lesion. Meanwhile, the draining lymphatics become thickened and cord-like. Multiple subcutaneous nodules and abscesses occur along the lymphatics.

Fixed sporotrichosis is a single nonlymphangitic nodule that is limited and less progressive. The fixed lesion is more common in endemic areas such as Mexico, where there is a high level of exposure and immunity in the population. Immunity limits the local spread of the infection.

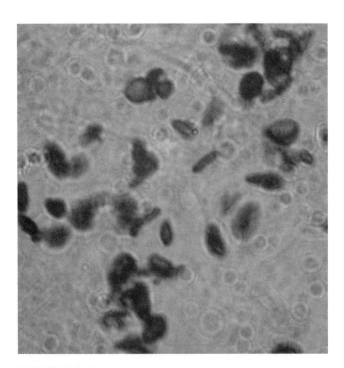

FIGURE 45–11 Sporotrichosis. Cutaneous tissue revealing the small spherical and elongated budding yeast cells (3–5 μm) of *Sporothrix schenckii,* which are stained black by the Gomori methenamine silver (GMS) stain. 400×.

There is usually little systemic illness associated with these lesions, but dissemination may occur, especially in debilitated patients. Rarely, primary pulmonary sporotrichosis results from inhalation of the conidia. This manifestation mimics chronic cavitary tuberculosis and tends to occur in patients with impaired cell-mediated immunity.

Diagnostic Laboratory Tests

A. Specimens

Specimens include biopsy material or exudate from granulous or ulcerative lesions.

B. Microscopic Examination

Although specimens can be examined directly with KOH or calcofluor white stain, the yeasts are rarely found. Even though they are sparse in tissue, the sensitivity of histopathologic sections is enhanced with routine fungal cell wall stains, such as Gomori methenamine silver, which stains the cell walls black, or the periodic acid-Schiff stain, which imparts a red color to the cell walls. Alternatively, they can be identified by fluorescent antibody staining. The yeasts are 3–5 μm in diameter and spherical to elongated.

Another structure termed an asteroid body is often seen in tissue, particularly in endemic areas such as Mexico, South Africa, and Japan. In hematoxylin and eosin-stained tissue, the asteroid body consists of a central basophilic yeast cell surrounded by radiating extensions of eosinophilic material, which are depositions of antigen–antibody complexes and complement.

C. Culture

The most reliable method of diagnosis is culture. Specimens are streaked on inhibitory mold agar or Sabouraud's agar containing antibacterial antibiotics and incubated at 25–30°C. The identification is confirmed by growth at 35°C and conversion to the yeast form.

D. Serology

Agglutination of yeast cell suspensions or of latex particles coated with antigen occurs in high titer with sera of infected patients but is not always diagnostic.

Treatment

In some cases, the infection is self-limited. Although the oral administration of saturated solution of potassium iodide in milk is quite effective, it is difficult for many patients to tolerate. Oral itraconazole or another of the azoles is the treatment of choice. For systemic disease, amphotericin B is given.

Epidemiology & Control

S schenckii occurs worldwide in close association with plants. For example, cases have been linked to contact with sphagnum moss, rose thorns, decaying wood, pine straw, prairie grass, and other vegetation. About 75% of cases occur in males, either because of increased exposure or because of an X-linked difference in susceptibility. The incidence is higher among agricultural workers, and sporotrichosis is considered an occupational risk for forest rangers, horticulturists, and workers in similar occupations. Prevention includes measures to minimize accidental inoculation and the use of fungicides, where appropriate, to treat wood. Animals are also susceptible to sporotrichosis.

CHROMOBLASTOMYCOSIS

Chromoblastomycosis (chromomycosis) is a subcutaneous mycotic infection caused by traumatic inoculation by any of five recognized fungal agents that reside in soil and vegetation. All are dematiaceous fungi, having melaninized cell walls: *Phialophora verrucosa, Fonsecaea pedrosoi, Rhinocladiella aquaspersa, Fonsecaea compacta,* and *Cladophialophora carrionii.* The infection is chronic and characterized by the slow development of progressive granulomatous lesions that in time induce hyperplasia of the epidermal tissue.

Morphology & Identification

The dematiaceous fungi are similar in their pigmentation, antigenic structure, morphology, and physiologic properties.

The colonies are compact, deep brown to black, and develop a velvety, often wrinkled surface. The agents of chromoblastomycosis are identified by their modes of conidiation. In tissue they appear the same, producing spherical brown cells (4–12 μm in diameter) termed muriform or sclerotic bodies that divide by transverse septation. Septation in different planes with delayed separation may give rise to a cluster of four to eight cells (Figure 45–12). Cells within superficial crusts or exudates may germinate into septate, branching hyphae.

A. Phialophora verrucosa

The conidia are produced from flask-shaped phialides with cup-shaped collarettes. Mature, spherical to oval conidia are extruded from the phialide and usually accumulate around it (Figure 45–13A).

B. Cladophialophora (Cladosporium) carrionii

Species of *Cladophialophora* and *Cladosporium* produce branching chains of conidia by distal (acropetalous) budding. The terminal conidium of a chain gives rise to the next conidium by a budding process. Species are identified based on differences in the length of the chains and the shape and size of the conidia. *C carrionii* produces elongated conidiophores with long, branching chains of oval conidia.

C. Rhinocladiella aquaspersa

This species produces lateral or terminal conidia from a lengthening conidiogenous cell—a sympodial process. The conidia are elliptical to clavate.

D. Fonsecaea pedrosoi

Fonsecaea is a polymorphic genus. Isolates may exhibit (1) phialides; (2) chains of blastoconidia, similar to *Cladosporium* species; or (3) sympodial, rhinocladiella-type conidiation.

Most strains of *F pedrosoi* form short branching chains of blastoconidia as well as sympodial conidia (Figure 45–13B).

E. Fonsecaea compacta

The blastoconidia produced by *F compacta* are almost spherical, with a broad base connecting the conidia. These structures are smaller and more compact than those of *F pedrosoi*.

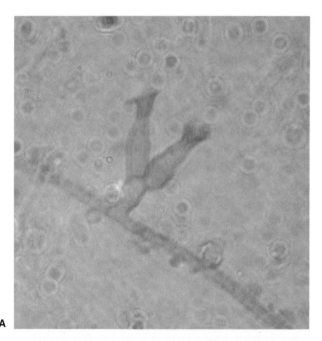

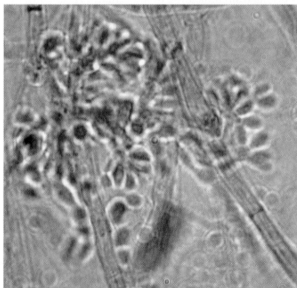

FIGURE 45–13 Identifying conidia produced in culture by the two most common agents of chromomycosis. **A**: *Phialophora verrucosa* produces conidia from these vase-shaped phialides with collarettes. 1000×. **B**: *Fonsecaea pedrosoi* usually displays short branching chains of blastoconidia, as well as other types of conidiogenesis. 1000×.

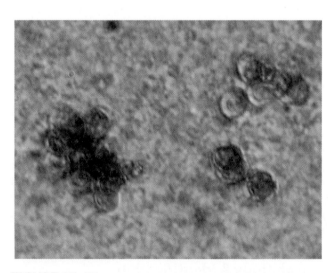

FIGURE 45–12 Chromomycosis. The diagnostic brownish, melanized sclerotic cells (4–12 μm diameter) are evident in this H&E-stained cutaneous biopsy. 400×.

Pathogenesis & Clinical Findings

The fungi are introduced into the skin by trauma, often of the exposed legs or feet. Over months to years, the primary lesion becomes verrucous and wart-like with extension along the draining lymphatics. Cauliflower-like nodules with crusting abscesses eventually cover the area. Small ulcerations or "black dots" of hemopurulent material are present on the warty surface. Rarely, elephantiasis may result from secondary infection, obstruction, and fibrosis of lymph channels. Dissemination to other parts of the body is very rare, though satellite lesions can occur due either to local lymphatic spread or to autoinoculation. Histologically, the lesions are granulomatous and the dark sclerotic bodies may be seen within leukocytes or giant cells.

Diagnostic Laboratory Tests

Specimens of scrapings or biopsies from lesions are placed in 10% KOH and examined microscopically for dark, spherical cells. Detection of the sclerotic bodies is diagnostic of chromoblastomycosis regardless of the etiologic agent. Tissue sections reveal granulomas and extensive hyperplasia of the dermal tissue.

Specimens should be cultured on inhibitory mold agar or Sabouraud's agar with antibiotics. The dematiaceous species is identified by its characteristic conidial structures, as described above. There are many similar saprophytic dematiaceous molds, but they differ from the pathogenic species in being unable to grow at 37°C and being able to digest gelatin.

Treatment

Surgical excision with wide margins is the therapy of choice for small lesions. Chemotherapy with flucytosine or itraconazole may be efficacious for larger lesions. Local applied heat is also beneficial. Relapse is common.

Epidemiology

Chromoblastomycosis occurs mainly in the tropics. The fungi are saprophytic in nature, probably occurring on vegetation and in soil. The disease occurs chiefly on the legs of barefoot agrarian workers following traumatic introduction of the fungus. Chromoblastomycosis is not communicable. Wearing shoes and protecting the legs probably would prevent infection.

PHAEOHYPHOMYCOSIS

Phaeohyphomycosis is a term applied to infections characterized by the presence of darkly pigmented septate hyphae in tissue. Both cutaneous and systemic infections have been described. The clinical forms vary from solitary encapsulated cysts in the subcutaneous tissue to sinusitis to brain abscesses. Over 100 species of dematiaceous molds have been associated with various types of phaeohyphomycotic infections. They are all exogenous molds that normally exist in nature. Some of the more common causes of subcutaneous phaeohyphomycosis are *Exophiala jeanselmei*, *Phialophora richardsiae*, *Bipolaris spicifera*, and *Wangiella dermatitidis*. These species and others (eg, *Exserohilum rostratum*, *Alternaria* species, and *Curvularia* species) may be implicated also in systemic phaeohyphomycosis. The incidence of phaeohyphomycosis and the range of pathogens have been increasing in recent years in both immunocompetent and compromised patients.

In tissue, the hyphae are large (5–10 μm in diameter), often distorted and may be accompanied by yeast cells, but these structures can be differentiated from other fungi by the melanin in their cell walls (Figure 45–14). Specimens are cultured on routine fungal media to identify the etiologic agent. In general, itraconazole or flucytosine is the drug of choice for subcutaneous phaeohyphomycosis. Brain abscesses are usually fatal, but when recognized they are managed with amphotericin B and surgery. The leading cause of cerebral phaeohyphomycosis is *Cladophialophora bantiana*.

MYCETOMA

Mycetoma is a chronic subcutaneous infection induced by traumatic inoculation with any of several saprophytic species of fungi or actinomycetous bacteria that are normally found in soil. The clinical features defining mycetoma are local swelling and interconnecting—often draining—sinuses

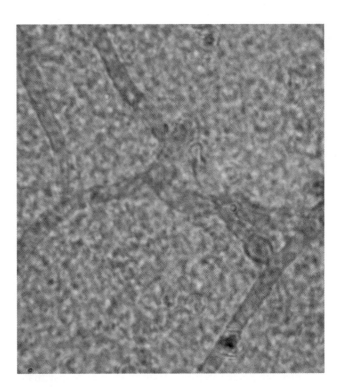

FIGURE 45–14 Phaeohyphomycosis. Melanized hyphae are observed in the tissue. 400×.

that contain granules, which are microcolonies of the agent embedded in tissue material. An **actinomycetoma** is a mycetoma caused by an actinomycete; a **eumycetoma** (maduromycosis, Madura foot) is a mycetoma caused by a fungus. The natural history and clinical features of both types of mycetoma are similar, but actinomycetomas may be more invasive, spreading from the subcutaneous tissue to the underlying muscle. Of course, the therapy is different. Mycetoma occurs worldwide but more often among impoverished people who do not wear shoes. Mycetomas occur only sporadically outside the tropics and are particularly prevalent in India, Africa, and Latin America. Actinomycetomas are discussed in Chapter 12.

Morphology & Identification

The fungal agents of mycetoma include, among others, *Pseudallescheria boydii* (anamorph, *Scedosporium apiospermum*), *Madurella mycetomatis*, *Madurella grisea*, *Exophiala jeanselmei*, and *Acremonium falciforme*. In the United States, the prevalent species is *P boydii*, which is homothallic and has the ability to produce ascospores in culture. *E jeanselmei* and the *Madurella* species are dematiaceous molds. These molds are identified primarily by their mode of conidiation. *P boydii* may also cause pseudallescheriasis, which is a systemic infection in compromised patients.

In tissue, the mycetoma granules may range up to 2 mm in size. The color of the granule may provide information about the agent. For example, the granules of mycetoma caused by *P boydii* and *A falciforme* are white; those of *M grisea* and *E jeanselmei* are black; and *M mycetomatis* produces a dark red to black granule. These granules are hard and contain intertwined, septate hyphae (3–5 μm in width). The hyphae are typically distorted and enlarged at the periphery of the granule.

Pathogenesis & Clinical Findings

Mycetoma develops after traumatic inoculation with soil contaminated with one of the agents. Subcutaneous tissues of the feet, lower extremities, hands, and exposed areas are most often involved. Regardless of the agent, the pathology is characterized by suppuration and abscess formation, granulomas, and the formation of draining sinuses containing the granules. This process may spread to contiguous muscle and bone. Untreated lesions persist for years and extend deeper and peripherally, causing deformation and loss of function.

Very rarely, *P boydii* may disseminate in an immunocompromised host or produces infection of a foreign body (eg, a cardiac pacemaker).

Diagnostic Laboratory Tests

Granules can be dissected out from the pus or biopsy material for examination and culture on appropriate media. The granule color, texture, and size and the presence of hyaline

or pigmented hyphae (or bacteria) are helpful in determining the causative agent. Draining mycetomas are often superinfected with staphylococci and streptococci.

Treatment

The management of eumycetoma is difficult, involving surgical debridement or excision and chemotherapy. *P boydii* is treated with topical nystatin or miconazole. Itraconazole, ketoconazole, and even amphotericin B can be recommended for *Madurella* infections and flucytosine for *E jeanselmei*. Chemotherapeutic agents must be given for long periods to adequately penetrate these lesions.

Epidemiology & Control

The organisms producing mycetoma occur in soil and on vegetation. Barefoot farm laborers are therefore commonly exposed. Properly cleaning wounds and wearing shoes are reasonable control measures.

ENDEMIC MYCOSES

Each of the four primary systemic (dimorphic) mycoses—coccidioidomycosis, histoplasmosis, blastomycosis, and paracoccidioidomycosis—is geographically restricted to specific areas of endemicity. The fungi that cause coccidioidomycosis and histoplasmosis exist in nature in dry soil or in soil mixed with guano, respectively. The agents of blastomycosis and paracoccidioidomycosis are presumed to reside in nature, but their habitats have not been clearly defined. Each of these four mycoses is caused by a thermally dimorphic fungus, and most infections are initiated in the lungs following inhalation of the respective conidia. Only a few infections lead to disease, which may involve dissemination from the lungs to other organs. With rare exceptions, these mycoses are not transmissible among humans or other animals. Table 45–3 summarizes and contrasts some of the fundamental features of these systemic or deep mycoses.

For all of these infections, the initial host defenses are provided by the alveolar macrophages, which are usually capable of inactivating the conidia and inducing a robust immune response. This process typically leads to granulomatous inflammation and the production of both antibodies and cell-mediated immunity. The induction of Th1 cytokines (eg, interleukin-12, interferon-γ, tumor necrosis factor α) will amplify the cellular defenses, activating macrophages and enhancing their fungicidal capacity. In an immunocompetent host, these responses lead to resolution of the inflammatory lesions. However, residual granulomata may retain dormant organisms with the potential for subsequent reactivation, constituting a latent form of the disease. Within the endemic areas for these fungi, most infections occur in immunocompetent individuals, but persons with impaired cellular immunity, such as patients with HIV/AIDS, have an increased risk of serious infection.

TABLE 45-3 **Summary of Endemic Mycoses**[a]

Mycosis	Etiology	Ecology	Geographic Distribution	Tissue Form
Histoplasmosis	*Histoplasma capsulatum*	Avian and bat habitats (guano); alkaline soil	Global; endemic in Ohio, Missouri and Mississippi River valleys; central Africa (var. *duboisii*)	Oval yeasts, 2 x 4 μm, intracellular in macrophages
Coccidioidomycosis	*Coccidioides posadasii* or *Coccidioides immitis*	Soil, rodents	Semiarid regions of southwestern United States, Mexico, Central and South America	Spherules, 10–80 μm, containing endospores, 2–4 μm
Blastomycosis	*Blastomyces dermatitidis*	Unknown (riverbanks?)	Mississippi, Ohio and St. Lawrence River valleys; southeastern United States	Thick-walled yeasts with broad-based, usually single, buds, 8–15 μm
Paracoccidioidomycosis	*Paracoccidioides brasiliensis*	Unknown (soil?)	Central and South America	Large, multiply budding yeasts, 15–30 μm

[a]All four endemic mycoses are caused by dimorphic fungi that reside in nature in the mold form producing hyaline septate hyphae and characteristic conidia. Infection is acquired by inhalation of the conidia. With the exception of blastomycosis, the evidence supports a high rate of infection within the endemic areas. Over 90% of infections occur in immunocompetent individuals, 75–90% in males, and 60–95% are asymptomatic and self-limited or latent. Symptomatic disease occurs frequently in immunocompromised patients, including those with HIV/AIDS.

COCCIDIOIDOMYCOSIS

Coccidioides posadasii and *C immitis* are phenotypically indistinguishable soil molds that cause **coccidioidomycosis**. The infection is endemic in well-circumscribed semiarid regions of the southwestern United States, Central America, and South America. Infection is usually self-limited; dissemination is rare but always serious, and it may be fatal.

Morphology & Identification

C posadasii was recently recognized by DNA-based analyses as a distinct species and a frequent cause of coccidioidomycosis. However, since it cannot be readily identified in the laboratory and since the clinical manifestations are the same with either *C immitis* or *C posadasii*, only the former, more familiar species name will be used in this chapter.

On most laboratory media, *C immitis* produces a white to tan cottony colony. The hyphae form chains of arthroconidia (arthrospores), which often develop in alternate cells of a hypha. These chains fragment into individual arthroconidia, which are readily airborne and highly resistant to adverse environmental conditions (Figure 45–16A). These small arthroconidia (3 × 6 μm) remain viable for years and are highly infectious. Following their inhalation, the arthroconidia become spherical and enlarge, forming spherules that contain endospores (Figure 45–16B). Spherules can also be produced in the laboratory by cultivation on a complex medium.

In histologic sections of tissue, sputum, or other specimens, the spherules are diagnostic of *C immitis*. At maturity, the spherules have a thick, doubly refractile wall and may attain a size of 80 μm in diameter. The spherule becomes packed with endospores (2–5 μm in size). Eventually, the wall ruptures to release the endospores, which may develop into new spherules (Figure 45–16B).

Antigenic Structure

Coccidioidin is a crude antigen preparation extracted from the filtrate of a liquid mycelial culture of *C immitis*. Spherulin is produced from a filtrate of a broth culture of spherules. In standardized doses, both antigens elicit positive delayed skin reactions in infected persons. They have also been used in a variety of serologic tests to measure serum antibodies to *C immitis*.

Pathogenesis & Clinical Findings

Inhalation of arthroconidia leads to a primary infection that is asymptomatic in 60% of individuals. The only evidence of infection is the development of serum precipitins and conversion to a positive skin test within 2–4 weeks. The precipitins will decline, but the skin test often remains positive for a lifetime. The other 40% of individuals develop a self-limited influenza-like illness with fever, malaise, cough, arthralgia, and headache. This condition is called **valley fever**, San Joaquin Valley fever, or desert rheumatism. After 1–2 weeks, about 15% of these patients develop hypersensitivity reactions, which present as a rash, erythema nodosum, or erythema multiforme. On radiographic examination, patients typically show hilar adenopathy along with pulmonary infiltrates, pneumonia, pleural effusions, or nodules. Pulmonary residua occur in about

5%, usually in the form of a solitary nodule or thin-walled cavity (Figure 45–15).

Less than 1% of persons infected with *C immitis* develop secondary or disseminated coccidioidomycosis, which is often debilitating and life-threatening. The risk factors for systemic coccidioidomycosis include heredity, sex, age, and compromised cell-mediated immunity. The disease occurs more frequently in certain racial groups. In decreasing order of risk, these are Filipinos, African Americans, Native Americans, Hispanics, and Asians. There is clearly a genetic component to the immune response to *C immitis*. Males are more susceptible than females, with the exception of women who are pregnant, which may relate to differences in the immune response or a direct effect of sex hormones on the fungus. For example, *C immitis* has estrogen-binding proteins, and elevated levels of estradiol and progesterone stimulate its growth. The young and the aged are also at greater risk. Because cell-mediated immune responses are required for adequate resistance, patients with AIDS and other conditions of cellular immunosuppression are at risk for disseminated coccidioidomycosis.

Some individuals develop a chronic but progressive pulmonary disease with multiplying or enlarging nodules or cavities. Dissemination will usually occur within a year after the primary infection. The spherules and endospores are spread by direct extension or hematogenously. A number of extrapulmonary sites may be involved, but the most frequent organs are the skin, the bones and joints, and the meninges. There are distinctive clinical manifestations associated with *C immitis* infections in each of these and other areas of the body.

Dissemination occurs when the immune response is inadequate to contain the pulmonary foci. In most persons, a positive skin test signifies a strong cell-mediated immune response and protection against reinfection. However, if such individuals become immunocompromised by taking cytotoxic drugs or by disease (eg, AIDS), dissemination can occur many years after primary infection (reactivation disease). Coccidioidomycosis in AIDS patients often presents with a rapidly fatal diffuse reticulonodular pneumonitis. Because of the radiologic overlap between this disease and *Pneumocystis* pneumonia and the different therapies for these two entities, it is important to be aware of the possibility of coccidioidal pneumonia in AIDS patients. Blood cultures are often positive for *C immitis*.

On histologic examination, the coccidioidal lesions contain typical granulomas with giant cells and interspersed suppuration. A diagnosis can be made by finding spherules and endospores. The clinical course is often characterized by remissions and relapses.

Diagnostic Laboratory Tests

A. Specimens

Specimens for culture include sputum, exudate from cutaneous lesions, spinal fluid, blood, urine, and tissue biopsies.

B. Microscopic Examination

Materials should be examined fresh (after centrifuging, if necessary) for typical spherules. KOH or calcofluor white stain will facilitate finding the spherules and endospores (Figure 45–16B). These structures are often found in histologic preparations.

C. Cultures

Cultures on inhibitory mold agar, Sabouraud's agar, or blood agar slants can be incubated at room temperature or at 37°C. The media can be prepared with or without antibacterial antibiotics and cycloheximide to inhibit contaminating bacteria or saprophytic molds, respectively. Because the arthroconidia are highly infectious, suspicious cultures are examined only in a biosafety cabinet (Figure 45–16A). Identification must be confirmed by detection of a *C immitis*-specific antigen, animal inoculation, or use of a specific DNA probe.

D. Serology

Within 2–4 weeks after infection, IgM antibodies to coccidioidin can be detected with a latex agglutination test. Specific IgG antibodies are detected by the immunodiffusion (ID) or complement fixation (CF) test. With resolution of the primary episode, these antibodies decline within a few months.

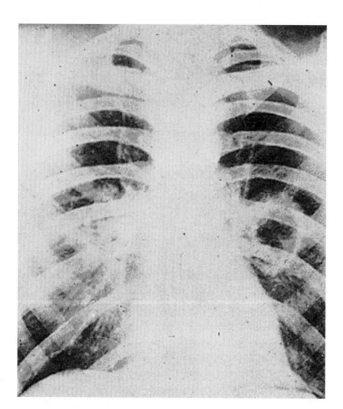

FIGURE 45–15 Chest radiograph of a patient with coccidioidomycosis revealing enlarged hilar lymph nodes and a cavity in the left lung.

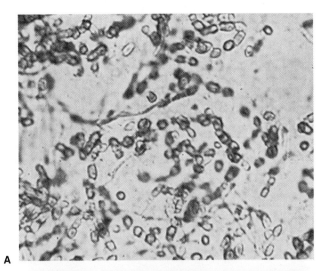

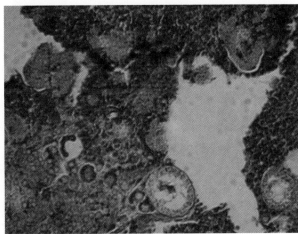

FIGURE 45–16 *Coccidioides* species and Coccidioidomycosis. **A**: In culture at ambient temperatures, *Coccidioides posadasii* produces hyaline, septate hyphae and arthroconidia. 400×. **B**: Large spherules containing endospores can be seen in this section of lung tissue. H&E. 200×.

In contrast, in disseminated coccidioidomycosis, the CF antibody titer continues to rise. Titers above 1:32 are indicative of dissemination, and their fall during treatment suggests improvement. However, CF titers <1:32 do not exclude coccidioidomycosis. Indeed, only half of the patients with coccidioidal meningitis have elevated serum antibodies, but antibody levels in the cerebrospinal fluid are usually high. In AIDS patients with coccidioidomycosis, these serologic tests are often negative.

E. Skin Test

The coccidioidin skin test reaches maximum induration (≥5 mm in diameter) between 24 and 48 hours after cutaneous injection of 0.1 mL of a standardized dilution. If patients with disseminated disease become anergic, the skin test will be negative, which implies a very poor prognosis. Cross-reactions

with antigens of other fungi may occur. Spherulin is more sensitive than coccidioidin in detecting reactors. Reactions to skin tests tend to diminish in size and intensity years after primary infection in persons residing in endemic areas, but skin testing exerts a booster effect. Following recovery from primary infection, there is usually immunity to reinfection.

Treatment

In most persons, symptomatic primary infection is self-limited and requires only supportive treatment, although itraconazole may reduce the symptoms. However, patients who have severe disease require treatment with amphotericin B, which is administered intravenously. This regimen may be followed by several months of oral therapy with itraconazole. Cases of coccidioidal meningitis have been treated with oral fluconazole, which has good penetration of the central nervous system; however, long-term therapy is required, and relapses have occurred. The azoles are not more efficacious than amphotericin B, but they are easier to administer and associated with fewer and less severe side effects. The newer lipid emulsions of amphotericin B promise to deliver higher doses with less toxicity. Surgical resection of pulmonary cavities is sometimes necessary and often curative.

Epidemiology & Control

The areas of endemicity for *C immitis* are semiarid regions, resembling the Lower Sonoran Life Zone. They include the southwestern states—particularly the San Joaquin and Sacramento Valleys of California, areas around Tucson and Phoenix in Arizona, the Rio Grand valley—and similar areas in Central and South America. Within these regions, *C immitis* can be isolated from the soil and indigenous rodents, and the level of skin test reactivity in the population indicates that many humans have been infected. The infection rate is highest during the dry months of summer and autumn, when dust is most prevalent. A high incidence of infection and disease may follow dust storms. During an epidemic of coccidioidomycosis in the San Joaquin Valley of California in 1991–1993, the rate of coccidioidomycosis increased more than 10-fold. Increased precipitation in the spring months of these years has been suggested as an environmental stimulus.

The disease is not communicable from person to person, and there is no evidence that infected rodents contribute to its spread. Some measure of control can be achieved by reducing dust, paving roads and airfields, planting grass or crops, and using oil sprays.

HISTOPLASMOSIS

H capsulatum is a dimorphic soil saprophyte that causes histoplasmosis, the most prevalent pulmonary mycotic infection in humans and animals. In nature, *H capsulatum* grows as a mold in association with soil and avian habitats, being enriched by alkaline nitrogenous substrates in guano. *H capsulatum* and histoplasmosis, which is initiated by inhalation of the conidia, occur worldwide. However, the incidence varies considerably, and most cases occur in the United States. *H capsulatum* received its name from the appearance of the yeast cells in histopathologic sections; however, it is neither a protozoan nor does it have a capsule.

Morphology & Identification

At temperatures below 37°C, primary isolates of *H capsulatum* often develop brown mold colonies, but the appearance varies. Many isolates grow slowly, and specimens require incubation for 4–12 weeks before colonies develop. The hyaline, septate hyphae produce microconidia (2–5 μm) and large, spherical thick-walled macroconidia with peripheral projections of cell wall material (8–16 μm) (Figure 45–17B). In tissue or in vitro on rich medium at 37°C, the hyphae and conidia convert to small, oval yeast cells (2 × 4 μm). In tissue, the yeasts are typically seen within macrophages, as *H capsulatum* is a facultative intracellular parasite (Figure 45–17A). In the laboratory, with appropriate mating strains, a sexual cycle can be demonstrated, yielding *Ajellomyces capsulatus,* a teleomorph that produces ascospores.

Antigenic Structure

Histoplasmin is a crude but standardized mycelial broth culture filtrate antigen. After initial infection, which is asymptomatic in over 95% of individuals, a positive delayed type skin test to histoplasmin is acquired. Antibodies to both yeast and mycelial antigens can be measured serologically (see Table 45–4).

Pathogenesis & Clinical Findings

After inhalation, the conidia develop into yeast cells and are engulfed by alveolar macrophages, where they are able to replicate. Within macrophages, the yeasts may disseminate to reticuloendothelial tissues such as the liver, spleen, bone marrow, and lymph nodes. The initial inflammatory reaction becomes granulomatous. In over 95% of cases, the resulting cell-mediated immune response leads to the secretion of cytokines that activate macrophages to inhibit the intracellular growth of the yeasts. Some individuals, such as immunocompetent persons who inhale a heavy inoculum, develop acute pulmonary histoplasmosis, which is a self-limited flu-like syndrome with fever, chills, myalgias, headaches, and nonproductive cough. On radiographic examination, most patients will have hilar lymphadenopathy and pulmonary infiltrates or nodules. These symptoms resolve spontaneously without therapy, and the granulomatous nodules in the lungs or other sites heal with calcification.

Chronic pulmonary histoplasmosis occurs most often in men and is usually a reactivation process, the breaking down of a dormant lesion that may have been acquired years before.

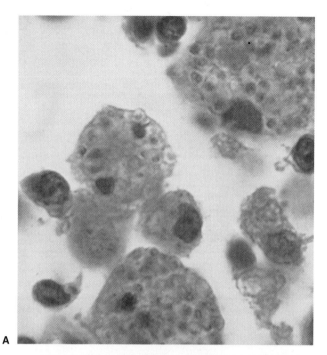

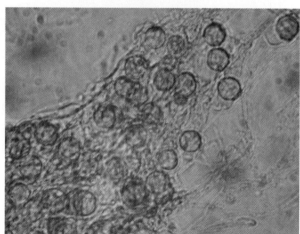

FIGURE 45–17 Histoplasmosis and *Histoplasma capsulatum.* **A**: Small, oval yeast cells (2–4 μm) packed within macrophages. Giemsa's stain. 1000×. **B**: In culture at ambient temperatures, *Histoplasma capsulatum* produces hyaline, septate hyphae bearing microconidia and large, spherical macroconidia. 400×.

This reactivation is usually precipitated by pulmonary damage such as emphysema.

Severe disseminated histoplasmosis develops in a small minority of infected individuals—particularly infants, the elderly, and the immunosuppressed, including AIDS patients. The reticuloendothelial system is especially apt to be involved, with lymphadenopathy, enlarged spleen and liver, high fever, anemia, and a high mortality rate without antifungal therapy. Mucocutaneous ulcers of the nose, mouth, tongue, and intestine can occur. In such individuals, histologic study reveals focal areas of necrosis within granulomas in many organs. The yeasts may be present in macrophages in the blood, liver, spleen, and bone marrow.

TABLE 45–4 Summary of Serologic Tests for Antibodies to Systemic Dimorphic Fungal Pathogens

Mycosis	Test[a]	Sensitivity and Value Antigen[b]	Diagnosis	Prognosis[c]	Comments
Coccidioidomycosis	TP	C	Early primary infection; 90% of cases positive	None	
	CF	C	Titer ≥ 1:32 = secondary disease	Titer reflects severity (except in meningeal disease)	Rarely cross-reactive with histoplasmin
	ID	C	>90% of cases positive, ie, F or HL band (or both)		More specific than CF test
Histoplasmosis	CF	H	≤84% of cases positive (titer ≥ 1:8)	Fourfold change in titer	Cross-reactions in patients with blastomycosis, cryptococcosis, aspergillosis; titer may be boosted by skin test with histoplasmin
	CF	Y	≤94% of cases positive (titer ≥ 1:8)	Fourfold change in titer	Less cross-reactivity than with histoplasmin
	ID	H	≥85% of cases positive, ie, m or m and h bands	Loss of h	Skin test with histoplasmin may boost m band; more specific than CF test
Blastomycosis	CF	By	<50% of cases are positive; reaction to homologous antigen only is diagnostic	Fourfold change in titer	Highly cross-reactive
	ID	Bcf	≤80% of cases are positive, ie, A band	Loss of A band	More specific and sensitive than CF test
	EIA	A	≤90% of cases are positive (titer ≥ 1:16)	Change in titer	92% specificity
Paracoccidioidomycosis	CF	P	80–95% of cases positive (titer ≥ 1:8)	Fourfold change in titer	Some cross-reactions at low titer with aspergillosis and candidiasis sera
	ID	P	98% of cases are positive (bands 1, 2, 3)	Loss of bands	Band 3 and band m (to histoplasmin) are identical

[a]Tests: CF, complement fixation; ID, immunodiffusion; TP, tube precipitin; EIA, enzyme immunoassay.

[b]Antigens: C, coccidioidin; H, histoplasmin; Y, yeast cells of *H capsulatum*; By, yeast cells of *B dermatitidis*; Bcf, culture filtrate of *B dermatitidis* yeast cells; A, antigen A of *B dermatitidis*; P, culture filtrate of *P brasiliensis* yeast cells. In the immunodiffusion tests, antibodies are detected to the following species-specific antigens: *C immitis*, F, HL; *H capsulatum*, m and h; *B dermatitidis*, A; and *P brasiliensis*, 1, 2, and 3.

[c]Fourfold changes in the complement fixation titer (eg, a fall from 1:32 to 1:8) are considered significant, as is the loss of specific immunodiffusion antibody (ie, becoming negative).

Diagnostic Laboratory Tests

A. Specimens

Specimens for culture include sputum, urine, scrapings from superficial lesions, bone marrow aspirates, and buffy coat blood cells. Blood films, bone marrow slides, and biopsy specimens may be examined microscopically. In disseminated histoplasmosis, bone marrow cultures are often positive.

B. Microscopic Examination

The small ovoid cells may be observed within macrophages in histologic sections stained with fungal stains (eg, Gomori methenamine silver, periodic acid Schiff, or calcofluor white) or in Giemsa-stained smears of bone marrow or blood (Figure 45–17A).

C. Culture

Specimens are cultured in rich media, such as glucosecysteine blood agar at 37°C and on Sabouraud's agar or inhibitory mold agar at 25–30°C. Cultures must be incubated for a minimum of 4 weeks. The laboratory should be alerted if histoplasmosis is suspected because special blood culture methods, such as lysis centrifugation or fungal broth medium, can be used to enhance the recovery of *H capsulatum*.

D. Serology

CF tests for antibodies to histoplasmin or the yeast cells become positive within 2–5 weeks after infection. CF titers rise during progressive disease and then decline to very low levels when the disease is inactive. With progressive disease, the CF titers are ≥ 1:32. Because cross-reactions may occur, antibodies to other fungal antigens are routinely tested. In the ID test, precipitins to two *H capsulatum*-specific antigens are detected: The presence of antibodies to the h antigen often signifies active histoplasmosis, while antibodies to the m antigen may arise from repeated skin testing or past exposure.

One of the most sensitive tests is a radioassay or enzyme immunoassay for circulating antigen of *H capsulatum*. Nearly all patients with disseminated histoplasmosis have a positive test for antigen in the serum or urine; the antigen level drops following successful treatment and recurs during relapse. Despite cross-reactions with other mycoses, this test for antigen is more sensitive than conventional antibody tests in AIDS patients with histoplasmosis.

E. Skin Test

The histoplasmin skin test becomes positive soon after infection and remains positive for years. It may become negative in progressive disseminated histoplasmosis. Repeated skin testing stimulates serum antibodies in sensitive individuals, interfering with the diagnostic interpretation of the serologic tests.

Immunity

Following initial infection, most persons appear to develop some degree of immunity. Immunosuppression may lead to reactivation and disseminated disease. AIDS patients may develop disseminated histoplasmosis through reactivation or new infection.

Treatment

Acute pulmonary histoplasmosis is managed with supportive therapy and rest. Itraconazole is the treatment for mild to moderate infection. In disseminated disease, systemic treatment with amphotericin B is often curative, though patients may need prolonged treatment and monitoring for relapses. Patients with AIDS typically relapse despite therapy that would be curative in other patients. Therefore, AIDS patients require maintenance therapy with itraconazole.

Epidemiology & Control

The incidence of histoplasmosis is highest in the United States, where the endemic areas include the central and eastern states and in particular the Ohio River Valley and portions of the Mississippi River Valley. Numerous outbreaks of acute histoplasmosis have resulted from exposure of many persons to large inocula of conidia. These occur when *H capsulatum* is disturbed in its natural habitat, ie, soil mixed with bird feces (eg, starling roosts, chicken houses) or bat guano (caves). Birds are not infected, but their excrement provides superb culture conditions for growth of the fungus. Conidia are also spread by wind and dust. The largest urban outbreak of histoplasmosis occurred in Indianapolis.

In some highly endemic areas, 80–90% of residents have a positive skin test by early adulthood. Many will have miliary calcifications in the lungs. Histoplasmosis is not communicable from person to person. Spraying formaldehyde on infected soil may destroy *H capsulatum*.

In Africa, in addition to the usual pathogen, there is a stable variant, *H capsulatum* var *duboisii*, which causes African histoplasmosis. This form differs from the usual disease by causing less pulmonary involvement and more skin and bone lesions with abundant giant cells that contain the yeasts, which are larger and more spherical.

BLASTOMYCOSIS

B dermatitidis is a thermally dimorphic fungus that grows as a mold in culture, producing hyaline, branching septate hyphae and conidia. At 37°C or in the host, it converts to a large, singly budding yeast cell (Figure 45–18). *B dermatitidis* causes **blastomycosis**, a chronic infection with granulomatous and suppurative lesions that is initiated in the lungs, whence dissemination may occur to any organ but preferentially to the skin and bones. The disease has been called North American blastomycosis because it is endemic and most cases occur in the United States and Canada. Despite this high prevalence in North America, blastomycosis has been documented in Africa, South America, and Asia. It is endemic for humans and dogs in the eastern United States.

Morphology & Identification

When *B dermatitidis* is grown on Sabouraud's agar at room temperature, a white or brownish colony develops, with branching hyphae bearing spherical, ovoid, or piriform conidia (3–5 μm in diameter) on slender terminal or lateral conidiophores (Figure 45–18B). Larger chlamydospores (7–18 μm) may also be produced. In tissue or culture at 37°C, *B dermatitidis* grows as a thick-walled, multinucleated, spherical yeast (8–15 μm) that usually produces single buds (Figure 45–18A). The bud and the parent yeast are attached with a broad base, and the bud often enlarges to the same size as the parent yeast before they become detached. The yeast colonies are wrinkled, waxy, and soft.

Antigenic Structure

Extracts of culture filtrates of *B dermatitidis* contain **blastomycin**, probably a mixture of antigens. As a skin test reagent, blastomycin lacks specificity and sensitivity. Patients are often negative or lose their reactivity, and false-positive

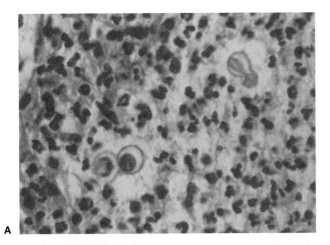

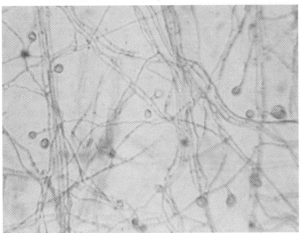

FIGURE 45–18 Blastomycosis and *Blastomyces dermatitidis.*
A: Note the large, spherical thick-walled yeast cells (8–15 μm in diameter) in this section of a cutaneous abscess. H&E. 400×. **B**: In culture at ambient temperatures, *Blastomyces dermatitidis* produces hyaline, septate hyphae and single conidia. 400×.

cross-reactions occur in people exposed to other fungi. Consequently, skin test surveys of the population to determine the level of exposure have not been conducted. The diagnostic value of blastomycin as an antigen in the CF test is also questionable because cross-reactions are common; however, many patients with widespread blastomycosis have high CF titers. In the ID test, using adsorbed reference antisera, antibodies can be detected to a specific *B dermatitidis* antigen, designated antigen A. More reliable is an enzyme immunoassay for antigen A. The immunodominant motif probably responsible for generating a protective cell-mediated immune response is part of a cell-surface and secreted protein, termed BAD.

Pathogenesis & Clinical Findings

Human infection is initiated in the lungs. Mild and self-limited cases have been documented, but their frequency is unknown because there is no adequate skin or serologic test with which to assess subclinical or resolved primary

infections. The most common clinical presentation is a pulmonary infiltrate in association with a variety of symptoms indistinguishable from other acute lower respiratory infections (fever, malaise, night sweats, cough, and myalgias). Patients can also present with chronic pneumonia. Histologic examination reveals a distinct pyogranulomatous reaction with neutrophils and noncaseating granulomas. When dissemination occurs, skin lesions on exposed surfaces are most common. They may evolve into ulcerated verrucous granulomas with an advancing border and central scarring. The border is filled with microabscesses and has a sharp, sloping edge. Lesions of bone, the genitalia (prostate, epididymis, and testis), and the central nervous system also occur; other sites are less frequently involved. Although immunosuppressed patients, including those with AIDS, may develop blastomycosis, it is not as common in these patients as are other systemic mycoses.

Diagnostic Laboratory Tests

A. Specimens

Specimens consist of sputum, pus, exudates, urine, and biopsies from lesions.

B. Microscopic Examination

Wet mounts of specimens may show broadly attached buds on thick-walled yeast cells. These may also be apparent in histologic sections (Figure 45–18A).

C. Culture

Colonies usually develop within 2 weeks on Sabouraud's or enriched blood agar at 30°C (Figure 45–18B). The identification is confirmed by conversion to the yeast form after cultivation on a rich medium at 37°C, by extraction and detection of the *B dermatitidis*-specific antigen A, or by a specific DNA probe.

D. Serology

As indicated in Table 45–4, antibodies can be measured by the CF and ID tests. In the EIA, high antibody titers to antigen A are associated with progressive pulmonary or disseminated infection. Overall, serologic tests are not as useful for the diagnosis of blastomycosis as they are in the case of the other endemic mycoses.

Treatment

Severe cases of blastomycosis are treated with amphotericin B. In patients with confined lesions, a 6-month course of itraconazole is very effective.

Epidemiology

Blastomycosis is a relatively common infection of dogs (and rarely other animals) in endemic areas. Blastomycosis cannot be transmitted by animals or humans. Unlike *C immitis* and *H capsulatum*, *B dermatitidis* has only rarely (and not reproducibly) been isolated from the environment, so its natural habitat is unknown. However, the occurrence of several small outbreaks has linked *B dermatitidis* to rural river banks.

PARACOCCIDIOIDOMYCOSIS

Paracoccidioides brasiliensis is the thermally dimorphic fungal agent of paracoccidioidomycosis (South American blastomycosis), which is confined to endemic regions of Central and South America.

Morphology & Identification

Cultures of the mold form of *P brasiliensis* grow very slowly and produce chlamydospores and conidia. The features are not distinctive. At 36°C, on rich medium, it forms large, multiply budding yeast cells (up to 30 μm). The yeasts are larger and have thinner walls than those of *B dermatitidis*. The buds are attached by a narrow connection (Figure 45–19).

Pathogenesis & Clinical Findings

P brasiliensis is inhaled, and initial lesions occur in the lung. After a period of dormancy that may last for decades, the pulmonary granulomas may become active, leading to chronic, progressive pulmonary disease or dissemination. Most patients are 30–60 years of age, and over 90% are men. A few patients (≤ 10%), typically less than 30 years of age, develop an acute or subacute progressive infection with a shorter incubation time. In the usual case of chronic paracoccidioidomycosis, the yeasts spread from the lung to other organs, particularly the skin and mucocutaneous tissue, lymph nodes, spleen, liver, adrenals, and other sites. Many patients present with painful sores involving the oral mucosa. Histology usually reveals either granulomas with central caseation or microabscesses. The yeasts are frequently observed in giant cells or directly in exudate from mucocutaneous lesions.

Skin test surveys have been conducted using an antigen extract, **paracoccidioidin**, which may cross-react with coccidioidin or histoplasmin.

Diagnostic Laboratory Tests

In sputum, exudates, biopsies, or other material from lesions, the yeasts are often apparent on direct microscopic examination with KOH or calcofluor white. Cultures on Sabouraud's or yeast extract agar are incubated at room temperature and confirmed by conversion to the yeast form by in vitro growth at 36°C. Serologic testing is most useful for diagnosis. Antibodies to paracoccidioidin can be measured by the CF or ID test (Table 45–4). Healthy persons in endemic areas do not have antibodies to *P brasiliensis*. In patients, titers tend to correlate with the severity of disease.

Treatment

Itraconazole appears to be most effective against paracoccidioidomycosis, but ketoconazole and trimethoprim-sulfamethoxazole are also efficacious. Severe disease can be treated with amphotericin B.

Epidemiology

Paracoccidioidomycosis occurs mainly in rural areas of Latin America, particularly among farmers. The disease manifestations are much more frequent in males than in females, but infection and skin test reactivity occur equally in both sexes. Since *P brasiliensis* has only rarely been isolated from nature, its natural habitat has not been defined. As with the other endemic mycoses, paracoccidioidomycosis is not communicable.

OPPORTUNISTIC MYCOSES

Patients with compromised host defenses are susceptible to ubiquitous fungi to which healthy people are exposed but usually resistant. In many cases, the type of fungus and the natural history of the mycotic infection are determined by the underlying predisposing condition of the host. As members of the normal microbial flora, *Candida* and related yeasts are endogenous opportunists. Other opportunistic mycoses are caused by exogenous fungi that are globally present in soil, water, and air. The more common pathogens will be discussed, but the incidence and the roster of fungal species causing serious mycotic infections in compromised individuals continue to increase.

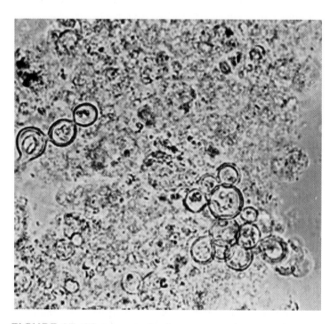

FIGURE 45–19 Paracoccidioidomycosis. Large, multiply budding yeast cells (15–30 μm) are observed in cutaneous lesion. KOH. 400×.

In patients with AIDS, the susceptibility and incidence of opportunistic mycoses are inversely correlated with the CD4 lymphocyte count.

CANDIDIASIS

Several species of the yeast genus *Candida* are capable of causing candidiasis. They are members of the normal flora of the skin, mucous membranes, and gastrointestinal tract. *Candida* species colonize the mucosal surfaces of all humans during or soon after birth, and the risk of endogenous infection is ever-present. Candidiasis is the most common systemic mycosis, and the most common agents are *C albicans*, *C tropicalis*, *C parapsilosis*, *C glabrata*, *C guilliermondii*, and *C dubliniensis*. The widespread use of fluconazole has precipitated the emergence of more azole-resistant species, such as *C krusei* and *C lusitaniae*.

Morphology & Identification

In culture or tissue, *Candida* species grow as oval, budding yeast cells (3–6 μm in size). They also form **pseudohyphae** when the buds continue to grow but fail to detach, producing chains of elongated cells that are pinched or constricted at the septations between cells. Unlike other species of *Candida*, *C albicans* is dimorphic; in addition to yeasts and pseudohyphae, it can also produce true hyphae (Figure 45–20). On agar media or within 24 hours at 37°C or room temperature, *Candida* species produce soft, cream-colored colonies with a yeasty odor. Pseudohyphae are apparent as submerged growth below the agar surface. Two simple morphologic tests distinguish *C albicans*, the most common pathogen, from other species of *Candida*: After incubation in serum for about 90 minutes at 37°C, yeast cells of *C albicans* will begin to form true hyphae or germ tubes (Figure 45–21), and on nutritionally deficient media *C albicans* produces large, spherical chlamydospores. Sugar fermentation and assimilation tests can be used to confirm the identification and speciate the more common *Candida* isolates, such as *C tropicalis*, *C parapsilosis*, *C guilliermondii*, *C kefyr*, *C krusei*, and *C lusitaniae*; *C glabrata* is unique among these pathogens because it produces only yeast cells and no pseudohyphal forms.

Antigenic Structure

The use of adsorbed antisera have defined two serotypes of *C albicans*: A (which includes *C tropicalis*) and B. Many other antigens have been characterized, including secreted proteases, an immunodominant enolase, and heat shock proteins.

Pathogenesis & Pathology

Superficial (cutaneous or mucosal) candidiasis is established by an increase in the local census of *Candida* and damage to the skin or epithelium that permits local invasion by the yeasts and pseudohyphae. Systemic candidiasis occurs when *Candida* enters the bloodstream and the phagocytic host defenses are inadequate to contain the growth and dissemination of the

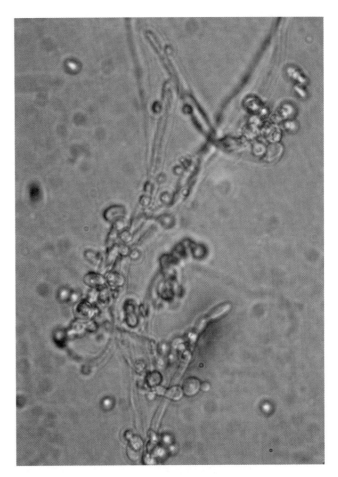

FIGURE 45–20 *Candida albicans.* Budding yeast cells (blastoconidia), hyphae and pseudohyphae. 400×.

yeasts. From the circulation, *Candida* can infect the kidneys, attach to prosthetic heart valves, or produce candidal infections almost anywhere (eg, arthritis, meningitis, endophthalmitis). The local histology of cutaneous or mucocutaneous lesions is characterized by inflammatory reactions varying from pyogenic abscesses to chronic granulomas. The lesions contain abundant budding yeast cells and pseudohyphae. Large increases of *Candida* in the intestinal tract often follow the administration of oral antibacterial antibiotics, and the yeasts can enter the circulation by crossing the intestinal mucosa.

Clinical Findings

A. Cutaneous & Mucosal Candidiasis

The risk factors associated with superficial candidiasis include AIDS, pregnancy, diabetes, young or old age, birth control pills, and trauma (burns, maceration of the skin). **Thrush** can occur on the tongue, lips, gums, or palate. It is a patchy to confluent, whitish pseudomembranous lesion composed of epithelial cells, yeasts, and pseudohyphae. Thrush develops in most patients with AIDS. Other risk factors include treatment with corticosteroids or antibiotics, high levels of glucose, and cellular immunodeficiency. Yeast invasion of

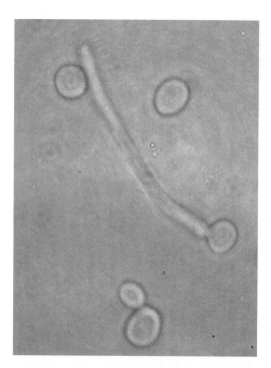

FIGURE 45–21 Germ tube. Unlike other species of *Candida*, *Candida albicans* produces true hyphae as well as budding yeast cells and pseudohyphae. After incubation in serum at 37°C for 60–90 min in the laboratory, clinical isolates of *Candida albicans* are stimulated to form hyphae, and this process is initiated by the production of germ tubes, which are thinner and more uniform than pseudohyphae (see Figure 45–20). 1000×.

the vaginal mucosa leads to **vulvovaginitis**, characterized by irritation, pruritus, and vaginal discharge. This condition is often preceded by factors such as diabetes, pregnancy, or antibacterial drugs that alter the microbial flora, local acidity, or secretions. Other forms of **cutaneous candidiasis** include invasion of the skin. This occurs when the skin is weakened by trauma, burns, or maceration. Intertriginous infection occurs in moist, warm parts of the body such as the axillae, groin, and intergluteal or inframammary folds; it is most common in obese and diabetic individuals. The infected areas become red and moist and may develop vesicles. Interdigital involvement between the fingers follows repeated prolonged immersion in water; it is most common in homemakers, bartenders, cooks, and vegetable and fish handlers. Candidal invasion of the nails and around the nail plate causes **onychomycosis**, a painful, erythematous swelling of the nail fold resembling a pyogenic paronychia, which may eventually destroy the nail.

B. Systemic Candidiasis

Candidemia can be caused by indwelling catheters, surgery, intravenous drug abuse, aspiration, or damage to the skin or gastrointestinal tract. In most patients with normal host defenses, the yeasts are eliminated and candidemia is transient. However, patients with compromised innate phagocytic defenses may develop occult lesions anywhere, especially the

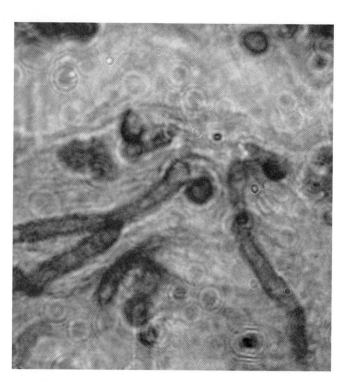

FIGURE 45–22 Candidiasis. Yeasts and pseudohyphae in tissue, stained with periodic acid Schiff. 1000×.

kidney, skin (maculonodular lesions), eye, heart, and meninges. Systemic candidiasis is most often associated with chronic administration of corticosteroids or other immunosuppressive agents; with hematologic diseases such as leukemia, lymphoma, and aplastic anemia; or with chronic granulomatous disease. Candidal endocarditis is frequently associated with deposition and growth of the yeasts and pseudohyphae on prosthetic heart valves or vegetations. Kidney infections are usually a systemic manifestation, whereas urinary tract infections are often associated with Foley catheters, diabetes, pregnancy, and antibacterial antibiotics.

C. Chronic Mucocutaneous Candidiasis

Most forms of this rare disease have onset in early childhood, are associated with cellular immunodeficiencies and endocrinopathies, and result in chronic superficial disfiguring infections of any or all areas of skin or mucosa.

Diagnostic Laboratory Tests

A. Specimens

Specimens include swabs and scrapings from superficial lesions, blood, spinal fluid, tissue biopsies, urine, exudates, and material from removed intravenous catheters.

B. Microscopic Examination

Tissue biopsies, centrifuged spinal fluid, and other specimens may be examined in Gram-stained smears or histopathological slides for pseudohyphae and budding cells (Figure 45–22). Skin or nail scrapings are first placed in a drop of 10% potassium hydroxide (KOH) and calcofluor white.

C. Culture

All specimens are cultured on fungal or bacteriologic media at room temperature or at 37°C. Yeast colonies are examined for the presence of pseudohyphae. *C albicans* is identified by the production of germ tubes or chlamydospores. Other *Candida* isolates are speciated with a battery of biochemical reactions. The interpretation of positive cultures varies with the specimen. Positive cultures from normally sterile body sites are significant. The diagnostic value of a quantitative urine culture depends on the integrity of the specimen and the yeast census. Contaminated Foley catheters may lead to "false-positive" urine cultures. Positive blood cultures may reflect systemic candidiasis or transient candidemia due to a contaminated intravenous line. Sputum cultures have no value because *Candida* species are part of the oral flora. Cultures of skin lesions are confirmatory.

D. Serology

In general, the currently available serologic tests have limited specificity or sensitivity. Serum antibodies and cell-mediated immunity are demonstrable in most people as a result of lifelong exposure to candida. In systemic candidiasis, antibody titers to various candidal antigens may be elevated, but there are no clear criteria for establishing a diagnosis serologically. The detection of circulating cell wall mannan, using a latex agglutination test or an enzyme immunoassay, is much more specific, but the test lacks sensitivity because many patients are only transiently positive or because they do not develop significant and detectable antigen titers until late in the disease. A promising new serological test for circulating β-glucan, which is found in the cell walls of many fungal species, is currently under evaluation.

Immunity

The basis of resistance to candidiasis is complex and incompletely understood. Cell-mediated immune responses, especially CD4 cells, are important in controlling mucocutaneous candidiasis, and the neutrophil is probably crucial for resistance to systemic candidiasis.

Treatment

Thrush and other mucocutaneous forms of candidiasis are usually treated with topical nystatin or oral ketoconazole or fluconazole. Systemic candidiasis is treated with amphotericin B, sometimes in conjunction with oral flucytosine, fluconazole, or caspofungin. The clearing of cutaneous lesions is accelerated by eliminating contributing factors such as excessive moisture or antibacterial drugs. Chronic mucocutaneous candidiasis responds well to oral ketoconazole and other azoles, but patients have a genetic cellular immune defect and often require lifelong treatment.

It is often difficult to establish an early diagnosis of systemic candidiasis—the clinical signs are not definitive, and cultures are often negative. Furthermore, there is no established prophylactic regimen for patients at risk, though treatment with an azole or with a short course of low-dose amphotericin B is often indicated for febrile or debilitated patients who are immunocompromised and do not respond to antibacterial therapy (see below).

Epidemiology & Control

The most important preventive measure is to avoid disturbing the normal balance of microbial flora and intact host defenses. Candidiasis is not communicable, since virtually all persons normally harbor the organism.

CRYPTOCOCCOSIS

Cryptococcus neoformans and *C gattii* are basidiomycetous yeasts with large polysaccharide capsules. *C neoformans* occurs worldwide in nature and is isolated readily from dry pigeon feces. *C gattii* is less common and typically associated with trees in tropical areas. Both species cause cryptococcosis, which follows inhalation of desiccated yeast cells or possibly the smaller basidiospores. From the lungs, these neurotropic yeasts typically migrate to the central nervous system where they cause meningoencephalitis. However, they also have the capacity to infect many other organs (eg, skin, eyes, prostate). *C neoformans* occurs in immunocompetent persons but more often in patients with HIV/AIDS, hematogenous malignancies, and other immunosuppressive conditions. Cryptococcosis due to *C gattii* is rarer and usually associated with apparently normal hosts.

Morphology & Identification

In culture, *Cryptococcus* species produce whitish mucoid colonies within 2–3 days. Microscopically, in culture or clinical material, the spherical budding yeast cells (5–10 μm in diameter) are surrounded by a thick nonstaining capsule (Figure 45–23). All species of *Cryptococcus*, including several nonpathogenic species, are encapsulated and possess urease. However, *C neoformans* and *C gattii* differ from nonpathogenic species by the abilities to grow at 37°C and the production of laccase, a phenol oxidase, which catalyzes the formation of melanin from appropriate phenolic substrates (eg, catecholamines). Both the capsule and laccase are well-characterized virulence factors. Clinical isolates are identified by demonstrating the production of laccase or a specific pattern of carbohydrate assimilations. Adsorbed antisera have defined five serotypes (A–D and AD); strains of *C neoformans* may possess serotype A, D, or AD, and isolates of *C gattii* may have serotype B or C. In addition to their capsular serotypes, the two species differ in their genotypes, ecology, some biochemical reactions, and clinical manifestations. Sexual reproduction can be demonstrated in the laboratory,

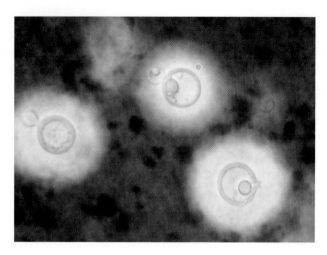

FIGURE 45–23 Cryptococcosis. The capsule of *Cryptococcus neoformans* is notably apparent in this pulmonary lavage specimen. Giemsa's stain. 1000×.

and successful mating results in the production of mycelia and basidiospores; the corresponding teleomorphs of the two varieties are *Filobasidiella neoformans* var *neoformans* (serotypes A and D) and *Filobasidiella neoformans* var *bacillispora* (serotypes B and C).

Antigenic Structure

The capsular polysaccharides, regardless of serotype, have a similar structure: They are long, unbranched polymers consisting of an α-1,3-linked polymannose backbone with β-linked monomeric branches of xylose and glucuronic acid. During infection, the capsular polysaccharide is solubilized in spinal fluid, serum, or urine and can be detected by an enzyme immunoassay or by the agglutination of latex particles coated with antibody to the polysaccharide. With proper controls, this test is diagnostic of cryptococcosis. Patient antibodies to the capsule can also be measured, but they are not used in diagnosis.

Pathogenesis

Infection is initiated by inhalation of the yeast cells, which in nature are dry, minimally encapsulated, and easily aerosolized. The primary pulmonary infection may be asymptomatic or may mimic an influenza-like respiratory infection, often resolving spontaneously. In patients who are compromised, the yeasts may multiply and disseminate to other parts of the body but preferentially to the central nervous system, causing cryptococcal meningoencephalitis. Other common sites of dissemination include the skin, adrenals, bone, eye, and prostate gland. The inflammatory reaction is usually minimal or granulomatous.

Clinical Findings

The major clinical manifestation is chronic meningitis, which can resemble a brain tumor, brain abscess, degenerative

central nervous system disease, or any mycobacterial or fungal meningitis. Cerebrospinal fluid pressure and protein may be increased and the cell count elevated, whereas the glucose is normal or low. Patients may complain of headache, neck stiffness, and disorientation. In addition, there may be lesions in skin, lungs, or other organs.

The course of cryptococcal meningitis may fluctuate over long periods, but all untreated cases are ultimately fatal. About 5–8% of patients with AIDS develop cryptococcal meningitis. The infection is not transmitted from person to person.

Diagnostic Laboratory Tests

A. Specimens, Microscopic Examination, & Culture

Specimens include cerebrospinal fluid, tissue, exudates, sputum, blood, cutaneous scrapings, and urine. Spinal fluid is centrifuged before microscopic examination and culture. For direct microscopy, specimens are often examined in wet mounts, both directly and after mixing with India ink, which delineates the capsule (Figure 45–23).

Colonies develop within a few days on most media at room temperature or 37°C. Media with cycloheximide inhibit *Cryptococcus* and should be avoided. Cultures can be identified by growth at 37°C and detection of urease. Alternatively, on an appropriate diphenolic substrate, the phenol oxidase (or laccase) of *C neoformans* and *C gattii* produces melanin in the cell walls and colonies develop a brown pigment.

B. Serology

Tests for capsular antigen can be performed on cerebrospinal fluid and serum. The latex slide agglutination test for cryptococcal antigen is positive in 90% of patients with cryptococcal meningitis. With effective treatment, the antigen titer drops—except in AIDS patients, who often maintain high antigen titers for long periods.

Treatment

Combination therapy of amphotericin B and flucytosine has been considered the standard treatment for cryptococcal meningitis, though the benefit from adding flucytosine remains controversial. Amphotericin B (with or without flucytosine) is curative in most patients. Since AIDS patients with cryptococcosis will almost always relapse when amphotericin B is withdrawn, they require suppressive therapy with fluconazole. Fluconazole offers excellent penetration of the central nervous system. HIV/AIDS patients treated with highly active antiretroviral therapy (HAART) have a lower incidence of cryptococcosis, and cases have a much better prognosis. However, a subset of these patients develop immune reconstitution inflammatory syndrome, which exacerbates their illness.

Epidemiology & Control

Bird droppings (particularly pigeon droppings) enrich for the growth of *C neoformans* and serve as a reservoir of infection. The organism grows luxuriantly in pigeon excreta, but the birds are not infected. In addition to patients with AIDS or hematologic malignancies, patients being maintained on corticosteroids are highly susceptible to cryptococcosis.

The vast majority of global cases of cryptococcosis are caused by *C neoformans* (serotype A, D or AD). However, the normally tropical species *C gattii* has emerged in the Pacific Northwest, where it has been isolated from several local species of trees, soil and water. Since 2000, human and veterinary cases have expanded from Vancouver Island to mainland British Columbia, Washington and Oregon.

ASPERGILLOSIS

Aspergillosis is a spectrum of diseases that may be caused by a number of *Aspergillus* species. *Aspergillus* species are ubiquitous saprobes in nature, and aspergillosis occurs worldwide. *A fumigatus* is the most common human pathogen, but many others, including *A flavus, A niger, A terreus,* and *A lentulus* may cause disease. This mold produces abundant small conidia that are easily aerosolized. Following inhalation of these conidia, atopic individuals often develop severe allergic reactions to the conidial antigens. In immunocompromised patients—especially those with leukemia, stem cell transplant patients, and individuals taking corticosteroids—the conidia may germinate to produce hyphae that invade the lungs and other tissues.

Morphology & Identification

Aspergillus species grow rapidly, producing aerial hyphae that bear characteristic conidial structures: long conidiophores with terminal vesicles on which phialides produce basipetal chains of conidia (Figure 45–6). The species are identified according to morphologic differences in the these structures, including the size, shape, texture, and color of the conidia.

Pathogenesis

In the lungs, alveolar macrophages are able to engulf and destroy the conidia. However, macrophages from corticosteroid-treated animals or immunocompromised patients have a diminished ability to contain the inoculum. In the lung, conidia swell and germinate to produce hyphae that have a tendency to invade preexisting cavities (aspergilloma or fungus ball) or blood vessels.

Clinical Findings

A. Allergic Forms

In some atopic individuals, development of IgE antibodies to the surface antigens of *Aspergillus* conidia elicits an immediate asthmatic reaction upon subsequent exposure. In others, the conidia germinate and hyphae colonize the bronchial tree

without invading the lung parenchyma. This phenomenon is characteristic of **allergic bronchopulmonary aspergillosis**, which is clinically defined as asthma, recurrent chest infiltrates, eosinophilia, and both type I (immediate) and type III (Arthus) skin test hypersensitivity to *Aspergillus* antigen. Many patients produce sputum with *Aspergillus* and serum precipitins. They have difficulty breathing and may develop permanent lung scarring. Normal hosts exposed to massive doses of conidia can develop **extrinsic allergic alveolitis**.

B. Aspergilloma & Extrapulmonary Colonization

Aspergilloma occurs when inhaled conidia enter an existing cavity, germinate, and produce abundant hyphae in the abnormal pulmonary space. Patients with previous cavitary disease (eg, tuberculosis, sarcoidosis, emphysema) are at risk. Some patients are asymptomatic; others develop cough, dyspnea, weight loss, fatigue, and hemoptysis. Cases of aspergilloma rarely become invasive. Localized, noninvasive infections (colonization) by *Aspergillus* species may involve the nasal sinuses, the ear canal, the cornea, or the nails.

C. Invasive Aspergillosis

Following inhalation and germination of the conidia, invasive disease develops as an acute pneumonic process with or without dissemination. Patients at risk are those with lymphocytic or myelogenous leukemia and lymphoma, stem cell transplant recipients, and especially individuals taking corticosteroids. The risk is much greater for patients receiving allogeneic (rather than autologous) hematopoietic stem cell transplants. In addition, AIDS patients with CD4 cell counts <50 CD4 cells/mm^3 are predisposed to invasive aspergillosis. Symptoms include fever, cough, dyspnea, and hemoptysis. Hyphae invade the lumens and walls of blood vessels, causing thrombosis, infarction, and necrosis. From the lungs, the disease may spread to the gastrointestinal tract, kidney, liver, brain, or other organs, producing abscesses and necrotic lesions. Without rapid treatment, the prognosis for patients with invasive aspergillosis is grave. Persons with less compromising underlying disease may develop chronic necrotizing pulmonary aspergillosis, which is a milder disease.

Diagnostic Laboratory Tests

A. Specimens, Microscopic Examination, & Culture

Sputum, other respiratory tract specimens, and lung biopsy tissue provide good specimens. Blood samples are rarely positive. On direct examination of sputum with KOH or calcofluor white or in histologic sections, the hyphae of *Aspergillus* species are hyaline, septate, and uniform in width (about 4 μm) and branch dichotomously (Figure 45–24). *Aspergillus* species grow within a few days on most media at room temperature. Species are identified according to the morphology of their conidial structures (Figure 45–6).

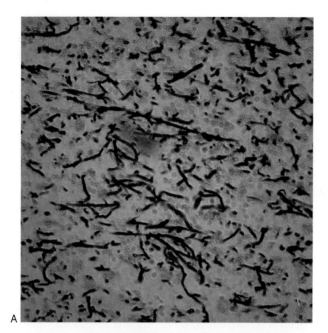

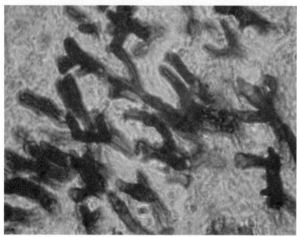

FIGURE 45–24 Invasive aspergillosis. **A:** Uniform, branching septate hyphae (ca. 4 μm in width) of *Aspergillus fumigatus* in lung tissue stained with Gomori methenamine silver. 400×. **B:** Similar preparation with Grocott stain. 1000×.

B. Serology

The ID test for precipitins to *A fumigatus* is positive in over 80% of patients with aspergilloma or allergic forms of aspergillosis, but antibody tests are not helpful in the diagnosis of invasive aspergillosis. However, a serologic test for circulating cell wall galactomannan is diagnostic.

Treatment

Aspergilloma is treated with itraconazole or amphotericin B and surgery. Invasive aspergillosis requires rapid administration of either the native or lipid formulation of amphotericin B or voriconazole, often supplemented with cytokine immunotherapy (eg, granulocyte-macrophage colony stimulating factor or interferon γ). Amphotericin B-resistant strains of *A. terreus* and other species, including *A flavus* and *A lentulus*, have emerged at several leukemia treatment centers, and the new triazole, posaconazole, may be more effective for these infections. The less severe chronic necrotizing pulmonary disease may be treatable with voriconazole or itraconazole. Allergic forms of aspergillosis are treated with corticosteroids or disodium chromoglycate.

Epidemiology & Control

For persons at risk for allergic disease or invasive aspergillosis, efforts are made to avoid exposure to the conidia of *Aspergillus* species. Most bone marrow transplant units employ filtered air-conditioning systems, monitor airborne contaminants in patients' rooms, reduce visiting, and institute other measures to isolate patients and minimize their risk of exposure to the conidia of *Aspergillus* and other molds. Some patients at risk for invasive aspergillosis are given prophylactic low-dose amphotericin B or itraconazole.

MUCORMYCOSIS

Mucormycosis (zygomycosis) is an opportunistic mycosis caused by a number of molds classified in the order Mucorales of the phylum Zygomycota. These fungi are ubiquitous thermotolerant saprobes. The leading pathogens among this group of fungi are species of the genera *Rhizopus* (Figure 45–2), *Rhizomucor, Absidia, Cunninghamella* (Figure 45–3), *Mucor* et al. However, the most prevalent agent is *Rhizopus oryzae*. The conditions that place patients at risk include acidosis—especially that associated with diabetes mellitus—leukemias, lymphoma, corticosteroid treatment, severe burns, immunodeficiencies, and other debilitating diseases as well as dialysis with the iron chelator deferoxamine.

The major clinical form is rhinocerebral mucormycosis, which results from germination of the sporangiospores in the nasal passages and invasion of the hyphae into the blood vessels, causing thrombosis, infarction, and necrosis. The disease can progress rapidly with invasion of the sinuses, eyes, cranial bones, and brain. Blood vessels and nerves are damaged, and patients develop edema of the involved facial area, a bloody nasal exudate, and orbital cellulitis. Thoracic mucormycosis follows inhalation of the sporangiospores with invasion of the lung parenchyma and vasculature. In both locations, ischemic necrosis causes massive tissue destruction. Less frequently, this process has been associated with contaminated wound dressings and other situations.

Direct examination or culture of nasal discharge, tissue, or sputum will reveal broad hyphae (10–15 μm) with uneven thickness, irregular branching, and sparse septations (Figure 45–25). These fungi grow rapidly on laboratory media, producing abundant cottony colonies. Identification is based on the sporangial structures. Treatment consists of

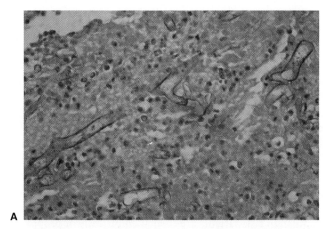
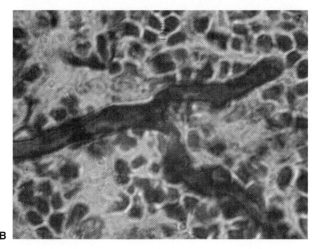

FIGURE 45–25 Mucormycosis (zygomycosis) **A**: Broad, ribbon-like sparsely septate hyphae (10–15 μm in width) of *Rhizopus oryzae* in lung tissue. H&E 400×. **B**: Similar histopathological specimen, stained with Gomori methenamine silver. 1000×.

aggressive surgical debridement, rapid administration of amphotericin B, and control of the underlying disease. Many patients survive, but there may be residual effects such as partial facial paralysis or loss of an eye.

PNEUMOCYSTIS PNEUMONIA

Pneumocystis jiroveci causes pneumonia in immunocompromised patients; dissemination is rare. Until recently, *P jiroveci* was thought to be a protozoan, but molecular biologic studies have proved that it is a fungus with a close relationship to ascomycetes. *Pneumocystis* species are present in the lungs of many animals (rats, mice, dogs, cats, ferrets, rabbits) but rarely cause disease unless the host is immunosuppressed. *P jiroveci* is the human species, and the more familiar *P carinii* is found only in rats. Until the AIDS epidemic, human disease was confined to interstitial plasma cell pneumonitis in malnourished infants and immunosuppressed patients (corticosteroid therapy, antineoplastic therapy, and transplant recipients).

Prior to the introduction of effective chemoprophylactic regimens, it was a major cause of death among AIDS patients. Chemoprophylaxis has resulted in a dramatic decrease in the incidence of pneumonia, but infections are increasing in other organs, primarily the spleen, lymph nodes, and bone marrow.

P jiroveci has morphologically distinct forms: thin-walled trophozoites and cysts, which are thick-walled, spherical to elliptical (4–6 μm), and contain four to eight nuclei. Cysts can be stained with silver stain, toluidine blue, and calcofluor white. In most clinical specimens, the trophozoites and cysts are present in a tight mass that probably reflects their mode of growth in the host. *P jiroveci* contains a surface glycoprotein that can be detected in sera from acutely ill or normal individuals.

P jiroveci is an extracellular pathogen. Growth in the lung is limited to the surfactant layer above alveolar epithelium. In non-AIDS patients, infiltration of the alveolar spaces with plasma cells leads to interstitial plasma cell pneumonitis. Plasma cells are absent in AIDS-related *Pneumocystis* pneumonia. Blockage of the oxygen exchange interface results in cyanosis.

To establish the diagnosis of *Pneumocystis* pneumonia, specimens of bronchoalveolar lavage, lung biopsy tissue, or induced sputum are stained and examined for the presence of cysts or trophozoites. Appropriate stains include Giemsa, toluidine blue, methenamine silver, and calcofluor white. A specific monoclonal antibody is available for direct fluorescent examination of specimens. *Pneumocystis* cannot be cultured. While not clinically useful, serologic testing has been used to establish the prevalence of infection.

In the absence of immunosuppression, *P jiroveci* does not cause disease. Serologic evidence suggests that most individuals are infected in early childhood, and the organism has worldwide distribution. Cell-mediated immunity presumably plays a dominant role in resistance to disease, as AIDS patients often have significant antibody titers, and *Pneumocystis* pneumonia is not usually seen until the CD4 lymphocyte count drops below 400/μL.

Acute cases of *Pneumocystis* pneumonia are treated with trimethoprim-sulfamethoxazole or pentamidine isethionate. Prophylaxis can be achieved with daily TMP-SMZ or aerosolized pentamidine. Other drugs are also available.

No natural reservoir has been demonstrated, and the agent may be an obligate member of the normal flora. Persons at risk are provided with chemoprophylaxis. The mode of infection is unclear, and transmission by aerosols may be possible.

PENICILLIOSIS

Only one of the numerous and ubiquitous environmental species of *Penicillium* is dimorphic, *Penicillium marneffei*, and this species has emerged as an endemic, opportunistic pathogen. *P marneffei* is found in several regions of southeast Asia, including southeastern China, Thailand, Vietnam, Indonesia, Hong Kong, Taiwan, and the Manipur state of India. Within these endemic areas, *P marneffei* has been isolated from soil and especially soil that is associated with bamboo rats and their habitats. At ambient temperatures, the mold form grows rapidly to develop a green-yellow colony with a diffusible reddish pigment. The septate, branching hyphae produce aerial conidiophores bearing phialides and basipetal chains of conidia, similar to the structures in Figure 45–4. In tissue, the hyphal forms convert to unicellular yeast-like cells (ca. 2×6 μm) that divide by fission. The major risk for infection is immunodeficiency due to HIV/AIDS, tuberculosis, corticosteroid treatment, or lymphoproliferative diseases. The clinical manifestations include fungemia, skin lesions, and systemic involvement of multiple organs, especially the reticuloendothelial system. Early signs and symptoms are nonspecific and may include cough, fever, fatigue, weight loss, and lymphadenopathy. However, 70% of patients, with or without AIDS, develop cutaneous or subcutaneous papules, pustules or rashes, which are often located on the face. From specimens of skin, blood or tissue biopsies, the diagnosis can be established by microscopic observation of the yeast-like cells and positive cultures. The treatment usually entails a defined course of amphotericin B followed by itraconazole. Without treatment, the mortality has exceeded 90%.

OTHER OPPORTUNISTIC MYCOSES

Individuals with compromised host defenses are susceptible to infections by many of the thousands of saprobic molds that exist in nature and produce airborne spores. Such opportunistic mycoses occur less frequently than candidiasis, aspergillosis, and mucormycosis because the fungi are less pathogenic. Advances in medicine have resulted in growing numbers of severely compromised patients in whom normally nonpathogenic fungi may become opportunistic pathogens. Devastating systemic infections have been caused by species of *Fusarium, Paecilomyces, Bipolaris, Curvularia, Alternaria,* and many others. Some opportunists are geographically restricted. For example, AIDS patients in Asia acquire systemic infections with *Penicillium marneffei,* which is a dimorphic pathogen endemic to the area. Another contributing factor is the increasing use of antifungal antibiotics, which has led to the selection of resistant fungal species and strains.

ANTIFUNGAL PROPHYLAXIS

Opportunistic mycoses are increasing among immunocompromised patients, especially in patients with hematological dyscrasias (eg, leukemia), hematopoietic stem cell recipients, and solid organ transplant patients, and others receiving cytotoxic and immunosuppressive drugs (eg, corticosteroids). For example, the incidence of systemic mycoses among patients with acute lymphocytic or myelogenous leukemia is 5–20%, and among allogeneic stem cell transplant patients, 5–10%. Many of these high risk patients have depressed innate host defenses, such as a reduction in the number and/or functionality of circulating neutrophils and monocytes. In addition, AIDS patients are highly susceptible to a variety of systemic mycoses when their CD4 cell counts drop below 200 cells per cubic milliliter.

The list of invasive opportunistic pathogens include species of *Candida, Cryptococcus, Saccharomyces,* and other yeasts; *Aspergillus* and other ascomycetous molds, such as *Fusarium, Paecilomyces,* and *Scopulariopsis*; dematiaceous molds (eg, species of *Bipolaris, Phialophora, Cladosporium*), and zygomycetes (*Rhizopus*). Because it is usually difficult to establish a definitive diagnosis early in the course of infection, many high risk patients are treated empirically or prophylactically with antifungal drugs. However, there is no universal consensus on the criteria for administering antifungal prophylaxis or the specific chemotherapy and regimen. Rather, most tertiary care hospitals have developed their own protocols for the administration of prophylactic antifungal chemotherapy to patients at high risk for invasive mycoses. Most hospitals will give oral fluconazole; others prescribe a short course of low-dose amphotericin B. Some of the criteria for administering antifungal prophylaxis to a patient with an underlying high risk disease or condition are persistent fever that is unresponsive to antibacterial antibiotics, neutropenia lasting more than 7 days, the observation of new and unexplained pulmonary infiltrates on radiographic examinations, or progressive, unexplained organ failure.

HYPERSENSITIVITY TO FUNGI

Throughout life, the respiratory tract is exposed to airborne conidia and spores of many saprophytic fungi. These particles often possess potent surface antigens capable of stimulating and eliciting strong allergic reactions. These hypersensitivity responses do not require growth or even viability of the inducing fungus, though in some cases (allergic bronchopulmonary aspergillosis) both infection and allergy may occur simultaneously. Depending on the site of deposition of the allergen, a patient may exhibit rhinitis, bronchial asthma, alveolitis, or generalized pneumonitis. Atopic persons are more susceptible. The diagnosis and range of a patient's hypersensitivity reactions can be determined by skin testing with fungal extracts. Management may entail avoidance of the offending allergen, treatment with corticosteroids, or attempts to desensitize patients.

Indoor air exposure to large numbers of fungal spores has led to the recognition of a condition described as "sick building syndrome" whereby moisture in construction materials, such as wood and fiberboard, allows contaminating molds to flourish. The production and contamination of the indoor air with large numbers of conidia have resulted in debilitating cases of systemic allergic or toxic reactions. Often the mold infestation is too extensive to eliminate with fungicides or filtration, and many such buildings have been demolished. The offending molds are usually noninfectious ascomycetes such as *Stachybotrys, Cladosporium, Fusarium,* and others.

MYCOTOXINS

Many fungi produce poisonous substances called mycotoxins that can cause acute or chronic intoxication and damage. The mycotoxins are secondary metabolites, and their effects are not dependent on fungal infection or viability. A variety of mycotoxins are produced by mushrooms (eg, *Amanita* species), and their ingestion results in a dose-related disease called **mycetismus**. Cooking has little effect on the potency of these toxins, which may cause severe or fatal damage to the liver and kidney. Other fungi produce mutagenic and carcinogenic compounds that can be extremely toxic for experimental animals. One of the most potent is **aflatoxin**, which is elaborated by *Aspergillus flavus* and related molds and is a frequent contaminant of peanuts, corn, grains, and other foods.

ANTIFUNGAL CHEMOTHERAPY

A limited but increasing number of antibiotics can be used to treat mycotic infections. Most have one or more limitations, such as profound side effects, a narrow antifungal spectrum, poor penetration of certain tissues, and the selection of resistant fungi. Promising new drugs are currently under development, and others are being evaluated in clinical trials. Finding suitable fungal targets is difficult because fungi, like humans, are eukaryotes. Many of the cellular and molecular processes are similar, and there is often extensive homology among the genes and proteins.

The classes of currently available drugs include the polyenes (amphotericin B and nystatin), which bind to ergosterol in the cell membrane; flucytosine, a pyrimidine analog; the azoles and other inhibitors of ergosterol synthesis, such as the allylamines; the echinocandins, which inhibit the synthesis of cell wall β-glucan; and griseofulvin, which interferes with microtubule assembly. Currently under investigation are inhibitors of cell wall synthesis, such as nikkomycin and pradimicin, and sordarin, which inhibits elongation factor 2.

In recent years, the number of antifungal drugs has increased, and additional compounds are currently under evaluation in clinical trials. Table 45–5 provides an abridged summary of the available drugs. Many of the newer chemotherapeutics are variations of the azole class of fungistatic drugs, such as the triazoles voriconazole and posaconazole. These drugs and the newer compounds were designed to improve the antifungal efficacy and pharmacokinetics, as well as to reduce the adverse side effects.

Amphotericin B

Description

The major polyene antibiotic is amphotericin B, a metabolite of streptomyces. Amphotericin B is the most effective drug for severe systemic mycoses. It has a broad spectrum, and the development of resistance is rare. The mechanism of action of the polyenes involves the formation of complexes with ergosterol in fungal cell membranes, resulting in membrane damage and leakage. Amphotericin B has greater affinity for ergosterol than cholesterol, the predominant sterol in mammalian cell membranes. Packaging of amphotericin B in liposomes and lipoidal emulsions has shown superb efficacy and excellent results in clinical studies. These formulations are currently available and may replace the conventional preparation. The lipid preparations are less toxic and permit higher concentrations of amphotericin B to be used.

Amphotericin B

Mechanism of Action

Amphotericin B is given intravenously as micelles with sodium deoxycholate dissolved in a dextrose solution. Though the drug is widely distributed in tissues, it penetrates poorly to the cerebrospinal fluid. Amphotericin B firmly binds to ergosterol in the cell membrane. This interaction alters the membrane fluidity and perhaps produces pores in the membrane through which ions and small molecules are lost. Unlike most other antifungals, amphotericin B is cidal. Mammalian cells lack ergosterol and are relatively resistant to these actions. Amphotericin B binds weakly to the cholesterol in mammalian membranes, and this interaction may explain its toxicity. At low levels, amphotericin B has an immunostimulatory effect.

Indications

Amphotericin B has a broad spectrum with demonstrated efficacy against most of the major systemic mycoses, including coccidioidomycosis, blastomycosis, histoplasmosis, sporotrichosis, cryptococcosis, aspergillosis, mucormycosis, and candidiasis. The response to amphotericin B is influenced by the dose and rate of administration, the site of the mycotic infection, the immune status of the patient, and the inherent susceptibility of the pathogen. Penetration of the joints and the central nervous system is poor, and intrathecal or intraarticular administration is recommended for some infections. Amphotericin B is used in combination with flucytosine to treat cryptococcosis. Some fungi, such as *Pseudallescheria boydii* and

TABLE 45–5   Comparison of Common Antifungal Drugs for the Treatment of Systemic Mycoses

Class and Mechanism	Drug	Route	Spectrum	Indications	Toxicity	Comments
Polyenes—bind ergosterol in fungal cell membrane; immune modulation	Amphotericin B	IV	Broad	Most Serious, invasive mycoses	Common: nephrotoxicity, acute infusion reactions, fever, chills, anemia, electrolyte disturbances, many others	Fungicidal; resistance is rare
	Amphotericin B lipid formulations[a]	IV	Broad	Most serious, invasive mycoses	Diminished nephrotoxicity fewer other side effects	Altered tissue distribution
Antimetabolite—converted to fluorouracil, perturbing synthesis of pyrimidines and RNA	Flucystosine	PO	Yeasts; dematiaceous molds	Candidiasis, cryptococcosis, phaeohyphomycosis	GI upset, (nausea, vomiting, and / or diarrhea), neuropathy, bone marrow	Resistance is common when used as monotherapy; high CSF and urine levels. Therapeutic drug levels are frequently monitored
Azoles[b]—inhibit ergosterol synthesis; block cytochrome P450-dependent 14-α-demethylation of lanosterol	Ketoconazole	PO, topical	Limited	Candidiasis, refractory dermatomycoses	Hormonal changes; hepatic toxicity, GI upset, neuropathy	Poor oral absorption
	Itraconazole	PO, IV	Broad	Endemic mycoses, aspergillosis, candidiasis, cryptococcosis, phaeohyphomycosis	Mild; GI upset, hepatic toxicity, neuropathy, bone marrow. Black box warning due to risk of cardiac toxicity.	Poor absorption, particularly with capsules. Absorption is better if given in solution, but diarrhea occurs more frequently. Blood levels must be monitored
	Fluconazole	PO, IV	Limited	Candidiasis, cryptococcosis	Comparatively safe; GI upset, dizziness, skin lesions, others	Excellent absorption; extensively used for prophylaxis and empiric therapy; resistance occurs, commonly with *Candida glabrata*, *Candida krusei*
	Voriconazole	PO, IV	Broad	Aspergillosis, candidiasis, rare molds, endemic mycoses, cryptococcosis, phaeohyphomycosis	Low; transient visual effects in ≈30%, hepatitic toxicity, GI upset, rash	Blood levels must be monitored
	Posaconazole	PO	Broad	Similar to voriconazole, plus zygomycetes	Comparatively safe, GI upset, headaches, somnolence, dizziness, fatigue, hepatic toxicity.	Absorption varies. Approved for prophylaxis in certain cancer patients
Echinocandins—perturb cell wall synthesis; inhibit 1,3-β-D-glucan synthase	Caspofungin	IV	Limited	Invasive candidiasis, refractory aspergillosis	Safe, minimal: GI upset, rash, headache	Used for empiric therapy
	Micafungin	IV	Limited	Esophageal candidiasis	Infrequent; fever	Used for prophylaxis
	Anidulafungin	IV	Limited	Invasive candidiasis	Infrequent	

[a]Amphotericin B Colloidal Dispersion (ABCD; Amphocil™ or Amphotec™), Amphotericin B Lipid Complex (ABLC; Abelcet™), and Liposomal Amphotericin B (L-AMB; Ambisome™).

[b]All of the azoles may inhibit host cytochrome P450 isoenzymes, and they may cause adverse interactions with many other drugs.

Aspergillus terreus, do not respond well to treatment with amphotericin B.

Side Effects

All patients have adverse reactions to amphotericin B, though these are greatly diminished with the new lipid preparations. Acute reactions that usually accompany the intravenous administration of amphotericin B include fever, chills, dyspnea, and hypotension. These effects can usually be alleviated by prior or concomitant administration of hydrocortisone or acetaminophen. Tolerance to the acute side effects develops during therapy.

Chronic side effects are usually the result of nephrotoxicity. Azotemia almost always occurs with amphotericin B therapy, and serum creatinine and ion levels must be closely monitored. Hypokalemia, anemia, renal tubular acidosis, headache, nausea, and vomiting are also frequently observed. While some of the nephrotoxicity is reversible, permanent reduction in glomerular and renal tubular function does occur. This damage can be correlated with the total dose of amphotericin B given. Toxicity is greatly diminished with the lipid formulations of amphotericin B (ie, Abelcet, Amphotec, and AmBisome).

Flucytosine

Description

Flucytosine (5-fluorocytosine) is a fluorinated derivative of cytosine. It is an oral antifungal compound used primarily in conjunction with amphotericin B to treat cryptococcosis or candidiasis. It is effective also against many dematiaceous fungal infections. It penetrates well into all tissues, including cerebrospinal fluid.

Flucytosine

Mechanism of Action

Flucytosine is actively transported into fungal cells by a permease. It is converted by the fungal enzyme cytosine deaminase to 5-fluorouracil and incorporated into 5-fluorodexoyuridylic acid monophosphate, which interferes with the activity of thymidylate synthetase and DNA synthesis. Mammalian cells lack cytosine deaminase and are therefore protected from the toxic effects of fluorouracil. Unfortunately, resistant mutants emerge rapidly, limiting the utility of flucytosine.

Indications

Flucytosine is used mainly in conjunction with amphotericin B for treatment of cryptococcosis and candidiasis. In vitro, it acts synergistically with amphotericin B against these organisms, and clinical trials suggest a beneficial effect of the combination, particularly in cryptococcal meningitis. The combination has also been shown to delay or limit the emergence of flucytosine-resistant mutants. By itself, flucytosine is effective against chromoblastomycosis and other dematiaceous fungal infections.

Side Effects

While flucytosine itself probably has little toxicity for mammalian cells and is relatively well tolerated, its conversion to fluorouracil results in a highly toxic compound that is probably responsible for the major side effects. Prolonged administration of flucytosine results in bone marrow suppression, hair loss, and abnormal liver function. The conversion of flucytosine to fluorouracil by enteric bacteria may cause colitis. Patients with AIDS may be more susceptible to bone marrow suppression by flucytosine, and serum levels should be closely monitored.

Azoles
Description

The antifungal imidazoles (eg, ketoconazole) and the triazoles (fluconazole, voriconazole, and itraconazole) are oral drugs used to treat a wide range of systemic and localized fungal infections. (See Figure 45–26.) The indications for their use are still being evaluated, but they have already supplanted amphotericin B in many less severe mycoses because they can be administered orally and are less toxic. Other imidazoles— miconazole and clotrimazole—are used topically and are discussed below.

Mechanism of Action

The azoles interfere with the synthesis of ergosterol. They block the cytochrome P450-dependent 14α-demethylation of lanosterol, which is a precursor of ergosterol in fungi and cholesterol in mammalian cells. However, the fungal cytochrome P450s are approximately 100–1000 times more sensitive to the azoles than mammalian systems. The various azoles are designed to improve their efficacy, availability, and pharmacokinetics and reduce their side effects. These are fungistatic drugs.

Indications

The indications for the use of antifungal azoles will broaden as the results of long-term studies—as well as new azoles— become available. Accepted indications for the use of antifungal azoles are listed below.

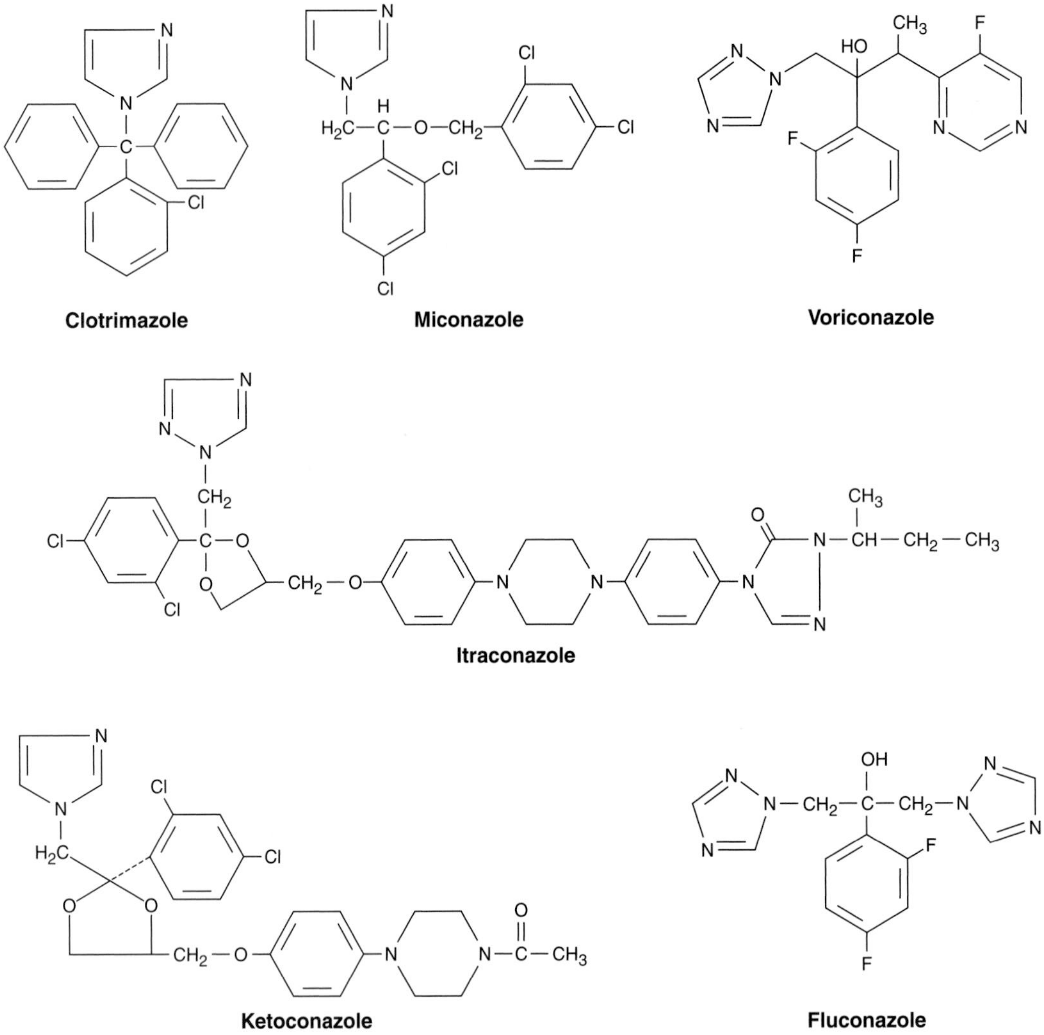

FIGURE 45–26 Structures of antifungal azoles. (Reproduced with permission from Katzung BG [editor]: *Basic and Clinical Pharmacology,* 11th ed. McGraw-Hill, 2009.)

Ketoconazole is useful in the treatment of chronic mucocutaneous candidiasis, dermatophytosis, and nonmeningeal blastomycosis, coccidioidomycosis, paracoccidioidomycosis, and histoplasmosis. Of the various azoles, fluconazole offers the best penetration of the central nervous system. It is used as maintenance therapy for cryptococcal and coccidioidal meningitis. Oropharyngeal candidiasis in AIDS patients and candidemia in immunocompetent patients can also be treated with fluconazole. Itraconazole is now the agent of first choice for histoplasmosis and blastomycosis as well as for certain cases of coccidioidomycosis, paracoccidioidomycosis, and aspergillosis. It has also been shown to be effective in the treatment of chromomycosis and onychomycosis due to dermatophytes and other molds. Voriconazole, which can be given orally or intravenously, exhibits a broad spectrum of activity against many molds and yeasts, especially aspergillosis, fusariosis, pseudallescheriasis, and

other less common systemic pathogens. The newest triazole is posaconazole (Figure 45–27A), which has a wide spectrum and demonstrated efficacy against fluconazole-resistant *Candida* species, aspergillosis, mucormycosis, and other opportunistic invasive molds. It is also well tolerated.

Side Effects

The adverse effects of the azoles are primarily related to their ability to inhibit mammalian cytochrome P450 enzymes. Ketoconazole is the most toxic, and therapeutic doses may inhibit the synthesis of testosterone and cortisol, which may cause a variety of reversible effects such as gynecomastia, decreased libido, impotence, menstrual irregularity, and occasionally adrenal insufficiency. Fluconazole and itraconazole at recommended therapeutic doses do not cause significant impairment of mammalian steroidogenesis. All the antifungal azoles can cause both asymptomatic

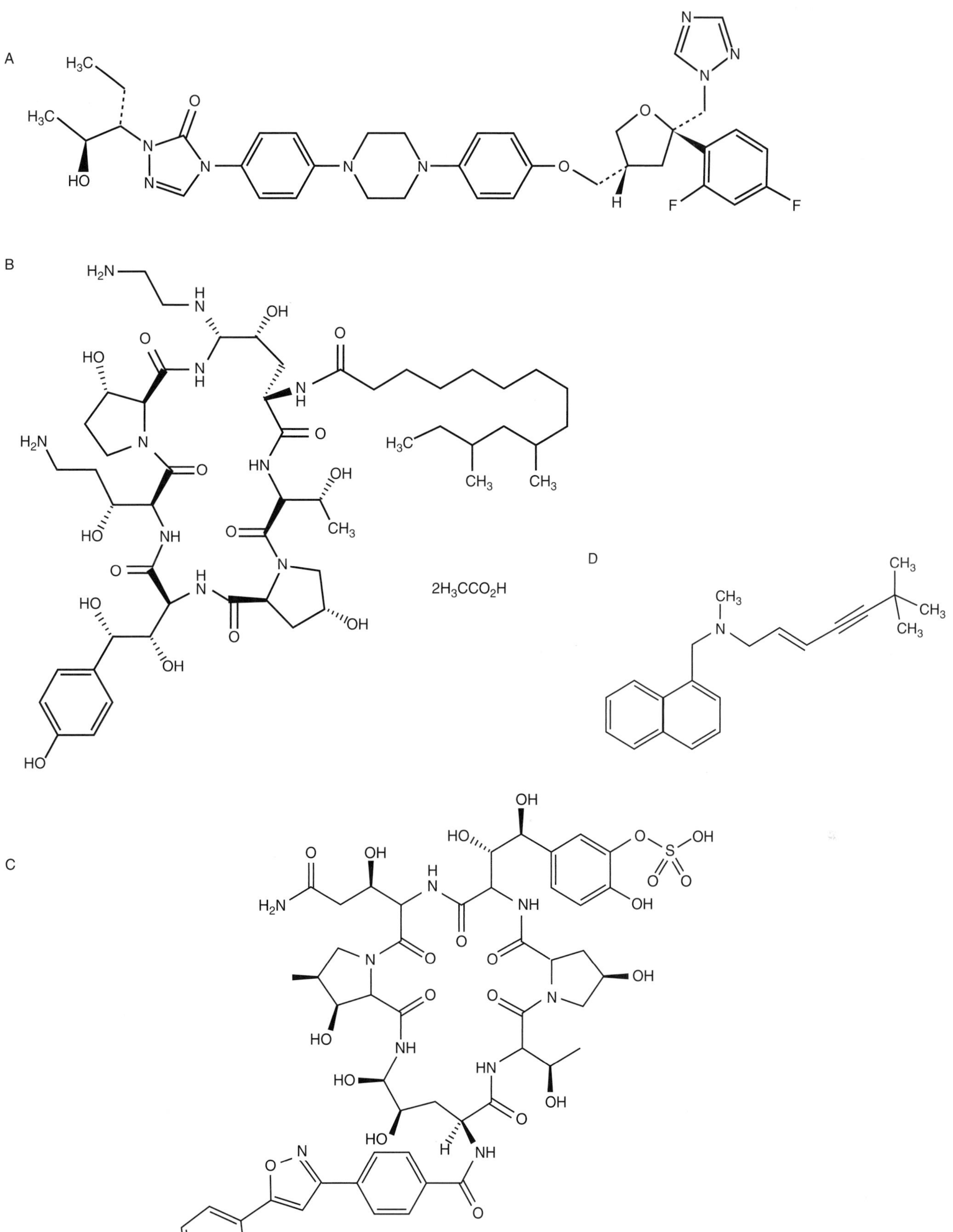

FIGURE 45–27 Newer antifungal drugs. **A**: Posaconazole. **B**: Caspofungin. **C**: Micafungin. **D**: Terbinafine.

elevations in liver function tests and rare cases of hepatitis. Voriconazole causes reversible visual impairment in about 30% of patients.

Since antifungal azoles interact with P450 enzymes that are also responsible for drug metabolism, some important drug interactions can occur. Increased antifungal azole concentrations can be seen when isoniazid, phenytoin, or rifampin is used. Antifungal azole therapy can also lead to higher than expected serum levels of cyclosporine, phenytoin, oral hypoglycemics, anticoagulants, digoxin, and probably many others. Serum monitoring of both drugs may be necessary to achieve a proper therapeutic range.

Echinocandins

The echinocandins are a new class of antifungal agents that perturb the synthesis of the pervasive cell wall polysaccharide β-glucan by inhibiting 1,3-β-glucan synthase and disrupting cell wall integrity. The first licensed drug, caspofungin has shown efficacy against invasive aspergillosis and systemic candidiasis due to a wide range of *Candida* species (Figure 45–27B). This intravenous agent may be especially indicated for refractory aspergillosis. Caspofungin is well tolerated.

Similar to caspofungin, two newly approved echinocandins, micafungin and anidulafungin, also inhibit the synthesis of β-glucan and have a similar spectrum of activity against species of *Candida* and *Aspergillus*, as well as several other molds. Micafungin (Figure 45–27C) and anidulafungin were recently licensed for the treatment of esophageal candidiasis and for the antifungal prophylaxis of hematopoietic stem cell transplant patients. Both seem to have better pharmacokinetics and in vivo stability than caspofungin. Clinical studies suggest that they will be useful in the treatment of mucosal and systemic candidiasis, refractory invasive aspergillosis, and in combination with amphotericin B or some of the newer triazoles.

Griseofulvin

Griseofulvin is an orally administered antibiotic derived from a species of penicillium. It is used to treat dermatophytoses and must be given for long periods. Griseofulvin is poorly absorbed and concentrated in the stratum corneum, where it inhibits hyphal growth. It has no effect on other fungi.

After oral administration, griseofulvin is distributed throughout the body but accumulates in the keratinized tissues. Within the fungus, griseofulvin interacts with microtubules and disrupts mitotic spindle function, resulting in inhibition of growth. Only actively growing hyphae are affected. Griseofulvin is clinically useful for the treatment of dermatophyte infections of the skin, hair, and nails. Oral therapy for weeks to months is usually required. Griseofulvin is generally well tolerated. The most common side effect is headache, which usually resolves without discontinuation of the drug. Less frequently observed side effects are gastrointestinal disturbances, drowsiness, and hepatotoxicity.

Terbinafine

Terbinafine is an allylamine drug; it blocks ergosterol synthesis by inhibiting squalene epoxidase (Figure 45–27D). Terbinafine is given orally to treat dermatophyte infections. It has proved quite effective in treating nail infections as well as other dermatophytoses. Side effects are not common but include gastrointestinal distress, headaches, skin reactions, and loss of sense of taste. For the long-term treatment of tinea unguium, terbinafine—as well as itraconazole and fluconazole—may be given intermittently, using a pulse treatment protocol.

TOPICAL ANTIFUNGAL AGENTS

Nystatin

Nystatin is a polyene antibiotic, structurally related to amphotericin B and having a similar mode of action. It can be used to treat local candidal infections of the mouth and vagina. Nystatin may also suppress subclinical esophageal candidiasis and gastrointestinal over-growth of candida. No systemic absorption occurs, and there are no side effects. However, nystatin is too toxic for parenteral administration.

Clotrimazole, Miconazole, & Other Azoles

A variety of antifungal azoles too toxic for systemic use are available for topical administration. Clotrimazole and miconazole are available in several formulations. Econazole, butaconazole, tioconazole, and terconazole are also available. All of these compounds seem to have comparable efficacy.

Topical azoles have a broad spectrum of activity. Tinea pedis, tinea corporis, tinea cruris, tinea versicolor, and cutaneous candidiasis respond well to local application of creams or powders. Vulvovaginal candidiasis can be treated with vaginal suppositories or creams. Clotrimazole is also available as an oral troche for treatment of oral and esophageal thrush in immunocompetent patients.

Griseofulvin

Other Topical Antifungal Agents

Tolnaftate and naftifine are topical antifungal agents used in the treatment of many dermatophyte infections and tinea versicolor. Formulations available include creams, powders, and sprays. Undecylenic acid is available in several formulations for the treatment of tinea pedis and tinea cruris. Although it is effective and well tolerated, antifungal azoles, naftifine, and tolnaftate are more effective. Haloprogin and ciclopirox are other topical agents commonly used in dermatophyte infections.

REVIEW QUESTIONS

1. Which statement regarding fungi is correct?
   (A) All fungi are able to grow as yeasts and molds
   (B) Although fungi are eukaryotes, they lack mitochondria
   (C) Fungi are photosynthetic
   (D) Fungi have one or more nuclei and chromosomes
   (E) Few fungi possess cell membranes

2. Which statement regarding fungal growth and morphology is correct?
   (A) Pseudohyphae are produced by all yeasts
   (B) Molds produce hyphae that may or may not be partitioned with cross-walls or septa
   (C) Conidia are produced by sexual reproduction
   (D) Most yeasts reproduce by budding and lack cell walls
   (E) Most pathogenic dimorphic molds produce hyphae in the host and yeasts at 30°C

3. Which statement regarding fungal cell walls is correct?
   (A) The major components of fungal cell walls are proteins such as chitin, glucans, and mannans
   (B) The cell wall is not essential for fungal viability or survival
   (C) Ligands associated with the cell walls of certain fungi mediate attachment to host cells
   (D) Fungal cell wall components are the targets for the major classes of antifungal antibiotics, such as the polyenes and azoles
   (E) Fungal cell wall components rarely stimulate an immune response

4. A 54-year-old man developed a slowly worsening headache followed by gradual, progressive weakness in his right arm. A brain scan revealed a left cerebral lesion. At surgery an abscess surrounded by granulomatous material was found. Sections of the tissue and subsequent culture showed darkly pigmented septate hyphae indicating phaeohyphomycosis. This infection may be caused by species of which genus below?
   (A) Aspergillus
   (B) Cladophialophora
   (C) Coccidioides
   (D) Malassezia
   (E) Sporothrix

5. A 35-year-old man is a farmer in a tropical area of West Africa. He developed a persistent scaly papule on his leg. Ten months later a new crop of wart-like purplish scaly lesions appeared. These lesions slowly progressed to a cauliflower-like appearance. Chromoblastomycosis (chromomycosis) was diagnosed. Which statement regarding this disease is most correct?
   (A) In tissue, the organisms convert to spherical cells that reproduce by fission and exhibit transverse septations
   (B) The etiologic agents are endogenous members of the mammalian flora and possess melaninized cell walls
   (C) The disease is caused by a single species
   (D) Most infections are systemic
   (E) Most infections are acute and clear spontaneously

6. A 42-year-old HIV-positive male, originally from Vietnam but now residing in Tucson, Arizona, presents with a painful ulcerative lesion on his upper lip (cheilitis). A biopsy was obtained, and the histopathologic slide (hematoxylin and eosin stain) revealed spherical structures (20–50 μm in diameter) with thick refractory cell walls. What is the likely disease consistent with this finding?
   (A) Infection with Penicillium marneffei
   (B) Cryptococcosis
   (C) Blastomycosis
   (D) Coccidioidomycosis
   (E) No diagnostic significance

7. A 47-year-old man with poorly controlled diabetes mellitus developed a bloody nasal discharge, facial edema, and necrosis of his nasal septum. Culture of his cloudy nasal secretions yielded Rhizopus species. What is the most important implication of this finding?
   (A) No diagnostic value because this mold is an airborne contaminant
   (B) Consider treatment for rhinocerebral mucormycosis (zygomycosis)
   (C) Strongly suggestive of ketoacidosis
   (D) Strongly suggestive of HIV infection
   (E) The patient has been exposed to indoor mold contamination

8. An 8-year-old boy develops a circular dry, scaly, and pruritic lesion on his leg. What is the diagnostic significance of observing branching, septate, nonpigmented hyphae in a potassium hydroxide/calcofluor white preparation of a scraping from this skin lesion?
   (A) Chromomycosis
   (B) Dermatophytosis
   (C) Phaeohyphomycosis
   (D) Sporotrichosis
   (E) No diagnostic significance

9. Which statement regarding the epidemiology of candidiasis is correct?
   (A) Patients receiving bone marrow transplants are not at risk for systemic candidiasis
   (B) Patients with impaired or low numbers of neutrophils and monocytes are not at risk for systemic candidiasis
   (C) Patients with any form of diabetes have enhanced resistance to candidiasis
   (D) Patients with AIDS frequently develop mucocutaneous candidiasis, such as thrush
   (E) Pregnancy lowers the risk of candidal vaginitis

10. Which statement regarding dermatophytosis is correct?
    (A) Chronic infections are associated with zoophilic dermatophytes, such as Microsporum canis

(B) Acute infections are associated with zoophilic dermatophytes, such as *Microsporum canis*

(C) Chronic infections are associated with anthropophilic dermatophytes, such as *Microsporum canis*

(D) Acute infections are associated with anthropophilic dermatophytes, such as *Microsporum canis*

11. Which statement regarding the laboratory identification of fungi is correct?

(A) *Histoplasma capsulatum* typically requires less than 48 hours of incubation to yield positive cultures from clinical specimens

(B) Since many saprobic (nonpathogenic) molds resemble dimorphic mycotic agents in culture at 30°C, the identification of putative dimorphic pathogenic fungi must be confirmed by conversion to the tissue form in vitro or by the detection of species-specific antigens or DNA sequence analysis

(C) Molds are routinely speciated by a battery of physiologic tests, such as the ability to assimilate various sugars

(D) A positive germ tube test provides a rapid presumptive identification of *Candida glabrata*

(E) Budding yeast cells and abundant pseudohyphae are typical of *Pneumocystis jiroveci*

12. A 28-year-old female sex worker from southern California complained of headaches, dizziness, and occasional episodes of "spacing out" during the past 2 weeks. A lumbar puncture revealed reduced sugar, elevated protein, and 450 mononuclear leukocytes per milliliter. She was seropositive for HIV. Her history is compatible with fungal meningitis due to *Cryptococcus neoformans, Coccidioides posadasii*, or a species of *Candida*. Which one of the following tests is confirmatory?

(A) Meningitis due to *Coccidioides posadasii* would be confirmed by a positive test of the CSF for cryptococcal capsular antigen

(B) Meningitis due to *Cryptococcus neoformans* would be confirmed by a positive test of the CSF for complement fixation antibodies to coccidioidin

(C) Meningitis due to a species of *Candida* would be confirmed by the microscopic observation of oval yeast cells and pseudohyphae in the CSF

(D) Meningitis due to *Coccidioides posadasii* would be confirmed by a positive skin test to coccidioidin

13. Which statement about phaeohyphomycosis is correct?

(A) The infection only occurs in immunocompetent patients

(B) Infected tissue reveals branching, septate nonpigmented hyphae

(C) The causative agents are members of the normal microbial flora and can be isolated readily from the skin and mucosa of healthy persons

(D) Phaeohyphomycosis may exhibit several clinical manifestations, including subcutaneous or systemic disease, as well as sinusitis

(E) Cases rarely respond to treatment with itraconazole

14. A 37-year-old male with AIDS, currently living in Indianapolis, Indiana, presented with osteomyelitis of the left hip. A needle biopsy of the bone marrow was obtained, and the calcofluor white smear revealed a variety of myelogenous cells, monocytes, and macrophages containing numerous intracellular yeast cells

that were elliptical and approximately 2 × 4 μm. What is the most likely diagnosis?

(A) Blastomycosis

(B) Candidiasis

(C) Cryptococcosis

(D) Histoplasmosis

(E) No diagnostic significance

15. The potassium hydroxide examination of sputum from a heart-transplant patient with fever and pulmonary infiltrates contains oval budding yeast cells and pseudohyphae. What is the diagnostic significance?

(A) Aspergillosis

(B) Candidiasis

(C) Hyalohyphomycosis

(D) Phaeohyphomycosis

(E) No diagnostic significance

16. A middle-aged male resident of southern California received a liver transplant. During the following months, he gradually developed fatigue, weight loss, cough, night sweats, dyspnea, and a nonhealing subcutaneous nodule on his nose. The chest radiograph revealed hilar lymphadenopathy and diffuse infiltrates. Direct examination and culture of a respiratory specimen were negative. Skin tests with PPD, blastomycin, coccidioidin, and histoplasmin were negative. Serological test results were as follows: Negative serum test for cryptococcal capsular antigen in blood, positive immunodiffusion test for serum precipitins to fungal antigen F and negative immunodiffusion tests for precipitins to antigens h, m, and A. Serum tests for fungal complement fixation antibodies were negative for *Blastomyces dermatitidis*, as well as both the mycelial and yeast antigens of *Histoplasma capsulatum* but yielded a titer of 1:32 to coccidioidin. Which interpretation of these data is the most tenable?

(A) Clinical and serologic findings are inconclusive

(B) Clinical and serologic findings are most consistent with active disseminated histoplasmosis

(C) Clinical and serologic findings are most consistent with active disseminated blastomycosis

(D) Clinical and serologic findings are consistent with active disseminated coccidioidomycosis

(E) Clinical and serologic findings exclude a diagnosis of blastomycosis, histoplasmosis, and coccidioidomycosis

17. Which statement regarding aspergillosis is correct?

(A) Patients with allergic bronchopulmonary aspergillosis rarely have eosinophilia

(B) Patients receiving parenteral corticosteroids are not at risk for invasive aspergillosis

(C) The diagnosis of pulmonary aspergillosis is frequently established by culturing *Aspergillus* from the sputum and blood

(D) The clinical manifestations of aspergillosis include local infections of the ear, cornea, nails, and sinuses

(E) Bone marrow transplant recipients are not at risk for invasive aspergillosis

18. Which statement regarding sporotrichosis is correct?

(A) The most common etiological agent is *Pseudallescheria boydii (Scedosporium apiospermum)*

(B) The etiologic agent is a dimorphic fungus

(C) The ecology of the etiologic agent is unknown

(D) Most cases are subcutaneous and nonlymphangitic

(E) Most patients are immunocompromised

19. A 24-year-old, HIV-negative migrant worker from Colombia presented with a painful ulcerative lesion on the tongue. The edge of the lesion was gently scraped and a calcofluor white-potassium hydroxide smear revealed tissue cells, debris, and several large, spherical, multiply budding yeast cells. Based on this observation, what is the most likely diagnosis?

    (A) Blastomycosis
    (B) Candidiasis
    (C) Coccidioidomycosis
    (D) Histoplasmosis
    (E) Paracoccidioidomycosis

20. Which statement about blastomycosis is correct?

    (A) Similar to other endemic mycoses, this infection occurs equally in men and women
    (B) Infection starts in the skin, and the organisms commonly disseminate to the lungs, bone, genitourinary tract, or other sites
    (C) The disease is endemic to certain areas of South America
    (D) In tissue, one finds large, thick-walled, single budding yeast cells with broad connections between the parent yeast and bud
    (E) All cases require treatment with amphotericin B

21. Which statement regarding dermatophytosis is correct?

    (A) Chronic infections are associated with zoophilic dermatophytes, such as *Trichophyton rubrum*
    (B) Acute infections are associated with zoophilic dermatophytes, such as *Trichophyton rubrum*
    (C) Chronic infections are associated with anthropophilic dermatophytes, such as *Trichophyton rubrum*
    (D) Acute infections are associated with anthropophilic dermatophytes, such as *Trichophyton rubrum*

22. Which statement regarding paracoccidiomycosis is not correct?

    (A) The etiological agent is a dimorphic fungus
    (B) Most patients acquired their infections in South America
    (C) Although the infection is acquired by inhalation and is initiated in the lungs, many patients develop cutaneous and mucocutaneous lesions
    (D) The vast majority of patients with active disease are males
    (E) The etiological agent is inherently resistant to amphotericin B

23. Your kidney transplant patient has developed nosocomial systemic candidiasis, but the patient's isolate of *Candida glabrata* is resistant to fluconazole. A reasonable alternative would be oral administration of:

    (A) Flucytosine
    (B) Posaconazole
    (C) Griseofulvin
    (D) Amphotericin B

24. Which one of the following antifungal drugs does not target the biosynthesis of ergosterol in the fungal membrane?

    (A) Voriconazole
    (B) Itraconazole
    (C) Terbinafine
    (D) Fluconazole
    (E) Micafungin

25. Which one of the following pathogenic yeasts is not a common member of the normal human flora or microbiota?

    (A) *Candida tropicalis*
    (B) *Malassezia globosa*
    (C) *Cryptococcus neoformans*
    (D) *Candida glabrata*
    (E) *Candida albicans*

Answers

1. D	8. B	15. E	22. E
2. B	9. D	16. D	23. B
3. C	10. B	17. D	24. E
4. B	11. B	18. B	25. C
5. A	12. C	19. E	
6. D	13. D	20. D	
7. B	14. D	21. C	

REFERENCES

Anaissie EJ, McGinnis MR, Pfaller MA: *Clinical Mycology.* Philadelphia, Churchill Livingstone, 2003.

Andes DR: Pharmacokinetics and pharmacodynamics of antifungals. Infect Dis Clin N Am 2006;20:679.

Ben-Ami R, Lewis RE, Raad II, Kontoyiannis DP: Phaeohyphomycosis in a tertiary care cancer center. Clin Infect Dis 2009;48:1033.

Borman AM, Campbell CK, Fraser M, Johnson EM: Analysis of the dermatophyte species isolated in the British Isles between 1980 and 2005 and review of worldwide dermatophyte trends over the last three decades. Med Mycol 2007;45:131.

Bradsher RW, Jr, Chapman SW, Pappas PG: Blastomycosis. Infect Dis Clin N Am 2003;17:21.

Calderone RA (editor): *Candida and Candidiasis.* Washington, DC: ASM Press, 2002.

Cappelletty D, Eiselstein-McKitrick K: The echinocandins. Pharmacother 2007;27:369.

Chayakulkeeree M, Perfect JR: Cryptococcosis. Infect Dis Clin N Am 2006;20:507.

Clemons KV et al (editors): Third Advances Against Aspergillosis. Med Mycol 2009;47(Suppl 1):S1.

Cohen J, Powderly WG (editors): Chapters 237–241. *Infectious Diseases,* 2nd ed, 2 vols. London, Mosby, pp 2341–2411, 2004.

Cooper CR Jr, Haycocks NG: *Penicillium marneffei:* An insurgent species among the penicillia. J Eukaryot Microbiol 2000;47:24.

da Rosa AC, Scroferneker ML, Vettorato R, Gervini RL, Vettorato G, Weber A: Epidemiology of sporotrichosis: A study of 304 cases in Brazil. J Am Acad Dermatol 2005;52:451.

de Camargo ZP, de Franco MF: Current knowledge on pathogenesis and immunodiagnosis of paracoccidioidomycosis. Rev Iberoam Micol 2000;17:41.

Dismukes WE, Pappas PG, Sobel JD: *Clinical Mycology.* New York, Oxford University Press, 2003.

Ferguson BJ (editor): Fungal rhinosinusitis: A spectrum of disease. Otolaryngol Clin N Am 2000;33:1.

Freifeld AG et al: Voriconazole use for endemic fungal infections. Antimicrob Agents Chemother 2009;53:1648.

Galgiani JN et al: Coccidioidomycosis. Clin Infect Dis 2005;41:1217.

Hope WW, Shoham S, Walsh TJ: The pharmacology and clinical use of caspofungin. Expert Opin Drug Metab Toxicol 2007;3:263.

Kauffman CA, Carver PL: Update on echinocandin antifungals. Semin Respir Crit Care Med 2008;29:211.

Kauffman CA: Endemic mycoses: Blastomycosis, histoplasmosis, and sporotrichosis. Infect Dis Clin N Am 2006;20:645.

Kim R, Khachikian D, Reboli AC: A comparative evaluation of properties and clinical efficacy of the echinocandins. Expert Opin Pharmacother 2007;8:1479.

Letsher-Bru V, Herbrecht R: Caspofungin: the first representative of a new antifungal class. Journal of Antimicrobial Chemotherapy 2003;51:513.

Lupi O, Tyring SK, McGinnis MR: Tropical dermatology: Fungal tropical diseases. J Am Acad Dermatol 2005;53:931.

Merz WG, Hay RJ (editors): *Topley & Wilson's Microbiology and Microbial Infections,* 10th ed, vol 4 *Medical Mycology.* London, Arnold, 2005.

Moen MD, Lyseng-Williamson KA, Scott LJ: Liposomal amphotericin B: A review of its use as empirical therapy in febrile neutropenia and in the treatment of invasive fungal infections. Drugs 2009;69:361.

Mohindra S, Mohindra S, Gupta R, Bakshi J, Gupta SK. Rhinocerebral mucormycosis: The disease spectrum in 27 patients. Mycoses 2007;50:290.

Nucci M, Anaissie EJ: *Fusarium* infections in immunocompromised patients. Clin Microbiol Rev 2007;20:695.

Perlroth J, Choi B, Spellberg BJ: Nosocomial fungal infections: Epidemiology, diagnosis, and treatment. Med Mycol 2007;45:321.

Pyrgos V, Shoham S, Walsh TJ: Pulmonary zygomycosis. Semin Respir Crit Care Med 2008;29:111.

Queiroz-Telles F, Esterre P, Perez-Blanco M, Vitale RG, Salgado CG, Bonifaz A: Chromoblastomycosis: An overview of clinical manifestations, diagnosis and treatment. Med Mycol 2009;47:3.

Revankar SG: Dematiaceous fungi. Mycoses 2007;50:91.

Ribes JA, Vanover-Sams CL, Baker DJ: Zygomycetes in human disease. Clin Microbiol Rev 2000;13:236.

Richardson MD, Lass-Flörl C: Changing epidemiology of systemic fungal infections. Clin Microbiol Infect 2008;14(Suppl.4):5.

Sobel JD. Vulvovaginal candidosis. Lancet 2007;369:1961.

Weitzman I, Summerbell RC: The dermatophytes. Clin MicrobiolRev 1995;8:240.

Wheat LJ, Kauffman CA: Histoplasmosis. Infect Dis Clin N Am 2003;17:1.

C H A P T E R

46

Medical Parasitology

Judy A. Sakanari, PhD and James H. McKerrow, MD, PhD

This chapter offers a brief survey of the protozoan and helminthic parasites of medical importance. A synopsis of each parasite is provided within tables that are organized by the organ system that is infected (eg, intestinal and blood/tissue protozoan infections and intestinal and blood/tissue helminthic infections). Key concepts are provided at the beginning of the protozoa and helminths sections to give the reader an overview of the paradigms in medical parasitology. Current updates to information provided in this chapter can be found at the Centers for Disease Control and Prevention Web site (www.cdc.gov/ncidod/dpd) under "Parasitic Diseases."

CLASSIFICATION OF PARASITES

The parasites covered in this chapter are categorized into two major groups: **parasitic protozoa** and **parasitic helminths**.

Protozoa are unicellular eukaryotes that form an entire kingdom. Classifying protozoan parasites into taxonomic groups is an ongoing process, and their status is often in a state of flux. For this reason, this chapter separates the parasitic protozoa into four traditional groups based on their means of locomotion and mode of reproduction: flagellates, amebae, sporozoa, and ciliates. Table 46–1 lists several medically important protozoan parasites by the organ system they infect, the mode of infection, diagnosis, treatment, and geographic location.

(1) Flagellates have one or more whiplike flagella and, in some cases, an undulating membrane (eg, trypanosomes). These include intestinal and genitourinary flagellates (*Giardia* and *Trichomonas*, respectively) and blood and tissue flagellates (*Trypanosoma* and *Leishmania*). **(2)**

Amebae are typically ameboid and use pseudopodia or protoplasmic flow to move. They are represented in humans by species of *Entamoeba*, *Naegleria*, and *Acanthamoeba*. **(3) Sporozoa** undergo a complex life cycle with alternating sexual and asexual reproductive phases. The human parasites *Cryptosporidium*, *Cyclospora*, and *Toxoplasma* and the malarial parasites (*Plasmodium* species) are all intracellular parasites. **(4) Ciliates** are complex protozoa bearing cilia distributed in rows or patches, with two kinds of nuclei in each individual. *Balantidium coli*, a giant intestinal ciliate of humans and pigs, is the only human parasite representative of this group, and because the disease is considered rare, it is not covered in this chapter.

Formerly listed with the sporozoa, because they possess polar filaments within a spore, **microsporidia** include more than 1000 species of intracellular parasites that infect invertebrates (mostly insects) and vertebrate hosts. In humans, microsporidians are opportunistic parasites of immunocompromised patients, including those undergoing chemotherapy and organ transplants.

Pneumocystis carinii was long considered a protozoan parasite but has been shown to be a member of the fungi rather than the protozoa. It causes interstitial plasma cell pneumonitis in immunosuppressed individuals and is considered an opportunistic pathogen.

Parasitic helminths, or worms of humans, belong to two phyla: Nematoda (roundworms) and Platyhelminthes (flatworms).

(1) Nematodes are among the most speciose and diverse animals. They are elongated and tapered at both ends, round in cross section, and unsegmented. They have only a set of longitudinal muscles, which allows them to move in a whiplike,

TABLE 46–1 Synopsis of Protozoan Infections by Organ System

Parasite/Disease	Site of Infection	Mechanism of Infection	Diagnosis	Treatment	Geographic Area
Intestinal protozoa					
Giardia lamblia (flagellate) Giardiasis	Small intestine	Ingest cysts in water, not killed by normal chlorination	Stool exam for O&P; EIA for antigens	Metronidazole or quinacrine	Ubiquitous: campers, ski resorts, dogs, wild animals, especially beavers
Entamoeba histolytica (amoeba) Amebiasis	Colon; liver; other organs	Ingest cysts from fecal contamination of water or food or oral/anal behaviors	Stool exam for O&P; EIA for antibodies and antigen	Iodoquinol, diloxanide furoate; metronidazole plus iodoquinol or paromomycin	Worldwide wherever fecal contamination occurs
Cryptosporidium (sporozoa) Cryptosporidiosis	Small intestine; respiratory tract	Ingest oocysts, fecal contamination	Stool exam/acid-fast staining; direct-fluorescent staining; EIA for antigens; PCR	Nitazoxanide for non-HIV infected	Ubiquitous, especially in cattle-raising areas
Cyclospora (sporozoa) Cyclosporiasis	Small intestine	Oocysts from fecal contamination of water, fresh produce	Stool exam—acid-fast staining, UV fluorescence microscopy	Trimethoprim/ sulfamethoxazole	Worldwide, tropics, subtropics
Sexually transmitted protozoa					
Trichomonas vaginalis (flagellate) Trichomoniasis	Vagina; males usually asymptomatic	Trophozoites passed from person to person through sexual intercourse	Microscopic exam of discharge, urine, tissue scraping	Metronidazole for both partners	Ubiquitous in sexually active populations
Blood and tissue flagellates					
Trypanosoma brucei rhodesiense East African trypanosomiasis, sleeping sickness	Blood, lymph	Tsetse bite (painful) lacerates skin and releases trypomastigotes	Trypomastigotes (extracellular) in blood smear, CSF or lymph node aspirate; serology (CATT)	Hemolytic stage: Suramin Late CNS involvement: Melarsoprol	East Africa; antelope, bushbuck are animal reservoirs for human infection
Trypanosoma brucei gambiense West African trypanosomiasis, sleeping sickness	Blood, lymph	Tsetse bite (painful) lacerates skin and releases trypomastigotes	Trypomastigotes (extracellular) in blood smear, CSF or lymph node aspirate; serology (CATT)	Hemolytic stage: Pentamidine Late CNS involvement: Eflornithine	West Africa: vegetation around rivers; humans only (not zoonotic)
Trypanosoma cruzi Chagas disease	Amastigotes intracellular; heart, parasympathetic ganglia	Kissing bug feces rubbed into bite or eye; blood transfusion; transplacental transmission	Trypomastigotes (extracellular) in blood smear; PCR; intracellular amastigotes in tissue bx	Nifurtimox	North, Central, and South America (bugs live in thatched roofs, mud cracks)
Leishmania major Leishmania tropica Old World cutaneous leishmaniasis	Skin; rolled edge ulceration	Sandfly injects promastigotes; amastigotes in macrophages, monocytes	Skin bx at edge of ulcer; histopathology; culture and PCR of organisms; intradermal leishmanin (Montenegro) skin test	Stibogluconate sodium, meglumine antimonate, pentamidine (all IM or IV)	Old World: Middle East, India, Africa, Russia

Organism / Disease	Clinical features	Transmission / Life cycle	Diagnosis	Treatment	Distribution
Leishmania mexicana complex New World cutaneous leishmaniasis	Skin; rolled edge ulceration	Sandfly injects promastigotes; amastigotes in macrophages, monocytes	Skin bx at edge of ulcer; histopathology; culture and PCR of organisms; intradermal leishmanin (Montenegro) skin test	Stibogluconate sodium, meglumine antimonate, pentamidine (all IM or IV)	New World: Mexico, Central and South America; chiclero ulcers on ears of chicle harvesters in Yucatan; disseminated leishmaniasis in Ethiopia and Venezuela induces specific anergy (distinctive syndrome)
Leishmania aethiopica, Leishmania mexicana pifano Disseminated or diffuse form of cutaneous leishmaniasis	Skin; anergy resulting in nonulcerating lesions over entire body	Sandfly injects promastigotes; amastigotes in macrophages, monocytes	Skin bx at edge of ulcer; histopathology; culture and PCR of organisms; intradermal leishmanin (Montenegro) skin test	Sodium stibogluconate, meglumine antimonate, pentamidine (all IM or IV)	Old World: Ethiopia New World: Venezuela
Leishmania brasiliensis complex Mucocutaneous leishmaniasis	Skin lesion; may destroy mucocutaneous tissues on face, mouth	Sandfly injects promastigotes; amastigotes in macrophages, monocytes	Skin bx at edge of ulcer; histopathology; culture and PCR of organisms; intradermal leishmanin (Montenegro) skin test	Sodium stibogluconate (IM or IV), meglumine antimonite (IM or IV), amphotericin B (IV)	Brazil, Peru, Bolivia
Leishmaniasis donovani Kala-azar, visceral leishmaniasis		Sandfly injects promastigotes; amastigotes in macrophages and monocytes of spleen, liver, bone marrow	Bx spleen, liver, bone marrow aspirate; histopathology; culture and PCR of organisms	Liposomal amphotericin B (IV), sodium stibogluconate (IM or IV), meglumine antimonite (IM or IV), amphotericin B (IV)	Post kala-azar dermal leishmaniasis 1–3 years after Rx in India, China, Mediterranean, Russia, Amazon Basin, Sudan, Kenya, South America
Tissue amebae					
Naegleria, Acanthamoeba, Balamuthia Primary amebic meningoencephalitis (*Entamoeba histolytica*—amebiasis, see intestinal protozoa)	Brain, spinal cord, eye	Swimming in warm freshwater lakes, ponds, rivers, hot springs; free-living amebae enter nasal membrane, pass to brain or via wound or penetration of eye (*Acanthamoeba*)	Trophozoite in CSF; clinical suspicion based on recent history of swimming or diving in warm waters	Amphotericin B: intrathecal+IV	Where free-living amebae survive in sediment of warm fresh waters
Blood and tissue sporozoa					
Plasmodium vivax Benign tertian malaria	Intracellular in RBCs; hypnozoites in liver can cause relapse	Female *Anopheles* mosquito releases sporozoites into bloodstream; parasites enter liver then blood; can relapse	Thick and thin blood smears; ring stage in RBC with Schüffner dots	¹Chloroquine (where no resistance), otherwise mefloquine or atovaquone/proguanil, either followed by primaquine for relapse	Tropics, Africa (rare in West Africa), Middle East, Asia, Central and South America

Continued

TABLE 46–1   Synopsis of Protozoan Infections by Organ System (Continued)

Parasite/Disease	Site of Infection	Mechanism of Infection	Diagnosis	Treatment	Geographic Area
Plasmodium falciparum Malignant tertian malaria	Intracellular in RBCs	Female *Anopheles* mosquito releases sporozoites into bloodstream; parasites enter liver then blood; no relapse	Thick and thin blood smears; banana-shaped gametocytes; double rings in RBCs	Chloroquine (where no resistance); quinine sulfate plus doxycycline or plus tetracycline or plus clindamycin; atovaquone/proguanil, mefloquine, ²artesunate plus doxycycline or clindamycin; coartemether/lumefantrine (coartem)	Predominant species, worldwide tropics, but especially sub-Saharan Africa
Plasmodium ovale Ovale malaria	Intracellular in RBCs; hypnozoites in liver can cause relapse	Female *Anopheles* mosquito releases sporozoites into bloodstream; parasites enter liver then blood; can relapse	Thick and thin blood smears	Chloroquine (where no resistance); primaquine for relapse	Tropical, sub-Saharan Africa
Plasmodium malariae Quartan or malariae malaria	Intracellular in RBCs; hypnozoites in liver can cause relapse	Enters liver from inoculation into bloodstream by infected mosquito; no relapse	Thick and thin blood smears	Chloroquine (where no resistance)	Tropical, Africa, Asia, South America
Babesia microti Babesiosis	Intracellular in RBCs	Tick bite; blood transfusions	Blood smears; tetrad forms ("Maltese Cross") inside RBCs	Clindamycin plus quinine; atovaquone plus azithromycin	USA (MA, NY, CT, NJ, WI, GA, CA); Europe
Toxoplasma gondii Toxoplasmosis	Intracellular in CNS, bone marrow	Ingestion of parasites in undercooked meat; ingestion of oocysts from cat feces; transplacental; blood transfusion	Serology (IgG and IgM)	Pyrimethamine plus sulfadiazine	Worldwide; areas where cats/felids live

CATT, card agglutination test for trypanosomes; CNS, central nervous system; CSF, cerebrospinal fluid; EIA, enzyme immunoassay; IM, intramuscular; IV, intravenous; O&P, ova and parasites; PCR, polymerase chain reaction; RBC, red blood cell.
ªRecommendations should be checked regularly (Phone: 877-FYI-TRIP; Internet: www.cdc.gov/travel/).
ᵇFor a review of malaria treatment, see Rosenthal PJ 2009.

penetrating fashion; a complete digestive system that is well adapted for ingestion of the host's gut contents, cells, blood, or cellular breakdown products; and a highly developed separate-sexed reproductive system. They shed their tough cuticles (molt) as they undergo development from larvae to adults, and the eggs and larval stages are well suited for survival in the external environment. Most human infections are acquired by ingestion of the egg or larval stage, but nematode infections can also be acquired from insect vectors and skin penetration. **(2) Platyhelminthes** are flatworms that are dorsoventrally flattened in cross section and are hermaphroditic, with a few exceptions. All medically important species belong to two classes: **Trematoda** (flukes) and **Cestoda** (tapeworms).

Trematodes are typically flattened and leaf shaped with two muscular suckers. They have a bifurcated gut and possess both circular and longitudinal muscles; they lack the cuticle characteristic of nematodes and instead have a syncytial epithelium. Trematodes are hermaphroditic, with the exception of the schistosomes (blood flukes), which have male and female worms that exist coupled together within small blood vessels of their hosts.

The life cycle of human trematodes is typically initiated when eggs are passed into fresh water via feces or urine. Eggs develop, hatch, and release a ciliated miracidium, which infects a snail host that is usually highly specific to the fluke species. Within the snail, the miracidium develops into a sporocyst, which contains germinal cells that ultimately develop into the final larval stage—the cercariae. These swim out of the snail and encyst as metacercariae in a second intermediate host or on vegetation, depending on the species. Most fluke infections are acquired by ingestion of the metacercariae. The cercariae of schistosomes, however, directly penetrate the skin of their hosts and do not encyst as metacercariae.

Cestodes, or tapeworms, are flat and have a ribbon-like chain of segments (proglottids) containing male and female reproductive structures. Adult tapeworms can reach lengths of 10 m and have hundreds of segments, with each segment releasing thousands of eggs. At the anterior end of an adult tapeworm is the scolex, which is often elaborated with muscular suckers, hooks, or structures that aid in its ability to attach to the intestinal wall. Adult tapeworms have no mouth or gut and absorb their nutrients directly from their host through their integument.

The life cycle of cestodes, like that of the trematodes, is usually indirect (involving one or more intermediate hosts and a final host). Eggs are excreted with the feces and ingested by an intermediate host (invertebrate, such as a flea, or vertebrate, such as a mammal); the larvae develop into certain forms that are peculiar to the specific species within the intermediate host (eg, cysticercus in the case of *Taenia solium* or hydatid cyst with *Echinococcus granulosus*). Cestode larvae are generally eaten, and the larva develops into an adult worm in the intestine of the final host.

INTESTINAL PROTOZOAN INFECTIONS

Key concepts pertaining to parasitic protozoa and the protozoa included in this chapter are listed in Tables 46–2 and 46–3. A synopsis of the parasitic protozoan infections is provided in Table 46–1.

GIARDIA LAMBLIA (INTESTINAL FLAGELLATE)

The Organism

Giardia lamblia (also referred to as *Giardia duodenalis* or *Giardia intestinalis*) is the causative agent of giardiasis and is the only common pathogenic protozoan found in the duodenum and jejunum of humans. *Giardia* exists in two forms: the trophozoite and the cyst forms. The trophozoite of *G lamblia* is a heart-shaped organism, has four pairs of flagella, and is approximately 15 μm in length (Figure 46–1A). A large concave sucking disk on the ventral surface helps the organism to adhere to intestinal villi. As the parasites pass into the colon, they typically encyst, and the cysts are passed in the stool (Figure 46–1B). They are ellipsoid, thick-walled, highly resistant, and 8–14 μm in length; they contain two nuclei as immature forms and four as mature cysts.

TABLE 46–2   Key Concepts: Parasitic Protozoa

Parasitic protozoa covered in this chapter are grouped into the flagellates, amebae, sporozoa, and ciliates.
Flagellates and amebae multiply by binary fission; sporozoans reproduce by a process known as merogony (also called schizogony) in which the nuclei replicate prior to cytokinesis.
Sporozoans also undergo sexual recombination, which leads to genomic and antigenic variation.
Protozoa can multiply quickly (on the order of several hours) in the host and can cause a rapid onset of symptoms.
Intestinal infections are acquired by ingestion of an environmentally resistant cyst (or oocyst) form; blood infections are vectorborne.
Infections by intracellular protozoa (*Trypanosoma cruzi*, *Leishmania* spp., *Cryptosporidium*, *Toxoplasma*, and *Plasmodium*) are difficult to treat because drugs must cross plasma membranes. No vaccines are available for any human parasitic disease.
Latent infections occur with *Toxoplasma* (parasites in tissue cysts are called bradyzoites) and *Plasmodium vivax* and *P ovale* (parasites in liver tissue are called hypnozoites).
In disseminated protozoal infections, fever and flu-like symptoms occur and are nonspecific.
Some parasitic protozoa are able to evade the host's immune response because they are intracellular and/or undergo antigenic variation.

TABLE 46–3   Parasitic Protozoa

Intestinal protozoa

 Giardia lamblia (Flagellate)

 Entamoeba histolytica (Ameba)

 Cryptosporidium hominis (Sporozoa)

 Cyclospora cayetanensis (Sporozoa)

Sexually transmitted protozoan infection

 Trichomonas vaginalis (Flagellate)

Blood and tissue protozoan infections

 Flagellates

  Trypanosoma brucei rhodesiense and *Trypanosoma brucei gambiense*

  Trypanosoma cruzi

  Leishmania donovani, Leishmania tropica, Leishmania mexicana

 Amebae

  Entamoeba histolytica (see intestinal protozoa)

  Naegleria fowleri and *Acanthamoeba castellanii*

 Sporozoa

  Plasmodium vivax, Plasmodium falciparum, Plasmodium ovale, and *Plasmodium malariae*

  Babesia microti

  Toxoplasma gondii

Microsporidia

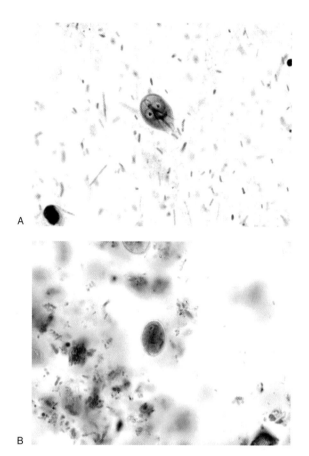

FIGURE 46–1 *Giardia lamblia.* **A**: Trophozoite (12–15 μm). **B**: Cyst (11–14 μm). (Used with permission from J. Sullivan, *A Color Atlas of Parasitology,* 8th ed., 2009.)

Pathology & Pathogenesis

G lamblia is usually only weakly pathogenic for humans. Cysts may be found in large numbers in the stools of entirely asymptomatic persons. In some persons, however, large numbers of parasites attached to the bowel wall may cause irritation and low-grade inflammation of the duodenal or jejunal mucosa, with consequent acute or chronic diarrhea associated with crypt hypertrophy, villous atrophy or flattening, and epithelial cell damage. Stools may be watery, semisolid, greasy, bulky, and foul smelling at various times during the course of the infection. Symptoms of malaise, weakness, weight loss, abdominal cramps, distention, and flatulence may continue for long periods. Collecting multiple stool samples over several days is recommended to increase the likelihood of microscopically detecting cysts in smears.

Epidemiology

G lamblia occurs worldwide. Humans are infected by ingestion of fecally contaminated water or food containing giardia cysts or by direct fecal contamination, as may occur in day care centers, refugee camps, and institutions, or during oral-anal sex. Epidemic outbreaks have been reported at ski resorts in the United States, where overloading of sewage facilities or contamination of the water supply has resulted in sudden outbreaks of giardiasis. Cysts can survive in water for up to 3 months. Outbreaks among campers in wilderness areas suggest that humans may be infected with various animal giardia harbored by rodents, deer, cattle, sheep, horses, or household pets.

ENTAMOEBA HISTOLYTICA (INTESTINAL & TISSUE AMEBA)

The Organism

Entamoeba histolytica cysts are present only in the lumen of the colon and in mushy or formed feces and range in size from 10 to 20 μm (Figure 46–2A). The cyst may contain a glycogen vacuole and chromatoid bodies (masses of ribonucleoprotein) with characteristic rounded ends (in contrast to splinter chromatoidals in developing cysts of *Entamoeba coli*). Nuclear division occurs within the cyst, resulting in a quadrinucleated cyst, and the chromatoid bodies and glycogen vacuoles disappear. Diagnosis in most cases rests on the characteristics of the cyst, as trophozoites usually appear only in diarrheic feces in active cases and survive for only a few hours.

The ameboid trophozoite is the only form present in tissues (Figure 46–2B). The cytoplasm has two zones, a hyaline outer margin and a granular inner region that may contain red blood cells (pathognomonic) but ordinarily contains no bacteria. The nuclear membrane is lined by fine, regular granules of chromatin with a small central body (endosome or karyosome).

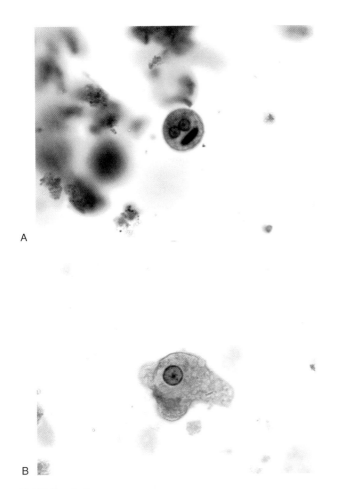

FIGURE 46–2 *Entamoeba histolytica.* **A**: Cyst (12–15 μm).
B: Trophozoite (10–20 μm). (Used with permission from J. Sullivan, *A Color Atlas of Parasitology,* 8th ed., 2009.)

Pathology & Pathogenesis of Invasive Amebiasis

It is estimated that approximately 50 million cases of invasive disease occur each year, with up to 100,000 deaths (Stanley, 2003). Disease results when the trophozoites of *E histolytica* invade the intestinal epithelium and form discrete ulcers with a pinhead-sized center and raised edges, from which mucus, necrotic cells, and amebae pass. The trophozoites multiply and accumulate above the muscularis mucosae, often spreading laterally. Rapid lateral spread of the multiplying amebae follows, undermining the mucosa and producing the characteristic "flask-shaped" ulcer of primary amebiasis: a small point of entry, leading via a narrow neck through the mucosa into an expanded necrotic area in the submucosa. Bacterial invasion usually does not occur at this time, cellular reaction is limited, and damage is by lytic necrosis.

Subsequent spread may coalesce colonies of amebae, undermining large areas of the mucosal surface. Trophozoites may penetrate the muscle layers and occasionally the serosa, leading to perforation into the peritoneal cavity. Subsequent

enlargement of the necrotic area produces gross changes in the ulcer, which may develop shaggy overhanging edges, secondary bacterial invasion, and accumulation of neutrophilic leukocytes. Secondary intestinal lesions may develop as extensions from the primary lesion (usually in the cecum, appendix, or nearby portion of the ascending colon). The organisms may travel to the ileocecal valve and terminal ileum, producing a chronic infection. The sigmoid colon and rectum are favored sites for later lesions. An amebic inflammatory or granulomatous tumor-like mass (ameboma) may form on the intestinal wall, sometimes growing sufficiently large to block the lumen.

Factors that determine invasion of amebae include the following: the number of amebae ingested, the pathogenic capacity of the parasite strain, host factors such as gut motility and immune competence, and the presence of suitable enteric bacteria that enhance amebic growth. Correct and prompt identification of the *Entamoeba* species remains a critical problem. Trophozoites, especially with red blood cells in the cytoplasm, found in liquid or semiformed stools are pathognomonic.

Symptoms vary greatly depending upon the site and intensity of lesions. Extreme abdominal tenderness, fulminating dysentery, dehydration, and incapacitation occur in serious disease. In less acute disease, onset of symptoms is usually gradual, and often includes episodes of diarrhea, abdominal cramps, nausea and vomiting, and an urgent desire to defecate. More frequently, there will be weeks of cramps and general discomfort, loss of appetite, and weight loss, with general malaise. Symptoms may develop within 4 days of exposure, may occur up to a year later, or may never occur.

Extraintestinal infection is metastatic and rarely occurs by direct extension from the bowel. By far the most common form is amebic hepatitis or liver abscess (4% or more of clinical infections), which is assumed to be due to microemboli, including trophozoites carried through the portal circulation. It is assumed that hepatic microembolism with trophozoites is a common accompaniment of bowel lesions but that these diffuse focal lesions rarely progress. A true amebic abscess is progressive, nonsuppurative (unless secondarily infected), and destructive without compression and formation of a wall. The contents are necrotic and bacteriologically sterile, active amebae being confined to the walls. A characteristic "anchovy paste" is produced in the abscess and seen on surgical drainage. More than half of patients with amebic liver abscess give no history of intestinal infection, and only one eighth of them pass cysts in their stools. Rarely, amebic abscesses also occur elsewhere (eg, lung, brain, spleen, or draining through the body wall). Any organ or tissue in contact with active trophozoites may become a site of invasion and abscess. Hepatic abscess, usually showing as an elevation of the right dome of the diaphragm, can be observed by ultrasonography, computerized tomography, magnetic resonance imaging, or radioisotope scanning. Serologic tests in these cases are usually strongly positive.

OTHER INTESTINAL AMEBAE

Invasive or pathogenic *E histolytica* is now considered a species distinct from the more common lumen-dwelling nonpathogenic commensal species, *E dispar,* with the name *E histolytica* reserved only for the pathogenic form. *E dispar* and the related *E moshkovskii* are, based on isoenzyme and genetic analyses, distinct species, even though they are microscopically identical. *E histolytica* must be distinguished not only from *E dispar* and *E moshkovskii* but also from four other ameba-like organisms that are also intestinal parasites of humans: (1) *Entamoeba coli,* which is very common; (2) *Dientamoeba fragilis* (a flagellate), the only intestinal parasite other than *E histolytica* that has been suspected of causing diarrhea and dyspepsia but is not invasive; (3) *Iodamoeba bütschlii;* and (4) *Endolimax nana.* Considerable experience is required to distinguish *E histolytica* from other forms, but it is necessary to do so because misdiagnosis often leads to unnecessary treatment, overtreatment, or a failure to treat.

Enzyme immunoassay (EIA) kits are available commercially for serodiagnosis of amebiasis when stools are often negative. EIA tests to detect amebic antigen in the stool are also sensitive and specific for *E histolytica* and can distinguish between pathogenic and nonpathogenic infections.

Epidemiology

E histolytica occurs worldwide, mostly in developing countries where sanitation and hygiene are poor. Infections are transmitted via the fecal-oral route; cysts are usually ingested through contaminated water, vegetables, and food; flies have also been linked to transmission in areas of fecal pollution. Most infections are asymptomatic, with the asymptomatic cyst passers being a source of contamination for outbreaks where sewage leaks into the water supply or breakdown of sanitation occurs (as in mental, geriatric, or children's institutions or prisons).

CRYPTOSPORIDIUM (INTESTINAL SPOROZOA)

The Organisms

Cryptosporidium species, typically *C hominis,* can infect the intestine in immunocompromised persons (eg, those with AIDS) and cause severe, intractable diarrhea. They have long been known as parasites of rodents, fowl, rhesus monkeys, cattle, and other herbivores and have probably been an unrecognized cause of self-limited, mild gastroenteritis and diarrhea in humans. Oocysts measuring 4–5 μm are passed in feces in enormous numbers and are immediately infectious. When oocysts in contaminated foods and water are ingested, sporozoites excyst and invade intestinal cells; the parasites multiply asexually within the apical portion of the intestinal cells, are released, and infect other intestinal cells to begin a new cycle. They also reproduce sexually, forming male microgamonts and female macrogamonts that fuse and develop into the oocysts.

Pathology & Pathogenesis

Cryptosporidium inhabits the brush border of mucosal epithelial cells of the gastrointestinal tract, especially the surface of villi of the lower small bowel (Figure 46–3A). The prominent clinical feature of cryptosporidiosis is watery diarrhea, which is mild and self-limited (1–2 weeks) in normal persons but may be severe and prolonged in immunocompromised or very young or old individuals. The small intestine is the most commonly infected site, but *Cryptosporidium* infections have also been found in other organs, including other digestive tract organs and the lungs.

Diagnosis depends on detection of oocysts in fresh stool samples. Stool concentration techniques using a modified acid-fast stain are usually necessary (Figure 46–3B). Monoclonal antibody-based testing can detect low-level infections, and

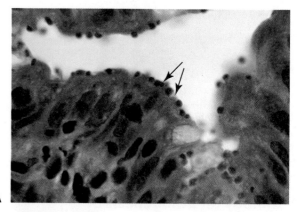

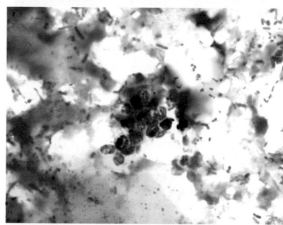

FIGURE 46–3 *Cryptosporidium.* **A**: Histological section of intestine with organisms (arrows) at the apical portion of the epithelial cells. (Courtesy of Pathology, UCSF.) **B**: Oocysts (4–5 μm) stain pink in stool samples stained with an acid-fast stain. (Used with permission from J. Sullivan, *A Color Atlas of Parasitology,* 8th ed., 2009.)

fluorescent microscopy with auramine staining is useful. EIA tests are now available for detection of fecal antigen.

Epidemiology

The incubation period for cryptosporidiosis is from 1 to 12 days, and the disease is acquired from infected animal or human feces or from fecally contaminated food or water. For those at high risk (immunocompromised and very young or old persons), avoidance of animal feces and careful attention to sanitation are required. The organisms are widespread and probably infect asymptomatically a significant proportion of the human population. Occasional outbreaks, such as the one that occurred in Milwaukee in early 1993, affecting more than 400,000 people, can result from inadequate protection, treatment, or filtration of water supplies for large urban centers. In this instance, cattle manure from large dairy farms apparently was the source of contamination of the water supply. The capacity of as few as 30 organisms to initiate an infection—and the ability of the parasite to complete its life cycle, including the sexual phase, within the same individual—makes possible the fulminating infections frequently observed in immunosuppressed individuals.

CYCLOSPORA (INTESTINAL SPOROZOA)

The Organism

The life cycle of *Cyclospora* is similar to that of *Cryptosporodium* and appears to involve only a single host. *Cyclospora*, however, differs from *Cryptosporidium* in that *Cyclospora* oocysts are not immediately infectious when passed in stools. Unlike *Cryptosporidium* oocysts, which are infectious in the feces, *Cyclospora* oocysts take days or weeks to become infectious, and because of this, direct person to person transmission through fecal exposure is unlikely to occur. Cyclosporiasis has been linked to waterborne and foodborne infections from various types of fresh produce, including raspberries, mesclun, and basil, since the 1990s (Herwaldt, 2000; Ho et al, 2002).

Pathology & Pathogenesis

Altered mucosal architecture with shortening of intestinal villi due to diffuse edema and infiltration of inflammatory cells leads to diarrhea, anorexia, fatigue, and weight loss. The duration of symptoms among untreated, nonimmune persons is often prolonged but ultimately self-limited, with remitting-relapsing symptoms lasting up to several weeks or months. The incubation period for *Cyclospora* infections is about 1 week, similar to infections with *Cryptosporidium*. Specific requests for laboratory testing of *Cyclospora* are necessary (same for *Cryptosporidium*) when examining stools for oocysts (8–10 µm), which are acid-fast positive (reddish). *Cyclospora* infections are treatable with trimethoprim-sulfamethoxazole (TMP-SMZ).

SEXUALLY TRANSMITTED PROTOZOAN INFECTION

TRICHOMONAS VAGINALIS (GENITOURINARY FLAGELLATE)

The Organism

Trichomonas vaginalis exists only as a trophozoite (no cyst stage); it has four free flagella that arise from a single stalk and a fifth flagellum, which forms an undulating membrane. It is pyriform and approximately 20 µm in length and 10 µm wide.

Pathology & Pathogenesis

T vaginalis is sexually transmitted, and most infections are asymptomatic or mild for both women and men. In women, the infection is normally limited to the vulva, vagina, and cervix; it does not usually extend to the uterus. The mucosal surfaces may be tender, inflamed, eroded, and covered with a frothy yellow or cream-colored discharge. In men, the prostate, seminal vesicles, and urethra may be infected. Signs and symptoms in females, in addition to profuse vaginal discharge, include local tenderness, vulval pruritus, and burning. About 10% of infected males have a thin, white urethral discharge. The incubation period is from around 5 to 28 days.

Epidemiology

T vaginalis is a common parasite of both males and females. Transmission is by sexual intercourse, but contaminated towels, douche equipment, examination instruments, and other objects may be responsible for some infections. Infants may be infected during birth. Control of *T vaginalis* infections always requires simultaneous treatment of both sexual partners. Mechanical protection (condoms) should be used during intercourse until the infection is eradicated in both partners.

BLOOD & TISSUE PROTOZOAN INFECTIONS

BLOOD FLAGELLATES

The hemoflagellates of humans include the genera *Trypanosoma* and *Leishmania* (Table 46–4). There are two distinct types of human trypanosomes: (1) African, which causes sleeping sickness and is transmitted by tsetse flies (eg, *Glossina*): *Trypanosoma brucei rhodesiense* and *Trypanosoma brucei gambiense*; and (2) American, which causes Chagas disease and is transmitted by kissing bugs (eg, *Triatoma*): *Trypanosoma cruzi*. The genus *Leishmania*, divided into a number of species infecting humans, causes

TABLE 46–4   Comparison of the *Trypanosoma* spp & *Leishmania* spp

Hemoflagellates	Disease	Vector	Stages in Humans
Trypanosoma brucei rhodesiense	African sleeping sickness (acute)	Tsetse fly	Trypomastigotes in blood
Trypanosoma brucei gambiense	African sleeping sickness (chronic)	Tsetse fly	Trypomastigotes in blood
Trypanosoma cruzi	Chagas disease	Kissing bug	Trypomastigotes in blood; amastigotes intracellular
Leishmania spp.	Cutaneous, mucocutaneous, visceral leishmaniasis	Sandfly	Amastigotes intracellular in macrophages and monocytes

cutaneous (Oriental sore), mucocutaneous (espundia), and visceral (kala-azar) leishmaniasis. All of these infections are transmitted by sandflies (*Phlebotomus* in the Old World and *Lutzomyia* in the New World).

TRYPANOSOMA BRUCEI RHODESIENSE & *T B GAMBIENSE* (BLOOD FLAGELLATES)

The Organisms

The genus *Trypanosoma* appears in the blood as trypomastigotes, with elongated bodies supporting a longitudinal lateral undulating membrane and a flagellum that borders the free edge of the membrane and emerges at the anterior end as a whiplike extension (Figure 46–4). The kinetoplast (circular DNA inside the single mitochondrion) is a darkly staining body lying immediately adjacent to the basal body from which the flagellum arises. ***Trypanosoma brucei rhodesiense, T b***

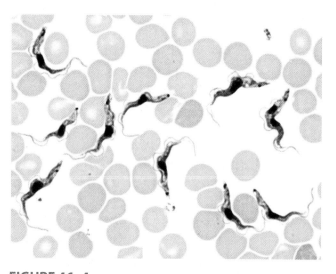

FIGURE 46–4 *Trypanosoma brucei gambiense* (or *Trypanosoma brucei rhodesiense*, indistinguishable in practice) trypomastigotes (14–35 μm) in a blood smear (RBCs = 10 μm). (Used with permission from J. Sullivan, *A Color Atlas of Parasitology*, 8th ed., 2009.)

gambiense, and T b brucei (which causes a sleeping sickness called nagana in livestock and game animals) are indistinguishable morphologically but differ biochemically, ecologically, and epidemiologically.

Pathology & Pathogenesis

Infective trypanosomes of *T b gambiense* and *T b rhodesiense* are introduced through the bite of the tsetse fly and multiply at the site of inoculation to cause variable induration and swelling (the primary lesion), which may progress to form a trypanosomal chancre. The African forms multiply extracellularly as trypomastigotes in the blood as well as in lymphoid tissues. They spread to lymph nodes, to the bloodstream, and, in terminal stages, to the central nervous system (CNS), where they produce the typical sleeping sickness syndrome: lassitude, inability to eat, tissue wasting, unconsciousness, and death.

CNS involvement is most characteristic of African trypanosomiasis. *T b rhodesiense* appears in the cerebrospinal fluid in about 1 month and *T b gambiense* in several months, but both are present in small numbers. *T b gambiense* infection is chronic and leads to progressive diffuse meningoencephalitis, with death from the sleeping syndrome usually following in 1–2 years. The more rapidly fatal *T b rhodesiense* produces somnolence and coma only during the final weeks of a terminal infection. The trypanosomes are transmissible through the placenta, and congenital infections occur in hyperendemic areas.

The African trypanosomes of the *T brucei* complex are remarkable in that they undergo antigenic variation through a series of genetically controlled surface glycoproteins that coat the surface of the organism (variant surface glycoproteins or VSGs). Successive waves of parasites in the host bloodstream are each covered with a distinct coat. This process is due to genetically induced changes of the surface glycoprotein. By producing different antigenic surface membranes, the parasite is able to evade the host's antibody response. Each population is reduced but is promptly replaced with another antigenic type before the preceding one is eliminated. Each trypanosome is thought to possess about 1000 VSG genes, an example of mosaic gene formation.

Epidemiology

African trypanosomiasis is restricted to recognized tsetse fly belts. *T b gambiense*, transmitted by the streamside tsetse *Glossina palpalis* and several other humid forest tsetse vectors, extends from West to Central Africa and produces a relatively chronic infection with progressive CNS involvement. *T b rhodesiense*, transmitted by the woodland-savanna *Glossina morsitans*, *Glossina pallidipes*, and *Glossina fuscipes*, occurs in the eastern and southeastern savannas of Africa, with foci west of Lake Victoria. It causes a smaller number of cases but is more virulent. Bushbuck and other antelopes may serve as reservoirs of *T b rhodesiense*, whereas humans are the principal reservoir of *T b gambiense*. Control depends upon searching for and then isolating and treating patients with the disease; controlling movement of people in and out of fly belts; using insecticides in vehicles; and instituting fly control, principally with aerial insecticides and by altering habitats. Contact with reservoir animals is difficult to control, and insect repellent is of little value against tsetse bites.

TRYPANOSOMA CRUZI (BLOOD FLAGELLATE)

The Organism

Trypanosoma cruzi has three developmental stages: epimastigotes in the vector, trypomastigotes (in the blood stream), and a rounded intracellular stage, the amastigote. The blood forms of *T cruzi* are present during the early acute stage and at intervals thereafter in smaller numbers. They are typical trypomastigotes with a large, rounded terminal kinetoplast in stained preparations but they are difficult to morphologically distinguish from African trypanosomes. The tissue forms, which are most common in heart muscle, liver, and brain, develop as amastigotes that multiply to form an intracellular colony after invasion of the host cell or phagocytosis of the parasite (Figure 46–5).

Pathology & Pathogenesis

Infective forms of *T cruzi* do not pass to humans by triatomine bug bites (which is the mode of entry of the nonpathogenic *T rangeli*); rather, they are introduced when infected bug feces are rubbed into the conjunctiva, the bite site, or a break in the skin. At the site of *T cruzi* entry, there may be a subcutaneous inflammatory nodule or chagoma. Unilateral swelling of the eyelids (Romaña's sign) is characteristic at onset, especially in children. The primary lesion is accompanied by fever, acute regional lymphadenitis, and dissemination to blood and tissues.

Interstitial myocarditis is the most common serious condition in Chagas disease. Other organs affected are the liver,

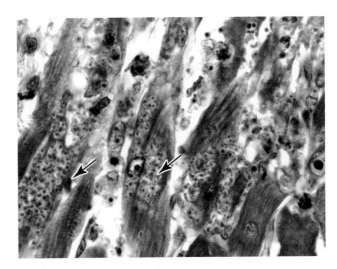

FIGURE 46–5 *Trypanosoma cruzi* amastigote colonies (arrows) in heart muscle. Amastigotes are 1–3 μm in diameter in tissue sections. (Used with permission from J. Sullivan, *A Color Atlas of Parasitology*, 8th ed., 2009.)

spleen, and bone marrow, especially with chronic *T cruzi* infection. Invasion or toxic destruction of nerve plexuses in the alimentary tract walls leads to megaesophagus and megacolon, especially in Brazilian Chagas disease. Megaesophagus and megacolon are absent in Colombian, Venezuelan, and Central American Chagas disease. *Trypanosoma rangeli* of South and Central America infects humans without causing disease and must therefore be carefully distinguished from the pathogenic species.

Epidemiology

American trypanosomiasis (Chagas disease) is especially important in Central and South America, although infection of animals extends much more widely—eg, to Maryland and southern California. A few autochthonous human cases have been reported in Texas and southern California. Since no effective treatment is known, it is particularly important to control the vectors with residual insecticides and habitat modification, such as replacement of mud-brick (adobe) houses with thatched roofs where the insects live, and to avoid contact with animal reservoirs. Chagas disease occurs largely among people in poor economic circumstances. An estimated 8–12 million people harbor the parasite, and many of these individuals sustain heart damage, with the result that their ability to work and their life expectancy are sharply reduced.

LEISHMANIA SPECIES (BLOOD FLAGELLATES)

The Organisms

The sandfly transmits the infective promastigotes during a bite. The promastigotes rapidly change to amastigotes after phagocytosis by macrophages or monocytes, and then multiply, filling the cytoplasm of the cell. The infected cells burst, and the released parasites are again phagocytosed. This process is repeated, producing a cutaneous lesion or visceral infection depending upon the species of parasite and the host response. The amastigotes are ovoid and approximately 2–3 μm in size. The nucleus and a dark-staining, rodlike kinetoplast can be seen as a "dot" and a "dash."

The genus *Leishmania*, widely distributed in nature, has a number of species that are nearly identical morphologically. Clinical characteristics of the disease are traditional differentiating characteristics, but many exceptions are now recognized. The different leishmanias present a range of clinical and epidemiologic characteristics that, for convenience only, are combined under three clinical groupings: (1) **cutaneous leishmaniasis** (Oriental sore, Baghdad boil, wet cutaneous sore, dry cutaneous sore, chiclero ulcer, uta, and other names), (2) **mucocutaneous leishmaniasis** (espundia), and (3) **visceral leishmaniasis** (kala-azar—Hindi for black fever).

There are strain differences in virulence, tissue tropism, and biologic and epidemiologic characteristics, as well as in the serologic and biochemical criteria. Some species can induce several disease syndromes (eg, visceral leishmaniasis from organisms of cutaneous leishmaniasis or cutaneous leishmaniasis from organisms of visceral leishmaniasis). Similarly, the same clinical condition can be caused by different agents.

Pathology & Pathogenesis

Leishmania tropica, *L major*, *L mexicana*, *L braziliensis*, and other **cutaneous forms** induce a dermal lesion at the site of inoculation by the sandfly (cutaneous leishmaniasis, Oriental sore, Delhi boil, etc.). The dermal layers are first affected, with cellular infiltration and proliferation of amastigotes intracellularly and spreading extracellularly, until the infection penetrates the epidermis and causes ulceration. Satellite lesions may be found (hypersensitivity or recidivans type of cutaneous leishmaniasis) that contain few or no parasites, do not readily respond to treatment, and induce a strong granulomatous scarring reaction. In Venezuela, a cutaneous disseminating form, caused by *L mexicana pifanoi,* is known. In Ethiopia, a form known as *L aethiopica* causes a similar nonulcerating, blistering, spreading cutaneous leishmaniasis. Both forms are typically anergic and nonreactive to skin test antigen and contain large numbers of parasites in the dermal blisters.

L braziliensis braziliensis causes **mucocutaneous** or **nasopharyngeal leishmaniasis** in Amazonian South America. It is known by many local names. The lesions are slow growing but extensive (sometimes 5–10 cm). From these sites, migration appears to occur rapidly to the nasopharyngeal or palatine mucosal surfaces, where no further growth may take place for years. After months to more than 20 years, relentless erosion may develop, destroying the nasal septum and surrounding regions. In some instances, death occurs

from asphyxiation due to blockage of the trachea, starvation, or respiratory infection. This is the classic clinical picture of espundia (Figure 46–6), most commonly found in the Amazon basin. At high altitudes in Peru, the clinical features (uta) resemble those of Oriental sore. *L braziliensis guyanensis* infection frequently spreads along lymphatic routes, where it appears as a linear chain of nonulcerating lesions. *L mexicana* infection is more typically confined to a single, indolent, ulcerative lesion that heals in about 1 year, leaving a characteristic depressed circular scar. In Mexico and Guatemala, the ears are frequently involved (chiclero ulcer), usually with a cartilage-attacking infection without ulceration and with few parasites.

L donovani, which causes **visceral leishmaniasis** or kala-azar, spreads from the site of inoculation to multiply in reticuloendothelial cells, especially macrophages in spleen, liver, lymph nodes, and bone marrow (Figure 46–7). This is accompanied by marked hyperplasia of the spleen. Progressive emaciation is accompanied by growing weakness. There is irregular fever, sometimes hectic. Untreated cases with symptoms of kala-azar usually are fatal. Some forms, especially in India, develop a postcure florid cutaneous resurgence, with abundant parasites in cutaneous vesicles, 1–2 years later (post-kala-azar dermal leishmanoid).

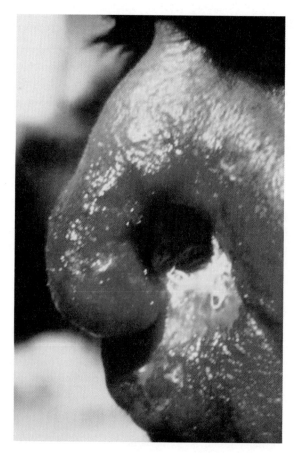

FIGURE 46–6 A patient with espundia caused by *Leishmania braziliensis*. (Reproduced with permission from WHO/TDR image library.)

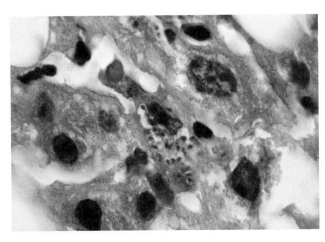

FIGURE 46–7 *Leishmania donovani* amastigotes (arrows) from a liver biopsy. (Courtesy of Pathology, UCSF.)

Epidemiology

According to the World Health Organization (WHO), 2 million new cases of leishmaniasis (1.5 million for cutaneous leishmaniasis and 500,000 for visceral leishmaniasis) are considered to occur annually, with an estimated 12 million people presently infected worldwide (WHO, 2009).

Oriental sore occurs mostly in the Mediterranean region, North Africa, and the Middle and Near East. The "wet" type, caused by *L major,* is rural, and burrowing rodents are the main reservoir; the "dry" type, caused by *L tropica,* is urban, and humans are presumably the only reservoir. For *L braziliensis,* there are a number of wild animal hosts but apparently there are no domestic animal reservoirs. Sandfly vectors are involved in all forms.

L donovani is found focally in most tropical and subtropical countries. Its local distribution is related to the prevalence of specific sandfly vectors. In the Mediterranean littoral and in middle Asia and South America, domestic and wild canids are reservoirs, and in the Sudan, various wild carnivores and rodents are reservoirs of endemic kala-azar. No animal reservoirs have been found for the forms from India and Kenya. Control is aimed at destroying breeding places and dogs, where appropriate, and protecting people from sandfly bites.

ENTAMOEBA HISTOLYTICA (TISSUE AMEBA)—See Intestinal Protozoan Infections Section

NAEGLERIA FOWLERI, ACANTHAMOEBA CASTELLANII, & BALAMUTHIA MANDRILLARIS (FREE-LIVING AMEBAE)

The Organisms

Primary amebic meningoencephalitis (PAM) and granulomatous amebic encephalitis (GAE) occurs in Europe and North America from amebic invasion of the brain. The free-living soil amebae *Naegleria fowleri, Acanthamoeba castellanii, Balamuthia mandrillaris,* and possibly species of *Hartmanella* have been implicated. Most cases have developed in children who were swimming and diving in warm, soil-contaminated water (eg, ponds and rivers).

Pathology & Pathogenesis

The amebae, primarily ***Naegleria fowleri,*** enter via the nose and the cribriform plate of the ethmoid bone, passing directly into brain tissue, where they rapidly form nests of amebae that cause extensive hemorrhage and damage, chiefly in the basilar portions of the cerebrum and the cerebellum (Figure 46–8).

The incubation period is from 1 to 14 days; early symptoms include headache, fever, lethargy, rhinitis, nausea, vomiting, and disorientation and resemble acute bacterial meningitis. In most cases, patients become comatose and die within a week. The key to diagnosis is clinical suspicion based on recent history of swimming or diving in warm waters.

Entry of ***Acanthamoeba*** into the CNS occurs from skin ulcers or traumatic penetration, such as keratitis from puncture of the corneal surface or ulceration from contaminated saline used with contact lenses. GAE is caused by *Acanthamoeba* and ***Balamuthia*** and is often associated with immunocompromised individuals. Infection of the CNS from the skin lesion may occur weeks or months later. It is termed granulomatous amebic encephalitis to distinguish it from the explosive, rapid brain infection from *Naegleria* (PAM). Treatment with amphotericin B has been successful in a few cases, chiefly in the rare instances when diagnosis can be made quickly.

PLASMODIUM SPECIES (BLOOD SPOROZOA)

Malaria is the number one killer of all the parasitic diseases. It is estimated that at least 1 million people die of malaria each

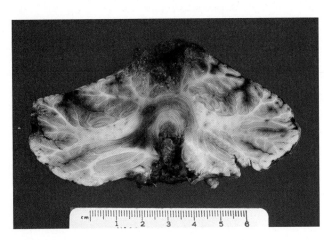

FIGURE 46–8 Dark areas of the cerebellum are regions of necrosis caused by *Naegleria fowleri* amoebae. (Courtesy of Pathology, UCSF.)

year, mostly children under 5 years of age (WHO Malaria Report, 2005). More than 80% of the deaths worldwide occur in sub-Saharan Africa.

The Organisms

Four species of *Plasmodium* cause malaria in humans: **Plasmodium vivax**, **P falciparum**, **P malariae**, and **P ovale**. The two most common species are *P vivax* and *P falciparum*, with falciparum being the most pathogenic of all. Transmission to humans is by the bloodsucking bite of female *Anopheles* mosquitoes (Figure 46–9). The morphology and other characteristics of these species are summarized in Table 46–5 and illustrated in Figures 46–10 and 46–11A–C.

Human infection results from the bite of an infected female *Anopheles* mosquito, through which the sporozoites are injected into the bloodstream. The sporozoites rapidly (usually within 1 hour) enter parenchymal cells of the liver, where the first stage of development in humans takes place (exoerythrocytic phase of the life cycle). Subsequently, numerous asexual progeny, the merozoites, rupture and leave the liver cells, enter the bloodstream, and invade erythrocytes. The merozoites do not return from red blood cells to liver cells.

Parasites in the red cells multiply in a species-characteristic fashion, breaking out of their host cells synchronously. This is the erythrocytic cycle, with successive broods of merozoites appearing at 48-hour intervals (*P vivax*, *P falciparum*, and *P ovale*) or every 72 hours (*P malariae*). During the erythrocytic cycles, certain merozoites enter red cells and become differentiated as male or female gametocytes. The sexual cycle therefore begins in the vertebrate host, but for its continuation into the sporogonic phase, the gametocytes must be taken up and ingested by bloodsucking female *Anopheles*.

P vivax and *P ovale* may persist as dormant forms, or hypnozoites, after the parasites have disappeared from the peripheral blood. Resurgence of an erythrocytic infection (relapse) occurs when merozoites from hypnozoites in the liver break out, are not phagocytosed in the bloodstream, and succeed in reestablishing a red cell infection (clinical malaria). Without treatment, *P vivax* and *P ovale* infections may persist as periodic relapses for up to 5 years. *P malariae* infections lasting 40 years have been reported; this is thought to be a cryptic erythrocytic rather than an exoerythrocytic infection and is therefore termed a recrudescence to distinguish it from a relapse.

Pathology & Pathogenesis

The incubation period for malaria is usually between 9 and 30 days, depending on the infecting species. For *P vivax* and *P falciparum*, this period is usually 10–15 days, but it may be weeks or months. The incubation period of *P malariae* averages about 28 days. Falciparum malaria, which can be fatal, must always be suspected if fever, with or without other

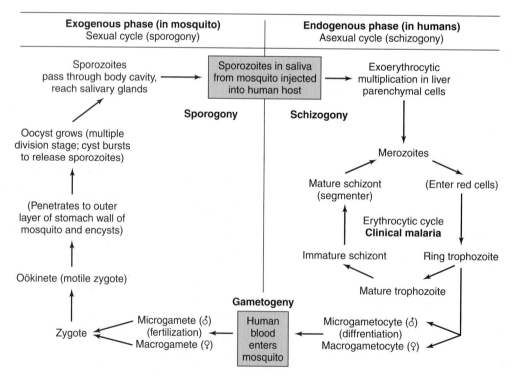

FIGURE 46–9 Life cycle of malaria parasites. Continuous cycling or delayed multiplication in the liver may cause periodic relapse over several years (1–2 years in *Plasmodium ovale*, 3–5 years in *Plasmodium vivax*). Relapse does not occur with *Plasmodium falciparum*, though a long prepatent period may occur, resulting in initial symptoms appearing up to 6 months or more after exposure.

TABLE 46–5 Some Characteristic Features of the Malarial Parasites of Humans (Romanowsky-Stained Preparations)

	Plasmodium vivax (Benign Tertian Malaria)	*P falciparum* (Malignant Tertian Malaria)	*P malariae* (Quartan Malaria)	*P ovale* (Ovale Malaria)
Parasitized red cells	Enlarged, pale. Fine stippling (Schüffner dots). Primarily invades reticulocytes, young red cells	Not enlarged. Coarse stippling (Maurer's clefts). Invades all red cells regardless of age	Not enlarged. No stippling (except with special stains). Primarily invades older red cells	Enlarged, pale. Schüffner dots conspicuous. Cells often oval, fimbriated, or crenated
Level of usual maximum parasitemia	Up to 30,000/μL of blood	May exceed 200,000/μL; commonly 50,000/μL	Fewer than 10,000/μL	Fewer than 10,000/μL
Ring stage trophozoites	Large rings (1/3–1/2 red cell diameter). Usually one chromatin granule; ring delicate	Small rings (1/5 red cell diameter). Often two granules; multiple infections common; ring delicate, may adhere to red cells	Large rings (1/3 red cell diameter). Usually one chromatin granule; ring thick	Large rings (1/3 red cell diameter). Usually one chromatin granule; ring thick
Pigment in developing trophozoites	Fine; light brown; scattered	Coarse; black; few clumps	Coarse; dark brown; scattered clumps; abundant	Coarse; dark yellow-brown; scattered
Older trophozoites	Very pleomorphic	Compact and rounded[a]	Occasional band forms	Compact and rounded
Mature schizonts (segmenters)	More than 12 merozoites (14–24)	Usually more than 12 merozoites (8–32). Very rare in peripheral blood[a]	Fewer than 12 large merozoites (6–12). Often in rosette	Fewer than 12 large merozoites (6–12). Often in rosette
Gametocytes	Round or oval	Crescentic	Round or oval	Round or oval
Distribution in peripheral blood	All forms	Only rings and crescents (gametocytes)[1]	All forms	All forms

[a]Ordinarily, only ring stages or gametocytes are seen in peripheral blood infected with *P falciparum*; post-ring stages make red cells sticky, and they tend to be retained in deep capillary beds, except in overwhelming, usually fatal, infections.

symptoms, develops at any time between 1 week after the first possible exposure to malaria and 2 months (or even longer) after the last possible exposure.

P vivax, P malariae, and *P ovale* parasitemias are relatively low grade, primarily because the parasites favor either young or old red cells but not both; *P falciparum* invades red cells of all ages, including the erythropoietic stem cells in bone marrow, so parasitemia may be very high. *P falciparum* also causes parasitized red cells to produce numerous projecting knobs that adhere to the endothelial lining of blood vessels, with resulting obstruction, thrombosis, and local ischemia. *P falciparum* infections are therefore far more serious than the others, with a much higher rate of severe and frequently fatal complications (cerebral malaria, malarial hyperpyrexia, gastrointestinal disorders, algid malaria, blackwater fever). Consideration of malaria in the differential diagnosis in patients with a suggestive presentation and history of travel to an endemic area is critical because delays in therapy can lead to severe illness or death with falciparum malaria.

Periodic paroxysms of malaria are closely related to events in the bloodstream. An initial chill, lasting from 15 minutes to 1 hour, begins as a synchronously dividing generation of parasites rupture their host red cells and escape into the blood.

Nausea, vomiting, and headache are common at this time. The succeeding febrile stage, lasting several hours, is characterized by a spiking fever that frequently reaches 40°C or more. During this stage, the parasites invade new red cells. The third, or sweating, stage concludes the episode. The fever subsides, and the patient falls asleep and later awakes feeling relatively well. In the early stages of infection, the cycles are frequently asynchronous and the fever pattern is irregular; later, paroxysms may recur at regular 48- or 72-hour intervals, although *P falciparum* pyrexia may last 8 hours or longer and may exceed 41°C. As the disease progresses, splenomegaly and, to a lesser extent, hepatomegaly appear. A normocytic anemia also develops, particularly in *P falciparum* infections.

Normocytic anemia of variable severity may be detected. During the paroxysms, there may be transient leukocytosis; subsequently, leukopenia develops, with a relative increase in large mononuclear cells. Liver function tests may give abnormal results during attacks but revert to normal with treatment or spontaneous recovery. The presence of protein and casts in the urine of children with *P malariae* is suggestive of quartan nephrosis. In severe *P falciparum* infections, renal damage may cause oliguria and the appearance of casts, protein, and red cells in the urine.

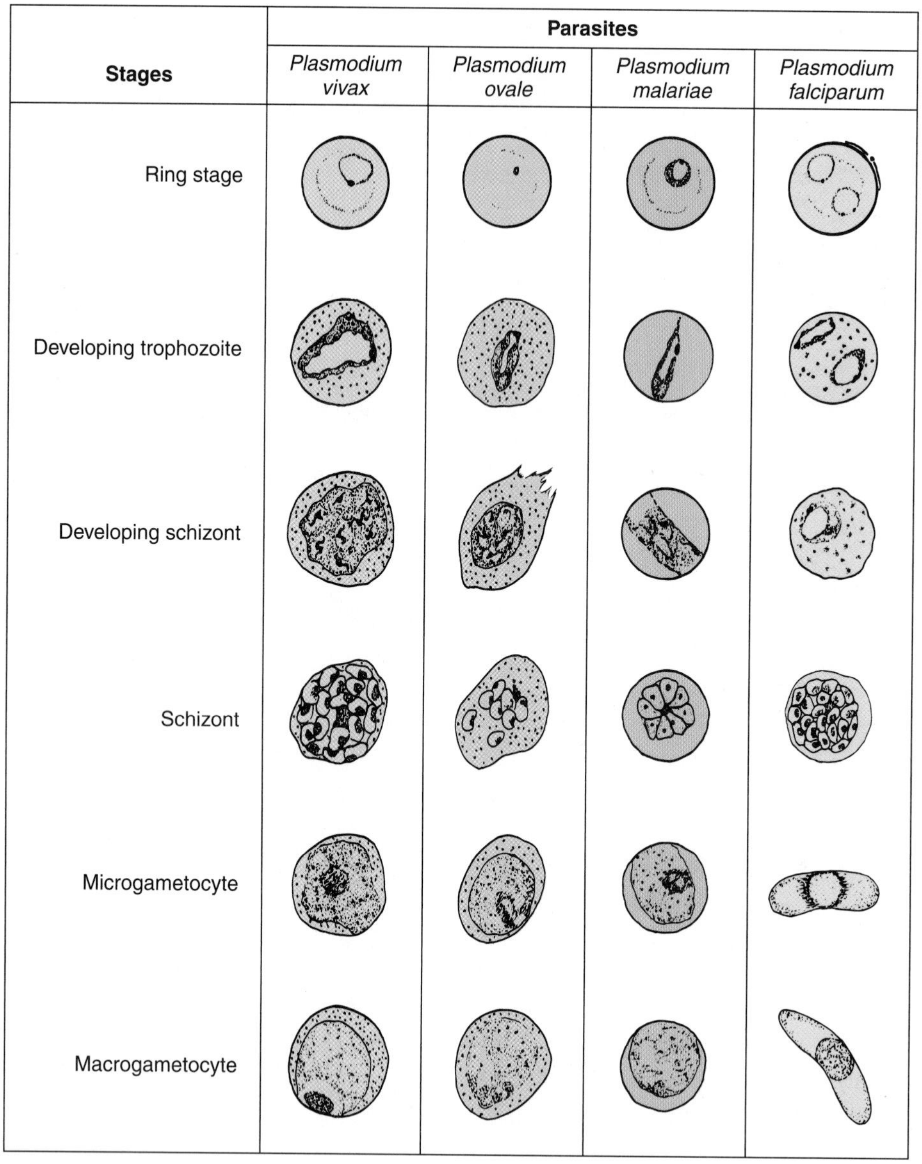

FIGURE 46–10 Morphologic characteristics of developmental stages of malarial parasites in the red blood cell. Note cytoplasmic Schüffner dots and enlarged host cells in *Plasmodium vivax* and *Plasmodium ovale* infections, the band-shaped trophozoite often seen in *Plasmodium malariae* infection, and the small, often multiply infected rings and the banana-shaped gametocytes in *Plasmodium falciparum* infections. Rings and gametocytes are typically seen in peripheral blood smears from patients with *Plasmodium falciparum* infections. (Reproduced with permission from Goldsmith R, Heyneman D: *Tropical Medicine and Parasitology.* McGraw-Hill, 1989.)

Epidemiology & Control

Malaria today is generally limited to the tropics and subtropics, although outbreaks in Turkey attest to the capacity of this disease to reappear in areas cleared of the agent. Malaria in the temperate zones is relatively uncommon, although severe epidemic outbreaks may occur when the largely nonimmune populations of these areas are exposed; it is usually unstable and relatively easy to control or eradicate. Tropical malaria is usually more stable, difficult to control, and far harder to eradicate. In the tropics, malaria generally disappears at altitudes above 6000 ft. *P vivax* and *P falciparum,* the most common species, are found throughout the malaria belt. *P malariae* is also broadly distributed but considerably less common. *P ovale* is rare except in West Africa, where it seems to replace *P vivax.* All forms of malaria can be transmitted transplacentally or by blood transfusion or by needles shared among drug misusers when one is infected. Such cases do not develop a liver infection; thus, relapse does not occur. Natural infection (other than transplacental transmission) takes place only through the bite of an infected female *Anopheles* mosquito.

Malaria control depends upon elimination of mosquito breeding places, personal protection against mosquitoes (eg, screens, pyrethrin-treated netting [see Figure 46–11D], protective clothing with sleeves and long trousers, and repellents), suppressive drugs for exposed persons, and adequate

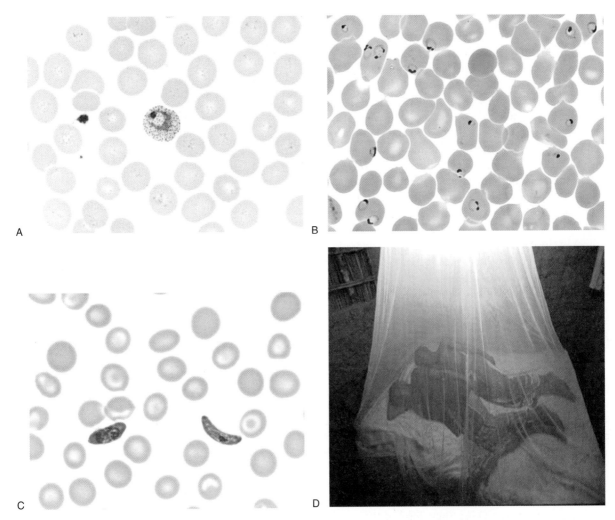

FIGURE 46–11 Distinguishing features between the two most common malarial parasites: **A**: *Plasmodium vivax* trophozoite inside a RBC with Schüffner dots. **B**: Double rings and **C**: banana-shaped gametocytes are typically seen in *Plasmodium falciparum* infections. **D**: Insecticide-impregnated bed nets are an important means of protection against mosquitoes transmitting malaria. (**A–C**: Used with permission from J. Sullivan, *A Color Atlas of Parasitology*, 8th ed., 2009. **D**: Used with permission from WHO/TDR image library/Crump.)

treatment of cases and carriers. Eradication requires prevention of biting contact between *Anopheles* mosquitoes and humans long enough to prevent transmission, with elimination of all active cases by treatment and by spontaneous cure. The results of massive efforts in highly endemic tropical areas have been unsuccessful. Costly eradication projects undertaken between 1955 and 1970 have been replaced with control programs specifically geared to the mosquito vector ecology and malaria epidemiology of each area. These programs must be continued as permanent public health responsibilities. A major WHO-led effort to "roll back malaria" is now under way.

Currently, there is no vaccine available for malaria. A sporozoite surface antigen has been tried as an antisporozoite vaccine, but its initial testing in humans was not successful. A synthetic tripeptide vaccine, SPf66, has been tested in Colombia and was found to be partially effective (<50%). A complete prophylactic vaccine would have to be active against both sporozoites and merozoites of the target species, with an antigametocyticidal effect to curb transmission. With increased reports of multidrug-resistant falciparum malaria and the complex and variable regimens suggested for different areas for both prophylaxis and treatment, referral to the Centers for Disease Control and Prevention for current recommendations is advised (http://wwwnc.cdc.gov/travel/; CDC Malaria Hot Line at 770-488-7788; CDC Voice Information System 1-888-CDC-FACT).

BABESIA MICROTI (BLOOD SPOROZOA)

Babesia species infect red blood cells and are widespread animal parasites that cause infectious jaundice in dogs and Texas cattle fever (red-water fever). Babesiosis, a tick-borne infection, is caused in the United States by ***Babesia microti***. It is considered an emerging infectious disease of humans and is increasing in numbers—more than 300 cases from Massachusetts with the primary focus being Nantucket Island. The great majority of infections in immunologically intact individuals are asymptomatic, but in affected persons the illness develops 7–10 days after the tick bite and is characterized by malaise, anorexia, nausea, fatigue, fever, sweats, myalgia, arthralgia, and depression. Human babesiosis is more severe in the elderly than in the young, in splenectomized individuals, and in AIDS patients. Babesiosis in these individuals may resemble falciparum malaria, with high fever, hemolytic anemia, hemoglobinuria, jaundice, and renal failure; infections are sometimes fatal. *Babesia* may be mistaken in humans for *P falciparum* in its ring form in red cells, though its "Maltese cross" form in the red cell without pigment or gametocytes is diagnostic.

TOXOPLASMA GONDII (TISSUE SPOROZOA)

The Organism

Toxoplasma gondii belongs to the group of sporozoans and has a worldwide distribution, infecting a wide range of animals and birds. The normal final hosts are strictly cats and members of the family Felidae. These are the only hosts in which the oocyst-producing sexual stage of *Toxoplasma* can develop.

Organisms (either sporozoites from oocysts or bradyzoites from tissue cysts) invade the mucosal cells of the cat's small intestine, where they form schizonts or gametocytes. After sexual fusion of the gametes, oocysts develop, exit from the host cell into the gut lumen of the cat, and pass out via the feces. In about 48 hours, the environmentally resistant oocysts are infective. When oocysts are ingested by the cat, the parasites repeat their asexual and sexual cycle. If oocysts are ingested by intermediate hosts such as certain birds, rodents, or mammals, including humans, the parasites can establish an infection but reproduce only asexually. In this latter case, the oocyst opens in the human's or animal's duodenum and releases the sporozoites, which pass through the gut wall, circulate in the body, and invade various cells, especially macrophages, where they form trophozoites, multiply, break out, and spread the infection to lymph nodes and other organs. These rapidly multiplying crescentic cells (**tachyzoites**) initiate the acute stage of disease. Subsequently, they penetrate nerve cells, especially of the brain and eye, where they multiply slowly (as bradyzoites) to form quiescent tissue cysts, initiating the chronic stage of disease. The tissue cysts (formerly called pseudocysts) are infective when ingested by cats (resulting in the intestinal sexual stage and oocyst production); when they are eaten by other animals, more tissue cysts are produced (asexually).

Pathology & Pathogenesis

The organism in humans produces either congenital or postnatal toxoplasmosis. Congenital infection, which develops only when nonimmune mothers are infected during pregnancy, is usually of great severity; postnatal toxoplasmosis is usually much less severe. Most human infections are asymptomatic. However, fulminating fatal infections may develop in patients with AIDS, presumably by alteration of a chronic infection to an acute one. Varying degrees of disease may occur in immunosuppressed individuals, resulting in retinitis or chorioretinitis, encephalitis, pneumonitis, or various other conditions.

The tachyzoite directly destroys cells and has a predilection for parenchymal cells and those of the reticuloendothelial system. Humans are relatively resistant, but a low-grade lymph node infection resembling infectious mononucleosis may occur. When a tissue cyst ruptures, releasing numerous bradyzoites, a local hypersensitivity reaction may cause inflammation, blockage of blood vessels, and cell death near the damaged cyst.

Congenital infection leads to stillbirths, chorioretinitis, intracerebral calcifications, psychomotor disturbances, and hydrocephaly or microcephaly. In these cases, the mother was infected for the first time during pregnancy. Prenatal toxoplasmosis is a major cause of blindness and other congenital defects. Infection during the first trimester generally results in stillbirth or major CNS anomalies. Second- and third-trimester infections induce less severe neurologic damage, though they are far more common. Clinical manifestations of these infections may be delayed until long after birth, even beyond childhood. Neurologic problems or learning difficulties may be caused by the long-delayed effects of late prenatal toxoplasmosis.

Epidemiology

Avoidance of human contact with cat feces is clearly important in control, particularly for pregnant women with negative serologic tests. Since oocysts usually take 48 hours to become infective, daily changing of cat litter (and its safe disposal) can prevent transmission. However, pregnant women should avoid all contact with cats, particularly kittens. An equally important source of human exposure is raw or undercooked meat, in which infective tissue cysts are frequently found. Humans (and other mammals) can become infected either from oocysts in cat feces or from tissue cysts in raw or undercooked meat. Freezing meat at –20°C for 48 hours or heating to 50°C for 4–6 minutes will provide sterilization. Kitchen cleanliness, hand washing after touching raw meat, and avoidance of cats and cat litter are essential during pregnancy. Periodic serologic screening for IgG and IgM antibodies to *Toxoplasma* is recommended. See Cook and Zumk (2003) and Guerrant et al (2006) for additional information.

MICROSPORIDIA

Microsporidia are a unique assemblage of intracellular parasites characterized by a unicellular spore containing a coiled spring-like tubular polar filament through which the sporoplasm is forcibly discharged into a host cell. Identification of species and genera is based upon electron microscopic morphology of the spore, nuclei, and coiled polar filament. A modified trichrome-blue stain may detect microsporidia in urine, stool, and nasopharyngeal specimens. All classes of vertebrates (especially fish) and many invertebrate groups (especially insects) are infected in essentially all tissues.

Transmission is chiefly by ingestion of spores in food or water. Transplacental transmission is common. Few cases were known among humans until intestinal, ophthalmic, and systemic infections were observed among AIDS patients. Microsporida is now increasingly recognized as a group of opportunistic parasites, probably widespread, abundant, and nonpathogenic in immunologically intact persons but a continuing threat to the immunocompromised. They often occur along with *Cryptosporidium* in AIDS patients.

The following microsporidial infections have been found among immunosuppressed individuals (mostly AIDS patients) (Guerrant et al, 2006). Ocular infections: *Encephalitozoon hellum, Vittaforma corneae (Nosema corneum),* and *Nosema ocularum.* Intestinal infections: *Enterocytozoon bieneusi* and *Encephalitozoon intestinalis.* There is no treatment for infections with *Encephalitozoon hellum, Encephalitozoon cuniculi, Pleistophora* sp., *Bracheola vesicularam, B (Nosema) algerae, B (Nosema) connori,* or *Trachipleistophora hominis,* which occur chiefly in AIDS patients.

INTESTINAL HELMINTHIC INFECTIONS

Key concepts pertaining to parasitic helminths and the helminthes included in this chapter are listed in Tables 46–6 and 46–7. A synopsis of the helminthic infections is provided in Table 46–8.

It is estimated that 1.5 billion people are infected with *Ascaris lumbricoides,* the giant roundworm of humans; 1.3 billion people are infected with hookworm (*Ancylostoma duodenale* or *Necator americanus*); and 800 million are infected with whipworm (*Trichuris trichiura*) (CDC Web site, www.cdc.gov/ncidod/dpd, "Parasitic Diseases").

Most intestinal helminthic infections are fairly benign, except when worm burdens are high and numbers of adult worms in the intestine reach the hundreds. In intestinal worm infections, the intestine usually harbors the adult stage of the parasite, except for *Strongyloides, Trichinella,* and *Taenia solium,* which not only reside in the intestine as adults but also have larvae capable of migrating throughout tissues.

Most nematode infections are acquired via the fecal-oral route, with human behaviors and poor sanitation and hygiene contributing to the transmission. In the case of the three most common intestinal infections (whipworm, hookworm, and ascariasis), the eggs require incubation in the soil for several days or weeks in warm, tropical climates.

Dietary habits of eating raw or lightly cooked food dishes contribute to most trematode and cestode infections. These infections can be acquired by eating improperly cooked intermediate hosts, including vegetables, fish, beef, and pork. Thorough cooking and freezing kill the parasites, thus preventing foodborne infections. Human behaviors and close associations with pets also are contributing factors for infection by *Dipylidium caninum* and *Echinococcus granulosus.*

ENTEROBIUS VERMICULARIS (PINWORM—INTESTINAL NEMATODE)

The Organism

Female **pinworms** (about 10 mm in length) have a slender, pointed posterior end. Males are approximately 3 mm in length and have a curved posterior end (Figure 46–12A and B). Pinworms are found worldwide but more commonly in temperate than tropical climates. They are the most common helminthic infection in the United States and infect mostly children.

Pathology & Pathogenesis

The main symptom associated with pinworm infections is perianal pruritus, especially at night, caused by a hypersensitivity reaction to the eggs that are laid around the perianal region by female worms, which migrate down from the colon at night. Scratching the anal region promotes transmission, as the eggs are highly infectious within hours of being laid (hand-to-mouth transmission). Irritability and fatigue from loss of sleep occur, but the infection is relatively benign.

Eggs are recovered using the "Scotch Tape" technique in the morning before a bowel movement. Transparent Scotch Tape is applied directly to the perianal area, and then placed on a microscope slide for examination. Eggs are football shaped, have a thin outer shell, and are approximately 50–60 μm in length (Figure 46–12C). Infectious larvae are often visible inside the egg. The small adult worms may be seen in a stool O&P (ova and parasites) test. Because the eggs are lightweight and highly infectious, it is important for bed linens, towels, and clothing to be washed in hot water to prevent reinfection.

TRICHURIS TRICHIURA (WHIPWORM—INTESTINAL NEMATODE)

The Organism

Adult female **whipworms** are approximately 30–50 mm in length; adult male worms are smaller (Figure 46–13A and B). The anterior end of the worms is slender, and the posterior end is thicker, giving it a "buggy whip" appearance, hence the name whipworm. Adult whipworms inhabit the colon, where male and female worms mate. Females release eggs (Figure 46–13C)

TABLE 46-6   Key Concepts: Parasitic Helminths

Parasitic helminths covered in this chapter are grouped into the nematodes, trematodes, and cestodes.
Most infections are acquired by ingestion of the egg or larval stage, with the exception of the hookworms, human threadworms, and schistosomes, whose larvae penetrate the skin, and the filarids, which are vectorborne.
Generally speaking, most intestinal nematode and cestode infections involve the adult stages and are not very pathogenic, except when there are large numbers of worms. Most of the pathology is associated with the larval stages (eg, microfilariae and trichinae in the case of nematodes; and cysticerci and hydatid cysts in the case of cestodes).
In trematode infections, the pathology is generally associated with the adult stage, because the adult worms are found in human tissues, eg, liver fluke and lung fluke (larval stages occur in animal hosts or on other sources).
Eosinophilia is a cardinal feature of a tissue infection by parasitic worms.
The pathologic features of the tissue-infecting nematodes are closely tied to the host response. Elephantiasis, a morbid gross enlargement of limbs, breasts, and genitalia, is an immunopathologic response to long-continued filarial infection by *Wuchereria* or *Brugia*.
Most helminths do not multiply by asexual multiplication in the human host: one egg or one larva yields one worm. The exception is *Echinococcus granulosus*, which multiplies asexually within hydatid cysts.
The only intracellular helminth is *Trichinella,* whose larval stage is intracellular within a muscle cell (known as a nurse cell).
Most worms that inhabit the intestinal lumen are easily treatable whereas worms inhabiting tissues are more difficult to treat with drugs.
The severity of disease and symptoms caused by helminthic infections are generally associated with heavy worm burdens (eg, hookworm disease and anemia).
Larva migrans is a term used when the larval stage of a nematode that normally infects an animal host migrates throughout human tissues (eg, skin, viscera, and central nervous system). A strong host immune response is elicited to the migrating worm and induces the pathology. Larva migrans is associated with zoonotic infections in which animals are the normal hosts and humans are accidentally infected.
The combination of poor sanitation, human behaviors, and tropical climates leads to a high prevalence of infection by the "soilborne" nematodes (*Ascaris*, whipworm, and hookworm).

that are passed in the feces, and eggs become infective after about 3 weeks of incubation in moist and shady soil. Humans acquire the infection by eating foods contaminated with infective eggs. Once eggs are swallowed, the larvae hatch in the small intestine, where they mature and migrate to the colon.

Pathology & Pathogenesis

The anterior ends of the worms lodge within the mucosa of the intestine, leading to small hemorrhages with mucosal cell destruction and infiltration of eosinophils, lymphs, and plasma cells. Infections with a low worm burden are usually asymptomatic, but infections of moderate to heavy worm loads present with lower abdominal pain, distention, and diarrhea. Severe infection may lead to profuse bloody diarrhea, cramps, tenesmus, urgency, and rectal prolapse. Occasionally worms migrate to the appendix, causing appendicitis.

ASCARIS LUMBRICOIDES (HUMAN ROUNDWORM—INTESTINAL NEMATODE)

The Organism

Adult *Ascaris* are large: females are 20–50 cm long, and males are 15–30 cm long (Figure 46–14). Humans acquire the infection after eggs are ingested; larvae hatch in the duodenum, penetrate through the mucosa, migrate in the circulatory

system, lodge in lung capillaries, penetrate the alveoli, and migrate from the bronchioles to the trachea and pharynx; larvae are swallowed and return to the intestine and mature into adults. After mating, females can release 200,000 eggs per day, which are passed in the feces. Eggs are infective after about 1 month in the soil and are infectious for several months (Figure 46–14B).

Pathology & Pathogenesis

If present in high numbers, adult worms may cause mechanical obstruction of the bowel and bile and pancreatic ducts. Worms tend to migrate if drugs such as anesthetics or steroids are given, leading to bowel perforation and peritonitis, anal passage of worms, vomiting, and abdominal pain. Larvae migrating through lungs induce an inflammatory response (pneumonitis), especially after second infection, leading to bronchial spasm, mucus production, and Löeffler syndrome (cough, eosinophilia, and pulmonary infiltrates).

ANCYLOSTOMA DUODENALE & NECATOR AMERICANUS (HUMAN HOOKWORMS—INTESTINAL NEMATODE)

The Organism

Female **hookworms** are approximately 10 mm in length; males are slightly smaller and have a taxonomically characteristic

TABLE 46–7   Parasitic Helminths

Intestinal helminthic infections
Nematodes
Enterobius vermicularis (pinworm)
Trichuris trichiura (whipworm)
Ascaris lumbricoides (human roundworm)
Ancylostoma duodenale and *Necator americanus* (human hookworms)
Strongyloides stercoralis (human threadworm)
Trichinella spiralis
Trematodes
Fasciolopsis buski (giant intestinal fluke)
Cestodes
Taenia saginata (beef tapeworm)
Taenia solium (pork tapeworm)
Diphyllobothrium latum (broad fish tapeworm)
Hymenolepis nana (dwarf tapeworm)
Dipylidium caninum (dog tapeworm)
Blood and tissue helminthic infections
Nematodes
Wuchereria bancrofti (lymphatic filariasis)
Brugia malayi (lymphatic filariasis)
Onchocerca volvulus (river blindness)
Dracunculus medinensis (Guinea worm)
Ancylostoma duodenale and *Necator americanus* (ground itch—see intestinal helminthic infections)
Strongyloides stercoralis (larva currens—see intestinal helminthic infections)
Trichinella spiralis (trichinellosis from larvae—see intestinal helminthic infections)
Larva migrans (Zoonotic infections by larval nematodes)
Ancylostoma caninum (dog hookworm)
Anisakis simplex (whaleworm)
Toxocara canis (dog roundworm)
Baylisascaris procyonis (raccoon roundworm)
Trematodes
Fasciola hepatica (sheep liver fluke)
Clonorchis sinensis (Chinese liver fluke)
Paragonimus westermani (lung fluke)
Schistosoma mansoni, Schistosoma japonicum, Schistosoma haematobium (blood flukes)
Cestodes (infections caused by the larval stages)
Taenia solium (cysticercosis/neurocysticosis—see intestinal helminthic infections)
Echinococcus granulosus (hydatid cyst)

copulatory bursa (broadened posterior end), which is used to mate with females. Females can release more than 10,000 eggs per day into the feces, where a larva hatches from the egg within a day or two (Figure 46–15A). Larvae can survive in moist soil for several weeks, waiting for an unsuspecting barefooted host to walk by. These larvae penetrate host skin and migrate throughout the host similarly to *Ascaris* and end up in the small intestine where they mature into adult worms.

Pathology & Pathogenesis

In the intestine, adult worms attach to intestinal villi with their buccal teeth (Figure 46–15B) and feed on blood and tissue with the aid of anticoagulants (Harrison et al, 2002). A few hundred worms in the intestine can cause hookworm disease, which is characterized by severe anemia and iron deficiency. Intestinal symptoms also include abdominal discomfort and diarrhea. The initial skin infection by the larvae causes a condition known as "ground itch," characterized by erythema and intense pruritus. Feet and ankles are common sites of infection due to exposure from walking barefoot.

STRONGYLOIDES STERCORALIS (HUMAN THREADWORM—INTESTINAL & TISSUE NEMATODE)

The Organism

Adult females (about 2 mm long) of **Strongyloides stercoralis** that inhabit the intestine are parthenogenic; that is, they do not need to mate with male worms to reproduce. They lay eggs within the intestine; larvae hatch from the eggs and are passed into the feces. These larvae can either develop into parasitic forms or develop into free-living male and female worms that mate and produce several generations of worms in the soil, a great example of an evolutionary adaptation to sustain a population. The larvae of these free-living forms, under certain environmental conditions such as temperature, can develop into parasitic forms. Hence *Strongyloides stercoralis* has a unique evolutionary adaptation that can greatly enhance its reproductive success.

Pathology & Pathogenesis

Of medical significance, *Strongyloides* can produce an internal reinfection or autoreinfection if newly hatched larvae never exit the host but, instead, undergo their molts within the intestine. These larvae penetrate the intestine, migrate throughout the circulatory system, enter the lungs (Figure 46–16) and heart (similar to the migration of hookworms upon penetrating skin), and develop into parasitic females in the intestine. These nematodes are able to sustain an infection for many years and, in the event of immunosuppression, produce a hyperinfection in which a fulminating, fatal infection occurs. In disseminated infections, clinical signs and symptoms primarily involve the gastrointestinal tract (severe diarrhea, abdominal pain, gastrointestinal bleeding, nausea, vomiting), lungs (coughing, wheezing, hemoptysis), and skin (rash, pruritus, larva currens). Larvae migrating from the intestine carrying enteric bacteria can cause local infections or sepsis, resulting in death.

TABLE 46–8 Synopsis of Helminthic Infections by Organ System

Parasite/Disease	Site of Infection	Mechanism of Infection	Diagnosis	Treatment	Geographic Area
Intestinal nematodes					
Enterobius vermicularis Pinworm	Lumen of cecum, colon	Ingestion of eggs; self-contamination anal–oral behavior	Scotch Tape test; microscopy for eggs	Pyrantel pamoate, mebendazole	Worldwide, temperate areas
Trichuris trichiura Whipworm	Cecum, colon	Ingestion of eggs from fecally contaminated soil or food	Stool exam for O&P (eggs)	Mebendazole, albendazole	Worldwide, very common
Ascaris lumbricoides Ascariasis, common roundworm	Small intestine; larvae through lungs	Ingestion of eggs from fecally contaminated soil or food	Stool exam for O&P (eggs)	Albendazole, mebendazole	Worldwide, very common
Ancylostoma duodenale, Necator americanus Human hookworms	Small intestine; larvae through skin, lungs	Larvae in soil penetrate skin	Stool exam for O&P (eggs)	Albendazole, mebendazole	Worldwide, tropics
Strongyloides stercoralis Strongyloidiasis, human threadworm	Small intestine; larvae through skin, lungs	Larvae in soil penetrate skin and (rarely) internal autoreinfection	Stool exam, sputum, bronchial lavage for O&P (larvae)	Ivermectin, albendazole	Worldwide, tropics, and subtropics
Trichinella spiralis Trichinosis	Adults in small intestine for 1–4 months; larvae encysted in muscle tissue	Eating undercooked, infected pork or other animal	Serology and muscle bx (larvae)	Albendazole (plus steroids for severe symptoms)	Worldwide
Intestinal trematode					
Fasciolopsis buski Giant intestinal fluke	Small intestine	Eating metacercariae encysted on aquatic vegetation	Stool exam for O&P (eggs)	Praziquantel	East and Southeast Asia
Intestinal cestodes					
Taenia saginata Beef tapeworm	Small intestine	Eating cysticerci encysted in undercooked beef	Stool exam for O&P (tapeworm segments)	Praziquantel	Africa, Mexico, United States, Argentina, Europe, where beef is eaten
Taenia solium Pork tapeworm (see also Cysticercosis)	Small intestine	Eating cysticerci encysted in undercooked pork	Stool exam for O&P (tapeworm segments)	Praziquantel	Worldwide, where pork is eaten, especially Mexico, Central and South America, the Philippines, Southeast Asia
Diphyllobothrium latum Broad fish tapeworm	Small intestine	Eating larvae encysted in undercooked fish	Stool exam for O&P (eggs, tapeworm segments)	Praziquantel	Worldwide, where fish is often eaten raw
Hymenolepsis nana Dwarf tapeworm	Small intestine	Eating eggs from feces or contaminated water; autoreinfection via fecal/oral route	Stool exam for O&P (eggs, tapeworm segments)	Praziquantel	Worldwide

Organism/Disease	Location	Transmission	Diagnosis	Treatment	Distribution
Dipylidium caninum Dog tapeworm	Small intestine	Eating larvae in fleas	Stool exam for O&P (tapeworm segments)	Praziquantel	Worldwide
Nematode tissue infections					
Wucheria bancrofti, Brugia malayi Filariasis	Adult worms in lymph nodes, lymphatic ducts	Bite of mosquitoes transmits larvae	Blood smear for microfilariae	Diethylcarbamazine	Tropics and subtropics, sub-Saharan Africa, Southeast Asia, Western Pacific, India, South America, Caribbean
Onchocerca volvulus Onchocerciasis African river blindness	Adults in skin nodules	Bite of black fly transmits larvae	Skin snips for microfilariae; subcutaneous nodules	Ivermectin	Tropical Africa, Central America
Dracunculus medinensis Guinea worm	Adults subcutaneous in lower legs, ankles, feet	Drinking water contaminated with infected copepods	Worm in skin blister	Slow removal of worm around stick; surgical removal; wound treatment	Almost eradicated with exception of few sub-Saharan African countries
Ancylostoma duodenale, Necator americanus, Trichinella spiralis, Strongyloides stercoralis	See intestinal nematodes				
Nematode larva migrans					
Ancylostoma caninum and other domestic hookworms Creeping eruption, CLM	Subcutaneous migrating larvae	Contact with soil contaminated by dog or cat feces	Physical exam and history	Albendazole, ivermectin, tropical thiabendazole	Tropical and subtropical, very local but widespread
Anisakis simplex Anisakiasis (VLM)	Gastrointestinal—larvae in stomach or intestinal wall, rarely penetrate viscera	Eating larvae in pickled, raw, or undercooked fish dishes	Endoscopy, radiology, eosinophilia	Surgical or endoscopic removal of worm	Pacific Basin (Japan, California, Hawaii), Scandinavian countries, where people eat fish
Toxocara species Dog and cat roundworms (VLM, OLM, NLM)	Larvae migrating in viscera, liver, lung, eye, brain	Ingesting eggs in soil contaminated by dog or cat feces	Serology; eosinophilia	Albendazole, Mebendazole	Worldwide, areas where dogs, cats defecate
Baylisascaris procyonis Raccoon roundworm (VLM, OLM, NLM)	Viscera, central nervous system—larvae migrate to eyes, brain	Ingesting eggs from raccoon feces	Serology (KR Kazacos, Dept Vet Med, Purdue Univ), eosinophilia, neuroimaging	Albendazole, Mebendazole, Corticosteroids	North America, Europe, areas where raccoons defecate (raccoon latrines)

Continued

TABLE 46–8 Synopsis of Helminthic Infections by Organ System (Continued)

Parasite/Disease	Site of Infection	Mechanism of Infection	Diagnosis	Treatment	Geographic Area
Trematode tissue infections					
Fasciola hepatica Fascioliasis, Sheep liver fluke	Adult worms in liver (bile duct, after migration through parenchyma)	Eating metacercariae on watercress, aquatic vegetation	Stool exam for O&P (eggs)	Triclabendazole, Bithionol	Worldwide, especially sheep-raising areas
Clonorchis sinensis Clonorchiasis, Chinese liver fluke	Adult worms in liver (bile ducts)	Eating metacercariae in undercooked fresh water fish	Stool exam for O&P (eggs)	Praziquantel	China, Korea, Indochina, Japan, Taiwan
Paragonimus westermani Paragonimiasis, Lung fluke	Adult worms in lungs	Eating metacercariae in raw crabs and other freshwater crustaceans	Stool exam for O&P, sputum, bronchial lavage (eggs)	Praziquantel	Asia, Central, South and North America, Africa
Schistosoma mansoni Bilharzia, Blood fluke, Schistosome	Adults in venous vessels of large intestine, liver	Cercariae (larvae) penetrate skin in snail-infested water	Stool exam for O&P (eggs)	Praziquantel	Africa to Near East, tropics and subtropics, South America, Caribbean
Schistosoma japonicum	Adults in venous vessels of small intestine, liver	Cercariae (larvae) penetrate skin in snail-infested water	Stool exam for O&P (eggs)	Praziquantel	China, Philippines, Japan
Schistosoma haematobium	Adults in venous vessels of urinary bladder	Cercariae (larvae) penetrate skin in snail-infested water	Urine for O&P (eggs)	Praziquantel	Africa, widely; Madagascar; Middle East
Cestode tissue infections					
Taenia solium (larval) Cysticercosis, Neurocysticercosis (CNS involvement)	Cysticerci in skin, liver, lungs, kidneys, muscles, eye, brain	Ingestion of eggs via human fecal–oral route	CT scans, MRI, x-rays, serology	Surgical excision, albendazole, praziquantel	Worldwide, especially pig-raising areas
Echinococcus granulosus (larval) Hydatid disease, Unilocular hydatid cyst	Hydatid cyst in liver, spleen, lungs, peritoneum, brain	Contact with dogs, foxes, other canids; eggs from feces	CT scans, MRI, x-rays, serology	Albendazole, surgical removal	Worldwide, especially sheep-raising areas

CLM, cutaneous larva migrans; CNS, central nervous system; CT, computerized tomography; MRI, magnetic resonance imaging; NLM, neural larva migrans; O&P, ova and parasites; OLM, ocular larva migrans; VLM, visceral larva migrans.

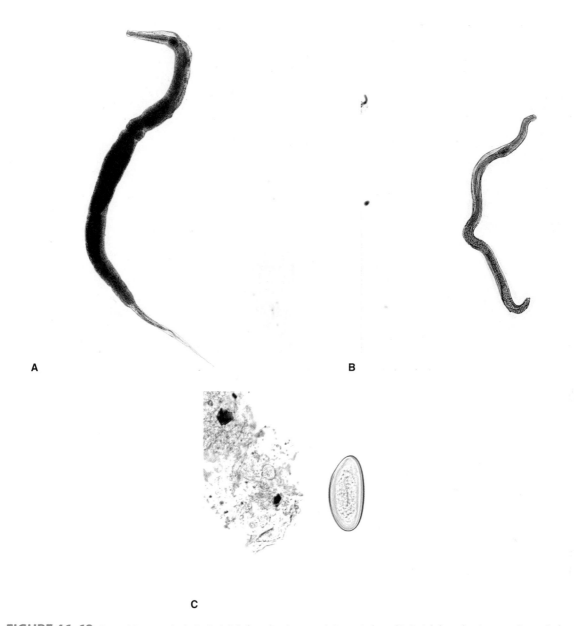

A

B

C

FIGURE 46–12 *Enterobius vermicularis*. **A**: Adult female pinworm (10 mm in length). **B**: Adult male pinworm (3 mm in length). **C**: A Scotch Tape test reveals a pinworm egg (50–60 μm in length) with an infectious larva inside. (Used with permission from J. Sullivan, *A Color Atlas of Parasitology*, 8th ed., 2009.)

TRICHINELLA SPIRALIS (INTESTINAL & TISSUE NEMATODE)

The Organism

Trichinella spiralis is acquired by eating raw or improperly cooked pork infected with the larval stage of these nematodes. In the small intestine, the larvae molt into adult worms, and, after mating with male worms, the female worms release live larvae. The larvae penetrate the intestine, circulate in the blood, and eventually encyst in muscle tissue. Adult female worms live for several weeks and after the first week of infection may cause diarrhea, abdominal pain, and nausea. Intestinal symptoms are mild to none and often go unnoticed.

Pathology & Pathogenesis

The main symptoms of trichinellosis are primarily caused by the larvae encysted in muscle tissue (Figure 46–17). The tissue migration phase lasts for about 1 month, with high fever, cough, and eosinophilia. As larvae encyst, edema occurs and inflammatory cells (PMNs and eosinophils) infiltrate the tissue. Calcification, which may or may not destroy the larvae, occurs within 5–6 months.

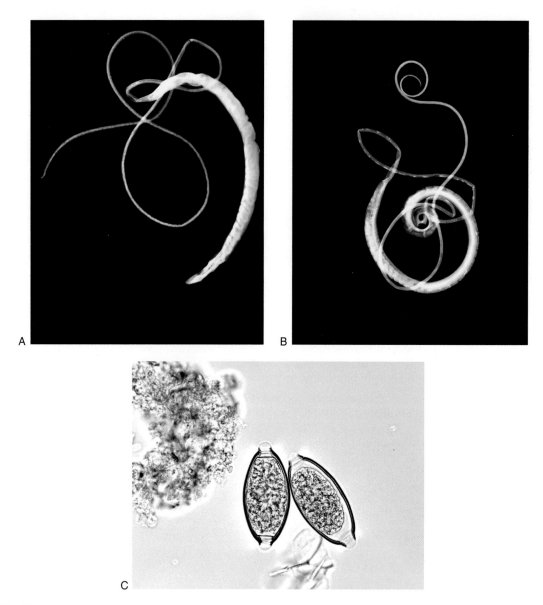

FIGURE 46–13 *Trichuris trichiura.* **A**: Adult female whipworm (30–50 mm in length). **B**: Adult male whipworm (30–45 mm). **C**: Whipworm eggs (50 µm) with distinct polar plugs. (Used with permission from J. Sullivan, *A Color Atlas of Parasitology,* 8th ed., 2009.)

Highly active muscle tissue such as the diaphragm, tongue, masseter, intercostals, and extraocular muscles are commonly infected. Individuals may suffer from myalgia and weakness, and eosinophilia may be increased for the first 6 months but then decreases.

Trichinellosis is a zoonotic disease; humans acquire the infection by eating raw or undercooked pork (eg, home-made sausages) but are a dead-end host for this infection. The lifecycle is maintained in wild animals such as boars and bears or in domestic animals, where pig-to-pig transmission occurs.

FASCIOLOPSIS BUSKI (GIANT INTESTINAL FLUKE—INTESTINAL TREMATODE)

Fasciolopsis buski, the giant intestinal fluke of humans (and pigs), is found in Asia and measures 20–75 mm in length. The larval metacercarial stage encysts on vegetation, such as water chestnuts or red caltrops. They are ingested with uncooked vegetation and then excyst and mature in the intestine. Most infections are light and asymptomatic, but heavy worm burdens cause ulceration, abscess of the intestinal wall, diarrhea, abdominal pain, and intestinal obstruction.

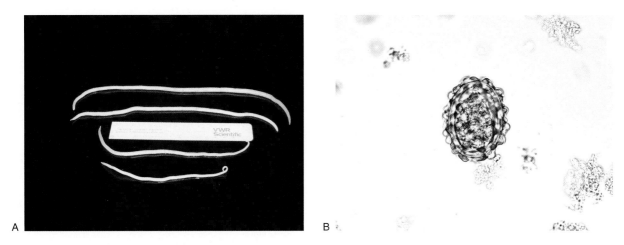

FIGURE 46-14 *Ascaris lumbricoides.* **A:** Adult females are larger than the adult male worms (length of ruler = 16 cm). **B:** An *Ascaris* egg (55–75 μm) with characteristic bumps (mammilated). (Used with permission from J. Sullivan, *A Color Atlas of Parasitology,* 8th ed., 2009.)

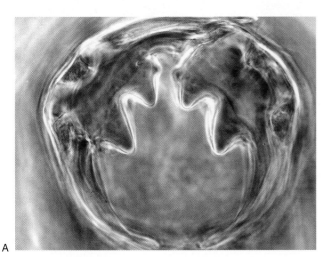

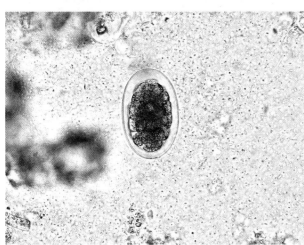

FIGURE 46-15 *Ancylostoma duodenale.* **A:** Adult hookworm with two pairs of teeth in the buccal capsule. **B:** The thin-shelled hookworm egg (60–75 μm) in early cleavage from an ova and parasite test. (Used with permission from J. Sullivan, *A Color Atlas of Parasitology,* 8th ed., 2009.)

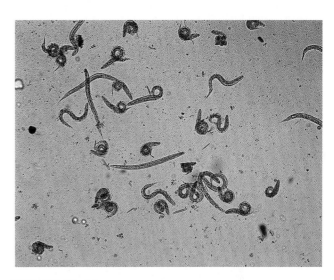

FIGURE 46-16 *Strongyloides stercoralis* larvae from a bronchiolar lavage. (Courtesy of Norman Setijono, UCSF.)

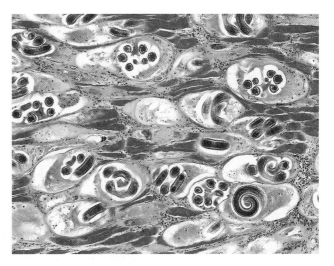

FIGURE 46-17 *Trichinella spiralis* larvae encyted in muscle tissue. (Used with permission from J. Sullivan, *A Color Atlas of Parasitology,* 8th ed., 2009.)

TAENIA SAGINATA (BEEF TAPEWORM— INTESTINAL CESTODE) & *TAENIA SOLIUM* (PORK TAPEWORM—INTESTINAL & TISSUE CESTODE)

The Organisms

If humans eat "measly beef" or "measly pork" containing the bladder-like larvae called cysticerci, they acquire infections of *T saginata* and *T solium*, respectively. The cysticerci, which are about the size of peas, develop into adult worms that can reach lengths of several meters in the intestine. Adult worms generally cause few problems, and most are asymptomatic; mild intestinal symptoms include diarrhea and abdominal pain.

In the intestine, egg-filled terminal segments break off from the adult worm and pass out with human feces. When the eggs from human feces are consumed by cows (*T saginata*) or pigs (*T solium*), larvae hatch from the eggs, migrate, and encyst as cysticerci in various tissues, including cow muscle (beef) or pig muscle (pork). Humans become infected when they eat raw or undercooked meats containing the cysticerci. These cysticerci then develop into adult worms in the human intestine.

Pathology & Pathogenesis

One medically significant difference between *T saginata* and *T solium* is that humans can be the intermediate host for *T solium*, similar to pigs. Thus, if humans ingest *T solium* eggs, the cysticerci encyst in various human tissues, including skin, muscle (Figure 46–18A), kidney, heart, liver, and brain (Figure 46–18B). This condition in humans is known as cysticercosis, and symptoms are associated with the involved tissues (eg, diminution of visual acuity with ophthalmocysticercosis; in neurocysticercosis, symptoms include headache, nausea, vomiting, mental disturbances, and seizures caused by encysted cysticerci in the brain). With the beef tapeworm *T saginata*, adult worms develop only in humans, and cysticerci of *T saginata* do not develop in humans (only in cattle or other herbivores).

DIPHYLLOBOTHRIUM LATUM (BROAD FISH TAPEWORM—INTESTINAL CESTODE)

The Organism

Diphyllobothrium latum, the broad fish tapeworm of humans (and many other fish-eating animals), reaches enormous size, sometimes exceeding 10 m in length. Humans acquire the infection when they eat improperly cooked or raw fish that is infected with the larvae known as plerocercoids, which look like white grains of rice in the fish flesh. In the intestine, the worm rapidly grows and develops a chain of segments capable of releasing more than 1 million eggs per day.

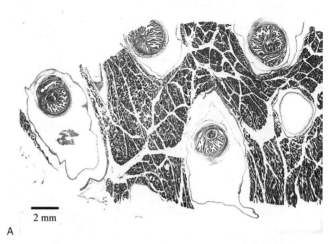

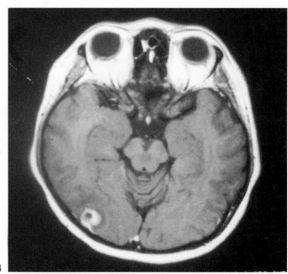

FIGURE 46–18 *Taenia solium* cysticercosis. **A**: Several cysticerci (larval forms) encysted in muscle. (Used with permission from J. Sullivan, *A Color Atlas of Parasitology,* 8th ed., 2009.) **B**: A single cysticercus seen in a magnetic resonance imaging scan of the brain. (Courtesy of Pathology, UCSF.)

Pathology & Pathogenesis

Disease caused by tapeworms is chiefly vague abdominal discomfort and loss of appetite, leading to weight loss. *D latum* has an unusual capacity to absorb vitamin B_{12}, and among some groups—especially Finns—a vitamin B_{12} deficiency leading to various levels of pernicious anemia may rarely develop.

HYMENOLEPIS NANA (DWARF TAPEWORM—INTESTINAL CESTODE)

The Organism

Hymenolepis nana, the dwarf tapeworm of humans (and rodents), is only about 4 cm in length. It is found worldwide and is one of the most common tapeworm infections in humans owing to the fact that the eggs can short-circuit the usual developmental phase in an insect and infect humans directly from eggs passed in feces of other humans (direct life cycle). Alternatively, if the insect that harbors the larval stage is inadvertently eaten, the larvae develop into adult worms in humans (indirect life cycle). Humans can be infected in both ways.

Pathology & Pathogenesis

Occasionally, massive infections, mostly in children, occur as a result of internal autoreinfection when the eggs hatch in the gut without leaving the intestine. Other than these instances of extremely heavy infection, disease caused by these worms is limited to minor intestinal disturbance.

DIPYLIDIUM CANINUM (DOG TAPEWORM—INTESTINAL CESTODE)

Dipylidium caninum is a cestode that commonly infects canids, felids, and pet owners, especially children. Adult worms inhabit the intestines and release characteristic double-pored segments containing egg clusters into the host's feces (Figure 46–19). Eggs are eaten by larval fleas, in which the parasite develops into its larval stage. Infected adult fleas

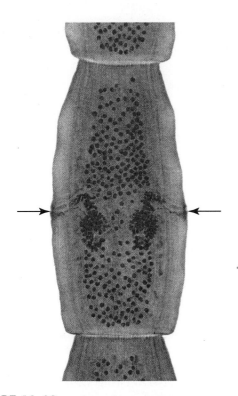

FIGURE 46–19 *Dipylidium caninum* tapeworm segment (23 mm long × 8 mm wide) is pumpkin seed-shaped and has characteristic genital pores (arrows) on both sides. (Used with permission from J. Sullivan, *A Color Atlas of Parasitology*, 8th ed., 2009.)

which still harbor the parasite are in turn eaten by dogs and cats when they lick the site where fleas are biting. Because of people's close association with their pets (and their fleas), humans can acquire the infection, but they are mostly asymptomatic. In children, infection can cause diarrhea and restlessness.

BLOOD & TISSUE HELMINTHIC INFECTIONS

WUCHERERIA BANCROFTI & BRUGIA MALAYI (LYMPHATIC FILARIASIS— TISSUE NEMATODES)

The Organisms

The filarid nematodes, **Wuchereria bancrofti**, **Brugia malayi**, and *Onchocerca volvulus*, are long, slender worms whose adult forms are found in tissues. Lymphatic filariasis is caused by the adult worms of *Wuchereria bancrofti* and *Brugia malayi* and affects more than 120 million people in 80 countries throughout the tropics and subtropics of Asia, Africa, the Western Pacific, and parts of the Caribbean and South America. Adult *W bancrofti* and *B malayi* (females 60–100 mm long; males 15–40 mm long) are found in the lymphatic vessels, where the female releases tiny larvae, called microfilariae, into the lymph. The microfilariae are swept into the peripheral blood and are found in the peripheral blood during specific times of the day, depending on the blood feeding habits of their insect vector (known as periodicity). With *W bancrofti* and *B malayi,* the infection is transmitted by mosquitoes; thus, prevention primarily involves protection against mosquito bites. Personal control measures include the use of insect repellent, mosquito netting, and protective clothing.

Pathology & Pathogenesis

Adult worms embedded in lymph tissues are the primary cause of inflammatory and fibrotic reactions. Signs and symptoms of acute infection include lymphangitis, with fever, painful lymph nodes, edema, and inflammation spreading from the affected lymph nodes. Elephantiasis is the name for the morbid gross enlargement of limbs, breasts, and genitalia that occurs in a chronic infection (Figure 46–20) and is an immunopathologic response to the mature or dying adult worms in the lymph tissues.

ONCHOCERCA VOLVULUS (RIVER BLINDNESS—TISSUE NEMATODE)

The WHO estimates that the global prevalence of onchocerciasis is more than 17 million, of whom 270,000 people are

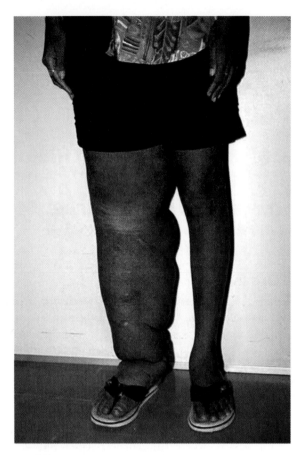

FIGURE 46–20 Woman with lymphatic filariasis. (Reproduced with permission from WHO/TDR image library/Crump.)

A

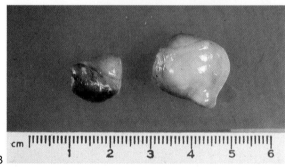

B

FIGURE 46–21 *Onchocera volvulus.* **A**: Palpation of a subcutaneous nodule. (Reproduced with permission from WHO/TDR image library/Crump.) **B**: Nodules that were surgically removed contain coiled adult worms. (Courtesy of Pathology, UCSF.)

blind and another 500,000 are visually impaired by the parasite. Most infected people live in Western and Central Africa, but the disease is also found in Yemen and six countries in the Americas.

The Organism

Onchocerca volvulus infections are transmitted when infected black flies of the genus *Simulium* feed on human skin. These flies do not pierce the blood vessels with thin, delicate mouthparts as do the mosquitoes. Instead, the infected black fly grinds the skin tissue and feeds upon the pool of blood and skin, where larval *Onchocerca* are released. The larvae develop into adult worms (females are 300–500 mm; males are 200–400 mm long) in the subcutaneous tissues, where they become encapsulated with host tissue to form a nodule (onchocercoma) that is 1–2 cm in diameter (Figure 46–21). Adult worms mate, and the female releases microfilariae, which migrate within skin. The black fly ingests the microfilariae during its bite, and the microfilaria becomes an infective larva in the black fly after about 1 week. Black flies require fast-flowing rivers and well-oxygenated waters to breed, hence the name for the disease: "river blindness."

Pathology & Pathogenesis

With *Onchocerca*, it is the microfilariae released from the female worms that cause the most severe damage. Migrating microfilariae, exclusively found in the interstitial fluids of the skin and subdermal tissues (*not* the bloodstream), cause changes in skin pigment and loss of elastic fibers, leading to "hanging groin," other skin changes, and severe pruritus, sometimes intractable and intolerable. Far more serious is the blindness that affects millions, mainly in Africa (primarily men). Visual loss develops over many years from an accumulation of microfilariae in the vitreous humor, since the microfilariae are not bloodborne and can concentrate and remain in the fluids of the eye. Visual clouding, photophobia, and ultimately retinal damage result in incurable blindness.

DRACUNCULUS MEDINENSIS (GUINEA WORM—TISSUE NEMATODE)

The Organism

The distantly related guinea worm, *Dracunculus medinensis*, has an aquatic cycle via copepods ("water fleas"—an abundant group of aquatic microcrustaceans). Copepods ingest larvae released from human skin blisters that burst when immersed in cold water, spewing great numbers of larvae. Infected copepods are inadvertently ingested by drinking unfiltered, infested water. After a year of systemic wandering in the body, the worms mature and mate. The females then travel to the skin—usually to the lower leg—where they induce blisters that form near the foot and ankle. What better way to soothe the pain and irritation from the blisters than to soak the afflicted leg in cool water? The cool water stimulates the female guinea worm to release her larvae, and the life cycle continues.

Pathology & Pathogenesis

D medinensis induces a broad range of pathologic changes depending on the site of adult infection and host response to the parasite's presence or to the worm's removal. Most disease caused by guinea worms is a result of secondary bacterial infections. These infections may be due to sepsis at the point of emergence of the anterior end of the worm from the cutaneous blisters. Killed adult worms (or pieces of them) in the skin may also initiate severe infection, leading to gangrene or anaphylaxis. These worms are important causes of debility and economic loss in Africa, where control efforts directed toward eradication are under way and complete eradication is a distinct possibility within a few years.

LARVA MIGRANS (ZOONOTIC LARVAL NEMATODE INFECTIONS)

The Organisms

Larva migrans occurs when humans are infected with nematodes that normally parasitize animal hosts. Humans are dead-end hosts; the larvae degenerate, inducing an immune response to the dead or dying worms, and they do not become reproductively mature in humans. Eosinophilia is a common feature, and stool exams for ova and parasites are not helpful in the diagnosis. There are several forms of larva migrans.

Pathology & Pathogenesis

Cutaneous larva migrans (CLM): also called creeping eruption, CLM is acquired when bare skin (often the hands and feet) contacts the soilborne larvae of **Ancylostoma caninum**, the **dog hookworm**; larvae migrate in the epithelial layers of the skin and leave red, itchy tracts on the skin. Signs of CLM are erythema and papules at the site of entry and serpiginous tracts of red inflammation.

Visceral larva migrans (VLM): marine mammals (eg, seals, dolphins, and whales) are the normal hosts of *Anisakis* (**whaleworm**). These larvae (about 15 mm in length) are found in intermediate hosts such as cod, herring, salmon, and rockfish, which, if accidentally eaten in raw or undercooked fish dishes, can invade the gastric mucosa or intestinal tissue and cause extreme abdominal pain that mimics appendicitis or small bowel obstruction. Eosinophilic granulomas form around the larva in stomach or intestinal tissues, and larvae can migrate to tissues outside the gastrointestinal tract.

Ocular larva migrans (OLM) and neural larva migrans (NLM): ingestion of eggs from the **dog roundworm** (*Toxocara canis*) and **raccoon roundworm** (*Baylisascaris procyonis*) can lead to VLM, OLM, and NLM. The larvae hatch out of the eggs in the intestine and migrate throughout the circulation. Larvae lodge in various tissues, which results in the formation of granulomas around the larvae. Symptoms of VLM include fever, hepatomegaly, and eosinophilia; OLM can lead to impaired vision and blindness in the affected eye. A single larva in the brain (NLM) can lead to serious motor dysfunction and blindness, and infections by the raccoon roundworm can be fatal (Gavin et al, 2006).

CLONORCHIS SINENSIS (CHINESE LIVER FLUKE), FASCIOLA HEPATICA (SHEEP LIVER FLUKE), & PARAGONIMUS WESTERMANI (LUNG FLUKE) —TISSUE TREMATODES

The Organisms

It is estimated that more than 980 million people from Southeast Asia and the Western Pacific region are at risk of acquiring a foodborne infection by *Clonorchis*, *Fasciola*, and *Paragonimus* (Keiser and Utzinger, 2005).

When humans eat uncooked or improperly cooked food items from endemic areas, they can acquire *Clonorchis* by eating the metacercariae encysted in freshwater fishes (eg, carp), *Fasciola* by eating metacercariae encysted on aquatic vegetation (eg, watercress), and *Paragonimus* by eating crustacean hosts such as a crayfish or freshwater crab (often as crushed crab in salad dressing).

Pathology & Pathogenesis

Metacercariae of the **Clonorchis sinensis** (Chinese liver fluke) excyst in the intestine and migrate up to the common bile ducts, where worm burdens of 500–1000 or more adult worms can be found. Chinese liver flukes cause mechanical irritation of the bile ducts that results in fibrosis and hyperplasia. In heavy infections, worms cause fever, chills, epigastric pain, and eosinophilia; chronic cholangitis may progress to atrophy of liver parenchyma, portal fibrosis, jaundice due to biliary obstruction, and cirrhosis of the liver.

Fasciola hepatica (sheep liver fluke), commonly found in the livers of sheep, cattle, and other herbivores, penetrates the intestinal wall, enters the coelom, invades the liver tissue, and resides in the bile ducts. Acute infection causes abdominal pain, intermittent fever, eosinophilia, malaise, and weight loss due to liver damage. Chronic infection may be asymptomatic or lead to intermittent biliary tract obstruction.

The metacercariae of the human lung fluke **Paragonimus westermani** excyst in the human gut, and young worms migrate to the lungs, where they become encapsulated in lung tissue (Figure 46–22). Eggs, released by the adult worms, move up the trachea to the pharynx, are expectorated or swallowed, and are then passed in the feces. Eggs in the lung induce an inflammatory response, forming granulomas around the eggs. Adult lung flukes appear as grayish-white nodules approximately 1 cm in size within the lung, but worms can be found in ectopic sites (brain, liver, and intestinal wall). Because pulmonary symptoms of pulmonary tuberculosis are similar to those of paragonimiasis (coughing and hemoptysis), it is

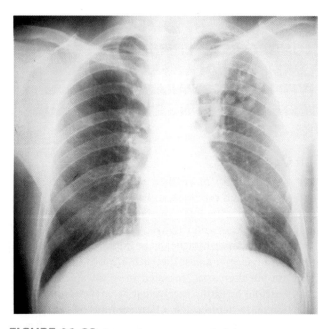

FIGURE 46–22 *Paragonimus westermani* adult worms are seen in the upper left quadrant of the lungs in a chest x-ray. (Courtesy of Radiology, UCSF.)

important to consider infection by the lung fluke in the differential diagnosis.

SCHISTOSOMA MANSONI, S JAPONICUM, & S HAEMATOBIUM (BLOOD FLUKES)

The Organisms

It is estimated that more than 200 million people are infected worldwide with *Schistosoma* species. The adult worms are long and slender (males are 6–12 mm in length; females are 7–17 mm in length) and can live in copula for 10–20 years within the venous system (Figure 46–23A): *S mansoni*: inferior mesenteric veins of large intestine; *S japonicum*: inferior and superior mesenteric veins of small intestine; *S haematobium*: veins of urinary bladder.

Humans acquire the infection when they contact water infested with the infectious cercariae. Cercariae are attracted to the warmth of a body and skin lipids and begin to burrow into exposed skin. Within 30 minutes, the cercariae have penetrated the epidermis and transformed into schistosomules, which enter the peripheral circulation, where they eventually become adults in the hepatoportal system or venus plexus surrounding the bladder. The female schistosomes begin releasing eggs approximately 5–8 weeks after infection.

Pathology & Pathogenesis

The most significant pathology is associated with the schistosome eggs, not the adult worms. Female schistosomes can lay hundreds or thousands of eggs per day within the venous system. When eggs are released, many are swept back into the circulation and lodge in the liver (*S mansoni* and *S japonicum*) or urinary bladder (*S haematobium*), while other eggs are able to reach the lumen of the intestine and pass out with the feces (*S mansoni* and *S japonicum*) or urine (*S haematobium*). A granulomatous reaction surrounds the eggs and leads to fibrosis of the liver with *S mansoni* and *S japonicum*. In chronic cases, blood flow to the liver is impeded, which leads to portal hypertension, accumulation of ascites in the abdominal cavity, hepatosplenomegaly, and esophageal varices.

With *S haematobium* infections, there is urinary tract involvement: urethral pain, increased urinary frequency, dysuria, hematuria, and bladder obstruction leading to secondary bacteria infections.

In travelers to endemic countries, clinical findings of acute schistosomiasis include an itchy rash (swimmer's itch) that occurs within an hour after cercariae penetrate the skin, followed by headache, chills, fever, diarrhea, and eosinophilia (known as snail fever or Katayama fever) 2–12 weeks after exposure (Salvana and King, 2008).

Diagnosis is by O&P: *S mansoni* (lateral spine) and *S japonicum* (barely visible nubby spine) eggs in stool; *S haematobium* (terminal spine) eggs in urine (Figure 46–23).

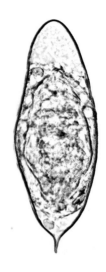

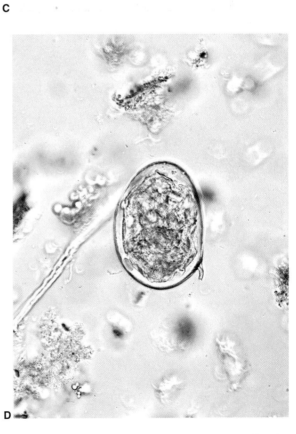

FIGURE 46–23 **A**: *Schistosoma mansoni* adult worms in copula. The female worm (arrows) is embraced within the male's gynecophoral canal. (Courtesy of Conor Caffrey, Sandler Center, UCSF.) **B**: *Schistosoma mansoni* egg with lateral spine (110–175 μm long × 45–70 μm wide). **C**: *Schistosoma haematobium* egg with terminal spine (110–170 μm long × 40–70 μm wide). **D**: *Schistosoma japonicum* egg with nubby spine (70–100 μm × 55–65 μm wide). (Used with permission from J. Sullivan, *A Color Atlas of Parasitology,* 8th ed., 2009.)

TISSUE CESTODE INFECTIONS (CAUSED BY THE LARVAL STAGES)

TAENIA SOLIUM—CYSTICERCOSIS/ NEUROCYSTICERCOSIS

See T Solium in the Intestinal Helminthic Infections section.

ECHINOCOCCUS GRANULOSUS (HYDATID CYST)

The Organism

Echinococcus granulosus is a small, three-segmented tapeworm found only in the intestine of dogs and other canids. The eggs leave these hosts and infect grazing animals. Similar to the beef and pork tapeworms, a larva hatches from the egg,

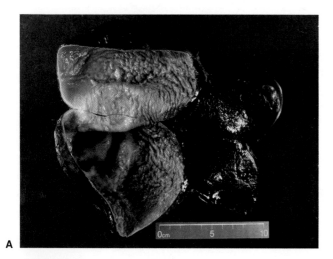

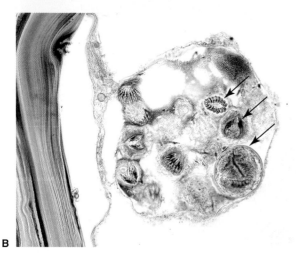

FIGURE 46–24 *Echinococcus granulosus.* **A**: A 14 cm hydatid cyst from a splenectomized patient. (Courtesy of Pathology, UCSF.) **B**: Histological section of a hydatid cyst showing several protoscolices (arrows) within a brood capsule. (Used with permission from J. Sullivan, *A Color Atlas of Parasitology,* 8th ed., 2009.)

penetrates the gut, and migrates to various tissues, especially liver, spleen (Figure 46–24A), muscle, and brain. Instead of a cysticercus developing, as in the case of the beef and pork tapeworms, the larva of *Echinococcus* develops into a fluid-filled cyst called a hydatid cyst. The cyst contains germinal epithelium in which thousands of future larvae (called protoscolices) develop (Figure 46–24B). Inside the hydatid cyst, the protoscolices are contained within brood capsules. If the hydatid cyst ruptures, the brood capsules can spill out of the cyst, metastasize to other sites, and develop into a hydatid cyst. Thus, ingestion of a single egg can give rise to several hydatid cysts, each containing several brood capsules.

Humans are infected only by ingesting *Echinococcus* eggs from dog feces. The dog, in turn, can acquire the infection only from an infected herbivore. Humans are only the intermediate and never the final host of this tapeworm.

Pathology & Pathogenesis

Hydatid cysts can grow about 1–7 cm per year, and the symptoms depend on the location of the cysts in the body. The liver is the most common site, where compression, atrophy, portal hypertension from mechanical obstruction, and cirrhosis can occur. Extreme care must be taken when removing the cyst. If the cyst ruptures, the highly immunogenic hydatid fluid can lead to anaphylactic shock and brood capsules can metastasize to form additional hydatid cysts.

REVIEW QUESTIONS

1. An outbreak of mild intestinal distress, sleeplessness, perianal itching, and anxiety has broken out among preschool children in a private home. The most likely cause of this condition is

    (A) *Trichomonas vaginalis*
    (B) *Enterobius vermicularis*
    (C) *Ascaris lumbricoides*
    (D) *Necator americanus*
    (E) *Entamoeba histolytica*

2. Chagas disease is especially feared in Latin America because of its threat to the heart and parasympathetic nervous system and the lack of an effective drug for the symptomatic later stages. Your patient is planning to reside in a Venezuelan village for 1–2 years. Which one of the following suggestions would be of special value for avoiding Chagas disease?

    (A) Boil or treat all of your drinking water.
    (B) Sleep under a bed net.
    (C) Do not keep domestic pets in your house.
    (D) Never walk barefoot in the village compound.
    (E) Do not eat lettuce or other raw vegetables or unpeeled fruit.

3. A 32-year-old male Peace Corps volunteer recently returned from a 2-year period working in a war zone in the southern Sudan in Central Africa. He presents with marked splenic enlargement, nonspecific hypergammaglobulinemia, and a negative leishmanin skin test (Montegro reaction). Mild fevers occur irregularly. The most likely parasitic disease is

    (A) Malaria
    (B) Cutaneous leishmaniasis

(C) Visceral leishmaniasis

(D) Trypanosomiasis

(E) Filariasis

4. A sexually active 24-year-old woman complains of vaginal itching and a malodorous purulent vaginal discharge. To verify your tentative diagnosis of trichomoniasis, you should include one of the following in your workup.

(A) Specific serologic test

(B) Ova and parasite fecal smear

(C) Wet mount of vaginal fluid

(D) ELISA test of serum

(E) Stool culture

5. You are working in a rural medical clinic in China and a 3-year-old girl is brought in by her mother. The child appears emaciated and, upon testing, is found to have a hemoglobin level of 5 g/dL. Her feet and ankles are swollen, and there is an extensive rash on her feet, ankles, and knees. The most likely parasitic infection that causes the child's condition is

(A) Schistosomiasis

(B) Cercarial dermatitis

(C) Cyclosporiasis

(D) Hookworm infection

(E) Trichuriasis

(F) Ascariasis

6. Pathologic effects of filariae in humans are caused by the adult worms in all but one species. In this case, the principal damage is caused by the microfilariae of

(A) *Brugia malayi*

(B) *Mansonella ozzardi*

(C) *Dracunculus medinensis*

(D) *Wuchereria bancrofti*

(E) *Onchocerca volvulus*

7. An 18-year-old male complains of abdominal pain, bloating, frequent loose stools, and loss of energy. He returned a month ago from a 3-week hiking and camping trek to the Mount Everest Base Camp in Nepal. The trek involved only high-elevation hiking, since he flew in and out of the 12,000-ft. starting point. Which of the following is an important consideration for the diagnosis?

(A) Exposure to high-level UV radiation

(B) The source and purification of water

(C) The use of insect repellents while hiking

(D) The presence of domestic animals en route

(E) The degree of contact with villagers en route

8. Which one of the following diagnostic tests should be conducted for the patient in Question 7?

(A) Blood and urine bacteriologic examination

(B) Series of ova and parasite tests and fecal smears

(C) ELISA or hemagglutination serologic tests for malaria

(D) Skin snip microfilarial test

(E) Endoscopic exam for whipworms

9. The parasite most likely to be responsible for the illness of the patient in Question 7 is

(A) *Entamoeba coli*

(B) *Plasmodium vivax*

(C) *Trichomonas vaginalis*

(D) *Naegleria gruberi*

(E) *Giardia lamblia*

10. Several Papua New Guinea villagers known to eat pork during celebrations were reported to be suffering from an outbreak of epileptiform seizures. One of the first things you should investigate is

(A) The prevalence of *Ascaris* infections in the population

(B) The presence of *Toxoplasma gondii* in cats

(C) The presence of *Trypanosoma brucei gambiense* in the villagers

(D) The presence of *Taenia* eggs in the drinking water

(E) The presence of adult *Taenia solium* in the pigs

11. A 32-year-old male tourist traveled to Senegal for 1 month. During the trip, he swam in the Gambia river. Two months after his return, he began complaining of intermittent lower abdominal pain with dysuria. Laboratory results of ova and parasites revealed eggs with a terminal spine. Which of the following parasites is the cause of the patient's symptoms?

(A) *Toxoplasma gondii*

(B) *Schistosoma mansoni*

(C) *Schistosoma haematobium*

(D) *Ascaris lumbricoides*

(E) *Taenia solium*

12. What type of specimen was collected for laboratory analysis based on the answer in the previous question?

(A) Thick blood smear

(B) Stool sample

(C) Urine sample

(D) Blood for serology

(E) Sputum sample

13. A previously healthy 23-year-old woman recently returned from her vacation after visiting friends in Arizona. She complained of severe headaches, saw "flashing lights," and had a purulent nasal discharge. She was admitted into the hospital with a diagnosis of bacterial meningitis and died 5 days later. Which of the following parasites should have been considered in the diagnosis? She had no prior history of travel outside of the United States.

(A) *Plasmodium falciparum*

(B) *Toxoplasma gondii*

(C) *Strongyloides stercoralis*

(D) *Entamoeba histolytica*

(E) *Naegleria fowleri*

14. How could the person have acquired the parasite in Question 13?

(A) Ingesting cysts from fecally contaminated drinking water

(B) Eating improperly cooked fish

(C) Eating improperly cooked beef

(D) Walking barefoot in the park

(E) Engaging in unprotected sexual intercourse

(F) Getting bitten by a sandfly

(G) Plunging into a natural hot spring

15. A 37-year-old sheep farmer from Australia presents with upper right quadrant pain and appears slightly jaundiced. A stool exam was negative for ova and parasites but a CT scan of the liver reveals a large 14 cm cyst that appears to contain fluid. Which of the following parasites should be considered?

(A) *Toxoplasma gondii*

(B) *Taenia solium*

(C) *Taenia saginata*

(D) *Clonorchis sinensis*

(E) *Schistosoma mansoni*

(F) *Echinococcus granulosus*

(G) *Paragonimus westermani*

16. An apparently run-down but alert 38-year-old woman has spent 6 months as a teacher in a rural Thailand village school. Her chief complaints include frequent headaches, occasional nausea and vomiting, and periodic fever. You suspect malaria and indeed find red blood cell parasites in a thin smear of finger-stick blood. To rule out the dangerous falciparum form of malaria, which one of the following choices would fit a diagnosis of *Plasmodium falciparum* malaria based on a microscopic examination of the blood smear?

    (A) Numerous large ovoid parasites in some of the red cells
    (B) Enlarged, somewhat misshapen parasitized red cells
    (C) Dividing parasites (schizonts) in red cells with 8–12 progeny
    (D) Dividing parasites in red cells with 16–24 progeny
    (E) Double ring forms found in the smear

17. Given a diagnosis of *Plasmodium falciparum* malaria for the patient in Question 16, which one of the following treatment regimens is appropriate?

    (A) Oral proguanil plus atovaquone
    (B) Oral chloroquine
    (C) Intravenous chloroquine
    (D) Oral proguanil
    (E) Intravenous quinidine

18. Given a diagnosis of *Plasmodium falciparum,* you should tell the patient in Question 16 that (select one)

    (A) There is little chance of a relapse in 1–3 years.
    (B) There is a strong likelihood that resistance will necessitate additional courses of treatment.
    (C) Returning to the tropics would be dangerous because hypersensitivity to the parasite may have developed.
    (D) She must avoid mosquito bites in this country, as they might induce a relapse of malaria.
    (E) A course of primaquine to clear the liver of hypnozoites would prevent recurrence of falciparum malaria.

19. The seriousness of *Plasmodium falciparum* infection compared with the other three forms of malaria is due to which one of the following?

    (A) Destruction of white blood cells compromises the immune reaction against malaria.
    (B) Stem cells in the marrow are largely destroyed.
    (C) Extensive damage to the liver can occur during the preerythrocytic phase of the parasite cycle.
    (D) Bloodstream parasites reinvade the liver and induce a more severe disease state.
    (E) Misshapen infected red cells adhere to the interior lining of blood vessels and block blood flow through these vessels.

20. A 52-year-old male, returning from a travel tour in India and Southeast Asia, was diagnosed with intestinal amebiasis and successfully treated with iodoquinol. A month later he returned to the clinic complaining of the following conditions, one of which is the most likely result of systemic amebiasis (even though the intestinal infection appears to be cured).

    (A) High periodic fever
    (B) Bloody urine
    (C) Tender, enlarged liver
    (D) Draining skin lesion
    (E) Enlarged painful spleen

Answers

1. B	6. E	11. C	16. E
2. B	7. B	12. C	17. A
3. C	8. B	13. E	18. A
4. C	9. E	14. G	19. E
5. D	10. D	15. F	20. C

REFERENCES

Abdalla SH, Pasvol G, Hoffman SL (editors): *Malaria: A Hematological Perspective.* Imperial College Press, 2004.

Ash LR, Orihel, TC: *Atlas of Human Parasitology,* 5th ed. American Society of Clinical Pathology Press, 2007.

Baird JK, Hoffman SL: Primaquine therapy for malaria. Clin Infect Dis 2004;39:1336.

Centers for Disease Control and Prevention: Parasitic diseases (www.cdc.gov/ncidod/dpd/).

Cook GC, Zumk A (editors): *Manson's Tropical Diseases,* 21st ed. Saunders, 2003.

Despommier DD, Gwadz RW, Hotez PJ, Knirsch CA (editors): *Parasitic Diseases,* 5th ed. Apple Trees Productions, 2005.

Drugs for parasitic infections. Treatment Guidelines from *The Medical Letter,* vol. 5, 2007.

Garcia LS: *Diagnostic Medical Parasitology,* 5th ed. American Society for Microbiology, 2007.

Gavin PJ, Kazacos KR, Shulman ST: Baylisascariasis. Clin Microbiol Rev 2005;18:703.

Goldsmith RS, Heyneman D (editors): *Tropical Medicine and Parasitology.* Appleton & Lange, 1989.

Goldsmith RS: Infectious diseases: Protozoal and helminthic. In: *Current Medical Diagnosis & Treatment 2006.* Tierney LM Jr, McPhee SJ, Papadakis MA (editors). McGraw-Hill, 2006.

Gomes ML, Galvao LMC, Macedo AM et al: Chagas' disease diagnosis: Comparative analysis of parasitologic, molecular, and serologic methods. Am J Trop Med Hyg 1999;60:205.

Guerrant RL, Walker DH, Weller PF (editors): *Tropical Infectious Diseases Principles, Pathogens, & Practice,* 2nd ed. Churchill Livingstone. Elsevier, 2 vols, 2006.

Haque R, Huston CD, Hughes M et al: Current concepts: Amebiasis. N Engl J Med 2003;348:1565.

Harrison LM, Nerlinger A, Bungiro RD et al: Molecular characterization of *Ancylostoma* inhibitors of coagulation factor Xa. J Biol Chem 2002;277:6223.

Herwaldt BL: *Cyclospora cayetanensis:* A review, focusing on the outbreaks of cyclosporiasis in the 1990s. Clin Micro Rev 2000;31:1040.

Keiser J, Utzinger J: Emerging foodborne trematodiasis. Emerg Infect Dis 2005;11:1507.

Long CA, Hoffman SL: Parasitology. Malaria—from infants to genomics to vaccines. Science 2002;297:345.

Mahmoud AAF (editor): *Schistosomiasis.* Imperial College Press, 2001.

McKerrow J, Parslow TG: Parasitic diseases. In: *Medical Immunology,* 10th ed. Parslow TG et al (editors). McGraw-Hill, 2001.

The Merck Manuals Online Medical Library (www.merck.com/mmpe/index.html).

Orihel TC, Ash LR: *Parasites in Human Tissues.* American Society of Clinical Pathology Press, 1995.

Peters W, Pasvol G: *Tropical Medicine & Parasitology,* 5th ed. Mosby, 2002.

Roberts LS, Janovy J Jr: *Foundations of Parasitology,* 8th ed. McGraw-Hill Higher Education, 2009.

Rosenthal PJ: Chap 52 Antiprotozoal drugs. In: *Basic and Clinical Pharmacology,* 11th ed. Katzung BG, Trevor A, Masters S (editors). McGraw-Hill, 2009.

Rowley HA, Uht RM, Kazacos KR et al: Radiologic-pathologic findings in raccoon roundworm (*Baylisascaris procyonis*) encephalitis. Am J Neuroradiol 2000;21:415.

Salvana EMT, King CH: Schistosomiasis in travelers and immigrants. Curr Infect Dis Reports 2008;10:42.

Stanley SL: Amoebiasis. Lancet 2003;361:1025.

Strickland GT: *Hunter's Tropical Medicine and Emerging Infectious Diseases.* Saunders, 2000.

Warrell DA, Gilles HM (eds): *Essential Microbiology,* 4th ed. Arnold Press, 2002.

Wilson WR, Sande MA (editors): *Current Diagnosis & Treatment in Infectious Diseases.* McGraw-Hill, 2001.

World Health Organization: World Malaria Report 2005 (http://rbm.who.int/wmr2005/).

World Health Organization: Leishmaniasis: Burden of disease 2009 (www.who.int/leishmaniasis/burden/en/).

Xiao L, Fayer R, Ryan U et al: *Cryptosporidium* taxonomy: Recent advances and implications for public health. Clin Microbiol Rev 2004;17:72.

Principles of Diagnostic Medical Microbiology

Diagnostic medical microbiology is concerned with the etiologic diagnosis of infection. Laboratory procedures used in the diagnosis of infectious disease in humans include the following:

1. Morphologic identification of the agent in stains of specimens or sections of tissues (light and electron microscopy).
2. Culture isolation and identification of the agent.
3. Detection of antigen from the agent by immunologic assay (latex agglutination, enzyme immunoassay [EIA], etc) or by fluorescein-labeled (or peroxidase-labeled) antibody stains.
4. DNA-DNA or DNA-RNA hybridization to detect pathogen-specific genes in patients' specimens.
5. Detection and amplification of organism nucleic acid in patients' specimens.
6. Demonstration of meaningful antibody or cell-mediated immune responses to an infectious agent.

In the field of infectious diseases, laboratory test results depend largely on the quality of the specimen, the timing and the care with which it is collected, and the technical proficiency and experience of laboratory personnel. Although physicians should be competent to perform a few simple, crucial microbiologic tests—make and stain a smear, examine it microscopically, and streak a culture plate—technical details of the more involved procedures are usually left to the bacteriologist or virologist and the technicians on the staff. Physicians who deal with infectious processes must know when and how to take specimens, what laboratory examinations to request, and how to interpret the results.

This chapter discusses diagnostic microbiology for bacterial, fungal, chlamydial, and viral diseases. The diagnosis of parasitic infections is discussed in Chapter 46.

COMMUNICATION BETWEEN PHYSICIAN & LABORATORY

Diagnostic microbiology encompasses the characterization of thousands of agents that cause or are associated with infectious diseases. The techniques used to characterize infectious agents vary greatly depending upon the clinical syndrome and the type of agent being considered, be it virus, bacterium, fungus, or other parasite. Because no single test will permit isolation or characterization of all potential pathogens, clinical information is much more important for diagnostic microbiology than it is for clinical chemistry or hematology. The clinician must make a tentative diagnosis rather than wait until laboratory results are available. When tests are requested, the physician should inform the laboratory staff of the tentative diagnosis (type of infection or infectious agent suspected). Proper labeling of specimens includes such clinical data as well as the patient's identifying data (at least two methods of definitive identification) and the requesting physician's name and pertinent contact information.

Many pathogenic microorganisms grow slowly, and days or even weeks may elapse before they are isolated and identified. Treatment cannot be deferred until this process is complete. After obtaining the proper specimens and informing the laboratory of the tentative clinical diagnosis, the physician should begin treatment with drugs aimed at the organism thought to be responsible for the patient's illness. As the laboratory staff begins to obtain results, they inform the physician, who can then reevaluate the diagnosis and clinical course of the patient and perhaps make changes in the therapeutic program. This "feedback" information from the laboratory consists of preliminary reports of the results of individual steps in the isolation and identification of the causative agent.

703

DIAGNOSIS OF BACTERIAL & FUNGAL INFECTIONS

Specimens

Laboratory examination usually includes microscopic study of fresh unstained and stained materials and preparation of cultures with conditions suitable for growth of a wide variety of microorganisms, including the type of organism most likely to be causative based on clinical evidence. If a microorganism is isolated, complete identification may then be pursued. Isolated microorganisms may be tested for susceptibility to antimicrobial drugs. When significant pathogens are isolated before treatment, follow-up laboratory examinations during and after treatment may be appropriate.

A properly collected specimen is the single most important step in the diagnosis of an infection, because the results of diagnostic tests for infectious diseases depend upon the selection, timing, and method of collection of specimens. Bacteria and fungi grow and die, are susceptible to many chemicals, and can be found at different anatomic sites and in different body fluids and tissues during the course of infectious diseases. Because isolation of the agent is so important in the formulation of a diagnosis, the specimen must be obtained from the site most likely to yield the agent at that particular stage of illness and must be handled in such a way as to favor the agent's survival and growth. For each type of specimen, suggestions for optimal handling are given in the following paragraphs and in the section on diagnosis by anatomic site, below.

Recovery of bacteria and fungi is most significant if the agent is isolated from a site normally devoid of microorganisms (a normally sterile area). Any type of microorganism cultured from blood, cerebrospinal fluid, joint fluid, or the pleural cavity is a significant diagnostic finding. Conversely, many parts of the body have normal microbiota (Chapter 10) that may be altered by endogenous or exogenous influences. Recovery of potential pathogens from the respiratory, gastrointestinal, or genitourinary tracts; from wounds; or from the skin must be considered in the context of the normal microbiota of each particular site. Microbiologic data must be correlated with clinical information in order to arrive at a meaningful interpretation of the results.

A few general rules apply to all specimens:

1. The quantity of material must be adequate.
2. The sample should be representative of the infectious process (eg, sputum, not saliva; pus from the underlying lesion, not from its sinus tract; a swab from the depth of the wound, not from its surface).
3. Contamination of the specimen must be avoided by using only sterile equipment and aseptic precautions.
4. The specimen must be taken to the laboratory and examined promptly. Special transport media may be helpful.
5. Meaningful specimens to diagnose bacterial and fungal infections must be secured before antimicrobial drugs are administered. If antimicrobial drugs are given before specimens are taken for microbiologic study, drug

therapy may have to be stopped and repeat specimens obtained several days later.

The type of specimen to be examined is determined by the presenting clinical picture. If symptoms or signs point to involvement of one organ system, specimens are obtained from that source. In the absence of localizing signs or symptoms, repeated blood samples for culturing are taken first, and specimens from other sites are then considered in sequence, depending in part upon the likelihood of involvement of a given organ system in a given patient and in part upon the ease of obtaining specimens.

Microscopy & Stains

Microscopic examination of stained or unstained specimens is a relatively simple and inexpensive but much less sensitive method than culture for detection of small numbers of bacteria. A specimen must contain at least 10^5 organisms per milliliter before it is likely that organisms will be seen on a smear. Liquid medium containing 10^5 organisms per milliliter does not appear turbid to the eye. Specimens containing 10^2–10^3 organisms per milliliter produce growth on solid media, and those containing ten or fewer bacteria per milliliter may produce growth in liquid media.

Gram staining is a very useful procedure in diagnostic microbiology. Most specimens submitted when bacterial infection is suspected should be smeared on glass slides, Gram-stained, and examined microscopically. The materials and method for Gram staining are outlined in Table 47–1. On microscopic examination, the Gram reaction (purple-blue indicates gram-positive organisms; red, gram-negative) and morphology (shape: cocci, rods, fusiform, or other; see Chapter 2) of bacteria should be noted. The appearance of bacteria on Gram-stained smears does not permit identification of species. Reports of gram-positive cocci in chains are suggestive of, but not definitive for, streptococcal species; gram-positive cocci in clusters suggest a staphylococcal species. Gram-negative rods can be large, small, or even coccobacillary. Some nonviable gram-positive bacteria can stain gram-negatively. Typically, bacterial morphology has been defined using organisms grown on agar. However, bacteria in body fluids or tissue can have highly variable morphology.

Specimens submitted for examination for mycobacteria should be stained for acid-fast organisms, using either **Ziehl-Neelsen stain** or **Kinyoun stain** (Table 47–1). An alternative fluorescent stain for mycobacteria, auramine-rhodamine stain, is more sensitive than other stains for acid-fast organisms but requires fluorescence microscopy and, if results are positive, confirmation of morphology with an acid-fast stain (Chapter 23).

Immunofluorescent antibody (IF) staining is useful in the identification of many microorganisms. Such procedures are more specific than other staining techniques but also more cumbersome to perform. The fluorescein-labeled antibodies in common use are made from antisera produced by injecting animals with whole organisms or complex antigen mixtures. The resultant **polyclonal antibodies** may react with multiple antigens

TABLE 47–1   Gram & Acid-Fast Staining Methods

Gram stain

(1) Fix smear by heat or using methanol.

(2) Cover with crystal violet.

(3) Wash with water. Do not blot.

(4) Cover with Gram's iodine.

(5) Wash with water. Do not blot.

(6) Decolorize for 10–30 seconds with gentle agitation in acetone (30 mL) and alcohol (70 mL).

(7) Wash with water. Do not blot.

(8) Cover for 10–30 seconds with safranin (2.5% solution in 95% alcohol).

(9) Wash with water and let dry.

Ziehl-Neelsen acid-fast stain

(1) Fix smear by heat.

(2) Cover with carbolfuchsin, steam gently for 5 minutes over direct flame (or for 20 minutes over a water bath). Do not permit slides to boil or dry out.

(3) Wash with deionized water.

(4) Decolorize in 3.0% acid-alcohol (95% ethanol and 3.0% hydrochloric acid) until only a faint pink color remains.

(5) Wash with water.

(6) Counterstain for 1 minute with Loeffler's methylene blue.

(7) Wash with deionized water and let dry.

Kinyoun carbolfuchsin acid-fast stain

(1) Formula: 4 g basic fuchsin, 8 g phenol, 20 mL 95% alcohol, 100 mL distilled water.

(2) Stain fixed smear for 3 minutes (no heat necessary) and continue as with Ziehl-Neelsen stain.

on the organism that was injected and may also cross-react with antigens of other microorganisms or possibly with human cells in the specimen. Quality control is important to minimize nonspecific IF staining. Use of **monoclonal antibodies** may circumvent the problem of nonspecific staining. IF staining is most useful in confirming the presence of specific organisms such as *Bordetella pertussis* or *Legionella pneumophila* in colonies isolated on culture media. The use of direct IF staining on specimens from patients is more difficult and less specific.

Stains such as calcofluor white, methenamine silver, and occasionally periodic acid-Schiff (PAS) and others are used for tissues and other specimens in which fungi or other parasites are present. Such stains are not specific for given microorganisms, but they may define structure so that morphologic criteria can be used for identification. Calcofluor white binds to cellulose and chitin in the cell walls of fungi and fluoresces under long-wavelength ultraviolet light. It may demonstrate morphology that is diagnostic of the species (eg, spherules with endospores in *Coccidioides immitis* infection). *Pneumocystis jiroveci* cysts are identified morphologically in silver-stained specimens. PAS is used to stain tissue sections when fungal infection is suspected. After primary isolation of fungi, stains such as lactophenol cotton blue are used to distinguish fungal growth and to identify organisms by their morphology.

Specimens to be examined for fungi can be examined unstained after treatment with a solution of 10% potassium hydroxide, which breaks down the tissue surrounding the fungal mycelia to allow a better view of the hyphal forms. Phase contrast microscopy is sometimes useful in unstained specimens. Darkfield microscopy is used to detect *Treponema pallidum* in material from primary or secondary syphilitic lesions or other spirochetes such as Leptospira.

Culture Systems

For diagnostic bacteriology, it is necessary to use several types of media for routine culture, particularly when the possible organisms include aerobic, facultatively anaerobic, and obligately anaerobic bacteria. The specimens and culture media used to diagnose the more common bacterial infections are listed in Table 47–2. The standard medium for specimens is blood agar, usually made with 5% sheep blood. Most aerobic and facultatively anaerobic organisms will grow on blood agar. Chocolate agar, a medium containing heated blood with or without supplements, is a second necessary medium; some organisms that do not grow on blood agar, including pathogenic *Neisseria* and *Haemophilus*, will grow on chocolate agar. A selective medium for enteric gram-negative rods (either MacConkey agar or eosin-methylene blue [EMB] agar) is a third type of medium used routinely. Specimens to be cultured for obligate anaerobes must be plated on at least two additional types of media, including a highly supplemented agar such as brucella agar with hemin and vitamin K and a selective medium containing substances that inhibit the growth of enteric gram-negative rods and facultatively anaerobic or anaerobic gram-positive cocci.

Many other specialized media are used in diagnostic bacteriology; choices depend on the clinical diagnosis and the organism under consideration. The laboratory staff selects the specific media on the basis of the information in the culture request. Thus, freshly made Bordet-Gengou or charcoal-containing medium is used to culture for *B pertussis* in the diagnosis of whooping cough, and other special media are used to culture for *Vibrio cholerae*, *Corynebacterium diphtheriae*, *Neisseria gonorrhoeae*, and *Campylobacter* species. For culture of mycobacteria, specialized solid and liquid media are commonly used. These media may contain inhibitors of other bacteria. Because many mycobacteria grow slowly, the cultures must be incubated and examined periodically for weeks (see Chapter 23).

Broth cultures in highly enriched media are important for back-up cultures of biopsy tissues and body fluids such as cerebrospinal fluid. Broth cultures may give positive results when there is no growth on solid media because of the small number of bacteria present in the inoculum (see above).

Many yeasts will grow on blood agar. Biphasic and mycelial phase fungi grow better on media designed specifically for fungi. Brain–heart infusion agar, with and without antibiotics, and inhibitory mold agar have largely replaced the traditional use of Sabouraud's dextrose agar to grow fungi. Media made with plant and vegetable materials, the natural habitats for many fungi, also grow many fungi that cause infections. Cultures for fungi are commonly done in paired sets, one set incubated at 25–30°C and the other at 35–37°C. Table 47–3

TABLE 47–2 Common Localized Bacterial Infections and Nocardiosis

Disease	Specimen	Common Causative Agents	Usual Microscopic Findings	Culture Media	Comments
Cellulitis of skin	Punch biopsy	Group A β-hemolytic streptococci, S aureus	Occasionally gram-positive cocci	Blood agar	Biopsy at leading edge of erythema may yield the organism
Impetigo	Swab	As for cellulitis (above)	As for cellulitis (above) and pharyngitis (below)		Culture rarely indicated
Skin ulcers	Punch biopsy; deep; tissue aspirate or biopsy	Mixed flora	Mixed flora	Blood, MacConkey, or EMB agar; anaerobe media	Skin ulcers below the waist often contain aerobes and anaerobes like gastrointestinal flora
Meningitis	CSF	Neisseria meningitidis	Gram-negative intracellular diplococci	Chocolate agar[a] and blood agar for CSF cultures	Latex agglutination (bacterial antigen detection) is a poor test to diagnose the cause of meningitis
		Haemophilus influenzae	Small gram-negative coccobacilli	Chocolate agar[s]	Latex agglutination (bacterial antigen detection) is a poor test to diagnose the cause of meningitis
		Streptococcus pneumoniae	Gram-positive cocci in pairs	Blood agar	Latex agglutination (bacterial antigen detection) is a poor test to diagnose the cause of meningitis; immunochromatographic assay is more sensitive
		Group B streptococci	Gram-positive cocci in pairs and chains	Blood agar	Latex agglutination (bacterial antigen detection) is a poor test to diagnose the cause of meningitis
		Escherichia coli and other Enterobacteriaceae	Gram-negative rods	Blood agar	Mainly in newborns; no need for selective media in CSF culture
		Listeria monocytogenes	Gram-positive rods	Blood agar	β-Hemolytic; may be confused with Group B Streptococci
Brain abscess	Pus	Mixed infection; anaerobic gram-positive and gram-negative cocci and rods, aerobic gram-positive cocci	Gram-positive cocci or mixed flora	Blood agar, chocolate agar,[a] anaerobe media	Specimen must be obtained surgically and transported under strict anaerobic conditions
Perioral abscess	Pus	Mixed flora of mouth and pharynx	Mixed flora	Blood agar, chocolate agar,[a] MacConkey or EMB agar; anaerobic media	Usually mixed bacterial infection; rarely, actinomycosis
Pharyngitis	Swab	Group A streptococci	Not recommended	Blood agar or selective medium	β-Hemolytic
		Corynebacterium diphtheriae	Not recommended	Loeffler's or Pai's medium, then cysteine-tellurite or Tinsdale's medium	Granular rods in "Chinese character" patterns in smears from culture; toxicity testing required

Condition	Specimen	Organism	Microscopy/smear	Media	Comments
Whooping cough (pertussis)	Swab	*Bordetella pertussis*	Not recommended	Regan-Lowe agar	Fluorescent antibody test identifies organisms from culture and occasionally in direct smears; PCR is more sensitive than culture
Epiglottitis	Swab	*Haemophilus influenzae*	Usually not helpful	Chocolate agar[a] (also use blood agar)	*Haemophilus influenzae* is part of normal flora in nasopharynx
Pneumonia	Sputum or other invasively obtained respiratory samples	*Streptococcus pneumoniae*	Many PMNs, gram-positive cocci in pairs or chains. Capsule swelling with omniserum (quellung test)	Blood agar; also MacConkey, EMB, and chocolate agars	*Streptococcus pneumoniae* is part of normal flora in nasopharynx. Blood cultures specific (positive) in 10–20%
		Staphylococcus aureus	Gram-positive cocci in pairs, tetrads, and clusters	Blood agar; also MacConkey, EMB, and chocolate agars	Uncommon cause of pneumonia. Usually β-hemolytic, coagulase-positive
		Enterobacteriaceae and other gram-negative rods	Gram-negative rods	Blood agar; MacConkey or EMB agar	Causes of hospital-associated pneumonia
		Mixed anaerobes and aerobes	Mixed respiratory tract flora; sometimes many PMNs	Blood, MacConkey, or EMB agar; anaerobe agar	Specimens must be obtained by bronchoscopy using a protected brush or transtracheal aspiration; expectorated sputum is unsatisfactory for anaerobes
Chest empyema	Pus	Same as pneumonia, or mixed flora infection	Mixed flora	Blood, MacConkey, or EMB agar; anaerobe media	Usually pneumonia; mixed aerobic and anaerobic flora derived from oropharynx
Liver abscess	Pus	*Escherichia coli*; *Bacteroides fragilis*; mixed aerobic or anaerobic flora	Gram-negative rods and mixed flora	Blood, MacConkey, or EMB agar; anaerobe media	Commonly enteric gram-negative aerobes and anaerobes; consider *Entamoeba histolytica* infection
Cholecystitis	Bile	Gram-negative enteric aerobes, also *Bacteroides fragilis*	Gram-negative rods	Blood, MacConkey, or EMB agar; anaerobic conditions	Usually gram-negative rods from gastrointestinal tract
Abdominal or perirectal abscess	Pus	Gastrointestinal flora	Mixed flora	Blood, MacConkey, or EMB agar; anaerobe media	Aerobic and anaerobic bowel flora; often more than five species grown
Enteric fever, typhoid	Blood, feces, urine	*Salmonella* Typhi	Not recommended	MacConkey, Hektoen, bismuth sulfite agars; others	Multiple specimens should be cultured; lactose negative. H_2S produced

(continued)

TABLE 47–2 Common Localized Bacterial Infections: Agents, Specimens, and Diagnostic Tests (Continued)

Disease	Specimen	Common Causative Agents	Usual Microscopic Findings	Culture Media	Comments
Enteritis, enterocolitis, bacterial diarrheas, "gastroenteritis"	Feces	Salmonella species other than Shigella typhi	Gram stain or methylene blue stain may show PMNs	MacConkey, Hektoen, bismuth sulfite agars; others	Nonlactose-fermenting colonies onto TSI[a] slants: Nontyphoid salmonellae produce acid and gas in butt, alkaline slant, and H$_2$S
		Shigella species	Gram stain or methylene blue stain may show PMNs	MacConkey, Hektoen, bismuth sulfite agars; others	Nonlactose-fermenting colonies onto TSI[a] slants: Shigellae produce alkaline slant, acid butt without gas
		Campylobacter jejuni	"Gull wing-shaped" gram-negative rods and often PMNs	Campy BAP or similar medium	Incubate at 42°C; colonies oxidase-positive; smear shows "gull wing-shaped" rods
		Vibrio cholerae	Not recommended	Thiosulfate citrate bile salts (TCBS) sucrose agar; others. Taurocholate-peptone broth for enrichment	Yellow colonies on TCBS. Vibrio cholerae is oxidase-positive
		Other vibrios	Not recommended	As for Vibrio cholerae	Differentiate from Vibrio cholerae by biochemical and culture tests
		Yersinia enterocolitica	Not recommended	MacConkey, CIN	Enrichment at 4°C helpful; incubate cultures at 25°C
Hemorrhagic colitis and hemolytic uremic syndrome	Feces	Escherichia coli O157:H7 and other serotypes	Not recommended	Sorbitol MacConkey medium	Look for sorbitol-negative colonies; type with antisera for O antigen 157 and flagellar antigen 7; EIA for shiga-like toxins are the preferred tests
Urinary tract infection	Urine (clean catch midstream specimen or one obtained by bladder catheterization or suprapubic aspiration)	Escherichia coli; Enterobacteriaceae; other gram-negative rods	Gram-negative rods seen on stained smear of uncentrifuged urine indicate more than 10^5 organisms/mL	Blood agar; MacConkey or EMB agar	Gray colonies that are β-hemolytic and give a positive spot indole test are usually Escherichia coli; others require further biochemical tests
Urethritis/cervicitis	Swab	Neisseria gonorrhoeae	Gram-negative diplococci in or on PMNs. Specific for urethral discharge in men; less reliable in women	Modified Thayer-Martin or similar antibiotic-containing selective medium	Positive stained smear diagnostic in men. Culture or nucleic acid amplification tests needed in women. Gonococci are oxidase-positive
		Chlamydia trachomatis	PMNs with no associated gram-negative diplococci	Culture in McCoy cells treated with cycloheximide	Crescent-shaped inclusions in epithelial cells by stains or immunofluorescence. Direct EIA or fluorescent antibody tests can be helpful; nucleic acid amplification tests are more sensitive

Condition	Specimen	Organism	Microscopy	Culture medium	Comments
Genital ulcers	Swab	Haemophilus ducreyi (chancroid)	Mixed flora	Chocolate agar with IsoVitaleX and vancomycin	Differential diagnosis of genital ulcers includes herpes simplex infection
		Treponema pallidum (syphilis)	Darkfield or fluorescent antibody examination shows spirochetes	None	
	Pus aspirated from suppurating lymph nodes	Chlamydia trachomatis (lymphogranuloma venereum)	PMNs with no associated gram-negative diplococci	Culture pus in cell culture (as for urethritis)	
Pelvic inflammatory disease	Cervical swab	Neisseria gonorrhoeae	PMNs with associated gram-negative diplococci; mixed flora may be present	Modified Thayer-Martin or similar antibiotic-containing selective medium; nucleic acid amplification test preferred	Causative organisms may be gonococci, anaerobes, others. Anaerobes always present in endocervix; thus, endocervical specimen not suitable for anaerobic culture
		Chlamydia trachomatis	See above; nucleic acid amplification test preferred	Cell culture (as for urethritis)	
	Aspirate from cul-de-sac or by laparoscope	Neisseria gonorrhoeae	Gram-negative diplococci in or on PMNs	Modified Thayer-Martin medium	
		Chlamydia trachomatis	See above	Cell culture (as for urethritis)	
		Mixed flora	Mixed flora	Blood, MacConkey, or EMB agar; anaerobic medium	Usually mixed anaerobic and aerobioc bacteria
Arthritis	Joint aspirate, blood	Staphylococcus aureus	Gram-positive cocci in pairs, tetrads, and clusters	Blood agar; chocolate agar[a]	Occurs in both children and adults; coagulase-positive; usually β-hemolytic
		Neisseria gonorrhoeae	Gram-negative diplococci in or on PMNs	Modified Thayer-Martin medium	
		Others	Morphology depends upon organisms	Blood agar, chocolate agar,[a] anaerobic medium	Includes streptococci, gram-negative rods, and anaerobes
Osteomyelitis	Pus or bone specimen obtained by aspiration or surgery	Multiple; often Staphylococcus aureus	Morphology depends upon organisms	Blood agar, MacConkey, EMB agar; anaerobic medium	Usually aerobic organisms; Staphylococcus aureus is most common; gram-negative rods frequent; anaerobes less common

TSI, triple sugar iron agar; CIN, Cefsulodin-Irgasan-Novobiocin medium; EIA, enzyme immunoassay.

[a] A chemical supplement such as IsoVitaleX enhances growth of Haemophilus and Neisseria species.

TABLE 47–3   Common Fungal Infections and Nocardiosis: Agents, Specimens, and Diagnostic Tests

	Specimen	Serologic and Other Tests	Comments
Invasive (deep-seated) mycoses			
Aspergillosis: *Aspergillus fumigatus*, other *Aspergillus* species			
Pulmonary	Respiratory secretions	Serum galactomannan assays are available—sensitivity and specificity are variable	Serology seldom useful
Disseminated	Biopsy specimen, blood	As above	Aspergillus is difficult to grow from blood of patients with disseminated infection.
Blastomycosis: *Blastomyces dermatiditis*			
Pulmonary	Respiratory secretions	CF; EIA	CF test usually negative and therefore not very useful. EIAs more sensitive but less specific. Culture is the best diagnostic test; serology seldom done. Urine antigen cross-reacts with other fungi.
Oral and cutaneous ulcers	Biopsy or swab specimen	CF, EIA, urine antigen	
Bone	Bone biopsy	CF, EIA, urine antigen	
Coccidioidomycosis: *Coccidioides immitis*			
Pulmonary	Respiratory secretions	CF, immunodiffusion, precipitation, latex agglutination, EIA, DNA probe (culture confirmation)	*Coccidioides immitis* will grow on routine blood agar cultures; positive cultures pose a serious hazard for laboratory workers. Culture confirmed by DNA probe. Serology often more useful than culture.
Disseminated	Biopsy specimen from site of infection, eg, skin, bone	As above except that skin test with coccidioidin may be negative	
Histoplasmosis: *Histoplasma capsulatum*			
Pulmonary	Respiratory secretions	CF, immunodiffusion, antibody tests, histoplasma antigen test on urine (most sensitive) serum and bronchoalveolar lavage fluid	Serology very useful, but less sensitive in immunosuppressed patients. Culture with DNA probe confirmation
Disseminated	Bone marrow, blood, biopsy specimen from site of infection	As above	
Nocardiosis: *Nocardia asteroides* complex			
Pulmonary	Respiratory secretions	Modified acid-fast stain	Nocardiae are bacteria that clinically behave like fungi. Weakly acid-fast, branching, filamentous gram-positive rods are nocardia.
Subcutaneous	Aspirate or biopsy of abscess	Will grow on standard laboratory media	
Brain	Material from brain abscess		
Paracoccidioidomycosis (South American blastomycosis): *Paracoccidioides brasiliensis*			
	Biopsy specimen from lesion	Immunodiffusion, CF, skin test (paracoccidioidin) is not reliable for diagnosis	Immunodiffusion test 95% sensitive and specific; CF test and skin test cross-react with histoplasmin. Positive skin test is of prognostic value.

Infection/Organism	Specimen	Serologic test	Comments
Sporotrichosis: *Sporothrix schenckii*			
Skin and subcutaneous nodules	Biopsy specimen	Agglutination	Culture more useful than serology
Disseminated	Biopsy specimen from infected site	As above	
Zygomycosis: *Rhizopus* species, *Mucor* species, others			
Rhinocerebral	Nasal-orbital tissue	None	Nonseptate hyphae seen in microscopic sections
Cutaneous; pulmonary and disseminated	Respiratory secretions, biopsy specimens	None	Culture useful
Yeast infections			
Candidiasis: *Candida albicans* and similar yeasts[a]			
Mucous membrane	Secretions	KOH wet mount useful for microscopy in localized infection	Usually easy to culture from clinical material
Skin	Swab specimen		
Systemic	Blood, biopsy specimen, urine	Immunodiffusion	As above. Serology seldom helpful.
Cryptococcosis: *Cryptococcus neoformans*			
Pulmonary	Respiratory secretions	Cryptococcal antigen rarely detected	Antibodies to *Cryptococcus neoformans* rarely found
Meningitis	CSF	Cryptococcal antigen detection is most useful	Repeated examination of CSF may be necessary to diagnose meningitis
Disseminated	Bone marrow, bone, blood, other	Cryptococcal antigen on serum	
Primary skin infections			
Dermatophytosis: *Microsporum* species, *Epidermophyton* species, *Trichophyton* species	Hair, skin, nails from infected sites	None	Can be cultured on dermatophyte agars

[a] *Candida tropicalis, Candida parapsilosis, Candida glabrata,* and other *Candida* species.

CF, complement fixation; EIA, enzyme immunoassays.

outlines specimens and other tests to be used for the diagnosis of fungal infections.

Antigen Detection

Immunologic systems designed to detect antigens of microorganisms can be used in the diagnosis of specific infections. IF tests (direct and indirect fluorescent antibody tests) are one form of antigen detection and are discussed in separate sections in this chapter on the diagnosis of bacterial, chlamydial, and viral infections and in the chapters on the specific microorganisms.

EIAs, including **enzyme-linked immunosorbent assays (ELISA),** and agglutination tests are used to detect antigens of infectious agents present in clinical specimens. The principles of these tests are reviewed briefly here.

There are many variations of EIAs to detect antigens. One commonly used format is to bind a capture antibody, specific for the antigen in question, to the wells of plastic microdilution trays. The specimen containing the antigen is incubated in the wells followed by washing of the wells. A second antibody for the antigen, labeled with enzyme, is used to detect the antigen. Addition of the substrate for the enzyme allows detection of the bound antigen by colorimetric reaction. A significant modification of EIAs is the development of immunochromatographic membrane formats for antigen detection. In this format, a nitrocellulose membrane is used to absorb the antigen from a specimen. A colored reaction appears directly on the membrane with sequential addition of conjugate followed by substrate. In some formats, the antigen is captured by bound antibody directed against the antigen. These assays have the advantage of being rapid and also frequently include a built-in positive control. An example of this type of assay is the Binax NOW *Streptococcus pneumoniae* urinary antigen test. In some EIAs, the initial antibody is not necessary, because the antigen will bind directly to the plastic of the wells. EIAs are used to detect viral, bacterial, chlamydial, protozoan, and fungal antigens in a variety of specimen types such as stool, cerebrospinal fluid, urine, and respiratory samples. Examples of these are discussed in the chapters on the specific etiological agents.

In latex agglutination tests, an antigen-specific antibody (either polyclonal or monoclonal) is fixed to latex beads. When the clinical specimen is added to a suspension of the latex beads, the antibodies bind to the antigens on the microorganism forming a lattice structure, and agglutination of the beads occurs. Coagglutination is similar to latex agglutination except that staphylococci rich in protein A (Cowan I strain) are used instead of latex particles; coagglutination is less useful for antigen detection compared with latex agglutination but is helpful when applied to identification of bacteria in cultures such as *S pneumoniae, Neisseria meningitidis, N gonorrhoeae,* and β-hemolytic streptococci.

Latex agglutination tests are primarily directed at the detection of carbohydrate antigens of encapsulated microorganisms. Antigen detection is used most often in the diagnosis of group A streptococcal pharyngitis. Detection of cryptococcal antigen is useful in the diagnosis of cryptococcal meningitis in patients with AIDS or other immunosuppressive diseases.

The sensitivity of latex agglutination tests in the diagnosis of bacterial meningitis may not be better than that of Gram stain, which is approximately 100,000 bacteria per milliliter. For that reason, the latex agglutination test is not recommended for direct specimen testing.

Western Blot Immunoassays

These assays are usually performed to detect antibodies against specific antigens of a particular organism. This method is based upon the electrophoretic separation of major proteins of the organism in question in a two-dimensional agarose gel. Organisms are mechanically or chemically disrupted and resultant solubilized antigen of the organism is placed in a polyacrylamide gel. An electric current is applied and major proteins are separated out on the basis of size (smaller proteins travel faster). The protein bands are transferred to strips of nitrocellulose paper. Following incubation of the strips with a patient's specimen containing antibody (usually serum), the antibodies bind to the proteins on the strip and are detected enzymatically in a fashion similar to the EIA methods described above. Western blot tests are used as specific tests for antibodies in HIV infection and Lyme disease.

Molecular Diagnostics

The principle behind early molecular assays is the hybridization of a characterized **nucleic acid probe** to a specific nucleic acid sequence in a test specimen followed by detection of the paired hybrid. For example, single-stranded probe DNA (or RNA) is used to detect complementary RNA or denatured DNA in a test specimen. The nucleic acid probe typically is labeled with enzymes, antigenic substrates, chemiluminescent molecules, or radioisotopes to facilitate detection of the hybridization product. By carefully selecting the probe or making a specific **oligonucleotide** and performing the hybridization under conditions of high stringency, detection of the nucleic acid in the test specimen can be extremely specific. Such assays are currently used primarily for rapid confirmation of a pathogen once growth is detected, eg, the identification of *Mycobacterium tuberculosis* in culture using the Gen-Probe Inc. (San Diego, CA) DNA probe. The Gen-Probe test is an example of a hybridization test format in which the probe and target are in solution. Most of the applications in use in clinical microbiology laboratories make use of solution hybridization formats. In situ hybridization involves the use of labeled DNA probes or labeled RNA probes to detect complementary nucleic acids in formalin-fixed paraffin-embedded tissues, frozen tissues, or cytologic preparations mounted on slides. Technically, this can be difficult and is usually performed in histology laboratories and not clinical microbiology laboratories. However, this technique has increased the knowledge of the biology of many infectious diseases, especially the hepatitides and oncogenic viruses, and is still useful in infectious diseases diagnosis. A novel technique that is somewhat of a modification of in situ hybridization makes use of peptide nucleic acid probes. Peptide nucleic acid probes are synthesized pieces of DNA in which the sugar phosphate backbone of DNA (normally negatively charged) is replaced by a polyamide of repetitive units (neutral charge). Individual nucleotide bases can be attached to the now neutral backbone, which allows for faster and more specific hybridization to complementary nucleic acids. Because the probes are synthetic, they are not subject to degradation by nucleases and other enzymes. A commercial company (AdvanDx, Woburn MA) has a number of FDA cleared assays for confirmation of *Staphylococcus aureus*, enterococci, certain *Candida* sp. and some gram-negative bacilli in positive blood culture bottles. The probe hybridization is detected by fluorescence and is called Peptide Nucleic Acid-Fluorescence In Situ Hybridization (PNA-FISH).

A. Identifying Bacteria Using 16s rRNA

The 16S rRNA of each species of bacteria has stable (conserved) portions of the sequence. Many copies are present in each organism. Labeled probes specific for the 16S rRNA of a species are added, and the amount of label on the double-stranded hybrid is measured. This technique is widely used for the rapid identification of many organisms. Examples include the most common and important *Mycobacterium* species, *C immitis, Histoplasma capsulatum,* and others.

Portions of the 16S rRNA are conserved across many species of microorganisms. Amplifying the 16S rRNA using primers to these conserved regions allows isolation and sequencing of the variable regions of the molecules. These variable sequences are genus- or species-specific markers that allow identification of microorganisms. Pathogens that are difficult or impossible to culture in the laboratory have been identified using this technique. One example is *Tropheryma whipplei*, the cause of Whipple disease.

Molecular diagnostic assays that use amplification of nucleic acid have become widely used and are evolving rapidly. These amplification systems fall into several basic categories as outlined below.

B. Target Amplification Systems

In these assays, the target DNA or RNA is amplified many times. The **polymerase chain reaction (PCR)** is used to amplify extremely small amounts of specific DNA present in a clinical specimen, making it possible to detect what were initially minute amounts of the DNA. PCR uses a thermostable DNA polymerase to produce a twofold amplification of target DNA with each temperature cycle. Conventional PCR utilizes three sequential reactions—denaturation, annealing,

and primer extension—as follows. The DNA extracted from the clinical specimen along with sequence-specific oligonucleotide primers, nucleotides, thermostable DNA polymerase, and buffer are heated to 90–95°C to denature (separate) the two strands of the target DNA. The temperature in the reaction is lowered, usually to 45–60°C depending upon the primers, to allow annealing of the primers to the target DNA. Each primer is then extended by the thermostable DNA polymerase by adding nucleotides complementary to the target DNA yielding the twofold amplification. The cycle is then repeated 30–40 times to yield amplification of the target DNA segment by as much as 10^5- to 10^6-fold. The amplified segment often can be seen in an electrophoretic gel or detected by Southern blot analysis using labeled DNA probes specific for the segment or by a variety of proprietary commercial techniques.

PCR can also be performed on RNA targets, which is called **reverse transcriptase PCR.** The enzyme reverse transcriptase is used to transcribe the RNA into complementary DNA for amplification.

PCR assays are available commercially for identification of a broad range of bacterial and viral pathogens such as *Chlamydia trachomatis, N gonorrhoeae, M tuberculosis*, cytomegalovirus, enteroviruses, and many others. An assay is available for HIV-1 viral load testing also. There are many other "in-house" PCRs that have been developed by individual laboratories to diagnose infections. Such assays are the tests of choice to diagnose many infections—especially when traditional culture and antigen detection techniques do not work well. Examples include testing of cerebrospinal fluid for herpes simplex virus to diagnose herpes encephalitis and testing of nasopharyngeal wash fluid to diagnose *B pertussis* infection (whooping cough).

A major consideration for laboratories that perform PCR assays is to prevent contamination of reagents or specimens with target DNA from the environment, which can obscure the distinction between truly positive results and falsely positive ones because of the contamination.

C. Probe Amplification Systems

The **ligase chain reaction (LCR)** is an amplification system different from PCR. LCR uses thermostable DNA polymerase and thermostable DNA ligase. LCR uses four oligonucleotide probes of 20–24 bases each. Each pair of oligonucleotides is designed to bind to the denatured target DNA only a few bases apart. The oligonucleotides are mixed with extracted target DNA from the specimen and other reagents and then heated to denature the target DNA. The reaction is then cooled to allow binding of the oligonucleotide probes to the target DNA. The short gap between the two probes is filled in by the DNA polymerase and linked by the DNA ligase, yielding double-stranded DNA molecules 40–50 bp in length. The cycle is repeated 30–40 times, yielding a large number of DNA molecules. This commercially available system includes automated detection of the amplified DNA. It can be used to detect *C trachomatis* and *N gonorrhoeae*. It is available only outside of the United States.

D. Signal Amplification Techniques

These assays strengthen the signal by amplifying the label (eg, fluorochromes, enzymes) that is attached to the target nucleic acid. The **branched DNA (bDNA)** system has a series of primary probes and a branched secondary probe labeled with enzyme. Multiple oligonucleotide probes specific for the target RNA (or DNA) are fixed to a solid surface such as a microdilution tray. These are the capture probes. The prepared specimen is added, and the RNA molecules are attached to the capture probes on the microdilution tray. Additional target probes bind to the target but not to the tray. The enzyme-labeled bDNA amplifier probes are added and attach to the target probes. A chemiluminescent substrate is added, and light emitted is measured to quantitate the amount of target RNA present. Examples of the use of this type of assay include the quantitative measurement of HIV-1, hepatitis C virus, and hepatitis B virus.

E. Amplification Methods: Non–PCR-Based

The **transcription-mediated amplification (TMA)** and the **nucleic acid sequence-based amplification (NASBA)** systems amplify large quantities of RNA in isothermal assays that coordinately use the enzymes reverse transcriptase, RNase H, and RNA polymerase. An oligonucleotide primer containing the RNA polymerase promoter is allowed to bind to the RNA target. The reverse transcriptase makes a single-stranded cDNA copy of the RNA. The RNase H destroys the RNA of the RNA-cDNA hybrid, and a second primer anneals to the segment of cDNA. The DNA-dependent DNA polymerase activity of reverse transcriptase extends the DNA from the second primer, producing a double-stranded DNA copy, with intact RNA polymerase. The RNA polymerase then produces many copies of the single-stranded RNA. Detection of *C trachomatis*, *N gonorrhoeae*, and *M tuberculosis* and quantitation of HIV-1 load are examples of the use of these types of assays.

Strand displacement assays (SDA) are isothermal amplification assays that employ use of restrictive endonuclease and DNA polymerase.

F. Real-Time PCR

Technologic advances, which have lead to "real-time amplification," have streamlined nucleic acid amplification platforms, improved the sensitivity of amplification tests, and have drastically reduced the potential for contamination. Real-time instruments have replaced the solid block used in conventional thermocyclers with fans that allow more rapid PCR cycling. Dramatic improvements in the chemistry of nucleic acid amplification reactions have resulted in homogeneous reaction mixtures in which fluorogenic compounds are present in the same reaction tube in which the amplification occurs. A variety of fluorogenic molecules are used. These include nonspecific dyes such as SYBR green, which binds to the minor groove of double-stranded DNA, and amplicon specific detection methods using fluorescently labeled oligonucleotide probes, which fall into three categories: TaqMan or hydrolysis probes; fluorescence energy transfer (FRET) probes; and molecular beacons. A complete discussion of these methods is beyond the scope of this chapter. The reader is referred to the book edited by Persing et al listed in the reference section. All of the methods allow for measurement of fluorescence with each amplification cycle, that is, "real-time" assessment of results. Since the reaction tube does not need to be opened to analyze the PCR products on a gel, there is much less risk of amplicon carry-over to the next reaction.

THE IMPORTANCE OF NORMAL BACTERIAL & FUNGAL FLORA

Organisms such as *M tuberculosis*, *Salmonella typhi*, and *Brucella* species are considered pathogens whenever they are found in patients. However, many infections are caused by organisms that are permanent or transient members of the normal flora. For example, *Escherichia coli* is part of the normal gastrointestinal flora and is also the most common cause of urinary tract infections. Similarly, the vast majority of mixed bacterial infections with anaerobes are caused by organisms that are members of the normal flora.

The relative numbers of specific organisms found in a culture are important when members of the normal flora are the cause of infection. When numerous gram-negative rods of species such as *Klebsiella pneumoniae* are found mixed with a few normal nasopharyngeal bacteria in a sputum culture, the gram-negative rods are strongly suspect as the cause of pneumonia because large numbers of gram-negative rods are not normally found in sputum or in the nasopharyngeal flora; the organisms should be identified and reported. In contrast, abdominal abscesses commonly contain a normal distribution of aerobic, facultatively anaerobic, and obligately anaerobic organisms representative of the gastrointestinal flora. In such cases, identification of all species present is not warranted; instead, it is appropriate to report "normal gastrointestinal flora."

Yeasts in small numbers are commonly part of the normal microbial flora. However, other fungi are not normally present and therefore should be identified and reported. Viruses usually are not part of the normal flora as detected in diagnostic microbiology laboratories. However, some latent viruses, eg, herpes simplex, or live vaccine viruses such as poliovirus occasionally appear in cultures for viruses. In some parts of the world, stool specimens commonly yield evidence of parasitic infection. In such cases, it is the relative number of parasites correlated with the clinical presentation that is important.

Members of the normal flora that are most commonly present in patient specimens and that may be reported as "normal flora" are discussed in Chapter 10.

LABORATORY AIDS IN THE SELECTION OF ANTIMICROBIAL THERAPY

The antimicrobial drug used initially in the treatment of an infection is chosen on the basis of clinical impression after the physician is convinced that an infection exists and has made a tentative etiologic diagnosis on clinical grounds. On the basis of this "best guess," a probable drug of choice can be selected (see Chapter 28). Before this drug is administered, specimens are obtained for laboratory isolation of the causative agent. The results of these examinations may necessitate selection of a different drug. The identification of certain microorganisms that are uniformly drug-susceptible eliminates the necessity for further testing and permits the selection of optimally effective drugs solely on the basis of experience. Under other circumstances, tests for drug susceptibility of isolated microorganisms may be helpful (see Chapter 28).

The commonly performed **disk diffusion susceptibility test** must be used judiciously and interpreted with restraint. In general, only one member of each major class of drugs is represented. For staphylococci, penicillin G, oxacillin, cefazolin, erythromycin, gentamicin, and vancomycin are used. For gram-negative rods, ampicillin, cefazolin and second- and third-generation cephalosporins, piperacillin and other "antipseudomonal penicillins," carbapenems, trimethoprim-sulfamethoxazole, fluoroquinolones, and the aminoglycosides (amikacin, tobramycin, gentamicin) are included. For urinary tract infections with gram-negative rods, nitrofurantoin, quinolones, and trimethoprim may be added. The choice of drugs to be included in a routine susceptibility test battery should be based on the susceptibility patterns of isolates in the laboratory, the type of infection (community-acquired or nosocomial), the source of the infection, and cost efficacy analysis for the patient population.

The sizes of zones of growth inhibition vary with the molecular characteristics of different drugs. Thus, the zone size of one drug cannot be compared to the zone size of another drug acting on the same organism. However, for any one drug the zone size can be compared to a standard, provided that media, inoculum size, and other conditions are carefully regulated. This makes it possible to define for each drug a minimum diameter of inhibition zone that denotes "susceptibility" of an isolate by the disk diffusion technique.

The disk test measures the ability of drugs to inhibit the growth of bacteria. The results correlate reasonably well with therapeutic response in those disease processes where body defenses can frequently eliminate infectious microorganisms.

In a few types of human infections, the results of disk tests are of little assistance (and may be misleading) because a bactericidal drug effect is required for cure. Outstanding examples are infective endocarditis, acute osteomyelitis, and severe infections in a host whose antibacterial defenses are inadequate, eg, persons with neoplastic diseases that have been treated with radiation and antineoplastic chemotherapy, or persons who are being given corticosteroids in high dosage and are immunosuppressed.

Instead of the disk test, a semiquantitative **minimum inhibitory concentration (MIC)** test procedure can be used. It measures more exactly the concentration of an antibiotic necessary to inhibit growth of a standardized inoculum under defined conditions. A semiautomated microdilution method is used in which defined amounts of drug are dissolved in a measured small volume of broth and inoculated with a standardized number of microorganisms. The end point, or minimum inhibitory concentration, is considered the last broth cup (lowest concentration of drug) remaining clear, ie, free from microbial growth. The minimum inhibitory concentration provides a better estimate of the probable amount of drug necessary to inhibit growth in vivo and thus helps in gauging the dosage regimen necessary for the patient.

Clinical microbiology laboratories perform disk diffusion tests and tests based upon determining the MIC and interpret their results using guidelines established by the Clinical Laboratory and Standards Institute (CLSI) located in Wayne, Pennsylvania. In addition, to help guide empiric therapy choices before the results of antimicrobial susceptibility tests are available, it is recommended by CLSI that laboratories publish an antibiogram annually that contains the results of susceptibility testing in aggregate for particular organism–drug combinations. For example, it may be important to know the most active β-lactam antimicrobial agent targeted against *Pseudomonas aeruginosa* among ICU patients in a particular hospital so that agent can be used when a patient develops an infection while in that unit.

There are other methods for assessing the efficacy of antimicrobial treatment. Bactericidal effects can be estimated by subculturing the clear broth onto antibiotic-free solid media. The result, eg, a reduction of colony-forming units by 99.9% below that of the control, is called the **minimal bactericidal concentration (MBC)**.

The selection of a bactericidal drug or drug combination for each patient can be guided by specialized laboratory tests. Such tests measure either the rate of killing (time-kill assay) or the proportion of the microbial population that is killed in a fixed time (serum bactericidal testing).

In urinary tract infections, the antibacterial activity of urine is far more important than that of serum.

DIAGNOSIS OF INFECTION BY ANATOMIC SITE

Wounds, Tissues, Bones, Abscesses, & Fluids

Microscopic study of smears and culture of specimens from wounds or abscesses often gives early and important indications of the nature of the infecting organism and thus helps in the choice of antimicrobial drugs. Specimens from diagnostic tissue biopsies should be submitted for bacteriologic as well as histologic examination. Such specimens for bacteriologic examination are kept away from fixatives and disinfectants, minced, and cultured by a variety of methods.

The pus in closed, undrained soft tissue abscesses frequently contains only one organism as the infecting agent; most commonly staphylococci, streptococci, or enteric gram-negative rods. The same is true in acute osteomyelitis, where the organisms can often be cultured from blood before the infection has become chronic. Multiple microorganisms are frequently encountered in abdominal abscesses and abscesses contiguous with mucosal surfaces as well as in open wounds. When deep suppurating lesions, such as chronic osteomyelitis, drain onto exterior surfaces through a sinus or fistula, the flora of the surface through which the lesion drains must not be mistaken for that of the deep lesion. Instead, specimens should be aspirated from the primary infection through uninfected tissue.

Bacteriologic examination of pus from closed or deep lesions must include culture by anaerobic methods. Anaerobic bacteria (*Bacteroides*, peptostreptococci) sometimes play an essential causative role, and mixtures of anaerobes are often present.

The methods used for cultures must be suitable for the semiquantitative recovery of common bacteria and also for recovery of specialized microorganisms, including mycobacteria and fungi. Eroded skin and mucous membranes are frequently the sites of yeast or fungus infections. *Candida*, *Aspergillus*, and other yeasts or fungi can be seen microscopically in smears or scrapings from suspicious areas and can be grown in cultures. Treatment of a specimen with KOH and calcofluor white greatly enhances the observation of yeasts and molds in the specimen.

Exudates that have collected in the pleural, peritoneal, pericardial, or synovial spaces must be aspirated with aseptic technique. If the material is frankly purulent, smears and cultures are made directly. If the fluid is clear, it can be centrifuged at high speed for 10 minutes and the sediment used for stained smears and cultures. The culture method used must be suitable for the growth of organisms suspected on clinical grounds—eg, mycobacteria, anaerobic organisms—as well as the commonly encountered pyogenic bacteria. Some fluid specimens clot, and culture of an anticoagulated specimen may be necessary. The following chemistry and hematology results are suggestive of infection: specific gravity > 1.018, protein content >3 g/dL (often resulting in clotting), and cell counts >500–1000/μL. Polymorphonuclear leukocytes predominate in acute untreated pyogenic infections; lymphocytes or monocytes predominate in chronic infections. Transudates resulting from neoplastic growth may grossly resemble infectious exudates by appearing bloody or purulent and by clotting on standing. Cytologic study of smears or of sections of centrifuged cells may prove the neoplastic nature of the process.

Blood

Since bacteremia frequently portends life-threatening illness, its early detection is essential. Blood culture is the single most important procedure to detect systemic infection due to bacteria. It provides valuable information for the management of febrile, acutely ill patients with or without localizing symptoms and signs and is essential in any patient in whom infective endocarditis is suspected even if the patient does not appear acutely or severely ill. In addition to its diagnostic significance, recovery of an infectious agent from the blood provides invaluable aid in determining antimicrobial therapy. Every effort should therefore be made to isolate the causative organisms in bacteremia.

In healthy persons, properly obtained blood specimens are sterile. Although microorganisms from the normal respiratory and gastrointestinal flora occasionally enter the blood, they are rapidly removed by the reticuloendothelial system. These transients rarely affect the interpretation of blood culture results. If a blood culture yields microorganisms, this fact is of great clinical significance provided that contamination can be excluded. Contamination of blood cultures with normal skin flora is most commonly due to errors in the blood collection procedure. Therefore, proper technique in performing a blood culture is essential.

The following rules, rigidly applied, yield reliable results:

1. Use strict aseptic technique. Wear gloves—they do not have to be sterile.
2. Apply a tourniquet and locate a fixed vein by touch. Release the tourniquet while the skin is being prepared.
3. Prepare the skin for venipuncture by cleansing it vigorously with 70–95% isopropyl alcohol. Using 2% tincture of iodine or 2% chlorhexidine, start at the venipuncture site and cleanse the skin in concentric circles of increasing diameter. Allow the antiseptic preparation to dry for at least 30 seconds. Do not touch the skin after it has been prepared.
4. Reapply the tourniquet, perform venipuncture, and (for adults) withdraw approximately 20 mL of blood.
5. Add the blood to labeled aerobic and anaerobic blood culture bottles.
6. Take specimens to the laboratory promptly, or place them in an incubator at 37°C.

Several factors determine whether blood cultures will yield positive results: the volume of blood cultured, the dilution of blood in the culture medium, the use of both aerobic and anaerobic culture media, and the duration of incubation. For adults, 20 to 30 mL per culture is usually obtained, and half is placed in an aerobic blood culture bottle and half in an anaerobic one, with one pair of bottles comprising a single blood culture. However, different volumes of blood may be required for the many different blood culture systems that exist. One widely used blood culture system uses bottles that hold 5 mL rather than 10 mL of blood. An optimal dilution of blood in a liquid culture medium is 1:300–1:150; this minimizes the effects of the antibody, complement, and white blood cell antibacterial systems that are present. Because such large dilutions are impractical in blood cultures, most such media contain 0.05% sodium polyanetholesulfonate (SPS), which inhibits the antibacterial systems. However, SPS also inhibits growth of some neisseriae and anaerobic gram-positive cocci and of *Gardnerella vaginalis*. If any of these organisms are suspected, alternative blood culture systems without SPS should be used.

Blood cultures are incubated for 5–7 days. Automated blood culture systems use a variety of methods to detect positive cultures. These automated methods allow frequent monitoring of the cultures—as often as every few minutes—and earlier detection of positive ones. The media in the automated blood culture systems are so enriched and the detection systems so sensitive that blood cultures using the automated systems do not need to be processed for more than 5 days. In general, subcultures are indicated only when the machine indicates that the culture is positive. Manual blood culture systems are obsolete and are likely to be used only in laboratories in developing countries that lack the resources to purchase automated blood culturing systems. In manual systems, the blood culture bottles are examined two or three times a day for the first 2 days and daily thereafter for 1 week. In the manual method, blind subcultures of all the blood culture bottles on days 2 and 7 may be necessary.

The number of blood specimens that should be drawn for cultures and the period of time over which this is done depend in part upon the severity of the clinical illness. In hyperacute infections, eg, gram-negative sepsis with shock or staphylococcal sepsis, it is appropriate to culture two blood specimens obtained from different anatomic sites over a period of 5–10 minutes. In other bacteremic infections, eg, subacute endocarditis, three blood specimens should be obtained over 24 hours. A total of three blood cultures yields the infecting bacteria in more than 95% of bacteremic patients. If the initial three cultures are negative and occult abscess, fever of unexplained origin, or some other obscure infection is suspected, additional blood specimens should be cultured when possible before antimicrobial therapy is started.

Several types of blood culture bottles are available that contain resins or other substances that absorb most antimicrobial drugs and some antimicrobial host factors as well. Indications for the use of the resin-containing bottles include the following: a clinically septic patient receiving antimicrobial therapy who already had negative sets of blood cultures; a patient with clinical evidence of endocarditis and negative blood cultures and who is receiving antimicrobial therapy; and a patient admitted to the hospital with sepsis who had been given antimicrobial therapy prior to admission. The resin-containing bottles should not be used to follow the effectiveness of therapy because the resin may absorb antimicrobials in the specimen and allow the culture to turn positive in spite of clinically efficacious therapy.

It is necessary to determine the significance of a positive blood culture. The following criteria may be helpful in differentiating "true positives" from contaminated specimens:

1. Growth of the same organism in repeated cultures obtained at different times from separate anatomic sites strongly suggests true bacteremia.
2. Growth of different organisms in different culture bottles suggests contamination but occasionally may follow clinical problems such as enterovascular fistulas.
3. Growth of normal skin flora, eg, coagulase-negative staphylococci, diphtheroids (corynebacteria and propionibacteria), or anaerobic gram-positive cocci, in only one of several cultures suggests contamination. Growth of such organisms in more than one culture or from specimens from a patient with a vascular prosthesis or central venous catheter enhances the likelihood that clinically significant bacteremia exists.
4. Organisms such as viridans streptococci or enterococci are likely to grow in blood cultures from patients suspected to have endocarditis, and gram-negative rods such as *E coli* in blood cultures from patients with clinical gram-negative sepsis. Therefore, when such "expected" organisms are found, they are more apt to be etiologically significant.

The following are the bacterial species most commonly recovered in positive blood cultures: staphylococci, including *S aureus*; viridans streptococci; enterococci, including *Enterococcus faecalis*; gram-negative enteric bacteria, including *E coli* and *K pneumoniae*; *P aeruginosa*; pneumococci; and *H influenzae*. *Candida* species, other yeasts, and some dimorphic fungi such as *H capsulatum* grow in blood cultures, but many fungi are rarely, if ever, isolated from blood. Cytomegalovirus and herpes simplex virus can occasionally be cultured from blood, but most viruses and rickettsiae and chlamydiae are not cultured from blood. Parasitic protozoa and helminths do not grow in blood cultures.

In most types of bacteremia, examination of direct blood smears is not useful. Diligent examination of Gram-stained smears of the buffy coat from anticoagulated blood will occasionally show bacteria in patients with *S aureus* infection, clostridial sepsis, or relapsing fever. In some microbial infections (eg, anthrax, plague, relapsing fever, rickettsiosis, leptospirosis, spirillosis, psittacosis), inoculation of blood into animals may give positive results more readily than does culture. In practicality, this is almost never done.

Urine

Bacteriologic examination of the urine is done mainly when signs or symptoms point to urinary tract infection, renal insufficiency, or hypertension. It should always be done in persons with suspected systemic infection or fever of unknown origin. It is desirable for women in the first trimester of pregnancy.

Urine secreted in the kidney is sterile unless the kidney is infected. Uncontaminated bladder urine is also normally sterile. The urethra, however, contains a normal flora, so that normal voided urine contains small numbers of bacteria. Because it is necessary to distinguish contaminating organisms from etiologically important organisms, only *quantitative* urine examination can yield meaningful results.

The following steps are essential in proper urine examination.

A. Proper Collection of Specimen

Proper collection of the specimen is the single most important step in a urine culture and the most difficult. Satisfactory specimens from females are problematic.

1. Have at hand a sterile, screw-cap specimen container and two to three gauze sponges soaked with nonbacteriostatic saline (antibacterial soaps for cleansing are not recommended).
2. Spread the labia with two fingers and keep them spread during the cleansing and collection process. Wipe the urethra area once from front to back with each of the saline gauzes.
3. Start the urine stream and, using the urine cup, collect a midstream specimen. Properly label the cup.

The same method is used to collect specimens from males; the foreskin should be kept retracted in uncircumcised males.

Catheterization carries a risk of introducing microorganisms into the bladder, but it is sometimes unavoidable. Separate specimens from the right and left kidneys and ureters can be obtained by the urologist using a catheter at cystoscopy. When an indwelling catheter and closed collection system are in place, urine should be obtained by sterile aspiration of the catheter with needle and syringe, not from the collection bag. To resolve diagnostic problems, urine can be aspirated aseptically directly from the full bladder by means of suprapubic puncture of the abdominal wall.

For most examinations, 0.5 mL of ureteral urine or 5 mL of voided urine is sufficient. Because many types of microorganisms multiply rapidly in urine at room or body temperature, urine specimens must be delivered to the laboratory rapidly or refrigerated not longer than overnight.

B. Microscopic Examination

Much can be learned from simple microscopic examination of urine. A drop of fresh uncentrifuged urine placed on a slide, covered with a coverglass, and examined with restricted light intensity under the high-dry objective of an ordinary clinical microscope can reveal leukocytes, epithelial cells, and bacteria if more than 10^5/mL are present. Finding 10^5 organisms per milliliter in a properly collected and examined urine specimen is strong evidence of active urinary tract infection. A Gram-stained smear of uncentrifuged midstream urine that shows gram-negative rods is diagnostic of a urinary tract infection.

Brief centrifugation of urine readily sediments pus cells, which may carry along bacteria and thus may help in microscopic diagnosis of infection. The presence of other formed elements in the sediments—or the presence of proteinuria—is of little direct aid in the specific identification of active urinary tract infection. Pus cells may be present without bacteria, and, conversely, bacteriuria may be present without pyuria. The presence of many squamous epithelial cells, lactobacilli, or mixed flora on culture suggests improper urine collection.

Some urine dipsticks contain leukocyte esterase and nitrite, measurements of polymorphonuclear cells and bacteria, respectively, in the urine. Positive reactions are strongly suggestive of bacterial urinary tract infection.

Although not readily embraced by clinical microbiology laboratories, many chemistry laboratories have implemented automated or semiautomated instruments for routine performance of urinalysis. A variety of techniques are used by these instruments to detect leukocytes and bacteria. The performance of these systems varies, but they bring a level of standardization for high volume testing that may not be accomplished using dipstick methods.

C. Culture

Culture of the urine, to be meaningful, must be performed quantitatively. Properly collected urine is cultured in measured amounts on solid media, and the colonies that appear after incubation are counted to indicate the number of bacteria per milliliter. The usual procedure is to spread 0.001–0.05 mL of undiluted urine on blood agar plates and other solid media for quantitative culture. All media are incubated overnight at 37°C; growth density is then compared with photographs of different densities of growth for similar bacteria, yielding semiquantitative data.

In active pyelonephritis, the number of bacteria in urine collected by ureteral catheter is relatively low. While accumulating in the bladder, bacteria multiply rapidly and soon reach numbers in excess of 10^5/mL—far more than could occur as a result of contamination by urethral or skin flora or from the air. Therefore, it is generally agreed that if more than 10^5 colonies/mL are cultivated from a properly collected and properly cultured urine specimen, this constitutes strong evidence of active urinary tract infection. The presence of more than 10^5 bacteria of the same type per milliliter in two consecutive specimens establishes a diagnosis of active infection of the urinary tract with 95% certainty. If fewer bacteria are cultivated, repeated examination of urine is indicated to establish the presence of infection.

The presence of fewer than 10^4 bacteria per milliliter, including several different types of bacteria, suggests that organisms come from the normal flora and are contaminants, usually from an improperly collected specimen. The presence of 10^4/mL of a single type of enteric gram-negative rod is strongly suggestive of urinary tract infection, especially in men. Occasionally, young women with acute dysuria and urinary tract infection will have 10^2–10^3/mL. If cultures are negative but clinical signs of urinary tract infection are present, "urethral syndrome," ureteral obstruction, tuberculosis of the bladder, gonococcal infection, or other disease must be considered.

Cerebrospinal Fluid

Meningitis ranks high among medical emergencies, and early, rapid, and precise diagnosis is essential. Diagnosis of meningitis depends upon maintaining a high index of suspicion, securing adequate specimens properly, and examining the specimens promptly. Because the risk of death or irreversible damage is great unless treatment is started immediately, there is rarely a second chance to obtain pretreatment specimens, which are essential for specific etiologic diagnosis and optimal management.

The most urgent diagnostic issue is the differentiation of acute purulent bacterial meningitis from "aseptic" and granulomatous meningitis. The immediate decision is usually based on the cell count, the glucose concentration and protein content of cerebrospinal fluid, and the results of microscopic search for microorganisms (see Case 1, Chapter 48). The initial impression is modified by the results of culture, serologic tests, nucleic acid amplification tests, and other laboratory procedures. In evaluating the results of cerebrospinal fluid glucose determinations, the simultaneous blood glucose level must be considered. In some central nervous system neoplasms, the cerebrospinal fluid glucose level is low.

A. Specimens

As soon as infection of the central nervous system is suspected, blood samples are taken for culture, and cerebrospinal fluid is obtained. To obtain cerebrospinal fluid, perform lumbar puncture with strict aseptic technique, taking care not to risk compression of the medulla by too rapid withdrawal of fluid when the intracranial pressure is markedly elevated. Cerebrospinal fluid is usually collected in three to four portions of 2–5 mL each, in sterile tubes. This permits the most convenient and reliable performance of tests to determine the several different values needed to plan a course of action.

B. Microscopic Examination

Smears are made from the sediment of centrifuged cerebrospinal fluid. Using a cytospin centrifuge to prepare the slides for staining is recommended because it concentrates cellular material and bacterial cells more effectively than standard centrifugation. Smears are stained with Gram stain. Study of stained smears under the oil immersion objective may reveal intracellular gram-negative diplococci (meningococci), intra- and extracellular lancet-shaped gram-positive diplococci (pneumococci), or small gram-negative rods (*H influenzae* or enteric gram-negative rods).

C. Antigen Detection

Cryptococcal antigen in cerebrospinal fluid may be detected by a latex agglutination or EIA test. *S pneumoniae* antigen may be detected by membrane immunoassay.

D. Culture

The culture methods used must favor the growth of microorganisms most commonly encountered in meningitis. Sheep blood and chocolate agar together grow almost all bacteria and fungi that cause meningitis. The diagnosis of tuberculous meningitis requires cultures on special media (see Table 47–2 and Chapter 23). Virus isolation can be attempted in aseptic meningitis or meningoencephalitis. The virus can be successfully isolated from the cerebrospinal fluid in infections caused by mumps virus, herpes simplex meningitis, and with some enteroviruses. Most viral central nervous system infections are best detected by nucleic acid amplification methods.

E. Follow-Up Examination of Cerebrospinal Fluid

The return of the cerebrospinal fluid glucose level and cell count toward normal is good evidence of adequate therapy. The clinical response is of paramount importance.

Respiratory Secretions

Symptoms or signs often point to involvement of a particular part of the respiratory tract, and specimens are chosen accordingly. In interpreting laboratory results, it is necessary to consider the normal microbial flora of the area from which the specimen was collected.

A. Specimens

1. Throat—Most "sore throats" are due to viral infection. Only 5–10% of "sore throats" in adults and 15–20% in children are associated with bacterial infections. The finding of a follicular yellowish exudate or a grayish membrane must arouse the suspicion that Lance-field group A β-hemolytic streptococcal, diphtherial, gonococcal, fusospirochetal, or candidal infection exists; such signs may also be present in infectious mononucleosis, adenovirus, and other virus infections.

Throat swabs are taken from each tonsillar area and from the posterior pharyngeal wall. The normal throat flora includes an abundance of viridans streptococci, neisseriae, diphtheroids, staphylococci, small gram-negative rods, and many other organisms. Microscopic examination of smears from throat swabs is of little value in streptococcal infections, because all throats harbor a predominance of streptococci.

Cultures of throat swabs are most reliable if inoculated promptly after collection. Media selective for streptococci can be used to culture for group A organisms. In streaking selective media for streptococci or blood agar culture plates, it is essential to spread a small inoculum thoroughly and avoid overgrowth by normal flora. This can be done readily by touching the throat swab to one small area of the plate and using a second, sterile applicator (or sterile bacteriologic loop) to streak the plate from that area. Detection of β-hemolytic colonies is facilitated by slashing the agar (to provide reduced oxygen tension) and incubating the plate for 2 days at 37°C.

Over the last two decades, a variety of antigen detection tests, probe methods, and nucleic acid amplification tests have been developed to enhance the detection of *Streptococcus pyogenes* from throat swabs in patients with acute streptococcal pharyngitis. In many circumstances, the rapid detection methods have been shown to be as sensitive as culture and have replaced culture in many laboratories. It is important that the user realizes that only *S pyogenes* will be detected or excluded by these tests, and thus, they cannot be relied upon to diagnose bacterial pharyngitis caused by other pathogens. One algorithm is to begin with a rapid test and send for culture those specimens that are negative by the rapid test.

2. Nasopharynx—Specimens from the nasopharynx are studied infrequently because special techniques must be used to obtain them. (See Diagnosis of Viral Infections, below.) Whooping cough is diagnosed by culture of *B pertussis* from nasopharyngeal or nasal washings or, by PCR amplification of *B pertussis* DNA in the specimen.

3. Middle ear—Specimens are rarely obtained from the middle ear because puncture of the drum is necessary. In acute otitis media, 30–50% of aspirated fluids are bacteriologically sterile. The most frequently isolated bacteria are pneumococci, *H influenzae, Moraxella catarrhalis,* and hemolytic streptococci.

4. Lower respiratory tract—Bronchial and pulmonary secretions of exudates are often studied by examining sputum. The most misleading aspect of sputum examination is the almost inevitable contamination with saliva and mouth flora. Thus, finding candida, *S aureus,* or even *S pneumoniae* in the sputum of a patient with pneumonitis has no etiologic significance unless supported by the clinical picture. Meaningful sputum specimens should be expectorated from the lower respiratory tract and should be grossly distinct from saliva. The presence of many squamous epithelial cells suggests heavy contamination with saliva; a large number of polymorphonuclear leukocytes (PMNs) suggests a purulent exudate. Sputum may be induced by the inhalation of heated hypertonic saline aerosol for several minutes. In pneumonia accompanied by a pleural effusion, the pleural fluid may yield the causative organisms more reliably than does sputum. In suspected tuberculosis, gastric washings (swallowed sputum) may yield organisms when expectorated material is not obtainable, eg, in the pediatric patient.

5. Transtracheal aspiration, bronchoscopy, lung biopsy, bronchoalveolar lavage—The flora in such specimens often reflects accurately the events in the lower respiratory tract. Specimens obtained by bronchoscopy may be necessary in the diagnosis of pneumocystis pneumonia or infection due to legionella or other organisms.

Bronchoalveolar lavage specimens are particularly useful in immunocompromised patients with diffuse pneumonia.

B. Microscopic Examination

Smears of purulent flecks or granules from sputum stained by Gram stain or acid-fast methods may reveal causative organisms and PMNs. A direct "quellung" test for pneumococci can be performed with polyvalent serum on fresh sputum.

C. Culture

The media used for sputum cultures must be suitable for the growth of bacteria (eg, pneumococci, *Klebsiella*), fungi (eg, *C immitis*), mycobacteria (eg, *M tuberculosis*), and other organisms. Specimens obtained by bronchoscopy and lung biopsy should also be cultured on other media (eg, for anaerobes, *Legionella*, and others). The relative prevalence of different organisms in the specimen must be estimated. Only a finding of one predominant organism or the simultaneous isolation of an organism from both sputum and blood can clearly establish its role in a pneumonic or suppurative process.

Gastrointestinal Tract Specimens

Acute symptoms referable to the gastrointestinal tract, particularly nausea, vomiting, and diarrhea, are commonly attributed to infection. In reality, most such attacks are caused by intolerance to food or drink, enterotoxins, drugs, or systemic illnesses.

Many cases of acute infectious diarrhea are due to viruses, which cannot be grown in tissue culture. On the other hand, many viruses that can be grown in culture (eg, adenoviruses, enteroviruses) can multiply in the gut without causing gastrointestinal symptoms. Similarly, some enteric bacterial pathogens may persist in the gut following an acute infection. Thus, it may be difficult to assign significance to a bacterial or viral agent cultured from the stool, especially in subacute or chronic illness.

These considerations should not discourage the physician from attempting laboratory isolation of enteric organisms but should constitute a warning of some common difficulties in interpreting the results.

The lower bowel has an exceedingly large normal bacterial flora. The most prevalent organisms are anaerobes (*Bacteroides*, gram-positive rods, and gram-positive cocci), gram-negative enteric organisms, and *E faecalis.* Any attempt to recover pathogenic bacteria from feces involves separation of pathogens from the normal flora, usually through the use of differential selective media and enrichment cultures. Important causes of acute gastroenteritis include viruses, toxins (of staphylococci, clostridia, vibrios, toxigenic *E coli*), invasive enteric gram-negative rods, slow lactose fermenters, shigellae and salmonellae, and campylobacters. The relative importance of these groups of organisms differs greatly in various parts of the world.

A. Specimens

Feces and rectal swabs are the most readily available specimens. Bile obtained by duodenal drainage may reveal infection of the biliary tract. The presence of blood, mucus, or helminths must be noted on gross inspection of the specimen. Leukocytes seen in suspensions of stool examined microscopically or detection of the leukocyte derived protein lactoferrin are useful means of differentiating invasive from noninvasive infectious diarrheas. However, it is important to note that leukocytes may be present in noninfectious, inflammatory conditions of the GI tract. Special techniques must be used to search for parasitic protozoa and helminths and their ova. Stained smears may reveal a prevalence of leukocytes and certain abnormal organisms, eg, candida or staphylococci, but they cannot be used to differentiate enteric bacterial pathogens from normal flora.

B. Culture

Specimens are suspended in broth and cultured on ordinary as well as differential media (eg, MacConkey agar, EMB agar) to permit separation of nonlactose fermenting gram-negative rods from other enteric bacteria. If salmonella infection is suspected, the specimen is also placed in an enrichment medium (eg, selenite F broth) for 18 hours before being plated on differential media (eg, Hektoen enteric or Shigella-Salmonella agar). *Yersinia enterocolitica* is more likely to be isolated after storage of fecal suspensions for 2 weeks at 4°C, but it can be isolated on yersinia or Shigella-Salmonella agar incubated at 25°C. Vibrios grow best on thiosulfate citrate bile salts sucrose agar. Thermophilic campylobacters are isolated on Campy agar or Skirrow's selective medium incubated at 40–42°C in 10% CO_2 with greatly reduced O_2 tension. Bacterial colonies are identified by standard bacteriologic methods. Agglutination of bacteria from suspect colonies by pooled specific antiserum is often the fastest way to establish the presence of salmonellae or shigellae in the intestinal tract.

C. Non–Culture-Based Methods

EIAs for detection of specific enteric pathogens, either directly in stool specimens or to confirm growth in broth or on plated media, are available. EIAs that detect Shiga toxins 1 and 2 in suspected cases of colitis caused by enterohemorrhagic *E coli* (also called Shiga toxin producing *E coli* or STEC) are available and are superior to culture. Also available are EIAs for direct detection of viral pathogens such as rotavirus, adenoviruses 40, 41 and noroviruses; bacterial pathogens such as *Campylobacter jejuni*; and the protozoan parasites *Giardia lamblia, Cryptosporidium parvum,* and *Entamoeba histolytica.* The performance of these assays is variable.

Intestinal parasites and their ova are discovered by repeated microscopic study of fresh fecal specimens. The specimens require special handling in the laboratory (see Chapter 46).

Sexually Transmitted Diseases

The causes of the genital discharge of urethritis in men are *N gonorrhoeae, C trachomatis,* and *Ureaplasma urealyticum.* Endocervicitis in women is caused by *N gonorrhoeae* and *C trachomatis.* The genital sores associated with diseases in both men and women are often herpes simplex, less commonly syphilis or chancroid, uncommonly lymphogranuloma venereum, and rarely granuloma inguinale. Each of these diseases has a characteristic natural history and evolution of lesions, but one can mimic another. The laboratory diagnosis of most of these infections is covered elsewhere in this book. A few diagnostic tests are listed below and outlined in Table 47–2.

A. Gonorrhea

A stained smear of a urethral or a cervical exudate that shows intracellular gram-negative diplococci strongly suggests gonorrhea. The sensitivity is about 90% for men and 50% for women—thus, culture or a nucleic acid amplification test is recommended for women. Exudate, rectal swab, or throat swab must be plated promptly on special media to yield *N gonorrhoeae*. Molecular methods to detect *N gonorrhoeae* DNA in urethral or cervical exudates or urine can be done, but falsely positive tests from detection of DNA sequences of nonpathogenic neisseriae may occur and varies with the type of commercial platforms. Serologic tests are not helpful.

B. Chlamydial Genital Infections

See section on the diagnosis of chlamydial infections, below.

C. Genital Herpes

See Chapter 33 and the section on the diagnosis of viral infections, below.

D. Syphilis

Darkfield or immunofluorescence examination of tissue fluid expressed from the base of the chancre may reveal typical *T pallidum.* Serologic tests for syphilis become positive 3–6 weeks after infection. A positive flocculation test (eg, VDRL or RPR) requires confirmation. A positive immunofluorescent treponemal antibody test (eg, FTA-ABS, *T pallidum* particle agglutination [TP-PA]—see Chapter 24) proves syphilitic infection.

E. Chancroid

Smears from a suppurating lesion usually show a mixed bacterial flora. Swabs from lesions should be cultured at 33°C on two or three media that are selective for *Haemophilus ducreyi*. Serologic tests are not helpful.

F. Granuloma Inguinale

Klebsiella (formerly *Calymmatobacterium*) *granulomatis,* the causative agent of this hard, granulomatous, proliferating lesion, can be grown in complex bacteriologic media, but this is rarely attempted and very difficult to perform successfully. Histologic demonstration of intracellular "Donovan bodies" in biopsy material most frequently supports the clinical impression. Serologic tests are not helpful.

G. Vaginosis/Vaginitis

Bacterial vaginosis associated with *G vaginalis* or *Mobiluncus* (see Chapter 21 and Case 13, Chapter 48) is diagnosed in the examining room by inspection of the vaginal discharge; the discharge (1) is grayish and sometimes frothy, (2) has a pH above 4.6, (3) has an amine ("fishy") odor when alkalinized with potassium hydroxide, and (4) contains "clue cells," large epithelial cells covered with gram-negative or gram-variable rods. Similar observations are used to diagnose *Trichomonas vaginalis* (see Chapter 46) infection; the motile organisms can be seen in wet-mount preparations or cultured from genital discharge. *Candida albicans* vaginitis is diagnosed by finding pseudohyphae in a potassium hydroxide preparation of the vaginal discharge or by culture.

ANAEROBIC INFECTIONS

A large majority of the bacteria that make up the normal human microbiota are anaerobes. When displaced from their normal sites into tissues or body spaces, anaerobes may produce disease. Certain characteristics are suggestive of anaerobic infections: (1) They are often contiguous with a mucosal surface. (2) They tend to involve mixtures of organisms. (3) They tend to form closed-space infections, either as discrete abscesses (lung, brain, pleura, peritoneum, pelvis) or by burrowing through tissue layers. (4) Pus from anaerobic infections often has a foul odor. (5) Most of the pathogenetically important anaerobes except *Bacteroides* and some *Prevotella* species are highly susceptible to penicillin G. (6) Anaerobic infections are favored by reduced blood supply, necrotic tissue, and a low oxidation-reduction potential, all of which also interfere with delivery of antimicrobial drugs. (7) It is essential to use special collection methods, transport media, and sensitive anaerobic techniques and media to isolate the organisms. Otherwise, bacteriologic examination may be negative or yield only incidental aerobes. (See also Chapter 21.)

The following are sites of important anaerobic infections.

Respiratory Tract

Periodontal infections, perioral abscesses, sinusitis, and mastoiditis may involve predominantly *Prevotella melaninogenica, Fusobacterium*, and peptostreptococci. Aspiration of saliva may result in necrotizing pneumonia, lung abscess, and empyema. Antimicrobial drugs and postural or surgical drainage are essential for treatment.

Central Nervous System

Anaerobes rarely produce meningitis but are common causes of brain abscess, subdural empyema, and septic thrombophlebitis. The organisms usually originate in the respiratory tract and spread to the brain via extension or hematogenously.

Intra-abdominal & Pelvic Infections

The flora of the colon consists predominantly of anaerobes, 10^{11} per gram of feces. *B fragilis,* clostridia, and peptostreptococci play a main role in abscess formation originating in perforation of the bowel. *Prevotella bivia* and *Prevotella disiens* are important in abscesses of the pelvis originating in the female genital organs. Like *B fragilis,* these species are often relatively resistant to penicillin; therefore, clindamycin, metronidazole, or another effective agent should be used.

Skin & Soft Tissue Infections

Anaerobes and aerobic bacteria often join to form synergistic infections (gangrene, necrotizing fasciitis, cellulitis). Surgical drainage, excision, and improved circulation are the most important forms of treatment, while antimicrobial drugs act as adjuncts. It is usually difficult to pinpoint one specific organism as being responsible for the progressive lesion, since mixtures of organisms are usually involved.

DIAGNOSIS OF CHLAMYDIAL INFECTIONS

Although *C trachomatis, Chlamydophila pneumoniae,* and *Chlamydophila psittaci* are bacteria, they are obligate intracellular parasites. Cultures and other diagnostic tests for chlamydia require procedures much like those used in diagnostic virology laboratories rather than those used in bacteriology and mycology laboratories. Thus, the diagnosis of chlamydial infections is discussed in a separate section of this chapter. The laboratory diagnosis of chlamydial infections also is discussed in Chapter 27.

Specimens

For *C trachomatis* ocular and genital infections, the specimens for direct examination or culture must be collected from infected sites by vigorous swabbing or scraping of the involved epithelial cell surface. Cultures or purulent discharges are not adequate, and purulent material should be cleaned away before the specimen is obtained. Thus, for inclusion conjunctivitis, a conjunctival scraping is obtained; for urethritis, a swab specimen is obtained from several centimeters into the urethra; and for cervicitis, a specimen is obtained from the columnar cell surface of the endocervical canal. Swab or urine samples may be used for nucleic acid amplification tests. When upper genital tract infection is suspected in women, scrapings of the endometrium provide a good sample. Fluid obtained by culdocentesis or aspiration of the uterine tube has a low yield for *C trachomatis* on culture.

For *C pneumoniae,* use nasopharyngeal (not throat) swab specimens.

For lymphogranuloma venereum, aspirates of buboes or fluctuant nodes provide the best specimen for culture.

For psittacosis, culture of sputum, blood, or biopsy material may yield *C psittaci.* This is not done routinely in clinical laboratories, both because it requires specialized methods and because of the risk to laboratory personnel.

Swabs, scrapings, and tissue specimens should be placed in transport medium. A useful medium has 0.2 mol/L sucrose in 0.02 M phosphate buffer, pH 7.0–7.2, with 5% fetal calf serum. Other transport media may be equally suitable. The transport medium should contain antibiotics to suppress bacteria other than chlamydia species. Gentamicin, 10 µg/mL, vancomycin, 100 µg/mL, and amphotericin B, 4 µg/mL, can be used in combination since they do not inhibit chlamydia. If specimens cannot be processed rapidly, they can be refrigerated for 24 hours; otherwise, they should be frozen at –60°C or colder until processed.

Microscopy & Stains

Cytologic examination is important and useful only in the examination of conjunctival scrapings to diagnose inclusion conjunctivitis and trachoma caused by *C trachomatis.* Typical intracytoplasmic inclusions can be seen, classically with Giemsa-stained specimens. Fluorescein-conjugated monoclonal antibodies can be used for direct examination of specimens from the genital tract and ocular specimens but are not as sensitive as chlamydial culture or molecular diagnostic tests.

Culture

When culture is required, cell culture techniques are recommended for the isolation of *Chlamydia* species. Cell culture for *C trachomatis* and *C psittaci* usually involves inoculation of the clinical specimens onto cycloheximide-treated McCoy cells, whereas *C pneumoniae* requires pretreated HL or HEP-2 cells. One technique uses a confluent growth of McCoy cells on 13-mm coverslips in small disposable vials. The inoculum is placed in duplicate vials and centrifuged onto the monolayers at approximately 3000 × *g* followed by incubation at 35°C for 48–72 hours and stained. To detect *C trachomatis,* immunofluorescence, Giemsa's stain, or iodine stain is used to search for intracytoplasmic inclusions. Immunofluorescent techniques are the most sensitive of the three stains but require special IF reagents and microscopy. Giemsa is more sensitive than iodine, but the microscopy is more difficult.

A second culture technique uses McCoy cells in 96-well microdilution plates and either iodine or fluorescent antibody staining. Because the surface area of the monolayer is less and

the inoculum is smaller, the microdilution plate method is less sensitive than the coverslip-vial technique.

Inclusions of *C trachomatis* stain with iodine, but inclusions of *C pneumoniae* and *C psittaci* do not (see Chapter 27). These two species are distinguished from *C trachomatis* by their different responses to iodine staining and by their susceptibility to sulfonamide. *C pneumoniae* in culture can be detected by using a genus-specific monoclonal antibody or, even better, a species-specific monoclonal antibody. Serologic techniques for species differentiation are not practical, though *C trachomatis* can be typed by the microimmunofluorescent method.

Antigen Detection & Nucleic Acid Hybridization

EIAs are used for detecting chlamydial antigens in genital tract specimens from patients with sexually transmitted disease. Compared with the more sensitive culture techniques for chlamydia (see above), the EIA has a sensitivity of about 90% and a specificity of about 97% when used in populations with a moderate to high prevalence of infection (5–20%). In this setting, the sensitivity, specificity, and positive predictive values are roughly comparable to those for the DFA test.

Commercial kits with nonradioisotopic molecular probes for *C trachomatis* 16S RNA sequences have been marketed for direct detection of *C trachomatis* in clinical specimens. The overall sensitivity and specificity of this method are about 85% and 98–99%, respectively. Nucleic acid amplification tests have also been developed and marketed. These assays are based on PCR, transcription-mediated amplification or strand displacement amplification. These tests are much more sensitive than culture and other nonamplification tests, which have required redefinition of sensitivity in the laboratory documentation of chlamydial infection. The specificity of the tests appears to be close to 100%. (See Chapter 27.)

Serology

The complement fixation (CF) test is widely used to diagnose psittacosis. The serologic diagnosis of chlamydial infections is discussed in Chapter 27.

The microimmunofluorescence method is more sensitive than CF for measuring antichlamydial antibodies. The titer of IgG antibodies can be diagnostic when fourfold titer rises are seen in acute and convalescent sera. However, it may be difficult to show a rise in the IgG titer because of high background titers in the sexually active population. The measurement of IgM antibodies is particularly useful in the diagnosis of *C trachomatis* pneumonia in neonates. Babies born to mothers with chlamydial infections have serum IgG antichlamydial antibodies from the maternal circulation. Babies with ocular or upper respiratory tract infections have low titers of antichlamydial IgM, whereas babies with chlamydial pneumonia have antichlamydial IgM titers of 1:32 or greater.

DIAGNOSIS OF VIRAL INFECTIONS

Diagnostic virology requires communication between the physician and the laboratory and depends on the quality of specimens and information supplied to the laboratory.

The choice of methods for laboratory confirmation of a viral infection depends upon the stage of the illness (Table 47–4). Antibody tests require samples taken at appropriate intervals, and the diagnosis often is not confirmed until convalescence. Virus isolation or antigen detection is required (1) when new epidemics occur, as with influenza; (2) when serologic tests are not useful; and (3) when the same clinical illness may be caused by many different agents. For example, aseptic (nonbacterial) meningitis may be caused by many different viruses; similarly, respiratory disease syndromes may be caused by many viruses as well as by mycoplasmas and other agents.

Diagnostic methods based on nucleic acid amplification techniques will soon replace some but not all virus culture approaches. However, the need for appropriate sample collection and test interpretation will not change. Furthermore, there will be times when recovery of the infectious agent is desired.

Isolation of a virus may not establish the cause of a given disease. Many other factors must be considered. Some viruses persist in human hosts for long periods of time, and therefore the isolation of herpesviruses, poliovirus, echoviruses, or coxsackieviruses from a patient with an undiagnosed illness does not prove that the virus is the cause of the disease. A consistent clinical and epidemiologic pattern must be established before it can be determined that a particular agent is responsible for a specific clinical picture.

Many viruses are most readily isolated during the first few days of illness. The specimens to be used in virus isolation attempts are listed in Table 47–5. A correlation of virus isolation and antibody presence helps in making the diagnosis, but it is rarely done.

Specimens can be refrigerated for up to 24 hours before virus cultures are done, with the exception of respiratory syncytial and certain other viruses. Otherwise, material should be frozen (preferably at –60°C or colder) if there is a delay in bringing it to the laboratory. Specimens that should not be frozen include (1) whole blood drawn for antibody determination, from which the serum must be separated before freezing; and (2) tissue for organ or cell culture, which should be kept at 4°C and taken to the laboratory promptly.

Virus is present in respiratory illnesses in pharyngeal or nasal secretions. Virus can be demonstrated in the throat fluid and scrapings from the base of vesicular rashes. In eye infections, virus is detectable in conjunctival swabs or scrapings and in tears. Encephalitides are usually diagnosed more readily by serologic means or nucleic acid amplification methods. Arboviruses and herpesviruses are not usually recovered from spinal fluid, but brain tissue from patients with viral encephalitis may yield the causative virus. In illnesses associated with enteroviruses, such as central nervous system disease, acute pericarditis, and myocarditis, viruses can be isolated from feces, throat swabs, or cerebrospinal fluid. However, as previously stated, nucleic acid amplification methods are the preferred methods for detecting enteroviruses in cerebrospinal fluid. Direct fluorescent antibody tests may be as sensitive as culture for detection of respiratory tract infections with respiratory syncytial virus, influenza viruses A and B, parainfluenza viruses, and adenoviruses. These tests provide answers within a few hours after collection of the specimen compared with days for virus culture, and for that reason they have become the tests of choice for etiologic diagnosis of respiratory tract viral infections.

Direct Examination of Clinical Material: Microscopy & Stains

Viral diseases for which direct microscopic examination of imprints or smears has proved useful include rabies and herpes simplex infection and varicella-zoster skin infections. Staining of viral antigens by immunofluorescence in a brain smear and corneal impressions from the rabid animal and from the skin of the nape of the neck of humans is the method of choice for routine diagnosis of rabies.

Virus Culture

A. Preparation of Inocula

Bacteria-free fluid materials such as cerebrospinal fluid, whole blood, plasma, or white blood cell buffy coat layer may be inoculated into cell cultures directly or after dilution with buffered phosphate solution (pH 7.6). Inoculation of embryonated eggs or animals for virus isolation is generally performed only in specialized laboratories.

Tissue is washed in media or sterile water, minced into small pieces with scissors, and ground to make a homogeneous paste. Diluent is added in amounts sufficient to make

TABLE 47–4   Relation of Stage of Illness to Presence of Virus in Test Materials & to Appearance of Specific Antibody

Stage or Period of Illness	Virus Detectable in Test Materials	Specific Antibody Demonstrable[a]
Incubation	Rarely	No
Prodrome	Occasionally	No
Onset	Frequently	Occasionally
Acute phase	Frequently	Frequently
Recovery	Rarely	Usually
Convalescence	Very rarely	Usually

[a]Antibody may be detected very early in previously vaccinated persons.

a concentration of 10–20% (weight/volume). This suspension can be centrifuged at low speed (not >2000 rpm) for 10 minutes to sediment insoluble cellular debris. The supernatant fluid may be inoculated; if bacteria are present, they are eliminated as discussed below.

Tissues may also be trypsinized, and the resulting cell suspension may be (1) inoculated on an existing tissue culture cell monolayer or (2) cocultivated with another cell suspension of cells known to be virus-free.

If the material to be tested contains bacteria (throat washings, stools, urine, infected tissue, or insects), they must be inactivated or removed before inoculation.

1. Bactericidal agents—Antibiotics are commonly employed in combination with differential centrifugation (see below).

2. Mechanical methods

a. *Filters*—Millipore-type membrane filters of cellulose acetate or similar inert material are preferred.

b. *Differential centrifugation*—This is a convenient method for removing many bacteria from heavily contaminated preparations of small viruses. Bacteria are sedimented at low speeds that do not sediment the virus, and high-speed centrifugation then sediments the virus. The virus-containing sediment is then resuspended in a small volume.

B. Cultivation in Cell Culture

Cell culture techniques are being replaced by antigen detection methods and nucleic acid amplification tests. However, they are still useful and are still practiced in clinical and public health virology laboratories. When viruses multiply in cell culture, they produce biologic effects (eg, cytopathic changes, viral interference, production of a hemagglutinin) that permit identification of the agent.

Test tube cultures are prepared by adding cells suspended in 1–2 mL of nutrient fluid that contains balanced salt solutions and various growth factors (usually serum, glucose, amino acids, and vitamins). Cells of fibroblastic or epithelial nature attach and grow on the wall of the test tube, where they may be examined with the aid of a low-power microscope.

With many viruses, growth of the agent is paralleled by degeneration of these cells (Figure 47–1). Some viruses produce a characteristic cytopathic effect (CPE) in cell culture, making a rapid presumptive diagnosis possible when the clinical syndrome is known. For example, respiratory syncytial virus characteristically produces multinucleated giant cells, whereas adenoviruses produce grape-like clusters of large round cells. Some viruses (eg, rubella virus) produce no direct cytopathic changes but can be detected by their interference with the CPE of a second challenge virus (viral interference). Influenza viruses and some paramyxoviruses may be detected within 24–48 hours if erythrocytes are added to infected cultures. Viruses maturing at the cell membrane produce a

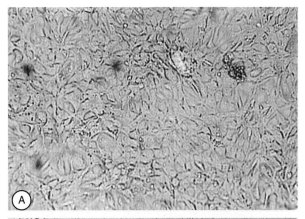

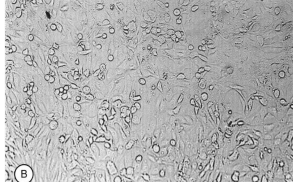

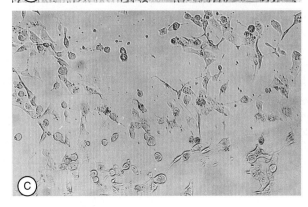

FIGURE 47–1 **A:** Monolayer of normal unstained monkey kidney cells in culture (120×). **B:** Unstained monkey kidney cell culture showing early stage of cytopathic effects typical of enterovirus infection (120×). Approximately 25% of the cells in the culture show cytopathic effects indicative of viral multiplication (1+ cytopathic effects). **C:** Unstained monkey kidney cell culture illustrating more advanced enteroviral cytopathic effect (3+ to 4+ cytopathic effects) (120×). Almost 100% of the cells are affected, and most of the cell sheet has come loose from the wall of the culture tube.

hemagglutinin that enables the erythrocytes to adsorb at the cell surface (hemadsorption). The identity of a virus isolate is established with type-specific antiserum, which inhibits virus growth or which reacts with viral antigens.

Some viruses can be grown in culture, but this is a very slow and arduous process. Alternative tests instead of culture are used to diagnose such infections (see below).

TABLE 47–5 Viral Infections: Agents, Specimens, & Diagnostic Tests

Syndrome and Virus	Specimen	Detection System	Comments
Respiratory diseases			
Influenza viruses	Nasopharyngeal washings or swab, sputum, invasively obtained respiratory specimens	Cell culture (PMK, MDCK), embryonated eggs, direct FA, EIA, nucleic acid amplification	Virus detected by hemadsorption of guinea pig erythrocytes in 2–4 days. HI or IF used to identify virus; nucleic acid amplification tests may be available in reference and public health labs
Parainfluenza viruses	Nasopharyngeal washings or swab, sputum	Cell culture (PMK, LLC-MK2), direct FA, nucleic acid amplification	Virus detected by hemadsorption of guinea pig erythrocytes in 4–7 days. HI, IF, and HAI used to identify the virus
Respiratory syncytial virus	Nasopharyngeal washings	Cell culture (HEL, HeLa, HEp-2), direct FA, nucleic acid amplification	CPE usually visible in 1–7 days. Detect antigen by EIA
Adenovirus	Nasopharyngeal washings or swab, feces, conjunctival swab	Cell culture (HEp-2, HEK), direct FA, nucleic acid amplification; EIA for enteric adenoviruses	CPE usually visible in 3–7 days. IF used to identify virus
Rhinoviruses	Nasopharyngeal washings or swab	Cell culture (HEL), nucleic acid amplification	
Enteroviruses	Nasopharyngeal washings or swab, feces	Cell culture (PMK, HEL), nucleic acid amplification	Coxsackievirus A rarely grows in tissue culture; nucleic acid amplification tests for central nervous system infections are preferred
Febrile diseases			
Dengue, other arboviruses	Serum, CSF, autopsy specimens, vector (*Aedes* mosquito)	Suckling mice, cell culture (Vero)	Many viruses in this group are highly infectious and easily transmissible to laboratory personnel. Some should only be studied in self-contained laboratories with controlled access. Use serology
Hemorrhagic fevers			
See Chapter 38.	Serum, blood	Suckling mice, cell culture (Vero)	See comment for febrile diseases
LCM			
LCM virus	Blood, CSF	Cell culture (Vero, BHK), suckling mice	IF and neutralization in mice used for identification of virus
Lassa fever			
Lassa virus	Blood, nasopharyngeal swab, exudates	Cell culture (Vero, BHK)	Lassa virus isolation is restricted to self-contained laboratories with controlled access
Encephalitis			
Arboviruses	Serum, CSF, nasopharyngeal swab	Suckling mice, cell culture (Vero)	See comment for febrile diseases
Enteroviruses	Feces, throat swab, CSF	Cell culture (PMK, HEL); nucleic acid amplification tests	
Rabies virus	Saliva, brain biopsy, skin biopsy (nape of neck)	Suckling mice, direct IF	Direct IF is preferable because speed of diagnosis is important for effective treatment
Herpesvirus	CSF	PCR	

Meningitis			
Enterovirus	CSF	Cell culture (PMK, HEL); nucleic acid amplification tests are preferred	
Mumps virus	CSF, nasopharyngeal swab, urine	Cell culture (PMK)	Virus detected by hemadsorption of guinea pig erythrocytes in 4–7 days. HAI and IF used to identify virus in culture; PCR is available in public health laboratories
Infectious mononucleosis			
Epstein-Barr (EB) virus	Blood, nasopharyngeal swab	Lymphoid cell culture, PCR; serologic testing	Culture of EB virus not performed routinely in clinical virology laboratories
Cytomegalovirus	Blood, urine, throat swab	Cell culture (HFF); shell vial culture, antigen detection, PCR	Tissue culture tubes should be held 4 weeks; shell vial 24 hours with early antigen staining by DFA
Hepatitis (see Chapter 37 for available and indicated tests)			
Hepatitis A virus	Serum, feces	IEM, EIA, RT-PCR	
Hepatitis B virus	Serum	EIA, PCR	
Hepatitis C virus	Serum	EIA, PCR	
Hepatitis D virus	Serum	EIA	
Enteritis			
Rotavirus	Feces	EIA	
Norwalk agent, caliciviruses, astroviruses	Feces	IEM	
Exanthems			
Varicella-zoster virus	Vesicle fluid	Cell culture (HEK). Direct fluorescent antibody is most sensitive.	CPE usually visible in 4 days to 2 weeks
Measles (rubeola) virus	Nasopharyngeal swab, blood, urine	Cell culture (PMK, HEK). Direct fluorescent antibody.	CPE usually visible in 2–3 weeks; serology
Rubella virus	Nasopharyngeal swab, blood, urine	Cell culture (AGMK, Vero)	Serology
Monkeypox, cowpox, vaccinia, and tanapoxviruses	Vesicle fluid	Embryonated eggs, electron microscopy	Testing performed only in public health laboratories
Herpes simplex virus	Vesicles, usually oral or genital	Cell culture (HFF, Vero); PCR	Cultures usually become positive in 24–72 hours; direct IF is rapid
Parvovirus	Blood	Serology, PCR	
Parotitis			
Mumps virus	Nasopharyngeal swab, urine	Cell culture (PMK)	See comment for Meningitis, above; serology

Continued

TABLE 47–5 Viral Infections: Agents, Specimens, & Diagnostic Tests (Continued)

Syndrome and Virus	Specimen	Detection System	Comments
Congenital anomalies			
Cytomegalovirus	Urine, throat swab	See Infectious Mononucleosis, above	
Rubella	Throat swab, CSF, blood	See Exanthems, above	Serology
Conjunctivitis			
Herpes simplex Herpes zoster	Conjunctival swabs, tears	See Exanthems, above	
Adenovirus Enterovirus		See Respiratory Diseases, above	
AIDS (acquired immunodeficiency syndrome)			
Human immunodeficiency virus	Blood, particularly leukocytes	Cell culture patient's PBC (not done in most clinical labs)	Antibody by EIA, confirmed by Western blot. Virus almost always detectable in PBC even when serum antibodies present; RT-PCR
Papovavirus infections			
Human papovavirus JC	Brain, urine, tissue specimens	Cell culture (HFF), EM	
Human papovavirus BK		Nucleic acid amplification	
Papillomaviruses	Biopsies, warts	DNA, IF	DNA hybridization performed on liquid Pap smears

AGMK, African green monkey kidney; BHK, baby hamster kidney; CF, complement fixation; CPE, cytopathic effect; EIA, enzyme immunoassay; FA, fluorescent antibody; HAI, hemadsorption inhibition; HEK, human embryonic kidney; HEL, human embryonic lung; HeLa, human epithelial cell line; HEp-2, human epithelial type 2 cells; HFF, human foreskin fibroblasts; HI, hemagglutination inhibition; IEM, immune electron microscopy; IF, immunofluorescent antibody; LLC-MK2, monkey kidney cell line; LCM, lymphocytic choriomeningitis; MDCK, dog kidney cell line; PBC, peripheral blood cells; PCR, polymerase chain reaction; PMK, primary monkey kidney; RIA, radioimmunoassay; RT-PCR, reverse transcriptase polymerase chain reaction; Vero, monkey kidney cell line.

[a]A variety of tests using patients' sera and known viral antigens are used for diagnosis. Similar tests using the virus isolated from the patient and known antisera are used to identify the infecting agent. These tests include neutralization of viral replication (inhibition of cytopathic effect), complement fixation, and others listed in the key.

C. Shell Vial Cultures

This method allows for rapid detection of viruses in clinical specimens. It has been adapted for several viruses, including cytomegalovirus and varicella-zoster virus. For example, CMV can be detected in 18–24 hours, compared with 2–4 weeks for classic cell culture; the sensitivities of shell vial and classic cell cultures for CMV are comparable. Monolayers of the appropriate cell line (eg, MRC-5 cells for CMV) are grown on coverslips in 15 × 45-mm shell vials. After inoculation with the specimen, the vials are centrifuged at $700 \times g$ for 40 minutes at room temperature. The vials are incubated at 37°C for 16–24 hours, fixed, and reacted with a monoclonal antibody specific for a CMV nuclear protein that is present very early in the culture; several such antibodies are commercially available. Direct or indirect antibody staining methods and fluorescence microscopy are used to determine positive shell vial cultures. Positive and negative control vials are included in each test run. A modification of the shell vial technique has been developed to allow the simultaneous recovery and detection of multiple respiratory viruses using R-Mix cells. This method typically is available through a commercial company Diagnostic Hybrids, Inc. Athens Ohio. One vial contains (mixes) two cell lines such as human lung carcinoma A549 and mink lung fibroblast Mv1Lu cells. The laboratory will typically inoculate two such vials. After 18–24 hours of incubation, one vial is stained using a "pooled" immunofluorescent antibody reagent that detects all of the common respiratory viruses. If the stain is positive, then the cells on the coverslip of the second vial are scraped, inoculated onto an eight-well slide, and then stained with individual monoclonal antibody reagents that detect the specific virus. Isolates are not obtained using the shell vial technique. If isolates are needed for susceptibility testing for antiviral drugs, the classic cell culture technique should be used.

Antigen Detection

The detection of viral antigens is widely used in diagnostic virology. Commercial kits are available to detect many viruses, including herpes simplex I and II, influenza A and B, respiratory syncytial virus, adenoviruses, parainfluenza viruses, rotavirus, and cytomegalovirus. Multiple types of assays are used as well: EIA, direct fluorescent antibody, indirect fluorescent antibody, latex agglutination, etc. The advantages of these procedures are that they allow detection of viruses that do not readily grow in cell culture (eg, rotaviruses, hepatitis A virus) or that grow very slowly (eg, cytomegalovirus). In general, the antigen detection assays for viruses are less sensitive than the viral culture and nucleic acid amplification methods.

Nucleic Acid Amplification & Detection

A wide variety of commercial assays are available to detect viral nucleic acid or to amplify and detect it. These procedures are rapidly becoming the standards for diagnostic virology, supplanting the traditional virus culture and antigen detection techniques. The methods include PCR, reverse transcriptase-PCR, and other proprietary methods. The procedures allow detection of viruses (eg, enteroviruses and many others) as well as quantitation of the viruses (eg, HIV-1, cytomegalovirus, Epstein-Barr virus, hepatitis viruses B, C, and HIV). Data from quantitative assays are used to guide antiviral therapy in multiple viral diseases. A good example is HIV/AIDS.

Nucleic Acid Hybridization

Nucleic acid hybridization to detect viruses is highly sensitive and specific. The specimen is spotted on a nitrocellulose membrane, and viral nucleic acid present in the sample is bound; it is then denatured with alkali in situ, hybridized with a labeled viral nucleic acid fragment, and the hybridized products detected. For rotavirus, which contains double-stranded RNA, the dot hybridization method is even more sensitive than EIA. RNA in heat-denatured fecal samples containing rotavirus is immobilized as above, and in situ hybridization is carried out with labeled single-stranded probes obtained by in vitro transcription of rotavirus.

Measuring the Immune Response to Virus Infection

Typically, a virus infection elicits immune responses directed against one or more viral antigens. Both cellular and humoral immune responses usually develop, and measurement of either may be used to diagnose a viral infection. Cellular immunity may be assessed by dermal hypersensitivity, lymphocyte transformation, and cytotoxicity tests. Humoral immune responses are of major diagnostic importance. Antibodies of the IgM class appear initially and are followed by IgG antibodies. The IgM antibodies disappear in several weeks, whereas the IgG antibodies persist for many years. Establishing the diagnosis of a viral infection is accomplished serologically by demonstrating a rise in antibody titer to the virus or by demonstrating antiviral antibodies of the IgM class (see Chapter 8). The methods used include the neutralization (Nt) test, the CF test, the hemagglutination inhibition (HI) test, and the IF test, passive hemagglutination, and immunodiffusion.

Measurement of antibodies by different methods does not necessarily give parallel results. Antibodies detected by the CF test are present during an enterovirus infection and in the convalescent period, but they do not persist. Antibodies detected by the Nt test also appear during infection and persist for many years. Assessment of antibodies by several methods in individuals or groups of individuals provides diagnostic information as well as information about epidemiologic features of the disease.

Serological tests for viral diagnosis are most useful when the virus has a long incubation period prior to the appearance of clinical manifestations. A partial list of such viruses includes Epstein-Barr virus, the hepatitis viruses, and HIV. Typically, testing for antibodies to these viruses is the first step in diagnosis and may be followed later by nucleic acid amplification that is used to assess the levels of circulating virus as an estimate of the level of infection and/or response to specific antiviral therapies. Another important utility of serological tests is to assess one's vulnerability or prior exposure to a virus and the potential for reactivation in the context of immunosuppression or organ transplantation.

Immune Electron Microscopy

Viruses not detectable by conventional techniques may be observed by immune electron microscopy (IEM). Antigen–antibody complexes or aggregates formed between virus particles in suspension are caused by the presence of antibodies in added antiserum and are detected more readily and with greater assurance than individual virus particles. IEM is used to detect viruses that cause enteritis and diarrhea; these viruses generally cannot be cultured by routine virus culture. Rotavirus is detected by EIA.

HIV

Human immunodeficiency virus-1 occurs worldwide, while HIV-2 occurs primarily in West Africa and a few other geographic areas. HIV infection presents a special case in diagnostic virology. The laboratory diagnosis must be established precisely, with little or no possibility of a falsely positive result. Once the diagnosis is established, laboratory tests are used to follow the progression of the infection and to help monitor the effectiveness of therapy. Blood banks use very sensitive tests to detect HIV-1 in donated blood and thus prevent transfusion-related HIV-1 infection.

Understanding of the diagnostic tests for HIV and those used to monitor infection requires understanding of HIV structure and replication as well as the immune response to infection. A brief summary of these topics is presented here. HIV is discussed in more detail in Chapter 44.

HIV-1 and HIV-2 are retroviruses. They are enveloped and have single strands of positive-sense RNA. Through the use of the viral enzyme reverse transcriptase, the RNA is transcribed into DNA, which is integrated into the host cell genome. The viral protein that is most commonly assayed directly is p24. Antibody responses can be found against a variety of other HIV gene products: *env* products, glycoprotein (gp) 160 (gp160, gp120, gp41); *gag*, p24, p17, p9, p7; and *pol*, p66, p51, p32, p11.

Two to six weeks after infection, 50% or more of patients develop an infectious mononucleosis-like syndrome. At this time there are high levels of HIV-1 in the blood, which can be detected by culture or by reverse transcriptase PCR. Antibodies to HIV-1 proteins become detectable 2–8 weeks after infection. There is an IgM response to *gag* gene products, which gradually shifts to an IgG response. Generally, IgG responses to p24 and gp120 occur initially, followed by responses to gp41 and other proteins. Viremia and blood p24 levels fall with the antibody response and may be undetectable during the asymptomatic period of infection, while p24 antibody levels remain high. Late in the course of the disease, the p24 antibody levels decrease while the p24 antigen increases. During the time soon after infection, when high levels of HIV-1 viremia occur, the CD4 T cell count decreases. Early in the asymptomatic period, it returns toward normal only to decrease gradually over time and more rapidly during the late stages of AIDS.

The tests used to diagnose HIV-1 infection are commercially available products that are highly developed and standardized. Several kits may be available for the same type of tests. These are outlined in a generic manner below.

Assays for Anti-HIV Antibodies

A. Enzyme-Linked Immunosorbent Assay (Enzyme Immunoassay)

ELISA (or EIA) is the primary screening test for diagnosis of HIV-1 infection. Generally, HIV-1 antigens are immobilized on a solid surface, commonly plastic wells or beads. The patient's serum and appropriate reagents are added. HIV-1 antibodies bound to the immobilized HIV-1 antigens are then detected with an enzyme-labeled antihuman IgG and a colorimetric reaction. The amount of color is proportionately higher with higher HIV-1 antibody concentrations. Color above a certain cutoff point is considered a positive test. ELISA for HIV-1 is over 99% sensitive and 99% specific. Infants born to HIV-1-infected mothers generally have positive ELISAs for HIV-1 because of transplacental transfer of antibodies. The test will gradually become negative if the infant is not truly infected with HIV-1.

The necessity to rapidly detect individuals who may be HIV positive, and to do so while they are still on-site in a clinic environment, has lead to several improvements in EIA testing. An assay has been developed to test oral secretions. In addition, several rapid immunoassays for use with whole blood and serum have also been approved for use. All rapid tests should be treated in the same fashion as conventional assays. Positive results should be confirmed by Western blot and a negative test in a person with strong clinical suspicion should have a repeat test if clinically indicated.

B. Western Blot

The Western blot test is used as a measure of specific HIV-1 antibodies to confirm a positive ELISA result. In the Western blot test, HIV-1 proteins are electrophoretically separated on a nitrocellulose strip. The strip is incubated with the patient's serum. The specific HIV-1 antibodies are subsequently detected using an enzyme-linked antihuman IgG. A positive colorimetric reaction forms bands on the nitrocellulose paper corresponding to the position of specific HIV-1 antigens. The criterion for a positive test is any two bands corresponding to p24, gp41, and gp120/160. The absence of bands is a negative result, while the presence of bands that do not meet the criterion for a positive test is an indeterminate result. False-positive and false-negative results are relatively uncommon. Patients with positive ELISAs and indeterminate Western blots need repeated testing and clinical evaluation. An infant born to an HIV-1-infected mother may have a positive Western blot, but this will gradually become negative if the infant is not truly infected with HIV-1.

Assays to Directly Detect HIV Infection

A. Detection of p 24 Antigen

ELISA is used to detect p24 antigen. Anti-p24 antibodies are immobilized on a solid surface and incubated with the patient's serum. The amount of p24 is detected using enzyme-linked anti-HIV-1 IgG and a colorimetric reaction. The p24 antigen is detectable during the acute viremic stage of infection and in the late stages of AIDS. A very small percentage of asymptomatic persons with HIV-1 infection are p24 antigen-positive.

B. Detection of HIV-1 RNA

A variety of commercial assays are available to detect and to quantitate HIV-1 RNA. These include PCR, NASBA, and bDNA assays for both detection and quantitation of HIV-1. (See the section on molecular diagnostics earlier in this chapter.) These assays can be used to detect HIV-1 infection prior to the time following infection when antibody tests become positive. They are also used to follow the effectiveness of anti-HIV-1 therapy.

C. Detection of HIV-1 Proviral DNA

DNA is extracted from mononuclear cells obtained from anticoagulated peripheral blood. Oligonucleotide primers specific to segments of the integrated HIV-1 proviral DNA are used in a PCR assay. (See the section on molecular diagnostics earlier in this chapter.) This type of assay can be used for an antibody-positive infant born to an HIV-1 infected mother to determine if the infant also is infected. This is the preferred test for the infant younger than 18 months of age.

D. HIV-1 Culture

Tissue culture was the first test developed to diagnose HIV-1 infection. It was used to establish HIV-1 as the cause of AIDS. Peripheral blood mononuclear cells from a potentially infected patient are cocultured with peripheral blood mononuclear cells from an uninfected person that have been stimulated with phytohemagglutinin and interleukin-2. The cultures are observed for formation of multinucleated giant cells, for HIV-1 reverse transcriptase activity, or for HIV-1 p24 antigen production. Quantitative cell culture and quantitative plasma culture can be performed as well. HIV-1 culture has a sensitivity of 95–99%. HIV-1 culture is time-consuming and expensive and thus not cost-effective for routine use.

Monitoring CD4 T Cell Counts

Absolute CD4 T cell counts are widely used to monitor the status of patients' HIV-1 infections. The counts generally are obtained using whole blood cells stained with anti-CD4 antibodies labeled with a fluorescent dye. The red blood cells are lysed and the CD4 cells counted using flow cytometry.

Prognostic Tests & Treatment Monitoring

A high viral load as measured by high HIV-1 RNA levels implies a poor prognosis. Similarly, a low CD4 T cell count indicates risk for an opportunistic infection and thus a poorer prognosis. Both the viral load and the CD4 T cell count are used to monitor the effectiveness of antiretroviral therapy.

Genotyping assays use the reverse transcriptase PCR to amplify HIV-1 RNA that codes for viral enzymes targeted by antiretroviral drugs. Analysis of the amplified sequences allows determination of mutations that code for resistance to the drugs. Such resistance testing is recommended in the settings of first-regimen or multiple-regimen failure or in pregnancy.

REVIEW QUESTIONS

1. A 47-year-old woman had a bone marrow transplant as part of her treatment for chronic myelogenous leukemia. While in the hospital she had a central venous catheter in place for administration of fluids. In the time following the transplant, but before it had engrafted, the patient had a very low white blood cell count. She developed a fever, and blood cultures were done. Which of the following scenarios suggests that the positive blood cultures resulted from a contaminant?

   (A) Two positive peripheral vein blood cultures with *Staphylococcus aureus*

   (B) Two positive peripheral vein blood cultures with *Staphylococcus epidermidis* along with two positive central line blood cultures with *Staphylococcus epidermidis*.

   (C) One positive peripheral vein blood culture and one positive central line blood culture with *Escherichia coli*

   (D) One positive central venous line blood culture with a *Corynebacterium* species and two negative peripheral vein blood cultures

   (E) Two positive central line blood cultures with *Candida albicans*

2. Two days ago a 22-year-old man returned from a 2-week trip to Mexico. Within 24 hours he developed diarrhea. Which of the following will not establish the etiology of his diarrhea?

   (A) Stool culture for *Salmonella*, *Shigella*, and *Campylobacter*

   (B) Stool culture for rotavirus and Norwalk-like virus

   (C) Stool enzyme immunoassay for *Giardia lamblia* antigen

   (D) Stool examination for *Entamoeba histolytica*

3. A 37-year-old man traveled to Peru during the time of the cholera epidemic. One day after returning home he developed severe

watery diarrhea. To enhance the isolation of *Vibrio cholerae* from his stool, the laboratory needs to include

(A) MacConkey agar
(B) Campylobacter blood agar
(C) Thiosulfate citrate bile salts sucrose agar
(D) Bismuth sulfite agar
(E) Hektoen agar

4. A 42-year-old man is known to have HIV/AIDS. Which of the following is the most appropriate method to follow the progress of his highly active antiretroviral therapy (HAART)?

(A) Determination of viral load
(B) Following anti HIV-1 antibody levels
(C) Using Western blot to assess his anti-P24 levels
(D) Repeated culture of his blood for HIV-1 to determine when the culture becomes negative
(E) Genotyping of his HIV-1 isolate to determine its antiretroviral susceptibility

5. A 2-year-old child develops diarrhea. Rotavirus infection is suspected. Which of the following would be most useful in diagnosing a rotavirus infection?

(A) Fluorescent antibody staining of the stool specimen
(B) Light microscopy to detect mucosal cells with cytopathic effect
(C) Detection of virus antigen in stool by enzyme-linked immunosorbent assay
(D) Virus culture

6. Which of the following is appropriate to determine the etiologic diagnosis of infection?

(A) Culture and identification of the agent
(B) DNA-DNA or DNA-RNA hybridization to detect pathogen-specific genes in patients' specimens
(C) Demonstration of a meaningful antibody or cell-mediated immune response to an infectious agent
(D) Morphologic identification of the agent in stains of specimens or sections of tissues by light or electron microscopy
(E) Detection of antigen from the agent by immunologic assay
(F) All of the above

7. A 45-year-old woman is admitted to the hospital because of fever, a 6-kg weight loss, and a new heart murmur. Probable endocarditis is diagnosed. How many blood cultures over what period of time should be done to provide evidence of specific bacterial infection in endocarditis?

(A) One
(B) Two over 10 minutes
(C) Three over 2 hours
(D) Three over 24 hours
(E) Six over 3 days

8. A 4-year-old boy develops bloody diarrhea. Hemorrhagic colitis due to *Escherichia coli* O157:H7 is suspected. What medium should be inoculated to help the laboratory staff make the diagnosis of this infection?

(A) Blood agar
(B) Sorbitol MacConkey agar
(C) Hektoen enteric agar
(D) CIN (cefsulodin, irgasan, novobiocin) agar
(E) Thiosulfate citrate bile salts sucrose agar

9. A 43-year-old black man frequently drove his long-haul 18-wheel truck through the Central Valley of California. Two months ago there was a major windstorm while he was driving through the

Valley. Two weeks after that he developed fever with a cough and pleuritic chest pain. An infiltrate was seen on chest x-ray. Pneumonia was diagnosed, and the patient was given erythromycin. The fever, cough, pleuritic pain, and infiltrate cleared over a 3-week period. Two weeks ago, he developed progressively severe headaches and for the past 2 days he has had vomiting. His cerebrospinal fluid contains 150 white blood cells per microliter, predominantly lymphocytes, and the glucose concentration is low. Meningitis due to *Coccidioides immitis* is suspected. Which of the following tests is the most sensitive and useful in confirming this diagnosis?

(A) Latex agglutination assay for coccidioidal antibodies performed on CSF
(B) Complement fixation test of cerebrospinal fluid for antibodies against *Coccidioides immitis*
(C) Immunodiffusion test of cerebrospinal fluid for antibodies against *Coccidioides immitis*
(D) Cerebrospinal fluid culture for *Coccidioides immitis*
(E) Complement fixation test of serum for antibodies against *Coccidioides immitis*

10. A 5-year-old kidney transplant patient being treated with cyclosporine develops a lymphoproliferative disorder. Which of the following viruses is most likely responsible for this disorder?

(A) Cytomegalovirus
(B) Herpes simplex virus
(C) Coxsackie B virus
(D) Hepatitis B virus
(E) Epstein-Barr virus

11. All of the following are appropriate indications for the use of serological tests for viruses *except*:

(A) As an indication of one's susceptibility to a particular viral infection
(B) For diagnosis when the virus has a long incubation period
(C) For screening purposes
(D) For confirmation of a viral infection
(E) To monitor the response to treatment

12. In August, a 2-year-old boy presents acutely with fever, signs of headache, decreased mental status, and neck stiffness. On physical examination the fever is confirmed, mild nuchal rigidity is present and although the child is irritable and mildly somnolent, he is arousable and is taking some oral fluids. The cerebrospinal fluid parameters reveal a protein of 60 μg/dL, glucose 40 μg/dL and a total of 200 WBCs predominately mononuclear. The most likely cause of this child's infection is:

(A) Bacterial
(B) Viral
(C) Protozoan
(D) Fungal
(E) Mycobacterial

13. In the case above, the most useful test for making a rapid definitive diagnosis of the most likely causative agent is:

(A) An antigen test for *Streptococcus pneumoniae*
(B) A latex agglutination test for Cryptococcal antigen
(C) A nucleic acid amplification test for viral RNA detection
(D) Culture on selective media combined with a probe test for confirmation
(E) Giemsa-stained smear of cerebrospinal fluid

14. Susceptibility testing using an MIC method, as opposed to disk diffusion, is preferred for all of the following types of infections *except*:
    (A) Urinary tract infections
    (B) Endocarditis
    (C) Osteomyelitis
    (D) Bacteremia in a neutropenic patient
    (E) Bacterial meningitis

15. Bacterial vaginosis is best diagnosed by all of the following *except*:
    (A) Measurement of the vaginal pH
    (B) Detection of a fishy odor when the discharge is alkalinized with KOH
    (C) Bacterial culture for aerobes and anaerobes
    (D) Examination of a Gram-stained smear for "clue cells"

Answers

1. D	5. D	9. B	13. C
2. B	6. F	10. E	14. A
3. C	7. D	11. E	15. C
4. A	8. B	12. B	

REFERENCES

Forbes BA, Sahm DF, Weissfeld AS (editors): *Bailey and Scott's Diagnostic Microbiology*, 12th ed. ASM Press, Washington, DC, 2007.

Murray PR et al (editors): *Manual of Clinical Microbiology,* 9th ed. ASM Press, 2007.

Persing D et al (editors): *Molecular Microbiology: Diagnostic Principles and Practice*, 2nd edition, ASM Press, (*In Press*).

Winn W et al (editors): *Koneman's Color Atlas and Textbook of Diagnostic Microbiology* 6th ed. Lippincott Williams and Wilkins, 2006.

Cases & Clinical Correlations

The management of infectious diseases requires an understanding of the presenting clinical manifestations and a knowledge of microbiology. Many infections present with constellations of focal and systemic signs and symptoms which in typical cases are highly suggestive of the diagnosis, though the disease might be caused by any of several different organisms. Making a clinical diagnosis with subsequent laboratory confirmation is part of the art of medicine. This chapter presents 19 cases and brief discussions of the differential diagnosis and management of those infections. An additional two cases and two outbreaks of naturally occurring diseases caused by agents associated with biologic warfare are presented.

The reader is referred to earlier chapters of this book for characterizations of the organisms; to Chapter 47 for information about diagnostic microbiology tests; and to textbooks of medicine and infectious diseases for more complete information about the clinical entities. One such book is McPhee SF, Papadakis MA (editors): *Current Medical Diagnosis & Treatment*, updated annually.

CENTRAL NERVOUS SYSTEM

CASE 1: MENINGITIS

A 3-year-old girl was brought to the emergency room by her parents because of fever and loss of appetite for the past 24 hours and difficulty in arousing her for the past 2 hours. The developmental history had been normal since birth. She attended a day care center and had a history of several episodes of presumed viral infections similar to those of other children at the center. Her childhood immunizations were current.

Clinical Features

Temperature was 39.5°C, pulse 130/min, and respirations 24/min. Blood pressure was 110/60 mm Hg.

Physical examination showed a well-developed and well-nourished child of normal height and weight who was somnolent. When her neck was passively flexed, her legs also flexed (positive Brudzinski sign, suggesting irritation of the meninges). Ophthalmoscopic examination showed no papilledema, indicating that there had been no long-term increase in intracranial pressure. The remainder of her physical examination was normal.

Laboratory Findings

Minutes later, blood was obtained for culture and other laboratory tests, and an intravenous line was placed. Lumbar puncture was performed less than 30 minutes after the patient arrived in the emergency room. The opening pressure was 350 mm of cerebrospinal fluid (CSF) (elevated). The fluid was cloudy. Several tubes of CSF were collected for culture, cell counts, and chemistry tests. One tube was taken immediately to the laboratory for Gram staining. The stain showed many polymorphonuclear (PMN) cells with cell-associated (intracellular) gram-negative diplococci suggestive of *Neisseria meningitidis* (Chapter 20).

Blood chemistry tests were normal. The hematocrit was normal. The white blood cell count was 25,000/μL (markedly elevated), with 88% PMN forms and an absolute PMN count of 22,000/μL (markedly elevated), 6% lymphocytes, and 6% monocytes. The CSF had 5000 PMNs/μL (normal, 0–5 lymphocytes/μL). The CSF protein was 100 mg/dL (elevated), and the glucose was 15 mg/dL (low, termed hypoglycorrhachia)—all consistent with bacterial meningitis. Cultures of blood and CSF grew serogroup B *N meningitidis*.

Treatment

Intravenous cefotaxime therapy was started within 35–40 minutes of the patient's arrival; dexamethasone was also given. The patient was treated with the antibiotic for 14 days and recovered without obvious sequelae. Further neurologic examinations and hearing tests were planned for the future. Rifampin prophylaxis was given to the other children who attended the day care center.

Comment

Clinical features of bacterial meningitis vary with the age of the patient. In the older child and the adult, bacterial meningitis usually presents with fever, headache, vomiting, photophobia, altered mental status ranging from sleepiness to coma, and neurologic signs ranging from abnormalities of cranial nerve function to seizures. However, subtle signs such as fever and lethargy are consistent with meningitis, particularly in infants. Meningitis is considered to be acute with signs and symptoms of less than 24 hours' duration and subacute when signs and symptoms have been present for 1–7 days. Lumbar puncture with examination of the CSF is indicated whenever there is any suspicion of meningitis.

Acute meningitis is most often caused by bacteria of a few species (Table 48–1): Lancefield serogroup B streptococci (*Streptococcus agalactiae*) (Chapter 14) and *Escherichia coli* (Chapter 15) in neonates; *Haemophilus influenzae* (Chapter 18) in unvaccinated children between the ages of 4–6 months and 6 years; *N meningitidis* in children and young adults; and *Streptococcus pneumoniae* (Chapter 14) occasionally in children and increasing in incidence in middle-aged and elderly persons. Many other species of microorganisms less commonly cause meningitis. *Listeria monocytogenes* (Chapter 12) causes meningitis in immunosuppressed patients and normal persons. The yeast *Cryptococcus neoformans* (Chapter 45) is the most common cause of meningitis in AIDS patients and can cause meningitis also in other immunosuppressed patients as well as in normal persons. Meningitis due to *Listeria* or *Cryptococcus* can be acute or insidious in onset. Gram-negative bacilli cause meningitis in acute head trauma and neurosurgical patients. *S pneumoniae* is found in recurrent meningitis in patients with basilar skull fractures. *Mycobacterium tuberculosis* (Chapter 23) can have a slow onset (chronic; >7 days) in immunologically normal persons but progresses more rapidly (subacute) in immunosuppressed persons such as AIDS patients. *Naegleria* species (Chapter 46), free-living amoebas, occasionally cause meningitis in persons with a recent history of swimming in warm fresh water. Viruses (Chapters 30, 33, 36) usually cause milder meningitis than bacteria. The viruses that most commonly cause meningitis are the enteroviruses (echoviruses and coxsackieviruses) and mumps virus.

The diagnosis of meningitis requires a high degree of suspicion when appropriate signs and symptoms are observed plus lumbar puncture without delay followed by examination of CSF. Findings in the spinal fluid typically include white blood cells in hundreds to thousands per microliter (PMNs for acute bacterial meningitis and lymphocytes for tuberculous and viral meningitis); glucose of <40 mg/dL, or less than 50% of the serum concentration; and protein of >100 mg/dL (see Table 48–2). In bacterial meningitis, Gram stain of cytocentrifuged sediment of CSF shows PMNs and bacterial morphology consistent with the species subsequently cultured: *N meningitidis*, intracellular gram-negative diplococci; *H influenzae*, small gram-negative coccobacilli; and serogroup B streptococci and pneumococci, gram-positive cocci in pairs and chains. Blood cultures should be done along with the CSF cultures.

Acute bacterial meningitis is fatal if untreated. Initial therapy for bacterial meningitis in infants < 1 month of age should consist of parenteral therapy known to be effective against the pathogens listed in Table 48–1 and including *L monocytogenes*. Ampicillin plus cefotaxime or ceftriaxone with or without gentamicin or ampicillin in combination with an aminoglycoside is recommended. For children between the ages of 1 month and 18 years of age and for

TABLE 48–1   Common Causes of Meningitis

Organism	Age Group	Comment	Chapter
Serogroup B streptococci (*S agalactiae*)	Neonates to age 3 months	As many as 25% of mothers have vaginal carriage of serogroup B streptococci. Ampicillin prophylaxis during labor of women at high risk (prolonged rupture of membranes, fever, etc) or of known carriers reduces the incidence of infection in babies.	14
Escherichia coli	Neonates	Commonly have the K1 antigen	15
Listeria monocytogenes	Neonates; elderly; immunocompromised children and adults	Not unusual in patients with cell mediated immune deficiency	12
Haemophilus influenzae	Children 6 months to 5 years	Widespread use of vaccine greatly reduces the incidence of *H influenzae* meningitis in children.	18
Neisseria meningitidis	Infants to 5 years and young adults	Polysaccharide vaccines against serogroups A, C, Y, and W135 are used in epidemic areas and in association with outbreaks.	20
Streptococcus pneumoniae	All age groups; highest incidence in the elderly	Often occurs with pneumonia; also with mastoiditis, sinusitis, and basilar skull fractures.	14
Cryptococcus neoformans	AIDS patients	Frequent cause of meningitis in AIDS patients.	45

TABLE 48-2 Typical Cerebrospinal Fluid Findings in Various Central Nervous System Diseases

Diagnosis	Cells (per μL)	Glucose (mg/dL)	Protein (mg/dL)	Opening Pressure
Normal[a]	0–5 lymphocytes	45–85	15–45	70–180 mm H_2O
Purulent meningitis (bacterial)[b]	200–20,000 PMNs	Low (<45)	High (>50)	++++
Granulomatous meningitis (mycobacterial, fungal)[b,c]	100–1000, mostly lymphocytes	Low (<45)	High (>50)	+++
Aseptic meningitis, viral or meningoencephalitis[c,d]	100–1000, mostly lymphocytes	Normal	Moderately high (>50)	Normal to +
Spirochetal meningitis (syphilis, leptospirosis)[c]	25–2000, mostly lymphocytes	Normal or low	High (>50)	+
"Neighborhood" reaction[e]	Variably increased	Normal	Normal or high	Variable

[a]CSF glucose level must be considered in relation to blood glucose level. Normally, CSF glucose level is 20–30 mg/dL lower than blood glucose level, or 50–70% of blood glucose normal value.

[b]Organisms in smear or culture of CSF.

[c]PMNs may predominate early.

[d]Virus isolation from CSF early; antibody titer rise in paired specimens of serum.

[e]May occur in mastoiditis, brain abscess, epidural abscess, sinusitis, septic thrombus, brain tumor, CSF culture usually negative.

the adult > 50 years of age, the recommended therapies are vancomycin plus a third-generation cephalosporin because of the prevalence of multidrug resistant *S pneumoniae*, reports of rising MICs to penicillin among meningococci, and the prevalence of β-lactamase production among *H influenzae*. Since adults over the age of 50 years are also susceptible to *L monocytogenes*, the addition of ampicillin to the regimen for older children and adults as listed above is recommended.

Available evidence supports administration of adjunctive dexamethasone 10–20 minutes prior to or concomitant with the first antimicrobial dose to children with *H influenzae* meningitis and in the adult with pneumococcal meningitis with continuation of steroids for the first 2–4 days of therapy.

Several vaccines are currently available and are recommended for the prevention of the more serious causes of bacterial meningitis. The *H influenzae* type B conjugate vaccine and the heptavalent conjugate pneumococcal vaccine are currently part of the routine vaccination series for infants and young children. The 23-valent polysaccharide pneumococcal vaccine is recommended for prevention of invasive pneumococcal disease in certain high risk groups over the age of 2 years. These include patients who are elderly, who have chronic underlying diseases such as cardiovascular disease, diabetes mellitus, chronic pulmonary problems, CSF leaks, and asplenia, among others. A quadrivalent conjugated meningococcal vaccine is currently recommended for all healthy 11- to 19-year-old and for 20- to 55-year-old persons at risk such as travelers to endemic areas. For at-risk children, 2–10 years of age, the conjugated vaccine may also be given. For adults over 55, the meningococcal polysaccharide vaccine is currently recommended pending evaluation of the conjugate vaccine in this age group.

REFERENCES

Chavez-Bueno S et al: Bacterial meningitis in children. Pediatr Clin North Am 2005;52:795.

Tunkel AR et al: Practice guidelines for the management of bacterial meningitis. Clin Infect Dis 2004;39:1267.

Van de Beek D et al: Community acquired bacterial meningitis in adults. N Engl J Med 2006;354:44.

CASE 2: BRAIN ABSCESS

A 57-year-old man presented to the hospital with seizures. Three weeks earlier he had developed bifrontal headaches that were relieved by aspirin. The headaches recurred several times, including the day prior to admission. On the morning of admission he was noted to have focal seizures with involuntary movements of the right side of his face and arm. While in the emergency room, he had a generalized seizure that was controlled by intravenous diazepam, phenytoin, and phenobarbital. Additional history from the patient's wife indicated that he had had a dental extraction and bridge work approximately 5 weeks earlier. He did not smoke, drank only socially, and took no medications. The remainder of his history was not helpful.

Clinical Features

The temperature was 37°C, the pulse 110/min, and respirations 18/min. The blood pressure was 140/80 mm Hg.

On physical examination, the patient was sleepy and had a decreased attention span. He moved all his extremities, though the right arm moved less than the left. There was slight blurring of the left optic disk, suggesting possible increased intracranial pressure. The remainder of his physical examination was normal.

Laboratory Findings & Imaging

Laboratory tests were all normal, including hemoglobin and hematocrit, white blood cell count and differential, serum electrolytes, blood urea nitrogen, serum creatinine, urinalysis, chest x-ray, and ECG. Lumbar puncture was not done and cerebrospinal fluid was not examined because of possible increased intracranial pressure due to a mass lesion. Blood cultures were negative. CT scan of the patient's head showed a 1.5-cm localized ring-enhancing lesion in the left parietal hemisphere suggestive of brain abscess.

Treatment

The patient had a neurosurgical procedure with drainage of the lesion, which was completely removed. Culture of necrotic material from the lesion yielded *Prevotella melaninogenica* (Chapter 21) and *Streptococcus anginosus* (Chapter 14). Pathologic examination of the tissue suggested that the lesion was several weeks old. The patient received antibiotic therapy for 4 weeks. He had no more seizures and no subsequent neurologic deficits. One year later, anticonvulsant medications were discontinued and a follow-up CT scan was negative.

Comment

A brain abscess is a localized pyogenic bacterial infection within the brain parenchyma. The major clinical manifestations are related to the presence of a space-occupying mass in the brain rather than the classic signs and symptoms of infection. Thus, patients commonly present with headache and a change in mental status from normal to lethargy or coma. Focal neurologic findings related to location of the abscess occur in less than half of patients; one-third have seizures, and less than half have fever. Occasionally, patients present with signs and symptoms suggesting acute meningitis. Initially, the clinician must differentiate brain abscess from other central nervous system processes, including primary or metastatic cancers, subdural or epidural abscesses, viral infections (herpes simplex encephalitis), meningitis, stroke, and a variety of other diseases.

Significant predisposing factors for brain abscess include distant site infections with bacteremia, such as endocarditis, lung infections, or other occult infections. Many patients have had relatively recent dental work. Brain abscess can also occur via spread from contiguous sites of infection such as in the middle ear, mastoid, or sinuses or following penetrating trauma. However, 20% of patients with brain abscesses have no discernible predisposing factors.

Brain abscess can be caused by a single species of bacteria, but more than one species are often isolated—in general, an average of two species. Of the facultative and aerobic bacteria, the viridans streptococci (including nonhemolytic and α- and β-hemolytic strains, the *S anginosus* group [milleri], *Streptococcus mitis*, etc; see Chapter 14) are most common, occurring in one-third to one-half of patients. *Staphylococcus aureus* (Chapter 13) is isolated in 10–15% and when present is often the only isolate found. Enteric gram-negative rods occur in about 25%, often in mixed cultures. Many other facultative or aerobic bacteria (eg, *S pneumoniae*, *Nocardia asteroides*, *M tuberculosis* and nontuberculous Mycobacteria) also occur in brain abscesses. Anaerobic bacteria are found in 50% or more of cases (Chapter 21). *Peptostreptococcus* is most common, followed by *Bacteroides* and *Prevotella* species. *Fusobacterium*, *Actinomyces*, and *Eubacterium* are less common, followed by other anaerobes. Fungi (Chapter 45) are seen almost exclusively in immunocompromised patients. *Candida* species are the most prevalent fungi, but opportunistic molds such as *Aspergillus* sp. and *Scedosporium apiospermum* are increasing in frequency. Dimorphic fungi such as *Coccidioides immitis* may also cause brain abscesses. *C neoformans* is an important pathogen in AIDS patients. Parasites (Chapter 46) responsible for brain abscesses include *Toxoplasma gondii*, the most common protozoal cause, particularly among AIDS patients, neurocystercercosis (larval form of *Taenia solium*), *Entamoeba histolytica*, *Schistosoma* sp., and *Paragonimus*.

Lumbar puncture to obtain CSF is generally not indicated in patients with brain abscess (or other mass lesions in the brain). The increased intracranial pressure makes the procedure life-threatening, because herniation of the brain through the tentorium cerebelli can result in midbrain compression. The findings in CSF are not specific for brain abscess: White blood cells, predominantly mononuclear cells, are often present; the glucose level may be moderately low and the protein concentration elevated. Thus, when fever and signs suggesting acute meningitis are absent and brain abscess is suspected, the clinician should obtain a CT scan. Brain abscesses typically show ring-enhanced uptake of contrast material on CT scan, though similar findings can be found in patients with brain tumors and other diseases. MRI may be helpful in differentiating brain abscesses from tumors. Definitive differentiation between brain abscess and tumor is done by pathologic examination and culture of tissue from the lesion obtained by a neurosurgical procedure.

Untreated brain abscesses are fatal. Surgical excision provides the initial therapy as well as the diagnosis of brain abscess. Needle aspiration using stereotactic technique is an alternative to surgical excision. Antibiotic therapy should be parenteral and should include high-dose penicillin G for streptococci and many anaerobes, metronidazole for anaerobes resistant to penicillin G, plus a third-generation cephalosporin for enteric gram-negative rods. Vancomycin or another drug specific for *S aureus* should be included in the initial therapy if the patient has endocarditis or is known to have staphylococcal bacteremia, or the abscess yields staphylococci. Initial therapy with antibiotics rather than surgery can be instituted in some patients whose brain abscesses are small (<2 cm), multiple, or difficult to reach surgically, but deteriorating neurologic functions indicate the need for surgery. Once culture results from the abscess material are known, initial antibiotic therapy should be modified to be specific for the bacteria isolated from the lesion. Antibiotic therapy should be continued for at least 3–4 weeks when surgical excision has been done or for 8 weeks or longer when there has been no surgery. Nonbacterial causes of brain abscesses generally require definitive diagnoses and specific therapy. Steroids to decrease swelling should be used only when there is mass effect.

REFERENCES

Bernardini GL: Diagnosis and management of brain abscess and subdural empyema. Curr Neurol and Neurosci Reports 2004;4:448.

Tunkel AR: Brain abscess. In *Mandell, Douglas, and Bennett's Principles and Practice of Infectious Diseases*, 7th ed. Mandell GL, Bennett JE, Dolin R (editors). Churchill Livingstone Elsevier , Philadelphia, PA, 2010.

Yogev R et al: Management of brain abscesses in children. Pediatr Infect Dis 2004;23:157.

CHEST

CASE 3: BACTERIAL PNEUMONIA

A 35-year-old man came to the emergency room because of fever and pain in his left chest when he coughed. Five days earlier he had developed signs of a viral upper respiratory infection with sore throat, runny nose, and increased cough. The day before presentation he developed left lateral chest pain when he coughed or took a deep breath. Twelve hours before coming to the emergency room he was awakened with a severe shaking chill and sweating. Further history taking disclosed that the patient drank moderate to heavy amounts of alcohol and had smoked one package of cigarettes daily for about 17 years. He worked as an automobile repair man. He had a history of two prior hospitalizations—4 years ago for alcohol withdrawal and 2 years ago for acute bronchitis.

Clinical Features

Temperature was 39°C, pulse 130/min and respirations 28/min. Blood pressure was 120/80 mm Hg.

Physical examination showed a slightly overweight man who was coughing frequently and holding his left chest when he coughed. He produced very little thick rusty-colored sputum. His chest examination showed normal movement of the diaphragm. There was dullness to percussion of the left lateral posterior chest, suggesting consolidation of the lung. Tubular (bronchial) breath sounds were heard in the same area along with dry crepitant sounds (rales), consistent with lung consolidation and viscous mucus in the airway. The remainder of his physical examination was normal.

Laboratory Findings & Imaging

Chest films showed a dense left lower lobe consolidation consistent with bacterial pneumonia. The hematocrit was 45% (normal). The white blood cell count was 16,000/μL (markedly elevated) with 80% PMN forms with an absolute PMN count of 12,800/μL (markedly elevated), 12% lymphocytes, and 8% monocytes. Blood chemistry tests, including electrolytes, were normal. Sputum was thick, yellow to rusty-colored, and purulent in appearance. Gram stain of the sputum showed many PMN cells and lancet-shaped gram-positive diplococci. Twenty-four hours later, the blood cultures were positive for *S pneumoniae* (Chapter 14). Cultures of sputum grew numerous *S pneumoniae* and a few colonies of *H influenzae* (Chapter 18).

Treatment

The initial diagnosis was bacterial pneumonia, probably pneumococcal. Parenteral aqueous penicillin G therapy was begun, and the patient was given parenteral fluids. Within 48 hours, his temperature was normal and he was coughing up large amounts of purulent sputum. Penicillin G was continued for 7 days. At follow-up 4 weeks after admission to the hospital, the lung consolidation had cleared.

CASE 4: VIRAL PNEUMONIA

A 31-year-old man presented with complaints of skin rash, cough, and shortness of breath. Four days previously he had begun to feel sick and developed a fever of 38°C. The next day he developed a skin rash that initially appeared as "bumps" but soon became vesicular. Several more crops of intensely pruritic skin lesions have subsequently appeared. Two hours before admission, the patient first experienced right-sided chest pain when he took a deep breath or coughed.

TABLE 48–3   **Characteristics and Treatment of Selected Pneumonias**

Organism	Clinical Setting	Gram-Stained Smears of Sputum	Chest Radiograph[a]	Laboratory Studies	Complications	Preferred Antimicrobial Therapy[b]	Chapter
Streptococcus pneumoniae	Chronic cardiopulmonary disease; follows upper respiratory tract infections	Gram-positive diplococci	Lobar consolidation	Gram-staining smear of sputum; culture of blood, pleural fluid; urinary antigen	Bacteremia, meningitis, endocarditis, pericarditis, empyema	Penicillin G (or V, oral); fluoroquinolones or vancomycin for highly penicillin resistant	14
Haemophilus influenzae	Chronic cardiopulmonary disease; follows upper respiratory tract infections	Small gram-negative coccobacilli	Lobar consolidation	Culture of sputum, blood, pleural fluid	Empyema, endocarditis	Ampicillin (or amoxicillin) if β-lactamase-negative; cefotaxime or ceftriaxone	18
Staphylococcus aureus	Influenza epidemics; nosocomial	Gram-positive cocci in clusters	Patchy infiltrates	Culture of sputum, blood, pleural fluid	Empyema, cavitation	Nafcillin[c]	13
Klebsiella pneumoniae	Alcohol abuse, diabetes mellitus; nosocomial	Gram-negative encapsulated rods	Lobar consolidation	Culture of sputum, blood, pleural fluid	Cavitation, empyema	A third or fourth generation cephalosporin; for severe infection[d], add gentamicin or tobramycin	15
Escherichia coli	Nosocomial; rarely, community-acquired	Gram-negative rods	Patchy infiltrates, pleural effusion	Culture of sputum, blood, pleural fluid	Empyema	A third-generation cephalosporin[d]	15
Pseudomonas aeruginosa	Nosocomial; cystic fibrosis	Gram-negative rods	Patchy infiltrates, cavitation	Culture of sputum, blood	Cavitation	Antipseudomonal cephalosporin or carbapenem or β-lactam/β-lactamase inhibitor, such as piperacillin/tazobactam plus an aminoglycoside	16
Anaerobes	Aspiration, periodontitis	Mixed flora	Patchy infiltrates in dependent lung zones	Culture of pleural fluid or of material obtained by transthoracic aspiration; bronchoscopy with protected specimen brush	Necrotizing pneumonia, abscess, empyema	Clindamycin	11, 20, 48
Mycoplasma pneumoniae	Young adults; summer and fall	PMNs and monocytes; no bacterial pathogens	Extensive patchy infiltrates	Complement fixation titer[e]; cold agglutinin serum titers are not helpful as they lack sensitivity and specificity; PCR	Skin rashes, bullous myringitis; hemolytic anemia	Erythromycin, azithromycin, or clarithromycin; doxycycline, fluoroquinolones	25

740

Organism	Page	Epidemiology / Clinical features	Gram stain	X-ray	Diagnosis	Complications	Treatment[b]
Legionella species	22	Summer and fall; exposure to contaminated construction site, water source, air conditioner; community-acquired or nosocomial	Few PMNs; no bacteria	Patchy or lobar consolidation	Immunofluorescent antibody titer[e]; culture of sputum or tissue[f]; *Legionella* urinary antigen (*L pneumophila* serogroup 1 only); PCR	Empyema, cavitation, endocarditis, pericarditis	Azithromycin, or clarithromycin, with or without rifampin; fluoroquinolones
Chlamydophila pneumoniae	27	Clinically similar to *M pneumoniae* pneumonia, but prodromal symptoms last longer (up to 2 weeks); sore throat with hoarseness common; mild pneumonia in teenagers and young adults	Nonspecific	Subsegmental infiltrate, less prominent than in *M pneumoniae* pneumonia; consolidation rare	Isolation very difficult; microimmunofluorescence with TWAR antigens is the recommended assay	Reinfection in older adults with underlying COPD or heart failure may be severe or even fatal	Doxycycline, erythromycin, clarithromycin; fluoroquinolones
Moraxella catarrhalis	20	Preexisting lung disease; elderly; corticosteroid or immunosuppressive therapy	Gram-negative diplococci	Patchy infiltrates; occasional lobar consolidation	Gram stain and culture of sputum or bronchial aspiration	Rarely, pleural effusions and bacteremia	Trimethoprim-sulfamethoxazole or amoxicillin-clavulanic acid or second or third-generation cephalosporins
Pneumocystis jiroveci	45	AIDS, immunosuppressive therapy	Not helpful in diagnosis	Diffuse interstitial and alveolar infiltrates; apical or upper lobe infiltrates on aerosolized pentamidine	Cysts and trophozoites of *P jiroveci* on methenamine silver or Giemsa's stains of sputum or BAL fluid; direct immunofluorescent antibody on BAL fluid	Pneumothorax, respiratory failure, ARDS, death	Trimethoprim-sulfamethoxazole, pentamidine isethionate

[a] X-ray findings lack specificity

[b] Microbial susceptibility testing should guide therapy.

[c] Nafcillin-resistant *S aureus* infections are treated with vancomycin.

[d] Extended spectrum β-lactamase-producing and carbapenemase-producing organisms may complicate therapy.

[e] Fourfold rise in titer is diagnostic.

[f] Selective media are required

Two weeks before admission, the patient's 8-year-old daughter had developed chickenpox (Chapter 33) and he had helped take care of her. The patient did not know if he had had chickenpox as a child.

Clinical Features

The temperature was 39°C, pulse 110/min, and respirations 30/min. Blood pressure was 115/70 mm Hg. The patient appeared to be acutely uncomfortable. He had a skin rash consisting of multiple crops or stages of lesions ranging from red maculopapules to vesicles that were broken and crusted over. His fingers and lips appeared to be slightly blue. Rales were heard bilaterally throughout both lung fields. The remainder of the physical examination was normal.

Laboratory Findings & Imaging

Chest films showed diffuse bilateral interstitial pulmonary infiltrates. Arterial blood gases showed a PO_2 of 60 mm Hg with 91% hemoglobin saturation. The hematocrit, white blood cell count, and serum electrolytes and liver tests were normal.

Treatment & Hospital Course

The patient was hospitalized and placed on oxygen therapy, which improved his hypoxia. He was given high-dose intravenous acyclovir. Over the next several days, his respiratory status improved, and on day 6 oxygen therapy was discontinued. The acyclovir was changed to oral therapy on day 3 and continued for a total of 10 days. The patient was discharged to home care on day 7.

Comment

Acute bacterial pneumonia commonly presents with an abrupt onset of chills and fever, cough, and often **pleuritic chest pain**. The cough frequently is productive of **purulent sputum**, but many patients with pneumonia are not adequately hydrated and do not produce sputum until they receive fluids, as in this case. Pleuritic chest pain occurs when the inflammatory process of the pneumonia involves the pleural lining of the lung and chest cavity; movement of the pleura, as occurs with coughing or deep breathing, yields localized pain. Patients with acute pneumonia appear ill and usually have tachypnea (rapid breathing) and tachycardia (rapid heart rate). Many patients with pneumonia have predisposing factors (congestive heart failure, chronic obstructive pulmonary disease, etc), which become exacerbated before or in association with the pneumonia.

The findings on physical examination are those associated with **consolidation of the lung tissue,** purulent mucus **(sputum)** in the airway, and, in some patients, fluid in the chest cavity. On percussion, there is dullness over the area of consolidation (or fluid). When consolidation occurs, the small airways are closed, leaving only the large airways open; on auscultation, there are tubular breath sounds over the area. If all the airways are blocked, no breath sounds are audible. Dry crepitant sounds (rales) or crackling sounds on auscultation indicate fluid or mucus in the airways; these sounds may change when the patient coughs.

Viral pneumonia is characterized by interstitial inflammation of the lung tissue and hyaline membrane formation in the alveolar spaces, often accompanied by bronchiolitis and sloughing of the ciliated cells of the small airways with peribronchial inflammation. The viruses that most commonly cause pneumonia are respiratory syncytial virus, parainfluenza viruses (typically type 3), influenza viruses, adenoviruses, measles virus, and varicella-zoster virus (Chapters 32, 39, 40). Cytomegalovirus (Chapter 33) causes pneumonia in allogeneic bone marrow and solid organ transplant patients; varicella-zoster virus may cause pneumonia in these patients as well. Emerging viral pathogens such as Metapneumovirus and newly discovered Coronaviruses may cause disease that mimics that of the more common viral respiratory pathogens (Chapters 40, 41). SARS Coronavirus was responsible for epidemic fatal respiratory disease in several countries. Many other infectious agents (and noninfectious agents also) can cause interstitial pneumonitis with or without focal consolidation in the lung. Examples include *Legionella pneumophila* (Chapter 22), *Mycoplasma pneumoniae* (Chapter 25), and *Pneumocystis jiroveci* (Chapter 45). The physical findings on chest examination in viral pneumonia frequently are limited; often only rales are heard on auscultation. Some of the viruses cause characteristic rashes that may serve as clues to diagnosis. Chest films show diffuse bilateral interstitial infiltrates. Focal areas of consolidation may be present. Supportive care such as oxygen therapy and specific antiviral chemotherapy, when possible, are important.

The most common causes of **community-acquired pneumonia** are *S pneumoniae*, *M pneumoniae* (**atypical pneumonia** in young persons), and *L pneumophila* (Table 48–3); collectively, these pathogens may cause as many as 75% of cases of community-acquired pneumonia. *H influenzae* pneumonia is common in association with chronic pulmonary disease. Other causes include *Chlamydophila pneumoniae* (Chapter 27) (as much as 10% of community-acquired pneumonia), *S aureus* (Chapter 13) in association with influenza virus infections, and *Klebsiella pneumoniae* (Chapter 15) in chronic alcoholics. Other gram-negative bacilli are uncommon causes of community-acquired pneumonia. **Pleural pulmonary infections with mixed anaerobic bacteria** are associated with predisposing factors such as periodontal disease, seizure disorders, stupor or coma, and aspiration of oropharyngeal bacteria into the lung. Pneumonia, lung abscesses, and infection of the pleural space (**empyema,** or pus in the chest cavity) occur with mixed anaerobic infections.

Hospital-acquired (nosocomial) pneumonia is frequently caused by enteric gram-negative bacilli such as *E coli* (Chapter 15), *Pseudomonas aeruginosa* (Chapter 16), and *S aureus* (Chapter 13), and *Legionella* may also cause hospital-acquired pneumonia. Fungi, including *Histoplasma capsulatum*, *C immitis*, and *C neoformans* (Chapter 45), cause **community-acquired pneumonia**; *Candida* and *Aspergillus* species (Chapter 45) are more likely to cause nosocomial infections.

Blood counts in patients with pneumonia usually show leukocytosis with increased PMN cells. Chest radiography shows segmental or lobar infiltrates. Cavities may be seen especially with mixed anaerobic infections or pneumonia due to *S aureus* or group A streptococci. Pleural effusions may also be found and, if present, may call for thoracentesis to obtain fluid for cell counts and culture. Blood cultures should be done in all patients with acute pneumonia. Sputum should be cultured as well.

Most patients with bacterial pneumonia and many patients with pneumonia due to other causes have mucopurulent sputum. Rusty-colored sputum suggests alveolar involvement and is associated with pneumococcal pneumonia but occurs with other organisms also. Foul-smelling sputum suggests mixed anaerobic infection. A purulent portion of the sputum should be chosen for Gram stain and microscopic examination; an adequate sputum specimen will have over 25 PMN cells and fewer than 10 epithelial cells per low-power field (100 × magnification). Traditionally, microscopic examination of the sputum has been used to help determine the cause of pneumonia; however, it may be difficult to differentiate organisms that are part of the normal oropharyngeal flora from those that are causing the pneumonia. The finding of numerous lancet-shaped gram-positive diplococci strongly suggests *S pneumoniae*, but streptococci that are part of the oropharyngeal flora can have the same appearance. The major value of stained sputum smears is when organisms that would not be expected are found (eg, numerous PMN cells along with numerous gram-negative bacilli suggesting enteric bacilli or pseudomonas, or numerous gram-positive cocci in clusters suggesting staphylococci). Sputum cultures have many of the same drawbacks as smears; it may be difficult to differentiate normal flora or colonizing bacteria from the cause of the pneumonia.

True demonstration of the cause of pneumonia comes from a limited set of specimens: a positive blood culture in a pneumonia patient with no confounding infections; a positive pleural fluid or direct lung aspirate culture; and detection of circulating antigen of a specific organism with no confounding infection (eg, *S pneumoniae* urinary antigen). A positive culture of sputum obtained by transtracheal aspiration is very helpful, but this procedure is rarely performed. Bronchoscopy is often used to obtain material for diagnostic studies in severely ill patients with pneumonia and is recommended for health care-associated pneumonia and pneumonia in the immunocompromised host. Quantitative bacterial culture performed on a carefully collected bronchoalveolar lavage (BAL) sample using 10^4 cfu/mL of a specific pathogen per sample as the cutoff for clinical significance is useful for establishing an etiology of bacterial pneumonia in patients not previously treated with antibiotics. Bronchoscopy with BAL may also yield a nonbacterial pathogen such as a filamentous mould or viral pathogen in the at-risk patient.

In the United States, several professional societies have established practice guidelines for the diagnosis and empirical and definitive treatment of community-acquired pneumonia and health care-associated and ventilator-associated pneumonia. For patients with community-acquired pneumonia, a macrolide, fluoroquinolone, or doxycycline is recommended as monotherapy for previously healthy outpatients. A macrolide plus a β-lactam or a fluoroquinolone alone is recommended for initial empiric treatment of outpatients in whom resistance is an issue and for patients who require hospitalization. These regimens should be modified in the event that an etiology is established and once the susceptibility of the causative agent is determined. In the case of hospital-acquired or health care-associated pneumonia, multidrug resistance is often a major problem and targeted antipseudomonal therapy with third-generation cephalosporins, carbapenems or β-lactam/β-lactamase inhibitor combinations in conjunction with an aminoglycoside may be required.

REFERENCES

ATS and IDSA Committees: Guidelines for the management of adults with hospital-acquired, ventilator-associated, and healthcare-associated pneumonia. Am J Respir Crit Care Med 2005;171:388.

Mandell LA et al: Infectious Diseases Society of America/American Thoracic Society consensus guidelines on the management of community-acquired pneumonia in adults. Clin Infect Dis 2007;44:527.

HEART

CASE 5: ENDOCARDITIS

A 45-year-old woman was admitted to the hospital because of fever, shortness of breath, and weight loss. Chills, sweats, and anorexia started 6 weeks before admission and increased in severity until admission. Persistent back pain developed 4 weeks prior to admission. Her shortness of breath on exertion increased to one block from her usual three blocks of walking. At the time of admission, she reported a 5-kg weight loss.

Rheumatic fever had been diagnosed in childhood, when she had swollen joints and fever and was confined to bed for 3 months. Subsequently, a heart murmur was heard.

Clinical Features

Temperature was 38°C, pulse 90/min, and respirations 18/min. Blood pressure was 130/80 mm Hg.

Physical examination showed a moderately overweight woman who was alert and oriented. She became short of breath while walking up two flights of stairs. Examination of her eyes showed a Roth spot (a round white spot surrounded by hemorrhage) in the retina of her right eye. Petechiae were seen in the conjunctiva of both eyes. Her head and neck were otherwise normal. Splinter hemorrhages were seen under two fingernails of her right hand and one finger of the left hand. Osler's nodes (tender, small, raised, red or purple lesions of the skin) were seen in the pads of one finger and one toe. Her heart size was normal to percussion. On auscultation, a low-pitched diastolic murmur consistent with mitral valve stenosis was heard at the apex; a loud mitral valve opening snap was heard over the left chest. Examination of her abdomen was difficult because of obesity; one observer felt an enlarged spleen. The remainder of her physical examination was normal.

Laboratory Findings & Imaging

The films from a chest x-ray showed a normal heart size and normal lungs. The ECG showed a normal sinus rhythm with broad P waves (atrial conduction). Echocardiography showed an enlarged left atrium, thickened mitral valve leaflets, and a vegetation on the posterior leaflet. The hematocrit was 29% (low). The white blood cell count was 9800/μL (high normal), with 68% PMNs (high), 24% lymphocytes, and 8% monocytes. The erythrocyte sedimentation rate was 68 mm/h (high). Blood chemistry tests, including electrolytes and tests of renal function, were normal. Three blood cultures were obtained on the day of admission; 1 day later, all three were positive for gram-positive cocci in chains that were viridans streptococci and subsequently identified as *Streptococcus sanguis* (Chapter 14).

Treatment

Endocarditis of the mitral valve was diagnosed. Intravenous penicillin G and gentamicin were begun and continued for 2 weeks. The patient was afebrile within 3 days after starting therapy. Following the successful treatment of her endocarditis, she was referred for long-term management of her heart disease.

Comment

The symptoms and signs of **endocarditis** are quite varied because any organ system can be secondarily (or primarily) involved. Fever occurs in 80–90% of patients, chills in 50%, anorexia and weight loss in about 25%, and skin lesions in about 25%. Nonspecific symptoms such as headache, backache, cough, and arthralgia are very common. Up to 25% of endocarditis patients present with neurologic signs or strokes secondary to emboli from **heart valve vegetations.** Backache, chest pain, and abdominal pain occur in 10–20% of the patients. Physical findings typically include fever in 90–95%, a **heart murmur** in 80–90% with a new or changing heart murmur in about 15%, and splenomegaly and skin lesions in about 50% of patients. Many other symptoms and physical

findings are directly related to the complications of metastatic infection and embolization from vegetations.

Streptococci cause about 70% of endocarditis cases. Viridans streptococci of several species (eg, *S sanguis, S salivarius, S mutans, S bovis* group; Chapter 14) are most common, followed by enterococci (eg, *E faecalis*) and other streptococci. The streptococci usually cause endocarditis on abnormal heart valves. *S aureus* causes 20–25% and *Staphylococcus epidermidis* about 5% of endocarditis (Chapter 13). *S aureus* can infect normal heart valves, is common in intravenous drug abusers, and produces more rapidly progressive disease than the streptococci. *S epidermidis* is a cause of endocarditis on prosthetic valves and only rarely infects native valves. Gram-negative bacilli (Chapters 15, 18) occur in about 5% and yeasts such as *Candida albicans* in about 3% of cases (Chapter 45). Emerging pathogens such as *Bartonella* sp. (Chapter 22) and *Tropheryma whipplei* (Chapter 22) have been reported with increasing frequency. Many other bacteria—indeed any species—can cause endocarditis; a small percentage are culture-negative.

The history and physical examination are important diagnostic procedures. The diagnosis is strongly suggested by repeatedly positive blood cultures with no other site of infection. Echocardiography can be a very helpful adjunctive procedure; the presence of vegetations in a patient with unexplained fever strongly suggests endocarditis.

Antibiotic therapy is essential because untreated endocarditis is fatal. Bactericidal drugs should be used. The choice of antibiotics depends upon the infecting organism: Penicillin G plus gentamicin for 2 weeks for viridans streptococci and for 4 weeks for susceptible enterococci is recommended. Vancomycin is the drug of choice for penicillin resistant strains. Multidrug resistance among enterococci may require the use of newer agents such as linezolide and daptomycin based upon susceptibility data. *S aureus* is treated with a penicillinase-resistant penicillin (eg, nafcillin) often with the addition of gentamicin for the first 5 days of therapy. Vancomycin is substituted for the β-lactam in cases of methicillin/oxacillin-resistant staphylococci. Treatment duration for staphylococcal endocarditis is 6 weeks. Bacteria other than the streptococci and staphylococci are treated with antibiotics of demonstrated activity. Surgery with valve replacement is sometimes necessary when valvular regurgitation (eg, aortic regurgitation) results in acute heart failure even when active infection is present. Surgery is required for fungal endocarditis and with failure of medical therapy; frequently necessary with gram-negative endocarditis; and important when the infection involves the sinus of Valsalva or produces septal abscesses and when embolization recurs.

REFERENCES

Baddour LM et al: Infective endocarditis: Diagnosis, antimicrobial therapy, and management of complications. A statement for healthcare professionals from the Committee on

Rheumatic Fever, Endocarditis, and Kawasaki Disease, Council on Cardiovascular Disease in the Young, and the Councils on Clinical Cardiology, Stroke, and Cardiovascular Surgery and Anesthesia, American Heart Association: Endorsed by the Infectious Diseases Society of America. Circulation 2005;111:e394; reference to these includes Correction, Circulation 2005; 112:2373. (Executive Summary, Circulation 2005;111:3167, Correction, Circulation 2005;112:2374). Accessed at http://circ.ahajournals.org/cgi/content/full/111/23/e394.

ABDOMEN

CASE 6: PERITONITIS & ABSCESSES

An 18-year-old male student was admitted to the hospital because of fever and abdominal pain. He had been well until 3 days prior to admission, when he developed diffuse abdominal pain and vomiting following the evening meal. The pain persisted through the night and was worse the following morning. He was seen in the emergency room, where abdominal tenderness was noted; x-rays of the chest and abdomen were normal; the white blood cell count was 24,000/μL; and other laboratory tests, including tests of liver, pancreas, and renal function, were normal. The patient returned home, but the abdominal pain and intermittent vomiting persisted and fever to 38°C developed. The patient was admitted to the hospital on the third day of illness.

There was no history of use of medication, drug or alcohol abuse, trauma, or infections, and the family history was negative.

Clinical Features

The temperature was 38°C, the pulse 100/min, respirations 24/min. The blood pressure was 110/70 mm Hg.

Physical examination showed a normally developed young man who appeared acutely ill and complained of diffuse abdominal pain. The chest and heart examinations were normal. The abdomen was slightly distended. There was diffuse periumbilical and right lower quadrant tenderness to palpation with guarding (muscle rigidity with palpation). There was a suggestion of a right lower quadrant mass. Bowel sounds were infrequent.

Laboratory Findings & Imaging

The hematocrit was 45% (normal), and the white blood cell count was 20,000/μL (markedly elevated) with 90% PMN cells (markedly elevated) and 12% lymphocytes. The serum amylase (a test for pancreatitis) was normal. Electrolytes and tests of liver and renal function were normal. X-ray films of the chest and abdomen were normal, though several distended loops of small bowel were seen. CT scan of the abdomen showed a fluid collection in the right lower quadrant with extension into the pelvis.

Treatment

The patient was taken to the operating room. At surgery, a perforated appendix with a large periappendiceal abscess extending into the pelvis was found. The appendix was removed, about 300 mL of foul-smelling abscess fluid was evacuated, and drains were placed. The patient was treated with gentamicin, ampicillin, and metronidazole for 2 weeks. The drains were advanced daily and totally removed 1 week after surgery. Culture of the abscess fluid revealed at least six species of bacteria, including E coli (Chapter 15), Bacteroides fragilis (Chapter 21), viridans streptococci, and enterococci (normal gastrointestinal flora). The patient recovered uneventfully.

Comment

Pain is the usual primary manifestation of **peritonitis** and **intra-abdominal abscess** formation. The localization and intensity of the pain are related to the primary disease of the abdominal viscera. Perforation of a peptic ulcer quickly yields epigastric pain that rapidly spreads throughout the abdomen with the spillage of gastric contents. A ruptured appendix or sigmoid colon diverticulum often produces more localized right or left lower quadrant pain, respectively, associated with the focal peritonitis and abscess formation. Nausea, vomiting, anorexia, and fever accompany the pain.

The signs and symptoms following acute spillage of bowel contents into the abdomen tend to take two phases. The first is the peritonitis stage, with acute pain associated with infection by E coli and other facultative anaerobic bacteria; this occurs over the first 1–2 days and if untreated is associated with a high mortality rate. The second stage is abscess formation associated with infection with B fragilis and other obligately anaerobic bacteria.

Physical examination during the acute phase shows abdominal rigidity and diffuse or local tenderness. Often the tenderness is pronounced when palpation of the abdomen is released, termed **rebound tenderness.** Later, abdominal distention and loss of bowel motility (**paralytic ileus**) occur.

The bacteria that make up the **normal gastrointestinal flora** (Chapter 10) are the causes of acute peritonitis and abscesses associated with bowel rupture: E coli and other enteric gram-negative rods, enterococci, viridans streptococci, B fragilis and other anaerobic gram-negative rods, and anaerobic gram-positive cocci and rods of many species.

The history and physical examination are the important initial steps in the diagnosis, to determine the acuteness and localization of the problem. Laboratory tests, such as white blood cell counts, yield nonspecific abnormal results or help rule out diseases such as pancreatitis, as in this case. X-ray films of the abdomen are very useful diagnostic adjuncts and may show gas and fluid collections in the large and small bowel. More definitive information indicating focal abnormalities is obtained using CT scans. When fluid is present, needle aspiration and culture yield a diagnosis of infection but do not define the underlying disease process.

Surgery may be necessary to obtain a definitive diagnosis, while at the same time it provides the definitive step in therapy. The underlying disease process, such as a gangrenous bowel or ruptured appendix, can be corrected and the localized infection drained. Antimicrobial drugs are important adjunctive therapy. The selection of drugs might include an antimicrobial active against the enteric gram-negative rods, one active against the enterococci and streptococci, and a third against the anaerobic gram-negative rods that are often resistant to penicillin G. Many regimens have been described; one regimen includes gentamicin, ampicillin, and metronidazole.

CASE 7: GASTROENTERITIS

Four members of a migrant farm worker family came to the hospital because of diarrhea and fever starting 6–12 hours earlier. The father was 28, the mother 24, and the children 6 and 4 years of age. The previous day, the family had a meal of mixed green salad, ground meat, beans, and tortillas prepared by another person in the encampment. Another child in the family, 8 months old, had not eaten the same meal and remained well. Approximately 24 hours after the meal, the children developed abdominal cramps, fever, and watery diarrhea. These symptoms had persisted for the preceding 12 hours, and in both children the diarrhea had become bloody. The parents had developed similar symptoms 6 and 8 hours earlier but did not have blood visible in their stools.

The parents stated that several other people in the camp had similar illnesses during the previous 2 weeks. The sanitation facilities in the camp were primitive.

Clinical Features

On physical examination, the children had temperatures of 39–39.5°C and the parents 38°C. All had tachycardia and appeared acutely ill. Both children appeared dehydrated.

White blood cell counts ranged from 12,000 to 16,000/μL, with 55–76% PMN cells. Multiple white blood cells were seen in the fecal wet mounts. Stools from the children were grossly bloody and mucoid. Cultures of the stools from each of the patients subsequently grew *Shigella flexneri* (Chapter 15).

Treatment

Both children were admitted to the hospital and given intravenous fluids and ampicillin. The parents were treated as outpatients, with oral fluids and oral ciprofloxacin. All recovered uneventfully. Public health follow-up led to improved sanitation conditions at the camp.

Comment

Nausea, vomiting, abdominal pain, diarrhea, and fever are the major clinical findings in gastrointestinal infections. The predominant symptoms are dependent upon the etiologic agent and whether it is toxigenic or invasive or both. When preformed toxins are in food, they often are associated with nausea and vomiting. For example, *S aureus* (Chapter 13) and *Bacillus cereus* (Chapter 11) produce **enterotoxins** in food; nausea and vomiting—and to a much lesser extent diarrhea—occur a few hours following ingestion of the food. Organisms that produce enterotoxins affect the proximal small bowel and tend to cause **watery diarrhea** (eg, enterotoxigenic *E coli* [Chapter 15], *Vibrio cholerae* [Chapter 17]). Agents such as rotaviruses, Norwalk virus (Chapter 37), and *Giardia lamblia* (Chapter 46) cause watery diarrhea through the mechanism of mucosal irritation or destruction. Invasive or cytotoxin-producing bacteria infect the colon and cause abdominal pain, frequent diarrhea, often with blood and mucus, fever, and dehydration, as in this case; this group of signs and symptoms is called **dysentery.** Organisms that cause dysentery include salmonellae of many serotypes, shigellae, *Campylobacter jejuni* (Chapter 17), enteroinvasive *E coli*, *Clostridium difficile* (Chapter 11), and *E histolytica* (Chapter 46). **Enteric fever** is a life-threatening infection characterized by fever, headache, and variable abdominal symptoms; *Salmonella typhi* (Chapter 15) (and also *Salmonella paratyphi* A and B, and *Salmonella choleraesuis*) and *Yersinia enterocolitica* (Chapter 19) cause enteric fever. The agents that commonly cause toxin-induced gastroenteritis, invasive, and noninvasive gastrointestinal infections are listed in Table 48–4.

Gastrointestinal infections are very common, especially in developing countries, where the associated mortality rate is high in infants and young children. Public health prevention through fostering good hygiene and providing sanitary water and food supplies is of the utmost importance.

In only a small percentage of cases is the etiologic agent demonstrated by means of stool culture or immunoassay.

TABLE 48–4   Agents that Commonly Cause Gastroenteritis

Organism	Typical Incubation Period	Signs and Symptoms	Epidemiology	Pathogenesis	Clinical Features	Chapter
Staphylococcus aureus	1–8 hours (rarely, up to 18)	Nausea and vomiting	Staphylococci grow in meats, dairy, and other foods and produce enterotoxin.	Enterotoxin acts on receptors in the gut that transmit impulse to medullary centers that control vomiting.	Very common, abrupt onset, intense vomiting for up to 24 hours, regular recovery in 24–48 hours. Occurs in persons eating the same food. No treatment usually necessary except to restore fluids and electrolytes	13
Bacillus cereus	2–16 hours	Vomiting or diarrhea	Reheated fried rice is common vehicle.	Enterotoxin formed in food or in gut from growth of *B cereus*.	With incubation period of 2–8 hours, mainly vomiting. With incubation period of 8–16 hours, mainly diarrhea	11
Clostridium perfringens	8–16 hours	Watery diarrhea	Clostridia grow in rewarmed meat dishes. Huge numbers ingested.	Enterotoxin produced during sporulation in gut, causes hypersecretion.	Abrupt onset of profuse diarrhea; vomiting occasionally. Recovery usual without treatment in 1–4 days. Many clostridia in cultures of food and feces of patients	11
Clostridium botulinum	18–24 hours	Paralysis	*C botulinum* grows in anaerobic food and produces toxin.	Toxin absorbed from gut blocks acetylcholine at neuromuscular junction.	Diplopia, dysphagia, dysphonia, difficulty breathing. Treatment requires ventilatory support and antitoxin. Diagnosis confirmed by finding toxin in blood or stool	11
Escherichia coli (enterotoxigenic; ETEC)	24–72 hours	Watery diarrhea	Most common cause of "traveler's diarrhea."	ETEC in the gut produces heat-labile (HL) or heat-stable (HS)enterotoxins. Toxins[a] cause hypersecretion in small intestine	Usually abrupt onset of diarrhea; vomiting rare. Serious infections in newborns. In adults, usually self-limiting in 1–3 days	9,15
Escherichia coli (enteroinvasive; EIEC)	48–72 hours	Dysentery	Occasional outbreaks of dysentery; infrequent cause of sporadic infection.	Inflammatory invasion of the colonic mucosa; similar to shigellosis. EIEC is closely related to *Shigella*	Acute bloody diarrhea with malaise, headache, high fever, and abdominal pain. Severe disease in poorly nourished children. WBC present in stool	9, 15
Escherichia coli (Shiga-toxin producing; STEC)	24–72 hours	Watery, bloody diarrhea	Bloody diarrhea associated with undercooked hamburgers in fast-food restaurants.	STEC produces shiga-like toxins. Often serotype 0157:H7	Causes bloody diarrhea, hemorrhagic colitis, and the majority of causes of hemolytic-uremic syndrome. Culture stool for sorbitol-negative *E coli* and serotype isolates with antisera for 0157:1–17. Other serotypes may be detected by toxin production using enzyme immunoassays that contain antibodies to the Shiga–like toxins	9, 15

(Continued)

TABLE 48-4 Agents that Commonly Cause Gastroenteritis (Continued)

Organism	Typical Incubation Period	Signs and Symptoms	Epidemiology	Pathogenesis	Clinical Features	Chapter
Escherichia coli (enteropathogenic; EPEC)	Slow onset	Watery diarrhea	Common cause of diarrhea in neonates in developing countries. Classically, cause of epidemic diarrhea in newborn nurseries with high mortality rates; less common now in developed countries.	EPEC attaches to mucosal epithelial cells and produces cytoskeletal changes; may invade cells. Different from other *E coli* that are enteroadherent or enteroaggregative and cause diarrhea	Insidious onset over 3–6 days with listlessness, poor feeding, and diarrhea. Usually lasts 5–15 days. Dehydration, electrolytic imbalance, and other complications may cause death. Antimicrobial therapy is important	9, 15
Vibrio parahaemolyticus	6–96 hours	Watery diarrhea	Organisms grow in seafood and in gut and produce toxin, or invade.	Toxin causes hypersecretion; vibrios invade epithelium; stools may be bloody	Abrupt onset of diarrhea in groups consuming same food, especially crabs and other seafood. Recovery is usually complete in 1–3 days. Food and stool cultures are positive	17
Vibrio cholerae	24–72 hours	Watery diarrhea	Organisms grow in gut and produce toxin.	Toxin causes hypersecretion in small intestine. Infective dose >10^5 organisms	Abrupt onset of liquid diarrhea in endemic area. Needs prompt replacement of fluid and electrolytes IV or orally. Stool cultures positive. Use selective media	9, 18
Shigella species (mild cases)	24–72 hours	Dysentery	Organisms grow in superficial gut epithelium.	Organisms invade epithelial cells; blood, mucus, and PMNs in stools. Infective dose <10^3 organisms	Abrupt onset of diarrhea; can have blood and pus in stools, cramps, tenesmus, and lethargy. WBC in stool. Stool cultures are positive. Often mild and self-limiting. Restore fluids	15
Shigella dysenteriae type 1 (Shiga bacillus)	24–72 hours	Dysentery, bloody diarrhea	Causes outbreaks in developing countries.	Produces cytotoxin and neurotoxin	Severe bloody diarrhea in children in developing countries; high fatality rate. Rare in the United States	15
Salmonella species	8–48 hours	Dysentery	Organisms grow in gut. Do not produce toxin.	Superficial infection of gut, little invasion. Infective dose >10^5 organisms	Gradual or abrupt onset of diarrhea and low-grade fever. WBC in stool. Stool cultures are positive. No antimicrobials unless systemic dissemination is suspected or patient is immuno-compromised. Prolonged carriage is frequent	15
Salmonella Typhi (*S paratyphi* A and B; *S choleraesuis*)	10–14 days	Enteric fever	Humans are the only reservoir for *S* typhi.	Invades intestinal mucosa and multiplies in macrophages in intestinal lymph follicles; enters mesenteric lymph glands to blood and dissemination	Insidious onset of malaise, anorexia, myalgias, and headache; high remittent fever; may have constipation or diarrhea. Hepatosplenomegaly in about 50% of patients. Diagnosis by culture of *S* typhi from blood, stool, or other site. Antibiotic therapy is important	15

Organism	Incubation period	Type	Epidemiology	Mechanism	Clinical features	Chapter
Yersinia enterocolitica	4–7 days	Enteric fever	Fecal-oral transmission. Food-borne. Animals infected.	Gastroenteritis or mesenteric adenitis. Occasional bacteremia. Toxin produced occasionally	Severe abdominal pain, diarrhea, fever; PMNS and blood in stool; polyarthritis, erythema nodosum, especially in children. Keep stool specimen at 4 °C before culture	19
Clostridium difficile	Days to weeks after antibiotic therapy	Dysentery	Antibiotic-associated pseudomembranous colitis.	Makes enterotoxin (toxin A) and cytotoxin (toxin B), which cause diarrhea and epithelial cell necrosis	Abrupt onset of bloody diarrhea and fever. Toxin in stool. Patient typically received antibiotics in previous days to weeks	11
Campylobacter jejuni	2–10 days	Dysentery	Infection via oral route from food, pets. Organisms grow in small intestine.	Invasion of mucous membrane. Toxin production uncertain	Fever, diarrhea; PMNs, and fresh blood in stools, especially in children. Usually self-limited. Special media needed for cultures at 42 °C. Patients usually recover in 5–8 days	17
Rotavirus	48–96 hours	Watery diarrhea, vomiting, mild fever	Virus is the major cause of diarrheal disease in infants and young children worldwide.	Induces histopathologic changes in intestinal mucosal cells	Fever and vomiting usually precede abdominal distress and diarrhea. Death in infants in developing countries follows dehydration, and electrolyte imbalance. Typical course is 3–9 days. Diagnosis by immunoassay detection of rotavirus antigen in stool	37
Norovirus	24–48 hours	Watery diarrhea, vomiting	Major cause of epidemic diarrhea especially in closed settings like cruise ships; high secondary attack rate	Induces histopathologic change in intestinal mucosa such as blunting of microvilli	Abrupt onset of abdominal pain followed by nausea, vomiting and diarrhea. Low-grade fever may occur; malaise, myalgias, and headache are described. Typical course is 2–3 days. Diagnosis requires RT-PCR or other assays not readily available	37
Giardia lamblia	1–2 weeks	Watery diarrhea	Most commonly identified intestinal parasite. Frequent pathogen in outbreaks of waterborne diarrhea.	Complex and poorly understood interaction of parasite with mucosal cells and patient's immune response	Diarrhea self-limiting in 1–3 weeks; chronic symptoms of intermittent diarrhea, malabsorption, and weight loss may last 6 months. Diagnosis by finding trophozoites or cysts in stool or duodenal contents, or by immunoassay detection of giardia antigen in stool	46
Entamoeba histolytica	Gradual onset 1–3 weeks	Dysentery	Highest prevalence in developing countries; 10% of world's population may be infected.	Invades colonic mucosa and lyses cells, including leukocytes	Diarrhea, abdominal pain, weight loss, and fever are common. Can give rise to many complications, including fulminant colitis, perforation, and liver abscess. Diagnosis by finding trophozoites or cysts in stool	46

aCholera toxin and *E coli* heat-labile toxin stimulate adenylyl cyclase activity, increasing cAMP concentration in gut, yielding secretion of chloride and water, and reduced reabsorption of sodium. *E coli* heat-stable toxin activates intestinal guanylyl cyclase and results in hypersecretion.

Finding white blood cells on fecal wet mounts is highly suggestive of infection with an invasive pathogen.

Maintaining adequate hydration is the most important feature of treatment, especially in infants and children. Antimicrobial therapy is necessary in treatment of enteric fever (**typhoid fever**) and shortens the duration of symptoms in *Shigella*, *Campylobacter*, and *V cholerae* infections, but it prolongs the symptoms and fecal shedding of *Salmonella*.

There is no specific therapy for infection due to rotaviruses, the most common viral cause of diarrhea.

REFERENCES

Clark B, McKendrick M: A review of viral gastroenteritis. Curr Opin Infect Dis 2004;17:461.

Guerrant RL et al: Practice guidelines for the management of infectious diarrhea. Clin Infect Dis 2001;32:331.

Guerrant RL, Bobak DA: Bacterial and protozoal gastroenteritis. N Engl J Med 1991;325:327.

Marcos LA, Dupont HL: Advances in defining etiology and new therapeutic approaches in acute diarrhea. J Infect 2007;55:385.

Patel MM, Hall AJ, Vinje J, Parashar UD: Noroviruses: A comprehensive review. J Clin Virol 2009;44:1.

URINARY TRACT

CASE 8: ACUTE UNCOMPLICATED BLADDER INFECTION

A 21-year-old woman presented to the university student health service with a 2-day history of increasing urinary frequency along with urgency and dysuria. Her urine had been pink or bloody for about 12 hours. She had no history of prior urinary tract infection. The patient had recently become sexually active and was using a diaphragm and spermicide.

Clinical Features

The temperature was 37.5°C, pulse 105/min, and respirations 18/min. The blood pressure was 105/70 mm Hg.

On physical examination, the only abnormal finding was mild tenderness to deep palpation in the suprapubic area.

Laboratory Findings

Laboratory tests showed a slightly elevated white blood cell count of 13,000/µL; 66% were PMNs, also elevated. Blood

urea nitrogen, serum creatinine and glucose, and serum electrolytes were normal. The urine sediment contained innumerable white cells, moderate numbers of red cells, and many bacteria suggestive of urinary tract infection. Culture yielded more than 10^5 colony-forming units (CFU)/mL of *E coli* (diagnostic of a urinary tract infection). Antimicrobial susceptibility tests were not done.

Treatment

The patient was cured by 3 days of oral sulfamethoxazole-trimethoprim therapy.

Comment

See below.

CASE 9: COMPLICATED URINARY TRACT INFECTION

A 67-year-old man developed fever and shock 3 days after a transurethral resection of his enlarged prostate gland. Two weeks earlier he had urinary obstruction with retention secondary to the enlargement; benign prostatic hypertrophy had been diagnosed. Urinary bladder catheterization had been necessary. Following the surgery, an indwelling urinary bladder catheter attached to a closed drainage system was left in place. Two days after surgery, the patient developed fever to 38°C; on the third postoperative day, he became confused and disoriented and had a shaking chill.

Clinical Features

The temperature was 39°C, the pulse was 120/min, and the respirations were 24/min. The blood pressure was 90/40 mm Hg.

On physical examination, the patient knew his name but was disoriented to time and place. His heart, lungs, and abdomen were normal. There was mild costovertebral tenderness over the area of the left kidney.

Laboratory Findings

Laboratory tests showed a normal hematocrit and hemoglobin but an elevated white blood cell count of 18,000/µL; 85% were PMNs (markedly elevated). Blood urea nitrogen, serum creatinine, serum glucose, and electrolytes were normal. Urine was obtained from the catheter port using a needle and syringe. The urine sediment contained innumerable white cells, a few red blood cells, and numerous bacteria, indicating a urinary tract infection. Urine culture yielded more than 10^5 CFU/mL of *K pneumoniae* (Chapter 15), confirming the diagnosis of urinary tract infection. Blood culture also yielded the *K pneumoniae*, which was susceptible to third-generation cephalosporins, gentamicin, and tobramycin

Treatment & Hospital Course

The patient had urinary tract infection associated with the bladder catheter. The left kidney was presumed to be involved based on the left costovertebral angle tenderness. He also had secondary bacteremia with shock (sometimes termed gram-negative sepsis and shock). He was treated with intravenous fluids and antibiotics and recovered. The same strain of *K pneumoniae* had been isolated from other patients in the hospital, indicating nosocomial spread of the bacteria.

Comment

Urinary tract infections may involve just the lower tract or both the lower and upper tracts. **Cystitis** is the term used to describe infection of the bladder with signs and symptoms including dysuria, urgency, and frequency, as in Case 8. **Pyelonephritis** is the term used to describe upper tract infection, often with flank pain and tenderness, and accompanying dysuria, urgency, and frequency, as in Case 9. Cystitis and pyelonephritis often present as acute diseases, but recurrent or chronic infections occur frequently.

It is generally accepted that 10^5 or more CFU/mL of urine is significant bacteriuria, though the patients may be symptomatic or asymptomatic. Some young women have dysuria and other symptoms of cystitis with less than 10^5CFU/mL of urine; in these women, as few as 10^3 CFU/mL of a gram-negative rod may be significant bacteriuria.

The prevalence of bacteriuria is 1–2% in school-age girls, 1–3% in nonpregnant women, and 3–8% during pregnancy. The prevalence of bacteriuria increases with age, and the sex ratio of infections becomes nearly equal. Over the age of 70 years, 20–30% or more of women and 10% or more of men have bacteriuria. Upper urinary tract infections routinely occur in patients with indwelling catheters even with optimal care and closed drainage systems: 50% after 4–5 days, 75% after 7–9 days, and 100% after 2 weeks. Sexual activity and use of spermicides increase the risk for UTIs in young women.

E coli (Chapter 15) causes 80–90% of acute uncomplicated bacterial lower tract infections (cystitis) in young women. Other enteric bacteria and *Staphylococcus saprophyticus* (Chapter 13) cause most of the other culture-positive bladder infections in this patient group. Some young women with acute dysuria suggesting cystitis have negative urine cultures for bacteria. In these patients, selective cultures for *Neisseria gonorrhoeae* (Chapter 20) and *Chlamydia trachomatis* (Chapter 27) and evaluation for herpes simplex infection should be considered.

In complicated upper tract infections, in the setting of anatomic abnormality or chronic catheterization, the spectrum in infecting bacteria is larger than in uncomplicated cases. *E coli* is frequently present, but other gram-negative rods of many species [eg, *Klebsiella*, *Proteus*, and *Enterobacter* (Chapter 15) and pseudomonads (Chapter 16)], enterococci, and staphylococci are also common. In many cases two or more species are present, and the bacteria are often resistant to antimicrobials given in association with prior therapy.

The presence of white blood cells in urine is highly suggestive but not specific for bacterial upper tract infections. White blood cells can be detected by microscopic examination of urine sediment or, indirectly, by dipstick detection of leukocyte esterase. The presence of red blood cells also is found on microscopy of the urine sediment, or indirectly by dipstick detection of hemoglobin. Proteinuria also is detected by dipstick. The presence of bacteria on Gram stain of noncentrifuged urine is strongly suggestive of 10^5 or more bacteria per milliliter of urine.

The presence of bacteriuria is confirmed by quantitative culture of the urine by any one of several methods. One frequently used method is to culture urine using a bacteriologic loop calibrated to deliver 0.01 or 0.001 mL followed by counting the number of colonies that grow.

Acute uncomplicated cystitis is usually caused by *E coli* susceptible to readily achievable urine concentrations of antibiotics appropriate for treatment of urinary tract infections. Thus, in the setting of the first such infection in a young woman, definitive identification and susceptibility testing of the bacteria are seldom necessary. Such cases can be treated by a single dose of appropriate antibiotic, but a 3-day course of therapy yields a lower relapse rate. Pyelonephritis is treated with 10–14 days of antibiotic therapy. Recurrent or complicated upper tract infections are best treated with antibiotics shown to be active against the infecting bacteria; definitive identification and susceptibility testing are indicated. Therapy for 14 days is appropriate and for 14–21 days if there is recurrence. Patients with complicated upper tract infections should have evaluations for anatomic abnormalities, stones, etc.

REFERENCES

Foster RT Sr: Uncomplicated urinary tract infections in women. Obstet Gynecol Clin North Am 2008;35:235.

Neal DE Jr: Complicated urinary tract infections. Urol Clin North Am 2008;35:13.

BONE & SOFT TISSUE

CASE 10: OSTEOMYELITIS

A 34-year-old man suffered an open fracture of the middle third of his tibia and fibula when his motorized three-wheel vehicle tipped over in a field and fell on him. He was taken to a hospital and promptly to the operating room. The wound was cleaned and debrided, the fracture was reduced, and the bone aligned. Metal plates were placed to span the fracture, align it, and hold it in place. Pins were placed through the skin and bone proximal and distal to the fracture to allow splinting and immobilization of the leg. One day after surgery, the leg remained markedly swollen; a moderate amount of serous drainage was present on the dressings. Two days later, the leg remained swollen and red, requiring opening of the surgical wound. Cultures of pus in the wound grew *S aureus* (Chapter 13) resistant to penicillin G but susceptible to nafcillin. The patient was treated with intravenous nafcillin for 10 days, and the swelling and redness decreased. Three weeks later, pus began to drain from a small opening in the wound. Cultures again grew *S aureus*. Exploration of the opening showed a sinus tract to the site of the fracture. An x-ray film of the leg showed poor alignment of the fracture. Osteomyelitis was diagnosed, and the patient was returned to the operating room, where the fracture site was debrided of necrotic soft tissue and dead bone; the pins and plates were removed. Bone grafts were placed. The fracture was immobilized by external fixation. Cultures obtained during surgery grew *S aureus*. The patient was treated with intravenous nafcillin for 1 month followed by oral dicloxacillin for 3 additional months. The wound and fracture slowly healed. After 6 months, there was no x-ray evidence of further osteomyelitis, and the patient was able to bear weight on the leg.

Comment

Osteomyelitis follows **hematogenous spread** of pathogenic bacteria from a distant site of infection to bone or, as in this case, direct inoculation of the bone and soft tissue, as can occur with an open fracture or from a contiguous site of soft tissue infection. The primary symptoms are fever and pain at the infected site; swelling, redness, and occasionally drainage can be seen, but the physical findings are highly dependent upon the anatomic location of the infection.

For example, osteomyelitis of the spine may present with fever, back pain, and signs of a paraspinous abscess; infection of the hip may show as fever with pain on movement and decreased range of motion. In children, the onset of osteomyelitis following hematogenous spread of bacteria can be very sudden, while in adults the presentation may be more indolent. Sometimes osteomyelitis is considered to be chronic or of long standing, but the clinical spectrum of osteomyelitis is broad, and the distinction between acute and chronic may not be clear either clinically or on morphologic examination of tissue.

S aureus (Chapter 13) is the primary agent of osteomyelitis in 60–70% of cases (90% in children). *S aureus* causes the infection after hematogenous spread or following direct inoculation. Community-acquired methicillin-resistant *S aureus* that contains the Panton-Valentine leukocidin causes acute hematogenous osteomyelitis affecting multiple sites, often in association with vascular complications. Streptococci cause osteomyelitis in about 10% of cases, and enteric gram-negative rods (eg, *E coli*) and other bacteria such as *P aeruginosa* (Chapter 16) in 20–30%. *Kingella kingae* (Chapter 16) is a common etiologic agent in infants and children. Anaerobic bacteria (eg, *Bacteroides* species [Chapter 21]) are also common, particularly in osteomyelitis of the bones of the feet associated with diabetes and foot ulcers. Any bacteria that cause infections in humans have been associated with osteomyelitis.

Definitive diagnosis of the etiology of osteomyelitis requires culture of a specimen obtained at surgery or by needle aspiration of bone or periosteum through uninfected soft tissue. Culture of pus from the opening of a draining sinus tract or superficial wound associated with the osteomyelitis commonly yields bacteria that are not present in the bone. Blood cultures are often positive when systemic symptoms and signs (fever, weight loss, elevated white blood cell count, high erythrocyte sedimentation rate) are present.

Early in the course of osteomyelitis, x-ray films of the infected site are negative. The initial findings noted radiologically usually are soft tissue swelling, loss of tissue planes, and demineralization of bone; 2–3 weeks after onset, bone erosions and evidence of periostitis appear. Bone scans with radionuclide imaging are about 90% sensitive. They become positive within a few days after onset and are particularly helpful in localizing the site of infection and determining if there are multiple sites of infection; however, bone scans do not differentiate between fractures, bone infarction (as occurs in sickle cell disease), and infection. CT and MRI also are sensitive and especially helpful in determining the extent of soft tissue involvement.

Antimicrobial therapy and surgical debridement are the mainstays of treatment of osteomyelitis. The specific antimicrobial should be selected after culture of a properly obtained specimen and susceptibility tests and continued for 6–8 weeks or longer, depending on the infection. Surgery should be done to remove any dead bone and sequestra that are present. Immobilization of infected limbs and fixation of fractures are important features of care.

REFERENCES

Calhoun JH, Manring MM: Adult Osteomyelitis. Infect Dis Clin N Amer 2005;19:265.

Kaplan SL: Osteomyelitis in children. Infect Dis Clin N Amer 2005;19:787.

CASE 11: GAS GANGRENE

A 22-year-old man fell while riding his new motorcycle and suffered an open fracture of his left femur and severe lacerations and crushing injury to the thigh and less extensive soft tissue injuries to other parts of his body. He was rapidly transported to the hospital and immediately taken to the operating room, where the fracture was reduced and the wounds debrided. At admission, results of his blood tests included a hematocrit of 45% and a hemoglobin of 15 g/dL. The immediate postoperative course was uneventful, but 24 hours later pain developed in the thigh. Fever was noted. Pain and swelling of the thigh increased rapidly.

Clinical Features & Course

The temperature was 40°C, the pulse 150/min, and respirations 28/min. The blood pressure was 80/40 mm Hg.

Physical examination showed an acutely ill young man who was in shock and delirious. The left thigh was markedly swollen and cool to touch. Large ecchymotic areas were present near the wound, and there was a serous discharge from the wound. Crepitus was felt, indicative of gas in the tissue of the thigh. An x-ray film also showed gas in the tissue planes of the thigh. Gas gangrene was diagnosed, and the patient was taken to the operating room for emergency extensive debridement of necrotic tissue. At the time of surgery, his hematocrit had fallen to 27% and his hemoglobin to 11 g/dL; his serum was red-brown in color, indicating hemolysis with free hemoglobin in his circulation. Anaerobic cultures of the specimen obtained at surgery grew *Clostridium perfringens* (Chapters 11, 21). The patient developed renal failure and heart failure, and died 3 days after his injury.

Comment

Case 11 illustrates a classic case of clostridial gas gangrene. *C perfringens* (or occasionally other *Clostridium* species) are inoculated into the traumatic wound from the environment; the clostridia are discussed in Chapters 11 and 21. The presence of necrotic tissue and foreign body material provides a suitable anaerobic environment for the organisms to multiply. After an incubation period usually of 2–3 days but sometimes only 8–12 hours, there is acute onset of pain, which rapidly

increases in intensity associated with shock and delirium. The extremity or wound shows tenderness, tense swelling, and a serosanguineous discharge. Crepitus is often present. The skin near the wound is pale but rapidly becomes discolored, and fluid-filled blebs form in the nearby skin. Skin areas of black necrosis appear. In severe cases, there is rapid progression.

In patients such as this one, Gram stain of fluid from a bleb or of a tissue aspirate shows large gram-positive rods with blunt ends and is highly suggestive of clostridial infection. PMN leukocytes are rare. Anaerobic culture provides the definitive laboratory confirmation. The differential diagnosis of clostridial gas gangrene includes anaerobic streptococcal myonecrosis, synergistic necrotizing myonecrosis, and necrotizing fasciitis. These clinically overlapping diseases can be differentiated from clostridial gas gangrene by Gram stain and cultures of appropriate specimens.

X-ray films of the infected site show gas in the fascial planes. Abnormal laboratory tests include a low hematocrit. The hemoglobin may be low or normal even when the hematocrit is low, consistent with hemolysis and cell-free circulating hemoglobin. Leukocytosis is usually present.

Extensive surgery with removal of all the dead and infected tissue is necessary as a lifesaving procedure. Penicillin G is the antibiotic of choice. Antitoxin is of no help. When shock and circulating free hemoglobin are present, renal failure and other complications are common and the prognosis is poor.

SEXUALLY TRANSMITTED DISEASES

CASE 12: URETHRITIS, ENDOCERVICITIS, & PELVIC INFLAMMATORY DISEASE

A 19-year-old woman came to the clinic because of lower abdominal pain of 2 days' duration and a yellowish vaginal discharge first seen 4 days previously on the day following the last day of her menstrual period. The patient had had intercourse with two partners in the previous month, including a new partner 10 days before presentation.

Clinical Features

Her temperature was 37.5°C; other vital signs were normal. Physical examination showed a yellowish mucopurulent discharge from the cervical os. Moderate left lower abdominal tenderness was present. The bimanual pelvic examination showed cervical motion tenderness and adnexal tenderness more severe on the left than on the right.

Laboratory Findings

Culture of the endocervix for *N gonorrhoeae* (Chapter 20) was negative. Culture for *C trachomatis* (Chapter 27) was positive.

Treatment

A diagnosis of pelvic inflammatory disease (PID) was made. The patient was treated as an outpatient with a single dose of ceftriaxone plus doxycycline for 2 weeks. Both of her partners came to the clinic and were treated.

Comment

In men, urethral discharge is classified as **gonococcal urethritis,** caused by *N gonorrhoeae,* or **nongonococcal urethritis,** caused usually by either *C trachomatis* (15–55% of cases) or *Ureaplasma urealyticum* (20–40% of cases) and infrequently by *Trichomonas vaginalis* (Chapter 46). The diagnosis is based on the presence or absence of gram-negative intracellular diplococci on stain of the urethral discharge. All patients with urethritis should be tested using nucleic acid amplification methods for both *C trachomatis* and *N gonorrhoeae.* Ceftriaxone is frequently used to treat gonococcal urethritis, but quinolones may be used in areas that report low resistance. Doxycycline or azithromycin is used to treat nongonococcal urethritis. It is highly recommended that men with gonococcal infection also be treated for chlamydial infection because of the likelihood that both infections may be present.

In women, the differential diagnosis of **endocervicitis (mucopurulent cervicitis)** is between gonorrhea and *C trachomatis* infection. The diagnosis is made by culture of the endocervical discharge or nucleic acid amplification tests for *N gonorrhoeae* and molecular diagnostic testing for *C trachomatis.* There are three major treatment options: (1) Treat for both *N gonorrhoeae* and *C trachomatis* before the culture results are available; (2) treat for *C trachomatis* only, if the prevalence of *N gonorrhoeae* infection is low but the likelihood of chlamydial infection is high; or (3) await culture results if the prevalence of both diseases is low and the likelihood of compliance with a recommendation for a return visit is high. Recommended treatments are the same as those mentioned above for urethritis.

Pelvic inflammatory disease (PID), also called **salpingitis,** is inflammation of the uterus, uterine tubes, and adnexal tissues that is not associated with surgery or pregnancy. PID is the major consequence of endocervical *N gonorrhoeae* and *C trachomatis* infections, and well over half of the cases are caused by one or both of these organisms. The incidence of gonococcal PID is high in inner city population, while chlamydial PID is more common in college students and more affluent population. Other common bacterial causes of PID are enteric organisms and anaerobic bacteria associated with bacterial vaginosis. Lower abdominal pain is the common presenting symptom. An abnormal vaginal discharge, uterine bleeding, dysuria, painful intercourse, nausea and vomiting, and fever also occur frequently. The major complication of PID is infertility due to uterine tubal occlusion. It is estimated that 8% of women become infertile after one episode of PID, 19.5% after two episodes, and 40% after three or more episodes. A clinical diagnosis of PID should be considered in any woman of childbearing age who has pelvic pain. Patients often have classic physical findings in addition to the presenting signs and symptoms, including lower abdominal, cervical motion, and adnexal tenderness. A clinical diagnosis can be confirmed by laparoscopic visualization of the uterus and uterine tubes, but this procedure is not practical and is infrequently performed; however, only about two-thirds of women with a clinical diagnosis of PID will have the disease when the uterine tubes and uterus are visualized. The differential diagnosis includes ectopic pregnancy and appendicitis as well as other diseases. In PID patients, hospitalization with intravenous therapy often is recommended to decrease the possibility of infertility. Inpatient drug regimens include cefoxitin and doxycycline or gentamicin and clindamycin. Outpatient regimens include cefoxitin or ceftriaxone in single doses plus doxycycline, or ofloxacin plus metronidazole.

REFERENCES

Centers for Disease Control and Prevention. Sexually transmitted diseases treatment guidelines. MMWR, Morb Mortal Wkly Rep 2002;51:1.

Lareau SM, Beigi RH: Pelvic inflammatory disease and tubo-ovarian abscess. Infect Dis Clin North Am 2008;22:693.

Trigg BG, Kerndt PR, Aynalem G: Sexually transmitted infections and pelvic inflammatory disease in women. Med Clin North Am 2008;92:1083.

CASE 13: VAGINOSIS & VAGINITIS

A 28-year-old woman came to the clinic because of a whitish-gray vaginal discharge with a bad odor, first noted 6 days previously. She had been sexually active with a single partner who was new to her in the past month.

Clinical Features

Physical examination showed a thin, homogeneous, whitish-gray discharge that was adherent to the vaginal wall. There was no discharge from the cervical os. The bimanual pelvic examination was normal, as was the remainder of the physical examination.

Laboratory Findings

The pH of the vaginal fluid was 5.5 (normal, <4.5). When KOH was added to vaginal fluid on a slide, an amine-like ("fishy") odor was perceived. A wet mount of the fluid showed many epithelial cells with adherent bacteria (clue cells). No PMN cells were seen. The diagnosis was bacterial vaginosis.

Treatment

Metronidazole twice daily for 7 days resulted in rapid clearing of the disorder. The decision was made not to treat her male partner unless she had a recurrence of vaginosis.

Comment

Bacterial vaginosis must be differentiated from a normal vaginal discharge and from *T vaginalis* vaginitis and *C albicans* vulvovaginitis. (see Table 48–5.) These diseases are very common, occurring in about one-fifth of women seeking gynecologic health care. Most women have at least one episode of vaginitis or vaginosis during their childbearing years.

Bacterial vaginosis is so named because no PMN cells are present in the vaginal discharge; ie, the disease is not an inflammatory process. In association with *Gardnerella vaginalis* (Chapter 22) infection, the lactobacilli of the normal vaginal flora decrease in number and the vaginal pH rises. Concomitantly, there is overgrowth of *G vaginalis* and vaginal anaerobic bacteria, producing the odorous amine-containing discharge. In addition to *G vaginalis,* curved gram-negative rods of the genus *Mobiluncus* have been associated with bacterial vaginosis. These curved bacteria can be seen on Gram stains of the vaginal discharge.

T vaginalis (Chapter 46) is a flagellated protozoan. *T vaginalis* **vaginitis** is best diagnosed by a wet mount of the vaginal fluid showing the motile trichomonads that are slightly larger than PMN cells. Because trichomonads lose their motility when cooled, it is best to use warm (37°C) saline, slides, and coverslips when making the wet mount preparations and to examine the preparations promptly.

Candidal vulvovaginitis frequently follows antibiotic therapy for a bacterial infection. The antibiotics decrease the normal genital flora, allowing the yeasts to proliferate and produce symptoms. Thus, candidal vulvovaginitis is not really a sexually transmitted disease.

REFERENCES

Wendel KA, Workowski KA: Trichomoniasis: Challenges to appropriate management. Clin Infect Dis 2007;44 Suppl 3:S123.

Johnston VJ, Mabey DC: Global epidemiology and control of *Trichomonas vaginalis.* Curr Opin Infect Dis 2008;21:56.

Nyirjesy P: Vulvovaginal candidiasis and bacterial vaginosis. Infect Dis Clin North Am 2008;22:637.

CASE 14: GENITAL SORES

A 21-year-old man came to the clinic with a chief complaint of a sore on his penis. The lesion began as a papule about 3 weeks earlier and slowly progressed to form the ulcer. It was painless, and the patient noticed no pus or discharge from the ulcer.

The patient was seen previously because of a sexually transmitted disease and was suspected of trading drugs for sex.

Clinical Features

The patient's temperature was 37°C, pulse 80/min, respirations 16/min, and blood pressure 110/80 mm Hg. There was a 1-cm ulcer on the left side of the penile shaft. The ulcer had

---

TABLE 48–5 Vaginitis and Bacterial Vaginosis

	Normal	Bacterial Vaginosis	*T vaginalis* Vaginitis	*C albicans* Vulvovaginitis
Primary symptoms	None	Discharge, bad odor, may have itching	Discharge, bad odor, may have itching	Discharge; itching and burning of vulvar skin
Vaginal discharge	Slight, white, flocculent	Increased, thin, homogeneous, white, gray, adherent	Increased, yellow, green, frothy, adherent; cervical petechiae often present	Increased, white, curdy like cottage cheese
pH	<4.5	>4.5	>4.5	≤4.5
Odor	None	Common, fishy	May be present, fishy	None
Microscopy	Epithelial cells with lactobacilli	Clue cells with adherent bacilli; no PMNs	Motile trichomonads; many PMNs	KOH preparation showing budding yeasts and pseudohyphae
Treatment	None	Metronidazole orally or topically	Metronidazole orally	Topical azole antifungal

a clean base and raised borders with moderate induration. There was little pain on palpation. Left inguinal lymph nodes 1–1.5 cm in diameter were palpable.

Laboratory Findings

The penile lesion was gently cleaned with saline and gauze. A small amount of clear exudate was then obtained from the base of the lesion, placed on a slide, and examined by dark field microscopy. Multiple spirochetes were seen. The rapid plasma reagin (RPR) screening serologic test for syphilis was positive at a 1:8 dilution. The confirmatory treponeme-specific fluorescent treponemal antibody-absorbed (FTA-ABS) test also was positive.

Treatment & Follow-Up

The patient was treated with a single dose of benzathine penicillin. Six months later, his RPR test had reverted to negative, but the FTA-ABS test was expected to stay positive for life.

The patient named five female sex partners for the month prior to his clinic visit. Three of these women were located by the public health investigators; two had positive serologic tests for syphilis and were treated. The two women who were not located had gone to unknown addresses in other cities.

Comment

The three major genital sore diseases are **syphilis, genital herpes,** and **chancroid.** (see Table 48–6.)

Two much less common genital sore diseases are the initial lesion of **lymphogranuloma venereum,** caused by *C trachomatis* (Chapter 27), and the rare disease **granuloma inguinale** (donovanosis), caused by *Klebsiella granulomatis.* Lymphogranuloma venereum is a systemic illness with fever, malaise, and lymphadenopathy; inguinal buboes may be present. The diagnosis usually is made by serologic tests, but culture of pus aspirated from an inguinal bubo may yield *C trachomatis.*

MYCOBACTERIUM TUBERCULOSIS INFECTIONS

CASE 15: PULMONARY TUBERCULOSIS

A 64-year-old man was admitted to the hospital with a 5-month history of progressive weakness and a weight loss of 13 kg. He also had fever, chills, and a chronic cough productive of yellowish sputum, occasionally streaked with blood.

The patient drank a lot of alcohol and lived in a boarding house next door to the tavern he frequented. He had smoked one pack of cigarettes a day for the past 45 years.

The patient had no history of tuberculosis, no record of prior skin tests for tuberculosis or abnormal chest radiographs, and no known exposure to tuberculosis.

Clinical Features

His temperature was 39°C, pulse 110/min, respirations 32/min, and blood pressure 120/80 mm Hg. He was a slender man. His dentition was poor, but the remainder of his head and neck examination was normal. On chest examination, many crackles were heard over the upper lung fields. The remainder of the physical examination was normal.

Laboratory Findings & Imaging

The hematocrit was 30% (low) and the white blood cell count was 9600/μL. Electrolyte concentrations and other blood tests were normal. The test for HIV-1 antibody was negative. A chest radiograph showed extensive cavitary infiltrates in both upper lobes. A tuberculin skin test was negative, as were skin tests with mumps and candida antigens, indicating anergy.

A sputum specimen was obtained immediately, and an acid-fast stain was done before the sputum concentration procedure. Numerous acid-fast bacteria were seen on the smear. Culture of the decontaminated and concentrated sputum was positive for acid-fast bacteria after 14 days' incubation; *M tuberculosis* was identified by molecular probe 2 days later. Susceptibility tests of the organisms showed susceptibility to isoniazid, rifampin, pyrazinamide, ethambutol, and streptomycin.

Hospital Course & Treatment

The patient was treated with isoniazid, rifampin, pyrazinamide, and ethambutol for 2 months, followed by directly observed twice-weekly administration of isoniazid and rifampin for 7 months. Follow-up sputum cultures were negative for *M tuberculosis.*

At hospitalization, the patient had been placed in isolation and asked to wear a mask at all times. However, before the mask and isolation were implemented, a medical student and a resident physician were exposed to the patient. The resident physician converted her tuberculin skin test and received isoniazid prophylaxis for 9 months.

An attempt was made to trace the patient's close contacts. A total of 34 persons were found to have positive tuberculin tests. Persons 35 years of age or younger were given isoniazid prophylaxis for 1 year; those older than 35 had periodic follow-up chest x-rays. Two cases of active tuberculosis also were diagnosed and treated. The *M tuberculosis* isolates from the two patients were identical to the index patient's isolate by DNA fingerprinting.

TABLE 48-6   The Major Genital Sore Diseases: Syphilis, Herpes, and Chancroid[a]

	Primary Syphilis	**Genital Herpes (Initial Lesions)**	**Chancroid**
Etiologic agent[b]	*Treponema pallidum*	Herpes simplex virus	*Haemophilus ducreyi*
Incubation period	3 weeks (10–90 days)	2–7 days	3–5 days
Usual clinical presentation	Slightly tender papule that ulcerates over 1 to several weeks	Marked pain in genital area; papules that ulcerate in 3–6 days; fever, headache, malaise, and inguinal adenopathy are common	Tender papule that ulcerates in 24 hours
Diagnostic tests	Dark-field examination of exudate from chancre; serologic tests	Virus culture of cells and fluid from chancre; serologic tests turn positive in 18–48 hours; fluorescent antibody stain of the same specimen	Culture of *Haemophilus ducreyi* on at least two kinds of enriched medium containing vancomycin and incubated at 33°C
Long-term sequelae	Secondary syphilis with mucocutaneous lesions; tertiary syphilis	Recurrent genital herpes	Inguinal bubo
Treatment	Benzathine penicillin G; doxycycline if penicillin allergy is present	Acyclovir or famciclovir or valacyclovir	Ceftriaxone, or azithromycin, or erythromycin, or ciprofloxacin

[a]*Source:* Sexually transmitted diseases treatment guidelines. MMWR Morb Mortal Wkly Rep 2006;6:1.

[b]HIV testing should be performed in patients with genital ulcer disease caused by these pathogens.

CASE 16: DISSEMINATED MILIARY TUBERCULOSIS

A 31-year-old Asian woman was admitted to the hospital with a history of 7 weeks of increasing malaise, myalgia, nonproductive cough, and shortness of breath. She had developed daily fevers of 38–39°C and had a recent 5-kg weight loss. She was given an oral cephalosporin with no benefit.

Her past medical history showed she had emigrated from the Philippines at age 24 and had had a negative chest radiograph at that time. The patient's grandmother had died of tuberculosis when the patient was an infant; the patient did not know if she had had contact with the grandmother. The patient was given BCG vaccine as a child. She was currently living with relatives who operated a boarding home for about 30 elderly persons.

Clinical Features

Her temperature was 39°C, pulse 100/min, respirations 20/min, and blood pressure 120/80 mm Hg. Her physical examination was entirely normal. The examiner was unable to palpate her spleen; the liver was of normal size to percussion; and there was no palpable lymphadenopathy.

Laboratory Findings & Imaging

The hemoglobin was 8.3 g/dL (normal, 12–15.5 g/dL), and the hematocrit was 27% (normal, 36–46%). The peripheral blood smear showed hypochromic, microcytic red blood cells compatible with chronic infection or iron deficiency anemia. The platelet count was 50,000/μL (normal, 140,000–450,000/μL). The white blood cell count was 7000/μL (normal), with a normal differential count. The prothrombin time was moderately prolonged and the partial thromboplastin time mildly prolonged, suggesting a coagulopathy of liver disease. The liver function tests were an aspartate aminotransferase (AST) of 140 units/L (normal, 10–40 units/L), alanine aminotransferase (ALT) 105 units/L (normal 5–35 units/L), bilirubin 2 mg/dL (twice normal), and alkaline phosphatase 100 units/L (normal 36–122 units/L). The serum albumin was 1.7 g/dL (normal, 3.4–5 g/dL). The creatinine, blood urea nitrogen, and electrolytes were normal. Urinalysis showed a few red and a few white blood cells. Two routine blood cultures were negative. Sputum and urine cultures grew small amounts of normal flora.

Serologic tests for HIV-1, hepatitis B virus antibody and antigen, coccidioidomycosis, leptospirosis, brucellosis, mycoplasmal infection, Lyme disease, and Q fever were negative. A tuberculin skin test was negative, as were skin tests with mumps and candida antigens, indicating anergy.

A chest radiograph was normal. Upper gastrointestinal and barium enema radiographs were negative. A CT scan of the abdomen was negative.

Hospital Course & Treatment

During the first few days of hospitalization the patient developed progressive shortness of breath and respiratory distress. Repeat chest radiography showed bilateral interstitial infiltrates. Adult respiratory distress syndrome was diagnosed. The hemoglobin was now 10.6 g/dL and the white blood cell count 4900/μL. Arterial blood gases showed a pH of 7.38, a PO_2 of 50 mm Hg (low), and a PCO_2 of 32 mm Hg. The patient was placed on oxygen therapy and intubated (for 4 days). BAL was performed. The lavage fluid was negative on routine culture, and an acid-fast stain was also negative. A second abdominal CT scan showed a normal-appearing liver, but periaortic lymphadenopathy and mild splenomegaly were present. The patient underwent laparoscopy with a liver biopsy and a bone marrow biopsy.

The liver and bone marrow biopsies both showed granulomas with giant cells; acid-fast bacilli were also present. (There were abundant iron stores, indicating that the anemia was due to chronic infection and not iron deficiency.) The patient was started on isoniazid, rifampin, pyrazinamide, and ethambutol. The chest radiographs continued to show diffuse infiltrates, but improvement was evident. The patient's fever decreased, and she showed general improvement.

Between 19 and 21 days of incubation, the liver and bone marrow biopsies and the lavage fluid all were culture-positive for acid-fast bacilli, identified as *M tuberculosis* by molecular probe. The mycobacteria were susceptible to all of the drugs the patient was receiving. The four-drug regimen was continued for 2 months until the susceptibility test results were obtained. The patient was then continued on isoniazid and rifampin for 10 more months for a total of 1 year of therapy.

The patient's relatives and the elderly persons who lived with them all had skin tests for tuberculosis. The persons with positive skin tests and those who were anergic or had recent histories of cough or weight loss also had chest radiographs. Three tuberculin-positive persons were found. No one had active tuberculosis. The three persons with positive skin tests were over 35 years of age and were not given prophylactic isoniazid because of the side effects of the drug in older persons.

The patient was thought to have had reactivation tuberculosis with hematogenous spread involving her lungs, liver, lymph nodes, and possibly her kidneys.

Comment

It is estimated that worldwide over 1.5 billion people, or approximately one-third of the world's population, have tuberculosis and that each year about 3 million people die of the disease. In the United States, a low incidence of tuberculosis of 9.4 cases per 100,000 population was reached in the mid-1980s. The rate increased slightly in the late 1980s, but since 1992 the rates have again declined.

The lowest (and most recently recorded) rate of 5.2 cases per 100,000 population (14,874 cases) was recorded in 2003. Tuberculosis in the United States occurs most commonly among lower socioeconomic populations: the urban poor, homeless persons, migrant farm workers, alcoholics, and intravenous drug users. Approximately half of the cases of tuberculosis occur in foreign-born individuals. The incidence of tuberculosis can be very high in selected groups and geographic areas (eg, HIV-positive intravenous drug abusers in the eastern United States, Haitian AIDS patients). Tuberculosis in elderly persons usually is due to reactivation of prior infection, while disease in children implies active transmission of *M tuberculosis*. About 80% of cases in children occur in ethnic minorities. However, active tuberculosis is most frequently diagnosed in young adults, often in association with HIV-1 infection. Concomitant tuberculosis and HIV-1 infections are especially important in developing countries; in Africa, millions of people have both infections. There is considerable concern about the spread of multidrug-resistant tuberculosis in Russia.

Spread of tuberculosis from a patient to another person occurs through infectious droplet nuclei generated during coughing, sneezing, or talking. The major factors in transmission of infection are the closeness and duration of contact and the infectiousness of the patient. Generally, <50% of contacts of active cases become infected as measured by conversion of tuberculin skin tests. Patients generally become noninfectious 2 weeks after beginning therapy. Once infected, 3–4% of persons develop active tuberculosis in the first year and about 10% at some later time. The ages when infection is most likely to yield active disease are infancy, age 15–25 years, and the elderly years.

The **tuberculin skin test** is performed by intracutaneous injection of 5 tuberculin units (TU) of purified protein derivative (PPD) using a number 26 or 27 needle. The reaction is read at 48–72 hours, and a positive test is induration of 10 mm or more; erythema is not considered in determining a positive test. Of persons with 10 mm induration, 90% have *M tuberculosis* infection while essentially all persons with more than 15 mm induration are infected. False-positive tests are caused by infection with nontuberculosis mycobacteria (eg, *Mycobacterium kansasii*). False-negative tests are due to generalized illness in tuberculosis patients or to immunosuppression. Additional skin tests with candida or mumps antigens, to which most immunologically normal persons react, can help determine if a patient is anergic.

Primary *M tuberculosis* infection in children includes mid or lower lung field infiltrates and hilar lymphadenopathy on chest films. Adolescents and adults may have a similar picture on primary infection, but infection will often quickly progress to **apical cavitary disease.** In the elderly, tuberculosis may present nonspecifically as a lower lobe pneumonia. When apical cavitary disease is present, it strongly suggests tuberculosis (the differential diagnosis includes histoplasmosis), but tuberculosis can mimic other diseases when parts of the lungs other than the apices are infected. Chronic pulmonary tuberculosis can be due to reactivation of endogenous infection or to exogenous reinfection.

Extrapulmonary tuberculosis occurs in less than 20% of cases, is more common in AIDS patients, and can be very serious and even life-threatening. The most common method of spread is by hematogenous dissemination at the time of primary infection or, less commonly, from chronic pulmonary or other foci. Direct extension of infection into the pleural, pericardial, or peritoneal spaces can occur, as can seeding of the gastrointestinal tract by swallowing infected secretions. In AIDS patients, unlike other patients, concurrent pulmonary and extrapulmonary disease is common. The major extrapulmonary forms of tuberculosis—in approximately descending order of frequency—are as follows: lymphatic, pleural, genitourinary, bones and joints, disseminated (miliary), meningeal, and peritoneal. However, any organ can be infected with *M tuberculosis,* and tuberculosis must be considered in the differential diagnosis of many other diseases.

The two major drugs used to treat tuberculosis are **isoniazid** and **rifampin**. The other first-line drugs are **pyrazinamide** and **ethambutol**. There are several second-line drugs that are more toxic or less effective, or both, and they should be used in therapy only when circumstances warrant their use (eg, treatment failure with standard drugs, multiple drug resistance). Several approved regimens exist for the treatment of susceptible *M tuberculosis* in children and adults. Most clinicians prefer 6-month regimens. The initial phase of a 6-month regimen for adults should consist of a 2-month period of INH, RIF, PZA, and EMB. Directly observed therapy 5 days per week is optimum. The continuation phase of treatment should consist of INH and RIF given for a minimum of 4 months. The continuation phase should be extended for an additional 3 months for patients who have cavitation on the initial or follow-up chest radiograph and are culture-positive at the time of completion of the initial phase of treatment (2 months).

Nine months of treatment is recommended if PZA cannot be included in the initial regimen, or if the isolate is determined to be resistant to PZA. A treatment course consisting of INH, RIF, and EMB should be given for the initial 2 months followed by INH and RIF for 7 months given either daily or twice weekly. Isoniazid and rifampin susceptibility or resistance are important factors in choosing appropriate drugs and establishing the duration of treatment. In noncompliant patients, directly observed therapy is important as well.

REFERENCES

American Thoracic Society, CDC and Infectious Diseases Society of America: Treatment of tuberculosis. MMWR, Morb Mortal Wkly Rep 2003;52(RR11):1.

LoBue P: Extensively drug-resistant tuberculosis. Curr Opin Infect Dis 2009;22:167.

Maartens G, Wilkinson RJ: Tuberculosis. Lancet 2007;370:2030.

HIV-1 & AIDS

CASE 17: DISSEMINATED *MYCOBACTERIUM AVIUM* COMPLEX (MAC) INFECTION

A 44-year-old man presented with a history of several weeks of intermittent fever accompanied at times by shaking chills. He had increased frequency of bowel movements without frank diarrhea but with occasional cramping and abdominal pain. There was no headache or cough. He had lost about 5 kg of body weight. The remainder of his medical history was negative.

Ten years prior to the present illness the patient's activities had put him at risk for acquiring HIV infection. He had never had laboratory tests to determine his HIV status.

Clinical Features

His temperature was 38°C, pulse 90/min, respirations 18/min, and blood pressure 110/70 mm Hg. He did not appear to be acutely ill. The tip of the spleen was palpable in the left upper abdominal quadrant 3 cm below the ribs (suggesting splenomegaly). Hepatomegaly and lymphadenopathy were not present, and there were no neurologic or meningeal signs. The balance of the physical examination was normal.

Laboratory Findings & Imaging

The patient's white count was stable at 3000/μL (below normal). The hematocrit was 29% (below normal). A CD4 T helper-inducer cell count was 75 cells/μL (normal, 425–1650/μL).

The chemistry panel was notable only for the liver enzyme alkaline phosphatase concentration of 210 units/L (normal, 36–122 units/L). Further evaluation of the cause of the patient's fever showed a normal urinalysis, negative routine blood cultures, and a normal chest radiograph. A serum cryptococcal antigen test was negative. Two blood cultures for mycobacteria were obtained. These turned positive 10 and 12 days after they were drawn. Three days later, the organism was identified by molecular probe as *M avium* complex (MAC).

A standard ELISA test for antibodies to HIV-1 was positive. Western blot analysis showed antibodies to each of the HIV-1 major antigen groups, Gag, Pol, and Env proteins. A branched-chain DNA assay to measure HIV-1 RNA was positive with 300,000 copies/mL.

TABLE 48-7   Summary of AIDS-Defining Infections, Their Treatment, and Prophylaxis

AIDS-Defining Infection	Infection Types	Treatment	Prophylaxis or Maintenance
Virus			
Cytomegalovirus	Retinitis, colitis, esophagitis, pneumonia, viremia	Valganciclovir orally and ganciclovir intraocular implant (retinitis); intravenous ganciclovir, foscarnet, famciclovir (oral and genital)	Oral or intravenous ganciclovir
Epstein-Barr virus	High-grade B-cell non-Hodgkin lymphomas	High dose cytotoxic therapy following HAART	
Herpes simplex	Cutaneous, oropharyngeal, or bronchial ulcers; proctitis	Acyclovir, foscarnet	Acyclovir, famciclovir, valacyclovir
JC virus	Progressive multifocal leukoencephalopathy		
Human herpesvirus 8 (Kaposi sarcoma-associated herpesvirus)	Kaposi sarcoma		
Bacteria			
Mycobacterium avium complex	Disseminated or extrapulmonary	Generally use two to four drugs: clarithromycin or azithromycin and ethambutol or rifabutin or ciprofloxacin or rifampin	Clarithromycin or azithromycin
Mycobacterium kansasii, other nontuberculous mycobacteria	Disseminated or extrapulmonary	According to established susceptibility patterns	
Mycobacterium tuberculosis	Any site: pulmonary, lymphadenitis, disseminated	Isoniazid, rifampin, pyrazinaminde, and ethambutol (others according to susceptibility test results) for 2 months; continue isoniazid and rifampin for at least 4 more months	Prevent transmission by good infection control practices; Isoniazid for tuberculin skin test positive ≥5 mm
Recurrent pyogenic bacterial infections	≥2 episodes within 2 years and <13 years of age; ≥2 episodes of pneumonia in 1 year and any age: *Streptococcus pneumoniae, Streptococcus pyogenes, Streptococcus agalactiae*, other streptococci, *Haemophilus influenzae, Staphylococcus aureus*	According to species	
Salmonella species	Bacteremia	Third-generation cephalosporin, ciprofloxacin	Ciprofloxacin
Pneumocystis jiroveci	Pneumonia	Trimethoprim-sulfamethoxazole; pentamidine isethionate; trimetrexate plus leucovorin with or without dapsone; clindamycin plus primaquine	Trimethoprim-sulfamethoxazole; dapsone with or without pyrimethamine plus leucovorin; aerosolized pentamidine isethionate, atovaquone
Fungi			
Candida albicans	Esophagitis, tracheo-bronchitis; also oropharyngeal, vaginitis	Amphotericin B, fluconazole, others	Fluconazole
Cryptococcus neoformans	Meningitis, disseminated; also pulmonary	Amphotericin B and flucytosine; fluconazole and flucytosine	Fluconazole
Histoplasma capsulatum	Extrapulmonary; also pulmonary	Amphotericin B, itraconazole	Itraconazole
Coccidioides immitis	Extrapulmonary; also pulmonary	Amphotericin B	Oral itraconazole or fluconazole
Protozoa			
Toxoplasma gondii	Encephalitis, disseminated	Pyrimethamine plus sulfadiazine and leucovorin; pyrimethamine and clindamycin plus folic acid	Trimethoprim-sulfamethoxazole or pyrimethamine-dapsone; atovaquone with or without pyrimethamine plus leucovorin
Cryptosporidium	Diarrhea for ≤1 month	Effective ART may result in clinical response; nitazoxanide, paromomycin	
Isospora species	Diarrhea for ≤1 month	Trimethoprim-sulfamethoxazole	Trimethoprim-sulfamethoxazole

TABLE 48–8 Common Complications in Patients with HIV Infection

Site	Complication and Etiology	Comment
General	Progressive generalized lymphadenopathy	Occurs in 50–70% of persons following primary HIV infection; must be differentiated from a large number of diseases that can cause lymphadenopathy
Nervous system	HIV encephalopathy; AIDS dementia	Short-term memory loss; difficulty organizing daily activities; inattention
	Cerebral toxoplasmosis; *Toxoplasma gondii*	Multifocal involvement of the brain is common and causes a wide spectrum of clinical disease: alteration of mental status, seizures, motor weakness, sensory abnormalities, cerebellar dysfunction, etc
	Cryptococcal meningitis; *Cryptococcus neoformans*	Often has an insidious onset with fever, headache, and malaise
	Progressive multifocal leukoencephalopathy; JC virus	Onset of focal neurologic deficits over a period of weeks
	Cytomegalovirus	Encephalitis, polyradiculopathy, mononeuritis multiplex
	Primary central nervous system lymphoma	Onset of focal neurologic deficits over a period of days to weeks
Eye	Cytomegalovirus	Retinitis
Skin	Kaposi sarcoma: human herpesvirus 8 (Kaposi sarcoma-associated herpesvirus)	Palpable firm cutaneous nodules 0.5–2 cm in diameter; initially may be smaller and later can be confluent, with large tumor masses; typically violaceous in color; may be hyperpigmented in dark-skinned persons; may involve many organ systems
	Staphylococcal folliculitis: *Staphylococcus aureus*	Infection of hair follicles of the central trunk, groin, or face
	Herpes zoster: varicella-zoster virus	Vesicles on an erythematous base in a dermatomal distribution
	Herpetic ulcers: herpes simplex virus	Grouped vesicles on an erythematous base that rapidly evolve into ulcers; usually on the face, hand, or genital areas
	Bacillary angiomatosis: *Bartonella henselae, Bartonella quintana*	Enlarging red papule with surrounding erythema; clinical appearance similar to that of Kaposi sarcoma but histologically very different
	Molluscum contagiosum	Discrete dome-shaped, pearly, flesh-colored papules or nodules that are often umbilicated. Usually appear along the beard line. Severe and prolonged infection can occur in patients with HIV
Mouth	Oral candidiasis: *Candida albicans*	Smooth red patches on the soft or hard palate; may form pseudomembranes
	Hairy leukoplakia: probably due to Epstein-Barr virus	Thickening of the oral mucosa, often with vertical folds or corrugations
	Gingivitis and periodontitis	Fiery red gingiva; necrotizing ulcers around the teeth
	Oral ulcers: herpes simplex, varicella-zoster virus, cytomegalovirus, and many other infectious agents	May present with recurrent vesicles that form ulcers
	Kaposi sarcoma	Purple-red lesions most often on the palate
Gastrointestinal	Esophagitis: *Candida albicans*, cytomegalovirus, herpes simplex virus	Presents with difficult and painful swallowing
	Gastritis: cytomegalovirus	Nausea, vomiting, early satiety, anorexia
	Enterocolitis: *Salmonella, Cryptosporidium, Isospora*, microsporidia, *Giardia, Entamoeba histolytica*, many others	Very common; diarrhea, abdominal cramping, and abdominal pain
	Proctocolitis: *Neisseria gonorrhoeae, Chlamydia trachomatis, Treponema pallidum, Campylobacter*, herpes simplex, cytomegalovirus	Rectal pain
Lung	Interstitial or consolidative pneumonia: Many tumors and many species of bacteria, fungi, viruses, and protozoa can cause pulmonary disease in HIV-infected patients	Onset may be slow or rapid, with fever, cough, and shortness of breath; diagnosis often made by bronchoscopy with bronchoalveolar lavage

(Continued)

TABLE 48–8   Common Complications in Patients with HIV Infection (Continued)

Site	Complication and Etiology	Comment
Genital tract	Vaginal candidiasis: *Candida albicans*	Abnormal curd-like discharge with vulvar redness and itching; common in HIV-infected women
	Genital warts: human papillomavirus	Can be severe in HIV-infected patients
	Invasive cervical carcinoma: human papillomavirus	Atypical cells on Pap smear up to and including carcinoma are common in HIV-infected women
	PID	More common and more severe in HIV-infected women than in other women
	Genital herpes: herpes simplex virus	Frequently recurrent and more severe in HIV-infected persons than other persons
	Syphilis: *Treponema pallidum*	Syphilis is a much more progressive disease in HIV-infected persons than in other persons; can yield accelerated development of neurologic syphilis

Treatment & Follow-Up

The patient was started on a three-drug regimen for MAC: clarithromycin, ethambutol, and ciprofloxacin. He noted an increased sense of well-being, a marked decrease in his fever and sweats, and an increased appetite. Concomitantly, the patient was started on **highly active antiretroviral therapy (HAART)**. The drugs used were two nucleoside reverse transcriptase inhibitors—Abacavir and lamivudine (3TC). At follow-up 4 months after initiating antiretroviral therapy, the patient's HIV-1 RNA viral load assay showed undetectable levels of the virus; the CD4 T cell count was 250 cells/μL.

Comment on HIV-1 Infection & AIDS

The incubation period from exposure to onset of acute HIV-1 disease is typically 2–4 weeks. Most persons develop acute illness that lasts 2–6 weeks. The common signs and symptoms are fever (97%), adenopathy (77%), pharyngitis (73%), rash (70%), and myalgia or arthralgia. The rash is erythematous, nonpruritic, and consists of maculopapular (slightly raised) lesions 5–10 mm in diameter, usually on the face or trunk—but the rash can be on the extremities or the palms and soles or may be generalized. Ulcers in the mouth are a distinctive feature of primary HIV infection. The acute illness has been described as "mononucleosis-like," but it truly is a distinct syndrome.

Anti-HIV-1 IgM antibodies appear within 2 weeks after the primary infection and precede the appearance of IgG antibodies, which are detectable within another few weeks. Detection of HIV-1 RNA early in the course of infection is a major concern for blood banks to prevent transfusion of antibody-negative HIV-1-positive blood.

AIDS is the major complication of HIV-1 infection. The syndrome is defined by the development of serious opportunistic infections, neoplasms, or other life-threatening manifestations resulting from progressive HIV-1-induced immunosuppression. AIDS is the most severe manifestation of several clinical illnesses following primary HIV infection. The first formal definition of AIDS as a syndrome was before HIV-1 had been characterized. The definition was modified in 1987 to include evidence of HIV-1 infection and again in 1993, when CD4 cell count criteria were added. The three CD4 cell count criteria are as follows: (1) >500 cells/μL, (2) 200–499/μL, and (3) <200/μL. The three clinical categories are as follows: (A) acute HIV-1 infection—persistent lymphadenopathy and asymptomatic disease; (B) patients with symptomatic conditions that are either attributed to HIV-1 infection or are complicated by HIV-1 infection (persistent oropharyngeal or vulvovaginal candidiasis, recurrent herpes zoster, bacillary angiomatosis, etc), and (C) AIDS-defining conditions (see below). The net result of the current HIV-1 infection classification is division into subtypes designated by letters and sub-subtypes denoted with numerals. The following subtypes and sub-subtypes are currently recognized: A1, A2, A3, A4, B, C, D, F1, F2, G, H, J, and K. The classification system is useful for studying the epidemiology, transmissibility, and possibly the response to antiretroviral therapy. However, the goal of helping to improve clinical and therapeutic management of AIDS patients in large part is still based on the CD4 cell count and viral load. Because CD4 cell counts are available in developed countries but not readily obtainable in much of the world, the value of the complex classification is limited in many geographic areas. The classification also does not provide for changing the status of patients who can improve dramatically with highly active antiretroviral therapy.

The AIDS-defining infections (clinical classification C, above) are listed in Table 48–7. AIDS-defining tumors include primary lymphoma of the brain, Burkitt or immunoblastic lymphoma, and invasive cervical carcinoma in women, in addition to Kaposi sarcoma. HIV-1 encephalopathy with disabling cognitive or motor functions and HIV-1 wasting disease (>10% weight loss and over 1 month of either diarrhea or weakness and fever) also are AIDS-defining.

HIV-1-infected patients may present with signs and symptoms referable to one or more organ systems. The common opportunistic infections are listed by anatomic site in Table 48–8. Typically, the evaluation of patients who may have HIV-1 infection or AIDS is based on a clinical and epidemiologic history of possible exposure coupled with a diagnostic evaluation of the presenting illness according to the site involved.

The status of knowledge about anti-HIV-1 drug therapy changes very rapidly, and for that reason anti-HIV-1 therapy recommendations should be considered interim ones. Only general guidelines are presented here. Postexposure prophylaxis with anti-HIV-1 drugs is effective, and treatment of primary HIV-1 infection may also have favorable prognostic implications. Many factors influence the decision to begin anti-HIV-1 treatment, including the rate of decrease of the CD4 cell count and the blood level of HIV-1 RNA. Early in the course of HIV-1 disease, when the CD4 cell count is >500 cells/μL, it is appropriate to monitor the clinical status. When the CD4 cell count is between 200 and 500 cells/μL, antiretroviral therapy may be indicated, depending upon the results of viral RNA testing. When the CD4 cell count falls to <200 cells/μL, therapy with drugs active against HIV-1 is generally recommended. The drugs used to treat HIV-1 infection are discussed in Chapter 30. When the CD4 cell count is <200/μL, treatment with two nucleoside analog reverse transcriptase inhibitors and a protease inhibitor or non-nucleoside reverse transcriptase inhibitor is recommended. This highly active antiretroviral therapy has significantly improved the lives and prognosis for many AIDS patients. Response to treatment should be monitored by following viral load measurements and for testing for resistance when clinical response is poor. When the CD4 cell count is <200/μL, prophylaxis for *P jiroveci* infection should be started. Prophylaxis for other opportunistic infections (Table 48–7) also may be appropriate.

REFERENCES

Hammer SM et al: International AIDS Society-USA. Antiretroviral treatment of adult HIV infection: 2008 recommendations of the International AIDS Society—USA panel. JAMA 2008;300:555.

Murphy RL (chair): Critical issues surrounding treatment in the era of active antiretroviral therapeutics. Clin Infect Dis 2000;30(2 Suppl). [Entire issue.]

Peiperl L, Coffey S, Bacon O, Volberding P (editors): HIV In Site Knowledge base [electronic resource], the comprehensive, on-line textbook of HIV disease from the University of California San Francisco and San Francisco General Hospital 2009; http://hivinsite.ucsf. edu/InSite.jsp?page=KB.

Taylor BS, Sobieszczyk ME, McCutchan FE, Hammer SM: The challenge of HIV-subtype diversity. N Engl J Med 2008;358:1590.

INFECTIONS IN TRANSPLANT PATIENTS

CASE 18: LIVER TRANSPLANTATION

A 61-year-old man underwent orthotopic liver transplantation for cirrhosis caused by chronic hepatitis C. He acquired hepatitis C from a transfusion of blood during coronary bypass surgery 10 years prior to his presentation with liver disease. Liver disease was diagnosed 2 years prior to orthotopic liver transplantation when he developed esophageal variceal bleeding. The bleeding was ultimately controlled, but the patient subsequently developed ascites and hepatic encephalopathy, only modestly controlled with medical therapy. He also suffered from insulin-dependent diabetes. At the time of his initial evaluation 4 months before the transplant, his liver function tests showed an AST of 43 units/L (normal, 10–40 units/L), ALT of 42 units/L (normal, 36–122 units/L), bilirubin of 2.9 mg/dL (normal, 0.1–1.2 mg/dL), albumin of 2.6 g/dL (normal 3.4–5 g/dL), and a prolonged prothrombin time of 1.8 International Normalized Ratio (INR). Anti-HCV was positive by the enzyme-linked immunoassay. The HCV genotype was type 1. The patient did not respond to interferon-α plus ribavirin therapy after 12 months. Viral load measurements were high at 500,000 IU/mL.

Orthotopic liver transplantation was accomplished without difficulty. Biliary reconstruction was by choledochocholedochostomy (primary anastomosis of the donor's to the recipient's common bile duct) with placement of a T-tube for external drainage of bile during healing of the anastomosis. A hepatocellular carcinoma was found incidentally on examination of the explant. The patient was started on intravenous tacrolimus (to reduce rejection) as a continuous infusion over 24 hours and corticosteroids for immunosuppression (also to help prevent rejection). The tacrolimus was changed to oral therapy on day 2. Intravenous ganciclovir was given on days 1–7 for prophylaxis against cytomegalovirus infection (hepatitis and pneumonia); after the ganciclovir was stopped, high-dose oral acyclovir was given four times daily for 3 months as continued prophylaxis against cytomegalovirus infection. Oral trimethoprim-sulfamethoxazole also was given twice weekly as prophylaxis against pneumocystis pneumonia.

Allograft function was established immediately after transplantation. On day 7 the AST was 40 units/L, alkaline phosphatase 138 units/L (normal, 36–122 units/L), and bilirubin 6.2 mg/dL. The differential diagnosis of the abnormal liver function was injury during liver preservation between donation and transplantation, hepatic artery thrombosis, and, rarely, herpes simplex hepatitis. Liver biopsy on day 7 showed injury during preservation.

The patient was discharged on day 12 on oral tacrolimus and prednisone to help prevent rejection. On day 21, a liver biopsy showed no evidence of cellular rejection and the liver tests were excellent: AST 18 units/L, alkaline phosphatase 96 units/L, and bilirubin 2 mg/dL. The serum creatinine was 2.2 mg/dL (normal, 0.5–1.4 mg/dL), and the dose of oral tacrolimus was decreased. On day 28, liver function tests rose to AST 296 units/L, alkaline phosphatase 497 units/L, and bilirubin 7 mg/dL. The differential diagnosis of abnormal liver function was acute cellular rejection and biliary obstruction. Cytomegalovirus hepatitis was possible, but this generally occurs after day 35, and the patient had been receiving prophylaxis for cytomegalovirus. A liver biopsy showed acute cellular rejection.

The patient was treated with two intravenous doses of methylprednisolone followed by oral prednisone. The tacrolimus blood level was in the therapeutic range. A follow-up liver biopsy 2 weeks later showed mild fatty change but no rejection. The AST was 15 units/L, alkaline phosphatase 245 units/L, and bilirubin 1.6 mg/dL.

One month later, 2.5 months posttransplantation, the AST again rose to 155 units/L but the alkaline phosphatase was unchanged at 178 units/L. Biopsy showed moderate fatty change, lobular hepatocyte necrosis, and mild portal inflammation consistent with posttransplant hepatitis C infection or resolving rejection. A polymerase chain reaction assay for HCV RNA was not done because it would have been positive and would have had limited prognostic value. The clinical impression was recurrent hepatitis C. The tacrolimus and prednisone were continued. Over the next month, liver function tests returned to normal.

At 6 months posttransplantation, the T-tube was removed from the bile drainage system. The patient immediately experienced severe diffuse abdominal pain. Culture of the bile grew *E coli* and *Enterococcus faecium*. The clinical impression was bile drainage into the abdomen. The patient was treated with ceftriaxone and vancomycin. Endoscopic retrograde cholangiopancreatography (ERCP) with sphincterotomy was performed to improve the bile flow. The patient was discharged 2 days later.

Eight months after the transplant, the patient presented with generalized subcutaneous edema (anasarca) and a lower extremity rash. His liver tests were mildly abnormal. The hematocrit and white blood cell count were normal. The blood urea nitrogen was 54 mg/dL (normal, 10–24 mg/dL), and serum creatinine was 2.8 mg/dL (normal, 0.6–1.2 mg/dL). Urinalysis showed 4+ protein and more than 50 red blood cells per high-power field. Skin biopsy showed a leukocytoclastic vasculitis. Cryoglobulinemia was diagnosed.

Four years posttransplant, the patient's liver tests have remained normal with the exception of intermittent mild AST and ALT elevations. Follow-up liver biopsies have shown moderate to severe fatty change with mild mononuclear cell portal inflammation. The patient remains an insulin-dependent diabetic. Renal function is mildly abnormal, with a serum creatinine of about 1.4 mg/dL. His quality of life is good. He is currently maintained on tacrolimus and prednisone. Compared with other liver transplant recipients, the patient is at increased risk for developing cirrhosis and suffering graft loss.

Comment

Transplant patients have their most important and life-threatening infections during the first few months following transplantation. Factors present prior to the transplant may be important. Underlying disease may contribute to susceptibility to infection. The patient may not have specific immunity—eg, may never have been exposed to cytomegalovirus—but the transplanted organ may be from a cytomegalovirus-positive donor or a blood transfusion may transmit the virus. The patient may have a latent infection that can become active during the period of immunosuppression following transplantation; examples include infections with herpes simplex virus, varicella-zoster virus, cytomegalovirus, and others, including tuberculosis. The patient may have received immunosuppressive drugs prior to transplantation.

A major factor determining infection is the type of transplantation: liver, heart, lung, kidney, etc. The duration and complexity of the surgical procedure also are important. Infections tend to involve the transplanted organ or to occur in association with the organ. In liver transplant patients, the surgery is complex and can take many hours. The type of biliary drainage that is established is an important determinant of abdominal infection. Direct connection of the donor biliary tract to the small bowel of the recipient (choledochojejunostomy) predisposes to biliary tract infection more so than does connection of the donor biliary tract to the recipient's existing biliary tract (choledochocholedochostomy). Liver transplant patients with surgery lasting 5–10 hours average one episode of infection posttransplant, while those whose surgery takes over 25 hours average three episodes. Liver transplant patients are prone to development of cytomegalovirus hepatitis and pneumonia. Heart–lung transplant recipients are prone to cytomegalovirus pneumonia. Ganciclovir given early in the posttransplant period is effective in reducing the impact of posttransplant cytomegalovirus disease. Other drugs often given as prophylaxis for posttransplant infection include the following: acyclovir for herpes simplex and varicella-zoster; trimethoprim-sulfamethoxazole for pneumocystis pneumonia; amphotericin B or other antifungal agent for fungal infections, primarily candidiasis and aspergillosis; isoniazid for tuberculosis; and a third-generation cephalosporin or other antibiotics for bacterial infections. The antibiotics often are given before, during, and shortly after operation to prevent wound infections and other infections directly associated with the procedure.

Immunosuppressive therapy in transplant patients also predisposes to infections. Corticosteroids in high doses, used to help prevent rejection or graft-versus-host disease, inhibit T cell proliferation, T cell-dependent immunity, and the expression of cytokine genes and thus have major effects on cellular immunity, antibody formation, and inflammation. Patients receiving high doses of corticosteroids are increasingly prone to fungal and other infections. Cyclosporine, a peptide, and tacrolimus, a macrolide, act on T cell function to prevent rejection. Other immunosuppressive drugs and antilymphocyte serum also are used. Collectively, the immunosuppressive agents can provide a setting where infections occur in transplant recipients.

Case 19 (below) presents a patient with bone marrow transplantation and includes comments on the infections that occur in that setting.

REFERENCES

Fishman JA, Rubin RH: Infection in organ transplant recipients. N Engl J Med 2007;357:2601. [PMID: 18094380]

Pizzo PA: Fever in immunocompromised patients. N Engl J Med 1999;341:893. [PMID: 10486422]

CASE 19: BONE MARROW TRANSPLANTATION

A 30-year-old man with chronic myelogenous leukemia underwent an allogeneic bone marrow transplant from an HLA-matched sibling donor. Prior to the transplant, the patient received total body radiation and high-dose cyclophosphamide to permanently destroy his leukemia, hematopoietic, and lymphoid cells.

The first infectious complication appeared at 10 days posttransplantation, before engraftment had occurred. The patient had mucositis, enteritis, and severe neutropenia with a white blood cell count of 100 cells/μL (normal, 3400–10,000 cells/μL). He was receiving prophylactic ceftazidime, low-dose amphotericin B, acyclovir, and trimethoprim-sulfamethoxazole. However, he became febrile to 39°C and looked sick. The clinical impression was probable bacterial sepsis related to the neutropenia, with the likely source being either his mouth or his gastrointestinal tract. Another possibility was infection of the central line used for his intravenous therapy. A fungal infection, either with Candida in the blood or Aspergillus pneumonia, would also be possible; however, these infections generally occur later following allogeneic bone marrow transplantation. The patient had been started on cyclosporine and low-dose prednisone therapy shortly after the bone marrow transplant to prevent graft-versus-host disease, which predisposed him to other opportunistic infections, but these also were less likely in the first few weeks following transplant.

When his condition worsened on posttransplant day 10, he was thought to have a bacterial infection. A blood culture was obtained, and the gram-negative antibiotic coverage was changed from ceftazidime to ciprofloxacin. Vancomycin was added pending the result of the blood culture. On day 12, the blood culture was reported positive for viridans streptococci. The patient was improved. The antibiotic therapy was continued until his white blood cell count increased to over 1000/μL.

On day 30 posttransplant, the patient was discharged to home care. He was engrafted and no longer neutropenic but was receiving cyclosporine and prednisone therapy for mild graft-versus-host disease.

On day 60 posttransplant, the patient developed fever, nausea, marked epigastric pain, and diarrhea. The clinical impression was cytomegalovirus enteritis or worsening graft-versus-host disease involving the gastrointestinal tract. Between day 30 and 60, the cyclosporine and prednisone therapy had gradually been decreased as his graft-versus-host disease had been stable. On day 60, the patient was admitted to hospital and examined by upper and lower gastrointestinal endoscopy. Mucosal lesions consistent with cytomegalovirus infection were seen and biopsied. On histologic examination, large intranuclear inclusion bodies consistent with cytomegalovirus infection were seen. Cultures were positive for cytomegalovirus. The patient was treated with ganciclovir and recovered.

The patient did well until day 120, when he developed abnormal liver function tests and diarrhea. Colonoscopy yielded a diagnosis of worsening graft-versus-host disease. His cyclosporine and prednisone dosages were increased.

On day 150 posttransplant, he developed fever and cough and was found to have multiple pulmonary infiltrates. The most likely diagnosis was fungal pneumonia, probably due to Aspergillus species, though P jiroveci and viral pneumonia were also possible. The patient underwent bronchoscopy with lavage and transbronchial biopsy. Cultures of the biopsy tissue grew Aspergillus fumigatus. The patient was treated with amphotericin B at the highest doses he could tolerate as determined by monitoring his renal function. This therapy was continued for 2 weeks in the hospital and then daily on an outpatient basis for 3 more weeks. The cyclosporine and prednisone dosages were decreased also.

By day 300, the patient was free of opportunistic infections. His graft-versus-host disease subsided, and his cyclosporine and prednisone dosages were tapered and then the drugs were discontinued. His chronic myelogenous leukemia remained in remission. He returned to work full-time 330 days after his bone marrow transplant.

Comment

Patients who undergo bone marrow transplantation receive ablative chemotherapy and radiation therapy to destroy their

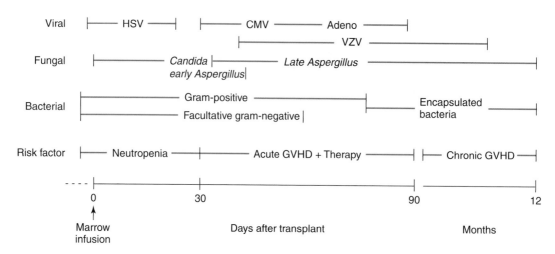

FIGURE 48–1 Predisposing risk factors and high incidence infections by time after human stem cell transplant. (GVHD, graft-versus-host disease.) (Modified with permission from Abeloff MD, Armitage JO, Niederhuber JE et al: *Clinical Oncology*, 3rd ed. Elsevier, 2004.)

hematopoietic and immune systems. The result is severe neutropenia and abnormal cellular immunity until the transplanted marrow engrafts. Because of the neutropenia, bone marrow transplantation patients are at especially high risk for infection compared with patients who receive solid organ transplants and are not neutropenic. Patients who have allogeneic bone marrow transplantation are also at risk for graft-versus-host disease, which does not occur in persons who have autologous bone marrow transplantation (ie, receive their own previously harvested bone marrow or stem cells). The immunosuppressive therapy used to control the graft-versus-host disease also helps provide a setting where patients are at high risk for infection.

The infections and the times they are likely to occur are shown in Figure 48–1. During the first month posttransplant, before engraftment occurs, there is severe neutropenia and damaged mucosal surfaces because of the pretransplant chemotherapy and radiation therapy. The patients are at greatest risk for infections caused by gram-negative and gram-positive bacteria that often are part of the normal flora of the skin, gastrointestinal tract, and respiratory tract. Recurrent herpes simplex virus infection may also occur at this time.

In the second and third months, after engraftment has occurred, the patients have continued impairment of humoral and cellular immunity. This impairment is more severe and persistent in patients with acute graft-versus-host disease. The major infections are interstitial pneumonia (about 50% caused by cytomegalovirus), aspergillus pneumonia, bacteremia, candidemia, and viral respiratory infections.

After 3 months posttransplant, there is gradual recovery of both humoral and cellular immunity. This reconstitution takes 1–2 years and can be significantly impaired by chronic graft-versus-host disease. Patients are at risk for varicella-

zoster infections and for respiratory tract infections, usually with encapsulated bacteria such as *S pneumoniae* (Chapter 14) and *H influenzae* (Chapter 18).

Prophylactic antimicrobial therapy is routinely used in bone marrow transplantation patients. Trimethoprim-sulfamethoxazole is given for 6 months or the duration of immunosuppression to prevent pneumocystis pneumonia. Acyclovir is given from the time of transplantation until engraftment occurs to prevent herpes simplex infection. Intravenous ganciclovir often is given early after transplantation and followed by oral acyclovir or oral ganciclovir to help prevent severe cytomegalovirus disease; the use of this prophylaxis varies depending upon whether the donor, the recipient, or both have evidence of prior cytomegalovirus infection. Fluoroquinolones or third-generation cephalosporins may be given during the engraftment period to help prevent bacterial infections. Antifungal agents—amphotericin B or fluconazole—may be used as prophylaxis for fungal disease. The use of vancomycin to prevent infections by gram-positive bacteria is controversial, in part because of potential selection for vancomycin-resistant enterococcal infection. After the immune system has returned to normal function, reimmunization with tetanus and diphtheria toxoids, pneumococcal and *H influenzae* polysaccharide vaccines, and killed viral vaccines (eg, polio, influenza) should be considered.

REFERENCE

Young JH, Weisdorf D: Infections in recipients of hematopoietic stem cell transplantation. In *Mandell, Douglas, and Bennett's Principles and Practice of Infectious Diseases,* 7th ed. Mandell GL, Bennett JE, Dolin R (editors). Churchill Livingstone Elsevier , 2010, p. 3821.

BIOLOGIC WARFARE & BIOTERRORISM

The grim specter of biologic warfare and bioterrorism is real. No people of the world are free from the potential consequences. The bioterrorism events with anthrax in the United States in the fall of 2001 are summarized below. A case of plague and outbreaks of smallpox and botulism, as they occur naturally, are described also. These cases and the accompanying discussion are intended to help the reader develop awareness of the importance and possible consequences of biologic warfare and bioterrorism.

CASE 20: BIOTERRORISM-RELATED ANTHRAX IN THE UNITED STATES, 2001

The Index Case in South Florida

On October 2, 2001, a 63-year-old man was admitted to the emergency department of a South Florida medical center because of fever, vomiting, and confusion. Four days earlier he had developed fever, myalgia, and malaise without specific focal symptoms. The patient had a history of mild heart disease but had otherwise been in good health.

On physical examination, the patient was lethargic and disoriented. His temperature was 39°C, blood pressure 150/80 mm Hg, pulse 110/min, and respirations 18/min. The only other potential abnormality on physical examination was the presence of rhonchi on auscultation of the chest. Rales were not heard. Nuchal rigidity was not present and Kernig and Brudzinski signs were not positive. Multiple blood cultures were obtained. Treatment with intravenous cefotaxime and vancomycin was begun prior to a spinal tap.

The hematocrit was 46%, the white blood cell count 9400/ μL with 77% PMN cells, 15% lymphocytes, and 8% monocytes. Blood chemistry tests were normal. The initial chest radiograph (see N Engl J Med 2001;354:1607) showed basilar infiltrates and a widened mediastinum. A spinal tap yielded cloudy CSF with a glucose of 57 mg/dL, a protein of 666 mg/ dL, and a white cell count of 4750/μL with 81% PMN cells and 19% monocytes. Red blood cells were also present in the CSF. Gram stain of the CSF revealed numerous PMNs and numerous large gram-positive rods, often in chains. A diagnosis of anthrax was considered, and high-dose intravenous penicillin G was added to the therapeutic regimen.

The medical center laboratory identified *Bacillus anthracis* (Chapter 11) 18 hours after the CSF had been cultured. *B anthracis* was also identified in the blood cultures.

On the second hospital day the patient had a generalized seizure. He was intubated, and assisted ventilation was begun. Low blood pressure and renal failure developed. The patient died on the third hospital day.

At autopsy there was generalized tissue edema with excessive fluid in the peritoneal cavity. Partial collapse of the lungs was present without consolidation of the lung tissue. There were areas of subpleural and perivascular hemorrhage.

There was gross blood in the mediastinum, and hemorrhagic lymph nodes were present. The heart, liver, and spleen were normal. The brain was not examined.

The patient worked as a photo editor for a large tabloid newspaper company in Palm Beach County, Florida. On September 19, 2001, the patient had examined a letter that coworkers described as suspicious and which contained a powder. The letter was never found. *B anthracis* (Chapter 11) was cultured from 20 of 136 investigation-directed environmental samples obtained October 8–10, 2001. The positive cultures included 2 of 21 from the patient's work area. Other positive cultures were obtained from the company mailroom, the mail van, and from the office of an asymptomatic mail carrier who had a nose culture-positive for *B anthracis*. Between October 25 and November 8, 2001, additional positive cultures were obtained from other areas of the workplace. Antibiotic prophylaxis was recommended for over 1000 people.

A second patient who worked at the publishing company and had extensive work-related mail exposure had onset of illness on September 28, 2001, and was reported as a possible case of inhalation anthrax on October 4. He was admitted to hospital on October 1 and started on antibiotic therapy. A nasal swab obtained on October 5, 2001, was positive for *B anthracis*, but cultures of blood, bronchial washings, and pleural fluid obtained after the start of antibiotic therapy were negative. Two pleural fluid samples were positive for *B anthracis* by the polymerase chain reaction. Thus, this patient was the second case of inhalation anthrax. He survived and was discharged from the hospital on October 17, 2001.

Washington DC Area, New York, New Jersey, & Connecticut

On September 18, 2001, a letter containing a large amount of *B anthracis* spores was mailed from Trenton, New Jersey, and addressed to "Editor, New York Post." An identical letter was mailed at the same time to the news anchorperson at NBC television. On October 9, 2001, two slightly different letters were mailed from Trenton to the District of Columbia offices of two U.S. senators. Each of the letters addressed to the senators contained a large quantity of highly refined weaponized *B anthracis* spores. These four letters contaminated Postal Service buildings and equipment, mailrooms, the offices to which they were addressed, and probably other pieces of mail as well.

A total of 22 cases of anthrax occurred, including the cases in Florida, as a result of contaminated letters. There were 11 cases of laboratory-confirmed inhalation anthrax, including 5 deaths. There were 11 cases of cutaneous anthrax, of which 9 were laboratory-confirmed and 2 were suspected. The cases occurred in South Florida (as described above), New Jersey, the District of Columbia area, New York, and Connecticut. Most cases occurred in persons who worked in postal facilities in New Jersey and the District of Columbia area where high-speed mail processing equipment was used and in persons who worked in mailrooms where letters were opened or handled. In several of the cases, including fatal inhalation anthrax cases in New York and Connecticut, exposure details have not been obtainable.

Many thousands of persons in Florida, the District of Columbia, New Jersey, and New York with potential exposure to *B anthracis* received antimicrobial prophylaxis. An experimental anthrax vaccine was offered to persons who were exposed. Contaminated sites were closed and decontaminated. There has been a nationwide and probably worldwide alert to the danger of anthrax, with investment of major resources by governments and local institutions. All of this has occurred at great cost. The alert status of the investigation and the need for investment of resources are likely to continue for some time.

REFERENCES

Bioterrorism-related anthrax. Emerg Infect Dis 2002;8(No. 10). Entire issue (31 articles).

Bush LM et al: Index case of fatal inhalation anthrax due to bioterrorism in the United States. N Engl J Med 2001;345:1607. [PMID: 11704685]

Kyriacou DN, Adamski A, Khardori N: Anthrax: from antiquity and obscurity to a front-runner in bioterrorism. Infect Dis Clin North Am 2006;20:227.

CASE 21: AN OUTBREAK OF SMALLPOX

The last previous case of smallpox in Yugoslavia was in 1927. Yugoslavia had continued population-wide vaccination to protect against imported cases. In 1972, a pilgrim returning from Mecca became ill with an undiagnosed febrile disease. Friends and relatives visited from a number of different areas; 2 weeks later, 11 of them became ill with high fever and rash. The patients were not aware of each other's illnesses, and their physicians, few of whom had seen a case of smallpox, failed to make a correct diagnosis.

One of the 11 patients was a 30-year-old teacher who quickly became critically ill with hemorrhagic smallpox, a form not readily diagnosed by experts. The teacher was first given penicillin at a local clinic. He became increasingly ill and was transferred to a dermatology ward in a city hospital, then to a similar ward in the capital city, and finally to a critical care unit because he was bleeding profusely and in shock. He died before a definitive diagnosis was made. He was buried 2 days before the first case of smallpox was recognized.

The first cases were correctly diagnosed 4 weeks after the first patient became ill, but by then 150 persons were already infected; of these, 38 (including two physicians, two nurses,

and four other hospital staff members) had been infected by the young teacher. The cases occurred in widely separate areas of the country. By the time of diagnosis, the 150 secondary cases had already begun to expose yet another generation of cases. Questions arose as to how many other undetected cases there were.

Health authorities launched a nationwide vaccination campaign. Mass vaccination clinics were held, and checkpoints along roads were established to examine vaccination certificates. Twenty million persons were vaccinated. Hotels and residential apartments were taken over, cordoned off by the military, and all known contacts of cases were forced into these centers under military guard. Some 10,000 persons spent 2 weeks or more in isolation. Meanwhile, neighboring countries closed their borders. Nine weeks after the first patient became ill, the outbreak stopped. In all, 175 patients contracted smallpox, and 35 died.

REFERENCE

Henderson DA: Bioterrorism as a public health threat. Emerg Infect Dis 1998; 4:488. (http://www.cdc.gov/ncidod/EID/vol4no3/hendrsn.htm)

CASE 22: PLAGUE

On August 2, 1996, an 18-year-old male resident of Flagstaff, Arizona, was taken to an outpatient clinic because of a 2-day history of fever, pain in his left groin, and diarrhea. On examination he was afebrile, had a pulse rate of 126/min, respiratory rate of 20/min, and a blood pressure of 130/80 mm Hg. Left groin swelling and tenderness were noted. A groin muscle strain was diagnosed and attributed to a fall 2 days earlier. He was treated with nonsteroidal anti-inflammatory agents, instructed about using a liquid diet, and released. On August 3, the patient reported feeling weak, had difficulty breathing, and collapsed while taking a shower. Emergency medical assistance was called, and the patient experienced cardiac arrest while emergency medical technicians were on site. He was transported to a hospital emergency department and pronounced dead shortly after arrival.

On August 8, cultures of blood samples obtained in the emergency department were presumptively positive for *Yersinia pestis* by fluorescent antibody staining and confirmed by specific bacteriophage lysis at the laboratory of the Arizona State Health Department. Additional isolates from postmortem brain, liver, lung, and vitreous fluid cultures were confirmed as *Y pestis* at the CDC. An epidemiologic investigation by public health officials indicated that the patient most likely became infected on July 27 as a result of bites by *Y pestis*-infected fleas while walking through a prairie dog *(Cynomys gunnisoni)* colony in Navajo County. High antibody titers to *Y pestis* were found in two of four pet dogs living in houses near the prairie dog colony. Dog owners were advised about the risk for plague and instructed to restrain their pets and periodically dust them with insecticide. Prairie dog burrows within one-half mile of the residences were dusted with insecticide to control flea population.

REFERENCES

Fatal human plague—Arizona and Colorado, 1996. MMWR, Morb Mortal Wkly Rep 1997;46:617.

Stenseth NC et al: Plague: Past, present, and future. PLoS Med 2008;15:5:e3.

CASE 23: AN OUTBREAK OF BOTULISM

In August and October 1993, public health officials in Italy were notified of seven cases of type B botulism from two apparently unrelated outbreaks in different communities. The following paragraphs summarize the investigations by the Regional Health Observatory of Campania and the Italian National Institute of Health. The illness was associated with eating commercially prepared roasted eggplant in oil.

Outbreak 1

On August 14, two waitresses working in a sandwich bar in Santa Maria di Castellabate were admitted to a local hospital with dysphagia, diplopia, and constipation; a clinical diagnosis of botulism was made. On August 12, the waitresses had prepared and eaten ham, cheese, and eggplant sandwiches. A third waitress also ate the sandwiches and developed dyspepsia, for which vomiting was induced; she did not have neurologic symptoms. The owner of the bar, who had tasted a small piece of eggplant from the same jar later on August 12, remained asymptomatic. The cook had initially opened the jar of commercially prepared sliced roasted eggplant in oil and had tasted its contents on August 11 and developed diarrhea. Both the cook and the owner reported that the eggplant tasted spoiled.

Botulism was presumptively diagnosed in the two hospitalized patients; both were treated with trivalent botulism antitoxin and gradually improved. No food samples were available for testing. No botulism toxin was detected in the serum of the two hospitalized patients. However, cultures of their stools subsequently yielded type B *Clostridium botulinum* (Chapters 11 and 21).

Outbreak 2

During October 5–6, four of nine members of an extended family who had dined together on October 2 were hospitalized in Naples with suspected botulism. The meal consisted of green olives, prosciutto, bean salad, green salad, mozzarella cheese, sausages, and commercially prepared roasted eggplant in oil. Based on an investigation and analysis of food histories, the eggplant was implicated as the probable source (relative risk, undefined; $P < 0.01$). All of the patients were treated with trivalent botulism antitoxin and gradually improved. Investigation indicated that on September 27, another family member had opened and dipped a fork into the implicated jar of eggplant; although he did not eat any eggplant, he used the fork for other food items. On September 28, he had developed vomiting, dysphagia, and double vision but was not hospitalized; his symptoms resolved spontaneously. On October 8, he was asymptomatic but was hospitalized and treated with trivalent botulism antitoxin after botulism was diagnosed in other family members.

One of the hospitalized patients developed respiratory muscle weakness and required mechanical ventilation. A serum specimen from one patient was negative for botulism toxin. Cultures of stool specimens from three patients yielded proteolytic type B *C botulinum*. No eggplant was available for testing.

Follow-Up Investigation

The commercially prepared eggplant suspected of causing both outbreaks was produced by one company and sold only in Italy. The company reported preparing the eggplant in the following manner: Eggplant slices were washed and soaked overnight in a solution of water, vinegar, and salt; roasted in an oven; and subsequently placed in glass jars. Garlic, peppers, oregano, and citric acid were added. The mixtures then were covered with sunflower oil and sealed with screw-on lids; after being filled, the jars were boiled in water for 30 minutes. The pH of the product was not consistently monitored. A total of 119 jars of eggplant from the same lot that caused the outbreaks were tested; neither *C botulinum* spores nor botulism toxin was detected. The pH of the product varied from 3.9 to 5.1; the pH was greater than 4.6 in 24 (20%) jars tested.

REFERENCE

Type B botulism associated with roasted eggplant in oil—Italy, 1993. MMWR, Morb Mortal Wkly Rep 1995;44:33.

Villar RG, Elliott SP, Davenport KM: Botulism: The many faces of botulinum toxin and its potential for bioterrorism. Infect Dis Clin North Am 2006;313.

A Brief History of Biologic Warfare

In August and October 1993, public health officials in Italy were notified of seven cases of type B botulism from two apparently unrelated outbreaks in different communities. The following paragraphs summarize the investigations by the Regional Health Observatory of Campania and the Italian National Institute of Health. The illness was associated with eating commercially prepared roasted eggplant in oil.

In the eighteenth century during the French and Indian War (1754–1767), smallpox was spread. On June 24, 1763, a British officer gave blankets containing fomites from a smallpox hospital to Native Americans of the Ohio River Valley. This was followed by an epidemic of smallpox among the indigenous people. Smallpox in European Americans may have contributed as well.

In World War I, Germany was thought to have used *B anthracis* and *Burkholderia mallei* (Chapter 16) to covertly contaminate animal feed and animals of neutral trading partners of the Allies.

Between 1932 and 1945, Japan had a biologic weapons development program in Harbin, China. There were 150 buildings, five satellite camps, and 3000 scientific staff. At least 11 Chinese cities were attacked. The organisms were *B anthracis, N meningitidis* (Chapter 20), *Shigella* species (Chapter 15), *V cholerae* (Chapter 17), and *Y pestis* (Chapter 19). Food and water were contaminated. Cultures were thrown into houses. Aerosols were sprayed from airplanes. Plague-infected fleas, 15 million per attack, were released from airplanes. Ten thousand prisoners died from experimental infections. Japanese troops had 10,000 casualties and 1700 deaths. An estimated 270,000 people in the villages died.

Between 1942 and 1969, the United States had a biologic weapons program at Fort Detrick, Maryland, with testing sites in Mississippi and Utah. Production sites were in Terre Haute, Indiana, and Pine Bluff, Arkansas. Weaponized and stockpiled biologic agents were *B anthracis*, botulinum toxin, *Francisella tularensis* (Chapter 18), *Brucella suis* (Chapter 18), *Coxiella burnetii* (Chapter 26), staphylococcal enterotoxin B, and Venezuelan equine encephalitis virus (Chapter 38). The biologic warfare capacity was destroyed between 1971 and 1973.

In the 1940s the Allies worked to develop biologic warfare capability with *B anthracis*. In bomb experiments, a small island—Gruinard Island in the North Sea off the coast of Scotland—was contaminated and unsafe. Viable *B anthracis* spores persisted on the island. In 1986, the island was decontaminated at great expense using formalin and seawater (Chapter 15).

In 1984, *Salmonella* Typhimurium was used by the Rajneeshee cult to contaminate the salad bars at ten restaurants in The Dalles, Oregon. There were 751 cases of enteritis and 45 hospitalizations.

In the 1990s, the Aum Shinrikyo cult allegedly launched three unsuccessful biologic attacks in Japan using *B anthracis*

and botulinum toxin. In 1992, they sent members to the former Zaire to obtain Ebola virus for weapons development. In March 1995, they released sarin into the Tokyo subway system, killing 12 people.

In 1996, a *Shigella dysenteriae* (Chapter 15) type 2 outbreak occurred in 12 of 45 hospital laboratory workers. The *Shigella* strain was identical to the one in the culture collection stored in the hospital's microbiology laboratory freezer. Donuts and muffins were contaminated and placed in the staff break room by an unknown person. Someone also announced the presence of the donuts and muffins by e-mail using the computer of a supervisor who was not present at the time.

Beginning in the late 1990s there were many hundreds, possibly thousands, of threats to transmit *B anthracis* through building ventilation systems or by mail. These events were investigated by several government agencies at great cost. All of the events proved to be hoaxes until the fall of 2001.

REFERENCE

Christopher GW et al: Biologic warfare: A historical perspective. JAMA 1997;278:412. [PMID: 9244333]

The Sverdlovsk Anthrax Outbreak of 1979

Sverdlovsk (Yekaterinburg) is a city of 1.2 million population 1400 km east of Moscow. It is located on the eastern foot of the Ural Mountains, on the border between Europe and Asia. It is the place where the last Tsar of Russia and his family were shot. It is the site of large armament factories that are now largely closed. In 1979, Boris Yeltsin was the senior Communist Party official in the region.

In April 1979, an outbreak of inhalation anthrax occurred in Sverdlovsk. The first case had onset on April 4. Over the subsequent 6 weeks, there were a total of 77 documented cases with 66 deaths—subsequently revised to 79 cases and 68 deaths. The disease occurred in 55 men with a mean age of 42 years; there was no man under age 24. There were 22 women with a mean age of 55; only two women, ages 24 and 32, were under age 40. No children were infected, which is not understood because children were at risk.

In order of frequency, the signs and symptoms were fever, dyspnea, cough, headache, vomiting, chills, weakness, abdominal pain, and chest pain.

Initially, patients were admitted to their local hospitals or clinics. Starting April 12, 1979, patients with high fever or other indications of possible anthrax were taken to city hospital Number 40.

In the region of the city where most of the patients resided, the building exteriors and trees were washed by fire brigades. Stray dogs were shot. Some unpaved streets were asphalted. Posters and newspapers warned against consumption of uninspected meat. (The outbreak was initially said to have been gastrointestinal anthrax from people eating tainted meat.)

Starting in mid April, a voluntary immunization program was implemented for healthy people 18–55 years old. A nonencapsulated spore vaccine was used. There were 59,000 people eligible, and 80% were immunized at least once.

Most of the 77 patients lived and worked in the southern area of the city. Some attended military reserve classes or had other reasons to be in that area. Other persons had occupations that might have taken them to that area of the city. Sixty of the 66 mapped cases were in a narrow zone approximately 4 km long extending south from the military microbiology facility to the southern city limit.

In an area up to 50 km south of Sverdlovsk, there were deaths or forced slaughter of sheep and cows. Involved animals were in six villages, all in a line south of the military microbiology facility. Animal anthrax has been enzootic in the Sverdlovsk region since before the Russian revolution, but the recorded cases in animals in 1979 were only in the downwind area to the south of the military microbiology facility.

Meteorologic records from the airport, 10 km east of the city, showed that only on Monday, April 2, 1979, were the winds from the north. All of this suggests that the human and animal anthrax cases resulted from release of *B anthracis* from the military microbiology facility on April 2, 1979.

An important conclusion from these observations is that the incubation period for inhalation anthrax is 2–43 days, with a mode of 9–10 days.

Autopsies were done on 42 of the persons who died. Hemorrhagic necrosis of thoracic lymph nodes and hemorrhagic mediastinitis were found.

In 1992, Boris Yeltsin, as President of Russia, stated that the air filters at the military microbiology facility had not been activated early in the morning of April 3, 1979, resulting in the unintentional release of *B anthracis* spores into the environment. Whether the contamination occurred on April 2, as indicated by the wind direction records, or on April 3, as indicated by Mr. Yeltsin statement, is not known.

PCR analysis was subsequently done on tissues from the autopsy specimens. Primers were used to detect the *vrrA* gene variable region of the *B anthracis* chromosome. The *vrrA* gene has unknown function. It has five categories of variable number of tandem repeats (VNTR). Only one category of VNTR is found in each *B anthracis* strain. The results of the PCR analysis showed that at least four of the five known strain categories were present in the tissue samples.

REFERENCES

Abramova FA et al: Pathology of inhalation anthrax in 42 cases from the Sverdlovsk outbreak of 1979. Proc Natl Acad Sci U S A 1993;90:2291. [PMID: 8460135]

Jackson PJ et al: PCR analysis of tissue samples from the 1979 Sverdlovsk anthrax victims: The presence of multiple *Bacillus anthracis* strains in different victims. Proc Natl Acad Sci U S A 1998;95:1224. [PMID: 9448313]

Meselson M et al: The Sverdlovsk anthrax outbreak of 1979. Science 1994;266:1202. [PMID: 7973702]

Iraq's Preparation for Biologic Warfare

The United Nations Special Commission (UNSCOM) and the International Atomic Energy Agency (IAEA) investigated Iraq's weapons of mass destruction from 1991 to 1998. Their open reports and additional personal experiences have been summarized and are outlined briefly below.

The biologic warfare program in Iraq began in the 1970s and was expanded in 1985. Two pathogenic bacteria were studied. For *B anthracis,* approximately 8000 L of solution with a spore and cell count of 10^9/mL were produced; 6000 L were used to fill weapons. The second pathogen was *C perfringens* (Chapters 11 and 20); 340 L of solution were produced.

Five viruses were studied. Two—yellow fever virus (Chapter 38) and Congo-Crimean hemorrhagic virus (Chapter 38)—were unsuitable because they required vectors. Enterovirus 17 (Chapter 36), human rotavirus (Chapter 37), and camelpox virus (Chapter 34) were studied in 1990, and development was stopped.

Toxins were studied also. For aflatoxin (hepatotoxin, nephrotoxin, carcinogen), 2200 L were produced. For botulinum toxin, 20,000 L were produced, and 11,500 L were put in Scud missile warheads. Ricin is a potent toxin that has been used for assassination; 10 L of a concentrated solution were produced. Tricothecene mycotoxins were studied, and 20 mL were produced.

Approximately 200 of the 400-lb bombs that could hold 85 L each were produced; 100 were filled with botulinum toxin, 50 with anthrax, and 7 with aflatoxin. The bombs were deployed at two sites. Iraq had 800 Scud missiles from Soviet bloc countries and an additional 80 manufactured by Iraq. They had a 300-km range with payload up to 1 metric ton. Some were modified for 600-km range and lower payload; 25 were fitted with biologic warheads, 13 with botulinum toxin, 10 with aflatoxin, and 2 with anthrax. All were deployed, 10 in a deep railroad tunnel and 15 in holes dug along the Tigris River.

Potential delivery systems included 122-mm rockets that apparently were not used. Several hundred Italian-made pesticide dispersal systems also apparently were not used.

In March 2003, the United States along with Great Britain—and with the concurrence of some other countries—invaded Iraq. The stated reason was because of Iraq's weapons of mass destruction. Iraq had previously stated they had destroyed these weapons. In the months before the invasion, United Nations inspectors had not found any such weapons. As of the time of this writing, no weapons of mass destruction had been found.

REFERENCE

Zilinskas RA: Iraq's biological weapons: The past as future? JAMA 1997;278:418. [PMID: 9244334]

Comment

Many agents have been suggested for use or used in acts of bioterrorism or warfare. Four agents are at the top of many lists: *B anthracis*, anthrax; smallpox; *Y pestis*, plague; and botulinum toxin. Plague usually requires a vector—the flea—and delivery of botulinum toxin could be difficult. However, anthrax and smallpox stand out as potential major problems, as illustrated by the accidental release of aerosolized anthrax from a Soviet Union military microbiology facility in 1979, the 22 cases of bioterrorism-induced anthrax in the United States in 2001, and by the smallpox outbreak in Yugoslavia in 1972.

The World Health Organization has estimated the number of casualties that might be produced by a hypothetical biologic warfare attack. It was assumed that there would be release of 50 kg of agent from an aircraft along a 2-km line upwind of a population center of 500,000 population. It was estimated that anthrax could reach over 20 km downwind and yield 125,000 incapacitated persons, including 95,000 deaths.

In 2001 in the United States, major efforts were implemented to deal with possible incidents involving weaponized biologic agents. It became readily apparent that meaningful present and future responses would require substantial resources from federal, state, and local governments. Education of the medical community is essential, as is education of the public and of the people who make public policy.

It is important to build an international consensus condemning the use of biologic weapons as agents of terrorism or warfare.

REFERENCES

Biological agents as weapons. JAMA 1997;278(5):347. [Entire issue.]

Cieslak TJ et al: Immunization against potential biologic agents. Clin Infect Dis 2000;30:843. [PMID: 10880299]

National symposium on medical and public health response to bioterrorism. Emerg Infect Dis 1999;5(4):491. [Entire issue.]

Index

Page numbers followed by f and t denote figures and tables, respectively.

SELECTED MEDICALLY IMPORTANT MICROORGANISMS (*Continued*)

(Continued from inside front cover)

Clostridium perfringens
Clostridium tetani
Clostridium species

GRAM-POSITIVE COCCI

Peptococcus niger
Peptostreptococcus species
Peptoniphilus species

II. VIRUSES

DNA VIRUSES

Adenoviridae
Mastadenovirus
Human adenoviruses
Hepadnaviridae
Orthohepadnavirus
Hepatitis B virus
Herpesviridae
Alphaherpesvirinae
Simplexvirus
Herpes B virus
Herpes simplex viruses 1 and 2
Varicellovirus
Varicella-zoster virus
Betaherpesvirinae
Cytomegalovirus
Cytomegalovirus
Roseolovirus
Human herpesviruses 6 and 7
Gammaherpesvirinae
Lymphocryptovirus
Epstein-Barr virus
Rhadinovirus
Human herpesvirus 8
Papillomaviridae
Papillomavirus
Human papillomaviruses
Parvoviridae
Bocavirus
Human bocavirus
Erythrovirus
Human parvovirus B19
Polyomaviridae
Polyomavirus
BK virus, JC virus, Merkel cell virus, SV40
Poxviridae
Molluscipoxvirus
Orthopoxvirus
Cowpox virus
Monkeypox virus
Smallpox virus (variola)
Vaccinia virus
Parapoxvirus
Orf virus
Pseudocowpox virus
Yatapoxvirus
Molluscum contagiosum virus

Yabapox and tanapox viruses

RNA VIRUSES

Arenaviridae
Arenavirus
Junin virus
Lymphocytic choriomeningitis virus
Lassa fever virus
Machupo virus
Astroviridae
Astrovirus
Human astroviruses
Bornaviridae
Bornavirus
Borna disease virus
Bunyaviridae
Hantavirus
Hantaan virus
Seoul virus
Sin Nombre virus
Nairovirus
Crimean-Congo hemorrhagic fever virus
Other serogroups
Orthobunyavirus
Bunyamwera serogroup
California serogroup
Other subgroups
Phlebovirus
Rift Valley fever virus
Sandfly fever viruses
Caliciviridae
Norovirus
Norwalk viruses
Sapovirus
Sapporo-like viruses
Coronaviridae
Coronavirus
Human coronaviruses
SARS coronavirus
Torovirus
Human toroviruses
Filoviridae
Ebolavirus
Ebola viruses
Marburgvirus
Marburg viruses
Flaviviridae
Flavivirus
Group B arboviruses, mosquito-borne viruses, encephalitis viruses, yellow fever, and dengue viruses
Tick-borne encephalitis viruses
Hepacivirus
Hepatitis C virus
Hepeviridae
Hepevirus
Hepatitis E virus

Orthomyxoviridae
Influenzavirus A, B
Influenza virus types A and B
Influenzavirus C
Influenza C virus
Paramyxoviridae
Respirovirus
Parainfluenza viruses
Rubulavirus
Mumps virus
Parainfluenza viruses
Morbillivirus
Measles virus
Pneumovirus
Respiratory syncytial virus
Henipavirus
Hendra virus
Nipah virus
Metapneumovirus
Human metapneumoviruses
Picornaviridae
Enterovirus
Coxsackie A viruses
Coxsackie B viruses
Echoviruses
Enteroviruses
Polioviruses
Hepatovirus
Hepatitis A virus
Parechovirus
Parechoviruses
Rhinovirus
Common cold viruses
Reoviridae
Coltivirus
Colorado tick fever virus
Rotavirus
Human rotaviruses
Retroviridae
Deltaretrovirus
Human T-lymphotropic viruses 1 and 2
Gammaretrovirus
XMRV retrovirus
Lentivirus
Human immunodeficiency viruses 1 and 2
Rhabdoviridae
Lyssavirus
Rabies virus
Vesiculovirus
Vesicular stomatitis virus
Togaviridae
Alphavirus
Group A arboviruses, mosquito- borne viruses, equine encephalitis viruses
Rubivirus
Rubella virus

UNCLASSIFIED HUMAN VIRUSES

Hepatitis D virus

UNCONVENTIONAL AGENTS (PRIONS)

Creutzfeldt-Jakob agent

III. FUNGI

DERMATOPHYTES

Epidermophyton floccosum
Microsporum canis
Microsporum gypseum
Microsporum species
Trichophyton mentagrophytes
Trichophyton rubrum
Trichophyton tonsurans
Trichophyton verrucosum
Trichophyton species

YEASTS AND YEAST-LIKE FUNGI

Candida albicans
Candida dubliniensis
Candida glabrata
Candida guilliermondii
Candida krusei
Candida lusitaniae
Candida parapsilosis
Candida tropicalis
Candida species
Cryptococcus gattii
Cryptococcus neoformans
Geotrichum species
Malassezia species
Pneumocystis jiroveci
Rhodotorula species
Saccharomyces species
Trichosporon species

DIMORPHIC FUNGI

Blastomyces dermatitidis
Coccidioides immitis
Coccidioides posadasii
Histoplasma capsulatum
Paracoccidioides brasiliensis
Penicillium marneffei
Sporothrix schenckii

HYALINE MOLDS

Acremonium species
Aspergillus flavus
Aspergillus fumigatus
Aspergillus lentulus
Aspergillus niger
Aspergillus terreus
Aspergillus species
Scedosporium apiospermum
Fusarium species
Paecilomyces species
Pseudallescheria boydii